3 ways to find your
MyPsychLab Ordering ISBN

MyPsychLab is available with texts for *Introductory Psychology*, *Biological Psychology*, *Social Psychology*, and *Abnormal Psychology* courses.

①

Look at the MyPsychLab ad
on the back cover of your text.

②

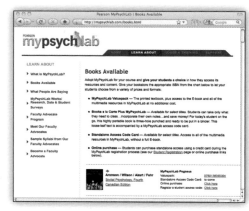

Visit the *Books Available* page of
MyPsychLab and scroll to your text.
www.mypsychlab.com/books.html

③

Visit the Pearson Psychology website
and search for your text.
www.pearsonhighered.com/psychology

More than 4 million students are now using Pearson MyLab products!

MyPsychLab This online, all-in-one study resource offers a dynamic, electronic version of your *Biopsychology* textbook with embedded video clips (close-captioned and with post-viewing activities) and embedded animations and simulations that dynamically illustrate chapter concepts. MyPsychLab also includes text-specific practice test questions for each chapter, which help prepare for exams. After completing a chapter pre-test, MyPsychLab generates a customized Study Plan for each individual student which helps to focus studying where it is needed most. Visit the site at **www.mypsychlab.com**.

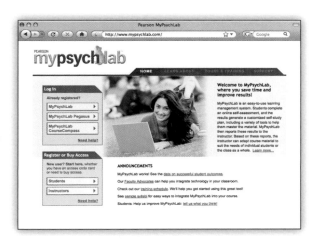

Virtual Brain within MyPsychLab

Students can immerse themselves in an interactive landscape of the human brain as it relates to *Biopsychology*. The Virtual Brain incorporates real-life scenarios as well as simulations, activities, quizzes, and more.

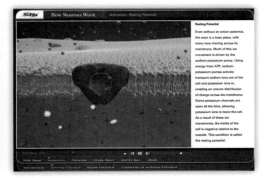

Bioflix within MyPsychLab

Bioflix within MyPsychLab: Highly visual interactive video on the toughest topics in biopsychology.

Pearson eText

Just like the printed text, students can highlight and add their own notes. Students save time and improve results by having access to their book online.

Save Time. Improve Results. **www.mypsychlab.com**

Why Do You Need This New Edition?

Seven new and updated features appear in this edition of *Biopsychology*.

1 New! **MyPsychLab**: This online, all-in-one study resource offers a dynamic, electronic version of the *Biopsychology* textbook with embedded video clips (close-captioned and with post-viewing activities) and embedded animations and simulations that dynamically illustrate chapter concepts. MyPsychLab also includes text-specific practice test questions for each chapter, which help students (or a student) master the concepts and prepare for exams. After a student completes a chapter pre-test, MyPsychLab generates a customized Study Plan for that student which helps her focus her studying where she needs it most. MyPsychLab is available in both course-management and website versions, and can be used as an instructor-driven assessment program and/or a student self-study learning program. Visit the site at www.mypsychlab.com.

2 New! **Virtual Brain**: Students can immerse themselves in an interactive landscape of the human brain as it relates to *Biopsychology*. The Virtual Brain incorporates real-life scenarios as well as simulations, activities, quizzes, and more. Available through MyPsychLab.

3 New! **Bioflix**: Interactive tutoring on the toughest topics in biopsychology. Available through MyPsychLab.

4 New! **Interactive PowerPoints**: Pearson has developed a set of **interactive PowerPoints** (available only on the Instructor's Resource DVD) with embedded animations, videos, and activities. Many of the slides include layered art, giving instructors the ability to highlight specific aspects of a figure—for example, identifying each part of the brain.

5 New! **748 new references**: Although *Biopsychology* often presents material within a historical context, it tries to provide carefully selected glimpses of cutting edge research and issues; substantial effort is required to maintain this cutting-edge focus, and this edition contains 748 new references.

6 New! **Special illustrations**: The eighth edition includes many new illustrations and brain scans, particularly in the later chapters, to stimulate student interest and clarify important points. The illustrations in *Biopsychology* were designed by Pinel and his artist/designer wife; thus they are the perfect complement to Pinel's written words.

7 New! **Study questions**: New study questions appear at the end of each chapter to promote study and review.

Biopsychology

Eighth Edition

John P. J. Pinel

University of British Columbia

PEARSON

Boston   Columbus   Indianapolis   New York   San Francisco   Upper Saddle River
Amsterdam   Cape Town   Dubai   London   Madrid   Milan   Munich   Paris   Montreal   Toronto
Delhi   Mexico City   Sao Paulo   Sydney   Hong Kong   Seoul   Singapore   Taipei   Tokyo

To Maggie, the love of my life.

Editor in Chief: Jessica Mosher
Executive Editor: Susan Hartman
Editorial Assistant: Laura Barry
Marketing Manager: Nicole Kunzmann
Senior Production Project Manager: Roberta Sherman
Manufacturing Buyer: Debbie Rossi
Cover Administrator: Joel Gendron
Editorial Production and Composition Service: Nesbitt
Interior Design: Nesbitt
Photo Researcher: Katherine S. Cebik
Developmental Editor: Erin K. L. Grelak
Production Editor: Jane Hoover

10 9 8 7 6 5 4 3 2 1 RRD-OH 14 13 12 11 10

Allyn & Bacon
is an imprint of

PEARSON

ISBN 10: 0-205-03099-8
ISBN 13: 978-0-205-03099-6

Brief Contents

Contents

Part Three
Sensory and Motor Systems

8 The Sensorimotor System   **191**
How You Move

Part Four
Brain Plasticity

Part Five

Biopsychology of Motivation

12 Hunger, Eating, and Health
Why Do Many People Eat Too Much? **298**

13 Hormones and Sex   327
What's Wrong with the Mamawawa?

14 Sleep, Dreaming, and Circadian Rhythms   355
How Much Do You Need to Sleep?

15 **Drug Addiction and the Brain's
Reward Circuits** **383**
Chemicals That Harm with Pleasure

Part Six
Disorders of Cognition and Emotion

16 Lateralization, Language, and the Split Brain 411
The Left Brain and the Right Brain of Language

Preface

Welcome to the Eighth Edition of *Biopsychology*! This edition builds on the strengths of its predecessors, but it also takes important new steps: In addition to covering many new cutting-edge research topics, it sharpens its focus on the human element of biopsychology and on promoting student thinking. Most importantly, this is the first edition of *Biopsychology* to focus on the role of creative thinking in scientific progress: It emphasizes instances in which progress in biopsychological science has been a product of creative thinking, and it encourages students to develop their own creative-thinking skills.

The Eighth Edition of *Biopsychology* is a clear, engaging introduction to current biopsychological theory and research. It is intended for use as a primary text in one- or two-semester courses in biopsychology—variously titled Biopsychology, Physiological Psychology, Brain and Behavior, Psychobiology, Behavioral Neuroscience, or Behavioral Neurobiology.

The defining feature of *Biopsychology* is its unique combination of biopsychological science and personal, reader-oriented discourse. It is a textbook that is "untextbooklike." Instead of presenting the concepts of biopsychology in the usual textbook fashion, it addresses students directly and interweaves the fundamentals of the field with clinical case studies, social issues, personal implications, useful metaphors, and memorable anecdotes.

Key Features Maintained in the Eighth Edition

The following are features that have characterized recent editions of *Biopsychology* and have been maintained or expanded in this edition.

Emphasis on Broad Themes   The emphasis of *Biopsychology* is "the big picture." Four broad themes are highlighted throughout the text by distinctive tabs: (1) thinking creatively, (2) clinical implications, (3) evolutionary perspective, and (4) neuroplasticity. A Themes Revisited section at the end of each chapter briefly summarizes how each theme was developed in that chapter. The four major themes provide excellent topics for essay assignments and exam questions.

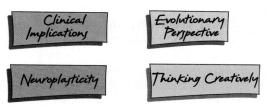

Effective Use of Case Studies   *Biopsychology* features many carefully selected case studies, which are highlighted in the text. These provocative cases stimulate interest, promote retention, and allow students to learn how biopsychological principles apply to the diagnosis and treatment of brain disorders.

Remarkable Illustrations   The illustrations in *Biopsychology* are special. Each one was conceptualized and meticulously designed to clarify and reinforce the text by a uniquely qualified scientist–artist team: Pinel and his artist/designer wife, Maggie Edwards.

Focus on Behavior   In some biopsychological textbooks, the coverage of neurophysiology, neurochemistry, and neuroanatomy subverts the coverage of behavioral research. *Biopsychology* gives top billing to behavior: It stresses that neuroscience is a team effort and that the unique contribution made by biopsychologists to this effort is their behavioral expertise.

Emphasis on the Scientific Method   *Biopsychology* emphasizes the scientific method. It portrays the scientific method as a means of answering questions that is as applicable in daily life as in the laboratory. And *Biopsychology* emphasizes that being a scientist is fun.

Discussion of Personal and Social Implications   Several chapters of *Biopsychology*—particularly those on eating, sleeping, sex, and drug addiction—carry strong personal and social messages. In these chapters, students are encouraged to consider the relevance of biopsychological research to their lives outside the classroom.

Engaging, Inspiring Voice   Arguably the strongest pedagogical feature of *Biopsychology* is its personal tone. Pinel addresses students directly and talks to them with warmth, enthusiasm, and good humor about recent advances in biopsychological science. Many students report being engaged and inspired by this approach.

Additions to the Eighth Edition

Six new or expanded features appear in the Eighth Edition of *Biopsychology*.

NEW! Thinking Creatively Theme   Many important advances in biopsychological research have been a product of creative thinking (thinking in productive, unconventional ways that are consistent with the evidence): Researchers who have sidestepped the constraints of convention and approached their problems from original perspectives have often made major breakthroughs. Although the role of *critical thinking* in science has been addressed by many textbooks, including this one, few have addressed the role of *creative thinking*. The role of creative thinking in scientific progress is one of the four major themes of the Eighth Edition of *Biopsychology*. Discoveries that are clearly attributable to creative thinking and points where Pinel challenges students to engage in creative thinking themselves are marked by thinking creatively tabs. The addition of this theme makes biopsychological research more interesting, teaches students an important lesson about the role of creativity in science, and fosters creative thinking in students.

NEW! Additional Check It Out Demonstrations
Biopsychology's Check It Out demonstrations encourage students to experience biopsychological concepts and phenomena for themselves. Several Check It Outs have been added to this edition. They are integrated into the flow of the text at points where their positive impact on understanding and retention is maximized.

NEW! MyPsychLab (www.mypsychlab.com)   MyPsychLab is an online study resource that offers a wealth of animations and practice tests, as well as additional study and research tools. There are over 150 assessments for students, a wealth of Web and video/media links, flashcards, and a fully interactive brain called the Virtual Brain.

NEW! More Illustrations and Brain Images   Building on *Biopsychology*'s strong art package, a number of new illustrations and brain images have been added. These have been carefully selected, designed, and positioned to support interest, clarity, and memorability.

NEW! End–of–Chapter Study Questions   Sample study questions are now included at the end of each chapter of *Biopsychology* to allow students to experience their usefulness and to encourage them to visit MyPsychLab for other learning and study resources.

NEW! Diagram-Based Test Questions   For the first time, the test bank for this edition of *Biopsychology* includes test questions referring to and accompanied by line drawings of the brain adapted from the text.

New Coverage in the Eighth Edition

Biopsychology remains one of the most rapidly progressing scientific fields. Like previous editions, the Eighth Edition of *Biopsychology* has meticulously incorporated recent developments in the field—it contains 748 citations of articles or books that did not appear in the preceding edition. These recent developments have dictated changes to many parts of the text. The following list presents some of the content changes to this edition, organized by chapter.

Chapter 1: Biopsychology as a Neuroscience

- Introduction of creative thinking as one of *Biopsychology*'s four major themes
- New list of Web sites on the ethics of animal research
- Discussion of the relationship between creative thinking and critical thinking in science
- Case of Howard Dully, the boy who was lobotomized at the insistence of his stepmother
- 10 new research citations

Chapter 2: Evolution, Genetics, and Experience

- Increased emphasis on epigenetic mechanisms
- Introduction of the term *enhancers*
- Emphasis of the fact that only a small proportion of chromosome segments contain protein-coding genes
- New discussions of microRNAs, alternative splicing, and monoallelic expression
- 56 new research citations

Chapter 3: Anatomy of the Nervous System

- Introduction of the reciprocal connections between thalamic nuclei and the neocortex
- Explanation of subcortical white matter
- 19 new research citations

Chapter 4: Neural Conduction and Synaptic Transmission

- Presentation of the Hodgkin-Huxley model from a modern perspective
- Increased emphasis on gap junctions
- Enhanced discussion of the diversity of neuron physiology
- 18 new research citations

Chapter 5: The Research Methods of Biopsychology

- Discussion of the weaknesses of the fMRI technique
- Added coverage of transcranial magnetic stimulation, with an interesting example
- Added coverage of genetic engineering, with illustrative examples

- Introduction of green fluorescent protein and its use as a neural stain
- Introduction of the brainbow neural staining technique
- Introduction of the fMRI default mode network
- Three striking new images: green fluorescent protein, brainbow, and default mode network
- 28 new research citations

Chapter 6: The Visual System

- Discussion of gap junctions in the retina
- Comparative analysis of color vision
- Description of transgenic mice that have an extra photopigment
- More systematic coverage of prosopagnosia
- Introduction of akinetopsia, with two new case studies
- Discussion and new illustration of area MT of the cortex
- 24 new research citations

Chapter 7: Mechanisms of Perception: Hearing, Touch, Smell, Taste, and Attention

- Reorganization of the coverage of the auditory system
- Updated descriptions of core, belt, and parabelt areas of auditory cortex
- Description of the discovery of a gene associated with the congenital absence of pain
- Discussion of the systematic layout of olfactory receptors
- Discussion of taste receptor proteins
- Introduction of olfactory glomeruli
- 46 new research citations

Chapter 8: The Sensorimotor System

- Description of the effects of posterior parietal cortex stimulation of conscious neurosurgical patients
- Discussion of the fact that some patients can stretch otherwise paralyzed limbs when they yawn
- 39 new research citations

Chapter 9: Development of the Nervous System

- Explanation of the role of gap junctions in neural migration and aggregation
- Description of various triggers for apoptosis
- Emphasis of the fact that experience fine-tunes normal neural development
- Introduction of autism spectrum disorders and Asperger's syndrome
- Review of studies of the brain damage commonly associated with autism
- 66 new research citations

Chapter 10: Brain Damage and Neuroplasticity

- Presentation of evidence that concussion can have lasting neurological consequences
- Explanation of how viruses can be used to map neural circuits

- Systematic discussion of the epidemiology of multiple sclerosis
- Description of various causal factors in multiple sclerosis
- Emphasis on epigenetic mechanisms in neuropsychological disorders
- Discussion of treatments for multiple sclerosis
- More systematic coverage of the role of neuroplastic responses in recovery from brain damage
- 74 new research citations

Chapter 11: Learning, Memory, and Amnesia

- Introduction of remote memory, grid cells, and smart drugs
- Systematic coverage of place cells, grid cells, and head-direction cells in the hippocampus and entorhinal cortex
- Explanation of the shortcomings of the cognitive map theory of hippocampal function
- Critical review of the evidence on the effectiveness of smart drugs
- 39 new research citations

Chapter 12: Hunger, Eating, and Health

- Integration of research on hunger and eating with the thinking creatively theme
- Surgical treatment of extreme obesity with illustrations of two common procedures
- Discussion of the increases in obesity typically associated with aging
- New important point: the reality of neuroplasticity is more compatible with theories of hunger that emphasize adaptation rather than set points
- 29 new research citations

Chapter 13: Hormones and Sex

- New section on hormones and sexual differentiation of the brain
- Systematic evaluation of the aromatization hypothesis
- Discussion of the limitations of the female default theory of brain development
- Examples of the independence of masculinization and defeminization and of feminization and demasculinization
- Up-to-date summary of research on the relation between menstrual cycles and variations in the female libido
- Discussion of the role of dopamine in male reproductive behavior
- Increased emphasis on human sexual identity
- 39 new references

Chapter 14: Sleep, Dreaming, and Circadian Rhythms

- Explanation of why the findings from the comparative study of sleep, which has been based almost entirely on animals in captivity, can be misleading
- Updated coverage of the comparative study of sleep

- Increased coverage of the treatment of sleep disorders
- Review of research on natural short sleepers and long sleepers
- Reordering of sections for clearer presentation
- 63 new research citations

Chapter 15: Drug Addiction and the Brain's Reward Circuits

- Discussion of the teratogenic effects of nicotine and the development of nicotine addiction and its treatment with nicotine patches
- Discussion of the risks of moderate alcohol drinking
- Critical evaluation of the evidence that moderate drinking is good for the heart
- Description of the relationship between marijuana use and schizophrenia
- New brain scans of the loss of cortex in methamphetamine users
- Description of the teratogenic effects of stimulants
- New brain scans of changes of dopamine binding observed in cocaine users
- New section on current issues in addiction research
- Review of research on deaths attributable to heroin overdose
- 80 new research citations

Chapter 16: Lateralization, Language, and the Split Brain

- New section on the evolutionary perspective of cerebral lateralization and language
- New illustration of the chimeric figures test used with split-brain patients
- Updated coverage of structural brain asymmetry
- Reduced coverage of the evidence against the Wernicke-Geschwind model
- Discussion of why functional brain imaging evidence does not prove causation
- 29 new research citations

Chapter 17: Biopsychology of Emotion, Stress, and Health

- New illustration of Ekman's six primary facial expressions programmed on a digitized face
- New illustration of a false smile
- Emphasis on the fact that stress can have both deleterious and beneficial effects on health
- New section on the cognitive neuroscience of emotion
- New brain scans showing that empathy is associated with activity in areas of sensorimotor cortex
- New brain scan showing that the recognition of facial expressions produces greater activity in the right hemisphere
- New brain scans showing calcification of the amygdalae in twins with Urbach-Wiethe disease
- 39 new research citations

Chapter 18: Biopsychology of Psychiatric Disorders

- Nine new figures, mostly brain images
- New discussion of the problems of diagnosis
- Discussion of the higher incidence of schizophrenia among people whose mothers suffered the effects of famine while pregnant
- Discussion of the relationships among hallucinogenic drugs, serotonin, and schizophrenia
- Description of the developmental course of brain pathology in schizophrenic patients
- Increased coverage of the concept of positive and negative symptoms of schizophrenia
- Introduction of the term *mood disorder*
- Description of the physical health problems associated with affective disorders
- Up-to-date discussion of SSRIs and suicide
- Improved description of mood stabilizers
- Up-to-date coverage on the effectiveness of antidepressants
- Description of the pathology of the amygdala and the cingulate gyrus associated with bipolar depression
- Description of the use of chronic stimulation of the brain through implanted electrodes to treat bipolar depression
- Discussion of the comorbidity of anxiety and affective disorders
- Clarification of a common misunderstanding about rebound following the suppression of Tourette tics
- Discussion of sensorimotor cortex involvement in Tourette syndrome
- A letter from Tourette patient, P.H.
- 50 new research citations

Pedagogical Learning Aids

Biopsychology has several features expressly designed to help students learn and remember the material:

- **Scan Your Brain** study exercises appear within chapters at key transition points, where students can benefit most from pausing to consolidate material before continuing.

- **Think about It** discussion questions at the end of each chapter challenge students to think critically and creatively about the content.
- **Check It Out** demonstrations apply biopsychological phenomena and concepts so that students can experience them themselves.

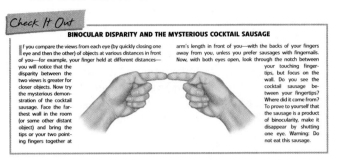

Check It Out

BINOCULAR DISPARITY AND THE MYSTERIOUS COCKTAIL SAUSAGE

If you compare the views from each eye (by quickly closing one eye and then the other) of objects at various distances in front of you—for example, your finger held at different distances— you will notice that the disparity between the two views is greater for closer objects. Now try the mysterious demonstration of the cocktail sausage. Face the farthest wall in the room (or some other distant object) and bring the tips or your two pointing fingers together at arm's length in front of you—with the backs of your fingers away from you, unless you prefer sausages with fingernails. Now, with both eyes open, look through the notch between your touching fingertips, but focus on the wall. Do you see the cocktail sausage between your fingertips? Where did it come from? To prove to yourself that the sausage is a product of binocularity, make it disappear by shutting one eye. Warning: Do not eat this sausage.

- **Themes Revisited** section at the end of each chapter summarizes the ways in which the book's four major themes relate to that chapter's subject matter.
- **Key Terms** appear in **boldface**, and other important terms of lesser significance appear in *italics*.
- **Appendixes** serve as convenient sources of additional information for students who want to expand their knowledge of selected biopsychology topics.

Ancillary Materials Available with *Biopsychology*

For Instructors

Pearson Education is pleased to offer the following supplements to qualified adopters.

Test Bank (0-205-03358-X) The test bank for the Eighth Edition of *Biopsychology* comprises more than 2,000 multiple-choice questions, including questions about accompanying brain images. The difficulty of each item is rated—easy (1), moderate (2), or difficult (3)—to assist instructors with test construction. Each item is also labeled with a topic and a page reference so that instructors can easily select appropriate questions for their tests. Textbook authors rarely prepare their own test banks; the fact that Pinel insists on preparing the *Biopsychology* test bank attests to its consistency with the text—and his commitment to helping students learn.

MyTest Test Bank (0-205-03359-8) This test bank is available in computerized format, which allows instructors to create and print quizzes and exams. Questions and tests can be authored online, allowing instructors maximum flexibility and the ability to efficiently manage assessments anytime, anywhere. Instructors can easily access existing questions and edit, create, and store questions using simple drag-and-drop controls. For more information, go to **www.PearsonMyTest.com**.

Instructor's Manual (0-205-03357-1) Skillfully prepared by Caroline Olko, Nassau Community College, the instructor's manual contains helpful teaching tools, including at-a-glance grids, activities and demonstrations for the classroom, handouts, lecture notes, chapter outlines, and other valuable course organization material for new and experienced instructors.

NEW! Interactive PowerPoint Slides These slides, available on the Instructor's DVD (0-205-03361-X), bring highlights of this edition of *Biopsychology* right into the classroom, drawing students into the lecture and providing engaging interactive activities, visuals, and videos.

Standard Lecture PowerPoint Slides (0-205-03639-2) Created by Jeffrey Grimm, Western Washington University, these slides have a more traditional format, with excerpts of the text material and artwork, and are also available on the Instructor's DVD (0-205-15055-1) as well as online at **www.pearsonhighered.com/irc**.

Films for the Humanities and Sciences Video for Biological Psychology (0-205-31913-0) This 60-minute biopsychology videotape is available to adopters of *Biopsychology*. Based on the *Films for the Humanities* series, this video provides students with glimpses of important biopsychological phenomena such as sleep recording, axon growth, memory testing in monkeys, the formation of synapses, gender differences in brain structure, human amnesic patients, rewarding brain stimulation, and brain scans.

For Students

NEW! MyPsychLab (www.mypsychlab.com) MyPsychLab, a new online student resource that replaces the CD-ROM available with previous editions of *Biopsychology*, provides a wealth of study tools for students looking to clarify and deepen their understanding of biopsychology concepts. MyPsychLab contains animations and videos, many of which were specifically designed to support, expand upon, and complement *Biopsychology*. Students can immerse themselves in interactive environments using BioFlix, a tutoring tool composed of 3-minute animations and accompanying PowerPoint® slideshows, as well as the Virtual Brain, an interactive diagram of the brain assembled with major content contributions from Deborah Carroll, PhD and Professor of Psychology at Southern Connecticut State University. The Virtual Brain contains a 3D interactive brain, assessments, and real-life scenarios. MyPsychLab also presents a set of self-scoring practice tests, interactive glossary flashcards, and access to MySearchLab, a resource that can help students with all parts of the writing process for research papers. MyPsychLab test questions authored by Ginger LeBlanc, Bakersfield College. Instructions on how to log onto MyPsychLab can be found at **www.mypsychlab.com**.

Study Card for Physiological Psychology (0-205-45346-5) Colorful, affordable, and packed with useful information, study cards make studying easier, more efficient, and more enjoyable. Course information is distilled down to the basics, helping students quickly master the fundamentals, review for understanding, or prepare for an exam.

A Colorful Introduction to the Anatomy of the Human Brain, Second Edition (0-205-54874-1) This book, written by John P. J. Pinel and Maggie Edwards, provides an easy and enjoyable means of learning or reviewing the fundamentals of human neuroanatomy through the acclaimed directed-coloring method.

Acknowledgments

I wrote *Biopsychology*, but Maggie Edwards took the responsibility for all other aspects of the manuscript and media preparation—Maggie is a talented artist and technical writer, and my partner in life. I am grateful for her encouragement and support and for her many contributions to this book. I also thank her on behalf of the many students who will benefit from her efforts.

And another special thank you goes to Steven Barnes. Several important additions to this edition were triggered by his insightful suggestions, and he stepped up and provided support and extra help when they were needed.

Allyn & Bacon did a remarkable job of producing this book. They shared my dream of a textbook that meets the highest standards of pedagogy but is also personal, attractive, and enjoyable. Thank you to Bill Barke, Stephen Frail, Susan Hartman, and other executives at Allyn & Bacon for having faith in *Biopsychology* and providing the financial and personal support necessary for it to stay at the forefront of its field. A special thank-you goes to Erin K. L. Grelak for her development assistance, her moral support, and her willingness to put up with our eccentricities. Special thanks also go to Roberta Sherman and Jane Hoover for coordinating the production—an excruciatingly difficult and often thankless job. Jane was also the copyeditor, making many improvements in the text and art, which were greatly appreciated. And many thanks to Maggie Edwards and Steven Barnes for compiling the reference list, the prototypical mind-numbing task.

I thank the following instructors for providing me with reviews of various editions of *Biopsychology*. Their comments have contributed substantially to the evolution of this edition:

L. Joseph Acher, Baylor University
Nelson Adams, Winston-Salem State University
Michael Babcock, Montana State University–Bozeman
Ronald Baenninger, College of St. Benedict
Carol Batt, Sacred Heart University
Noel Jay Bean, Vassar College

Danny Benbasset, George Washington University
Thomas Bennett, Colorado State University
Linda Brannon, McNeese State University
Peter Brunjes, University of Virginia
Michelle Butler, United States Air Force Academy
Donald Peter Cain, University of Western Ontario
Deborah A. Carroll, Southern Connecticut State University
John Conklin, Camosun College
Michael A. Dowdle, Mt. San Antonio College
Doug Engwall, Central Connecticut State University
Gregory Ervin, Brigham Young University
Robert B. Fischer, Ball State University
Allison Fox, University of Wollongong
Ed Fox, Purdue University
Thomas Goettsche, SAS Institute, Inc.
Arnold M. Golub, California State University--Sacramento
Mary Gotch, Solano College
Jeffrey Grimm, Western Washington University
Kenneth Guttman, Citrus College
Melody Smith Harrington, St. Gregory's University
Theresa D. Hernandez, University of Colorado
Cindy Ellen Herzog, Frostburg State University
Peter Hickmott, University of California–Riverside
Tony Jelsma, Atlantic Baptist University
Roger Johnson, Ramapo College
John Jonides, University of Michigan
Jon Kahane, Springfield College
Craig Kinsley, University of Richmond
Ora Kofman, Ben-Gurion University of the Negev
Louis Koppel, Utah State University
Maria J. Lavooy, University of Central Florida
Victoria Littlefield, Augsburg College
Linda Lockwood, Metropolitan State College of Denver
Charles Malsbury, Memorial University
Michael R. Markham, Florida International University
Michael P. Matthews, Drury College
Lin Meyers, California State University–Stanislaus
Russ Morgan, Western Illinois University
Henry Morlock, SUNY–Plattsburgh
Caroline Olko, Nassau Community College
Lauretta Park, Clemson University
Ted Parsons, University of Wisconsin–Platteville
Jim H. Patton, Baylor University
Edison Perdorno, Minnesota State University
Michael Peters, University of Guelph
Michelle Pilati, Rio Hondo College
Joseph H. Porter, Virginia Commonwealth University
David Robbins, Ohio Wesleyan University
Dennis Rodriguez, Indiana University–South Bend
Margaret G. Ruddy, College of New Jersey
Jeanne P. Ryan, SUNY–Plattsburgh
Jerome Siegel, David Geffen School of Medicine, UCLA

Patti Simone, Santa Clara University
Ken Sobel, University of Central Arkansas
David Soderquist, University of North Carolina at
    Greensboro
Michael Stoloff, James Madison University
Stuart Tousman, Rockford College
Dallas Treit, University of Alberta
Margaret Upchurch, Transylvania University

Dennis Vincenzi, University of Central Florida
Ashkat Vyas, Hunter College
Charles Weaver, Baylor University
Linda Walsh, University of Northern Iowa
David Widman, Juniata College
Jon Williams, Kenyon College
David Yager, University of Maryland
H.P. Ziegler, Hunter College

To the Student

In the 1960s, I was, in the parlance of the times, "turned on" by an undergraduate course in biopsychology. I could not imagine anything more interesting than a field of science dedicated to studying the relation between psychological processes and the brain. My initial fascination led to a long career as a student, researcher, teacher, and writer of biopsychological science. *Biopsychology* is my attempt to share my fascination with you.

I have tried to make *Biopsychology* a different kind of textbook, a textbook that includes clear, concise, and well-organized explanations of the key points but is still interesting to read—a book from which you might suggest a suitable chapter to an interested friend or relative. To accomplish this goal, I thought about what kind of textbook I would have liked when I was a student, and I decided immediately to avoid the stern formality and ponderous style of conventional textbook writing.

I wanted *Biopsychology* to have a relaxed and personal style. In order to accomplish this, I imagined that you and I were chatting as I wrote, and that I was telling you— usually over a glass of something—about the interesting things that go on in the field of biopsychology. Imagining these chats kept my writing from drifting back into conventional "textbookese," and it never let me forget that I was writing this book for you.

I am particularly excited, and a bit nervous, about this edition of *Biopsychology*. This edition marks the first time that I am trying to share something about biopsychological research that has fascinated me throughout my career: Often science and creativity are considered to be opposites, but in my experience many of the major advances in biopsychological science have resulted from creative thinking. These major advances have been made by biopsychologists who have recognized that there are alternatives to the conventional ways of thinking about biopsychological issues that have been engrained in them by their culture and training and have adopted creative new approaches. Two things in particular have fascinated me about the interplay between creative thinking and biopsychological science: how difficult it is to identify and shed conventional approaches even when they clearly haven't been working, and how often solutions to long-standing problems become apparent when they are approached from a new perspective. I hope that my focus on creative thinking makes the study of biopsychology more interesting for you and that it helps you become a more creative thinker.

I hope that *Biopsychology* teaches you much, and that reading it generates in you the same positive feelings that writing it did in me. If you are so inclined, I welcome your comments and suggestions. You can contact me at the Department of Psychology, University of British Columbia, Vancouver, BC, Canada, V6T 1Z4, or at the following e-mail address:

jpinel@psych.ubc.ca

About the Author

John Pinel, the author of *Biopsychology*, obtained his Ph.D. from McGill University in Montreal and worked briefly at the Massachusetts Institute of Technology before taking a faculty position at the University of British Columbia in Vancouver, where he is currently Professor Emeritus. Professor Pinel is an award-winning teacher and the author of over 200 scientific papers. However, he feels that *Biopsychology* is his major career-related accomplishment: "It ties together everything I love about my job: students, teaching, writing, and research."

Pinel attributes much of his success to his wife, Maggie, who is an artist and professional designer. Over the years, they have collaborated on many projects, and the high quality of *Biopsychology*'s illustrations is largely attributable to her skill and effort.

Pinel is an enthusiastic West African drummer who performs at local clubs, festivals, and drum circles with Nigerian drum master Kwasi Iruoje. For relaxation, his favorite pastime is cuddling his three cats: Rastaman, Sambala, and Squeak.

1

Biopsychology as a Neuroscience
What Is Biopsychology, Anyway?

The appearance of the human brain is far from impressive (see Figure 1.1). The human brain is a squishy, wrinkled, walnut-shaped hunk of tissue weighing about 1.3 kilograms. It looks more like something that you might find washed up on a beach than like one of the wonders of the world—which it surely is. Despite its disagreeable external appearance, the human brain is an amazingly intricate network of **neurons** (cells that receive and transmit electrochemical signals). Contemplate for a moment the complexity of your own brain's neural circuits. Consider the 100 billion neurons in complex array (see Azevedo et al., 2009), the estimated 100 trillion connections among them, and the almost infinite number of paths that neural signals can follow through this morass. The complexity of the human brain is hardly surprising, considering what it can do. An organ capable of creating a *Mona Lisa,* an artificial limb, and a supersonic aircraft; of

> **Simulate**
> How Neurons Work
> **www.mypsychlab.com**

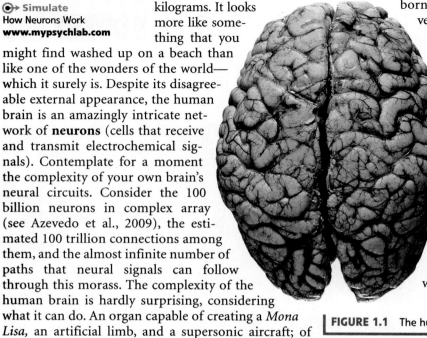

FIGURE 1.1   The human brain.

traveling to the moon and to the depths of the sea; and of experiencing the wonders of an alpine sunset, a newborn infant, and a reverse slam dunk *must* be complex.

> **Simulate**
> Visit the Virtual Brain, an interactive tool that allows you to explore Biopsychology through simulations, videos, and quizzes.
> **www.mypsychlab.com**

Paradoxically, **neuroscience** (the scientific study of the nervous system) may prove to be the brain's ultimate challenge: Does the brain have the capacity to understand something as complex as itself?

Neuroscience comprises several related disciplines. The primary purpose of this chapter is to introduce you to one of them: biopsychology. Each of this chapter's seven sections characterizes the neuroscience of biopsychology in a different way.

Before you proceed to the body of this chapter, I would like to tell you about two things: (1) the case of Jimmie G., which will give you a taste of the interesting things that lie ahead, and (2) the major themes of this book.

The Case of Jimmie G., the Man Frozen in Time

Jimmie G. was a good-looking, friendly 49-year-old. He liked to talk about his school days and his experiences in the navy, which he was able to describe in detail. Jimmie was an intelligent man with superior abilities in math and science. In fact, it was not readily apparent why he was a resident of a neurological ward.

When Jimmie talked about his past, there was a hint of his problem. When he talked about his school days, he used the past tense; when he recounted his early experiences in the navy, however, he switched to the present tense. More worrisome was that he never talked about anything that happened to him after his time in the navy.

Jimmie G. was tested by eminent neurologist Oliver Sacks, and a few simple questions revealed a curious fact: The 49-year-old patient believed that he was 19. When he was asked to describe what he saw in a mirror, Jimmie became so frantic and confused that Dr. Sacks immediately took the mirror out of the room.

Returning a few minutes later, Dr. Sacks was greeted by a once-again cheerful Jimmie, who acted as if he had never seen Sacks before. Indeed, even when Sacks suggested that they had met recently, Jimmie was certain that they had not.

Then Dr. Sacks asked where Jimmie thought he was. Jimmie replied that all the beds and patients made him think that the place was a hospital. But he couldn't understand why he would be in a hospital. He was afraid that he might have been admitted because he was sick, but didn't know it.

Further testing confirmed what Dr. Sacks feared. Although Jimmie had good sensory, motor, and cognitive abilities, he had one terrible problem: He forgot everything that was said or shown to him within a few seconds. Basically, Jimmie could not remember anything that had happened to him since his early 20s, and he was not going to remember anything that happened to him for the rest of his life. Sacks was stunned by the implications of Jimmie's condition.

Jimmie G.'s situation was heart-wrenching. Unable to form new lasting memories, he was, in effect, a man frozen in time, a man without a recent past and no prospects for a future, stuck in a continuous present, lacking any context or meaning.

("The Case of Jimmie G., the Man Frozen in Time," reprinted with the permission of Simon & Schuster, Inc. and Pan Macmillan, London from *The Man Who Mistook His Wife for a Hat and Other Clinical Tales* by Oliver Sacks. Copyright © 1970, 1981, 1983, 1984, 1986 by Oliver Sacks. Electronic rights with permission of the Wylie Agency.)

Remember Jimmie G.; you will encounter him again, later in this chapter.

Four Major Themes of This Book

You will learn many new facts in this book—new findings, concepts, terms, and the like. But more importantly, many years from now, long after you have forgotten most of those facts, you will still be carrying with you productive new ways of thinking. I have selected four of these for special emphasis: They are the major themes of this book.

◉ Watch
Watch the author explain how and why the themes of *Biopsychology* have been highlighted for your convenience.
www.mypsychlab.com

To help you give these themes the special attention they deserve and to help you follow their development as you progress though the book, I have marked relevant passages with tabs. The following are the four major themes and their related tabs.

Thinking Creatively about Biopsychology   Because many biopsychological topics are so interesting (as you have already seen in the case of Jimmie G.) and relevant to everyday life, we are fed a steady diet of biopsychological information and opinion—by television, newspapers, the Internet, friends, relatives, books, teachers, etc. As a result, you almost certainly hold strong views, based on conventional wisdom, about many of the topics that you are going to encounter in this book. Because these preconceptions are shared by many biopsychological researchers, they have often impeded scientific progress. But, some of the most important advances in biopsychological science have been made by researchers who have managed to overcome the restrictive effects of conventional thinking and have taken creative new approaches that are consistent with the evidence. Indeed, **thinking creatively** (thinking in productive, unconventional ways) is the cornerstone of any science. The thinking creatively

Thinking Creatively

tab marks points in the text where I describe research that involves thinking "outside the box," where I have tried to be creative myself in the analysis of the research that I am presenting, and where I encourage you to base your thinking on the evidence rather than on widely accepted views.

Clinical Implications   Clinical (pertaining to illness or treatment) considerations are woven through the fabric of biopsychology. There are two aspects to clinical impli-

Clinical Implications

cations: Much of what biopsychologists learn about the functioning of the normal brain comes from studying the diseased or damaged brain; and, conversely, much of what biopsychologists discover has relevance for the treatment of brain disorders.

This book focuses on the interplay between brain dysfunction and biopsychology, and each major example of that interplay is highlighted by a clinical implications tab.

The Evolutionary Perspective   Although the events that led to the evolution of the human species can never be determined with certainty, thinking of the environmental pressures that likely led to the evolution of our brains and behavior often leads to important biopsychological insights. This approach is called the **evolutionary perspective**. An important component of the evolutionary

Evolutionary Perspective

perspective is the *comparative approach* (trying to understand biological phenomena by comparing them in different species). You will learn throughout the text that we humans have learned much about ourselves by studying species that are related to us through evolution. The evolutionary approach has proven to be one of the cornerstones of modern biopsychological inquiry. Each discussion that relates to the evolutionary perspective is marked by an evolutionary perspective tab. A closely related topic, the genetics of behavior, is covered in Chapter 2.

Neuroplasticity   Until the early 1990s, most neuroscientists thought of the brain as a three-dimensional array of neural elements "wired" together in a massive network of circuits. The complexity of this "wiring diagram" of the brain was staggering, but it failed to capture one of the brain's most important features. In the last two decades, research has clearly demonstrated that the adult brain is not

Neuroplasticity

a static network of neurons: It is a *plastic* (changeable) organ that continuously grows and changes in response to the individual's genes and experiences. The discovery of neuroplasticity, arguably the single most influential discovery in modem neuroscience, is currently influencing many areas of biopsychological research. Each example of such research is marked in this book by a neuroplasticity tab.

1.1

What Is Biopsychology?

Biopsychology is the scientific study of the biology of behavior—see Dewsbury (1991). Some refer to this field as *psychobiology*, *behavioral biology*, or *behavioral neuroscience;* but I prefer the term biopsychology because it denotes a biological approach to the study of psychology rather than a psychological approach to the study of biology: Psychology commands center stage in this text. *Psychology* is the scientific study of behavior—the scientific study of all overt activities of the organism as well as all the internal processes that are presumed to underlie them (e.g., learning, memory, motivation, perception, and emotion).

The study of the biology of behavior has a long history, but biopsychology did not develop into a major neuroscientific discipline until the 20th century. Although it is not possible to specify the exact date of biopsychology's birth, the publication of *The Organization of Behavior* in 1949 by D. O. Hebb played a key role in its emergence (see Brown & Milner, 2003; Cooper, 2005; Milner, 1993). In his book, Hebb developed the first comprehensive theory of how complex psychological phenomena, such as perceptions, emotions, thoughts, and memories, might be produced by brain activity. Hebb's theory did much to discredit the view that psychological functioning is too complex to have its roots in the physiology and chemistry of the brain. Hebb based his theory on experiments involving both humans and laboratory animals, on clinical case studies, and on logical arguments developed from his own insightful observations of daily life. This eclectic approach has become a hallmark of biopsychological inquiry.

In comparison to physics, chemistry, and biology, biopsychology is an infant—a healthy, rapidly growing infant, but an infant nonetheless. In this book, you will reap the benefits of biopsychology's youth. Because biopsychology does not have a long and complex history, you will be able to move directly to the excitement of current research.

1.2

What Is the Relation between Biopsychology and the Other Disciplines of Neuroscience?

Neuroscience is a team effort, and biopsychologists are important members of the team (see Albright, Kandel, & Posner, 2000; Kandel & Squire, 2000). Biopsychology can be further defined by its relation to other neuroscientific disciplines.

Biopsychologists are neuroscientists who bring to their research a knowledge of behavior and of the methods of behavioral research. It is their behavioral orientation and expertise that make their contribution to neuroscience unique (see Cacioppo & Decety, 2009). You will be able to better appreciate the importance of this contribution if you consider that the ultimate purpose of the nervous system is to produce and control behavior (see Grillner & Dickinson, 2002).

Biopsychology is an integrative discipline. Biopsychologists draw together knowledge from the other neuroscientific disciplines and apply it to the study of behavior. The following are a few of the disciplines of neuroscience that are particularly relevant to biopsychology:

Neuroanatomy. The study of the structure of the nervous system (see Chapter 3).

Neurochemistry. The study of the chemical bases of neural activity (see Chapter 4).

Neuroendocrinology. The study of interactions between the nervous system and the endocrine system (see Chapters 13 and 17).

Neuropathology. The study of nervous system disorders (see Chapter 10).

Neuropharmacology. The study of the effects of drugs on neural activity (see Chapters 4, 15, and 18).

Neurophysiology. The study of the functions and activities of the nervous system (see Chapter 4).

1.3

What Types of Research Characterize the Biopsychological Approach?

Although biopsychology is only one of many disciplines that contribute to neuroscience, it is itself broad and diverse. Biopsychologists study many different phenomena, and they approach their research in many different ways. In order to characterize biopsychological research, this section discusses three major dimensions along which approaches to biopsychological research vary. Biopsychological research can involve either human or nonhuman subjects; it can take the form of either formal experiments or nonexperimental studies; and it can be either pure or applied.

Human and Nonhuman Subjects

Both human and nonhuman animals are the subject of biopsychological research. Of the nonhumans, rats are the most common subjects; however, mice, cats, dogs, and nonhuman primates are also widely studied.

Humans have several advantages over other animals as experimental subjects of biopsychological research: They can follow instructions, they can report their subjective experiences, and their cages are easier to clean. Of course, I am joking about the cages, but the joke does serve to draw attention to one advantage that humans have over other species of experimental subjects: Humans are often cheaper. Because only the highest standards of animal care are acceptable, the cost of maintaining an animal laboratory can be prohibitive for all but the most well-funded researchers.

Of course, the greatest advantage that humans have as subjects in a field aimed at understanding the intricacies of human brain function is that they have human brains. In fact, you might wonder why biopsychologists would bother studying nonhuman subjects at all. The answer lies in the evolutionary continuity of the brain. The brains

Evolutionary Perspective

of humans differ from the brains of other mammals primarily in their overall size and the extent of their cortical development. In other words, the differences between the brains of humans and those of related species are more quantitative than qualitative, and thus many of the principles of human brain function can be clarified by the study of nonhumans (see Nakahara et al., 2002; Passingham, 2009; Platt & Spelke, 2009).

Conversely, nonhuman animals have three advantages over humans as subjects in biopsychological research. The first is that the brains and behavior of nonhuman subjects are simpler than those of human subjects. Hence, the study of nonhuman species is more likely to reveal fundamental brain–behavior interactions. The second advantage is that insights frequently arise from the **comparative approach**, the study of biological processes by comparing different species. For example, comparing the behavior of species that do not have a cerebral cortex with the behavior of species that do can provide valuable clues about cortical function. The third advantage is that it is possible to conduct research on laboratory animals that, for ethical reasons, is not possible with human subjects. This is not to say that the study of nonhuman animals is not governed by a strict code of ethics (see Demers et al., 2006; Goldberg & Hartung, 2006)—it is. However, there are fewer ethical constraints on the study of laboratory species than on the study of humans.

In my experience, most biopsychologists display considerable concern for their subjects, whether they are of their own species or not; however, ethical issues are not left to the discretion of the individual researcher. All biopsychological research, whether it involves human or nonhuman subjects, is regulated by independent committees according to strict ethical guidelines: "Researchers cannot escape the logic that if the animals we observe are reasonable models of our own most intricate actions, then they must be respected as we would respect our own sensibilities" (Ulrich, 1991, p. 197).

If you are concerned about the ethics of biopsychological research on nonhuman species, be sure to read the Check It Out feature.

Experiments and Nonexperiments

Biopsychological research involves both experiments and nonexperimental studies. Two common types of nonexperimental studies are quasiexperimental studies and case studies.

Experiments The experiment is the method used by scientists to study causation, that is, to find out what causes what. As such, it has been almost single-handedly responsible for the knowledge that is the basis for our modern way of life. It is paradoxical that a method capable of such complex feats is itself so simple. To conduct an experiment involving living subjects, the experimenter

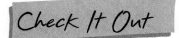

THE ETHICS OF BIOPSYCHOLOGICAL RESEARCH ON NONHUMAN ANIMALS

Ethical issues are never straightforward. Is the stress on nonhuman animals caused by being research subjects outweighed by the potential benefits of the research? Web sites addressing this question fall into one of two categories: those that argue that the ethics of research on nonhuman animals is a complex issue that needs careful consideration, and those that oppose the use of nonhuman animals as research subjects under any circumstance.

I do not want to influence your position on this question; I am still wrestling with aspects of it myself. However, I think that it is important to warn you against people who take extreme positions on difficult and complex issues. Often their positions are based on emotion and misconception rather than on a careful consideration of the evidence. Please check out sites from the following two lists.

For the Responsible Use of Nonhuman Animals in Research

American Psychological Association (APA): Guidelines for Ethical Conduct in the Care and Use of Animals
www.apa.org/science/leadership/care/guidelines.aspx

Understanding Animal Research
www.understandinganimalresearch.org.uk

Animal Research for Life
www.animalresearchforlife.eu

American Physiological Society (APS): Animal Research
www.the-aps.org/pa/policy/animals/intro.htm

Against the Use of Nonhuman Animals in Research Under Any Circumstance

Uncaged: Protecting Animals
www.uncaged.co.uk

National Anti-Vivisection Society (NAVS)
www.navs.org

Stop Animal Tests
www.stopanimaltests.com

first designs two or more conditions under which the subjects will be tested. Usually, a different group of subjects is tested under each condition (**between-subjects design**), but sometimes it is possible to test the same group of subjects under each condition (**within-subjects design**). The experimenter assigns the subjects to conditions, administers the treatments, and measures the outcome in such a way that there is only one relevant difference between the conditions that are being compared. This difference between the conditions is called the **independent variable**. The variable that is measured by the experimenter to assess the effect of the independent variable is called the **dependent variable**. If the experiment is done correctly, any differences in the dependent variable between the conditions must have been caused by the independent variable.

Why is it critical that there be no differences between conditions other than the independent variable? The reason is that when there is more than one difference that could affect the dependent variable, it is difficult to determine whether it was the independent variable or the unintended difference—called a **confounded variable**—that led to the observed effects on the dependent variable. Although the experimental method is conceptually simple, eliminating all confounded variables can be quite difficult. Readers of research papers must be constantly on the alert for confounded variables that have gone unnoticed by the experimenters themselves.

An experiment by Lester and Gorzalka (1988) illustrates the experimental method in action. The experiment was a demonstration of the **Coolidge effect**. The Coolidge effect is the fact that a copulating male who becomes incapable of continuing to copulate with one sex partner can often recommence copulating with a new sex partner (see Figure 1.2). Before your imagination starts running wild, I should mention that the subjects in Lester and Gorzalka's experiment were hamsters, not students from the undergraduate subject pool.

Lester and Gorzalka argued that the Coolidge effect had not been demonstrated in females because it is more difficult to conduct well-controlled Coolidge-effect experiments with females—not because females do not display a Coolidge effect. The confusion, according to Lester and Gorzalka, stemmed from the fact that the males of most mammalian species become sexually fatigued more readily than do the females. As a result, attempts to demonstrate the Coolidge effect in females are often confounded by the fatigue of the males. When, in the midst of copulation, a female is provided with a new sex partner, the increase in her sexual receptivity could be either a legitimate Coolidge effect or a reaction to the greater vigor of the new male. Because female mammals usually display little sexual fatigue, this confounded variable is not a serious problem in demonstrations of the Coolidge effect in males.

Lester and Gorzalka devised a clever new procedure to control for this confounded variable. At the same time that a female subject was copulating with one male (the familiar male), the other male to be used in the test (the *Thinking Creatively* unfamiliar male) was copulating with another female. Then, both males were given a rest while the female was copulating with a third male. Finally, the female subject was tested with either the familiar male or the unfamiliar male. The dependent variable was the amount of

FIGURE 1.2 President Calvin Coolidge and Mrs. Grace Coolidge. Many students think that the Coolidge effect is named after a biopsychologist named Coolidge. In fact, it is named after President Calvin Coolidge, of whom the following story is told. (If the story isn't true, it should be.) During a tour of a poultry farm, Mrs. Coolidge inquired of the farmer how his farm managed to produce so many eggs with such a small number of roosters. The farmer proudly explained that his roosters performed their duty dozens of times each day.

"Perhaps you could point that out to Mr. Coolidge," replied the First Lady in a pointedly loud voice.

The President, overhearing the remark, asked the farmer, "Does each rooster service the same hen each time?"

"No," replied the farmer, "there are many hens for each rooster."

"Perhaps you could point that out to Mrs. Coolidge," replied the President.

time that the female displayed **lordosis** (the arched-back, rump-up, tail-diverted posture of female rodent sexual receptivity) during each sex test. As Figure 1.3 illustrates, the females responded more vigorously to the unfamiliar males than they did to the familiar males during the third test, despite the fact that both the unfamiliar and familiar males were equally fatigued and both mounted the females with equal vigor. This experiment illustrates the importance of good experimental design as well as making a point that you will encounter in Chapter 13: that males and females are more similar than many people appreciate.

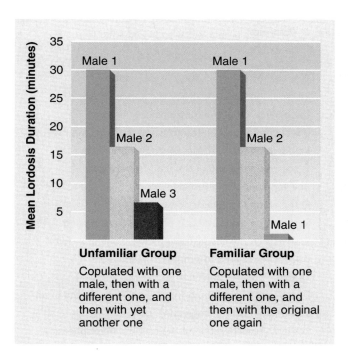

FIGURE 1.3 The experimental design and results of Lester and Gorzalka (1988). On the third test, the female hamsters were more sexually receptive to an unfamiliar male than they were to the male with which they had copulated on the first test.

Quasiexperimental Studies It is not possible for biopsychologists to bring the experimental method to bear on all problems of interest to them. There are frequently physical or ethical impediments that make it impossible to assign subjects to particular conditions or to administer the conditions once the subjects have been assigned to them. For example, experiments on the causes of brain damage in human alcoholics are not feasible because it would not be ethical to assign a subject to a condition that involves years of alcohol consumption. (Some of you may be more concerned about the ethics of assigning subjects to a control condition that involves years of sobriety.) In such prohibitive situations, biopsychologists sometimes conduct **quasiexperimental studies**—studies of groups of subjects who have been exposed to the conditions of interest in the real world. These studies have the appearance of experiments, but they are not true experiments because potential confounded variables have not been controlled—for example, by the random assignment of subjects to conditions.

In one quasiexperimental study, a team of researchers compared 100 detoxified male alcoholics from an alcoholism treatment unit with 50 male nondrinkers obtained from various sources (Acker et al., 1984). The alcoholics as a group performed more poorly on various tests of perceptual, motor, and cognitive ability, and their brain scans revealed extensive

brain damage. Although this quasiexperimental study seems like an experiment, it is not. Because the participants themselves decided which group they would be in—by drinking alcohol or not—the researchers had no means of ensuring that exposure to alcohol was the only variable that distinguished the two groups. Can you think of differences other than exposure to alcohol that could reasonably be expected to exist between a group of alcoholics and a group of abstainers—differences that could have contributed to the neuroanatomical or intellectual differences that were observed between them? There are several. For example, alcoholics as a group tend to be more poorly educated, more prone to accidental head injury, more likely to use other drugs, and more likely to have poor diets. Accordingly, quasiexperimental studies have revealed that alcoholics tend to have more brain damage than nonalcoholics, but such studies have not indicated why.

Have you forgotten Jimmie G.? His condition was a product of long-term alcohol consumption.

Case Studies Studies that focus on a single case or subject are called **case studies**. Because they focus on a single case, they often provide a more in-depth picture than that provided by an experiment or a quasiexperimental study, and they are an excellent source of testable hypotheses. However, there is a major problem with all case studies: their **generalizability**—the degree to which their results can be applied to other cases. Because humans differ from one another in both brain function and behavior, it is important to be skeptical of any biopsychological theory based entirely on a few case studies.

Pure and Applied Research

Biopsychological research can be either pure or applied. Pure research and applied research differ in a number of respects, but they are distinguished less by their own attributes than by the motives of the individuals involved in their pursuit. **Pure research** is research motivated primarily by the curiosity of the researcher—it is done solely for the purpose of acquiring knowledge. In contrast, **applied research** is research intended to bring about some direct benefit to humankind.

Many scientists believe that pure research will ultimately prove to be of more practical benefit than applied research. Their view is that applications flow readily from an understanding of basic principles and that attempts to move directly to application without first gaining a basic understanding are shortsighted. Of course, it is not necessary for a research project to be completely pure or completely applied; many research programs have elements of both approaches.

One important difference between pure and applied research is that pure research is more vulnerable to the vagaries of political regulation because politicians and the

TABLE 1.1   Nobel Prizes Specifically Related to the Nervous System or Behavior

Nobel Winner	Date	Accomplishment
Ivan Pavlov	1904	Research on the physiology of digestion
Camillo Golgi and Santiago Romón y Cajal	1906	Research on the structure of the nervous system
Charles Sherrington and Edgar Adrian	1932	Discoveries about the functions of neurons
Henry Dale and Otto Loewi	1936	Discoveries about the transmission of nerve impulses
Joseph Erlanger and Herbert Gasser	1944	Research on the functions of single nerve fibers
Walter Hess	1949	Research on the role of the brain in behavior
Egas Moniz	1949	Development of prefrontal lobotomy
Georg von Békésy	1961	Research on the auditory system
John Eccles, Alan Hodgkin, and Andrew Huxley	1963	Research on the ionic basis of neural transmission
Ragnor Granit, Haldan Hartline, and George Wald	1967	Research on the chemistry and physiology of vision
Bernard Katz, Ulf von Euler, and Julius Axelrod	1970	Discoveries related to synaptic transmission
Karl Von Frisch, Konrad Lorenz, and Nikolass Tinbergen	1973	Studies of animal behavior
Roger Guillemin and Andrew Schally	1977	Discoveries related to hormone production by the brain
Herbert Simon	1979	Research on human cognition
Roger Sperry	1981	Research on separation of the cerebral hemispheres
David Hubel and Torsten Wiesel	1981	Research on neurons of the visual system
Rita Levi-Montalcini and Stanley Cohen	1986	Discovery and study of nerve growth factors
Erwin Neher and Bert Sakmann	1991	Research on ion channels
Alfred Gilman and Martin Rodbell	1994	Discovery of G-protein–coupled receptors
Arvid Carlsson, Paul Greengard, and Eric Kandel	2000	Discoveries related to synaptic transmission
Linda Buck and Richard Axel	2004	Research on the olfactory system

voting public have difficulty understanding why research of no immediate practical benefit should be supported. If the decision were yours, would you be willing to grant hundreds of thousands of dollars to support the study of squid *motor neurons* (neurons that control muscles), learning in recently hatched geese, the activity of single nerve cells in the visual systems of monkeys, the hormones released by the *hypothalamus* (a small neural structure at the base of the brain) of pigs and sheep, or the function of the *corpus callosum* (the large neural pathway that connects the left and right halves of the brain)? Which, if any, of these projects would you consider worthy of support? Each of these seemingly esoteric projects was supported, and each earned a Nobel Prize for the scientist(s) involved.

Table 1.1 lists some of the Nobel Prizes awarded for research related to the brain and behavior (see Benjamin, 2003). The purpose of this list is to give you a general sense of the official recognition that behavioral and brain research has received, not to have you memorize the list. You will learn later in the chapter that, when it comes to evaluating science, the Nobel Committee has not been infallible.

⊙→ Simulate
History of the Brain
www.mypsychlab.com

1.4

What Are the Divisions of Biopsychology?

As you have just learned, biopsychologists conduct their research in a variety of fundamentally different ways. Biopsychologists who take the same approaches to their research tend to publish their research in the same journals, attend the same scientific meetings, and belong to the same professional societies. The particular approaches to biopsychology that have flourished and grown have gained wide recognition as separate divisions of biopsychological research. The purpose of this section of the chapter is to give you a clearer sense of biopsychology and its diversity by describing six of its major divisions: (1) physiological psychology, (2) psychopharmacology, (3) neuropsychology, (4) psychophysiology, (5) cognitive neuroscience, and (6) comparative psychology. For simplicity, they are presented as distinct approaches; but there is much overlap among them, and many biopsychologists regularly follow more than one approach.

Physiological Psychology

Physiological psychology is the division of biopsychology that studies the neural mechanisms of behavior through the direct manipulation of the brain in controlled experiments—surgical and electrical methods of brain manipulation are most common. The subjects of physiological psychology research are almost always laboratory animals, because the focus on direct brain manipulation and controlled experiments precludes the use of human subjects in most instances. There is also a tradition of pure research in physiological psychology; the emphasis is usually on research that contributes to the development of theories of the neural control of behavior rather than on research that is of immediate practical benefit.

Psychopharmacology

Psychopharmacology is similar to physiological psychology, except that it focuses on the manipulation of neural activity and behavior with drugs. In fact, many of the early psychopharmacologists were simply physiological psychologists who moved into drug research, and many of today's biopsychologists identify closely with both approaches. However, the study of the effects of drugs on the brain and behavior has become so specialized that psychopharmacology is regarded as a separate discipline. A substantial portion of psychopharmacological research is applied. Although drugs are sometimes used by psychopharmacologists to study the basic principles of brain–behavior interaction, the purpose of many psychopharmacological experiments is to develop therapeutic drugs (see Chapter 18) or to reduce drug abuse (see Chapter 15). Psychopharmacologists study the effects of drugs on laboratory species—and on humans, if the ethics of the situation permits it.

Clinical Implications

Neuropsychology

Neuropsychology is the study of the psychological effects of brain damage in human patients. Obviously, human subjects cannot ethically be exposed to experimental treatments that endanger normal brain function. Consequently, neuropsychology deals almost exclusively with case studies and quasiexperimental studies of patients with brain damage resulting from disease, accident, or neurosurgery. The outer layer of the cerebral hemispheres—the **cerebral cortex**—is most likely to be damaged by accident or surgery; this is one reason why neuropsychology has focused on this important part of the human brain.

Clinical Implications

Neuropsychology is the most applied of the biopsychological subdisciplines; the neuropsychological assessment of human patients, even when part of a program of pure research, is always done with an eye toward benefiting them in some way. Neuropsychological tests facilitate diagnosis and thus help the attending physician prescribe effective treatment (see Benton, 1994). They can also be an important basis for patient care and counseling; Kolb and Whishaw (1990) described such an application.

The Case of Mr. R., the Brain-Damaged Student Who Switched to Architecture

Mr. R., a 21-year-old left-handed man, struck his head on the dashboard in a car accident. . . . Prior to his accident Mr. R. was an honor student at a university. . . . However, a year after the accident he had become a mediocre student who had particular trouble completing his term papers. . . . He was referred to us for neuropsychological assessment, which revealed several interesting facts.

First, Mr. R. was one of about one-third of left-handers whose language functions are represented in the right rather than left hemisphere. . . . In addition, although Mr. R. had a superior IQ, his verbal memory and reading speed were only low-average, which is highly unusual for a person of his intelligence and education. These deficits indicated that his right temporal lobe may have been slightly damaged in the car accident, resulting in an impairment of his language skills. On the basis of our neuropsychological investigation we were able to recommend vocations to Mr. R. that did not require superior verbal memory skills, and he is currently studying architecture.

("The Case of Mr. R., the Brain-Damaged Student Who Switched to Architecture" from *Fundamentals of Human Neuropsychology*, 3/e, by Bryan Kolb and Ian Q. Whishaw. © 1980, 1985, 1990 by W. H. Freeman and Company. Used with permission of Worth Publishers.)

Psychophysiology

Psychophysiology is the division of biopsychology that studies the relation between physiological activity and psychological processes in human subjects. Because the subjects of psychophysiological research are human, psychophysiological recording procedures are typically noninvasive; that is, the physiological activity is recorded from the surface of the body. The usual measure of brain activity is the scalp **electroencephalogram (EEG)** (see Chapter 5). Other common psychophysiological measures are muscle tension, eye movement, and several indicators of autonomic nervous system activity (e.g., heart rate, blood pressure, pupil dilation, and electrical conductance of the skin). The **autonomic nervous system (ANS)** is the division of the nervous system that regulates the body's inner environment (see Chapter 3).

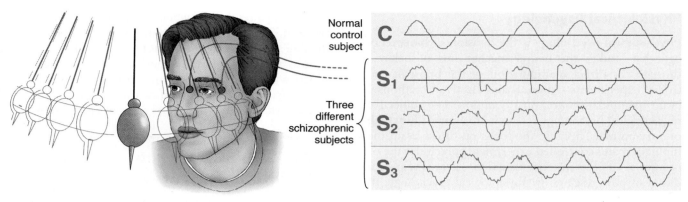

Normal control subject **C**

Three different schizophrenic subjects **S₁** **S₂** **S₃**

FIGURE 1.4   Visual tracking of a pendulum by a normal control subject (top) and three schizophrenics. (Adapted from Iacono & Koenig, 1983.)

Most psychophysiological research focuses on understanding the physiology of psychological processes, such as attention, emotion, and information processing, but there have also been a number of interesting clinical applications of the psychophysiological method. For example, psychophysiological experiments have indicated that schizophrenics have difficulty smoothly tracking a moving object such as a pendulum (e.g., Chen et al., 2008)—see Figure 1.4.

Clinical Implications

Cognitive Neuroscience

Cognitive neuroscience is the youngest division of biopsychology, but it is currently among the most active and exciting. Cognitive neuroscientists study the neural bases of **cognition**, a term that generally refers to higher intellectual processes such as thought, memory, attention, and complex perceptual processes (see Cabeza & Kingston, 2002; Raichle, 2008). Because of its focus on cognition, most cognitive neuroscience research involves human subjects; and because of its focus on human subjects, its methods tend to be noninvasive, rather than involving penetration or direct manipulation of the brain.

The major method of cognitive neuroscience is *functional brain imaging* (recording images of the activity of the living human brain; see Chapter 5) while a subject is engaged in a particular cognitive activity. For example, Figure 1.5 shows that the visual areas of the left and right cerebral cortex at the back of the brain became active when the subject viewed a flashing light.

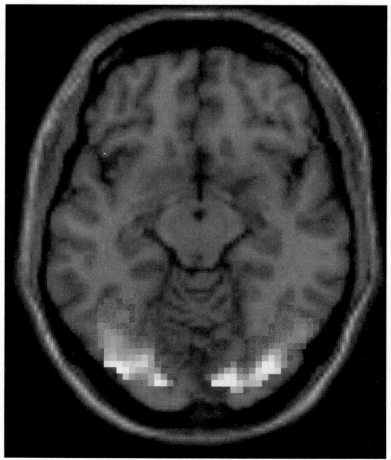

FIGURE 1.5   Functional brain imaging is the major method of cognitive neuroscience. This image—taken from the top of the head with the subject lying on her back—reveals the locations of high levels of neural activity at one level of the brain as the subject views a flashing light. The red and yellow areas indicate high levels of activity in the visual cortex at the back of the brain. (Courtesy of Todd Handy, Department of Psychology, University of British Columbia.)

Because the theory and methods of cognitive neuroscience are so complex and pertinent to so many fields, most cognitive neuroscientific publications result from interdisciplinary collaboration among many individuals with different types of training. For example, biopsychologists, cognitive psychologists, social psychologists, economists, computing and mathematics experts, and various types of neuroscientists commonly contribute to the field. Cognitive neuroscience research sometimes involves noninvasive electrophysiological recording, and it sometimes focuses on subjects with brain pathology; in these cases, the boundaries between cognitive neuroscience and psychophysiology and neuropsychology, respectively, are blurred.

Comparative Psychology

Although most biopsychologists study the neural mechanisms of behavior, there is more to biopsychology than this. As Dewsbury (1991) asserted:

The "biology" in "psychobiology" should include the whole-animal approaches of ethology, ecology, evolution . . . as well as the latest in physiological methods and thought. . . . The "compleat psychobiologist" should use whatever explanatory power can be found with modern physiological techniques, but never lose sight of the problems that got us going in the first place: the integrated behavior of whole, functioning, adapted organisms. (p. 98)

The division of biopsychology that deals generally with the biology of behavior, rather than specifically with the neural mechanisms of behavior, is **comparative psychology**. Comparative psychologists compare the behavior of different species in order to understand the evolution, genetics, and adaptiveness of behavior. Some comparative psychologists study behavior in the laboratory; others engage in **ethological research** —the study of animal behavior in its natural environment.

⊙→ Simulate
Tour of an Animal Cell
www.mypsychlab.com

Because two important areas of biopsychological research often employ comparative analysis, I have included them as part of comparative psychology. One of these is *evolutionary psychology* (a subfield that focuses on understanding behavior by considering its likely evolutionary origins; see Caporael, 2001; Duchaine, Cosmides, & Tooby, 2001; Kenrick, 2001). The other is behavioral genetics (the study of genetic influences on behavior; see Carson & Rothstein, 1999; Plomin et al., 2002).

In case you have forgotten, the purpose of this section has been to demonstrate the diversity of biopsychology by describing its six major divisions. These are summarized for you in Table 1.2. You will learn about the progress being made in each of these divisions in subsequent chapters.

TABLE 1.2 The Six Major Divisions of Biopsychology, with Examples of How They Have Approached the Study of Memory	
The Six Divisions of Biopsychology	*Examples of How the Six Approaches Have Pursued the Study of Memory*
Physiological psychology: study of the neural mechanisms of behavior by manipulating the nervous systems of nonhuman animals in controlled experiments.	**Physiological psychologists** have studied the contributions of the hippocampus to memory by surgically removing the hippocampus in rats and assessing their ability to perform various memory tasks.
Psychopharmacology: study of the effects of drugs on the brain and behavior.	**Psychopharmacologists** have tried to improve the memory of Alzheimer's patients by administering drugs that increase the levels of the neurotransmitter acetylcholine.
Neuropsychology: study of the psychological effects of brain damage in human patients.	**Neuropsychologists** have shown that patients with alcohol-produced brain damage have particular difficulty in remembering recent events.
Psychophysiology: study of the relation between physiological activity and psychological processes in human subjects by noninvasive physiological recording.	**Psychophysiologists** have shown that familiar faces elicit the usual changes in autonomic nervous system activity even when patients with brain damage report that they do not recognize a face.
Cognitive neuroscience: study of the neural mechanisms of human cognition, largely through the use of functional brain imaging.	**Cognitive neuroscientists** have used brain-imaging technology to observe the changes that occur in various parts of the brain while human volunteers perform memory tasks.
Comparative psychology: study of the evolution, genetics, and adaptiveness of behavior, largely through the use of the comparative method.	**Comparative psychologists** have shown that species of birds that cache their seeds tend to have big hippocampi, confirming that the hippocampus is involved in memory for location.

Scan Your Brain

To see if you are ready to proceed to the next section of the chapter, scan your brain by filling in each of the following blanks with one of the six divisions of biopsychology. The correct answers are provided at the end of the exercise. Before proceeding, review material related to your errors and omissions.

1. A biopsychologist who studies the memory deficits of human patients with brain damage would likely identify with the division of biopsychology termed _____.
2. Biopsychologists who study the physiological correlates of psychological processes by recording physiological signals from the surface of the human body are often referred to as _____.
3. The biopsychological research of _____ frequently involves the direct manipulation or recording of the neural activity of laboratory animals by various invasive surgical, electrical, and chemical means.
4. The division of biopsychology that focuses on the study of the effects of drugs on behavior is often referred to as _____.
5. _____ is a division of biopsychology that investigates the neural bases of human cognition; its major method is functional brain imaging.
6. _____ are biopsychologists who study the genetics, evolution, and adaptiveness of behavior, often by using the comparative approach.

Scan Your Brain answers: (1) neuropsychology, (2) psychophysiologists, (3) physiological psychologists, (4) psychopharmacology, (5) Cognitive neuroscience, (6) Comparative psychologists.

1.5

Converging Operations: How Do Biopsychologists Work Together?

Because none of the six biopsychological approaches to research is without its shortcomings and because of the complexity of the brain and its role in psychological processes, major biopsychological issues are rarely resolved by a single experiment or even by a single series of experiments taking the same general approach. Progress is most likely when different approaches are focused on a single problem in such a way that the strengths of one approach compensate for the weaknesses of the others; this combined approach is called **converging operations** (see Thompson, 2005).

Consider, for example, the relative strengths and weaknesses of neuropsychology and physiological psychology in the study of the psychological effects of damage to the human cerebral cortex. In this instance, the strength of the neuropsychological approach is that it deals directly with human patients; its weakness is that its focus on human patients precludes experiments. In contrast, the strength of the physiological psychology approach is that it can bring the power of the experimental method and neuroscientific technology to bear through research on nonhuman animals; its weakness is that the relevance of research on laboratory animals to human neuropsychological deficits is always open to question. Clearly these two approaches complement each other well; together they can answer questions that neither can answer individually.

To examine converging operations in action, let's return to the case of Jimmie G. The neuropsychological disorder from which Jimmie G. suffered *Clinical Implications* was first described in the late 19th century by S. S. Korsakoff, a Russian physician, and subsequently became known as **Korsakoff's syndrome**. The primary symptom of Korsakoff's syndrome is severe memory loss, which is made all the more heartbreaking—as you have seen in Jimmie G.'s case—by the fact that its sufferers are often otherwise quite capable. Because Korsakoff's syndrome commonly occurs in alcoholics, it was initially believed to be a direct consequence of the toxic effects *Thinking Creatively* of alcohol on the brain. This conclusion proved to be a good illustration of the inadvisability of basing causal conclusions on quasiexperimental research. Subsequent research showed that Korsakoff's syndrome is largely caused by the brain damage associated with *thiamine* (vitamin B_1) deficiency.

The first support for the thiamine-deficiency interpretation of Korsakoff's syndrome came from the discovery of the syndrome in malnourished persons who consumed little or no alcohol. Additional support came from experiments in which thiamine-deficient rats were compared with otherwise identical groups of control rats. The thiamine-deficient rats displayed memory deficits and patterns of brain damage similar to those observed in human alcoholics (see Mumby, Cameli, & Glenn, 1999). Alcoholics often develop Korsakoff's syndrome because most of their caloric intake comes in the form of alcohol, which lacks vitamins, and because alcohol interferes with the metabolism of what little thiamine they do consume. However, alcohol has been shown to accelerate the development of brain damage in thiamine-deficient rats, so it may have a direct toxic effect on the brain as well (Zimitat et al., 1990).

The point of all this (in case you have forgotten) is that progress in biopsychology typically comes from converging operations—in this case, from the convergence of neuropsychological case studies (case studies of Korsakoff patients), quasiexperiments with human subjects (comparisons of alcoholics with people who do not drink alcohol), and controlled experiments on laboratory animals (comparison of thiamine-deficient and control rats). The strength of biopsychology lies in the diversity of its methods and approaches. This means that, in evaluating biopsychological claims, it is rarely sufficient to consider the results of one study or even of one line of experiments using the same method or approach.

So what has all the research on Korsakoff's syndrome done for Jimmie G. and others like him? Today, alcoholics are counseled to stop drinking and are treated with massive doses of thiamine. The thiamine limits the development of further brain damage and often leads to a slight improvement in the patient's condition; but, unfortunately, the brain damage that has already occurred is largely permanent.

1.6

Scientific Inference: How Do Biopsychologists Study the Unobservable Workings of the Brain?

Scientific inference is the fundamental method of biopsychology and of most other sciences—it is what makes being a scientist fun. This section provides further insight into the nature of biopsychology by defining, illustrating, and discussing scientific inference.

The scientific method is a system for finding things out by careful observation, but many of the processes studied by scientists cannot be observed. For example, scientists use empirical (observational) methods to study ice ages, gravity, evaporation, electricity, and nuclear fission—none of which can be directly observed; their effects can be observed, but the processes themselves cannot. Biopsychology is no different from other sciences in this respect. One of its main goals is to characterize, through empirical methods, the unobservable processes by which the nervous system controls behavior.

The empirical method that biopsychologists and other scientists use to study the unobservable is called **scientific inference**. The scientists carefully measure key events that they can observe and then use these measures as a basis for logically inferring the nature of events that they cannot observe. Like a detective carefully gathering clues from which to recreate an unwitnessed crime, a biopsychologist carefully gathers relevant measures of behavior and neural activity from which to infer the nature of the neural processes that regulate behavior. The fact that the neural mechanisms of behavior cannot be directly observed and must be studied through scientific inference is what makes biopsychological research such a challenge—and, as I said before, so much fun.

To illustrate scientific inference, I have selected a research project in which you can participate. By making a few simple observations about your own visual abilities under different conditions, you will be able to discover the principle by which your brain translates the movement of images on your retinas into perceptions of movement (see Figure 1.6 on page 14). One feature of the mechanism is immediately obvious. Hold your hand in front of your face, and then move its image across your retinas by moving your eyes, by moving your hand, or by moving both at once. You will notice that only those movements of the retinal image that are produced by the movement of your hand are translated into the sight of motion; movements of the retinal image that are produced by your own eye movements are not. Obviously, there must be a part of your brain that monitors the movements of your retinal image and subtracts from the total those image movements that are produced by your own eye movements, leaving the remainder to be perceived as motion.

Now, let's try to characterize the nature of the information about your eye movements that is used by your brain in its perception of motion. Try the following. Shut one eye, then rotate your other eye slightly upward by gently pressing on your lower eyelid with your fingertip. What do you see? You see all of the objects in your visual field moving downward. Why? It seems that the brain mechanism that is responsible for the perception of motion does not consider eye movement per se. It considers only those eye movements that are actively produced by neural signals from the brain to the eye muscles, not those that are passively produced by external means (e.g., by your finger). Thus, when your eye was moved passively, your brain assumed that it had remained still and attributed the movement of your retinal image to the movement of objects in your visual field.

It is possible to trick the visual system in the opposite way; instead of the eyes being moved when no active signals have been sent to the eye muscles, the eyes can be held stationary despite the brain's attempts to move them. Because this experiment involves paralyzing the eye muscles, you cannot participate. Hammond, Merton, and Sutton (1956) injected a *paralytic* (movement-inhibiting) substance into the eye muscles of their subject—who was Merton himself. This paralytic substance was the active ingredient of *curare,* with which some South American natives coat their blow darts. What do you think Merton saw when he then tried to move his eyes? He saw the stationary visual world moving in the same direction as his attempted eye movements. If a visual object is focused on part of your retina, and it stays focused there despite the fact that you have moved your eyes to the right, it too must have

Simulate
Perception of Motion
www.mypsychlab.com

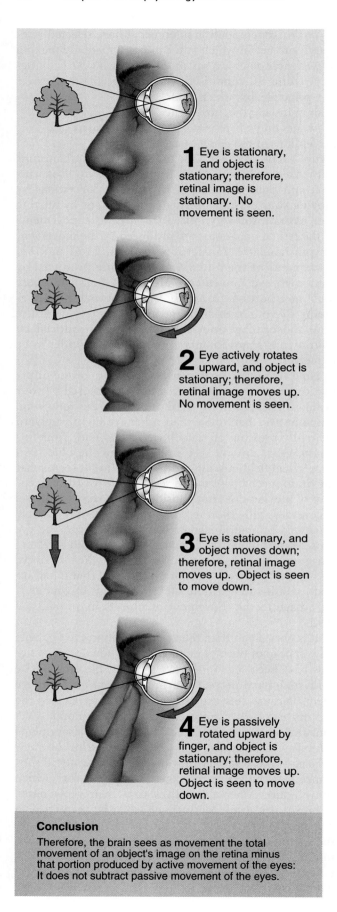

1 Eye is stationary, and object is stationary; therefore, retinal image is stationary. No movement is seen.

2 Eye actively rotates upward, and object is stationary; therefore, retinal image moves up. No movement is seen.

3 Eye is stationary, and object moves down; therefore, retinal image moves up. Object is seen to move down.

4 Eye is passively rotated upward by finger, and object is stationary; therefore, retinal image moves up. Object is seen to move down.

Conclusion

Therefore, the brain sees as movement the total movement of an object's image on the retina minus that portion produced by active movement of the eyes: It does not subtract passive movement of the eyes.

FIGURE 1.6   The perception of motion under four different conditions.

moved to the right. Consequently, when Merton sent signals to his eye muscles to move his eyes to the right, his brain assumed that the movement had been carried out, and it perceived stationary objects as moving to the right.

The point of the eye-movement example is that biopsychologists can learn much about the activities of the brain through scientific inference, without directly observing them—and so can you. By the way, neuroscientists are still interested in the kind of feedback mechanisms inferred from the demonstrations of Hammond and colleagues, and they are finding a lot of direct evidence for such mechanisms using modern neural recording techniques (see Lindner et al., 2006; Munoz, 2006).

1.7
Critical Thinking about Biopsychological Claims

We have all heard or read that we use only a small portion of our brains, that it is important to eat three meals a day, that intelligence is inherited, that everybody needs at least 8 hours of sleep per night, that there is a gene for schizophrenia, that morphine is a particularly dangerous (hard) drug, that neurological diseases can now be cured by genetic engineering, and that homosexuality is caused by inappropriate upbringing—to note just a few claims about biopsychological phenomena that have been widely disseminated. You may believe some of these claims. But are they true? How does one find out? And if they are not true, why do so many people believe them?

As you have already learned, one of the major goals of this book is to teach you how to think creatively (to think in productive, unconventional ways) about biopsychological information. Often the first step in creative thinking is spotting the weaknesses of existing ideas and the evidence on which they are based—the process by which these weaknesses are recognized is called **critical thinking**. The identification of weaknesses in existing beliefs is one of the major stimuli for scientists to adopt creative new approaches. The purpose of this final section of the chapter is to begin the development of your creative thinking ability by describing two claims that were once widely accepted but *Thinking Creatively* were subsequently shown to be unfounded. Notice that if you keep your wits about you, you do not have to be an expert to spot the weaknesses.

The first step in judging the validity of any scientific claim is to determine whether the claim and the research on which it is based were published in a reputable scientific journal (Rensberger, 2000). The reason is that, in order to be published in a reputable scientific journal, an article must first be reviewed by experts in the field—usually three or four of them—and judged to be of good quality. Indeed, the best scientific journals publish only a small proportion of the manuscripts submitted to them. You should be particularly skeptical of scientific claims that have not gone through this review process, but, as you are about to learn, the review process is not a guarantee that scientific papers are free of unrecognized flaws.

The first case that follows deals with an unpublished claim that was largely dispensed through the news media. The second deals with a claim that was initially supported by published research. Because both of these cases are part of the history of biopsychology, we have the advantage of 20/20 hindsight in evaluating their claims.

Case 1: José and the Bull

José Delgado, a particularly charismatic neuroscientist, demonstrated to a group of newspaper reporters a remarkable new procedure for controlling aggression (see Horgan, 2005). Delgado strode into a Spanish bull ring carrying only a red cape and a small radio transmitter. With the transmitter, he could activate a battery-powered stimulator that had previously been mounted on the horns of the other inhabitant of the ring. As the raging bull charged, Delgado calmly activated the stimulator and sent a weak electrical current from the stimulator through an electrode that had been implanted in the caudate nucleus (see Chapter 3), a structure deep in the bull's brain. The bull immediately veered from its charge. After a few such interrupted charges, the bull stood tamely as Delgado swaggered about the ring. According to Delgado, this demonstration marked a significant scientific breakthrough—the discovery of a caudate taming center and the fact that stimulation of this structure could eliminate aggressive behavior, even in bulls specially bred for their ferocity.

To those present at this carefully orchestrated event and to most of the millions who subsequently read about it, Delgado's conclusion was compelling. Surely, if caudate stimulation could stop the charge of a raging bull, the caudate must be a taming center. It was even suggested that caudate stimulation through implanted electrodes might be an effective treatment for human psychopaths. What do you think?

Analysis of Case 1 The fact of the matter is that Delgado's demonstration provided little or no support for his conclusion. It should have been obvious to anyone who

did not get caught up in the provocative nature of Delgado's media event that there are numerous ways in which brain stimulation can abort a bull's charge, most of which are simpler or more direct, and thus more probable, than the one suggested by Delgado. For example, the stimulation may have simply rendered the bull confused, dizzy, nauseous, sleepy, or temporarily blind rather than nonaggressive; or the stimulation could have been painful. Clearly, any observation that can be interpreted in so many different ways provides little support for any one interpretation. When there are several possible interpretations for a behavioral observation, the rule is to give precedence to the simplest one; this rule is called **Morgan's Canon**. The following comments of Valenstein (1973) provide a reasoned view of Delgado's demonstration:

> Actually there is no good reason for believing that the stimulation had any direct effect on the bull's aggressive tendencies. An examination of the film record makes it apparent that the charging bull was stopped because as long as the stimulation was on it was forced to turn around in the same direction continuously. After examining the film, any scientist with knowledge in this field could conclude only that the stimulation had been activating a neural pathway controlling movement. (p. 98) . . . he [Delgado] seems to capitalize on every individual effect his electrodes happen to produce and presents little, if any, experimental evidence that his impression of the underlying cause is correct. (p. 103) . . . his propensity for dramatic, albeit ambiguous, demonstrations has been a constant source of material for those whose purposes are served by exaggerating the omnipotence of brain stimulation. (p. 99)

Case 2: Becky, Moniz, and Prefrontal Lobotomy

In 1949, Dr. Egas Moniz was awarded the Nobel Prize in Physiology and Medicine for the development of **prefrontal lobotomy**—a surgical procedure in which the connections between the prefrontal lobes and the rest of the brain are cut as a treatment for mental illness. The **prefrontal lobes** are the large areas, left and right, at the very front of the brain (see Figure 1.7 on page 16). Moniz's discovery was based on the report that Becky, a chimpanzee that frequently became upset when she made errors during the performance of a food-rewarded task, did not do so following the creation of a large bilateral lesion (an area of damage to both sides of the brain) of her prefrontal lobes. After hearing about this isolated observation at a scientific meeting in 1935, Moniz persuaded neurosurgeon Almeida Lima to operate on a series of psychiatric patients; Lima

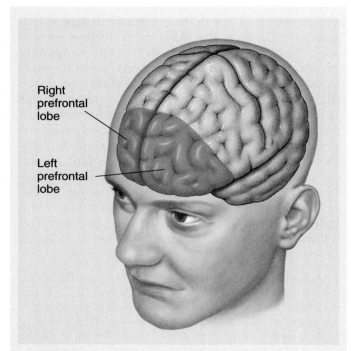

FIGURE 1.7   The right and left prefrontal lobes, whose connections to the rest of the brain are disrupted by prefrontal lobotomy.

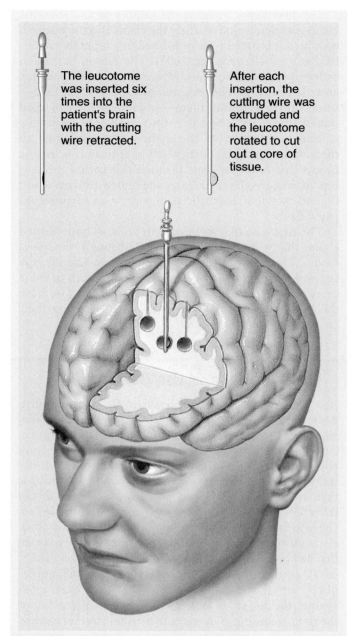

FIGURE 1.8   The prefrontal lobotomy procedure developed by Moniz and Lima.

cut out six large cores of prefrontal tissue with a surgical device called a **leucotome** (see Figure 1.8).

Following Moniz's claims that prefrontal surgery was therapeutically successful and had no significant side effects, there was a rapid proliferation of various forms of prefrontal psychosurgery (see Valenstein, 1980, 1986). One such variation was **transorbital lobotomy**, which was developed in Italy and then popularized in the United States by Walter Freeman in the late 1940s. It involved inserting an ice-pick-like device under the eyelid, driving it through the orbit (the eye socket) with a few taps of a mallet, and pushing it into the frontal lobes, where it was waved back and forth to sever the connections between the prefrontal lobes and the rest of the brain (see Figure 1.9). This operation was frequently performed in the surgeon's office.

Analysis of Case 2   Incredible as it may seem, Moniz's program of **psychosurgery** (any brain surgery, such as prefrontal lobotomy, performed for the treatment of a psychological problem) was largely based on the observation of a single chimpanzee in a single situation. Thus, Moniz displayed a complete lack of appreciation for the diversity of brain and behavior, both within and between species. No program of psychosurgery should ever be initiated without a thorough assessment of the effects of the surgery on a large sample of subjects from various nonhuman mammalian species. To do so is not only unwise, it is unethical.

A second major weakness in the scientific case for prefrontal lobotomy was the failure of Moniz and others to carefully evaluate the consequences of the surgery in the first patients to undergo the operation (see Mashour, Walker, & Martuza, 2005; Singh, Hallmayer, & Illes, 2007). The early reports that the operation was therapeutically successful were based on the impressions of the individuals

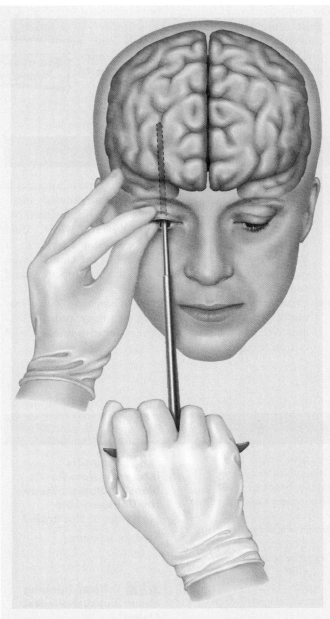

FIGURE 1.9 The transorbital procedure for performing prefrontal lobotomy

who were the least objective—the physicians who had prescribed the surgery and their colleagues. Patients were frequently judged as improved if they were more manageable, and little effort was made to evaluate more important aspects of their psychological adjustment or to document the existence of adverse side effects.

Eventually, it became clear that prefrontal lobotomies are of little therapeutic benefit and that they can produce a wide range of undesirable side effects, such as amorality, lack of foresight, emotional unresponsiveness, epilepsy, and urinary incontinence. This led to the abandonment of prefrontal lobotomy in many parts of the world—but not before over 40,000 patients had been lobotomized in the United States alone. And, prefrontal lobotomies still continue to be performed in some countries.

A particularly troubling aspect of the use of prefrontal lobotomy is that not only informed, consenting adults received this "treatment." In his recent memoir, Howard Dully described how he had been lobotomized at the age of 12 (Dully & Fleming, 2007). The lobotomy was arranged by Dully's stepmother, agreed to by his father, and performed in 10 minutes by Walter Freeman. Dully spent most of the rest of his life in asylums, jails, and halfway houses, wondering what he had done to deserve the lobotomy and how much it had been responsible for his troubled life. Investigation of the medical documents and interviews with some of those involved in the case have indicated that Dully was a normal child whose stepmother was obsessed by her hatred for him. Tragically, neither his father nor the medical profession intervened to protect him from Freeman's ice-pick.

Some regard sound scientific methods as unnecessary obstacles in the paths of patients seeking treatment and therapists striving to provide it. However, the unforeseen consequences of prefrontal lobotomy should caution us against abandoning science for expediency. Only by observing the rules of science can scientists protect the public from bogus scientific claims.

You are about to enter a world of amazing discovery and intriguing ideas: the world of biopsychology. I hope that your brain enjoys learning about itself.

Themes Revisited

The seeds of three of the major themes of this book were planted in this chapter, but the thinking creatively theme predominated. First, you saw the creative approach that Lester and Gorzalka took in their research on the Coolidge effect in females. Then, you learned three important new ideas that will help you think about biopsychological claims: (1) the experimental method, (2) converging operations, and (3) scientific inference. Finally, you were introduced to two biopsychological claims that were once

Thinking Creatively

widely believed and saw how critical thinking identified their weaknesses and replaced them with creative new interpretations.

You also learned that two of the other major themes of the book—clinical implications and the evolutionary perspective—tend to be associated with particular divisions

of biopsychology. Clinical implications most commonly emerge from neuropsychological, psychopharmacological, and psychophysiological research; the evolutionary perspective is a defining feature of comparative psychology.

Think about It

1. This chapter tells you in general conceptual terms what biopsychology is. Another, and perhaps better, way of defining biopsychology is to describe what biopsychologists do. Ask your instructor what she or he did to become a biopsychologist and what she or he does each workday. I think that you will be surprised. Is your instructor predominantly a physiological psychologist, a psychopharmacologist, a neuropsychologist, a psychophysiologist, a cognitive neuroscientist, or a comparative psychologist?

2. What ethical considerations should guide biopsychological research on nonhuman animals? How should these ethical considerations differ from those guiding biopsychological research on humans?

3. In retrospect, the entire story of prefrontal lobotomies is shocking. How could physicians, who are generally intelligent, highly educated, and dedicated to helping their patients, participate in such a travesty? How could somebody win a Nobel Prize for developing a form of surgery that left over 40,000 people in the United States alone mentally crippled? Why did this happen? Could something like this happen today?

4. Creative thinking is as important in biopsychological laboratories as it is in the life of biopsychology students: Discuss. What is the relation between creative thinking and critical thinking?

Key Terms

Neurons (p. 2)
Neuroscience (p. 2)
Thinking creatively (p. 3)
Clinical (p. 3)
Evolutionary perspective (p. 3)

1.1 What Is Biopsychology?

Biopsychology (p. 3)

1.2 What Is the Relation between Biopsychology and the Other Disciplines of Neuroscience?

Neuroanatomy (p. 4)
Neurochemistry (p. 4)
Neuroendocrinology (p. 4)
Neuropathology (p. 4)
Neuropharmacology (p. 4)
Neurophysiology (p. 4)

1.3 What Types of Research Characterize the Biopsychological Approach?

Comparative approach (p. 5)
Between-subjects design (p. 5)
Within-subjects design (p. 5)
Independent variable (p. 6)
Dependent variable (p. 6)
Confounded variable (p. 6)
Coolidge effect (p. 6)
Lordosis (p. 6)
Quasiexperimental studies (p. 7)
Case studies (p. 7)
Generalizability (p. 7)
Pure research (p. 7)
Applied research (p. 7)

1.4 What Are the Divisions of Biopsychology?

Physiological psychology (p. 9)
Psychopharmacology (p. 9)

Neuropsychology (p. 9)
Cerebral cortex (p. 9)
Psychophysiology (p. 9)
Electroencephalogram (EEG) (p. 9)
Autonomic nervous system (ANS) (p. 9)
Cognitive neuroscience (p. 10)
Cognition (p. 10)
Comparative psychology (p. 11)
Ethological research (p. 11)

1.5 Converging Operations: How Do Biopsychologists Work Together?

Converging operations (p. 12)
Korsakoff's syndrome (p. 12)

1.6 Scientific Inference: How Do Biopsychologists Study the Unobservable Workings of the Brain?

Scientific inference (p. 13)

1.7 Critical Thinking about Biopsychological Claims

Critical thinking (p. 14)
Morgan's Canon (p. 15)
Prefrontal lobotomy (p. 15)
Prefrontal lobes (p. 15)
Leucotome (p. 16)
Transorbital lobotomy (p. 16)
Psychosurgery (p. 16)

✓•─Quick Review Test your comprehension of the chapter with this brief practice test. You can find the answers to these questions as well as more practice tests, activities, and other study resources at www.mypsychlab.com.

1. According to the text, creative thinking about biopsychology is thinking
   a. in new ways.
   b. in productive ways.
   c. in ways that are consistent with the evidence rather than widely accepted views.
   d. "outside the box."
   e. all of the above

2. The field that focuses on the study of the structure of the nervous system is
   a. neurophysiology.
   b. behavioral neuroscience.
   c. neurochemistry.
   d. neuropharmacology.
   e. none of the above

3. Which division of biopsychology relies on functional brain imaging as its major research method?
   a. cognitive neuroscience
   b. neuropsychology
   c. psychophysiology
   d. behavioral neuroscience
   e. physiological psychology

4. Korsakoff's syndrome is
   a. most commonly observed in males of Russian descent.
   b. caused in large part by thiamine deficiency.
   c. often associated with chronic alcoholism.
   d. all of the above
   e. both b and c

5. Who was awarded a Nobel Prize for the development of pre-frontal lobotomy as a treatment for psychiatric disorders?
   a. Lima
   b. Valenstein
   c. Moniz
   d. Freeman
   e. Delgado

2 Evolution, Genetics, and Experience
Thinking about the Biology of Behavior

We all tend to think about things in ways that have been ingrained in us by our **Zeitgeist** (pronounced "ZYTE-gyste"), the general intellectual climate of our culture. That is why this is a particularly important chapter for you. You see, you are the intellectual product of a Zeitgeist that promotes ways of thinking about the biological bases of behavior that are inconsistent with the facts. The primary purpose of this chapter is to help you bring your thinking about the biology of behavior in line with modern biopsychological science.

2.1

Thinking about the Biology of Behavior: From Dichotomies to Interactions

We tend to ignore the subtleties, inconsistencies, and complexities of our existence and to think in terms of simple, mutually exclusive dichotomies: right–wrong, good–bad, attractive–unattractive, and so on. The allure of this way of thinking is its simplicity.

The tendency to think about behavior in terms of dichotomies is illustrated by two kinds of questions that people commonly ask about behavior: (1) Is it physiological, or is it psychological? (2) Is it inherited, or is it learned? Both questions have proved to be misguided; yet they are among the most common kinds of questions asked in biopsychology classrooms. That is why I am dwelling on them here.

Is It Physiological, or Is It Psychological?

The idea that human processes fall into one of two categories, physiological or psychological, has a long history in many cultures. In Western cultures, it rose to prominence following the Dark Ages, in response to a 17th-century conflict between science and the Roman Church. For much of the history of Western civilization, truth was whatever was decreed to be true by the Church. Then, in about 1400, things started to change. The famines, plagues, and marauding armies that had repeatedly swept Europe during the Dark Ages subsided, and interest turned to art, commerce, and scholarship—this was the period of the Renaissance, or rebirth (1400 to 1700). Some Renaissance scholars were not content to follow the dictates of the Church; instead, they started to study things directly by observing them—and so it was that modern science was born.

Much of the scientific knowledge that accumulated during the Renaissance was at odds with Church dictates. However, the conflict was resolved by the prominent French philosopher René Descartes (pronounced "day-CART"). Descartes (1596–1650) advocated a philosophy that, in a sense, gave one part of the universe to science and the other part to the Church. He argued that the universe is composed of two elements: (1) physical matter, which behaves according to the laws of nature and is thus a suitable object of scientific investigation; and (2) the human mind (soul, self, or spirit), which lacks physical substance, controls human behavior, obeys no natural laws, and is thus the appropriate purview of the Church. The human body, including the brain, was assumed to be entirely physical, and so were nonhuman animals.

Cartesian dualism, as Descartes's philosophy became known, was sanctioned by the Roman Church, and so the idea that the human brain and the mind are separate entities became even more widely accepted. It has survived to this day, despite the intervening centuries of scientific progress. Most people now understand that human behavior has a physiological basis, but many still cling to the dualistic assumption that there is a category of human activity that somehow transcends the human brain (Bloom & Weisberg, 2007).

Is It Inherited, or Is It Learned?

The tendency to think in terms of dichotomies extends to the way people think about the development of behavioral capacities. For centuries, scholars have debated whether humans and other animals inherit their behavioral capacities or acquire them through learning. This debate is commonly referred to as the **nature–nurture issue**.

◉ **Watch**
Twin Studies
www.mypsychlab.com

Most of the early North American experimental psychologists were totally committed to the nurture (learning) side of the nature–nurture issue (de Waal, 1999). The degree of this commitment is illustrated by the oft-cited words of John B. Watson, the father of *behaviorism*:

> We have no real evidence of the inheritance of [behavioral] traits. I would feel perfectly confident in the ultimately favorable outcome of careful upbringing of a healthy, well-formed baby born of a long line of crooks, murderers and thieves, and prostitutes. Who has any evidence to the contrary?
> . . . Give me a dozen healthy infants, well-formed, and my own specified world to bring them up in and I'll guarantee to take any one at random and train him to become any type of specialist I might select—doctor, lawyer, artist, merchant-chief and, yes even beggar-man and thief. (Watson, 1930, pp. 103–104)

At the same time that experimental psychology was taking root in North America, **ethology** (the study of animal behavior in the wild) was becoming the dominant approach to the study of behavior in Europe. European ethology, in contrast to North American experimental psychology, focused on the study of **instinctive behaviors** (behaviors that occur in all like members of a species, even when there seems to have been no opportunity for them to have been learned), and it emphasized the role of nature, or inherited factors, in behavioral development (Burkhardt, 2005). Because instinctive behaviors do not seem to be learned, the early ethologists assumed that they are entirely inherited. They were wrong, but then so were the early experimental psychologists.

Problems with Thinking about the Biology of Behavior in Terms of Traditional Dichotomies

The physiological-or-psychological debate and the nature-or-nurture debate are based on incorrect ways of thinking about the biology of behavior, and a new generation of questions is directing the current boom in biopsychological research (Churchland, 2002). What is wrong with these old ways of thinking about the biology of behavior, and what are the new ways?

Physiological-or-Psychological Thinking Runs into Difficulty   Not long after Descartes's mind–brain dualism was officially sanctioned by the Roman Church, it started to come under public attack.

> In 1747, Julien Offroy de la Mettrie anonymously published a pamphlet that scandalized Europe. . . . La Mettrie fled to Berlin, where he was forced to live in exile for the rest of his life. His crime? He had argued that thought was produced by the brain—a dangerous assault, in the eyes of his contemporaries. (Corsi, 1991, cover)

There are two lines of evidence against *physiological-or-psychological thinking* (the assumption that some aspects of human psychological functioning are so complex that they could not possibly be the product of a physical brain). The first line is composed of the many demonstrations that even the most complex psychological changes (e.g., changes in self-awareness, memory, or emotion) can be produced by damage to, or stimulation of, parts of the brain (see Farah & Murphy, 2009). The second line of evidence is composed of demonstrations that some nonhuman species, particularly *primate* species, possess abilities that were once assumed to be purely psychological and thus purely human (see Huffman, Nahallage, & Leca, 2008; Kornell, 2009; Okamoto-Barth, Call, & Tomasello, 2007; Warneken et al., 2007; Wood et al., 2007). The following two cases illustrate these two kinds of evidence. Both cases deal with self-awareness, which is widely regarded as one hallmark of the human mind (see Damasio, 1999).

The first case is Oliver Sacks's (1985) account of "the man who fell out of bed." This patient was suffering from **asomatognosia**, a deficiency in the awareness of parts of one's own body. Asomatognosia typically involves the left side of the body and usually results from damage to the *right parietal lobe* (see Figure 2.1). The point here is that

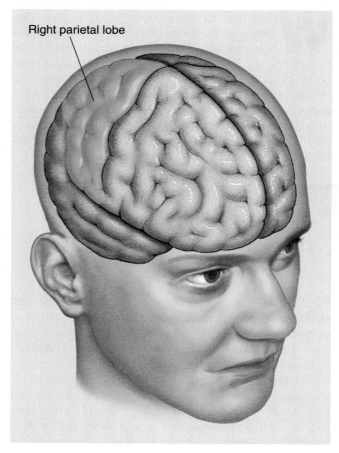

Right parietal lobe

FIGURE 2.1 Asomatognosia typically involves damage to the right parietal lobe.

although the changes in self-awareness displayed by the patient were very complex, they were clearly the result of brain damage: Indeed, the full range of human experience can be produced by manipulations of the brain.

The Case of the Man Who Fell Out of Bed

When he awoke, Dr. Sacks's patient felt fine—that is, until he touched the thing in bed next to him. It was a severed human leg, all hairy and still warm! At first, the patient was confused. Then, he figured it out. One of the nurses must have taken it from the autopsy department and put it in his bed as a joke. Some joke; it was disgusting. So, he threw the leg out of the bed, but somehow he landed on the floor with it attached to him.

The patient became agitated and desperate, and Dr. Sacks tried to comfort him and help him back into the bed. Making one last effort to reduce the patient's confusion, Sacks asked him where his left leg was, if the one attached to him wasn't it. Turning pale and looking like he was about to pass out, the patient replied that he had no idea where his own leg was—it had disappeared.

("The Case of the Man Who Fell Out of Bed," reprinted with the permission of Simon & Schuster, Inc. and Pan Macmillan, London from *The Man Who Mistook His Wife for a Hat and Other Clinical Tales* by Oliver Sacks. Copyright © 1970, 1981, 1983, 1984, 1986 by Oliver Sacks. Electronic rights with permission of the Wylie Agency.)

The second case describes G. G. Gallup's research on self-awareness in chimpanzees (see Gallup, 1983; Parker, Mitchell, & Boccia, 1994). The point of this case is that even nonhumans, which are assumed to have no mind, are capable of considerable psychological complexity—in this case, self-awareness. Although their brains are less complex than the brains of humans, some species are capable of levels of psychological complexity (e.g., self-awareness) that were once believed to imply the existence of a human mind.

The Case of the Chimps and the Mirrors

An organism is self-aware to the extent that it can be shown capable of becoming the object of its own attention. . . . One way to assess an organism's capacity to become the object of its own attention is to confront it with a mirror.

. . . I gave a number of group-reared, preadolescent chimpanzees individual exposure to themselves in mirrors. . . . Invariably, their first reaction to the mirror was to respond as if they were seeing another chimpanzee. . . . After about two days, however, . . . they . . . started to use the mirror to groom and inspect parts of their bodies they had not seen before, and progressively began to experiment with the reflection by making faces, looking at themselves upside down, and assuming unusual postures while monitoring the results in the mirror. . . .

So in an attempt to provide a more convincing demonstration of self-recognition, I devised an unobtrusive and more rigorous test. . . . Each chimpanzee was anesthetized. . . . I carefully painted the uppermost portion of an eyebrow ridge and the top half of the opposite ear with a bright red, odorless, alcohol soluble dye. . . .

Following recovery from anesthesia . . . the mirror was then reintroduced as an explicit test of self-recognition. Upon seeing their painted faces in the mirror, all the chimpanzees showed repeated mark-directed responses, consisting of attempts to touch and inspect marked areas on their eyebrow and ear while watching the image. [See Figure 2.2.] In addition, there was over a three-fold increase in viewing time. . . . Several chimpanzees also showed noteworthy attempts to visually examine and smell the fingers which had been used to touch these facial marks. I suspect that you would respond pretty much the

FIGURE 2.2   The reactions of chimpanzees to their *own* images indicate that they are self-aware. In this photo, the chimpanzee is reacting to the bright red, odorless dye that was painted on its eyebrow ridge while it was anesthetized. (Photograph by Donna Bierschwale, courtesy of the New Iberia Research Center.)

same way, if upon awakening one morning you saw yourself in the mirror with red spots on your face.

(From "Toward a Comparative Psychology of Mind" by G. G. Gallup, Jr., *American Journal of Primatology* 2:237–248, 1983. Copyright © 1983 John Wiley & Sons, Inc. Reprinted with permission of Wiley-Liss, Inc., a subsidiary of John Wiley & Sons, Inc.)

Nature-or-Nurture Thinking Runs into Difficulty

The history of nature-or-nurture thinking can be summed up by paraphrasing Mark Twain: "Reports of its death have been greatly exaggerated." Each time it has been discredited, it has resurfaced in a slightly modified form. First, factors other than genetics and learning were shown to influence behavioral development; factors such as the fetal environment, nutrition, stress, and sensory stimulation also proved to be influential. This led to a broadening of the concept of nurture to include a variety of experiential factors in addition to learning. In effect, it changed the nature-or-nurture dichotomy from "genetic factors or learning" to "genetic factors or experience."

○ **Watch**
Separated Twins Reunited
www.mypsychlab.com

Next, it was argued convincingly that behavior always develops under the combined control of both nature and nurture (see Johnston, 1987; Rutter, 1997), not under the control of one or the other. Faced with this point, many people merely substituted one kind of nature-or-nurture thinking for another. They stopped asking, "Is it genetic, or is it the result of experience?" and started asking, "How much of it is genetic, and how much of it is the result of experience?"

Like earlier versions of the nature-or-nurture question, the how-much-of-it-is-genetic-and-how-much-of-it-is-the-result-of-experience version is fundamentally flawed. The problem is that it is based on the premise that genetic factors and experiential factors combine in an additive fashion— that a behavioral capacity, such as intelligence, is created through the combination or mixture of so many parts of genetics and so many parts of experience, rather than through the interaction of genetics and experience. Once you learn more about how genetic factors and experience interact, you will better appreciate the folly of this assumption. For the time being, however, let me illustrate its weakness with a metaphor embedded in an anecdote.

The Case of the Thinking Student

One of my students told me that she had read that intelligence was one-third genetic and two-thirds experience, and she wondered whether this was true. I responded by asking her the following question: "If I wanted to get a better understanding of music, would it be reasonable for me to begin by asking how much of it came from the musician and how much of it came from the instrument?"

Thinking Creatively

"That would be dumb," she said. "The music comes from both; it makes no sense to ask how much comes from the musician and how much comes from the instrument. Somehow the music results from the interaction of the two together. You would have to ask about the interaction."

"That's exactly right," I said. "Now, do you see why . . ."

"Don't say any more," she interrupted. "I see what you're getting at. Intelligence is the product of the interaction of genes and experience, and it is dumb to try to find how much comes from genes and how much comes from experience."

"And the same is true of any other behavioral trait," I added.

The point of this metaphor, in case you have forgotten, is to illustrate why it is nonsensical to try to understand interactions between two factors by asking how much each factor contributes. We would not ask how much musicians and how much instruments contribute to music; we would not ask how much the water and how much the temperature contributes to evaporation; and we would not ask how much males and how much females contribute to copulation. Similarly, we shouldn't ask how much genetic and experiential factors contribute to behavioral development. In each case, the answers lie in understanding the nature of the interactions (see Jasny, Kelner, & Pennisi, 2008; Robinson, Fernald, & Clayton, 2008).

The importance of thinking in an interactive way about development will become obvious to you in Chapter 9, which focuses on the mechanisms of neural development. At this point, however, it is sufficient for you to appreciate three general points: (1) neurons become active long before they are fully developed; (2) the subsequent course of their development (e.g., the number of connections they form or whether or not they survive) depends greatly on their activity, much of which is triggered by external experience; and (3) experience continuously modifies genetic expression. Please stop and think about these three points: You may find them counterintuitive.

A Model of the Biology of Behavior So far in this section, you have learned why people tend to think about the biology of behavior in terms of dichotomies, and 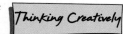 you have learned some of the reasons why this way of thinking is inappropriate. Now, let's look at the way of thinking about the biology of behavior that has been adopted by many biopsychologists (see Kimble, 1989). It is illustrated in Figure 2.3. Like other powerful ideas, it is simple and logical. This model boils down to the single premise that all behavior is the product of interactions among three factors: (1) the organism's genetic endowment, which is a product of its evolution; (2) its experience; and (3) its perception of the current situation. Please examine the model carefully, and consider its implications.

The next three sections of this chapter deal with three elements of this model of behavior: evolution, genetics, and the interaction of genetics and experience in behavioral development. The final section of the chapter deals with the genetics of human psychological differences.

2.2

Human Evolution

Modern biology began in 1859 with the publication of Charles Darwin's *On the Origin of Species*. In this monumental work, Darwin described his theory of evolution— the single most influential theory in the biological sciences. Darwin was not the first to suggest that species **evolve** (undergo gradual orderly change) from preexisting species, but he was the first to amass a large body of supporting evidence and the first to suggest how evolution occurs (see Bowler, 2009).

Darwin presented three kinds of evidence to support his assertion that species evolve: (1) He documented the evolution of fossil records through progressively more recent geological layers. (2) He described striking structural similarities among living species (e.g., a human's hand, a bird's wing, and a cat's paw), which suggested that they had evolved from common ancestors.

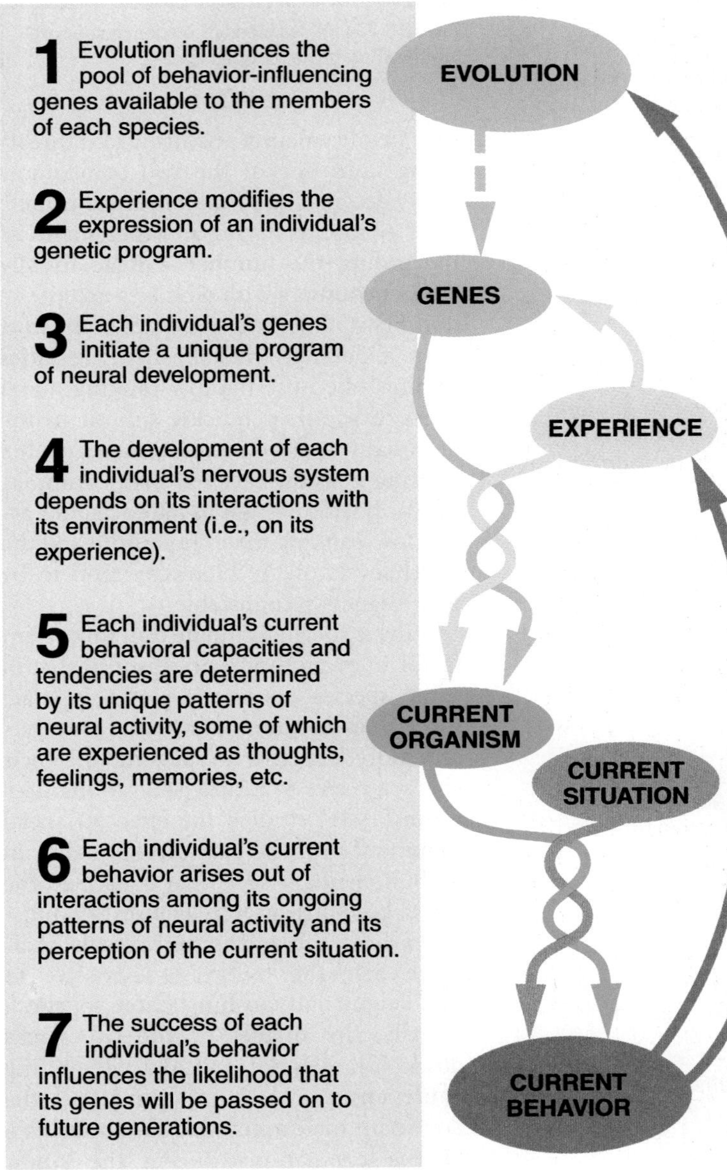

1 Evolution influences the pool of behavior-influencing genes available to the members of each species.

2 Experience modifies the expression of an individual's genetic program.

3 Each individual's genes initiate a unique program of neural development.

4 The development of each individual's nervous system depends on its interactions with its environment (i.e., on its experience).

5 Each individual's current behavioral capacities and tendencies are determined by its unique patterns of neural activity, some of which are experienced as thoughts, feelings, memories, etc.

6 Each individual's current behavior arises out of interactions among its ongoing patterns of neural activity and its perception of the current situation.

7 The success of each individual's behavior influences the likelihood that its genes will be passed on to future generations.

EVOLUTION

GENES

EXPERIENCE

CURRENT ORGANISM

CURRENT SITUATION

CURRENT BEHAVIOR

FIGURE 2.3 A schematic illustration of the way in which many biopsychologists think about the biology of behavior.

tions (see Kingsley, 2009). He argued that natural selection, when repeated for generation after generation, leads to the evolution of species that are better adapted to surviving and reproducing in their particular environmental niche. Darwin called this process *natural selection* to emphasize its similarity to the artificial selective breeding practices employed by breeders of domestic animals. Just as horse breeders create faster horses by selectively breeding the fastest of their existing stock, nature creates fitter animals by selectively breeding the fittest. **Fitness**, in the Darwinian sense, is the ability of an organism to survive and contribute its genes to the next generation.

The theory of evolution was at odds with the various dogmatic views that were embedded in the 19th-century Zeitgeist, so it met with initial resistance. Although resistance still exists, virtually none comes from people who understand the evidence (see Mayr, 2000). Many critics of evolution say that it is *only* a theory, but anybody who makes, or even accepts, such a statement understands neither evolution nor science (see Branch & Scott, 2009). True, evolution is a theory, but that does not mean that it is a vague, unreliable speculation: A *scientific theory* is an explanation that provides the best current account of some phenomenon based on the available evidence. In the case of evolution, that evidence is extensive, diverse, and totally convincing—and it increases daily. Like evidence for the theories of gravity, electricity, and earth's orbit around the sun, evidence for the theory of evolution is so strong that almost all biologists regard it as fact.

Evolution is both a beautiful concept and an important one, more crucial nowadays to human welfare, to medical science, and to our understanding of the world than ever before [see Mindell, 2009]. It's also deeply persuasive—a theory you can take to the bank . . . the supporting evidence is abundant, various, ever increasing, and easily available in museums, popular books, textbooks, and a mountainous accumulation of scientific studies. No one needs to, and no one should, accept evolution merely as a matter of faith. (Quammen, 2004, p. 8)

(3) He pointed to the major changes that had been brought about in domestic plants and animals by programs of selective breeding. However, the most convincing evidence of evolution comes from direct observations of rapid evolution in progress (see Orr, 2009). For example, Grant (1991) observed evolution of the finches of the Galápagos Islands—a population studied by Darwin himself—after only a single season of drought. Figure 2.4 on page 26 illustrates these four kinds of evidence.

Darwin argued that evolution occurs through **natural selection**. He pointed out that the members of each species vary greatly in their structure, physiology, and behavior, and that the heritable traits that are associated with high rates of survival and reproduction are the most likely ones to be passed on to future genera-

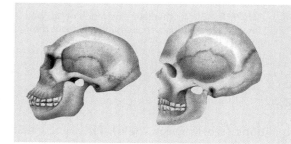

Fossil records change systematically through geological layers. Illustrated here is the evolution of the hominid skull.

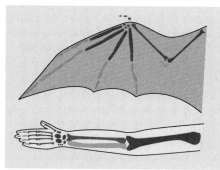

There are striking structural similarities among diverse living species (e.g., between a human arm and a bat's wing).

Major changes have been created in domestic plants and animals by programs of selective breeding.

Evolution has been observed in progress. For example, an 18-month drought on one of the Galápagos Islands left only large, difficult-to-eat seeds and increased the beak size in one species of finch.

FIGURE 2.4   Four kinds of evidence supporting the theory that species evolve.

they involve mainly posturing and threatening until one of the two combatants backs down. The dominant male usually wins encounters with all other males of the group; the number 2 male usually wins encounters with all males except the dominant male; and so on down the line. Once a hierarchy is established, hostilities diminish because the low-ranking males learn to avoid or quickly submit to the dominant males. Because most of the fighting goes on between males competing for positions high in the social hierarchy, low-ranking males fight little and the lower levels of the hierarchy tend to be only vaguely recognizable.

Why is social dominance an important factor in evolution? One reason is that in some species dominant males copulate more than nondominant males and thus are more effective in passing on their characteristics to future generations. McCann (1981) studied the effect of social dominance on the rate of copulation in 10 bull elephant seals that cohabited the same breeding beach. Figure 2.5 illustrates how these massive animals challenge each other by raising themselves to full height and pushing chest to chest. Usually, the smaller of the two backs down; if it does not, a vicious neck-biting battle ensues. McCann found that the dominant male accounted for about 37% of the copulations during the study, whereas poor number 10 accounted for only about 1% (see Figure 2.5).

Another reason why social dominance is an important factor in evolution is that in some species dominant females are more likely to produce more, and more healthy, offspring. For example, Pusey, Williams, and Goodall (1997) found that high-ranking female chimpanzees produced more offspring and that these offspring were more likely to survive to sexual maturity. They attributed these advantages to the fact that high-ranking female chimpanzees are more likely to maintain access to productive food foraging areas.

Evolution and Behavior

Some behaviors play an obvious role in evolution. For example, the ability to find food, avoid predation, or defend one's young obviously increases an animal's ability to pass on its genes to future generations. Other behaviors play a role that is less obvious but no less important (e.g., Bergman et al., 2003; Dunbar, 2003; Silk, Alberts, & Altmann, 2003). Two examples are social dominance and courtship display.

Social Dominance   The males of many species establish a stable *hierarchy of social dominance* through combative encounters with other males. In some species, these encounters often involve physical damage; in others,

Courtship Display   An intricate series of courtship displays precedes copulation in many species. The male approaches the female and signals his interest. His signal (which may be olfactory, visual, auditory, or tactual) may

FIGURE 2.5 Two massive bull elephant seals challenge one another. Dominant bull elephant seals copulate more frequently than those that are lower in the dominance hierarchy. (Adapted from McCann, 1981.)

Y-axis: Percentage of Total Observed Copulations
X-axis: Rank of Bull Elephant Seal (1 2 3 4 5 6 7 8 9 10 Others)

elicit a signal in the female, which may elicit another response in the male, and so on until copulation ensues. But copulation is unlikely to occur if one of the pair fails to react appropriately to the signals of the other.

Simulate
Recognizing Facial Expressions of Emotion
www.mypsychlab.com

Courtship displays are thought to promote the evolution of new species. Let me explain. A **species** is a group of organisms that is reproductively isolated from other organisms; that is, the members of a species can produce fertile offspring only by mating with members of the same species (Zimmer, 2008). A new species begins to branch off from an existing species when some barrier discourages breeding between a subpopulation of the existing species and the remainder of the species. Once such a reproductive barrier forms, the subpopulation evolves independently of the remainder of the species until cross-fertilization becomes impossible (see Willis, 2009).

The reproductive barrier may be geographic; for example, a few birds may fly together to an isolated island, where many generations of their offspring breed among themselves and evolve into a separate species. Alternatively—to get back to the main point—the reproductive barrier may be behavioral. A few members of a species may develop different courtship displays, and these may form a reproductive barrier between themselves and the rest of their **conspecifics** (members of the same species): Only the suitable exchange of displays between a courting couple will lead to reproduction.

Course of Human Evolution

Simulate
The Complexity of Humans: Phil Zimbardo
www.mypsychlab.com

By studying fossil records and comparing current species, we humans have looked back in time and pieced together the evolutionary history of our species—although some of the details are still controversial. The course of human evolution, as it is currently understood, is summarized in this section.

Evolution of Vertebrates Complex multicellular water-dwelling organisms first appeared on earth about 600 million years ago (Bottjer, 2005). About 150 million years later, the first chordates evolved. **Chordates** (pronounced "KOR-dates") are animals with dorsal nerve cords (large nerves that run along the center of the back, or *dorsum*); they are 1 of the 20 or so large categories, or *phyla* (pronounced "FY-la"), into which zoologists group animal species. The first chordates with spinal bones to protect their dorsal nerve cords evolved about 25 million years later. The spinal bones are called *vertebrae* (pronounced "VERT-eh-bray"), and the chordates that possess them are called **vertebrates**. The first vertebrates were primitive bony fishes. Today, there are seven classes of vertebrates: three classes of fishes, plus amphibians, reptiles, birds, and mammals.

Recently an important fossil was discovered in northern Canada (Daeschler, Shubin, & Jenkins, 2006; Shubin, Daeschler, & Jenkins, 2006). The fossil is about 375 million years old, from a time when some fish were starting to evolve into four-legged land vertebrates. This fossilized creature had been a little of each: Along with the scales, teeth, and gills of a fish, it had several anatomical features found only in land animals (such as primitive wrists and finger bones). In short, this is just the type of link between fish and land vertebrates predicted by the theory of evolution. See Figure 2.6 on page 28.

Evolution of Amphibians About 410 million years ago, the first bony fishes started to venture out of the water. Fishes that could survive on land for brief periods of time had two great advantages: They could escape from stagnant pools to nearby fresh water, and they could take advantage of terrestrial food sources. The advantages of life on land were so great that natural selection transformed the fins and gills of bony fishes to legs and lungs, respectively—and so it was that the first **amphibians** evolved about 400 million years ago. Amphibians (e.g., frogs, toads, and salamanders) in their

FIGURE 2.6 A recently discovered fossil of a missing evolutionary link is shown on the right, and a reconstruction of the creature is shown on the left. It had scales, teeth, and gills like a fish and primitive wrist and finger bones similar to those of land animals.

larval form must live in the water; only adult amphibians can survive on land.

Evolution of Reptiles About 300 million years ago, reptiles (e.g., lizards, snakes, and turtles) evolved from a branch of amphibians. Reptiles were the first vertebrates to lay shell-covered eggs and to be covered by dry scales. Both of these adaptations reduced the reliance of reptiles on watery habitats. A reptile does not have to spend the first stage of its life in the watery environment of a pond or lake; instead, it spends the first stage of its life in the watery environment of a shell-covered egg. And once hatched, a reptile can live far from water, because its dry scales greatly reduce water loss through its water-permeable skin.

Evolution of Mammals About 180 million years ago, during the height of the age of dinosaurs, a new class of vertebrates evolved from one line of small reptiles. The females of this new class fed their young with secretions from special glands called *mammary glands*, and the members of the class are called **mammals** after these glands. Eventually, mammals stopped laying eggs; instead, the females nurtured their young in the watery environment of their bodies until the young were mature enough to be born. The duck-billed platypus is one surviving mammalian species that lays eggs.

Spending the first stage of life inside one's mother proved to have considerable survival value; it provided the long-term security and environmental stability necessary for complex programs of development to unfold. Today, most classification systems recognize about 20 different orders of mammals. The order to which we belong is the order **primates**. We humans—in our usual humble way—named our order using the Latin term *primus*, which means "first" or "foremost."

Primates have proven particularly difficult to categorize because there is no single characteristic that is possessed by all primates but no other animals. Still most experts agree that there are about a dozen families of primates. Members of five of them appear in Figure 2.7.

Apes (gibbons, orangutans, gorillas, and chimpanzees) are thought to have evolved from a line of Old-World monkeys. Like Old-World monkeys, apes have long arms and grasping hind feet that are specialized for arboreal (treetop) travel, and they have opposable thumbs that are not long enough to be of much use for precise manipulation (see Figure 2.8 on page 30). Unlike Old-World monkeys, though, apes have no tails and can walk upright for short distances. Chimpanzees are the closest living relatives of humans; almost 99% of genes are identical in the two species (Chimpanzee Sequencing and Analysis Consortium, 2005; Pollard, 2009)—but see Cohen (2007). However, the actual ape ancestor of humans is likely long extinct (Jaeger & Marivaux, 2005).

Emergence of Humankind Primates of the family that includes humans are the **hominins**. According to the simplest view, this family is composed of two *genera* (the plural of *genus*): *Australopithecus* and *Homo*. *Homo* is thought to be composed of two species: *Homo erectus*, which is extinct, and *Homo sapiens* (humans), which is not. One version of the *taxonomy* (biological classification) of the human species is presented in Figure 2.9 on page 30.

It is extremely difficult to reconstruct the events of human evolution because the evidence is so sparse. Only a few hominin fossils dating from the critical period have been discovered, and they are only fragments (a jawbone and a few teeth). There have, however, been some exciting recent findings. For example, fossil evidence indicates that a population of 3-foot-tall hominins inhabited an Indonesian Island (Flores) as recently as 13,000 years ago (see Diamond, 2004; Wong, 2005a). Also, an uncommonly complete fossil of a 3-year-old Australopithecus girl was recently found in Ethiopia (see Figure 2.10 on page 30).

Most experts believe that the australopithecines evolved about 6 million years ago in Africa (Lovejoy et al., 2009; White et al., 2009), from a line of apes (*australo*

APE
Silver-Backed Lowland Gorilla

FIGURE 2.7   Examples of the five different families of primates.

PROSIMIAN
Tarsus Monkey

OLD-WORLD MONKEY
Hussar Monkey

NEW-WORLD MONKEY
Squirrel Monkey

HOMININ
Human

record by modern humans (*Homo sapiens*). Then, about 50,000 years ago, modern humans began to migrate out of Africa (Anikovich et al., 2007; Goebel, 2007; Grine et al., 2007).

Paradoxically, although the big three human attributes—large brain, upright posture, and free hands with an opposable thumb—have been evident for hundreds of thousands of years, most human accomplishments are of recent origin. Artistic products (e.g., wall paintings and carvings) did not appear until about 40,000 years ago, ranching and farming were not established until about 10,000 years ago (e.g., Dillehay et al., 2007), and writing was not invented until about 3,500 years ago.

Thinking about Human Evolution

Figure 2.12 on page 32 illustrates the main branches of vertebrate evolution. As you examine it, consider the following commonly misunderstood points about evolution. They should provide you with a new perspective from which to consider your own origins.

means "southern," and *pithecus* means "ape"). Several species of *Australopithecus* are thought to have roamed the African plains for about 5 million years before becoming extinct. Australopithecines were only about 1.3 meters (4 feet) tall, and they had small brains; but analysis of their pelvis and leg bones indicates that their posture was as upright as yours or mine. Any doubts about their upright posture were erased by the discovery of the fossilized footprints pictured in Figure 2.11 on page 31 (Agnew & Demas, 1998).

The first *Homo* species are thought to have evolved from one species of *Australopithecus* about 2 million years ago (Spoor et al., 2007). One distinctive feature of the early *Homo* species was the large size of their brain cavity, larger than that of *Australopithecus,* but smaller than that of modern humans. The early *Homo* species used fire and tools (see Ambrose, 2001) and coexisted in Africa with various species of *Australopithecus* for about a half-million years, until the australopithecines died out.

About 200,000 years ago (Pääbo, 1995), early *Homo* species were gradually replaced in the African fossil

- Evolution does not proceed in a single line. Although it is common to think of an evolutionary ladder or scale, a far better metaphor for evolution is a dense bush.
- We humans have little reason to claim evolutionary supremacy. We are the last surviving species of a family (i.e., hominins) that has existed for only a blip of evolutionary time.
- Evolution does not always proceed slowly and gradually. Rapid evolutionary changes (i.e., in a few generations) can be triggered by sudden changes in the environment or by adaptive genetic mutations. Whether human evolution occurred gradually or suddenly is still a matter of intense debate among *paleontologists* (those who scientifically study fossils). About the time that hominins evolved, there was a sudden cooling of the earth, leading to a decrease in African forests and an increase in African grasslands (see Behrensmeyer, 2006). This may have accelerated human evolution.

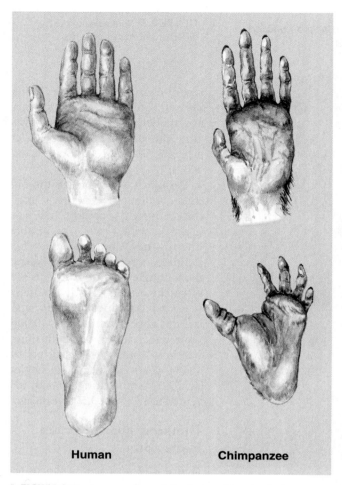

FIGURE 2.8 A comparison of the feet and hands of a human and a chimpanzee.

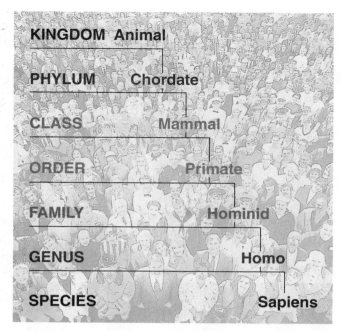

FIGURE 2.9 A taxonomy of the human species.

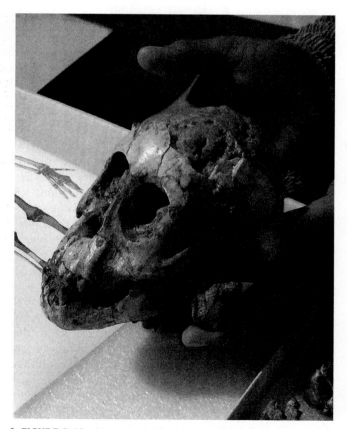

FIGURE 2.10 The remarkably complete skull of a 3-year-old *Australopithecus* girl. The fossil is 3.3 million years old.

- Few products of evolution have survived to the present day—only the tips of the branches of the evolutionary bush have survived. Fewer than 1% of all known species are still in existence.

- Evolution does not progress to preordained perfection—evolution is a tinkerer, not an architect. Increases in adaptation occur through changes to existing programs of development; and although the results are improvements in their particular environmental context, they are never perfect designs. For example, the fact that mammalian sperm do not develop effectively at body temperature led to the evolution of the scrotum—hardly a perfect solution to any design problem (see Shubin, 2009).

- Not all existing behaviors or structures are adaptive. Evolution often occurs through changes in developmental programs that lead to several related characteristics, only one of which might be adaptive—the incidental nonadaptive evolutionary by-products are called **spandrels**.

The human belly button is a spandrel; it serves no adaptive function and is merely the by-product of the umbilical cord. Also, behaviors or structures that were once adaptive might become nonadaptive, or even maladaptive, if the environment changes.

- Not all existing adaptive characteristics evolved to perform their current function. Some characteristics, called **exaptations**, evolved to perform one function and were later co-opted to perform another. For example, bird wings are exaptations—they are limbs that first evolved for the purpose of walking.
- Similarities among species do not necessarily mean that the species have common evolutionary origins. Structures that are similar because they have a common evolutionary origin are termed **homologous**; structures that are similar but do not have a common evolutionary origin are termed **analogous**. The similarities between analogous structures result from **convergent evolution**, the evolution in unrelated species of similar solutions to the same environmental demands. Deciding whether a structural similarity is analogous or homologous requires careful analysis of the similarity. For example, a bird's wing and a human's arm have a basic underlying commonality of skeletal structure that suggests a common ancestor; in contrast, a bird's wing and a bee's wing have few structural similarities, although they do serve the same function.

Evolution of the Human Brain

Early research on the evolution of the human brain focused on size. This research was stimulated by the assumption that brain size and intellectual capacity are closely related—an assumption that quickly ran into two problems. First, it was shown that modern humans, whom modern humans believe to be the most intelligent of all creatures, do not have the biggest brains. With brains weighing about 1,350 grams, humans rank far behind whales and elephants, whose brains weigh between 5,000 and 8,000 grams (Harvey & Krebs, 1990). Second, the sizes of the brains of acclaimed intellectuals (e.g., Einstein) were found to be unremarkable, certainly no match for their gigantic intellects. It is now clear that, although healthy adult human brains vary greatly in size—between about 1,000 and 2,000 grams—there is no clear relationship between overall human brain size and intelligence.

One obvious problem in relating brain size to intelligence is the fact that larger animals tend to have larger brains, presumably because larger bodies require more brain tissue to control and regulate them. Thus, the facts that large men tend to have larger brains than small men, that men tend to have larger brains than women, and that elephants have larger brains than humans do not suggest anything about the relative intelligence of these populations. This problem led to the proposal that brain weight expressed as a percentage of total body weight might be a better measure of intellectual capacity. This measure allows humans (2.33%) to take their rightful place ahead of elephants (0.20%), but it also allows both humans and elephants to be surpassed by that

FIGURE 2.11 Fossilized footprints of *Australopithecus* hominins who strode across African volcanic ash about 3.6 million years ago. They left a 70-meter trail. There were two adults and a child; the child often walked in the footsteps of the adults.

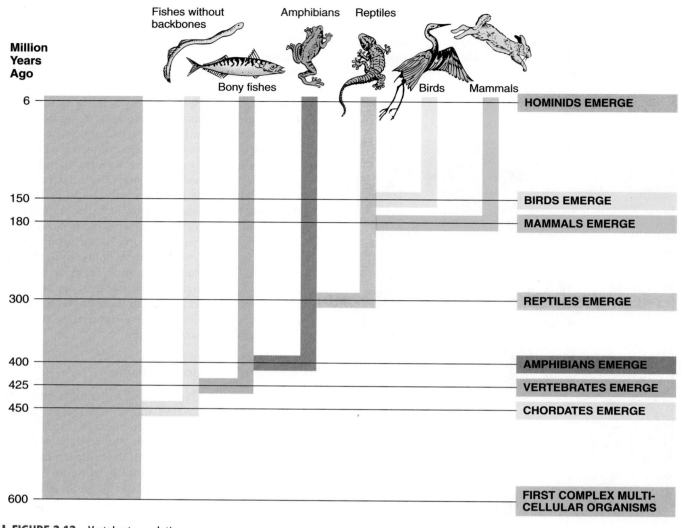

FIGURE 2.12 Vertebrate evolution.

intellectual giant of the animal kingdom, the shrew (3.33%).

A more reasonable approach to the study of brain evolution has been to compare the evolution of different brain regions (Finlay & Darlington, 1995; Killacky, 1995). For example, it has been informative to consider the evolution of the **brain stem** separately from the evolution of the **cerebrum** (cerebral hemispheres). In general, the brain stem regulates reflex activities that are critical for survival (e.g., heart rate, respiration, and blood glucose level), whereas the cerebrum is involved in more complex adaptive processes such as learning, perception, and motivation.

Figure 2.13 is a schematic representation of the relative size of the brain stems and cerebrums of several species that are living descendants of species from which humans evolved. This figure makes three important points about the evolution of the human brain:

- It has increased in size during evolution.
- Most of the increase in size has occurred in the cerebrum.
- An increase in the number of **convolutions**—folds on the cerebral surface—has greatly increased the volume of the *cerebral cortex,* the outermost layer of cerebral tissue (Hilgetag & Barbas, 2009).

Although there are differences among the brains of related species, there is a fundamental similarity: All brains are constructed of neurons, and the neural structures in the brains of one species can usually be found in the brains of related species (see Passingham, 2009). For example, the brains of humans, monkeys, rats, and mice contain the same major structures connected in the same ways, and similar structures tend to perform similar functions (see Cole et al., 2009). Human abilities appear to result from the modification of abilities found in our closest evolutionary relatives (see Herrmann et al., 2007; Landau & Lakusta, 2009).

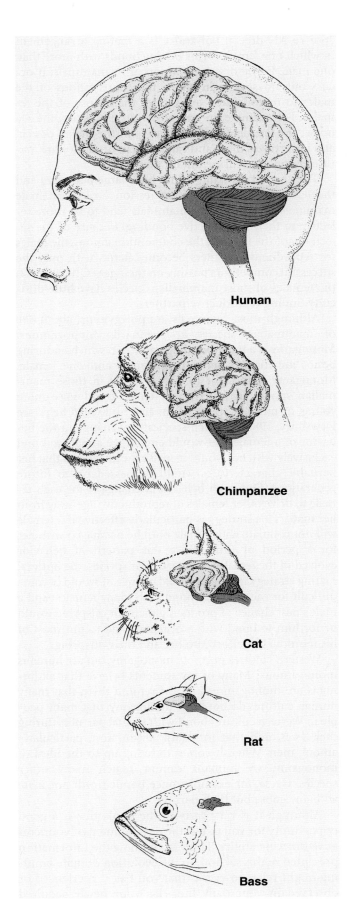

Human

Chimpanzee

Cat

Rat

Bass

Evolutionary Psychology: Understanding Mate Bonding

The evolutionary approach has been embraced by many psychologists. Indeed, a new field of psychology, termed *evolutionary psychology*, has coalesced around it. Evolutionary psychologists try to understand human behaviors through a consideration of the pressures that led to their evolution (see Schmitt & Pilcher, 2004). Some of the most interesting and controversial work in this field has focused on questions of sex differences in mate bonding, questions you may be dealing with in your own life.

◉ **Watch**
Dating and Finding a Mate:
Ralf and Stephani
www.mypsychlab.com

In most species, mating is totally promiscuous. *Promiscuity* is a mating arrangement in which the members of both sexes indiscriminately copulate with many different partners during each mating period. Although such indiscriminate copulation is the predominant mode of reproduction, the males and females of some species form *mating bonds* (enduring mating relationships) with members of the other sex.

Most mammals tend to form mating bonds. Why? An influential theory, proposed by Trivers (1972), attributes the evolution of mate bonding in many mammalian species to the fact that female mammals give birth to relatively small numbers of helpless, slow-developing young. As a result, it is adaptive for the males of many mammalian species to stay with the females who are carrying their offspring and to promote the successful development of those offspring. A male mammal that behaves in this way may be more likely to pass on his heritable characteristics to future generations. Thus, natural selection has promoted the evolution in mammalian males of the tendency to bond with the females with which they have copulated. Similarly, there is selection pressure on female mammals to behave in ways that will induce males to bond to them because this improves their ability to pass on their own heritable characteristics to future generations. In many species, mating bonds last a lifetime. But what kind of mating bonds do mammals tend to form, and why?

The pattern of mate bonding that is most prevalent in mammals is **polygyny** (pronounced "pol-IG-in-ee"), an arrangement in which one male forms mating bonds with more than one female. Why did polygyny evolve in so many mammalian species? The evidence suggests that polygyny evolved as the predominant pattern of mate bonding in mammals because female mammals make a far greater contribution to the rearing of their young than

FIGURE 2.13 The brains of animals of different evolutionary ages. Cerebrums are shown in yellow; brain stems are shown in purple.

do males (Trivers, 1972). Mammalian mothers carry their developing young in their bodies, sometimes for many months, and then suckle and care for them after they are born. In contrast, mammalian fathers often contribute little more to reproduction than sperm. One major consequence of this common one-sided mammalian parenting arrangement is that the females of most mammalian species can produce only a few offspring during their lifetimes, whereas males have the capacity to sire many offspring.

Because each female mammal can produce only a few offspring, she must make the best of her chances if her heritable characteristics are going to be passed on to future generations in significant numbers. In particular, it is important that she mate with particularly fit males. Mating with fit males increases the likelihood that her offspring will be fit and will pass on her genes, along with those of her mate, to the next generation; it also increases the likelihood that what little parental support her offspring will receive from their father will be effective. Thus, according to current theory, the tendency to establish mating bonds with only the fittest males evolved in females of many mammalian species. In contrast, because male mammals can sire so many offspring, there has been little evolutionary pressure on them to become selective in their bonding—the males of most mammalian species will form mating bonds with as many females as possible. The inevitable consequence of the selective bonding of female mammals and the nonselective bonding of male mammals is polygyny—see Figure 2.14.

The strongest evidence in support of the theory that polygyny evolves when females make a far greater contribution to reproduction and parenting than males do comes from the studies of **polyandry** (pronounced

FIGURE 2.14 Horses, like most mammals, are polygynous. The stallion breeds with all the mares in the herd by virtue of his victories over other males.

"pol-ee-AN-dree"). Polyandry is a mating arrangement in which one female forms mating bonds with more than one male. Polyandry does not occur in mammals; it occurs only in species in which the contributions of the males to reproduction are greater than those of the females. For example, in one polyandrous species, the sea horse, the female deposits her eggs in the male's pouch, and he fertilizes them and carries them until they are mature enough to venture out on their own.

The current thinking is that both large body size and the tendency to engage in aggression evolved in male mammals because female mammals tend to be more selective in their reproductive bonding. Because of the selectivity of the females, the competition among the males for reproductive partners becomes fierce, with only the successful competitors passing on their genes. In contrast, the females of most mammalian species have little difficulty finding reproductive partners.

Although most mammals are polygynous, about 4% of mammalian species are primarily monogamous. **Monogamy** is a mate-bonding pattern in which enduring bonds are formed between one male and one female. Monogamy is thought to have evolved in those mammalian species in which each female could raise more young, or more fit young, if she had undivided help (see Dewsbury, 1988). In such species, any change in the behavior of a female that would encourage a male to bond exclusively with her would increase the likelihood that her heritable characteristics would be passed on to future generations. One such behavioral change is for each female to drive other females of reproductive age away from her mate. This strategy is particularly effective if a female will not copulate with a male until he has stayed with her for a period of time. Once this pattern of behavior evolved in the females of a particular species, the optimal mating strategy for males would change. It would become difficult for each male to bond with many females, and a male's best chance of producing many fit offspring would be for him to bond with a fit female and to put most of his reproductive effort into her and their offspring.

Western cultures promote monogamy, but are humans monogamous? Many of my students believe that all humans are monogamous, until I remind them that many human cultures do not favor monogamy, that many people in Western cultures bond with several partners during their lives, and that infidelity is common, particularly among men. When it comes to living up to the ideal of monogamy, we humans cannot match many other species. Geese, for example, once bonded, will not mate with any goose but their partner.

Although it is very early in the book, this is a good opportunity for you to test the development of your creative thinking ability. Think about how the information presented in this section on the evolution of mate bonding might relate to events that you have experienced or observed in your daily life. Has your newly acquired

evolutionary perspective enabled you to think about these events in new ways?

Thinking about Evolutionary Psychology

It is important not to lose sight of the fact that the significance of evolutionary psychology does not lie in the many theories it has generated. It is easy to speculate about how particular human behaviors evolved without ever having one's theories disproved, because it is not possible to know for sure how an existing behavior evolved. Good theories of behavioral evolution have predictions about current behaviors built into them so that the predictions—and thus the theory—can be tested. Theories that cannot be tested have little use.

The foregoing evolutionary theory of mate bonding has led to several predictions about current aspects of human mate selection. Buss (1992) has confirmed several of them, for example:

- Men in most cultures value youth and attractiveness (both indicators of fertility) in their mates more than women do; in contrast, women value power and earning capacity more than men do.
- Physical attractiveness best predicts which women will bond with men of high occupational status.
- The major mate-attraction strategy of women is increasing their physical attractiveness; in men, it is displaying their power and resources.
- Men are more likely than women to commit adultery.

It is important to appreciate that behavioral tendencies shaped by evolution exist in humans without any need for our awareness of them or their evolutionary origins. It's also important to remember that all inherited tendencies are modulated by experience.

Scan Your Brain

This is a good place for you to pause to scan your brain to see if you are ready to proceed: Do you remember what you have read about misleading dichotomies and evolution? Fill in the following blanks with the most appropriate terms from the first two sections of the chapter. The correct answers are provided at the end of the exercise. Before proceeding, review material related to your errors and omissions.

1. The _____ side of the nature–nurture controversy is that all behavior is learned.

2. Physiological-or-psychological thinking was given official recognition in the 17th century when the Roman Church officially supported _____.
3. In the Darwinian sense, _____ refers to the ability of an organism to survive and produce large numbers of fertile offspring.
4. A _____ is a group of reproductively isolated organisms.
5. Mammals are thought to have evolved from _____ about 180 million years ago.
6. _____ are the closest living relatives of humans; they have about 99% of the same genetic material as humans.
7. The first hominins were the _____.
8. The best metaphor for evolution is not a ladder; it is a dense _____.
9. Fewer than _____ % of all known species still exist.
10. Nonadaptive structures or behaviors that evolved because they were linked to a characteristic that was adaptive are called _____.
11. Structures or behaviors that evolved to perform one function but were later co-opted to perform another are called _____.
12. Structures that are similar because they have a common evolutionary origin are called _____ structures.
13. _____ structures are similar because of convergent evolution.

Scan Your Brain answers: (1) nurture, (2) Cartesian dualism, (3) fitness, (4) species, (5) reptiles, (6) Chimpanzees, (7) australopithecines, (8) bush, (9) 1, (10) spandrels, (11) exaptations, (12) homologous, (13) Analogous.

2.3
Fundamental Genetics

Darwin did not understand two of the key facts on which his theory of evolution was based. He did not understand why conspecifics differ from one another, and he did not understand how anatomical, physiological, and behavioral characteristics are passed from parent to offspring. While Darwin puzzled over these questions, there was an unread manuscript in his files that contained the answers. It had been sent to him by an unknown Augustinian monk, Gregor Mendel. Unfortunately for Darwin (1809–1882) and for Mendel (1822–1884), the significance of Mendel's research was not recognized until the early part of the 20th century, well after both of their deaths.

Mendelian Genetics

Mendel studied inheritance in pea plants. In designing his experiments, he made two wise decisions. He decided to study dichotomous traits, and he decided to begin his experiments by crossing the offspring of true-breeding lines. **Dichotomous traits** are traits that occur in one form or the other, never in combination. For example, seed color is a dichotomous pea plant trait: Every pea plant has either brown seeds or white seeds. **True-breeding lines** are breeding lines in which interbred members always produce offspring with the same trait (e.g., brown seeds), generation after generation.

In one of his early experiments, Mendel studied the inheritance of seed color: brown or white. He began by cross breeding the offspring of a line of pea plants that had bred true for brown seeds with the offspring of a line of pea plants that had bred true for white seeds. The offspring of this cross all had brown seeds. Then, Mendel bred these first-generation offspring with one another, and he found that about three-quarters of the resulting second-generation offspring had brown seeds and about one-quarter had white seeds. Mendel repeated this experiment many times with various pairs of dichotomous pea plant traits, and each time the result was the same. One trait, which Mendel called the **dominant trait**, appeared in all of the first-generation offspring; the other trait, which he called the **recessive trait**, appeared in about one-quarter of the second-generation offspring. Mendel would have obtained a similar result if he had conducted an experiment with true-breeding lines of brown-eyed (dominant) and blue-eyed (recessive) humans.

The results of Mendel's experiment challenged the central premise on which all previous ideas about inheritance had rested: that offspring inherit the traits of their parents. Somehow, the recessive trait (white seeds) was passed on to one-quarter of the second-generation pea plants by first-generation pea plants that did not themselves possess it. An organism's observable traits are referred to as its **phenotype**; the traits that it can pass on to its offspring through its genetic material are referred to as its **genotype**.

Mendel devised a theory to explain his results. It comprised four ideas. First, Mendel proposed that there are two kinds of inherited factors for each dichotomous trait—for example, that a brown-seed factor and a white-seed factor control seed color. Today, we call each inherited factor a **gene**. Second, Mendel proposed that each organism possesses two genes for each of its dichotomous traits; for example, each pea plant possesses either two brown-seed genes, two white-seed genes, or one of each. The two genes that control the same trait are called **alleles** (pronounced "a-LEELZ"). Organisms that possess two identical genes for a trait are said to be **homozygous** for that trait; those that possess two different genes for a trait are said to be **heterozygous** for that trait. Third, Mendel proposed that one of the two kinds of genes for each dichotomous trait dominates the other in heterozygous organisms. For example, pea plants with a brown-seed gene and a white-seed gene always have brown seeds because the brown-seed gene always dominates the white-seed gene. And fourth, Mendel

proposed that for each dichotomous trait, each organism randomly inherits one of its "father's" two factors and one of its "mother's" two factors. Figure 2.15 illustrates how Mendel's theory accounts for the result of his experiment on the inheritance of seed color in pea plants.

Chromosomes: Reproduction and Recombination

It was not until the early 20th century that genes were found to be located on **chromosomes**—the threadlike structures in the *nucleus* of each cell. Chromosomes occur in matched pairs, and each species has a characteristic number of pairs in each of its body cells; humans have 23 pairs. The two genes (alleles) that control each trait are situated at the same location, one on each chromosome of a particular pair.

The process of cell division that produces **gametes** (egg cells and sperm cells) is called **meiosis** (pronounced "my-OH-sis")—see Sluder and McCollum (2000). In meiosis, the chromosomes divide, and one chromosome of each pair goes to each of the two gametes that results from the cell division. As a result, each gamete has only half the usual number of chromosomes (23 in humans); and when a sperm cell and an egg cell combine during fertilization (see Figure 2.16), a **zygote** (a fertilized egg cell) with the full complement of chromosomes is produced.

The random division of the pairs of chromosomes into two gametes is not the only way meiosis contributes to genetic diversity. Let me explain. During the first stage of meiosis, the chromosomes line up in their pairs. Then, the members of each pair cross over one another at random points, break apart at the points of contact, and exchange sections of themselves. As a result of this **genetic recombination**, each of the gametes that formed the zygote that developed into you contained chromosomes that were unique, spliced-together recombinations of chromosomes from your mother and father.

In contrast to the meiotic creation of the gametes, all other cell division in the body occurs by **mitosis** (pronounced "my-TOE-sis"). Just prior to mitotic division, the number of chromosomes doubles so that, when the cell divides, both daughter cells end up with the full complement of chromosomes.

Chromosomes: Structure and Replication

Each chromosome is a double-stranded molecule of **deoxyribonucleic acid (DNA)**. Each strand is a sequence of **nucleotide bases** attached to a chain of *phosphate* and *deoxyribose*; there are four nucleotide bases: *adenine*, *thymine*, *guanine*, and *cytosine*. It is the sequence of these bases on each chromosome that constitutes the genetic code—just as sequences of letters constitute the code of our language.

The two strands that compose each chromosome are coiled around each other and bonded together by the

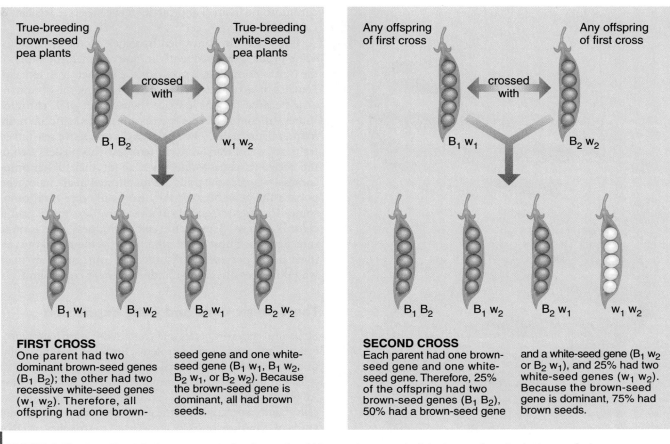

FIRST CROSS
One parent had two dominant brown-seed genes (B_1 B_2); the other had two recessive white-seed genes (w_1 w_2). Therefore, all offspring had one brown-seed gene and one white-seed gene (B_1 w_1, B_1 w_2, B_2 w_1, or B_2 w_2). Because the brown-seed gene is dominant, all had brown seeds.

SECOND CROSS
Each parent had one brown-seed gene and one white-seed gene. Therefore, 25% of the offspring had two brown-seed genes (B_1 B_2), 50% had a brown-seed gene and a white-seed gene (B_1 w_2 or B_2 w_1), and 25% had two white-seed genes (w_1 w_2). Because the brown-seed gene is dominant, 75% had brown seeds.

FIGURE 2.15   How Mendel's theory accounts for the results of his experiment on the inheritance of seed color in pea plants.

attraction of adenine for thymine and guanine for cytosine. This specific bonding pattern has an important consequence: The two strands that compose each chromosome are exact complements of each other. For example, the sequence of adenine, guanine, thymine, cytosine, and guanine on one strand is always attached to the complementary sequence of thymine, cytosine, adenine, guanine, and cytosine on the other. Figure 2.17 on page 38 illustrates the structure of DNA.

Replication is a critical process of the DNA molecule. Without it, mitotic cell division would not be possible. Figure 2.18 on page 39 illustrates how DNA replication is thought to work. The two strands of DNA start to unwind. Then the exposed nucleotide bases on each of the two strands attract their complementary bases, which are floating in the fluid of the nucleus. Thus, when the unwinding is complete, two double-stranded DNA molecules, both of which are identical to the original, have been created.

Chromosome replication does not always go according to plan; there may be errors. Sometimes, these errors are gross errors. For example, in *Down syndrome*, which you will learn about in Chapter 10, there is an extra chromosome in each cell. But more commonly, errors in duplication take the form of **mutations**—accidental alterations in individual genes. In most cases, mutations disappear from the gene pool within a few generations because the organisms that inherit them are less fit. However, in rare instances, mutations increase fitness and in so doing contribute to rapid evolution.

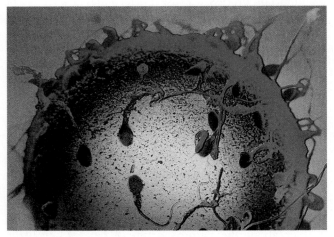

FIGURE 2.16   During fertilization, sperm cells attach themselves to the surface of an egg cell; only one will enter the egg cell and fertilize it.

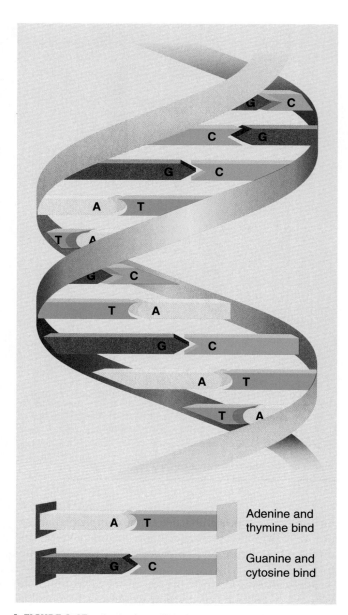

FIGURE 2.17 A schematic illustration of the structure of a DNA molecule. Notice the complementary pairings of nucleotide bases: thymine with adenine, and guanine with cytosine.

Adenine and thymine bind

Guanine and cytosine bind

Sex Chromosomes and Sex-Linked Traits

There is one exception to the rule that chromosomes always come in matched pairs. The typical chromosomes, which come in matched pairs, are called **autosomal chromosomes;** the one exception is the pair of **sex chromosomes**—the pair of chromosomes that determines an individual's sex. There are two types of sex chromosomes, X and Y, and the two look different and carry different genes. Female mammals have two X chromosomes, and male mammals have one X chromosome and one Y chromosome. Traits that are influenced by genes on the sex chromosomes are referred to as **sex-linked traits.** Virtually all sex-linked traits are controlled by genes on the X chromosome because the Y

chromosome is small and carries few genes (see Jegalian & Lahn, 2001).

Traits that are controlled by genes on the X chromosome occur more frequently in one sex than the other. If the trait is dominant, it occurs more frequently in females. Females have twice the chance of inheriting the dominant gene because they have twice the number of X chromosomes. In contrast, recessive sex-linked traits occur more frequently in males. The reason is that recessive sex-linked traits are manifested only in females who possess two of the recessive genes—one on each of their X chromosomes—whereas the traits are manifested in all males who possess the gene because they have only one X chromosome. The classic example of a recessive sex-linked trait is color blindness. Because the color-blindness gene is quite rare, females almost never inherit two of them and thus almost never possess the disorder; in contrast, every male who possesses one color-blindness gene is color blind.

The Genetic Code and Gene Expression

Structural genes are genes that contain the information necessary for the synthesis of proteins. **Proteins** are long chains of **amino acids;** they control the physiological activities of cells and are important components of cellular structure. All the cells in the body (e.g., brain cells, hair cells, and bone cells) contain exactly the same genes. How then do different kinds of cells develop? The answer lies in stretches of DNA that lack structural genes—indeed, although all genes were once assumed to be structural genes, those genes compose only a small portion of each chromosome.

The stretches of DNA that lack structural genes are not well understood, but it is clear that they include portions called *enhancers* (or *promoters*). **Enhancers** are stretches of DNA whose function is to determine whether particular structural genes initiate the synthesis of proteins and at what rate. The control of **gene expression** by enhancers is an important process, because it determines how a cell will develop and how it will function once it reaches maturity. Enhancers are like switches, and like switches, they can be regulated in two ways: They can be turned up or they can be turned down. Proteins that bind to DNA and influence the extent to which genes are expressed are called **transcription factors.** Many of the transcription factors that control enhancers are influenced by signals received by the cell from its environment (see West, Griffith, & Greenberg, 2002). If it has not already occurred to you, this is a major mechanism by which experience can interact with genes to influence development (Carroll, Prud'homme, & Gompel, 2008; Flavell & Greenberg, 2008; Wray & Babbitt, 2008). Please pause and consider the relevance of this point to what you learned earlier in this chapter about the nature–nurture issue. Has it given you new insight?   *Thinking Creatively*

The expression of a structural gene is illustrated in Figure 2.19 on page 40. First, the small section of the chromosome that contains the gene unravels, and the unraveled

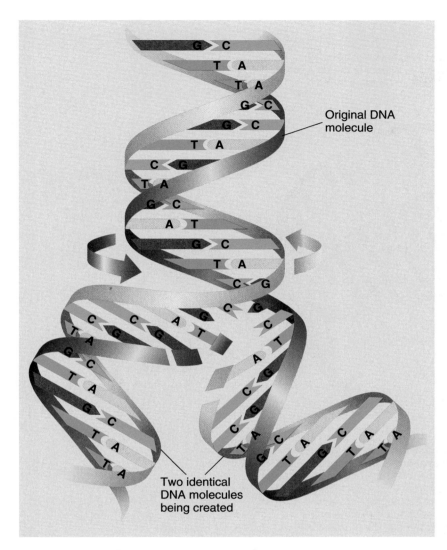

Original DNA molecule

Two identical DNA molecules being created

FIGURE 2.18 DNA replication. As the two strands of the original DNA molecule unwind, the nucleotide bases on each strand attract free-floating complementary bases. Once the unwinding is complete, two DNA molecules, each identical to the first, will have been created.

transfer RNA molecule that is attached to the appropriate amino acid. The ribosome reads codon after codon and adds amino acid after amino acid until it reaches a codon that tells it the protein is complete, whereupon the completed protein is released into the cytoplasm. Thus, the process of gene expression involves two phases: the *transcription* of the DNA base-sequence code to an RNA base-sequence code and the *translation* of the RNA base-sequence code into a sequence of amino acids.

Mitochondrial DNA

So far, we have discussed only the DNA that composes the chromosomes in the cell nucleus. Indeed, you may have the impression that all the DNA is in the nucleus. It isn't. The cells' mitochondria also contain DNA, called *mitochondrial DNA*. **Mitochondria** are the energy-generating structures located in the cytoplasm of every cell, including neurons (see Chapter 3). Human mitochondrial genes are inherited solely from one's mother.

Mitochondrial DNA is of great interest to evolutionary biologists because mutations develop in mitochondrial DNA at a reasonably consistent rate. As a result, mitochondrial DNA can be used as an evolutionary clock (see Kaessmann & Pääbo, 2002). Analysis of mutations of mitochondrial DNA in human populations has added to the substantial evidence—from anthropological, archeological, linguistic, and other genetic analyses—that hominins evolved in Africa and gradually spread over the earth (Goebel, Waters, & O'Rourke, 2008; Stix, 2008; Wallace, 1997)—see Figure 2.20 on page 41.

Evolutionary Perspective

Modern Genetics

Arguably, the most ambitious scientific project of all time began in 1990. Known as the **human genome project**, it was a loosely knit collaboration of major research institutions and individual research teams in several countries. The purpose of this collaboration was to compile a map of the sequence of all 3 billion bases that compose human

section of one of the DNA strands serves as a template for the transcription of a short strand of **ribonucleic acid (RNA)**. RNA is like DNA except that it contains the nucleotide base uracil instead of thymine and has a phosphate and ribose backbone instead of a phosphate and deoxyribose backbone. The strand of transcribed RNA is called **messenger RNA** because it carries the genetic code out of the nucleus of the cell. Once it has left the nucleus, the messenger RNA attaches itself to one of the many **ribosomes** in the cell's *cytoplasm* (the clear fluid within the cell). The ribosome then moves along the strand of messenger RNA, translating the genetic code as it proceeds.

Each group of three consecutive nucleotide bases along the messenger RNA strand is called a **codon**. Each codon instructs the ribosome to add 1 of the 20 different kinds of amino acids to the protein that it is constructing; for example, the sequence guanine-guanine-adenine instructs the ribosome to add the amino acid glycine. Each kind of amino acid is carried to the ribosome by molecules of **transfer RNA**; as the ribosome reads a codon, it attracts a

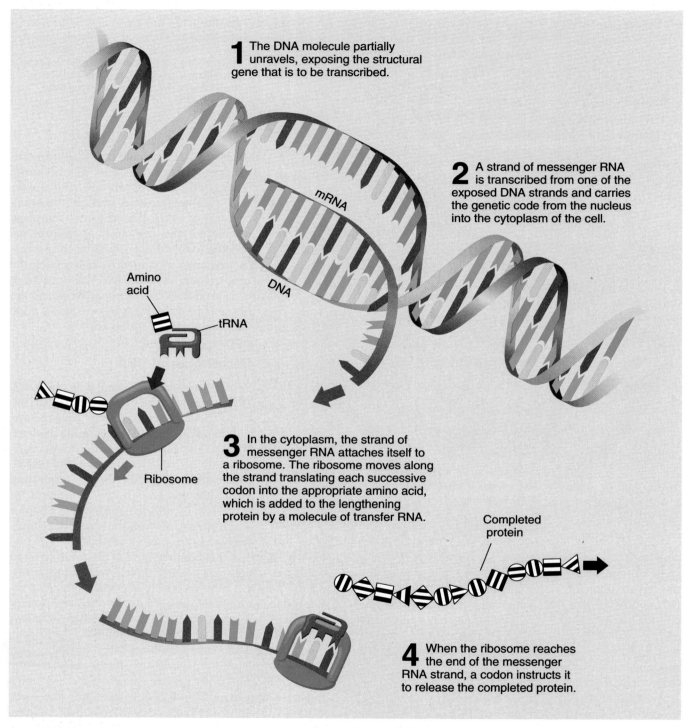

1 The DNA molecule partially unravels, exposing the structural gene that is to be transcribed.

2 A strand of messenger RNA is transcribed from one of the exposed DNA strands and carries the genetic code from the nucleus into the cytoplasm of the cell.

mRNA

DNA

Amino acid

tRNA

Ribosome

3 In the cytoplasm, the strand of messenger RNA attaches itself to a ribosome. The ribosome moves along the strand translating each successive codon into the appropriate amino acid, which is added to the lengthening protein by a molecule of transfer RNA.

Completed protein

4 When the ribosome reaches the end of the messenger RNA strand, a codon instructs it to release the completed protein.

FIGURE 2.19 Gene expression. Transcription of a section of DNA into a complementary strand of messenger RNA is followed by the translation of the messenger RNA strand into a protein.

chromosomes. This ambitious task was completed in 2001, marking the beginning of the modern era of genetics research.

During the compilation of the human genome, many technical advances were implemented to speed up the process. Many researchers began to use this new technol-ogy to compile the genomes of other species. Genomes have been established for many species, and many more will soon be complete.

Undoubtedly, the most surprising result to emerge from research on genomes is the fact that we humans have a relatively small number of genes. Humans have about

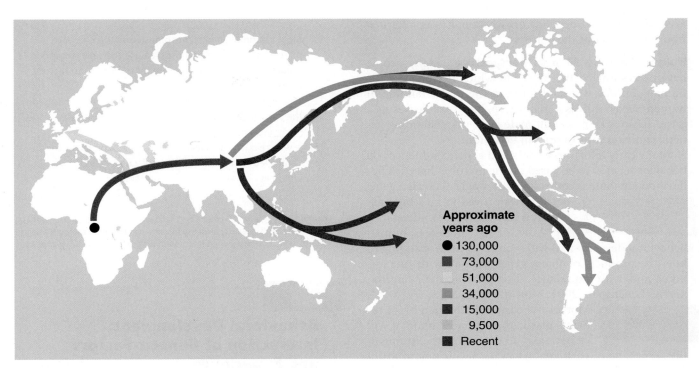

FIGURE 2.20 The analysis of mitochondrial DNA indicates that hominins evolved in Africa and spread over the earth in a series of migrations. (Adapted from Wallace, 1997.)

20,000 genes; mice have about the same number, and corn has many more (Ast, 2005; Lee, Hughes, & Frey, 2006). The unexpectedly small number of human genes in relation to the complexity of the human body suggests that knowledge of how DNA works is far from complete—indeed, protein-encoding (i.e., structural) genes constitute only about 2% of human DNA.

The discovery that genes compose only 2% of human DNA led to the rapid growth of a new field of research: epigenetics. **Epigenetics** focuses on mechanisms that influence the expression of genes without changing the genes themselves (see Bird, 2007). Epigenetic mechanisms are assumed to be the means by which a small number of genes are able to orchestrate the development of humans in all their complexity. The following are four currently influential lines of inquiry in epigenetics.

Active Nongene DNA It had long been assumed that the primary, if not only, function of DNA was the synthesis of proteins. Consequently, those portions of DNA that did not directly participate in the synthesis of proteins were thought to be nonfunctioning evolutionary remnants and were often referred to as *pseudogenes* or *junk DNA*. Today, many areas of *active nongene DNA* are being discovered. Many of these areas control structural gene expression, thus their ability to influence human development and behavior is immense (Gerstein & Zheng, 2006).

MicroRNAs MicroNRAs are short single strands of RNA. Until recently, their function was a mystery, and they were largely ignored by researchers. However, it has been established that there are hundreds of types of microRNAs and that they have major effects on gene expression through their actions on enhancers and messenger RNA. In so doing, they influence brain development (Coolen & Bally-Cuif, 2009; Kosik, 2009) and synapse function (Schratt, 2009), and their disruption has been associated with neurodegenerative disorders (Eaker, Dawson, & Dawson, 2009; Hébert & De Strooper, 2009).

Alternative Splicing It had been a "law" in genetics that one gene encodes one protein. However, the discovery of alternative splicing necessitated revision of this "law." **Alternative splicing** occurs when some strands of messenger RNA are broken apart and the pieces are spliced to new segments. This allows a single gene to encode more than one protein (Li, Lee, & Black, 2007). Alternative splicing is particularly prevalent in neural tissue.

Monoallelic Expression As you have learned, body cells normally have two copies (alleles) of each gene (except for genes on the Y chromosome), and which of the two is expressed depends on their dominance and recessiveness. Recently, it has become apparent that there are many exceptions to this generalization, particularly in the nervous system. In some cases, one of the two alleles is inactivated by as yet unidentified epigenetic mechanisms, and the other is expressed—a phenomenon called **monoallelic expression** (Gimelbrant et al., 2007; Ohlsson, 2007; Wilkinson, Davies, & Isles, 2007). Sometimes, which of two alleles is expressed depends on whether it was inherited from the mother or the father.

Human Genome Map in Perspective Many people have overestimated the degree to which deciphering the human genome will contribute to the understanding of human behavior. It is a major step, but it still leaves us a great distance from the ultimate goal: understanding genetic contributions to human behavior.

Many early efforts to understand genetic influences on behavior seemed to be based on the premise that each behavioral attribute is controlled by a single dedicated gene,

Thinking Creatively

but this has proved to be a big mistake. It is now clear that understanding how genes influence human behavioral development will require an understanding of how the products of many genes interact with one another and with experience through epigenetic mechanisms (Greenspan, 2004; Moffitt, Caspi, & Rutter, 2006).

Neuroplasticity

Moreover, because the human brain remains plastic through adulthood, it will be necessary to determine the timing of the expression of particular genes in specific neural structures throughout the entire life span (McConkey & Varki, 2005). This will not happen soon.

Scan Your Brain

Do you remember what you have just read about genetics so that you can move on to the next section with confidence? To find out, fill in the following blanks with the most appropriate terms. The correct answers are provided at the end of the exercise. Before proceeding, review material related to your errors and omissions.

1. In his ground-breaking experiments, Mendel studied _____ traits in true-breeding lines of pea plants.
2. An organism's observable traits form its _____; the traits that it can pass on to its offspring through its genetic material constitute its _____.
3. The two genes that control each trait are called _____.
4. Organisms that possess two identical alleles for a particular trait are _____ for that trait.
5. Egg cells and sperm cells are _____.
6. All body cells except sperm cells and egg cells are created by _____.
7. Genetic recombination contributes to genetic _____.
8. Each strand of DNA is a sequence of _____ bases.
9. Because organisms that inherit them are less fit, _____ usually disappear from the gene pool within a few generations.

10. Most mammalian chromosomes come in matched pairs; the _____ chromosomes are the only exception.
11. Genes can be turned off or on by transcription factors acting on _____.
12. The massive international research effort that mapped the sequence of bases in human chromosomes was the _____ project.
13. It was recently discovered that gene expression can be controlled by a class of RNA molecules called _____, which act on enhancers and messenger RNA.

Scan Your Brain answers: (1) dichotomous, (2) phenotype, genotype, (3) alleles, (4) homozygous, (5) gametes, (6) mitosis, (7) diversity, (8) nucleotide, (9) mutations, (10) sex, (11) enhancers, (12) human genome, (13) microRNAs.

2.4 Behavioral Development: Interaction of Genetic Factors and Experience

This section of the chapter provides three classic examples of how genetic factors and experience interact to direct behavioral ontogeny. (**Ontogeny** is the development of individuals over their life span; **phylogeny**, in contrast, is the evolutionary development of species through the ages.) These three examples have been particularly influential in shaping modern views of behavioral ontogenetic development. In each example, you will see that this development is a product of gene–experience interaction.

Selective Breeding of "Maze-Bright" and "Maze-Dull" Rats

You have already learned in this chapter that most early psychologists assumed that behavior develops largely through learning. Tryon (1934) undermined this assumption by showing that behavioral traits can be selectively bred.

Tryon focused his selective-breeding experiments on the behavior that had been the focus of early psychologists in their investigations of learning: the maze running of laboratory rats. Tryon began by training a large heterogeneous group of laboratory rats to run a complex maze; the rats received a food reward when they reached the goal box. Tryon then mated the females and males that least frequently entered incorrect alleys during training—he referred to these rats as *maze-bright*. And he bred the females and males that most frequently entered incorrect alleys during training—he referred to these rats as *maze-dull*.

When the offspring of both the maze-bright and the maze-dull rats matured, their maze-learning performance was assessed. Then, the brightest of the maze-bright offspring were mated with one another, as were the dullest of

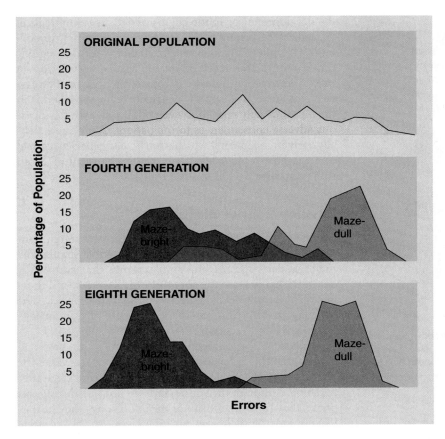

FIGURE 2.21 Selective breeding of maze-bright and maze-dull strains of rats by Tryon (1934)

by the breeding. Indeed, Searle (1949) compared maze-dull and maze-bright rats on 30 different behavioral tests and found that they differed on many of them. The pattern of differences suggested that the maze-bright rats were superior maze learners not because they were more intelligent but because they were less fearful—a trait that is not adaptive in many natural environments.

Selective-breeding studies have proved that genes influence the development of behavior. This conclusion in no way implies that experience does not. This point was driven home by Cooper and Zubek (1958) in a classic study of maze-bright and maze-dull rats. The researchers reared maze-bright and maze-dull rats in one of two environments: (1) an impoverished environment (a barren wire-mesh group cage) or (2) an enriched environment (a wire-mesh group cage that contained tunnels, ramps, visual displays, and other objects designed to stimulate interest). When the maze-dull rats reached maturity, they made significantly more errors than the maze-bright rats only if they had been reared in the impoverished environment (see Figure 2.22 on page 44).

Phenylketonuria: A Single-Gene Metabolic Disorder

It is often easier to understand the genetics of a behavioral disorder than it is to understand the genetics of normal behavior. The reason is that many genes influence the development of a normal behavioral trait, but it sometimes takes only one abnormal gene to screw it up. A good example of this point is the neurological disorder **phenylketonuria (PKU)**.

PKU was discovered in 1934 when a Norwegian dentist, Asbjörn Fölling, noticed a peculiar odor in the urine of his two mentally retarded children. He correctly assumed that the odor was related to their disorder, and he had their urine analyzed. High levels of **phenylpyruvic acid** were found in both samples. Spurred on by his discovery, Fölling identified other retarded children who had abnormally high levels of urinary phenylpyruvic acid, and he concluded that this subpopulation of retarded

the maze-dull offspring. This selective breeding procedure was continued for 21 generations (and the descendants of Tryon's original strains are still available today). By the eighth generation, there was almost no overlap in the maze-learning performance of the two strains. With a few exceptions, the worst of the maze-bright strain made fewer errors than the best of the maze-dull strain (see Figure 2.21).

To control for the possibility that good maze-running performance was somehow being passed from parent to offspring through learning, Tryon used a *cross-fostering control procedure*: He tested maze-bright offspring that had been reared by maze-dull parents and maze-dull offspring that had been reared by maze-bright parents. However, the offspring of maze-bright rats made few errors even when they were reared by maze-dull rats, and the offspring of maze-dull rats made many errors even when they were reared by maze-bright rats.

Since Tryon's seminal selective-breeding experiments, many behavioral traits have been selectively bred. Indeed, it appears that any measurable behavioral trait that varies among members of a species can be selectively bred.

An important general point made by studies of selective breeding is that selective breeding based on one behavioral trait usually brings a host of other behavioral traits along with it. This indicates that the behavioral trait used as the criterion for selective breeding is not the only behavioral trait that is influenced by the genes segregated

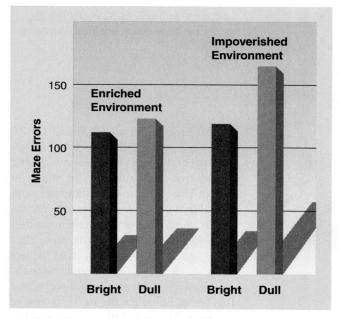

FIGURE 2.22 Maze-dull rats did not make significantly more errors than maze-bright rats when both groups were reared in an enriched environment. (Adapted from Cooper & Zubek, 1958.)

children was suffering from the same disorder. In addition to mental retardation, the symptoms of PKU include vomiting, seizures, hyperactivity, hyperirritability, and brain damage (Antshel & Waisbren, 2003; Sener, 2003).

The pattern of transmission of PKU through the family trees of afflicted individuals indicates that it is transmitted by a single gene mutation. About 1 in 100 people of European descent carry the PKU gene; but because the gene is recessive, PKU develops only in homozygous individuals (those who inherit a PKU gene from both their mother and their father). In the United States, about 1 in 10,000 white infants is born with PKU; the incidence is much lower among infants of African heritage.

The biochemistry of PKU turned out to be reasonably straightforward. PKU homozygotes lack *phenylalanine hydroxylase*, an enzyme that is required for the conversion of the amino acid *phenylalanine* to *tyrosine*. As a result, phenylalanine accumulates in the body; and levels of *dopamine*, a neurotransmitter normally synthesized from tyrosine, are low. The consequence is abnormal brain development.

Like other behavioral traits, the behavioral symptoms of PKU result from an interaction between genetic and environmental factors: between the PKU gene and diet (see Widaman, 2009). Accordingly, in most modern hospitals, the blood of newborn infants is routinely screened for high phenylalanine levels (Saxena, 2003; Wall et al., 2003). If the level is high, the infant is immediately placed on a special phenylalanine-restricted diet; this diet reduces both the amount of phenylalanine in the blood and the development of mental retardation—however, it does not prevent

the development of subtle cognitive deficits (Huijbregts et al., 2002). The timing of this treatment is extremely important. The phenylalanine-restricted diet does not significantly reduce the development of mental retardation in PKU homozygotes unless it is initiated within the first few weeks of life; conversely, the restriction of phenylalanine in the diet is usually relaxed in late childhood, with few obvious adverse consequences to the patient. The period, usually early in life, during which a particular experience must occur to have a major effect on the development of a trait is the **sensitive period** for that trait.

Development of Birdsong

In the spring, the songs of male songbirds threaten conspecific male trespassers and attract potential mates. The males of each species sing similar songs that are readily distinguishable from the songs of other species, and there are recognizable local dialects within each species. The learning of birdsong has many parallels to human language learning (Iacoboni, 2009; Mooney, 2009).

Studies of the ontogenetic development of birdsong suggest that this behavior develops in two phases. The first phase, called the **sensory phase**, begins several days after hatching. Although the young birds do not sing during this phase, they form memories of the adult songs they hear—usually sung by their own male relatives—that later guide the development of their own singing (Aronov, Andalman, & Fee, 2008; London & Clayton, 2008). The young males of many songbird species are genetically prepared to acquire the songs of their own species during the sensory phase. They cannot readily acquire the songs of other species; nor can they acquire the songs of their own species if they do not hear them during the sensory phase. Males who do not hear the songs of their own species early in their lives may later develop a song, but it is likely to be abnormal.

The second phase of birdsong development, the **sensorimotor phase**, begins when the juvenile males begin to twitter *subsongs* (the immature songs of young birds), usually when they are several months old. During this phase, the rambling vocalizations of subsongs are gradually refined until they resemble the songs of the birds' earlier adult tutors. Auditory feedback is necessary for the development of singing during the sensorimotor phase; unless the young birds are able to hear themselves sing, their subsongs do not develop into adult songs (Doupe et al., 2004). However, once stable adult song has crystallized, songbirds are much less dependent on hearing for normal song production (Lombardino & Nottebohm, 2000).

When it comes to the retention of their initial crystallized adult songs, there are two common patterns among songbird species. Most songbird species, such as the widely studied zebra finches and white-crowned sparrows, are *age-limited learners*; in these species, adult songs, once crystallized, remain unchanged for the rest of the birds' lives. In contrast, some species are *open-ended*

FIGURE 2.23 Male zebra finches (age-limited song learners) and male canaries (open-ended song learners) are common subjects of research on birdsong development. (Illustration kindly provided by *Trends in Neuroscience;* original photograph by Arturo Alvarez-Buylla.)

learners; they are able to add new songs to their repertoire throughout their lives. For example, at the end of each mating season, male canaries return from a period of stable song to a period of plastic song—a period during which they can add new songs for the next mating season. Male zebra finches (age-limited learners) and male canaries (open-ended learners) are shown in Figure 2.23.

Figure 2.24 is a simplified version of the neural circuit that controls birdsong in the canary. It has two major components: the descending motor pathway and the anterior forebrain pathway. The *descending motor pathway* descends from the high vocal center on each side of the brain to the syrinx (voice box) on the same side; it mediates song production. The *anterior forebrain pathway* mediates song learning (Doupe, 1993; Vicario, 1991).

The canary song neural circuit is remarkable in four respects. First, the left descending motor pathway plays a more important role in singing than the right descending motor pathway (which duplicates the left-hemisphere dominance for language in humans). Second, the high vocal center is four times larger in male canaries than in females (see MacDougall-Shackleton & Ball, 1999). Third, each spring, as the male canary prepares its new repertoire of songs for the summer seduction, the song-control structures of its brain double in size, only to shrink back in the fall; this springtime burst of brain growth and singing is triggered by elevated levels of the hormone testosterone that result from the increasing daylight (Brenowitz, 2004; Van der Linden et al., 2009). Fourth, the *Neuroplasticity* seasonal increase in size of the song-control brain structures results from the growth of new neurons, not from an increase in the size of existing ones (Tramontin, Hartman, & Brenowitz, 2000)—this finding was one of the first documented examples of adult *neurogenesis* (growth of new neurons).

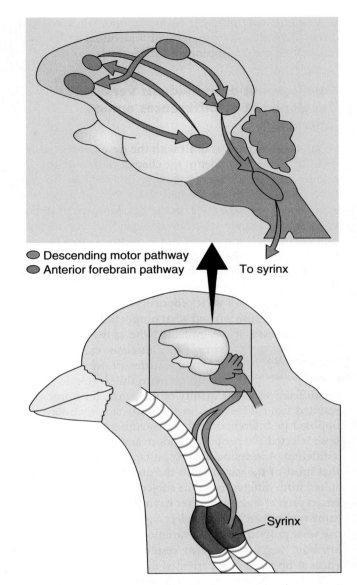

○ Descending motor pathway
○ Anterior forebrain pathway

To syrinx

Syrinx

FIGURE 2.24 The neural pathway responsible for the production and learning of song in the male canary.

2.5

Genetics of Human Psychological Differences

This chapter has focused on three topics—human evolution, genetics, and the interaction of genetics and experience. All three topics converge on one fundamental question: Why are we the way we are? You have learned that each of us is a product of gene–experience interactions and that the effects of genes and experience on individual development are inseparable This final section of the chapter continues to look at the effects of gene–experience interactions, but it focuses on a developmental issue that is fundamentally different from the ones we have been discussing—the development of individual differences rather than the development of individuals.

Development of Individuals versus Development of Differences among Individuals

So far, this chapter has dealt with the development of individuals. The remainder of the chapter deals with the development of differences among individuals. In the development of the individual, the effects of genes and experience are inseparable. In the development of differences among individuals, they are separable. This distinction is extremely important, but it confuses many people. Let me return to the musician metaphor to explain it.

The music of an individual musician is the product of the interaction of the musician and the instrument, and it is nonsensical to ask what proportion of the music is produced by the musician and what proportion by the instrument. However, if we evaluated the playing of a large

Thinking Creatively

sample of musicians, each playing a different instrument, we could statistically estimate the degree to which the differences in the quality of the music they produced resulted from differences in the musicians themselves as opposed to differences in their instruments. For example, if we selected 100 people at random and had each one play a different professional-quality guitar, we would likely find that most of the variation in the quality of the music resulted from differences in the subjects, some being experienced players and some never having played before. In the same way, researchers can select a group of volunteers and ask what proportion of the variation among them in some attribute (e.g., intelligence) results from genetic differences as opposed to experiential differences.

To assess the relative contributions of genes and experience to the development of differences in psychological attributes, behavioral geneticists study individuals of known genetic similarity. For example, they often compare **monozygotic twins** (identical twins), who developed from the same zygote and thus are genetically identical, with **dizygotic twins** (fraternal twins), who developed from two zygotes and thus are no more similar than any pair of siblings. Studies of pairs of monozygotic and dizygotic twins who have been separated at infancy by adoption are particularly informative about the relative contributions of genetics and experience to differences in human psychological development. The most extensive of such adoption studies is the Minnesota Study of Twins Reared Apart (see Bouchard & Pedersen, 1998).

Minnesota Study of Twins Reared Apart

The Minnesota Study of Twins Reared Apart involved 59 pairs of identical twins and 47 pairs of fraternal twins who

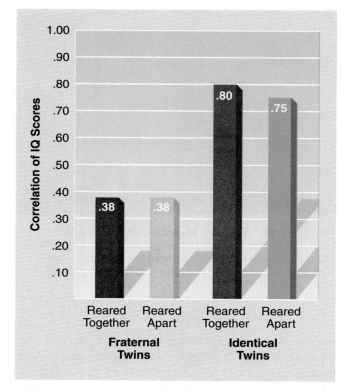

FIGURE 2.25 The correlations of the intelligence quotients (IQs) of identical and fraternal twins, reared together or apart.

had been reared apart, as well as many pairs of identical and fraternal twins who had been reared together. Their ages ranged from 19 to 68 years. Each twin was brought to the University of Minnesota for approximately 50 hours of testing, which focused on the assessment of intelligence and personality. Would the adult identical twins reared apart prove to be similar because they were genetically identical, or would they prove to be different because they had been brought up in different family environments?

The results of the Minnesota Study of Twins Reared Apart proved to be remarkably consistent—both internally, between the various cognitive and personality dimensions that were studied, and externally, with the findings of other, similar studies. In general, adult identical twins were substantially more similar to one another on all psychological dimensions than were adult fraternal twins, whether or not both twins of a pair were raised in the same family environment (see Turkheimer, 2000). General intelligence (as measured by the Wechsler Adult Intelligence Scale) has been the most widely studied psychological attribute of twins; Figure 2.25 illustrates the general pattern of findings (see Bouchard, 1998).

The results of the Minnesota study have been widely disseminated by the popular press. Unfortunately, the meaning of the results has often been distorted. Sometimes, the misrepresentation of science by the popular press does not matter—at least not much. This is not one of those times. People's misbeliefs about the origins of human intelligence and personality are often translated into inappropriate and discriminatory social attitudes and practices (see McGuffin, Riley, & Plomin, 2001). The accompanying newspaper story illustrates how the results of the Minnesota study have been misrepresented to the public.

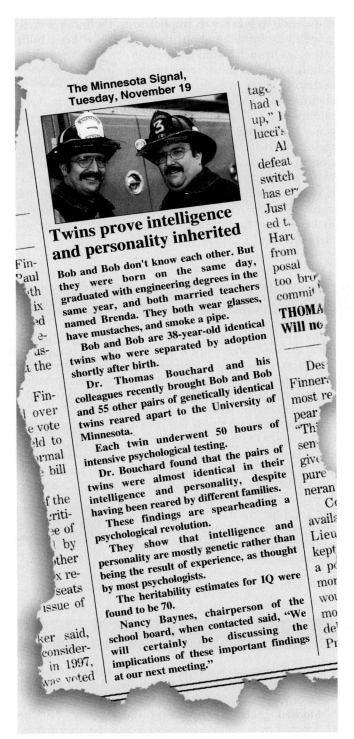

This story is misleading in four ways. You should have no difficulty spotting the first: It oozes nature-or-nurture thinking and all of the misconceptions associated with it. Second, by focusing on the similarities of Bob and Bob, the story creates the impression that Bob and Bob (and the other monozygotic pairs of twins reared apart) are cognitively identical. They are similar, but they are not even close to being identical. It is easy to come up with a long list of similarities between any two people if one asks them enough questions and ignores the dissimilarities. Third, the story creates the impression that the results of the Minnesota study are revolutionary. On the contrary, the importance of the Minnesota study lies mainly in the fact that it constitutes a particularly thorough confirmation of the results of previous adoption studies. Fourth, and most important, the story creates the false impression that the results of the Minnesota study make some general point about the relative contributions of genes and experience to the development of intelligence and personality in individuals. They do not, and neither do the results of any other adoption study. True, Bouchard and his colleagues estimated the heritability of IQ to be .70, but they did not conclude that IQ is 70% genetic. A **heritability estimate** is not about individual development; it is a numerical estimate of the proportion of variability that occurred in a particular trait in a particular study as a result of the genetic variation in that study (see Plomin & DeFries, 1998). Thus, heritability estimates tell us about the contribution of genetic differences to phenotypic differences among the subjects in a study; they have nothing to say about the relative contributions of genes and experience to the development of individuals.

The concept of heritability estimates can be quite confusing. I suggest that you pause here and carefully think through the definition. *Thinking Creatively* The musician metaphor will help.

The magnitude of a study's heritability estimate depends on the amount of genetic and environmental variation from which it was calculated, and it cannot be applied to other kinds of situations. For example, in the Minnesota study, there was relatively little environmental variation. All subjects were raised in industrialized countries (Great Britain, Canada, and the United States) by parents who could meet the strict standards required for adoption. Accordingly, most of the variation in the subjects' intelligence and personality resulted from genetic variation. If the twins had been separately adopted by European royalty, African Bushmen, Hungarian Gypsies, Los Angeles rap stars, London advertising executives, and Argentinian army officers, the resulting heritability estimates for IQ and personality would likely have been lower. Bouchard and his colleagues emphasize this point in their papers.

A commonly overlooked point about the role of genetic factors in the development of human psychological differences is that genetic differences may promote psychological differences by influencing experience (see Plomin & Neiderhiser, 1992). At first, this statement seems paradoxical because we have been conditioned to think of genes and experience as separate developmental influences. However, there is now ample evidence that individuals of similar genetic endowment tend to seek out similar environments and experiences. For example, individuals whose genetic endowments promote aggression are likely to become involved in aggressive activities (e.g., football or competitive fighting), and these experiences are likely to further promote the development of aggressive tendencies. When a particular gene encourages a developing individual to select experiences that increase the behavioral effects of the gene, the gene is said to have a **multiplier effect**.

In an influential paper, Eric Turkheimer (2000) identified three findings that have been supported, almost unanimously, by the results of adoption studies on the heritability of human behavioral traits—in particular, adoption studies that separately evaluated the effects on the development of individual differences of two classes of experience: (1) the effects of the particular family environment in which a person was raised, and (2) the effects of all experiences other than particular family environments. Here are Turkheimer's three consistent findings:

- All human behavioral traits are highly heritable—values of heritability estimates typically range from .40 to .70.

- Being raised in different family environments contributes little to the diversity of behavioral traits.
- Experiences other than the family environment contribute significantly to behavioral diversity.

In thinking about heritability estimates, it is paramount that you remember that heritability estimates depend on the particular subjects in a given study. This point is driven home by the important study of Turkheimer and colleagues (2003). I hope that you take some time to ponder its implications. Its findings challenge fuzzy thinking about the meaning of heritability estimates in such a simple and compelling way that considering them is certain to sharpen your own understanding. Turkheimer and colleagues studied the heritability of IQ in a sample of 7-year-old twins. Unlike the other studies that you have encountered in this chapter, this study focused on the heritability of a trait as a function of socioeconomic status. Remarkably, among the twins in the sample whose families were very poor, the heritability estimate for IQ was low, whereas among the twins from affluent families, it was high. What do you make of these findings and of their implications for social policy?

Thinking Creatively

Themes Revisited

This chapter introduced the topics of evolution, genetics, and development, but its unifying focus was thinking creatively about the biology of behavior. Not surprisingly, then, of this book's four major themes, the thinking creatively theme received the most attention. This chapter challenged you to think about important biopsychological phenomena in new ways. Thinking creatively tabs marked points in the chapter where you were encouraged to sharpen your thinking about the nature–nurture issue, the physiological-or-psychological dichotomy, human evolution, the biopsychological implications of the human genome project, the genetics of human psychological differences, the meaning of heritabiliity estimates, and the important study of Turkheimer and colleagues.

The other three themes also received coverage in this chapter, and it was marked by the appropriate tab in each

Thinking Creatively

case. The evolutionary perspective was illustrated by comparative research on self-awareness in chimps, by consideration of the evolutionary significance of social dominance and courtship displays, by efforts to understand mate bonding, and by the use of mitochondrial DNA to study human evolution. The clinical implications theme was illustrated by the case of the man who fell out of bed and the discussion of phenylketonuria (PKU). The neuroplasticity theme arose at two points: when you learned that the brain's ability to change and develop through adulthood constitutes a major challenge for the field of genetics, and when you learned that brain growth occurs in male songbirds prior to each breeding season.

Evolutionary Perspective

Clinical Implications

Neuroplasticity

Think about It

1. Nature-or-nurture thinking about intelligence is sometimes used as an excuse for racial discrimination. How can the interactionist view, which has been championed in this chapter, be used as a basis for arguing against discriminatory practices?

2. Imagine that you are a biopsychology instructor. One of your students asks you whether depression is physiological or psychological. What would you say?
3. Modern genetics can prevent the tragedy of a life doomed by heredity; embryos can now be screened for

some genetic diseases. But what constitutes a disease? Should genetic testing be used to select a child's characteristics? If so, what characteristics?

4. In the year 2030, a major company demands that all prospective executives take a gene test. As a result, some lose their jobs, and others fail to qualify for health insurance. Discuss.

5. "All men are created equal." Discuss.

6. The field of epigenetics is changing conventional views about the role of genes in human development. Discuss.

Key Terms

Zeitgeist (p. 21)

2.1 Thinking about the Biology of Behavior: From Dichotomies to Relations and Interactions

Cartesian dualism (p. 21)
Nature–nurture issue (p. 21)
Ethology (p. 21)
Instinctive behaviors (p. 21)
Asomatognosia (p. 22)

2.2 Human Evolution

Evolve (p. 24)
Natural selection (p. 25)
Fitness (p. 25)
Species (p. 27)
Conspecifics (p. 27)
Chordates (p. 27)
Vertebrates (p. 27)
Amphibians (p. 27)
Mammals (p. 28)
Primates (p. 28)
Hominins (p. 28)
Spandrels (p. 30)

Exaptations (p. 31)
Homologous (p. 31)
Analogous (p. 31)
Convergent evolution (p. 32)
Brain stem (p. 32)
Cerebrum (p. 32)
Convolutions (p. 32)
Polygyny (p. 33)
Polyandry (p. 34)
Monogamy (p. 34)

2.3 Fundamental Genetics

Dichotomous traits (p. 36)
True-breeding lines (p. 36)
Dominant trait (p. 36)
Recessive trait (p. 36)
Phenotype (p. 36)
Genotype (p. 36)
Gene (p. 36)
Alleles (p. 36)
Homozygous (p. 36)
Heterozygous (p. 36)
Chromosomes (p. 36)
Gametes (p. 36)

Meiosis (p. 36)
Zygote (p. 36)
Genetic recombination (p. 36)
Mitosis (p. 36)
Deoxyribonucleic acid (DNA) (p. 36)
Nucleotide bases (p. 36)
Replication (p. 37)
Mutations (p. 37)
Autosomal chromosomes (p. 38)
Sex chromosomes (p. 38)
Sex-linked traits (p. 38)
Proteins (p. 38)
Amino acids (p. 38)
Enhancers (p. 38)
Gene expression (p. 38)
Transcription factors (p. 38)
Ribonucleic acid (RNA) (p. 39)
Messenger RNA (p. 39)
Ribosomes (p. 39)
Codon (p. 39)
Transfer RNA (p. 39)
Mitochondria (p. 39)
Human genome project (p. 39)

Epigenetics (p. 41)
MicroRNAs (p. 41)
Alternative splicing (p. 41)
Monoallelic expression (p. 41)

2.4 Behavioral Development: The Interaction of Genetic Factors and Experience

Ontogeny (p. 42)
Phylogeny (p. 42)
Phenylketonuria (PKU) (p. 43)
Phenylpyruvic acid (p. 43)
Sensitive period (p. 44)
Sensory phase (p. 44)
Sensorimotor phase (p. 44)

2.5 The Genetics of Human Psychological Differences

Monozygotic twins (p. 46)
Dizygotic twins (p. 46)
Heritability estimate (p. 47)
Multiplier effect (p. 48)

✓ Quick Review   Test your comprehension of the chapter with this brief practice test. You can find the answers to these questions as well as more practice tests, activities, and other study resources at www.mypsychlab.com.

1. A wolf is a conspecific of a
   a. dog.
   b. wolf.
   c. cat.
   d. hyena.
   e. coyote.

2. All humans are
   a. mammals.
   b. vertebrates.
   c. primates.
   d. Homo sapiens.
   e. all of the above

3. The pattern of mate bonding that is most prevalent in mammals is
   a. polygyny.
   b. monogamy.

   c. polygamy.
   d. polyandry.
   e. promiscuity.

4. Factors that influence the expression of genes without changing the genes themselves are often referred to as
   a. epigenetic mechanisms.
   b. gene maps.
   c. monoallelic expressions
   d. mitochondrial factors.
   e. genetic recombination factors.

5. Phenylketonuria commonly develops in individuals who
   a. have PKU in their urine.
   b. are homozygous for the PKU gene.
   c. eat a phenylalanine-free diet.
   d. are the children of Norwegian dentists.
   e. eat a tyrosine-free diet.

3

Anatomy of the Nervous System

Systems, Structures, and Cells That Make Up Your Nervous System

In order to understand what the brain does, it is first necessary to understand what it is—to know the names and locations of its major parts and how they are connected to one another. This chapter introduces you to these fundamentals of brain anatomy.

Before you begin this chapter, I want to apologize for the lack of foresight displayed by early neuroanatomists in their choice of names for neuroanatomical structures—but, then, how could they have anticipated that Latin and Greek, universal languages of the educated in their day, would not be compulsory university fare in our time? To help you, I have provided the literal English meanings of many of the neuroanatomical terms, and I have kept this chapter as brief, clear, and to the point as possible, covering only the most important structures. The payoff for your effort will be a fundamental understanding of the structure of the human brain and a new vocabulary to discuss it.

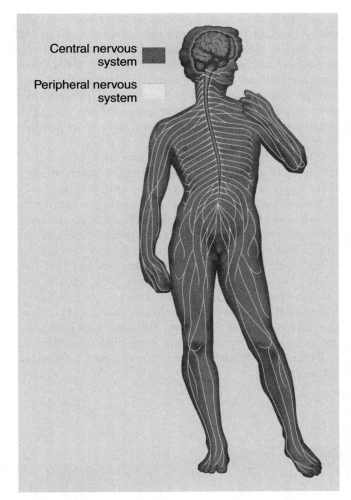

FIGURE 3.1 The human central nervous system (CNS) and peripheral nervous system (PNS). The CNS is represented in red; the PNS in yellow. Notice that even those portions of nerves that are within the spinal cord are considered to be part of the PNS.

3.1

General Layout of the Nervous System

Divisions of the Nervous System

The vertebrate nervous system is composed of two divisions: the central nervous system and the peripheral nervous system (see Figure 3.1). Roughly speaking, the **central nervous system (CNS)** is the division of the nervous system that is located within the skull and spine; the **peripheral nervous system (PNS)** is the division that is located outside the skull and spine.

The central nervous system is composed of two divisions: the brain and the spinal cord. The *brain* is the part of the CNS that is located in the skull; the *spinal cord* is the part that is located in the spine.

The peripheral nervous system is also composed of two divisions: the somatic nervous system and the autonomic nervous system. The **somatic nervous system (SNS)** is the part of the PNS that interacts with the external environment. It is composed of **afferent nerves** that carry sensory signals from the skin, skeletal muscles, joints, eyes, ears, and so on, to the central nervous system, and **efferent nerves** that carry motor signals from the central nervous system to the skeletal muscles. The **autonomic nervous system (ANS)** is the part of the peripheral nervous system that regulates the body's internal environment. It is composed of afferent nerves that carry sensory signals from internal organs to the CNS and efferent nerves that carry motor signals from the CNS to internal organs. You will not confuse the terms *afferent* and *efferent* if you remember that many words that involve the idea of going toward something—in this case, going toward the CNS—begin with an *a* (e.g., *advance, approach, arrive*) and that many words that involve the idea of going away from something begin with an *e* (e.g., *exit, embark, escape*).

The autonomic nervous system has two kinds of efferent nerves: sympathetic nerves and parasympathetic nerves. The **sympathetic nerves** are those autonomic motor nerves that project from the CNS in the *lumbar* (small of the back) and *thoracic* (chest area) regions of the spinal cord. The **parasympathetic nerves** are those autonomic motor nerves that project from the brain and *sacral* (lower back) region of the spinal cord. See Appendix I. (Ask your instructor to specify the degree to which you are responsible for material in the appendices.) All sympathetic and parasympathetic nerves are two-stage neural paths: The sympathetic and parasympathetic neurons project from the CNS and go only part of the way to the target organs before they *synapse on* other neurons (second-stage neurons) that carry the signals the rest of the way. However, the sympathetic and parasympathetic systems differ in that the sympathetic neurons that project from the CNS synapse on second-stage neurons at a substantial distance from their target organs, whereas the parasympathetic neurons that project from the CNS synapse near their target organs on very short second-stage neurons (see Appendix I).

The conventional view of the respective functions of the sympathetic and parasympathetic systems stresses three important principles: (1) that sympathetic nerves stimulate, organize, and mobilize energy resources in threatening situations, whereas parasympathetic nerves act to conserve energy; (2) that each autonomic target organ receives opposing sympathetic and parasympathetic input, and its activity is thus controlled by relative levels of sympathetic and parasympathetic activity; and (3) that sympathetic changes are indicative of psychological arousal, whereas parasympathetic changes are indicative of psychological relaxation. Although these principles are generally correct, there are significant qualifications and exceptions to each of them (see Guyenet, 2006)—see Appendix II.

Most of the nerves of the peripheral nervous system project from the spinal cord, but there are 12 pairs of exceptions: the 12 pairs of **cranial nerves**, which project from the brain. They are numbered in sequence from front to back. The cranial nerves include purely sensory nerves such as the olfactory nerves (I) and the optic nerves (II), but most contain both sensory and motor fibers. The longest cranial nerves are the vagus nerves (X), which contain motor and sensory fibers traveling to and from the gut. The 12 pairs of cranial nerves and their targets are illustrated in Appendix III; the functions of these nerves are listed in Appendix IV. The autonomic motor fibers of the cranial nerves are parasympathetic.

The functions of the various cranial nerves are commonly assessed by neurologists as a basis for diagnosis. Because the functions and locations of the cranial nerves are specific, disruptions of particular cranial nerve functions provide excellent clues about the location and extent of tumors and other kinds of brain pathology.

Figure 3.2 summarizes the major divisions of the nervous system. Notice that the nervous system is a "system of twos."

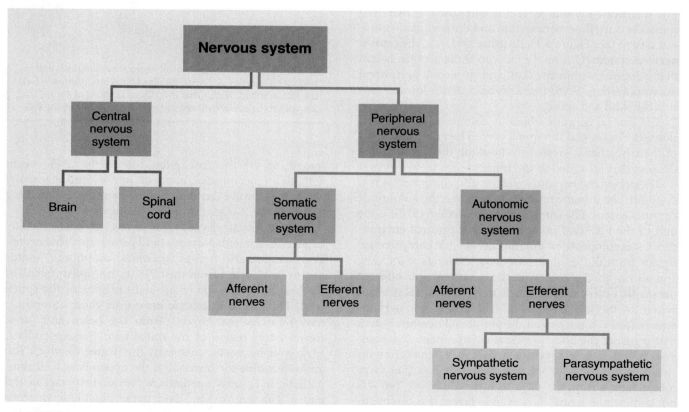

FIGURE 3.2   The major divisions of the nervous system.

Meninges, Ventricles, and Cerebrospinal Fluid

The brain and spinal cord (the CNS) are the most protected organs in the body. They are encased in bone and covered by three protective membranes, the three **meninges** (pronounced "men-IN-gees"). The outer *meninx* (which, believe it or not, is the singular of *meninges*) is a tough membrane called the **dura mater** (tough mother). Immediately inside the dura mater is the fine **arachnoid membrane** (spiderweblike membrane). Beneath the arachnoid membrane is a space called the **subarachnoid space**, which contains many large blood vessels and cerebrospinal fluid; then comes the innermost meninx, the delicate **pia mater** (pious mother), which adheres to the surface of the CNS.

Also protecting the CNS is the **cerebrospinal fluid (CSF)**, which fills the subarachnoid space, the central canal of the spinal cord, and the cerebral ventricles of the brain. The **central canal** is a small central channel that runs the length of the spinal cord; the **cerebral ventricles** are the four large internal chambers of the brain: the two lateral ventricles, the third ventricle, and the fourth ventricle (see Figure 3.3). The subarachnoid space, central canal, and cerebral ventricles are interconnected by a series of openings and thus form a single reservoir.

The cerebrospinal fluid supports and cushions the brain. Patients who have had some of their cerebrospinal fluid drained away often suffer raging headaches and experience stabbing pain each time they jerk their heads.

Cerebrospinal fluid is continuously produced by the **choroid plexuses**—networks of capillaries (small blood vessels) that protrude into the ventricles from the pia mater. The excess cerebrospinal fluid is continuously absorbed from the subarachnoid space into large blood-filled spaces, or *dural sinuses*, which run through the dura mater and drain into the large jugular veins of the neck. Figure 3.4 on page 54 illustrates the absorption of cerebrospinal fluid from the subarachnoid space into the large sinus that runs along the top of the brain between the two cerebral hemispheres.

Occasionally, the flow of cerebrospinal fluid is blocked by a tumor near one of the narrow channels that link the ventricles—for example, near the *cerebral aqueduct*, which connects the third and fourth ventricles. The resulting buildup of fluid in the ventricles causes the walls of the ventricles, and thus the entire brain, to expand, producing a condition called *hydrocephalus* (water head). Hydrocephalus is treated by draining the excess fluid from the ventricles and trying to remove the obstruction.

Blood–Brain Barrier

The brain is a finely tuned electrochemical organ whose function can be severely disturbed by the introduction of certain kinds of chemicals. Fortunately, there is a mechanism that impedes the passage of many toxic substances from the blood into the brain: the **blood–brain barrier**

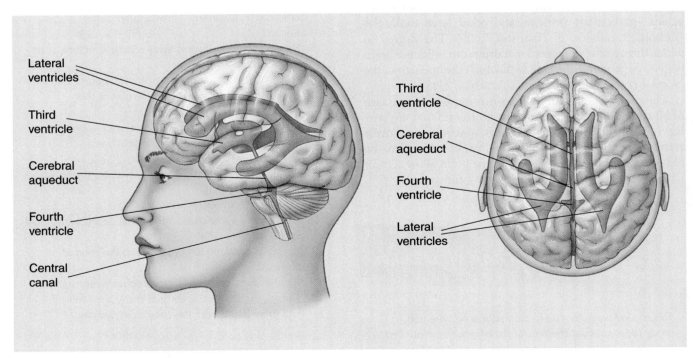

FIGURE 3.3　The cerebral ventricles.

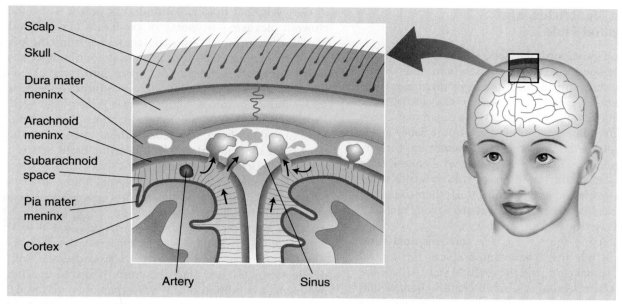

Scalp
Skull
Dura mater meninx
Arachnoid meninx
Subarachnoid space
Pia mater meninx
Cortex

Artery   Sinus

FIGURE 3.4   The absorption of cerebrospinal fluid from the subarachnoid space (blue) into a major sinus. Note the three meninges.

(see Banerjee & Bhat, 2007). This barrier is a consequence of the special structure of cerebral blood vessels. In the rest of the body, the cells that compose the walls of blood vessels are loosely packed; as a result, most molecules pass readily through them into surrounding tissue. In the brain, however, the cells of the blood vessel walls are tightly packed, thus forming a barrier to the passage of many molecules—particularly proteins and other large molecules (Abbott, Rönnbäck, & Hannson, 2005). The degree to which therapeutic or recreational drugs can influence brain activity depends on the ease with which they penetrate the blood–brain barrier (Löscher & Potschka, 2005).

The blood–brain barrier does not impede the passage of all large molecules. Some large molecules that are critical for normal brain function (e.g., glucose) are actively transported through cerebral blood vessel walls. Also, the blood vessel walls in some areas of the brain allow certain large molecules to pass through them unimpeded.

Scan Your Brain

This is good place for you to scan your brain: Are you ready to learn about the cells of the nervous system? Test your grasp of the first section of this chapter by filling in the following blanks with the most appropriate terms. The correct answers are provided at the end of the exercise.

Before proceeding, review material related to your errors and omissions.

1. The _____ system is composed of the brain and the spinal cord.
2. The part of the peripheral nervous system that regulates the body's internal environment is the _____ system.
3. Nerves that carry signals away from a structure, such as the CNS, are _____ nerves.
4. The ANS nerves that project from the thoracic and lumbar regions of the spinal cord are part of the _____ system.
5. _____ nerves stimulate, organize, and mobilize energy resources in threatening situations.
6. The vagus nerves are the longest _____.
7. The olfactory nerves and optic nerves are the only two purely sensory _____.
8. The innermost meninx is the _____.
9. The cerebral ventricles, central canal, and subarachnoid space are filled with _____.
10. _____ is continuously produced by the choroid plexuses.
11. A tumor near the _____ can produce hydrocephalus.
12. The _____ blocks the entry of many large molecules into brain tissue from the circulatory system.

Scan Your Brain answers: (1) central nervous, (2) autonomic nervous, (3) efferent, (4) sympathetic nervous, (5) Sympathetic, (6) cranial nerves, (7) cranial nerves, (8) pia mater, (9) cerebrospinal fluid, (10) Cerebrospinal fluid, (11) cerebral aqueduct, (12) blood–brain barrier.

Cells of the Nervous System

Most of the cells of the nervous system are of two funda-mentally different types: neurons and glial cells. Their anatomy is discussed in the following two subsections.

Anatomy of Neurons

As you learned in Chapter 1, **neurons** are cells that are specialized for the reception, conduction, and transmis-sion of electrochemical signals. They come in an incredi-ble variety of shapes and sizes (see Nelson, Sugino, & Hempel, 2006); however, many are similar to the one illustrated in Figures 3.5 and 3.6 (on page 56).

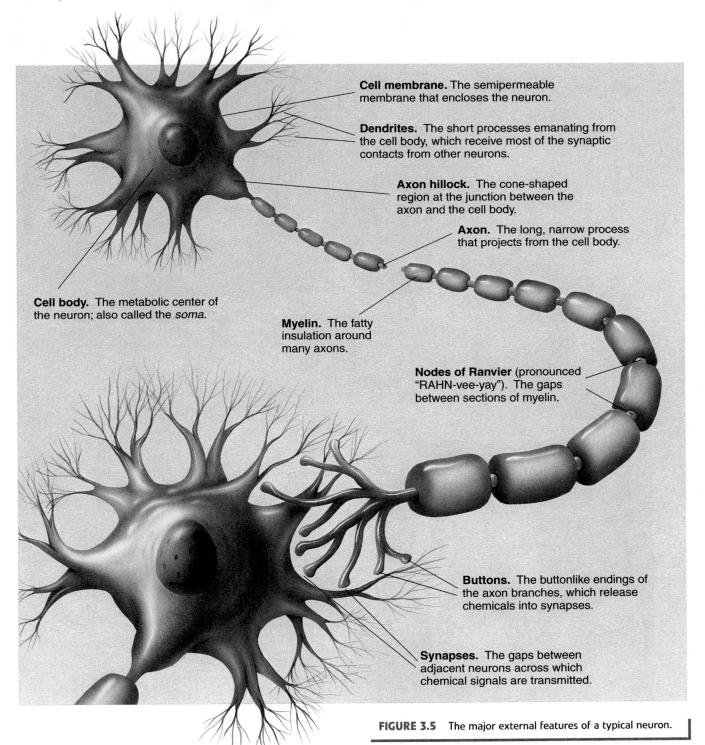

Cell membrane. The semipermeable membrane that encloses the neuron.

Dendrites. The short processes emanating from the cell body, which receive most of the synaptic contacts from other neurons.

Axon hillock. The cone-shaped region at the junction between the axon and the cell body.

Axon. The long, narrow process that projects from the cell body.

Cell body. The metabolic center of the neuron; also called the *soma*.

Myelin. The fatty insulation around many axons.

Nodes of Ranvier (pronounced "RAHN-vee-yay"). The gaps between sections of myelin.

Buttons. The buttonlike endings of the axon branches, which release chemicals into synapses.

Synapses. The gaps between adjacent neurons across which chemical signals are transmitted.

FIGURE 3.5 The major external features of a typical neuron.

Endoplasmic reticulum. A system of folded membranes in the cell body; rough portions (those with ribosomes) play a role in the synthesis of proteins; smooth portions (those without ribosomes) play a role in the synthesis of fats.

Cytoplasm. The clear internal fluid of the cell.

Ribosomes. Internal cellular structures on which proteins are synthesized; they are located on the endoplasmic reticulum.

Golgi complex. A connected system of membranes that packages molecules in vesicles.

Nucleus. The spherical DNA-containing structure of the cell body.

Mitochondria. Sites of aerobic (oxygen-consuming) energy release.

Microtubules. Tubules responsible for the rapid transport of material throughout neurons.

Synaptic vesicles. Spherical membrane packages that store neurotransmitter molecules ready for release near synapses.

Neurotransmitters. Molecules that are released from active neurons and influence the activity of other cells.

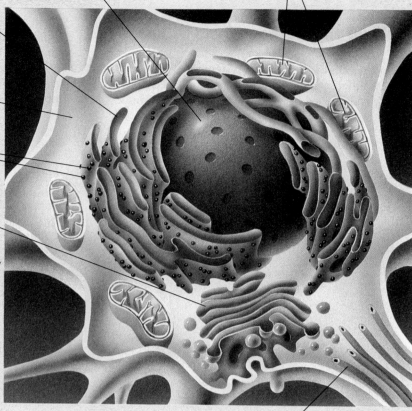
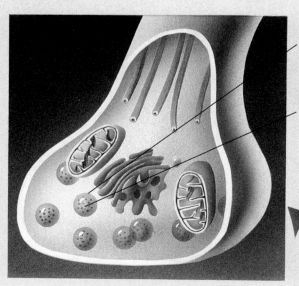

FIGURE 3.6 The major internal features of a typical neuron.

External Anatomy of Neurons   Figure 3.5 is an illustration of the major external features of one type of neuron. For your convenience, the definition of each feature is included in the illustration.

Internal Anatomy of Neurons   Figure 3.6 is an illustration of the major internal features of one type of neuron. Again, the definition of each feature is included in the illustration.

Neuron Cell Membrane   The neuron cell membrane is composed of a *lipid bilayer* (Piomelli, Astarita, & Rapaka, 2007), or two layers of fat molecules (see Figure 3.7). Embedded in the lipid bilayer are numerous protein molecules that are the basis of many of the cell membrane's functional properties. Some membrane proteins are *channel proteins*, through which certain molecules can pass; others are *signal proteins*, which transfer a signal to the inside of the neuron when particular molecules bind to them on the outside of the membrane.

Classes of Neurons   Figure 3.8 on page 58 illustrates a way of classifying neurons that is based on the number of processes (projections) emanating from their cell bodies. A neuron with more than two processes extending from its cell body is classified as a **multipolar neuron**; most neurons are multipolar. A neuron with one process extending from its cell body is classified as a **unipolar neuron**, and a neuron with two processes extending from its cell body is classified as a **bipolar neuron**. Neurons with a short axon or no axon at all are called **interneurons**; their function is to integrate the neural activity within a single brain structure, not to conduct signals from one structure to another.

Neurons and Neuroanatomical Structure   In general, there are two kinds of gross neural structures in the nervous system: those composed primarily of cell bodies and those composed primarily of axons. In the central nervous system, clusters of cell bodies are called **nuclei** (singular *nucleus*); in the peripheral nervous system, they are called **ganglia** (singular *ganglion*). (Note that the word *nucleus* has two different neuroanatomical meanings; it is a

structure in the neuron cell body and a cluster of cell bodies in the CNS.) In the central nervous system, bundles of axons are called **tracts**; in the peripheral nervous system, they are called **nerves**.

Glial Cells: The Forgotten Cells

Neurons are not the only cells in the nervous system; **glial cells** are found throughout the system. Although they have been widely reported to outnumber neurons 10 to 1, this view has been challenged by recent research. Glial cells do predominate in some brain structures, but overall the numbers of glial cells and neural cells are approximately equal (Azevedo et al., 2009).

There are several kinds of glial cells (Fields & Stevens-Graham, 2002). **Oligodendrocytes**, for example, are glial cells with extensions that wrap around the axons of some neurons of the central nervous system. These extensions are rich in **myelin**, a fatty insulating substance, and the **myelin sheaths** that they form increase the speed and efficiency of axonal conduction. A similar function is performed in the peripheral nervous system by **Schwann cells**, a second class of glial cells. Oligodendrocytes and Schwann cells are illustrated in Figure 3.9 on page 58. Notice that each Schwann cell constitutes one myelin segment, whereas each oligodendrocyte provides several myelin segments, often on more than one axon. Another important difference between Schwann cells and oligodendrocytes is that only Schwann cells can guide axonal *regeneration* (regrowth) after damage. That is why effective axonal regeneration in the mammalian nervous system is restricted to the PNS.

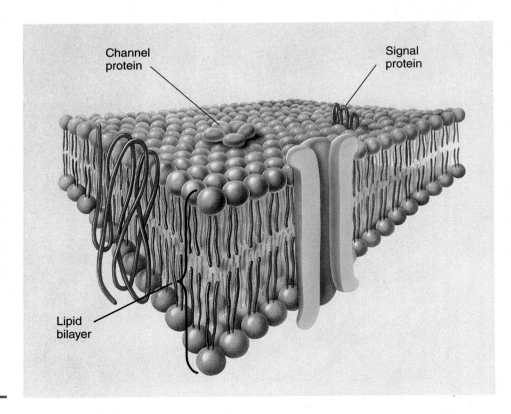

Channel protein

Signal protein

Lipid bilayer

FIGURE 3.7   The cell membrane is a lipid bilayer with signal proteins and channel proteins embedded in it.

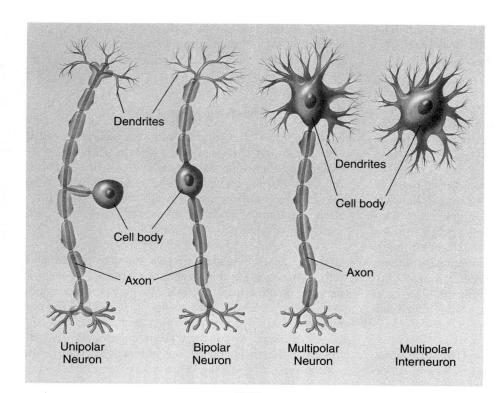

Dendrites

Cell body

Axon

Unipolar
Neuron

Bipolar
Neuron

Dendrites

Cell body

Axon

Multipolar
Neuron

Multipolar
Interneuron

FIGURE 3.8   A unipolar neuron, a bipolar neuron, a multipolar neuron, and an interneuron.

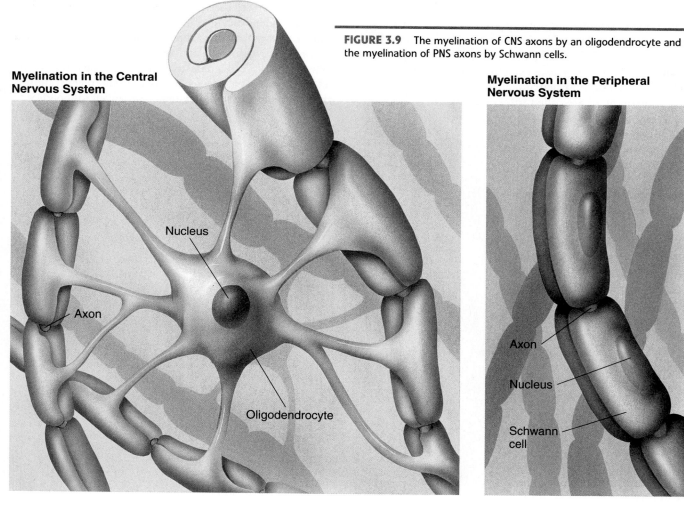

FIGURE 3.9   The myelination of CNS axons by an oligodendrocyte and the myelination of PNS axons by Schwann cells.

Myelination in the Central Nervous System

Nucleus

Axon

Oligodendrocyte

Myelination in the Peripheral Nervous System

Axon

Nucleus

Schwann cell

Microglia make up a third class of glial cells. Microglia are smaller than other glia—thus their name. They respond to injury or disease by multiplying, engulfing cellular debris, and triggering inflammatory responses (Nimmerjahn, Kirchhoff, & Helmchen, 2005).

Astrocytes constitute a fourth class of glial cells. They are the largest glial cells and they are so named because they are star-shaped (*astron* means "star"). The extensions of some astrocytes cover the outer surfaces of blood vessels that course through the brain; they also make contact with neuron cell bodies (see Figure 3.10). These particular astrocytes play a role in allowing the passage of some chemicals from the blood into CNS neurons and in blocking other chemicals (Abbott, Rönnbäck, & Hannson, 2006).

For decades, it was assumed that the function of astrocytes was merely to provide support for neurons—providing them with nutrition, clearing waste, and forming a physical matrix to hold neural circuits together (*glia*

means "glue"). But this limited view of the role of astrocytes is rapidly changing, thanks to a series of remarkable findings (see Kettenmann & Verkhratsky, 2008). For example, astrocytes have been shown to send and receive signals from neurons and other glial cells, to control the establishment and maintenance of synapses between neurons (Jourdain et al., 2007), to modulate neural activity (Rouach et al., 2008), to maintain the function of axons (Edgar & Nave, 2009), and to participate in glial circuits (Giaume et al., 2010). Now that the first wave of discoveries has focused neuroscientists' attention on astrocytes and other glial cells, appreciation of their role in nervous system function is growing rapidly. And their role in various nervous system disorders is currently being investigated intensively.

3.3
Neuroanatomical Techniques and Directions

This section of the chapter first describes a few of the most common neuroanatomical techniques. Then, it explains the system of directions that neuroanatomists use to describe the location of structures in vertebrate nervous systems.

Neuroanatomical Techniques

The major problem in visualizing neurons is not their minuteness. The major problem is that neurons are so tightly packed and their axons and dendrites so intricately intertwined that looking through a microscope at unprepared neural tissue reveals almost nothing about them. The key to the study of neuroanatomy lies in preparing neural tissue in a variety of ways, each of which permits a clear view of a different aspect of neuronal structure, and then combining the knowledge obtained from each of the preparations. This point is illustrated by the following widely used neuroanatomical techniques.

Golgi Stain   The greatest blessing to befall neuroscience in its early years was the accidental discovery of the **Golgi stain** by Camillo Golgi (pronounced "GOLE-jee"), an Italian physician, in the early 1870s; see Rapport (2005). Golgi was trying to stain the meninges, by exposing a block of neural tissue to potassium dichromate and silver nitrate, when he noticed an amazing thing. For some unknown reason, the silver chromate created by the chemical reaction of the two substances Golgi was using invaded a few neurons in each slice of tissue and stained each invaded neuron entirely black. This discovery made it possible to see individual neurons for the first time, although only in silhouette (see Figure 3.11). Golgi stains are commonly used when the overall shape of neurons is of interest.

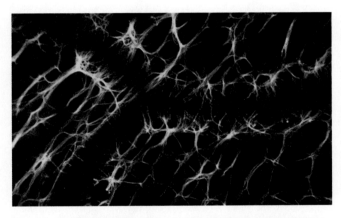

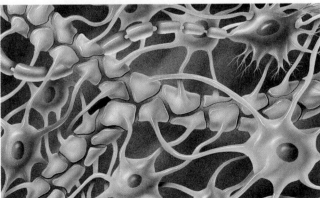

FIGURE 3.10 Astrocytes have an affinity for blood vessels, and they form a supportive matrix for neurons. The photograph on the top is of a slice of brain tissue stained with a glial stain; the unstained channels are blood vessels. The illustration on the bottom is a three-dimensional representation of the image on the top showing how the feet of astrocytes cover blood vessels and contact neurons. Compare the two panels. (Photograph courtesy of T. Chan-Ling.)

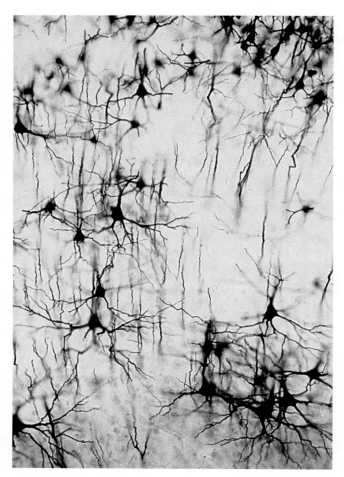

FIGURE 3.11   Neural tissue that has been stained by the Golgi method. Because only a few neurons take up the stain, their silhouettes are revealed in great detail, but their internal details are invisible. Usually, only part of a neuron is captured in a single slice. (Ed Reschke © Peter Arnold, Inc.)

Nissl Stain   Although the Golgi stain permits an excellent view of the silhouettes of the few neurons that take up the stain, it provides no indication of the number of neurons in an area or the nature of their inner structure. The first neural staining procedure to overcome these shortcomings was the **Nissl stain**, which was developed by Franz Nissl, a German psychiatrist, in the 1880s. The most common dye used in the Nissl method is cresyl violet. Cresyl violet and other Nissl dyes penetrate all cells on a slide, but they bind effectively only to structures in neuron cell bodies. Thus, they often are used to estimate the number of cell bodies in an area, by counting the number of Nissl-stained dots. Figure 3.12 is a photograph of a slice of brain tissue stained with cresyl violet. Notice that only the layers composed mainly of neuron cell bodies are densely stained.

Electron Microscopy   A neuroanatomical technique that provides information about the details of neuronal structure is **electron microscopy** (pronounced "my-CROSS-

cuh-pee"). Because of the nature of light, the limit of magnification in light microscopy is about 1,500 times, a level of magnification that is insufficient to reveal the fine anatomical details of neurons. Greater detail can be obtained by first coating thin slices of neural tissue with an electron-absorbing substance that is taken up by different parts of neurons to different degrees, then passing a beam of electrons through the tissue onto a photographic film. The result is an *electron micrograph*, which captures neuronal structure in exquisite detail (see Figure 4.11 on page 88). A *scanning electron microscope* provides spectacular electron micrographs in three dimensions (see Figure 3.13), but it is not capable of as much magnification as a conventional electron microscope. The strength of electron microscopy is also a weakness: Because the images are so detailed, they can make it difficult to visualize general aspects of neuroanatomical structure.

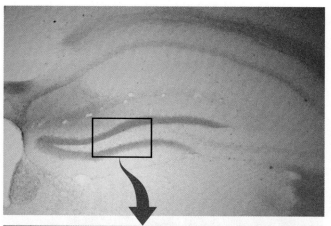

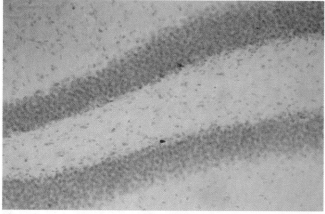

FIGURE 3.12   The Nissl stain. Presented here is a Nissl-stained coronal section through the rat hippocampus, at two levels of magnification to illustrate two uses of Nissl stains. Under low magnification (top panel), Nissl stains provide a gross indication of brain structure by selectively staining groups of neural cell bodies—in this case, the layers of the hippocampus. Under higher magnification (bottom panel), one can distinguish individual neural cell bodies and thus count the number of neurons in various areas. (Courtesy of my good friends Carl Ernst and Brian Christie, Department of Psychology, University of British Columbia.)

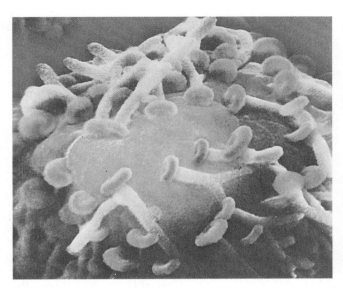

FIGURE 3.13 A color-enhanced scanning electron micrograph of a neuron cell body (green) studded with terminal buttons (orange). Each neuron receives numerous synaptic contacts. (Courtesy of Jerold J. M. Chun, M.D., Ph.D.)

Neuroanatomical Tracing Techniques Neuroanatomical tracing techniques are of two types: anterograde (forward) tracing methods and retrograde (backward) tracing methods. *Anterograde tracing methods* are used when an investigator wants to trace the paths of axons projecting away from cell bodies located in a particular area. The investigator injects into the area one of several chemicals commonly used for anterograde tracing—chemicals that are taken up by cell bodies and then transported forward along their axons to their terminal buttons. After a few days, the brain is removed and sliced; the slices are then treated to reveal the locations of the injected chemical. *Retrograde tracing methods* work in reverse; they are used when an investigator wants to trace the paths of axons projecting into a particular area. The investigator injects into the area one of several chemicals commonly used for retrograde tracing—chemicals that are taken up by terminal buttons and then transported backward along their axons to their cell bodies. After a few days, the brain is removed and sliced; the

FIGURE 3.14 Anatomical directions in representative vertebrates, my cats Sambala and Rastaman.

slices are then treated to reveal the locations of the injected chemical.

Directions in the Vertebrate Nervous System

It would be difficult for you to develop an understanding of the layout of an unfamiliar city without a system of directional coordinates: north–south, east–west. The same goes for the nervous system. Thus, before introducing you to the locations of major nervous system structures, I will describe the three-dimensional system of directional coordinates used by neuroanatomists.

Directions in the vertebrate nervous system are described in relation to the orientation of the spinal cord. This system is straightforward for most vertebrates, as Figure 3.14 indicates. The vertebrate nervous system has three axes: anterior–posterior, dorsal–ventral, and medial–lateral. First, **anterior** means toward the nose end (the anterior end), and **posterior** means toward the tail end (the posterior end); these same directions are sometimes referred to as *rostral* and *caudal*, respectively. Second, **dorsal** means toward the surface of the back or the top of the head (the dorsal surface), and **ventral** means toward the surface of the chest or the bottom of the head (the ventral surface). Third, **medial** means toward the midline of the body, and **lateral** means away from the midline toward the body's lateral surfaces.

We humans complicate this simple three-axis (anterior–posterior, ventral–dorsal, medial–lateral) system of neuroanatomical directions by insisting on walking around on our hind legs. This changes the orientation of our cerebral hemispheres in relation to our spines and brain stems.

You can save yourself a lot of confusion if you remember that the system of vertebrate neuroanatomical directions was adapted for use in humans in such a way that the terms used to describe the positions of various body surfaces are the same in humans as they are in more typical, non-upright vertebrates. Specifically, notice that the top

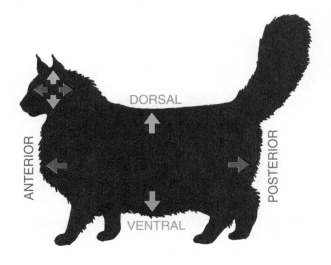

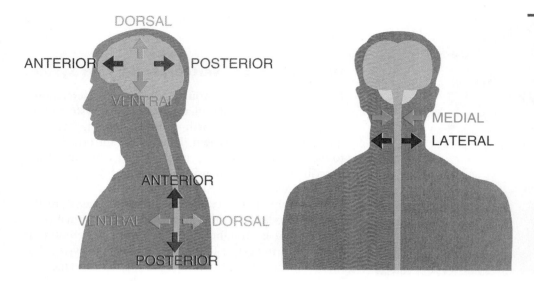

FIGURE 3.15 Anatomical directions in a human. Notice that the directions in the cerebral hemispheres are rotated by 90° in comparison to those in the spinal cord and brain stem because of the unusual upright posture of humans.

planes: **horizontal sections**, **frontal sections** (also termed *coronal sections*), and **sagittal sections**. These three planes are illustrated in Figure 3.16. A section cut down the center of the brain, between the two hemispheres, is called a *midsagittal section*. A section cut at a right angle to any long, narrow structure, such as the spinal cord or a nerve, is called a **cross section**.

of the human head and the back of the human body are both referred to as *dorsal* even though they are in different directions, and the bottom of the human head and the front of the human body are both referred to as *ventral* even though they are in different directions (see Figure 3.15). To circumvent this complication, the terms **superior** and **inferior** are often used to refer to the top and bottom of the primate head, respectively.

Proximal and distal are two other common directional terms. In general, **proximal** means "close," and **distal** means "far." Specifically, with regard to the peripheral nervous system, *proximal* means closer to the CNS, and *distal* means farther from the CNS. Your shoulders are proximal to your elbows, and your elbows are proximal to your fingers.

In the next few pages, you will be seeing drawings of sections (slices) of the brain cut in one of three different

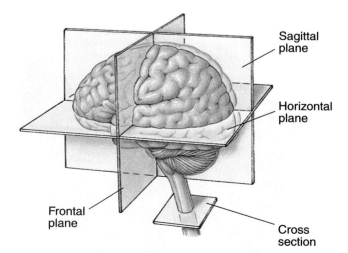

FIGURE 3.16 Horizontal, frontal (coronal), and sagittal planes in the human brain and a cross section of the human spinal cord.

Scan Your Brain

This is a good place for you to pause to scan your brain. Are you ready to proceed to the structures of the brain and spinal cord? Test your grasp of the preceding sections of this chapter by drawing a line between each term in the left column and the appropriate word or phrase in the right column. The correct answers are provided at the end of the exercise. Before proceeding, review material related to your incorrect answers.

1. myelin	a. gaps
2. soma	b. cone-shaped region
3. axon hillock	c. packaging membranes
4. Golgi complex	d. fatty substance
5. ribosomes	e. neurotransmitter storage
6. synapses	f. cell body
7. glial cells	g. PNS clusters of cell bodies
8. synaptic vesicles	h. protein synthesis
9. astrocytes	i. the forgotten cells
10. ganglia	j. CNS myelinators
11. oligodendrocytes	k. black
12. Golgi stain	l. largest glial cells
13. dorsal	m. caudal
14. posterior	n. top of head

Scan Your Brain answers: (1) d, (2) f, (3) b, (4) c, (5) h, (6) a, (7) i, (8) e, (9) l, (10) g, (11) j, (12) k, (13) n, (14) m.

3.4
Spinal Cord

In the first three sections of this chapter, you learned about the divisions of the nervous system, the cells that compose it, and some of the neuroanatomical techniques that are used to study it. This section begins your ascent of the human CNS by focusing on the spinal cord. The final two sections of the chapter focus on the brain.

In cross section, it is apparent that the spinal cord comprises two different areas (see Figure 3.17): an inner H-shaped core of gray matter and a surrounding area of white matter. **Gray matter** is composed largely of cell bodies and unmyelinated interneurons, whereas **white matter** is composed largely of myelinated axons. (It is the myelin that gives the white matter its glossy white sheen.) The two dorsal arms of the spinal gray matter are called the **dorsal horns**, and the two ventral arms are called the **ventral horns**.

Pairs of *spinal nerves* are attached to the spinal cord—one on the left and one on the right—at 31 different levels of the spine. Each of these 62 spinal nerves divides as it nears the cord (see Figure 3.17), and its axons are joined to the cord via one of two roots: the *dorsal root* or the *ventral root*.

All dorsal root axons, whether somatic or autonomic, are sensory (afferent) unipolar neurons with their cell bodies grouped together just outside the cord to form the **dorsal root ganglia** (see Figure 3.17). Many of their synaptic terminals are in the dorsal horns of the spinal gray matter (see Figure 3.18). In contrast, the neurons of the ventral root are motor (efferent) multipolar neurons with their cell bodies in the ventral horns. Those that are part of the somatic nervous system project to skeletal muscles; those that are part of the autonomic nervous

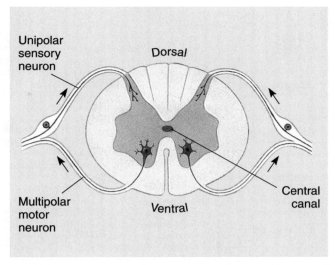

FIGURE 3.18   A schematic cross section of the spinal cord.

system project to ganglia, where they synapse on neurons that in turn project to internal organs (heart, stomach, liver, etc.). See Appendix I.

3.5
Five Major Divisions of the Brain

A necessary step in learning to live in an unfamiliar city is learning the names and locations of its major neighborhoods or districts. Those who possess this information can easily communicate the general location of any destination in the city. This section of the chapter introduces you to the five "neighborhoods," or divisions, of the brain—for much the same reason.

To understand why the brain is considered to be composed of five divisions, it is necessary to understand its early development (see Holland, 2009). In the vertebrate embryo, the tissue that eventually develops into the CNS is recognizable as a fluid-filled tube (see Figure 3.19 on page 64). The first indications of the developing brain are three swellings that occur at the anterior end of this tube. These three swellings eventually develop into the adult *forebrain*, *midbrain*, and *hindbrain*.

⊙ **Watch**
The Forebrain; The Midbrain; The Hindbrain
www.mypsychlab.com

Before birth, the initial three swellings in the neural tube become five (see Figure 3.19). This occurs because the forebrain swelling grows into two different swellings, and so does the hindbrain swelling. From anterior to posterior, the five swellings that compose the developing brain at birth are the *telencephalon*, the *diencephalon*, the

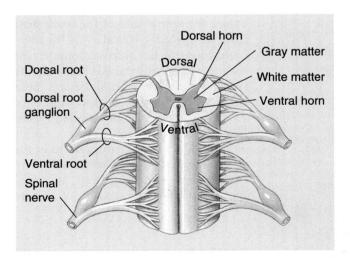

FIGURE 3.17   The dorsal and ventral roots of the spinal cord.

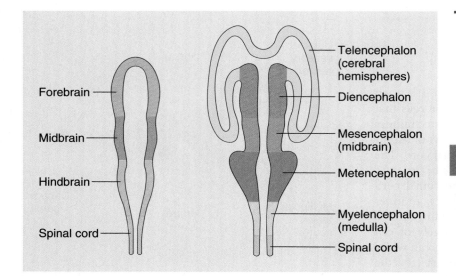

Forebrain

Midbrain

Hindbrain

Spinal cord

Telencephalon (cerebral hemispheres)

Diencephalon

Mesencephalon (midbrain)

Metencephalon

Myelencephalon (medulla)

Spinal cord

FIGURE 3.19 The early development of the mammalian brain illustrated in schematic horizontal sections. Compare with the adult human brain in Figure 3.20

mesencephalon (or midbrain), the *metencephalon*, and the *myelencephalon* (*encephalon* means "within the head"). These swellings ultimately develop into the five divisions of the adult brain. As a student, I memorized their order by remembering that the *telencephalon* is on the *top* and the other four divisions are arrayed below it in alphabetical order.

Figure 3.20 illustrates the locations of the telencephalon, diencephalon, mesencephalon, metencephalon, and myelencephalon in the adult human brain. Notice that in humans, as in other higher vertebrates, the telencephalon (the left and right *cerebral hemispheres*) undergoes the greatest growth during development. The other four divisions of the brain are often referred to collectively as the **brain stem**—the stem on which the cerebral hemispheres sit. The myelencephalon is often referred to as the *medulla*.

3.6
Major Structures of the Brain

Now that you have learned the five major divisions of the brain, it is time to introduce you to their major structures. This section of the chapter begins its survey of brain structures in the myelencephalon, then ascends through the other divisions to the telencephalon. The brain structures introduced and defined in this section are boldfaced but are not included in the Key Terms list at the end of the chapter. Rather, they are arranged according to their locations in the brain in Figure 3.30 on page 71.

Here is a reminder before you delve into the anatomy of the brain: The directional coordinates are the same for the brain stem as for the spinal cord, but they are rotated by 90° for the forebrain.

Myelencephalon

Not surprisingly, the **myelencephalon** (or **medulla**), the most posterior division of the brain, is composed largely of tracts carrying signals between the rest of the brain and the body. An interesting part of the myelencephalon from a psychological perspective is the **reticular formation** (see Figure 3.21). It is a complex network of about 100 tiny nuclei that occupies the central core of the brain stem from the posterior boundary of the myelencephalon to the anterior boundary of the midbrain. It is so named because of its netlike appearance (*reticulum* means "little net"). Sometimes, the reticular formation is referred to as the *reticular activating system* because parts of it seem to play a role in arousal. However, the various nuclei of the reticular formation are involved in a variety of functions—including sleep, attention, movement, the maintenance of muscle

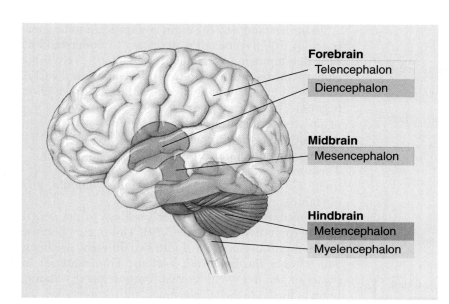

Forebrain
Telencephalon
Diencephalon

Midbrain
Mesencephalon

Hindbrain
Metencephalon
Myelencephalon

FIGURE 3.20 The divisions of the adult human brain.

tone, and various cardiac, circulatory, and respiratory reflexes. Accordingly, referring to this collection of nuclei as an
activating system can be misleading.

Metencephalon

The **metencephalon**, like the myelencephalon, houses
many ascending and descending tracts and part of the
reticular formation. These structures create a bulge, called
the **pons**, on the brain stem's ventral surface. The pons is
one major division of the metencephalon; the other is the
cerebellum (little brain)—see Figure 3.21. The **cerebellum**
is the large, convoluted structure on the brain stem's dorsal surface. It is an important sensorimotor structure;
cerebellar damage eliminates the ability to precisely control one's movements and to adapt them to changing conditions. However, the fact that cerebellar damage also
produces a variety of cognitive deficits (e.g., deficits in decision making and in the use of language suggests that the
functions of the cerebellum are not restricted to sensorimotor control.

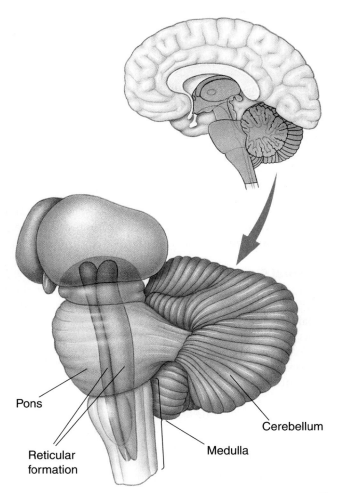

FIGURE 3.21 Structures of the human myelencephalon
(medulla) and metencephalon.

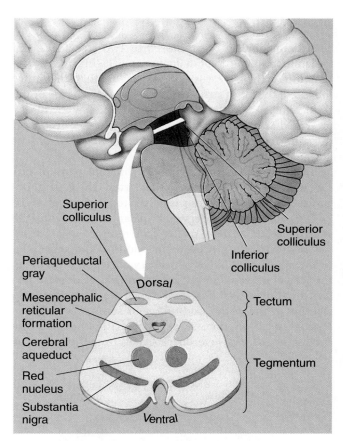

FIGURE 3.22 The human mesencephalon (midbrain).

Mesencephalon

The **mesencephalon**, like the metencephalon, has two divisions. The two divisions of the mesencephalon are the
tectum and the tegmentum (see Figure 3.22). The **tectum**
(roof) is the dorsal surface of the midbrain. In mammals,
the tectum is composed of two pairs of bumps, the
colliculi (little hills). The posterior pair, called the **inferior
colliculi**, have an auditory function; the anterior pair,
called the **superior colliculi**, have a visual function. In
lower vertebrates, the function of the tectum is entirely
visual; thus, the tectum is sometimes referred to as the
optic tectum.

The **tegmentum** is the division of the mesencephalon
ventral to the tectum. In addition to the reticular formation
and tracts of passage, the tegmentum contains three colorful structures that are of particular interest to biopsychologists: the periaqueductal gray, the substantia nigra, and the
red nucleus (see Figure 3.22). The **periaqueductal gray** is
the gray matter situated around the **cerebral aqueduct**, the
duct connecting the third and fourth ventricles; it is of special interest because of its role in mediating the analgesic
(pain-reducing) effects of opiate drugs. The **substantia
nigra** (black substance) and the **red nucleus** are both important components of the sensorimotor system.

Diencephalon

The **diencephalon** is composed of two structures: the thalamus and the hypothalamus (see Figure 3.23). The **thalamus** is the large, two-lobed structure that constitutes the top of the brain stem. One lobe sits on each side of the third ventricle, and the two lobes are joined by the **massa intermedia**, which runs through the ventricle. Visible on the surface of the thalamus are white *lamina* (layers) that are composed of myelinated axons.

The thalamus comprises many different pairs of nuclei, most of which project to the cortex. The general organization of the thalamus is illustrated in Appendix V.

The most well understood thalamic nuclei are the **sensory relay nuclei**—nuclei that receive signals from sensory receptors, process them, and then transmit them to the appropriate areas of sensory cortex. For example, the **lateral geniculate nuclei**, the **medial geniculate nuclei**, and the **ventral posterior nuclei** are important relay stations in the visual, auditory, and somatosensory systems, respectively. Sensory relay nuclei are not one-way streets; they all receive feedback signals from the very areas

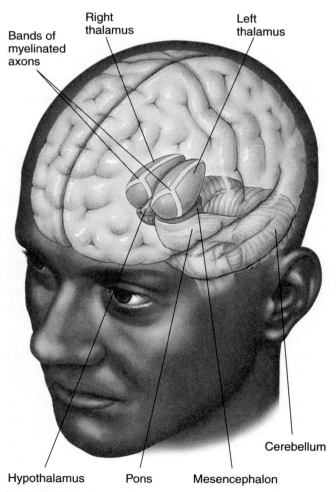

FIGURE 3.23 The human diencephalon.

(Labels: Bands of myelinated axons; Right thalamus; Left thalamus; Hypothalamus; Pons; Mesencephalon; Cerebellum)

of cortex to which they project (Cudeiro & Sillito, 2006). Although less is known about the other thalamic nuclei, the majority of them receive input from areas of the cortex and project to other areas of the cortex (Sherman, 2007).

The **hypothalamus** is located just below the anterior thalamus (*hypo* means "below")—see Figure 3.24. It plays an important role in the regulation of several motivated behaviors (e.g., eating, sleep, and sexual behavior). It exerts its effects in part by regulating the release of hormones from the **pituitary gland**, which dangles from it on the ventral surface of the brain. The literal meaning of *pituitary gland* is "snot gland"; it was discovered in a gelatinous state behind the nose of an unembalmed cadaver and was incorrectly assumed to be the main source of nasal mucus.

In addition to the pituitary gland, two other structures appear on the inferior surface of the hypothalamus: the optic chiasm and the mammillary bodies (see Figure 3.24). The **optic chiasm** is the point at which the *optic nerves* from each eye come together. The X shape is created because some of the axons of the optic nerve **decussate** (cross over to the other side of the brain) via the optic chiasm. The decussating fibers are said to be **contralateral** (projecting from one side of the body to the other), and the nondecussating fibers are said to be **ipsilateral** (staying on the same side of the body). The **mammillary bodies**, which are often considered to be part of the hypothalamus, are a pair of spherical nuclei located on the inferior surface of the hypothalamus, just behind the pituitary. The mammillary bodies and the other nuclei of the hypothalamus are illustrated in Appendix VI.

Telencephalon

The **telencephalon**, the largest division of the human brain, mediates the brain's most complex functions. It initiates voluntary movement, interprets sensory input, and mediates complex cognitive processes such as learning, speaking, and problem solving.

Cerebral Cortex   The cerebral hemispheres are covered by a layer of tissue called the **cerebral cortex** (cerebral bark). Because the cerebral cortex is mainly composed of small, unmyelinated neurons, it is gray and is often referred to as the *gray matter*. In contrast, the layer beneath the cortex is mainly composed of large myelinated axons, which are white and often referred to as the *white matter* (Fields, 2008). In humans, the cerebral cortex is deeply convoluted (furrowed)—see Figure 3.25. The *convolutions* have the effect of increasing the amount of cerebral cortex without increasing the overall volume of the brain. Not all mammals have convoluted cortexes; most mammals are *lissencephalic* (smooth-brained). It was once believed that the number and size of cortical

> ◉ **Watch**
> Major Brain Structures and Functions: The Brain; The Cerebral Cortex
> **www.mypsychlab.com**

are the **central fissure** and the **lateral fissure**. These fissures partially divide each hemisphere into four lobes: the **frontal lobe**, the **parietal lobe** (pronounced "pa-RYE-e-tal"), the **temporal lobe**, and the **occipital lobe** (pronounced "ok-SIP-i-tal"). Among the largest gyri are the **precentral gyri**, the **postcentral gyri**, and the **superior temporal gyri** in the frontal, parietal, and temporal lobes, respectively.

It is important to understand that the cerebral lobes are not functional units. It is best to think of the cerebral cortex as a flat sheet of cells that just happens to be divided into lobes because pressure causes it to be folded in on itself at certain places during development. Thus, it is incorrect to think that a lobe is a functional unit, having one set of functions. Still, it is useful at this early stage of your biopsychological education to get a general idea of various functions of areas within each lobe. More thorough discussions of the cerebral localization of brain functions are presented in later chapters.

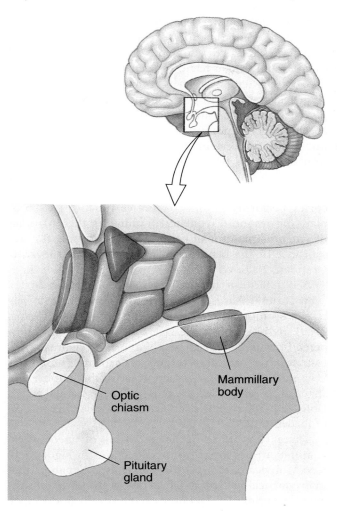

FIGURE 3.24 The human hypothalamus (in color) in relation to the optic chiasm and the pituitary gland.

Evolutionary Perspective convolutions determined a species' intellectual capacities; however, the number and size of cortical convolutions appear to be related more to body size. Every large mammal has an extremely convoluted cortex.

The large furrows in a convoluted cortex are called **fissures**, and the small ones are called **sulci** (singular *sulcus*). The ridges between fissures and sulci are called **gyri** (singular *gyrus*). It is apparent in Figure 3.25 that the cerebral hemispheres are almost completely separated by the largest of the fissures: the **longitudinal fissure**.

◉ **Watch**
Hemispheric Specialization
www.mypsychlab.com

The cerebral hemispheres are directly connected by a few tracts spanning the longitudinal fissure; these hemisphere-connecting tracts are called **cerebral commissures**. The largest cerebral commissure, the **corpus callosum**, is clearly visible in Figure 3.25.

As Figures 3.25 and 3.26 (on page 68) indicate, the two major landmarks on the lateral surface of each hemisphere

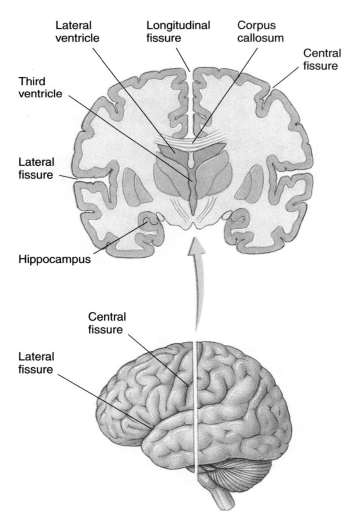

FIGURE 3.25 The major fissures of the human cerebral cortex.

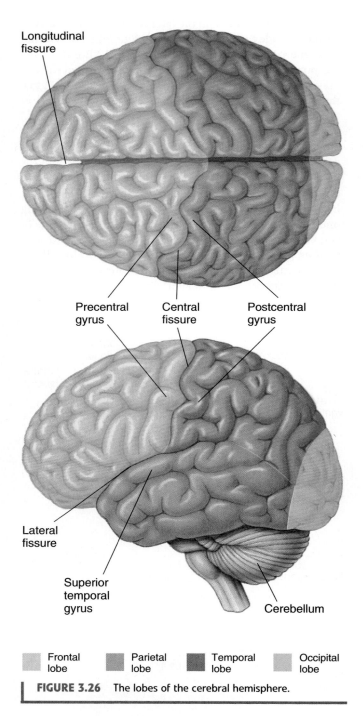

Longitudinal
fissure

Precentral   Central   Postcentral
gyrus   fissure   gyrus

Lateral
fissure

Superior
temporal
gyrus   Cerebellum

| | Frontal lobe | | Parietal lobe | | Temporal lobe | | Occipital lobe |

FIGURE 3.26   The lobes of the cerebral hemisphere.

The main function of the occipital lobes is quite straightforward: We humans rely heavily on the analysis of visual input to guide our behavior, and the occipital cortex and large areas of adjacent cortex perform this function. There are two large functional areas in each parietal lobe: The postcentral gyrus analyzes sensations from the body (e.g., touch), whereas the remaining areas

of cortex in the posterior parts of the parietal lobes play roles in perceiving the location of both objects and our own bodies and in directing our attention. The cortex of each temporal lobe has three general functional areas: the superior temporal gyrus is involved in hearing and language; the inferior temporal cortex identifies complex visual patterns; and the medial portion of temporal cortex (which is not visible from the usual side view) is important for certain kinds of memory. Lastly, each frontal lobe has two distinct functional areas: the precentral gyrus and adjacent frontal cortex have a motor function, whereas the frontal cortex anterior to motor cortex performs complex cognitive functions, such as planning response sequences, evaluating the outcomes of potential patterns of behavior, and assessing the significance of the behavior of others (Huey, Krueger, & Grafman, 2006; Wise, 2008).

About 90% of human cerebral cortex is **neocortex** (new cortex); that is, it is six-layered cortex of relatively recent evolution (see Douglas & Martin, 2004; Rakic, 2009). By convention, the layers of neocortex are numbered I through VI, starting at the surface. Figure 3.27 illustrates two adjacent sections of neocortex. One has been stained with a Nissl stain to reveal the number and shape of its cell bodies; the other has been stained with a Golgi stain to reveal the silhouettes of a small proportion of its neurons.

Three important characteristics of neocortical anatomy are apparent from the sections in Figure 3.27 (see Molyneaux et al., 2007). First, it is apparent that many cortical neurons fall into one of two different categories: pyramidal (pyramid-shaped) cells and stellate (star-shaped) cells. **Pyramidal cells** are large multipolar neurons with pyramid-shaped cell bodies, a large dendrite called an *apical dendrite* that extends from the apex of the pyramid straight toward the cortex surface, and a very long axon (Spruston, 2008). In contrast, **stellate cells** are small star-shaped interneurons (neurons with a short axon or no axon). Second, it is apparent that the six layers of neocortex differ from one another in terms of the size and density of their cell bodies and the relative proportion of pyramidal and stellate cell bodies that they contain. Third, it is apparent that many long axons and dendrites course vertically (i.e., at right angles to the cortical layers) through the neocortex. This vertical flow of information is the basis of the neocortex's **columnar organization**; neurons in a given vertical column of neocortex often form a mini-circuit that performs a single function (Laughlin & Sejnowski, 2003).

A fourth important characteristic of neocortical anatomy is not apparent in Figure 3.27: Although neocortex is six-layered, there are variations in the thickness of the respective layers from area to area (see Zilles & Amunts, 2010). For example, because the stellate cells of layer IV are specialized for receiving sensory signals from the thalamus, layer IV is extremely thick in areas of sensory cortex. Conversely, because the pyramidal cells of layer V conduct signals from the neocortex to the brain stem and spinal cord, layer V is extremely thick in areas of motor cortex.

The **hippocampus** is one important area of cortex that is not neocortex—it has only three major layers (see Förster, Ahao, & Frotscher, 2006). The hippocampus is located at the medial edge of the cerebral cortex as it folds back on itself in the medial temporal lobe (see Figure 3.25 on page 67). This folding produces a shape that is, in cross section, somewhat reminiscent of a sea horse (*hippocampus* means "sea horse"). The hippocampus plays a major role in some kinds of memory, particularly memory for spatial location (see Chapter 11).

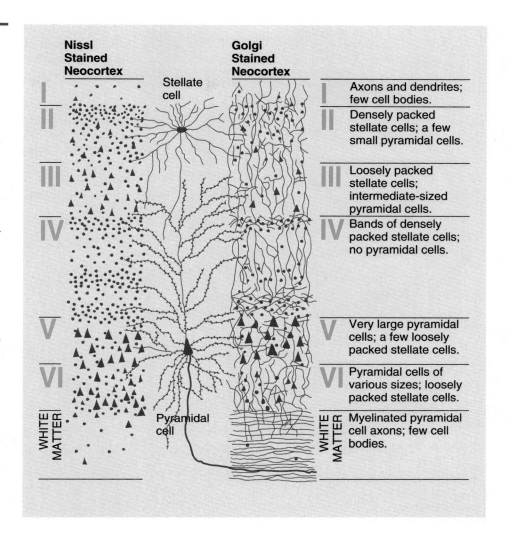

FIGURE 3.27 The six layers of neocortex. (Adapted from Rakic, 1979.)

Nissl Stained Neocortex

Golgi Stained Neocortex

Stellate cell

Pyramidal cell

WHITE MATTER

I — Axons and dendrites; few cell bodies.

II — Densely packed stellate cells; a few small pyramidal cells.

III — Loosely packed stellate cells; intermediate-sized pyramidal cells.

IV — Bands of densely packed stellate cells; no pyramidal cells.

V — Very large pyramidal cells; a few loosely packed stellate cells.

VI — Pyramidal cells of various sizes; loosely packed stellate cells.

WHITE MATTER — Myelinated pyramidal cell axons; few cell bodies.

The Limbic System and the Basal Ganglia

Although much of the subcortical portion of the telencephalon is taken up by axons projecting to and from the neocortex, there are several large subcortical nuclear groups. Some of them are considered to be part of either the *limbic system* or the *basal ganglia motor system*. Don't be misled by the word *system* in these contexts; it implies a level of certainty that is unwarranted. It is not entirely clear exactly what these hypothetical systems do, exactly which structures should be included in them, or even whether it is appropriate to view them as unitary systems. Nevertheless, if not taken too literally, the concepts of *limbic system* and *basal ganglia motor system* provide a useful means of conceptualizing the organization of several subcortical structures.

The **limbic system** is a circuit of midline structures that circle the thalamus (*limbic* means "ring"). The limbic system is involved in the regulation of motivated behaviors—including the four *F*s of motivation: fleeing, feeding, fighting, and sexual behavior. (This joke is as old as biopsychology

itself, but it is a good one.) In addition to the structures about which you have already read (the mammillary bodies and the hippocampus), major structures of the limbic system include the amygdala, the fornix, the cingulate cortex, and the septum.

Let's begin tracing the limbic circuit (see Figure 3.28 on page 70) at the **amygdala**—the almond-shaped nucleus in the anterior temporal lobe (*amygdala* means "almond" and is pronounced "a-MIG-dah-lah")—see Swanson & Petrovich (1998). Posterior to the amygdala is the hippocampus, which runs beneath the thalamus in the medial temporal lobe. Next in the ring are the cingulate cortex and the fornix. The **cingulate cortex** is the large strip of cortex in the **cingulate gyrus** on the medial surface of the cerebral hemispheres, just superior to the corpus callosum; it encircles the dorsal thalamus (*cingulate* means "encircling"). The **fornix**, the major tract of the limbic system, also encircles the dorsal thalamus; it leaves the dorsal end of the hippocampus and sweeps forward in an arc coursing along the superior

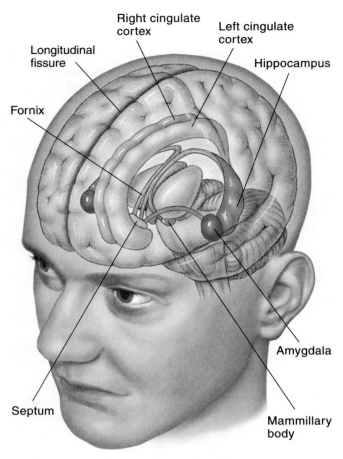

FIGURE 3.28   The major structures of the limbic system: amygdala, hippocampus, cingulate cortex, fornix, septum, and mammillary body.

surface of the third ventricle and terminating in the septum and the mammillary bodies (*fornix* means "arc"). The **septum** is a midline nucleus that is located at the anterior tip of the cingulate cortex. Several tracts connect the septum and mammillary bodies with the amygdala and hippocampus, thereby completing the limbic ring.

The functions of the hypothalamus and the amygdala have been investigated more than those of the other limbic structures. As stated previously, the hypothalamus is involved in a variety of motivated behaviors such as eating, sleep, and sexual behavior. The amygdala, on the other hand, is involved in emotion, particularly fear—you will learn much more about these structures in Chapters 12, 13, 14, and 17.

The **basal ganglia** are illustrated in Figure 3.29. As we did with the limbic system, let's begin our examination of the basal ganglia with the amygdala, which is considered to be part of both systems. Sweeping out of each amygdala, first in a posterior direction and then in an anterior

direction, is the long tail-like **caudate** (*caudate* means "tail-like"). Each caudate forms an almost complete circle; in its center, connected to it by a series of fiber bridges, is the **putamen** (pronounced "pew-TAY-men"). Together, the caudate and the putamen, which both have a striped appearance, are known as the **striatum** (striped structure). The remaining structure of the basal ganglia is the pale circular structure known as the **globus pallidus** (pale globe). The globus pallidus is located medial to the putamen, between the putamen and the thalamus.

The basal ganglia play a role in the performance of voluntary motor responses. Of particular interest is a pathway that projects to the striatum from the substantia nigra of the midbrain. *Parkinson's disease*, a disorder that

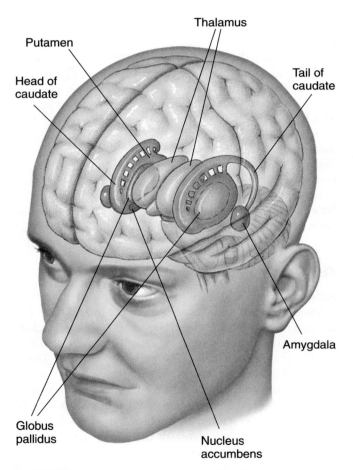

FIGURE 3.29   The basal ganglia: amygdala, striatum (caudate plus putamen), and globus pallidus, Notice that, in this view, the right globus pallidus is largely hidden behind the right thalamus, and the left globus pallidus is totally hidden behind the left putamen. Although the globus pallidus is usually considered to be a telencephalic structure, it actually originates from diencephalic tissue that migrates into its telencephalic location during the course of prenatal development.

is characterized by rigidity, tremors,

Clinical Implications and poverty of voluntary movement, is associated with the deterioration of this pathway. Another part of the basal ganglia that is currently of particular interest to biopsychologists is the *nucleus accumbens*, which is in the medial portion of the ventral striatum (see Figure 3.29). The nucleus accumbens is thought to play a role in the rewarding effects of addictive drugs and other reinforcers.

Figure 3.30 summarizes the major brain divisions and structures—whose names have appeared in boldface in this section.

FIGURE 3.30 Summary of major brain structures. This display contains all the brain anatomy key terms that appear in boldface in Section 3.6.

Telencephalon	Cerebral cortex	Neocortex Hippocampus
	Major fissures	Central fissure Lateral fissure Longitudinal fissure
	Major gyri	Precentral gyrus Postcentral gyrus Superior temporal gyrus Cingulate gyrus
	Four lobes	Frontal lobe Temporal lobe Parietal lobe Occipital lobe
	Limbic system	Amygdala Hippocampus Fornix Cingulate cortex Septum Mammillary bodies
	Basal ganglia	Amygdala Caudate } Striatum Putamen } Globus pallidus
	Cerebral commissures	Corpus callosum
Diencephalon	Thalamus	Massa intermedia Lateral geniculate nuclei Medial geniculate nuclei Ventral posterior nuclei
	Hypothalamus	Mammillary bodies
	Optic chiasm	
	Pituitary gland	
Mesencephalon	Tectum	Superior colliculi Inferior colliculi
	Tegmentum	Reticular formation Cerebral aqueduct Periaqueductal gray Substantia nigra Red nucleus
Metencephalon	Reticular formation Pons Cerebellum	
Myelencephalon or Medulla	Reticular formation	

Scan Your Brain

If you have not previously studied the gross anatomy of the brain, your own brain is probably straining under the burden of new terms. To determine whether you are ready to proceed, scan your brain by labeling the following midsagittal view of a real human brain. You may find it challenging to switch from color-coded diagrams to a photograph of a real brain.

The correct answers are provided at the end of the exercise. Before proceeding, review material related to your errors and omissions. Notice that Figure 3.30 includes all the brain anatomy terms that have appeared in bold type in this section and thus is an excellent review tool.

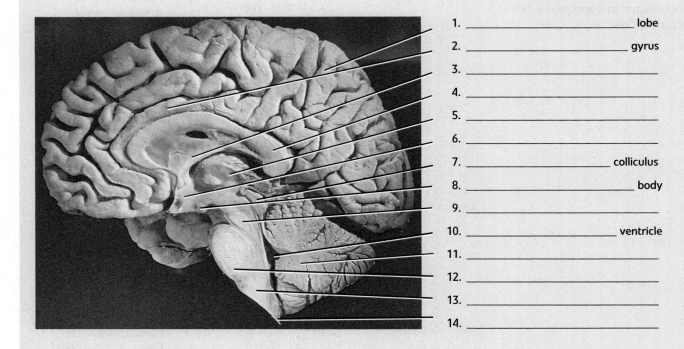

1. _____ lobe
2. _____ gyrus
3. _____
4. _____
5. _____
6. _____
7. _____ colliculus
8. _____ body
9. _____
10. _____ ventricle
11. _____
12. _____
13. _____
14. _____

Scan Your Brain answers: (1) parietal, (2) cingulate, (3) fornix, (4) corpus callosum, (5) thalamus, (6) hypothalamus, (7) superior, (8) mammillary, (9) tegmentum, (10) fourth, (11) cerebellum, (12) pons, (13) medulla, or myelencephalon, (14) spinal cord.

Figure 3.31 concludes this chapter, for reasons that too often get lost in the shuffle of neuroanatomical terms and technology. I have included it here to illustrate the beauty of the brain and the art of those who study its structure. I hope you are inspired by it. I wonder what thoughts its neural circuits once contained.

FIGURE 3.31 The art of neuroanatomical staining. This slide was stained with both a Golgi stain and a Nissl stain. Clearly visible on the Golgi-stained pyramidal neurons are the pyramid-shaped cell bodies, the large apical dendrites, and numerous dendritic spines. Less obvious here is the long, narrow axon that projects from each pyramidal cell body off the bottom of this slide. (Courtesy of Miles Herkenham, Unit of Functional Neuroanatomy, National Institute of Mental Health, Bethesda, MD.)

Themes Revisited

Clinical Implications This chapter contributed relatively little to the development of the book's themes; that development was temporarily slowed while you were being introduced to the key areas and structures of the human brain. A knowledge of fundamental neuroanatomy will serve as the foundation of discussions of brain function in subsequent chapters. However, the clinical implications theme did arise three times: in discussions of the importance of the cranial nerves in neurological diagnosis, the role of blockage of cerebral aqueducts in hydrocephalus, and the involvement of damage to the pathway from the substantia nigra to the striatum in Parkinson's disease. Also, the evolutionary perspective was evident when the text noted interspecies differences in cortical convolutions.

Think about It

1. Which of the following extreme positions do you think is closer to the truth? (a) The primary goal of all psychological research should be to relate psychological phenomena to the anatomy of neural circuits. (b) Psychologists should leave the study of neuroanatomy to neuroanatomists.

2. Perhaps the most famous mistake in the history of biopsychology was made by Olds and Milner (see Chapter 15). They botched an electrode implantation in the brain of a rat, and the tip of the stimulation electrode ended up in an unknown structure. When they subsequently tested the effects of electrical stimulation of this unknown structure, they made a fantastic discovery: The rat seemed to find the brain stimulation extremely pleasurable. In fact, the rat would press a lever for hours at an extremely high rate if every press produced a brief stimulation to its brain through the electrode. If you had accidentally stumbled on this intracranial self-stimulation phenomenon, what neuroanatomical procedures would you have used to identify the stimulation site and the neural circuits involved in the pleasurable effects of the stimulation?

Key Terms

3.1 General Layout of the Nervous System

Central nervous system (CNS) (p. 51)
Peripheral nervous system (PNS) (p. 51)
Somatic nervous system (SNS) (p. 51)
Afferent nerves (p. 51)
Efferent nerves (p. 51)
Autonomic nervous system (ANS) (p. 51)
Sympathetic nerves (p. 51)
Parasympathetic nerves (p. 51)
Cranial nerves (p. 52)
Meninges (p. 53)
Dura mater (p. 53)
Arachnoid membrane (p. 53)
Subarachnoid space (p. 53)
Pia mater (p. 53)
Cerebrospinal fluid (CSF) (p. 53)

Central canal (p. 53)
Cerebral ventricles (p. 53)
Choroid plexuses (p. 53)
Blood–brain barrier (p. 53)

3.2 Cells of the Nervous System

Neuron (p. 55)
Multipolar neuron (p. 57)
Unipolar neuron (p. 57)
Bipolar neuron (p. 57)
Interneurons (p. 57)
Nuclei (p. 57)
Ganglia (p. 57)
Tracts (p. 57)
Nerves (p. 57)
Glial cells (p. 57)
Oligodendrocytes (p. 57)
Myelin (p. 57)
Myelin sheaths (p. 57)
Schwann cells (p. 57)

Microglia (p. 59)
Astrocytes (p. 59)

3.3 Neuroanatomical Techniques and Directions

Golgi stain (p. 59)
Nissl stain (p. 60)
Electron microscopy (p. 60)
Anterior (p. 61)
Posterior (p. 61)
Dorsal (p. 61)
Ventral (p. 61)
Medial (p. 61)
Lateral (p. 61)
Superior (p. 62)
Inferior (p. 62)
Proximal (p. 62)
Distal (p. 62)
Horizontal sections (p. 62)
Frontal sections (p. 62)
Sagittal sections (p. 62)
Cross section (p. 62)

3.4 The Spinal Cord

Gray matter (p. 63)
White matter (p. 63)
Dorsal horns (p. 63)
Ventral horns (p. 63)
Dorsal root ganglia (p. 63)

3.5 The Five Major Divisions of the Brain

Brain stem (p. 64)

3.6 Major Structures of the Brain

Sensory relay nuclei (p. 66)
Decussate (p. 66)
Contralateral (p. 66)
Ipsilateral (p. 66)
Sulci (p. 67)
Pyramidal cells (p. 68)
Stellate cells (p. 68)
Columnar organization (p. 69)

✔•—Quick Review Test your comprehension of the chapter with this brief practice test. You can find the answers to these questions as well as more practice tests, activities, and other study resources at www.mypsychlab.com.

1. The sympathetic nervous system is a component of the
   a. peripheral nervous system.
   b. parasympathetic nervous system.
   c. autonomic nervous system.
   d. all of the above
   e. both a and c

2. In a typical multipolar neuron, emanating from the cell body are many
   a. axons.
   b. microglia.
   c. dendrites.
   d. nuclei.
   e. astrocytes.

3. If a researcher wished to count the number of neurons in a slice of cortical tissue, she should stain the slice using
   a. a Nissl stain.

   b. a Golgi stain.
   c. an electron stain.
   d. a tell-tale stain.
   e. a Weigert stain.

4. The pons and the cerebellum compose the
   a. mesencephalon.
   b. hypothalamus.
   c. telencephalon.
   d. metencephalon.
   e. reticular formation.

5. Which of the following structures does not belong in the list?
   a. striatum
   b. hippocampus
   c. caudate
   d. globus pallidus
   e. putamen

4 Neural Conduction and Synaptic Transmission

How Neurons Send and Receive Signals

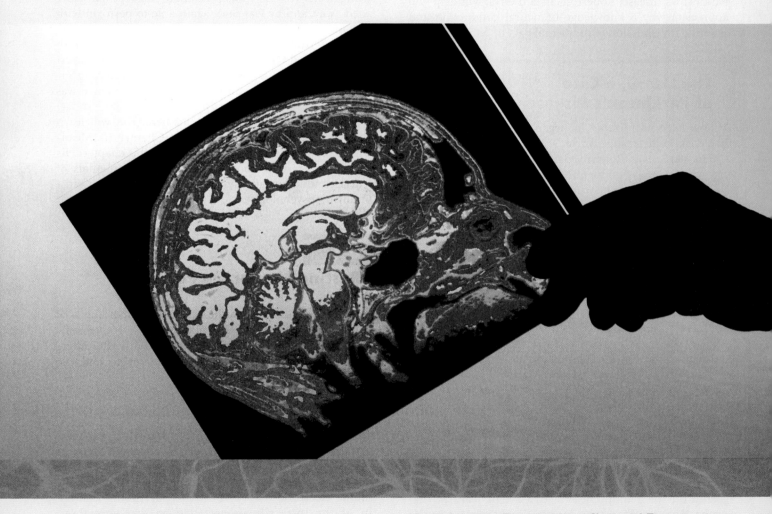

hapter 3 introduced you to the anatomy of neurons. This chapter introduces you to their function—how neurons conduct and transmit electrochemical signals through your nervous system. It begins with a description of how signals are generated in resting neurons; then, it follows the signals as they are conducted through neurons and transmitted across synapses to other neurons. It concludes with a discussion of how drugs are used to study the relation between synaptic transmission and behavior. "The Lizard," a case study of a patient with Parkinson's disease, Roberto Garcia d'Orta, will help you appreciate why a knowledge of neural conduction and synaptic transmission is an integral part of biopsychology.

The Lizard, a Case of Parkinson's Disease

"I have become a lizard," he began. "A great lizard frozen in a dark, cold, strange world."

His name was Roberto Garcia d'Orta. He was a tall thin man in his sixties, but like most patients with Parkinson's disease, he appeared to be much older than his actual age. Not many years before, he had been an active, vigorous business man. Then it happened—not all at once, not suddenly, but slowly, subtly, insidiously. Now he turned like a piece of granite, walked in slow shuffling steps, and spoke in a monotonous whisper.

Clinical Implications

What had been his first symptom?

A tremor.

Had his tremor been disabling?

"No," he said. "My hands shake worse when they are doing nothing at all"—a symptom called *tremor-at-rest*.

The other symptoms of Parkinson's disease are not quite so benign. They can change a vigorous man into a lizard. These include rigid muscles, a marked poverty of spontaneous movements, difficulty in starting to move, and slowness in executing voluntary movements once they have been initiated.

The term "reptilian stare" is often used to describe the characteristic lack of blinking and the widely opened eyes gazing out of a motionless face, a set of features that seems more reptilian than human. Truly a lizard in the eyes of the world.

What was happening in Mr. d'Orta's brain? A small group of nerve cells called the *substantia nigra* (black substance) were unaccountably dying. These neurons make a particular chemical called dopamine, which they deliver to another part of the brain, known as the *striatum*. As the cells of the substantia nigra die, the amount of dopamine they can deliver goes down. The striatum helps control movement, and to do that normally, it needs dopamine.

(Adapted from NEWTON'S MADNESS by Harold Klawans (Harper & Row 1990). Reprinted by permission of Jet Literary Associates, Inc.)

Although dopamine levels are low in Parkinson's disease, dopamine is not an effective treatment because it does not readily penetrate the blood–brain barrier. However, knowledge of dopaminergic transmission has led to the development of an effective treatment: L-*dopa*, the chemical precursor of dopamine, which readily penetrates the blood–brain barrier and is converted to dopamine once inside the brain.

Mr. d'Orta's neurologist prescribed L-dopa, and it worked. He still had a bit of tremor; but his voice became stronger, his feet no longer shuffled, his reptilian stare faded away, and he was once again able to perform with ease many of the activities of daily life (e.g., eating, bathing, writing, speaking, and even making love with his wife). Mr. d'Orta had been destined to spend the rest of his life trapped inside a body that was becoming increasingly difficult to control, but his life sentence was repealed—at least temporarily.

Mr. d'Orta's story does not end here. You will learn what ultimately happened to him in Chapter 10. Meanwhile, keep him in mind while you read this chapter: His case illustrates why knowledge of the fundamentals of neural conduction and synaptic transmission is a must for any biopsychologist.

4.1
Resting Membrane Potential

As you are about to learn, the key to understanding how neurons work—and how they malfunction—is the membrane potential. The **membrane potential** is the difference in electrical charge between the inside and the outside of a cell.

Recording the Membrane Potential

To record a neuron's membrane potential, it is necessary to position the tip of one electrode inside the neuron and the tip of another electrode outside the neuron in the extracellular fluid. Although the size of the extracellular electrode is not critical, it is paramount that the tip of the intracellular electrode be fine enough to pierce the neural membrane without severely damaging it. The intracellular electrodes are called **microelectrodes**; their tips are less than one-thousandth of a millimeter in diameter—much too small to be seen by the naked eye.

Resting Membrane Potential

When both electrode tips are in the extracellular fluid, the voltage difference between them is zero. However, when the tip of the intracellular electrode is inserted into a neuron, a steady potential of about –70 millivolts (mV) is recorded. This indicates that the potential inside the resting neuron is about 70 mV less than that outside the neuron. This steady membrane potential of about –70 mV is called the neuron's **resting potential**. In its resting state, with the –70 mV charge built up across its membrane, a neuron is said to be *polarized*.

Ionic Basis of the Resting Potential

Why are resting neurons polarized? Like all salts in solution, the salts in neural tissue separate into positively and negatively charged particles called **ions**. The resting potential results from the fact that the ratio of negative to positive charges is greater inside the neuron than outside. Why this unequal distribution of charges occurs can be understood in terms of the interaction of four factors: two factors that act to distribute ions equally throughout the intracellular and extracellular fluids of the nervous system and two features of the neural membrane that counteract these homogenizing effects.

The first of the two homogenizing factors is *random motion*. The ions in neural tissue are in constant random motion, and particles in random motion tend to become evenly distributed because they are more likely to move down their *concentration gradients* than up them; that is, they are more likely to move from areas of high concentration to areas of low concentration than vice versa. The second factor that promotes the even distribution of ions is *electrostatic pressure*. Any accumulation of charges, positive or negative, in one area tends to be dispersed by the repulsion among the like charges in the vicinity and the attraction of opposite charges concentrated elsewhere.

Despite the continuous homogenizing effects of random movement and electrostatic pressure, no single class of ions is distributed equally on the two sides of the neural membrane. Four kinds of ions contribute significantly to the resting potential: sodium ions (Na^+), potassium ions (K^+), chloride ions (Cl^-), and various negatively charged protein ions. The concentrations of both Na^+ and Cl^- ions are greater outside a resting neuron than inside, whereas K^+ ions are more concentrated on the inside. The negatively charged protein ions are synthesized inside the neuron and, for the most part, stay there (see Figure 4.1). By the way, the symbols for sodium and potassium were derived from their Latin names: *natrium* (Na) and *kalium* (K), respectively.

Two properties of the neural membrane are responsible for the unequal distribution of Na^+, K^+, Cl^-, and protein ions in resting neurons. One of these properties is passive; that is, it does not involve the consumption of energy. The other is active and does involve the consumption of energy. The passive property of the neural membrane that contributes to the unequal disposition of Na^+, K^+, Cl^-, and protein ions is its differential permeability to those ions. In resting neurons, K^+ and Cl^- ions pass readily through the neural membrane, Na^+ ions pass through it with difficulty, and the negatively charged protein ions do not pass through it at all. Ions pass through the neural membrane at specialized pores called **ion channels**, each

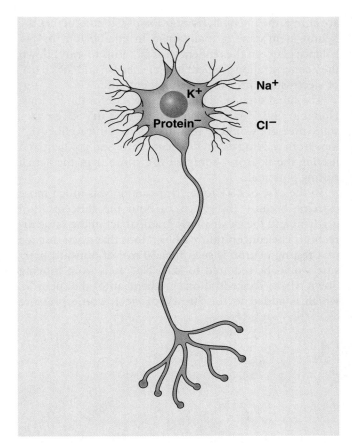

FIGURE 4.1 In its resting state, more Na^+ and Cl^- ions are outside the neuron than inside, and more K^+ ions and negatively charged protein ions are inside the neuron than outside.

type of which is specialized for the passage of particular ions.

In the 1950s, the classic experiments of neurophysiologists Alan Hodgkin and Andrew Huxley provided the first evidence that an energy-consuming process is involved in the maintenance of the resting potential. Hodgkin and Huxley began by wondering why the high extracellular concentrations of Na^+ and Cl^- ions and the high intracellular concentration of K^+ ions were not eliminated by the tendency for them to move down their concentration gradients to the side of lesser concentration. Could the electrostatic pressure of –70 mV across the membrane be the counteracting force that maintained *Thinking Creatively* the unequal distribution of ions? To answer this question, Hodgkin and Huxley took a creative approach for which they received a Nobel Prize.

First, they calculated for each of the three ions the electrostatic charge that would be required to offset the tendency for them to move down their concentration gradients. For Cl^- ions, this calculated electrostatic charge was –70 mV, the same as the actual resting potential. Hodgkin and Huxley thus concluded that when neurons are at rest, the unequal distribution of Cl⁻ ions across the neural membrane is maintained in equilibrium by the balance between the tendency for Cl⁻ ions to move down their concentration gradient into the neuron and the 70 mV of electrostatic pressure driving them out.

The situation turned out to be different for the K⁺ ions. Hodgkin and Huxley calculated that 90 mV of electrostatic pressure would be required to keep intracellular K⁺ ions from moving down their concentration gradient and leaving the neuron—some 20 mV more than the actual resting potential.

In the case of Na⁺ ions, the situation was much more extreme because the effects of both the concentration gradient and the electrostatic gradient act in the same direction. The concentration of Na⁺ ions that exists outside of a resting neuron is such that 50 mV of outward pressure would be required to keep Na⁺ ions from moving down their concentration gradient into the neuron, which is added to the 70 mV of electrostatic pressure

acting to move them in the same direction. Thus, the equivalent of a whopping 120 mV of pressure is acting to force Na⁺ ions into resting neurons.

Subsequent experiments confirmed Hodgkin and Huxley's calculations. They showed that K⁺ ions are continuously being driven out of resting neurons by 20 mV of pressure and that, despite the high resistance of the cell membrane to the passage of Na⁺ ions, those ions are continuously being driven in by the 120 mV of pressure. Why, then, do the intracellular and extracellular concentrations of Na⁺ and K⁺ remain constant in resting neurons? Hodgkin and Huxley discovered that there are active mechanisms in the cell membrane to counteract the *influx* (inflow) of Na⁺ ions by pumping Na⁺ ions out as rapidly as they pass in and to counteract the *efflux* (outflow) of K⁺ ions by pumping K⁺ ions in as rapidly as they pass out. Figure 4.2 summarizes Hodgkin and Huxley's findings and conclusions.

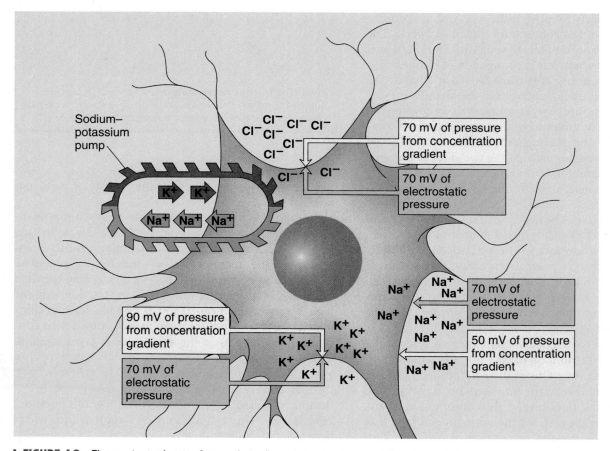

FIGURE 4.2   The passive and active factors that influence the distribution of Na⁺, K⁺, and Cl⁻ ions across the neural membrane. Passive factors continuously drive K⁺ ions out of the resting neuron and Na⁺ ions in; therefore, K⁺ ions must be actively pumped in and Na⁺ ions must be actively pumped out to maintain the resting equilibrium.

It was subsequently discovered that the transport of Na$^+$ ions out of neurons and the transport of K$^+$ ions into them are not independent processes. Such ion transport is performed by energy-consuming mechanisms in the cell membrane that continually exchange three Na$^+$ ions inside the neuron for two K$^+$ ions outside. These transporters are commonly referred to as **sodium–potassium pumps**.

Since the discovery of sodium–potassium pumps, several other classes of **transporters** (mechanisms in the membrane of a cell that actively transport ions or molecules across the membrane) have been discovered (e.g., Tzingounis & Wadiche, 2007). You will encounter more of them later in this chapter.

Table 4.1 summarizes the major factors that are responsible for maintaining the differences between the intracellular and extracellular concentrations of Na$^+$, K$^+$, and Cl$^-$ ions in resting neurons. These differences plus the negative charges of the various protein ions, which are trapped inside the neuron, are largely responsible for the resting membrane potential.

Now that you understand these basic properties of the resting neuron, you are prepared to consider how neurons respond to input.

4.2

Generation and Conduction of Postsynaptic Potentials

When neurons fire, they release from their terminal buttons chemicals called *neurotransmitters*, which diffuse across the synaptic clefts and interact with specialized receptor molecules on the receptive membranes of the next neurons in the circuit. When neurotransmitter molecules bind to postsynaptic receptors, they typically have one of two effects, depending on the structure of both the neurotransmitter and the receptor in question. They may **depolarize** the receptive membrane (decrease the resting membrane potential, from −70 to −67 mV, for example) or they may **hyperpolarize** it (increase the resting membrane potential, from −70 to −72 mV, for example).

Postsynaptic depolarizations are called **excitatory postsynaptic potentials (EPSPs)** because, as you will soon learn, they increase the likelihood that the neuron will fire. Postsynaptic hyperpolarizations are called **inhibitory postsynaptic potentials (IPSPs)** because they decrease the likelihood that the neuron will fire. Both EPSPs and IPSPs are **graded responses**. This means that the amplitudes of EPSPs and IPSPs are proportional to the intensity of the signals that elicit them: Weak signals elicit small postsynaptic potentials, and strong signals elicit large ones.

EPSPs and IPSPs travel passively from their sites of generation at synapses, usually on the dendrites or cell body, in much the same way that electrical signals travel through a cable. Accordingly, the transmission of postsynaptic potentials has two important characteristics. First, it is rapid—so rapid that it can be assumed to be instantaneous for most purposes. It is important not to confuse the duration of EPSPs and IPSPs with their rate of transmission; although the duration of EPSPs and IPSPs varies considerably, all postsynaptic potentials, whether brief or enduring, are transmitted at great speed. Second, the transmission of EPSPs and IPSPs is *decremental*: EPSPs and IPSPs decrease in amplitude as they travel through the neuron, just as a sound wave loses amplitude (the sound grows fainter) as it travels through air. Most EPSPs and IPSPs do not travel more than a couple of millimeters from their site of generation before they fade out; thus, they never travel very far along an axon.

TABLE 4.1	Factors Responsible for Maintaining the Differences in the Intracellular and Extracellular Concentrations of Na$^+$, K$^+$, and Cl$^-$ Ions in Resting Neurons
Na$^+$	Na$^+$ ions tend to be driven into the neurons by both the high concentration of Na$^+$ ions outside the neuron and the negative internal resting potential of −70 mv. However, the membrane is resistant to the passive diffusion of Na$^+$, and the sodium–potassium pumps are thus able to maintain the high external concentration of Na$^+$ ions by pumping them out at the same slow rate as they move in.
K$^+$	K$^+$ ions tend to move out of the neuron because of their high internal concentration, although this tendency is partially offset by the internal negative potential. Despite the tendency for the K$^+$ ions to leave the neuron, they do so at a substantial rate because the membrane offers little resistance to their passage. To maintain the high internal concentration of K$^+$ ions, the sodium–potassium pumps in the cell membrane pump K$^+$ ions into neurons at the same rate as they move out.
Cl$^-$	There is little resistance in the neural membrane to the passage of Cl$^-$ ions. Thus, Cl$^-$ ions are readily forced out of the neuron by the negative internal potential. As chloride ions begin to accumulate on the outside, there is an increased tendency for them to move down their concentration gradient back into the neuron. When the point is reached where the electrostatic pressure for Cl$^-$ ions to move out of the neuron is equal to the tendency for them to move back in, the distribution of Cl$^-$ ions is held in equilibrium. This point of equilibrium occurs at −70 mV.

4.3

Integration of Postsynaptic Potentials and Generation of Action Potentials

The postsynaptic potentials created at a single synapse typically have little effect on the firing of the postsynaptic neuron (Bruno & Sakmann, 2006). The receptive areas of most neurons are covered with thousands of synapses, and whether or not a neuron fires is determined by the net effect of their activity. More specifically, whether or not a neuron fires depends on the balance between the excitatory and inhibitory signals reaching its axon. Until recently, it was believed that action potentials were generated at the **axon hillock** (the conical structure at the junction between the cell body and the axon), but they are actually generated in the adjacent section of the axon (Palmer & Stuart, 2006).

The graded EPSPs and IPSPs created by the action of neurotransmitters at particular receptive sites on a neuron's membrane are conducted instantly and decrementally to the axon hillock. If the sum of the depolarizations and hyperpolarizations reaching the section of the axon adjacent to the axon hillock at any time is sufficient to depolarize the membrane to a level referred to as its **threshold of excitation**—usually about –65 mV—an action potential is generated near the axon hillock. The **action potential (AP)** is a massive but momentary—lasting for 1 millisecond—reversal of the membrane potential from about –70 to about +50 mV. Unlike postsynaptic potentials, action potentials are not graded responses; their magnitude is not related in any way to the intensity of the stimuli that elicit them. To the contrary, they are **all-or-none responses**; that is, they either occur to their full extent or do not occur at all. See Figure 4.3 for an illustration of an EPSP, an IPSP, and an AP. Although many neurons display APs of the type illustrated in Figure 4.3, others do not—for example, some neurons display APs that are longer, that have lower amplitude, or that involve multiple spikes.

In effect, each multipolar neuron adds together all the graded excitatory and inhibitory postsynaptic potentials reaching its axon and decides to fire or not to fire on the basis of their sum. Adding or combining a number of individual signals into one overall signal is called **integration**. Neurons integrate incoming signals in two ways: over space and over time.

Figure 4.4 shows the three possible combinations of **spatial summation**. It shows how local EPSPs that are produced simultaneously on different parts of the receptive membrane sum to form a greater EPSP, how simultaneous IPSPs sum to form a greater IPSP, and how simultaneous EPSPs and IPSPs sum to cancel each other out.

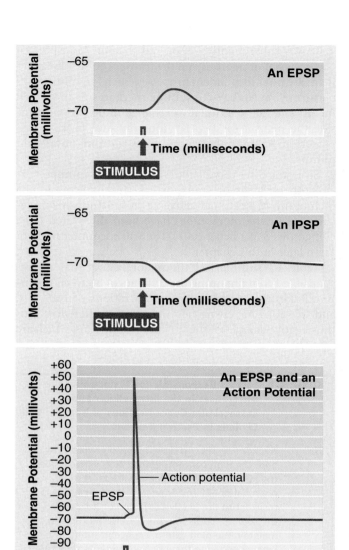

FIGURE 4.3   An EPSP, and IPSP, and an EPSP followed by a typical AP.

Figure 4.5 on page 82 illustrates **temporal summation**. It shows how postsynaptic potentials produced in rapid succession at the same synapse sum to form a greater signal. The reason that stimulations of a neuron can add together over time is that the postsynaptic potentials they produce often outlast them. Thus, if a particular synapse is activated and then activated again before the original postsynaptic potential has completely dissipated, the effect of the second stimulus will be superimposed on the lingering postsynaptic potential produced by the first. Accordingly, it is possible for a brief subthreshold excitatory stimulus to fire a neuron if it is administered twice in rapid succession. In the same way,

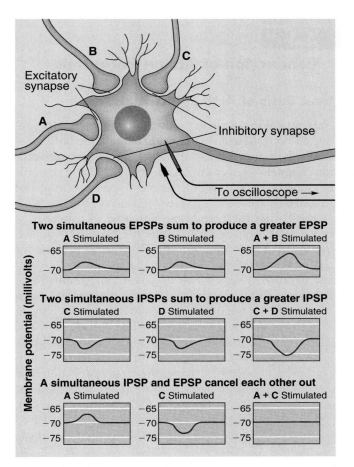

FIGURE 4.4 The three possible combinations of spatial summation.

regardless of where they originate (Williams & Stuart, 2002, 2003).

Before you learn how action potentials are conducted along the axon, pause here to make sure that you understand how action potentials are created. Fill in each blank with the most appropriate term. The correct answers are provided at the end of the exercise. Before proceeding, review material related to your errors and omissions.

1. Roberto Garcia d'Orta referred to himself as "a great lizard frozen in a dark, cold, strange world." He suffered from _____.
2. Tremor-at-rest is a symptom of _____.
3. Microelectrodes are required to record a neuron's resting _____.
4. The _____ is about −70 mV.
5. In its resting state, a neuron is said to be _____.
6. Two factors promote the even distribution of ions across neural membranes: _____ and electrostatic pressure.
7. In the resting state, there is a greater concentration of Na⁺ ions _____ the neural membrane than _____ the neural membrane.
8. *Natrium* is Latin for _____.
9. Ions pass through neural membranes via specialized pores called _____.
10. From their calculations, Hodgkin and Huxley inferred the existence of _____ in neural membranes.
11. Neurotransmitters typically have one of two effects on postsynaptic neurons: They either depolarize them or _____ them.
12. Postsynaptic depolarizations are commonly referred to in their abbreviated form: _____.
13. Action potentials are generated near, but not at, the _____.
14. An action potential is elicited when the depolarization of the neuron reaches the _____.
15. Unlike postsynaptic potentials, which are graded, action potentials are _____ responses.
16. Neurons integrate postsynaptic potentials in two ways: through spatial summation and through _____ summation.

an inhibitory synapse activated twice in rapid succession can produce a greater IPSP than that produced by a single stimulation.

Each neuron continuously integrates signals over both time and space as it is continually bombarded with stimuli through the thousands of synapses covering its dendrites and cell body. Remember that, although schematic diagrams of neural circuitry rarely show neurons with more than a few representative synaptic contacts, most neurons receive thousands of such contacts.

The location of a synapse on a neuron's receptive membrane has long been assumed to be an important factor in determining its potential to influence the neuron's firing. Because EPSPs and IPSPs are transmitted decrementally, synapses near the axon trigger zone have been assumed to have the most influence on the firing of the neuron (see Mel, 2002). However, it has been demonstrated that some neurons have a mechanism for amplifying dendritic signals that originate far from their cell bodies; thus, in these neurons, all dendritic signals reaching the cell body have a similar amplitude,

Thinking Creatively

In some ways, the firing of a neuron is like the firing of a gun. Both reactions are triggered by graded responses. As a trigger is squeezed, it gradually moves back until it causes the gun to fire; as a neuron is stimulated, it becomes less polarized until the threshold of excitation is reached and firing occurs. Furthermore, the firing of a gun and neural firing are both all-or-none events. Just as squeezing a trigger harder does not make the bullet travel faster or farther, stimulating a neuron more intensely does not increase the speed or amplitude of the resulting action potential.

4.4
Conduction of Action Potentials

Ionic Basis of Action Potentials

How are action potentials produced, and how are they conducted along the axon? The answer to both questions is basically the same: through the action of **voltage-activated ion channels**—ion channels that open or close in response to changes in the level of the membrane potential (see Armstrong, 2007).

Recall that the membrane potential of a neuron at rest is relatively constant despite the high pressure acting to drive Na^+ ions into the cell. This is because the resting membrane is relatively impermeable to Na^+ ions and because those few that do pass in are pumped out. But things suddenly change when the membrane potential of the axon is reduced to the threshold of excitation. The voltage-activated sodium channels in the axon membrane open wide, and Na^+ ions rush in, suddenly driving the membrane potential from about -70 to about $+50$ mV. The rapid change in the membrane potential that is associated with the *influx* of Na^+ ions then triggers the opening of voltage-activated potassium channels. At this point, K^+ ions near the membrane are driven out of the cell through these channels—first by their relatively high internal concentration and then, when the action potential is near its peak, by the positive internal charge. After about 1 millisecond, the sodium channels close. This marks the end of the *rising phase* of the action potential and the beginning of *repolarization* by the continued efflux of K^+ ions. Once repolarization has been achieved, the potassium channels gradually close. Because they close gradually, too many K^+ ions flow out of the neuron, and it is left hyperpolarized for a brief period of time. Figure 4.6 illustrates the timing of the opening and closing of the sodium and potassium channels during an action potential.

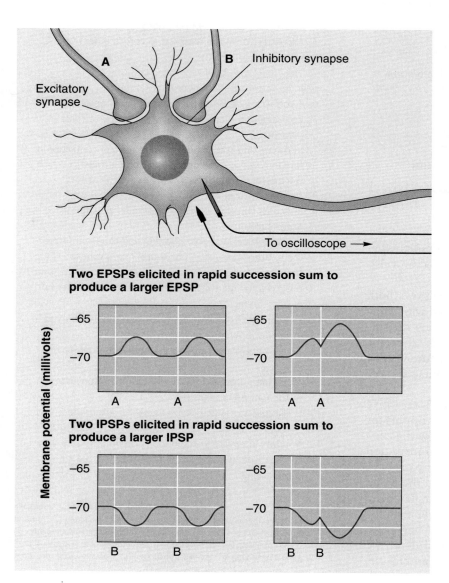

Two EPSPs elicited in rapid succession sum to produce a larger EPSP

Two IPSPs elicited in rapid succession sum to produce a larger IPSP

Membrane potential (millivolts)

FIGURE 4.5 The two possible combinations of temporal summation.

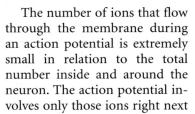

FIGURE 4.6 The opening and closing of voltage-activated sodium and potassium channels during the three phases of the action potential: rising phase, repolarization, and hyperpolarization.

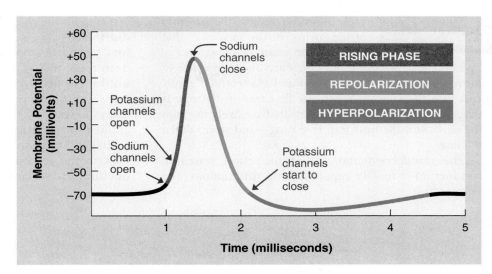

The number of ions that flow through the membrane during an action potential is extremely small in relation to the total number inside and around the neuron. The action potential involves only those ions right next to the membrane. Therefore, a single action potential has little effect on the relative concentrations of various ions inside and outside the neuron, and the resting ion concentrations next to the membrane are rapidly reestablished by the random movement of ions. The sodium–potassium pumps play only a minor role in the reestablishment of the resting potential.

Refractory Periods

There is a brief period of about 1 to 2 milliseconds after the initiation of an action potential during which it is impossible to elicit a second one. This period is called the **absolute refractory period**. The absolute refractory period is followed by the **relative refractory period**—the period during which it is possible to fire the neuron again, but only by applying higher-than-normal levels of stimulation. The end of the relative refractory period is the point at which the amount of stimulation necessary to fire a neuron returns to baseline.

The refractory period is responsible for two important characteristics of neural activity. First, it is responsible for the fact that action potentials normally travel along axons in only one direction. Because the portions of an axon over which an action potential has just traveled are left momentarily refractory, an action potential cannot reverse direction. Second, the refractory period is responsible for the fact that the rate of neural firing is related to the intensity of the stimulation. If a neuron is subjected to a high level of continual stimulation, it fires and then fires again as soon as its absolute refractory period is over—a maximum of about 1,000 times per second. However, if the level of stimulation is of an intensity just sufficient to fire the neuron when it is at rest, the neuron does not fire again until both the absolute and the relative refractory periods have run their course. Intermediate levels of stimulation produce intermediate rates of neural firing.

Axonal Conduction of Action Potentials

The conduction of action potentials along an axon differs from the conduction of EPSPs and IPSPs in two important ways. First, the conduction of action potentials along an axon is *nondecremental*; action potentials do not grow weaker as they travel along the axonal membrane. Second, action potentials are conducted more slowly than postsynaptic potentials.

The reason for these two differences is that the conduction of EPSPs and IPSPs is passive, whereas the axonal conduction of action potentials is largely active. Once an action potential has been generated, it travels passively along the axonal membrane to the adjacent voltage-activated sodium channels, which have yet to open. The arrival of the electrical signal opens these channels, thereby allowing Na+ ions to rush into the neuron and generate a full-blown action potential on this portion of the membrane. This signal is then conducted passively to the next sodium channels, where another action potential is actively triggered. These events are repeated again and again until a full-blown action potential is triggered in all the terminal buttons (Huguenard, 2000). However, because there are so many ion channels on the axonal membrane and they are so close together, it is usual to think of axonal conduction as a single wave of excitation spreading actively at a constant speed along the axon, rather than as a series of discrete events.

The wave of excitation triggered by the generation of an action potential near the axon hillock always spreads passively back through the cell body and dendrites of the neuron. Although little is yet known about the functions of these backward action potentials, they are currently the subject of intensive investigation.

The following analogy may help you appreciate the major characteristics of axonal conduction. Consider a row of mouse traps on a wobbly shelf, all of them set and ready to be triggered. Each trap stores energy by holding back its striker against the pressure of the spring, in the same way that each sodium channel stores energy by holding back Na⁺ ions, which are under pressure to move down their concentration and electrostatic gradients into the neuron. When the first trap in the row is triggered, the vibration is transmitted passively through the shelf, and the next trap is sprung—and so on down the line.

The nondecremental nature of action potential conduction is readily apparent from this analogy; the

last trap on the shelf strikes with no less intensity than did the first. This analogy also illustrates the refractory period: A trap cannot respond again until it has been reset, just as a section of axon cannot fire again until it has been repolarized. Furthermore, the row of traps can transmit in either direction, just like an axon. If electrical stimulation of sufficient intensity is applied to the terminal end of an axon, an action potential will be generated and will travel along the axon back to the cell body; this is called **antidromic conduction**. Axonal conduction in the natural direction—from cell body to terminal buttons—is called **orthodromic conduction**. The elicitation of an action potential and the direction of orthodromic conduction are summarized in Figure 4.7.

Conduction in Myelinated Axons

In Chapter 3, you learned that the axons of many neurons are insulated from the extracellular fluid by segments of fatty tissue called *myelin*. In myelinated axons, ions can pass through the axonal membrane only at the **nodes of Ranvier**—the gaps between adjacent myelin segments. Indeed, in myelinated axons, axonal sodium channels are concentrated at the nodes of Ranvier (Salzer, 2002). How, then, are action potentials transmitted in myelinated axons?

When an action potential is generated in a myelinated axon, the signal is conducted passively—that is, instantly and decrementally—along the first segment of myelin to the next node of Ranvier. Although the signal is somewhat diminished

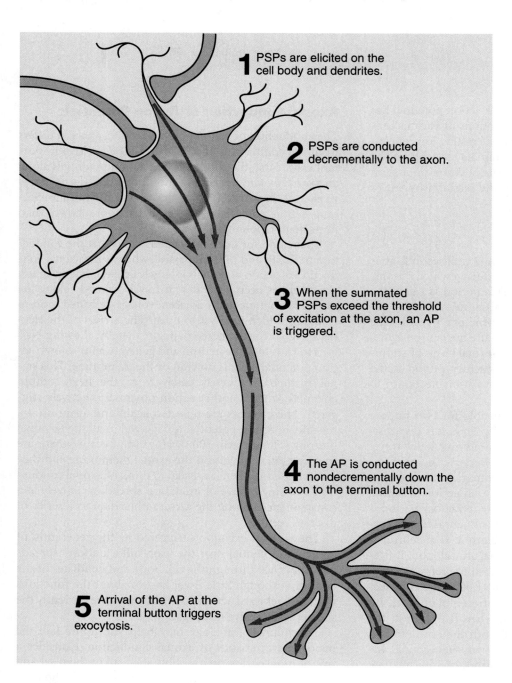

1 PSPs are elicited on the cell body and dendrites.

2 PSPs are conducted decrementally to the axon.

3 When the summated PSPs exceed the threshold of excitation at the axon, an AP is triggered.

4 The AP is conducted nondecrementally down the axon to the terminal button.

5 Arrival of the AP at the terminal button triggers exocytosis.

FIGURE 4.7 The direction of signals conducted orthodromically through a typical multipolar neuron.

by the time it reaches that node, it is still strong enough to open the voltage-activated sodium channels at the node and to generate another full-blown action potential. This action potential is then conducted passively along the axon to the next node, where another full-blown action potential is elicited, and so on.

Myelination increases the speed of axonal conduction. Because conduction along the myelinated segments of the axon is passive, it occurs instantly, and the signal thus "jumps" along the axon from node to node. There is, of course, a slight delay at each node of Ranvier while the action potential is actively generated, but conduction is still much faster in myelinated axons than in unmyelinated axons, in which passive conduction plays a less prominent role (see Poliak & Peles, 2003). The transmission of action potentials in myelinated axons is called **saltatory conduction** (*saltare* means "to skip or jump"). Given the important role of myelin in neural conduction, it is hardly surprising that the *neurodegenerative diseases* (diseases that damage the nervous system) that attack myelin have devastating effects on neural activity and behavior—see the discussion of multiple sclerosis in Chapter 10.

The Velocity of Axonal Conduction

At what speed are action potentials conducted along an axon? The answer to this question depends on two properties of the axon (see ffrench-Constant, Colognato, & Franklin, 2004). Conduction is faster in large-diameter axons, and—as you have just learned—it is faster in those that are myelinated. Mammalian *motor neurons* (neurons that synapse on skeletal muscles) are large and myelinated; thus, some can conduct at speeds of 100 meters per second (about 224 miles per hour). In contrast, small, unmyelinated axons conduct action potentials at about 1 meter per second.

There is a misconception about the velocity of motor neuron action potentials in humans. The maximum velocity of motor neuron action potentials was found to be about 100 meters per second in cats and was then assumed to be the same in humans: It is not. The maximum velocity of conduction in human motor neurons is about 60 meters per second (Peters & Brooke, 1998).

Conduction in Neurons without Axons

Action potentials are the means by which axons conduct all-or-none signals nondecrementally over relatively long distances. Thus, to keep what you have just learned about action potentials in perspective, it is important for you to remember that many neurons in mammalian brains either do not have axons or have very short ones, and many of these neurons do not normally display action potentials. Conduction in these *interneurons* is typically passive and decremental (Juusola et al., 1996).

The Hodgkin–Huxley Model in Perspective

The preceding account of neural conduction is based heavily on the *Hodgkin-Huxley model*, the theory first proposed by Hodgkin and Huxley in the early 1950s (see Huxley, 2002). Perhaps you have previously encountered some of this information about neural conduction in introductory biology and psychology courses, where it is often presented as a factual account of neural conduction and its mechanisms, rather than as a theory. The Hodgkin-Huxley model was a major advance in our understanding of neural conduction (Armstrong, 2007). Fully deserving of the 1963 Nobel Prize, the model provided a simple effective introduction to what we now understand about the general ways in which neurons conduct signals. The problem is that the simple neurons and mechanisms of the Hodgkin-Huxley model are not representative of the variety, complexity, and plasticity of many of the neurons in the mammalian brain.

The Hodgkin-Huxley model was based on the study of squid motor neurons. Motor neurons are simple, large, and readily accessible in the PNS—squid motor neurons are particularly large. The simplicity, size, and accessibility of squid motor neurons contributed to the initial success of Hodgkin and Huxley's research, but these same properties make it difficult to apply the model directly to the mammalian brain. Hundreds of different kinds of neurons are found in the mammalian brain, and many of these have actions not found in motor neurons (see Debanne, 2004; Markram et al., 2004; Nusser, 2009). Thus, the Hodgkin-Huxley model must be applied to cerebral neurons with caution. The following are some properties of cerebral neurons that are not shared by motor neurons:

- Many cerebral neurons fire continually even when they receive no input (Lisman, Raghavachari, & Tsien, 2007; Schultz, 2007; Surmeier, Mercer, & Chan, 2005).
- The axons of some cerebral neurons can actively conduct both graded signals and action potentials (Alle & Geiger, 2006, 2008).
- Action potentials of all motor neurons are the same, but action potentials of different classes of cerebral neurons vary greatly in duration, amplitude, and frequency (Bean, 2007).
- Many cerebral neurons have no axons and do not display action potentials.
- The dendrites of some cerebral neurons can actively conduct action potentials (Chen, Midtgaard, & Shepherd, 1997).

Clearly, cerebral neurons are far more complex than motor neurons, which have traditionally been the focus of neurophysiological research, and thus, results of studies of motor neurons should be applied to the brain with caution.

Synaptic Transmission: Chemical Transmission of Signals among Neurons

You have learned in this chapter how postsynaptic potentials are generated on the receptive membrane of a resting neuron, how these graded potentials are conducted passively to the axon, how the sum of these graded potentials can trigger action potentials, and how these all-or-none potentials are actively conducted down the axon to the terminal buttons. In the remaining sections of this chapter, you will learn how action potentials arriving at terminal buttons trigger the release of neurotransmitters into synapses and how neurotransmitters carry signals to other cells. This section provides an overview of five aspects of synaptic transmission: (1) the structure of synapses; (2) the synthesis, packaging, and transport of neurotransmitter molecules; (3) the release of neurotransmitter molecules; (4) the activation of receptors by neurotransmitter molecules; and (5) the reuptake, enzymatic degradation, and recycling of neurotransmitter molecules.

⊙→ Simulate
Synaptic Transmission
www.mypsychlab.com

Structure of Synapses

Some communication among neurons occurs across synapses such as the one illustrated in Figure 4.8. Neurotransmitter molecules are released from buttons into synaptic clefts, where they induce EPSPs or IPSPs in other neurons by binding to receptors on their postsynaptic membranes. The synapses featured in Figure 4.8 are *axodendritic synapses*—synapses of axon terminal buttons on dendrites. Notice that many axodendritic synapses terminate on **dendritic spines** (nodules of various shapes that are located on the surfaces of many dendrites)—see Figure 3.31 on page 73. Also common are *axosomatic synapses*—synapses of axon terminal buttons on *somas* (cell bodies).

Although axodendritic and axosomatic synapses are the most common synaptic arrangements, there are several others (Shepherd & Erulkar, 1997). For example, there are *dendrodendritic synapses*, which are interesting because they are often capable of transmission in either direction. *Axoaxonic synapses* are particularly important because they can mediate *presynaptic facilitation and inhibition*. As illustrated in Figure 4.9, an axoaxonic synapse on, or near, a terminal button can selectively facilitate or inhibit the effects of that button on the postsynaptic neuron. The advantage of presynaptic facilitation and inhibition (compared to EPSPs and IPSPs, which you have already learned about) is that they can selectively influence one particular synapse rather than the entire presynaptic neuron.

The synapses depicted in Figure 4.9 are **directed synapses**—synapses at which the site of neurotransmitter release and the site of neurotransmitter reception are in close proximity. This is a common arrangement, but there are also many nondirected synapses in the mammalian nervous system. **Nondirected synapses** are synapses at which the site of release is at some distance from the site of reception. One type of nondirected synapse is depicted in Figure 4.10. In this type of arrangement, neurotransmitter molecules are released from a

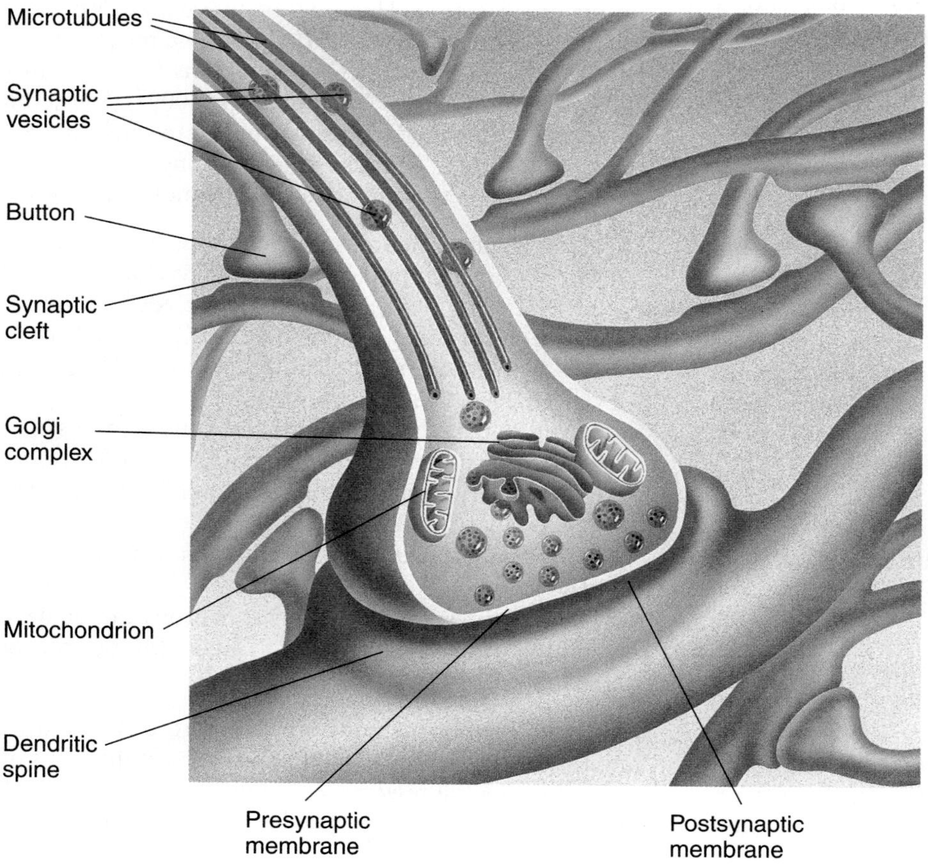

Microtubules
Synaptic vesicles
Button
Synaptic cleft
Golgi complex
Mitochondrion
Dendritic spine
Presynaptic membrane
Postsynaptic membrane

FIGURE 4.8 The anatomy of a typical synapse.

FIGURE 4.9 Presynaptic facilitation and inhibition.

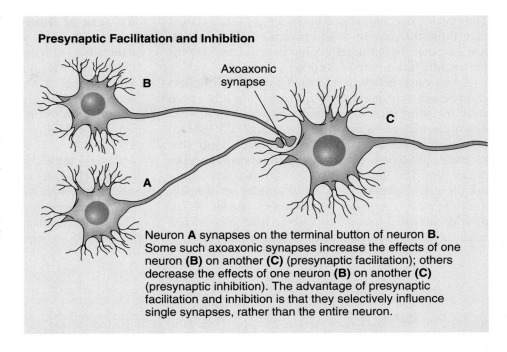

Presynaptic Facilitation and Inhibition

Neuron **A** synapses on the terminal button of neuron **B**. Some such axoaxonic synapses increase the effects of one neuron (**B**) on another (**C**) (presynaptic facilitation); others decrease the effects of one neuron (**B**) on another (**C**) (presynaptic inhibition). The advantage of presynaptic facilitation and inhibition is that they selectively influence single synapses, rather than the entire neuron.

series of *varicosities* (bulges or swellings) along the axon and its branches and thus are widely dispersed to surrounding targets. Because of their appearance, these synapses are often referred to as *string-of-beads synapses*.

Synthesis, Packaging, and Transport of Neurotransmitter Molecules

👁 **Watch** Neurotransmitters: Communicators between Neurons **www.mypsychlab.com**

There are two basic categories of neurotransmitter molecules: small and large. The small neurotransmitters are of several types; large neurotransmitters are all neuropeptides. **Neuropeptides** are short amino acid chains comprising between 3 and 36 amino acids; in effect, they are short proteins.

Small-molecule neurotransmitters are typically synthesized in the cytoplasm of the terminal button and packaged in **synaptic vesicles** by the button's **Golgi complex** (see Brittle & Waters, 2000). (This may be a good point at which to review the internal structures of neurons in Figure 3.6 on page 56.) Once filled with neurotransmitter, the vesicles are stored in clusters next to the presynaptic membrane. In

contrast, neuropeptides, like other proteins, are assembled in the cytoplasm of the cell body on *ribosomes*; they are then packaged in vesicles by the cell body's Golgi complex and transported by *microtubules* to the terminal buttons at a rate of about 40 centimeters per day.

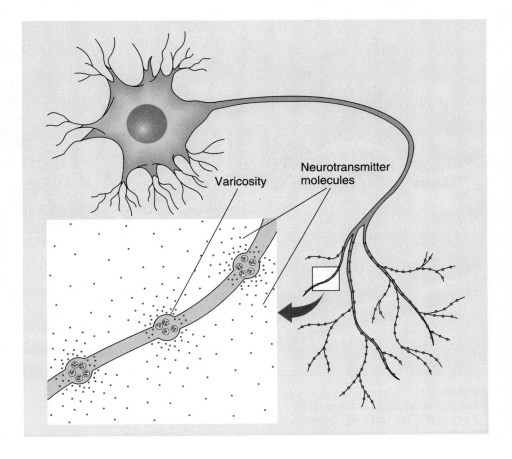

Varicosity

Neurotransmitter molecules

FIGURE 4.10 Nondirected neurotransmitter release. Some neurons release neurotransmitter molecules diffusely from varicosities along the axon and its branches.

The vesicles that contain neuropeptides are usually larger than those that contain small-molecule neurotransmitters, and they do not usually congregate as closely to the presynaptic membrane as the other vesicles do.

It was once believed that each neuron synthesizes and releases only one neurotransmitter, but it has been clear for some time that many neurons contain two neurotransmitters—a situation that is generally referred to as **coexistence**. It may have escaped your notice that the button illustrated in Figure 4.8 contains synaptic vesicles of two sizes. This suggests that it contains two neurotransmitters: a neuropeptide in the larger vesicles and a small-molecule neurotransmitter in the smaller vesicles. So far, most documented cases of coexistence have involved one small-molecule neurotransmitter and one neuropeptide.

Release of Neurotransmitter Molecules

Exocytosis—the process of neurotransmitter release—is illustrated in Figure 4.11 (see Schweizer & Ryan, 2006). When a neuron is at rest, synaptic vesicles that contain small-molecule neurotransmitters tend to congregate near sections of the presynaptic membrane that are particularly rich in *voltage-activated calcium channels* (see Rizzoli & Betz, 2004, 2005). When stimulated by action potentials, these channels open, and Ca^{2+} ions enter the button. The entry of the Ca^{2+} ions causes synaptic vesicles to fuse with the presynaptic membrane and empty their contents into the synaptic cleft (see Collin, Marty, & Llano, 2005; Schneggenburger & Neher, 2005). At many—but not all—synapses, one action potential causes the

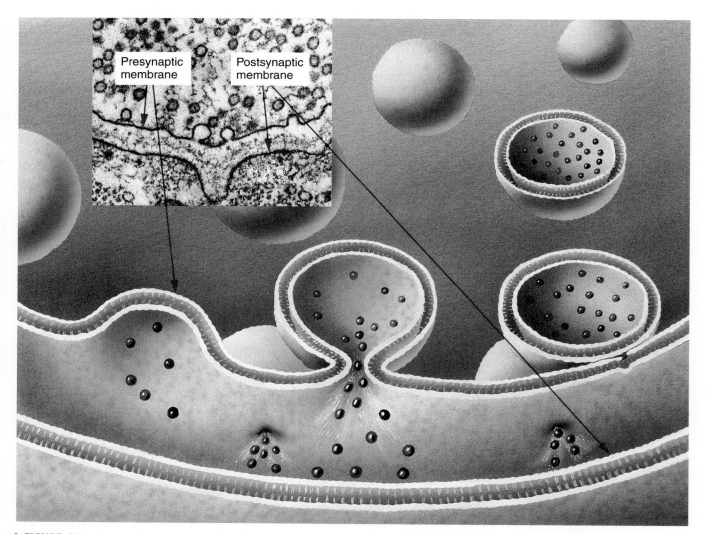

Presynaptic membrane

Postsynaptic membrane

FIGURE 4.11 Schematic and photographic illustrations of exocytosis. (The photomicrograph was reproduced from J. E. Heuser et al., *Journal of Cell Biology*, 1979, *81*, 275–300, by copyright permission of The Rockefeller University Press.)

release of neurotransmitter molecules from one vesicle (Matsui & Jahr, 2006).

The exocytosis of small-molecule neurotransmitters differs from the exocytosis of neuropeptides. Small-molecule neurotransmitters are typically released in a pulse each time an action potential triggers a momentary influx of Ca^{2+} ions through the presynaptic membrane; in contrast, neuropeptides are typically released gradually in response to general increases in the level of intracellular Ca^{2+} ions, such as might occur during a general increase in the rate of neuron firing.

Activation of Receptors by Neurotransmitter Molecules

Once released, neurotransmitter molecules produce signals in postsynaptic neurons by binding to **receptors** in the postsynaptic membrane. Each receptor is a protein that contains binding sites for only particular neurotransmitters; thus, a neurotransmitter can influence only those cells that have receptors for it. Any molecule that binds to another is referred to as its **ligand**, and a neurotransmitter is thus said to be a ligand of its receptor.

👁 **Watch**
Synaptic Chemical Messengers
www.mypsychlab.com

It was initially assumed that there is only one type of receptor for each neurotransmitter, but this has not proved to be the case. As more receptors have been identified, it has become clear that most neurotransmitters bind to several different types of receptors. The different types of receptors to which a particular neurotransmitter can bind are called the **receptor subtypes** for that neurotransmitter. The various receptor subtypes for a neurotransmitter are typically located in different brain areas, and they typically respond to the neurotransmitter in different ways (see Darlison & Richter, 1999). Thus, one advantage of receptor subtypes is that they enable one neurotransmitter to transmit different kinds of messages to different parts of the brain.

The binding of a neurotransmitter to one of its receptor subtypes can influence a postsynaptic neuron in one of two fundamentally different ways, depending on whether the receptor is ionotropic or metabotropic (Heuss & Gerber, 2000; Waxham, 1999). **Ionotropic receptors** are those receptors that are associated with ligand-activated ion channels; **metabotropic receptors** are those receptors that are associated with signal proteins and **G proteins** (*guanosine-triphosphate–sensitive proteins*); see Figure 4.12.

When a neurotransmitter molecule binds to an ionotropic receptor, the associated ion channel usually opens or closes immediately, thereby inducing an immediate postsynaptic potential. For example, in some neurons, EPSPs (depolarizations) occur because the neurotransmitter opens sodium channels, thereby increasing the flow of Na^+

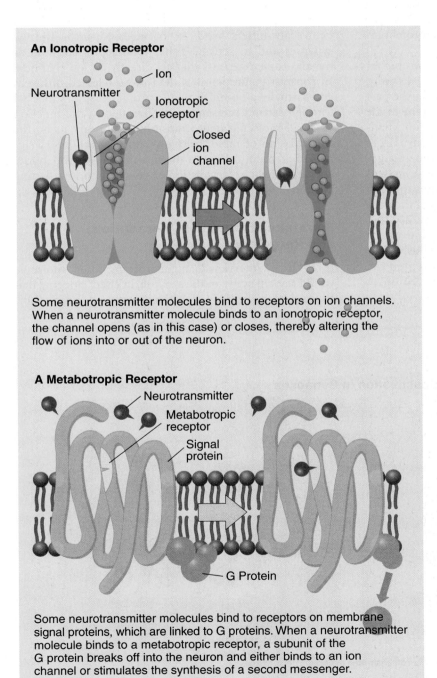

An Ionotropic Receptor

Neurotransmitter — Ion
Ionotropic receptor
Closed ion channel

Some neurotransmitter molecules bind to receptors on ion channels. When a neurotransmitter molecule binds to an ionotropic receptor, the channel opens (as in this case) or closes, thereby altering the flow of ions into or out of the neuron.

A Metabotropic Receptor

Neurotransmitter
Metabotropic receptor
Signal protein
G Protein

Some neurotransmitter molecules bind to receptors on membrane signal proteins, which are linked to G proteins. When a neurotransmitter molecule binds to a metabotropic receptor, a subunit of the G protein breaks off into the neuron and either binds to an ion channel or stimulates the synthesis of a second messenger.

FIGURE 4.12 Ionotropic and metabotropic receptors.

ions into the neuron. In contrast, IPSPs (hyperpolarizations) often occur because the neurotransmitter opens potassium channels or chloride channels, thereby increasing the flow of K$^+$ ions out of the neuron or the flow of Cl$^-$ ions into it, respectively.

Metabotropic receptors are more prevalent than ionotropic receptors, and their effects are slower to develop, longer-lasting, more diffuse, and more varied. There are many different kinds of metabotropic receptors, but each is attached to a serpentine signal protein that winds its way back and forth through the cell membrane seven times. The metabotropic receptor is attached to a portion of the signal protein outside the neuron; the G protein is attached to a portion of the signal protein inside the neuron.

When a neurotransmitter binds to a metabotropic receptor, a subunit of the associated G protein breaks away. Then, one of two things happens, depending on the particular G protein. The subunit may move along the inside surface of the membrane and bind to a nearby ion channel, thereby inducing an EPSP or IPSP; or it may trigger the synthesis of a chemical called a **second messenger** (neurotransmitters are considered to be the *first messengers*). Once created, a second messenger diffuses through the cytoplasm and may influence the activities of the neuron in a variety of ways (Neves, Ram, & Iyengar, 2002)—for example, it may enter the nucleus and bind to the DNA, thereby influencing genetic expression Thus, a neurotransmitter's binding to a metabotropic receptor can have radical, long-lasting effects—see the discussion of *epigenetics* in Chapter 2.

One type of metabotropic receptor—autoreceptors—warrants special mention. **Autoreceptors** are metabotropic receptors that have two unconventional characteristics: They bind to their neuron's own neurotransmitter molecules; and they are located on the presynaptic, rather than the postsynaptic, membrane. Their usual function is to monitor the number of neurotransmitter molecules in the synapse, to reduce subsequent release when the levels are high, and to increase subsequent release when they are low.

Differences between small-molecule and peptide neurotransmitters in patterns of release and receptor binding suggest that they serve different functions. Small-molecule neurotransmitters tend to be released into directed synapses and to activate either ionotropic receptors or metabotropic receptors that act directly on ion channels. In contrast, neuropeptides tend to be released diffusely, and virtually all bind to metabotropic receptors that act through second messengers. Consequently, the function of small-molecule neurotransmitters appears to be the transmission of rapid, brief excitatory or inhibitory signals to adjacent cells; and the function of neuropeptides appears to be the transmission of slow, diffuse, long-lasting signals.

Reuptake, Enzymatic Degradation, and Recycling

If nothing intervened, a neurotransmitter molecule would remain active in the synapse, in effect clogging that channel of communication. However, two mechanisms

Two Mechanisms of Neurotransmitter Deactivation in Synapses

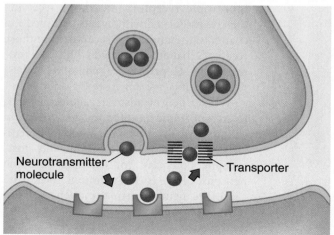

Reuptake

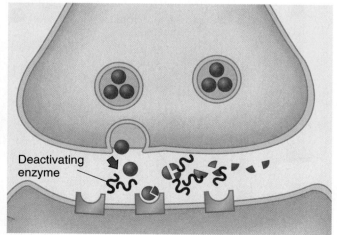

Enzymatic Degradation

FIGURE 4.13   The two mechanisms for terminating neurotransmitter action in the synapse: reuptake and enzymatic degradation.

terminate synaptic messages and keep that from happening. These two message-terminating mechanisms are **reuptake** by transporters and **enzymatic degradation** (see Figure 4.13).

Reuptake is the more common of the two deactivating mechanisms. The majority of neurotransmitters, once released, are almost immediately drawn back into the presynaptic buttons by transporter mechanisms.

In contrast, other neurotransmitters are degraded (broken apart) in the synapse by the action of **enzymes**—proteins that stimulate or inhibit biochemical reactions without being affected by them. For example, *acetylcholine*, one of the few neurotransmitters for which enzymatic degradation is the main mechanism of synaptic deactivation, is broken down by the enzyme **acetylcholinesterase**.

Terminal buttons are models of efficiency. Once released, neurotransmitter molecules or their breakdown products are drawn back into the button and recycled, regardless of the mechanism of their deactivation. Even the vesicles, once they have done their job, are drawn back into the neuron from the presynaptic membrane and are used to create new vesicles (Südhof, 2004).

Glial Function and Synaptic Transmission

Once overlooked as playing merely supportive roles in the nervous system, glial cells have been thrust to center stage by a wave of remarkable findings. For example, astrocytes have been shown to release chemical transmitters, to contain receptors for neurotransmitters, to conduct signals, and to participate in neurotransmitter reuptake (see Fields & Burnstock, 2006; Miller & Cleveland, 2005). Indeed, it is becoming inappropriate to think of brain function solely in terms of neuron–neuron connections. Neurons are only part of the story.

The importance of glial cells in brain function is suggested by the greater prevalence of these cells in intelligent organisms. Will *neuroscience* prove to be a misnomer? Anybody for "gliascience"?

Evolutionary Perspective

Gap Junctions   Interest in gap junctions has recently been rekindled. **Gap junctions** are narrow spaces between adjacent neurons that are bridged by fine tubular channels, called *connexins,* that contain cytoplasm. Consequently, the cytoplasm of the two neurons is continuous, allowing electrical signals and small molecules to pass from one neuron to the next (see Figure 4.14). Gap junctions are sometimes called *electrical synapses.*

Gap junctions are commonplace in invertebrate nervous systems, but their existence was more difficult to establish in mammals (see Bennett, 2000). They were first demonstrated in mammals in the 1970s, but few mammalian examples accumulated over the ensuing 30 years. Then technological developments led to the discovery of gap junctions throughout the mammalian brain; they seem to be an integral feature of local neural inhibitory circuits (Hestrin & Galarreta, 2005). In addition, astrocytes have been shown to communicate with each other, neurons, and other cells through gap junctions (Bennett et al., 2003). Thus, the focus on glial function is reviving interest in gap junctions.

The role of gap junctions in nervous system activity is both underappreciated (Conners & Long, 2004) and poorly understood (Nagy, Dudek, & Rash, 2004). Although they are less selective than synapses, gap junctions have two advantages. One is that communication across them is very fast because it does not involve active mechanisms. The other advantage is that gap junctions permit communication in either direction.

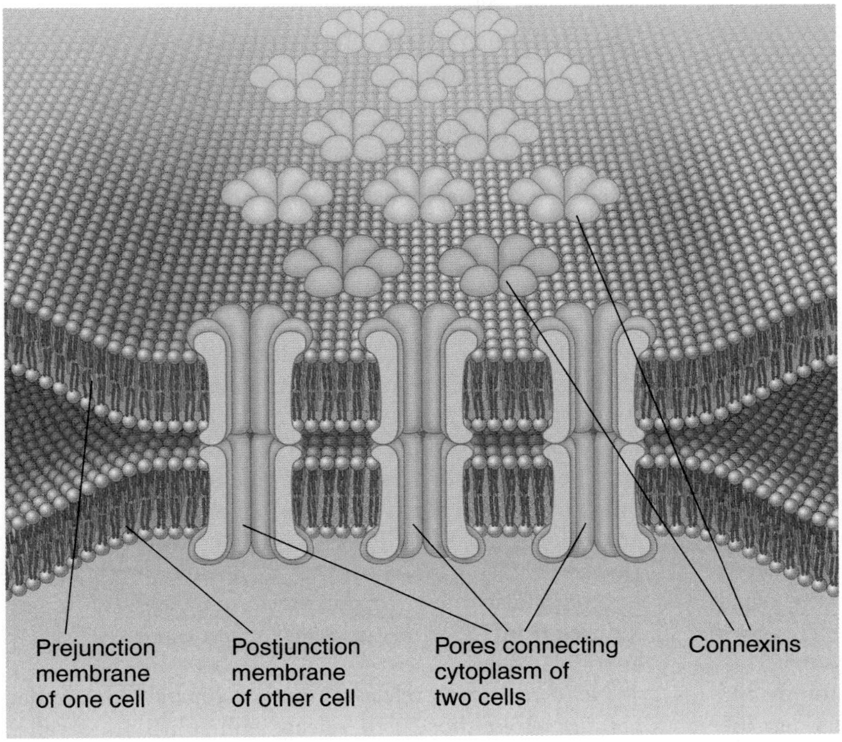

Prejunction membrane of one cell   Postjunction membrane of other cell   Pores connecting cytoplasm of two cells   Connexins

FIGURE 4.14   Gap junctions. Gap junctions connect the cytoplasm of two cells.

Scan Your Brain

Before moving on to the discussion of specific neurotransmitters, review the general principles of axon conduction and synaptic transmission. Draw a line to connect each term in the left column with the appropriate word or phrase in the right column. The correct answers are provided at the end of the exercise. Before proceeding, review material related to your errors and omissions.

1. fatty
2. sclerosis
3. cell bodies
4. dendritic spines
5. nondecremental
6. presynaptic facilitation
7. nondirected synapses
8. synaptic vesicles
9. from cell body to terminal buttons
10. acetylcholinesterase
11. short amino acid chains
12. saltatory
13. metabotropic receptors
14. electrical synapses
15. spines

a. axonal conduction of action potentials
b. orthodromic
c. myelin
d. nodes of Ranvier
e. multiple
f. dendritic
g. compartmentalize dendrites
h. somas
i. axoaxonic synapses
j. string-of-beads
k. neuropeptides
l. store neurotransmitters
m. G proteins
n. enzymatic degradation
o. gap junctions

Scan Your Brain answers: (1) c, (2) e, (3) h, (4) g, (5) a, (6) i, (7) j, (8) l, (9) b, (10) n, (11) k, (12) d, (13) m, (14) o, (15) f.

4.6
Neurotransmitters

Now that you understand the basics of neurotransmitter function, let's take a closer look at some of the well over 100 neurotransmitter substances that have been identified (see Purves et al., 2004). The following are three classes of conventional small-molecule neurotransmitters: the *amino acids*, the *monoamines*, and *acetylcholine*. Also, there is a fourth group of various small-molecule neurotransmitters, which are often referred to as *unconventional neurotransmitters* because their mechanisms of action are unusual. In contrast to the small-molecule neurotransmitters, there is only one class of large-molecule neurotransmitters: the *neuropeptides*. Most neurotransmitters produce either excitation or inhibition, not both; but a few produce excitation when they

bind to some of their receptor subtypes and inhibition when they bind to others. All of the neurotransmitter classes and individual neurotransmitters that appear in this section in boldface type are presented in Figure 4.17 at the end of this section.

Amino Acid Neurotransmitters

The neurotransmitters in the vast majority of fast-acting, directed synapses in the central nervous system are amino acids—the molecular building blocks of proteins. The four most widely studied **amino acid neurotransmitters** are **glutamate**, **aspartate**, **glycine**, and **gamma-aminobutyric acid (GABA)**. The first three are common in the proteins we consume, whereas GABA is synthesized by a simple modification of the structure of glutamate. Glutamate is the most prevalent excitatory neurotransmitter in the mammalian central nervous system. GABA is the most prevalent inhibitory neurotransmitter (see Jacob, Moss, & Jurd, 2008; Orser, 2007); however, it has excitatory effects at some synapses (Szabadics et al., 2006).

Monoamine Neurotransmitters

Monoamines are another class of small-molecule neurotransmitters. Each is synthesized from a single amino acid—hence the name monoamine (one amine). **Monoamine neurotransmitters** are slightly larger than amino acid neurotransmitters, and their effects tend to be more diffuse (see Bunin & Wightman, 1999). The monoamines are present in small groups of neurons whose cell bodies are, for the most part, located in the brain stem. These neurons often have highly branched axons with many varicosities (string-of-beads synapses), from which monoamine neurotransmitters are diffusely released into the extracellular fluid (see Figures 4.10 and 4.15).

There are four monoamine neurotransmitters: **dopamine**, **epinephrine**, **norepinephrine**, and **serotonin**. They are subdivided into two groups, **catecholamines** and **indolamines**, on the basis of their structures. Dopamine, norepinephrine, and epinephrine are catecholamines. Each is synthesized from the amino acid *tyrosine*. Tyrosine is converted to L-dopa, which in turn is converted to dopamine. Neurons that release norepinephrine have an extra enzyme (one that is not present in dopaminergic neurons), which converts the dopamine in them to norepinephrine. Similarly, neurons that release epinephrine have all the enzymes present in neurons that release norepinephrine, along with an extra enzyme that converts norepinephrine to epinephrine (see Figure 4.16). In contrast to the other monoamines, serotonin (also called *5-hydroxytryptamine*, or *5-HT*) is synthesized from the amino acid *tryptophan* and is classified as an indolamine.

Neurons that release norepinephrine are called *noradrenergic*; those that release epinephrine are called

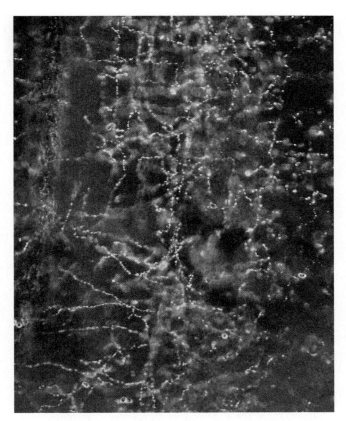

FIGURE 4.15   String-of-beads noradrenergic nerve fibers. The bright, beaded structures represent sites in these multiple-branched axons where the monoamine neurotransmitter norepinephrine is stored in high concentration and released into the surrounding extracellular fluid. (Courtesy of Floyd E. Bloom, M.D., The Scripps Research Institute, La Jolla, California.)

adrenergic. There are two reasons for this naming. One is that epinephrine and norepinephrine used to be called *adrenaline* and *noradrenaline*, respectively, by many scientists, until a drug company registered *Adrenalin* as a brand name. The other reason will become apparent to you if you try to say *norepinephrinergic*.

Acetylcholine

Acetylcholine (abbreviated Ach) is a small-molecule neurotransmitter that is in one major respect like a professor who is late for a lecture: It is in a class by itself. It is created by adding an *acetyl* group to a *choline* molecule. Acetylcholine is the neurotransmitter at neuromuscular junctions, at many of the synapses in the autonomic nervous system, and at synapses in several parts of the central nervous system. As you learned in the last section, acetylcholine is broken down in the synapse by the enzyme *acetylcholinesterase*. Neurons that release acetylcholine are said to be *cholinergic*.

Unconventional Neurotransmitters

The unconventional neurotransmitters act in ways that are different from those that neuroscientists have come to think of as typical for such substances. One class of unconventional neurotransmitters, the **soluble-gas neurotransmitters**, includes **nitric oxide** and **carbon monoxide** (Boehning & Snyder, 2003). These neurotransmitters are produced in the neural cytoplasm and immediately diffuse through the cell membrane into the extracellular fluid and then into nearby cells. They easily pass through cell membranes because they are soluble in lipids. Once inside another cell, they stimulate the production of a second messenger and in a few seconds are deactivated by being converted to other molecules. They are difficult to study because they exist for only a few seconds.

Soluble-gas neurotransmitters have been shown to be involved in *retrograde transmission*. At some synapses, they transmit feedback signals from the postsynaptic neuron back to the presynaptic neuron. The function of retrograde transmission seems to be to regulate the activity of presynaptic neurons (Ludwig & Pittman, 2003).

Another class of unconventional neurotransmitters, the endocannabinoids, has only recently been discovered. **Endocannabinoids** are neurotransmitters that are

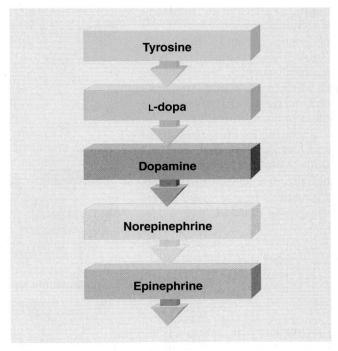

FIGURE 4.16   The steps in the synthesis of catecholamines from tyrosine.

similar to *delta-9-tetrahydrocannabinol* (THC), the main *psychoactive* (producing psychological effects) constituent of marijuana (see Chapter 15). So far, two endocannabinoids have been discovered (Van Sickle et al., 2005). The most widely studied is **anandamide** (from the Sanskrit word *ananda*, which means "eternal bliss"). Like the soluble gases, the endocannabinoids are produced immediately before they are released. Endocannabinoids are synthesized from fatty compounds in the cell membrane; they tend to be released from the dendrites and cell body; and they tend to have most of their effects on presynaptic neurons, inhibiting subsequent synaptic transmission (see Glickfield & Scanziani, 2005).

Neuropeptides

Over 100 neuropeptides have been identified (see Ludwig & Leng, 2006). The actions of each neuropeptide depend on its amino acid sequence.

It is usual to loosely group **neuropeptide transmitters** into five categories. Three of these categories acknowledge that neuropeptides often function in multiple capacities, not just as neurotransmitters: One category (**pituitary peptides**) contains neuropeptides that were first identified as hormones released by the pituitary; a second category (**hypothalamic peptides**) contains neuropeptides that were first identified as hormones released by the hypothalamus; and a third category (**brain–gut peptides**) contains neuropeptides that were first discovered in the gut. The fourth category (**opioid peptides**) contains neuropeptides that are similar in structure to the active ingredients of opium, and the fifth (**miscellaneous peptides**) is a catch-all category that contains all of the neuropeptide transmitters that do not fit into one of the other four categories.

Figure 4.17 summarizes all the neurotransmitters that were introduced in this section. If it has not already occurred to you, this table should be very useful for reviewing the material.

Scan Your Brain

This is a good place for you to pause to scan your brain to see if you are ready to proceed. Are you familiar with the neurotransmitters to which you have just been introduced? Find out by filling in the blanks. The correct answers are provided at the end of the exercise. Before proceeding, review material related to your errors and omissions.

Amino acids are the neurotransmitters in the vast majority of (1) _____ acting, directed synapses. Four amino acids are widely recognized neurotransmitters: (2) _____, (3) _____, (4) _____, and (5) _____. In contrast to the amino acid neurotransmitters, the (6) _____ are small-molecule neurotransmitters with slower, more diffuse effects; they belong to one of two categories: (7) _____ or indolamines. In the former category are epinephrine, (8) _____, and (9) _____; (10) _____ is the only neurotransmitter in the latter category. (11) _____, the neurotransmitter at neuromuscular junctions, is a neurotransmitter in a class by itself. There are also unconventional neurotransmitters: the (12) _____ neurotransmitters, such as nitric oxide

Small-Molecule Neurotransmitters

Amino acids		Glutamate Aspartate Glycine GABA
Monoamines	Catecholamines	Dopamine Epinephrine Norepinephrine
	Indolamines	Serotonin
Acetylcholine		Acetylcholine
Unconventional neurotransmitters	Soluble gases	Nitric oxide Carbon monoxide
	Endocannabinoids	Anandamide

Large-Molecule Neurotransmitters

Neuropeptides	Pituitary peptides Hypothalamic peptides Brain–gut peptides Opioid peptides Miscellaneous peptides

FIGURE 4.17 Classes of neurotransmitters and the particular neurotransmitters that were discussed (and appeared in boldface) in this section.

and carbon monoxide, and the endocannabinoids. Finally, the neuropeptides, which are short chains of (13) _____, are the only large-molecule neurotransmitters. They are usually grouped into five categories: the pituitary peptides, the hypothalamic peptides, the brain–gut peptides, the (14) _____ peptides, and the miscellaneous peptides.

Scan Your Brain answers: (1) fast, (2, 3, 4, 5) glutamate, aspartate, glycine, and GABA, in any order, (6) monoamines, (7) catecholamines, (8, 9) norepinephrine and dopamine, in either order, (10) serotonin, (11) Acetylcholine, (12) soluble-gas, (13) amino acids, (14) opioid.

4.7

Pharmacology of Synaptic Transmission and Behavior

In case you have forgotten, the reason I have asked you to invest so much effort in learning about the neurotransmitters is that they play a key role in how the brain works. We began this chapter on a behavioral note by considering the pathological behavior of Roberto Garcia d'Orta, which resulted from a Parkinson's disease–related disruption of his dopamine function. Now, let's return to behavior.

Most of the methods that biopsychologists use to study the behavioral effects of neurotransmitters are *pharmacological* (involving drugs). To study neurotransmitters and behavior, researchers administer to human or nonhuman subjects drugs that have particular effects on particular neurotransmitters and then assess the effects of the drugs on behavior.

Drugs have two fundamentally different kinds of effects on synaptic transmission: They facilitate it or they inhibit it. Drugs that facilitate the effects of a particular neurotransmitter are said to be **agonists** of that neurotransmitter. Drugs that inhibit the effects of a particular neurotransmitter are said to be its **antagonists**.

◉ Simulate
Psychoactive Drugs
www.mypsychlab.com

How Drugs Influence Synaptic Transmission

Although synthesis, release, and action vary from neurotransmitter to neurotransmitter, the following seven general steps are common to most neurotransmitters: (1) synthesis of the neurotransmitter, (2) storage in vesicles, (3) breakdown in the cytoplasm of any neurotransmitter that leaks from the vesicles, (4) exocytosis, (5) inhibitory feedback via autoreceptors, (6) activation of postsynaptic receptors, and (7) deactivation. Figure 4.18 on page 96 illustrates these seven steps, and Figure 4.19 on page 97 illustrates some ways that agonistic and antagonistic drugs influence them.

For example, some agonists of a particular neurotransmitter bind to postsynaptic receptors and activate them, whereas some antagonistic drugs, called **receptor blockers**, bind to postsynaptic receptors without activating them and, in so doing, block the access of the usual neurotransmitter.

Behavioral Pharmacology: Three Influential Lines of Research

You will encounter discussions of the *putative* (hypothetical) behavioral functions of various neurotransmitters in subsequent chapters. However, this chapter ends with descriptions of three particularly influential lines of research on neurotransmitters and behavior. Each line of research led to the discovery of an important principle of neurotransmitter function, and each illustrates how drugs are used to study the nervous system and behavior.

Wrinkles and Darts: Discovery of Receptor Subtypes

It was originally assumed that there was one kind of receptor for each neurotransmitter, but this notion was dispelled by research on acetylcholine receptors (see Changeux & Edelstein, 2005). Some acetylcholine receptors bind to *nicotine* (a CNS stimulant and major psychoactive ingredient of tobacco), whereas other acetylcholine receptors bind to *muscarine* (a poisonous substance found in some mushrooms). These two kinds of acetylcholine receptors thus became known as *nicotinic receptors* and *muscarinic receptors*.

Next, it was discovered that nicotinic and muscarinic receptors are distributed differently in the nervous system, have different modes of action, and consequently have different behavioral effects. Both nicotinic and muscarinic receptors are found in the CNS and the PNS. In the PNS, many nicotinic receptors occur at the junctions between motor neurons and muscle fibers, whereas many muscarinic receptors are located in the autonomic nervous system (ANS). Nicotinic and muscarinic receptors are ionotropic and metabotropic, respectively.

Many of the drugs that are used in research and in medicine are extracts of plants that have long been used for medicinal and recreational purposes. The cholinergic agonists and antagonists illustrate this point well. For example, the ancient Greeks consumed extracts of the belladonna plant to treat stomach ailments and to make themselves more attractive. Greek women believed that the pupil-dilating effects of these extracts enhanced their beauty (*belladonna* means "beautiful lady"). **Atropine**, which is the main active ingredient of belladonna, is a receptor blocker that exerts its antagonist effect by binding to muscarinic receptors, thereby blocking the effects of acetylcholine on them. The pupil-dilating effects of atropine are mediated by its antagonist actions on muscarinic receptors in the ANS. In contrast, the disruptive effects of large doses of atropine on memory is mediated by its antagonistic effect on muscarinic receptors in the

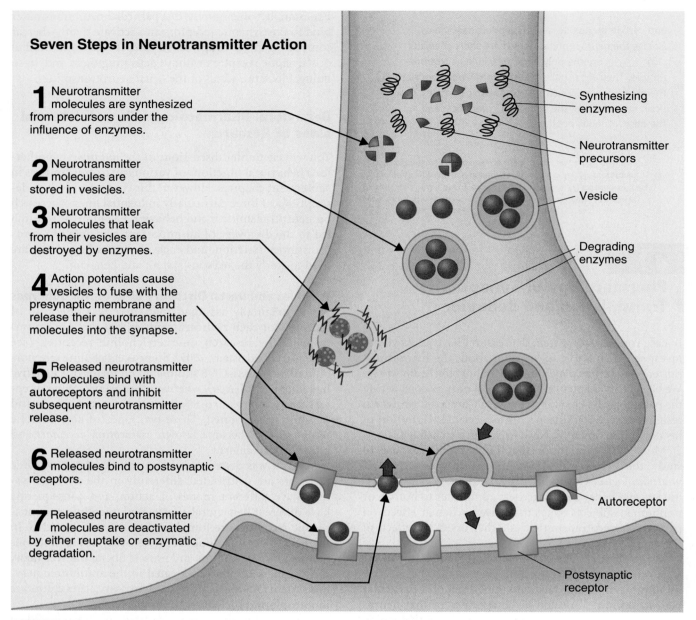

Seven Steps in Neurotransmitter Action

1 Neurotransmitter molecules are synthesized from precursors under the influence of enzymes.

2 Neurotransmitter molecules are stored in vesicles.

3 Neurotransmitter molecules that leak from their vesicles are destroyed by enzymes.

4 Action potentials cause vesicles to fuse with the presynaptic membrane and release their neurotransmitter molecules into the synapse.

5 Released neurotransmitter molecules bind with autoreceptors and inhibit subsequent neurotransmitter release.

6 Released neurotransmitter molecules bind to postsynaptic receptors.

7 Released neurotransmitter molecules are deactivated by either reuptake or enzymatic degradation.

Synthesizing enzymes

Neurotransmitter precursors

Vesicle

Degrading enzymes

Autoreceptor

Postsynaptic receptor

FIGURE 4.18 Seven steps in neurotransmitter action: (1) synthesis, (2) storage in vesicles, (3) breakdown of any neurotransmitter leaking from the vesicles, (4) exocytosis, (5) inhibitory feedback via autoreceptors, (6) activation of postsynaptic receptors, and (7) deactivation.

CNS. The disruptive effect of high doses of atropine on memory was one of the earliest clues that cholinergic mechanisms may play a role in memory (see Chapter 11).

South American natives have long used *curare*—an extract of a certain class of woody vines—on the tips of darts they use to kill their game and occasionally their enemies. Like atropine, curare is a receptor blocker at cholinergic synapses, but it acts at nicotinic receptors. By binding to nicotinic receptors, curare blocks transmission at neuromuscular junctions, thus paralyzing its recipients and killing them by blocking their respiration. You may be surprised, then, to learn that the active ingredient of curare

is sometimes administered to human patients during surgery to ensure that their muscles do not contract during an incision. When curare is used for this purpose, the patient's breathing must be artificially maintained by a respirator.

Botox (short for *Botulinium toxin*), a neurotoxin released by a bacterium often found in spoiled food, is another nicotinic antagonist, but its mechanism of action is different: It blocks the release of acetylcholine at neuromuscular junctions and is thus a deadly poison. However, injected in minute doses at specific sites, it has applications

 Clinical Implications

in medicine (e.g., reduction of tremors) and cosmetics (e.g., reduction of wrinkles; see Figure 4.20 on page 98).

Pleasure and Pain: Discovery of Endogenous Opioids
Opium, the sticky resin obtained from the seed pods of the opium poppy, has been used by humans since prehistoric times for its pleasurable effects. Morphine, its major psychoactive ingredient, is highly addictive. But morphine also has its good side: It is an effective *analgesic* (painkiller)—see Chapters 7 and 15.

In the 1970s, it was discovered that opiate drugs such as morphine bind effectively to receptors in the brain. These receptors were generally found in the hypothalamus and other limbic areas, but they were most concentrated in the area of the brain stem around the cerebral aqueduct, which connects the third and fourth ventricles; this part of the

Clinical Implications

brain stem is called the **periaqueductal gray (PAG)**. Microinjection of morphine into the PAG, or even electrical stimulation of the PAG, produces strong analgesia.

The existence of selective opiate receptors in the brain raised an interesting question: Why are they there? They are certainly not there so that once humans discovered opium, opiates would have a place to bind. The existence of opiate receptors suggested that *opioid* (opiate-like) chemicals occur naturally in the brain, and that possibility triggered an intensive search for them.

Several families of **endogenous** (occurring naturally within the body) opioids have been discovered. First discovered were the **enkephalins** (meaning "in the head"). Another major family of endogenous opioids are the **endorphins** (a contraction of "endogenous morphine"). All endogenous opioid neurotransmitters are neuropeptides, and their receptors are metabotropic.

Tremors and Insanity: Discovery of Antischizophrenic Drugs   Arguably, the most important event in the treatment of mental illness has been the development of drugs for the treatment of schizophrenia (see Chapter 18). Surprisingly, Parkinson's disease, the disease from which

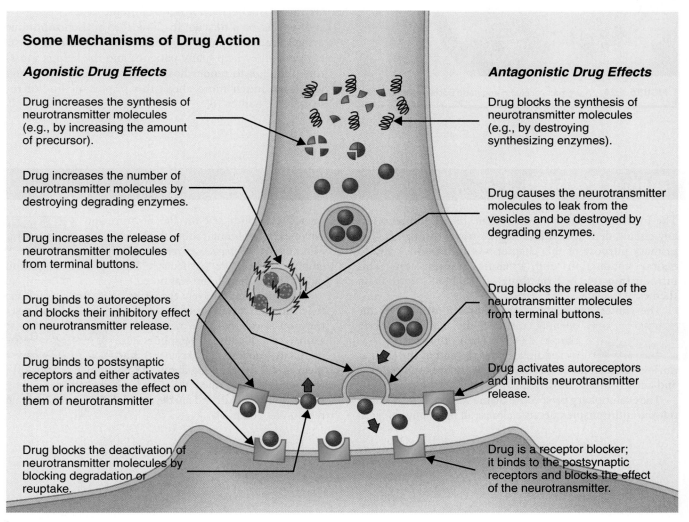

FIGURE 4.19   Some mechanisms of agonistic and antagonistic drug effects.

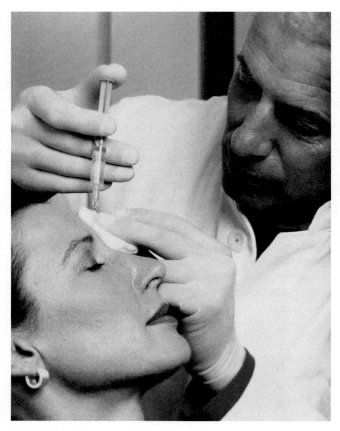

FIGURE 4.20   A woman receiving cosmetic Botox injections.

Roberto Garcia d'Orta suffered, played a major role in their discovery.

In the 1950s, largely by chance, two drugs were found to have antischizophrenic effects. Although these two drugs were not related structurally, they both produced a curious pattern of effects: Neither drug appeared to have any antischizophrenic activity until patients had been taking it for about 3 weeks, at which point the drug also started to produce mild Parkinsonian symptoms (e.g., tremor-at-rest). Researchers put this result together with two then-recent findings: (1) that Parkinson's disease is associated with the degeneration of the main *dopamine* pathway of the brain, and (2) that dopamine agonists—*cocaine* and *amphetamines*—produce a temporary disorder that resembles schizophrenia. Together, these findings suggested that schizophrenia is caused by excessive activity at dopamine synapses, and thus that potent dopamine antagonists would be effective in its treatment.

It was ultimately discovered that one particular dopamine receptor, the D_2 receptor, plays a key role in schizophrenia and that drugs that most effectively block it are the most effective antischizophrenic drugs.

It would be a mistake to think that antischizophrenic drugs cure schizophrenia or that they help in every case. However, they help many patients, and the help is sometimes enough to render hospitalization unnecessary. You will learn much more about this important line of research in Chapter 18.

Themes Revisited

The function of the nervous system, like the function of any circuit, depends on how signals travel through it. The primary purpose of this chapter was to introduce you to neural conduction and synaptic transmission. This introduction touched on three of the book's four main themes.

The clinical implications theme was illustrated by the opening case of the Lizard, Roberto Garcia d'Orta. Then, this theme was picked up again at the end of the chapter during discussions of curare, Botox, endogenous opioids, and antischizophrenic drugs.

The evolutionary perspective theme was implicit throughout the entire chapter, because almost all neurophysiological research is conducted on the neurons and  synapses of nonhuman subjects. However, the evolutionary perspective received explicit emphasis when the particularly high glial-cell-to-neuron ratio of the human brain was noted.

The thinking creatively theme arose in two metaphors: the firing-gun metaphor of action potentials and the mouse-traps-on-a-wobbly-shelf metaphor of axonal conduction. Metaphors are useful in teaching, and scientists find them useful for thinking about the phenomena they study. The text also described the creative Nobel-Prize–winning research of Hodgkin and Huxley on the ionic bases of resting membrane potentials.

Think about It

1. Just as computers operate on binary (yes-no) signals, the all-or-none action potential is the basis of neural communication. The human brain is thus nothing more than a particularly complex computer. Discuss.
2. How have the findings described in this chapter changed your understanding of brain function?
3. Why is it important for biopsychologists to understand neural conduction and synaptic transmission? Is it important for all psychologists to have such knowledge? Discuss.
4. The discovery that neurotransmitters can act directly on DNA via G proteins uncovered a mechanism through which experience and genes can interact (see Chapter 2). Discuss.
5. Dendrites and glial cells are currently "hot" subjects of neuroscientific research. Describe the findings that have generated such interest, and explain how they have changed our conception of brain function.

Key Terms

4.1 Resting Membrane Potential

Membrane potential (p. 76)
Microelectrodes (p. 76)
Resting potential (p. 77)
Ions (p. 77)
Ion channels (p. 77)
Sodium–potassium pumps (p. 79)
Transporters (p. 79)

4.2 Generation and Conduction of Postsynaptic Potentials

Depolarize (p. 79)
Hyperpolarize (p. 79)
Excitatory postsynaptic potentials (EPSPs) (p. 79)
Inhibitory postsynaptic potentials (IPSPs) (p. 79)
Graded responses (p. 79)

4.3 Integration of Postsynaptic Potentials and Generation of Action Potentials

Axon hillock (p. 80)
Threshold of excitation (p. 80)
Action potential (AP) (p. 80)
All-or-none responses (p. 80)
Integration (p. 80)
Spatial summation (p. 80)
Temporal summation (p. 80)

4.4 Conduction of Action Potentials

Voltage-activated ion channels (p. 82)
Absolute refractory period (p. 83)
Relative refractory period (p. 83)
Antidromic conduction (p. 84)
Orthodromic conduction (p. 84)
Nodes of Ranvier (p. 84)
Saltatory conduction (p. 85)

4.5 Synaptic Transmission: Chemical Transmission of Signals among Neurons

Dendritic spines (p. 86)
Directed synapses (p. 86)
Nondirected synapses (p. 86)
Neuropeptides (p. 87)
Synaptic vesicles (p. 87)
Golgi complex (p. 87)
Coexistence (p. 88)

Exocytosis (p. 88)
Receptors (p. 89)
Ligand (p. 89)
Receptor subtypes (p. 89)
Ionotropic receptors (p. 89)
Metabotropic receptors (p. 89)
G proteins (p. 89)
Second messenger (p. 90)
Autoreceptors (p. 90)
Reuptake (p. 91)
Enzymatic degradation (p. 91)
Enzymes (p. 91)
Acetylcholinesterase (p. 91)
Gap junctions (p. 91)

4.6 Neurotransmitters

Amino acid neurotransmitters (p. 92)
Glutamate (p. 92)
Aspartate (p. 92)
Glycine (p. 92)
Gamma-aminobutyric acid (GABA) (p. 92)
Monoamine neurotransmitters (p. 92)
Dopamine (p. 92)
Epinephrine (p. 92)
Norepinephrine (p. 92)
Serotonin (p. 92)
Catecholamines (p. 92)

Indolamines (p. 92)
Acetylcholine (p. 93)
Soluble-gas neurotransmitters (p. 93)
Nitric oxide (p. 93)
Carbon monoxide (p. 93)
Endocannabinoids (p. 93)
Anandamide (p. 94)
Neuropeptide transmitters (p. 94)
Pituitary peptides (p. 94)
Hypothalamic peptides (p. 94)
Brain–gut peptides (p. 94)
Opioid peptides (p. 94)
Miscellaneous peptides (p. 94)

4.7 Pharmacology of Synaptic Transmission and Behavior

Agonists (p. 95)
Antagonists (p. 95)
Receptor blockers (p. 95)
Atropine (p. 95)
Botox (p. 96)
Periaqueductal gray (PAG) (p. 97)
Endogenous (p. 97)
Enkephalins (p. 97)
Endorphins (p. 97)

1. IPSPs are
   a. inhibitory.
   b. graded.
   c. all-or-none.
   d. all of the above
   e. both a and b

2. Which of the following ions triggers exocytosis by its influx into terminal buttons?
   a. Cl^-
   b. Ca^{2+}
   c. glutamate
   d. glycine
   e. Na^+

3. Which of the following is the most common mechanism of deactivating neurotransmitter molecules in synapses?
   a. enzymatic degradation
   b. acetylcholinesterase
   c. reuptake by transporters
   d. all of the above
   e. both a and b

4. All of the following are monoamine neurotransmitters except
   a. epinephrine.
   b. serotonin.
   c. norepinephrine.
   d. dopamine.
   e. acetylcholine.

5. Botox is a
   a. nicotinic agonist
   b. nicotinic antagonist.
   c. cholinergic agonist.
   d. cholinergic antagonist.
   e. poison used by some South American natives on their darts.

5 The Research Methods of Biopsychology

Understanding What Biopsychologists Do

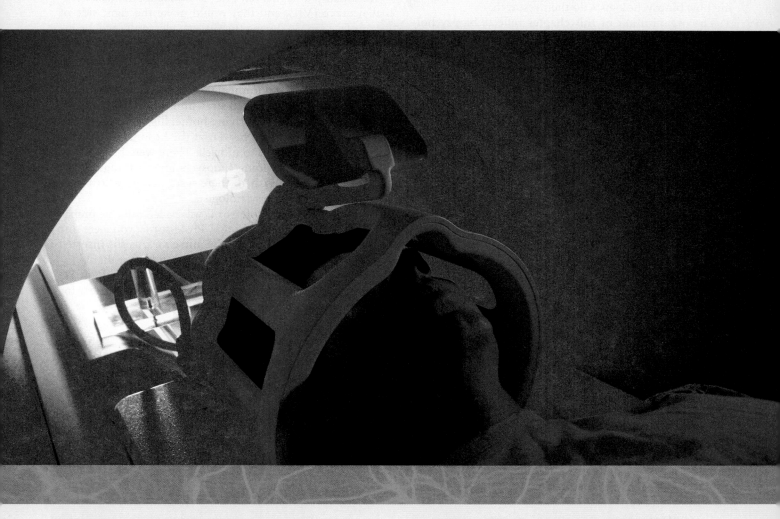

Chapters 1 and 2 introduced you to the general interests, ideas, and approaches that characterize biopsychology. In Chapters 3 and 4, your introduction to biopsychology was temporarily curtailed while background material in neuroanatomy, neurophysiology, and neurochemistry was presented. This chapter gets down to the nitty-gritty of biopsychology; it describes the specific day-to-day activities of the biopsychology laboratory. It is intended to prepare you for later chapters and to sharpen your understanding of biopsychology by describing how biopsychologists do their research.

The organization of this chapter reflects biopsychology's intrinsic duality. The chapter has two major parts: One deals with methods of studying the nervous system, and the other deals with methods of studying behavior.

As you read through this chapter, you should keep in mind that most of the methods that are used to study the human brain are also used for clinical purposes, for either diagnosis or treatment. The case of Professor P. makes this point.

The Ironic Case of Professor P.

Two weeks before his brain surgery, Professor P. reported to the hospital for a series of tests. What amazed Professor P. most about these tests was how familiar they seemed. No, *Clinical Implications* Professor P. was not a psychic; he was a behavioral neuroscientist, and he was struck by how similar the tests performed on him were to the tests he had seen in his department.

Professor P. had a brain tumor on his right auditory-vestibular cranial nerve (cranial nerve VIII; see Appendices III and IV), and he had to have it *excised* (cut out). First, Professor P.'s auditory abilities were assessed by measuring his ability to detect sounds of various volumes and pitches and then by measuring the magnitude of the EEG signals evoked in his auditory cortex by clicks in his right ear.

Next, Professor P.'s vestibular function (balance) was tested by injecting cold water into his ear.

"Do you feel anything, Professor P.?"

"Well, a cold ear."

"Nothing else?"

"No."

So colder and colder water was tried with no effect until the final, coldest test was conducted. "Ah, that feels weird," said Professor P. "It's kind of like the bed is tipping."

The results of the tests were bad, or good, depending on your perspective. Professor P.'s hearing in his right ear was poor, and his right vestibular nerve was barely functioning. "At the temperatures we flushed down there, most people would have been on their hands and knees puking their guts out," said the medical technician. Professor P. smiled at the technical terminology.

Of course, he was upset that his brain had deteriorated so badly, but he sensed that his neurosurgeon was secretly pleased: "We won't have to try to save the nerve; we'll just cut it."

There was one last test. The skin of his right cheek was lightly pricked while the EEG responses of his somatosensory cortex were recorded from his scalp. "This is just to establish a baseline for the surgery," it was explained. "One main risk of removing tumors on the auditory-vestibular cranial nerve (VIII) is damaging the facial cranial nerve (VII), and that would make the right side of your face sag. So during the surgery, electrodes will be inserted in your cheek, and your cheek will be repeatedly stimulated with tiny electrical pulses. The cortical responses will be recorded and fed into a loudspeaker so that the surgeon can immediately hear changes in the activity if his scalpel starts to stray into the area."

As Professor P. was driving home, his mind wandered from his own plight to his day at the hospital. "Quite interesting," he thought to himself. There were biopsychologists everywhere, doing biopsychological things. In all three labs he had visited, there were people who began their training as biopsychologists.

Two weeks later, Professor P. was rolled into the preparation room. "Sorry to do this, Professor P., you were one of my favorite instructors," the nurse said, as she inserted a large needle into Professor P.'s face and left it there.

Professor P. didn't mind; he was barely conscious. He did not know that he wouldn't regain consciousness for several days—at which point he would be incapable of talking, eating, or even breathing.

Don't forget Professor P.; you will learn more about his case in Chapter 10. For now, this case has demonstrated to you that many of the research methods of biopsychology are also used in clinical settings (see Matthews, Honey, & Bullmore, 2006). Let's move on to the methods themselves.

PART ONE
METHODS OF STUDYING THE NERVOUS SYSTEM

5.1
Methods of Visualizing and Stimulating the Living Human Brain

Prior to the early 1970s, biopsychological research was impeded by the inability to obtain images of the organ of primary interest: the living human brain. Conventional

X-ray photography is next to useless for this purpose. When an X-ray photograph is taken, an X-ray beam is passed through an object and then onto a photographic plate. Each of the molecules through which the beam passes absorbs some of the radiation; thus, only the unabsorbed portions of the beam reach the photographic plate. X-ray photography is therefore effective in characterizing internal structures that differ substantially from their surroundings in the degree to which they absorb X-rays—for example, a revolver in a suitcase full of clothes or a bone in flesh. However, by the time an X-ray beam has passed through the numerous overlapping structures of the brain, which differ only slightly from one another in their ability to absorb X-rays, it carries little information about the structures through which it has passed.

👁 **Watch**
Brain Imaging
www.mypsychlab.com

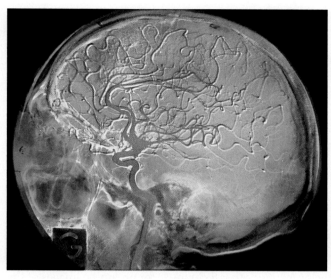

FIGURE 5.1   A cerebral angiogram of a healthy subject.

Contrast X-Rays

Although conventional X-ray photography is not useful for visualizing the brain, contrast X-ray techniques are. **Contrast X-ray techniques** involve injecting into one compartment of the body a substance that absorbs X-rays either less than or more than the surrounding tissue. The injected substance then heightens the contrast between the compartment and the surrounding tissue during X-ray photography.

One contrast X-ray technique, **cerebral angiography**, uses the infusion of a radio-opaque dye into a cerebral artery to visualize the cerebral circulatory system during X-ray photography (see Figure 5.1). Cerebral angiograms are most useful for localizing vascular damage, but the displacement of blood vessels from their normal position also can indicate the location of a tumor.

Clinical Implications

X-Ray Computed Tomography

In the early 1970s, the study of the living human brain was revolutionized by the introduction of computed tomography. **Computed tomography (CT)** is a computer-assisted X-ray procedure that can be used to visualize the brain and other internal structures of the living body. During cerebral computed tomography, the neurological patient lies with his or her head positioned in the center of a large cylinder, as depicted in Figure 5.2.

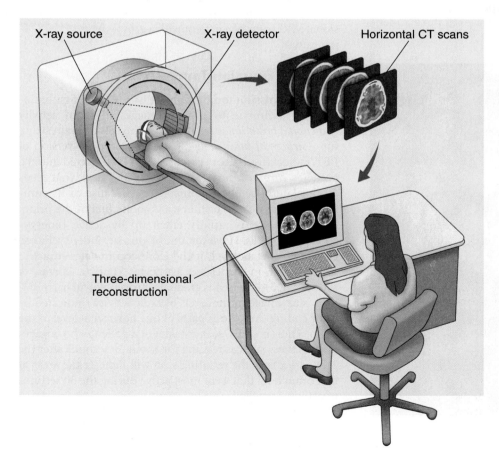

X-ray source   X-ray detector   Horizontal CT scans

Three-dimensional reconstruction

FIGURE 5.2   Computed tomography (CT) uses X-rays to create a CT scan of the brain.

On one side of the cylinder is an X-ray tube that projects an X-ray beam through the head to an X-ray detector mounted on the other side. The X-ray tube and detector automatically rotate around the head of the patient at one level of the brain, taking many individual X-ray photographs as they rotate. The meager information in each X-ray photograph is combined by a computer to generate a CT scan of one horizontal section of the brain. Then, the X-ray tube and detector are moved along the axis of the patient's body to another level of the brain, and the process is repeated. Scans of eight or nine horizontal brain sections are typically obtained from a patient; combined, they provide a three-dimensional representation of the brain.

Magnetic Resonance Imaging

The success of computed tomography stimulated the development of other techniques for obtaining images of the inside of the living body. Among these techniques is **magnetic resonance imaging (MRI)**—a procedure in which high-resolution images are constructed from the measurement of waves that hydrogen atoms emit when they are activated by radio-frequency waves in a magnetic field. MRI provides clearer images of the brain than does CT. A color-coded two-dimensional MRI scan of the midsagittal brain is presented in Figure 5.3.

In addition to providing relatively high **spatial resolution** (the ability to detect and represent differences in spatial location), MRI can produce images in three dimen-

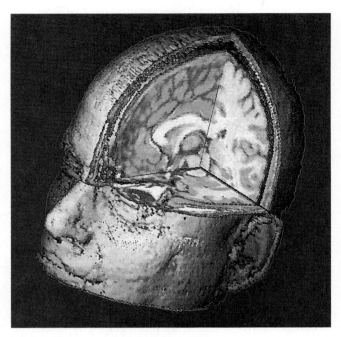

FIGURE 5.4 Structural MRI can be used to provide three-dimensional images of the entire brain. (Courtesy of Bruce Foster and Robert Hare, University of British Columbia.)

sions. Figure 5.4 is a three-dimensional MRI scan. Figure 5.5 shows two-dimensional MRI scans of a patient with a growing tumor.

Positron Emission Tomography

Positron emission tomography (PET) was the first brain-imaging technique to provide images of brain activity (*functional brain images*) rather than images of brain structure (*structural brain images*). In one common version of PET, radioactive **2-deoxyglucose (2-DG)** is injected into the patient's *carotid artery* (an artery of the neck that feeds the ipsilateral cerebral hemisphere). Because of its similarity to glucose, the primary metabolic fuel of the brain, 2-deoxyglucose is rapidly taken up by active (energy-consuming) cells. However, unlike glucose, 2-deoxyglucose cannot be metabolized; it therefore accumulates in active neurons—or in associated astrocytes (Barros, Porras, & Bittner, 2005)—until it is gradually broken down. Each PET scan is an image of the levels of radioactivity (indicated by color coding) in various parts of one horizontal level of the brain. Thus, if a PET scan is taken of a patient who engages in an activity such as reading for about 30 seconds after the 2-DG injection, the resulting scan will indicate the areas at that brain level that were most active during the 30 seconds of activity (see Figure 5.6).

Notice from Figure 5.6 that PET scans are not really images of the brain. Each PET scan is merely a colored map of the amount of radioactivity in each of the tiny

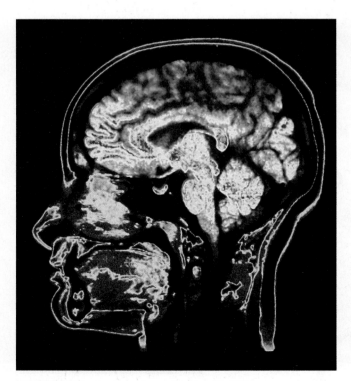

FIGURE 5.3 A color-enhanced midsagittal MRI scan.

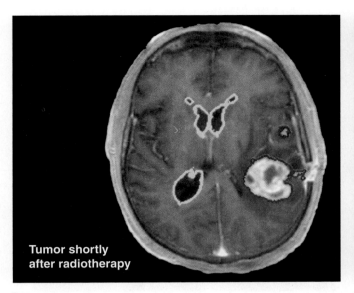

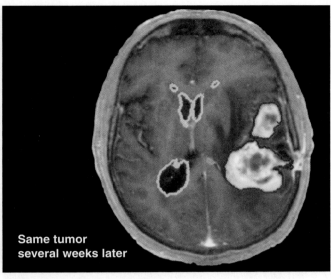

Tumor shortly
after radiotherapy

Same tumor
several weeks later

FIGURE 5.5   Structural MRI can also be used to provide two-dimensional images of brain slices. The MRI scan on the left shows a tumor shortly after radiotherapy, and the MRI scan on the right shows the same tumor several weeks later—clearly, the tumor has continued to grow. Ventricles are outlined in yellow; the tumor is outlined in red. (Based on Calmon et al., 1998; courtesy of Neil Roberts, University of Liverpool.)

cubic voxels (volume pixels) that compose the scan. Exactly how each voxel maps onto a particular brain structure can be estimated only by superimposing the scan on a brain image.

Functional MRI

MRI technology has been used to produce functional images of the brain. Indeed, functional MRI has become the most influential tool of cognitive neuroscience (Poldrack, 2008) and is now widely used for medical diagnosis (Holdsworth & Bammer, 2008).

Functional MRI (fMRI) produce images representing the increase in oxygen flow in the blood to active areas of the brain. Functional MRI is possible because of two attributes of oxygenated blood (see Raichle & Mintun, 2006). First, active areas of the brain take up more oxygenated blood than they need for their energy requirements, and thus oxygenated blood accumulates in active areas of the brain. Second, oxygenated blood has magnetic properties (oxygen influences the effect of magnetic fields on iron in the blood). The signal recorded by fMRI is called the **BOLD signal** (the blood-oxygen-level-dependent signal).

Functional MRI has four advantages over PET: (1) Nothing has to be injected into the subject; (2) it provides both structural and functional information in the same image; (3) its spatial resolution is better; and (4) it can be used to produce three-dimensional images of activity over the entire brain. Functional MRIs are shown in Figure 5.7.

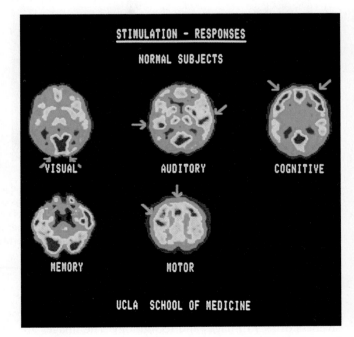

FIGURE 5.6   A series of PET scans. Each scan is a horizontal section recorded during a different cognitive or perceptual task. Areas of high activity are indicated by reds and yellows. For example, notice the high level of activity in the visual cortex of the occipital lobe when the subject scanned a visual display. (From "Positron Tomography: Human Brain Function and Biochemistry" by Michael E. Phelps and John C. Mazziotta, *Science*, 228 [9701], May 17, 1985, p. 804. Copyright 1985 by the AAAS. Reprinted by permission. Courtesy of Drs. Michael E. Phelps and John Mazziotta, UCLA School of Medicine.)

Right hemisphere lateral surface

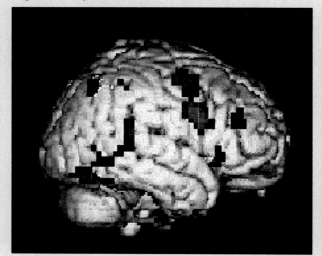

Left hemisphere lateral surface

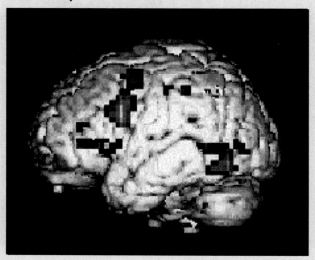

Left hemisphere medial surface

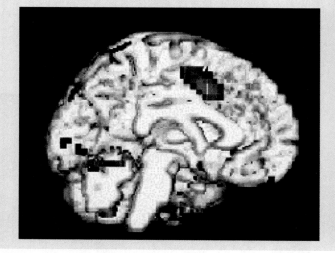

Right hemisphere medial surface

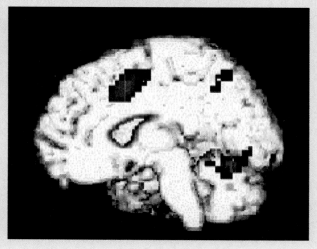

FIGURE 5.7 Functional magnetic resonance images (fMRIs). These images illustrate the areas of cortex that became more active when the subjects observed strings of letters and were asked to specify which strings were words; in the control condition, subjects viewed strings of asterisks (Kiehl et al., 1999). Thesis fMRIs illustrate surface activity; but images of sections through the brain can also be displayed. (Courtesy of Kent Kiehl and Peter Liddle, Department of Psychiatry, University of British Columbia.)

It is important not to be unduly swayed by the impressiveness of fMRI images and technology. The images are often presented—particularly in the popular press or general textbooks—as if they are pictures of human neural activity. They aren't: They are images of the BOLD signal, and the relation between the BOLD signal and neural activity is proving to be complex and variable (see Bartels, Logothetis, & Moutoussis, 2008; Ekstrom et al., 2009; Goense & Logothetis, 2008; Shmuel & Leopold, 2009; Zhang et al., 2009). Furthermore, fMRI technology is too slow to capture many neural responses—it takes 2 or 3 seconds to create an fMRI image, and many

neural responses, such as action potentials, occur in milliseconds (see Dobbs, 2005; Poldrack, 2008).

Magnetoencephalography

Another technique that is used to monitor the brain activity of human subjects is **magnetoencephalography (MEG)**. MEG measures changes in magnetic fields on the surface of the scalp that are produced by changes in underlying patterns of neural activity. Its major advantage over fMRI is its **temporal resolution**; it can record fast changes in neural activity.

Transcranial Magnetic Stimulation

PET, fMRI, and magnetoencephalography have allowed cognitive neuroscientists to create images of the activity of the human brain. But these methods all have the same weakness: They can be used to show a correlation between brain activity and cognitive activity, but they can't prove that the brain activity caused the cognitive activity (Rorden & Karnath, 2004; Sack, 2006). For example, a brain-imaging technique may show that the cingulate cortex becomes active when subjects view disturbing photographs, but it can't prove that the cingulate activity causes the emotional experience—there are many other explanations. One way of supporting the hypothesis that the cingulate cortex is an area for emotional experience would be to assess emotional experience in people lacking a functional cingulate cortex—for example, by studying patients with cingulate damage or studying healthy patients whose cingulate cortex has somehow been "turned off." Transcranial magnetic stimulation is a way of accomplishing this.

Transcranial magnetic stimulation (TMS) is a technique for affecting the activity in an area of the cortex by creating a magnetic field under a coil positioned next to the skull (see Fitzpatrick & Rothman, 2000; Pascual-Leone, Walsh, & Rothwell, 2000). In effect, the magnetic stimulation temporarily turns off part of the brain while the effects of the disruption on cognition and behavior are assessed. Although there are still fundamental questions about safety, depth of effect, and mechanisms of neural disruption (see Allen et al., 2007; Bestmann, 2007; Wagner, Valero-Cabre, & Pascual-Leone, 2007), TMS is often employed to circumvent the difficulty that brain-imaging studies have in determining causation.

5.2

Recording Human Psychophysiological Activity

The preceding section introduced you to functional brain imaging, the cornerstone of cognitive neuroscience research. This section deals with *psychophysiological recording methods* (methods of recording physiological activity from the surface of the human body). Five of the most widely studied psychophysiological measures are described: one measure of brain activity (the scalp EEG), two measures of somatic nervous system activity (muscle tension and eye movement), and two measures of autonomic nervous system activity (skin conductance and cardiovascular activity).

Scalp Electroencephalography

The *electroencephalogram (EEG)* is a measure of the gross electrical activity of the brain. It is recorded through large electrodes by a device called an *electroencephalograph (EEG machine)*, and the technique is called **electroencephalography**. In EEG studies of human subjects, each

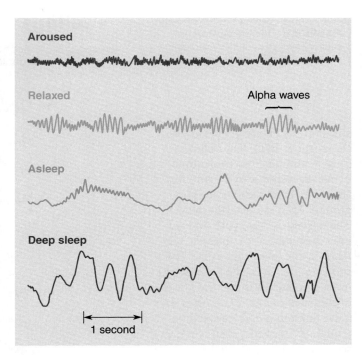

FIGURE 5.8 Some typical electroencephalograms and their psychological correlates.

channel of EEG activity is usually recorded from disk-shaped electrodes, about half the size of a dime, which are taped to the scalp.

The scalp EEG signal reflects the sum of electrical events throughout the head. These events include action potentials and postsynaptic potentials, as well as electrical signals from the skin, muscles, blood, and eyes. Thus, the utility of the scalp EEG does not lie in its ability to provide an unclouded view of neural activity. Its value as a research and diagnostic tool rests on the fact that some EEG wave forms are associated with particular states of consciousness or particular types of cerebral pathology (e.g., epilepsy). For example, **alpha waves** are regular, 8- to 12-per-second, high-amplitude waves that are associated with relaxed wakefulness. A few examples of EEG wave forms and their psychological correlates are presented in Figure 5.8.

Watch
Visit to a Cognitive Neuroscience Laboratory www.mypsychlab.com

Because EEG signals decrease in amplitude as they spread from their source, a comparison of signals recorded from various sites on the scalp can sometimes indicate the origin of particular waves. This is why it is usual to record EEG activity from many sites simultaneously.

Psychophysiologists are often more interested in the EEG waves that accompany certain psychological events than they are in the background EEG signal. These accompanying EEG waves are generally referred to as **event-related potentials (ERPs)**. One commonly studied type of event-related potential is the **sensory evoked potential**—the change in the cortical EEG signal that is elicited by the momentary presentation of a sensory

FIGURE 5.9 The averaging of an auditory evoked potential. Averaging increases the signal-to-noise ratio.

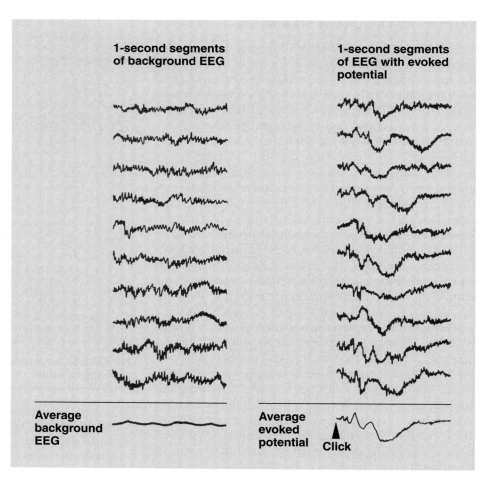

1-second segments of background EEG

1-second segments of EEG with evoked potential

Average background EEG

Average evoked potential **Click**

stimulus. As Figure 5.9 illustrates, the cortical EEG that follows a sensory stimulus has two components: the response to the stimulus (the signal) and the ongoing background EEG activity (the noise). The *signal* is the part of any recording that is of interest; the *noise* is the part that isn't. The problem in recording sensory evoked potentials is that the noise of the background EEG is often so great that the sensory evoked potential is masked. Measuring a sensory evoked potential can be like detecting a whisper at a rock concert.

A method used to reduce the noise of the background EEG is **signal averaging**. First, a subject's response to a stimulus, such as a click, is recorded many—let's say 1,000—times. Then, a computer identifies the millivolt value of each of the 1,000 traces at its starting point (i.e., at the click) and calculates the mean of these 1,000 scores. Next, it considers the value of each of the 1,000 traces 1 millisecond (msec) from its start, for example, and calculates the mean of these values. It repeats this process at the 2-msec mark, the 3-msec mark, and so on. When these averages are plotted, the average response evoked by the click is more apparent, because the random background EEG is canceled out by the averaging. See Figure 5.9, which illustrates the averaging of an auditory evoked potential.

The analysis of *average evoked potentials (AEPs)* focuses on the various waves in the averaged signal. Each wave is characterized by its direction, positive or negative, and by its latency. For example, the **P300 wave** illustrated in Figure 5.10 is the positive wave that occurs about 300 milliseconds after a momentary stimulus that has meaning for the subject (e.g., a stimulus to which the subject must respond)—see Friedman, Cycowicz, and Gaeta (2001). In contrast, the portions of an evoked potential recorded in the first few milliseconds after a stimulus are not influenced by the meaning of the stimulus for the subject. These small waves are called **far-field potentials** because, although they are recorded from the scalp, they originate far away in the sensory nuclei of the brain stem.

Although electroencephalography scores high on temporal resolution, it initially failed miserably on spatial resolution. With conventional electroencephalographic procedures, one can only roughly estimate the source of a particular signal. However, newer techniques employing

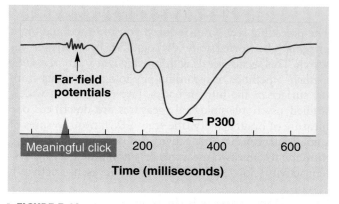

Far-field potentials

P300

Meaningful click   200   400   600

Time (milliseconds)

FIGURE 5.10 An average auditory evoked potential. Notice the P300 wave. This wave occurs only if the stimulus has meaning for the subject; in this case, the click signals the imminent delivery of a reward. By convention, positive EEG waves are always shown as downward deflections.

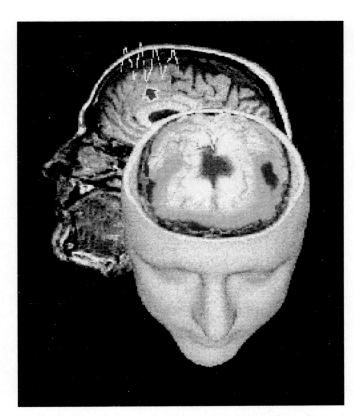

FIGURE 5.11 The marriage of electroencephalography and magnetic resonance imaging: The distribution of EEG signals can be represented on a structural cerebral MRI. Plotted in this illustration is the distribution of theta waves recorded while the subjects worked on a memory task. The highest incidence of theta waves (indicated by red in the three-dimensional MRI of the dorsal brain surface and by blue on the midsagittal section) occurred in the anterior cingulate cortex. (Courtesy of Alan Gevins, EEG Systems Laboratory & SAM Technology, San Francisco.)

sophisticated computer software and many electrodes can accurately locate the source of signals. The spatial resolution of these techniques is sufficient to enable the amplitude of evoked EEG signals recorded on the cortex to be color-coded and plotted on the surface of a three-dimensional MRI scan (Gevins et al., 1995). This useful marriage of techniques is illustrated in Figure 5.11.

Muscle Tension

Each skeletal muscle is composed of millions of threadlike muscle fibers. Each muscle fiber contracts in an all-or-none fashion when activated by the motor neuron that innervates it. At any given time, a few fibers in each resting muscle are likely to be contracting, thus maintaining the overall tone (tension) of the muscle. Movement results when a large number of fibers contract at the same time.

In everyday language, anxious people are commonly referred to as "tense." This usage acknowledges the fact that anxious, or otherwise aroused, individuals typically display high resting levels of tension in their muscles. This is why psychophysiologists are interested in this measure; they use it as an indicator of psychological arousal.

Electromyography is the usual procedure for measuring muscle tension. The resulting record is called an *electromyogram (EMG)*. EMG activity is usually recorded between two electrodes taped to the surface of the skin over the muscle of interest. An EMG record is presented in Figure 5.12. You will notice from this figure that the main correlate of an increase in muscle contraction is an increase in the amplitude of the raw EMG signal, which reflects the number of muscle fibers contracting at any one time.

Most psychophysiologists do not work with raw EMG signals; they convert them to a more workable form. The raw signal is fed into a computer that calculates the total amount of EMG spiking per unit of time—in consecutive 0.1-second intervals, for example. The integrated signal (i.e., the total EMG activity per unit of time) is then plotted. The result is a smooth curve, the amplitude of which is a simple, continuous measure of the level of muscle tension (see Figure 5.12).

Eye Movement

The electrophysiological technique for recording eye movements is called **electrooculography**, and the resulting record is called an *electrooculogram (EOG)*. Electrooculography is

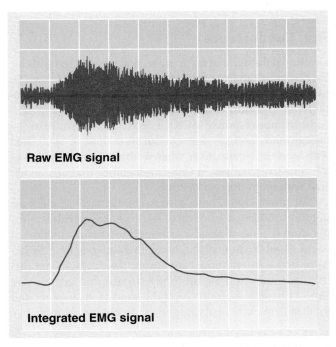

Raw EMG signal

Integrated EMG signal

FIGURE 5.12 The relation between a raw EMG signal and its integrated version. The subject tensed the muscle beneath the electrodes and then gradually relaxed it.

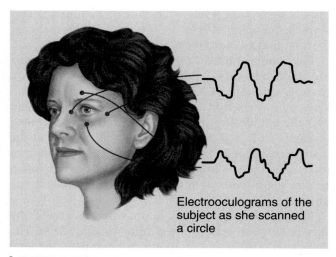

Electrooculograms of the subject as she scanned a circle

FIGURE 5.13   The typical placement of electrodes around the eye for electrooculography. The two electrooculogram traces were recorded as the subject scanned a circle.

based on the fact that there is a steady potential difference between the front (positive) and back (negative) of the eyeball. Because of this steady potential, when the eye moves, a change in the electrical potential between electrodes placed around the eye can be recorded. It is usual to record EOG activity between two electrodes placed on each side of the eye to measure its horizontal movements and between two electrodes placed above and below the eye to measure its vertical movements (see Figure 5.13).

Skin Conductance

Emotional thoughts and experiences are associated with increases in the ability of the skin to conduct electricity. The two most commonly employed indexes of *electrodermal activity* are the **skin conductance level (SCL)** and the **skin conductance response (SCR)**. The SCL is a measure of the background level of skin conductance that is associated with a particular situation, whereas the SCR is a measure of the transient changes in skin conductance that are associated with discrete experiences.

The physiological bases of skin conductance changes are not fully understood, but there is considerable evidence implicating the sweat glands. Although the main function of sweat glands is to cool the body, these glands tend to become active in emotional situations. Sweat glands are distributed over most of the body surface; but, as you are almost certainly aware, those of the hands, feet, armpits, and forehead are particularly responsive to emotional stimuli.

Cardiovascular Activity

The presence in our language of phrases such as *chicken-hearted*, *white with fear*, and *blushing bride* indicates that

modern psychophysiologists were not the first to recognize the relationship between *cardiovascular activity* and emotion. The cardiovascular system has two parts: the blood vessels and the heart. It is a system for distributing oxygen and nutrients to the tissues of the body, removing metabolic wastes, and transmitting chemical messages. Three different measures of cardiovascular activity are frequently employed in psychophysiological research: heart rate, arterial blood pressure, and local blood volume.

Heart Rate   The electrical signal that is associated with each heartbeat can be recorded through electrodes placed on the chest. The recording is called an **electrocardiogram** (abbreviated either **ECG**, for obvious reasons, or **EKG**, from the original German). The average resting heart rate of a healthy adult is about 70 beats per minute, but it increases abruptly at the sound, or thought, of a dental drill.

Blood Pressure   Measuring arterial blood pressure involves two independent measurements: a measurement of the peak pressure during the periods of heart contraction, the *systoles*, and a measurement of the minimum pressure during the periods of relaxation, the *diastoles*. Blood pressure is usually expressed as a ratio of systolic over diastolic blood pressure in millimeters of mercury (mmHg). The normal resting blood pressure for an adult is about 130/70 mmHg. A chronic blood pressure of more than 140/90 mmHg is viewed as a serious health hazard and is called **hypertension**.

You have likely had your blood pressure measured with a *sphygmomanometer*—a crude device composed of a hollow cuff, a rubber bulb for inflating it, and a pressure gauge for measuring the pressure in the cuff (*sphygmos* means "pulse"). More reliable, fully automated methods are used in research.

Blood Volume   Changes in the volume of blood in particular parts of the body are associated with psychological events. The best-known example of such a change is the engorgement of the genitals that is associated with sexual arousal in both males and females. **Plethysmography** refers to the various techniques for measuring changes in the volume of blood in a particular part of the body (*plethysmos* means "an enlargement").

One method of measuring these changes is to record the volume of the target tissue by wrapping a strain gauge around it. Although this method has utility in measuring blood flow in fingers or similarly shaped organs, the possibilities for employing it are somewhat limited. Another plethysmographic method is to shine a light through the tissue under investigation and to measure the amount of the light that is absorbed by it. The more blood there is in a structure, the more light it will absorb.

5.3

Invasive Physiological Research Methods

We turn now from a consideration of the noninvasive techniques employed in research on living human brains to a consideration of more direct techniques, which are commonly employed in biopsychological studies of laboratory animals. Most physiological techniques used in biopsychological research on laboratory animals fall into one of three categories: lesion methods, electrical stimulation methods, and invasive recording methods. Each of these three methods is discussed in this section of the chapter, but we begin with a description of *stereotaxic surgery*.

Stereotaxic Surgery

Stereotaxic surgery is the first step in many biopsychological experiments. *Stereotaxic surgery* is the means by which experimental devices are precisely positioned in the depths of the brain. Two things are required in stereotaxic surgery: an atlas to provide directions to the target site and an instrument for getting there.

The **stereotaxic atlas** is used to locate brain structures in much the same way that a geographic atlas is used to locate geographic landmarks. There is, however, one important difference. In contrast to the surface of the earth, which has only two dimensions, the brain has three. Accordingly, the brain is represented in a stereotaxic atlas by a series of individual maps, one per page, each representing the structure of a single, two-dimensional frontal brain slice. In stereotaxic atlases, all distances are given in millimeters from a designated reference point. In some rat atlases, the reference point is **bregma**—the point on the top of the skull where two of the major *sutures* (seams in the skull) intersect.

The **stereotaxic instrument** has two parts: a *head holder*, which firmly holds each subject's brain in the prescribed position and orientation; and an *electrode holder*, which holds the device to be inserted. A system of precision gears allows the electrode holder to be moved in three dimensions: anterior–posterior, dorsal–ventral, and lateral–medial. The implantation by stereotaxic surgery of an electrode in the amygdala of a rat is illustrated in Figure 5.14.

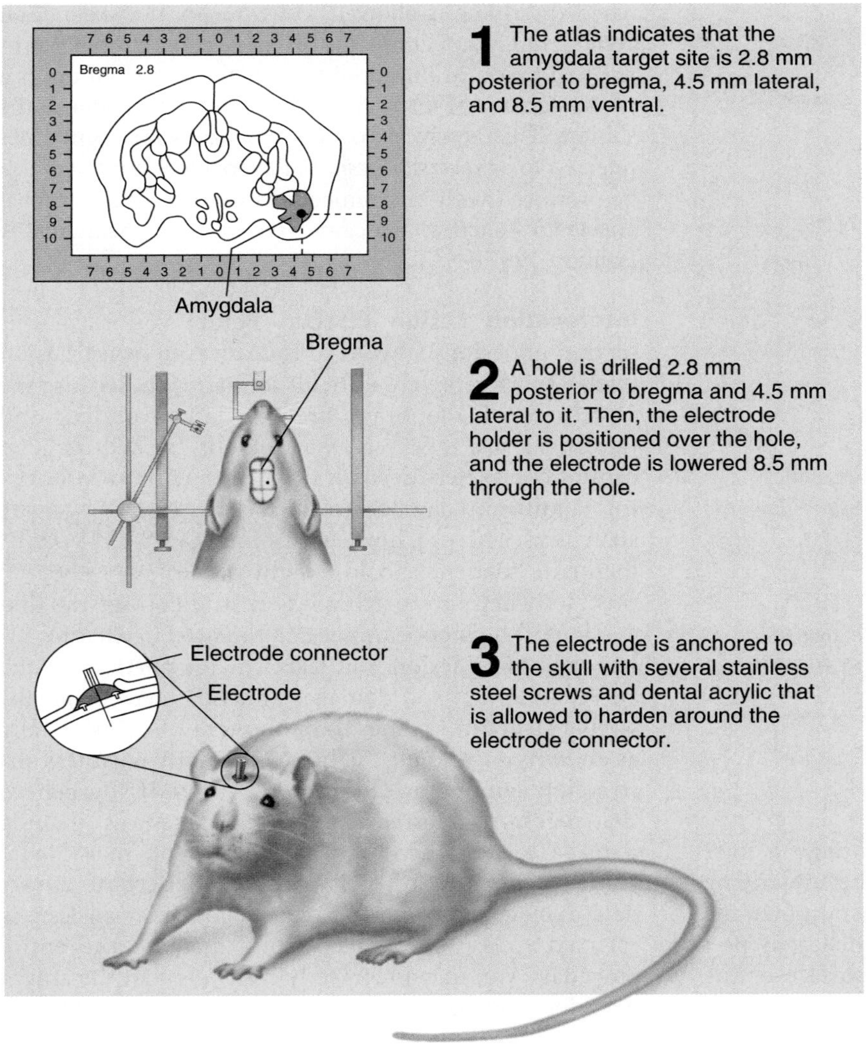

Lesion Methods

Those of you with an unrelenting drive to dismantle objects to see how they work will appreciate the lesion methods. In those methods, a part of the brain is removed, damaged, or destroyed; then, the behavior of the subject is carefully assessed in an effort to determine the functions of the lesioned structure. Four types of lesions are discussed here: aspiration lesions, radio-frequency lesions, knife cuts, and cryogenic blockade.

Aspiration Lesions   When a lesion is to be made in an area of cortical tissue that is accessible to the eyes and instruments of the surgeon, **aspiration** is frequently the method of choice. The

1 The atlas indicates that the amygdala target site is 2.8 mm posterior to bregma, 4.5 mm lateral, and 8.5 mm ventral.

2 A hole is drilled 2.8 mm posterior to bregma and 4.5 mm lateral to it. Then, the electrode holder is positioned over the hole, and the electrode is lowered 8.5 mm through the hole.

3 The electrode is anchored to the skull with several stainless steel screws and dental acrylic that is allowed to harden around the electrode connector.

FIGURE 5.14 Stereotaxic surgery: implanting an electrode in the rat amygdala.

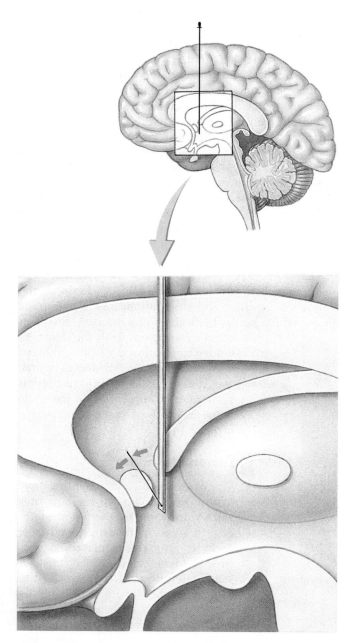

FIGURE 5.15 A device for performing subcortical knife cuts. The device is stereotaxically positioned in the brain; then, the blade swings out to make the cut, Here, the anterior commissure is being sectioned.

cortical tissue is drawn off by suction through a fine-tipped handheld glass pipette. Because the underlying white matter is slightly more resistant to suction than the cortical tissue itself, a skilled surgeon can delicately peel off the layers of cortical tissue from the surface of the brain, leaving the underlying white matter and major blood vessels undamaged.

Radio-Frequency Lesions   Small subcortical lesions are commonly made by passing *radio-frequency current* (high-frequency current) through the target tissue from the tip of a stereotaxically positioned electrode. The heat from the current destroys the tissue. The size and shape of the lesion are determined by the duration and intensity of the current and the configuration of the electrode tip.

Knife Cuts   *Sectioning* (cutting) is used to eliminate conduction in a nerve or tract. A tiny, well-placed cut can unambiguously accomplish this task without producing extensive damage to surrounding tissue. How does one insert a knife into the brain to make a cut without severely damaging the overlying tissue? One method is depicted in Figure 5.15.

Cryogenic Blockade   An alternative to destructive lesions is **cryogenic blockade**. When coolant is pumped through an implanted *cryoprobe*, such as the one depicted in Figure 5.16, neurons near the tip are cooled until they stop firing. The temperature is maintained above the freezing level, so there is no structural damage. Then, when the tissue is allowed to warm up, normal neural activity returns. A cryogenic blockade is functionally similar to a lesion in that it eliminates the contribution of a particular area of the brain to the ongoing behavior of the subject. This is why cryogenic blockades are sometimes referred to as *reversible lesions*. Reversible lesions can also be produced with microinjections into the brain of local anesthetics such as lidocaine (see Floresco, Seamans, & Phillips, 1997).

Interpreting Lesion Effects   Before you leave this section on lesions, a word of caution is in order. Lesion effects are deceptively difficult to interpret. Because the structures of the brain are small, convoluted, and tightly packed together, even a highly skilled surgeon cannot completely destroy a structure without producing significant damage to adjacent structures. There is, however, an unfortunate tendency to lose sight of *Thinking Creatively* this fact when interpreting lesion studies—see the discussion of the hippocampus and memory in Chapter 11. For example, a lesion that leaves major portions of the amygdala intact and damages an assortment of neighboring structures comes to be thought of simplistically as an *amygdala lesion*. Such an apparently harmless abstraction can be misleading in two ways. If you believe that all lesions referred to as "amygdala lesions" include damage to no other brain structure, you may incorrectly attribute all of their behavioral effects to amygdala damage; conversely, if you believe that all lesions referred to as "amygdala lesions" include the entire amygdala, you may incorrectly conclude that the amygdala does not participate in behaviors uninfluenced by the lesion.

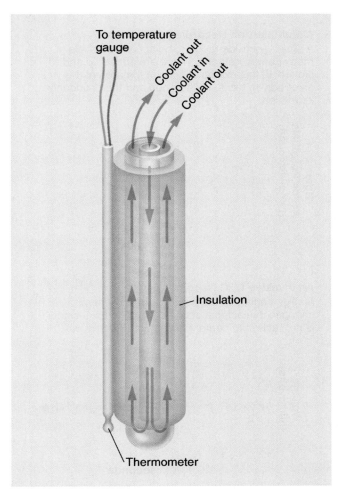

To temperature gauge

Coolant out

Coolant in

Coolant out

Insulation

Thermometer

FIGURE 5.16 A cryoprobe. The cryoprobe is implanted in the brain; then the brain area at the uninsulated tip of the cryoprobe is cooled while the effects on behavior are assessed. Cryoprobes are slender so that they can be implanted in the brain without causing substantial damage; they are typically constructed of hypodermic tubing of two gauges.

Bilateral and Unilateral Lesions As a general principle—but one with several notable exceptions—the behavioral effects of *unilateral lesions* (lesions restricted to one half of the brain) are much milder than those of symmetrical *bilateral lesions* (lesions involving both sides of the brain), particularly in nonhuman species. Indeed, behavioral effects of unilateral lesions to some brain structures can be difficult to detect. As a result, most experimental studies of lesion effects are studies of bilateral, rather than unilateral, lesions.

Electrical Stimulation

Clues about the function of a neural structure can be obtained by stimulating it electrically. Electrical brain stimulation is usually delivered across the two tips of a *bipolar electrode*—two insulated wires wound tightly together and cut at the end. Weak pulses of current produce an immediate increase in the firing of neurons near the tip of the electrode.

Electrical stimulation of the brain is an important biopsychological research tool because it often has behavioral effects, usually opposite to those produced by a lesion to the same site. It can elicit a number of behavioral sequences, including eating, drinking, attacking, copulating, and sleeping. The particular behavioral response that is elicited depends on the location of the electrode tip, the parameters of the current, and the test environment in which the stimulation is administered.

Invasive Electrophysiological Recording Methods

This section describes four invasive electrophysiological recording methods: intracellular unit recording, extracellular unit recording, multiple-unit recording, and invasive EEG recording. See Figure 5.17 on page 114 for an example of each method.

Intracellular Unit Recording A method whose findings were discussed at length in Chapter 4, *intracellular unit recording*, provides a moment-by-moment record of the graded fluctuations in one neuron's membrane potential. Most experiments using this recording procedure are performed on chemically immobilized animals because it is next to impossible to keep the tip of a microelectrode positioned inside a neuron of a freely moving animal.

Extracellular Unit Recording It is possible to record the action potentials of a neuron through a microelectrode whose tip is positioned in the extracellular fluid next to it—each time the neuron fires, there is an electrical disturbance and a blip is recorded at the electrode tip. Accordingly, *extracellular unit recording* provides a record of the firing of a neuron but no information about the neuron's membrane potential. It is difficult to record extracellularly from a single neuron in a freely moving animal without the electrode tip shifting away from the neuron, but it can be accomplished with special flexible microelectrodes that can shift slightly with the brain. Initially, extracellular unit recording involved recording from one neuron at a time, each at the tip of a separately implanted electrode. However, it is now possible to simultaneously record extracellular signals from up to about 100 neurons by analyzing the correlations among the signals picked up through several different electrodes implanted in the same general area (e.g., Nicolelis & Ribeiro, 2006).

Multiple-Unit Recording In *multiple-unit recording*, the electrode tip is much larger than that of a microelectrode; thus, it picks up signals from many neurons, and slight shifts in its position due to movement of the subject have little

An Intracellular Unit Recording
An intracellular microelectrode records the membrane potential from one neuron as it fires.

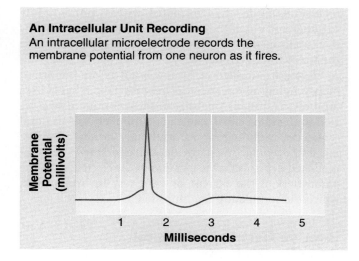

A Multiple-Unit Recording
A small electrode records the action potentials of many nearby neurons. These are added up and plotted. In this example, firing in the area of the electrode tip gradually declined and then suddenly increased.

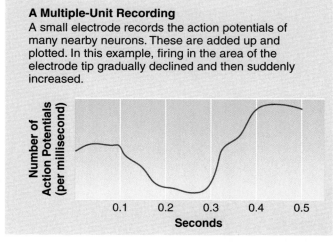

An Extracellular Unit Recording
An extracellular microelectrode records the electrical disturbance that is created each time an adjacent neuron fires.

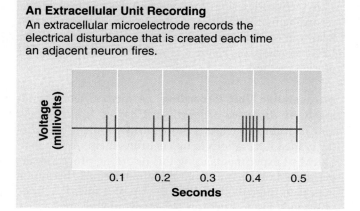

An Invasive EEG Recording
A large implanted electrode picks up general changes in electrical brain activity. The EEG signal is not related to neural firing in any obvious way.

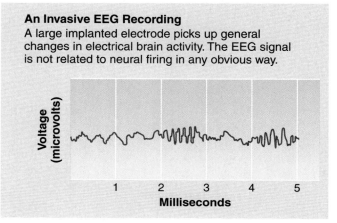

FIGURE 5.17 Four methods of recording electrical activity of the nervous system.

effect on the overall signal. The many action potentials picked up by the electrode are fed into an integrating circuit, which adds them together. A multiple-unit recording is a graph of the total number of recorded action potentials per unit of time (e.g., per 0.1 second).

Invasive EEG Recording In laboratory animals, EEG signals are recorded through large implanted electrodes rather than through scalp electrodes. Cortical EEG signals are frequently recorded through stainless steel skull screws, whereas subcortical EEG signals are typically recorded through stereotaxically implanted wire electrodes.

5.4
Pharmacological Research Methods

In the preceding section, you learned how physiological psychologists study the brain by manipulating it and

recording from it using surgical and electrical methods. In this section, you will learn how psychopharmacologists manipulate and record from the brain using chemical methods.

The major research strategy of psychopharmacology is to administer drugs that either increase or decrease the effects of particular neurotransmitters and to observe the behavioral consequences. You learned in Chapter 4 how *agonists* and *antagonists* affect neurotransmitter systems. Described here are routes of drug administration, methods of using chemicals to make selective brain lesions, methods of measuring the chemical activity of the brain that are particularly useful in biopsychological research, and methods for locating neurotransmitter systems.

Routes of Drug Administration

In most psychopharmacological experiments, drugs are administered in one of the following ways: (1) They are fed to the subject; (2) they are injected through a tube into the stomach (*intragastrically*); or (3) they are injected hypodermically into the peritoneal cavity of the abdomen (*intraperitoneally, IP*), into a large muscle (*intramuscularly, IM*), into the fatty tissue beneath the skin (*subcutaneously, SC*), or into a large surface vein (*intravenously, IV*). A problem with these peripheral routes of administration is that many drugs do not readily pass through the blood–brain barrier. To overcome this problem, drugs can be administered in small amounts through a fine, hollow tube, called a **cannula**, that has been stereotaxically implanted in the brain.

Selective Chemical Lesions

The effects of surgical, electrolytic, and cryogenic lesions are frequently difficult to interpret because they affect all neurons in the target area. In some cases, it is possible to make more selective lesions by injecting **neurotoxins** (neural poisons) that have an affinity for certain components of the nervous system. There are many selective neurotoxins. For example, when either *kainic acid* or *ibotenic acid* is administered by microinjection, it is preferentially taken up by cell bodies at the tip of the cannula and destroys those neurons, while leaving neurons with axons passing through the area largely unscathed.

Another widely used selective neurotoxin is *6-hydroxydopamine (6-OHDA)*. It is taken up by only those neurons that release the neurotransmitter *norepinephrine* or *dopamine*, and it leaves other neurons at the injection site undamaged.

Measuring Chemical Activity of the Brain

There are many procedures for measuring the chemical activity of the brains of laboratory animals. Two techniques that have proved particularly useful in biopsychological research are the 2-deoxyglucose technique and cerebral dialysis.

The 2-Deoxyglucose Technique   The *2-deoxyglucose (2-DG) technique* entails placing an animal that has been injected with radioactive 2-DG in a test situation in which it engages in the activity of interest. Because 2-DG is similar in structure to glucose—the brain's main source of energy—neurons active during the test absorb it at a high rate but do not metabolize it. Then the subject is killed, and its brain is removed and sliced. The slices are then subjected to **autoradiography**; they are coated with a photographic emulsion, stored in the dark for a few days, and then developed much like film. Areas of the brain that absorbed high levels of the radioactive 2-DG during the test appear as black spots on the slides. The density of the spots in various regions of the brain can then be color-coded (see Figure 5.18).

Cerebral Dialysis   **Cerebral dialysis** is a method of measuring the extracellular concentration of specific neurochemicals in behaving animals (see Robinson & Justice, 1991)—most other techniques for measuring neurochemicals require that the animals be killed so that tissue can be extracted. Cerebral dialysis involves the implantation in the brain of a fine tube with a short semipermeable section. The semipermeable section is positioned in the brain structure of interest so that extracellular chemicals from the structure will diffuse into the tube. Once in the tube, they can be collected for freezing, storage, and later analysis; or they can be carried in solution directly to a *chromatograph* (a device for measuring the chemical constituents of liquids or gases).

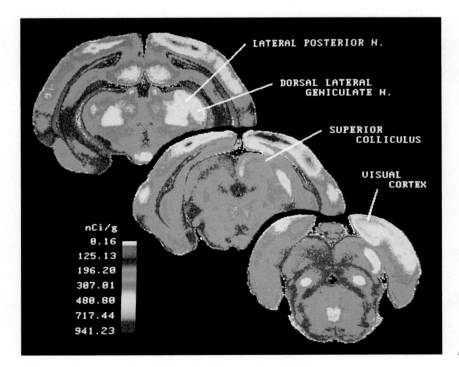

FIGURE 5.18   The 2-deoxyglucose technique. The accumulation of radioactivity is shown in three frontal sections taken from the brain of a Richardson's ground squirrel. The subject was injected with radioactive 2-deoxyglucose; then, for 45 minutes, it viewed brightly illuminated black and white stripes through its left eye while its right eye was covered. Because the ground squirrel visual system is largely crossed, most of the radioactivity accumulated in the visual structures of the right hemisphere (the hemisphere on your right). (Courtesy of Rod Cooper, Department of Psychology, University of Calgary.)

Locating Neurotransmitters and Receptors in the Brain

A key step in trying to understand the psychological function of a particular neurotransmitter or receptor is finding out where it is located in the brain. Two of the techniques available for this purpose are immunocytochemistry and in situ hybridization. Each involves exposing brain slices to a labeled *ligand* of the molecule under investigation (the ligand of a molecule is another molecule that binds to it).

Immunocytochemistry   When a foreign protein (an *antigen*) is injected into an animal, the animal's body creates *antibodies* that bind to it and help the body remove or destroy it; this is known as the body's *immune reaction*. Neurochemists have created stocks of antibodies to the brain's peptide neurotransmitters (neuropeptides; see Chapter 4) and their receptors. **Immunocytochemistry** is a procedure for locating particular neuroproteins in the brain by labeling their antibodies with a dye or radioactive element and then exposing slices of brain tissue to the labeled antibodies. Regions of dye or radioactivity accumulation in the brain slices mark the locations of the target neuroprotein.

Because all enzymes are proteins and because only those neurons that release a particular neurotransmitter are likely to contain all the enzymes required for its synthesis, immunocytochemistry can be used to locate neurotransmitters by binding to their enzymes. This is done by exposing brain slices to labeled antibodies that bind to enzymes located in only those neurons that contain the neurotransmitter of interest (see Figure 5.19).

In Situ Hybridization   Another technique for locating peptides and other proteins in the brain is **in situ hybridization**. This technique takes advantage of the fact that all peptides and proteins are transcribed from sequences of nucleotide bases on strands of messenger RNA (see Chapter 2).

The nucleotide base sequences that direct the synthesis of many neuroproteins have been identified, and hybrid strands of mRNA with the complementary base sequences have been artificially created. In situ hybridization (see Figure 5.20) involves the following steps. First, hybrid RNA strands with the base sequence complementary to that of the mRNA that directs the synthesis of the target neuroprotein are obtained. Next, the hybrid RNA strands are labeled with a dye or radioactive element. Finally, the brain slices are exposed to the labeled hybrid RNA strands; they bind to the complementary mRNA strands, marking the location of neurons that release the target neuroprotein.

5.5
Genetic Engineering

Genetics is a science that has made amazing progress in the last two decades, and biopsychologists are reaping the benefits. Modern genetic methods are now widely used in biopsychological research, which just a few years ago would have seemed like science fiction.

Gene Knockout Techniques

Gene knockout techniques are procedures for creating organisms that lack a particular gene under investigation (see Eisener-Dorman, Lawrence, & Bolivar, 2008). Mice (the favored mammalian subjects of genetic research) that are the products of gene knockout techniques are referred to as *knockout mice*. (This term often makes me

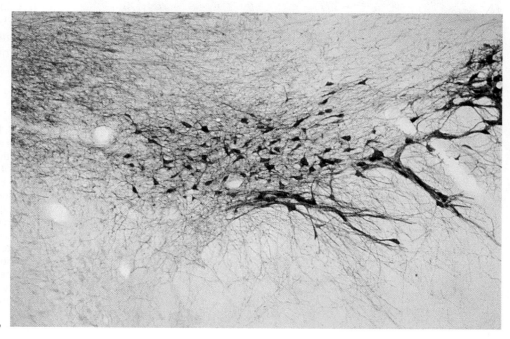

FIGURE 5.19   Immunocyto-chemistry. This section through a rat's substantia nigra reveals dopaminergic neurons that have taken up the antibody for tyrosine hydroxylase, the enzyme that converts tyrosine to L-dopa. (Courtesy of Mark Klitenick and Chris Fibiger, Department of Psychiatry, University of British Columbia.)

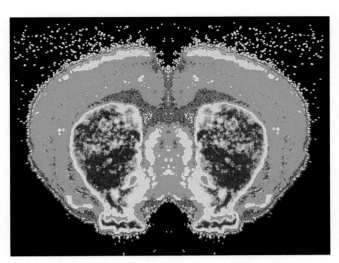

FIGURE 5.20 In situ hybridization. This color-coded frontal section through a rat brain reveals high concentrations of mRNA expression for an endorphin in the striatum (in red and yellow). (Courtesy of Ningning Guo and Chris Fibiger, Department of Psychiatry, University of British Columbia.)

smile, as images of little mice with boxing gloves flit through my mind.)

Many gene knockout studies have been conducted to clarify the neural mechanisms of behavior. For example, Ruby and colleagues (2002) and Hattar and colleagues (2003) used *melanopsin knockout mice* (mice in whom the gene for the synthesis of melanopsin has been deleted) to study the role of melanopsin in regulating the light–dark cycles that control circadian (about 24 hours) rhythms of bodily function—for example, daily cycles of sleep, eating, and body temperature. *Melanopsin* is a protein found in some neurons in the mammalian *retina* (the receptive layer of the eye), and it had been implicated in the control of circadian ryhythms by light because many of the neurons containing melanopsin project to the circadian clock mechanism in the hypothalamus. Knockout of the gene for synthesizing melanopsin reduced, but did not eliminate, the responses of the clock mechanism to changes in light, and it impaired, but did not eliminate, the ability of mice to adjust their circadian rhythms in response to changes in the light–dark cycle. Thus, melansopsin appears to contribute to the control of circadian rhythms by light, but it is not the only factor.

This type of result is typical of gene knockout studies of behavior: Many genes have been discovered that contribute to particular behaviors, but invariably other mechanisms are involved. It may be *Thinking Creatively* tempting to think that each behavior is controlled by a single gene, but the reality is much more complex. Each behavior is controlled by many genes interacting with one another and with experience.

Gene Replacement Techniques

It is now possible to replace one gene with another. **Gene replacement techniques** have created interesting possibilities for research and therapy. Pathological genes from human cells can be inserted in other animals such as mice—mice that contain the genetic material of another species are called **transgenic mice.** For example, Shen and colleagues (2008) created transgenic mice by inserting a defective human gene that had been found to be associated with schizophrenia in a Scottish family with a particularly high incidence of the disorder. The transgenic mice displayed a variety of cerebral abnormalities (e.g., reduced cerebral cortex and enlarged ventricles) and abnormal behaviors reminiscent of human schizophrenia, confirming that the defective gene was a causal factor in the familial schizophrenia of the Scottish family.

In another gene replacement technique, a gene is replaced with one that is identical except for the addition of a few bases that can act as a switch, turning the gene off or on in response to particular chemicals. The chemicals can then be used to activate or suppress the gene at a particular point *Clinical Implications* in development. Treating neurological disease by replacing faulty genes in patients suffering from genetic disorders is an exciting, but as yet unrealized, goal.

Fantastic Fluorescence and the Brainbow

Green fluorescent protein (GFP) is a protein that exhibits bright green fluorescence when exposed to blue light. First isolated by Shimomura, Johnson, and Saiga (1962), from a species of jellyfish *Evolutionary Perspective* found off the west coast of North America, GFP is currently stimulating advances in many fields of biological research. Martin Chalfie, Osamu Shimomura, and Roger Y. Tsien were awarded the 2008 Nobel Prize in chemistry for its discovery and study.

The utility of GFP as a research tool in the biological sciences could not be realized until its gene was identified and cloned in the early 1990s. The general strategy is to activate the GFP gene in only the particular cells under investigation so that they can readily be visualized. This can be accomplished in two ways: by inserting the GFP gene in only the target cells or by introducing the GFP gene in all cells of the subject but expressing the gene in only the target cells. Chalfie and colleagues (1994) were the first to use GFP to visualize neurons. They introduced the GFP gene into a small transparent roundworm, *Caenorhabditis elegans,* in an area of its chromosomes that controls the development of touch receptor neurons. Figure 5.21 on page 118 shows the glowing touch receptor neurons. The GFP gene has now been expressed in the cells of many plant and animal species, including humans.

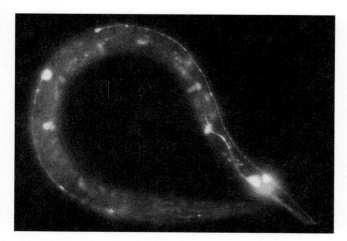

FIGURE 5.21 Touch receptor neurons of the transparent *Caenorhabditis elegans* labeled by green fluorescent protein.

Livet and colleagues took the very useful GFP technique one step further—one big step. First, Tsien (1998) found that making minor alterations to the GFP gene resulted in the synthesis of proteins that fluoresced in different colors. Livet and colleagues (2007) then introduced the mutated genes for cyan, yellow, and blue fluorescent proteins into the genomes of developing mice in such a way that they were expressed in developing neurons. Each neuron produced different amounts of the three proteins, giving it a distinctive color—in the same way that a color printer can make any color by mixing only three colored inks in differing proportions. Because each neuron was labeled with its own distinctive color, the pathways of neural axons could be traced to their destinations through the cellular morass. This technique has been dubbed **brainbow** for obvious reasons—see Figure 5.22.

Thinking Creatively

Scan Your Brain

The research methods of biopsychology illustrate a psychological disorder suffered by many scientists. I call it "unabbreviaphobia"—the fear of leaving any term unabbreviated. To determine whether you have mastered Part One of this chapter and are ready for Part Two, supply the full term for each of the following abbreviations. The correct answers are provided at the end of the exercise. Before proceeding, review material related to your incorrect answers and omissions.

1. CT: _____
2. MRI: _____
3. PET: _____
4. 2-DG: _____
5. fMRI: _____
6. MEG: _____
7. TMS: _____
8. EEG: _____
9. ERP: _____
10. AEP: _____
11. EMG: _____
12. EOG: _____
13. SCL: _____
14. SCR: _____
15. ECG: _____
16. EKG: _____
17. IP: _____
18. IM: _____
19. IV: _____
20. SC: _____
21. 6-OHDA: _____
22. GFP: _____

Scan Your Brain answers: (1) computed tomography, (2) magnetic resonance imaging, (3) positron emission tomography, (4) 2-deoxyglucose, (5) functional MRI, (6) magnetoencephalography, (7) transcranial magnetic stimulation, (8) electroencephalogram, (9) event-related potential, (10) average evoked potential, (11) electromyogram, (12) electrooculogram, (13) skin conductance level, (14) skin conductance response, (15) electrocardiogram, (16) electrocardiogram, (17) intraperitoneal, (18) intramuscular, (19) intravenous, (20) subcutaneous, (21) 6-hydroxydopamine, (22) green fluorescent protein.

PART TWO
BEHAVIORAL RESEARCH METHODS OF BIOPSYCHOLOGY

We turn now from methods used by biopsychologists to study the nervous system to those that deal with the behavioral side of biopsychology. Because of the inherent invisibility of neural activity, the primary objective of the methods used in its investigation is to render the unobservable observable. In contrast, the major objectives of behavioral research methods are to control, to simplify, and to objectify.

A single set of procedures developed for the investigation of a particular behavioral phenomenon is commonly referred to as a **behavioral paradigm**. Each behavioral paradigm normally comprises a method for producing the behavioral phenomenon under investigation and a method for objectively measuring it.

FIGURE 5.22 With the research technique called *brainbow*, each neuron is labeled with a different color, facilitating neuron tracing.

modern *customized-test-battery approach.*

Single-Test Approach Before the 1950s, the few existing neuropsychological tests were designed to detect the presence of brain damage; in particular, the goal of these early tests was to discriminate between patients with psychological problems resulting from structural brain damage and those with psychological problems resulting from functional, rather than structural, changes to the brain. This approach proved unsuccessful, in large part because no single test could be developed that would be sensitive to all the varied and complex psychological symptoms that could potentially occur in a brain-damaged patient.

5.6
Neuropsychological Testing

A patient suspected of suffering from some sort of nervous system dysfunction is usually referred to a *neurologist*, who assesses simple sensory and motor functions. More subtle changes in perceptual, emotional, motivational, or cognitive functions are the domain of the *neuropsychologist.*

Clinical Implications

Because neuropsychological testing is so time consuming, it is typically prescribed for only a small portion of brain-damaged patients. This is unfortunate; the results of neuropsychological testing can help brain-damaged patients in three important ways: (1) by assisting in the diagnosis of neural disorders, particularly in cases in which brain imaging, EEG, and neurological testing have proved equivocal; (2) by serving as a basis for counseling and caring for the patients; and (3) by providing a basis for objectively evaluating the effectiveness of the treatment and the seriousness of its side effects.

Modern Approach to Neuropsychological Testing

The nature of neuropsychological testing has changed radically since the 1950s (see Stuss & Levine, 2002). Indeed, the dominant approach to psychological testing has evolved through three distinct phases: the *single-test approach*, the *standardized-test-battery approach*, and the

Standardized-Test-Battery Approach The standardized-test-battery approach to neuropsychological testing grew out of the failures of the single-test approach, and by the 1960s, it was predominant. The objective stayed the same—to identify brain-damaged patients—but the testing involved standardized batteries (sets) of tests rather than a single test. The most widely used standardized test battery has been the *Halstead-Reitan Neuropsychological Test Battery.* The Halstead-Reitan is a set of tests that tend to be performed poorly by brain-damaged patients in relation to other patients or healthy control subjects; the scores on each test are added together to form a single aggregate score. An aggregate score below the designated cutoff leads to a diagnosis of brain damage. The standardized-test-battery approach has proved only marginally successful; standardized test batteries discriminate effectively between neurological patients and healthy patients, but they are not so good at discriminating between neurological patients and psychiatric patients.

The Customized-Test-Battery Approach The customized-test-battery approach began to be used routinely in a few elite neuropsychological research institutions in the 1960s. This approach proved highly successful in research, and it soon spread to clinical practice. It now predominates in both the research laboratory and the neurological ward (see Lezak, 1997; Strub & Black, 1997).

The objective of current neuropsychological testing is not merely to identify patients with brain damage; the objective is to characterize the nature of the psychological deficits of each brain-damaged patient. So how does the customized-test-battery approach to neuropsychological testing work? It usually begins in the same way for all patients: with a common battery of tests selected by the neuropsychologist to provide an indication of the general nature of the neuropsychological symptoms. Then, depending on the results of the common test battery, the neuropsychologist selects a series of tests customized to each patient in an effort to characterize in more detail the general symptoms revealed by the common battery. For example, if the results of the test battery indicated that a patient had a memory problem, subsequent tests would include those designed to reveal the specific nature of the memory problem.

The tests used in the customized-test-battery approach differ in three respects from earlier tests. First, the newer tests are specifically designed to measure aspects of psychological function that have been spotlighted by modern theories and data. For example, modern theories, and the evidence on which they are based, suggest that the mechanisms of short-term and long-term memory are totally different; thus, the testing of patients with memory problems virtually always involves specific tests of both short-term and long-term memory. Second, the interpretation of the test results often does not rest entirely on how well the patient does; unlike early neuropsychological tests, currently used tests often require the neuropsychologist to assess the cognitive strategy that the patient employs in performing the test. Brain damage often changes the strategy that a neuropsychological patient uses to perform a test without lowering the overall score. Third, the customized-test-battery approach requires more skill and knowledge on the part of the neuropsychologist to select just the right battery of tests to expose a patient's deficits and to identify qualitative differences in cognitive strategy.

Tests of the Common Neuropsychological Test Battery

Because the customized-test-battery approach to neuropsychological testing typically involves two phases—a battery of general tests given to all patients followed by a series of specific tests customized to each patient—the following examples of neurological tests are presented in two subsections. First are some tests that are often administered as part of the initial common test battery, and second are some tests that might be used by a neuropsychologist to investigate in more depth particular problems revealed by the common battery.

Clinical Implications

Intelligence Although the overall *intelligence quotient* (*IQ*) is a notoriously poor measure of brain damage, a test of general intelligence is nearly always included in the battery of neuropsychological tests routinely given to all patients. Many neuropsychological assessments begin with the **Wechsler Adult Intelligence Scale (WAIS)**, first published in 1955 and standardized in 1981 on a sample of 1,880 U.S. citizens between 16 and 71. The WAIS is often the first test because knowing a patient's IQ can help a neuropsychologist interpret the results of subsequent tests. Also, a skilled neuropsychologist can sometimes draw inferences about a patient's neuropsychological dysfunction from the pattern of deficits on the 15 subtests of the WAIS. For example, low scores on subtests of verbal ability tend to be associated with left hemisphere damage, whereas right hemisphere damage tends to reduce scores on performance subtests. The 11 original subtests of the WAIS are described in Table 5.1.

👁 **Watch**
Robert Sternberg on Intelligence
www.mypsychlab.com

Memory One weakness of the WAIS is that it often fails to detect memory deficits, despite including subtests specifically designed to test memory function. For example, the information subtest of the WAIS assesses memory for general knowledge (e.g., "Who is Queen Elizabeth?"), and the **digit span** subtest (the most widely used test of short-term memory) identifies the longest sequence of random digits that a patient can repeat correctly 50% of the time; most people have a digit span of 7. However, these two forms of memory are among the least likely to be disrupted by brain damage—patients with seriously disturbed memories often show no deficits on either the information or the digit span subtest. Be that as it may, memory problems rarely escape unnoticed; they are often reported by the patient or the family of the patient.

⏵ **Simulate**
Digit Span
www.mypsychlab.com

Language If a neuropsychological patient has taken the WAIS, deficits in the use of language can be inferred from a low aggregate score on the verbal subtests. A patient who has not taken the WAIS can be quickly screened for language-related deficits with the **token test**. Twenty tokens of two different shapes (squares and circles), two different sizes (large and small), and five different colors (white, black, yellow, green, and red) are placed on a table in front of the subject. The test begins with the examiner reading simple instructions—for example, "Touch a red square"—and the subject trying to follow them. Then, the test progresses to more difficult instructions, such as "Touch the small, red circle and then the large, green square." Finally, the subject is asked to read the instructions aloud and follow them.

Language Lateralization It is usual for one hemisphere to participate more than the other in language-related activities. In most people, the left hemisphere is dominant for language, but in some, the right hemisphere

TABLE 5.1	The 11 Original Subtests of the Wechsler Adult Intelligence Scale (WAIS)

Verbal Subtests

Information Read to the subject are 29 questions of general information—for example "Who is the president of the United States?"

Digit Span Three digits are read to the subject at 1-second intervals, and the subject is asked to repeat them in the same order. Two trials are given at three digits, four digits, five digits, and so on until the subject fails both trials at one level.

Vocabulary The subject is asked to define a list of 35 words that range in difficulty.

Arithmetic The subject is presented with 14 arithmetic questions and must answer them without the benefit of pencil and paper.

Comprehension The subject is asked 16 questions that test the ability to understand general principles—for example, why should people vote?

Similarities The subject is presented with pairs of items and is asked to explain how the items in each pair are similar.

Performance Subtests

Picture Completion The subject must identify the important part missing from 20 drawings—for example, a drawing of a squirrel with no tail.

Picture Arrangement The subject is presented with 10 sets of cartoon drawings and is asked to arrange each set so that it tells a sensible story.

Block Design The subject is presented with blocks that are red on two sides, white on two sides, and half red and half white on the other two. The subject is shown pictures of nine patterns and is asked to duplicate them by arranging the blocks appropriately.

Object Assembly The subject is asked to put together the pieces of four simple jigsaw puzzles to form familiar objects.

Digit Symbol The subject is presented with a key that matches each of a series of symbols with a different digit. On the same page is a series of digits and the subject is given 90 seconds to write the correct symbol, according to the key, next to as many digits as possible.

is dominant (see Chapter 16). A test of language lateralization is often included in the common test battery because knowing which hemisphere is dominant for language is often useful in interpreting the results of other tests. Furthermore, a test of language lateralization is virtually always given to patients before any surgery that might encroach on the cortical language areas. The results are used to plan the surgery, trying to avoid the language areas if possible.

There are two widely used tests of language lateralization. The sodium amytal test (Wada, 1949) is one, and the dichotic listening test (Kimura, 1973) is the other.

The **sodium amytal test** involves injecting the anesthetic sodium amytal into either the left or right carotid artery in the neck. This temporarily anesthetizes the *ipsilateral* (same-side) hemisphere while leaving the *contralateral* (opposite-side) hemisphere largely unaffected. Several tests of language function are quickly administered while the ipsilateral hemisphere is anesthetized. Later, the process is repeated for the other side of the brain. When the injection is on the side dominant for language, the patient is completely mute for about 2 minutes. When the injection is on the nondominant side, there are only a few minor speech problems. Because the sodium amytal test is invasive, it can be administered only for medical reasons—usually to determine the dominant language hemisphere prior to brain surgery.

In the standard version of the **dichotic listening test**, sequences of spoken digits are presented to subjects through stereo headphones. Three digits are presented to one ear at the same time that three different digits are presented to the other ear. Then the subjects are asked to report as many of the six digits as they can. Kimura (1973) found that subjects correctly report more of the digits heard by the ear contralateral to their dominant hemisphere for language, as determined by the sodium amytal test.

Tests of Specific Neuropsychological Function

Following analysis of the results of a neuropsychological patient's performance on the common test battery, the neuropsychologist selects a series of specific tests to clarify the nature of the general problems exposed by the common battery. There are thousands of tests that might be selected. This section describes a few of them and mentions some of the considerations that might influence their selection.

Clinical Implications

Memory Following the discovery of memory impairment by the common test battery, at least four fundamental questions about the memory impairment must be answered (see Chapter 11): (1) Does the memory impairment involve *short-term memory*, *long-term memory*, or both? (2) Are any deficits in long-term memory *anterograde* (affecting the retention of things learned after the damage), *retrograde* (affecting the retention of things learned before the damage), or both? (3) Do any deficits in long-term memory involve *semantic memory* (memory for knowledge of the world) or *episodic memory* (memory for personal experiences)? (4) Are any deficits in long-term memory deficits of *explicit memory* (memories of which the patient is aware and can thus express verbally), *implicit memory* (memories that are demonstrated by the improved performance of the patient without the patient being conscious of them), or both?

Many amnesic patients display severe deficits in explicit memory with no deficits at all in implicit memory (Curran & Schacter, 1997). **Repetition priming tests** have proven instrumental in the assessment and study of this pattern. Patients are first shown a list of words and asked to study them; they are not asked to remember them. Then, at a later time, they are asked to complete a list of word fragments, many of which are fragments of words from the initial list. For example, if "purple" had been in the initial test, "pu_p_ _" could be one of the test word fragments. Amnesic patients often complete the fragments as well as healthy control subjects. But—and this is the really important part—they often have no conscious memory of any of the words in the initial list or even of ever having seen the list. In other words, they display good implicit memory of experiences without explicit memories of them.

Language   If a neuropsychological patient turns out to have language-related deficits on the common test battery, a complex series of tests is administered to clarify the nature of the problem (see Chapter 16). For example, if a patient has a speech problem, it may be one of three fundamentally different problems: problems of *phonology* (the rules governing the sounds of the language), problems of *syntax* (the grammar of the language), or problems of *semantics* (the meaning of the language). Because brain-damaged patients may have one of these problems but not the others, it is imperative that the testing of all neuropsychological patients with speech problems include tests of each of these three capacities (Saffran, 1997).

Reading aloud can be disrupted in different ways by brain damage, and follow-up tests must be employed that can differentiate between the different patterns of disruption (Coslett, 1997). Some *dyslexic* patients (those with reading problems) remember the rules of pronunciation but have difficulties pronouncing words that do not follow these rules, words such as *come* and *tongue*, whose pronunciation must be remembered. Other dyslexic patients pronounce simple familiar words based on memory but have lost the ability to apply the rules of pronunciation—they cannot pronounce nonwords such as *trapple* or *fleeming*.

Frontal-Lobe Function

Injuries to the frontal lobes are common, and the **Wisconsin Card Sorting Test** (see Figure 5.23) is a component of many customized test batteries because performance on it is sensitive to frontal-lobe damage (see Eling, Derckx, & Maes, 2008). On each Wisconsin card is either one symbol or two, three, or four identical symbols. The symbols are all either triangles, stars, circles, or crosses; and they are all either red, green, yellow, or blue.

At the beginning of the test, the patient is confronted with four stimulus cards that differ from one another in the form, color, and number of symbols they display. The task is to correctly sort cards from a deck into piles in front of the stimulus cards. However, the patient does not know whether to sort by form, by color, or by number. The patient begins by guessing and is told after each card has been sorted whether it was sorted correctly or incorrectly. At first, the task is to learn to sort by color. But as soon as the patient makes several consecutive correct responses, the sorting principle is changed to shape or number without any indication other than the fact that responses based on color become incorrect. Thereafter, each time the patient learns a new sorting principle, the principle is changed.

Patients with damage to their frontal lobes often continue to sort on the basis of one sorting principle for 100 or more trials after it has become incorrect (Demakis, 2003). They seem to have great difficulty learning and remembering that previously appropriate guidelines for

FIGURE 5.23   The Wisconsin Card Sorting Test. This woman is just starting the test. If she places the first card in front of the stimulus card with the three green circles, she is sorting on the basis of color. She must guess until she can learn which principle—color, shape, or number—should guide her sorting. After she has placed a card she is told whether or not her placement is correct.

effective behavior are no longer appropriate, a problem called *perseveration*.

5.7

Behavioral Methods of Cognitive Neuroscience

Cognitive neuroscience is predicated on two related assumptions. The first premise is that each complex cognitive process results from the combined activity of simple cognitive processes called **constituent cognitive processes**. The second premise is that each constituent cognitive process is mediated by neural activity in a particular area of the brain. One of the main goals of cognitive neuroscience is to identify the parts of the brain that mediate various constituent cognitive processes.

With the central role played by PET and fMRI in cutting-edge cognitive neuroscience research, the **paired-image subtraction technique** has become one of the key behavioral research methods in such research (see Posner &

Raichle, 1994). Let me illustrate this technique with an example from a PET study of single-word processing by Petersen and colleagues (1988). Petersen and his colleagues were interested in locating the parts of the brain that enable a subject to make a word association (to respond to a printed word by saying a related word). You might think this would be an easy task to accomplish by having a subject perform a word-association task while a PET image of the subject's brain is recorded. The problem with this approach is that many parts of the brain that would be active during the test period would have nothing to do with the constituent cognitive process of forming a word association; much of the activity recorded would be associated with other processes such as seeing the words, reading the words, and speaking. The paired-image subtraction technique was developed to deal with this problem.

The paired-image subtraction technique involves obtaining PET or fMRI images during several different cognitive tasks. Ideally, the tasks are designed so that pairs of them differ from each other in terms of only a single constituent cognitive process. Then, the brain activity associated with that process can be estimated by subtracting the activity in the image associated with one of the two tasks from the activity in the image associated with the other. For example, in one of the tasks in the study by Petersen and colleagues, subjects spent a minute reading aloud printed nouns as they appeared on a screen; in another, they observed the same nouns on the screen but responded to each of them by saying aloud an associated verb (e.g., *truck— drive*). Then, Petersen and his colleagues subtracted the activity in the images that they recorded during the two tasks to obtain a *difference image*. The difference image illustrated the areas of the brain that were specifically involved in the constituent cognitive process of forming the word association; the activity associated with fixating on the screen, seeing the nouns, saying the words, and so on was eliminated by the subtraction (see Figure 5.24).

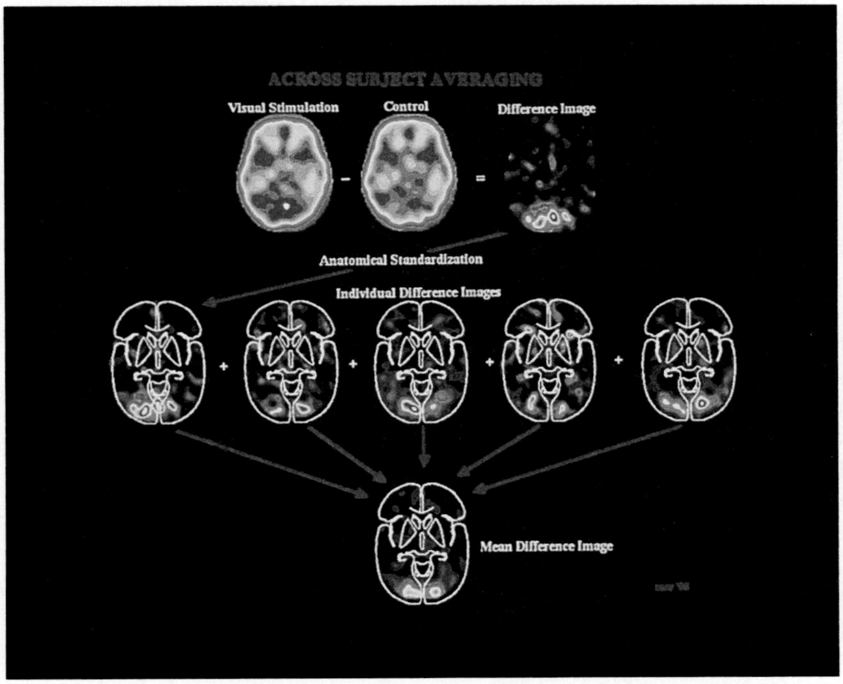

Interpretation of difference images is complicated by the fact that there is substantial brain activity when human subjects sit quietly and let their minds wander—this level of activity has been termed the brain's **default mode** (Raichle, 2010). The brain structures that are typically active in the default mode are collectively referred to as the **default mode network**, which comprises medial and

FIGURE 5.24 The paired-image subtraction technique, which is commonly employed in cognitive neuroscience. Here we see that the brain of a subject is generally active when the subject looks at a flickering checkerboard pattern (visual stimulation condition). However, if the activity that occurred when the subject stared at a blank screen (control situation) is subtracted, it becomes apparent that the perception of the flashing checkerboard pattern was associated with an increase in activity that was largely restricted to the occipital lobe. The individual difference images of five subjects were averaged to produce the mean difference image. (PET scans courtesy of Marcus Raichle, Mallinckrodt Institute of Radiology, Washington University Medical Center.)

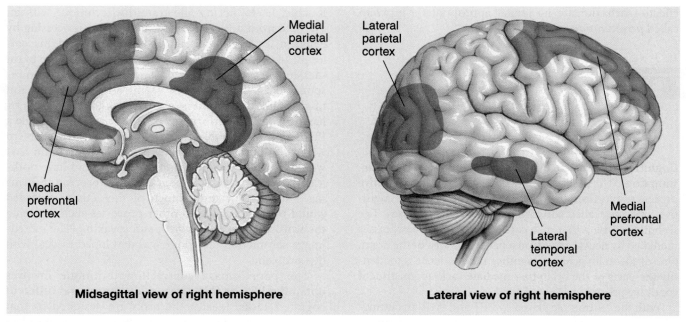

FIGURE 5.25   The default mode network: areas of the brain in which activity is commonly recorded by functional brain imaging techniques when the mind wanders.

lateral parietal cortex, medial frontal cortex, and lateral temporal cortex. See Figure 5.25.

Another difficulty in using PET and fMRI to locate constituent cognitive processes results from the *noise* associated with random cerebral events that occur during the test—for example, thinking about a sudden pang of hunger, noticing a fly on the screen, or wondering whether the test will last much longer (see Mason et al., 2007). The noise created by such events can be significantly reduced with a technique discussed earlier in this chapter: *signal averaging*. By averaging the difference images obtained from repetitions of the same tests, the researchers can greatly increase the *signal-to-noise ratio*. It is standard practice to average the images obtained from several subjects; the resulting mean (averaged) difference image emphasizes areas of activity that are common to most of the subjects and deemphasizes areas of activity that are peculiar to a few of them (see Figure 5.24). However, this averaging procedure can lead to a serious problem: If two subjects had specific but different patterns of cortical activity, the average image derived from the two would reveal little about either. Because people differ substantially from one another in the cortical localization of cognitive abilities, this is a serious problem (see Haynes & Rees, 2006). Moreover, the area of cortex that controls a particular ability can change in an individual as a result of experience.

Thinking Creatively

Neuroplasticity

5.8
Biopsychological Paradigms of Animal Behavior

Noteworthy examples of the behavioral paradigms used to study the biopsychology of laboratory species are provided here under three headings: (1) paradigms for the assessment of species-common behaviors, (2) traditional conditioning paradigms, and (3) seminatural animal learning paradigms. In each case, the focus is on methods used to study the behavior of the laboratory rat, the most common subject of biopsychological research.

Paradigms for Assessment of Species-Common Behaviors

Many of the behavioral paradigms that are used in biopsychological research are used to study species-common behaviors. **Species-common behaviors** are those that are displayed by virtually all members of a species, or at least by all those of the same age and sex. Commonly studied species-common behaviors include grooming, swimming, eating, drinking, copulating, fighting, and nest building. Described here are the open-field test, tests of aggressive and defensive behavior, and tests of sexual behavior.

Open-Field Test In the **open-field test**, the subject is placed in a large, barren chamber, and its activity is recorded (see Brooks & Dunnett, 2009). It is usual to measure general activity either with an automated activity recorder or by drawing lines on the floor of the chamber and counting the number of line-crossings during the test. It is also common in the open-field test to count the number of *boluses* (pieces of excrement) that were dropped by an animal during the test. Low activity scores and high bolus counts are frequently used as indicators of fearfulness. Fearful rats are highly **thigmotaxic**; that is, they rarely venture away from the walls of the test chamber and rarely engage in such activities as rearing and grooming. Rats are often fearful when they are first placed in a strange open field, but this fearfulness usually declines with repeated exposure to the same open field.

Tests of Aggressive and Defensive Behavior Typical patterns of aggressive and defensive behavior can be observed and measured during combative encounters between the dominant male rat of an established colony and a smaller male intruder (see Blanchard & Blanchard, 1988). This is called the **colony-intruder paradigm**. The behaviors of the dominant male are considered to be aggressive and those of the hapless intruder defensive. The dominant male of the colony (the *alpha male*) moves sideways toward the intruder, with its hair erect. When it nears the intruder, it tries to push the intruder off balance and to deliver bites to its back and flanks. The defender tries to protect its back and flanks by rearing up on its hind legs and pushing the attacker away with its forepaws or by rolling onto its back. Thus, piloerection, lateral approach, and flank- and back-biting indicate conspecific aggression in the rat; freezing, boxing (rearing and pushing away), and rolling over indicate defensiveness.

Some tests of rat defensive behavior assess reactivity to the experimenter rather than to another rat. For example, it is common to rate the resistance of a rat to being picked up—no resistance being the lowest category and biting the highest—and to use the score as one measure of defensiveness (Kalynchuk et al., 1997).

The **elevated plus maze**, a four-armed, plus-sign-shaped maze that is typically mounted 50 centimeters above the floor, is a test of defensiveness that is commonly used to study in rats the *anxiolytic* (anxiety-reducing) effects of

Clinical Implications

drugs. Two of the arms of the maze have sides, and two do not. The measure of defensiveness, or anxiety, is the proportion of time the rats spend in the protected closed arms rather than on the exposed arms. Many established anxiolytic drugs significantly increase the proportion of time that rats spend on the open arms (see Pellow et al., 1985), and, conversely, many new drugs that prove to be effective in reducing rats' defensiveness on the maze often turn out to be effective in the treatment of human anxiety.

Tests of Sexual Behavior Most attempts to study the physiological bases of rat sexual behavior have focused on the copulatory act itself. The male mounts the female from behind and clasps her hindquarters. If the female is receptive, she responds by assuming the posture called **lordosis**; that is, she sticks her hindquarters in the air, she bends her back in a U, and she deflects her tail to the side. During some mounts, the male inserts his penis into the female's vagina; this act is called **intromission**. After intromission, the male dismounts by jumping backwards. He then returns a few seconds later to mount and intromit once again. Following about 10 such cycles of mounting, intromitting, and dismounting, the male mounts, intromits, and **ejaculates** (ejects his sperm).

Three common measures of male rat sexual behavior are the number of mounts required to achieve intromission, the number of intromissions required to achieve ejaculation, and the interval between ejaculation and the reinitiation of mounting. The most common measure of female rat sexual behavior is the **lordosis quotient** (the proportion of mounts that elicit lordosis).

Traditional Conditioning Paradigms

Learning paradigms play a major role in biopsychological research for three reasons. The first is that learning is a phenomenon of primary interest to psychologists. The second is that learning paradigms provide an effective technology for producing and controlling animal behavior. Because animals cannot follow instructions from the experimenter, it is often necessary to train them to behave in a fashion consistent with the goals of the experiment. The third reason is that it is possible to infer much about the sensory, motor, motivational, and cognitive state of an animal from its ability to learn and perform various responses.

If you have taken a previous course in psychology, you will likely be familiar with the Pavlovian and operant conditioning paradigms. In the **Pavlovian conditioning paradigm**, the experimenter pairs an initially neutral stimulus called a *conditional stimulus* (e.g., a tone or a light) with an *unconditional stimulus* (e.g., meat powder)—a stimulus that elicits an *unconditional* (reflexive) *response* (e.g., salivation). As a result of these pairings, the conditional stimulus eventually acquires the capacity, when administered alone, to elicit a *conditional response* (e.g., salivation)—a response that is often, but not always, similar to the unconditional response.

In the **operant conditioning paradigm**, the rate at which a particular voluntary response (such as a lever press) is emitted is increased by *reinforcement* or decreased by *punishment*. One of the most widely used operant conditioning paradigms in biopsychology is the self-stimulation paradigm. In the **self-stimulation paradigm**, animals press a lever to deliver electrical stimulation to particular sites in their own brains;

◉ Watch
Stimulus Generalization and
Stimulus Discrimination in
Operant Conditioning
www.mypsychlab.com

those structures in the brain that support self-stimulation are often called *pleasure centers*.

Seminatural Animal Learning Paradigms

In addition to Pavlovian and operant conditioning paradigms, biopsychologists use animal learning paradigms that have been specifically designed to mimic situations that an animal might encounter in its natural environment (see Gerlai & Clayton, 1999). Development of these paradigms stemmed in part from the reasonable assump-

tion that forms of learning tending to benefit an animal's survival in the wild are likely to be more highly developed and more directly related to innate neural mechanisms. The following are four common seminatural learning paradigms: the conditioned taste aversion, radial arm maze, Morris water maze, and conditioned defensive burying.

Conditioned Taste Aversion A **conditioned taste aversion** is the avoidance response that develops to tastes of food whose consumption has been followed by illness (see Garcia & Koelling, 1966). In the standard conditioned taste aversion experiment, rats receive an *emetic* (a nausea-inducing drug) after they consume a food with an unfamiliar taste. On the basis of this single conditioning trial, the rats learn to avoid the taste.

The ability of rats to readily learn the relationship between a particular taste and subsequent illness unquestionably increases their chances of survival in their natural environment, where potentially edible substances are not routinely screened by government agencies. Rats and many other animals are *neophobic* (afraid of new things); thus, when they first encounter a new food, they consume it in only small quantities. If they subsequently become ill, they will not consume it again. Conditioned aversions also develop to familiar tastes, but these typically require more than a single trial to be learned.

Humans also develop conditioned taste aversions. Cancer patients have been reported to develop aversions to foods consumed before nausea-inducing chemotherapy (Bernstein & Webster, 1980). Many of you will be able to testify on the basis of personal experience about the effectiveness of conditioned taste aversions. I still have vivid memories of a long-ago batch of red laboratory punch that I overzealously consumed after eating two pieces of blueberry pie. But that is another story—albeit a particularly colorful one.

The discovery of conditioned taste aversion challenged three widely accepted principles of learning (see Revusky & Garcia, 1970) that had grown out of research on traditional operant and Pavlovian conditioning paradigms. First, it challenged the view that animal conditioning is always a gradual step-by-step process; robust taste aversions can be established in only a single trial. Second, it showed that *temporal contiguity* is not essential for

conditioning; rats acquire taste aversions even when they do not become ill until several hours after eating. Third, it challenged the *principle of equipotentiality*—the view that conditioning proceeds in basically the same manner regardless of the particular stimuli and responses under investigation. Rats appear to have evolved to readily learn associations between tastes and illness; it is only with great difficulty that they learn relations between the color of food and nausea or between taste and footshock.

Radial Arm Maze The radial arm maze taps the well-developed spatial abilities of rodents. The survival of rats in the wild depends on their ability to navigate quickly and accurately through their environment and to learn which locations in it are likely to contain food and water. This task is much more complex for a rodent than it is for us. Most of us obtain food from locations where the supply is continually replenished; we go to the market confident that we will find enough food to satisfy our needs. In contrast, the foraging rat must learn, and retain, a complex pattern of spatially coded details. It must not only learn where morsels of food are likely to be found but must also remember which of these sites it has recently stripped of their booty so as not to revisit them too soon. Designed by Olton and Samuelson (1976) to study these spatial abilities, the **radial arm maze** (see Figure 5.26) is an array of arms—usually eight or more—radiating from a central starting area. At the end of each arm is a food cup, which may or may not be baited, depending on the purpose of the experiment.

In one version of the radial arm maze paradigm, rats are placed each day in a maze that has the same arms baited each day. After a few days of experience, rats rarely visit unbaited arms at all, and they rarely visit baited arms more than once in the same day—even when control procedures make it impossible for them to recognize odors left during previous visits to an arm or to make their visits in a systematic sequence. Because the arms are identical, rats must orient themselves in the maze with reference to external room

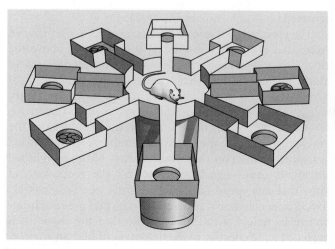

FIGURE 5.26 A radial arm maze.

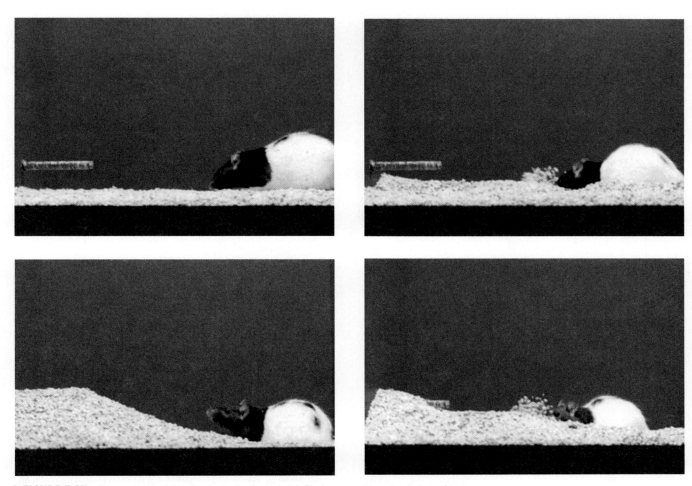

FIGURE 5.27 These photos (viewed clockwise from top left) show a rat burying a test object from which it has just received a single mild shock. (Photographs by Jack Wong.)

cues; thus, their performance can be disrupted by rotation of the maze or by changes in the appearance of the room.

Morris Water Maze Another seminatural learning paradigm that has been designed to study the spatial abilities of rats is the **Morris water maze** (Morris, 1981). The rats are placed in a circular, featureless pool of cool milky water, in which they must swim until they discover the escape platform—which is invisible just beneath the surface of the water. The rats are allowed to rest on the platform before being returned to the water for another trial. Despite the fact that the starting point is varied from trial to trial, the rats learn after only a few trials to swim directly to the platform, presumably by using spatial cues from the room as a reference. The Morris water maze is useful for assessing the navigational skills of brain-lesioned or drugged animals.

Conditioned Defensive Burying Yet another seminatural learning paradigm that is useful in biopsychological research is conditioned defensive burying (e.g., Pinel &

Mana, 1989; Pinel & Treit, 1978). In studies of **conditioned defensive burying**, rats receive a single aversive stimulus (e.g., a shock, air blast, or noxious odor) from an object mounted on the wall of the chamber just above the floor, which is littered with bedding material. After a single trial, almost every rat learns that the test object is a threat and responds by flinging bedding material at the test object with its head and forepaws (see Figure 5.27). Antianxiety drugs reduce the amount of conditioned defensive burying, and thus the paradigm is used to study the neurochemistry of anxiety (e.g., Treit, 1987).

Before moving on to the next chapter, you need to appreciate that to be effective these research methods must be used together. Seldom, if ever, is an important biopsychological issue resolved by use of a single method. The reason for this is that neither the methods used to manipulate the brain nor the methods used to assess the behavioral consequences of these manipulations are totally selective;

Thinking Creatively

there are no methods of manipulating the brain that change only a single aspect of brain function, and there are no measures of behavior that reflect only a single psychological process. Accordingly, lines of research that use a single method can usually be interpreted in more than one way and thus cannot provide unequivocal evidence for any one interpretation. Typically, important research questions are resolved only when several methods are brought to bear on a single problem. This general approach, as you learned in Chapter 1, is called *converging operations*.

Scan Your Brain

Scan your brain to see how well you remember the behavioral research methods of biopsychology. In each blank, write the name of a behavioral test or paradigm. The correct answers are provided at the end of the exercise. Before proceeding, review material related to your incorrect answers and omissions.

1. Many neuropsychological assessments begin with the _____.
2. The most common test of short-term memory is the _____.
3. The most common invasive test of language lateralization is the _____.
4. The most common tests of explicit memory are the _____.

5. A common test of frontal-lobe damage is the _____.
6. PET and fMRI studies almost always employ the _____.
7. A commonly used test of rat fearfulness is the _____.
8. Male rats' aggressive and defensive behavior is readily assessed by the _____.
9. The most commonly used test of anxiolytic drug effects is the _____.
10. The most common measure of the sexual receptivity of a female rat is the _____.
11. Animals press a lever to deliver stimulation of their own brains in the _____.
12. The spatial abilities of foraging rodents are often assessed with a _____.
13. The ability of a rat to find an invisible safety platform can be assessed in a _____.

Scan Your Brain answers: (1) WAIS, (2) digit-span test, (3) sodium amytal test, (4) repetition priming tests, (5) Wisconsin Card Sorting Test, (6) paired-image subtraction technique, (7) open-field test, (8) colony-intruder paradigm, (9) elevated plus maze, (10) lordosis quotient, (11) self-stimulation paradigm, (12) radial arm maze, (13) Morris water maze.

Themes Revisited

This chapter introduced you to the two kinds of research methods used by biopsychologists: methods of studying the brain and methods of studying behavior. In the descriptions of these methods, all four of the main themes of the book were apparent.

The chapter-opening case of Professor P. alerted you to the fact that many of the methods used by biopsychologists to study the human brain are also used clinically, in either diagnosis or treatment. The clinical implications theme came up again during discussions of brain imaging, genetic engineering, neuropsychological testing, and use of the elevated plus maze to test anxiolytic drugs.

Clinical Implications

The neuroplasticity theme arose during the discussion of the methods of cognitive neuroscience. Experience can produce

Neuroplasticity

changes in brain organization that can complicate the interpretation of functional brain images.

The evolutionary perspective theme arose in the discussion of green fluorescent protein, first isolated from jellyfish, and again during the discussion of the rationale for using seminatural animal learning paradigms, which assess animal behavior in environments similar to those in which it evolved.

Evolutionary Perspective

The thinking creatively theme came up several times. The development of new research methods often requires considerable creativity, and understanding the particular weaknesses and strengths of each research method is the foundation on which creative thinking rests.

Thinking Creatively

Think about It

1. The current rate of progress in the development of new and better brain-scanning devices will soon render behavioral tests of brain damage obsolete. Discuss.

2. You are taking a physiological psychology course, and your laboratory instructor gives you two rats: one rat with a lesion in an unknown brain structure and one normal rat. How would you test the rats to determine which one has the lesion? How would you determine the behavioral effects of the lesion? How would your approach differ from one that you might use to test a human patient suspected of having brain damage?

3. The search for the neural mechanisms of learning should focus on forms of learning necessary for survival in the wild. Discuss.

4. All patients should complete a battery of neuropsychological tests both before and after neurosurgery. Discuss.

5. The methods that biopsychologists use to study behavior are fundamentally different from the methods that they use to study the brain, and these fundamental differences lead to an under appreciation of behavioral methods by those who lack expertise in their use. Discuss.

6. Functional brain-imaging techniques are impressive and widely used, but they are far from perfect. Critically evaluate them.

Key Terms

PART ONE Methods of Studying the Nervous System

5.1 Methods of Visualizing and Stimulating the Living Human Brain

Contrast X-ray techniques (p. 103)
Cerebral angiography (p. 103)
Computed tomography (CT) (p. 103)
Magnetic resonance imaging (MRI) (p. 104)
Spatial resolution (p. 104)
Positron emission tomography (PET) (p. 104)
2-Deoxyglucose (2-DG) (p. 104)
Functional MRI (fMRI) (p. 105)
BOLD signal (p. 105)
Magnetoencephalography (MEG) (p. 106)
Temporal resolution (p. 106)
Transcranial magnetic stimulation (TMS) (p. 107)

5.2 Recording Human Psychophysiological Activity

Electroencephalography (p. 107)

Alpha waves (p. 107)
Event-related potentials (ERPs) (p. 107)
Sensory evoked potential (p. 107)
Signal averaging (p. 108)
P300 wave (p. 108)
Far-field potentials (p. 108)
Electromyography (p. 109)
Electrooculography (p. 109)
Skin conductance level (SCL) (p. 110)
Skin conductance response (SCR) (p. 110)
Electrocardiogram (ECG or EKG) (p. 110)
Hypertension (p. 110)
Plethysmography (p. 110)

5.3 Invasive Physiological Research Methods

Stereotaxic atlas (p. 111)
Bregma (p. 111)
Stereotaxic instrument (p. 111)
Aspiration (p. 111)
Cryogenic blockade (p. 112)

5.4 Pharmacological Research Methods

Cannula (p. 115)
Neurotoxins (p. 115)
Autoradiography (p. 115)
Cerebral dialysis (p. 115)

Immunocytochemistry (p. 116)
In situ hybridization (p. 116)

5.5 Genetic Engineering

Gene knockout techniques (p. 116)
Gene replacement techniques (p. 117)
Transgenic mice (p. 117)
Green fluorescent protein (GFP) (p. 117)
Brainbow (p. 118)

PART TWO Behavioral Research Methods of Biopsychology

Behavioral paradigm (p. 118)

5.6 Neuropsychological Testing

Wechsler Adult Intelligence Scale (WAIS) (p. 120)
Digit span (p. 120)
Token test (p. 120)
Sodium amytal test (p. 121)
Dichotic listening test (p. 121)
Repetition priming tests (p. 122)
Wisconsin Card Sorting Test (p. 122)

5.7 Behavioral Methods of Cognitive Neuroscience

Cognitive neuroscience (p. 123)
Constituent cognitive processes (p. 123)

Paired-image subtraction technique (p. 123)
Default mode (p. 123)
Default mode network (p. 123)

5.8 Biopsychological Paradigms of Animal Behavior

Species-common behaviors (p. 124)
Open-field test (p. 125)
Thigmotaxic (p. 125)
Colony-intruder paradigm (p. 125)
Elevated plus maze (p. 125)
Lordosis (p. 125)
Intromission (p. 125)
Ejaculate (p. 125)
Lordosis quotient (p. 125)
Pavlovian conditioning paradigm (p. 125)
Operant conditioning paradigm (p. 125)
Self-stimulation paradigm (p. 125)
Conditioned taste aversion (p. 126)
Radial arm maze (p. 126)
Morris water maze (p. 127)
Conditioned defensive burying (p. 127)

✔•─Quick Review Test your comprehension of the chapter with this brief practice test. You can find the answers to these questions as well as more practice tests, activities, and other study resources at www.mypsychlab.com.

1. A method of measuring the extracellular concentration of particular neurochemicals in the brain is
   a. cerebral dialysis.
   b. immunocytochemistry.
   c. extracellular unit recording.
   d. intracellular unit recording.
   e. the 2-deoxyglucose technique.

2. Mice that have had genetic material of another species (e.g., a pathological human gene) inserted into their genome are called
   a. knockout mice.
   b. transgenic mice.
   c. homozygous.
   d. heterozygous.
   e. both a and d

3. The most widely used test of short-term memory is the
   a. token test.
   b. WAIS.
   c. sodium amytal test.
   d. digit span test.
   e. repetition priming test.

4. The colony-intruder paradigm is commonly used to study
   a. natural environment.
   b. spatial perception.
   c. reproduction.
   d. epigenetics.
   e. aggressive and defensive behavior.

5. A seminatural animal learning paradigm that is often used to study spatial ability is
   a. self-stimulation paradigm.
   b. conditioned defensive burying paradigm.
   c. radial arm maze.
   d. Morris water maze.
   e. both c and d

6 The Visual System
How We See

This chapter is about your visual system. Most people think that their visual system has evolved to respond as accurately as possible to the patterns of light that enter their eyes. They recognize the obvious limitations in the accuracy of their visual system, of course; and they appreciate those curious instances, termed *visual illusions*, in which it is "tricked" into seeing things the way they aren't. But such shortcomings are generally regarded as minor imperfections in a system that responds as faithfully as possible to the external world.

But, despite the intuitive appeal of thinking about it in this way, this is not how the visual system works. The visual system does not produce an accurate internal copy of the external world. It does much more. From the tiny, distorted, upside-down, two-dimensional retinal images projected on the visual receptors that line the backs of the eyes, the visual system creates an accurate, richly detailed, three-dimensional perception that is—and this is the really important part—in some respects even better than the external reality from which it was created. My primary goal in this chapter is to help you appreciate the inherent creativity of your own visual system.

Thinking Creatively

You will learn in this chapter that understanding the visual system requires the integration of two types of research: (1) research that probes the visual system with sophisticated neuroanatomical, neurochemical, and neurophysiological techniques; and (2) research that focuses on the assessment of what we see. Both types of research receive substantial coverage in this chapter, but it is the second type that provides you with a unique educational opportunity: the opportunity to participate in the very research you are studying. Throughout this chapter, you will be encouraged to participate in a series of Check It Out demonstrations designed to illustrate the relevance of what you are learning in this text to life outside its pages.

This chapter is composed of six sections. The first three sections take you on a journey from the external visual world to the visual receptors of the retina and from there over the major visual pathway to the primary visual cortex. The next two sections describe how the neurons of this pathway mediate the perception of two particularly important features of the visual world: edges and color. The final section deals with the flow of visual signals from the primary visual cortex to other parts of cortex that participate in the complex process of vision.

Before you begin the first section of the chapter, I'd like you to consider an interesting clinical case. Have you ever wondered whether one person's subjective experiences are like those of others? This case provides evidence that at least some of them are. It was reported by Whitman Richards (1971), and his subject was his wife. Mrs. Richards suffered from migraine headaches, and like 20% of migraine sufferers, she often experienced visual displays, called *fortification illusions*, prior to her attacks

(see Dodick & Gargus, 2008; Pietrobon & Striessnig, 2003).

The Case of Mrs. Richards: Fortification Illusions and the Astronomer

Each fortification illusion began with a gray area of blindness near the center of her visual field—see Figure 6.1. During the next few minutes, the gray area would begin to expand into a horseshoe shape, with a zigzag pattern of flickering lines at its advancing edge (this pattern reminded people of the plans for a fortification, hence the name of the illusions).

It normally took about 20 minutes for the lines and the trailing area of blindness to reach the periphery of her visual field. At this point, her headache would usually begin.

Clinical Implications

Because the illusion expanded so slowly, Mrs. Richards was able to stare at a point on the center of a blank sheet of paper and periodically trace on the sheet the details of her illusion. This method made it apparent that the lines became thicker and the expansion of the area of blindness occurred faster as the illusion spread into the periphery.

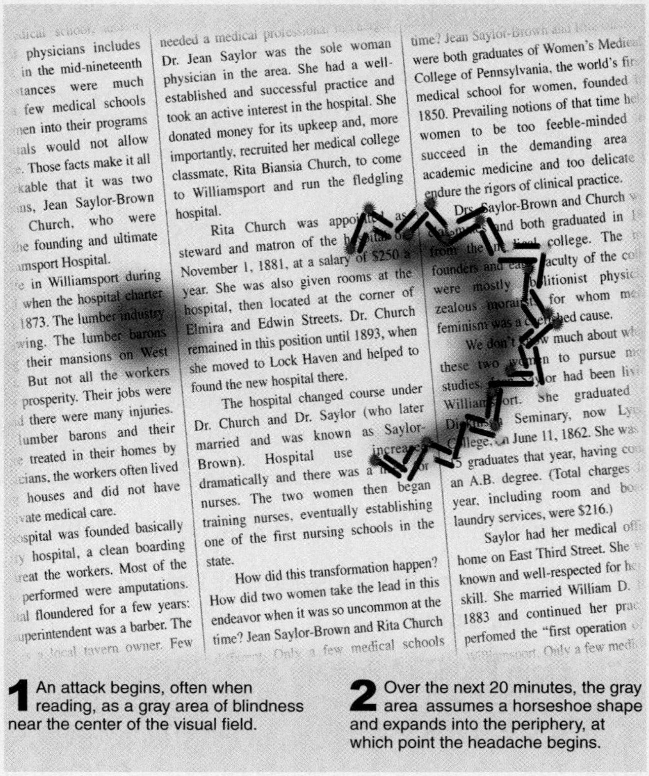

1 An attack begins, often when reading, as a gray area of blindness near the center of the visual field.

2 Over the next 20 minutes, the gray area assumes a horseshoe shape and expands into the periphery, at which point the headache begins.

FIGURE 6.1   The fortification illusions associated with migraine headaches.

Interestingly, Dr. Richards discovered that a similar set of drawings was published in 1870 by the famous British astronomer George Biddell Airy. They were virtually identical to those done by Mrs. Richards.

We will return to fortification illusions after you have learned a bit about the visual system. At that point, you will be able to appreciate the significance of their features.

6.1

Light Enters the Eye and Reaches the Retina

Everybody knows that cats, owls, and other nocturnal animals can see in the dark. Right? Wrong! Some animals have special adaptations that allow them to see under very dim illumination, but no animal can see in complete darkness. The light reflected into your eyes from the objects around you is the basis for your ability to see them; if there is no light, there is no vision.

You may recall from high-school physics that light can be thought of in two different ways: as discrete particles of energy, called photons, traveling through space at about 300,000 kilometers (186,000 miles) per second, or as waves of energy. Both theories are useful; in some ways light behaves like particles, and in others it behaves like waves. Physicists have learned to live with this nagging inconsistency, and we must do the same.

Light is sometimes defined as waves of electromagnetic energy that are between 380 and 760 *nanometers* (billionths of a meter) in length (see Figure 6.2). There is nothing

special about these wavelengths except that the human visual system responds to them. In fact, some animals can see wavelengths that we cannot (see Fernald, 2000). For example, rattlesnakes can see *infrared waves*, which are too long for humans to see; as a result, they can see warm-blooded prey in what for us would be complete darkness. So, if I were writing this book for rattlesnakes, I would be forced to provide a different definition of light for them.

Evolutionary Perspective

Wavelength and intensity are two properties of light that are of particular interest—wavelength because it plays an important role in the perception of color, and intensity because it plays an important role in the perception of brightness. In everyday language, the concepts of *wavelength* and *color* are often used interchangeably, and so are *intensity* and *brightness*. For example, we commonly refer to an intense light with a wavelength of 700 nanometers as being a bright red light (see Figure 6.2), when in fact it is our perception of the light, not the light itself, that is bright and red. I know that these distinctions may seem trivial to you now, but by the end of the chapter you will appreciate their importance.

The Pupil and the Lens

The amount of light reaching the *retinas* is regulated by the donut-shaped bands of contractile tissue, the *irises*, which give our eyes their characteristic color (see Figure 6.3 on page 134). Light enters the eye through the *pupil*, the hole in the iris. The adjustment of pupil size in response to changes in illumination represents a compromise between **sensitivity** (the ability to detect the presence of dimly lit objects) and **acuity** (the ability to see the details of objects). When the level of illumination is high and sensitivity is thus not important, the visual system takes advantage of the situation by constricting the pupils. When the pupils are constricted, the image falling on each retina is sharper and there is a greater *depth of focus*; that is, a greater range of depths are simultaneously kept in focus on the retinas. However, when the level of illumination is too low to adequately activate the receptors, the pupils dilate to let in more light, thereby sacrificing acuity and depth of focus.

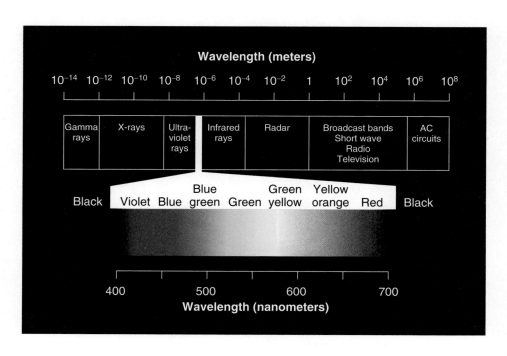

FIGURE 6.2 The electromagnetic spectrum and the colors associated with the wavelengths that are visible to humans.

FIGURE 6.3 The human eye. Light enters the eye through the pupil, whose size is regulated by the iris. The iris gives the eye its characteristic color—blue, brown, or other.

Behind each pupil is a *lens*, which focuses incoming light on the retina (see Figure 6.4). When we direct our gaze at something near, the tension on the ligaments holding each lens in place is adjusted by the **ciliary muscles**, and the lens assumes its natural cylindrical shape.

⊙→ Simulate The Structure of the Human Eye
www.mypsychlab.com

This increases the ability of the lens to *refract* (bend) light and thus brings close objects into sharp focus. When we focus on a distant object, the lens is flattened. The process of adjusting the configuration of the lenses to bring images into focus on the retina is called **accommodation**.

Eye Position and Binocular Disparity

No description of the eyes of vertebrates would be complete without a discussion of their most obvious feature: the fact that they come in pairs. One reason vertebrates have two eyes is that vertebrates have two sides: left and right. By having one eye on each side, which is by far the most common arrangement, vertebrates can see in almost every direction without moving their heads. But then why do some vertebrates, including humans, have their eyes mounted side by side on the front of their heads? This arrangement sacrifices the ability to see behind so that what is in front can be viewed through both eyes simultaneously—an arrangement that is an important basis for our visual system's ability to create three-dimensional perceptions (to see depth) from two-dimensional retinal images. Why do you think the two-eyes-on-the-front arrangement has evolved in some species but not in others? (The following Check It Out demonstration answers this question.)

Evolutionary Perspective

The movements of your eyes are coordinated so that each point in your visual world is projected to corresponding

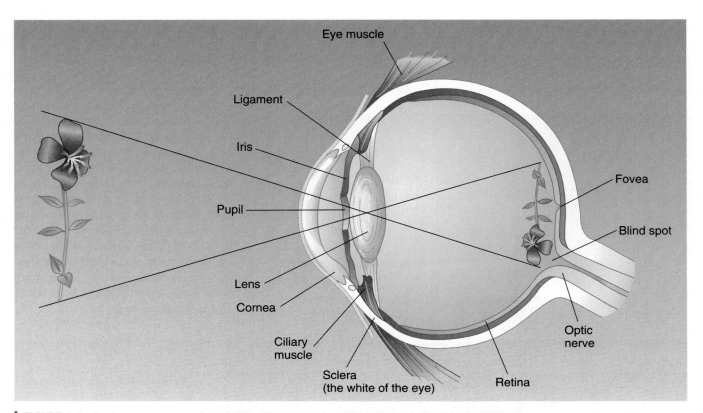

FIGURE 6.4 The human eye, a product of 600 million years of evolution (Lamb, Collin, & Pugh, 2007).

Check It Out

THE POSITION OF EYES

Here you see three animals whose eyes are on the front of their heads (a human, an owl, and a lion) and three whose eyes are on the sides of their heads (an antelope, a canary, and a squirrel). Why do a few vertebrate species have their eyes side by side on the front of the head while most species have one eye on each side?

In general, predators tend to have front facing eyes because this enables them to accurately perceive how far away prey animals are; prey animals tend to have side-facing eyes because this gives them a larger field of vision and the ability to see predators approaching from most directions.

points on your two retinas. To accomplish this, your eyes must *converge* (turn slightly inward); convergence is greatest when you are inspecting things that are close. But the positions of the images on your two retinas can never correspond exactly because your two eyes do not view the world from exactly the same position. **Binocular disparity**—

the difference in the position of the same image on the two retinas—is greater for close objects than for distant objects; therefore, your visual system can use the degree of binocular disparity to construct one three-dimensional perception from two two-dimensional retinal images (see Parker, 2007). (Look at the Check It Out demonstration below.)

Check It Out

BINOCULAR DISPARITY AND THE MYSTERIOUS COCKTAIL SAUSAGE

If you compare the views from each eye (by quickly closing one eye and then the other) of objects at various distances in front of you—for example, your finger held at different distances—you will notice that the disparity between the two views is greater for closer objects. Now try the mysterious demonstration of the cocktail sausage. Face the farthest wall in the room (or some other distant object) and bring the tips or your two pointing fingers together at

arm's length in front of you—with the backs of your fingers away from you, unless you prefer sausages with fingernails. Now, with both eyes open, look through the notch between your touching fingertips, but focus on the wall. Do you see the cocktail sausage between your fingertips? Where did it come from? To prove to yourself that the sausage is a product of binocularity, make it disappear by shutting one eye. Warning: Do not eat this sausage.

6.2

The Retina and Translation of Light into Neural Signals

After light passes through the pupil and the lens, it reaches the retina. The retina converts light to neural signals, conducts them toward the CNS, and participates in the processing of the signals (Field & Chichilnisky, 2007; Werblin & Roska, 2007).

Simulate Can You Spot the Mistake? www.mypsychlab.com

Figure 6.5 illustrates the fundamental cellular structure of the retina. The retina is composed of five layers of different types of neurons: **receptors, horizontal cells, bipolar cells, amacrine cells,** and **retinal ganglion cells.** Each of these five types of retinal neurons comes in a variety of subtypes: Over 50 different kinds of retinal neurons have been identified (Dacey, 2004; Masland, 2001; Wässle, 2004). Notice that the amacrine cells and the horizontal cells are specialized for *lateral communication* (communication across the major channels of sensory input). Retinal neurons communicate both chemically via synapses and electrically via gap junctions (Bloomfield & Völgyi, 2009).

Also notice in Figure 6.5 that the retina is in a sense inside-out: Light reaches the receptor layer only after passing through the other four layers. Then, once the receptors have been activated, the neural message is transmitted back out through the retinal layers to the retinal ganglion cells, whose axons project across the inside of the retina before gathering together in a bundle and exiting the eyeball. This inside-out arrangement creates two visual problems: One is that the incoming light is distorted by the retinal tissue through which it must pass before reaching the receptors. The other is that for the bundle of retinal ganglion cell axons to leave the eye, there must be a gap in the receptor layer; this gap is called the **blind spot.**

The first of these two problems is minimized by the fovea (see Figure 6.6). The **fovea** is an indentation, about 0.33 centimeter in diameter, at the center of the retina; it is the area of the retina that is specialized for high-acuity vision (for seeing fine details). The thinning of the retinal ganglion cell layer at the fovea reduces the distortion of incoming light. The blind spot, the second of the two visual problems created by the inside-out structure of the

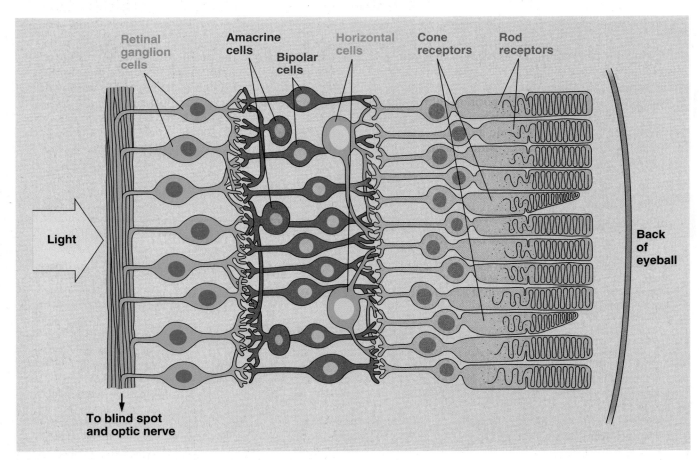

FIGURE 6.5 The cellular structure of the mammalian retina.

It is a mistake to think that completion is merely a response to blind spots. Indeed, completion is a fundamental visual system function (see Komatsu, 2006; Lleras & Moore, 2006). When you look at an object, your visual system does not conduct an image of that object from your retina to your cortex. Instead, it extracts key information about the object—primarily information about its edges and their location—and conducts that information to the cortex, where a perception of the entire object is created from that partial information. For example, the color and brightness of large unpatterned surfaces are not perceived directly but are filled in (completed) by a completion process called **surface interpolation** (the process by which we perceive surfaces; the visual system extracts information about edges and from it infers the appearance of large surfaces). The central role of surface interpolation in vision is an extremely important but counterintuitive concept. I suggest that you read this paragraph again and think about it. Are your creative thinking skills developed enough to feel comfortable with this new way of thinking about your own visual system?

> ◉→ **Simulate** Demonstration of Surface Interpolation **www.mypsychlab.com**

Thinking Creatively

Cone and Rod Vision

You likely noticed in Figure 6.5 that there are two different types of receptors in the human retina: cone-shaped receptors called **cones**, and rod-shaped receptors called **rods** (see Figure 6.7). The existence of these two types of receptors puzzled researchers until 1866, when it was first noticed that species active only in the day tend to have cone-only retinas and species active only at night tend to have rod-only retinas.

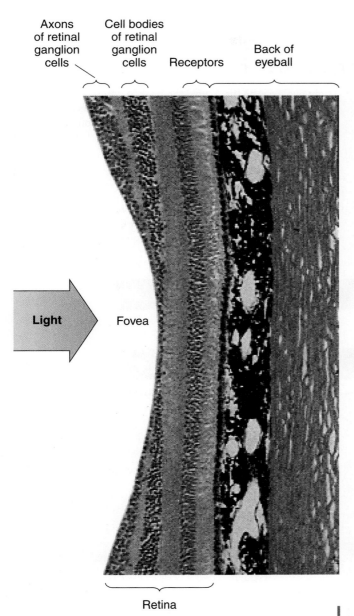

Axons of retinal ganglion cells   Cell bodies of retinal ganglion cells   Receptors   Back of eyeball

Light   Fovea

Retina

FIGURE 6.6   A section of the retina. The fovea is the indentation at the center of the retina; it is specialized for high-acuity vision.

retina, requires a more creative solution—which is illustrated in the Check It Out demonstration on page 138.

In the Check It Out demonstration, you will experience **completion** (or *filling in*). The visual system uses information provided by the receptors around the blind spot to fill in the gaps in your retinal images. When the visual system detects a straight bar going into one side of the blind spot and another straight bar leaving the other side, it fills in the missing bit for you; and what you see is a continuous straight bar, regardless of what is actually there. The completion phenomenon is one of the most compelling demonstrations that the visual system does much more than create a faithful copy of the external world.

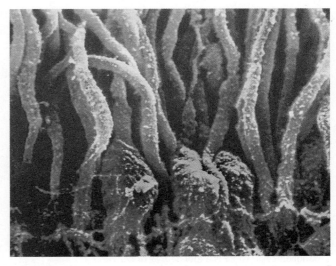

FIGURE 6.7   Cones and rods. The smaller, conical cells are cones; the larger, cylindrical cells are rods.

Check It Out

YOUR BLIND SPOT AND COMPLETION

First, prove to yourself that you do have areas of blindness that correspond to your retinal blind spots. Close your left eye and stare directly at the A below, trying as hard as you can to not shift your gaze. While keeping the gaze of your

right eye fixed on the A, hold the book at different distances from you until the black dot to the right of the A becomes focused on your blind spot and disappears at about 20 centimeters (8 inches).

If each eye has a blind spot, why is there not a black hole in your perception of the world when you look at it with one eye? You will discover the answer by focusing on B with your right eye while holding the book at the same

distance as before. Suddenly, the broken line to the right of B will become whole. Now focus on C at the same distance with your right eye. What do you see?

From this observation emerged the **duplexity theory of vision**—the theory that cones and rods mediate different kinds of vision. **Photopic vision** (cone-mediated vision) predominates in good lighting and provides high-acuity (finely detailed) colored perceptions of the world. In dim illumination, there is not enough light to reliably excite the cones, and the more sensitive **scotopic vision** (rod-mediated vision) predominates. However, the sensitivity of scotopic vision is not achieved without cost: Scotopic vision lacks both the detail and the color of photopic vision.

The differences between photopic and scotopic vision result in part from a difference in the way the two systems are "wired." As Figure 6.8 illustrates, there is a large difference in convergence between the two systems. In the scotopic system,

the output of several hundred rods converge on a single retinal ganglion cell, whereas in the photopic system, only a few cones converge on each retinal ganglion cell to

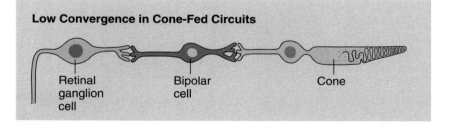

FIGURE 6.8 A schematic representation of the convergence of cones and rods on retinal ganglion cells. There is a low degree of convergence in cone-fed pathways and a high degree of convergence in rod-fed pathways.

receive input from only a few cones. As a result, the effects of dim light simultaneously stimulating many rods can summate (add) to influence the firing of the retinal ganglion cell onto which the output of the stimulated rods converges, whereas the effects of the same dim light applied to a sheet of cones cannot summate to the same degree, and the retinal ganglion cells may not respond at all to the light.

The convergent scotopic system pays for its high degree of sensitivity with a low level of acuity. When a retinal ganglion cell that receives input from hundreds of rods changes its firing, the brain has no way of knowing which portion of the rods contributed to the change. Although a more intense light is required to change the firing of a retinal ganglion cell that receives signals from cones, when such a retinal ganglion cell does react, there is less ambiguity about the location of the stimulus that triggered the reaction.

Cones and rods differ in their distribution on the retina. As Figure 6.9 illustrates, there are no rods at all in the fovea, only cones. At the boundaries of the foveal indentation, the proportion of cones declines markedly, and there is an increase in the number of rods. The density of rods reaches a maximum at 20° from the center of the fovea. Notice that there are more rods in the *nasal hemiretina* (the half of each retina next to the nose) than in the *temporal hemiretina* (the half next to the temples).

Spectral Sensitivity

Generally speaking, more intense lights appear brighter. However, wavelength also has a substantial effect on the perception of brightness. Because our visual systems are not equally sensitive to all wavelengths in the visible spectrum, lights of the same intensity but of different wavelengths can differ markedly in brightness. A graph of the relative brightness of lights of the same intensity presented at different wavelengths is called a *spectral sensitivity curve*.

By far the most important thing to remember about spectral sensitivity curves is that humans and other animals with both cones and rods have two of them: a **photopic spectral sensitivity curve** and a **scotopic spectral sensitivity curve**. The photopic spectral sensitivity of humans can be determined by having subjects judge the relative brightness of different wavelengths of light shone on the fovea. Their scotopic spectral sensitivity can be determined by asking subjects to judge the relative brightness of different wavelengths of light shone on the periphery of the retina at an intensity too low to activate the few peripheral cones that are located there.

The photopic and scotopic spectral sensitivity curves of human subjects are plotted in Figure 6.10. Notice that under photopic conditions, the visual system is maximally sensitive to wavelengths of about 560 nanometers; thus, under photopic conditions, a light at 500 nanometers would have to be much more intense than one at 560 nanometers to be seen as equally bright. In contrast, under scotopic conditions, the visual system is maximally sensitive to wavelengths of about 500 nanometers; thus, under scotopic conditions, a light of 560 nanometers would have to be much more intense than one at 500 nanometers to be seen as equally bright.

Because of the difference in photopic and scotopic spectral sensitivity, an interesting visual effect can be observed during the transition from photopic to scotopic vision. In 1825, Jan Purkinje described the following occurrence, which has become known as the **Purkinje effect** (pronounced "pur-KIN-jee"). One evening, just before dusk, while Purkinje was walking in his garden, he noticed how bright most of his yellow and red flowers appeared in relation to his blue ones. What amazed him was that just a

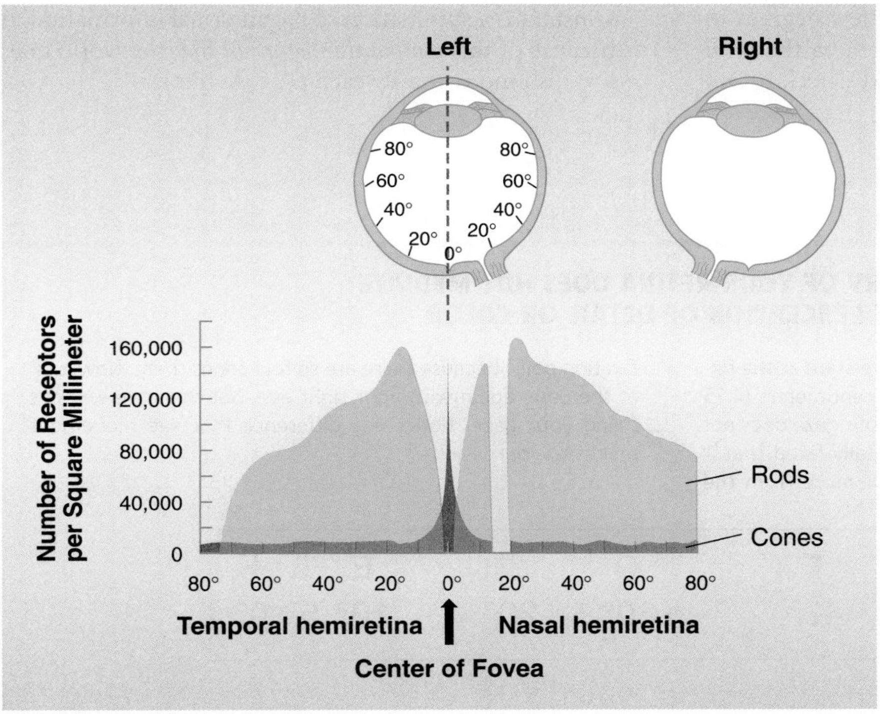

FIGURE 6.9 The distribution of cones and rods over the human retina. The figure illustrates the number of cones and rods per square millimeter as a function of distance from the center of the fovea. (Based on Lindsay & Norman, 1977.)

FIGURE 6.10 Human photopic (cone) and scotopic (rod) spectral sensitivity curves. The peak of each curve has been arbitrarily set at 100%.

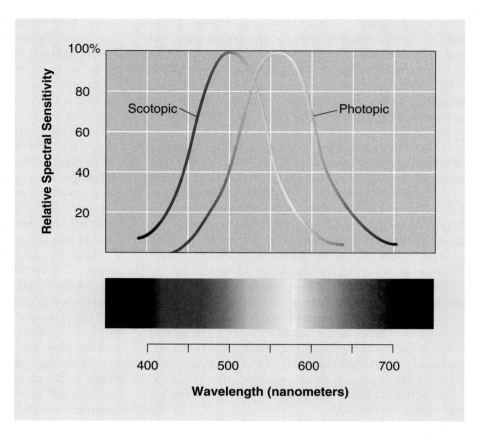

few minutes later the relative brightness of his flowers had somehow been reversed; the entire scene, when viewed at night, appeared *Thinking Creatively* completely in shades of gray, but most of the blue flowers appeared as brighter grays than did. the yellow and red ones. Can you explain this shift in relative brightness by referring to the photopic and scotopic spectral sensitivity curves in Figure 6.10?

Eye Movement

If cones are in fact responsible for mediating high-acuity color vision under photopic conditions, how can they accomplish their task when most of them are crammed into the fovea? Look around you. What you see is not a few colored details at the center of a grayish scene. You seem to see an expansive, richly detailed, lavishly colored world. How can such a perception be the product of a photopic system that, for the most part, is restricted to a few degrees in the center of your *visual field* (the entire area that you can see at a particular moment)? The next Check It Out

demonstration provides a clue. It shows that what we see is determined not just by what is projected on the retina at that instant. Although we are not aware of it, the eyes continually scan the visual field, and our visual perception at any instant is a summation of recent visual information. It is because of this *temporal integration* that the world does not vanish momentarily each time we blink.

Check It Out

PERIPHERY OF YOUR RETINA DOES NOT MEDIATE THE PERCEPTION OF DETAIL OR COLOR

Close your left eye, and with your right eye stare at the fixation point (+) at a distance of about 12 centimeters (4.75 inches) from the page. Be very careful that your gaze does not shift. You will notice when your gaze is totally fixed that it is difficult to see detail and color at 20° or more from the

fixation point because there are so few cones there. Now look at the page again with your right eye, but this time without fixing your gaze. Notice the difference that eye movement makes to your vision.

W	F	D	M	E	A	**+**
50°	40°	30°	20°	10°	5°	0°

Even when we fix our gaze on an object, our eyes continuously move. These involuntary **fixational eye movements** are of three kinds: tremor, drifts, and **saccades** (small jerky movements, or flicks; pronounced "sah-KAHDS"). Although we are normally unaware of fixational eye movements, they have a critical visual function (Martinez et al., 2005; Trommershäuser, Glimcher, & Gegenfurtner, 2009). We must fix our gaze to perceive the minute details of our world, but, ironically, if we were to fixate perfectly, our world would fade and disappear. This would happen because visual neurons respond to change; if retinal images are artificially stabilized (kept from moving on the retina), the images start to disappear and reappear. Thus, fixational eye movements enable us to see during fixation by keeping the images moving on the retina.

Visual Transduction: The Conversion of Light to Neural Signals

Transduction is the conversion of one form of energy to another. *Visual transduction* is the conversion of light to neural signals by the visual receptors. A breakthrough in the study of visual transduction came in 1876, when a red *pigment* (a pigment is any substance that absorbs light) was extracted from rods. This pigment had a curious property. When the pigment—which became known as **rhodopsin**—was exposed to continuous intense light, it was *bleached* (lost its color), and it lost its ability to absorb light; but when it was returned to the dark, it regained both its redness and its light-absorbing capacity.

It is now clear that rhodopsin's absorption of light (and the accompanying bleaching) is the first step in rod-mediated vision. Evidence comes from demonstrations that the degree to which rhodopsin absorbs light in various situations predicts how humans see under the very same conditions. For example, it has been shown that the degree to which rhodopsin absorbs lights of different wavelengths is related to the ability of humans and other animals with rods to detect the presence of different wavelengths of light under scotopic conditions. Figure 6.11 illustrates the relationship between the **absorption spectrum** of rhodopsin and the human scotopic spectral sensitivity curve. The goodness of the fit leaves little doubt that, in dim light, our sensitivity to various wavelengths is a direct consequence of rhodopsin's ability to absorb them.

Rhodopsin is a G-protein–coupled receptor that responds to light rather than to neurotransmitter molecules (see Koutalos & Yau, 1993; Molday & Hsu, 1995). Rhodopsin receptors, like other G-protein–coupled receptors, initiate a cascade of intracellular chemical events when they are activated (see Figure 6.12 on page 142). When rods are in darkness, their sodium channels are partially open, thus keeping the rods slightly depolarized and allowing a steady flow of excitatory glutamate neurotransmitter molecules to emanate from them. However, when rhodopsin receptors are bleached by light, the resulting cascade of intracellular chemical events

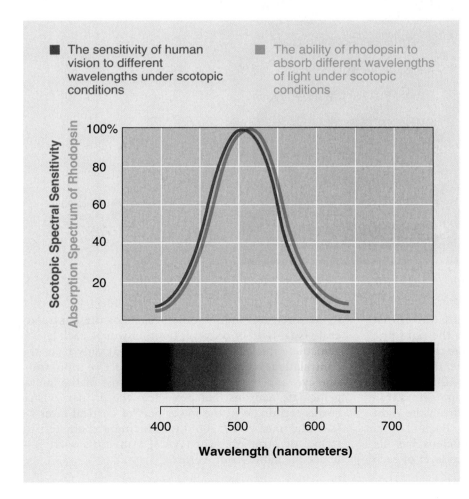

■ The sensitivity of human vision to different wavelengths under scotopic conditions

■ The ability of rhodopsin to absorb different wavelengths of light under scotopic conditions

FIGURE 6.11 The absorption spectrum of rhodopsin compared with the human scotopic spectral sensitivity curve.

FIGURE 6.12 The inhibitory response of rods to light. When light bleaches rhodopsin molecules, the rods' sodium channels close; as a result, the rods become hyperpolarized and release less glutamate.

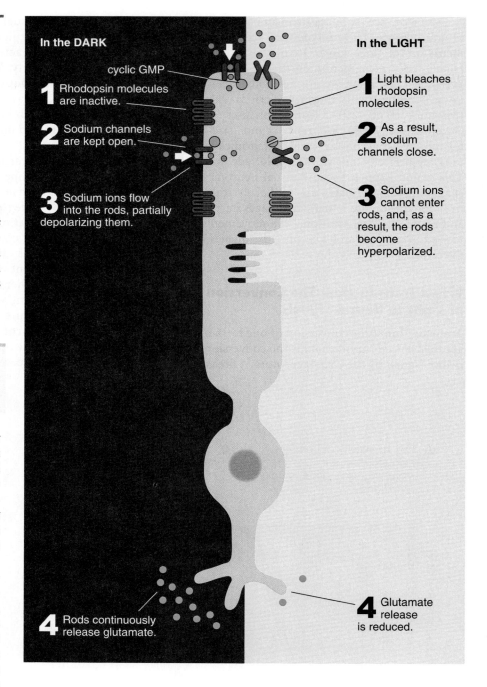

In the DARK

cyclic GMP

1 Rhodopsin molecules are inactive.

2 Sodium channels are kept open.

3 Sodium ions flow into the rods, partially depolarizing them.

4 Rods continuously release glutamate.

In the LIGHT

1 Light bleaches rhodopsin molecules.

2 As a result, sodium channels close.

3 Sodium ions cannot enter rods, and, as a result, the rods become hyperpolarized.

4 Glutamate release is reduced.

closes the sodium channels, hyperpolarizes the rods, and reduces the release of glutamate. The transduction of light by rods exemplifies an important point: Signals are often transmitted through neural systems by inhibition.

6.3

From Retina to Primary Visual Cortex

Many pathways in the brain carry visual information. By far the largest and most thoroughly studied visual pathways are the **retina-geniculate-striate pathways**, which conduct signals from each retina to the **primary visual cortex**, or striate cortex, via the **lateral geniculate nuclei** of the thalamus.

About 90% of axons of retinal ganglion cells become part of the retina-geniculate-striate pathways (see Tong, 2003). No other sensory system has such a predominant pair (left and right) of pathways to the cortex. The organization of these visual pathways is illustrated in Figure 6.13. Examine it carefully.

The main idea to take away from Figure 6.13 is that all signals from the left visual field reach the right primary visual cortex, either ipsilaterally from the *temporal hemiretina* of the right eye or contralaterally (via the *optic chiasm*) from the *nasal hemiretina* of the left eye—

Simulate The Primary Visual Pathways from Retina to Visual Cortex
www.mypsychlab.com

and that the opposite is true of all signals from the right visual field. Each lateral geniculate nucleus has six layers, and each layer of each

nucleus receives input from all parts of the contralateral visual field of one eye. In other words, each lateral geniculate nucleus receives visual input only from the contralateral visual field; three layers receive input from one eye, and three from the other. Most of the lateral geniculate neurons that project to the primary visual cortex terminate in the lower part of cortical layer IV (see Martinez et al., 2005), producing a characteristic stripe, or striation, when viewed in cross section—hence the name *striate cortex*.

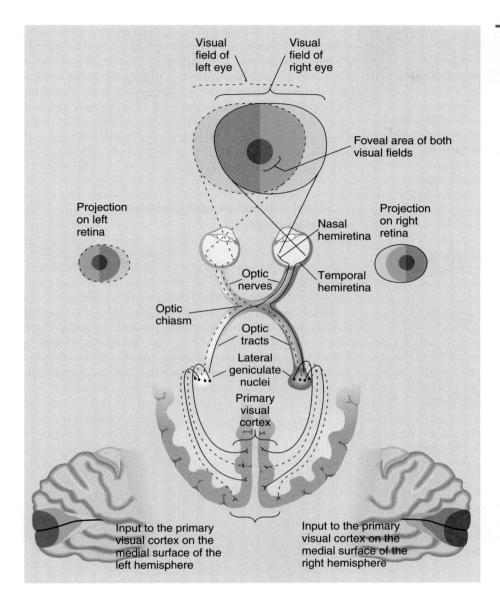

Visual field of left eye
Visual field of right eye

Foveal area of both visual fields

Projection on left retina

Nasal hemiretina

Projection on right retina

Optic nerves

Temporal hemiretina

Optic chiasm

Optic tracts

Lateral geniculate nuclei

Primary visual cortex

Input to the primary visual cortex on the medial surface of the left hemisphere

Input to the primary visual cortex on the medial surface of the right hemisphere

FIGURE 6.13 The retina-geniculate-striate system: the neural projections from the retinas through the lateral geniculate nuclei to the left and right primary visual cortex (striate cortex). The colors indicate the flow of information from various parts of the receptive fields of each eye to various parts of the visual system. (Based on Netter, 1962.)

an array of electrodes in the primary visual cortex of patients who were blind because of damage to their eyes. If electrical current was administered simultaneously through an array of electrodes forming a shape, such as a cross, on the surface of a patient's cortex, the patient reported "seeing" a glowing image of that shape; see also Merabet and colleagues (2005).

Clinical Implications

The M and P Channels

Not apparent in Figure 6.13 is the fact that at least two parallel channels of communication flow through each lateral geniculate nucleus (see Nassi, Lyon, & Callaway, 2006). One channel runs through the top four layers. These layers are called the **parvocellular layers** (or *P layers*) because they are composed of neurons with small cell bodies (*parvo* means "small"). The other channel runs through the bottom two layers, which are called the **magnocellular layers** (or *M layers*) because they are composed of neurons with large cell bodies (*magno* means "large").

The parvocellular neurons are particularly responsive to color, to fine pattern details, and to stationary or slowly moving objects. In contrast, the magnocellular neurons are particularly responsive to movement. Cones provide the majority of the input to the P layers, whereas rods provide the majority of the input to the M layers.

The parvocellular and magnocellular neurons project to different sites in the lower part of layer IV of the striate cortex. In turn, these M and P portions of lower layer IV project to different parts of visual cortex (Levitt, 2001; Yabuta, Sawatari, & Callaway, 2001).

Retinotopic Organization

The retina-geniculate-striate system is **retinotopic**; each level of the system is organized like a map of the retina. This means that two stimuli presented to adjacent areas of the retina excite adjacent neurons at all levels of the system. The retinotopic layout of the primary visual cortex has a disproportionate representation of the fovea; although the fovea is only a small part of the retina, a relatively large proportion of the primary visual cortex (about 25%) is dedicated to the analysis of its input.

A dramatic demonstration of the retinotopic organization of the primary visual cortex was provided by Dobelle, Mladejovsky, and Girvin (1974). They implanted

Scan your Brain answers: (1) retinal ganglion, (2) blind spot, (3) fovea, (4) photopic, (5) optic chiasm, (6) rhodopsin, (7) retinotopically, (8) absorption, (9) convergence.

6.4
Seeing Edges

Edge perception (seeing edges) does not sound like a particularly important topic, but it is. Edges are the most informative features of any visual display because they *Evolutionary Perspective* define the extent and position of the various objects in it. Given the importance of perceiving visual edges and the unrelenting pressure of natural selection, it is not surprising that the visual systems of many species are particularly good at edge perception.

Before considering the visual mechanisms underlying edge perception, it is important to appreciate exactly what a visual edge is. In a sense, a visual edge is nothing: It is merely the place where two different areas of a visual image meet. Accordingly, the perception of an edge is really the perception of a contrast between two adjacent areas of the visual field. This section of the chapter reviews the perception of edges (the perception of contrast) between areas that differ from one another in brightness (i.e., show brightness contrast).

Lateral Inhibition and Contrast Enhancement

Carefully examine the stripes in Figure 6.14. The intensity graph in the figure indicates what is there—a series of homogeneous stripes of different intensity. But this is not exactly what you see, is it? What you see is indicated in the brightness graph. Adjacent to each edge, the brighter stripe looks brighter than it really is and the darker stripe looks darker than it really is. The nonexistent stripes of brightness and darkness running adjacent to the edges are called *Mach bands*; they enhance the contrast at each edge and make the edge easier to see.

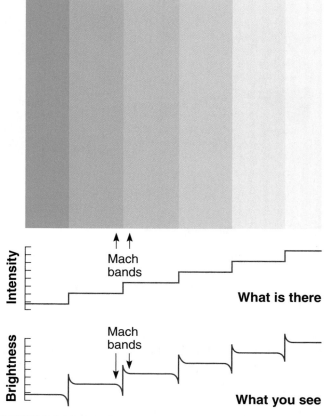

FIGURE 6.14 The illusory bands visible in this figure are often called Mach bands, although Mach used a different figure to generate them in his studies (see Eagleman, 2001).

It is important to appreciate that **contrast enhancement** is not something that occurs just in books. Although

we are normally unaware of it, every edge we look at is highlighted for us by the contrast-enhancing mechanisms of our nervous systems. In effect, our perception of edges is better than the real thing (as determined by measurements of the physical properties of the light entering our eyes).

The classic studies of the physiological basis of contrast enhancement were conducted on the eyes of an unlikely subject: the *horseshoe crab* (e.g., Ratliff, 1972). The lateral

Evolutionary Perspective eyes of the horseshoe crab are ideal for certain types of neurophysiological research. Unlike mammalian eyes, they are composed of very large receptors, called *ommatidia*, each with its own large axon. The axons of the ommatidia are interconnected by a lateral neural network.

In order to understand the physiological basis of contrast enhancement in the horseshoe crab, you must know two things. The first is that if a single ommatidium is illuminated, it fires at a rate that is proportional to the intensity of the light striking it; more intense lights produce more firing. The second is that when a receptor fires, it inhibits its neighbors via the lateral neural network; this inhibition is called **lateral inhibition** because it spreads laterally across the array of receptors. The amount of lateral inhibition produced by a receptor is greatest when the

receptor is most intensely illuminated, and the inhibition has its greatest effect on the receptor's immediate neighbors.

The neural basis of contrast enhancement can be understood in terms of the firing rates of the receptors on each side of an edge, as indicated in Figure 6.15. Notice that the receptor adjacent to the edge on the more intense side (receptor D) fires more than the other intensely illuminated receptors (A, B, C), while the receptor adjacent to the edge on the less well-illuminated side (receptor E) fires less than the other receptors on that side (F, G, H). Lateral inhibition accounts for these differences. Receptors A, B, and C all fire at the same rate, because they are all receiving the same high level of stimulation and the same high degree of lateral inhibition from all their highly stimulated neighbors. Receptor D fires more than A, B, and C, because it receives as much stimulation as they do but less inhibition from its neighbors, many of which are on the dimmer side of the edge. Now consider the receptors on the dimmer side. Receptors F, G, and H fire at the same rate, because they are all being stimulated by the same low level of light and receiving the same low level of inhibition from their neighbors. However, receptor E fires even less, because it is receiving the same excitation but more inhibition from its neighbors, many of which are on the more intense side of the edge. Now that you understand the neural basis of contrast enhancement, take another look at Figure 6.14. Also, if you are still having a hard time believing that Mach bands are created by your own visual system, look at the following Check It Out demonstration on page 146.

Receptive Fields of Visual Neurons

The Nobel Prize–winning research of David Hubel and Torsten Wiesel (see Hubel & Wiesel, 2004) is the fitting climax to this discussion of brightness contrast. Their research has revealed much about the neural mechanisms of vision, and their method has been adopted by subsequent generations of sensory neurophysiologists.

Hubel and Wiesel's influential method is a technique for studying single neurons in the visual systems of laboratory animals—their research subjects were cats and monkeys. First, the tip of a microelectrode is

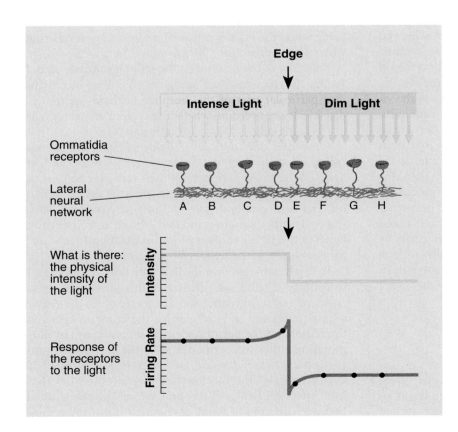

FIGURE 6.15 How lateral inhibition produces contrast enhancement. (Based on Ratliff, 1972.)

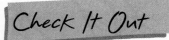

CONTRAST ENHANCEMENT AND MACH BANDS

The Mach band demonstration is so compelling that you may be confused by it. You may think that the Mach bands below have been created by the printers of the book, rather than by your own visual system. To prove to yourself that the Mach bands are a creation of your visual system, view each stripe individually by covering the adjacent ones with two pieces of paper. You will see at once that each stripe is completely homogeneous. Then, take the paper away, and the Mach bands will suddenly reappear along the edges of the stripe.

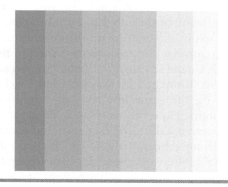

positioned near a single neuron in the part of the visual system that is under investigation. During testing, eye movements are blocked by paralyzing the eye muscles, and the images on a screen in front of the subject are focused sharply on the retina by an adjustable lens. The next step in the procedure is to identify the receptive field of the neuron. The **receptive field** of a visual neuron is the area of the visual field within which it is possible for a visual stimulus to influence the firing of that neuron. The final step in the method is to record the responses of the neuron to various stimuli within its receptive field in order to characterize the types of stimuli that most influence its activity. Then, the electrode is advanced slightly, and the entire process of identifying and characterizing the receptive field properties is repeated for another neuron, and then for another, and another, and so on. The general strategy is to begin by studying neurons near the receptors and gradually working up through "higher" and "higher" levels of the system in an effort to understand the increasing complexity of the neural responses at each level.

Receptive Fields: Neurons of the Retina–Geniculate–Striate System

Hubel and Wiesel (1979) began their studies of visual system neurons by recording from the three levels of the retina-geniculate-striate system: first from retinal

ganglion cells, then from lateral geniculate neurons, and finally from the striate neurons of lower layer IV, the terminus of the system. They found little change in the receptive fields as they worked through the levels.

When Hubel and Wiesel compared the receptive fields recorded from retinal ganglion cells, lateral geniculate nuclei, and lower layer IV neurons, four commonalties were readily apparent:

- At each level, the receptive fields in the foveal area of the retina were smaller than those at the periphery; this is consistent with the fact that the fovea mediates fine-grained (high-acuity) vision.
- All the neurons (retinal ganglion cells, lateral geniculate neurons, and lower layer IV neurons) had receptive fields that were circular.
- All the neurons were **monocular**; that is, each neuron had a receptive field in one eye but not the other.
- Many neurons at each of the three levels of the retina-geniculate-striate system had receptive fields that comprised an excitatory area and an inhibitory area separated by a circular boundary.

Let me explain this last point—it is important. When Hubel and Wiesel shone a spot of white light onto the various parts of the receptive fields of a neuron in the retina-geniculate-striate pathway, they discovered two different responses. The neuron responded with either "on" firing or "off" firing, depending on the location of the spot of light in the receptive field. That is, the neuron either displayed a burst of firing when the light was turned on ("*on*" *firing*), or it displayed an inhibition of firing when the light was turned on and a burst of firing when it was turned off ("*off*" *firing*).

For most of the neurons in the retina-geniculate-striate system, the reaction—"on" firing or "off" firing—to a light in a particular part of the receptive field was quite predictable. It depended on whether they were on-center cells or off-center cells, as illustrated in Figure 6.16.

On-center cells respond to lights shone in the central region of their receptive fields with "on" firing and to lights shone in the periphery of their receptive fields with inhibition, followed by "off" firing when the light is turned off. **Off-center cells** display the opposite pattern: They respond with inhibition and "off" firing in response to lights in the center of their receptive fields and with "on" firing to lights in the periphery of their receptive fields.

In effect, on-center and off-center cells respond best to contrast. Figure 6.17 illustrates this point. The most effective way to influence the firing rate of an on-center or off-center cell is to maximize the contrast between the center and the periphery of its receptive field by illuminating either the entire center or the entire surround (periphery), while leaving the other region completely dark. Diffusely illuminating the entire receptive field has little effect on firing. Hubel and Wiesel thus concluded that one function of many of the neurons in the retina-

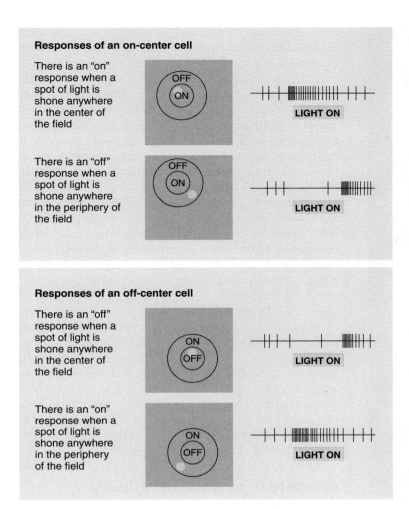

Responses of an on-center cell

There is an "on" response when a spot of light is shone anywhere in the center of the field

There is an "off" response when a spot of light is shone anywhere in the periphery of the field

Responses of an off-center cell

There is an "off" response when a spot of light is shone anywhere in the center of the field

There is an "on" response when a spot of light is shone anywhere in the periphery of the field

FIGURE 6.16 The receptive fields of an on-center cell and an off-center cell.

Receptive Fields: Simple Cortical Cells

The striate cortex neurons that you just read about—that is, the neurons of lower layer IV—are exceptions. Their receptive fields are unlike those of the vast majority of striate neurons. The receptive fields of most primary visual cortex neurons fall into one of two classes: simple or complex. Neither of these classes includes the neurons of lower layer IV.

Simple cells, like lower layer IV neurons, have receptive fields that can be divided into antagonistic "on" and "off" regions and are thus unresponsive to diffuse light. And like lower layer IV neurons, they are all monocular. The main difference is that the borders between the "on" and "off" regions of the cortical receptive fields of simple cells are straight lines rather than circles. Several examples of receptive fields of simple cortical cells are presented in Figure 6.18 on page 148. Notice that simple cells respond best to bars of light in a dark field, dark bars in a light field, or single straight edges between dark and light areas;

geniculate-striate system is to respond to the degree of brightness contrast between the two areas of their receptive fields (see Livingstone & Hubel, 1988).

Before moving on, notice one important thing from Figures 6.16 and 6.17 about visual system neurons: Most are continually active, even when there is no visual input (Tsodyks et al., 1999). Indeed, spontaneous activity is a characteristic of most cerebral neurons, and responses to external stimuli consume only a small portion of the constant energy required by ongoing brain activity (Raichle, 2006). Arieli and his colleagues (1996) have shown that the level of activity of visual cortical neurons at the time that a visual stimulus is presented influences how the cells respond to the stimulus—this may be one means by which cognition influences perception.

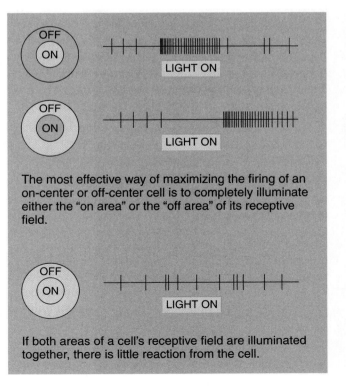

The most effective way of maximizing the firing of an on-center or off-center cell is to completely illuminate either the "on area" or the "off area" of its receptive field.

If both areas of a cell's receptive field are illuminated together, there is little reaction from the cell.

FIGURE 6.17 The responses of an on-center cell to contrast.

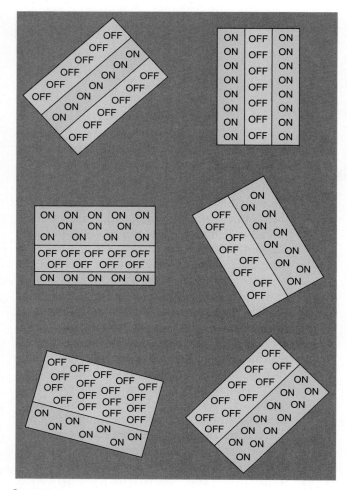

FIGURE 6.18 Examples of visual fields of simple cortical cells.

that each simple cell responds maximally only when its preferred straight-edge stimulus is in a particular position and in a particular orientation; and that the receptive fields of simple cortical cells are rectangular rather than circular.

Receptive Fields: Complex Cortical Cells

Complex cells are more numerous than simple cells. Like simple cells, complex cells have rectangular receptive fields, respond best to straight-line stimuli in a specific orientation, and are unresponsive to diffuse light. However, complex cells differ from simple cells in three important ways. First, they have larger receptive fields. Second, it is not possible to divide the receptive fields of complex cells into static "on" and "off" regions: A complex cell responds to a particular straight-edge stimulus of a particular orientation regardless of its position within the receptive field of that cell. Thus, if a stimulus (e.g., a 45° bar of light) that produces "on" firing in a particular complex cell is swept across that cell's receptive field, the

cell will respond continuously to it as it moves across the field. Many complex cells respond more robustly to the movement of a straight line across their receptive fields in a particular direction. Third, unlike simple cortical cells, which are all monocular (respond to stimulation of only one of the eyes), many complex cells are **binocular** (respond to stimulation of either eye). Indeed, in monkeys, over half the complex cortical cells are binocular.

If the receptive field of a binocular complex cell is measured through one eye and then through the other, the receptive fields in each eye turn out to have almost exactly the same position in the visual field, as well as the same orientation preference. In other words, what you learn about the cell by stimulating one eye is confirmed by stimulating the other. What is more, if the appropriate stimulation is applied through both eyes simultaneously, a binocular cell usually fires more robustly than if only one eye is stimulated.

Most of the binocular cells in the primary visual cortex of monkeys display some degree of *ocular dominance;* that is, they respond more robustly to stimulation of one eye than they do to the same stimulation of the other. In addition, some binocular cells fire best when the preferred stimulus is presented to both eyes at the same time but in slightly different positions on the two retinas (e.g., Ohzawa, 1998). In other words, these cells respond best to *retinal disparity* and thus are likely to play a role in depth perception (e.g., Livingstone & Tsao, 1999).

Columnar Organization of Primary Visual Cortex

The study of the receptive fields of primary visual cortex neurons has led to two important conclusions. The first conclusion is that the characteristics of the receptive fields of visual cortex neurons are attributable to the flow of signals from neurons with simpler receptive fields to those with more complex fields (see Reid & Alonso, 1996). Specifically, it seems that signals flow from on-center and off-center cells in lower layer IV to simple cells and from simple cells to complex cells (see Hirsch & Martinez, 2006).

The second conclusion is that primary visual cortex neurons are grouped in functional vertical columns (in this context, *vertical* means at right angles to the cortical layers). Much of the evidence for this conclusion comes from studies of the receptive fields of neurons along various vertical and horizontal electrode tracks (see Figure 6.19). If an electrode is advanced vertically through the layers of the visual cortex, with stops to plot the receptive fields of many neurons along the way, the results show that each cell in the column has a receptive field in the same area of the visual field. In addition, all the cells in a column respond best to straight lines in the very same orientation, and those neurons in a column that are either monocular or binocular with ocular dominance are all most sensitive to light in the same eye, left or right.

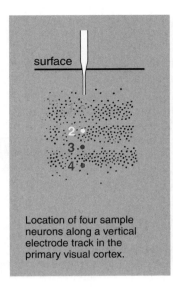

Location of four sample neurons along a vertical electrode track in the primary visual cortex.

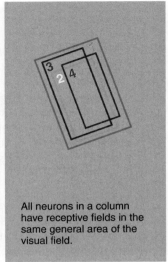

All neurons in a column have receptive fields in the same general area of the visual field.

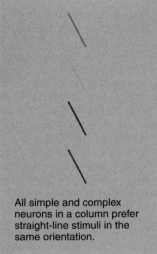

All simple and complex neurons in a column prefer straight-line stimuli in the same orientation.

1 right-eye dominant

2 right-eye dominant

3 right-eye dominant

4 right-eye dominant

In a given column, all monocular neurons and all binocular neurons that display dominance are dominated by the same eye.

Location of four sample neurons along a horizontal electrode track in the primary visual cortex.

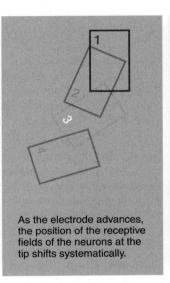

As the electrode advances, the position of the receptive fields of the neurons at the tip shifts systematically.

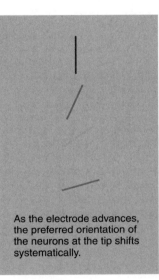

As the electrode advances, the preferred orientation of the neurons at the tip shifts systematically.

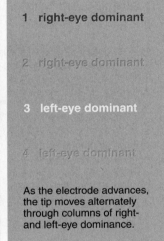

1 right-eye dominant

2 right-eye dominant

3 left-eye dominant

4 left-eye dominant

As the electrode advances, the tip moves alternately through columns of right- and left-eye dominance.

FIGURE 6.19   The organization of the primary visual cortex: the receptive-field properties of cells encountered along typical vertical and horizontal electrode tracks in the primary visual cortex.

In contrast, if an electrode is advanced horizontally through the tissue of the primary visual cortex, each successive cell encountered is likely to have a receptive field in a slightly different location and to be maximally responsive to straight lines of a slightly different orientation. And during a horizontal electrode pass, the tip passes alternately through areas of left-eye dominance and right-eye dominance—commonly referred to as *ocular dominance columns*.

All of the functional columns in the primary visual cortex that analyze input from one area of the retina are clustered together. Figure 6.20 on page 150 illustrates the organization of such a cluster as proposed by Hubel and Wiesel. Half of a cluster is thought to receive input from

the left eye, and the other half from the right eye. Furthermore, each cluster is thought to include neurons with preferences for straight-line stimuli of various orientations.

Now that you understand how the visual cortex is organized, you are in a better position to think constructively about Mrs. Richards's fortification illusions.

The Case of Mrs. Richards, Revisited

There was obviously a disturbance in Mrs. Richards's visual system: But where? And what kind of disturbance? And why the straight lines? A simple test located the disturbance. Mrs. Richards was asked to shut one eye and then the other and to report what happened to her illusion. The answer was, "Nothing." This suggested that the disturbance was cortical, because the visual cortex is the first part of the retina-geniculate-striate system that contains neurons that receive input from both eyes.

This hypothesis was confirmed by a few simple calculations: The gradual acceleration of the illusion as it spread out to the periphery is consistent with a wave of disturbance expanding from the "foveal area" of the primary visual cortex to its boundaries at a constant rate of about 3 millimeters per minute—the illusion accelerated because proportionally less visual cortex is dedicated to receiving signals from the periphery of the visual field.

And why the lines? Would you expect anything else from an area of the cortex whose elements appear to be specialized for coding straight-line stimuli?

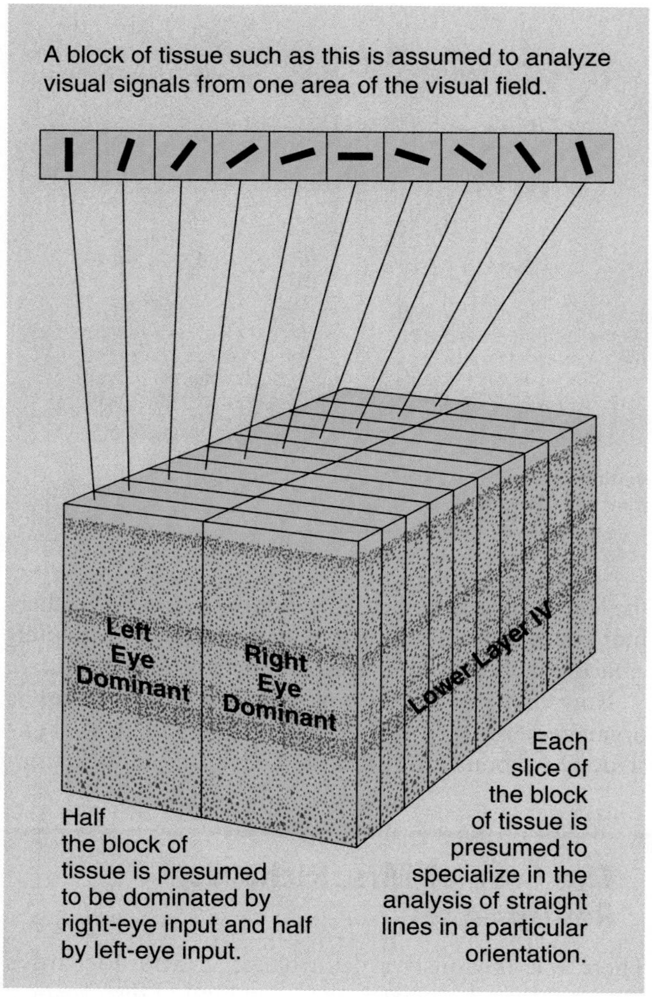

Plasticity of Receptive Fields of Neurons in the Visual Cortex

Most neuroscientific research on the visual system is based on two implicit assumptions. One is that the mechanisms of visual processing can be identified by studies using simplified, artificial stimuli. The second assumption is that the receptive field properties of each neuron are static, unchanging properties of that neuron. Research that has employed video clips of real scenes involving natural movement suggests that neither of these assumptions is correct (see Felsen & Dan, 2006).

Studies of the responses of visual cortex to natural scenes—just the type of scenes the visual system has evolved to perceive—indicate that the response of a visual cortex neuron depends not only on the stimuli in its receptive field, but on the larger scene in which these stimuli are embedded (see Bair, 2005; Kayser, Körding, & König, 2004). This *plasticity*, or ability to adapt to change, which has been largely ignored, appears to be a fundamental property of visual cortex function. It means that research based solely on the study of reaction to simple stimuli (e.g., spots and bars of light) cannot provide a complete explanation of how the visual system works; receptive field properties depend on the scene in which the stimuli to its field are embedded.

6.5
Seeing Color

Color is one of the most obvious qualities of human visual experience. So far in this chapter, we have largely limited our discussion of vision to black, white, and gray. Black is experienced when there is an absence of light; the perception of white is produced by an intense mixture of a wide range of wavelengths in roughly equal proportions; and the perception of gray is produced by the same mixture at lower intensities. In this section, we deal with the perception of colors such as blue, green, and yellow. The correct term for colors is *hues*, but in everyday language they are referred to as colors; and for the sake of simplicity, I will do the same.

What is there about a visual stimulus that determines the color we perceive? To a large degree, the perception of an object's color depends on the wavelengths of light that it reflects into the eye. Figure 6.2 on page 133 is an illustration of the colors associated with individual wavelengths; however, outside the laboratory, one never encounters objects that reflect single wavelengths. Sunlight and most sources of artificial light contain complex mixtures of most visible wavelengths. Most objects absorb the different wavelengths of light that strike them to

A block of tissue such as this is assumed to analyze visual signals from one area of the visual field.

Left Eye Dominant

Right Eye Dominant

Lower Layer IV

Half the block of tissue is presumed to be dominated by right-eye input and half by left-eye input.

Each slice of the block of tissue is presumed to specialize in the analysis of straight lines in a particular orientation.

FIGURE 6.20 Hubel and Wiesel's model of the organization of a cluster of functional columns in the primary visual cortex.

varying degrees and reflect the rest. The mixture of wavelengths that objects reflect influences our perception of their color, but it is not the entire story—as you are about to learn.

Component and Opponent Processing

The **component theory** (*trichromatic theory*) of color vision was proposed by Thomas Young in 1802 and refined by Hermann von Helmholtz in 1852. According to this theory, there are three different kinds of color receptors (cones), each with a different spectral sensitivity, and the color of a particular stimulus is presumed to be encoded by the ratio of activity in the three kinds of receptors. Young and Helmholtz derived their theory from the observation that any color of the visible spectrum can be matched by a mixing together of three different wavelengths of light in different proportions. This can be accomplished with any three wavelengths, provided that the color of any one of them cannot be matched by a mixing of the other two. The fact that three is normally the minimum number of different wavelengths necessary to match every color suggested that there were three types of receptors.

Another theory of color vision, the **opponent-process theory** of color vision, was proposed by Ewald Hering in 1878. He suggested that there are two different classes of cells in the visual system for encoding color and another class for encoding brightness. Hering hypothesized that each of the three classes of cells encoded two complementary color perceptions. One class of color-coding cells signaled red by changing its activity in one direction (e.g., hyperpolarization) and signaled red's complementary

color, green, by changing its activity in the other direction (e.g., hypopolarization). Another class of color-coding cells was hypothesized to signal blue and its complement, yellow, in the same opponent fashion; and a class of brightness-coding cells was hypothesized to similarly signal both black and white. **Complementary colors** are pairs of colors (e.g., green light and red light) that produce white or gray when combined in equal measure.

Hering based his opponent-process theory of color vision on several behavioral observations. One was that complementary colors cannot exist together: There is no such thing as bluish yellow or reddish green (see Billock & Tsou, 2010). Another was that the afterimage produced by staring at red is green and vice versa, and the afterimage produced by staring at yellow is blue and vice versa (try the Check It Out demonstration).

A somewhat misguided debate raged for many years between supporters of the component (trichromatic) and opponent theories of color vision. I say "misguided" because it was fueled more by the adversarial predisposition of scientists than by the incompatibility of the two theories. In fact, research subsequently proved that both color-coding mechanisms coexist in our visual systems (see DeValois et al., 2000).

It was the development in the early 1960s of a technique for measuring the absorption spectrum of the photopigment contained in a single cone that allowed researchers (e.g., Wald, 1964) to confirm the conclusion that Young had reached over a century and a half before. They found that there are indeed three different kinds of cones in the retinas of those vertebrates with good color vision, and they found that each of the three has a different photopigment with its own characteristic absorption spectrum. As Figure 6.21 on page 152 illustrates, some cones are most sensitive to short wavelengths, some are most sensitive to medium wavelengths, and some are most sensitive to long wavelengths.

Although the coding of color by cones seems to operate on a purely component basis (see Jameson, Highnote, & Wasserman, 2001), there is evidence of opponent processing of color at all subsequent levels of the retina-geniculate-striate system. That is, at all subsequent levels, there are cells that respond in one direction (e.g., increased firing) to one color and in the opposite direction (e.g., decreased firing) to its complementary color (see Chatterjee & Callaway, 2003; Gegenfurtner & Kiper, 2003).

Most primates are *trichromats* (possessing three color vision photopigments); see Jacobs & Nathans (2009). Most other mammals are *dichromats* (possessing two color vision photopigments)—they lack the photopigment sensitive to long wavelengths and thus have difficulty seeing light at the red end of the visible spectrum (see Figure 6.2). In contrast, some birds, fish, and reptiles have four

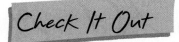

COMPLEMENTARY AFTERIMAGES

Have you ever noticed complementary afterimages? To see them, stare at the fixation point (x) in the left panel for 1 minute without moving your eyes, then quickly shift your gaze to the fixation point in the right panel. In the right panel, you will see four squares whose colors are complementary to those in the left panel.

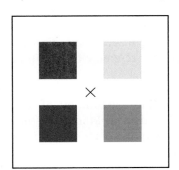
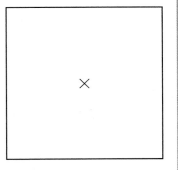

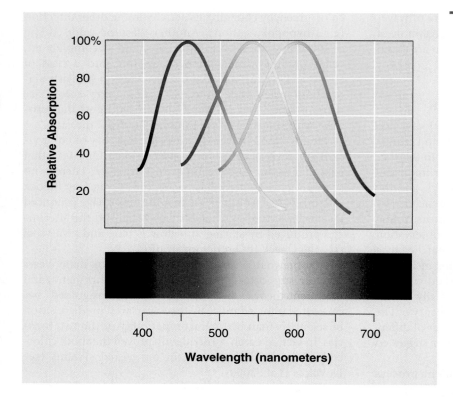

FIGURE 6.21 The absorption spectra of the three classes of cones.

during the course of the day. However, although the wavelengths reflected by my shirt change markedly, its color does not. My shirt will be just as blue in midmorning and in late afternoon as it is now. Color constancy is the tendency for an object to stay the same color despite major changes in the wavelengths of light that it reflects.

Although the phenomenon of color constancy is counterintuitive, its advantage is obvious. Color constancy improves our ability to tell objects apart in a memorable way so that we can respond appropriately to them; our ability to recognize objects would be greatly lessened if their color changed every time there was a change in illumination (see Spence et al., 2006). In essence, if it were not for color constancy, color vision would have little survival value.

Although color constancy is an important feature of our vision, we are normally unaware of it. Under everyday conditions, we have no way of appreciating just how much the wavelengths reflected by an object can change without the object changing its color. It is only in the controlled environment of the laboratory that one can fully appreciate that color constancy is more than an important factor in color vision: It is the essence of color vision.

Edwin Land (1977) developed several dramatic laboratory demonstrations of color constancy. In these demonstrations, Land used three adjustable projectors. Each projector emitted only one wavelength of light: one a short-wavelength light, one a medium-wavelength light, and one a long-wavelength light. Thus, it was clear that only three wavelengths of light were involved in the demonstrations. Land shone the three projectors on a test display like the one in Figure 6.22. (These displays are called *Mondrians* because they resemble the paintings of the Dutch artist Piet Mondrian.)

Land found that adjusting the amount of light emitted from each projector—and thus the amount of light of each wavelength being reflected by the Mondrian—had no effect at all on the perception of its colors. For example, in one demonstration Land used a photometer to measure the amounts of the three wavelengths being reflected by a rectangle judged to be pure blue by his subjects. He then

photopigments; the fourth allows them to detect ultraviolet light, which is invisible to humans.

In a remarkable study, Jacobs and colleagues (2007) introduced into mice the gene for the long-wavelength photopigment, thus converting them from dichromats to trichromats. Behavioral tests indicated that the transgenic mice had acquired the ability to perceive long wavelengths and to make color discriminations involving light at that end of the spectrum.

Color Constancy and the Retinex Theory

Neither component nor opponent processing can account for the single most important characteristic of color vision: color constancy. **Color constancy** refers to the fact that the perceived color of an object is not a simple function of the wavelengths reflected by it.

Color constancy is an important, but much misunderstood, concept. Let me explain it with an example. As I write this at 7:15 on a December morning, it is dark outside, and I am working in my office by the light of a tiny incandescent desk lamp. Later in the morning, when students start to arrive, I will turn on my nasty fluorescent office lights; and then, in the afternoon, when the sun has shifted to my side of the building, I will turn off the lights and work by natural light. The point is that because these light sources differ markedly in the wavelengths they contain, the wavelengths reflected by various objects in my office—my blue shirt, for example—change substantially

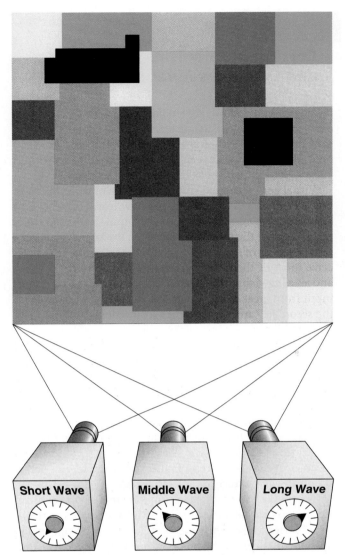

FIGURE 6.22 The method of Land's (1977) color-vision experiments. Subjects viewed Mondrians that were illuminated by various proportions of three different wavelengths: a short wavelength, a middle wavelength, and a long wavelength.

lighting) and as long as the object is viewed as part of a scene, not in isolation.

According to Land's **retinex theory** of color vision, the color of an object is determined by its *reflectance*—the proportion of light of different wavelengths that a surface reflects. Although the wavelengths of light reflected by a surface change dramatically with changes in illumination, the efficiency with which a surface absorbs each wavelength and reflects the unabsorbed portion does not change. According to the retinex theory, the visual system calculates the reflectance of surfaces, and thus perceives their colors, by comparing the light reflected by adjacent surfaces in at least three different wavelength bands (short, medium, and long)—see Hurlbert and Wolf (2004).

Why is Land's research so critical for neuroscientists trying to discover the neural mechanisms of color vision? It is important because it suggests one type of cortical neuron that is likely to be involved in color vision (see Shapely & Hawken, 2002). If the perception of color depends on the analysis of contrast between adjacent areas of the visual field, then the critical neurons should be responsive to color contrast (see Hurlbert, 2003). And they are. For example, **dual-opponent color cells** in the monkey visual cortex respond with vigorous "on" firing when the center of their circular receptive field is illuminated with one wavelength, such as green, and the surround (periphery) is simultaneously illuminated with another wavelength, such as red. And the same cells display vigorous "off" firing when the pattern of illumination is reversed—for example, red in the center and green in the surround. In essence, dual-opponent color cells respond to the contrast between wavelengths reflected by adjacent areas of their receptive field.

A major breakthrough in the understanding of the organization of the primary visual cortex came with the discovery that dual-opponent color cells are not distributed evenly throughout the primary visual cortex of monkeys (see Zeki, 1993a). Livingstone and Hubel (1984) found that these neurons are concentrated in the primary visual cortex in peglike columns that penetrate the layers of the monkey primary visual cortex, with the exception of lower layer IV. Many neurons in these peglike columns are particularly rich in the mitochondrial enzyme **cytochrome oxidase**; thus, their distribution in the primary visual cortex can be visualized if one stains slices of tissue with stains that have an affinity for this enzyme.

When a section of monkey striate tissue is cut parallel to the cortical layers and stained in this way, the pegs are seen as "blobs" of stain scattered over the cortex (unless the section is cut from lower layer IV). To the relief of instructors and students alike, the term **blobs** has become the accepted scientific label for peglike, cytochrome oxidase–rich, dual-opponent color columns. The blobs were found to be located in the midst of ocular dominance

adjusted the emittance of the projectors, and he measured the wavelengths reflected by a red rectangle on a different Mondrian, until the wavelengths were exactly the same as those that had been reflected by the blue rectangle on the original. When he showed this new Mondrian to his subjects, the red rectangle looked—you guessed it—red, even though it reflected exactly the same wavelengths as had the blue rectangle on the original Mondrian.

The point of Land's demonstration is that blue objects stay blue, green objects stay green, and so forth, regardless of the wavelengths they reflect. This color constancy occurs as long as the object is illuminated with light that contains some short, medium, and long wavelengths (such as daylight, firelight, and virtually all manufactured

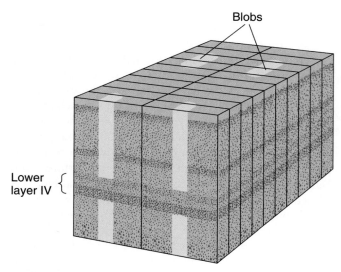

Blobs

Lower layer IV

FIGURE 6.23 Hubel and Livingstone's model of primary visual cortex organization. The blobs are peglike columns that contain dual-opponent color cells.

columns (compare Figure 6.23 with Figure 6.20). Functional MRI studies have provided evidence of dual-opponent color cells in the human visual cortex (Engel, 1999).

Scan Your Brain

The striate cortex is the main entrance point of visual signals to the cortex. In the upcoming section, we will follow visual signals to other parts of the cortex. This is a good point to pause and review what you have learned. Draw a line to connect each term in the first column with the closely related word or phrase in the second column. Each term should be linked to only one item in the second column. The correct answers are provided at the end of this exercise. Before proceeding, review material related to your errors and omissions.

1. contrast enhancement	a. many are binocular
2. simple cortical cells	b. complementary afterimages
3. complex cortical cells	c. blobs
4. ocular dominance columns	d. reflectance
5. component	e. static on and off areas
6. opponent	f. ommatidia
7. retinex	g. Mach bands
8. cytochrome oxidase	h. striate cortex
9. horseshoe crab	i. three

Scan Your Brain answers: (1) g, (2) e, (3) a, (4) h, (5) i, (6) b, (7) d, (8) c, (9) f.

6.6

Cortical Mechanisms of Vision and Conscious Awareness

So far, you have followed the major visual pathways from the eyes to the primary visual cortex, but there is much more to the human visual system—we are visual animals. The entire occipital cortex as well as large areas of temporal cortex and parietal cortex are involved in vision (see Figure 6.24).

Visual cortex is often considered to be of three different types. *Primary visual cortex*, as you have learned, is that area of cortex that receives most of its input from the visual relay nuclei of the thalamus (i.e., from the lateral geniculate nuclei). Areas of **secondary visual cortex** are those that receive most of their input from the primary visual cortex, and areas of **visual association cortex** are those that receive input from areas of secondary visual cortex as well as from the secondary areas of other sensory systems.

The primary visual cortex is located in the posterior region of the occipital lobes, much of it hidden from view in the longitudinal fissure. Most areas of secondary visual cortex are located in two general regions: in the prestriate cortex and in the inferotemporal cortex. The **prestriate cortex** is the band of tissue in the occipital lobe that surrounds the primary visual cortex. The **inferotemporal cortex** is the cortex of the inferior temporal lobe. Areas of association cortex that receive visual input are located in several parts of the cerebral cortex, but the largest single area is in the **posterior parietal cortex**.

The major flow of visual information in the cortex is from the primary visual cortex to the various areas of secondary visual cortex to the areas of association cortex. As one moves up this visual hierarchy, the neurons have larger receptive fields and the stimuli to which the neurons respond are more specific and more complex (see Zeki, 1993b).

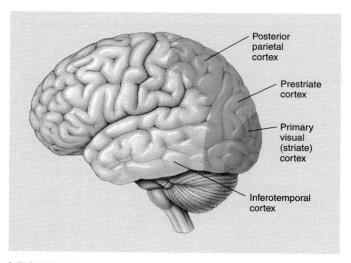

Posterior parietal cortex

Prestriate cortex

Primary visual (striate) cortex

Inferotemporal cortex

FIGURE 6.24 The visual areas of the human cerebral cortex.

Damage to Primary Visual Cortex: Scotomas and Completion

Damage to an area of the primary visual cortex produces a **scotoma**—an area of blindness—in the corresponding area of the contralateral visual field of both eyes (see Figure 6.13). Neurological patients with suspected damage to the primary visual cortex are usually given a **perimetry test**. While the patient's head is held motionless on a chin rest, the patient stares with one eye at a fixation point on a screen. A small dot of light is then flashed on various parts of the screen, and the patient presses a button to record when the dot is seen. Then, the entire process is repeated for the other eye. The result is a map of the visual field of each eye, which indicates any areas of blindness. Figure 6.25 illustrates the perimetric maps of the visual fields of a man with a bullet wound in his left primary visual cortex. Notice the massive scotoma in the right visual field of each eye.

Clinical Implications

Many patients with extensive scotomas are not consciously aware of their deficits. One of the factors that contributes to this lack of awareness is completion. A patient with a scotoma who looks at a complex figure, part of which lies in the scotoma, often reports seeing a complete image (Zur & Ullman, 2003). In some cases, this completion may depend on residual visual capacities in the scotoma; however, completion also occurs in cases in which this explanation can be ruled out. For example, patients who are **hemianopsic** (having a scotoma covering half of the visual field) may see an entire face when they focus on a person's nose, even when the side of the face in the scotoma has been covered by a blank card.

Consider the completion phenomenon experienced by the esteemed physiological psychologist Karl Lashley (1941). He often developed a large scotoma next to his fovea during a migraine attack (see Figure 6.26 on page 156).

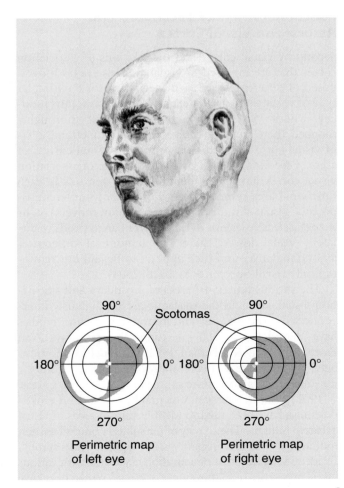

FIGURE 6.25 The perimetric maps of a subject with a bullet wound in his left primary visual cortex. The scotomas (areas of blindness) are indicated in gray. (Based on Teuber, Battersby, & Bender, 1960.)

The Case of the Physiological Psychologist Who Made Faces Disappear

Talking with a friend I glanced just to the right of his face wherein his head disappeared. His shoulders and necktie were still visible but the vertical stripes on the wallpaper behind him seemed to extend down to the necktie. It was impossible to see this as a blank area when projected on the striped wallpaper of uniformly patterned surface although any intervening object failed to be seen. (Lashley, 1941, p. 338)

Clinical Implications

Damage to Primary Visual Cortex: Scotomas, Blindsight, and Conscious Awareness

You probably equate perception with **conscious awareness**; that is, you assume that if a person sees something, he or she will be consciously aware of seeing it. In everyday thinking, perceiving and being aware are inseparable processes: We assume that someone who has seen something will always be able to acknowledge that he or she has seen it and be able to describe it—in humans, conscious awareness is usually inferred from the ability to verbally describe the object of awareness. In the following pages, you will encounter examples of phenomena for which this is not the case: People see things but have no conscious awareness of them. Blindsight is the first example.

Blindsight is another phenomenon displayed by patients with scotomas resulting from damage to primary visual cortex. **Blindsight** is the ability of such patients to respond to visual stimuli in their scotomas even though they have no conscious awareness of the stimuli (Danckert & Rossetti, 2005; Weiskrantz, 2004). Of all visual abilities, perception of motion is most likely to survive damage to primary

Clinical Implications

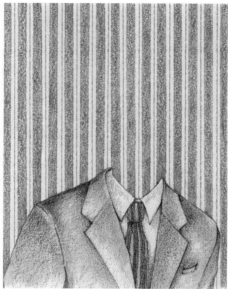

Lashley's scotoma   What Lashley saw

FIGURE 6.26 The completion of a migraine-induced scotoma as described by Karl Lashley (1941).

visual cortex (Intriligator, Xie, & Barton, 2002). For example, a subject might reach out and grab a moving object in her scotoma, all the while claiming not to see it.

If blindsight confuses you, imagine how it confuses people who experience it. Consider, for example, the reactions to blindsight of D.B., a patient who was blind in his left visual field following surgical removal of his right occipital lobe (Weiskrantz, 2004; Weiskrantz et al., 1974).

The Case of D.B., the Man Confused by His Own Blindsight

Even though the patient had no awareness of "seeing" in his blind [left] field, evidence was obtained that he (a) could reach for visual stimuli [in his left field] with considerable accuracy; (b) could differentiate the orientation of a vertical line from a horizontal or diagonal line; (c) could differentiate the letters "X" and "O."

Needless to say, he was questioned repeatedly about his vision in his left half-field, and his most common response was that he saw nothing at all. . . . When he was shown his results [through his good, right-half field] he expressed surprise and insisted several times that he thought he was just "guessing." When he was shown a video film of his reaching and judging orientation of lines, he was openly astonished. (Weiskrantz et al., 1974, pp. 721, 726)

Two neurological interpretations of blindsight have been proposed. One is that the striate cortex is not completely destroyed and the remaining islands of functional cells are capable of mediating some visual abilities in the absence of conscious awareness (see Wüst, Kasten, & Sabel, 2002). The other is that those visual pathways that ascend directly to the secondary visual cortex from subcortical visual structures without passing through the primary visual cortex are capable of maintaining some visual abilities in the absence of cognitive awareness (see Kentridge, Heywood, & Weiskrantz, 1997). There is some support for both theories, but the evidence is not conclusive for either (see Gross, Moore, & Rodman, 2004; Rosa, Tweedale, & Elston, 2000; Schärli, Harman, & Hogben, 1999a, 1999b). Indeed, it is possible that both mechanisms contribute to the phenomenon.

Functional Areas of Secondary and Association Visual Cortex

Secondary visual cortex and the portions of association cortex that are involved in visual analysis are both composed of different areas, each specialized for a particular type of visual analysis. For example, in the macaque monkey, whose visual cortex has been most thoroughly mapped, there are more than 30 different functional areas of visual cortex; in addition to primary visual cortex, 24 areas of secondary visual cortex and 7 areas of association visual cortex have been identified. The neurons in each functional area respond most vigorously to different aspects of visual stimuli (e.g., to their color, movement, or shape); selective lesions to the different areas produce different visual losses; there are anatomical differences among the areas; and each appears to be laid out retinotopically (Grill-Spector & Mallach, 2004).

The various functional areas of secondary and association visual cortex in the macaque are prodigiously interconnected. Anterograde and retrograde tracing studies have identified over 300 interconnecting pathways (Van Essen, Anderson, & Felleman, 1992). Although connections between areas are virtually always reciprocal, the major flow of signals is from more simple to more complex areas.

PET (positron emission tomography), fMRI, and evoked potentials have been used to identify various areas of visual cortex in humans. The activity of the subjects' brains has been monitored while they inspect various types of visual stimuli. By identifying the areas of activation associated with various visual properties (e.g., movement or color), researchers have so far delineated about a dozen different functional areas of human visual cortex (see Grill-Spector & Mallach, 2004). A map of these areas is shown in Figure 6.27. Most are similar in terms of location, anatomical characteristics, and function to areas already identified in the macaque.

- V3A
- V3
- V2
- V1/Primary
- V4
- MT/V5

Dorsal and Ventral Streams

As you have already learned, most visual information enters the primary visual cortex via the lateral geniculate nuclei. The information from the two lateral geniculate nuclei is received in the primary visual cortex, combined, and then segregated into multiple pathways that project separately to the various functional areas of secondary, and then association, visual cortex (see Horton & Sincich, 2004).

Many pathways that conduct information from the primary visual cortex through various specialized areas of secondary and association cortex are parts of two major streams: the dorsal stream and the ventral stream (Ungerleider & Mishkin, 1982). The **dorsal stream** flows from the primary visual cortex to the dorsal prestriate cortex to the posterior parietal cortex, and the **ventral stream** flows from the primary visual cortex to the ventral prestriate cortex to the inferotemporal cortex—see Figure 6.28 on page 158.

⊙→ Simulate The Dorsal and Ventral Visual Pathways **www.mypsychlab.com**

Most visual cortex neurons in the dorsal stream respond most robustly to spatial stimuli, such as those indicating the location of objects or their direction of movement. In contrast, most neurons in the ventral stream respond to the characteristics of objects, such as color and shape. Indeed, several lines of evidence suggest that there are clusters of visual neurons in the ventral stream, each of which responds specifically to a particular class of objects—for example, to faces, bodies, letters, animals, or tools (Haxby, 2006; Reddy & Kanwisher, 2006).

Ungerleider and Mishkin (1982) proposed that the dorsal and ventral visual streams perform different visual functions. They suggested that the dorsal stream is involved in the perception of "where" objects are and the ventral stream is involved in the perception of "what" objects are.

A major implication of the **"where" versus "what" theory** of vision is that damage to some areas of cortex may abolish certain aspects of vision while leaving others unaffected. Indeed, the most convincing support for the influential "where" versus "what" theory has come from the comparison of the specific effects of damage to the dorsal and ventral streams (see Ungerleider & Haxby, 1994). Patients with damage to the posterior parietal cortex often have difficulty reaching accurately for objects that they have no difficulty describing; conversely, patients with damage to the inferotemporal cortex often have no difficulty reaching accurately for objects that they have difficulty describing.

Although the "where" versus "what" theory has many advocates, there is an alternative interpretation for the same evidence (Goodale, 2004). Goodale and Milner (1992) argued that the key difference between the dorsal and ventral streams is not the kinds of information they carry but the use to which that information is put. They suggested that the function of the dorsal stream is to direct behavioral interactions with objects, whereas the function of the ventral stream is to mediate the conscious perception of objects; this

FIGURE 6.28 Information about particular aspects of a visual display flow out of the primary visual cortex over many pathways. The pathways can be grouped into two general streams: dorsal and ventral.

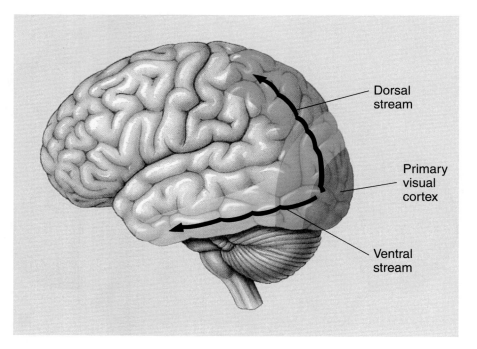

Dorsal stream

Primary visual cortex

Ventral stream

is the **"control of behavior"** versus **"conscious perception" theory** (see Logothetis & Sheinberg, 1996). One of the most interesting aspects of this theory is its evolutionary implication. Goodale (2004) has suggested that the conscious awareness mediated by the ventral stream is one thing that distinguishes humans and their close relatives from their evolutionary ancestors.

The "control of behavior" versus "conscious perception" theory can readily explain the two major neuropsychological findings that are the foundation of the "where" versus "what" theory. Namely, the "control of behavior" versus "conscious perception" theory suggests that patients with dorsal stream damage may do poorly on tests of location and movement because most tests of location and movement involve performance measures, and that patients with ventral stream damage may do poorly on tests of visual recognition because most tests of visual recognition involve verbal, and thus conscious, awareness.

The major support for the "control of behavior" versus "conscious perception" theory is the confirmation of its two primary assertions: (1) that some patients with bilateral lesions to the ventral stream have no conscious experience of seeing and yet are able to interact with objects under visual guidance, and (2) that some patients with bilateral lesions to the dorsal stream can consciously see objects but cannot interact with them under visual guidance (see Figure 6.29). Following are two such cases.

The Case of D.F., the Woman Who Could Grasp Objects She Did Not Consciously See

D.F. has bilateral damage to her ventral prestriate cortex, thus interrupting the flow of the ventral stream; her case is described in an interesting book by Goodale and Milner (2004). Amazingly, she can respond accurately to visual stimuli that she does not consciously see. Goodale and Milner (1992) describe her thusly:

> Despite her profound inability to recognize the size, shape and orientation of visual objects, D.F. showed

strikingly accurate guidance of hand and finger movements directed at the very same objects. Thus, when she was presented with a pair of rectangular blocks of the same or different dimensions, she was unable to distinguish between them. When she was asked to indicate the width of a single block by means of her index finger and thumb, her matches bore no relationship to the dimensions of the object and showed considerable trial to trial variability. However, when she was asked simply to reach out and pick up the block, the aperture between her index finger and thumb changed systematically with the width of the object, just as in normal subjects. In other words, D.F. scaled her grip to the dimensions of the objects she was about to pick up, even though she appeared to be unable to [consciously] "perceive" those dimensions.

> A similar dissociation was seen in her responses to the orientation of stimuli. Thus, when presented with a large slot that could be placed in one of a number of different orientations, she showed great difficulty in indicating the orientation either verbally or manually (i.e., by rotating her hand or a hand-held card). Nevertheless, she was as good as normal subjects at reaching out and placing her hand or the card into the slot, turning her hand appropriately from the very onset of the movement. (p. 22)

The Case of A.T., the Woman Who Could Not Accurately Grasp Unfamiliar Objects That She Saw

The case of A.T. is in major respects complementary to that of D.F. The patient A.T. is a woman with a lesion of the occipitoparietal region, which likely interrupts her dorsal route.

Dorsal and Ventral Streams: Two Theories and What They Predict

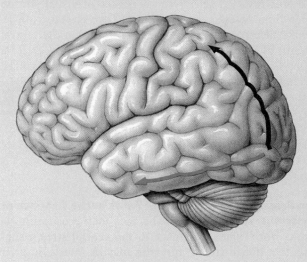

"Where" vs. "What" Theory

Dorsal stream specializes in visual spatial perception
Ventral stream specializes in visual pattern recognition

Predicts
• Damage to **dorsal stream** disrupts visual spatial perception
• Damage to ventral stream disrupts visual pattern recognition

"Control of Behavior" vs. "Conscious Perception" Theory

Dorsal stream specializes in visually guided behavior
Ventral stream specializes in conscious visual perception

Predicts
• Damage to **dorsal stream** disrupts visually guided behavior but not conscious visual perception
• Damage to ventral stream disrupts conscious visual perception but not visually guided behavior

FIGURE 6.29 The "where" versus "what" and the "control of behavior" versus "conscious perception" theories make different predictions.

A.T. was able to recognize objects, and was also able to demonstrate their size with her fingers. By contrast, pre-shape of the hand during object-directed movements was incorrect. Correlation between object size and maximum grip size was lacking, with the consequence that objects could not be grasped between the fingertips; instead, the patient made awkward palmar grasps. The schema framework offers a compelling explanation for this deficit. Because the grasp schemas were destroyed by the lesion, or disconnected from visual input, the grip aperture did not stop at the required size, grip closure was delayed and the transport was prolonged in order to remain coordinated with the grasp.

A.T. cannot preshape her hand for neutral objects like plastic cylinders, yet, when faced with a familiar object whose size is a semantic property, like a lipstick, she can grasp it with reasonable accuracy. This interaction reflects the role of the abundant anatomical interconnections between the two cortical systems. (Jeannerod et al., 1995, p. 320)

The characterization of functional differences between the dorsal and ventral streams is far from complete. For example, the streams certainly do not function in isolation from one another (Konen & Kastner, 2008; Nassi & Calloway, 2009), but little attention has been paid to how they might interact.

Now that you have been introduced to dorsal and ventral streams, you are prepared to proceed to the end of the chapter. The chapter concludes with a discussion of two neuropsychological disorders of vision: one (prosopagnosia)

associated with damage to an area of the ventral stream and the other (akinetopsia) associated with damage to an area of the dorsal stream.

Prosopagnosia

Prosopagnosia, briefly put, is visual agnosia for faces. Let me explain. **Agnosia** is a failure of recognition (*gnosis* means "to know") that is not attributable to a sensory deficit or to verbal or intellectual impairment; **visual agnosia** is a specific agnosia for visual stimuli. Visual agnosics can see visual stimuli, but they don't know what they are.

Visual agnosias themselves are often specific to a particular aspect of visual input and are named accordingly; for example, *movement agnosia, object agnosia,* and *color agnosia* are difficulties in recognizing movement, objects, and color, respectively. It is presumed that each specific visual agnosia results from damage to an area of secondary visual cortex that mediates the recognition of that particular attribute. Prosopagnosics are visual agnosics with a specific difficulty in recognizing faces.

Prosopagnosics can usually recognize a face as a face, but they have problems recognizing whose face it is. They often report seeing a jumble of individual facial parts (e.g., eyes, nose, chin, cheeks) that for some reason are never fused, or bound, into an easy-to-recognize whole (see Stephan & Caine, 2009). In extreme cases, prosopagnosics cannot recognize themselves: Imagine what it would be like to stare in the mirror every morning and not recognize the face that is looking back.

Is Prosopagnosia Specific to Faces? The belief that prosopagnosia is a deficit specific to the recognition of faces has been challenged. To understand this challenge, you need to know that the diagnosis of prosopagnosia is typically applied to neuropsychological patients who have difficulty recognizing particular faces, but can readily identify other test objects (e.g., a chair, a dog, or a tree). Surely, this is powerful evidence that prosopagnosics have recognition difficulties specific to faces. Not so. Pause for a moment, and think about this evidence: It is seriously flawed.

Because prosopagnosics have no difficulty recognizing faces as faces, the fact that they can recognize chairs as chairs, pencils as pencils, and doors as doors is not relevant. The critical question is whether they can recognize which chair, which pencil, and which door. Careful testing of prosopagnosics usually reveals that their recognition deficits are not restricted to faces: For example, a farmer lost his ability to recognize particular cows when he lost his ability to recognize faces. This suggests that some prosopagnosics have a general problem recognizing specific objects that belong to complex classes of objects (e.g., particular automobiles or particular houses), not a specific problem recognizing faces (see Behrmann et al., 2005)—although the facial-recognition problems are likely to be the most problematic. On the other hand, several thorough case studies of prosopagnosia have failed to detect recognition deficits unrelated to faces (De Renzi, 1997; Duchaine & Nakayama, 2005; Farah, 1990). It seems likely that prosopagnosia is not a unitary disorder (Duchaine & Nakayama, 2006); that is, it appears that only some patients have pattern recognition deficits restricted to facial recognition.

Thinking Creatively

R.P., a Typical Case of Prosopagnosia

R.P. is a typical prosopagnosic. With routine testing, he displayed a severe deficit in recognizing faces and in identifying facial expressions (Laeng & Caviness, 2001) but no other recognition problems. If testing had stopped there, it would have been concluded that R.P. is an agnosic with recognition problems specific to human faces. However, more thorough testing suggested that R.P. is deficient in recognizing all objects with complex curved surfaces (e.g., amoeboid shapes), not just faces.

What Brain Pathology Is Associated with Prosopagnosia? The diagnosis of prosopagnosia is often associated with damage to the ventral stream in the area of the boundary between the occipital and temporal lobes. This area of human cortex has become known as the *fusiform face area*, and parts of it are selectively activated by

human faces (Kanwisher, 2006; Pourtois et al., 2005)—it should be emphasized however that parts of the fusiform face area respond selectively to classes of visual stimuli other than faces (Grill-Spector, Sayres, & Ress, 2006). Similar face-specific areas have been found in the ventral streams of macaque monkeys (Moeller, Freiwald, & Tsao, 2008; Tsao et al., 2006). It makes sense that specialized mechanisms to perceive faces have evolved in the human brain because face perception plays such a major role in human social behavior (Pascalis & Kelly, 2009; Tsao & Livingstone, 2008). The extent to which the development of the fusiform face area depends on a person's early experience with faces is still unclear (Peissig & Tarr, 2007).

Evolutionary Perspective

Can Prosopagnosics Perceive Faces in the Absence of Conscious Awareness? The fact that prosopagnosia results from bilateral damage to the ventral stream suggests that dorsal-stream function may be intact. In other words, it suggests that prosopagnosics may be able to unconsciously recognize faces that they cannot recognize consciously. This is, indeed, the case.

Tranel and Damasio (1985) were the first to demonstrate that prosopagnosics can recognize faces in the absence of conscious awareness. They presented a series of photographs to several patients, some familiar to the patients, some not. The subjects claimed not to recognize any of the faces. However, when familiar faces were presented, the subjects displayed a large skin conductance response, which did not occur with unfamiliar faces, thus indicating that the faces were being unconsciously recognized by undamaged portions of the brain.

Akinetopsia

Akinetopsia is a deficiency in the ability to see movement progress in a normal smooth fashion. Akinetopsia can be triggered by high doses of certain antidepressants, as the following two cases illustrate (Horton, 2009).

Two Cases of Drug-Induced Akinetopsia

A 47-year-old depressed male receiving 100 mg nefazodone twice daily reported a bizarre derangement of motion perception. Each moving object was followed by a trail of multiple freeze-frame images, which disappeared once the motion ceased. A 48-year-old female receiving 400 mg of nefazodone once daily at bedtime reported similar symptoms, with persistent multiple strobelike trails following moving objects. In both cases, stationary elements were perceived normally, indicating a selective impairment of the visual perception of motion. Vision returned to normal in both patients once the dosage was reduced.

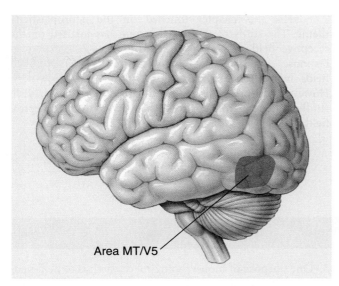

FIGURE 6.30 The location of MT: Damage to this middle temporal area of the human brain is associated with akinetopsia.

Akinetopsia is often associated with damage to the middle temporal (MT) area of the cortex. The location of MT—near the junction of the temporal, parietal, and occipital lobes—is illustrated in Figure 6.30. Sometimes MT is called V5, or MT/V5. This is because researchers studying the visual system in different primate species used different systems of neuroanatomical classification. All three terms appear to refer to comparable areas.

The function of MT appears to be the perception of motion. Given the importance of the perception of motion in primate survival, it is reasonable that an area of the visual system is dedicated to it. Some neurons at lower levels of the visual hierarchy (e.g., in the primary visual cortex) respond to movement, as well as color and shape; however, they provide little information about the direction of movement because their receptive fields are so small. In contrast, 95% of the neurons of MT respond to specific directions of movement and

little else. Also, each MT neuron has a large binocular receptive field, allowing it to track movement over a wide range.

The following four lines of research implicate MT in the visual perception of motion and damage to MT as a cause of akinetopsia:

- Patients with akinetopsia tend to have unilateral or bilateral damage to MT (Schenk & Zihl, 1997; Shipp et al., 1994; Vaina et al., 2001).
- As measured by fMRI, activity in MT increases when humans view movement (Grossman et al., 2000).
- Blocking activity in MT with transcranial magnetic stimulation (TMS) produces motion blindness (Beckers & Hömberg, 1992; Beckers & Zeki, 1995).
- Electrical stimulation of MT in human patients induces the visual perception of motion (Blanke et al., 2002; Lee et al., 2000; Richer et al., 1991).

Conclusion

A key goal of this chapter was to help you understand that vision is a creative process. Your visual system does not transmit complete and intact visual images of the world to the cortex. It carries information about a few critical features of the visual field—for example, information about location, movement, brightness contrast, and color contrast—and from these bits of information, it creates a perception that is far better than the retinal image in all respects and better than the external reality in some. Another main point is that your visual system can perceive things without your conscious awareness of them.

The Check It Out demonstrations in this chapter offered you many opportunities to experience firsthand important aspects of the visual process. I hope that you did take the time to check them out and that your experience made you more aware of the amazing abilities of your own visual system and the relevance of what you have learned in this chapter to your everyday life.

Themes Revisited

This chapter developed all four of the book's major themes. First, the evolutionary perspective theme was emphasized, largely because the majority of research on the neural mechanisms of human vision has been comparative and because thinking about the adaptiveness of various aspects of vision (e.g., color vision) has led to important insights.

Second, the thinking creatively theme was emphasized because the main point of the chapter was that we tend to think about our own visual systems in a way that is

fundamentally incorrect: The visual system does not passively provide 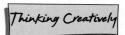 images of the external world; it extracts some features of the external world, and from these it creates our visual perceptions. Once you learn to think in this unconventional way, you will be able to better appreciate the amazingness of your own visual system.

Third, the clinical implications theme was developed through a series of clinical case studies: Mrs. Richards, who experienced fortification illusions before her migraine attacks; Karl Lashley, the physiological psychologist who used his scotoma to turn a friend's head into a wallpaper pattern; D.B., the man with blindsight; D.F., who showed by her accurate reaching that she perceived the size, shape, and orientation of objects that she could not describe; A.T., who could describe the size and shape of objects that she could not accurately reach for; R.P., a typical prosopagnosic, and two patients with akinetopsia induced by an antidepressant.

Clinical Implications

Fourth, this chapter touched on the neuroplasticity theme. The study of the visual system has focused on the receptive field properties of neurons in response to simple stimuli, and receptive fields have been assumed to be static. However, when natural visual scenes have been used in such studies, it has become apparent that each neuron's receptive field changes depending on the visual context.

Neuroplasticity

Think about It

1. It is difficult to define the term *illusion* rigorously, because in a sense, all of what we see is an illusion. Explain and discuss.
2. Some sensory pathways control behavior directly without conscious awareness, whereas others control behavior consciously. Discuss the evolutionary implications of this arrangement.
3. One purpose of biopsychological research is to help neuropsychological patients, but these patients also help biopsychologists understand the brain mechanisms of psychological processes. Discuss and give some examples.
4. How do most people think color vision works? What does the phenomenon of color constancy suggest?
5. Why should natural scenes be used to study visual neurons?

Key Terms

6.1 Light Enters the Eye and Reaches the Retina

Sensitivity (p. 133)
Acuity (p. 133)
Ciliary muscles (p. 134)
Accommodation (p. 134)
Binocular disparity (p. 135)

6.2 The Retina and Translation of Light into Neural Signals

Receptors (p. 136)
Horizontal cells (p. 136)
Bipolar cells (p. 136)
Amacrine cells (p. 136)
Retinal ganglion cells (p. 136)
Blind spot (p. 136)
Fovea (p. 136)
Completion (p. 137)
Surface interpolation (p. 137)
Cones (p. 137)
Rods (p. 137)
Duplexity theory (p. 138)
Photopic vision (p. 138)
Scotopic vision (p. 138)

Photopic spectral sensitivity curve (p. 139)
Scotopic spectral sensitivity curve (p. 139)
Purkinje effect (p. 139)
Fixational eye movements (p. 141)
Saccades (p. 141)
Transduction (p. 141)
Rhodopsin (p. 141)
Absorption spectrum (p. 141)

6.3 From Retina to Primary Visual Cortex

Retina-geniculate-striate pathways (p. 142)
Primary visual cortex (p. 142)
Lateral geniculate nuclei (p. 142)
Retinotopic (p. 143)
Parvocellular layers (p. 143)
Magnocellular layers (p. 143)

6.4 Seeing Edges

Contrast enhancement (p. 145)
Lateral inhibition (p. 145)

Receptive field (p. 146)
Monocular (p. 146)
On-center cells (p. 146)
Off-center cells (p. 146)
Simple cells (p. 147)
Complex cells (p. 148)
Binocular (p. 148)

6.5 Seeing Color

Component theory (p. 151)
Opponent-process theory (p. 151)
Complementary colors (p. 151)
Color constancy (p. 152)
Retinex theory (p. 153)
Dual-opponent color cells (p. 153)
Cytochrome oxidase (p. 153)
Blobs (p. 153)

6.6 Cortical Mechanisms of Vision and Conscious Awareness

Secondary visual cortex (p. 154)
Visual association cortex (p. 154)

Prestriate cortex (p. 154)
Inferotemporal cortex (p. 154)
Posterior parietal cortex (p. 154)
Scotoma (p. 155)
Perimetry test (p. 155)
Hemianopsic (p. 155)
Conscious awareness (p. 155)
Blindsight (p. 155)
Dorsal stream (p. 157)
Ventral stream (p. 157)
"Where" versus "what" theory (p. 157)
"Control of behavior" versus "conscious perception" theory (p. 158)
Prosopagnosia (p. 159)
Agnosia (p. 159)
Visual agnosia (p. 159)
Akinetopsia (p. 160)

✓•⟶Quick Review Test your comprehension of the chapter with this brief practice test. You can find the answers to these questions as well as more practice tests, activities, and other study resources at www.mypsychlab.com.

1. Fortification illusions are often associated with
   a. acuity.
   b. migraines.
   c. binocular disparity.
   d. epilepsy.
   e. surface interpolation.

2. Photopic vision is
   a. achromatic.
   b. rod-mediated.
   c. limited to the periphery of the retina.
   d. all of the above
   e. none of the above

3. The major advantage of the retinex theory over the classic component and opponent-process theories of color vision is that the retinex theory
   a. can explain color constancy.
   b. is newer.
   c. can explain Mondrians.
   d. can explain the perception of blobs.
   e. all of the above

4. Damage to the fusiform face area is often associated with
   a. akinetopsia.
   b. blindsight.
   c. prosopagnosia.
   d. blockage of the dorsal stream.
   e. hemianopsia.

5. The middle temporal (MT) area of human cortex appears to play an important role in the perception of
   a. motion.
   b. faces.
   c. illusions.
   d. akinetopsia.
   e. color.

7

Mechanisms of Perception: Hearing, Touch, Smell, Taste, and Attention

How You Know the World

Two chapters in this text focus primarily on sensory systems. Chapter 6 was the first, and this is the second. Chapter 6 introduced the visual system; this chapter focuses on the remaining four of the five **exteroceptive sensory systems**: the *auditory* (hearing), *somatosensory* (touch), *olfactory* (smell), and *gustatory* (taste) systems. In addition, this chapter describes the mechanisms of attention: how our brains manage to attend to a small number of sensory stimuli despite being continuously bombarded by thousands of them.

Before you begin the first section of this chapter, consider the following case (Williams, 1970). As you read the chapter, think about this patient, about the nature of his deficit, and the likely location of his brain damage. By the time you have reached the final section of this chapter, you will better understand this patient's problem.

The Case of the Man Who Could See Only One Thing at a Time

A 68-year-old patient was referred because he had difficulty finding his way around—even around his own home. The patient attributed his problems to his "inability to see properly." It was found that if two objects (e.g., two pencils) were held in front of him at the same time, he could see only one of them, whether they were held side by side, one above the other, or even one partially behind the other. Pictures of single objects or faces could be identified, even when quite complex; but if a picture included two objects, only one object could be identified at one time, though that one would sometimes fade, whereupon the other would enter the patient's perception. If a sentence were presented in a line, only the rightmost word could be read, but if one word were presented spread over the entire area previously covered by the sentence, the word could be read in its entirety. If the patient was shown overlapping drawings (i.e., one drawn on top of another), he would see one but deny the existence of the other.

Clinical Implications

7.1

Principles of Sensory System Organization

The visual system, which you learned about in Chapter 6, is by far the most thoroughly studied sensory system and, as a result, the most well understood. As more has been discovered about the other sensory systems, it has become increasingly clear that they are organized in a way similar to the visual system.

The sensory areas of the cortex are, by convention, considered to be of three fundamentally different types: primary, secondary, and association. The **primary sensory cortex** of a system is the area of sensory cortex that receives most of its input directly from the thalamic relay nuclei of that system. For example, as you learned in Chapter 6, the primary visual cortex is the area of the cerebral cortex that receives most of its input from the lateral geniculate nucleus of the thalamus. The **secondary sensory cortex** of a system comprises the areas of the sensory cortex that receive most of their input from the primary sensory cortex of that system or from other areas of the secondary sensory cortex of the same system. **Association cortex** is any area of cortex that receives input from more than one sensory system. Most input to areas of association cortex comes via areas of secondary sensory cortex.

The interactions among these three types of sensory cortex and among other sensory structures are characterized by three major principles: hierarchical organization, functional segregation, and parallel processing.

Hierarchical Organization

Sensory systems are characterized by **hierarchical organization**. A hierarchy is a system whose members can be assigned to specific levels or ranks in relation to one another. For example, an army is a hierarchical system because all soldiers are ranked with respect to their authority. In the same way, sensory structures are organized in a hierarchy on the basis of the specificity and complexity of their function (see Figure 7.1). As one moves through a sensory system from receptors, to thalamic nuclei, to

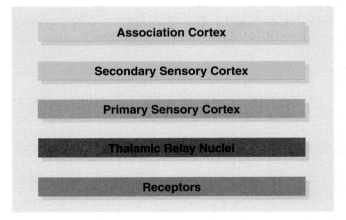

FIGURE 7.1   The hierarchical organization of the sensory systems. The receptors perform the simplest and most general analyses, and the association cortex performs the most complex and specific analyses.

primary sensory cortex, to secondary sensory cortex, to association cortex, one finds neurons that respond optimally to stimuli of greater and greater specificity and complexity. Each level of a sensory hierarchy receives most of its input from lower levels and adds another layer of analysis before passing it on up the hierarchy (see Rees, Kreiman, & Koch, 2002).

The hierarchical organization of sensory systems is apparent from a comparison of the effects of damage to various levels: The higher the level of damage, the more specific and complex the deficit. For example, destruction of a sensory system's receptors produces a complete loss of ability to perceive in that sensory modality (e.g., total blindness or deafness); in contrast, destruction of an area of association or secondary sensory cortex typically produces complex and specific sensory deficits, while leaving fundamental sensory abilities intact. Dr. P., the man who mistook his wife for a hat (Sacks, 1985), displayed such a pattern of deficits.

The Case of the Man Who Mistook His Wife for a Hat

Dr. P. was a highly respected musician and teacher—a charming and intelligent man. He had been referred to eminent neurologist Oliver Sacks for help with a vision *Clinical* problem. At least, as Dr. P. explained to *Implications* the neurologist, other people seemed to think that he had a vision problem, and he did admit that he sometimes made odd errors.

Dr. Sacks tested Dr. P.'s vision and found his visual acuity to be excellent—Dr. P. could easily spot a pin on the floor. The first sign of a problem appeared when Dr. P. needed to put his shoe back on following a standard reflex test. Gazing at his foot, he asked Sacks if it was his shoe.

Continuing the examination, Dr. Sacks showed Dr. P. a glove and asked him what it was. Taking the glove and puzzling over it, Dr. P. could only guess that it was a container divided into five compartments for some reason. Even when Sacks asked whether the glove might fit on some part of the body, Dr. P. displayed no signs of recognition.

At that point, Dr. P. seemed to conclude that the examination was over and, from the expression on his face, that he had done rather well. Preparing to leave, he turned and grasped his wife's head and tried to put it on his own. Apparently, he thought it was his hat.

Mrs. P. showed little surprise. That kind of thing happened a lot.

(Reprinted with the permission of Simon & Schuster, Inc. and Pan Macmillan, London from *The Man Who Mistook His Wife for a Hat and Other Clinical Tales* by Oliver Sacks. Copyright © 1970, 1981, 1983, 1984, 1986 by Oliver Sacks. Electronic rights with permission of the Wylie Agency.)

In recognition of the hierarchical organization of sensory systems, psychologists divide the general process of perceiving into two general phases: sensation and perception. **Sensation** is the process of detecting the presence of stimuli, and **perception** is the higher-order process of integrating, recognizing, and interpreting complete patterns of sensations. Dr. P.'s problem was clearly one of visual perception, not visual sensation.

Functional Segregation

It was once assumed that the primary, secondary, and association areas of a sensory system were each *functionally homogeneous*. That is, it was assumed that all areas of cortex at any given level of a sensory hierarchy acted together to perform the same function. However, research has shown that **functional segregation**, rather than functional homogeneity, characterizes the organization of sensory systems. It is now clear that each of the three levels of cerebral cortex—primary, secondary, and association—in each sensory system contains functionally distinct areas that specialize in different kinds of analysis.

Parallel Processing

It was once believed that the different levels of a sensory hierarchy were connected in a serial fashion. A *serial system* is a system in which information flows among the components over just one pathway, like a string through a strand of beads. However, there is now evidence that sensory systems are *parallel systems*—systems in which information flows through the components over multiple pathways. Parallel systems feature **parallel processing**—the simultaneous analysis of a signal in different ways by the multiple parallel pathways of a neural network.

There appear to be two fundamentally different kinds of parallel streams of analysis in our sensory systems: one that is capable of influencing our behavior without our conscious awareness and one that influences our behavior by engaging our conscious awareness.

Summary Model of Sensory System Organization

Figure 7.2 summarizes the information in this section of the chapter by illustrating how thinking about the organization of sensory systems has changed. In the 1960s, sensory systems were believed to be hierarchical, functionally homogeneous, and serial. However, subsequent research has established that sensory systems are hierarchical, functionally segregated, and parallel (see Tong, 2003).

Sensory systems are characterized by a division of labor: Multiple specialized areas, at multiple levels, are interconnected by multiple parallel pathways. For example, each area of the visual system is specialized for perceiving specific aspects of visual scenes (e.g., shape, color,

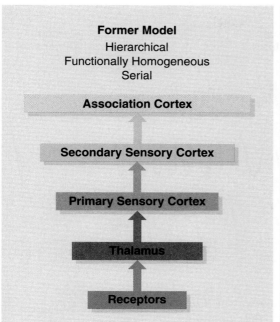
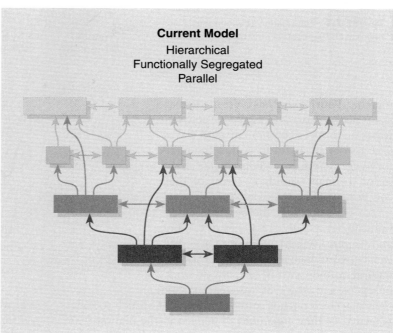

FIGURE 7.2   Two models of sensory system organization: The former model was hierarchical, functionally homogeneous, and serial; the current model, which is more consistent with the evidence, is hierarchical, functionally segregated, and parallel. Not shown in the current model are the many descending pathways that are means by which higher levels of sensory systems can influence sensory input.

movement). Yet, complex stimuli are normally perceived as integrated wholes, not as combinations of independent attributes. How does the brain combine individual sensory attributes to produce integrated perceptions? This is called the *binding problem* (see Billock & Tsou, 2004; Botly & De Rosa, 2008).

One possible solution to the binding problem is that there is a single area of the cortex at the top of the sensory hierarchy that receives signals from all other areas of the sensory system and puts them together to form perceptions; however, there are no areas of cortex to which all areas of a single sensory system report. It seems, then, that perceptions must be a product of the combined activity of different interconnected cortical areas.

Not shown in Figure 7.2 are the many neurons that descend through the sensory hierarchies. Although most sensory neurons carry information from lower to higher levels of their respective sensory hierarchies, some conduct in the opposite direction (from higher to lower levels). These are said to carry top-down signals (see Saalmann, Pigarev, & Vidyasagar, 2007; Sillito, Cudeiro, & Jones, 2006; Yantis, 2008).

Now that you have an understanding of the general principles of sensory system organization, let's take a look at the auditory system, the somatosensory system, and the chemical sensory systems (smell and taste).

7.2
Auditory System

The function of the auditory system is the perception of sound—or, more accurately, the perception of objects and events through the sounds that they make. Sounds are vibrations of air molecules that stimulate the auditory system; humans hear only those molecular vibrations between about 20 and 20,000 hertz (cycles per second). Figure 7.3 on page 168 illustrates how sounds are commonly recorded in the form of waves and the relation between the physical dimensions of sound vibrations and our perceptions of them. The *amplitude, frequency,* and *complexity* of the molecular vibrations are most closely linked to perceptions of *loudness, pitch,* and *timbre,* respectively.

Pure tones (sine wave vibrations) exist only in laboratories and sound recording studios; in real life, sound is always associated with complex patterns of vibrations. For example, Figure 7.4 on page 168 illustrates the complex sound wave associated with one note of a clarinet. The figure also illustrates that any complex sound wave can be broken down mathematically into a series of sine waves of various frequencies and amplitudes; these component sine waves produce the original sound when they are added

FIGURE 7.3 The relation between the physical and perceptual dimensions of sound.

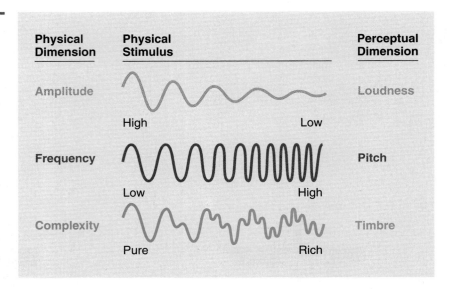

together. **Fourier analysis** is the mathematical procedure for breaking down complex waves into their component sine waves. One theory of audition is that the auditory system performs a Fourier-like analysis of complex sounds in terms of their component sine waves.

For any pure tone, there is a close relationship between the frequency of the tone and its perceived pitch; however, the relation between the frequencies that make up natural sounds (which are always composed of a mixture of frequencies) and their perceived pitch is complex: The pitch of such sounds is related to their *fundamental frequency* (the highest frequency of which the various component frequencies of a sound are multiples). For example, a sound that is a mixture of 100, 200, and 300 Hz frequencies normally has a pitch related to 100 Hz, because 100 Hz is the highest frequency of which the three components are multiples. An extremely important characteristic of pitch perception is the fact that the pitch of a complex sound may not be directly related to the frequency of any of the sound's components (see Bendor & Wang, 2006). For example, a mixture of pure tones with frequencies of 200, 300, and 400 Hz would be perceived as having the same pitch as a pure tone of 100 Hz—because 100 Hz is the fundamental frequency of 200, 300, and 400 Hz. This important aspect of pitch perception is referred to as the *missing fundamental.*

The Ear

The ear is illustrated in Figure 7.5. Sound waves travel from the outer ear down the auditory canal and cause the **tympanic membrane** (the eardrum) to vibrate. These vibrations are then transferred to the three **ossicles**—the small bones of the middle ear: the *malleus* (the hammer), the *incus* (the anvil), and the stapes (the stirrup). The vibrations of the stapes trigger vibrations of the membrane called the

oval window, which in turn transfers the vibrations to the fluid of the snail-shaped **cochlea** (*kokhlos* means "land snail"). The cochlea is a long, coiled tube with an internal membrane running almost to its tip. This internal membrane is the auditory receptor organ, the **organ of Corti**.

⊙➔ **Simulate** Structures of the Human Ear **www.mypsychlab.com**

Each pressure change at the oval window travels along the organ of Corti as a wave. The organ of Corti is composed of two membranes: the basilar membrane and the tectorial membrane. The auditory receptors, the **hair cells**, are mounted in the **basilar membrane**, and the **tectorial membrane** rests on the hair cells (Kelly & Chen, 2009). Accordingly, a deflection of the organ of Corti at any point along its length produces a shearing force on the hair cells at the same point (Kelley, 2006). This force stimulates the hair cells, which in turn increase firing in axons

FIGURE 7.4 The breaking down of a sound—in this case, the sound of a clarinet—into its component sine waves by Fourier analysis. When added together, the component sine waves produce the complex sound wave.

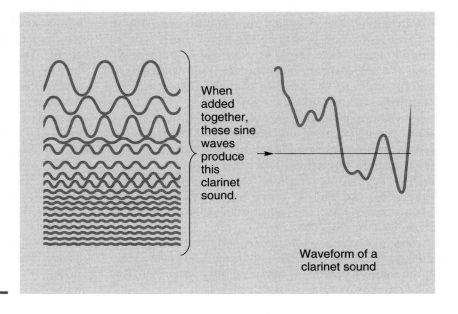

When added together, these sine waves produce this clarinet sound.

Waveform of a clarinet sound

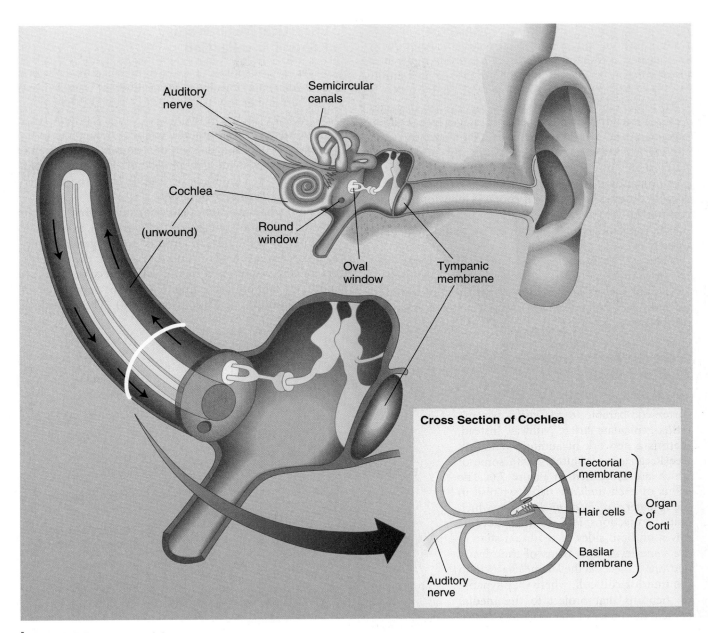

FIGURE 7.5   Anatomy of the ear.

of the **auditory nerve**—a branch of cranial nerve VIII (the *auditory-vestibular nerve*). The vibrations of the cochlear fluid are ultimately dissipated by the *round window*, an elastic membrane in the cochlea wall.

Simulate   Sound Waves in the Human Ear
www.mypsychlab.com

The cochlea is remarkably sensitive (see Dallos, 2008; Ren & Gillespie, 2007). Humans can hear differences in pure tones that differ in frequency by only 0.2%. The major principle of cochlear coding is that different frequencies produce maximal stimulation of hair cells at different points along the basilar membrane—with higher frequencies producing greater activation closer to the windows and lower frequencies producing greater

activation at the tip of the basilar membrane. Thus, the many component frequencies that compose each complex sound activate hair cells at many different points along the basilar membrane, and the many signals created by a single complex sound are carried out of the ear by many different auditory neurons. Like the cochlea, most other structures of the auditory system are arrayed according to frequency. Thus, in the same way that the organization of the visual system is primarily **retinotopic**, the organization of the auditory system is primarily **tonotopic**.

This brings us to the major unsolved mystery of auditory processing. Imagine yourself in a complex acoustic environment such as a party. The music is playing; people are dancing, eating, and drinking; and numerous conversations are going on around you. Because the component frequencies in each individual sound activate many sites along your basilar membrane, the number of sites simultaneously activated at any one time by the party noises is enormous. But somehow your auditory system manages to sort these individual frequency messages into separate categories and combine them so that you hear each source of complex sounds independently (see Feng & Ratnam, 2000). For example, you hear the speech of the person standing next to you as a separate sequence of sounds, despite the fact that it contains many of the same component frequencies coming from other sources. The mechanism underlying this important ability remains a mystery.

Figure 7.5 also shows the **semicircular canals**—the receptive organs of the **vestibular system**. The vestibular system carries information about the direction and intensity of head movements, which helps us maintain our balance.

From the Ear to the Primary Auditory Cortex

There is no major auditory pathway to the cortex comparable to the visual system's retina-geniculate-striate pathway. Instead, there is a network of auditory pathways (see Recanzone & Sutter, 2008), some of which are illustrated in Figure 7.6. The axons of each *auditory nerve* synapse in the ipsilateral *cochlear nuclei*, from which many projections lead to the **superior olives** on both sides of the brain stem at the same level. The axons of the olivary neurons project via the *lateral lemniscus* to the **inferior colliculi**, where they synapse on neurons that project to the **medial geniculate nuclei** of the thalamus, which in turn project to the *primary auditory cortex*. Notice that signals from each ear are combined at a very low level (in the superior olives) and are transmitted to both ipsilateral and contralateral auditory cortex.

Because of the complexity of the subcortical auditory pathways, their analysis has been difficult. However, there is one function of the subcortical auditory system that is well understood: the localization of sounds in space.

FIGURE 7.6 Some of the pathways of the auditory system that lead from one ear to the cortex.

Subcortical Mechanisms of Sound Localization

Localization of sounds in space is mediated by the lateral and medial superior olives, but in different ways. When a sound originates to a person's left, it reaches the left ear first, and it is louder at the left ear. Some neurons in the *medial superior olives* respond to slight differences in the time of arrival of signals from the two ears, whereas some neurons in the *lateral superior olives* respond to slight differences in the amplitude of sounds from the two ears (see Heffner & Masterton, 1990).

The medial and lateral superior olives project to the *superior colliculus* (not shown in Figure 7.6), as well as to the inferior colliculus. In contrast to the general tonotopic organization of the auditory system, the deep layers of the superior colliculi, which receive auditory input, are laid out according to a map of auditory space (King, Schnupp, & Thompson, 1998). The superficial

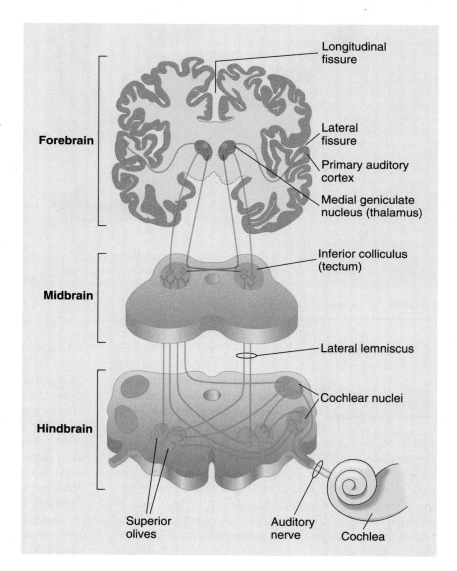

layers of the superior colliculi, which receive visual input, are organized retinotopically. Thus, it appears that the general function of the superior colliculi is locating sources of sensory input in space.

Many researchers interested in sound localization have studied barn owls because these owls can locate sources of sounds better than any other animal whose hearing has been tested (see Konishi, 2003). They are nocturnal hunters and must be able to locate field mice solely by the rustling sounds the mice make in the dark. Not surprisingly, the auditory neurons of the barn owl's superior colliculus region are very finely tuned; that is, each neuron responds to sounds from only a particular location in the range of the owl's hearing (see Cohen & Knudsen, 1999).

Auditory Cortex

Recent progress in the study of human auditory cortex has resulted from the convergence of functional brain-imaging studies in humans and invasive neural recording studies in monkeys (Cohen, Russ, & Gifford, 2005). Still, primate auditory cortex is far from being well understood—for example, our understanding of it lags far behind our current understanding of visual cortex.

In primates, the primary auditory cortex, which receives the majority of its input from the medial geniculate nucleus, is located in the temporal lobe, hidden from view within the lateral fissure (see Figure 7.7). Adjacent to the primary auditory area in each hemisphere are two other areas: Together these three areas are referred to as the *core region*. Surrounding the core region is a band—often called the *belt*—of areas of secondary cortex. Areas of secondary auditory cortex outside the belt are called *parabelt*

areas. There about 10 separate areas of secondary auditory cortex in primates (see Bendor & Wang, 2006).

Organization of Primate Auditory Cortex   Two important principles of organization of primary auditory cortex have been identified. First, like the primary visual cortex, the primary auditory cortex is organized in functional columns (see Schreiner, 1992): All of the neurons encountered during a vertical microelectrode penetration of primary auditory cortex (i.e., a penetration at right angles to the cortical layers) tend to respond optimally to sounds in the same frequency range. Second, like the cochlea, auditory cortex is organized tonotopically (see Schreiner, Read, & Sutter, 2000): Each area of primary and secondary auditory cortex appears to be organized on the basis of frequency.

What Sounds Should Be Used to Study Auditory Cortex?   Why has research on auditory cortex lagged behind research on visual cortex? There are several reasons, but a major one is a lack of clear understanding of the dimensions along which auditory cortex evaluates sound. You may recall from Chapter 6 that research on the visual cortex did not start to progress rapidly until Hubel and Weisel discovered that most visual neurons respond to contrast. There is clear evidence of a hierarchical organization in auditory cortex—the neural responses of secondary auditory cortex tend to be more complex and varied than those of primary auditory cortex (see Scott, 2005). However, the responses at even the lowest cortical level of auditory analysis, the primary auditory cortex, are so complex and varied (see Nelken, 2008) that progress has been limited.

Many neurons in auditory cortex respond only weakly to simple stimuli such as pure tones, but these stimuli have been widely employed in electrophysiological studies of auditory cortex. This practice is changing, partly in response to the discovery that many auditory cortex neurons in monkeys respond robustly to monkey calls (see Romanski & Averbeck, 2009). Our lack of understanding of auditory cortex has created a Catch-22 situation: Because we have no idea what auditory cortex does, it is difficult to ask the right research questions or to probe the system with appropriate acoustic stimuli (Griffiths et al., 2004; Hromádka & Zador, 2009).

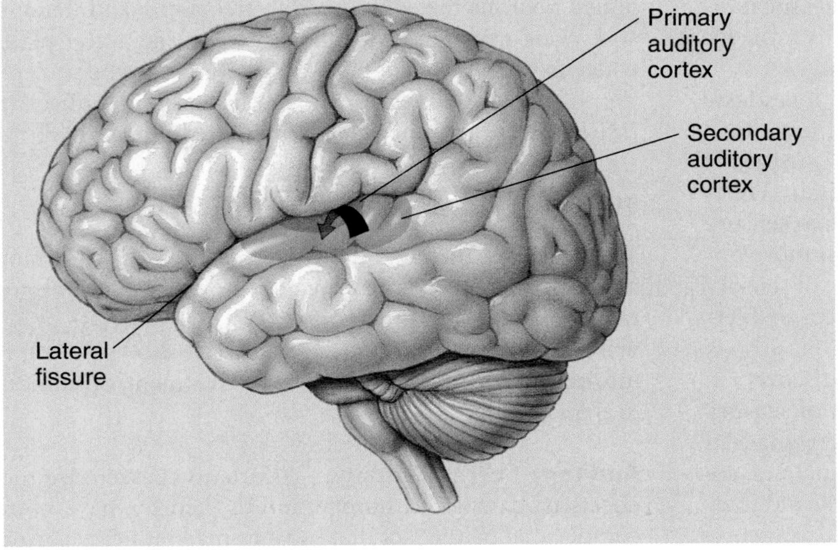

Primary
auditory
cortex

Secondary
auditory
cortex

Lateral
fissure

FIGURE 7.7   General location of the primary auditory cortex and areas of secondary auditory cortex. Most auditory cortex is hidden from view in the temporal cortex of the lateral fissure.

FIGURE 7.8 The hypothesized anterior and posterior auditory pathways.

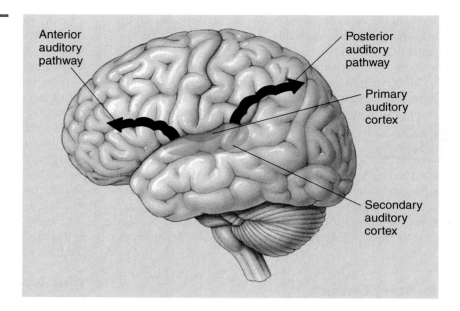

Two Streams of Auditory Cortex

Thinking about the general organization of auditory cortex has been inspired by research on visual cortex. Researchers have proposed that, just as there are two main cortical streams of visual analysis (dorsal and ventral), there are two main cortical streams of auditory analysis. Auditory signals are ultimately conducted to two large areas of association cortex: prefrontal cortex and posterior parietal cortex. It has been hypothesized that the *anterior auditory pathway* is more involved in identifying sounds (what), whereas the *posterior auditory pathway* is more involved in locating sounds (where)—see Hackett and Kaas (2004), Scott (2005), and Lomber and Malhotra (2008). This hypothesis is illustrated in Figure 7.8.

Auditory–Visual Interactions Sensory systems have traditionally been assumed to interact in association cortex. Indeed, as you have already learned, association cortex is usually defined as areas of cortex where such interactions, or associations, take place. Much of the research on sensory system interactions has focused on interactions between the auditory and visual systems, particularly on those that occur in the posterior parietal cortex (see Bulkin & Groh, 2006; Cohen, Russ, & Gifford, 2005). In one study of monkeys (Mulette-Gillman, Cohen, & Groh, 2005), some posterior parietal neurons were found to have visual receptive fields, some were found to have auditory receptive fields, and some were found to have both. Those that had both visual and auditory receptive fields had both fields covering the same location of the subject's immediate environment.

Functional brain imaging is widely used to investigate sensory system interactions. One advantage of functional brain imaging is that it does not focus on any one part of the brain; it records activity throughout the brain. Functional brain-imaging studies have confirmed that sensory interactions do occur in association cortex, but more importantly, they have repeatedly found evidence of sensory interactions at the lowest level of the sensory cortex hierarchy, in areas of primary sensory cortex (see Macaluso & Driver, 2005; Schroeder & Foxe, 2005). This discovery is changing how we think about the interaction of sensory systems: Sensory system interaction is not merely tagged on after *unimodal* (involving one system) analyses are complete; sensory system interaction is an early and integral part of sensory processing.

Where Does the Perception of Pitch Occur? Recent research has answered one fundamental question about auditory cortex: Where does the perception of pitch likely occur? This seemed like a simple question to answer because most areas of auditory cortex have a clear tonotopic organization. However, when experimenters used sound stimuli in which frequency and pitch were different—for example, by using the missing fundamental technique—most auditory neurons responded to changes in frequency rather than pitch. This information led Bendor and Wang (2005) to probe primary and secondary areas of monkey auditory cortex with microelectrodes to assess the responses of individual neurons to missing fundamental stimuli. They discovered one small area just anterior to primary auditory cortex that contained many neurons that responded to pitch rather than frequency, regardless of the quality of the sound. The same small area also contained neurons that responded to frequency, and Bendor and Wang suggested that this area was likely the place where frequencies of sound were converted to the perception of pitch. A similar pitch area has been identified by fMRI studies in a similar location in the human brain.

Effects of Damage to the Auditory System

The study of damage to the auditory system is important for two reasons. First, it provides information about how the auditory system works. Second, it can serve as a source of *Clinical Implications* information about the causes and treatment of clinical deafness.

Auditory Cortex Damage Efforts to characterize the effects of damage to human auditory cortex have been complicated by the fact that most human auditory cortex is in the lateral fissure. Consequently, it is rarely destroyed in its entirety; and if it is, there is almost always extensive damage to surrounding tissue. As a result, efforts to understand the effects of auditory cortex damage have relied largely on the study of surgically placed lesions in nonhumans.

Most studies of the effects of auditory cortex lesions have assessed the effects of large lesions that involve the core region and most of the belt and parabelt areas. Given the large size of the lesions in most studies, the lack of severe permanent deficits is surprising, suggesting that the subcortical circuits serve more complex and important auditory functions than was once assumed. For example, few permanent hearing deficits of any kind have been reliably detected in rats following auditory cortex lesions.

Although the effects of auditory cortex lesions depend somewhat on the species, the effects in humans and monkeys appear to be quite similar (see Heffner & Heffner, 2003), as far as we can tell, given the small number of relevant human cases. Following bilateral lesions, there is often a complete loss of hearing, which presumably results from the shock of the lesion, because hearing recovers in the ensuing weeks. The major permanent effects are loss of the ability to localize sounds and impairment of the ability to discriminate frequencies (see Heffner & Heffner, 2003).

Evolutionary Perspective

The effects of unilateral auditory cortex lesions suggest that the system is partially contralateral. A unilateral lesion disrupts the ability to localize sounds in space contralateral, but not ipsilateral, to the lesion. However, other auditory deficits produced by unilateral auditory cortex lesions tend to be only slightly greater for contralateral sounds.

Deafness in Humans Deafness is one of the most prevalent human disabilities: An estimated 250 million people currently suffer from disabling hearing impairments (Taylor & Forge, 2005). Total deafness is rare, occurring in only 1% of hearing-impaired individuals. The rarity of total deafness is likely a consequence of the diffuse, parallel network of auditory pathways: If one auditory brain structure is destroyed, alternative pathways over which auditory information can flow remain.

Because of the parallel organization of the auditory system, severe hearing problems typically result from damage to the inner ear or the middle ear or to the nerves leading from them, rather than from more central damage. There are two common classes of hearing impairments: those associated with damage to the ossicles (*conductive deafness*) and those associated with damage to the cochlea or auditory nerve (*nerve deafness*). In this book on the nervous system, we are more concerned with nerve deafness. The major common cause of nerve deafness is a loss of hair cell receptors (Taylor & Forge, 2005).

If only part of the cochlea is damaged, individuals may have nerve deafness for some frequencies but not others. This is a characteristic of age-related hearing loss. The first age-related hearing loss to develop is usually a specific deficit in perceiving high frequencies. That is why elderly people often have difficulty distinguishing "s," "f," and "t" sounds: They can hear people speaking to them but often have difficulty understanding what people are saying. Often, relatives and friends do not realize that much of the confusion displayed by elderly people stems from difficulty discriminating sounds (see Wingfield, Tun, & McCoy, 2005).

Hearing loss is sometimes associated with **tinnitus** (ringing of the ears). When only one ear is damaged, the ringing is perceived as coming from that ear; however, cutting the nerve from the ringing ear has no effect on the ringing. *Neuroplasticity* This suggests that changes to the central auditory system that were caused by the deafness are the cause of tinnitus (Eggermont & Roberts, 2004).

Some people with nerve deafness benefit from cochlear implants (see Figure 7.9 on page 174). *Cochlear implants* bypass damage to the auditory hair cells by converting sounds picked up by a microphone on the patient's ear to electrical signals, which are then carried into the cochlea by a bundle of electrodes. These signals excite the auditory nerve. Although cochlear implants can provide major benefits, they do not restore normal hearing. The sooner a person receives a cochlear implant after becoming deaf, the more likely he or she is to benefit, because disuse leads to degeneration of the auditory neural pathways (see Ryugo, Kretzmer, & Niparko, 2005).

Scan Your Brain

Before we go on to discuss the other sensory systems, pause and test your knowledge of what you have learned in this chapter so far. The correct answers are provided at the end of the exercise. Before proceeding, review material related to your errors and omissions.

1. Areas of cortex that receive input from more than one sensory system are called _____ cortex.
2. The three principles of sensory system organization are hierarchical organization, _____, and parallel processing.
3. Fourier analysis breaks down complex sounds into component _____ waves.
4. The highest frequency of which the various component frequencies of a sound are multiples is their _____ frequency.
5. The middle _____ is the incus.
6. The auditory nerve is a branch of cranial nerve VIII, the _____ nerve.
7. The layout of the auditory system tends to be _____.

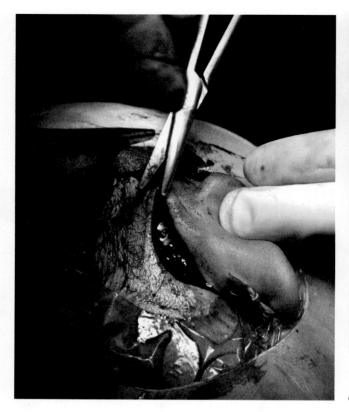

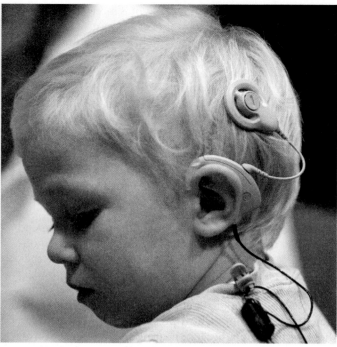

FIGURE 7.9 Cochlear implant: The surgical implantation is shown on the left, and a child with an implant is shown on the right.

8. The axons of the auditory nerves synapse in the ipsilateral _____ nuclei.

9. One function of the superior olives is sound _____.

10. Some areas of secondary auditory cortex are located in a band that is adjacent to and surrounds the core region. These areas are often called the _____ areas.

11. Although _____ have been widely used in electrophysiological studies, auditory neurons often respond only weakly to them.

12. Many studies of auditory-visual interactions have focused on association cortex in the posterior _____ cortex.

Scan Your Brain answers: (1) association, (2) functional segregation, (3) sine, (4) fundamental, (5) ossicle, (6) auditory-vestibular, (7) tonotopic, (8) cochlear, (9) localization, (10) belt, (11) pure tones, (12) parietal.

7.3

Somatosensory System: Touch and Pain

Sensations from your body are referred to as *somatosensations*. The system that mediates these bodily sensations—the *somatosensory system*—is, in fact, three separate but interacting systems: (1) an *exteroceptive system*, which

senses external stimuli that are applied to the skin; (2) a *proprioceptive system*, which monitors information about the position of the body that comes from receptors in the muscles, joints, and organs of balance; and (3) an *interoceptive system*, which provides general information about conditions within the body (e.g., temperature and blood pressure). This discussion deals almost exclusively with the exteroceptive system, which itself comprises three somewhat distinct divisions: a division for perceiving *mechanical stimuli* (touch), one for *thermal stimuli* (temperature), and one for *nociceptive stimuli* (pain).

Cutaneous Receptors

There are many kinds of receptors in the skin (see Johnson, 2001). Figure 7.10 illustrates four of them. The simplest cutaneous receptors are the **free nerve endings** (neuron endings with no specialized structures on them), which are particularly sensitive to temperature change and pain. The largest and deepest cutaneous receptors are the onionlike **Pacinian corpuscles**; because they adapt rapidly, they respond to sudden displacements of the skin but not to constant pressure. In contrast, *Merkel's disks* and *Ruffini endings* both adapt slowly and respond to gradual skin indentation and skin stretch, respectively.

To appreciate the functional significance of fast and slow receptor adaptation, consider what happens when a constant pressure is applied to the skin. The pressure evokes a burst of firing in all receptors, which corresponds

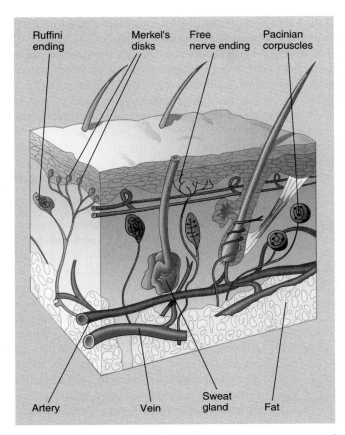

FIGURE 7.10 Four cutaneous receptors that occur in human skin.

Dermatomes

The neural fibers that carry information from cutaneous receptors and other somatosensory receptors gather together in nerves and enter the spinal cord via the *dorsal roots*. The area of the body that is innervated by the left and right dorsal roots of a given segment of the spinal cord is called a **dermatome**. Figure 7.11 is a dermatomal map of the human body. Because there is considerable overlap between adjacent dermatomes, destruction of a single dorsal root typically produces little somatosensory loss.

Two Major Somatosensory Pathways

Somatosensory information ascends from each side of the body to the human cortex over two major pathways: the dorsal-column medial-lemniscus system and the anterolateral system. The **dorsal-column medial-lemniscus system** tends to carry information about touch and proprioception, and the **anterolateral system** tends to

to the sensation of being touched; however, after a few hundred milliseconds, only the slowly adapting receptors remain active, and the quality of the sensation changes. In fact, you are often totally unaware of constant skin pressure; for example, you are usually unaware of the feeling of your clothes against your body until you focus attention on it. As a consequence, when you try to identify objects by touch, you manipulate them in your hands so that the pattern of stimulation continually changes. (The identification of objects by touch is called **stereognosis**.) Having some receptors that adapt quickly and some that adapt slowly provides information about both the dynamic and static qualities of tactual stimuli.

The structure and physiology of each type of somatosensory receptor are specialized, allowing the receptor to be sensitive to a particular type of tactual stimulation. However, in general, the various receptors tend to function in the same way: Stimuli applied to the skin deform or change the chemistry of the receptor, and this in turn changes the permeability of the receptor cell membrane to various ions (see Lumpkin & Bautista, 2005; Tsunozaki & Bautista, 2009). The result is a neural signal.

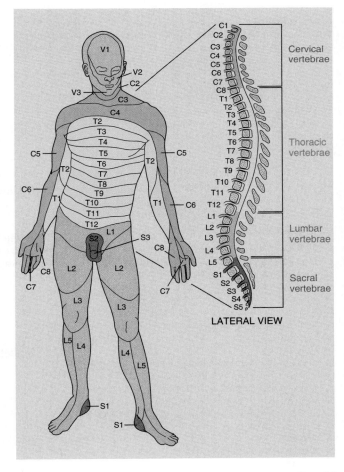

FIGURE 7.11 The dermatomes of the human body. S, L, T, and C refer respectively to the sacral, lumbar, thoracic, and cervical regions of the spinal cord. V1, V2, and V3 stand for the three branches of the trigeminal nerve.

carry information about pain and temperature. The key words in the preceding sentence are "tends to": The separation of function in the two pathways is far from complete. Accordingly, lesions of the dorsal-column medial-lemniscus system do not eliminate touch perception or proprioception, and lesions of the anterolateral system do not eliminate perception of pain or temperature.

The dorsal-column medial-lemniscus system is illustrated in Figure 7.12. The sensory neurons of this system enter the spinal cord via a dorsal root, ascend ipsilaterally in the **dorsal columns**, and synapse in the *dorsal column nuclei* of the medulla. The axons of dorsal column nuclei neurons *decussate* (cross over to the other side of the brain) and then ascend in the **medial lemniscus** to the contralateral **ventral posterior nucleus** of the thalamus. The ventral posterior nuclei also receive input via the three branches of the trigeminal nerve, which carry somatosensory information from the contralateral areas of the face. Most neurons of the ventral posterior nucleus project to the *primary somatosensory cortex (SI)*; others project to the *secondary somatosensory cortex (SII)* or the posterior parietal cortex. Neuroscience trivia buffs will almost certainly want to add to their collection the fact that the dorsal column neurons that originate in the toes are the longest neurons in the human body.

The anterolateral system is illustrated in Figure 7.13. Most dorsal root neurons of the anterolateral system synapse as soon as they enter the spinal cord. The axons of most of the second-order neurons decussate but then ascend to the brain in the contralateral anterolateral portion of the spinal cord; however, some do not decussate but ascend ipsilaterally. The anterolateral system comprises three different tracts: the *spinothalamic tract*, which projects to the ventral posterior nucleus of the thalamus (as does the dorsal-column medial-lemniscus system); the *spinoreticular tract*, which projects to the *reticular formation* (and then to the *parafascicular nuclei* and *intralaminar nuclei* of the thalamus); and the *spinotectal tract*, which projects to the *tectum* (colliculi). The three branches of the trigeminal nerve carry pain and temperature information from the face to the same thalamic sites. The pain and temperature information that reaches the thalamus is then distributed to SI, SII, posterior parietal cortex, and other parts of the brain.

If both ascending somatosensory paths are completely transected by a spinal injury, the patient can feel no body sensation from below the level of the cut. Clearly, when it comes to spinal injuries, lower is better.

Mark, Ervin, and Yakolev (1962) assessed the effects of lesions to the thalamus on the chronic pain of patients in the

Clinical Implications

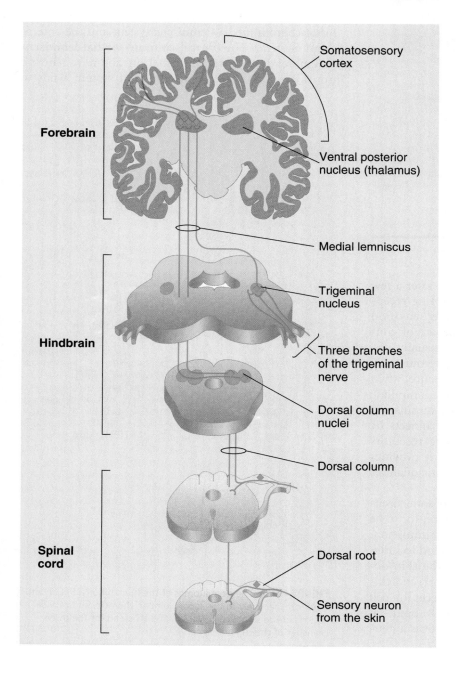

Somatosensory cortex

Ventral posterior nucleus (thalamus)

Medial lemniscus

Trigeminal nucleus

Three branches of the trigeminal nerve

Dorsal column nuclei

Dorsal column

Dorsal root

Sensory neuron from the skin

Forebrain

Hindbrain

Spinal cord

FIGURE 7.12 The dorsal-column medial-lemniscus system. The pathways from only one side of the body are shown.

FIGURE 7.13 The anterolateral system. The pathways from only one side of the body are shown.

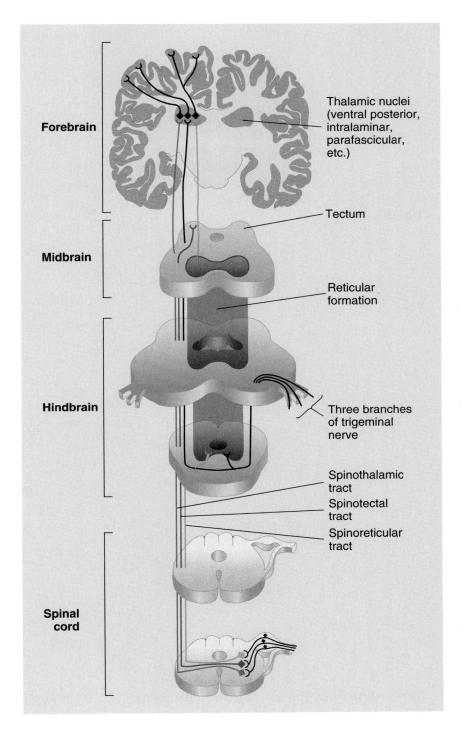

Clinical Implications

advanced stages of cancer. Lesions to the ventral posterior nuclei, which receive input from both the spinothalamic tract and the dorsal-column medial-lemniscus system, produced some loss of cutaneous sensitivity to touch, to temperature change, and to sharp pain; but the lesions had no effect on deep, chronic pain. In contrast, lesions of the parafascicular and intralaminar nuclei, both of which receive input from the spinoreticular tract, reduced deep chronic pain without disrupting cutaneous sensitivity.

Cortical Areas of Somatosensation

In 1937, Penfield and his colleagues mapped the primary somatosensory cortex of patients during neurosurgery (see Figure 7.14 on page 178). Penfield applied electrical stimulation to various sites on the cortical surface, and the patients, who were fully conscious under a local anesthetic, described what they felt. When stimulation was applied to the *postcentral gyrus*, the patients reported somatosensory sensations in various parts of their bodies. When Penfield mapped the relation between each site of stimulation and the part of the body in which the sensation was felt, he discovered that the human primary somatosensory cortex (SI) is **somatotopic**—organized according to a map of the body surface. This somatotopic map is commonly referred to as the **somatosensory homunculus** (*homunculus* means "little man").

Notice in Figure 7.14 that the somatosensory homunculus is distorted; the greatest proportion of SI is dedicated to receiving input from the parts of the body that we use to make tactile discriminations (e.g., hands, lips, and tongue). In contrast, only small areas of SI receive input from large areas of the body, such as the back, that are not usually used to make somatosensory discriminations. The demonstration in the Check It Out on page 179 allows you to experience the impact this organization has on your ability to perceive touches.

A second somatotopically organized area, SII, lies just ventral to SI in the postcentral gyrus, and much of it extends into the lateral fissure. SII receives most of its input from SI and is thus regarded as secondary somatosensory cortex. In contrast to SI, whose input is largely contralateral, SII receives substantial input from both sides of the body. Much of the output of SI and SII goes to the association cortex of the *posterior parietal lobe*.

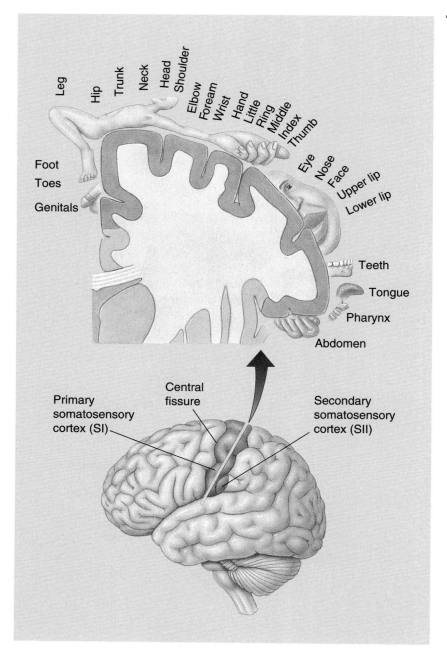

FIGURE 7.14 The locations of human primary somatosensory cortex (SI) and one area of secondary somatosensory cortex (SII) with the conventional portrayal of the somatosensory homunculus. Something has always confused me about this portrayal of the somatosensory homunculus: The body is upside-down, while the head is right side up. It now appears that this conventional portrayal is wrong. The results of an fMRI study suggest that the face representation is also inverted (Servos et al., 1999).

particular column of primary somatosensory cortex had a receptive field on the same part of the body and responded most robustly to the same type of tactile stimuli (e.g., light touch or heat). Moreover, unit recordings suggested that primary somatosensory cortex is composed of four functional strips, each with a similar, but separate, somatotopic organization. Each strip of primary somatosensory cortex is most sensitive to a different kind of somatosensory input (e.g., to light touch or pressure). Thus, if one were to record from neurons in a horizontal line across the four strips, one would find neurons that "preferred" four different kinds of tactile stimulation, all to the same part of the body. Also, one would find that as one moved from anterior to posterior, the preferences of the neurons would tend to become more complex and specific (see Caselli, 1997), suggesting an anterior-to-posterior hierarchical organization (Iwamura, 1998).

Although there is general consensus about the localization of SI and SII, there is still considerable debate about the full extent and organization of somatosensory cortex (see Hsiao, 2008). For example, there is evidence of two narrow bands of secondary cortex, one on each side of SI, and of another band next to SII. Furthermore, SII itself may comprise two or three different areas of secondary cortex. Finally, it has been proposed that two streams of analysis proceed from SI: a dorsal stream that projects to posterior parietal cortex and participates in multisensory integration and direction of attention, and a ventral stream that projects to SII and participates in the perception of objects' shapes (Hsiao, 2008).

The receptive fields of many neurons in the primary somatosensory cortex, like those of visual system neurons, can be divided into antagonistic excitatory and inhibitory areas (DiCarlo & Johnson, 2000; DiCarlo, Johnson, & Hsaio, 1998). Figure 7.15 on page 180 illustrates the receptive field of a neuron of the primary somatosensory cortex that is responsive to light touch (Mountcastle & Powell, 1959).

Studies that systematically explored single neurons in primary somatosensory cortex (e.g., Kaas et al., 1981) found the same columnar organization characteristic of other areas of primary sensory cortex. Each cell in a

TOUCHING A BACK

Because only a small portion of human primary somatosensory cortex receives input from the entire back, people have difficulty recognizing objects that touch their backs. You may not have noticed this tactile deficiency—unless, of course, you often try to identify objects by feeling them with your back. You will need one thing to demonstrate the recognition deficiencies of the human back: a friend. Touch your friend on the back with one, two, or three fingers, and ask your friend how many fingers he or she feels. When using two or three fingers, be sure they touch the back simultaneously because temporal cues invalidate this test of tactile discrimination. Repeat the test many times, adjusting the distance between the touches on each trial. Record the results. What you should begin to notice is that the back is incapable of discriminating between separate touches unless the distance between the touches is considerable. In contrast, fingertips can distinguish the number of simultaneous touches even when the touches are very close.

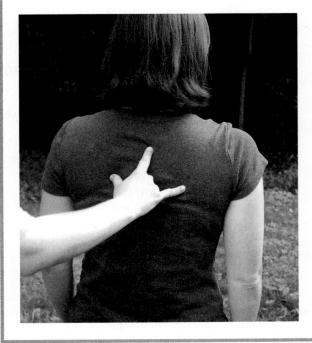

Effects of Damage to the Primary Somatosensory Cortex

Like the effects of damage to the primary auditory cortex, the effects of damage to the primary somatosensory cortex are often remarkably mild—presumably because, like the auditory system, the somatosensory system features numerous parallel pathways. Corkin, Milner, and

Rasmussen (1970) assessed the somatosensory abilities of epileptic patients before and after a unilateral excision that included SI. Following the surgery, the patients displayed two minor contralateral deficits: a reduced ability to detect light touch and a reduced ability to identify objects by touch (i.e., a deficit in stereognosis). These deficits were bilateral only in those cases in which the unilateral lesion encroached on SII.

Somatosensory System and Association Cortex

Somatosensory signals are ultimately conducted to the highest level of the sensory hierarchy, to areas of association cortex in prefrontal and posterior parietal cortex.

Posterior parietal cortex contains *bimodal neurons* (neurons that respond to activation of two different sensory systems) that respond to both somatosensory and visual stimuli. The visual and somatosensory receptive fields of each neuron are spatially related; for example, if a neuron has a somatosensory receptive field centered in the left hand, its visual field is adjacent to the left hand. Remarkably, as the left hand moves, the visual receptive field of the neuron moves with it. The existence of these bimodal neurons motivated the following interesting case study by Schendel and Robertson (2004).

The Case of W.M., Who Reduced His Scotoma with His Hand

W.M. suffered a stroke in his right posterior cerebral artery. The stroke affected a large area of his right occipital and parietal lobes and left him with severe left *hemianopsia* (a condition in which a scotoma covers half the visual field). When tested with his left hand in his lap, W.M. detected 97.8% of the stimuli presented in his *Clinical Implications* right visual field and only 13.6% of those presented in his left visual field. However, when he was tested with his left hand extended into his left visual field, his ability to detect stimuli in his left visual field improved significantly. Further analysis showed that this general improvement resulted from W.M.'s greatly improved ability to see those objects in the left visual field that were near his left hand. *Neuroplasticity* Remarkably, this area of improved performance around his left hand was expanded even further when he held a tennis racket in his extended left hand.

Somatosensory Agnosias

There are two major types of somatosensory agnosia. One is **astereognosia**—the inability to recognize objects by

FIGURE 7.15 The receptive field of a neuron of the primary somatosensory cortex. Notice the antagonistic excitatory and inhibitory areas.

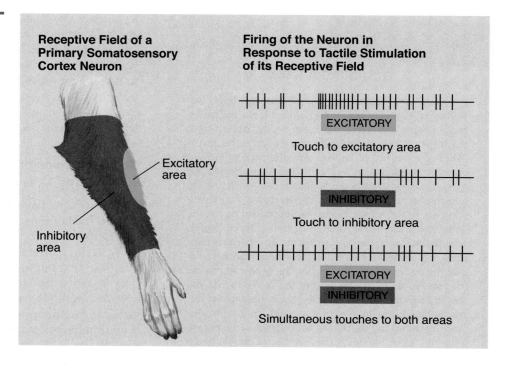

Receptive Field of a Primary Somatosensory Cortex Neuron

Excitatory area

Inhibitory area

Firing of the Neuron in Response to Tactile Stimulation of its Receptive Field

EXCITATORY
Touch to excitatory area

INHIBITORY
Touch to inhibitory area

EXCITATORY
INHIBITORY
Simultaneous touches to both areas

Clinical Implications

touch. Cases of pure astereognosia—those that occur in the absence of simple sensory deficits—are rare (Corkin, Milner, & Rasmussen, 1970). The other type of somatosensory agnosia is **asomatognosia**—the failure to recognize parts of one's own body. Asomatognosia is usually unilateral, affecting only the left side of the body; and it is usually associated with extensive damage to the right posterior parietal lobe. The case of Aunt Betty is an example.

The Case of Aunt Betty, Who Lost Half of Her Body

It was time to see Aunt Betty—she wasn't really my aunt, but I grew up thinking that she was. She was my mother's best friend. She had had a stroke in her right hemisphere.

As we walked to her room, one of the medical students described the case. "Left hemiplegia [left-side paralysis]," I was told.

Aunt Betty was lying on her back with her head and eyes turned to the right. "Betty," I called out. Not Aunt Betty, but Betty. I was 37; I'd dropped the "Aunt" long ago—at least 2 years earlier.

I approached her bed from the left, but Aunt Betty did not turn her head or even her eyes to look towards me.

"Hal," she called out. "Where are you?"

I turned her head gently toward me. We talked. It was clear that she had no speech problems, no memory loss, and no confusion. She was as bright as ever. But her eyes still looked to the right as if the left side of her world did not exist.

I picked up her right hand and held it in front of her eyes. "What's this?" I asked.

"My hand, of course," she said with an intonation that suggested what she thought of my question.

"Well then, what's this?" I said, as I held up her limp left hand where she could see it.

"A hand."

"Whose hand?"

"Your hand, I guess," she replied. She seemed genuinely puzzled. I carefully placed her hand on the bed.

"Why have you come to this hospital?" I asked.

"To see you," she replied hesitantly. I could tell that she didn't really know the answer.

"Is there anything wrong with you?"

"No."

"How about your left hand and leg?"

"They're fine," she said. "How are yours?"

"They're fine too," I replied. There was nothing else to do. Aunt Betty was in trouble.

(Adapted from NEWTON'S MADNESS by Harold Klawans (Harper & Row 1990). Reprinted by permission of Jet Literary Associates, Inc.)

As in the case of Aunt Betty, asomatognosia is often accompanied by **anosognosia**—the failure of neuropsychological patients to recognize their own symptoms. Indeed, anosognosia is a common, but curious, symptom of many neurological disorders.

Asomatognosia is commonly a component of **contralateral neglect**—the tendency not to respond to stimuli that are contralateral to a right-hemisphere injury. You will learn more about contralateral neglect in Chapter 8.

Perception of Pain

A paradox is a logical contradiction. The perception of pain is paradoxical in three important respects, which are explained in the following three subsections.

Adaptiveness of Pain   One paradox of pain is that an experience that seems in every respect to be so bad is in fact extremely important for our survival. There is no

Evolutionary Perspective

special stimulus for pain; it is a response to potentially harmful stimulation of any type (see Craig, 2003). It warns us to stop engaging in potentially harmful activities or to seek treatment (see Basbaum & Julius, 2006).

The value of pain is best illustrated by the cases of people who experience no pain. The way that I think about

Thinking Creatively

pain was forever changed by the case study of Miss C.

The Case of Miss C., the Woman Who Felt No Pain

Miss C., a young Canadian girl who was a student at McGill University in Montreal. . . . The young lady was highly intelligent and seemed normal in every way except that she had never felt pain. As a child, she had bitten off

Clinical Implications

the tip of her tongue while chewing food, and had suffered third-degree burns after kneeling on a radiator to

look out of the window. . . . She felt no pain when parts of her body were subjected to strong electric shock, to hot water at temperatures that usually produce reports of burning pain, or to a prolonged ice-bath. Equally astonishing was the fact that she showed no changes in blood pressure, heart rate, or respiration when these stimuli were presented. Furthermore, she could not remember ever sneezing or coughing, the gag reflex could be elicited only with great difficulty, and corneal reflexes (to protect the eyes) were absent. A variety of other stimuli, such as inserting a stick up through the nostrils, pinching tendons, or injections of histamine under the skin—which are normally considered as forms of torture—also failed to produce pain.

Miss C. had severe medical problems. She exhibited pathological changes in her knees, hip, and spine, and underwent several orthopaedic operations. The surgeon attributed these changes to the lack of protection to joints usually given by pain sensation. She apparently failed to shift her weight when standing, to turn over in her sleep, or to avoid certain postures, which normally prevent inflammation of joints. . . .

Miss C. died at the age of twenty-nine of massive infections . . . and extensive skin and bone trauma.

(From *The Challenge of Pain*, pp. 16–17, by Ronald Melzack and Patrick D. Wall, 1982, London: Penguin Books Ltd. Copyright © Ronald Melzack and Patrick D. Wall, 1982.)

Cox and colleagues (2006) studied six cases of congenital insensitivity to pain among members of a family from Pakistan. They were able to identify the gene abnormality underlying the disorder in these six individuals: a problem with a gene that in-

Clinical Implications

fluences synthesis of the sodium ion channels. This gene could prove to be the target of a new generation of analgesics.

Lack of Clear Cortical Representation of Pain   The second paradox of pain is that it has no obvious cortical representation (Rainville, 2002). Painful stimuli activate many areas of cortex, but the particular areas of activation vary from study to study and from person to person (see Apkarian, 2008; Tracey, 2005). However, none of those areas is necessary for the perception of pain. For example, painful stimuli usually elicit responses in SI and SII (see Zhuo, 2008). However, removal of SI and SII in humans is not associated with any change in the threshold for pain. Indeed, *hemispherectomized* patients (those with one cerebral hemisphere removed) can still perceive pain from both sides of their bodies.

The cortical area that has been most frequently linked to the experience of pain is the **anterior cingulate cortex** (the cortex of the anterior cingulate gyrus; see Figure 7.16). The anterior cingulate cortex appears to be involved in the emotional reaction to pain rather than to the perception of pain itself (Panksepp, 2003; Price, 2000).

Descending Pain Control   The third paradox of pain is that this most compelling of all sensory experiences can be so effectively suppressed by cognitive and emotional

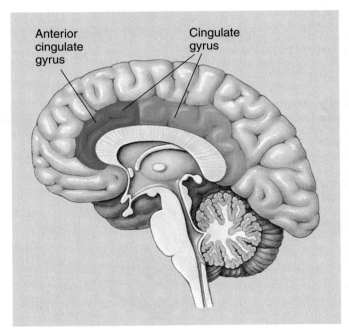

FIGURE 7.16   Location of anterior cingulate cortex in the cingulate gyrus.

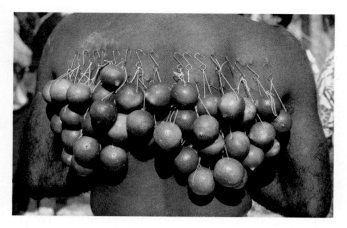

FIGURE 7.17 When experienced as part of a religious rite, normally excruciating conditions often produce little pain. Limes are used here because of the caustic effects of lime juice.

factors (Wager, 2005; Wiech, Ploner, & Tracey, 2008). For example, men participating in a certain religious ceremony suspend objects from hooks embedded in their backs with little evidence of pain (see Figure 7.17); severe wounds suffered by soldiers in battle are often associated with little pain; and people injured in life-threatening situations frequently feel no pain until the threat is over.

Melzack and Wall (1965) proposed the **gate-control theory** to account for the ability of cognitive and emotional factors to block pain. They theorized that signals descending from the brain can activate neural gating circuits in the spinal cord to block incoming pain signals.

Three discoveries led to the identification of a descending pain-control circuit. First was the discovery that electrical stimulation of the **periaqueductal gray (PAG)** has analgesic (pain-blocking) effects: Reynolds (1969) was able to perform surgery on rats with no analgesia other than that provided by PAG stimulation. Second was the discovery that the PAG and other areas of the brain contain specialized receptors for opiate analgesic drugs such as morphine. And third was the isolation of several endogenous (internally produced) opiate analgesics, the **endorphins**, which you learned about in Chapter 4. These three findings together suggested that analgesic drugs and psychological factors might block pain through an endorphin-sensitive circuit that descends from the PAG.

Figure 7.18 illustrates the descending analgesia circuit first hypothesized by Basbaum and Fields (1978). They proposed that the output of the PAG excites the serotonergic neurons of the *raphé nuclei* (a cluster of serotonergic nuclei in the core of the medulla), which in turn project down the dorsal columns of the spinal cord and excite interneurons that block incoming pain signals in the dorsal horn.

Descending analgesia pathways have been the subject of intensive investigation since the first model was proposed by Basbaum and Fields in 1978. In order to incorporate the mass of accumulated data, models of the descending analgesia circuits have grown much more complex.

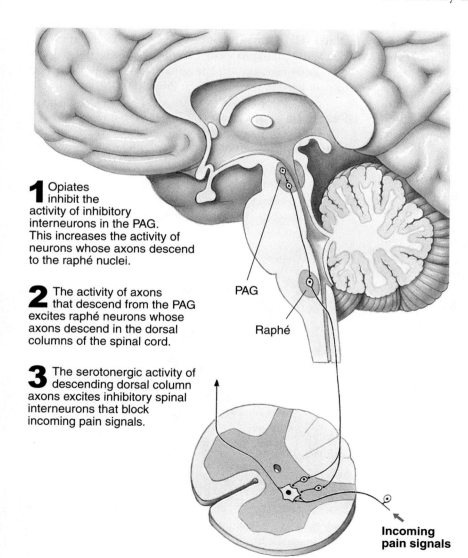

1 Opiates inhibit the activity of inhibitory interneurons in the PAG. This increases the activity of neurons whose axons descend to the raphé nuclei.

2 The activity of axons that descend from the PAG excites raphé neurons whose axons descend in the dorsal columns of the spinal cord.

3 The serotonergic activity of descending dorsal column axons excites inhibitory spinal interneurons that block incoming pain signals.

PAG

Raphé

Incoming pain signals

FIGURE 7.18 Basbaum and Fields's (1978) model of the descending analgesia circuit.

Still, a descending component involving opiate activity in the PAG and serotonergic activity in the raphé nuclei remains a key part of most of these models. One important addition to the original descending analgesia model is the discovery that some descending circuits can increase, rather than reduce, the perception of pain (Fields, 2004; Gebhart, 2004; Vanegas & Schaible, 2004).

Neuropathic Pain

In most cases, plasticity of the human nervous system helps it function more effectively. In the case of neuropathic pain, just the opposite is true. **Neuropathic pain** is severe chronic pain in the absence of a recognizable pain stimulus. A typical case of neuropathic

Neuroplasticity

pain develops after an injury: The injury heals and there seems to be no reason for further pain, but the patient experiences chronic excruciating pain. In many cases, neuropathic pain can be triggered by an innocuous stimulus, such as a gentle touch.

Although the exact mechanisms of neuropathic pain are unknown, it is somehow caused by pathological changes in the nervous system induced by the original injury (see Reichling & Levine, 2009). Recent research has implicated signals from aberrant glial cells in neuropathic pain; these signals are thought to trigger hyperactivity in neural pain pathways (Fields, 2009).

Although the neuropathic pain may be perceived to be in a limb—even in an amputated limb (see Chapter 10)—it is caused by abnormal activity in the CNS. Thus, cutting nerves from the perceived location of the pain often brings little or no comfort. And, unfortunately, medications that have been developed to treat the pain associated with injury often prove to be ineffective against neuropathic pain.

7.4
Chemical Senses: Smell and Taste

Olfaction (smell) and *gustation* (taste) are referred to as the chemical senses because their function is to monitor the chemical content of the environment. Smell is the response of the olfactory system to airborne chemicals that are drawn by inhalation over receptors in the nasal passages, and taste is the response of the gustatory system to chemicals in solution in the oral cavity.

When we are eating, smell and taste act in concert. Molecules of food excite both smell and taste receptors and produce an integrated sensory impression termed **flavor**. The contribution of olfaction to flavor is often underestimated, but you won't make this mistake if you remember that people with no sense of smell have difficulty distinguishing the flavors of apples and onions.

In humans, the main adaptive role of the chemical senses is flavor recognition. However, in many other species, the chemical senses also play a significant role in regulating social interactions (e.g., Zufall & Leinders-Zufall, 2007). The members of many species release **pheromones**—chemicals that influence the physiology and behavior of *conspecifics* (others of the same species). For example, Murphy and Schneider (1970) showed that the sexual and aggressive behavior of hamsters is under pheromonal control. Normal male hamsters attack and kill unfamiliar males that are placed in their colonies, whereas they mount and impregnate unfamiliar sexually receptive females. However, male hamsters that are unable to smell the intruders engage in neither aggressive nor sexual behavior. Murphy and Schneider confirmed the olfactory basis of hamsters' aggressive and sexual behavior in a particularly devious fashion. They swabbed a male intruder with the vaginal secretions of a sexually receptive female before placing it in an unfamiliar colony; in so doing, they converted it from an object of hamster assassination to an object of hamster lust.

The possibility that humans may release sexual pheromones has received considerable attention because of its financial and recreational potential. There have been many suggestive findings. For example, (1) the olfactory sensitivity of women is greatest when they are ovulating or pregnant; (2) the menstrual cycles of women living together tend to become synchronized; (3) humans—particularly women—can tell the sex of a person from the breath or the underarm odor; and (4) men can judge the stage of a woman's menstrual cycle on the basis of her vaginal odor. However, there is still no direct evidence that human odors can serve as sex attractants. Most subjects do not find the aforementioned body odors to be particularly attractive.

Olfactory System

The olfactory system is illustrated in Figure 7.19 on page 184. The olfactory receptor cells are located in the upper part of the nose, embedded in a layer of mucus-covered tissue called the **olfactory mucosa**. Their dendrites are located in the nasal passages, and their axons pass through a porous portion of the skull (the *cribriform plate*) and enter the **olfactory bulbs,** where they synapse on neurons that project via the *olfactory* tracts to the brain. There are about 10 million olfactory receptor cells in mice (Imai & Sakano, 2007).

➤ **Simulate** Investigating Olfaction: The Nose Knows
www.mypsychlab.com

For decades, it was assumed that there were only a few types of olfactory receptors. Different profiles of activity in a small number of receptor types were thought to lead to the perception of various smells—in the same way that the profiles of activity in three types of cones lead to the perception of colors. Then, at the turn of the 21st century, it was discovered that rats and mice have about 1,500 different kinds of receptor proteins and that humans have almost 1,000 (see Keller & Vosshall, 2008).

In mammals, each olfactory receptor cell contains only one type of receptor protein molecule (see Imai & Sakano, 2007). Olfactory receptor proteins are in the membranes of the dendrites of the olfactory receptor cells, where they can be stimulated by circulating airborne chemicals in the nasal passages. Researchers have attempted to discover the functional principle by which the various receptors are distributed through the olfactory mucosa. If there is such a principle, it has not yet been discovered: All of the types of receptor appear to be scattered throughout the mucosa, providing no clue about the organization of the system. Because each type of receptor responds in varying degrees to a wide variety of odors, each odor seems to be encoded by component processing—that is, by the pattern of activity across receptor types (Gottfried, 2009; Zou & Buck, 2006).

Despite the fact that olfactory receptors of each kind are scattered throughout the olfactory mucosa, somehow all of the receptors that contain the same receptor protein project to the same general location in the olfactory bulb (see Wilson, 2008).

The olfactory receptor axons terminate in the discrete clusters of neurons near the surface of the ofactory bulbs—these clusters are **olfactory glomeruli.** Each glomerulus receives input from several thousand olfactory receptor cells, all with the same receptor protein (Zou, Chesler, & Firestein, 2009). There appear to be two glomeruli in each olfactory bulb for each receptor protein (see Schoppa, 2009).

Because systematic topographic organization is apparent in other sensory systems (i.e., retinotopic and tonotopic layouts), researchers have been trying to discover the principle by which the glomeruli sensitive to particular odors are arrayed on the surfaces of the olfactory bulbs. There is evidence of a systematic layout (see Soucy et al., 2009):

- There is mirror symmetry between the left and right bulbs—glomeruli sensitive to particular odors tend to be located at the same sites on the two bulbs.
- The glomeruli sensitive to particular odors are arrayed on the bulbs in the same way in different members of the same species (i.e., mice).
- The layout of the glomeruli is similar in related species (i.e., rats and mice).

However, the principle according to which the glomeruli are arrayed has yet to be discovered (Schoppa, 2009). Soucy and colleagues (2009) found that there is only a slight tendency for glomeruli sensitive to similar odors to be located near one another.

The olfactory receptor cells differ from the receptor cells of other sensory systems in one important way. New olfactory receptor cells are created throughout each individual's life, to replace those that have deteriorated (Doty, 2001). Once created, the new receptor cells develop axons,

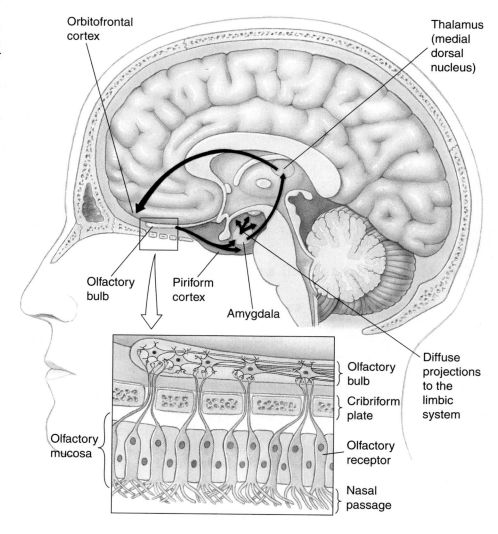

FIGURE 7.19   The human olfactory system.

which grow until they reach appropriate sites in the olfactory bulb. Each new olfactory receptor cell survives only a few weeks before being replaced.

Each olfactory tract projects to several structures of the medial temporal lobes, including the amygdala and the **piriform cortex**—an area of medial temporal cortex adjacent to the amygdala. The piriform cortex is considered to be primary olfactory cortex, but this designation is somewhat arbitrary (Bensafi et al., 2004). The olfactory system is the only sensory system whose major sensory pathway reaches the cerebral cortex without first passing through the thalamus (see Wilson & Mainen, 2006).

Two major olfactory pathways leave the amygdala-piriform area. One projects diffusely to the limbic system, and the other projects via the **medial dorsal nuclei** of the thalamus to the **orbitofrontal cortex**—the area of cortex on the inferior surface of the frontal lobes, next to the *orbits* (eye sockets)—see Gottfried & Zald (2005). The limbic projection is thought to mediate the emotional response to odors; the thalamic-orbitofrontal projection is

thought to mediate the conscious perception of odors. Little is known about how neurons receptive to different odors are organized in the cortex (see Savic, 2002).

Gustatory System

Taste receptors are found on the tongue and in parts of the oral cavity; they typically occur in clusters of about 50, called **taste buds**. On the tongue, taste buds are often located around small protuberances called *papillae* (singular *papilla*). The relation between taste receptors, taste buds, and papillae is illustrated in Figure 7.20 (see Gilbertson, Damak, & Margolskee, 2000). Unlike olfactory receptors, taste receptors do not have their own axons; each neuron that carries signals away from a taste bud receives input from many receptors.

⊙→ Simulate
The Four Basic Tastes
www.mypsychlab.com

According to the conventional view of taste perception, there are receptors for each of five primary tastes—sweet, sour, bitter, salty, and *unami* (meaty)—and each taste we experience is produced by a different combination of activity in these five kinds of receptors. One major problem with this conventional view is that many tastes cannot be created by combinations of the primary tastes (Schiffman & Erickson, 1980). Although it has become apparent from this finding and others that a component-processing explanation of taste perception is not correct, a comprehensive explanation that is more consistent with the evidence has not yet emerged (Simon et al., 2006; Spector & Travers, 2005).

Although the mechanisms of taste perception remain a mystery, a significant step toward solving this mystery was the identification of 33 gustatory receptor proteins: 1 unami, 2 sweet, and 30 bitter receptor proteins (see Scott, 2004). Sour and salty appear to have no receptor proteins; instead, substances with these tastes appear to influence taste receptor cells by acting directly on their ion channels (see Spector & Glendinning, 2009). So far, it appears that only one type of receptor protein appears in each taste receptor cell.

Surface of Tongue

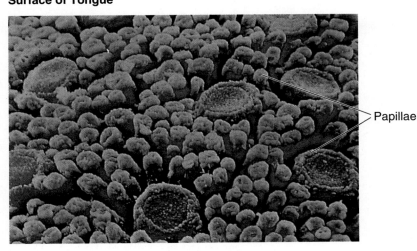

Papillae

Cross Section of a Papilla

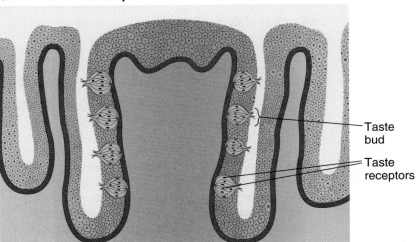

Taste bud

Taste receptors

FIGURE 7.20 Taste receptors, taste buds, and papillae on the surface of the tongue. Two sizes of papillae are visible in the photograph; only the larger papillae contain taste buds and receptors.

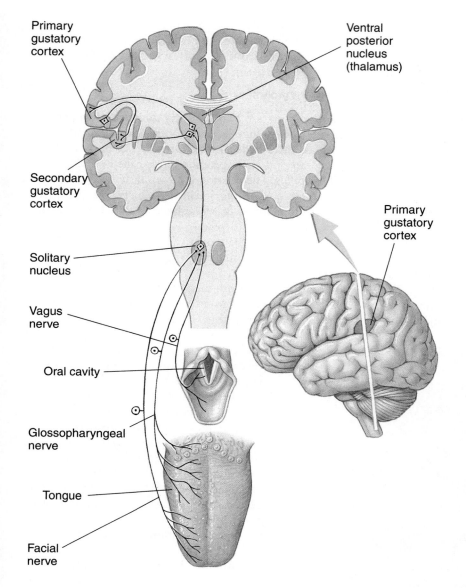

FIGURE 7.21 The human gustatory system.

system are primarily ipsilateral. Thus, particular tastes seem to be encoded in the brain by profiles of activity in groups of neurons (e.g., high activity in some and low in others).

Brain Damage and the Chemical Senses

The inability to smell is called **anosmia**; the inability to taste is called **ageusia**. The most common neurological cause of anosmia is a blow to the head that causes a displacement of the brain within the skull and shears the ol-factory nerves where they pass through the cribriform plate. Less complete deficits in olfaction have been linked to a wide variety of neurological disorders including Alzheimer's disease, Down syndrome, epilepsy, multiple sclerosis, Korsakoff's syndrome, and Parkinson's disease (see Doty, 2001).

Clinical Implications

Ageusia is rare, presumably because sensory signals from the mouth are carried via three separate pathways. However, partial ageusia, limited to the anterior two-thirds of the tongue on one side, is sometimes observed after damage to the ear on the same side of the body. This is because the branch of the facial nerve (VII) that carries gustatory information from the anterior two-thirds of the tongue passes through the middle ear.

The major pathways over which gustatory signals are conducted to the cortex are illustrated in Figure 7.21. Gustatory afferent neurons leave the mouth as part of the *facial* (VII), *glossopharyngeal* (IX), and *vagus* (X) *cranial nerves*, which carry information from the front of the tongue, the back of the tongue, and the back of the oral cavity, respectively. These fibers all terminate in the **solitary nucleus** of the medulla, where they synapse on neurons that project to the *ventral posterior nucleus* of the thalamus. The gustatory axons of the ventral posterior nucleus project to the *primary gustatory cortex*, which is near the face area of the somatosensory homunculus, on the superior lip of the lateral fissure, and to the *secondary gustatory cortex*, which is hidden from view in the lateral fissure (Sewards & Sewards, 2001). Unlike the projections of other sensory systems, the projections of the gustatory

Scan Your Brain

Now that you have reached the threshold of this chapter's final section, a section that focuses on attention, you should scan your brain to test your knowledge of the sensory systems covered in the preceding sections. Complete each sentence with the name of the appropriate system. The correct answers are provided at the end of the exercise. Before proceeding, review material related to your incorrect answers and omissions.

1. The primary _____ cortex is organized tonotopically.
2. The inferior colliculi and medial geniculate nuclei are components of the _____ system.
3. The dorsal-column medial-lemniscus system and the anterolateral system are pathways of the _____ system.
4. The ventral posterior nuclei, the intralaminar nuclei, and the parafascicular nuclei are all thalamic nuclei of the _____ system.
5. The periaqueductal gray and the raphé nuclei are involved in blocking the perception of _____.
6. One pathway of the _____ system projects from the amygdala and piriform cortex to the orbitofrontal cortex.
7. Parts of the ventral posterior nuclei are thalamic relay nuclei of both the somatosensory system and the _____ system.
8. Unlike the neuronal projections of all other sensory systems, those of the _____ system are primarily ipsilateral.
9. Anosmia is caused by damage to the _____ system.
10. Ageusia is caused by damage to the _____ system.

Scan Your Brain answers: (1) auditory, (2) auditory, (3) somatosensory, (4) somatosensory, (5) pain, (6) olfactory, (7) gustatory, (8) gustatory, (9) olfactory, (10) gustatory.

7.5

Selective Attention

We consciously perceive only a small subset of the many stimuli that excite our sensory organs at any one time and largely ignore the rest (Bays & Husain, 2008; Huang, Treisman, & Pashler, 2007). The process by which this occurs is **selective attention**.

Selective attention has two characteristics: It improves the perception of the stimuli that are its focus, and it interferes with the perception of the stimuli that are not its focus. For example, if you focus your attention on a potentially important announcement in a noisy airport, your chances of understanding it increase; but your chances of understanding a simultaneous comment from a traveling companion decrease.

Attention can be focused in two different ways: by internal cognitive processes (*endogenous attention*) or by external events (*exogenous attention*)—see Knudsen (2007). For example, your attention can be focused on a table top because you are searching for your keys (endogenous attention), or it can be drawn there because your cat tipped over a lamp (exogenous attention). Endogenous attention

is thought to be mediated by top-down (from higher to lower levels) neural mechanisms, whereas exogenous attention is thought to be mediated by bottom-up (from lower to higher levels) neural mechanisms (see Fritz et al., 2007; Moore, 2006).

Where do top-down attentional influences on sensory systems originate? There is a general consensus that both prefrontal cortex and posterior parietal cortex play major roles in directing top-down attention (Morishima et al., 2009; Saalmann, Pigarev, & Viayasagar, 2007; Yantis, 2008).

Eye movements often play an important role in visual attention, but it is important to realize that visual attention can be shifted without shifting the direction of visual focus (Rees et al., 1999). To prove this to yourself, look at the following Check it Out demonstration.

One other important characteristic of selective attention is the cocktail-party phenomenon (see Feng & Ratnam, 2000). The **cocktail-party phenomenon** is the fact that even when you are focusing so intently on one conversation that you are totally unaware of the content of other conversations going on around you, the mention of your

Check It Out

SHIFTING VISUAL ATTENTION WITHOUT SHIFTING VISUAL FOCUS

Fix your gaze on the +, concentrate on it. Next, shift your attention to one of the letters without shifting your gaze from +. Now, shift your attention to other letters, again without shifting your gaze from the +. You have experienced *covert attention*—a shift of visual attention without corresponding eye movement. A change in visual attention that involves a shift in gaze is called *overt attention*.

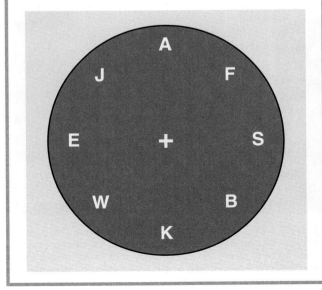

name in one of the other conversations will immediately gain access to your consciousness. This phenomenon suggests that your brain can block from conscious awareness all stimuli except those of a particular kind while still unconsciously monitoring the blocked-out stimuli just in case something comes up that requires attention.

Change Blindness

There is no better illustration of the importance of attention than the phenomenon of change blindness (Simons & Ambinder, 2005; Simons & Rensink, 2005). To study

○→ Simulate
Change Blindness
www.mypsychlab.com

change blindness, a subject is shown a photographic image on a computer screen and is asked to report any change in the image as soon as it is noticed. In fact, the image is composed of two images that alternate with a delay of less than 0.1 second between them. The two photographic images are identical except for one gross feature. For example, the two images in Figure 7.22 are identical except that the picture in the center of the wall is missing from one. You might think that any subject would immediately notice the picture disappearing and reappearing. But this is not what happens. Most subjects spend many seconds staring at the image—searching, as instructed, for some change—before they notice the disappearing and reappearing picture. When this finally happens, they wonder in amazement why it took them so long.

Why does change blindness occur? It occurs because, contrary to our impression, when we view a scene, we have absolutely no memory for parts of the scene that are not the focus of our attention. When viewing the scene in Figure 7.22, most subjects attend to the two people and do not notice when the picture disappears from the wall between them. Because they have no memory of the parts of the image to which they did not attend, subjects are not aware when those parts change.

The change blindness phenomenon does not occur without the brief (i.e., less than 0.1 second) intervals between images, although they barely produce a flicker. Without the intervals, no memory is required and the changes are perceived immediately.

Neural Mechanisms of Attention

Moran and Desimone (1985) first demonstrated the effects of attention on neural activity. They trained monkeys to stare at a fixation point on a screen while they recorded the activity of neurons in a prestriate area that was part of the ventral stream and particularly sensitive to color. In one experiment, they recorded from individual neurons that responded to either red or green bars of light in their receptive fields. When the monkey was trained to perform a task that required attention to the red cue, the response to the red cue was increased, and the response to the green cue was reduced. The opposite happened when the monkey attended to green.

Experiments paralleling those in monkeys have been conducted in humans using functional brain-imaging techniques. For example, Corbetta and colleagues (1990) presented a collection of moving, colored stimuli of various shapes and asked their subjects to discriminate among the stimuli based on their movement, color, or

<0.1 second

FIGURE 7.22   The change blindness phenomenon. These two illustrations were continually alternated, with a brief (less than 0.1 second) interval between each presentation, and the subjects were asked to report any changes they noticed. Amazingly, it took most of them many seconds to notice the disappearing and reappearing picture in the center of the wall.

shape. Attention to shape or color produced increased activity in areas of the ventral stream; attention to movement produced increased activity in an area of the dorsal stream (see Chapter 6).

In another study of attention in human subjects, Ungerleider and Haxby (1994) showed subjects a series of faces. The subjects were asked whether the faces belonged to the same person or whether they were located in the same position relative to the frame. When the subjects were attending to identity, regions of the ventral stream were more active; when the subjects were attending to position, regions of the dorsal stream were more active.

The preceding studies indicate how the neural mechanisms of selective attention work. Selective attention is thought to work by strengthening the neural responses to attended-to aspects and by weakening the responses to others. In general, anticipation of a stimulus increases neural activity in the same circuits affected by the stimulus itself (Carlsson et al., 2000).

The demonstration of one neural mechanism of attention surprised a lot of biopsychologists, not because it is inconsistent with current neural theories of attention, but because it involves a surprising degree of neural plasticity (see Connor, 2006; Jääskeläinen et al., 2007). The location of the receptive fields of visual neurons has usually been assumed to be a static property of the neurons; however, Wommelsdorf and colleagues (2006) found that spatial attention can shift the location of receptive fields. Recording from neurons in an area of monkey secondary visual cortex in the dorsal stream, they found that the receptive fields of many of the neurons shifted toward points in the visual field to which the subjects were attending.

Neuroplasticity

Simultanagnosia

I have not forgotten that I asked you to think about the patient whose case opened this chapter. He could identify objects in any part of his visual field if they were presented individually; thus, he was not suffering from blindness or other visual field defects. His was a disorder of attention called **simultanagnosia**. Specifically, he suffered from *visual simultanagnosia*, a difficulty in attending visually to more than one object at a time. Because the dorsal stream of the posterior parietal cortex is responsible for visually localizing objects in space, you may have hypothesized that the patient's problem was associated with damage to this area. If you did, you were correct. The damage associated with simultanagnosia is typically bilateral.

Clinical Implications

Themes Revisited

The clinical implications theme was prominent in this chapter, but you saw it in a different light. Previous chapters discussed how biopsychological research is leading to the development of new treatments; this chapter focused exclusively on what particular clinical cases have revealed about the organization of healthy sensory systems. The following cases played a key role in this chapter: the patient with visual simultanagnosia; Dr. P., the visual agnosic who mistook his wife for a hat; Aunt Betty, the asomatognosic who lost the left side of her body; Miss C., the student who felt no pain and died as a result; and W.M., the man who reduced his scotoma with his hand.

Two of the other major themes were also developed in this chapter: the neuroplasticity theme and the evolutionary perspective theme. Although this chapter did not systematically discuss the plasticity of sensory systems—upcoming chapters focus on this topic—three important examples of sensory system plasticity were mentioned: the effects of tinnitus on the auditory system, partial recovery of vision in a scotoma by placing a hand in the scotoma, and the movement of the receptive fields of visual neurons toward a location that is the focus of attention. The value of the evolutionary perspective was illustrated by comparative studies of some species that have proven particularly informative because of their evolutionary specializations (e.g., the auditory localization abilities of the barn owl and the tendency of the secondary auditory cortex of monkeys to respond to monkey calls).

The thinking creatively theme came up once. The case of Miss C. taught us that pain is a positive sensation that we can't live without.

Clinical Implications

Neuroplasticity

Evolutionary Perspective

Thinking Creatively

Think about It

1. How has this chapter changed your concept of perception?
2. Why are most people amazed by the change blindness phenomenon? Why does change blindness occur, and what does it indicate about attentional mechanisms?

3. Which sensory system would you study if you were a biopsychologist who studied sensory systems? Why?
4. Bob went through a stop sign and hit a car near a busy school crossing. If you were his lawyer, how could you use the change blindness phenomenon to plead for leniency?

5. Discuss the paradoxes of pain.
6. Design an experiment to demonstrate that flavor is the product of both taste and smell. Try it on yourself.
7. Damage to areas of primary sensory cortex has surprisingly little impact on perception. Discuss with respect to parallel pathways.

8. What is simultanagnosia? What brain damage is often associated with it, and what evidence is there that it is a disorder of selective attention?

Key Terms

Exteroceptive sensory systems (p. 165)

7.1 Principles of Sensory System Organization

Primary sensory cortex (p. 165)
Secondary sensory cortex (p. 165)
Association cortex (p. 165)
Hierarchical organization (p. 165)
Sensation (p. 166)
Perception (p. 166)
Functional segregation (p. 166)
Parallel processing (p. 166)

7.2 Auditory System

Fourier analysis (p. 168)
Tympanic membrane (p. 168)
Ossicles (p. 168)

Oval window (p. 168)
Cochlea (p. 168)
Organ of Corti (p. 168)
Hair cells (p. 168)
Basilar membrane (p. 168)
Tectorial membrane (p. 168)
Auditory nerve (p. 169)
Retinotopic (p. 169)
Tonotopic (p. 169)
Semicircular canals (p. 170)
Vestibular system (p. 170)
Superior olives (p. 170)
Inferior colliculi (p. 170)
Medial geniculate nuclei (p. 170)
Tinnitus (p. 173)

7.3 Somatosensory System: Touch and Pain

Free nerve endings (p. 174)
Pacinian corpuscles (p. 174)
Stereognosis (p. 175)

Dermatome (p. 175)
Dorsal-column medial-lemniscus system (p. 175)
Anterolateral system (p. 175)
Dorsal columns (p. 176)
Medial lemniscus (p. 176)
Ventral posterior nucleus (p. 176)
Somatotopic (p. 177)
Somatosensory homunculus (p. 177)
Astereognosia (p. 179)
Asomatognosia (p. 180)
Anosognosia (p. 180)
Contralateral neglect (p. 180)
Anterior cingulate cortex (p. 181)
Gate-control theory (p. 182)
Periaqueductal gray (PAG) (p. 182)
Endorphins (p. 182)
Neuropathic pain (p. 183)

7.4 Chemical Senses: Smell and Taste

Flavor (p. 183)
Pheromones (p. 183)
Olfactory mucosa (p. 183)
Olfactory bulbs (p. 183)
Olfactory glomeruli (p. 184)
Piriform cortex (p. 185)
Medial dorsal nuclei (p. 185)
Orbitofrontal cortex (p. 185)
Taste buds (p. 185)
Solitary nucleus (p. 186)
Anosmia (p. 186)
Ageusia (p. 186)

7.5 Selective Attention

Selective attention (p. 187)
Cocktail-party phenomenon (p. 187)
Change blindness (p. 188)
Simultanagnosia (p. 189)

✓●[Quick Review] Test your comprehension of the chapter with this brief practice test. You can find the answers to these questions as well as more practice tests, activities, and other study resources at www.mypsychlab.com.

1. Primate auditory cortex is organized
   a. retinotopically.
   b. somatotopically.
   c. tonotopically.
   d. none of the above
   e. both b and c

2. The somatosensory homunculus is in the
   a. primary somatosensory cortex.
   b. postcentral gyrus.
   c. parietal cortex.
   d. all of the above
   e. none of the above

3. The inability to recognize objects by touch is
   a. astereognosia.
   b. stereognosis.
   c. asomatognosia.
   d. running gnosia.
   e. prosopagnosia.

4. Which of the following kinds of damage is often the recognizable cause of neuropathic pain?
   a. scar tissue
   b. unhealed tissue damage
   c. unhealed puncture wounds
   d. unhealed burns
   e. none of the above

5. Bob heard a noise behind him and turned just in time to see a cat pushing the door open. This is an example of
   a. nonselective attention.
   b. exogenous attention.
   c. endogenous attention.
   d. ageusia.
   e. a top-down neural mechanism.

8 The Sensorimotor System
How You Move

The evening before I started to write this chapter, I was standing in a checkout line at the local market. As I waited, I furtively scanned the headlines on the prominently displayed magazines—WOMAN GIVES BIRTH TO CAT; FLYING SAUCER LANDS IN CLEVELAND SHOPPING MALL; HOW TO LOSE 20 POUNDS IN 2 DAYS. Then, my mind began to wander, and I started to think about beginning to write this chapter. That is when I began to watch Rhonda's movements and to wonder about the neural system that controlled them. Rhonda is a cashier—the best in the place.

The Case of Rhonda, the Dexterous Cashier

I was struck by the complexity of even Rhonda's simplest movements. As she deftly transferred a bag of tomatoes to the scale, there was a coordinated adjustment in almost every part of her body. In addition to her obvious finger, hand, arm, and shoulder movements, coordinated movements of her head and eyes tracked her hand to the tomatoes; and there were adjustments in the muscles of her feet, legs, trunk, and other arm, which kept her from lurching forward. The accuracy of these responses suggested that they were guided in part by the patterns of visual, somatosensory, and vestibular changes they produced. The term *sensorimotor* in the title of this chapter formally recognizes the critical contribution of sensory input to guiding motor output.

As my purchases flowed through her left hand, Rhonda registered the prices with her right hand and bantered with Rick, the bagger. I was intrigued by how little of what Rhonda was doing appeared to be under her conscious control. She made general decisions about which items to pick up and where to put them, but she seemed to give no thought to the exact means by which these decisions were carried out. Each of her responses could have been made with an infinite number of different combinations of finger, wrist, elbow, shoulder, and body adjustments; but somehow she unconsciously picked one. The higher parts of her sensorimotor system—perhaps her cortex—seemed to issue conscious general commands to other parts of the system, which unconsciously produced a specific pattern of muscular responses that carried them out.

The automaticity of Rhonda's performance was a far cry from the slow, effortful responses that had characterized her first days at the market. Somehow, experience had integrated her individual movements into smooth sequences, and it seemed to have transferred the movements' control from a mode that involved conscious effort to one that did not.

I was suddenly jarred from my contemplations by a voice. "Sir, excuse me, sir, that will be $18.65," Rhonda said, with just a hint of delight at catching me in mid-daydream. I hastily paid my bill, muttered "thank you," and scurried out of the market.

As I write this, I am smiling both at my own embarrassment and at the thought that Rhonda has unknowingly introduced you to three principles of sensorimotor control that are the foundations of this chapter: (1) The sensorimotor system is hierarchically organized. (2) Motor output is guided by sensory input. (3) Learning can change the nature and the locus of sensorimotor control.

8.1
Three Principles of Sensorimotor Function

Before getting into the details of the sensorimotor system, let's take a closer look at the three principles of sensorimotor function introduced by Rhonda. You will better appreciate these principles if you recognize that they also govern the operation of any large, efficient company—perhaps because that is another system for controlling output that has evolved in a competitive environment. You may find this metaphor useful in helping you understand the principles of sensorimotor system organization—many scientists find that metaphors help them think creatively about their subject matter.

The Sensorimotor System Is Hierarchically Organized

The operation of both the sensorimotor system and a large, efficient company is directed by commands that cascade down through the levels of a hierarchy (see Graziano, 2009)—from the association cortex or the company president (the highest levels) to the muscles or the workers (the lowest levels). Like the orders that are issued from the office of a company president, the commands that emerge from the association cortex specify general goals rather than specific plans of action. Neither the association cortex nor the company president routinely gets involved in the details. The main advantage of this *hierarchical organization* is that the higher levels of the hierarchy are left free to perform more complex functions.

Thinking Creatively

Both the sensorimotor system and a large, efficient company are parallel hierarchical systems; that is, they are hierarchical systems in which signals flow between levels over multiple paths (see Darian-Smith, Burman, & Darien-Smith, 1999). This parallel structure enables the association cortex or company president to exert control over the lower levels of the hierarchy in more than one way. For example, the association cortex can directly inhibit an eye blink reflex to allow the insertion of a contact lens, just as a company president can personally organize a delivery to an important customer.

The sensorimotor and company hierarchies are also characterized by *functional segregation*. That is, each level of the sensorimotor and company hierarchies tends to be composed of different units (neural structures or departments), each of which performs a different function.

In summary, the sensorimotor system—like the sensory systems you read about in Chapter 7—is a parallel, functionally segregated, hierarchical system. The main difference between the sensory systems and the sensorimotor system is the primary direction of information flow. In sensory systems, information mainly flows up through the hierarchy; in the sensorimotor system, information mainly flows down.

Motor Output Is Guided by Sensory Input

Efficient companies are flexible. They continuously monitor the effects of their own activities, and they use this information to fine-tune their activities. The sensorimotor system does the same (Gomi, 2008; Sommer & Wurtz, 2008).

Neuroplasticity

The eyes, the organs of balance, and the receptors in skin, muscles, and joints all monitor the body's responses, and they feed their information back into sensorimotor circuits. In most instances, this **sensory feedback** plays an important role in directing the continuation of the responses that produced it. The only responses that are not normally influenced by sensory feedback are *ballistic movements*—brief, all-or-none, high-speed movements, such as swatting a fly.

Behavior in the absence of just one kind of sensory feedback—the feedback that is carried by the somatosensory nerves of the arms—was studied in G.O., a former darts champion.

The Case of G.O., the Man with Too Little Feedback

An infection had selectively destroyed the somatosensory nerves of G.O.'s arms. He had great difficulty performing intricate responses such as doing up his buttons or picking up coins, even under visual guidance. Other difficulties resulted

Clinical Implications

from his inability to adjust his motor output in the light of unanticipated external disturbances; for example, he could not keep from spilling a cup of coffee if somebody brushed against him. However, G.O.'s greatest problem was his inability to maintain a constant level of muscle contraction:

> The result of this deficit was that even in the simplest of tasks requiring a constant motor output to the hand, G.O. would have to keep a visual check on his progress. For example, when carrying a suitcase, he would frequently glance at it to reassure himself that he had not dropped it some paces back. However, even visual feedback was of little use to him in many tasks. These tended to be those requiring a constant force output such as grasping a pen while writing or holding a cup. Here, visual information

> was insufficient for him to be able to correct any errors that were developing in the output since, after a period, he had no indication of the pressure that he was exerting on an object; all he saw was either the pen or cup slipping from his grasp. (Rothwell et al., 1982, p. 539)

Many adjustments in motor output that occur in response to sensory feedback are controlled unconsciously by the lower levels of the sensorimotor hierarchy without the involvement of the higher levels (see Poppele & Bosco, 2003). In the same way, large companies run more efficiently if the clerks do not have to check with the company president each time they encounter a minor problem.

Learning Changes the Nature and Locus of Sensorimotor Control

When a company is just starting up, each individual decision is made by the company president after careful consideration. However, as the company develops, many individual actions are coordinated into sequences of prescribed procedures that are routinely carried out by personnel at lower levels of the hierarchy.

Similar changes occur during sensorimotor learning (see Ashe et al., 2006; Kübler, Dixon, & Garavan, 2006). During the initial stages of motor learning, each individual response is performed under conscious control; then, after much practice, individual responses become organized into continuous integrated sequences of action that flow smoothly and

Neuroplasticity

are adjusted by sensory feedback without conscious regulation. If you think for a moment about the sensorimotor skills you have acquired (e.g., typing, swimming, knitting, basketball playing, dancing, piano playing), you will appreciate that the organization of individual responses into continuous motor programs and the transfer of their control to lower levels of the nervous system characterize most sensorimotor learning.

A General Model of Sensorimotor System Function

Figure 8.1 on page 194 is a model that illustrates several principles of sensorimotor system organization; it is the framework of this chapter. Notice its hierarchical structure, the functional segregation of the levels (e.g., of secondary motor cortex), the parallel connections between levels, and the numerous feedback pathways.

This chapter focuses on the neural structures that play important roles in the control of voluntary behavior (e.g., picking up an apple). It begins at the level of association cortex and traces major motor signals as they descend the sensorimotor hierarchy to the skeletal muscles that ultimately perform the movements.

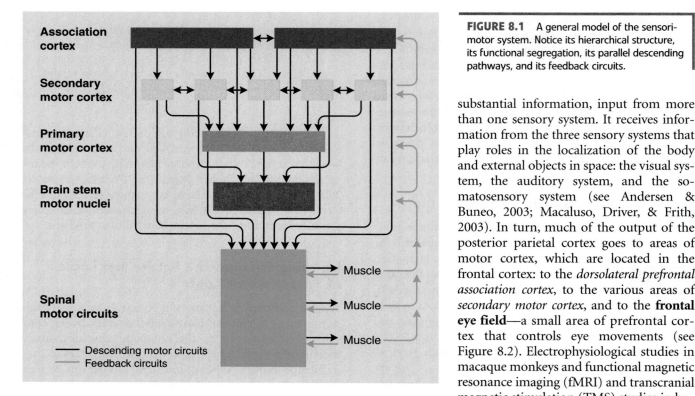

FIGURE 8.1 A general model of the sensorimotor system. Notice its hierarchical structure, its functional segregation, its parallel descending pathways, and its feedback circuits.

substantial information, input from more than one sensory system. It receives information from the three sensory systems that play roles in the localization of the body and external objects in space: the visual system, the auditory system, and the somatosensory system (see Andersen & Buneo, 2003; Macaluso, Driver, & Frith, 2003). In turn, much of the output of the posterior parietal cortex goes to areas of motor cortex, which are located in the frontal cortex: to the *dorsolateral prefrontal association cortex*, to the various areas of *secondary motor cortex*, and to the **frontal eye field**—a small area of prefrontal cortex that controls eye movements (see Figure 8.2). Electrophysiological studies in macaque monkeys and functional magnetic resonance imaging (fMRI) and transcranial magnetic stimulation (TMS) studies in humans indicate that the posterior parietal cortex comprises a mosaic of small areas, each specialized for guiding particular movements of eyes, head, arms, or hands (Culham & Valyear, 2006; Fogassi & Luppino, 2005).

In one important study (Desmurget et al., 2009), electrical stimulation was applied to the inferior portions of the posterior parietal cortexes of conscious neurosurgical patients. At low current levels, the patients experienced an intention to perform a particular action and, at high levels, felt that they had actually performed it. However, in neither case did the action actually occur.

Damage to the posterior parietal cortex can produce a variety of deficits, including deficits in the perception and memory of spatial relationships, in accurate reaching and grasping, in the control of eye movement, and in attention (Freund, 2003). However, apraxia and contralateral neglect are the two most striking consequences of posterior parietal cortex damage.

Apraxia is a disorder of voluntary movement that is not attributable to a simple motor deficit (e.g., not to paralysis or weakness) or to any deficit in comprehension or motivation (see Heilman, Watson, & Rothi, 1997). Remarkably, apraxic patients have difficulty making specific movements when they are requested to do so, particularly when the movements are out of context; however, they can often readily perform the very same movements under natural conditions, when they are not thinking about doing so. For example, an apraxic carpenter who has no difficulty at all hammering a nail during the course of her work might not be able to

Clinical Implications

8.2

Sensorimotor Association Cortex

Association cortex is at the top of your sensorimotor hierarchy. There are two major areas of sensorimotor association cortex: the posterior parietal association cortex and the dorsolateral prefrontal association cortex (see Brochier & Umiltà, 2007; Haggard, 2008). Posterior parietal cortex and the dorsolateral prefrontal cortex are each composed of several different areas, each with different functions (see Culham & Kanwisher, 2001; Fuster, 2000). However, there is no general consensus on how best to divide either of them for analysis or even how comparable the areas are in humans, monkeys, and rats (Calton & Taube, 2009; Husain & Nachev, 2006).

Posterior Parietal Association Cortex

Before an effective movement can be initiated, certain information is required. The nervous system must know the original positions of the parts of the body that are to be moved, and it must know the positions of any external objects with which the body is going to interact. The **posterior parietal association cortex** (the portion of parietal neocortex posterior to the primary somatosensory cortex) plays an important role in integrating these two kinds of information, in directing behavior by providing spatial information, and in directing attention (Britten, 2008; Husain & Nachev, 2006).

You learned in Chapter 7 that the posterior parietal cortex is classified as *association cortex* because it receives

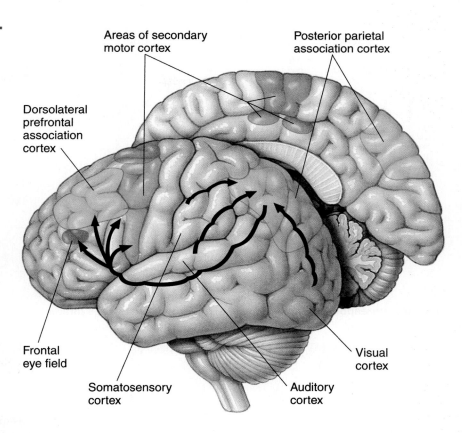

FIGURE 8.2 The major cortical input and output pathways of the posterior parietal association cortex. Shown are the lateral surface of the left hemisphere and the medial surface of the right hemisphere.

Areas of secondary motor cortex

Posterior parietal association cortex

Dorsolateral prefrontal association cortex

Frontal eye field

Somatosensory cortex

Auditory cortex

Visual cortex

demonstrate hammering movements when requested to make them, particularly in the absence of a hammer. Although its symptoms are bilateral, apraxia is often caused by unilateral damage to the left posterior parietal lobe or its connections (Jax, Buxbaum, & Moll, 2006).

Contralateral neglect, the other striking consequence of posterior parietal cortex damage, is a disturbance of a patient's ability to respond to stimuli on the side of the body opposite (contralateral) to the side of a brain lesion, in the absence of simple sensory or motor deficits (see Driver, Vuilleumier, & Husain, 2004; Luauté et al. 2006; Pisella & Mattingley, 2004). Most patients with contralateral neglect often behave as if the left side of their world does not exist, and they often fail to appreciate that they have a problem (Berti et al., 2005). The disturbance is often associated with large lesions of the right posterior parietal lobe (see Mort et al., 2003; Van Vleet & Robertson, 2006). For example, Mrs. S. suffered from contralateral neglect after a massive stroke to the posterior portions of her right hemisphere.

The Case of Mrs. S., the Woman Who Turned in Circles

Mrs. S.'s stroke had left her unable to recognize or respond to things to the left—including external objects as well as parts of her own body. For example, Mrs. S. often put makeup on the right side of her face but ignored the left.

Mrs. S.'s left-side contralateral neglect created many problems for her, but a particularly bothersome one was that she had difficulty getting enough to eat. When a plate of food was put in front of her, she could see only the food on the right half of the plate and thus ate only that much, even if she was very hungry.

After asking for and receiving a wheelchair capable of turning in place, Mrs. S. developed an effective way of getting more food if she was still hungry after completing a meal. She turns her wheelchair around to the right in a full circle until she sees the remaining half of her meal. Then, she eats that food, or more precisely, she eats the right half of that food. If she is still hungry after that, she turns once again to the right until she discovers the remaining quarter of her meal and eats half of that . . . and so on.

("The Case of Mrs. S., the Woman Who Turned in Circles," reprinted with the permission of Simon & Schuster, Inc. and Pan Macmillan, London from *The Man Who Mistook His Wife for a Hat and Other Clinical Tales* by Oliver Sacks. Copyright © 1970, 1981, 1983, 1984, 1986 by Oliver Sacks. Electronic rights with permission of the Wylie Agency.)

Most patients with contralateral neglect have difficulty responding to things to the left. But to the left of what? For most patients with contralateral neglect, the deficits in responding occur for stimuli to the left of their own bodies, referred to as *egocentric left*. Egocentric left is partially defined by gravitational coordinates: When patients tilt their heads, their field of neglect is not normally tilted with it (see the top panel of Figure 8.3 on page 196).

In addition to failing to respond to objects on their egocentric left, many patients tend not to respond to the left sides of objects, regardless of where the objects are in their visual fields (Kleinman et al., 2007). Neurons that have egocentric receptive fields and others with object-based receptive fields have been found in primate parietal cortex (Olson, 2003; Pouget & Driver, 2000). Object-based

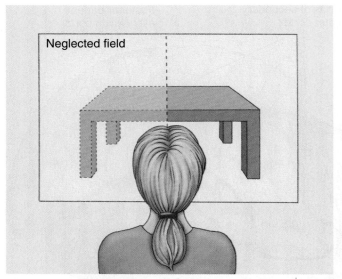
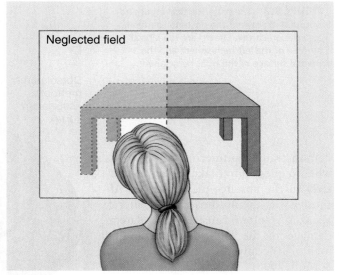

Patient is unresponsive to things to the left, even if the head is tilted.

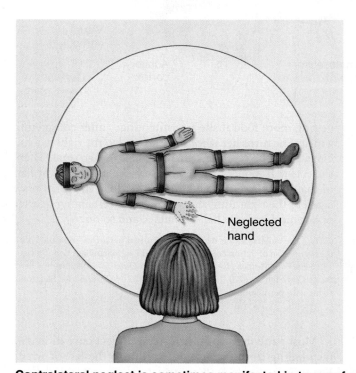
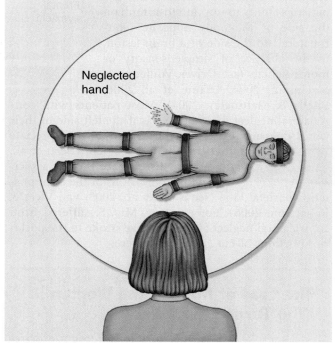

Contralateral neglect is sometimes manifested in terms of object-based coordinates.

FIGURE 8.3 Contralateral neglect is sometimes manifested in terms of gravitational coordinates, sometimes in terms of object-based coordinates.

contralateral neglect is illustrated in the lower panel of Figure 8.3. In this demonstration, patients with contralateral neglect had deficits in responding to the right hand of an experimenter who was facing them, regardless of its specific location in the patients' visual field.

You have learned in the preceding chapters that the failure to perceive an object consciously does not necessarily

mean that the object is not perceived. Indeed, two types of evidence suggest that information about objects that are not noticed by patients with contralateral neglect may be unconsciously perceived. First, when objects were repeatedly presented at the same spot to the left of patients with contralateral neglect, they tended to look to the same spot on future trials, although they were unaware of the objects

(Geng & Behrmann, 2002). Second, patients could more readily identify fragmented (partial) drawings viewed to their right if complete versions of the drawings had previously been presented to the left, where they were not consciously perceived (Vuilleumier et al., 2002).

Dorsolateral Prefrontal Association Cortex

The other large area of association cortex that has important sensorimotor functions is the **dorsolateral prefrontal association cortex**. It receives projections from the posterior parietal cortex, and it sends projections to areas of *secondary motor cortex*, to *primary motor cortex*, and to the *frontal eye field*. These projections are shown in Figure 8.4. Not shown are the major projections back from dorsolateral prefrontal cortex to posterior parietal cortex.

Dorsolateral prefrontal cortex seems to play a role in the evaluation of external stimuli and the initiation of voluntary reactions to them (Matsumoto & Tanaka, 2004; Ohbayashi, Ohki, & Miyashita, 2003). This view is supported by the response characteristics of neurons in this area of association cortex. Several studies have

Evolutionary Perspective

characterized the activity of monkey dorsolateral prefrontal neurons as the monkeys identify and respond to objects (e.g., Rao, Rainer, & Miller, 1997). The activity of some neurons depends on the characteristics of objects; the activity of others depends on the locations of objects; and the activity of still others depends on a combination of both. The activity of other dorsolateral prefrontal neurons is related to the response, rather than to the object. These neurons typically begin to fire before the response and continue to fire until the response is complete. There are neurons in all cortical motor areas that begin to fire in anticipation of a motor activity, but those in the dorsolateral prefrontal association cortex fire first.

The response properties of dorsolateral prefrontal neurons suggest that decisions to initiate voluntary movements may be made in this area of cortex (Rowe et al., 2000; Tanji & Hoshi, 2001), but these decisions depend on critical interactions with posterior parietal cortex (Brass et al., 2005; Connolly, Andersen, & Goodale, 2003; de Lange, Hagoort, & Toni, 2005).

8.3
Secondary Motor Cortex

Areas of **secondary motor cortex** are those that receive much of their input from association cortex (i.e., posterior parietal cortex and dorsolateral prefrontal cortex) and send much of their output to primary motor cortex (see Figure 8.5 on page 198). For many years, only two areas of secondary motor cortex were known: the supplementary motor area and the premotor cortex. Both of these large areas are clearly visible on the lateral surface of the frontal lobe, just anterior to the *primary motor cortex*. The **supplementary motor area** wraps over the top of the frontal lobe and extends down its medial surface into the longitudinal fissure, and the **premotor cortex** runs in a strip from the supplementary motor area to the lateral fissure.

Identifying the Areas of Secondary Motor Cortex

The simple two-area conception of secondary motor cortex has become more complex. Neuroanatomical and neurophysiological research with monkeys has

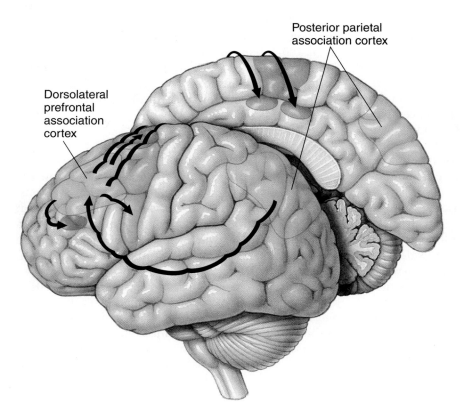

Posterior parietal association cortex

Dorsolateral prefrontal association cortex

FIGURE 8.4 The major cortical input and output pathways of the dorsolateral prefrontal association cortex. Shown are the lateral surface of the left hemisphere and the medial surface of the right hemisphere.

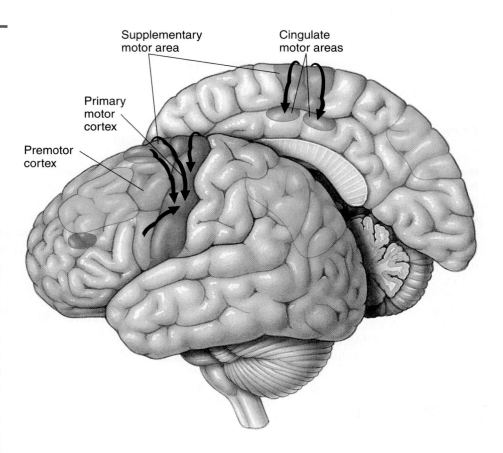

FIGURE 8.5 Four areas of secondary motor cortex—the supplementary motor area, the premotor cortex, and two cingulate motor areas—and their output to the primary motor cortex. Shown are the lateral surface of the left hemisphere and the medial surface of the right hemisphere.

made a case for at least eight areas of secondary motor cortex in each hemisphere, each with its own subdivisions (see Graziano, 2006; Nachev, Kennard, & Husain, 2008): three different supplementary motor areas (SMA, preSMA, and supplementary eye field), two premotor areas (dorsal and ventral), and three small areas—the **cingulate motor areas**—in the cortex of the cingulate gyrus. Although most of the research on secondary motor cortex has been done in monkeys, functional brain-imaging studies have suggested that human secondary motor cortex is similar to that of other primates (see Rizzolatti, Fogassi, & Gallese, 2002).

To qualify as secondary motor cortex, an area must be appropriately connected with association and secondary motor areas (see Figure 8.5). Electrical stimulation of an area of secondary motor cortex typically elicits complex movements, often involving both sides of the body. Neurons in an area of secondary motor cortex often become more active just prior to the initiation of a voluntary movement and continue to be active throughout the movement.

In general, areas of secondary motor cortex are thought to be involved in the programming of specific patterns of movements after taking general instructions from dorsolateral prefrontal cortex (see Hoshi & Tanji, 2007). Evidence of such a function comes from brain-imaging studies in which the patterns of activity in the brain have been measured while the subject is either imagining his or her own performance of a particular series of movements or planning the performance of the same movements (see Kosslyn, Ganis, & Thompson, 2001; Sirigu & Duhamel, 2001).

Despite evidence of similarities among areas of secondary motor cortex, substantial effort has been put into discovering their differences. Until recently, this research

Evolutionary Perspective

⊙► Simulate The Motor Areas of the Cerebral Cortex **www.mypsychlab.com**

has focused on differences between the supplementary motor area and the premotor cortex as originally defined. Although several theories have been proposed to explain functional differences between these areas (e.g., Hoshi & Tanji, 2007), none has received consistent support. As the boundaries between various areas of secondary motor cortex become more accurately characterized, the task of determining the function of each area will become easier—and vice versa.

Mirror Neurons

Few discoveries have captured the interest of neuroscientists as much as the discovery of mirror neurons (Lyons, Santos, & Keil, 2006). **Mirror neurons** are neurons that fire when an individual performs a particular goal-directed hand movement or when she or he observes the same goal-directed movement performed by another (Fadiga, Craighero, & Olivier, 2005).

Mirror neurons were discovered in the early 1990s in the laboratory of Giacomo Rizzolatti (see Rizzolatti, Fogassi, & Gallese, 2006). Rizzolatti and his colleagues had been studying a class of macaque ventral premotor neurons that seemed to encode particular goal objects; that is, these neurons fired when the monkey reached for one object (e.g., a toy) but not when the monkey reached for another object. Then, the researchers noticed something

FIGURE 8.6 Responses of a mirror neuron of a monkey.

strange: Some of these neurons, later termed *mirror neurons*, fired just as robustly when the monkey watched the experimenter pick up the same object, but not any other—see Figure 8.6.

Why did the discovery of mirror neurons in the ventral premotor cortex create such a stir? The reason is that they provide a possible mechanism for *social cognition* (knowledge of the perceptions, ideas, and intentions of others). Mapping the actions of others onto one's own action repertoire would facilitate social understanding, cooperation, and imitation (Iacoboni, 2005; Jackson & Decety, 2004; Knoblick & Sebanz, 2006).

Support for the idea that mirror neurons might play a role in social cognition has come from demonstrations that these neurons respond to the *understanding* of an action, not to some superficial aspect of it (Rizzolatti et al., 2006). For example, mirror neurons that reacted to the sight of an action that made a sound (e.g., cracking a peanut) were found to respond just as robustly to the sound alone—in other words, they responded fully to the par-

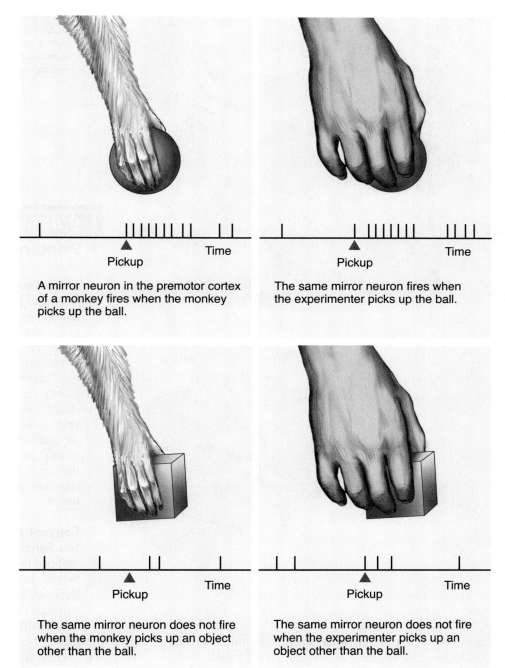

A mirror neuron in the premotor cortex of a monkey fires when the monkey picks up the ball.

The same mirror neuron fires when the experimenter picks up the ball.

The same mirror neuron does not fire when the monkey picks up an object other than the ball.

The same mirror neuron does not fire when the experimenter picks up an object other than the ball.

ticular action regardless of how it was detected. Indeed, many ventral premotor mirror neurons fire when a monkey does not perceive the key action but has enough clues to create a mental representation of it. The researchers identified mirror neurons that fired when the experimenter reached and grasped an object on a table. Then, a screen was placed in front of the object before the experimenter reached for it. Half the neurons still fired robustly, even though the monkeys could only have imagined what was happening.

Mirror neurons have also been found in the inferior portion of the posterior parietal lobe. Some of these respond to the purpose of an action rather than to the action itself. For example, some posterior parietal mirror neurons fired robustly when a monkey grasped a piece of food only if it was clear that the food would be subsequently eaten—if the food was repeatedly grasped and then placed in a bowl, the grasping was associated with little firing. The same neurons fired strongly when the monkey watched the experimenter picking up pieces of food to eat them, but not when the experimenter picked up food and placed it in a bowl.

The existence of mirror neurons has not yet been confirmed in humans because there are few opportunities to

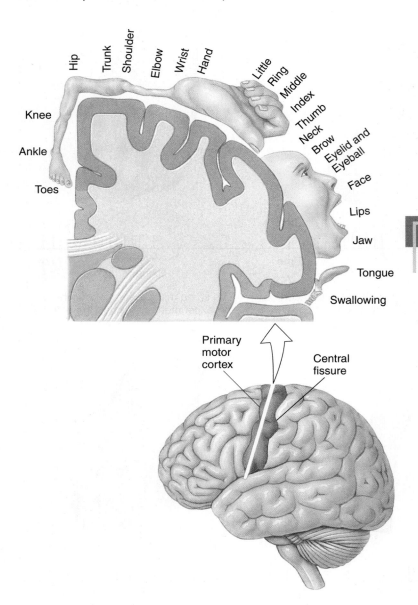

Primary motor cortex

Central fissure

FIGURE 8.7 The motor homunculus: the somatotopic map of the human primary motor cortex. Stimulation of sites in the primary motor cortex elicits simple movements in the indicated parts of the body. (Based on Penfield & Rasmussen, 1950.)

8.4
Primary Motor Cortex

The **primary motor cortex** is located in the *precentral gyrus* of the frontal lobe (see Figure 8.5 and Figure 8.7). It is the major point of convergence of cortical sensorimotor signals, and it is the major, but not the only, point of departure of sensorimotor signals from the cerebral cortex. Understanding of the function of primary motor cortex has recently undergone radical changes—see Graziano (2006). The following two subsections describe this evolution: First, we consider the conventional view of primary motor cortex function, and second, we learn the current view of primary motor cortex function and some of the evidence on which it is based.

Conventional View of Primary Motor Cortex Function In 1937, Penfield and Boldrey mapped the primary motor cortex of conscious human patients during neurosurgery by applying brief, low-intensity electrical stimulations to various points on the cortical surface and noting which part of the body moved in response to each stimulation. They found that the stimulation of each particular cortical site activated a particular contralateral muscle and produced a simple movement. When they mapped out the relation between each cortical site and the muscle that was activated by its stimulation, they found that the primary motor cortex is organized somatotopically—that is, according to a map of the body. The **somatotopic** layout of the human primary motor cortex is commonly referred to as the **motor homunculus** (see Figure 8.7). Notice that most of the primary motor cortex is dedicated to controlling parts of the body that are capable of intricate movements, such as the hands and mouth.

It is important to appreciate that each site in the primary motor cortex receives sensory feedback from receptors in the muscles and joints that the site influences. One interesting exception to this general pattern of feedback

record the firing of individual neurons in humans while conducting the required behavioral tests (Turella et al., 2007). Although it has not been possible to directly demonstrate the existence of human mirror neurons, indirect lines of evidence suggest that mirror neurons do exist in the human brain. For example, functional brain-imaging studies have found areas of human motor cortex that are active when a person performs, watches, or imagines a particular action (e.g., Rizzolatti & Fabbri-Destro, 2008; Rodriguez et al., 2008). Indeed, many researchers believe that the evidence for human mirror neurons is so strong that they have started to consider the possible role that pathology of these neurons might play in human neuropsychological disorders (see Ramachandran & Oberman, 2006).

Evolutionary Perspective

has been described in monkeys: Monkeys have at least two different hand areas in the primary motor cortex of each hemisphere, and one receives input from receptors in the skin rather than from receptors in the muscles and joints. Presumably, this adaptation facilitates **stereognosis**—the process of identifying objects by touch. Close your eyes and explore an object with your hands; notice how stereognosis depends on a complex interplay between motor responses and the somatosensory stimulation produced by them (see Johansson & Flanagan, 2009).

What is the function of each primary motor cortex neuron? Until recently, each neuron was thought to encode the direction of movement. The main evidence for this was the finding that each neuron in the arm area of the primary motor cortex fires maximally when the arm reaches in a particular direction; that is, each neuron has a different preferred direction.

Current View of Primary Motor Cortex Function

Recent efforts to map the primary motor cortex have used a new stimulation technique—see Graziano (2006). Rather than stimulating with brief pulses of current that are just above the threshold to produce a reaction, investigators have used longer bursts of current (e.g., 0.5 second), which are more similar to the duration of a motor response, at slightly higher intensities. The results were amazing: Rather than eliciting the contractions of individual muscles, these currents elicited complex natural-looking response sequences. For example, stimulation at one site reliably produced a feeding response: The arm reached forward, the hand closed as if clasping some food, the closed hand was brought to the mouth, and finally the mouth opened. These recent studies have revealed a crude somatotopic organization—that is, stimulation in the face area tended to elicit face movements. However, the elicited responses were complex species-typical movements, which often involved several parts of the body (e.g., hand, shoulder, and mouth), rather than individual muscle contractions. Also, sites that moved a particular body part overlapped greatly with sites that moved other body parts (Sanes et al., 1995). That is why small lesions in the hand area of the primary motor cortex of humans (Scheiber, 1999) or monkeys (Scheiber & Poliakov, 1998) never selectively disrupt the activity of a single finger.

The conventional view that many primary motor cortex neurons are tuned to movement in a particular direction has also been challenged. In the studies that have supported this view, the monkey subjects were trained to make arm movements from a central starting point so that the relation between neural firing and the direction of movement could be assessed. However, there is another reasonable interpretation of the results of these studies. Graziano (2006) recognized this alternative interpretation when he recorded from individual primary

Thinking Creatively

motor cortex neurons as monkeys moved about freely, rather than as they performed simple, learned arm movements from a set starting point. The firing of many primary motor cortex neurons was most closely related to the end point of a movement, not to the direction of the movement. If a monkey reached toward a particular location, primary motor cortex neurons sensitive to that target location tended to become active regardless of the direction of the movement that was needed to get to the target.

The importance of the target of a movement, rather than the direction of a movement, for the function of primary motor cortex was also apparent in Graziano's stimulation studies. For example, if stimulation of a particular cortical site causes the left elbow to bend to a 90° angle, opposite responses will be elicited if the arm was initially straight (180° angle) and if it was bent halfway (45° angle)—but the end point will always be the same. Stop for a moment and consider the implications of this finding, because they are as important as they are counterintuitive. First, the finding means that the signals from every site in the primary motor cortex diverge greatly, so each particular site has the ability to get a body part (e.g., an arm) to a target location regardless of the starting position. Second, it means that the sensorimotor system is inherently plastic. So far, I have not said much about neuroplasticity, which is a major theme of this book, but it will soon start to play a central role. Here you have learned that each location in the primary motor cortex can produce innumerable patterns of muscle contraction required to get a body part from any starting point to a target location. The

Neuroplasticity

key point is that the route that neural signals follow from a given area of primary motor cortex is extremely plastic and is presumably determined at any point in time by somatosensory feedback (Davidson et al., 2007).

The neurons of the primary motor cortex play a major role in initiating body movements. With an appropriate interface, could they control the movements of a machine (see Craelius, 2002; König & Verschure, 2002; Taylor, Tillery, & Schwartz, 2002)? Belle says, "yes."

Belle: The Monkey That Controlled a Robot with Her Mind

In the laboratory of Miguel Nicolesis and John Chapin (2002), a tiny owl monkey called Belle watched a series of lights on a control panel. Belle had learned that if she moved the joystick in her right hand in the direction of a light, she would be rewarded with a drop of fruit juice. On this particular day, Nicolesis and Chapin demonstrated an amazing feat. As a light flashed on the panel, 100 microelectrodes recorded extracellular unit activity from neurons in Belle's primary motor cortex. This activity moved Belle's arm toward the light, but at the same time, the signals were analyzed by a computer, which fed the output to a laboratory several hundred kilometers away, at the Massachusetts Institute of Technology. At MIT, the signals from Belle's brain entered the circuits of a robotic arm. On each trial, the activity of Belle's primary motor cortex moved her arm toward the test light, and it moved the robotic arm in the same direction. Belle's neural signals were directing the activity of a robot.

Belle's remarkable feat raises a possibility. Perhaps one day injured people will routinely control wheelchairs, prosthetic limbs, or even their own paralyzed limbs through the power of their thoughts (Lebedev & Nicolesis, 2006; Scherberger, 2009). Indeed, some neuroprosthetic devices for human patients are currently being evaluated (see Pancrazio & Peckham, 2009; Patil, 2009).

Effects of Primary Motor Cortex Lesions Extensive damage to the human primary motor cortex has less effect than you might expect, given that this cortex is the major point of departure of motor fibers from the cerebral cortex. Large lesions to the primary motor cortex may disrupt a patient's ability to move one body part (e.g., one finger) independently of others, may produce **astereognosia** (deficits in stereognosis), and may reduce the speed, accuracy, and force of a patient's movements. Such lesions do not, however, eliminate voluntary movement, presumably because there are parallel pathways that descend directly from secondary motor areas to subcortical motor circuits without passing through primary motor cortex.

8.5
Cerebellum and Basal Ganglia

The cerebellum and the basal ganglia (see Figure 3.21 on page 65 and Figure 3.29 on page 70) are both important sensorimotor structures, but neither is a major part of the pathway by which signals descend through the sensorimotor hierarchy. Instead, both the cerebellum and the basal ganglia interact with different levels of the sensorimotor hierarchy and, in so doing, coordinate and modulate its activities. The interconnections between sensory and motor areas via the cerebellum and basal ganglia are thought to be one reason why damage to cortical connections between visual cortex and frontal motor areas does not abolish visually guided responses (Glickstein, 2000).

Cerebellum

The functional complexity of the cerebellum is suggested by its structure (see Apps & Hawkes, 2009). For example, although it constitutes only 10% of the mass of the brain, it contains more than half of the brain's neurons (Azevedo et al., 2009).

The cerebellum receives information from primary and secondary motor cortex, information about descending motor signals from brain stem motor nuclei, and feedback from motor responses via the somatosensory and vestibular systems. The cerebellum is thought to compare these three sources of input and correct ongoing movements that deviate from their intended course (see Bastian, 2006; Bell, Han, & Sawtell, 2008). By performing this function, it is believed to play a major role in motor learning, particularly in the learning of sequences of movements in which timing is a critical factor (D'Angelo & De Zeeuw, 2008; Jacobson, Rokni, & Yarom, 2008).

The consequences of diffuse cerebellar damage for motor function are devastating. The patient loses the ability to control precisely the direction, force, velocity, and amplitude of movements and the ability to adapt patterns of motor output to changing conditions. It is difficult to maintain steady postures (e.g., standing), and attempts to do so frequently lead to tremor. There are also severe disturbances in balance, gait, speech, and the control of eye movement. Learning new motor sequences is particularly difficult (Shin & Ivry, 2003; Thach & Bastian, 2004).

The traditional view that the function of the cerebellum is limited to the fine-tuning (see Apps & Garwicz, 2005) and learning of motor responses has been challenged. This challenge has been based on functional brain images of activity in the cerebellums of healthy volunteers recorded while they performed a variety of non-motor cognitive tasks (see Strick, Dum, & Fiez, 2009), from the documentation of cognitive deficits in patients with cerebellar damage (e.g., Fabbro et al., 2004; Hoppenbrouwers et al., 2008), and from the demonstrated connection of the cerebellum with cognitive areas such as the prefrontal cortex (Ramnani, 2006). Various alternative theories to the traditional view have been proposed, but the most parsimonious of them tend to argue that the cerebellum functions in the fine-tuning and learning of cognitive responses in the same way that it functions in the fine-tuning and learning of motor responses (e.g., Doya, 2000).

Basal Ganglia

The basal ganglia do not contain as many neurons as the cerebellum, but in one sense they are more complex. Unlike the cerebellum, which is organized systematically in lobes, columns, and layers, the basal ganglia are a complex heterogeneous collection of interconnected nuclei.

The anatomy of the basal ganglia suggests that, like the cerebellum, they perform a modulatory function (see Kreitzer, 2009). They contribute few fibers to descending motor pathways; instead, they are part of neural loops

Simulate Major Pathways of the Basal Ganglia www.mypsychlab.com

that receive cortical input from various cortical areas and transmit it back to the cortex via the thalamus (see McHaffie et al., 2005; Smith et al., 2004). Many of these loops carry signals to and from the motor areas of the cortex (see Nambu, 2008).

Theories of basal ganglia function have evolved in much the same way that theories of cerebellar function have changed. The traditional view of the basal ganglia was that they, like the cerebellum, play a role in the modulation of motor output. Now, the basal ganglia are thought to be involved in a variety of cognitive functions in addition to their role in the modulation of motor output (see Graybiel, 2005; Graybiel & Saka, 2004; Strick, 2004). This expanded view of the function of the basal ganglia is consistent with the fact that they project to cortical areas known to have cognitive functions (e.g., prefrontal lobes).

In experiments on laboratory animals, the basal ganglia have been shown to participate in learning to respond correctly in order to obtain reward and avoid punishment, a type of response learning that is often acquired gradually, trial by trial (see Joshua, Adler, & Bergman, 2009; Surmeier, Plotkin, & Shen, 2009). However, the basal ganglia's cognitive functions do not appear to be limited to this form of response learning (e.g., Ravizza & Ivry, 2001).

Scan Your Brain

Are you ready to continue your descent into the sensorimotor circuits of the spinal cord? This is a good place for you to pause to scan your brain to evaluate your knowledge of the sensorimotor circuits of the cortex, cerebellum, and basal ganglia by completing the following statements. The correct answers are provided at the end of the exercise. Before proceeding, review material related to your incorrect answers and omissions.

1. Visual, auditory, and somatosensory input converges on the _____ association cortex.
2. A small area of frontal cortex called the frontal _____ plays a major role in the control of eye movement.
3. Contralateral neglect is often associated with large lesions of the right _____ lobe.
4. The _____ prefrontal cortex seems to play an important role in initiating complex voluntary responses.

5. The secondary motor area that is just dorsal to the premotor cortex and is largely hidden from view on the medial surface of each hemisphere is the _____.
6. Most of the direct sensory input to the supplementary motor area comes from the _____ system.
7. Most of the direct sensory input to the premotor cortex comes from the _____ system.
8. The _____ cortex is the main point of departure of motor signals from the cerebral cortex to lower levels of the sensorimotor hierarchy.
9. The foot area of the motor homunculus is in the _____ fissure.
10. Although the _____ constitutes only 10% of the mass of the brain, it contains more than half of the brain's neurons.
11. The _____ are part of neural loops that receive input from various cortical areas and transmit it back to various areas of motor cortex via the thalamus.
12. Although both are considered to be motor structures, damage to the _____ or the _____ produces cognitive deficits.

Scan Your Brain answers: (1) posterior parietal, (2) eye field, (3) parietal, (4) dorsolateral, (5) supplementary motor area, (6) somatosensory, (7) visual, (8) primary motor, (9) longitudinal, (10) cerebellum, (11) basal ganglia, (12) cerebellum; basal ganglia.

8.6
Descending Motor Pathways

Neural signals are conducted from the primary motor cortex to the motor neurons of the spinal cord over four different pathways. Two pathways descend in the *dorsolateral* region of the spinal cord, and two descend in the *ventromedial* region of the spinal cord. Signals conducted over these pathways act together in the control of voluntary movement (see Iwaniuk & Whishaw, 2000). Like a large company, the sensorimotor system does not work well unless there are good lines of communication from the executive level (the cortex) to the office personnel (the spinal motor circuits) and workers (the muscles).

Dorsolateral Corticospinal Tract and Dorsolateral Corticorubrospinal Tract

One group of axons that descends from the primary motor cortex does so through the *medullary pyramids*—two bulges on the ventral surface of the medulla—then decussates and continues to descend in the contralateral dorsolateral spinal white matter. This group of axons constitutes the **dorsolateral corticospinal tract**. Most notable among its neurons are the **Betz cells**—extremely large pyramidal neurons of the primary motor cortex.

Most axons of the dorsolateral corticospinal tract synapse on small interneurons of the spinal gray matter, which synapse on the motor neurons of distal muscles of the wrist, hands, fingers, and toes. Primates and the few other mammals (e.g., hamsters and raccoons) that are capable of moving their digits independently have dorsolateral corticospinal tract neurons that synapse directly on digit motor neurons (see Porter & Lemon, 1993).

A second group of axons that descends from the primary motor cortex synapses in the *red nucleus* of the midbrain. The axons of neurons in the red nucleus then decussate and descend through the medulla, where some of them terminate in the nuclei of the cranial nerves that control the muscles of the face. The rest continue to descend in the dorsolateral portion of the spinal cord. This pathway is called the **dorsolateral corticorubrospinal tract** (*rubro* refers to the red nucleus). The axons of the dorsolateral corticorubrospinal tract synapse on interneurons that in turn synapse on motor neurons that project to the distal muscles of the arms and legs.

The two divisions of the dorsolateral motor pathway—the direct dorsolateral corticospinal tract and the indirect dorsolateral corticorubrospinal tract—are illustrated schematically in Figure 8.8.

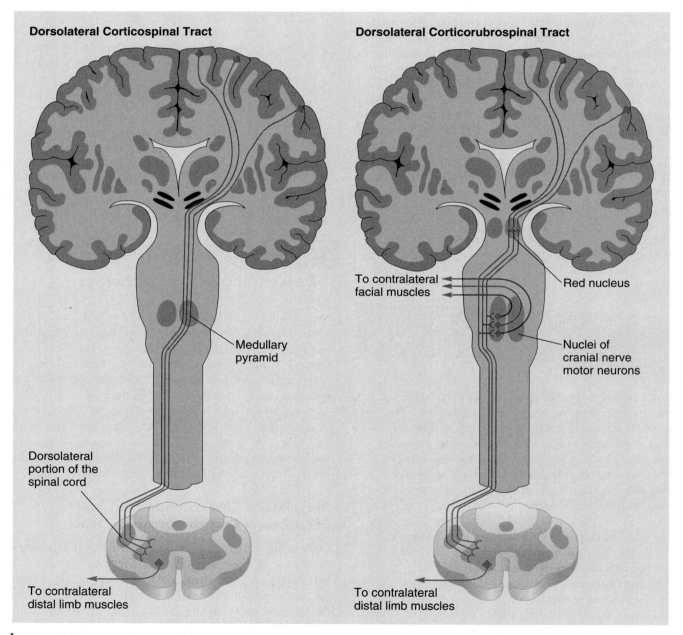

FIGURE 8.8 The two divisions of the dorsolateral motor pathway: the dorsolateral corticospinal tract and the dorsolateral corticorubrospinal tract. The projections from only one hemisphere are shown.

Ventromedial Corticospinal Tract and Ventromedial Cortico–brainstem–spinal Tract

Just as there are two major divisions of the dorsolateral motor pathway, one direct (the corticospinal tract) and one indirect (the corticorubrospinal tract), there are two major divisions of the ventromedial motor pathway, one direct and one indirect. The direct ventromedial pathway is the **ventromedial corticospinal tract**, and the indirect one—as you might infer from its cumbersome but descriptive name—is the **ventromedial cortico-brainstem-spinal tract**.

The long axons of the ventromedial corticospinal tract descend ipsilaterally from the primary motor cortex directly into the ventromedial areas of the spinal white matter. As each axon of the ventromedial corticospinal tract descends, it branches diffusely and innervates the interneuron circuits in several different spinal segments on both sides of the spinal gray matter.

The ventromedial cortico-brainstem-spinal tract comprises motor cortex axons that feed into a complex network of brain stem structures. The axons of some of the neurons in this complex brain stem motor network then descend bilaterally in the ventromedial portion of the spinal cord. Each side carries signals from both hemispheres, and each neuron synapses on the interneurons of several different spinal cord segments that control the proximal muscles of the trunk and limbs.

Which brain stem structures interact with the ventromedial cortico-brainstem-spinal tract? There are four major ones: (1) the **tectum**, which receives auditory and visual information about spatial location; (2) the **vestibular nucleus**, which receives information about balance from receptors in the semicircular canals of the inner ear; (3) the **reticular formation**, which, among other things, contains motor programs that regulate complex species-typical movements such as walking, swimming, and jumping; and (4) the motor nuclei of the cranial nerves that control the muscles of the face.

The two divisions of the descending ventromedial pathway—the direct ventromedial corticospinal tract and the indirect ventromedial cortico-brainstem-spinal tract—are illustrated in Figure 8.9 on page 206.

Comparison of the Two Dorsolateral Motor Pathways and the Two Ventromedial Motor Pathways

The descending dorsolateral and ventromedial pathways are similar in that each is composed of two major tracts, one whose axons descend directly to the spinal cord and another whose axons synapse in the brain stem on neurons that in turn descend to the spinal cord. However, the two dorsolateral tracts differ from the two ventromedial tracts in two major respects:

- The two ventromedial tracts are much more diffuse. Many of their axons innervate interneurons on both sides of the spinal gray matter and in several different segments, whereas the axons of the two dorsolateral tracts terminate in the contralateral half of one spinal cord segment, sometimes directly on a motor neuron.
- The motor neurons that are activated by the two ventromedial tracts project to proximal muscles of the trunk and limbs (e.g., shoulder muscles), whereas the motor neurons that are activated by the two dorsolateral tracts project to distal muscles (e.g., finger muscles).

Because all four of the descending motor tracts originate in the cerebral cortex, all are presumed to mediate voluntary movement; however, major differences in their routes and destinations suggest that they have different functions. This difference was first demonstrated in two experiments on monkeys that were reported by Lawrence and Kuypers in 1968.

Evolutionary Perspective

In their first experiment, Lawrence and Kuypers (1968a) transected (cut through) the left and right dorsolateral corticospinal tracts of their subjects in the medullary pyramids, just above the decussation of the tracts. Following surgery, these monkeys could stand, walk, and climb quite normally; however, their ability to use their limbs for other activities was impaired. For example, their reaching movements were weak and poorly directed, particularly in the first few days following the surgery. Although there was substantial improvement in the monkeys' reaching ability over the ensuing weeks, two other deficits remained unabated. First, the monkeys never regained the ability to move their fingers independently of one another; when they picked up pieces of food, they did so by using all of their fingers as a unit, as if they were glued together. And second, they never regained the ability to release objects from their grasp; as a result, once they picked up a piece of food, they often had to root for it in their hand like a pig rooting for truffles in the ground. In view of this latter problem, it is remarkable that they had no difficulty releasing their grasp on the bars of their cage when they were climbing. This point is important because it shows that the same response performed in different contexts can be controlled by different parts of the central nervous system. The point is underlined by the finding that some patients can stretch otherwise paralyzed limbs when they yawn (Provine, 2005).

In their second experiment, Lawrence and Kuypers (1968b) made additional transections in the monkeys whose dorsolateral corticospinal tracts had already been transected in the first experiment. The dorsolateral corticorubrospinal tract was transected in one group of these monkeys. The monkeys could stand, walk, and climb after this second transection; but when they were sitting, their arms hung limply by their sides (remember that monkeys normally use their arms for standing and walking). In those few instances in which the monkeys did use an arm for reaching, they used it like a rubber-handled rake—throwing it out from the shoulder and using it to draw small objects of interest back along the floor.

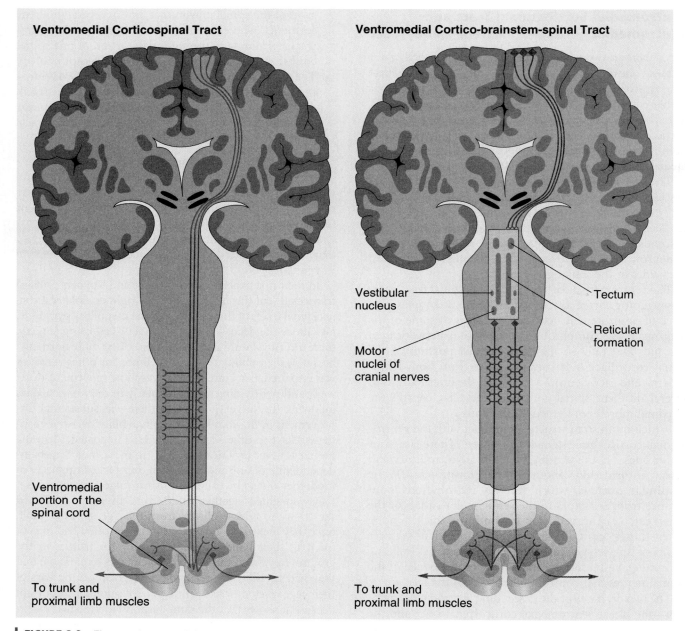

Ventromedial Corticospinal Tract

Ventromedial Cortico-brainstem-spinal Tract

Vestibular nucleus

Tectum

Reticular formation

Motor nuclei of cranial nerves

Ventromedial portion of the spinal cord

To trunk and proximal limb muscles

To trunk and proximal limb muscles

FIGURE 8.9 The two divisions of the ventromedial motor pathway: the ventromedial corticospinal tract and the ventromedial cortico-brainstem-spinal tract. The projections from only one hemisphere are shown.

The other group of monkeys in the second experiment had both of their ventromedial tracts transected. In contrast to the first group, these subjects had severe postural abnormalities: They had great difficulty walking or sitting. If they did manage to sit or stand without clinging to the bars of their cages, the slightest disturbance, such as a loud noise, frequently made them fall. Although they had some use of their arms, the additional transection of the two ventromedial tracts eliminated their ability to control their shoulders. When they fed, they did so with elbow and whole-hand movements while their upper arms hung limply by their sides.

What do these experiments tell us about the roles of the various descending sensorimotor tracts in the control of primate movement? They suggest that the two ventromedial tracts are involved in the control of posture and whole-body movements (e.g., walking and climbing) and that they can exert control over the limb movements involved in such activities. In contrast, both dorsolateral tracts—the corticospinal tract and the corticorubrospinal tract—control the movements of the limbs. This redundancy was presumably the basis for the good recovery of limb movement after the initial lesions of the corticospinal dorsolateral tract. However, only the corticospinal division of the dorsolateral system is capable of mediating independent movements of the digits.

Sensorimotor Spinal Circuits

We have descended the sensorimotor hierarchy to its lowest level: the spinal circuits and the muscles they control. Psychologists, including me, tend to be brain-oriented, and they often think of the spinal cord motor circuits as mere cables that carry instructions from the brain to the muscles. If you think this way, you will be surprised: The motor circuits of the spinal cord show considerable complexity in their functioning, independent of signals from the brain (see Grillner & Jessell, 2009). Again, the business metaphor helps put this in perspective: Can the office workers (spinal circuits) and workers (muscles) of a company function effectively when all of the executives and branch managers are at a convention in Hawaii? Of course they can—and the sensorimotor spinal circuits are also capable of independent functioning.

Muscles

Motor units are the smallest units of motor activity. Each motor unit comprises a single motor neuron and all of the individual skeletal muscle fibers that it innervates (see Figure 8.10). When the motor neuron fires, all the muscle fibers of its unit contract together. Motor units differ appreciably in the number of muscle fibers they contain; the units with the fewest fibers—those of the fingers and face—permit the highest degree of selective motor control.

A skeletal muscle comprises hundreds of thousands of threadlike muscle fibers bound together in a tough membrane and attached to a bone by a *tendon. Acetylcholine*, which is released by motor neurons at *neuromuscular junctions*, activates the **motor end-plate** on each muscle fiber and causes the fiber to contract. Contraction is the only method that muscles have for generating force, thus any muscle can generate force in only one direction. All of the motor neurons that innervate the fibers of a single muscle are called its **motor pool**.

◉ Watch
Muscle Contraction
www.mypsychlab.com

Although it is an oversimplification (see Gollnick & Hodgson, 1986), skeletal muscle fibers are often considered to be of two basic types: fast and slow. *Fast muscle fibers*, as you might guess, are those that contract and relax quickly. Although they are capable of generating great force, they fatigue quickly because they are *poorly vascularized* (have few blood vessels, which gives them a pale color). In contrast, *slow muscle fibers*, although slower and weaker, are capable of more sustained contraction because they are more richly vascularized (and

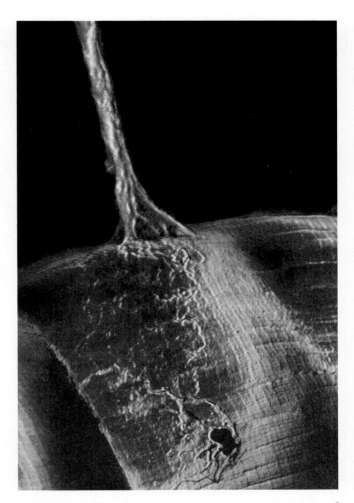

FIGURE 8.10   An electron micrograph of a motor unit: a motor neuron (pink) and the muscle fibers that it innervates.

hence much redder). Each muscle has both fast and slow fibers—the fast muscle fibers participate in quick movements such as jumping, whereas the slow muscle fibers participate in gradual movements such as walking. Because each muscle can apply force in only one direction, joints that move in more than one direction must be controlled by more than one muscle. Many skeletal muscles belong unambiguously to one of two categories: flexors or extensors. **Flexors** act to bend or flex a joint, and **extensors** act to straighten or extend it. Figure 8.11 on page 208 illustrates the *biceps* and *triceps*—the flexor and extensor, respectively, of the elbow joint. Any two muscles whose contraction produces the same movement, be it flexion or extension, are said to be **synergistic muscles**; those that act in opposition, like the biceps and the triceps, are said to be **antagonistic muscles**.

To understand how muscles work, it is important to realize that they are elastic, rather than inflexible and cablelike. If you think of an increase in muscle tension as being analogous to an increase in the tension of an elastic band joining two bones, you will appreciate that

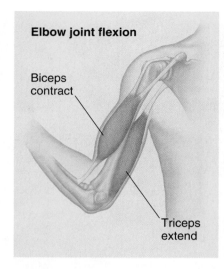

Elbow joint flexion

Biceps
contract

Triceps
extend

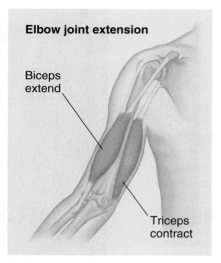

Elbow joint extension

Biceps
extend

Triceps
contract

FIGURE 8.11   The biceps and triceps, which are the flexor and extensor, respectively, of the elbow joint.

increasing the number of neurons in its motor pool that are firing, by increasing the firing rates of those that are already firing, or more commonly by a combination of these two changes.

Receptor Organs of Tendons and Muscles

The activity of skeletal muscles is monitored by two kinds of receptors: Golgi tendon organs and muscle spindles. **Golgi tendon organs** are embedded in the *tendons*, which connect each skeletal muscle to bone; **muscle spindles** are embedded in the muscle tissue itself. Because of their different locations, Golgi tendon organs and muscle spindles respond to different aspects of muscle contraction. Golgi tendon organs respond to increases in muscle tension (i.e., to the pull of the muscle on the tendon), but they are completely insensitive to changes in muscle length. In contrast, muscle spindles respond to changes in muscle length, but they do not respond to changes in muscle tension.

Under normal conditions, the function of Golgi tendon organs is to provide the central nervous system with information about muscle tension, but they also serve a protective function. When the contraction of a muscle is so extreme that there is a risk of damage, the Golgi tendon organs excite inhibitory interneurons in the spinal cord that cause the muscle to relax.

Figure 8.12 is a schematic diagram of the *muscle-spindle feedback*

muscle contraction can be of two types. Activation of a muscle can increase the tension that it exerts on two bones without shortening and pulling them together; this is termed **isometric contraction**. Or it can shorten and pull them together; this is termed **dynamic contraction**. The tension in a muscle can be increased by

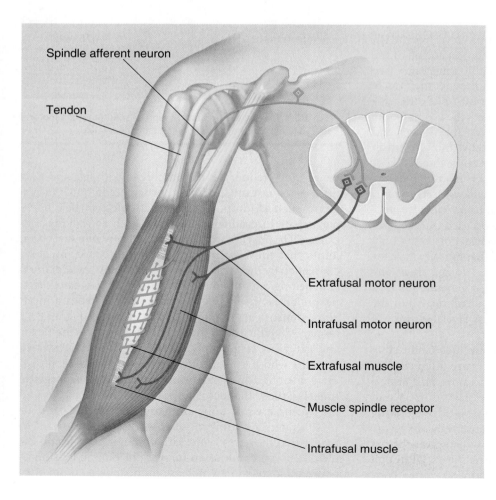

Spindle afferent neuron

Tendon

Extrafusal motor neuron

Intrafusal motor neuron

Extrafusal muscle

Muscle spindle receptor

Intrafusal muscle

FIGURE 8.12   The muscle-spindle feedback circuit. There are many muscle spindles in each muscle; for clarity, only one much-enlarged muscle spindle is illustrated here.

circuit. Examine it carefully. Notice that each muscle spindle has its own threadlike **intrafusal muscle**, which is innervated by its own **intrafusal motor neuron**. Why

would a receptor have its own muscle and motor neuron? The reason becomes apparent when you consider what would happen to a muscle spindle without them. Without its intrafusal motor input, a muscle spindle would fall slack each time its **skeletal muscle (extrafusal muscle)** contracted. In this slack state, the muscle spindle could not do its job, which is to respond to slight changes in extrafusal muscle length. As Figure 8.13 illustrates, the intrafusal motor

neuron solves this problem by shortening the intrafusal muscle each time the extrafusal muscle becomes shorter, thus keeping enough tension on the middle, stretch-sensitive portion of the muscle spindle to keep it responsive to slight changes in the length of the extrafusal muscle.

Stretch Reflex

When the word *reflex* is mentioned, many people think of themselves sitting on the edge of their doctor's examination table having their knees rapped with a little rubber-headed hammer. The resulting leg extension is called the **patellar tendon reflex** (*patella* means "knee"). This reflex

is a **stretch reflex**—a reflex that is elicited by a sudden external stretching force on a muscle.

When your doctor strikes the tendon of your knee, the extensor muscle running along your thigh is stretched. This initiates the chain of events that is depicted in Figure 8.14 on page 210. The sudden stretch of the thigh muscle stretches its muscle-spindle stretch receptors, which in turn initiate a volley of action potentials that are carried from the stretch receptors into the spinal cord by **spindle afferent neurons** via the *dorsal root*. This volley of action potentials excites motor neurons in the *ventral horn* of the spinal cord, which respond by sending action potentials back to the muscle whose stretch originally excited them (see Illert & Kummel, 1999). The arrival of these impulses back at the starting point results in a compensatory muscle contraction and a sudden leg extension.

The method by which the patellar tendon reflex is typically elicited in a doctor's office—that is, by a sharp blow to the tendon of a completely relaxed muscle—is designed to make the reflex readily observable. However, it does little to communicate its functional significance. In real-life situations, the function of the stretch reflex is to keep external forces from altering the intended position of the body. When an external force, such as a push on your arm while you are holding a cup of coffee, causes an unanticipated extrafusal muscle stretch, the muscle-spindle feedback circuit produces an immediate compensatory contraction of the muscle that counteracts the force and keeps you from spilling the coffee—unless, of course, you are wearing your best clothes.

The mechanism by which the stretch reflex maintains limb stability is illustrated in Figure 8.15 on page 211. Examine it carefully because it illustrates two of the principles of sensorimotor system function that are the focus of this chapter: the important role played by sensory feedback in the regulation of motor output and the ability of lower circuits in the motor hierarchy to take care of "business details" without the involvement of higher levels.

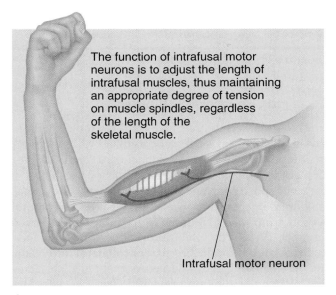

Without intrafusal motor neurons, the spindles of a skeletal muscle would become slack and unresponsive to stretch during a muscle contraction.

Extrafusal muscle   Intrafusal muscle

The function of intrafusal motor neurons is to adjust the length of intrafusal muscles, thus maintaining an appropriate degree of tension on muscle spindles, regardless of the length of the skeletal muscle.

Intrafusal motor neuron

FIGURE 8.13   The function of intrafusal motor neurons.

Withdrawal Reflex

I am sure that, at one time or another, you have touched something painful—a hot pot, for example—and suddenly pulled back your hand. This is a **withdrawal reflex**. Unlike the stretch reflex, the withdrawal reflex is *not monosynaptic*. When a painful stimulus is applied to the hand, the first responses are recorded in the motor neurons of the arm flexor muscles about 1.6 milliseconds later, about the time it takes a neural signal to cross two synapses. Thus, the shortest route in the withdrawal-reflex circuit involves one interneuron. Other responses are recorded in the motor neurons of the arm flexor muscles after the initial volley; these responses are triggered by signals that have traveled over multisynaptic pathways—some involving the cortex. See Figure 8.16 on page 212.

Simulate Monosynaptic and Polysynaptic Reflex www.mypsychlab.com

Reciprocal Innervation

Reciprocal innervation is an important principle of spinal cord circuitry. It refers to the fact that antagonistic muscles are innervated in a way that permits a smooth, unimpeded motor response: When one is contracted, the other relaxes. Figure 8.16 illustrates the role of reciprocal innervation in the withdrawal reflex. "Bad news" of a sudden painful event in the hand arrives in the dorsal horn of the spinal cord and has two effects: The signals excite both excitatory and inhibitory interneurons. The excitatory interneurons excite the motor neurons of the elbow flexor; the inhibitory interneurons inhibit the motor neurons of the elbow extensor. Thus, a single sensory input produces a coordinated pattern of motor output; the activities of agonists and antagonists are automatically coordinated by the internal circuitry of the spinal cord.

Movements are quickest when there is simultaneous excitation of all agonists and complete inhibition of all antagonists; however, this is not the way voluntary movement is normally produced. Most muscles are always contracted to some degree, and movements are produced by adjustment in the level of relative cocontraction between antagonists. Movements that are produced by **cocontraction** are smooth, and they can be stopped with precision by a slight increase in the contraction of the antagonistic muscles. Moreover, cocontraction insulates us from the effects of unexpected external forces.

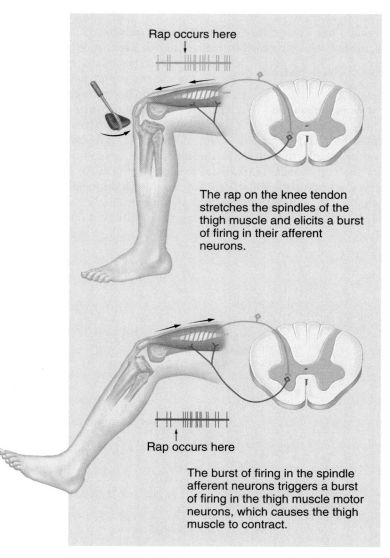

Rap occurs here

The rap on the knee tendon stretches the spindles of the thigh muscle and elicits a burst of firing in their afferent neurons.

Rap occurs here

The burst of firing in the spindle afferent neurons triggers a burst of firing in the thigh muscle motor neurons, which causes the thigh muscle to contract.

FIGURE 8.14 The elicitation of a stretch reflex. All of the muscle spindles in a muscle are activated during a stretch reflex, but only a single muscle spindle is depicted here.

Recurrent Collateral Inhibition

Like most workers, muscle fibers and the motor neurons that innervate them need an occasional break, and there are inhibitory neurons in the spinal cord that make sure they get it. Each motor neuron branches just before it leaves the spinal cord, and the branch synapses on a small inhibitory interneuron, which inhibits the very motor neuron from which it receives its input (see Illert & Kummel, 1999). The inhibition produced by these local feedback circuits is called **recurrent collateral inhibition**, and the small inhibitory interneurons that mediate recurrent collateral inhibition are called *Renshaw cells*. As a consequence of recurrent collateral inhibition, each time

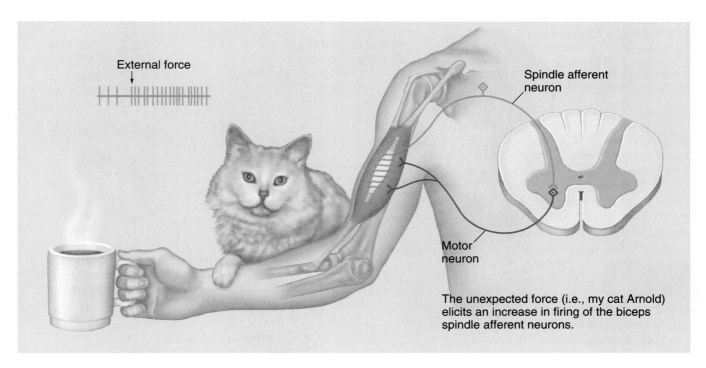

External force

Spindle afferent
neuron

Motor
neuron

The unexpected force (i.e., my cat Arnold)
elicits an increase in firing of the biceps
spindle afferent neurons.

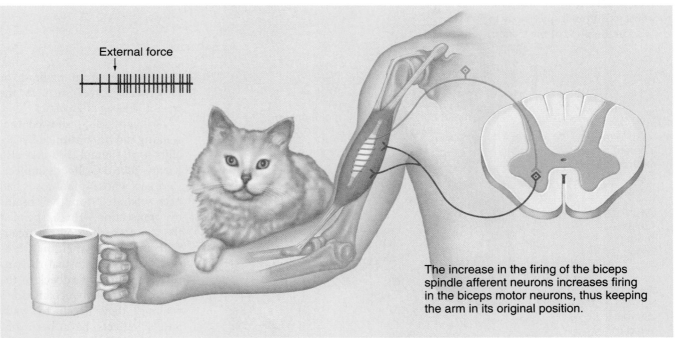

External force

The increase in the firing of the biceps
spindle afferent neurons increases firing
in the biceps motor neurons, thus keeping
the arm in its original position.

FIGURE 8.15 The automatic maintenance of limb position by the muscle-spindle feedback system.

a motor neuron fires, it momentarily inhibits itself and shifts the responsibility for the contraction of a particular muscle to other members of the muscle's motor pool.

Figure 8.17 on page 212 provides a summary; it illustrates recurrent collateral inhibition and other factors that directly excite or inhibit motor neurons.

Walking: A Complex Sensorimotor Reflex

Most reflexes are much more complex than withdrawal and stretch reflexes. Think for a moment about the complexity of the program of reflexes that is needed to control an activity such as walking (see Capaday, 2002; Dietz, 2002;

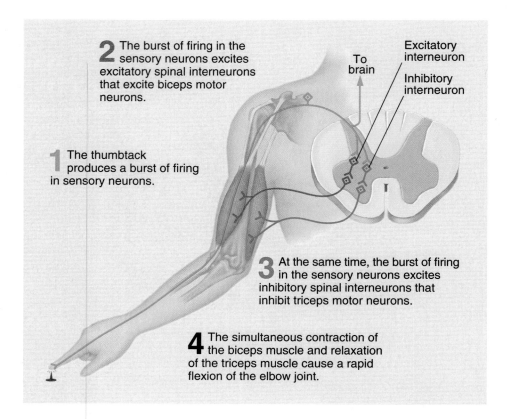

FIGURE 8.16 The reciprocal innervation of antagonistic muscles in the arm. During a withdrawal reflex, elbow flexors are excited, and elbow extensors are inhibited.

2 The burst of firing in the sensory neurons excites excitatory spinal interneurons that excite biceps motor neurons.

1 The thumbtack produces a burst of firing in sensory neurons.

To brain

Excitatory interneuron

Inhibitory interneuron

3 At the same time, the burst of firing in the sensory neurons excites inhibitory spinal interneurons that inhibit triceps motor neurons.

4 The simultaneous contraction of the biceps muscle and relaxation of the triceps muscle cause a rapid flexion of the elbow joint.

Nielsen, 2002). Such a program must integrate visual information from the eyes; somatosensory information from the feet, knees, hips, arms, and so on; and information about balance from the semicircular canals of the inner ears. And it must produce, *Neuroplasticity* on the basis of this information, an integrated series of movements that involves the muscles of the trunk, legs, feet, and upper arms. This program of reflexes must also be incredibly plastic; it must be able to adjust its output immediately to changes in the slope of the terrain,

to instructions from the brain, or to external forces such as the weight of a bag of groceries.

Grillner (1985) showed that walking can be controlled by circuits in the spinal cord (see also Kiehn, 2006). Grillner's subjects were cats whose spinal cords had been separated from their brains by transection. He suspended the cats in a sling over a treadmill; amazingly, when the treadmill was started so that the cats received sensory feedback of the sort that normally accompanies walking, they began to walk. Similar effects have been observed in other species, but in humans the descending motor pathways seem to play a greater role in walking (see Drew, Jiang, & Widajewicz, 2002).

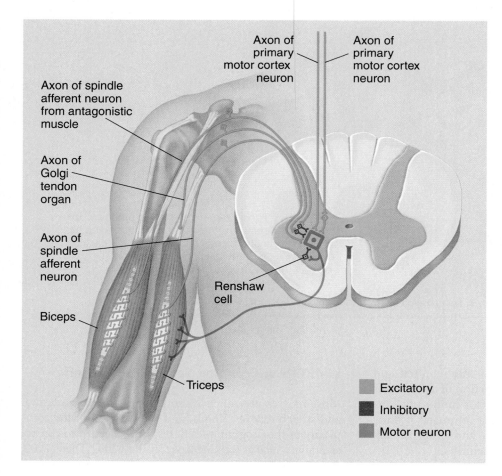

Axon of primary motor cortex neuron

Axon of primary motor cortex neuron

Axon of spindle afferent neuron from antagonistic muscle

Axon of Golgi tendon organ

Axon of spindle afferent neuron

Renshaw cell

Biceps

Triceps

■ Excitatory
■ Inhibitory
■ Motor neuron

FIGURE 8.17 The excitatory and inhibitory signals that directly influence the activity of a motor neuron.

Scan Your Brain

Before beginning the important section on central sensorimotor programs, test your knowledge of descending sensorimotor pathways and spinal circuits. Before each item on the left, write the letter for the most related item on the right. The correct answers are provided at the end of the exercise. Before proceeding, review material related to your incorrect answers and omissions.

____ 1. cerebellum	a. very large
____ 2. medullary pyramids	b. acetylcholine
____ 3. Betz cells	c. biceps and triceps
____ 4. rubro	d. Golgi tendon organs
____ 5. motor end-plate	e. extrafusal
____ 6. triceps	f. learning and fine-tuning
____ 7. antagonistic	of responses
____ 8. muscle tension	g. red nucleus
____ 9. muscle length	h. two synapses
____ 10. skeletal muscle	i. complex reflex
____ 11. patellar tendon reflex	j. inhibitory interneurons
____ 12. withdrawal reflex	k. bulges
____ 13. Renshaw cells	l. stretch reflex
____ 14. walking	m. muscle spindles
	n. extensor

Scan Your Brain answers: (1) f, (2) k, (3) a, (4) g, (5) b, (6) n, (7) c, (8) d, (9) m, (10) e, (11) l, (12) h, (13) j, (14) i.

8.8
Central Sensorimotor Programs

In this chapter, you have learned that the sensorimotor system is like the hierarchy of a large efficient company. You have learned how the executives—the dorsolateral prefrontal cortex, the supplementary motor area, and the premotor cortex—issue commands based on information supplied to them in part by the posterior parietal cortex. And you have learned how these commands are forwarded to the director of operations (the primary motor cortex) for distribution over four main channels of communication (the two dorsolateral and the two ventromedial spinal motor pathways) to the metaphoric office managers of the sensorimotor hierarchy (the spinal sensorimotor circuits). Finally, you have learned how spinal sensorimotor circuits direct the activities of the workers (the muscles). As you have seen, recent research has taught us that the lower levels of the sensorimotor hierarchy (e.g., primary motor cortex and spinal cord) are much more complex than once assumed, but that the functions of the higher levels are even more complex.

One theory of sensorimotor function is that the sensorimotor system comprises a hierarchy of **central sensorimotor programs** (see Brooks, 1986; Georgopoulos, 1991). The central sensorimotor program theory suggests that all but the highest levels of the sensorimotor system have certain patterns of activity programmed into them and that complex movements are produced by activating the appropriate combinations of these programs (see Swinnen, 2002; Tresch et al., 2002). Accordingly, if you decide that you want to look at a magazine, your association cortex will activate high-level cortical programs that in turn will activate lower-level programs—perhaps in your brain stem—for walking, bending over, picking up, and thumbing through. These programs in turn will activate spinal programs that control the various elements of the sequences and cause your muscles to complete the objective (Grillner & Jessell, 2009).

Once activated, each level of the sensorimotor system is capable of operating on the basis of current sensory feedback, without the direct control of higher levels. Thus, although the highest levels of your sensorimotor system retain the option of directly controlling your activities, most of the individual responses that you make are performed without direct cortical involvement, and you are barely aware of them.

👁️**Watch**
Drumming without Direct
Cortical Involvement
www.mypsychlab.com

In much the same way, a company president who wishes to open a new branch office simply issues the command to one of the executives, and the executive responds in the usual fashion by issuing a series of commands to the appropriate people lower in the hierarchy, who in turn do the same. Each of the executives and workers of the company knows how to complete many different tasks and executes them in the light of current conditions when instructed to do so. Good companies have mechanisms for ensuring that the programs of action at different levels of the hierarchy are well coordinated and effective. In the sensorimotor system, these mechanisms seem to be the responsibility of the cerebellum and basal ganglia.

Central Sensorimotor Programs Are Capable of Motor Equivalence

Like a large, efficient company, the sensorimotor system does not always accomplish a particular task in exactly the same way. The fact that the same basic movement can be carried out in different ways involving different muscles is called **motor equivalence**. For example, you have learned to sign your name with stereotypical finger and hand movements, yet if you wrote your name with your toe on a sandy beach, your signature would still retain many of its typical characteristics.

Motor equivalence illustrates the inherent plasticity of the sensorimotor system. It suggests that specific central sensorimotor programs for signing your name are not stored in the neural circuits that directly control your preferred hand; general programs are stored higher in your sensorimotor hierarchy and then are adapted to the situation as required. In an fMRI study, Rijntjes and others (1999) showed that the central sensorimotor programs for signing one's name seem to be stored in areas of secondary motor cortex that control the preferred hand. Remarkably, these same hand areas were also activated when the signature was made with a toe.

Sensory Information That Controls Central Sensorimotor Programs Is Not Necessarily Conscious

In Chapter 6, you learned that the neural mechanisms of conscious visual perception (ventral stream) are not necessarily the same as those that mediate the visual control of behavior (dorsal stream). Initial evidence for this theory came from neuropsychological patients who could respond to visual stimuli of which they had no conscious awareness and from others who could not effectively interact with objects that they consciously perceived.

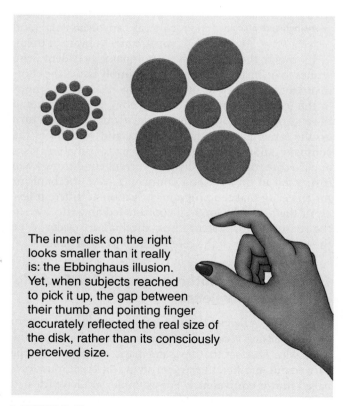

The inner disk on the right looks smaller than it really is: the Ebbinghaus illusion. Yet, when subjects reached to pick it up, the gap between their thumb and pointing finger accurately reflected the real size of the disk, rather than its consciously perceived size.

FIGURE 8.18 The Ebbinghaus illusion. Notice that the central disk on the left appears larger than the one on the right. Haffenden and Goodale (1998) found that when subjects reached out to pick up either of the central disks, the position of their fingers as they approached the disks indicated that their responses were being controlled by the actual sizes of the disks, not their consciously perceived sizes.

Is there evidence for the separation of conscious perception and sensory control of behavior in intact humans? Haffenden and Goodale (1998) supplied such evidence (see also Ganel, Tanzer, & Goodale, 2008; Goodale & Westwood, 2004). They showed healthy volunteers a three-dimensional version of the visual illusion in Figure 8.18—notice that the two central disks appear to be different sizes, even though they are identical. Remarkably, when the volunteers were asked to indicate the size of each central disk with their right thumb and pointing finger, they judged the disk on the left to be bigger than the one on the right; however, when they were asked to reach out and pick up the disks with the same two digits, the preparatory gap between the digits was a function of the actual size of each disk rather than its perceived size.

Central Sensorimotor Programs Can Develop without Practice

What type of experience is necessary for the normal development of central sensorimotor programs? In particular, is it necessary to practice a particular behavior for its central sensorimotor program to develop?

Although central sensorimotor programs for some behaviors can be established by practicing the behaviors, the central sensorimotor programs for many species-typical behaviors are established without explicit practice of the behaviors. This point was made clear by the classic study of Fentress (1973). Fentress showed that adult mice raised from birth without forelimbs still made the patterns of shoulder movements typical of grooming in their species—and that these movements were well coordinated with normal tongue, head, and eye movements. For example, the mice blinked each time they made the shoulder movements that would have swept their forepaws across their eyes. Fentress's study also demonstrated the importance of sensory feedback in the operation of central sensorimotor programs. The forelimbless mice, deprived of normal tongue-forepaw contact during face grooming, would often interrupt ostensible grooming sequences to lick a cage-mate or even the floor.

Practice Can Create Central Sensorimotor Programs

Although central sensorimotor programs for many species-typical behaviors develop without practice, practice is a certain way to generate or modify such programs (Sanes, 2003). Theories of sensorimotor learning emphasize two kinds of processes that influence the learning of central sensorimotor programs: response chunking and shifting control to lower levels of the sensorimotor system.

Response Chunking According to the **response-chunking hypothesis**, practice combines the central sensorimotor programs that control individual response into programs that control sequences (chunks) of behavior. In a novice typist, each response necessary to type a word is individually triggered and controlled; in a skilled typist, sequences of letters are activated as a unit, with a marked increase in speed and continuity.

An important principle of chunking is that chunks can themselves be combined into higher-order chunks. For example, the responses needed to type the individual letters and digits of one's address may be chunked into longer sequences necessary to produce the individual words and numbers, and these chunks may in turn be combined so that the entire address can be typed as a unit.

Shifting Control to Lower Levels In the process of learning a central sensorimotor program, control is shifted from higher levels of the sensorimotor hierarchy to lower levels (see Ashe et al., 2006). Shifting the level of control to lower levels of the sensorimotor system during training (see Seitz et al., 1990) has two advantages. One is that it frees up the higher levels of the system to deal with more esoteric aspects of performance. For example, skilled pianists can concentrate on interpreting a piece of music because they do not have to consciously focus on pressing the right keys. The other advantage of shifting the level of control is that it permits great speed because different circuits at the lower levels of the hierarchy can act simultaneously, without interfering with one another. It is possible to type 120 words per minute only because the circuits responsible for activating each individual key press can become active before the preceding response has been completed.

Functional Brain Imaging of Sensorimotor Learning

Functional brain-imaging techniques have provided opportunities for studying the neural correlates of sensorimotor learning. By recording the brain activity of human subjects as they learn to perform new motor sequences, researchers can develop hypotheses about the roles of various structures in sensorimotor learning. A good example of this approach is the PET study of Jenkins and colleagues (1994). These researchers recorded PET activity from human subjects who performed two different sequences of key presses. There were four different keys, and each sequence was four presses long. The presses were performed with the right hand, one every 3 seconds, and tones indicated when to press and whether or not a press was correct. There were three conditions: a rest control condition, a condition in which the subjects performed a newly learned sequence, and a condition in which the subjects performed a well-practiced sequence.

The following are six major findings of this study (see Figure 8.19 on page 216). They recapitulate important points that have already been made in this chapter.

- **Finding 1:** Posterior parietal cortex was activated during the performance of both the newly learned sequence and the well-practiced sequence, but it was more active during the newly learned sequence. This finding is consistent with the hypothesis that the posterior parietal cortex integrates sensory stimuli (in this case, the tones) that are used to guide motor sequences, and it is consistent with the finding that the posterior parietal cortex is more active when subjects are attending more to the stimuli, as is often the case during the early stages of motor learning.

- **Finding 2:** Dorsolateral prefrontal cortex was activated during the performance of the newly learned sequence but not the well-practiced sequence. This suggests that the dorsolateral prefrontal cortex plays a particularly important role when motor sequences are being performed largely under conscious control, as is often the case during the early stages of motor learning.

- **Finding 3:** The areas of secondary motor cortex responded differently. The contralateral premotor cortex was more active during the performance of the newly learned sequence, whereas the supplementary motor area was more active bilaterally during the well-practiced sequence. This finding is consistent with the hypothesis that the premotor cortex plays a more prominent role when performance is being guided largely by sensory stimuli, as is often the case in the early stages of motor learning, and that the supplementary motor area plays a more prominent role when performance is largely independent of sensory stimuli, as is often the case for well-practiced motor sequences, which can be run off automatically with little sensory feedback.

- **Finding 4:** Contralateral primary motor and somatosensory cortexes were equally activated during the performance of both the newly learned and the well-practiced motor sequences. This finding is consistent with the fact that the motor elements were the same during both sequences.

- **Finding 5:** The contralateral basal ganglia were equally activated during the performance of the newly learned sequence and the well-practiced sequence. Jenkins and colleagues speculated that different subpopulations of basal ganglia neurons may have been active during the two conditions, but this could not be detected because of the poor spatial resolution of PET.

- **Finding 6:** The cerebellum was activated bilaterally during the performance of both the newly learned and

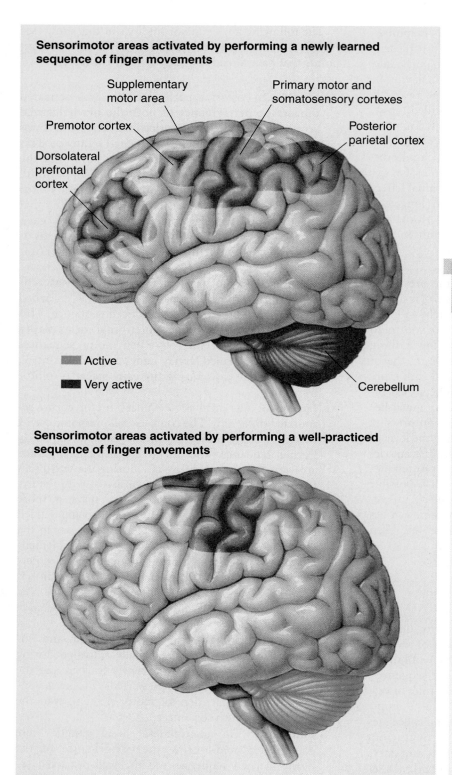

Sensorimotor areas activated by performing a newly learned sequence of finger movements

Supplementary motor area

Premotor cortex

Dorsolateral prefrontal cortex

Primary motor and somatosensory cortexes

Posterior parietal cortex

■ Active

■ Very active

Cerebellum

Sensorimotor areas activated by performing a well-practiced sequence of finger movements

FIGURE 8.19   The activity recorded by PET scans during the performance of newly learned and well-practiced sequences of finger movements. (Based on Jenkins et al., 1994.)

the well-practiced sequences, but it was more active during the newly learned sequence. This is consistent with the idea that the cerebellum plays a prominent role in motor learning.

The Case of Rhonda, Revisited

A few days after I finished writing this chapter, I stopped off to pick up a few fresh vegetables and some fish for dinner, and I once again found myself waiting in Rhonda's line. It was the longest line, but I am a creature of habit. This time, I felt rather smug as I watched her. All of the reading and thinking that had gone into the preparation of this chapter had provided me with some new insights into what she was doing and how she was doing it. I wondered whether she appreciated her own finely tuned sensorimotor system as much as I did. Then I hatched my plot—a little test of Rhonda's muscle-spindle feedback system. How would Rhonda's finely tuned sensorimotor system react to a bag that looked heavy but was in fact extremely light? Next time, I would get one of those paper bags at the mushroom counter, blow it up, drop one mushroom in it, and then fold the top so it looked completely full. I smiled at the thought. But I wasn't the only one smiling. My daydreaming ended abruptly, and the smile melted from my face, as I noticed Rhonda's extended hand and her amused grin. Will I never learn?

Themes Revisited

All four of this book's major themes were addressed in this chapter. Most prominent was the clinical implications theme. You learned how research with neuropsychological patients with sensorimotor deficits, as well as with normal human subjects, has contributed to current theories of sensorimotor functioning.

The evolutionary perspective theme was evident in the discussion of several comparative experiments on the sensorimotor system, largely in nonhuman primates. An important point to keep in mind is that although the sensorimotor functions of nonhuman primates are similar to those of humans, they are not identical (e.g., monkeys walk on both their hands and feet).

You learned how metaphors can be used to think productively about science—in particular, how a large, efficient company can serve as a useful metaphor for the sensorimotor system. You also leaned how recent analyses have suggested  that primary motor cortex encodes the end point of movements rather than the movements themselves.

Finally, you learned that the sensorimotor system is fundamentally plastic. General commands to act are issued by cortical circuits, but exactly how an act is actually completed depends on the current situation (e.g., body position). Moreover, the sensorimotor system maintains the ability to change itself in response to practice.

Think about It

1. Both sensorimotor systems and large businesses are complex systems trying to survive in a competitive milieu. It is no accident that they function in similar ways. Discuss.
2. We humans tend to view cortical mechanisms as preeminent, presumably because we are the species with the largest cortexes. However, one might argue from several perspectives that the lower sensorimotor functions are more important. Discuss.
3. This chapter has presented more evidence of parallel processing: Some neural circuits perform their function under conscious awareness, and other circuits perform the same function in the absence of conscious control. Discuss.
4. Belle, the owl monkey, controlled a robotic arm with her brain. How might the technology that allowed her to do this be used to improve the lives of paralyzed patients?
5. In order to be efficient, the sensorimotor system must be plastic. Discuss.

Key Terms

8.1 Three Principles of Sensorimotor Function

Sensory feedback (p. 193)

8.2 Sensorimotor Association Cortex

Posterior parietal association cortex (p. 194)
Frontal eye field (p. 194)
Apraxia (p. 194)
Contralateral neglect (p. 195)
Dorsolateral prefrontal association cortex (p. 197)

8.3 Secondary Motor Cortex

Secondary motor cortex (p. 197)
Supplementary motor area (p. 197)
Premotor cortex (p. 197)

Cingulate motor areas (p. 198)
Mirror neurons (p. 198)

8.4 Primary Motor Cortex

Primary motor cortex (p. 200)
Somatotopic (p. 200)
Motor homunculus (p. 200)
Stereognosis (p. 201)
Astereognosia (p. 202)

8.6 Descending Motor Pathways

Dorsolateral corticospinal tract (p. 203)
Betz cells (p. 203)
Dorsolateral corticorubrospinal tract (p. 204)
Ventromedial corticospinal tract (p. 205)

Ventromedial cortico-brainstem-spinal tract (p. 205)
Tectum (p. 205)
Vestibular nucleus (p. 205)
Reticular formation (p. 205)

8.7 Sensorimotor Spinal Circuits

Motor units (p. 207)
Motor end-plate (p. 207)
Motor pool (p. 207)
Flexors (p. 207)
Extensors (p. 207)
Synergistic muscles (p. 207)
Antagonistic muscles (p. 207)
Isometric contraction (p. 208)
Dynamic contraction (p. 208)
Golgi tendon organs (p. 208)
Muscle spindles (p. 208)
Intrafusal muscle (p. 209)

Intrafusal motor neuron (p. 209)
Skeletal muscle (extrafusal muscle) (p. 209)
Patellar tendon reflex (p. 209)
Stretch reflex (p. 209)
Spindle afferent neurons (p. 209)
Withdrawal reflex (p. 210)
Reciprocal innervation (p. 210)
Cocontraction (p. 210)
Recurrent collateral inhibition (p. 210)

8.8 Central Sensorimotor Programs

Central sensorimotor programs (p. 213)
Motor equivalence (p. 213)
Response-chunking hypothesis (p. 215)

✓•─ **Quick Review**   Test your comprehension of the chapter with this brief practice test. You can find the answers to these questions as well as more practice tests, activities and other study resources at www.mypsychlab.com.

1. A striking consequence of posterior parietal cortex damage is
   a. apraxia.
   b. contralateral neglect.
   c. stereognosis.
   d. both a and b
   e. both b and c

2. A neuron that fires just as robustly when a monkey sees a particular action being performed as when the monkey performs the action itself is a
   a. multipolar neuron.
   b. bipolar neuron.
   c. mirror neuron.
   d. motor neuron.
   e. sensory neuron.

3. Primary motor cortex is located in the
   a. parietal lobe.
   b. frontal lobe.
   c. precentral gyrus.
   d. postcentral gyrus.
   e. both b and c

4. Primary motor cortex is laid out
   a. ectopically.
   b. somatotopically.
   c. tonotopically.
   d. retinotopically.
   e. randomly.

5. The cell bodies of Betz cells are located in the
   a. fingertips.
   b. primary motor cortex.
   c. premotor cortex.
   d. dorsal horn.
   e. ventral horn.

9 Development of the Nervous System
From Fertilized Egg to You

Most of us tend to think of the brain as a three-dimensional array of neural elements "wired" together in a massive network of circuits. However, the brain is not a static network of interconnected elements.

Neuroplasticity It is a *plastic* (changeable), living organ that continuously changes in response to its genetic programs and environment.

This chapter focuses on the incredible process of *neuro-development* (neural development), which begins with a single fertilized egg cell and ends with a functional adult brain. Three general ideas are emphasized: (1) the amazing nature of neurodevelopment, (2) the important role of experience in neurodevelopment, and (3) the dire consequences of neurodevelopmental errors. The chapter culminates in a discussion of two devastating disorders of human neurodevelopment: autism and Williams disorder.

⊙ **Watch**
The Central Nervous System
www.mypsychlab.com

But first, a case study. Many of us are reared in similar circumstances—we live in warm, safe, stimulating environments with supportive families and communities and plenty to eat and drink. Because

Thinking Creatively there is so little variation in most people's early experience, the critical role of experience in human cerebral and psychological development is not always obvious. In order to appreciate the critical role played by experience in neurodevelopment, it is important to consider cases in which children have been reared in grossly abnormal environments. Genie is such a case (Curtiss, 1977; Rymer, 1993).

The Case of Genie

When Genie was admitted to the hospital at the age of 13, she was only 1.35 meters (4 feet, 5 inches) tall and weighed only 28.1 kilograms (62 pounds). She could not stand erect, chew solid food, or control her bladder or bowels.

Clinical Implications Since the age of 20 months, Genie had spent most days tied to a potty in a small, dark, closed room. Her only clothing was a cloth harness, which kept her from moving anything other than her feet and hands. In the evening, Genie was transferred to a covered crib and a straitjacket. Her father was intolerant of noise, and he beat Genie if she made any sound whatsoever. According to her mother, who was almost totally blind, Genie's father and brother rarely spoke to Genie, although they sometimes barked at her like dogs. The mother was permitted only a few minutes with Genie each day, during which time she fed Genie cereal or baby food—Genie was allowed no solid food. Genie's severe childhood deprivation left her seriously scarred. When she was admitted to hospital, she made almost no sounds and was totally incapable of speech.

After Genie's discovery, a major effort was made to get her development back on track and to document her problems and improvements; however, after a few years,

Genie "disappeared" in a series of legal proceedings, foster homes, and institutions. (Rymer, 1993)

Genie received special care and training after her rescue, but her behavior never became normal. The following were a few of her continuing problems: She did not react to extremes of warmth and cold; she tended to have silent tantrums during which she would flail, spit, scratch, urinate, and rub her own "snot" on herself; she was easily terrified (e.g., of dogs and men wearing khaki); she could not chew; she could speak only short, poorly pronounced phrases. Genie is currently living in a home for retarded adults. Clearly, experience plays a major role in the processes of neurodevelopment, processes to which you are about to be introduced.

9.1
Phases of Neurodevelopment

In the beginning, there is a *zygote*, a single cell formed by the amalgamation of an *ovum* and a *sperm*. The zygote divides to form two daughter cells. These two divide to form four, the four divide to form eight, and so on, until a mature organism is produced. Of course, there must be more to development than this; if there were not, each of us would have ended up like a bowl of rice pudding: an amorphous mass of homogeneous cells.

To save us from this fate, three things other than cell multiplication must occur. First, cells must *differentiate*; some must become muscle cells, some must become multipolar neurons, some must become glial cells, and so on. Second, cells must make their way to appropriate sites and align themselves with the cells around them to form particular structures. And third, cells must establish appropriate functional relations with other cells. This section describes how developing neurons accomplish these things in five phases: (1) induction of the neural plate, (2) neural proliferation, (3) migration and aggregation, (4) axon growth and synapse formation, and (5) neuron death and synapse rearrangement.

Induction of the Neural Plate

Three weeks after conception, the tissue that is destined to develop into the human nervous system becomes recognizable as the **neural plate**—a small patch of ectodermal tissue on the dorsal *Neuroplasticity* surface of the developing embryo. The ectoderm is the outermost of the three layers of embryonic cells: *ectoderm*, *mesoderm*, and *endoderm*. The development of the neural plate is the first major stage of neurodevelopment in all vertebrates.

The development of the neural plate seems to be *induced* by chemical signals from an area of the underlying mesoderm layer—an area that is consequently referred to as an *organizer* (see Placzek & Briscoe, 2005). Tissue taken from the dorsal mesoderm of one embryo (i.e., the *donor*) and implanted beneath the ventral ectoderm of another embryo (i.e., the *host*) induces the development of an extra neural plate on the ventral surface of the host. The search for the particular substances that are released by the organizer and induce the development of the neural plate is in full swing (see Muñoz-Sanjuán & Brivanlou, 2002).

An important change occurs to the cells of the developing nervous system at about the time that the neural plate becomes visible (see Rhinn, Picker, & Brand, 2006). The earliest cells of the human embryo are **totipotent**—that is, they have the ability to develop into any type of cell in the body if transplanted to the appropriate site. However, as the embryo develops, the destiny of various cells becomes more *specified* (Molyneaux et al., 2007; Okano & Temple, 2009). When the neural plate develops, its cells lose some of their potential to become different kinds of cells. Each cell of the early neural plate still has the potential to develop into most types of mature nervous system cell, but it cannot normally develop into other kinds of cells. Cells like these are said to be **multipotent**, rather than totipotent.

The cells of the neural plate are often referred to as embryonic stem cells. **Stem cells** are cells that meet two specific criteria (see Brivanlou et al., 2003; Seaberg & van der Kooy, 2003): (1) They have a seemingly unlimited capacity for self-renewal if maintained in an appropriate cell culture, and (2) they have the ability to develop into different types of mature cells. However, as the neural plate develops into the neural tube, some of its cells become specified as future glial cells of various types, and others become specified as future neurons of various types (see Placzek & Briscoe, 2005; Rowitch, 2004). Because these cells still have the capacity for unlimited self-renewal and are still multipotent, these cells are termed *glial stem cells* and *neural stem cells*, respectively. Other neural tube stem cells do not become specified until later (see Kriegstein & Alvarez-Buylla, 2009).

Why do stem cells maintained in a cell culture have an almost unlimited capacity for self-renewal? This capacity results from the fact that when a stem cell divides, two different daughter cells are created: one that eventually develops into some type of body cell and one that develops into another stem cell (Spradling & Zheng, 2007). In theory, stem cells can keep dividing forever through this process of mitosis, but eventually errors accumulate, which can disrupt the process. That is why stem cell cultures do not last forever.

Because of the ability of embryonic stem cells to develop into different types of mature cells, their therapeutic potential is currently under intensive investigation. Will embryonic stem cells injected into a damaged part of a mature brain develop into the appropriate brain structure and improve function? You will learn about the potential of stem-cell therapy in Chapter 10.

As Figure 9.1 illustrates, the neural plate folds to form the *neural groove*, and then the lips of the neural groove fuse to form the **neural tube**. The inside of the

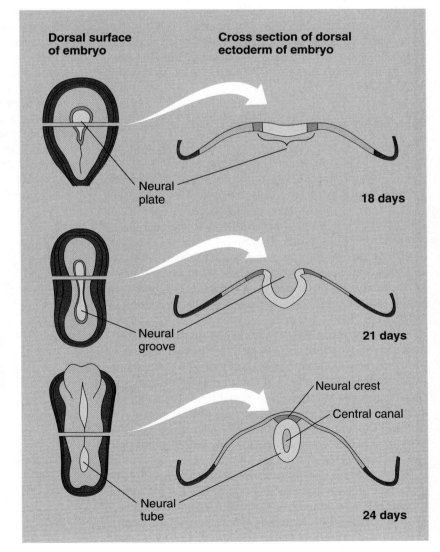

FIGURE 9.1 How the neural plate develops into the neural tube during the third and fourth weeks of human embryological development. (Based on Cowan, 1979.)

neural tube eventually becomes the *cerebral ventricles* and *spinal canal*. By 40 days after conception, three swellings are visible at the anterior end of the human neural tube; these swellings ultimately develop into the *forebrain, midbrain,* and *hindbrain* (see Figure 3.19 on page 64).

Neural Proliferation

Once the lips of the neural groove have fused to create the neural tube, the cells of the tube begin to *proliferate* (increase greatly in number). This **neural proliferation** does not occur simultaneously or equally in all parts of the tube. Most cell division in the neural tube occurs in the *ventricular zone*—the region adjacent to the *ventricle* (the fluid-filled center of the tube). In each species, the cells in different parts of the neural tube proliferate in a particular sequence that is responsible for the pattern of swelling and folding that gives the brain of each member of that species the characteristic shape. The complex pattern of proliferation is in part controlled by chemical signals from two organizer areas in the neural tube: the *floor plate*, which runs along the midline of the anterior surface of the tube, and the *roof plate*, which runs along the midline of the dorsal surface of the tube (see Chizhikou & Millen, 2004).

Neuroplasticity

Migration and Aggregation

Migration   Once cells have been created through cell division in the ventricular zone of the neural tube, they migrate to the appropriate target location. During this period of **migration**, the cells are still in an immature form, lacking the processes (i.e., axons and dendrites) that characterize mature neurons. Two major factors govern migration in the developing neural tube: time and location. In a given region of the tube, subtypes of neurons arise on a precise and predictable schedule and then migrate together to their prescribed destinations (Okano & Temple, 2009).

Neuroplasticity

Cell migration in the developing neural tube is considered to be of two kinds (see Figure 9.2): **Radial migration** proceeds from the ventricular zone in a straight line outward toward the outer wall of the tube; **tangential migration** occurs at a right angle to radial migration—that is, parallel to the tube's walls. Most cells engage in both radial and tangential migration to get from their point of origin in the ventricular zone to their target destination (see Hatten, 2002).

There are two methods by which developing cells migrate (see Figure 9.3; Liu & Roo, 2004). One is somal translocation. In **somal translocation**, an extension grows from the developing cell in the general direction of the migration; the extension seems to explore the immediate environment for attractive and repulsive cues as it grows. Then, the cell body itself moves into and along the extending process, and trailing processes are retracted (see Marin, Valdeolmillos, & Moya, 2006).

The second method of migration is **glia-mediated migration** (see Figure 9.3). Once the period of neural proliferation is well underway and the walls of the neural tube are thickening, a temporary network of glial cells, called **radial glial cells**, appears in the developing neural tube (Campbell & Gotz, 2002). At this point, most cells engaging in radial migration do so by moving along the radial glial network (see Nadarajah & Parnavelas, 2002).

Most research on migration in the developing neural tube has focused on the cortex. This research has revealed orderly waves of migrating cells, progressing from deeper to more superficial layers. Because each wave of cortical cells migrates through the already formed lower layers of cortex before reaching its destination, this radial pattern of cortical development is referred to as an **inside-out pattern** (see Cooper, 2008). Cortical migration patterns are more complex than was first thought (Kriegstein & Noctor, 2004): Many cortical cells engage in long tangential migrations to reach their final destinations, and the patterns of proliferation and migration are different for different areas of the cortex (Rash & Groves, 2006). The migration of interneurons is particularly complex (Wonders & Anderson, 2006).

The **neural crest** is a structure that is situated just dorsal to the neural tube (see Figure 9.1). It is formed from cells that break off from the neural tube as it is being formed. Neural crest cells develop into the neurons and glial cells of the peripheral nervous system, and thus many of them must migrate over considerable distances (see Jessen & Mirsky, 2005).

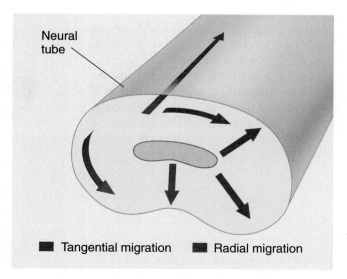

Neural tube

■ Tangential migration   ■ Radial migration

FIGURE 9.2   Two types of neural migration: radial migration and tangential migration.

Somal Translocation (Radial or Tangential)

Glia-Mediated Migration (Radial Only)

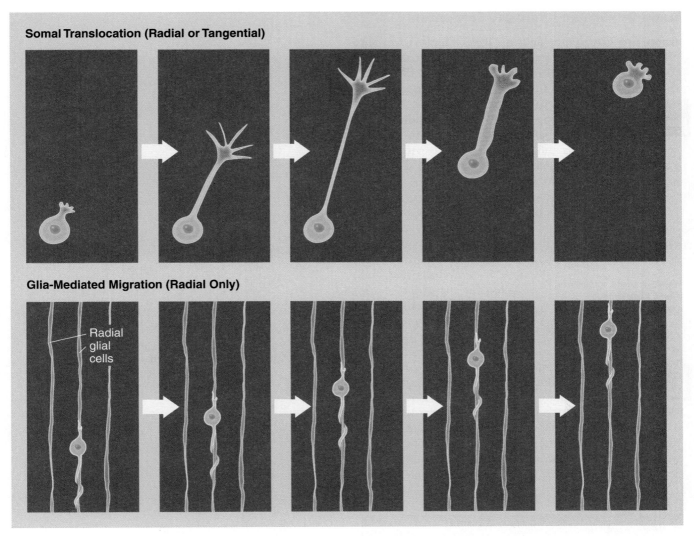

Radial glial cells

FIGURE 9.3 Two methods by which cells migrate in the developing neural tube: somal translocation and glia-mediated migration.

Numerous chemicals that guide migrating neurons by either attracting or repelling them have been discovered (Marin & Rubenstein, 2003; Neary & Zimmerman, 2009). Some of these guidance molecules are released by glial cells (see Auld, 2001; Marin et al., 2001).

Aggregation Once developing neurons have migrated, they must align themselves with other developing neurons that have migrated to the same area to form the structures of the nervous system. This process is called **aggregation**.

Both migration and aggregation are thought to be mediated by **cell-adhesion molecules (CAMs)**, which are located on the surfaces of neurons and other cells (Dalva, McClelland, & Kayser, 2007; Fuerst & Burgess, 2009; Shapiro, Love, & Colman, 2007). Cell-adhesion molecules have the ability to recognize molecules on other cells and adhere to them. Elimination of just one type of CAM in a *knockout mouse* (see Chapter 5) has been shown to have a devastating effect on brain development (DiCicco-Bloom, 2006; Lien et al., 2006). This finding suggests that abnormalities of CAM function may be causal factors in some neurological disorders.

Clinical Implications

Gap junctions between adjacent cells have been found to be particularly prevalent during brain development. You may recall from Chapter 4 that *gap junctions* are points of connection between adjacent neurons; gap junctions are not as wide as synapses, and the gaps are bridged by narrow tubes called *connexins*, through which cells can exchange cytoplasm. There is increasing evidence that gap junctions play a role in migration and aggregation (see Elias & Kriegstein, 2008; Montoro & Yuste, 2004).

Axon Growth and Synapse Formation

Axon Growth   Once neurons have migrated to their appropriate positions and aggregated into neural structures, axons and dendrites begin to grow from them. For the nervous system to function, these projections must

Neuroplasticity

grow to appropriate targets. At each growing tip of an axon or dendrite is an amoebalike structure called a **growth cone**, which extends and retracts fingerlike cytoplasmic extensions called *filopodia* (see Figure 9.4), as if searching for the correct route (Wen & Zheng, 2006).

Remarkably, most growth cones reach their correct targets. A series of studies of neural regeneration by Roger Sperry in the early 1940s first demonstrated that axons are capable of precise growth and suggested how it occurs.

In one study, Sperry cut the optic nerves of frogs, ro-

⊙ **Simulate** Roger Sperry's
Axon Regeneration
www.mypsychlab.com

tated their eyeballs 180°, and waited for the axons of the **retinal ganglion cells**, which compose the optic nerve, to

regenerate (grow again). (Frogs, unlike mammals, have retinal ganglion cells that regenerate.) Once regeneration

Evolutionary Perspective

was complete, Sperry used a convenient behavioral test to assess the frogs' visual capacities (see Figure 9.5). When he dan-

gled a lure behind the frogs, they struck forward, thus indicating that their visual world, like their eyes, had been rotated 180°. Frogs whose eyes had been rotated, but whose optic nerves had not been cut, responded in exactly the same way. This was strong behavioral evidence

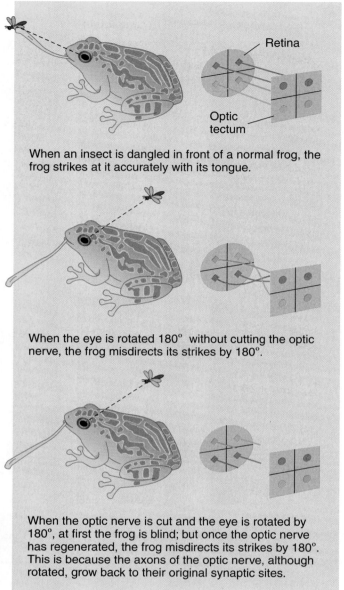

When an insect is dangled in front of a normal frog, the frog strikes at it accurately with its tongue.

When the eye is rotated 180° without cutting the optic nerve, the frog misdirects its strikes by 180°.

When the optic nerve is cut and the eye is rotated by 180°, at first the frog is blind; but once the optic nerve has regenerated, the frog misdirects its strikes by 180°. This is because the axons of the optic nerve, although rotated, grow back to their original synaptic sites.

FIGURE 9.5   Sperry's classic study of eye rotation and regeneration.

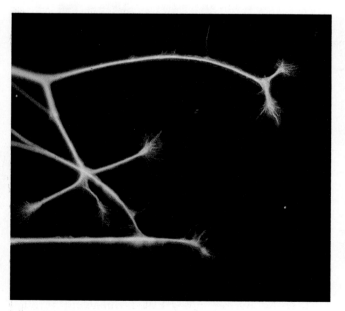

FIGURE 9.4   Growth cones. The cytoplasmic fingers (the filopodia) of growth cones seem to grope for the correct route. (Courtesy of Naweed I. Syed, Ph.D., Departments of Anatomy and Medical Physiology, the University of Calgary.)

that each retinal ganglion cell had grown back to the same point of the **optic tectum** (called the *superior colliculus* in mammals) to which it had originally been connected. Neuroanatomical investigations have confirmed that this is exactly what happens (see Guo & Udin, 2000).

On the basis of his studies of regeneration, Sperry proposed the **chemoaffinity hypothesis** of axonal development (see Sperry, 1963). He hypothesized that each postsynaptic surface in the nervous system releases a specific chemical label and that each growing axon is attracted by the label to its postsynaptic target during both neural development and regeneration. Indeed, it is difficult to imagine another mechanism by which an axon growing out from a rotated eyeball could find its precise target on the optic tectum. Several guidance molecules for axon growth have been identified (e.g., Guthrie, 2007; O'Donnell, Chance, & Bashaw, 2009; Round & Stein, 2007).

The chemoaffinity hypothesis fails to account for the discovery that some growing axons follow the same circuitous route to reach their target in every member of a species, rather than growing directly to it (see Araújo & Tear, 2003). This discovery led to a revised notion of how growing axons reach their specific targets. According to this revised hypothesis, a growing axon is not attracted to its target by a single specific attractant released by the target, as Sperry thought. Instead, growth cones seem to be influenced by a series of chemical signals along the route (see Imai et al., 2009; Miyamichi & Luo, 2009; Vanderhaegen & Polleux, 2004). These guidance molecules are similar to those that guide neural migration in the sense that some attract and others repel the growing axons.

Guidance molecules are not the only signals that guide growing axons to their targets (Lindwall, Fothergill, & Richards, 2007). Other signals come from adjacent growing axons (see Murai & Pasquale, 2008). **Pioneer growth cones**—the first growth cones to travel along a particular route in a developing nervous system—are presumed to follow the correct trail by interacting with guidance molecules along the route. Then, subsequent growth cones embarking on the same journey follow the routes blazed by the pioneers. The tendency of developing axons to grow along the paths established by preceding axons is called **fasciculation**.

Much of the axonal development in complex nervous systems involves growth from one topographic array of neurons to another. The neurons on one array project to another, maintaining the same topographic relation they had on the first array; for example, the topographic map of the retina is maintained on the optic tectum. It has been suggested that topographic maps evolved as a means of minimizing the volume of neural connections in the brain. There is evolutionary pressure for richly connected neurons to be as close as possible to each other (Chklovskii & Koulakov, 2004).

At first, it was assumed that the integrity of topographical relations in the developing nervous system was maintained by a point-to-point chemoaffinity, with each retinal ganglion cell growing toward a specific chemical label. However, evidence indicates that the mechanism must be more complex. In most species, the synaptic connections between retina and optic tectum are established long before either reaches full size. Then, as the retinas and the optic tectum grow at different rates, the initial synaptic connections shift to other tectal neurons so that each retina is precisely mapped onto the tectum, regardless of their relative sizes.

Studies of the regeneration (rather than the development) of retinal-tectum projections tell a similar story. In one informative series of studies, the optic nerves of mature frogs or fish were cut and their pattern of regeneration was assessed after parts of either the retina or the optic tectum had been destroyed. In both cases, the axons did not grow out to their original points of connection (as the chemoaffinity hypothesis predicted they would); instead, they grew out to fill the available space in an orderly fashion. These results are illustrated schematically in Figure 9.6 on page 226.

The **topographic gradient hypothesis** has been proposed to explain accurate axonal growth involving topographic mapping in the developing brain (see Flanagan, 2006; McLaughlin & O'Leary, 2005; Stoeckli, 2006). According to this hypothesis, axons growing from one topographic surface (e.g., the retina) to another (e.g., the optic tectum) are guided to specific targets that are arranged on the terminal surface in the same way as the axons' cell bodies are arranged on the original surface (Rakic et al., 2009). The key part of this hypothesis is that the growing axons are guided to their destinations by two intersecting signal gradients (e.g., an anterior-posterior gradient and a medial-lateral gradient).

Several guidance molecules have been implicated in signal gradients; however, the evidence for such involvement is strongest for a family of molecules called *ephrins* (e.g., Clandinin & Feldheim, 2009; Polleux, Ince-Dunn, & Gosh, 2007). Figure 9.7 on page 227 illustrates how gradients of ephrin-A and ephrin-B define locations on the vertebrate retina.

Synapse Formation Once axons have reached their intended sites, they must establish an appropriate pattern of synapses. A single neuron can grow an axon on its own, but it takes coordinated activity in at least two neurons to create a synapse between them (see Arikkath & Reichardt, 2008; Lai & Ip, 2009). This is one reason why our understanding of how axons connect to their targets has lagged behind our understanding of how they reach them. Still, some exciting breakthroughs have been made—for example, some of the chemical signals that play a role in the location and formation of synapses have been identified (see Chen & Cheng, 2009; Inestrosa & Arenas, 2010).

Perhaps the most exciting recent discovery about **synaptogenesis** (the formation of new synapses) is that it depends on the presence of glial cells, particularly astrocytes (see Allen & Barres, 2005; Shors, 2009). Retinal ganglion cells maintained in culture formed seven times more synapses when astrocytes were present. Moreover, synapses formed in the presence of astrocytes were quickly lost when the astrocytes were removed. Early theories about the contribution of astrocytes to synaptogenesis emphasized a nutritional role: Developing neurons need high levels of cholesterol during synapse formation,

⊕ Simulate Brain Building www.mypsychlab.com

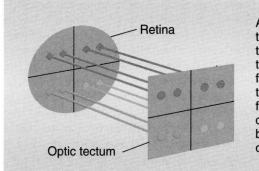

Axons normally grow from the frog retina and terminate on the optic tectum in an orderly fashion. The assumption that this orderliness results from point-to-point chemoaffinity is challenged by the following two observations.

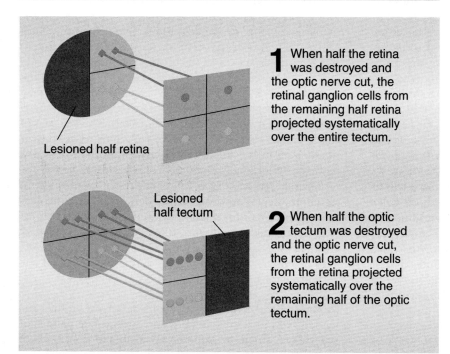

1 When half the retina was destroyed and the optic nerve cut, the retinal ganglion cells from the remaining half retina projected systematically over the entire tectum.

2 When half the optic tectum was destroyed and the optic nerve cut, the retinal ganglion cells from the retina projected systematically over the remaining half of the optic tectum.

FIGURE 9.6 The regeneration of the optic nerve of the frog after portions of either the retina or the optic tectum have been destroyed. These phenomena support the topographic gradient hypothesis.

Neuron Death and Synapse Rearrangement

Neuron Death Neuron death is a normal and important part of neurodevelopment. Many more neurons—about 50% more—are produced than are required, and large-scale neuron death occurs in waves in various parts of the brain throughout development.

Neuron death during development was initially assumed to be a passive process. It was assumed that developing neurons died when they failed to get adequate nutrition. However, it is now clear that cell death during development is usually active. Genetic programs inside neurons are triggered and cause them to actively commit suicide. Passive cell death is called **necrosis** (ne-KROE-sis); active cell death is called **apoptosis** (A-poe-toe-sis).

Apoptosis is safer than necrosis. Necrotic cells break apart and spill their contents into extracellular fluid, and the consequence is potentially harmful inflammation. In contrast, in apoptotic cell death, DNA and other internal structures are cleaved apart and packaged in membranes before the cell breaks apart. These membranes contain molecules that attract scavenger microglia and other molecules that prevent inflammation (Freeman, 2006; Savill, Gregory, & Haslett, 2003). Apoptosis removes excess neurons in a safe, neat, and orderly way. But apoptosis has a dark side as well. If genetic programs for apoptotic cell death are blocked, the consequence can be cancer; if the programs are inappropriately activated, the consequence can be neurodegenerative disease (see Haase et al., 2008).

What triggers the genetic programs that cause apoptosis in developing neurons? There appear to be two kinds of triggers (Miguel-Aliaga & Thor, 2009). First, some developing neurons appear to be genetically programmed for an early death—once they have fulfilled their functions, groups of neurons die together, in the absence of any obvious external stimulus (Chao, Ma, & Shen, 2009). Second, some developing neurons seem to die because they fail to obtain the life-preserving chemicals that are supplied by their targets (see Deppmann et al., 2008; Innocenti & Price, 2005). Evidence that life-preserving chemicals are

and the extra cholesterol is supplied by astrocytes (Mauch et al., 2001; Pfrieger, 2002). However, current evidence suggests that astrocytes play a much more extensive role in synaptogenesis by processing, transferring, and storing information supplied by neurons (Perea, Navarrete, & Araque, 2009).

Most current research on synaptogenesis is focusing on elucidating the chemical signals that must be exchanged between presynaptic and postsynaptic neurons for a synapse to be created (see Chih, Engelman, & Scheiffele, 2005; Dean & Dresbach, 2006; Hussain & Sheng, 2005; Takeichi, 2007). One complication researchers face is the promiscuity that developing neurons display when it comes to synaptogenesis. Although the brain must be "wired" according to a specific plan in order to function, in-vitro studies suggest that any type of neuron will form synapses with any other type. However, once established, synapses that do not function appropriately tend to be eliminated (Waites, Craig, & Garner, 2005).

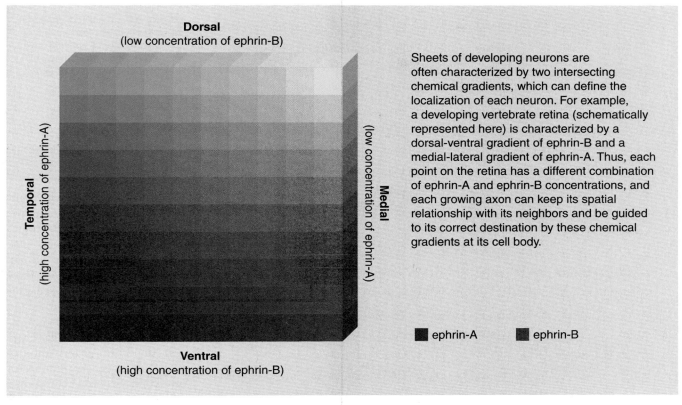

Dorsal
(low concentration of ephrin-B)

Temporal
(high concentration of ephrin-A)

Medial
(low concentration of ephrin-A)

Ventral
(high concentration of ephrin-B)

Sheets of developing neurons are often characterized by two intersecting chemical gradients, which can define the localization of each neuron. For example, a developing vertebrate retina (schematically represented here) is characterized by a dorsal-ventral gradient of ephrin-B and a medial-lateral gradient of ephrin-A. Thus, each point on the retina has a different combination of ephrin-A and ephrin-B concentrations, and each growing axon can keep its spatial relationship with its neighbors and be guided to its correct destination by these chemical gradients at its cell body.

■ ephrin-A   ■ ephrin-B

FIGURE 9.7   The topographic gradient hypothesis. Gradients of ephrin-A and ephrin-B on the developing retina (see McLaughlin, Hindges, & O'Leary, 2003).

supplied to developing neurons by their postsynaptic targets comes from two kinds of observations: Grafting an extra target structure (e.g, an extra limb) to an embryo before the period of synaptogenesis reduces the death of neurons growing into the area, and destroying some of the neurons growing into an area before the period of cell death increases the survival rate of the remaining neurons.

Several life-preserving chemicals that are supplied to developing neurons by their targets have been identified. The most prominent class of these chemicals is the **neurotrophins**. **Nerve growth factor (NGF)** was the first neurotrophin to be isolated (see Levi-Montalcini, 1952, 1975). The neurotrophins promote the growth and survival of neurons, function as axon guidance molecules, and stimulate synaptogenesis.

Synapse Rearrangement   During the period of cell death, neurons that have established incorrect connections are particularly likely to die. As they die, the space they leave vacant on postsynaptic membranes is filled by the sprouting axon terminals of surviving neurons. Thus, cell death results in a massive rearrangement of synaptic connections. This phase of synapse rearrangement tends to focus the output of each neuron on a smaller number of postsynaptic cells, thus increasing the selectivity of transmission (see Figure 9.8 on page 228).

Scan Your Brain

Are you ready to focus on the continuing development of the human brain after birth? To find out if you are prepared to proceed, scan your brain by filling in the blanks in the following chronological list of stages of neurodevelopment. The correct answers are provided at the end of the exercise. Before proceeding, review material related to your errors and omissions.

1. Induction of the neural _____
2. Formation of the _____ tube
3. Neural _____
4. Neural _____
5. _____ aggregation
6. Growth of neural _____
7. Formation of _____
8. Neuron _____ and synapse _____

Scan Your Brain answers: (1) plate, (2) neural, (3) proliferation, (4) migration, (5) Neural, (6) processes (axons and dendrites), (7) synapses, (8) death; rearrangement.

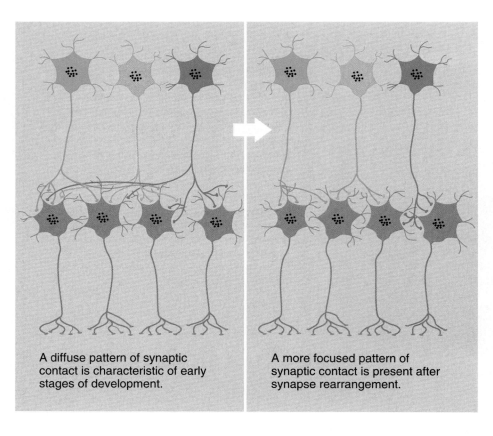

A diffuse pattern of synaptic contact is characteristic of early stages of development.

A more focused pattern of synaptic contact is present after synapse rearrangement.

FIGURE 9.8 The effect of neuron death and synapse rearrangement on the selectivity of synaptic transmission. The synaptic contacts of each axon become focused on a smaller number of cells.

9.2

Postnatal Cerebral Development in Human Infants

Most of our knowledge of the development of the human brain comes from the study of nonhuman species (see

Evolutionary Perspective

Bystron, Blakemore, & Rakic, 2008). This fact emphasizes the value of the evolutionary perspective. There is, however, one way in which the development of the human brain is unique: The human brain develops more slowly than those of other species, not achieving full maturity until late adolescence (see Blakemore, 2008).

This section deals with the part of cerebral development that occurs after birth. It focuses on the development of the *prefrontal cortex* (see Figure 1.7 on page 16) because the prefrontal cortex is the last part of the human brain to reach maturity (Casey, Giedd, & Thomas, 2000).

Postnatal Growth of the Human Brain

The human brain grows substantially after birth: Its volume quadruples between birth and adulthood (see Johnson, 2001). This increase in size does not, however, result from the development of additional neurons. With the exception of two structures (i.e., the olfactory bulb and the hippocampus) in which many new neurons

continue to be created during *Neuroplasticity* the adult years, all of the neurons that will compose the adult human brain have developed by the seventh month of prenatal development. The postnatal growth of the human brain seems to result from three other kinds of growth: synaptogenesis, myelination of axons, and increased branching of dendrites.

There has been particular interest in the postnatal formation of synapses because the number of connections between neurons in a particular region of the brain is assumed to be an indicator of its analytic ability. There is a general increase in *synaptogenesis* in the human cortex shortly after birth, but there are differences among the cortical regions. For example, in the primary visual and auditory cortexes, there is a major burst of synaptogenesis in the fourth postnatal month, and maximum synapse density (150% of adult levels) is achieved in the seventh or eighth postnatal month; whereas synaptogenesis in the prefrontal cortex occurs at a relatively steady rate, reaching maximum synapse density in the second year.

Myelination increases the speed of axonal conduction, and the myelination of various areas of the human brain during development roughly parallels their functional development (Nagy, Westerberg, & Klinsberg, 2004; Sherman & Brophy, 2005). Myelination of sensory areas occurs in the first few months after birth, and myelination of the motor areas follows soon after that, whereas myelination of prefrontal cortex continues into adulthood (Fields, 2008a, 2008b).

In general, the pattern of *dendritic branching* duplicates the original pattern of neural migration—dendritic branching progresses from deeper to more superficial layers. Technical advances in imaging live neurons in culture are leading to insights into how dendrites can reconfigure themselves (see Wong & Wong, 2000). Most surprising is the speed with which even mature dendrites can change their shape—changes can be observed during a few seconds (Alvarez & Sabatini, 2007).

⊙➔ **Simulate** The Brain and the Brain Stem **www.mypsychlab.com**

Postnatal human brain development is not a one-way street; there are regressive changes as well as growth. For example, once maximum synaptic density has been achieved, there are periods of synaptic loss. Like periods of synaptogenesis, periods of synaptic loss occur at different times in different parts of the brain. For example, synaptic density in primary visual cortex declines to adult levels by about 3 years of age, whereas its decline to adult levels in prefrontal cortex is not achieved until adolescence. It has been suggested that the overproduction of synapses may underlie the greater plasticity of the young brain.

Several structural MRI studies of the developing human brain have focused on the relationship between the growth of cortical gray matter and the growth of cortical white matter. Cortical white matter grows slowly and steadily until early adulthood. In contrast, the growth pattern of cortical gray matter is an inverted U. The gray matter grows until it is larger than it will be in the adult brain; then it decreases in size (Amso & Casey, 2006). The achievement of the adult level of gray matter in a particular cortical area is correlated with that area's reaching functional maturity—sensory and motor areas reach their mature form before cognitive areas (Casey et al., 2005).

Development of the Prefrontal Cortex

As you have just learned, the prefrontal cortex displays the most prolonged period of development of any brain region. Its development is believed to be largely responsible for the course of human cognitive development, which occurs over the same period (see Yurgelun-Todd, 2007).

Given the size, complexity, and heterogeneity of the prefrontal cortex, it is hardly surprising that no single theory can explain its function. Nevertheless, four types of cognitive functions have often been linked to this area in studies of adults with extensive prefrontal damage. Various parts of the adult prefrontal cortex seem to play roles in (1) *working memory*, that is, keeping relevant information accessible for short periods of time while a task is being completed; (2) planning and carrying out sequences of actions; (3) inhibiting responses that are inappropriate in the current context but not in others; and (4) following rules for social behavior (see Bunge & Zelazo, 2006; Kagan & Baird, 2004). Young humans do not begin to demonstrate these cognitive functions until prefrontal development has progressed.

One interesting line of research on prefrontal cortex development is based on Piaget's classic studies of psychological development in human babies. In his studies of 7-month-old children, Piaget noticed an intriguing error. A small toy was shown to an infant; then, as the child watched, it was placed behind one of two screens, left or right (see Figure 9.9). After a brief delay, the infant was allowed to reach for the toy. Piaget found that almost all 7-month-old infants reached for the screen behind which they had seen the toy being placed. However, if, after being placed behind the same screen on several consecutive trials, the toy was placed behind the other screen (as the infant watched), most of the 7-month-old infants kept reaching for the previously correct screen, rather than the screen that currently hid the toy. Children tend to make this *perseverative error* between about 7 and 12 months, but not thereafter (Diamond, 1985). **Perseveration** is the tendency to continue making a formerly correct response when it is currently incorrect.

Diamond (1991) hypothesized that this perseverative error occurred in infants between 7 and 12 months because the neural circuitry of the prefrontal cortex is not yet fully developed during that period. Synaptogenesis in the prefrontal cortex is not maximal until early in the second year, and correct performance of the task involved two of

FIGURE 9.9 Testing object permanence. This infant, like other infants younger than 12 months, has not yet developed object permanence. Once the stuffed toy is hidden by the screen, the toy stops existing for the infant. Older infants try to reach around the screen to get the toy and often get upset if they can't reach it.

the major functions of this brain area: holding information in working memory and suppressing previously correct, but currently incorrect, responses.

In support of her hypothesis, Diamond conducted a series of comparative experiments. First, she showed that infant, but not adult, monkeys make the same perseverative error on Piaget's test as 7-to-12-month-old human infants do. Then, she tested adult monkeys with bilateral lesions to their dorsolateral prefrontal cortex (see Figure 8.2 on page 195 for the location of this area of cortex), and she found that the lesioned adult monkeys made perseverative errors similar to those made by the infant monkeys. Control monkeys with lesions in the hippocampus or posterior parietal cortex did not make such errors.

Evolutionary Perspective

9.3

Effects of Experience on the Early Development, Maintenance, and Reorganization of Neural Circuits

Genetic programs of neurodevelopment do not act in a vacuum. Neurodevelopment unfolds through interactions between neurons and their environment—a neuron's environment includes all of the structures and events both inside and outside the organism that can affect the neuron. You learned in the first section of this chapter how factors (e.g., neurotrophins and CAMs) in a neuron's internal environment can influence its migration, aggregation, and growth. This section focuses on how the external environment (i.e., the experiences of the developing organism) can influence the development and maintenance of neural circuits.

One question that developmental biopsychologists ask about the effects of experience on development is whether the experiences are permissive or instructive. **Permissive experiences** are those that are necessary for information in genetic programs to be manifested; **instructive experiences** are those that contribute to the direction of development (see Chalupa & Huberman, 2004).

An important feature of the effects of experience on development is that they are time-dependent: The effect of a given experience on development depends on when it occurs during development. In most cases, there is a window of opportunity in which a particular experience can influence development. If it is absolutely essential (i.e., critical) for an experience to occur within a particular interval to influence development, the interval is called a **critical period**. If an experience has a great effect on development when it occurs during a particular interval but can still have weak effects outside the interval, the interval is called a **sensitive period**. Although the term *critical period* is widely used, the vast majority of experiential effects

on development have sensitive periods (see Morishita & Hensch, 2008; Thomas & Johnson, 2008).

The effects on development of experiences within a sensitive period are often complex and specific to the neural circuits under investigation (see Hensch, 2004), but there is one simple, general principle that seems to characterize the effects of experience on the development of virtually all neural circuits: If neural circuits, once formed, are not used, they do not survive and function normally (see Hockfield & Kalb, 1993; Kalil, 1989). That is, use it or lose it. One advantage of the slowness of human brain development is that it provides many opportunities for experience to fine-tune development (Johnson, 2001).

Evolutionary Perspective

Early Studies of Experience and Neurodevelopment: Deprivation and Enrichment

Most research on the effects of experience on the development of the brain has focused on sensory and motor systems—which lend themselves to experiential manipulation. Much of the early research focused on two general manipulations of experience: sensory deprivation and enrichment.

Neuroplasticity

The first studies of sensory deprivation assessed the effects of rearing animals in the dark. Rats reared from birth in the dark were found to have fewer synapses and fewer dendritic spines in their primary visual cortexes, and as adults they were found to have deficits in depth and pattern vision. In contrast, the first studies of early exposure to enriched environments found that enrichment had beneficial effects. For example, rats that were raised in enriched (complex) group cages rather than by themselves in barren cages were found to have thicker cortexes with more dendritic spines and more synapses per neuron (see Sale, Berardi, & Maffei, 2008).

Competitive Nature of Experience and Neurodevelopment: Ocular Dominance Columns

Recent research on the effects of experience on brain development has mostly progressed beyond assessing general sensory deprivation or enrichment. Manipulations of early experience have become more selective. Many of these selective manipulations of early experience have revealed a competitive aspect to the effects of experience on neurodevelopment. This competitive aspect is clearly illustrated by the disruptive effects of early monocular deprivation on the development of ocular dominance columns in primary visual cortex (see Chapter 6).

Neuroplasticity

Depriving one eye of input for a few days early in life has a lasting adverse effect on vision in the deprived eye, but this does not happen if the other eye is also blindfolded.

When only one eye is blindfolded, the ability of that eye to activate the visual cortex is reduced, whereas the ability of the other eye is increased. Both of these effects occur because early monocular deprivation changes the pattern of synaptic input into layer IV of the primary visual cortex.

In many species, ocular dominance columns (see Figure 6.20 on page 150) in layer IV of the primary visual cortex are almost fully developed at birth (see Feller & Scanziani, 2005; Grubb & Thompson, 2004). However, if just one eye is deprived of light for several days at some point during the first few months of life, the system is reorganized: The width of the columns of input from the deprived eye is decreased, and the width of the columns of input from the nondeprived eye is increased (Hata & Stryker, 1994; Hubel, Wiesel, & LeVay, 1977). The exact timing of the sensitive period for this effect is specific to each species—and it is a sensitive period rather than a critical period, because monocular deprivation in adulthood can have modest effects on ocular dominance columns (Hofer et al., 2006).

Evolutionary Perspective

Because the adverse effects of early monocular deprivation manifest themselves so quickly (i.e., in a few days), it was believed that they could not be mediated by structural changes. However, Antonini and Stryker (1993) found that a few days of monocular deprivation produce a massive decrease in the axonal branching of the lateral geniculate nucleus neurons that normally carry signals from the deprived eye to layer IV of the primary visual cortex (see Figure 9.10).

Effects of Experience on Topographic Sensory Cortex Maps

Some of the most remarkable demonstrations of the effects of experience on the organization of the nervous system come from research on sensory topographic maps (see Sur & Rubenstein, 2005). The following are three such demonstrations:

Neuroplasticity

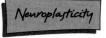

- Roe and colleagues (1990) surgically altered the course of developing axons of ferrets' retinal ganglion cells so that the axons synapsed in the medial geniculate nucleus of the auditory system instead of in the lateral geniculate nucleus of the visual system. Remarkably, the experience of visual input caused the auditory cortex of the ferrets to become organized retinotopically (laid out like a map of the retina). Typically, surgically attaching the inputs of one sensory system to cortex that would normally develop into the primary cortex of another system leads that cortex to develop many, but not all, characteristics typical of the newly attached system (see Majewska & Sur, 2006).

- Knudsen and Brainard (1991) raised barn owls with vision-displacing prisms over their eyes. This led to a corresponding change in the auditory spatial map in the tectum. For example, an owl that was raised wearing prisms that shifted its visual world 23° to the right had an auditory map that was also shifted 23° to the right, so that objects were heard to be where they were seen to be.

Evolutionary Perspective

- Several studies have shown that early music training influences the organization of human cortex (see Münte, Altenmüller, & Jänke, 2002). For example, early musical training expands the area of auditory cortex that responds to complex musical tones.

Experience Fine-Tunes Neurodevelopment

Don't be misled by the way studies of the effects of experience on neurodevelopment are typically conducted. In their studies, biopsychologists often alter the normal course of neurodevelopment by exposing their subjects to pathological early experiences such as blindness or isolation. Consequently, you may come to think of the role of experience in neurodevelopment as pathological, rather than normal. The truth is that long before the nervous system is fully developed, neurons begin to fire spontaneously (Blankenship & Feller, 2010) and begin to interact

Thinking Creatively

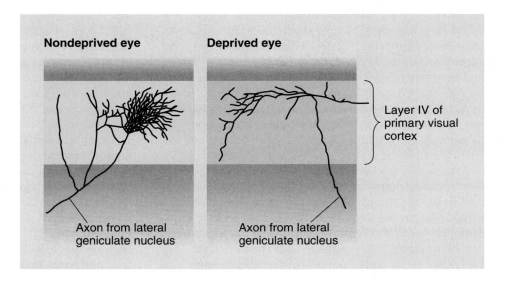

FIGURE 9.10 The effect of a few days of early monocular deprivation on the structure of axons projecting from the lateral geniculate nucleus into layer IV of the primary visual cortex. Axons carrying information from the deprived eye displayed substantially less branching. (Based on Antonini & Stryker, 1993.)

Nondeprived eye   Deprived eye

Layer IV of primary visual cortex

Axon from lateral geniculate nucleus   Axon from lateral geniculate nucleus with the environment. The resulting patterns of neural activity fine-tune subsequent stages of neurodevelopment (see Huberman, Feller, & Chapman, 2008; Learney, Van Wart, & Sur, 2009; Sanes & Bao, 2009). This fine-tuning constitutes the critical, final phase of normal development.

Experience clearly has major effects on the development and maintenance of neural circuits, but the mechanisms through which experience exerts these effects are not well understood. The problem is not the lack of possible mechanisms, but rather that there are so many (see Feldman & Brecht, 2005; Gottlieb, 2000). For example, it is now well established that experience can influence gene expression (e.g., Fagiolini, Jensen, & Champagne, 2009; Majdan & Shatz, 2006), and experience can alter any of the phases of development in innumerable ways by this one mechanism alone.

Neuroplasticity

9.4
Neuroplasticity in Adults

If this book were a road trip that you and I were taking together, at this point, we would spot the following highway sign: SLOW, IMPORTANT VIEWPOINT AHEAD. You see, you are about to encounter an idea that has changed how neuroscientists think about the human brain.

Neuroplasticity was once thought to be restricted to the developmental period. Mature brains were considered to be set in their ways, incapable of substantial reorganization. Now, the accumulation of evidence has made clear that mature brains are continually changing and adapting (Abraham, 2008; Gogolla, Galimberti, & Caroni, 2007; McCoy, Huang, & Philpot, 2009). Many lines of research are contributing to this changing view. For now, consider the following two. You will encounter more in the next two chapters.

Neurogenesis in Adult Mammals

When I was a student, I learned two important principles of brain development. The first I learned through experience: The human brain starts to function in the womb and never stops working until one stands up to speak in public. The second I learned in a course on brain development: **Neurogenesis** (the growth of new neurons) does not occur in adults. The first principle appears to be fundamentally correct, at least when applied to me, but the second has proved wrong.

Prior to the early 1980s, brain development after the early developmental period was seen as a downhill slope: Neurons continually die throughout a person's life, and it was assumed that the lost cells are never replaced by new ones. Although researchers began to chip away at this misconception in the early 1980s, it persisted until the turn of the century as one of the central principles of neurodevelopment.

The first serious challenge to the assumption that neurogenesis is restricted to early stages of development came with the discovery of the growth of new neurons in the brains of adult birds. Nottebohm and colleagues (e.g., Goldman & Nottebohm, 1983) found that brain structures involved in singing begin to grow in songbirds just before each mating season and that this growth results from an increase in the number of neurons. This finding stimulated the re-examination of earlier unconfirmed claims that new neurons are created in the adult rat hippocampus.

Evolutionary Perspective

Then, in the 1990s, researchers, armed with newly developed immunohistochemical markers that have a selective affinity for recently created neurons, convincingly showed that adult neurogenesis does indeed occur in the rat hippocampus (Cameron et al., 1993)—see Figure 9.11. And shortly thereafter, it was discovered that new neurons are also continually added to adult rat olfactory bulbs. Subsequently, it was reported that new neurons are added to the cortex (Gould et al., 1999), but apparently this report was erroneous (Au & Fishell, 2006; Bhardwaj et al., 2006; Rakic, 2006). In adult mammals, substantial neurogenesis seems to be restricted to the olfactory bulbs and hippocampuses—claims of adult neurogenesis in other parts of the mammalian brain remain controversial (see Fowler, Liu, & Wang, 2007).

Neuroplasticity

At first, reports of adult neurogenesis were not embraced by a generation of neuroscientists who had been trained to think of the adult brain as fixed, but acceptance grew as confirmatory reports accumulated. Particularly influential were reports that new neurons are added to the hippocampuses of primates (e.g., Kornack & Rakic, 1999), including humans (Erikkson et al., 1998), and that the number of new neurons added to the adult hippocampus is substantial, an estimated 2,000 per hour (West, Slomianka, & Gunderson, 1991).

Where do the neurons created during adult neurogenesis come from? New olfactory bulb neurons and new hippocampal neurons come from different places (Hagg, 2006). New olfactory bulb neurons are created from *adult neural stem cells* at certain sites in the subventricular zone of the lateral ventricles (Lledo & Saghatelyan, 2006; Sawamoto et al., 2006) and then migrate to the olfactory bulbs (Curtis et al., 2007). In contrast, new hippocampal cells are created near their final location in the dentate gyrus of the hippocampus (Kempermann et al., 2004).

One particularly promising line of research on adult neurogenesis began with a study of the effects on adult rodents of living in *enriched environments* (variable environments, with toys, running wheels, and other rats). It turned out that adult rats living in enriched environments produced 60% more new hippocampal neurons than did adult rats living in nonenriched environments (Kempermann & Gage, 1999). However, before you start enriching your apartment, you should be aware that the

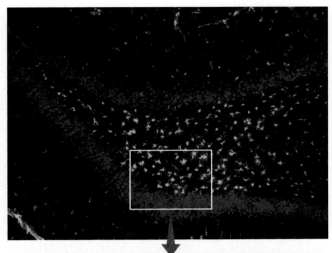

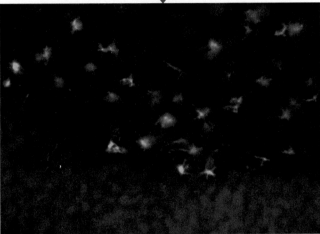

FIGURE 9.11 Adult neurogenesis. The top panel shows new cells in the dentate gyrus of the hippocampus—the cell bodies of neurons are stained blue, mature glial cells are stained green, and new cells are stained red. The bottom panel shows the new cells from the top panel under higher magnification, which makes it apparent that the new cells have taken up both blue and red stain and are thus new neurons. (Courtesy of Carl Ernst and Brian Christie, Department of Psychology, University of British Columbia.)

integrated into neural circuits, and begin to conduct neural signals (Doetsch & Hen, 2005; Kempermann, Wiskott, & Gage, 2004; Toni et al., 2008). Adult-generated olfactory neurons become *interneurons*, and adult-generated hippocampal neurons become *granule cells* in the *dentate gyrus*—see Figure 9.11—see Duan et al., 2008; Lledo, Alonso, & Grubb, 2006. However, although there has been some progress in understanding the anatomy and physiological functions of adult-generated neurons, understanding of their behavioral functions has proven to be elusive (see Ming & Song, 2005; Van Praag, Zhao, & Gage, 2004).

Effects of Experience on the Reorganization of the Adult Cortex

I said that we would consider two lines of current research on adult neuroplasticity, and you have just learned about adult neurogenesis. The second line of research on adult neuroplasticity *Neuroplasticity* deals with the effects of experience on the reorganization of adult cortex (see Elbert & Rockstroh, 2004; Steven & Blakemore, 2004).

Experience in adulthood can lead to reorganization of sensory and motor cortical maps. For example, Mühlnickel and colleagues (1998) found that *tinnitus* (ringing in the ears) produces a major reorganization of primary auditory cortex; Elbert and colleagues (1995) showed that adult musicians who play stringed instruments that are fingered with the left hand (e.g., the violin) have an enlarged hand-representation area in their right somatosensory cortex; and Rossini and colleagues (1994) showed that anesthetizing the second and fourth fingers reduced their representation in contralateral somatosensory cortex.

One study of adult neuroplasticity warrants special attention because it demonstrates an important aspect of this plasticity. Hofer and colleagues (2005) showed that elimination of visual input to one eye of adult mice reduced the size of the ocular dominance columns for that eye in layer IV of primary visual cortex. More importantly, they showed that the reductions in size of the ocular dominance columns occurred more quickly and were more enduring if the adult mice had previously experienced visual deprivation in the same eye. Thus, once the brain has adapted to abnormal environmental conditions, it acquires the ability to adapt more effectively if it encounters the same conditions again.

Although the cellular mechanisms underlying changes to adult sensory cortex are unknown, there are numerous possibilities. Experience has been shown to increase, decrease, or otherwise modify cortical synapses, buttons, and dendritic spines (see Holtmaat & Svoboda, 2009).

The discovery of adult neuroplasticity is changing the way that we humans think about ourselves. More importantly for those with brain damage, it has suggested some promising new treatment options. You will learn about these in Chapter 10.

observed positive effect on neurogenesis in the adult rat hippocampus depends largely, if not entirely, on the increases in exercise that typically occur in enriched environments (Farmer et al., 2004; Van Praag et al., 1999). *Clinical Implications* This finding has a provocative implication: Since the hippocampus is involved in some kinds of memory (see Balu & Lucki, 2008; Frankland & Miller, 2008), perhaps exercise can be used as a treatment for memory problems. Indeed, aerobic exercise has been shown to improve cognitive function in older adults (Hertzog et al., 2008).

What are the functions of the newly created hippocampal and olfactory neurons? It is now established that neurons generated during adulthood survive, become

Scan Your Brain

Before you delve into the last section of the chapter, on disorders of neurodevelopment, review what you have learned since the first section. Fill in each of the blanks with the most appropriate term. The correct answers are provided at the end of the exercise. Before proceeding, review material related to your errors and omissions.

1. The development of the human brain is unique in one respect: The human brain develops far more _____ than the brains of other species.
2. The last part of the human brain to reach maturity is the _____ cortex.
3. The postnatal growth of the human brain results from myelination, dendritic branching, and _____.
4. Human cortical _____ matter grows slowly and steadily until early adulthood.
5. In general, the sensory and motor areas of cerebral cortex achieve maturity before the _____ areas.
6. _____ is the tendency to continue making a formerly correct response when it is currently incorrect.
7. The competitive nature of neurodevelopment has been demonstrated by the effects of early monocular deprivation on the size of _____.
8. Many studies have shown that early experience influences the _____ maps of sensory cortex.
9. Adult _____ occurs in two areas of the mammalian brain.
10. Substantial neurogenesis occurs in adult mammalian hippocampuses and _____.
11. New neurons that are added to the hippocampus of an adult mammal are created near the _____ gyrus.
12. Adult rats living in enriched environments produced 60% more _____ neurons.

Scan Your Brain answers: (1) slowly, (2) prefrontal, (3) synaptogenesis, (4) white, (5) cognitive, (6) perseveration, (7) ocular dominance columns, (8) topographic, (9) neurogenesis, (10) olfactory bulbs, (11) dentate, (12) hippocampal.

9.5

Disorders of Neurodevelopment: Autism and Williams Syndrome

Like all complex processes, neurodevelopment is easily thrown off track; and, unfortunately, one tiny misstep can have far-reaching consequences because it can disrupt subsequent stages. Ironically, much of what we have learned about normal development has come from studying disturbances in normal developmental programs—as you have already seen with the case of Genie. This will become even more apparent to you in this final section of the chapter, which focuses on two disorders of neurodevelopment: autism and Williams syndrome. It is informative to consider these two disorders together because, as you will soon learn, they are similar in some respects and opposite in others.

Autism

Autism is a complex neurodevelopmental disorder. It is a difficult disorder to define because cases differ so greatly; however, there are some commonalities. For example, in almost all cases, the disorder is apparent before the age of 3 and does not increase in severity after that age. Despite considerable between-patient variation in symptoms, three are considered to be *core symptoms* because they are displayed in some form in most cases of the disorder: (1) a reduced ability to interpret the emotions and intentions of others, (2) a reduced capacity for social interaction and communication, and (3) a preoccupation with a single subject or activity (Baron-Cohen & Belmonte, 2005). Although the diagnosis of autism is based on these three core symptoms, all three are not present in every case (see Happé, Ronald, & Plomin, 2008). Also, there are other characteristics that tend to occur in many cases of the disorder—80% of those with autism are male, 50% suffer mental retardation (see Dawson et al., 2007), and 35% have seizures. Older mothers are more likely to give birth to a child with autism, but the probability of a young mother (under 30) giving birth to a child with autism increases if the father is over 40 (Shelton, Tancredi, & Hertz-Picciotto, 2010).

Alex suffers from autism.

The Case of Alex: Are You Ready to Rock?

Clinical Implications

Alex cried so hard when he was a baby that he would sometimes vomit. It was the first sign of his autism. Alex is 7. He spends much of each day scampering around the house.

When Alex sees the delivery boy, he yells, "Are you ready to rock?" because the delivery boy once said that in passing to him. Alex is obsessed by spaghetti, chocolate ice cream, and trucks. He can spot trucks in a magazine that are so small that most people would not immediately recognize them, and he has all his toy trucks lined up in the main hallway of his house.

Despite his severe mental retardation, Alex has little difficulty with computers. He recently went on the Internet and ordered a video.

Alex has echolalia; that is, he repeats almost everything that he hears. He recently told his mother that he loves her, but she doubts very much that he understands what this means—he is just repeating what she has said to him. He loves music and knows the words to many popular songs.

Because the variability of autism is so great, many researchers who study it regard it as a *spectrum disorder*—that is, as a group of related disorders—and refer to **autism spectrum disorders**. You may have heard of **Asperger's syndrome**, a mild autism spectrum disorder in which cognitive and linguistic functions are well preserved.

Autism spectrum disorders are common enough that everybody should be alert for their two main early signs: delayed development of language (e.g., no babbling by 12 months, no words by 16 months, no self-initiated meaningful phrases by 24 months) and delayed development of social interaction (e.g., no smiles or happy expressions by 9 months, no communicative gestures such as pointing or waving by 12 months). Also, many infants with autism spectrum disorders display minor anomalies of ear structure: square-shaped ears positioned too low on the head and rotated slightly backward with the tops flopped over (see Figure 9.12).

Autism spectrum disorders are the most prevalent childhood neurological disorders. Until the early 1990s, most *epidemiological studies* (studies of the incidence and distribution of disease in the general population) reported that autism in the United States occurred in fewer than 1 in 1,000 births; however, the Centers for Disease Control and Prevention now estimate the incidence to be 1 in 166 births. This large increase in the incidence of autism is cause for concern, although it may, in part, reflect recent broadening of the diagnostic criteria, increasing public awareness of the disorder, and improved methods of identifying cases (Hertz-Picciotto & Delwiche, 2009).

Autism is a difficult disorder to treat. Intensive behavioral therapy can improve the lives of some individuals, but it is often difficult for a person with autism to live independently. A person who cannot understand the feelings and motivations of others and who has difficulty communicating will certainly struggle to cope with the demands of living in today's complex society.

Autism Is a Heterogeneous Disorder   Autism is *heterogeneous* in the sense that afflicted individuals may be severely impaired in some respects but may be normal, or even superior, in others. For example, autistic patients who suffer from mental retardation often perform well on tests involving rote memory, jigsaw puzzles, music, and art. Even for the single category of speech disability, there may be a heterogeneous pattern of deficits. Many individuals with autism have sizable vocabularies, are good spellers, and can read aloud textual material they do not understand. However, the same individuals are often unable to use intonation to communicate emotion, to coordinate facial expression with speech, and to speak metaphorically.

Autistic Savants   Perhaps the single most remarkable aspect of autism is the tendency for some persons with autism to be savants. **Savants** are intellectually handicapped individuals who nevertheless display amazing and specific cognitive or artistic abilities. About 10% of individuals with autism display some savant abilities; conversely, about 50% of savants are diagnosed with autism. Savant abilities can take many forms: feats of memory, naming the day of the week for any future or past date, identifying prime numbers (any number divisible only by itself and 1), drawing, and playing musical instruments (see Bonnel et al., 2003). Consider the following savants (Ramachandran & Blakeslee, 1998; Sacks, 1985).

FIGURE 9.12   A boy with autism shows the typical anomalies of ear structure.

Cases of Amazing Savant Abilities

- One savant could tell the time of day to the exact second without ever referring to his watch. Even when he was asleep, he would mumble the correct time.
- Tom was blind and could not tie his own shoes. He had never had any musical training, but he could

play the most difficult piano piece after hearing it just once, even if he was playing with his back to the piano.

- One pair of autistic twins had difficulty doing simple addition and subtraction and could not even comprehend multiplication and division. Yet, if given any date in the last or next 40,000 years, they could specify the day of week that it fell on.

Savant abilities remain a mystery. These abilities do not appear to develop through learning or practice; they seem to emerge spontaneously. Even savants with language abilities cannot explain their own feats. They seem to recognize patterns and relations that escape others. Several investigators have speculated that somehow damage to certain parts of their brains has led to compensatory overdevelopment in other parts (Treffert & Christensen, 2005; Treffert & Wallace, 2002).

Genetic Basis of Autism   Genetic factors influence the development of autism (see Rodier, 2000). Siblings of people with autism have about a 5% chance of being diagnosed with the disorder. This is well above the rate in the general population, but well below the 50% chance that would be expected if autism were caused solely by a single dominant gene or the 25% chance that would be expected if autism were caused solely by a single recessive gene. Also, if one monozygotic twin is diagnosed with autism, the other has a 60% chance of receiving the same diagnosis. These findings suggest that autism is triggered by several genes interacting with the environment, and several genes have already been implicated (Bourgeron, 2009; Burbach & van der Zwaag, 2008).

Neural Mechanisms of Autism   The heterogeneity of the symptoms of autism spectrum disorders—that is, severe deficits in some behavioral functions but not others—suggest underlying damage to some neural structures but not others. Moreover, the fact that the pattern of behavioral deficits varies so markedly from patient to patient suggests similar variability in the underlying neural pathology. Given this situation, it is clear that large, systematic studies will be required to identify the neuropathology underlying the various symptoms of autism spectrum disorders. Unfortunately, such studies have not been conducted (Amaral, Schumann, & Nordahl, 2008). There have been numerous postmortem and structural MRI studies that involve only a few autistic patients and focus on particular parts of the brain. These studies suggest pathology in the cerebellum, amygdala (Schulkin, 2007), and frontal cortex (Bachevalier & Loveland, 2006), but there is little agreement on the nature of the pathology (Amaral et al., 2008).

Two lines of research on the neural mechanisms of autism spectrum disorders warrant mention. Both lines were initially inspired by the severe deficits in social interaction displayed by most children with such disorders. The first line has focused on the abnormal reaction of individuals with autism to faces: They spend less time than normal looking at faces, particularly at the eyes. Research has shown that the *fusiform face area* (see Chapter 6) of autistic patients displays less fMRI activity in response to the presentation of faces (e.g., Bölte et al., 2006; Dalton et al., 2005). The second line of research focuses on *mirror neurons*. You may recall from Chapter 8 that mirror neurons are cells that fire when a monkey performs a particular goal-directed action or observes the same action being performed by another. There is indirect evidence of mirror neurons in humans and they are believed to be part of a system that helps one understand the intentions of others. On this basis, it was hypothesized that children with autism might be deficient in mirror neuron function. Indeed, areas of the cortex that normally display fMRI activity when healthy volunteers observe others performing tasks do not become active in children with autism, even in those with no mental retardation (e.g., Dapretto et al., 2006).

Williams Syndrome

Williams syndrome, like autism, is a neurodevelopmental disorder associated with mental retardation and a heterogeneous pattern of abilities and disabilities, However, in contrast to the withdrawn, emotionally insensitive, uncommunicative individuals who have autism, people with Williams syndrome are sociable, empathetic, and talkative. In many respects, autism and Williams syndrome are opposites, which is why they can be fruitfully studied together.

Williams syndrome occurs in approximately 1 of 7,500 births (Strømme, Bjornstad, & Ramstad, 2002). Anne Louise McGarrah has Williams syndrome (Finn, 1991).

The Case of Anne Louise McGarrah: Uneven Abilities of Those with Williams Syndrome

Anne Louise McGarrah can't add 15 and 25. Yet, she is an avid reader and quite articulate about it. "I love to read," she says. "Biographies, fiction, novels, different articles in newspapers, articles in magazine, just about anything . . . ."

At 42, McGarrah has difficulty telling left from right. Yet she plays the piano and recorder and appreciates classical music: "I love listening to music. I like a little bit of Beethoven, but I especially like Mozart and Chopin and Bach. I like the way they develop their music—it's very light, it's very airy, and it's very cheerful music. . . ."

If McGarrah is asked to get several items from a cupboard, she'll get confused and come back with only one.

Yet she is one of the most sensitive people you're likely to meet—and she's very aware of her own condition. . . . "One time I had a very weird experience. I was in the store, and I was shopping, minding my own business, doing my usual, regular shopping. A woman came up to me, and she was staring and staring and staring at me, and the next thing you know she ran away. It was very odd. I felt really, really bewildered by it. I wanted to talk to her and I wanted to ask her, 'Well why are you staring at me? Is there something wrong? Can I help you to understand that I have a disability?'" (Finn, 1991).

It is the language abilities of Williams syndrome patients that have attracted the most attention. Although they display a delay in language development and language deficits in adulthood (Bishop, 1999; Paterson et al., 1999), their language skills are remarkable considering their characteristically low IQs—which average around 60. For example, in one test, children with Williams syndrome were asked to name as many animals as they could in 60 seconds. Answers included koala, yak, ibex, condor, Chihuahua, brontosaurus, and hippopotamus. When asked to look at a picture and tell a story about it, children with Williams syndrome often produced an animated narrative. As they told the story, the children altered the pitch, volume, rhythm, and vocabulary of their speech to engage the audience. Sadly, the verbal and social skills of these children often lead teachers to overestimate their cognitive abilities, and thus they do not always receive the extra academic support they need.

Clinical Implications

Persons with Williams syndrome have other cognitive strengths, several of which involve music (Lennoff et al., 1997). Although most cannot learn to read music, some have perfect or near-perfect pitch and an uncanny sense of rhythm. Many retain melodies for years, and some are professional musicians. As a group, people with Williams syndrome show more interest in, and emotional reaction to, music than does the general population. One child with Williams syndrome said, "Music is my favorite way of thinking." Yet another cognitive strength of individuals with Williams syndrome is their remarkable ability to recognize faces.

On the other hand, like any group of individuals with an average IQ of 60, persons with Williams syndrome display many severe cognitive deficits. For example, their spatial abilities are even worse than those of people with comparable IQs—they have great difficulty remembering the location of a few blocks placed on a test board, their space-related speech is poor, and their ability to draw objects is almost nonexistent (Jordan et al., 2002).

Williams syndrome is also associated with a variety of health problems, including several involving the heart. One heart disorder was found to result from a mutation in a gene on chromosome 7 that controls the synthesis of *elastin*, a protein that imparts elasticity to many organs and tissues, including the heart. Aware that the same cardiac problem is prevalent in people with Williams syndrome, investigators assessed the status of this gene in that group. Remarkably, they found that the gene on one of the two copies of chromosome 7 was absent in 95% of the individuals with Williams syndrome (see Howald et al., 2006). Other genes were missing as well: Through an accident of reproduction, an entire region of chromosome 7 had been deleted. Once the functions of the other genes in this region have been determined, researchers will have a much fuller understanding of the etiology of Williams syndrome.

People with Williams syndrome show a general thinning of the cortex and underlying white matter (Meyer-Lindenberg, Mervis, & Berman, 2006; Toga, Thompson, & Sowell, 2006). The cortical thinning is greatest in two areas: at the boundary of parietal and occipital cortex and in the **orbitofrontal cortex** (the inferior area of frontal cortex near the orbits, or eye sockets)—see Figure 9.13. Lack

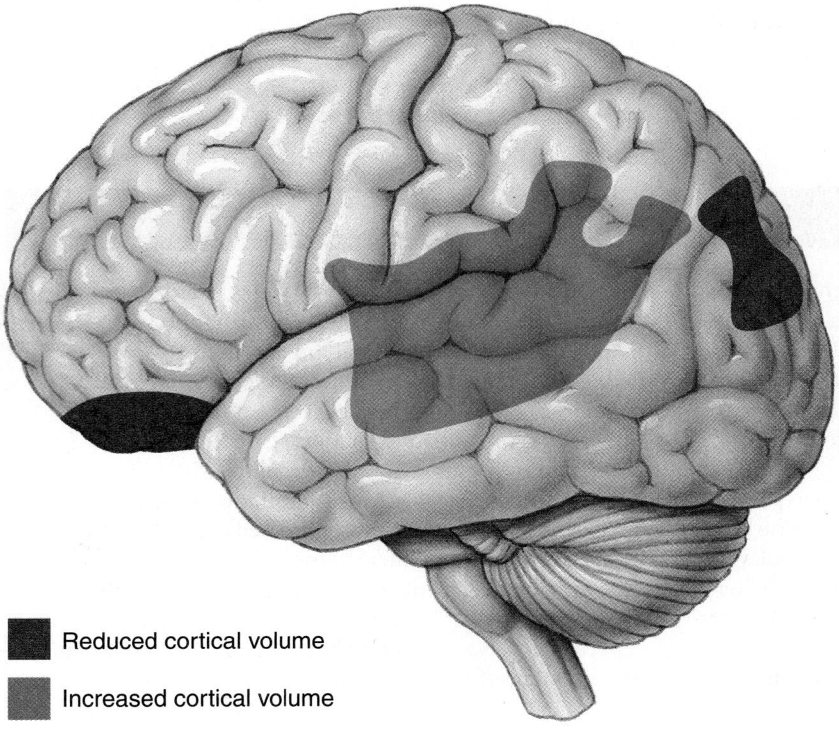

■ Reduced cortical volume

■ Increased cortical volume

FIGURE 9.13 Two areas of reduced cortical volume and one area of increased cortical volume observed in people with Williams syndrome. (See Meyer-Lindenberg et al., 2006; Toga & Thompson, 2005.)

FIGURE 9.14   People with Williams syndrome are characterized by their elfin appearance.

of cortical development in these two areas may be related to two of the major symptoms of Williams syndrome: profound impairment of spatial cognition and remarkable hypersociability, respectively. Conversely, the thickness of the cortex in one area in people with Williams syndrome is often normal: the **superior temporal gyrus**, which includes primary and secondary auditory cortex (refer to Figure 9.13). The normalcy of this area may be related to the relatively high levels of

language and music processing in those with Williams syndrome.

Many cultures feature tales involving magical little people (pixies, elves, leprechauns, etc). Descriptions of these creatures portray them as virtually identical to persons with Williams syndrome: short with small upturned noses, oval ears, broad mouths, full lips, puffy eyes, and small chins (see Figure 9.14). Even the typical behavioral characteristics of elves—engaging storytellers, talented musicians, loving, trusting, and sensitive to the feelings of others—match those of individuals with Williams syndrome. These similarities suggest that folk tales about elves may have originally been based on persons with Williams syndrome.

To conclude this final section, I would like you to think about the similarities and differences between autism and Williams syndrome. Why do you think that some researchers have suggested that investigating them together could prove fruitful? *Thinking Creatively*

Epilogue

As a student, I was amazed to learn that—starting as one cell—my brain had constructed itself through the interaction of its genetic programs and experience. Decades later, I was astounded to learn that my adult brain was still plastic. I hope that this chapter has communicated my enthusiasm for this field of research and helped your brain appreciate its own development.

Also, I hope that the case of Genie, which began the chapter, and the cases about people with autism and Williams syndrome, which completed it, have helped you appreciate the devastating consequences when complex programs of neurodevelopment go wrong.

Themes Revisited

This chapter was all about neurodevelopment, with a particular emphasis on how the brain continues to develop throughout an individual's life span. I *Neuroplasticity* hope you now fully appreciate that your own brain is constantly changing in response to interactions between your genetic programs and experience.

The clinical implications and evolutionary perspective themes were also emphasized in this chapter. One of the best ways to understand the principles of normal neurodevelopment is to consider what happens when *Clinical Implications* it goes wrong: Accordingly, the chapter began with the tragic case of Genie and ended with a discussion of autism and Williams syndrome.

The evolutionary perspective theme was *Evolutionary Perspective* emphasized because much of the information that we have about normal human neurodevelopment and human neurodevelopmental disorders has come from studying other species.

The thinking creatively tab appeared infrequently in this chapter, but the theme was pervasive. Thinking creatively tabs highlighted discussions that made two general points. First, neurodevelopment always proceeds from gene–experience interactions rather *Thinking Creatively* than from either genetics or environment alone. Second, it is important to realize that "normal" experience plays an important role in fine-tuning in the development of neural function.

Think about It

1. What does the case of Genie teach us about normal development?
2. Neuron death is a necessary stage in neurodevelopment. Discuss.
3. Even the adult brain displays plasticity. What do you think is the evolutionary significance of this ability?
4. Discuss the evolutionary significance of the slow development of the human brain.
5. Autism spectrum disorders and Williams syndrome are similar, but they are opposites in several ways. Thus, their neural mechanisms can be fruitfully studied together. Discuss.

Key Terms

9.1 Phases of Neurodevelopment

Neural plate (p. 220)
Mesoderm layer (p. 221)
Totipotent (p. 221)
Multipotent (p. 221)
Stem cells (p. 221)
Neural tube (p. 221)
Neural proliferation (p. 222)
Ventricular zone (p. 222)
Migration (p. 222)
Radial migration (p. 222)
Tangential migration (p. 222)
Somal translocation (p. 222)
Glia-mediated migration (p. 222)
Radial glial cells (p. 222)

Inside-out pattern (p. 222)
Neural crest (p. 222)
Aggregation (p. 223)
Cell-adhesion molecules (CAMs) (p. 223)
Growth cone (p. 224)
Retinal ganglion cells (p. 224)
Optic tectum (p. 224)
Chemoaffinity hypothesis (p. 224)
Pioneer growth cones (p. 225)
Fasciculation (p. 225)
Topographic gradient hypothesis (p. 225)
Synaptogenesis (p. 225)
Necrosis (p. 226)
Apoptosis (p. 226)
Neurotrophins (p. 227)

Nerve growth factor (NGF) (p. 227)

9.2 Postnatal Cerebral Development in Human Infants

Perseveration (p. 229)

9.3 Effects of Experience on the Early Development, Maintenance, and Reorganization of Neural Circuits

Permissive experiences (p. 230)
Instructive experiences (p. 230)
Critical period (p. 230)
Sensitive period (p. 230)

9.4 Neuroplasticity in Adults

Neurogenesis (p. 232)

9.5 Disorders of Neurodevelopment: Autism and Williams Syndrome

Autism (p. 234)
Autism spectrum disorders (p. 235)
Asperger's syndrome (p. 235)
Savants (p. 235)
Williams syndrome (p. 236)
Orbitofrontal cortex (p. 237)
Superior temporal gyrus (p. 238)

✔●─**Quick Review** Test your comprehension of the chapter with this brief practice test. You can find the answers to these questions as well as more practice tests, activities and other study resources at www.mypsychlab.com.

1. Radial migration in the developing neural tube normally occurs along
   a. radial axons.
   b. radial dendrites.
   c. the central canal.
   d. radial glial cells.
   e. both a and b

2. At the tip of a growing axon is a
   a. hillock.
   b. CAM.
   c. growth cone.
   d. neural crest cell.
   e. radial glial cell.

3. Fasciculation is
   a. against the law in some states.
   b. the tendency of developing axons to grow along paths established by preceding axons.
   c. the main method used by pioneer growth cones to reach their targets.
   d. a form of synaptogenesis.
   e. both b and c

4. Which part of the human brain seems to be the last to reach maturity?
   a. prefrontal cortex
   b. hippocampus
   c. amygdala
   d. sensory cortex
   e. both a and c

5. In mammals, substantial adult neurogenesis occurs in the
   a. hippocampus.
   b. neocortex.
   c. olfactory bulbs.
   d. both a and b
   e. both a and c

10 Brain Damage and Neuroplasticity

Can the Brain Recover from Damage?

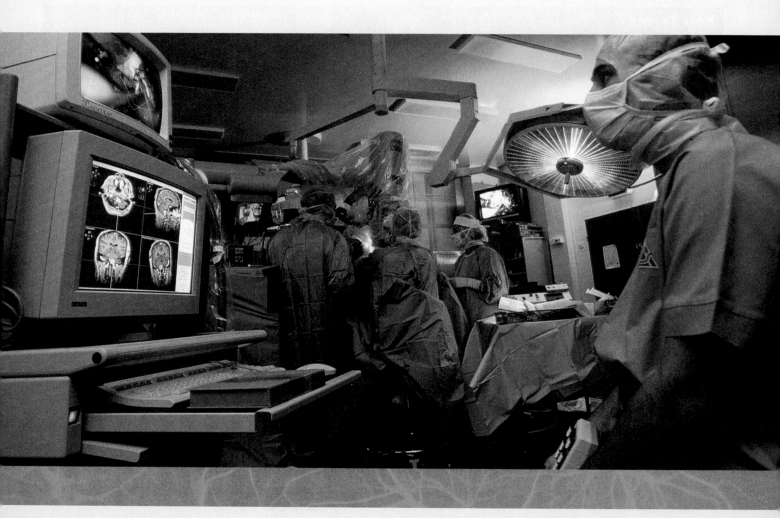

The study of human brain damage serves two purposes: It increases our understanding of the healthy brain, and it serves as a basis for the development of new treatments. The first three sections of this chapter focus on brain damage itself. The last two sections continue the neuroplasticity theme that was the focus of Chapter 9: The fourth section focuses on the recovery and reorganization of the brain after damage, and the fifth discusses exciting new treatments that promote neuroplasticity. But first, the continuation of the ironic case of Professor P., which you first encountered in Chapter 5, provides a personal view of brain damage.

The Ironic Case of Professor P.

One night Professor P. sat at his desk staring at a drawing of the cranial nerves, much like the one in Appendix III of

Clinical Implications

this book. As he mulled over the location and function of each cranial nerve (see Appendix IV), the painful truth became impossible for him to deny. The irony of the situation was that Professor P. was a neuroscientist, all too familiar with what he was experiencing.

His symptoms started subtly, with slight deficits in balance. . . . Professor P. chalked these occasional lurches up to aging—after all, he thought to himself, he was past his prime. Similarly, his doctor didn't seem to think that it was a problem, but Professor P. monitored his symptoms nevertheless. Three years later, his balance problems unabated, Professor P. started to worry. He was trying to talk on the phone but was having trouble hearing until he changed the phone to his left ear and the faint voice on the other end became louder. Professor P. was going deaf in his right ear.

Professor P. made an appointment with his doctor, who referred him to a specialist. After a cursory and poorly controlled hearing test, the specialist gave him good news. "You're fine, Professor P.; lots of people experience a little hearing loss when they reach middle age; don't worry about it." To this day, Professor P. regrets that he did not insist on a second opinion.

It was about a year later that Professor P. sat staring at the illustration of the cranial nerves. By then he had begun to experience numbness on the right side of his mouth; he was having problems swallowing; and his right tear ducts were not releasing enough tears. He sat staring at his own illustration, focused on the point where the auditory and vestibular nerves come together to form cranial nerve VIII (the auditory-vestibular nerve). He knew it was there, and he knew that it was large enough to be affecting cranial nerves V through X as well. It was something slow-growing, perhaps a tumor? Was he going to die? Was his death going to be terrible and lingering?

He didn't see his doctor right away. A friend of his was conducting a brain MRI study, and Professor P. volunteered to be a control subject, knowing that his problem would show up on the scan. It did: a large tumor sitting, as predicted, on the right cranial nerve VIII.

Then, MRI in hand, Professor P. went back to his doctor, who referred him to a neurologist, who in turn referred him to a neurosurgeon. Several stressful weeks later, Professor P. found himself on life support in the intensive care unit of his local hospital, tubes emanating seemingly from every part of his body. During the 6-hour surgery, Professor P. had stopped breathing.

In the intensive care unit, near death and hallucinating from the morphine, Professor P. thought he heard his wife, Maggie, calling for help, and he tried to go to her assistance. But one gentle morphine-steeped professor was no match for five burly nurses intent on saving his life. They quickly turned up his medication, and the next time he regained consciousness, he was tied to the bed.

Professor P.'s auditory-vestibular nerve was transected during his surgery, which has left him permanently deaf and without vestibular function on the right side. He was also left with partial hemifacial paralysis, including serious blinking and tearing problems.

Professor P. has now returned to his students, his research, and his writing, hoping that the tumor was completely removed. Indeed, at the very moment that I am writing these words, Professor P. is working on the forthcoming edition of his textbook. . . . If it has not yet occurred to you, I am Professor P. This chapter has come to have special meaning for me.

10.1
Causes of Brain Damage

This section provides an introduction to six causes of brain damage: brain tumors, cerebrovascular disorders, closed-head injuries, infections of the brain, neurotoxins, and genetic factors. It concludes with a discussion of programmed cell death, which mediates many forms of brain damage.

👁 **Watch**
My Tumor and Welcome to It
www.mypsychlab.com

Brain Tumors

A **tumor**, or **neoplasm** (literally, "new growth"), is a mass of cells that grows independently of the rest of the body (see Wechsler-Reya & Scott, 2001). In other words, it is a cancer.

Clinical Implications

About 20% of tumors found in the human brain are **meningiomas** (see Figure 10.1)—tumors that grow between the *meninges*, the three membranes that cover the central nervous system. All meningiomas are **encapsulated**

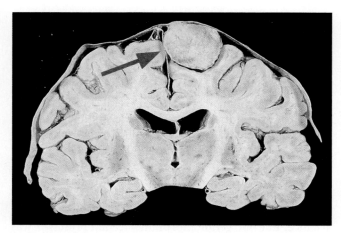

FIGURE 10.1 A meningioma. (Courtesy of Kenneth Berry, Head of Neuropathology, Vancouver General Hospital.)

tumors—tumors that grow within their own membrane. As a result, they are particularly easy to identify on a CT scan, they can influence the function of the brain only by the pressure they exert on surrounding tissue, and they are almost always **benign tumors**—tumors that are surgically removable with little risk of further growth in the body (see Grimson et al., 1999).

Unfortunately, encapsulation is the exception rather than the rule when it comes to brain tumors. Aside from meningiomas, most brain tumors are infiltrating. **Infil-**

trating tumors are those that grow diffusely through surrounding tissue. As a result, they are usually **malignant tumors**; that is, it is difficult to remove or destroy them completely, and any cancerous tissue that remains after surgery continues to grow.

About 10% of brain tumors do not originate in the brain. They grow from infiltrating cells that are carried to the brain by the bloodstream from some other part of the body (see Klein, 2008). These tumors are called **metastatic tumors**; *metastasis* refers to the transmission of disease from one organ to another. Many metastatic brain tumors originate as cancers of the lungs. Figure 10.2 illustrates the ravages of metastasis. Currently, the chance of recovering from a cancer that has already attacked two or more separate sites is slim.

Fortunately, my tumor was encapsulated. Encapsulated tumors that grow on cranial nerve VIII are referred to as *acoustic neuromas* (neuromas are tumors that grow on nerves or tracts). Figure 10.3 is an MRI scan of my acoustic neuroma, the very same scan that I took to my doctor.

Cerebrovascular Disorders: Strokes

Strokes are sudden-onset cerebrovascular disorders that cause brain damage. In the United States, stroke is the third leading cause of death, the major cause of neurological dysfunction, and the most common cause of adult disability (see Janardhan & Qureshi, 2004; Wang, 2006). The symptoms of a

Clinical Implications

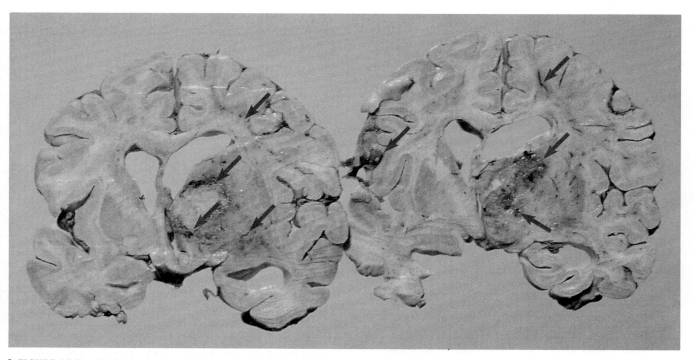

FIGURE 10.2 Multiple metastatic brain tumors. The arrows indicate some of the more advanced areas of metastatic tumor development.

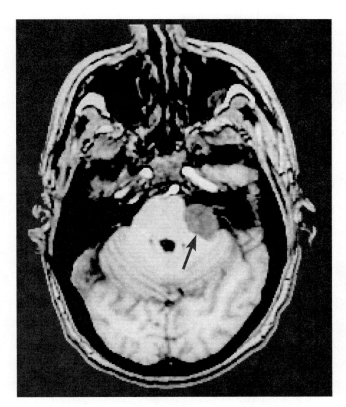

FIGURE 10.3 An MRI of Professor P.'s acoustic neuroma. The arrow indicates the tumor.

elasticity of the artery wall is defective. Although aneurysms of the brain are particularly problematic, aneurysms can occur in any part of the body (Elefteriades, 2005). Aneurysms can be **congenital** (present at birth) or can result from exposure to vascular poisons or infection (see Kalaria, 2001). Individuals who have aneurysms should make every effort to avoid high blood pressure or any strenuous activity, such as weight lifting (see Galluzzi, Blomgren, & Kroemer, 2009).

Cerebral Ischemia   **Cerebral ischemia** is a disruption of the blood supply to an area of the brain. The three main causes of cerebral ischemia are thrombosis, embolism, and arteriosclerosis. In **thrombosis**, a plug called a *thrombus* is formed and blocks blood flow at the site of its formation. A thrombus may be composed of a blood clot, fat, oil, an air bubble, tumor cells, or any combination thereof. **Embolism** is similar, except that the plug, called an *embolus* in this case, is carried by the blood from a larger vessel, where it was formed, to a smaller one, where it becomes lodged; in essence, an embolus is just a thrombus that has taken a trip. In **arteriosclerosis**, the walls of blood vessels thicken and the channels narrow, usually as the result of fat deposits; this narrowing can eventually lead to complete blockage of the blood vessels (Libby, 2002). The *angiogram* in Figure 10.4 illustrates partial blockage of one carotid artery.

Much of the damage produced by cerebral ischemia takes a day or two to develop fully, and, paradoxically, some of the brain's own neurotransmitters play a key role in its development (Wahlgren & Ahmed, 2004). Much of the brain damage associated with stroke is a consequence of excessive release of excitatory amino acid neurotransmitters, in particular **glutamate**, the brain's most prevalent excitatory neurotransmitter.

stroke depend on the area of the brain that is affected, but common consequences of stroke are amnesia, aphasia (language difficulties), paralysis, and coma.

The area of dead or dying tissue produced by a stroke is called an *infarct*. Surrounding the infarct is a dysfunctional area called the **penumbra**. The tissue in the penumbra may recover or die, depending on a variety of factors, and the goal of treatment following stroke is to save it (see Murphy & Corbett, 2009).

There are two major types of strokes: those resulting from cerebral hemorrhage and those resulting from cerebral ischemia (pronounced "iss-KEEM-ee-a").

Cerebral Hemorrhage   **Cerebral hemorrhage** (bleeding in the brain) occurs when a cerebral blood vessel ruptures and blood seeps into the surrounding neural tissue and damages it. Bursting aneurysms are a common cause of intracerebral hemorrhage. An **aneurysm** is a pathological balloonlike dilation that forms in the wall of an artery at a point where the

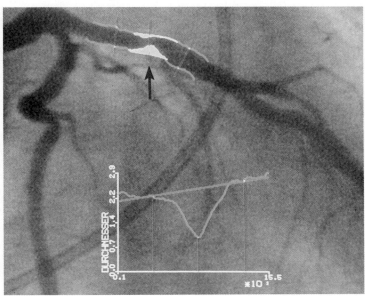

FIGURE 10.4 An angiogram that illustrates narrowing of the carotid artery (see arrow), the main pathway of blood to the brain. Compare this angiogram with the normal angiogram in Figure 5.1 on page 103.

Here is how this mechanism is thought to work (see Greer, 2006). After a blood vessel becomes blocked, many of the blood-deprived neurons become overactive and release excessive quantities of glutamate. The glutamate in turn overactivates glutamate receptors in the membranes of postsynaptic neurons; the glutamate receptors that are most involved in this reaction are the **NMDA (N-methyl-D-aspartate) receptors**. As a result, large numbers of Na^+ and Ca^{2+} ions enter the postsynaptic neurons.

The excessive internal concentrations of Na^+ and Ca^{2+} ions affect the postsynaptic neurons in two ways: They trigger the release of excessive amounts of glutamate from the neurons, thus spreading the toxic cascade to yet other neurons; and they trigger a sequence of internal reactions that ultimately kill the postsynaptic neurons. (See Figure 10.5.)

Ischemia-induced brain damage has three important properties (Krieglstein, 1997). First, it takes a while to develop. Soon after a temporary cerebral ischemic episode, say, one 10 minutes in duration, there usually is little or no evidence of brain damage; however, substantial neuron loss can often be detected a day or two later. Second, ischemia-induced brain damage does not occur equally in all parts of the brain; particularly susceptible are neurons in certain areas of the hippocampus (Ohtaki et al., 2003). Third, the mechanisms of ischemia-induced damage vary somewhat from structure to structure within the brain, and in at least some areas, astrocytes have been implicated (see Seifert, Schilling, & Steinhäuser, 2006).

An implication of the discovery that excessive glutamate release causes much of the brain damage associated with stroke is the possibility of preventing stroke-related brain damage by blocking the glutaminergic cascade. The search is on for a glutamate antagonist that is effective and safe for use in human stroke victims (Lo, Dalkara, & Moskowitz, 2003). Several have proved to be effective in laboratory animals, but so far none has been shown to limit brain damage from strokes in humans (see Greer, 2006). Wahlgren and Ahmed (2004) have argued that if such treatments are to be effective, they need to be initiated soon after the stroke, not hours or days later (see Suwanwela & Koroshetz, 2007).

Closed-Head Injuries

It is not necessary for the skull to be penetrated for the brain to be seriously damaged. In fact, any blow to the

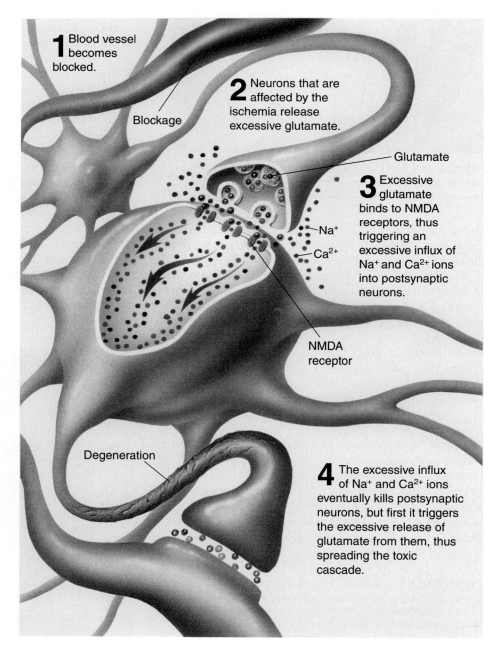

1 Blood vessel becomes blocked.

Blockage

2 Neurons that are affected by the ischemia release excessive glutamate.

Glutamate

3 Excessive glutamate binds to NMDA receptors, thus triggering an excessive influx of Na^+ and Ca^{2+} ions into postsynaptic neurons.

Na^+

Ca^{2+}

NMDA receptor

Degeneration

4 The excessive influx of Na^+ and Ca^{2+} ions eventually kills postsynaptic neurons, but first it triggers the excessive release of glutamate from them, thus spreading the toxic cascade.

FIGURE 10.5 The cascade of events by which the stroke-induced release of glutamate kills neurons.

head should be treated with extreme caution, particularly when confusion, sensorimotor disturbances, or loss of consciousness ensues. Brain injuries produced by blows that do not penetrate the skull are called *closed-head injuries.*

Contusions are closed-head injuries that involve damage to the cerebral circulatory system. Such damage produces internal hemorrhaging, which results in a hematoma. A **hematoma** is a localized collection of clotted blood in an organ or tissue—in other words, a bruise.

It is paradoxical that the very hardness of the skull, which protects the brain from penetrating injuries, is the major factor in the development of contusions. Contusions from closed-head injuries occur when the brain slams against the inside of the skull. As Figure 10.6 illustrates, blood from such injuries can accumulate in the *subdural space*—the space between the dura mater and arachnoid membrane—and severely distort the surrounding neural tissue.

It may surprise you to learn that contusions frequently occur on the side of the brain opposite the side struck by a blow. The reason for such so-called **contrecoup injuries** is that the blow causes the brain to strike the inside of the skull on the other side of the head.

When there is a disturbance of consciousness following a blow to the head and there is no evidence of a contusion or other structural damage, the diagnosis is **concussion**. It was commonly assumed that a concussion entails a temporary disruption of normal cerebral function with no long-term damage. However, there is now substantial evidence that the cognitive, motor, and neurological effects of concussion can last many years (De Beaumont et al., 2009).

The **punch-drunk syndrome** is the **dementia** (general intellectual deterioration) and cerebral scarring observed in boxers and other individuals who experience repeated concussions. The case of Jerry Quarry is particularly severe.

The Case of Jerry Quarry, Ex-Boxer

Jerry Quarry thumps his hard belly with both fists. Smiles at the sound. Like a stone against a tree.

"Feel it," he says proudly, punching himself again and again.

He pounds big, gnarled fists into meaty palms. Cocks his head. Stares. Vacant blue eyes. Punch-drunk at 50. Medical name: *Dementia pugilistic* [punch-drunk syndrome]. Cause: Thousands of punches to the head.

A top heavyweight contender in the 1960s and '70s, Quarry now needs help shaving, showering, putting on shoes and socks. Soon, probably, diapers. His older brother, James, cuts meat into little pieces so he won't choke. Jerry smiles like a kid. Shuffles like an old man.

Slow, slurred speech. Random thoughts snagged on branches in a dying brain. Memories twisted. Voices no one else hears. (Steve Wiltstein, Associated Press, 1995)

Infections of the Brain

An invasion of the brain by microorganisms is a *brain infection*, and the resulting inflammation is **encephalitis**. There are two common types of brain infections: bacterial infections and viral infections.

Bacterial Infections   When bacteria infect the brain, they often lead to the formation of *cerebral abscesses*—pockets of pus in the brain. Bacteria are also the major cause of **meningitis** (inflammation of the meninges), which is fatal in 25% of adults (Nau & Brück, 2002). Penicillin and other antibiotics sometimes eliminate bacterial infections of the brain, but they cannot reverse brain damage that has already been produced.

Syphilis is one bacterial brain infection you have likely heard about (see Zimmer, 2008). Syphilis bacteria are passed from infected to noninfected individuals through contact with genital sores. The infecting bacteria then go into a dormant stage for several years before they become virulent and attack many parts of the body, including the brain. The syndrome of insanity and dementia that results from a syphilitic infection is called **general paresis**.

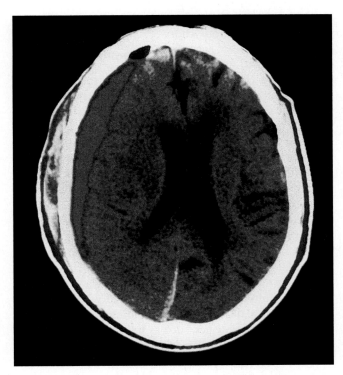

FIGURE 10.6   A CT scan of a subdural hematoma. Notice that the subdural hematoma has displaced the left lateral ventricle.

Viral Infections There are two types of viral infections of the nervous system: those that have a particular affinity for neural tissue and those that attack neural tissue but have no greater affinity for it than for other tissues.

Rabies, which is usually transmitted through the bite of a rabid animal, is a well-known example of a virus that has a particular affinity for the nervous system. The fits of rage caused by the virus's effects on the brain increase the probability that rabid animals that normally attack by biting (e.g., dogs, cats, raccoons, bats, and mice) will spread the disorder. Although the effects of the rabies virus on the brain are almost always lethal (Willoughby, 2007), the virus does have one redeeming feature: It does not usually attack the brain for at least a month after it has been contracted, thus allowing time for a preventive vaccination.

The *mumps* and *herpes* viruses are common examples of viruses that can attack the nervous system but have no special affinity for it. Although these viruses sometimes spread into the brain, they typically attack other tissues of the body.

Viruses may play a far greater role in neuropsychological disorders than is currently thought (see van den Pol, 2006). Their involvement in the *etiology* (cause) of disorders is often difficult to recognize because they can lie dormant for many years before producing symptoms.

Neurotoxins

The nervous system can be damaged by exposure to any one of a variety of toxic chemicals, which can enter general circulation from the gastrointestinal tract, from the lungs, or through the skin. For example, heavy metals such as mercury and lead can accumulate in the brain and permanently damage it, producing a **toxic psychosis** (chronic insanity produced by a neurotoxin). Have you ever wondered why *Alice in Wonderland*'s Mad Hatter was a mad hatter and not a mad something else? In 18th- and 19th-century England, hat makers were commonly driven mad by the mercury employed in the preparation of the felt used to make hats. In a similar vein, the word *crackpot* originally referred to the toxic psychosis observed in some people in England—primarily the poor—who steeped their tea in cracked ceramic pots with lead cores.

Sometimes, the very drugs used to treat neurological disorders prove to have toxic effects. For example, some of the antipsychotic drugs introduced in the early 1950s produced effects of distressing scope. By the late 1950s, millions of psychotic patients were being maintained on these new drugs. However, after several years of treatment, many of the patients developed a motor disorder termed **tardive dyskinesia (TD)**. Its primary symptoms are involuntary smacking and sucking movements of the lips, thrusting and rolling of the tongue, lateral jaw movements, and puffing of the cheeks. Safer antipsychotic drugs have since been developed.

Some neurotoxins are *endogenous* (produced by the patient's own body). For example, the body can produce antibodies that attack particular components of the nervous system (see Newsom-Davis & Vincent, 1991). Also, you have just learned that excessive release of glutamate causes much of the brain damage following stroke.

Genetic Factors

Normal human cells have 23 pairs of chromosomes; however, sometimes accidents of cell division occur, and the fertilized egg ends up with an abnormal chromosome or with an abnormal number of normal chromosomes. Then, as the fertilized egg divides and redivides, these chromosomal anomalies are duplicated in every cell of the body.

 Listen Genetics
www.mypsychlab.com

Most neuropsychological diseases of genetic origin are caused by abnormal recessive genes that are passed from parent to offspring. (In Chapter 2, you learned about one such disorder, *phenylketonuria*, or *PKU*.) Inherited neuropsychological disorders are rarely associated with dominant genes because dominant genes that disturb neuropsychological function tend to be eliminated from the gene pool—every individual who carries one has major survival and reproductive disadvantages. In contrast, individuals who inherit one abnormal recessive gene do not develop the disorder, and the gene is passed on to future generations.

Genetic accident is another major cause of neuropsychological disorders of genetic origin. **Down syndrome**, which occurs in about 0.15% of births, is such a disorder. The genetic accident associated with Down syndrome occurs in the mother during ovulation, when an extra chromosome 21 is created in the egg. Thus, when the egg is fertilized, there are three of this chromosome, rather than two, in the zygote. The consequences tend to be characteristic disfigurement (flattened skull and nose, folds of skin over the inner corners of the eyes, and short fingers; see Figure 10.7), intellectual impairment, and troublesome medical complications. The probability of giving birth to a child with Down syndrome increases markedly with advancing maternal age: For example, the probability goes from 1 in 1,667 at maternal age 20 to 1 in 11 at maternal age 49 (Egan, 2004).

Rapid progress is being made in locating and characterizing the faulty genes that are associated with some neuropsychological disorders. This is leading to new treatment and prevention strategies, such as splicing in healthy genes to replace faulty ones and using specific DNA-binding proteins to block the expression of faulty genes. Also, determining which proteins are encoded by these genes and how the proteins influence programs of neurodevelopment suggests methods for early diagnosis, prevention, and treatment (Ben-Ari, 2008; Orr & Zoghbi, 2007).

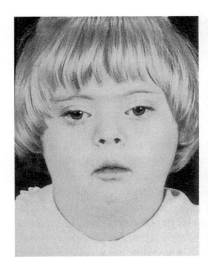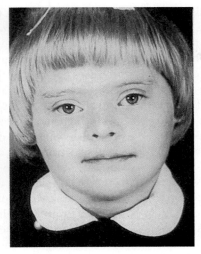

FIGURE 10.7 A child with Down syndrome before and after plastic surgery. The purpose of these photographs is not to promote cosmetic surgery but to challenge our culture's reaction to people with Down syndrome. The little girl on the left and the little girl on the right are the same girl; they deserve the same respect and consideration. (Courtesy of Kenneth E. Salyer, Director, International Craniofacial Institute.)

Programmed Cell Death

You learned in Chapter 9 that neurons and other cells have genetic programs for destroying themselves by a process called **apoptosis** (pronounced "A-poe-toe-sis"). Apoptosis plays a critical role in early development by eliminating extra neurons. It also plays a role in brain damage. Indeed, all of the six causes of brain damage that have been discussed in this chapter (tumors, cerebrovascular disorders, closed-head injuries, infections, toxins, and genetic factors) produce neural damage, in part, by activating apoptotic programs of self-destruction (Allsop & Fazakerley, 2000; Dirnagl, Simon, & Hallenbeck, 2003; Nijhawan, Honarpour, & Wang, 2000).

Neuroplasticity

Clinical Implications

It was once assumed that the death of neurons following brain damage was totally necrotic—*necrosis* is passive cell death resulting from injury. It now seems that if cells are not damaged too severely, they will attempt to marshal enough resources to "commit suicide." However, cell death is not an either-or situation: Some dying cells display signs of both necrosis and apoptosis (see Elibol et al., 2001).

It is easy to understand why apoptotic mechanisms have evolved: Apoptosis is clearly more adaptive than necrosis. In necrosis, the damaged neuron swells and breaks apart, beginning in the axons and dendrites and ending in the cell body. This fragmentation leads to inflammation, which can damage other cells in the vicinity. Necrotic cell death is quick—it is typically complete in a few hours. In contrast, apoptotic cell death is slow, typically requiring a day or two. Apoptosis of a neuron proceeds gradually, starting with shrinkage of the cell body. Then, as parts of the neuron die, the resulting debris is packaged in vesicles. As a result, there is no inflammation, and damage to nearby cells is kept to a minimum.

Evolutionary Perspective

Neuropsychological Diseases

The preceding section focused on the causes of human brain damage. This section considers five diseases that are associated with brain damage: epilepsy, Parkinson's disease, Huntington's disease, multiple sclerosis, and Alzheimer's disease.

Epilepsy

The primary symptom of **epilepsy** is the epileptic *seizure*, but not all persons who suffer seizures are considered to have epilepsy. Sometimes, an otherwise healthy person may have one seizure and never have another (Haut & Shinnar, 2008)—such a one-time convulsion could be triggered by exposure to a convulsive toxin or by a high fever. The diagnosis of *epilepsy* is applied to only those patients whose seizures are repeatedly generated by their own chronic brain dysfunction. About 1% of the population are diagnosed as epileptic at some point in their lives.

Clinical Implications

In view of the fact that epilepsy is characterized by epileptic seizures—or, more accurately, by spontaneously recurring epileptic seizures—you might think that the task of diagnosing this disorder would be an easy one. But you would be wrong. The task is made difficult by the diversity and complexity of epileptic seizures. You are probably familiar with seizures that take the form of **convulsions** (motor seizures); these often involve tremors (*clonus*), rigidity (*tonus*), and loss of both balance and consciousness. But many seizures do not take this form; instead, they involve subtle changes of thought, mood, or behavior that are not easily distinguishable from normal ongoing activity.

There are many causes of epilepsy. Indeed, all of the causes of brain damage that have been described in this chapter—including viruses, neurotoxins, tumors, and blows to the head—can cause epilepsy, and over 70 different faulty genes have been linked to it (Berkovic et al., 2006). Many cases of epilepsy appear to be associated with faults at inhibitory synapses (e.g., GABAergic synapses) that cause many neurons in a particular area to fire in synchronous bursts (see Cossart, Bernard, & Ben-Ari, 2005), a pattern of firing that is rare in the normal brain (Ecker et al., 2010).

The diagnosis of epilepsy rests heavily on evidence from electroencephalography (EEG). The value of scalp electroencephalography in confirming suspected cases of epilepsy stems from the fact that epileptic seizures are associated with bursts of high-amplitude EEG spikes, which are often apparent in the scalp EEG during an attack (see Figure 10.8), and from the fact that individual spikes often punctuate the scalp EEGs of epileptics between attacks (Cohen et al., 2002).

Some epileptics experience peculiar psychological changes just before a convulsion. These changes, called **epileptic auras**, may take many different forms—for example, a bad smell, a specific thought, a vague feeling of familiarity, a hallucination, or a tightness of the chest. Epileptic auras are important for two reasons. First, the nature of the auras provides clues concerning the location of the epileptic focus. Second, epileptic auras can warn the patient of an impending convulsion.

Once an individual has been diagnosed as epileptic, it is usual to assign the epilepsy to one of two general categories—*partial epilepsy* or *generalized epilepsy*—and then to one of their respective subcategories (see Tuxhorn & Kotagal, 2008). The various seizure types are so different from one another that epilepsy is best viewed not as a single disease but as a number of different, but related, diseases.

Partial Seizures   A **partial seizure** is a seizure that does not involve the entire brain. The epileptic neurons at a focus begin to discharge together in bursts, and it is this synchronous bursting of neurons (see Figure 10.9) that produces epileptic spiking in the EEG. The synchronous activity tends to spread to other areas of the brain—but, in the case of partial seizures, not to the entire brain. The specific behavioral symptoms of a partial epileptic seizure depend on where the disruptive discharges begin and into what structures they spread. Because partial seizures do not involve the entire brain, they are not usually accompanied by a total loss of consciousness or equilibrium.

There are two major categories of partial seizures: simple and complex. **Simple partial seizures** are partial seizures whose symptoms are primarily sensory or motor or both; they are sometimes called *Jacksonian seizures* after the famous 19th-century neurologist Hughlings Jackson. As the epileptic discharges spread through the sensory or motor areas of the brain, the symptoms spread systematically through the body.

In contrast, **complex partial seizures** are often restricted to the temporal lobes, and those who experience them are often said to have *temporal lobe epilepsy*. During a complex partial seizure, the patient engages in compulsive, repetitive, simple behaviors commonly referred to as *automatisms* (e.g., doing and undoing a button) and in more complex behaviors that appear almost normal. The diversity of complex partial seizures is illustrated by the following two cases.

The Subtlety of Complex Partial Seizures: Two Cases

One morning a doctor left home to answer an emergency call from the hospital and returned several hours later, a trifle confused, feeling as though he had experienced a bad dream. At the hospital he had performed a difficult . . . [operation] with his usual competence, but later had done and said things deemed inappropriate.

A young man, a music teacher, when listening to a concert, walked down the aisle and onto the platform, circled the piano, jumped to the floor, did a hop, skip, and jump up the aisle, and regained his senses when part way home. He often found himself on a trolley [bus] far from his destination. (Lennox, 1960, pp. 237–238)

Although patients appear to be conscious throughout their complex partial seizures, they usually have little or no subsequent recollection of them. About half of all cases of epilepsy in adults are of the complex partial variety—the temporal lobes are particularly susceptible to epileptic discharges (Bernard et al., 2004).

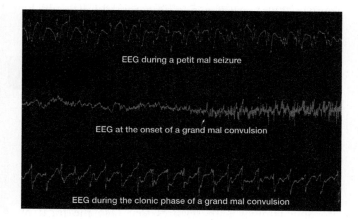

FIGURE 10.8 Cortical EEG recorded during epileptic attacks. Notice that each trace is characterized by epileptic spikes (sudden, high amplitude EEG signals that accompany epileptic attacks).

FIGURE 10.9 The bursting of an epileptic neuron, recorded by extracellular unit recording.

Generalized Seizures **Generalized seizures** involve the entire brain. Some begin as focal discharges that gradually spread through the entire brain. In other cases, the discharges seem to begin almost simultaneously in all parts of the brain. Such sudden-onset generalized seizures may result from diffuse pathology or may begin focally in a structure, such as the thalamus, that projects to many parts of the brain.

Like partial seizures, generalized seizures occur in many forms. One is the **grand mal** (literally, "big trouble") **seizure**. The primary symptoms of a grand mal seizure are loss of consciousness, loss of equilibrium, and a violent *tonic-clonic convulsion*—a convulsion involving both tonus and clonus. Tongue biting, urinary incontinence, and *cyanosis* (turning blue from excessive extraction of oxygen from the blood during the convulsion) are common manifestations of grand mal convulsions. The **hypoxia** (shortage of oxygen supply to tissue, for example, to the brain) that accompanies a grand mal seizure can itself cause brain damage.

A second major category of generalized seizure is the **petit mal** (literally, "small trouble") **seizure** (see Crunelli & Leresche, 2002). Petit mal seizures are not associated with convulsions; their primary behavioral symptom is the *petit mal absence*—a disruption of consciousness that is associated with a cessation of ongoing behavior, a vacant look, and sometimes fluttering eyelids. The EEG of a petit mal seizure is different from that of other seizures; it is a bilaterally symmetrical **3-per-second spike-and-wave discharge** (see Figure 10.10). Petit mal seizures are most common in children, and they frequently cease at puberty. They often go undiagnosed; thus, children with petit mal epilepsy are sometimes considered to be "daydreamers" by their parents and teachers.

Although there is no cure for epilepsy, the frequency and severity of seizures can often be reduced by antiepileptic medication (see Rogawski & Löscher, 2004). Unfortunately, these drugs often have adverse side effects (Toledano & Gil-Nagel, 2008). Brain surgery is sometimes performed, but only in grave situations (see Guerrini, Dobyns, & Barkovitch, 2008).

Parkinson's Disease

Parkinson's disease is a movement disorder of middle and old age that affects 1–2% of the elderly population

Clinical Implications

(see Strickland & Bertoni, 2004). It is about 2.5 times more prevalent in males than in females (see Lorincz, 2006).

The initial symptoms of Parkinson's disease are mild—perhaps no more than a slight stiffness or tremor of the fingers—but they inevitably increase in severity with advancing years. The most common symptoms of the full-blown disorder are a tremor that is pronounced during inactivity but not during voluntary movement or sleep, muscular rigidity, difficulty initiating movement, slowness of movement, and a masklike face. Pain and depression often develop before the motor symptoms become severe.

Although Parkinson's patients often display cognitive deficits, dementia is not typically associated with the disorder (Poliakoff & Smith-Spark, 2008). In essence, Parkinson's disease victims are thinking people trapped inside bodies they cannot control. Do you remember the case of "The Lizard"—Roberto Garcia d'Orta from Chapter 4?

Like epilepsy, Parkinson's disease seems to have no single cause; faulty genes, brain infections, strokes, tumors, traumatic brain injury, and neurotoxins have all been implicated in specific cases (see Klein & Schlossmacher, 2006; Sulzer, 2007). However, in the majority of cases, no cause is obvious, and there is no family history of the disorder (see Calne et al., 1987).

Parkinson's disease is associated with widespread degeneration, but it is particularly severe in the **substantia nigra**—the midbrain nucleus whose neurons project via the **nigrostriatal pathway** to the **striatum** of the basal

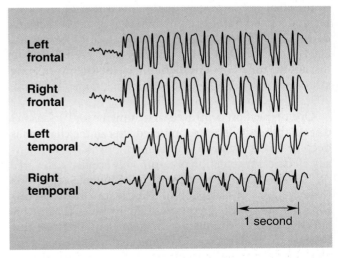

FIGURE 10.10 The bilaterally symmetrical, 3-per-second spike-and-wave EEG discharge that is associated with petit mal epileptic seizures.

ganglia. Although *dopamine* is normally the major neurotransmitter released by most neurons of the substantia nigra, there is little dopamine in the substantia nigra and striatum of long-term Parkinson's patients. Autopsy often reveals clumps of proteins in the surviving dopaminergic neurons of the substantia nigra—the clumps are called **Lewy bodies**, after the German pathologist who first reported them in 1912 (see Lozano & Kalia, 2005).

As you saw in the case of d'Orta, the symptoms of Parkinson's disease can be alleviated by injections of L-**dopa**—the chemical from which the body synthesizes

Neuroplasticity

dopamine. However, L-dopa is not a permanent solution; it typically becomes less and less effective with continued use, until its side effects (e.g., involuntary movements) outweigh its benefits (Jenner, 2008). This is what happened to d'Orta. L-Dopa therapy gave him a 3-year respite from his disease, but ultimately it became ineffective. His prescription was then changed to another dopamine agonist, and again his condition improved—but again the improvement was only temporary. There is currently no drug that will block the progressive development of Parkinson's disease (see Biglan & Ravina, 2007) or permanently reduce the severity of its symptoms (see Hermanowicz, 2007). We will return to d'Orta's roller-coaster case later in this chapter.

About 10 different gene mutations have been linked to Parkinson's disease (see Lorincz, 2006). This has led many people to believe that a cure is just around the corner. However, it is important to realize that each of these gene mutations has been discovered in a different family, each of which had members suffering from a rare *familial*

Thinking Creatively

(running in families) form of early-onset Parkinson's disease. Thus, these mutations are unlikely to be significant factors in typical forms of the disease. Still, the study of the effects of these gene mutations is leading to a better understanding of the physiological changes that underlie the symptoms of the disorder (see Moore et al., 2005). Indeed, all of the mutations have been found to disrupt the function of **mitochondria**, the energy-creating structures in each cell (Abou-Sleiman, Muqit, & Wood, 2006; Chan, Gertler, & Surmeir, 2009).

One of the most controversial treatments for Parkinson's disease (others will be encountered later in the chapter) is **deep brain stimulation**, a treatment in which low-intensity electrical stimulation is continually applied to an area of the brain through a stereotaxically implanted electrode (Montgomery & Gale, 2008). The treatment of Parkinson's disease by this method usually involves chronic bilateral electrical stimulation of a nucleus that lies just beneath the thalamus and is connected to the basal ganglia: the **subthalamic nucleus**. High-frequency electrical stimulation is employed, which blocks the function of the target structure, much as a lesion would. Once the current is turned on, symptoms are alleviated within

minutes. Although the effectiveness of deep brain stimulation slowly declines, improvement relative to the patient's pretreatment status is often still apparent 2 years after stimulation begins (Kleiner-Fisman et al., 2003). If the stimulation is, for any reason, turned off, the therapeutic improvements dissipate within an hour or two. Unfortunately, deep brain stimulation can cause side effects such as cognitive, speech, and gait problems (see Frank et al., 2007; Uc & Follett, 2007).

Huntington's Disease

Like Parkinson's disease, **Huntington's disease** is a progressive motor disorder of middle and old age; but, unlike Parkinson's disease, it is rare (1 in 10,000), it has a strong genetic basis, and it is associated with severe dementia.

Clinical Implications

The first sign of Huntington's disease is often increased fidgetiness. As the disorder develops, rapid, complex, jerky movements of entire limbs (rather than individual muscles) begin to predominate. Eventually, motor and intellectual deterioration become so severe that sufferers are incapable of feeding themselves, controlling their bowels, or recognizing their own children. There is no cure; death typically occurs about 15 years after the appearance of the first symptoms.

Huntington's disease is passed from generation to generation by a single dominant gene, called **huntingtin**. The protein it codes for is known as the **huntingtin protein**. Because the gene is dominant, all individuals carrying the gene develop the disorder, as do about half their offspring. The Huntington's gene is often passed from parent to child because the first symptoms of the disease do not appear until after the peak reproductive years (at about age 40). We still do not know how the huntingtin protein damages the brain (see Bossy-Wetzel, Petrilli, & Knott, 2008).

If one of your parents were to develop Huntington's disease, the chance would be 50/50 that you too would develop it. If you were in such a

⊙ Watch Genetic Counseling
www.mypsychlab.com

situation, would you want to know whether you would suffer the same fate? Medical geneticists have developed a test that can tell relatives of Huntington's patients whether they are carrying the gene (Gilliam, Gusella, &

⊙→ Simulate Some Common Types of Genetic Disorders and Their Defects
www.mypsychlab.com

Lehrach, 1987; Martin, 1987). Some choose to take the test, and some do not. One advantage of the test is that it permits the relatives of Huntington's patients who have not inherited the gene to have children without the fear of passing the disorder on to them.

Multiple Sclerosis

Multiple sclerosis (MS) is a progressive disease that attacks the myelin of axons in the CNS. It is particularly disturbing

Clinical Implications

because it typically attacks young people just as they are beginning their adult life. First, there are microscopic areas of degeneration on myelin sheaths; but eventually damage to the myelin is so severe that the associated axons become dysfunctional and degenerate (see Trapp & Nave, 2008). Ultimately, many areas of hard scar tissue develop in the CNS (*sclerosis* means "hardening"). Figure 10.11 illustrates degeneration of the white matter of a patient with multiple sclerosis.

Multiple sclerosis is an *autoimmune disorder*—a disorder in which the body's immune system attacks part of the body, as if it were a foreign substance. In multiple sclerosis, myelin is the focus of the faulty immune reaction. Indeed, an animal model of multiple sclerosis, termed *experimental autoimmune encephalomyelitis*, can be induced by injecting laboratory animals with myelin and a preparation that stimulates the immune system.

Diagnosing multiple sclerosis is difficult because the nature and severity of the disorder depend on a variety of factors including the number, size, and position of the sclerotic lesions (see Kantarci, 2008). Furthermore, in some cases, there are periods of remission (up to 2 years), during which the patient seems almost normal; however, these are usually just oases in the progression of the disorder, which eventually becomes continuous and severe. Common symptoms of advanced multiple sclerosis are visual disturbances, muscular weakness, numbness, tremor, and **ataxia** (loss of motor coordination). In addition, cognitive deficits and emotional changes occur in some patients (Benedict & Bobholz, 2007).

Epidemiological studies of multiple sclerosis have provided evidence of both genetic and environmental factors in its etiology (causes)—see Ascherio and Munger (2008) and Ramagopalan, Dyment, and Ebers (2008). **Epidemiology** is the study of the various factors, such as diet, geographic location, age, sex, and race that influence the distribution of a disease in the general population. The following are three epidemiological findings that implicate genetic factors in the etiology of multiple sclerosis:

- higher concordance rate in monozygotic (25%) than in dizygotic (5%) twins,
- three times higher incidence in females than in males, and
- substantially higher incidence in Caucasians (0.15%) than in other ethnic groups, such as native Asians and Africans.

In contrast, the following are three epidemiological findings that implicate environmental factors:

- incidence is higher among populations living in colder climates, as opposed to near the equator;
- individuals who migrate from a high-incidence region to a low-incidence region, particularly at a young age, reduce their susceptibility (and vice versa); and
- cigarette smokers are at greater risk.

Although epidemiological evidence has implicated genetic factors in the etiology of multiple sclerosis, the effects are weak—the concordance rate in identical twins is only 25%, and only one chromosomal locus (region on a chromosome) has been linked to multiple sclerosis with any certainty (see Ascherio & Munger, 2008). Consequently, researchers searching for the causes of multiple sclerosis have started to focus on epigenetic mechanisms, that is, on interactions between environmental factors and genetic predispositions. One good example of such research is the study of the role of vitamin D in multiple sclerosis: The incidence of multiple sclerosis is highest in regions far from the equator, which are exposed to little strong sunshine, a major source of vitamin D (see Ascherio & Munger, 2008), and vitamin D has been shown to influence a chromosomal locus implicated in multiple sclerosis (Ramagopalan et al., 2009).

A number of treatments have been shown to be effective with the animal model of multiple sclerosis, and some of these have proven beneficial to human patients (see De Jager & Hafler, 2007; Franklin & ffrench-Constant, 2008; Wingerchuk, 2008). However, none of these treatments constitutes a cure: The beneficial effects of each treatment tend to be temporary and to act on only some symptoms in some patients.

Alzheimer's Disease

Alzheimer's disease is the most common cause of *dementia* (Yaari & Corey-Bloom, 2007). It sometimes appears in individuals as young as 40, but the likelihood of its development becomes greater with advancing years. *Clinical Implications* About 10% of the general population over the age of 65 suffer from the disease, and the proportion is about 35% in those over 85 (see St. George-Hyslop, 2000; Turner, 2006).

Alzheimer's disease is progressive. Its early stages are often characterized by a selective decline in memory,

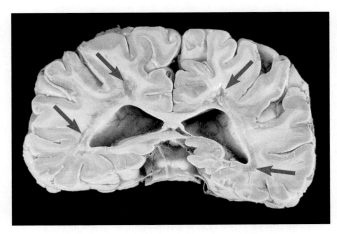

FIGURE 10.11 Areas of sclerosis (see arrows) in the white matter of a patient with MS.

FIGURE 10.12 Amyloid plaques (stained blue) in the brain of a deceased patient who had Alzheimer's disease.

neurons and a protein called **amyloid**, which is present in normal brains in only very small amounts. In addition, there is substantial neuron loss. The presence of amyloid plaques in the brain of a patient who died of Alzheimer's disease is illustrated in Figure 10.12.

Although neurofibrillary tangles, amyloid plaques, and neuron loss tend to occur throughout the brains of Alzheimer's patients, they are more prevalent in some areas than in others. For example, they are particularly prevalent in medial temporal lobe structures such as the *entorhinal cortex*, *amygdala*, and *hippocampus*—all structures that are involved in various aspects of memory (see Collie & Maruff, 2000; Selkoe, 2002). They are also prevalent in the inferior temporal cortex, posterior parietal cortex, and prefrontal cortex—all areas that mediate complex cognitive functions. (See Figure 10.13.)

Alzheimer's disease has a major genetic component. People with an Alzheimer's victim in their immediate family have a 50% chance of being stricken by the disease if they survive into their 80s (Breitner, 1990). For practical reasons, much of the early research on the genetics of Alzheimer's disease has focused on the rare early-onset familial form of the disease, and three specific gene mutations have been found to be associated with that form. So far, about 20 different chromosomal loci have been implicated in the common, late-onset form, but each has only a slight effect (see Bertram & Tanzi, 2008; Rademakers & Rovelet-Lecrux, 2009). This has led many investigators to assume that the cause of Alzheimer's disease lies in epigenetic mechanisms (Gatz, 2007). There is currently no cure for Alzheimer's disease (Brody & Holtzman, 2008).

One factor complicating the search for a treatment or cure for Alzheimer's disease is that it is not clear which

●) Watch Alzheimer's Disease
www.mypsychlab.com

deficits in attention, and personality changes; its intermediate stages are marked by confusion, irritability, anxiety, and deterioration of speech; and in its advanced stages, the patient deteriorates to the point that even simple responses such as swallowing and controlling the bladder are difficult (see Salmon & Bondi, 2009; Storandt, 2008). Alzheimer's disease is terminal.

Because Alzheimer's disease is not the only cause of dementia, it cannot be diagnosed with certainty on the basis of its behavioral symptoms—definitive diagnosis of Alzheimer's disease must await autopsy. The two defining characteristics of the disease are neurofibrillary tangles and amyloid plaques. *Neurofibrillary tangles* are threadlike tangles of protein in the neural cytoplasm, and *amyloid plaques* are clumps of scar tissue composed of degenerating

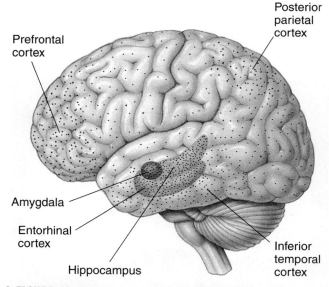

FIGURE 10.13 The typical distribution of neurofibrillary tangles and amyloid plaques in the brains of patients with advanced Alzheimer's disease. (Based on Goedert, 1993, and Selkoe, 1991.)

Thinking Creatively

symptom is primary (see Lee, 2001; Mudher & Lovestone, 2002). This is a key issue because an effective treatment will most likely be developed only by research focusing on the primary symptom. The *amyloid hypothesis* is currently the dominant view. It proposes that amyloid plaques are the primary symptom of the disorder and cause all the other symptoms (see Hardy & Selkoe, 2002). However, there is some support for the hypothesis that neurofibrillary tangles are the primary symptom (see Ballatore, Lee, & Trojanowski, 2007; Spires-Jones et al., 2009).

The main support for the amyloid hypothesis has come from the genetic analysis of families with early-onset Alzheimer's disease (see Wolfe, 2006). All three different gene mutations that cause early-onset Alzheimer's disease influence the synthesis of amyloid. Also, amyloid has been found to be toxic to neurons that have been artificially maintained in tissue culture.

The first efforts to develop treatments for Alzheimer's disease focused on the fact that declines in acetylcholine levels were among the earliest neurochemical changes appearing in patients. Cholinergic agonists are still sometimes prescribed, but, except for a few minor benefits early in the disorder, they have proven ineffective. Several other treatments are currently under development. Arguably, the most promising of these is the *immunotherapeutic approach* (see Brody & Holtzman, 2008), which involves administering an amyloid vaccine to reduce plaque deposits. This vaccine initially proved effective in tests on transgenic mice carrying human Alzheimer genes. However, tests on human patients were curtailed because of serious side effects. The search is on for safer immunotherapeutic alternatives.

Scan Your Brain

This is a good place for you to pause to scan your brain. Are you ready to progress to the following section, which discusses animal models of some of the disorders that you have just learned about? Fill in the following blanks. The correct answers are provided at the end of the exercise. Before proceeding, review material related to your errors and omissions.

1. The two major categories of epileptic seizures are _____ and _____.
2. _____ are simple repetitive responses that occur during complex partial seizures.
3. The disorder characterized by tremor at rest is _____ disease.
4. Parkinson's disease is associated with degeneration in the _____ dopamine pathway.
5. _____ disease is passed from generation to generation by a single dominant gene.

6. Genetic studies of Parkinson's disease and Alzheimer's disease have focused on early-onset _____ forms of the disorders.
7. Experimental autoimmune encephalomyelitis is an animal model of _____.
8. The most common cause of dementia is _____ disease.
9. Two major neuropathological symptoms of Alzheimer's disease are _____ tangles and _____ plaques.

Scan Your Brain answers: (1) partial and generalized (in either order), (2) Automatisms, (3) Parkinson's, (4) nigrostriatal, (5) Huntington's, (6) familial, (7) multiple sclerosis, (8) Alzheimer's, (9) neurofibrillary, amyloid.

10.3
Animal Models of Human Neuropsychological Diseases

The first two sections of this chapter focused on neuropsychological diseases and their causes, but they also provided some glimpses into the ways in which researchers attempt to solve the puzzles of neurological dysfunction. This section focuses on one of these ways: the experimental investigation of animal models.

Because identifying the neuropathological bases of human neuropsychological diseases is seldom possible based on research on the patients themselves, research on animal models of the diseases often plays an important role. Unfortunately, using and interpreting animal models is far from straightforward: Even the best animal models of neuropsychological diseases display only some of the features of the diseases they are modeling (see Maries et al., 2003), though researchers often treat animal models as if they duplicate the human conditions that they are claimed to model in every respect. That is why I have included this section on animal models in this chapter: I do not want you to fall into the trap of thinking about them in this way. This section discusses three widely used animal models: the kindling model of epilepsy, the transgenic mouse model of Alzheimer's disease, and the MPTP model of Parkinson's disease.

Evolutionary Perspective

Thinking Creatively

Kindling Model of Epilepsy

In the late 1960s, Goddard, McIntyre, and Leech (1969) delivered one mild electrical stimulation per day to rats through implanted amygdalar electrodes. There was no behavioral response to the first few stimulations, but soon each stimulation began to elicit a convulsive response. The first convulsions were mild, involving only a slight tremor of the face. However, with each subsequent stimulation, the elicited convulsions became more generalized, until each convulsion involved the entire body. The progressive development and intensification of convulsions elicited by a series of periodic brain stimulations became known as the **kindling phenomenon**, one of the first neuroplastic phenomena to be widely studied.

Although kindling is most frequently studied in rats subjected to repeated amygdalar stimulation, it is a remarkably general phenomenon. For example, kindling has been reported in mice, rabbits, cats, dogs, and various primates. Moreover, kindling can be produced by the repeated stimulation of many brain sites other than the amygdala, and it can be produced by the repeated application of initially subconvulsive doses of convulsive chemicals.

Evolutionary Perspective

There are many interesting features of kindling, but two warrant emphasis. The first is that the neuroplastic changes underlying kindling are permanent. A subject that has been kindled and then left unstimulated for several months still responds to each low-intensity stimulation with a generalized convulsion. The second is that kindling is produced by distributed, as opposed to massed, stimulations. If the intervals between successive stimulations are shorter than an hour or two, it usually requires many more stimulations to kindle a subject; and under normal circumstances, no kindling at all occurs at intervals of less than about 20 minutes.

Neuroplasticity

Much of the interest in kindling stems from the fact that it models epilepsy in two ways (see Morimoto, Fahnestock, & Racine, 2004). First, the convulsions elicited in kindled animals are similar in many respects to those observed in some types of human epilepsy. Second, the kindling phenomenon itself is comparable to the **epileptogenesis** (the development, or genesis, of epilepsy) that can follow a head injury: Some individuals who at first appear to have escaped serious injury after a blow to the head begin to experience convulsions a few weeks later, and these convulsions sometimes begin to recur more and more frequently and with greater and greater intensity.

It must be stressed that the kindling model as it is applied in most laboratories does not model epilepsy in one important respect. You will recall from earlier in this chapter that epilepsy is a disease in which epileptic attacks recur spontaneously; in contrast, kindled convulsions are elicited. However, a model that overcomes this shortcoming has been developed in several species. If subjects are kindled for a very long time—about 300 stimulations in rats—a syndrome can be induced that is truly epileptic, in the sense that the subjects begin to display spontaneous seizures and continue to display them even after the regimen of stimulation is curtailed.

Thinking Creatively

One interesting and potentially important development in the study of kindling is that some researchers have started to use it to model *interictal behavior* (behavior that occurs in epileptics between their seizures). For some human epileptics, particularly those who suffer from complex partial seizures, pathological changes in interictal behavior are more distressing and more difficult to treat than

the seizures themselves (Leung, Ma, & McLachlan, 2000). Several studies of kindling have shown that kindled subjects display a variety of changes in interictal emotional behavior that are similar to those observed in human epileptics (Kalynchuk, 2000; Wintink et al., 2003).

Transgenic Mouse Models of Alzheimer's Disease

Perhaps the most promising step forward in the study of Alzheimer's disease has been the development of several transgenic models of the disorder. **Transgenic** refers to animals into which genes of another species have been introduced (see Götz & Ittner, 2008).

In one transgenic mouse model of Alzheimer's disease (Hsiao et al., 1996), genes that accelerate the synthesis of human amyloid are injected into newly fertilized mouse eggs, which are then injected into a foster mother to develop. When the transgenic mice mature, their brains contain many amyloid plaques like those of human Alzheimer's patients. Moreover, the distribution of the amyloid plaques is comparable to that observed in human Alzheimer's patients, with the highest concentrations occurring in structures of the medial temporal lobes (e.g., hippocampus, amygdala, and entorhinal cortex). The mice whose brains contain amyloid plaques display neural loss and memory disturbances (see Lansbury, 2006).

Although transgenic mouse models are arguably the best current animal models of Alzheimer's disease, they are not without problems. For example, most of these models do not display neurofibrillary tangles, which is a serious problem if neurofibrillary tangles prove to be the primary symptom of Alzheimer's disease. However, a *triple transgenic mouse model* (a mouse into which three different human Alzheimer genes have been inserted), which displays both amyloid plaques and neurofibrillary tangles, is currently being tested (Pietropaolo, Feldon, & Yee, 2008).

MPTP Model of Parkinson's Disease

The preeminent animal model of Parkinson's disease grew out of an unfortunate accident, which resulted in the following cases.

Evolutionary Perspective

The Cases of the Frozen Addicts

Parkinson's disease . . . rarely occurs before the age of 50. It was somewhat of a surprise then to see a group of young drug addicts at our hospital in 1982 who had developed symptoms of severe and what proved to be irreversible parkinsonism. The only link between these patients was the recent use of a new "synthetic heroin." They exhibited virtually all of the typical motor features of Parkinson's disease, including the classic triad of bradykinesia (slowness

Clinical Implications

of movement), tremor and rigidity of their muscles. Even the subtle features, such as seborrhea (oiliness of the skin) and micrographia (small handwriting), that are typical of Parkinson's disease were present. After tracking down samples of this substance, the offending agent was tentatively identified as 1-methyl-4-phenyl-1,2,3,6-tetrahydropyridine or **MPTP**. . . . There has been no sign of remission, and most are becoming increasingly severe management problems. (Langston, 1985, p. 79)

Researchers immediately turned the misfortune of these few to the advantage of many by developing a much-needed animal model of Parkinson's disease (Langston, 1986). It was quickly established that nonhuman primates respond to MPTP the same way humans do. The brains of nonhuman primates exposed to MPTP have cell loss in the substantia nigra similar to that observed in the brains of Parkinson's patients. Considering that the substantia nigra is the major source of the brain's dopamine, it is not surprising that the level of dopamine is greatly reduced in both the MPTP model and in the naturally occurring disorder. However, it is curious that in a few monkeys MPTP produces a major depletion of dopamine without producing any gross motor symptoms (Taylor et al., 1990).

The MPTP animal model has benefited patients with Parkinson's disease. For example, it was discovered that **deprenyl**, a monoamine agonist, blocks the effects of

MPTP in an animal model, and it was subsequently shown that deprenyl administered to early Parkinson's patients retards the progression of the disease (Tetrud & Langston, 1989)—see Figure 10.14. Still, it is important to remember that the MPTP model does not model etiological factors in Parkinson's disease.

10.4

Neuroplastic Responses to Nervous System Damage: Degeneration, Regeneration, Reorganization, and Recovery

In the first three sections of this chapter, you have learned about some of the ways brain damage occurs, some of the neuropsychological disorders such damage can produce, and some of the animal models used to study these disorders. This section focuses on four neuroplastic responses of the brain to damage: degeneration, regeneration, reorganization, and recovery of function.

Neural Degeneration

Neural degeneration (neural deterioration) is a common component of both brain development and disease (see Coleman, 2005; Low & Cheng, 2005; Luo & O'Leary, 2005). A widely used method for the controlled study of neural degeneration is to cut the axons of neurons. Two kinds *Neuroplasticity* of neural degeneration ensue: anterograde degeneration and retrograde degeneration (see Vargas & Barres, 2007). **Anterograde degeneration** is the degeneration of the **distal segment**—the segment of a cut axon between the cut and the synaptic terminals. **Retrograde degeneration** is the degeneration of the **proximal segment**—the segment of a cut axon between the cut and the cell body.

Anterograde degeneration occurs quickly following *axotomy*, because the cut separates the distal segment of the axon from the cell body, which is the metabolic center of the neuron. The entire distal segment becomes badly swollen within a few hours, and it breaks into fragments within a few days.

The course of retrograde degeneration is different; it progresses gradually back from the cut to the cell body. In about 2 or 3 days, major changes become apparent in the cell bodies of most axotomized neurons. These early cell body changes are either degenerative or regenerative in nature. Early degenerative changes to the cell body (e.g., a decrease in size) suggest that the neuron will ultimately die—usually by apoptosis but sometimes by necrosis or a combination of both (Syntichaki & Tavernarakis, 2003). Early regenerative changes (e.g., an increase in size) indicate that the cell body is involved in a massive synthesis of the proteins that will be used to replace the degenerated

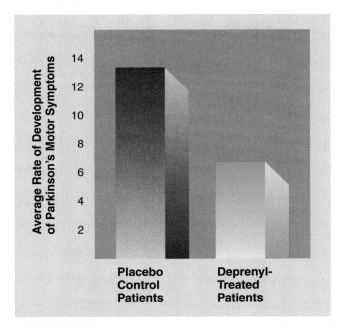

FIGURE 10.14 Average rate of motor symptom development in early Parkinson's patients treated with deprenyl (a monoamine oxidase inhibitor) or with a placebo. Deprenyl slowed the progression of the disease by 50%. (Based on Tetrud & Langston, 1989.)

axon. But early regenerative changes in the cell body do not guarantee the long-term survival of the neuron; if the regenerating axon does not manage to make synaptic contact with an appropriate target, the neuron eventually dies.

Sometimes, degeneration spreads from damaged neurons to neurons that are linked to them by synapses; this is called **transneuronal degeneration**. In some cases, transneuronal degeneration spreads from damaged neurons to the neurons on which they synapse; this is called *anterograde transneuronal degeneration*. And in some cases, it spreads from damaged neurons to the neurons that synapse on them; this is called *retrograde transneuronal degeneration*. Neural and transneuronal degeneration are illustrated in Figure 10.15.

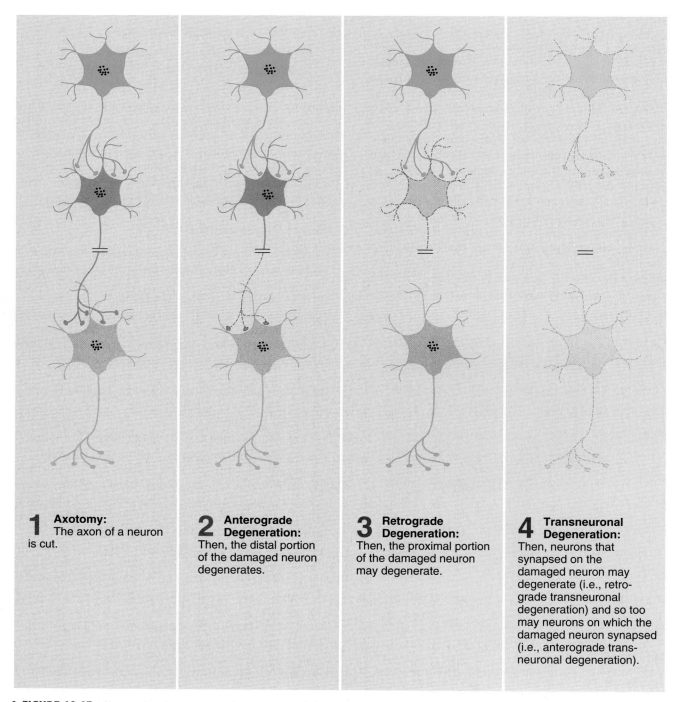

1 Axotomy:
The axon of a neuron is cut.

2 Anterograde Degeneration:
Then, the distal portion of the damaged neuron degenerates.

3 Retrograde Degeneration:
Then, the proximal portion of the damaged neuron may degenerate.

4 Transneuronal Degeneration:
Then, neurons that synapsed on the damaged neuron may degenerate (i.e., retrograde transneuronal degeneration) and so too may neurons on which the damaged neuron synapsed (i.e., anterograde transneuronal degeneration).

FIGURE 10.15   Neuronal and transneuronal degeneration following axotomy.

Neural Regeneration

Neural regeneration—the regrowth of damaged neurons—does not proceed as successfully in mammals and other higher vertebrates as it does in most invertebrates and  lower vertebrates. For example, in Chapter 9, you learned about accurate regeneration in the frog visual system as demonstrated by Sperry's eye-rotation experiments. The capacity for accurate axonal growth, which higher vertebrates possess during their early development, is lost once they reach maturity. Regeneration is virtually nonexistent in the CNS of adult mammals, and is at best a hit-or-miss affair in the PNS.

In the mammalian PNS, regrowth from the proximal stump of a damaged nerve usually begins 2 or 3 days after axonal damage. What happens next depends on the nature of the injury (see Chen, Yu, & Strickland, 2007); there are three possibilities. First, if the original Schwann cell myelin sheaths remain intact, the regenerating peripheral axons grow through them to their original targets at a rate of a few millimeters per day. Second, if the peripheral nerve is severed and the cut ends become separated by a few millimeters, regenerating axon tips often grow into incorrect sheaths and are guided by them to incorrect destinations; that is why it is often difficult to regain the coordinated use of a limb affected by nerve damage even if there has been substantial regeneration. And third, if the cut ends of a severed mammalian peripheral nerve become widely separated or if a lengthy section of the nerve is damaged, there may be no meaningful regeneration at all; regenerating axon tips grow in a tangled mass around the proximal stump, and the neurons ultimately die. These three patterns of mammalian peripheral nerve regeneration are illustrated in Figure 10.16.

Why do mammalian PNS neurons sometimes regenerate, but mammalian CNS neurons do not? The obvious answer is that PNS neurons are inherently capable of regeneration, whereas CNS neurons are not. However, this answer has proved to be incorrect. Some CNS neurons are capable of regeneration if they are transplanted to the PNS, whereas some PNS neurons are not capable of regeneration if they are transplanted to the CNS (see Dusart et al., 2005). Clearly, there is something about the environment of the PNS that promotes regeneration and something about the environment of the CNS that does not (Goldberg & Barres, 2000). Schwann cells seem to be the key.

Schwann cells, which myelinate PNS axons, clear the debris resulting from neural degeneration and promote

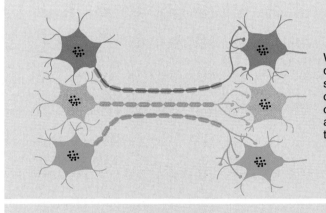

When a nerve is damaged without severing the Schwann cell sheaths (e.g., by crushing), individual axons regenerate to their correct targets.

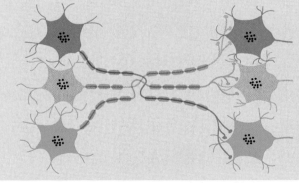

When a nerve is damaged and the severed ends of the Schwann cell sheaths are slightly separated, individual axons often regenerate up incorrect sheaths and reach incorrect targets.

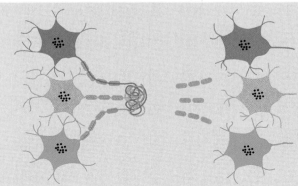

When a nerve is damaged and the severed ends of the Schwann cell sheaths are widely separated, there is typically no functional regeneration.

FIGURE 10.16 Three patterns of axonal regeneration in mammalian peripheral nerves.

regeneration in the mammalian PNS by producing both neurotrophic factors and cell-adhesion molecules (CAMs). The neurotrophic factors released by Schwann cells stimulate the growth of new axons, and the cell-adhesion molecules on the cell membranes of Schwann cells mark the paths along which regenerating PNS axons grow. In contrast, **oligodendroglia**, which myelinate CNS axons, do not clear debris or stimulate or guide regeneration; indeed, they release factors that actively block regeneration (see Yiu & He, 2006). Furthermore, oligodendroglia tend to survive for long periods of time (e.g., months) after nerve damage, thus chronically inhibiting regeneration of the axons (see Vargas & Barres, 2007).

When an axon degenerates, axon branches grow out from adjacent healthy axons and synapse at the sites vacated by the degenerating axon; this is called **collateral sprouting**. Collateral sprouts may grow out from the axon terminal branches or the nodes of Ranvier on adjacent neurons. Collateral sprouting is illustrated in Figure 10.17.

In contrast to neural regeneration in mammals, that in lower vertebrates is extremely accurate. It is accurate in both the CNS and the PNS, and it is accurate even when the regenerating axons do not grow into remnant Schwann cell myelin sheaths. The accuracy of regeneration in lower vertebrates offers hope of a medical breakthrough:

If the factors that promote accurate regeneration in lower vertebrates can be identified and applied to the human brain, it might be possible to cure currently untreatable brain injuries. Remarkably, when invertebrates lose an entire limb, the regenerating axons release a factor that promotes regeneration of that limb (Kumar et al., 2007; Muneoka, Han, & Gardiner, 2008; Stocum, 2007).

Neural Reorganization

You learned in Chapter 9 that adult mammalian brains have the ability to reorganize themselves in response to experience. You will learn in this section that they can also reorganize themselves in response to damage.

Cortical Reorganization Following Damage in Laboratory Animals

Most studies of neural reorganization following damage have focused on the sensory and motor cortex of laboratory animals. Sensory and motor cortex are ideally suited to the study of neural reorganization because of their topographic layout. The damage-induced reorganization of the primary sensory and motor cortex has been studied under two conditions: following damage to peripheral nerves and following damage to the cortical areas themselves (Buonomano & Merzenich, 1998).

Demonstrations of cortical reorganization following neural damage in laboratory animals started to be reported in substantial numbers in the early 1990s. The following three studies have been particularly influential:

- Kaas and colleagues (1990) assessed the effect of making a small lesion in one retina and removing the other. Several months after the retinal lesion was made, primary visual cortex neurons that originally had receptive fields in the lesioned area of the retina were found to have receptive fields in the area of the retina next to the lesion; remarkably, this change began within minutes of the lesion (Gilbert & Wiesel, 1992).
- Pons and colleagues (1991) mapped the primary somatosensory cortex of monkeys whose contralateral arm sensory neurons had been cut 10 years before. They found that the cortical face representation had systematically expanded into the original arm area. This study created a stir because the scale of the reorganization was far greater than had been assumed to be possible: The primary somatosensory cortex face area had expanded its border by well over a centimeter, likely as a consequence of the particularly long (10-year) interval between surgery and testing.
- Sanes, Suner, and Donoghue (1990) transected the motor neurons that controlled the muscles of rats' *vibrissae* (whiskers). A few weeks later, stimulation of

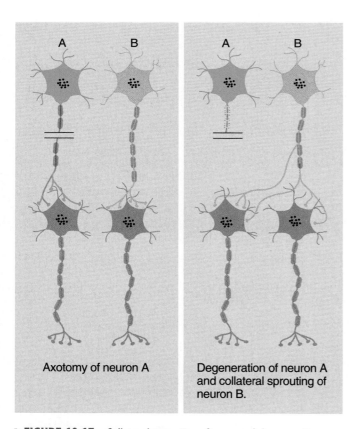

| Axotomy of neuron A | Degeneration of neuron A and collateral sprouting of neuron B. |

FIGURE 10.17 Collateral sprouting after neural degeneration.

the area of motor cortex that had previously elicited vibrissae movement now activated other muscles of the face. This result is illustrated in Figure 10.18.

Cortical Reorganization Following Damage in Humans

Demonstrations of cortical reorganization in controlled experiments on nonhumans provided an incentive to search for similar effects in human clinical populations. One such line of research has used brain-imaging technology to study the cortices of blind individuals. The findings are consistent with the hypothesis that there is continuous competition for cortical space by functional circuits. Without visual input to the cortex, there is an expansion of auditory and somatosensory cortex (see Elbert et al., 2002), and auditory and somatosensory input

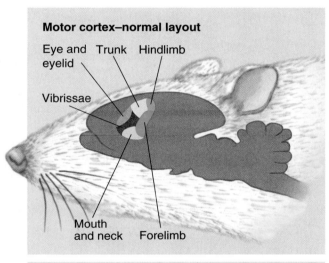

Motor cortex–normal layout

Eye and eyelid
Trunk
Hindlimb
Vibrissae
Mouth and neck
Forelimb

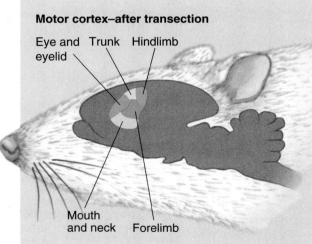

Motor cortex–after transection

Eye and eyelid
Trunk
Hindlimb
Mouth and neck
Forelimb

FIGURE 10.18 Reorganization of the rat motor cortex following transection of the motor neurons that control movements of the vibrissae. The motor cortex was mapped by brain stimulation before transection and then again a few weeks after. (Based on Sanes et al., 1990)

is processed in formerly visual areas (see Amedi et al., 2005). There seems to be a functional consequence to this reorganization: Blind volunteers have demonstrated skills superior to those of sighted controls on a variety of auditory and somatosensory tasks (see Gougoux et al., 2005).

Mechanisms of Neural Reorganization

Two kinds of mechanisms have been proposed to account for the reorganization of neural circuits: a strengthening of existing connections, possibly through release from inhibition, and the establishment of new connections by collateral sprouting (see Cafferty, McGee, & Strittmatter, 2008). Indirect support for the first mechanism comes from two observations: Reorganization often occurs too quickly to be explained by neural growth, and rapid reorganization never involves changes of more than 2 millimeters of cortical surface. Indirect support for the second mechanism comes from the observation that the magnitude of long-term reorganization can be too great to be explained by changes in existing connections. Figure 10.19 on page 260 illustrates how these two mechanisms might account for the reorganization that occurs after damage to a peripheral somatosensory nerve.

Although sprouting and release from inhibition are considered to be the likely mechanisms of cortical reorganization following damage, these are not the only possibilities. For example, neural degeneration, adjustment of dendritic trees, and adult neurogenesis may all be involved. It is also important to appreciate that cortical reorganization following damage is not necessarily mediated by changes to the cortex itself: Changes to the cortex can be produced by adjustments to subcortical structures, such as the thalamus (Fox, Glazewski, & Schulze, 2000).

Recovery of Function after Brain Damage

Recovery of function in humans after nervous system damage is a poorly understood phenomenon. Nevertheless, there is a general consensus that recovery of function is most likely when lesions are small and the patient is young; see Figure 10.20 on page 260 (Payne & Lomber, 2001). Neuroplastic phenomena, like those that you have just been reading about, are presumed to underlie recovery of function. However, it has proven difficult to provide strong evidence for this assumption.

Clinical Implications

Neuroplasticity

Recovery of function after nervous system damage is difficult to study because there are other compensatory changes that can easily be confused with it. For example, any improvement in the week or two after damage could reflect a decline in *cerebral edema* (brain swelling) rather than a recovery from the neural damage itself, and any gradual improvement in the months after damage could

FIGURE 10.19 The two-stage model of neural reorganization: (1) strengthening of existing connections through release from inhibition and (2) establishment of new connections by collateral sprouting.

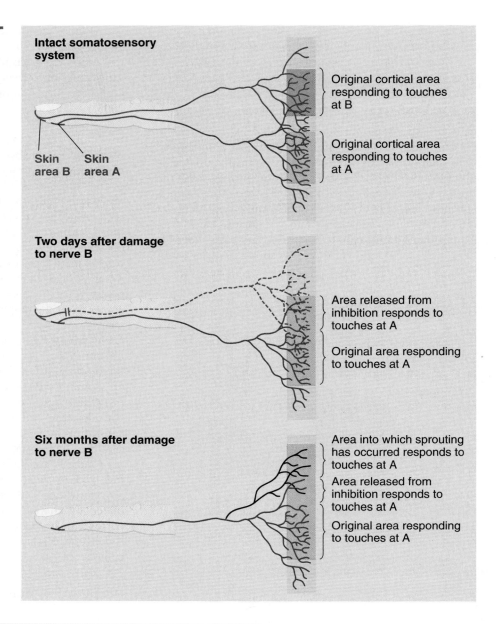

reflect the learning of new cognitive and behavioral strategies (i.e., substitution of functions) rather than the return of lost functions (see Wilson, 1998).

Cognitive reserve (roughly equivalent to education and intelligence) is thought to play a role in the improvements observed after brain damage that do not result from true recovery of brain function. Let me explain. Kapur (1997) conducted a biographical study of doctors and neuroscientists with brain damage, and he observed a surprising degree of what appeared to be cognitive recovery. His results suggested, however, that the observed improvement did not occur because these patients had actually recovered lost brain function but because their cognitive reserve allowed them to accomplish tasks in alternative ways. Cognitive reserve has also been used to explain why educated people are less susceptible to the effects of aging-related brain deterioration (Reuter-Lorenz & Cappell, 2008).

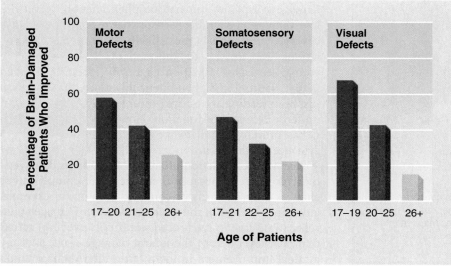

FIGURE 10.20 Percentage of patients showing improvement following brain injury. Teuber (1975) assessed the deficits of brain-damaged soldiers within a week of their injury and again 20 years later. This graph shows the percentage of patients who displayed improvement over the intervening 20 years as a function of age at the time of injury and the particular defect.

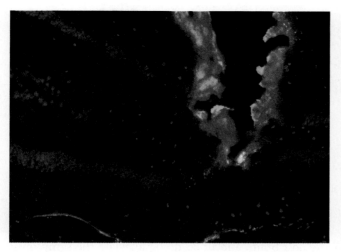

FIGURE 10.21   Increased neurogenesis in the dentate gyrus following damage. The left panel shows (1) an electrolytic lesion in the dentate gyrus (damaged neurons are stained turquoise) and (2) the resulting increase in the formation of new cells (stained red), many of which develop into mature neurons (stained dark blue). The right panel displays the comparable control area in the unlesioned hemisphere, showing the normal number of new cells (stained red). (These images are courtesy of my friends Carl Ernst and Brian Christie, Department of Psychology, University of British Columbia.)

The search for the neuroplastic responses that contribute to the recovery of function following brain damage is further complicated by the fact that damage to one part of the brain can induce widespread changes (Nudo, 2006). In some cases, neuroplastic changes to undamaged circuits have been shown to contribute to recovery after CNS damage (Nishimura et al., 2007)—for example, mice that have had their spinal cords transected often recover some motor control in the absence of any neural regrowth across the cut (Courtine et al., 2008).

There has been substantial interest in the possibility that adult neurogenesis might contribute to recovery from brain damage. Evidence in favor of this possibility is the finding that stem cells tend to migrate short distances into areas of brain damage in adult laboratory animals (Zhou et al., 2003)—for example, cerebral ischemia increases neurogenesis in the hippocampus (see Figure 10.21). However, there is no evidence that stem cells can migrate from their sites of genesis in the hippocampus and olfactory bulbs to distant areas of damage in the adult human brain (Ghashghaei, Lai, & Anton, 2007).

Maximizing recovery from brain damage will require discovering the most effective ways of administering treatments. For example, several recent experiments in laboratory animals have shown that various treatments are maximally effective only if initiated within a few days of the brain damage (see Kolb et al., 2006; Nudo, 2006). This finding has a potentially vital implication: It suggests that rehabilitation of human patients with brain damage should begin as soon as possible after the damage occurs.

10.5

Neuroplasticity and the Treatment of Nervous System Damage

This section reveals one reason for all the excitement about the phenomenon of neuroplasticity: the dream that recent discoveries about neuroplasticity can be applied to the treatment of brain damage in human patients. The following four subsections describe research on some major new treatment approaches. Most of this research has focused on animal models, but some of it has progressed to clinical trials with human patients. Although I discuss each treatment individually, they are best used in combination (Boulenguez & Vinay, 2009).

Clinical Implications

Evolutionary Perspective

Reducing Brain Damage by Blocking Neurodegeneration

Several studies in animals have shown that it may be possible to reduce brain damage by blocking neural degeneration in human patients. For example, in one study, Xu and colleagues (1999) induced cerebral ischemia in rats by limiting blood flow to the brain. This had two major effects on rats in the control group: It produced damage to the hippocampus, a structure that is particularly susceptible to ischemic damage, and it produced deficits in the rats'

Neuroplasticity

performance in the Morris water maze (see Chapter 5). The hippocampuses of rats in the experimental group were treated with viruses genetically engineered to release *apoptosis inhibitor protein*. Amazingly, the apoptosis inhibitor protein reduced both the loss of hippocampal neurons and the deficits in Morris water maze performance.

In addition to apoptosis inhibitor protein, several other neurochemicals have been shown to block the degeneration of damaged neurons. The most widely studied of these is *nerve growth factor* (see Sofroniew, Howe, & Mobley, 2001). You may be surprised to learn that estrogens have a similar effect (see Behl, 2002; Sawada & Shimohama, 2000; Stein, 2001; Wise et al., 2001). **Estrogens** are a class of steroid hormones that are released in large amounts by the *ovaries* (the female gonads). These hormones have several important effects on the maturation of the female body, which you will learn about in Chapter 13, but they also have been shown to limit or delay neuron death in animal models. These neuroprotective effects of estrogens may explain why several brain disorders (e.g., Parkinson's disease) are more prevalent in males than in females.

In general, molecules that limit neural degeneration also promote regeneration. This point leads us to the next subsection.

Promoting Recovery from CNS Damage by Promoting Regeneration

Although regeneration does not normally occur in the mammalian CNS (see Tanaka & Ferretti, 2009), several

Neuroplasticity

studies in laboratory animals have shown that it can be induced. The following two studies are particularly promising because they have shown that such regeneration can be associated with functional recovery.

Cheng, Cao, and Olson (1996) transected the spinal cords of rats, thus rendering them *paraplegic* (paralyzed in the posterior portion of their bodies). The researchers then transplanted sections of myelinated peripheral nerve across the transection. As a result, spinal cord neurons regenerated through the implanted Schwann cell myelin sheaths, and the regeneration allowed the rats to regain use of their hindquarters.

A similar study involved transplanting *olfactory ensheathing cells* rather than Schwann cells. Olfactory ensheathing cells, which are similar to Schwann cells, were selected because the olfactory system is unique in its ability to support continual growth of axons from new PNS neurons into the CNS (i.e., into the olfactory bulbs). Raisman and Li (2007) made unilateral lesions in the corticospinal tract of rats, which produced partial paralysis on the same side of the body. Then they implanted bridges of olfactory ensheathing cells across the lesion. Axons grew through the lesion, and the motor function of the affected paw was partially restored.

Promoting Recovery from CNS Damage by Neurotransplantation

Efforts to treat CNS damage by neurotransplantation have taken two different approaches (see Björklund & Lindvall, 2000). The first is to transplant fetal tissue into the damaged area; the second is to transplant cultures of stem cells.

Clinical Implications

Transplanting Fetal Tissue The first approach to neurotransplantation was to replace a damaged structure with fetal tissue that would develop into the same structure. Could the *donor* tissue develop and become integrated into the *host* brain, and in so doing alleviate the symptoms? This approach focused on Parkinson's disease. Parkinson's patients lack the dopamine-releasing cells of the nigrostriatal pathway: Could they be cured by transplanting the appropriate fetal tissue into the site?

Neuroplasticity

Early signs were positive. Bilateral transplantation of fetal substantia nigra cells was successful in treating the MPTP monkey model of Parkinson's disease (Bankiewicz et al., 1990; Sladek et al., 1987). Fetal substantia nigra transplants survived in the MPTP-treated monkeys; the transplanted cells innervated adjacent striatal tissue, released dopamine, and, most importantly, alleviated the severe poverty of movement, tremor, and rigidity produced by the MPTP.

Soon after the favorable effects of neurotransplants in the MPTP monkey model were reported, neurotransplantation was offered as a treatment for Parkinson's disease at major research hospitals. The results of the first case studies were promising. The fetal substantia nigra implants survived, and they released dopamine into the host striatum (see Sawle & Myers, 1993). More importantly, some of the patients improved.

The results of these case studies triggered a large-scale double-blind evaluation study of patients suffering from advanced Parkinson's disease. The study was extremely thorough; it even included placebo controls—patients who received surgery but no implants. The initial results were encouraging: Although control patients showed no improvement, the implants survived in the experimental patients, and some displayed a modest improvement. Unfortunately, however, about 15% of these patients started to display a variety of uncontrollable writhing and chewing movements about a year after the surgery (Greene et al., 1999).

The results of this first double-blind placebo-controlled clinical trial of the effectiveness of fetal tissue transplants created widespread debate (see Dunnett, Björklund, & Lindvall, 2001). The incidence of adverse motor side effects is likely to stifle further attempts to develop neurotransplantation as a treatment for Parkinson's disease. However, many still believe that this is an extremely promising therapeutic approach, but that the large-scale clinical trial was premature. Researchers do not yet know how to maximize the survival and growth of neurotransplants and how to minimize their side effects (see Winkler, Kirik, & Björklund, 2004). It is important to achieve a balance between the pressure to develop new treatments quickly and the need to base treatments on a carefully constructed foundation of scientific understanding (see Döbrössy & Dunnett, 2001).

In Chapter 4, you were introduced to Roberto Garcia d'Orta—the Lizard. D'Orta, who suffered from Parkinson's disease, initially responded to L-dopa therapy; but, after 3 years of therapy, his condition worsened. Then he responded to treatment with a dopamine agonist, but again the improvement was only temporary. D'Orta was in a desperate state when he heard about *adrenal medulla autotransplantation* (transplanting a patient's own adrenal medulla cells into her or his striatum, usually for the treatment of Parkinson's disease). Adrenal medulla cells release small amounts of dopamine, and there were some early indications that adrenal medulla autotransplantation might alleviate the symptoms of Parkinson's disease.

D'Orta demanded adrenal medulla autotransplantation from his doctor. When his doctor refused, on the grounds that the effectiveness of the treatment was unproven, d'Orta found himself another doctor—a neurosurgeon who was not nearly so responsible. Subsequent research found this surgical procedure to be ineffective, and its study has been abandoned.

Clinical Implications

The Case of Roberto Garcia d'Orta: The Lizard Gets an Autotransplant

Roberto flew to Juarez. The neurosurgeon there greeted him with open arms. As long as Roberto could afford the cost, he'd be happy to do an adrenal implant on him. . . .

Were there any dangers?

The neurosurgeon seemed insulted by the question. If Señor d'Orta didn't trust him, he could go elsewhere. . . .

Roberto underwent the procedure.

He flew back home two weeks later. He was no better. He was told that it took time for the cells to grow and make the needed chemicals. . . .

Then I received an unexpected call from Roberto's wife. Roberto was dead. . . .

He'd died of a stroke. . . . Had the stroke been a complication of his surgery? It was more than a mere possibility. (Klawans, 1990, pp. 63–64)

(Adapted from NEWTON'S MADNESS by Harold Klawans (Harper & Row 1990). Reprinted by permission of Jet Literary Associates, Inc.)

Transplanting Stem Cells   In Chapter 9, you learned about *embryonic stem cells*, which are *multipotent* (have the capacity to develop into many types of mature cells).

Investigators are trying to develop procedures for repairing brain damage by injecting embryonic stem cells into the damaged site (Pluchino et al., 2005). Once injected, the stem cells could develop and replace the damaged neurons or myelin, under guidance from surrounding tissue (Keirstead, 2005). A study by McDonald and colleagues (1999) illustrates the potential of this method.

Neuroplasticity

McDonald and colleagues injected embryonic neural stem cells into an area of spinal damage. Their subjects were rats that had been rendered paraplegic by a blow. The stem cells migrated to different areas around the damaged area, where they developed into mature neurons. Remarkably, the rats receiving the implants became capable of supporting their weight with their hindlimbs and walking, albeit awkwardly.

The study by McDonald and colleagues and several similar ones triggered a frenzy of research activity. Effective treatment for severe CNS damage appeared to be within reach. However, it quickly became apparent that much research still needs to be done (Li et al., 2008). First, techniques for promoting the survival, maturation, and the establishment of correct connections with surviving cells need to be identified. Second, methods for encouraging functional recovery have to be developed. For example, little attention has been paid to the behavioral treatment of patients with neural stem cell implants, which is likely to be an important factor in their recovery. And third, methods for keeping the implanted cells from themselves becoming pathological need to be developed (Amariglio et al., 2009; Brundin et al., 2008). In short, although therapeutic neural stem cell transplantation is one of the most exciting areas of investigation in all of neuroscience, the ultimate goal is an ambitious one and will take longer than once thought to achieve (see Lanza & Rosenthal, 2004).

Promoting Recovery from CNS Damage by Rehabilitative Training

Several demonstrations of the important role of experience in the organization of the developing and adult brain kindled a renewed interest in the use of rehabilitative training to promote recovery from CNS damage. The following innovative rehabilitative training programs were derived from such findings.

Neuroplasticity

Clinical Implications

Strokes   Small strokes produce a core of brain damage, which is often followed by a gradually expanding loss of neural function in the surrounding penumbra. Nudo and colleagues (1996) produced small *ischemic lesions* in the hand area of the motor cortex of monkeys. Then, 5 days later, a program of hand training and practice was initiated. During the ensuing 3 or 4 weeks, the monkeys plucked hundreds of tiny food pellets from food wells of different sizes. This practice substantially reduced the expansion of cortical damage into the surrounding penumbra. The monkeys that received the rehabilitative training also showed greater recovery in the use of their affected hand.

One of the principles that has emerged from the study of neurodevelopment is that neurons seem to be in a competitive situation: They compete with other neurons for synaptic sites and neurotrophins, and the losers die. Weiller and Rijntjes (1999) designed a rehabilitative program based on this principle. Their procedure, called *constraint-induced therapy* (Taub, Uswatte, & Elbert, 2002), was to tie down the functioning arm for 2 weeks while the affected arm received intensive training. Performance with the affected arm improved markedly over the 2 weeks, and there was an increase in the area of motor cortex controlling that arm.

Spinal Injury   In one approach to treating spinal injuries (see Wolpaw & Tennissen, 2001), patients who were incapable of walking were supported by a harness over a moving treadmill. With most of their weight supported and the treadmill providing feedback, the patients gradually learned to make walking movements. Then, as they improved, the amount of support was gradually reduced. In one study using this technique, over 90% of the trained patients eventually became independent walkers, compared with only 50% of those receiving conventional physiotherapy. The effectiveness of this treatment has been confirmed and extended in human patients (e.g., Herman et al., 2002) and in nonhuman subjects (Frigon & Rossignol, 2008).

Benefits of Cognitive and Physical Exercise   There are numerous studies linking cognitive and physical activity to beneficial neurological outcomes in human patients. Individuals who are cognitively and physically active are less likely to contract neurological disorders; and if they do, their symptoms tend to be less severe and their recovery better (see Särkämo et al., 2008). However, in such correlational studies, there are always problems of causal interpretation: Do more active individuals tend to have better neurological outcomes because they are more active, or do they tend to be more active because they are less ill? Because of these problems of causal interpretation, research in this area has relied heavily on controlled experiments using animal models (see Nithianantharajah & Hannan, 2006).

One experimental approach to studying the benefits of cognitive and physical activity has been to assess the neurological benefits of housing animals in enriched environments. **Enriched environments** are those that are designed to promote cognitive and physical activity—they typically involve group housing, toys, activity wheels, and changing stimulation (see Figure 10.22). The health-promoting effects of enriched environments have already been demon-

strated in animal models of epilepsy, Huntington's disease, Alzheimer's disease, Parkinson's disease, Down syndrome, and various forms of stroke and traumatic brain injury (see Lazarov et al., 2005; Nithianantharajah & Hannan, 2006). Although the mechanisms underlying the neurological benefits of enriched environments are unclear, there are many possibilities: Enriched environments have been shown to increase dendritic branching, the size and number of dendritic spines, the size of synapses, the rate of adult neurogenesis, and the levels of various neurotrophic factors.

Physical exercise has also been shown to have a variety of beneficial effects on the rodent brain (Cotman, Berchtold, & Christie, 2007). For example, Van Praag and colleagues found that wheel running can increase adult neurogenesis in the hippocampus (2002), reduce age-related declines in the number of neurons in the hippocampus (2005), and improve performance on tests of memory and navigation, two abilities that have been linked to the hippocampus. Also, Adlard and colleagues (2005) found that wheel running reduced the development of amyloid plaques in mice genetically predisposed to develop a model of Alzheimer's disease (see also Woodlee and Schallert, 2006).

Phantom Limbs: Neuroplastic Phenomena   Most amputees continue to experience the limbs that have been amputated—a condition referred to as **phantom limb**. As you are about to learn, phantom limbs are the product of neuroplasticity.

The most striking feature of phantom limbs is their reality. Their existence is so compelling that a patient may try to jump out of bed onto a nonexistent leg or to lift a cup with a nonexistent hand. In most cases, the amputated limb behaves like a normal limb; for example, as an amputee walks, a phantom arm seems to swing back and

FIGURE 10.22   A rodent in an enriched laboratory environment.

forth in perfect coordination with the intact arm. However, sometimes an amputee feels that the amputated limb is stuck in a peculiar position. For example, one amputee felt that his phantom arm extended straight out from the shoulder, and as a result, he turned sideways whenever he passed through doorways (Melzack, 1992).

About 50% of amputees experience chronic severe pain in their phantom limbs. A typical complaint is that an amputated hand is clenched so tightly that the fingernails are digging into the palm of the hand. Phantom limb pain can occasionally be treated by having the amputee concentrate on opening the amputated hand, but often surgical treatments are attempted. Based on the premise that phantom limb pain results from irritation at the stump, surgical efforts to control it have often involved cutting off the stump or various parts of the neural pathway between the stump and the cortex. Unfortunately, these treatments haven't worked (see Melzack, 1992).

Tom and Philip experienced phantom limbs. Their neuropsychologist was the esteemed V. S. Ramachandran.

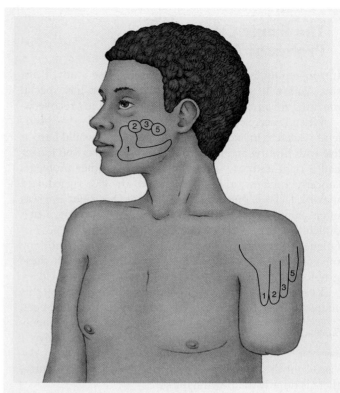

FIGURE 10.23 The places on Tom's body where touches elicited sensations in his phantom hand. (Based on Ramachandran & Blakeslee, 1998.)

The Cases of Tom and Philip: Phantom Limbs and Ramachandran

Dr. Ramachandran read about the study of Pons and colleagues (1991), which you have already encountered in this chapter. In this study, severing the sensory neurons in the arms of monkeys led to a reorganization of so-

Clinical Implications

matosensory cortex: The area of the somatosensory cortex that originally received input from the damaged arm now received input from areas of the body normally mapped onto adjacent areas of somatosensory cortex.

Thinking Creatively

Ramachandran was struck by a sudden insight: Perhaps phantom limbs were not in the stump at all, but in the brain; perhaps the perception of a phantom arm originated from parts of the body that now innervated the original arm area of the somatosensory cortex (see Ramachandran & Blakeslee, 1998).

Excited by his hypothesis, Dr. Ramachandran asked one of his patients, Tom, if he would participate in a simple test. He touched various parts of Tom's body and asked Tom what he felt. Remarkably, when he touched the

Neuroplasticity

side of Tom's face on the same side as his amputated arm, Tom felt sensations from various parts of his phantom hand as well as his face. Indeed, when some warm water was dropped on his face, he felt it running down his phantom hand. A second map of his hand was found on his shoulder (see Figure 10.23).

Philip, another patient of Dr. Ramachandran, suffered from severe chronic pain in his phantom arm. For a

decade, Philip's phantom arm had been frozen in an awkward position (Ramachandran & Rogers-Ramachandran, 2000), and Philip suffered great pain in his elbow.

Could Philip's pain be relieved by teaching him to move his phantom arm? Knowing how important feedback is in movement (see Chapter 8), Dr. Ramachandran constructed a special feedback apparatus for Philip. This

Thinking Creatively

was a box divided in two by a vertical mirror. Philip was instructed to put his good right hand into the box through a hole in the front and view it through a hole in the top. When he looked at his hand, he could see it and its mirror image. He was instructed to put his phantom limb in the box and try to position it, as best he could, so that it corresponded to the mirror image of his good hand. Then, he was instructed to make synchronous, bilaterally symmetrical movements of his arms—his actual right arm and his phantom left arm—while viewing his good arm and its mirror image. Ramachandran sent Philip home with the box and instructed him to use it frequently. Three weeks later, Philip phoned.

"Doctor," he exclaimed, "it's gone. . . . You know the excruciating pain that I always had in my elbow? . . . Well, now I don't have an elbow and I don't have that pain anymore." (Ramachandran & Blakeslee, 1998, p. 49)

The Ironic Case of Professor P.: Recovery

If you remember the chapter-opening case study, I am sure you will appreciate why this chapter has special meaning for me. Writing it played a part in my recovery.

When I was released from the hospital, I had many problems related to my neurosurgery. I knew that I would have to live with hearing and balance problems because I no longer had a right auditory-vestibular nerve. My other problems concerned me more. The right side of my face sagged, and making facial expressions was difficult. My right eye was often painful—likely because of inadequate tearing. I had difficulty talking, and I experienced debilitating attacks of fatigue. Unfortunately, neither my neurosurgeon nor my medical doctor seemed to know how to deal with these problems, and I was pretty much left to fend for myself.

I used information that I had learned from writing this chapter. Little about recovery from brain damage had been proven, but the results of experiments on animal models were suggestive. I like to think that the program I devised contributed to my current good health, but, of course, there is no way of knowing for sure.

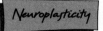 *Thinking Creatively*

I based my program of recovery on evidence that cognitive and physical exercise promotes recovery and other forms of neuroplasticity. My job constituted the cognitive part of my recovery program. Because I was influenced by recent evidence that the beneficial effects of exercise are greatest soon after the brain trauma, I returned to work 2 weeks after leaving the hospital.

Once back at work I got more mental, oral, and facial exercise than I had anticipated. A few conversations were enough to make my throat and face ache and to totally exhaust me, at which point I would retreat to my office until I was fit to emerge once again for more "treatment."

Being a university professor is not physically demanding—perhaps you've noticed. I needed some physical exercise, but my balance problems limited my options. I turned to African hand drumming. I love the rhythms, and I found that learning and playing them could be a serious cognitive and physical challenge—particularly for somebody as enthusiastic and inept as I. So I began to practice, take lessons, and drum with my new friends at every opportunity. Gradually, I worked, talked, smiled, and drummed myself to recovery. Today, my face is reasonably symmetrical, my speech is good, I am fit, and my balance has improved.

 Clinical Implications

Themes Revisited

This is the second chapter of the book to focus on neuroplasticity. It covered the neuroplastic changes associated with neurological disease and brain damage and the efforts to maximize various neuroplastic changes to promote recovery.

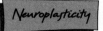 *Neuroplasticity*

Because this entire chapter dealt with clinical issues, the clinical implications tab made numerous appearances. In particular, it drew attention to the many cases in the chapter: the ironic case of Professor P.; Jerry Quarry, the punch-drunk ex-boxer; the cases of complex partial epilepsy; the cases of MPTP poisoning; and Tom and Philip, the amputees with phantom limbs.

 Clinical Implications

The chapter stressed creative thinking in several places. Attention was drawn to thinking about the relation between genes and Parkinson's disease, about the need to identify the primary symptom of Alzheimer's disease, about the applicability of animal models to humans, and about the correlation between exercise and recovery of function after nervous system damage. Particularly interesting were the creative approaches that Dr. Ramachandran took in treating Tom and Philip, who suffered from phantom limb pain.

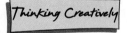 *Thinking Creatively*

The evolutionary perspective theme was also highlighted at several points. You were introduced to the concept of animal models, which is based on the comparative approach, and you learned that most of the research on neural regeneration and reorganization following brain damage has been done with animal models. Finally, you learned that research into the mechanisms of neural regeneration has been stimulated by the fact that this process occurs accurately in some species.

Evolutionary Perspective

Think about It

1. An epileptic is brought to trial for assault. The lawyer argues that her client is not a criminal and that the assaults in question were psychomotor attacks. She points out that her client takes her medication faithfully, but that it does not help. The prosecution lawyer argues that the defendant has a long history of violent assault and must be locked up. What do you think the judge should do?

2. Describe a bizarre incident you have observed that you think in retrospect might have been a complex partial or petit mal seizure.

3. The more that is known about a disease, the easier it is to diagnose; and the more accurately it can be diagnosed, the easier it is to find things out about it. Explain and discuss.

4. Total dementia often creates less suffering than partial dementia. Discuss.

5. Major breakthroughs in the treatment of CNS damage are on the horizon. Discuss.

6. The first evaluation of the effectiveness of neurotransplantation in the treatment of Parkinson's disease suggested that the treatment, as administered, was not effective. What do you think should be the next step?

Key Terms

10.1 Causes of Brain Damage

Tumor (neoplasm) (p. 241)
Meningiomas (p. 241)
Encapsulated tumors (p. 241)
Benign tumors (p. 242)
Infiltrating tumors (p. 242)
Malignant tumors (p. 242)
Metastatic tumors (p. 242)
Strokes (p. 242)
Penumbra (p. 243)
Cerebral hemorrhage (p. 243)
Aneurysm (p. 243)
Congenital (p. 243)
Cerebral ischemia (p. 243)
Thrombosis (p. 243)
Embolism (p. 243)
Arteriosclerosis (p. 243)
Glutamate (p. 243)
NMDA (N-methyl-D-aspartate) receptors (p. 244)
Contusions (p. 245)
Hematoma (p. 245)
Contrecoup injuries (p. 245)
Concussion (p. 245)
Punch-drunk syndrome (p. 245)

Dementia (p. 245)
Encephalitis (p. 245)
Meningitis (p. 245)
General paresis (p. 245)
Toxic psychosis (p. 246)
Tardive dyskinesia (TD) (p. 246)
Down syndrome (p. 246)
Apoptosis (p. 247)

10.2 Neuropsychological Diseases

Epilepsy (p. 247)
Convulsions (p. 247)
Epileptic auras (p. 248)
Partial seizure (p. 248)
Simple partial seizures (p. 248)
Complex partial seizures (p. 248)
Generalized seizures (p. 249)
Grand mal seizure (p. 249)
Hypoxia (p. 249)
Petit mal seizure (p. 249)
3-per-second spike-and-wave discharge (p. 249)
Parkinson's disease (p. 249)
Substantia nigra (p. 249)

Nigrostriatal pathway (p. 249)
Striatum (p. 249)
Lewy bodies (p. 250)
L-dopa (p. 250)
Mitochondria (p. 250)
Deep brain stimulation (p. 250)
Subthalamic nucleus (p. 250)
Huntington's disease (p. 250)
Huntingtin (p. 250)
Huntingtin protein (p. 250)
Multiple sclerosis (MS) (p. 250)
Ataxia (p. 251)
Epidemiology (p. 251)
Alzheimer's disease (p. 251)
Amyloid (p. 252)

10.3 Animal Models of Human Neuropsychological Diseases

Kindling phenomenon (p. 254)
Epileptogenesis (p. 254)
Transgenic (p. 254)
MPTP (p. 255)
Deprenyl (p. 255)

10.4 Neuroplastic Responses to Nervous System Damage: Degeneration, Regeneration, Reorganization, and Recovery

Anterograde degeneration (p. 255)
Distal segment (p. 255)
Retrograde degeneration (p. 255)
Proximal segment (p. 255)
Transneuronal degeneration (p. 256)
Neural regeneration (p. 257)
Schwann cells (p. 257)
Oligodendroglia (p. 258)
Collateral sprouting (p. 258)

10.5 Neuroplasticity and the Treatment of Nervous System Damage

Estrogens (p. 262)
Enriched environments (p. 264)
Phantom limb (p. 264)

✔•ⵌQuick Review Test your comprehension of the chapter with this brief practice test. You can find the answers to these questions as well as more practice tests, activities, and other study resources at www.mypsychlab.com.

1. All meningiomas are
   a. malignant.
   b. infiltrating.
   c. encapsulated.
   d. both a and b
   e. none of the above

2. Cerebral ischemia is caused by
   a. thrombosis.
   b. embolism.
   c. arteriosclerosis.
   d. all of the above
   e. both a and b

3. The Mad Hatter suffered from
   a. lead poisoning.
   b. mercury poisoning.

   c. HIV infection.
   d. syphilis.
   e. tardive dyskinesia.

4. A 3-per-second spike-and-wave discharge is a sign of
   a. a complex partial seizure.
   b. Parkinson's disease.
   c. Huntington's disease.
   d. multiple sclerosis.
   e. a petit mal seizure.

5. The MPTP model is a model of
   a. kindling.
   b. Alzheimer's disease.
   c. epilepsy.
   d. multiple sclerosis.
   e. none of the above

11

Learning, Memory, and Amnesia
How Your Brain Stores Information

Learning and memory are two ways of thinking about the same thing: Both are neuroplastic processes; they deal with the ability of the brain to change its functioning in response to experience. **Learning** deals with how experience changes the brain, and **memory**

Neuroplasticity

deals with how these changes are stored and subsequently reactivated. Without the ability to learn and remember, we would experience every moment as if waking from a life-long sleep—each person would be a stranger, each act a new challenge, and each word incomprehensible.

This chapter focuses on the roles played by various brain structures in the processes of learning and memory. Our knowledge of these roles has come to a great extent

Watch
Memory Deficits
www.mypsychlab.com

from the study of neuropsychological patients with brain-damage–produced **amnesia** (any pathological loss of memory) and from research on animal models of the same memory problems.

11.1
Amnesic Effects of Bilateral Medial Temporal Lobectomy

Ironically, the person who contributed more than any other to our understanding of the neuropsychology of memory was not a neuropsychologist. In fact, although he collabo-

Clinical Implications

rated on dozens of studies of memory, he had no formal research training and not a single degree to his name. He was H.M., a man who in 1953, at the age of 27, had the medial portions of his temporal lobes removed for the treatment of a severe case of epilepsy. Just as the Rosetta Stone provided archaeologists with important clues to the meaning of Egyptian hieroglyphics, H.M.'s memory deficits were instrumental in the achievement of our current understanding of the neural bases of memory (see Corkin, 2002).

The Case of H.M., the Man Who Changed the Study of Memory

During the 11 years preceding his surgery, H.M. suffered an average of one generalized seizure each week and many partial seizures each day, despite massive doses of anticonvulsant medication. Electroencephalography suggested that H.M.'s convulsions arose from foci in the medial portions of both his left and right temporal lobes. Because the removal of one medial temporal lobe had proved to be an effective treatment for patients with a unilateral temporal lobe focus, the decision was made to perform a **bilateral medial temporal lobectomy**—the removal of the medial portions of

both temporal lobes, including most of the **hippocampus**, **amygdala**, and adjacent cortex (see Figure 11.1). (A **lobectomy** is an operation in which a lobe, or a major part of one, is removed from the brain; a **lobotomy** is an operation in which a lobe, or a major part of one, is separated from the rest of the brain by a large cut but is not removed.)

In several respects, H.M.'s bilateral medial temporal lobectomy was an unqualified success. His generalized seizures were all but eliminated, and the incidence of partial seizures was reduced to one or two per day, even though the level of his anticonvulsant medication was substantially reduced. Furthermore, H.M. entered surgery a reasonably well-adjusted individual with normal perceptual and motor abilities and superior intelligence, and he left it in the same condition. Indeed, H.M.'s IQ increased from 104 to 118 as a result of his surgery, presumably because of the decline in the incidence of his seizures. Be that as it may, H.M. was the last patient to receive a bilateral medial temporal lobectomy—because of its devastating amnesic effects.

In assessing the amnesic effects of brain surgery, it is usual to administer tests of the patient's ability to remember things learned before the surgery and tests of the patient's ability to remember things learned after the surgery. Deficits on the former tests lead to a diagnosis of **retrograde** (backward-acting) **amnesia**; those on the latter tests lead to a diagnosis of **anterograde** (forward-acting) **amnesia**. If a patient is found to have anterograde amnesia, the next step is usually to determine whether the difficulty

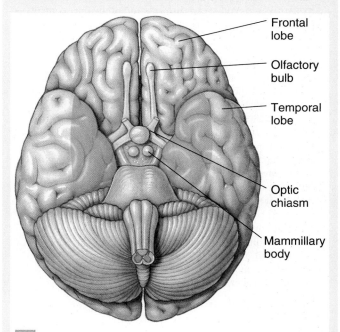

Frontal lobe

Olfactory bulb

Temporal lobe

Optic chiasm

Mammillary body

Tissue typically excised in medial temporal lobectomy

FIGURE 11.1 Medial temporal lobectomy. The portions of the medial temporal lobes that were removed from H.M.'s brain are illustrated in a view of the inferior surface of the brain.

in storing new memories influences **short-term memory** (storage of new information for brief periods of time while a person attends to it), **long-term memory** (storage of new information once the person stops attending to it), or both.

Like his intellectual abilities, H.M.'s memory for events predating his surgery remained largely intact. Although he had a mild retrograde amnesia for those events that occurred in the 2 years before his surgery, his memory for more remote events (e.g., for the events of his childhood) was reasonably normal.

H.M.'s short-term anterograde memory also remained normal: For example, his **digit span,** the classic test of short-term memory (see Chapter 5), was six digits (Wickelgren, 1968)—this means that if a list of six digits was read to him, he could usually repeat the list correctly, but he would have difficulty repeating longer lists.

In contrast, H.M. had an almost total inability to form new long-term memories: Once he stopped thinking about a new experience, it was lost forever. In effect, H.M. became suspended in time on that day in 1953 when he regained his health but lost his future:

> As far as we can tell, this man has retained little if anything of events subsequent to the operation. . . . Ten months before I examined him, his family had moved from their old house to a new one a few blocks away on the same street. He still had not learned the new address (though remembering the old one perfectly), nor could he be trusted to find his way home alone. He did not know where objects in constant use were kept, and his mother stated that he would read the same magazines over and over again without finding their contents familiar. . . . [F]orgetting occurred the instant the patient's focus of attention shifted. (Milner, 1965, pp. 104–105)
>
> During three of the nights at the Clinical Research Center, the patient rang for the night nurse, asking her, with many apologies, if she would tell him where he was and how he came to be there. He clearly realized that he was in a hospital but seemed unable to reconstruct any of the events of the previous day. On another occasion he remarked "Every day is alone in itself, whatever enjoyment I've had, and whatever sorrow I've had." Our own impression is that . . . events fade for him long before the day is over. . . . His experience seems to be that of a person who is just becoming aware of his surroundings without fully comprehending the situation, because he does not remember what went before.
>
> He still fails to recognize people who are close neighbours or family friends but who got to know him only after the operation. . . . Although he gives his date of birth unhesitatingly and accurately, he always underestimates his own age and can only make wild guesses as to the date. (Milner, Corkin, & Teuber, 1968, pp. 216–217)

H.M. lived in a nursing home for many years. He spent much of each day doing crossword puzzles—his progress on a crossword puzzle was never lost because it was written down. H.M. died in 2008.

(From "Further analysis of the Hippocampal amnesic syndrome: 14 year follow-up study of H.M.," by Brenda Milner et al., *Neuropsychologia*, 6(3), Sept. 1968, 215–234, 8. Copyright © 1968. Reprinted by permission of Elsevier Ltd.)

Formal Assessment of H.M.'s Anterograde Amnesia: Discovery of Unconscious Memories

In order to characterize H.M.'s anterograde memory problems, researchers began by measuring his performance on objective tests of various kinds of memory. This subsection describes seven tests that were used to assess H.M.'s long-term memory. The results of the first two tests documented H.M.'s severe deficits in long-term memory, whereas the results of the last five indicated that H.M.'s brain was capable of storing long-term memories but that H.M. had no conscious awareness of those memories. This finding changed the way biopsychologists think about the brain and memory.

Digit Span + 1 Test   H.M.'s inability to form certain long-term memories was objectively illustrated by his performance on the *digit span + 1 test,* a classic test of verbal long-term memory. H.M. was asked to repeat 5 digits that were read to him at 1-second intervals. He repeated the sequence correctly. On the next trial, the same 5 digits were presented in the same sequence with 1 new digit added on the end. This same 6-digit sequence was presented a few times until he got it right, and then another digit was added to the end of it, and so on. After 25 trials, H.M. had not managed to repeat the 8-digit sequence. Normal subjects can correctly repeat about 15 digits after 25 trials of the digit span + 1 test (Drachman & Arbit, 1966).

◉ Watch
Digit Span Test
www.mypsychlab.com

Block-Tapping Memory-Span Test   H.M. had **global amnesia**—amnesia for information presented in all sensory modalities. Milner (1971) demonstrated that H.M.'s amnesia was not restricted to verbal material by assessing his performance on the + 1 version of the *block-tapping memory-span test.* An array of 9 blocks was spread out on a board in front of H.M., and he was asked to watch the neuropsychologist touch a sequence of them and then to repeat the same sequence of touches. H.M. had a *block-tapping span* of 5 blocks, which is in the normal range; but he could not learn to correctly touch a sequence of 6 blocks, even when the same sequence was repeated 12 times.

Mirror-Drawing Test   The first indication that H.M.'s anterograde amnesia did not involve all long-term memories came from the results of a *mirror-drawing test* (Milner, 1965). H.M.'s task was to draw a line within the boundaries of a star-shaped target by watching his hand in a mirror. H.M. was asked to trace the star 10 times on each of 3 consecutive days, and the number of times he went outside the boundaries on each trial was recorded. As Figure 11.2 shows, H.M.'s performance improved over the 3 days, which indicates retention of the task. However, despite his improved performance, H.M. could not recall ever having seen the task before.

FIGURE 11.2 The learning and retention of the mirror-drawing task by H.M. Despite his good retention of the task, H.M. had no conscious recollection of having performed it before. (Based on Milner, 1965.)

Rotary-Pursuit Test In the *rotary-pursuit test* (see Figure 11.3), the subject tries to keep the tip of a stylus in contact with a target that rotates on a revolving turntable. Corkin (1968) found that H.M.'s performance on the rotary-pursuit test improved significantly over 9 daily practice sessions, despite the fact that H.M. claimed each day that he had never seen the pursuit rotor before. His improved performance was retained over a 7-day retention interval.

Incomplete-Pictures Test The discovery that H.M. was capable of forming long-term memories for mirror drawing and rotary pursuit suggested that sensorimotor tasks were the one exception to his inability to form long-term memories. However, this view was challenged by the demonstration that H.M. could also form new long-term memories for the **incomplete-pictures test** (Gollin, 1960)—a nonsensorimotor test of memory that employs five sets of fragmented drawings. Each set contains drawings of the same 20 objects, but the sets differ in their degree of sketchiness: Set 1 contains the most fragmented drawings, and set 5 contains the complete drawings. The subject is asked to identify the 20 objects

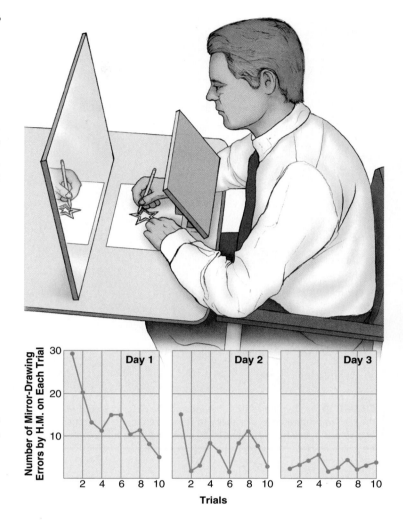

from the sketchiest set (set 1); then, those objects that go unrecognized are presented in their set 2 versions, and so on, until all 20 items have been identified. Figure 11.4 on page 272 illustrates the performance of H.M. on this test and his improved performance 1 hour later (Milner et al., 1968). Despite his improved performance, H.M. could not recall previously performing the task.

Pavlovian Conditioning H.M. learned an eye-blink Pavlovian conditioning task, albeit at a retarded rate (Woodruff-Pak, 1993). A tone was sounded just before a puff of air was administered to his eye; these trials were repeated until the tone alone elicited an eye blink. Two years later, H.M. retained this conditioned response almost perfectly, although he had no conscious recollection of the training.

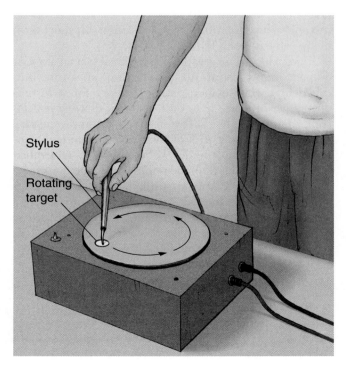

FIGURE 11.3 The rotary-pursuit task. The subject tries to keep the stylus in contact with the rotating target, and time-on-target is automatically recorded. H.M. learned and retained this task, although he had no conscious recollection of the learning trials.

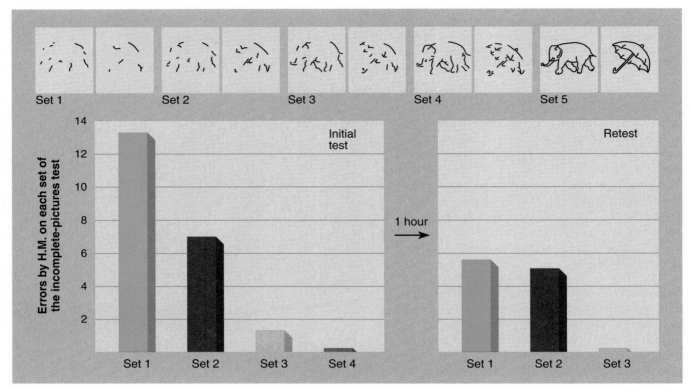

FIGURE 11.4 Two items from the incomplete-pictures test. H.M.'s memory for the 20 items on the test was indicated by his ability to recognize the more fragmented versions of them when he was retested. Nevertheless, he had no conscious awareness of having previously seen the items.

Three Major Scientific Contributions of H.M.'s Case

H.M.'s case is a story of personal tragedy, but his contributions to the study of the neural basis of memory were immense. The following three contributions proved to be particularly influential.

First, by showing that the medial temporal lobes play an especially important role in memory, H.M.'s case challenged the then prevalent view that memory functions are diffusely and equivalently distributed throughout the brain. In so doing, H.M.'s case renewed efforts to relate individual brain structures to specific *mnemonic* (memory-related) processes; in particular, H.M.'s case spawned a massive research effort aimed at clarifying the mnemonic functions of the hippocampus and other medial temporal lobe structures.

Second, the discovery that bilateral medial temporal lobectomy abolished H.M.'s ability to form certain kinds of long-term memories without disrupting his performance on tests of short-term memory or his **remote memory** (memory for experiences in the distant past) supported the theory that there are different modes of storage for short-term, long-term, and remote memory (see Nee et al.,

2008; Squire & Bayley, 2007). H.M.'s specific problem appeared to be a difficulty in **memory consolidation** (the translation of short-term memories into long-term memories).

⊙→ **Simulate** Memory **www.mypsychlab.com**

Third, H.M.'s case was the first to reveal that an amnesic patient might claim no recollection of a previous experience, while demonstrating memory for it by improved performance (e.g., on the mirror-drawing and incomplete-pictures tests). This discovery led to the creation of two distinct categories of long-term memories:. Conscious long-term memories became known as **explicit memories**, and long-term memories that are demonstrated by improved test performance without conscious awareness became known as **implicit memories**. As you will soon learn, this distinction is of general relevance: Many people with amnesia lose their ability to form explicit memories while maintaining their ability to form implicit memories.

Medial Temporal Lobe Amnesia

Neuropsychological patients with a profile of mnemonic deficits similar to those of H.M., with preserved intellectual

functioning, and with evidence of medial temporal lobe damage are said to suffer from **medial temporal lobe amnesia**.

Research on medial temporal lobe amnesia has shown that H.M.'s difficulty in forming explicit long-term memories while retaining the ability to form implicit long-term memories of the same experiences is not unique to him (see Eichenbaum, 1999). This problem has proved to be a symptom of medial temporal lobe amnesia, as well as many other amnesic disorders. As a result, the assessment of implicit long-term memories now plays an important role in the study of human memory (see Schacter, Dobbins, & Schnyer, 2004).

Tests that have been developed to assess implicit memory are called **repetition priming tests**. The incomplete-pictures test is an example, but repetition priming tests that involve memory for words are more common. First, the participants are asked to examine a list of words; they are not asked to learn or remember anything. Later, they are shown a series of fragments (e.g., _ O B _ _ E R) of words from the original list and are simply asked to complete them. Controls who have seen the original words perform well. Surprisingly, participants with amnesia often perform equally well, even though they have no explicit memory of seeing the original list. (By the way, the correct answer to the repetition priming example is "lobster.")

The discovery that there are two memory systems—explicit and implicit—raises an important question: Why do we have two parallel memory systems, one conscious (explicit) and one unconscious (implicit)? *Evolutionary Perspective* Presumably, the implicit system was the first to evolve because it is more simple (it does not involve consciousness), so the question is actually this: What advantage is there in having a second, conscious system?

Two experiments, one with amnesic patients (Reber, Knowlton, & Squire, 1996) and one with amnesic monkeys with medial temporal lobe lesions (Buckley & Gaffan, 1998), suggest that the answer is "flexibility." In both experiments, the amnesic subjects learned an implicit learning task as well as control subjects did; however, if they were asked to use their implicit knowledge in a different way or in a different context, they failed miserably. Presumably, the evolution of explicit memory systems provided for the flexible use of information.

Semantic and Episodic Memories

H.M. was able to form very few new explicit memories. However, most people with medial temporal lobe amnesia display memory deficits that are less complete. The study of these amnesics has found that explicit memories fall into two categories and that many of these amnesics tend to have far greater difficulties with one category than the other.

Explicit long-term memories come in two varieties: semantic and episodic (see Hampton & Schwartz, 2004; Rubin, 2006). **Semantic memories** are explicit memories for general facts or information; **episodic memories** are explicit memories for the particular events (i.e., episodes) of one's life (see Patterson, Nestor, & Rogers, 2007). People with medial temporal lobe amnesia have particular difficulty with episodic memories. In other words, they have difficulty remembering specific events from their lives, even though their memory for general information is often normal. Although they can't remember having lunch, going to a movie, chatting with a friend, or attending a lecture, they often remember what their friends are like, a movie they have seen, a language they learned, writing, world events, and the sorts of things learned at school.

Endel Tulving has been a major force in research on the semantic-episodic dichotomy (Tulving, 2002). Following is a description of Tulving's patient K.C. Episodic memory (also called *autobiographical memory*) has been likened to traveling back in time mentally and experiencing one's past.

The Case of K.C., the Man Who Can't Time Travel

K.C. had a motorcycle accident in 1981. He suffered diffuse brain damage, including damage to the medial temporal lobes. Despite severe amnesia, K.C.'s other cognitive abilities remain remarkably normal. His general intelligence and use of language are normal; he has no difficulty concentrating; he plays the organ, chess, and various card games; and his reasoning abilities are good. His knowledge of mathematics, history, science, geography, and other school subjects is good.

Similarly, K.C. has good retention of many of the facts of his early life. He knows his birth date, where he lived as a youth, where his parents' summer cottage was located, the names of schools he attended, the makes and colors of cars that he has owned.

Still, in the midst of these normal memories, K.C. has severe amnesia for personal experiences. He cannot recall a single personal event for more than a minute or two. This inability to recall any episodes (events) at which he was present covers his entire life. Despite these serious memory problems, K.C. has no difficulty having a conversation, and his memory problems are far less obvious to others than one would expect. Basically, he does quite well using his semantic memory.

K.C. understands time but cannot "time travel," into either the past or the future. He cannot imagine his future any better than he can recall his past: He can't imagine what he will be doing for the rest of the day, the week, or his life.

Vargha-Khadem and colleagues (1997) followed the maturation of three patients with medial temporal lobe amnesia who experienced bilateral medial temporal lobe damage early in life. Remarkably, although they could remember few of the experiences they had during their daily lives (episodic memory), they progressed through mainstream schools and acquired reasonable levels of language ability and factual knowledge (semantic memory). However, despite their academic success, their episodic memory did not improve (de Haan et al., 2006).

It is difficult to spot episodic memory deficits, even when the deficits are extreme. This occurs in part because neuropsychologists usually have no way of knowing the true events of a patient's life and in part because the patients become very effective at providing semantic answers to episodic questions. The following paraphrased exchange illustrates why neuropsychologists have difficulty spotting episodic memory problems.

The Case of the Clever Neuropsychologist: Spotting Episodic Memory Deficits

Neuropsychologist: I understand that you were a teacher.
Patient: That's right, I taught history.
Neuropsychologist: You must have given some good lectures in your time. Can you recall one of them that stands out?
Patient: Sure. I have given thousands of lectures. I especially liked Greek history.
Neuropsychologist: Was there any particular lecture that stood out—perhaps because it was very good or because something funny happened?
Patient: Oh, yes. Many stand out. My students liked my lectures—at least some of them—and sometimes I was quite funny.
Neuropsychologist: But is there one—just one—that you remember? And can you tell me something about it?
Patient: Oh yes, no problem. I didn't understand what you wanted. I can remember giving lectures and all my students were there watching and smiling.
Neuropsychologist: But can you describe a lecture where something happened that never happened in any other lecture? Perhaps something funny or disturbing.
Patient: That's hard.
Neuropsychologist: Before I go, I have some news for you that I think you will like. I understand that you are a hockey fan and follow the Toronto Maple Leafs.
Patient: Jeez, you guys know everything.
Neuropsychologist: Last night was a great night for Toronto. They beat New York 6–0. Do you think that you can remember that score for me? I will ask you about it a bit later.
Patient: That's great news. I will have no problem remembering that.
[Neuropsychologist leaves the room and returns an hour later.]
Neuropsychologist: I asked you to remember something the last time we chatted. Do you remember it?
Patient: I don't think so. I seem to have forgotten. It must have been a long time ago.
Neuropsychologist: That's strange. Do you remember anything specific about our last meeting, or even when it was?
Patient: Yes, we chatted for a time, I think about my memory.
Neuropsychologist: I understand that you are a Toronto Maple Leafs fan. Are they a good team?
Patient: Yes, they are very good. I used to go to every game with my father when I was a kid. They had great players; they were fast skaters and worked very hard. Did you know that they beat the New York Rangers 6–0? Now that's good.

Effects of Cerebral Ischemia on the Hippocampus and Memory

Patients who have experienced **cerebral ischemia**—that is, have experienced an interruption of blood supply to their brains—often suffer from medial temporal lobe amnesia. R.B. is one such individual (Zola-Morgan, Squire, & Amaral, 1986).

Clinical Implications

The Case of R.B., Product of a Bungled Operation

At the age of 52, R.B. underwent cardiac bypass surgery. The surgery was bungled, and, as a consequence, R.B. suffered brain damage. The pump that was circulating R.B.'s blood to his body while his heart was disconnected broke down, and it was several minutes before a replacement arrived from another part of the hospital. R.B. lived, but the resulting ischemic brain damage left him amnesic.

Although R.B.'s amnesia was not as severe as H.M.'s, it was comparable in many aspects. R.B. died in 1983 of a heart attack, and a detailed postmortem examination of his brain was carried out with the permission of his family.

Obvious brain damage was restricted largely to the **pyramidal cell layer** of just one part of the hippocampus—the **CA1 subfield** (see Figure 11.5).

R.B.'s case suggested that hippocampal damage by itself can produce medial temporal lobe amnesia. However, this conclusion has been challenged—as you will learn later in this chapter.

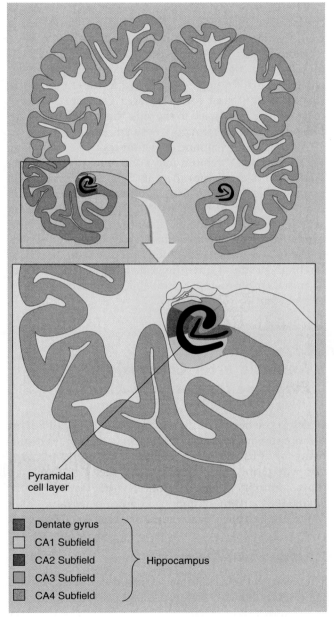

Pyramidal
cell layer

- ■ Dentate gyrus
- □ CA1 Subfield
- ■ CA2 Subfield   ⎫
- □ CA3 Subfield   ⎬ Hippocampus
- ▨ CA4 Subfield   ⎭

FIGURE 11.5 The major components of the hippocampus: CA1, CA2, CA3, and CA4 subfields and the dentate gyrus. R.B.'s brain damage appeared to be restricted largely to the pyramidal cell layer of the CA1 subfield. (CA stands for *cornu ammonis*, another name for hippocampus.)

11.2
Amnesia of Korsakoff's Syndrome

As you learned in Chapter 1, **Korsakoff's syndrome** is a disorder of memory that is common in people who have consumed large amounts of alcohol; the disorder is largely attributable to the brain damage associated with the thiamine deficiency that often accompanies heavy alcohol consumption. In its advanced stages, it is characterized by a variety of sensory and motor problems, extreme confusion, personality changes, and a risk of death from liver, gastrointestinal, or heart disorders. Postmortem examination typically reveals lesions to the *medial diencephalon* (the medial thalamus and the medial hypothalamus) and diffuse damage to several other brain structures, most notably the neocortex, hippocampus, and cerebellum (e.g., Sullivan & Marsh, 2003).

The amnesia of Korsakoff's syndrome is similar to medial temporal lobe amnesia in some respects. For example, during the early stages of the disorder, anterograde amnesia for explicit episodic memories is the most prominent symptom. However, as the disorder progresses, severe retrograde amnesia, which can extend back into childhood, also develops.

The gradual, insidious onset and progressive development of Korsakoff's syndrome complicate the study of the resulting retrograde amnesia. It is never entirely clear to what extent Korsakoff amnesia for recent events reflects the ret-

Clinical Implications

rograde disruption of existing memories or the gradually increasing anterograde blockage of the formation of new ones.

Because the brain damage associated with Korsakoff's syndrome is diffuse, it has not been easy to identify which part of it is specifically responsible for the amnesia. The first hypothesis, which was based on several small postmortem studies, was that damage to the *mammillary bodies* of the hypothalamus was responsible for the memory deficits of Korsakoff patients; however, subsequent studies revealed cases of Korsakoff amnesia with no mammillary body damage. But in all of these exceptional cases, there was damage to another pair of medial diencephalic nuclei: the **mediodorsal nuclei** of the thalamus. However, it is unlikely that the memory deficits of Korsakoff patients are attributable to the damage of any single diencephalic structure (see Tsivilis et al., 2008; Vann & Aggleton, 2004).

N.A. is a particularly well-known patient with **medial diencephalic amnesia** (amnesia, such as Korsakoff amnesia, associated with damage to the medial diencephalon). Although his memory deficits were conventional, their cause was not.

The Up-Your-Nose Case of N.A.

After a year of junior college, N.A. joined the U.S. Air Force; he served as a radar technician until his accident in December of 1960. On that fateful day, N.A.

was assembling a model airplane in his barracks room. His roommate had removed a miniature fencing foil from the wall and was making thrusts behind N.A.'s chair. N.A. turned suddenly and was stabbed through the right nostril. The foil penetrated the cribriform plate [the thin bone around the base of the frontal lobes], taking an upward course to the left into the forebrain. (Squire, 1987, p. 177)

Clinical Implications

The examiners . . . noted that at first he seemed to be unable to recall any significant personal, national or international events for the two years preceding the accident, but this extensive retrograde amnesia appeared to shrink. . . . Two-and-a-half years after the accident, the retrograde amnesia was said to involve a span of perhaps two weeks immediately preceding the injury, but the exact extent of this retrograde loss was (and remains) impossible to determine. . . .

During . . . convalescence (for the first six to eight months after the accident), the patient's recall of day-to-day events was described as extremely poor, but "occasionally some items sprang forth uncontrollably; he suddenly recalled something he seemed to have no business recalling."

Since his injury, he has been unable to return to any gainful employment, although his memory has continued to improve, albeit slowly. (Teuber, Milner, & Vaughan, 1968, pp. 268–269)

An MRI of N.A.'s brain was taken in the late 1980s (Squire et al., 1989). It revealed extensive medial diencephalic damage, including damage to the mediodorsal nuclei and mammillary bodies.

11.3
Amnesia of Alzheimer's Disease

Alzheimer's disease is another major cause of amnesia. The first sign of Alzheimer's disease is often a mild deterioration of memory. However, the disorder is progressive: Eventually, *dementia* develops and becomes so severe that the patient is incapable of even simple activities (e.g., eating, speaking, recognizing a spouse, or bladder control). Alzheimer's disease is terminal.

Clinical Implications

Efforts to understand the neural basis of Alzheimer's amnesia have focused on *predementia Alzheimer's patients* (Alzheimer's patients who have yet to develop dementia). The memory deficits of these patients are

◉ **Watch**
Alzheimer's Smell Test
www.mypsychlab.com

more general than those associated with medial temporal lobe damage, medial diencephalic damage, or Korsakoff's syndrome (see Butters & Delis, 1995). In addition to major anterograde and retrograde deficits in tests of explicit memory, predementia Alzheimer's patients often display deficits in short-term memory and in some types of implicit memory: Their implicit memory for verbal and perceptual material is often deficient, whereas their implicit memory for sensorimotor learning is not (see Gabrieli et al., 1993; Postle, Corkin, & Growdon, 1996).

The level of acetylcholine is greatly reduced in the brains of Alzheimer's patients. This reduction results from the degeneration of the **basal forebrain** (a midline area located just above the hypothalamus; see Figure 11.17 on page 288), which is the brain's main source of acetylcholine. This finding, coupled with the finding that strokes in the basal forebrain area can cause amnesia (Morris et al., 1992), led to the view that acetylcholine depletion is the cause of Alzheimer's amnesia.

Although acetylcholine depletion resulting from damage to the basal forebrain may contribute to Alzheimer's amnesia, it is clearly not the only factor. The brain damage associated with Alzheimer's disease is extremely diffuse (see Figure 10.13 on page 252) and involves many areas, including the medial temporal lobe and the prefrontal cortex, which play major roles in memory. Furthermore, damage to some structures of the basal forebrain produces attentional deficits, which can easily be mistaken for memory problems (see Baxter & Chiba, 1999; Everitt & Robbins, 1997).

11.4
Amnesia after Concussion: Evidence for Consolidation

Blows to the head that do not penetrate the skull but are severe enough to produce *concussion* (a temporary disturbance of consciousness produced by a nonpenetrating head injury) are the most common causes of amnesia (see Levin, 1989). Amnesia following a nonpenetrating blow to the head is called **posttraumatic amnesia (PTA)**.

Clinical Implications

Posttraumatic Amnesia

The *coma* (pathological state of unconsciousness) following a severe blow to the head usually lasts a few seconds or minutes, but in severe cases it can last weeks. Once the patient regains consciousness, he or she experiences a period of confusion. Victims of concussion are typically not tested by a neuropsychologist until after the period of confusion—if they are tested at all. Testing usually reveals that the patient has permanent retrograde amnesia for the events that led up to the blow and permanent anterograde

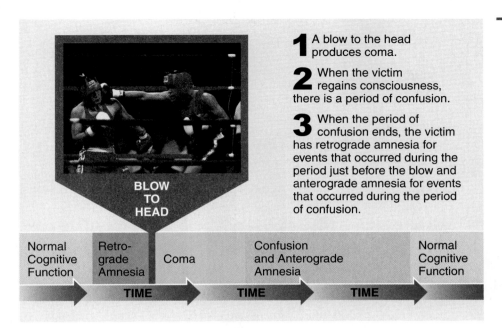

1. A blow to the head produces coma.

2. When the victim regains consciousness, there is a period of confusion.

3. When the period of confusion ends, the victim has retrograde amnesia for events that occurred during the period just before the blow and anterograde amnesia for events that occurred during the period of confusion.

| Normal Cognitive Function | Retro-grade Amnesia | Coma | Confusion and Anterograde Amnesia | Normal Cognitive Function |

TIME → TIME → TIME →

FIGURE 11.6 The retrograde amnesia and anterograde amnesia that are associated with a concussion-producing blow to the head.

amnesia for many of the events that occurred during the subsequent period of confusion.

The anterograde memory deficits that follow a non-penetrating head injury are often quite puzzling to the friends and relatives who have talked to the patient during the period of confusion—for example, during a hospital visit. The patient may seem reasonably lucid at the time, because short-term memory is normal, but later may have no recollection whatsoever of the conversation.

Figure 11.6 summarizes the effects of a closed-head injury on memory. Note that the duration of the period of confusion and anterograde amnesia is typically longer than that of the coma, which is typically longer than the period of retrograde amnesia. More severe blows to the head tend to produce longer comas, longer periods of confusion, and longer periods of amnesia (Levin, Papanicolaou, & Eisenberg, 1984). Not illustrated in Figure 11.6 are *islands of memory*—surviving memories for isolated events that occurred during periods for which other memories have been wiped out.

Gradients of Retrograde Amnesia and Memory Consolidation

Gradients of retrograde amnesia after concussion seem to provide evidence for *memory consolidation* (see Riccio, Millin, & Gisquet-Verrier, 2003). The fact

Neuroplasticity

that concussions preferentially disrupt recent memories suggests that the storage of older memories has been strengthened (i.e., consolidated).

The most prominent theory of memory consolidation is Hebb's theory. He argued that memories of experiences are stored in the short term by neural activity *reverberating* (circulating) in closed circuits. These reverberating patterns of neural activity are susceptible to disruption—for example, by

a blow to the head—but eventually they induce structural changes in the involved synapses, which provide stable long-term storage.

Electroconvulsive shock seemed to provide a controlled method of studying memory consolidation. **Electroconvulsive shock (ECS)** is an intense, brief, diffuse, seizure-inducing current that is administered to the brain through large electrodes attached to the scalp. The rationale for using ECS to study memory consolidation was that by disrupting neural activity, ECS would erase from storage only those memories that had not yet been converted to structural synaptic changes; the length of the period of retrograde amnesia produced by an ECS would thus provide an estimate of the amount of time needed for memory consolidation.

Many studies have employed ECS to study consolidation. Some studies have been conducted on human patients, who receive ECS for the treatment of depression. However, the most well-controlled studies have been conducted with laboratory animals.

Evolutionary Perspective

In one such study, thirsty rats were placed for 10 minutes on each of 5 consecutive days in a test box that contained a small niche. By the fifth of these habituation sessions, most rats explored the niche only 1 or 2 times per session. On the sixth day, a water spout was placed in the niche, and each rat was allowed to drink for 15 seconds after it discovered the spout. This was the learning trial. Then, 10 seconds, 1 minute, 10 minutes, 1 hour, or 3 hours later, each experimental rat received a single ECS. The next day, the retention of all subjects was assessed on the basis of how many times each explored the niche when the water spout was not present. The control rats that experienced the learning trial but received no ECS explored the empty niche an average of 10 times during the 10-minute test session, thereby indicating that they remembered their discovery of water the previous day. The rats that had received ECS 1 hour or 3 hours after the learning trial also explored the niche about 10 times. In contrast, the rats that received the ECS 10 seconds, 1 minute, or 10 minutes after the learning trial explored the empty niche significantly less on the test day. This result suggested that the consolidation of the memory of the learning trial took between 10 minutes and 1 hour (see Figure 11.7).

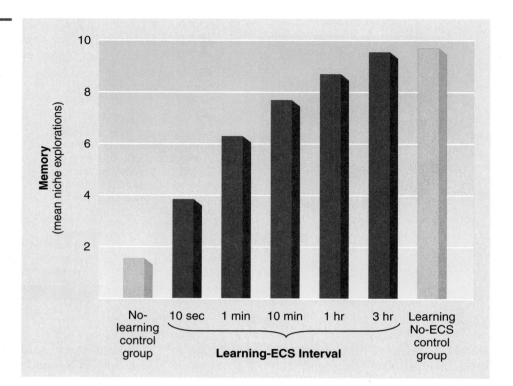

FIGURE 11.7   A short gradient of ECS-produced retrograde amnesia. Retention of one-trial learning by a control group of rats and by groups of rats that received ECS at various intervals after the learning trial. The rats that received no ECS displayed memory by exploring a niche where they had encountered water. Only the rats that received ECS within 10 minutes of the learning trial displayed significant retrograde amnesia for it. (Based on Pinel, 1969.)

Numerous variations of this experiment were conducted in the 1950s and 1960s, with different learning tasks, different species, and different numbers and intensities of electroconvulsive shocks. Initially, there was some consistency in the findings: Most seemed to suggest a rather brief consolidation time of a few minutes or less (e.g., Chorover & Schiller, 1965). But some researchers observed very long gradients of ECS-produced retrograde amnesia. For example, Squire, Slater, and Chace (1975) measured the memory of a group of ECS-treated patients for television shows that had played for only one season in different years prior to their electroconvulsive therapy. They tested each subject twice on different forms of the test: once before they received a series of five electroconvulsive shocks and once after. The difference between the before-and-after scores served as an estimate of memory loss for the events of each year. Figure 11.8 illustrates that five electroconvulsive shocks disrupted the retention of television shows that had played in the

3 years prior to treatment but not those that had played earlier.

Long gradients of retrograde amnesia are incompatible with Hebb's theory of consolidation. It is reasonable to think of the neural activity resulting from an experience reverberating through the brain for a few seconds or even

FIGURE 11.8   Demonstration of a long gradient of ECS-produced retrograde amnesia. A series of five electroconvulsive shocks produced retrograde amnesia for television shows that played for only one season in the 3 years before the shocks; however, the shocks did not produce amnesia for one-season shows that had played prior to that. (Based on Squire, Slater, & Chace, 1975.)

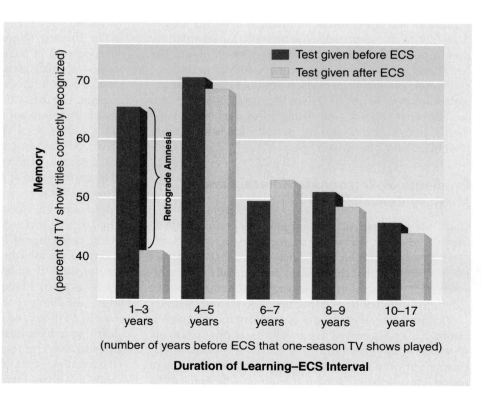

a few minutes; but gradients of retrograde amnesia covering days, weeks, or years cannot be easily accounted for by the disruption of reverberatory neural activity (e.g., Squire & Spanis, 1984). Long gradients of retrograde amnesia indicate that memory consolidation can continue for a very long time after learning, perhaps indefinitely.

Hippocampus and Consolidation

The discovery that H.M. seemed to be suffering from a temporally graded retrograde amnesia led Scoville and Milner (1957) to conclude that the hippocampus and related structures play a role in consolidation. To account for the fact that the bilateral medial temporal lobectomy disrupted only those retrograde memories acquired in the period just before H.M.'s surgery, they suggested that memories are temporarily stored in the hippocampus until they can be transferred to a more stable cortical storage system. This theory has become known as the **standard consolidation theory** (Moscovitch et al., 2006) and has been supported by several demonstrations that medial temporal lobe lesions produce temporally graded retrograde amnesia in experimental animals (e.g., Haist, Bowden, & Mao, 2001; Hanson, Bunsey, & Riccio, 2002; Squire, Clark, & Knowlton, 2001).

Several alternative theories of memory consolidation have been proposed (see James & MacKay, 2001). The **multiple-trace theory** of Nadel and Moscovitch (1997) is particularly compatible with the finding that gradients of retrograde amnesia are often very long. Nadel and Moscovitch proposed that the hippocampus and other structures involved in memory storage store memories for as long as they exist—not just during the period immediately after learning. When a conscious experience occurs, it is rapidly and sparsely encoded in a distributed fashion throughout the hippocampus and other involved structures. According to Nadel and Moscovitch, retained memories become progressively more resistant to disruption by hippocampal damage because each time a similar experience occurs or the original memory is recalled, a new **engram** (a change in the brain that stores a memory) is established and linked to the original engram, making the memory easier to recall and the original engram more difficult to disrupt.

Reconsolidation

One theoretical construct that has recently attracted researchers' attention is *reconsolidation* (see Lee, 2009; Nader & Hardt, 2009; Tronson & Taylor, 2007). The hypothesis is that each time a memory is retrieved from long-term storage, it is temporarily held in *labile* (changeable or unstable) short-term memory, where it is once again susceptible to posttraumatic amnesia until it is reconsolidated.

Interest in the process of reconsolidation originated with several studies in the 1960s, but then faded until a

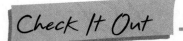

Memory Tests

Now that you have completed four sections about memory loss, why don't you assess your own memory capacities? Go to www.memorylossonline.com, where you will be able to assess your verbal and visual long-term memory. H.M. would score close to zero on the tests you will be taking.

key study by Nader, Schafe, and LeDoux (2000) rekindled it. These researchers infused the protein-synthesis inhibitor *anisomycin* into the amygdalae of rats shortly after the rats had been required to recall a fear conditioning trial. The infusion produced retrograde amnesia for the fear conditioning, even though the original conditioning trial had occurred days before. Most research on reconsolidation has involved fear conditioning, and some evidence suggests that not all kinds of memories are susceptible to reconsolidation (Biedenkapp & Rudy, 2004).

Scan Your Brain

This chapter is about to move from discussion of human memory disorders to consideration of animal models of human memory disorders. Are you ready? Scan your brain to assess your knowledge of human memory disorders by filling in the blanks in the following sentences. The correct answers are provided at the end of the exercise. Before proceeding, review the material related to your errors and omissions.

1. H.M. had his _____ temporal lobes removed.
2. The mirror-drawing test, the rotary-pursuit test, the incomplete-pictures test, and the repetition priming test are all tests of _____ memory.
3. H.M. was virtually incapable of forming new long-term _____ memories.
4. Support for the view that hippocampal damage can by itself cause amnesia comes from the study of R.B., who suffered _____ damage to the pyramidal cells of his CA1 hippocampal subfield.
5. The current view is that damage to the _____ diencephalon is responsible for most of the memory deficits of people with Korsakoff's disease.

6. The gradual onset of Korsakoff's syndrome complicates the study of the resulting _____ amnesia.

7. The _____ nuclei are the structures in the medial diencephalon that have been most frequently implicated in memory.

8. Alzheimer's disease is associated with degeneration of the basal forebrain and resulting depletion of _____.

9. Posttraumatic amnesia can be induced with _____ shock, which is used in the treatment of depression.

10. The transfer of a memory from short-term storage to long-term storage is termed _____.

11. Because some gradients of retrograde amnesia are extremely long, it is unlikely that memory consolidation is mediated by _____ neural activity, as hypothesized by Hebb.

12. The changes in the brain that store memories are called _____.

Scan Your Brain answers: (1) medial, (2) implicit, (3) explicit, (4) ischemic, (5) medial, (6) retrograde, (7) mediodorsal, (8) acetylcholine, (9) electroconvulsive, (10) consolidation, (11) reverberating, (12) engrams.

11.5

Neuroanatomy of Object–Recognition Memory

As interesting and informative as the study of patients with amnesia can be, it has major limitations. Many important questions about the neural bases of amnesia can be answered only by controlled experiments. For example, in order to identify the particular structures of the brain that participate in various kinds of memory, it is necessary to make precise lesions in various structures and to control what and when the subjects learn, and how and when their retention is tested. Because such experiments are not feasible with humans, there has been a major effort to develop animal models of human brain-damaged–produced amnesia.

The first reports of H.M.'s case in the 1950s triggered a massive effort to develop an animal model of his disorder so that it could be subjected to experimental analysis. In its early years, this effort was a dismal failure; lesions of medial temporal lobe structures did not produce severe anterograde amnesia in rats, monkeys, or other nonhuman species.

In retrospect, there were two reasons for the initial difficulty in developing an animal model of medial temporal lobe amnesia. First, it was not initially apparent that H.M.'s anterograde amnesia did not extend to all kinds of long-term memory—that is, that it was specific to explicit long-term memories—and most animal memory tests that were widely used in the 1950s and

1960s were tests of implicit memory (e.g., Pavlovian and operant conditioning). Second, it was incorrectly assumed that the amnesic effects of medial temporal lobe lesions were largely, if not entirely, attributable to hippocampal damage; and most efforts to develop animal models of medial temporal lobe amnesia thus focused on hippocampal lesions.

Thinking Creatively

Monkey Model of Object–Recognition Amnesia: The Delayed Nonmatching-to-Sample Test

Finally, in the mid 1970s, over two decades after the first reports of H.M.'s remarkable case, an animal model of his disorder was developed. It was hailed as a major breakthrough because it opened up the neuroanatomy of medial temporal lobe amnesia to experimental investigation.

Evolutionary Perspective

In separate laboratories, Gaffan (1974) and Mishkin and Delacour (1975) showed that monkeys with bilateral medial temporal lobectomies have major problems forming long-term memories for objects encountered in the **delayed nonmatching-to-sample test**. In this test, a monkey is presented with a distinctive object (the *sample object*), under which it finds food (e.g., a banana pellet). Then, after a delay, the monkey is presented with two test objects: the sample object and an unfamiliar object. The monkey must remember the sample object so that it can select the unfamiliar object to obtain food concealed beneath it. The correct performance of a trial is illustrated in Figure 11.9.

Intact, well-trained monkeys performed correctly on about 90% of delayed nonmatching-to-sample trials when the retention intervals were a few minutes or less. In contrast, monkeys with bilateral medial temporal lobe lesions had major object-recognition deficits (see Figure 11.10 on page 282). These deficits modeled those of H.M. in key respects. For example, the monkeys' performance was normal at delays of a few seconds but fell off to near chance levels at delays of several minutes, and their performance was extremely susceptible to the disruptive effects of distraction (Squire & Zola-Morgan, 1985). In fact, humans with medial temporal lobe amnesia have been tested on the delayed nonmatching-to-sample test—their rewards were coins rather than banana pellets—and their performance mirrored that of monkeys with similar brain damage.

The development of the delayed nonmatching-to-sample test for monkeys provided a means of testing the assumption that the amnesia resulting from medial temporal lobe damage is entirely the consequence of hippocampal damage—Figure 11.11 on page 282 illustrates the locations in the monkey brain of the three major temporal lobe structures: hippocampus, amygdala, and adjacent **rhinal cortex**. But before we consider this important

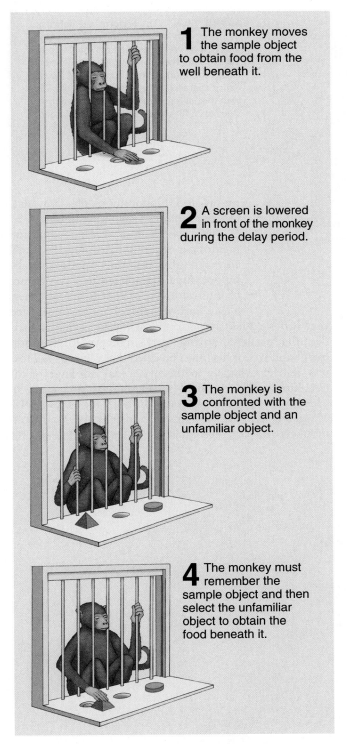

1 The monkey moves the sample object to obtain food from the well beneath it.

2 A screen is lowered in front of the monkey during the delay period.

3 The monkey is confronted with the sample object and an unfamiliar object.

4 The monkey must remember the sample object and then select the unfamiliar object to obtain the food beneath it.

FIGURE 11.9 The correct performance of a delayed nonmatching-to-sample trial. (Based on Mishkin & Appenzeller, 1987.)

line of research, we need to look at another important methodological development: the rat version of the delayed nonmatching-to-sample test.

Delayed Nonmatching–to–Sample Test for Rats

In order to understand why the development of the rat version of the delayed nonmatching-to-sample test played an important role in assessing the specific role of hippocampal damage in medial temporal lobe amnesia, examine Figure 11.12 on page 283, which illustrates the usual methods of making hippocampal lesions in monkeys and rats. Because of the size and location of the hippocampus, almost all studies of hippocampal lesions in monkeys have involved *aspiration* (suction) of large portions of the rhinal cortex in addition to the hippocampus. However, in rats, the extraneous damage associated with aspiration lesions of the hippocampus is typically limited to a small area of parietal neocortex. Furthermore, the rat hippocampus is small enough that it can be lesioned electrolytically or with intracerebral neurotoxin injections; in either case, there is little extraneous damage.

The version of the delayed nonmatching-to-sample test for rats that most closely resembles that for monkeys was developed by David Mumby using an apparatus that has become known as the **Mumby box**. This rat version of the test is illustrated in Figure 11.13 on page 284.

It was once assumed that rats could not perform a task as complex as that required for the delayed nonmatching-to-sample test; Figure 11.14 on page 285 indicates otherwise. Rats perform almost as well as monkeys with delays of up to 1 minute (Mumby, Pinel, & Wood, 1989).

The validity of the rat version of the delayed nonmatching-to-sample test has been established by studies of the effects of medial temporal lobe lesions. Combined bilateral lesions of rats' hippocampus, amygdala, and rhinal cortex produce major retention deficits at all but the shortest retention intervals (Mumby, Wood, & Pinel, 1992).

Neuroanatomical Basis of the Object-Recognition Deficits Resulting from Medial Temporal Lobectomy

To what extent are the object-recognition deficits following bilateral medial temporal lobectomy a consequence of hippocampal damage? In the early 1990s, researchers began assessing the relative effects on performance in the delayed nonmatching-to-sample test of lesions to various medial temporal lobe structures in both monkeys and rats. Challenges to the view that hippocampal damage is the critical factor in medial temporal amnesia quickly accumulated (e.g., Meunier et al., 1990; Mumby, Wood, & Pinel, 1992; Zola-Morgan et al., 1989). Most reviewers of the relevant research (Brown & Aggleton, 2001; Bussey & Saksida, 2005; Duva, Kornecook, & Pinel, 2000; Mumby, 2001; Murray, 1996; Murray & Richmond, 2001) have reached

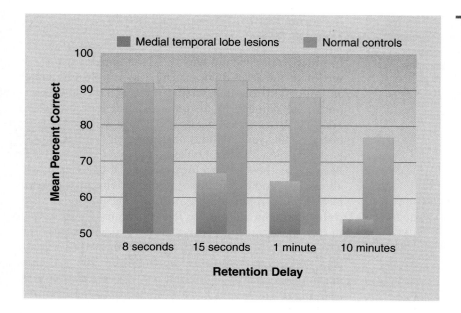

FIGURE 11.10 The performance deficits of monkeys with large bilateral medial temporal lobe lesions on the delayed nonmatching-to-sample test. There were significant deficits at all but the shortest retention interval. These deficits parallel the memory deficits of humans with medial temporal lobe amnesia on the same task. (Based on Squire & Zola-Morgan, 1991.)

similar conclusions: Bilateral surgical removal of the rhinal cortex consistently produces severe and permanent deficits in performance on the delayed nonmatching-to-sample test and other tests of object recognition. In contrast, bilateral surgical removal of the hippocampus produces either moderate deficits or none at all, and bilateral destruction of the amygdala has no effect. Figure 11.15 on page 285 compares the effects of rhinal cortex lesions on object recog-

nition and hippocampus-plus-amygdala lesions on object recognition in rats.

The reports that object-recognition memory is severely disrupted by rhinal cortex lesions but only moderately by hippocampal lesions led to a resurgence of interest in the case of R.B. and others like it. Earlier in this chapter, you learned that R.B. was left amnesic following an ischemic accident that occurred during heart surgery and that subsequent analysis of his brain revealed that obvious cell loss was restricted largely to the pyramidal cell layer of his CA1 hippocampal subfield (see Figure 11.5 on page 275). This result has been replicated in both monkeys (Zola-Morgan et al., 1992) and rats (Wood et al., 1993). In both monkeys and rats, cerebral ischemia leads to a loss of CA1

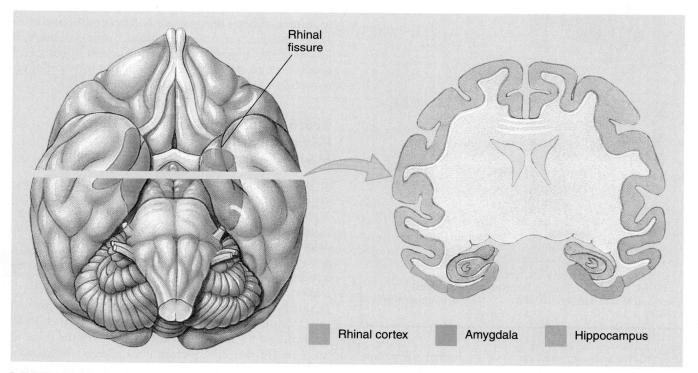

FIGURE 11.11 The three major structures of the medial temporal lobe, illustrated in the monkey brain: the hippocampus, the amygdala, and the rhinal cortex.

FIGURE 11.12 Aspiration lesions of the hippocampus in monkeys and rats. Because of differences in the size and location of the hippocampus in monkeys and in rats, hippocampectomy typically involves the removal of large amounts of rhinal cortex in monkeys, but not in rats.

The Location of the Hippocampus

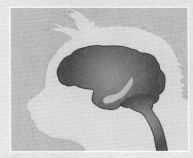
Monkeys

Rats

In monkeys, the hippocampus is usually removed by aspiration via the inferior surface of the brain, thus destroying substantial amounts of rhinal cortex.

In rats, aspiration of the hippocampus is usually performed via the dorsal surface of the brain, thus destroying small amounts of parietal neocortex.

hippocampal pyramidal cells and severe deficits in performance on the delayed nonmatching-to-sample test.

In summary, the relation between ischemia-produced hippocampal damage and object-recognition deficits in humans, monkeys, and rats seems to provide strong support for the theory that the hippocampus plays a key role in object-recognition memory. Although this conclusion is widely accepted, there is a gnawing problem with it: How can ischemia-produced lesions to one small part of the hippocampus be associated with severe deficits in performance on the delayed nonmatching-to-sample test when the deficits associated with total removal of the hippocampus are only slight?

Thinking Creatively

Mumby and colleagues (1996) conducted an experiment that appears to resolve this paradox. They hypothesized that: (1) the ischemia-produced hyperactivity of CA1 pyramidal cells damages neurons outside the hippocampus, possibly through the excessive release of excitatory amino acid neurotransmitters; (2) this extrahippocampal damage is not readily detectable by conventional histological analysis (i.e., it does not involve concentrated cell loss); and (3) this extrahippocampal damage is largely responsible for the object-recognition deficits that are produced by cerebral ischemia. Mumby and colleagues supported their hypothesis by showing that bilateral hippocampectomy actually blocks the development of ischemia-produced deficits in performance on the delayed nonmatching-to-sample test. First, they produced cerebral ischemia in rats by temporarily tying off their carotid arteries. Then, one group of the ischemic rats received a bilateral hippocampectomy 1 hour later, a second group received a bilateral hippocampectomy 1 week later, and a third group received no bilateral hippocampectomy. Following recovery, the latter two groups of ischemic rats displayed severe object-recognition deficits, whereas the rats whose hippocampus had been removed 1 hour after ischemia did not. Explaining how hippocampectomy can prevent the development of the object-recognition

Thinking Creatively

deficits normally produced by cerebral ischemia is a major problem for the theory that the hippocampus plays a critical role in object-recognition memory.

The most obvious damage following cerebral ischemia is in the CA1 subfield of the hippocampus, but there is substantial damage to other areas that is more diffuse and thus more difficult to quantify (see Fujioka et al., 1997; Katsumata et al., 2006; van Groen et al., 2005). Therefore, the presence of object-recognition deficits following cerebral ischemia does not constitute strong evidence for the involvement of the hippocampus in object-recognition memory. A study by Allen and colleagues (2006) made a similar point. They found that those patients with a greatly reduced hippocampal volume were much more likely to suffer from anterograde amnesia; however, these same patients also tended to have extensive neocortical damage.

Rhinal cortex comprises two areas: **entorhinal cortex** (the cortex within the rhinal fissure) and **perirhinal cortex** (the cortex around the rhinal fissure). Evidence indicates that it is the perirhinal cortex that plays the role in object recognition. For example, reviews by Bussey and Saksida (2005) and Murray, Bussey, and Saksida (2007) concluded that the hippocampus plays a role in spatial cognition (as you will learn in the next section) and that

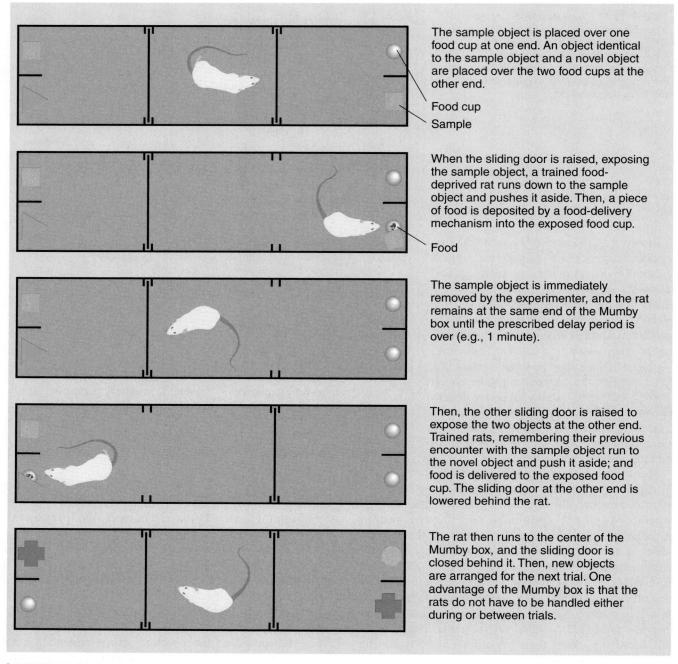

The sample object is placed over one food cup at one end. An object identical to the sample object and a novel object are placed over the two food cups at the other end.

Food cup

Sample

When the sliding door is raised, exposing the sample object, a trained food-deprived rat runs down to the sample object and pushes it aside. Then, a piece of food is deposited by a food-delivery mechanism into the exposed food cup.

Food

The sample object is immediately removed by the experimenter, and the rat remains at the same end of the Mumby box until the prescribed delay period is over (e.g., 1 minute).

Then, the other sliding door is raised to expose the two objects at the other end. Trained rats, remembering their previous encounter with the sample object run to the novel object and push it aside; and food is delivered to the exposed food cup. The sliding door at the other end is lowered behind the rat.

The rat then runs to the center of the Mumby box, and the sliding door is closed behind it. Then, new objects are arranged for the next trial. One advantage of the Mumby box is that the rats do not have to be handled either during or between trials.

FIGURE 11.13   The Mumby box and the rat version of the delayed nonmatching-to-sample test.

the perirhinal cortex plays a role in object recognition (see Albasser et al., 2009; Eichenbaum, Yonelinas, & Ranganath, 2007; Good et al., 2007; Norman & Eacott, 2005). The reviews further concluded that the function of perirhinal cortex is to construct and maintain complex representations of objects that are used in a variety of processes (e.g., perception and categorization) in addition to memory (see Eacott, Machin, & Gaffan, 2001; Hampton & Murray, 2002).

11.6
Hippocampus and Memory for Spatial Location

As you have just read, the perirhinal cortex, rather than the hippocampus, plays a major role in object recognition. This discovery has led to a re-evaluation of the contributions of

FIGURE 11.14 A comparison of the performance of intact monkeys (Zola-Morgan, Squire, & Mishkin, 1982) and intact rats (Mumby, Pinel, & Wood, 1989) on the delayed nonmatching-to-sample test.

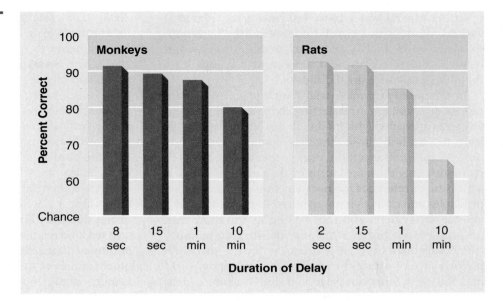

the hippocampus to memory, which has focused on the substantial evidence implicating the hippocampus in memory for spatial location.

Hippocampal Lesions Disrupt the Performance of Spatial Tasks

Bilateral lesions of the hippocampus in laboratory animals often have little or no effect on performance on memory

tests. But there is one exception to this general result: Hippocampal lesions consistently disrupt the performance of tasks that involve the memory for spatial location (e.g., Kaut & Bunsey, 2001; McDonald & White, 1993; O'Keefe, 1993). For example, hippocampal lesions disrupt performance on the Morris water maze test and the radial arm maze test.

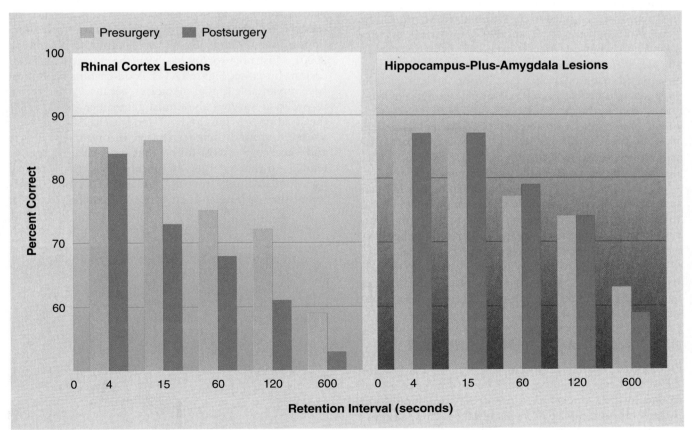

FIGURE 11.15 Effects of rhinal cortex lesions and hippocampus-plus-amygdala lesions in rats. Lesions of the rhinal cortex, but not of the hippocampus and amygdala combined, produced severe deficits in performance of the delayed nonmatching-to-sample test in rats. (Based on Mumby & Pinel, 1994; Mumby, Wood, & Pinel, 1992.)

In the **Morris water maze test**, intact rats placed at various locations in a circular pool of murky water rapidly learn to swim to a stationary platform hidden just below the surface. Rats with hippocampal lesions learn this simple task with great difficulty.

In the **radial arm maze test**, several (e.g., eight) arms radiate out from a central starting chamber, and the same few arms are baited with food each day. Intact rats readily learn to visit only those arms that contain food and do not visit the same arm more than once each day. The ability to visit only the baited arms of the radial arm maze is a measure of **reference memory** (memory for the general principles and skills that are required to perform a task), and the ability to refrain from visiting an arm more than once in a given day is a measure of **working memory** (temporary memory that is necessary for the successful performance of a task on which one is currently working). Rats with hippocampal lesions display major deficits on both the reference memory and the working memory measures of radial arm maze performance.

Hippocampal Place Cells

Consistent with the view that the hippocampus plays a role in spatial processing is the fact that many hippocampal neurons are **place cells** (Ahmed & Mehta, 2009; Fenton, 2007; Leutgeb et al., 2007)—neurons that respond only when a subject is in specific locations (i.e., in the *place fields* of the neurons). For example, when a rat is first placed in an unfamiliar test environment, none of its hippocampal neurons have a place field in that environment; then, as the rat familiarizes itself with the environment, many hippocampal neurons acquire a place field in it—that is, each fires only when the rat is in a particular part of the test environment. Each place cell has a place field in a different part of the environment.

By placing a rat in an ambiguous situation in a familiar test environment, it is possible to determine where the rat thinks it is from the route that it takes to get to the location in the environment where it has previously been rewarded. Using this strategy, researchers (Kubie et al., 2007; O'Keefe & Speakman, 1987; Wilson & McNaughton, 1993) have shown that the firing of a rat's place cells indicates where the rat "thinks" it is in the test environment, not necessarily where it actually is.

Hippocampal and Entorhinal Grid Cells

One line of research on hippocampal place cells has focused on the entorhinal cortex, a major source of neural signals to the hippocampus. An answer to the question of how hippocampal place cells obtain their spatial information came from the discovery of so-called grid cells in the entorhinal cortex. **Grid cells** are entorhinal neurons that each have an extensive array of evenly spaced place fields, producing a pattern reminiscent of graph paper (see Jeffrey, 2007; Moser, Kropff, & Moser,

2008). The distance between the evenly spaced place fields is flexible; in experimental animals kept in smaller environments, the fields are closer together (Barry et al., 2007). The even spacing of the place fields in the grid cells could enable spatial computations in hippocampal place cells. There are other types of neurons in the entorhinal cortex associated with spatial location: For example, *head-direction cells* are tuned to the direction of head orientation, and *border cells* fire when the subject is near the borders of its immediate environment (Solstad et al., 2008).

What is the relation between entorhinal grid cells and hippocampal place cells? There is no unequivocal answer to this question; however, it appears that entorhinal grid cells respond relatively reflexively to location, whereas hippocampal place cells respond to place in combination with other features of the test environment (see Buzsáki, 2005; Leutgeb et al., 2004; Poucet & Save, 2005; Wills et al., 2005).

Comparative Studies of the Hippocampus and Spatial Memory

Although most of the evidence that the hippocampus plays a role in spatial memory comes from research on rats, the hippocampus seems to perform a similar function in many other species (see Colombo & Broadbent, 2000). Most noteworthy has been the research on food-caching birds (see Figure 11.16). Food-caching birds must have remarkable spatial memories, because in order to survive, they must remember the locations of hundreds of food caches *Evolutionary Perspective* scattered around their territories. In one study, Sherry and Vaccarino (1989) found that food-caching species tended to have larger hippocampuses than related non–food-caching species. Indeed, Clayton (2001) found that caching and retrieving are required to trigger

FIGURE 11.16 A female acorn woodpecker caching an acorn beneath roof shingles.

hippocampal growth and maintain its size in mountain chickadees.

Although research on a variety of species indicates that the hippocampus does play a role in spatial memory, the evidence from primate studies has been less consistent. The hippocampal pyramidal cells of primates do have place fields (Rolls, Robertson, & Georges-François, 1995), but the effects of hippocampal damage on the performance of spatial memory tasks have not been consistent (e.g., Henke et al., 1999; Kessels et al., 2001; Maguire et al., 1998). The problem may be that spatial memory in humans and monkeys is often tested while they remain stationary and make judgments of locations on computer screens, whereas spatial memory in rats, mice, and birds is typically studied as subjects navigate through test environments (see Suzuki & Clayton, 2000).

Theories of Hippocampal Function

There are many theories of hippocampal function (see Bird & Burgess, 2008; Moscovitch et al., 2006), most of which acknowledge the important role of the hippocampus in spatial memory. The most influential of these has been the **cognitive map theory** of hippocampal function of O'Keefe and Nadel (1978). According to this theory, there are several systems in the brain that specialize in the memory for different kinds of information, and the specialization of the hippocampus is memory for spatial location. Specifically, O'Keefe and Nadel proposed that the hippocampus constructs and maintains allocentric maps of the external world from the sensory input that it receives. (*Allocentric* refers to representations of space based on relations among external objects and landmarks; in contrast, *egocentric* refers to representations of space based on relations of objects and landmarks to one's own position.) According to the cognitive map theory, the hippocampus plays a role in episodic memory because the spatial context plays a critical role in acquiring and recalling the memory of any episode.

Most subsequent theories of hippocampal function have focused on perceived shortcomings of the cognitive map theory. Here are three of the criticisms of the cognitive map theory (see Bannerman et al., 2004; Bird & Burgess, 2008; Shapiro, Kennedy, & Ferbinteanu, 2006):

- The firing of place cells has been found to depend on more than spatial location; it also depends on recent or pending behavior in their place fields.
- Hippocampal damage sometimes impairs the performance of tasks that have no obvious spatial component.
- The hippocampus is a large, complex structure, and thus unlikely to act as a unit with a single function. The functions of each part of the hippocampus need to be addressed.

11.7
Where Are Memories Stored?

So far, this chapter has focused on the involvement of medial temporal lobe structures in memory, and you have learned that structures such as the hippocampus and rhinal cortex appear to play major roles. This does not mean, however, that memories are stored in these structures.

The most straightforward method of identifying sites of memory storage in the brain is by assessing the ability of various brain lesions to produce retrograde amnesia. For example, if a particular structure were the storage site for all memories of a particular type, then destruction of that structure should eliminate all memories of that type that were acquired prior to the lesion. No brain structure has shown this result: Lesions of particular structures tend to produce either no retrograde amnesia at all or retrograde amnesia for only the experiences that occurred in the days or weeks just before the surgery. These findings have led to two major conclusions: (1) Memories are stored diffusely in the brain and thus can survive destruction of any single structure; and (2) memories become more resistant to disruption over time.

Another aspect of memory storage has made it difficult to use the lesion method to study this process. Each memory appears to be stored in the structures of the brain that participated in the original experience that created the memory (see Martin, 2006; Pasternak & Greenlee, 2005; Weinberger, 2004). Thus, substantial sensory and motor loss is often associated with lesion-produced retrograde amnesia, which can make it difficult to assess the memory loss. As a result, evidence implicating various structures in memory storage has tended to rely on recording and stimulation studies rather than on lesion studies.

So far, this chapter has focused on four neural structures that appear to play some role in the storage of memories: (1) The hippocampus and (2) the perirhinal cortex have roles in spatial and object memory, respectively; and (3) the mediodorsal nucleus and (4) the basal forebrain have been implicated in memory by Korsakoff's and Alzheimer's disease, respectively. In this section, we take a brief look at five other areas of the brain that have been implicated in memory storage: inferotemporal cortex, amygdala, prefrontal cortex, cerebellum, and striatum. See Figure 11.17 on page 288.

Inferotemporal Cortex

Areas of secondary sensory cortex are presumed to play an important role in storing sensory memories. For example, because the **inferotemporal cortex** (the cortex of the inferior temporal lobe) is involved in the visual perception of objects, it is thought to participate, in concert with the perirhinal cortex, in storing memories of visual patterns (see Bussey & Saksida, 2005). In support of this view, Naya,

FIGURE 11.17   The structures of the brain that have been shown to play a role in memory. Because it would have blocked the view of other structures, the striatum is not included. (See Figure 3.29 on page 70.)

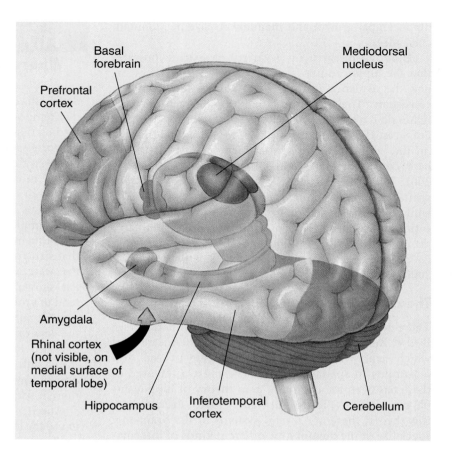

Yoshida, and Miyashita (2001) recorded the responses of neurons in inferotemporal cortex and perirhinal cortex while monkeys learned the relation between the two items in pairs of visual images. When a pair was presented, responses were first recorded in inferotemporal neurons and then in perirhinal neurons; however, when the monkeys were required to recall that pair, activity was recorded in perirhinal neurons before inferotemporal neurons. Naya and colleagues concluded that this reversed pattern of activity reflected the retrieval of visual memories from inferotemporal cortex.

Amygdala

The amygdala is thought to play a special role in memory for the emotional significance of experiences (Erlich et al., 2009; LaBar & Cabeza, 2006; Reijmers et al., 2007). Rats with amygdalar lesions, unlike intact rats, do not respond with fear to a neutral stimulus that has previously been followed by electric foot shock (see McGaugh, 2002; Medina et al., 2002). Also, Bechara and colleagues (1995) reported the case of a neuropsychological patient with bilateral damage to the amygdala who could not acquire conditioned autonomic startle responses to various visual or auditory stimuli but had good explicit memory for the training. However, there is little evidence that the amygdala stores memories; it appears to be involved in strengthening emotionally significant memories stored in other structures (Paz et al., 2006; Roozendaal, McEwen, & Chattarji, 2009). This involvement of the amygdala accounts for the fact that emotion-provoking events are remembered better than neutral events.

Prefrontal Cortex

Patients with damage to the **prefrontal cortex** (the area of frontal cortex anterior to motor cortex) are not grossly amnesic; they often display no deficits at all on conventional tests of memory. Stuss and Alexander (2005) have argued that different parts of the prefrontal cortex play different roles in memory, and that lumping all patients with prefrontal cortex damage together seriously clouds researchers' thinking.

Two episodic memory abilities are often lost by patients with large prefrontal lesions. Patients with large prefrontal lesions often display both anterograde and retrograde deficits in memory for the temporal order of events, even when they can remember the events themselves. They also display deficits in *working memory* (the ability to maintain relevant memories while a task is being completed)—see Kimberg, D'Esposito, and Farah (1998) and Smith (2000). As a result of these two deficits, patients with prefrontal cortex damage often have difficulty performing tasks that involve a series of responses (see Colvin, Dunbar, & Grafman, 2001).

The Case of the Cook Who Couldn't

The story of one patient with prefrontal cortex damage is very well known because she was the sister of Wilder Penfield, the famous Montreal neurosurgeon. Before her brain damage, she had been an excellent cook; and afterward, she retained all the requisite knowledge. She remembered her favorite recipes, and she remembered how to perform each individual cooking technique. However, she was incapable of preparing even simple meals because she could not carry out the various steps in proper sequence (Penfield & Evans, 1935).

The prefrontal cortex is a large structure that is composed of many anatomically distinct areas that have different connections and, presumably, different functions. Until recently, neuropsychological studies have tended to ignore this fact: The location of prefrontal damage has often not been determined accurately, and patients with different areas of damage have often been lumped together. However, the study of prefrontal cortex is changing. Functional brain-imaging studies are finding that specific complex patterns of prefrontal activity are associated with various memory functions. Some regions of prefrontal cortex perform fundamental cognitive processes (e.g., attention and task management) during working memory tasks, and other regions of prefrontal cortex seem to participate in other memory processes (Mitchell et al., 2004; Wig et al., 2005).

Cerebellum and Striatum

Just as explicit memories of experiences are presumed to be stored in the circuits of the brain that mediated their original perception, implicit memories of sensorimotor learning are presumed to be stored in sensorimotor circuits (see Graybiel, 2008). Most research on the neural mechanisms of memory for sensorimotor tasks have focused on two structures: the cerebellum and the striatum.

The **cerebellum** is thought to participate in the storage of memories of learned sensorimotor skills through its various neuroplastic mechanisms (Boyden, Katoh, & Raymond, 2004). Its role in the Pavlovian conditioning of the eye-blink response of rabbits has been most intensively investigated (see Delgado-García & Gruart, 2006). In this paradigm, a tone (conditional stimulus) is sounded just before a puff of air (unconditional stimulus) is delivered to the eye. After several trials, the tone comes to elicit an eye blink. The convergence of evidence from stimulation, recording, and lesion studies suggests that the effects of this conditioning are stored in the form of changes in the way that cerebellar neurons respond to the tone (see De Zeeuw & Yeo, 2005; Koekkoek et al., 2003).

The **striatum** is thought to store memories for consistent relationships between stimuli and responses—the type of memories that develop incrementally over many trials (see Laubach, 2005; White, 1997). Sometimes this striatum-based form of learning is referred to as *habit formation* (Schultz, Tremblay, & Hollermar, 2003).

Although few would disagree that the cerebellum and the striatum play a role in sensorimotor memory, there is growing evidence that these structures also play a role in certain types of memory with no obvious motor component (e.g., Maddox et al., 2005). For example, Knowlton, Mangels, and Squire (1996) found that the Parkinson's patients, who had striatal damage, could not solve a probabilistic discrimination problem.

The problem was a computer "weather forecasting" game, and the task was to correctly predict the weather by pressing one of two keys, rain or shine. The patients based their predictions on stimulus cards presented on the screen—each card had a different probability of leading to sunshine, which the patients had to learn and remember. The Parkinson's patients did not improve over 50 trials, although they displayed normal explicit (conscious) memory for the training episodes. In contrast, amnesic patients with medial temporal lobe or medial diencephalic damage displayed marked improvement in performance but had no explicit memory of their training.

Scan Your Brain

The preceding sections of this chapter have dealt with the gross neuroanatomy of memory—with the structures of the brain that are involved in various aspects. Before you proceed to the next section, which deals with the synaptic mechanisms of learning and memory, test your knowledge by writing the name of the relevant brain structure in each of the following blanks. The correct answers are provided at the end of the exercise. Before proceeding, review the material related to your errors and omissions.

1. Medial temporal lobe structure involved in spatial memory: _____
2. Area of rhinal cortex that participates in spatial memory through its hippocampal projections: _____
3. Area of rhinal cortex that plays a role in object recognition: _____.
4. Medial diencephalic nucleus that has been linked to Korsakoff's amnesia: _____.
5. Cholinergic area that has been linked to the memory problems of Alzheimer's disease: _____
6. Large area of neocortex that participates in the visual perception and memory of objects: _____.
7. Medial temporal lobe structure involved in the emotional significance of memories: _____.
8. Cortical area involved in memory for temporal order and in working memory: _____.
9. Mesencephalic structure that has been implicated in the retention of eye-blink conditioning and other learned sensorimotor skills: _____.
10. Subcortical structure that has been linked to habit formation: _____.

Scan Your Brain answers: (1) hippocampus, (2) entorhinal cortex, (3) perirhinal cortex, (4) mediodorsal nucleus, (5) basal forebrain, (6) inferotemporal cortex, (7) amygdala, (8) prefrontal cortex, (9) cerebellum, (10) striatum.

11.8

Synaptic Mechanisms of Learning and Memory

So far, this chapter has focused on the particular structures of the human brain that are involved in learning and memory and on what happens when these structures are damaged. In this section, the level of analysis changes: The focus shifts to the neuroplastic mechanisms within these structures that are thought to be the fundamental bases of learning and memory.

◉ Watch
Key Processes in the Stages of Memory
www.mypsychlab.com

Most modern thinking about the neural mechanisms of memory began with Hebb (1949). Hebb argued so convincingly that enduring changes in the efficiency of synaptic transmission were the basis of long-term memory that the search for the neural bases of learning and memory has focused almost exclusively on the synapse (see Kandel, 2001; Malenka, 2003).

Long-Term Potentiation

Because Hebb's hypothesis that enduring facilitations of synaptic transmission are the neural bases of learning and memory was so influential, there was great excitement when such an effect was discovered. In 1973, Bliss and Lømø showed that there is a facilitation of synaptic transmission following high-frequency electrical stimulation applied to presynaptic neurons. This phenomenon has been termed **long-term potentiation (LTP)**.

LTP has been demonstrated in many species and in many parts of their brains, but it has been most frequently studied in the rat hippocampus. Figure 11.18 illustrates three hippocampal synapses at which LTP has been commonly studied.

Figure 11.19 illustrates LTP in the granule cell layer of the rat hippocampal dentate gyrus. First, a single low-intensity pulse of current was delivered to the perforant path (the major input to the dentate gyrus), and the response was recorded through an extracellular multiple-unit electrode in the granule cell layer of the hippocampal dentate gyrus; the purpose of this initial stimulation was to determine the initial response baseline. Second, high-intensity, high-frequency stimulation lasting 10 seconds was delivered to the perforant path to induce the LTP. Third, the granule cells' responses to single pulses of low-intensity current were measured again after various delays. Figure 11.19 shows that transmission at the granule

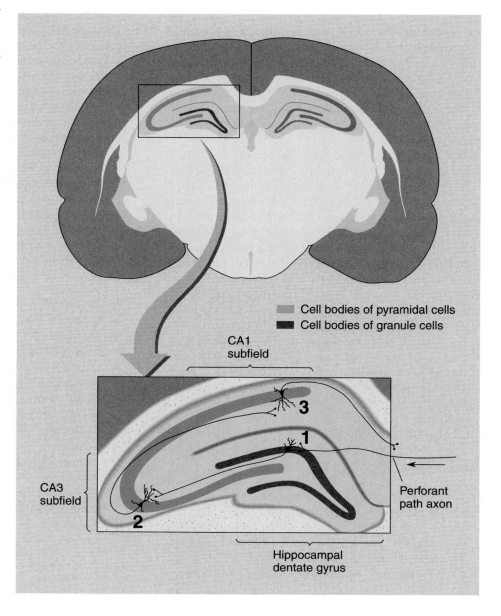

Cell bodies of pyramidal cells
Cell bodies of granule cells

CA1 subfield

3

1

CA3 subfield

2

Perforant path axon

Hippocampal dentate gyrus

FIGURE 11.18 A slice of rat hippocampal tissue that illustrates the three synapses at which LTP is most commonly studied: (1) the dentate granule cell synapse, (2) the CA3 pyramidal cell synapse, and (3) the CA1 pyramidal cell synapse.

FIGURE 11.19 Long-term potentiation in the granule cell layer of the rat hippocampal dentate gyrus. (Traces courtesy of Michael Corcoran, Department of Psychology, University of Saskatchewan.)

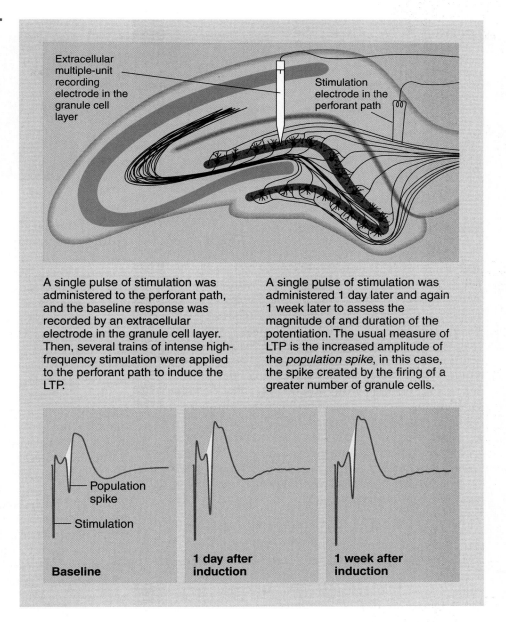

Extracellular multiple-unit recording electrode in the granule cell layer

Stimulation electrode in the perforant path

A single pulse of stimulation was administered to the perforant path, and the baseline response was recorded by an extracellular electrode in the granule cell layer. Then, several trains of intense high-frequency stimulation were applied to the perforant path to induce the LTP.

A single pulse of stimulation was administered 1 day later and again 1 week later to assess the magnitude of and duration of the potentiation. The usual measure of LTP is the increased amplitude of the *population spike*, in this case, the spike created by the firing of a greater number of granule cells.

— Population spike

— Stimulation

Baseline

1 day after induction

1 week after induction

cells' synapses was still potentiated 1 week after the high-frequency stimulation.

LTP is among the most widely studied neuroscientific phenomena. Why? The reason goes back to 1949 and Hebb's influential theory of memory. The synaptic changes that Hebb hypothesized as underlying long-term memory seemed to be the same kind of changes that underlie LTP (see Cooper, 2005).

LTP has two key properties that Hebb proposed as characteristics of the physiological mechanisms of learning and memory. First, LTP can last for a long time—for several months after multiple stimulations (see Abraham, 2006). Second, LTP develops only if the firing of the presynaptic neuron is followed by the firing of the postsynaptic neuron; it does not develop when the presynaptic neuron fires and the postsynaptic neuron does not, and it does not develop when the presynaptic neuron does not fire and the postsynaptic neuron does (see Bi & Poo, 2001). The *co-occurrence* of firing in presynaptic and postsynaptic cells is now recognized as the critical factor in LTP, and the assumption that co-occurrence is a physiological necessity for learning and memory is often referred to as *Hebb's postulate for learning.*

Additional support for the idea that LTP is related to the neural mechanisms of learning and memory has come from several observations (see Lisman, Lichtman, & Sanes, 2003; Lynch, 2004; Morris et al., 2003): (1) LTP can be elicited by low levels of stimulation that mimic normal neural activity; (2) LTP effects are most prominent in structures that have been implicated in learning and memory, such as the hippocampus; (3) behavioral conditioning can produce LTP-like changes in the hippocampus; (4) many drugs that influence learning and memory have parallel effects on LTP; (5) the induction of maximal LTP blocks the learning of a Morris water maze until the LTP has subsided; (6) mutant mice that display little hippocampal LTP have difficulty learning the Morris water maze; and (7) LTP occurs at specific synapses that have been shown to participate in learning and memory in simple invertebrate nervous systems. Still, it is important to keep in mind that all of this evidence is indirect and that LTP as induced in the laboratory by electrical stimulation is at best a caricature of the subtle cellular events that underlie learning and memory (see Cain, 1997; Eichenbaum, 1996; Shors & Matzel, 1997).

Conceiving of LTP as a three-part process, many researchers are investigating the mechanisms of *induction, maintenance,* and *expression*—that is, the processes by which high-frequency stimulations induce LTP (learning), the changes responsible for storing LTP (memory), and the changes that allow it to be expressed during the test (recall).

Resting   **Firing**   More calcium

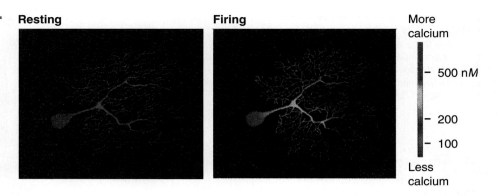

FIGURE 11.20 The influx of calcium ions into active neurons. This influx can be visualized with microfluorometric techniques. Notice that the greatest influx (measured in nanomolars) occurs in the axon terminal branches. (Courtesy of Tank et al., 1988.)

500 n*M*

200

100

Less calcium

Induction of LTP: Learning

The NMDA (or N-methyl-D-aspartate) receptor is prominent at the synapses at which LTP is commonly studied. The **NMDA receptor** is a receptor for **glutamate**—the main excitatory neurotransmitter of the brain, as you learned in Chapter 4. An NMDA receptor does not respond maximally unless two events occur simultaneously: Glutamate must bind to it, and the postsynaptic neuron must already be partially depolarized. This dual requirement stems from the fact that the calcium channels that are associated with NMDA receptors allow only small numbers of calcium ions to enter the neuron unless the neuron is already depolarized when glutamate binds to the receptors; it is the influx of calcium ions that triggers action potentials and the cascade of events in the postsynaptic neuron that induces LTP. The study of calcium influx has been greatly facilitated by the development of *optical imaging techniques* for visualizing it (see Figure 11.20).

An important characteristic of the induction of LTP at glutaminergic synapses stems from the nature of the NMDA receptor and the requirement for co-occurrence of firing for LTP to occur. This characteristic is not obvious under the usual, but unnatural, experimental condition in which LTP is induced by high-intensity, high-frequency stimulation, which always activates the postsynaptic neurons through massive temporal and spatial summation. However, when a more natural, low-intensity stimulation is applied, the postsynaptic neurons do not fire, and thus LTP is not induced—unless the postsynaptic neurons are already partially depolarized so that their calcium channels open wide when glutamate binds to their NMDA receptors.

The requirement for the postsynaptic neurons to be partially depolarized when the glutamate binds to them is an extremely important characteristic of conventional LTP because it permits neural networks to learn associations. Let me explain. If one glutaminergic neuron were to fire by itself and release its glutamate neurotransmitter across a synapse onto the NMDA receptors of a postsynaptic neuron, there would be no potentiation of transmission at that synapse because the postsynaptic cell would not fire.

However, if the postsynaptic neuron were partially depolarized by input from other neurons when the presynaptic neuron fired, the binding of the glutamate to the NMDA receptors would open wide the calcium channels, calcium ions would flow into the postsynaptic neuron, and transmission across the synapses between the presynaptic and postsynaptic neuron would be potentiated. Accordingly, the requirement for co-occurrence and the dependence of NMDA receptors on simultaneous binding and partial depolarization mean that, under natural conditions, synaptic facilitation records the fact that there has been simultaneous activity in at least two converging inputs to the postsynaptic neuron—as would be produced by the "simultaneous" presentation of a conditional stimulus and an unconditional stimulus.

The exact mechanisms by which calcium influx induces LTP are complex (see Lisman, 2003); however, there is substantial evidence that calcium exerts some of its effects by activating *protein kinases* (a class of enzymes that influence many chemical reactions of the cell) in the neural cytoplasm (see Kind & Neumann, 2001). A consistent finding has been that protein kinase inhibitors block the induction of LTP (see Bashir & Collingridge, 1992).

Figure 11.21 summarizes the induction of NMDA-receptor–mediated LTP. Although it is well established that the induction of LTP at synapses with NMDA receptors depends on the influx of calcium ions into the postsynaptic neuron, the next stages of the induction process are not well understood. This is probably because several mechanisms are involved (see Sheng & Kim, 2002).

Maintenance and Expression of LTP: Storage and Recall

The search for the mechanisms underlying the maintenance and expression of LTP began with attempts to determine whether these mechanisms occur in presynaptic or postsynaptic neurons. This question has been answered: The maintenance and expression of LTP involve changes in both presynaptic and postsynaptic neurons. This discovery indicated that the mechanisms underlying the maintenance and expression of LTP are complex. Indeed, after more than three decades of research, we still

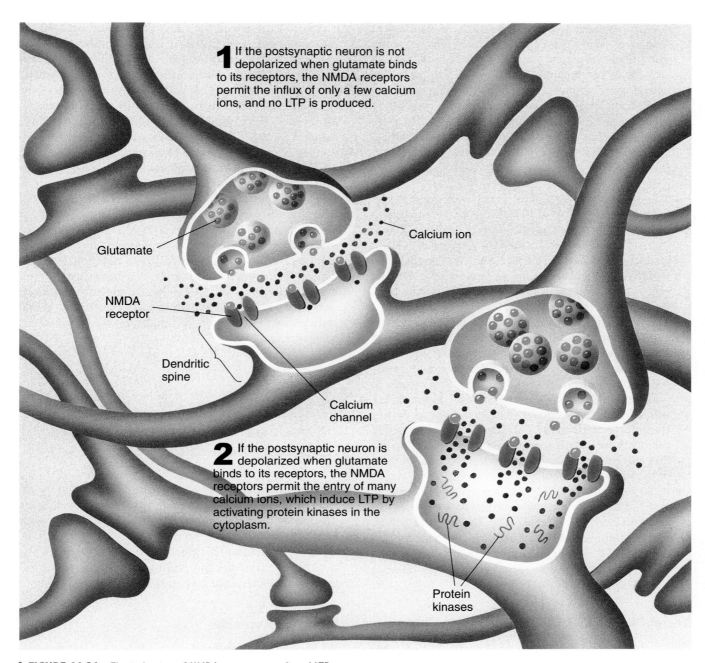

1 If the postsynaptic neuron is not depolarized when glutamate binds to its receptors, the NMDA receptors permit the influx of only a few calcium ions, and no LTP is produced.

Calcium ion

Glutamate

NMDA receptor

Dendritic spine

Calcium channel

2 If the postsynaptic neuron is depolarized when glutamate binds to its receptors, the NMDA receptors permit the entry of many calcium ions, which induce LTP by activating protein kinases in the cytoplasm.

Protein kinases

FIGURE 11.21   The induction of NMDA-receptor–mediated LTP.

do not have satisfactory answers. However, the quest for the neural mechanisms of LTP has contributed to many important discoveries that have changed the face of neuroscience. The following five have completely changed our understanding of the mechanisms of neuroplasticity:

- Once it became apparent that only those synapses that were depolarized before the high-frequency stimulation were involved in LTP (other synapses on the same postsynaptic neurons were unaffected), it was clear

that there must be a mechanism for keeping the events at one set of synapses on a postsynaptic neuron from affecting other synapses on the same neuron. This specificity is due to the **dendritic spines**; the calcium ions that enter a dendritic spine do not readily pass out of it, and thus they exert their effects locally.

- It became apparent that maintenance of LTP involves structural changes, which depend on protein synthesis. The discovery that LTP causes structural changes was exciting because the structure of neurons and neural circuits had been assumed to be static. Many kinds of structural changes have been described (e.g., increases in number and size of synapses, increases in number and size of postsynaptic spines, changes in presynaptic and postsynaptic membranes, and changes in dendritic branching), and the changes have turned out to occur far more rapidly and more frequently than was once assumed (e.g., Kopec & Malinow, 2006; Segal, 2005). This knowledge has completely changed our conception of how neurons store experiences (see Raymond, 2007).

- The discovery of structural changes in neurons following the induction of LTP stimulated a search for a mechanism by which a neuron's activity could change its structure. This led to the discovery of numerous **transcription factors** (intracellular proteins that bind to DNA and influence the operation of particular genes) that were activated by neural activity (e.g., Collingridge, Isaac, & Wang, 2004; Fields, 2005).

- The facts that the induction of LTP begins in postsynaptic neurons and that its maintenance and expression involve presynaptic changes indicate that there must be some type of signal that passes from the postsynaptic neurons back to the presynaptic neurons. At synapses where NMDA receptors are predominant, this signal takes the form of the soluble-gas neurotransmitter **nitric oxide**. Nitric oxide is synthesized in the postsynaptic neurons in response to calcium influx and then diffuses back into the terminal buttons of the presynaptic neurons.

- It has become apparent that astrocytes play a role in synaptic transmission and that they have a substantial effect on LTP (e.g., Ge et al., 2006; Pascual et al., 2005). These findings mean that figuring out the mechanisms of LTP is going to require looking at more than presynaptic and postsynaptic neurons (see Bains & Oliet, 2007).

Variability of LTP

When I first started reading about research on LTP, I was excited at the possibilities. If LTP were the key to understanding the neural bases of learning and memory, then important discoveries would soon be forthcoming—largely because LTP seemed like such a simple model. A generation of neuroscientists has shared my view, and LTP has become the most researched topic in all of neuroscience. With such a massive effort, surely identifying the mechanisms underlying the induction, maintenance, and expression of LTP should be relatively straightforward. However, it seems that researchers are further from ultimate answers than I naively thought they were about a quarter-century ago. What has happened? Many important discoveries have been made, but rather than leading to solutions, they have revealed that LTP is far more complex than first thought and have led to many more questions.

Most of the research on LTP has focused on NMDA-receptor–mediated LTP in the hippocampus. It is now clear that NMDA-receptor–mediated LTP involves a complex array of changes that are difficult to sort out. In addition, LTP has been documented in many other parts of the CNS, where it tends to be mediated by different mechanisms (e.g., Ikeda et al., 2003; Kullman & Lamsa, 2007; Lamsa et al., 2007). And then there is LTD (long-term depression), which is the flip side of LTP and occurs in response to prolonged low-frequency stimulation of presynaptic neurons (Bliss & Schoepfer, 2004; Liu et al., 2004). Presumably, a full understanding of LTP and its role in memory will require an understanding of LTD (Kemp & Manahan-Vaughan, 2006; Massey & Bashir, 2007).

The dream of discovering the neural basis of learning and memory is what has attracted so many neuroscientists to focus on LTP. Although this dream has not yet been fulfilled, the study of LTP has led to many important discoveries about the function and plasticity of neural systems. By this criterion, this massive research effort has been a success.

11.9
Conclusion: Biopsychology of Memory and You

Because this chapter has so far focused on the amnesic affects of brain damage, you may have been left with the impression that the biopsychological study of memory has little direct relevance to individuals, like you, with intact healthy brains. This final section shows that such is not the case. It makes this point by describing two interesting lines of research and one provocative case study.

Infantile Amnesia

We all experience **infantile amnesia**; that is, we remember virtually nothing of the events of our infancy (Hayne, 2003). Newcombe and her colleagues (2000) addressed the following question: Do normal children who fail explicitly to recall or recognize things from their early childhood display preserved implicit memories for these things? The results of two experiments indicate that the answer is "yes."

In one study of infantile amnesia (Newcombe & Fox, 1994), children were shown a series of photographs of preschool-aged children, some of whom had been their preschool classmates. The subjects recognized a few of their former preschool classmates. However, whether they explicitly remembered a former classmate or not, they consistently displayed a large skin conductance response to the photographs of those classmates.

In a second study of infantile amnesia, Drummey and Newcombe (1995) used a version of the incomplete-pictures test. First, they showed a series of drawings to 3-year-olds, 5-year-olds, and adults. Three months later, the researchers assessed the implicit memories for these drawings by asking each participant to identify them ("It's a car," "It's a chair," etc.) and some control drawings as quickly as they could. During the test, the drawings were first presented badly out of focus, but became progressively sharper over time. Following this test of implicit memory, explicit memory was assessed by asking the participants which of the drawings they remembered seeing before. The 5-year-olds and adults showed better explicit memory than the 3-year-olds did; that is, they were more likely to recall seeing drawings from the original series. However, all three groups displayed substantial implicit memory: All participants were able to identify the drawings they had seen before, even when they had no conscious recollection of having seen them, sooner than they could identify control drawings.

Smart Drugs: Do They Work?

Nootropics, or **smart drugs**, are substances (drugs, supplements, herbal extracts, etc.) that are thought to improve memory. The shelves of health food stores are full of them, and even more are available on the Internet. Perhaps you have heard of, or even tried, some of them: *Ginkgo biloba* extracts, ginseng extracts, multivitamins, glucose, cholinergic agonists, Piracetam, antioxidants, phospholipids, and many more. Those offering nootropics for sale claim that scientific evidence has proven that these substances improve the memories of healthy children and adults and block the adverse effects of aging on memory. Are these claims really supported by valid scientific evidence?

The evidence that nootropics enhance memory has been reviewed several times by independent scientists (e.g., Gold, Cahill, & Wenk, 2002; McDaniel, Maier, & Einstein, 2002; Rose, 2002). The following pattern of conclusions has emerged from these reviews:

- Although nootropics are often marketed to healthy adults wanting to improve their memories, most research has been done with either nonhumans or humans with memory difficulties (i.e., the elderly).
- The relevant research with humans tends to be of low quality, with few participants, poor controls, and little effort to differentiate among various kinds of memory.
- For each purported nootropic, there are typically a few positive findings, on which the vendors focus; however, these findings have typically been difficult to replicate or to generalize to other conditions.

In short, no purported nootropic has been convincingly shown to have memory-enhancing effects. There may be enough positive evidence to warrant continued investigation of some potential nootropics, but there  is not nearly enough to justify the various claims that are made in advertisements for these substances. Why do you think there is such a huge gulf between the evidence and the claims? What are the ethical implications for the vendors and users of purported nootropics?

Posttraumatic Amnesia and Episodic Memory

This chapter began with the case of H.M.; it ends with the case of R.M. The case of R.M. is one of the most ironic that I have encountered. R.M. is a biopsychologist, and, as you will learn, his vocation played an important role in one of his symptoms.

The Case of R.M., the Biopsychologist Who Remembered H.M.

R.M. fell on his head while skiing; when he regained consciousness, he was suffering from both retrograde and anterograde amnesia. For several hours, he could recall few of the events of his previous life. He could not remember if he was married, where he lived, or where he worked. He had lost most of his episodic memory.

Also, many of the things that happened to him in the hours after his accident were forgotten as soon as his attention was diverted from them. For example, in the car on his way to the hospital, R.M. chatted with the person sitting next to him—a friend of a friend with whom he had skied all day. But each time his attention was drawn elsewhere—for example, by the mountain scenery—he completely forgot this person and reintroduced himself.

This was a classic case of posttraumatic amnesia. Like H.M., R.M. was trapped in the present, with only a cloudy past and seemingly no future. The irony of the situation was that during those few hours, when R.M. could recall few of the events of his own life, his thoughts repeatedly drifted to one semantic memory—his memory of a person he remembered learning about somewhere in his muddled past. Through the haze, he remembered H.M., his fellow prisoner of the present and wondered if the same fate lay in store for him.

R.M. recovered fully and looks back on what he can recall of his experience with relief and a feeling of empathy for H.M. Unlike H.M., R.M. received a reprieve, but his experience left him with a better appreciation for the situation of those with amnesia, like H.M., who are serving life sentences.

Themes Revisited

All four of the book's themes played major roles in this chapter. The biopsychology of memory is a neuroplastic phenomenon—it focuses on the changes in neural function that store experiences. Thus, the neuroplasticity theme was implicit throughout the chapter.

Neuroplasticity

Because the study of the neural mechanisms of memory is based largely on the study of humans with amnesia, the clinical implications theme played a significant role. The study of memory disorders has so far been a one-way street: We have learned much about memory and its neural mechanisms from studying amnesic patients, but we have not yet learned enough to treat their memory problems.

Clinical Implications

Animal models have also played a major role in the study of memory disorders. Only so much progress can be made studying human clinical cases; questions of causation must be addressed with animal models. The study of medial temporal lobe amnesia illustrates the comparative approach at its best.

Evolutionary Perspective

Finally, the thinking creatively tab marked four points in the chapter where you were encouraged to think in unconventional ways. You were encouraged to consider (1) why it proved so difficult to develop an animal model of medial temporal lobe amnesia, (2) why cases of cerebral ischemia do not provide strong evidence for involvement of the hippocampus in object recognition memory, (3) the clever demonstration by Mumby and colleagues that ischemia-induced object recognition deficits can be blocked by appropriately timed hippocampal lesions, and (4) the gulf between the actual evidence for smart drugs and the advertised claims.

Thinking Creatively

Think about It

1. The study of the anatomy of memory has come a long way since H.M.'s misfortune. What kind of advances do you think will be made in the next decade?
2. Using examples from your own experience, compare implicit and explicit memory.
3. What are the advantages and shortcomings of animal models of amnesia? Compare the usefulness of monkey and rat models.
4. LTP is one of the most intensely studied of all neuroscientific phenomena. Why? Has the effort been successful?
5. Case studies have played a particularly important role in the study of memory. Discuss.

Key Terms

Learning (p. 269)
Memory (p. 269)
Amnesia (p. 269)

11.1 Amnesic Effects of Bilateral Medial Temporal Lobectomy

Bilateral medial temporal lobectomy (p. 269)
Hippocampus (p. 269)
Amygdala (p. 269)
Lobectomy (p. 269)
Lobotomy (p. 269)
Retrograde amnesia (p. 269)
Anterograde amnesia (p. 269)
Short-term memory (p. 269)
Long-term memory (p. 270)
Digit span (p. 270)
Global amnesia (p. 270)

Incomplete-pictures test (p. 271)
Remote memory (p. 272)
Memory consolidation (p. 272)
Explicit memories (p. 272)
Implicit memories (p. 272)
Medial temporal lobe amnesia (p. 273)
Repetition priming tests (p. 273)
Semantic memories (p. 273)
Episodic memories (p. 273)
Cerebral ischemia (p. 274)
Pyramidal cell layer (p. 275)
CA1 subfield (p. 275)

11.2 Amnesia of Korsakoff's Syndrome

Korsakoff's syndrome (p. 275)
Mediodorsal nuclei (p. 275)

Medial diencephalic amnesia (p. 275)

11.3 Amnesia of Alzheimer's Disease

Alzheimer's disease (p. 276)
Basal forebrain (p. 276)

11.4 Amnesia after Concussion: Evidence for Consolidation

Posttraumatic amnesia (PTA) (p. 276)
Electroconvulsive shock (ECS) (p. 277)
Standard consolidation theory (p. 279)
Multiple-trace theory (p. 279)
Engram (p. 279)

11.5 Neuroanatomy of Object–Recognition Memory

Delayed nonmatching-to-sample test (p. 280)
Rhinal cortex (p. 280)
Mumby box (p. 281)
Entorhinal cortex (p. 283)
Perirhinal cortex (p. 283)

11.6 Hippocampus and Memory for Spatial Location

Morris water maze test (p. 286)
Radial arm maze test (p. 286)
Reference memory (p. 286)
Working memory (p. 286)
Place cells (p. 286)

✓● **Quick Review**   Test your comprehension of the chapter with this brief practice test. You can find the answers to these questions as well as more practice tests, activities, and other study resources at www.mypsychlab.com.

1. H.M.'s digit span was
   a. six.
   b. in the normal range.
   c. only 32%.
   d. less than 5%.
   e. both a and b

2. Korsakoff's syndrome has been associated with damage to the area around the
   a. hippocampus.
   b. medial temporal lobes.
   c. mediodorsal nucleus.
   d. both a and b
   e. both b and c

3. Posttraumatic amnesia can be induced by
   a. electroconvulsive shock.
   b. thiamine.
   c. alcohol.
   d. repetition priming.
   e. both b and c

4. Amnesia resulting from medial temporal lobectomy appears to be largely a product of damage to the
   a. medial diencephalon.
   b. hippocampus.
   c. amygdala.
   d. rhinal cortex.
   e. both b and c

5. The hippocampus appears to play a major role in memory for
   a. faces.
   b. spatial location.
   c. emotional significance.
   d. words.
   e. place cells.

12 Hunger, Eating, and Health

Why Do Many People Eat Too Much?

Eating is a behavior that is of interest to virtually everyone. We all do it, and most of us derive great pleasure from it. But for many of us, it becomes a source of serious personal and health problems.

Most eating-related health problems in industrialized nations are associated with eating too much—the average American consumes 3,800 calories per day, about twice the average daily requirement (see Kopelman, 2000). For example, it is estimated that 65% of the adult U.S. population is either overweight or clinically obese, qualifying this problem for epidemic status (see Abelson & Kennedy, 2004; Arnold, 2009). The resulting financial and personal costs are huge. Each year in the United States, about $100 billion is spent treating obesity-related disorders (see Olshansky et al., 2005). Moreover, each year, an estimated 300,000 U.S. citizens die from disorders caused by their excessive eating (e.g., diabetes, hypertension, cardiovascular diseases, and some cancers). Although the United States is the trend-setter when it comes to overeating and obesity, many other countries are not far behind (Sofsian, 2007). Ironically, as overeating and obesity have reached epidemic proportions, there has been a related increase in disorders associated with eating too little (see Polivy & Herman, 2002). For example, almost 3% of American adolescents currently suffer from *anorexia* or *bulimia*, which can be life-threatening in extreme cases.

Watch
You Are What You Eat
www.mypsychlab.com

The massive increases in obesity and other eating-related disorders that have occurred over the last few decades in many countries stand in direct opposition to most people's thinking about hunger and eating. Many people—and I assume that this includes you—believe that hunger and eating are normally triggered when the body's energy resources fall below a prescribed optimal level, or **set point**. They appreciate that many factors influence hunger and eating, but they assume that the hunger and eating system has evolved to supply the body with just the right amount of energy.

Watch
Thinking about Hunger
www.mypsychlab.com

This chapter explores the incompatibility of the set-point assumption with the current epidemic of eating disorders. If we all have hunger and eating systems whose primary function is to maintain energy resources at optimal levels, then eating disorders should be rare. The fact that they are so prevalent suggests that hunger and eating are regulated in some other way. This chapter will repeatedly challenge you to think in new ways about issues that impact your health and longevity and will provide new insights of great personal relevance—I guarantee it.

Thinking Creatively

Before you move on to the body of the chapter, I would like you to pause to consider a case study. What would a severely amnesic patient do if offered a meal shortly after finishing one? If his hunger and eating were controlled by energy set points, he would refuse the second meal. Did he?

The Case of the Man Who Forgot Not to Eat

R.H. was a 48-year-old male whose progress in graduate school was interrupted by the development of severe amnesia for long-term explicit memory. His amnesia was similar in pattern and severity to that of H.M., whom you met in Chapter 11, and an MRI examination revealed bilateral damage to the medial temporal lobes.

Clinical Implications

The meals offered to R.H. were selected on the basis of interviews with him about the foods he liked: veal parmigiana (about 750 calories) plus all the apple juice he wanted. On one occasion, he was offered a second meal about 15 minutes after he had eaten the first, and he ate it. When offered a third meal 15 minutes later, he ate that, too. When offered a fourth meal he rejected it, claiming that his "stomach was a little tight."

Then, a few minutes later, R.H. announced that he was going out for a good walk and a meal. When asked what he was going to eat, his answer was "veal parmigiana."

Clearly, R.H.'s hunger (i.e., motivation to eat) did not result from an energy deficit (Rozin et al., 1998). Other cases like that of R.H. have been reported by Higgs and colleagues (2008).

12.1
Digestion, Energy Storage, and Energy Utilization

The primary purpose of hunger is to increase the probability of eating, and the primary purpose of eating is to supply the body with the molecular building blocks and energy it needs to survive and function (see Blackburn, 2001). This section provides the foundation for our consideration of hunger and eating by providing a brief overview of the processes by which food is digested, stored, and converted to energy.

Digestion

The *gastrointestinal tract* and the process of digestion are illustrated in Figure 12.1 on page 300. **Digestion** is the gastrointestinal process of breaking down food and absorbing its constituents into the body. In order to appreciate the basics of digestion, it is useful to consider the body without its protuberances, as a simple living tube

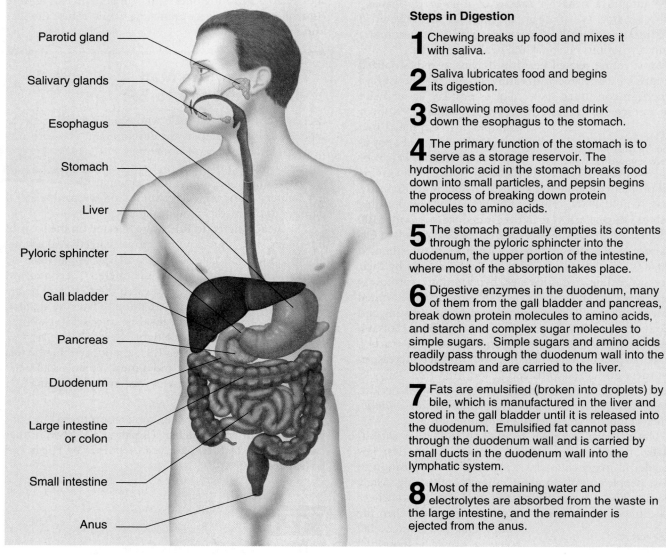

Steps in Digestion

1 Chewing breaks up food and mixes it with saliva.

2 Saliva lubricates food and begins its digestion.

3 Swallowing moves food and drink down the esophagus to the stomach.

4 The primary function of the stomach is to serve as a storage reservoir. The hydrochloric acid in the stomach breaks food down into small particles, and pepsin begins the process of breaking down protein molecules to amino acids.

5 The stomach gradually empties its contents through the pyloric sphincter into the duodenum, the upper portion of the intestine, where most of the absorption takes place.

6 Digestive enzymes in the duodenum, many of them from the gall bladder and pancreas, break down protein molecules to amino acids, and starch and complex sugar molecules to simple sugars. Simple sugars and amino acids readily pass through the duodenum wall into the bloodstream and are carried to the liver.

7 Fats are emulsified (broken into droplets) by bile, which is manufactured in the liver and stored in the gall bladder until it is released into the duodenum. Emulsified fat cannot pass through the duodenum wall and is carried by small ducts in the duodenum wall into the lymphatic system.

8 Most of the remaining water and electrolytes are absorbed from the waste in the large intestine, and the remainder is ejected from the anus.

FIGURE 12.1 The gastrointestinal tract and the process of digestion.

with a hole at each end. To supply itself with energy and other nutrients, the tube puts food into one of its two holes—the one with teeth—and passes the food along its internal canal so that the food can be broken down and partially absorbed from the canal into the body. The leftovers are jettisoned from the other end. Although this is not a particularly appetizing description of eating, it does serve to illustrate that, strictly speaking, food has not been consumed until it has been digested.

Energy Storage in the Body

As a consequence of digestion, energy is delivered to the body in three forms: (1) **lipids** (fats), (2) **amino acids** (the breakdown products of proteins), and (3) **glucose**

(a simple sugar that is the breakdown product of complex *carbohydrates*, that is, starches and sugars).

The body uses energy continuously, but its consumption is intermittent; therefore, it must store energy for use in the intervals between meals. Energy is stored in three forms: *fats, glycogen,* and *proteins*. Most of the body's energy reserves are stored as fats, relatively little as glycogen and proteins (see Figure 12.2). Thus, changes in the body weights of adult humans are largely a consequence of changes in the amount of their stored body fat.

Why is fat the body's preferred way of storing energy? Glycogen, which is largely stored in the liver and muscles, might be expected to be the body's preferred mode of energy storage because it is so readily converted to glucose—the body's main directly utilizable source of energy. But there

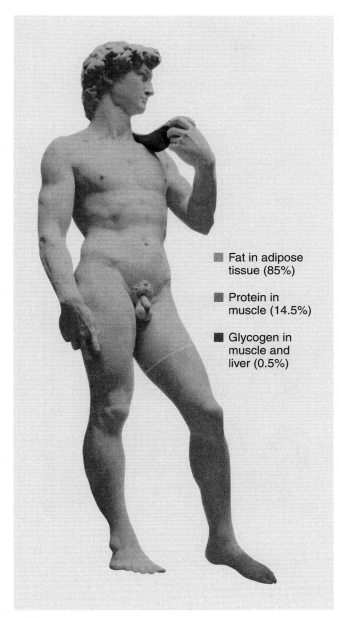

- ■ Fat in adipose tissue (85%)
- ■ Protein in muscle (14.5%)
- ■ Glycogen in muscle and liver (0.5%)

FIGURE 12.2 Distribution of stored energy in an average person.

are two reasons why fat, rather than glycogen, is the primary mode of energy storage: One is that a gram of fat can store almost twice as much energy as a gram of glycogen; the other is that glycogen, unlike fat, attracts and holds substantial quantities of water. Consequently, if all your fat calories were stored as glycogen, you would likely weigh well over 275 kilograms (600 pounds).

Three Phases of Energy Metabolism

There are three phases of *energy metabolism* (the chemical changes by which energy is made available for an organism's use): the cephalic phase, the absorptive phase, and the fasting phase. The **cephalic phase** is the preparatory phase; it often begins with the sight, smell, or even just the thought of food, and it ends when the food starts to be absorbed into the bloodstream. The **absorptive phase** is the period during which the energy absorbed into the bloodstream from the meal is meeting the body's immediate energy needs. The **fasting phase** is the period during which all of the unstored energy from the previous meal has been used and the body is withdrawing energy from its reserves to meet its immediate energy requirements; it ends with the beginning of the next cephalic phase. During periods of rapid weight gain, people often go directly from one absorptive phase into the next cephalic phase, without experiencing an intervening fasting phase.

The flow of energy during the three phases of energy metabolism is controlled by two pancreatic hormones: insulin and glucagon. During the cephalic and absorptive phases, the pancreas releases a great deal of insulin into the bloodstream and very little glucagon. **Insulin** does three things: (1) It promotes the use of glucose as the primary source of energy by the body. (2) It promotes the conversion of bloodborne fuels to forms that can be stored: glucose to glycogen and fat, and amino acids to proteins. (3) It promotes the storage of glycogen in liver and muscle, fat in adipose tissue, and proteins in muscle. In short, the function of insulin during the cephalic phase is to lower the levels of bloodborne fuels, primarily glucose, in anticipation of the impending influx; and its function during the absorptive phase is to minimize the increasing levels of bloodborne fuels by utilizing and storing them.

In contrast to the cephalic and absorptive phases, the fasting phase is characterized by high blood levels of **glucagon** and low levels of insulin. Without high levels of insulin, glucose has difficulty entering most body cells; thus, glucose stops being the body's primary fuel. In effect, this saves the body's glucose for the brain, because insulin is not required for glucose to enter most brain cells. The low levels of insulin also promote the conversion of glycogen and protein to glucose. (The conversion of protein to glucose is called **gluconeogenesis**.)

On the other hand, the high levels of fasting-phase glucagon promote the release of **free fatty acids** from adipose tissue and their use as the body's primary fuel. The high glucagon levels also stimulate the conversion of free fatty acids to **ketones**, which are used by muscles as a source of energy during the fasting phase. After a prolonged period without food, however, the brain also starts to use ketones, thus further conserving the body's resources of glucose.

Figure 12.3 summarizes the major metabolic events associated with the three phases of energy metabolism.

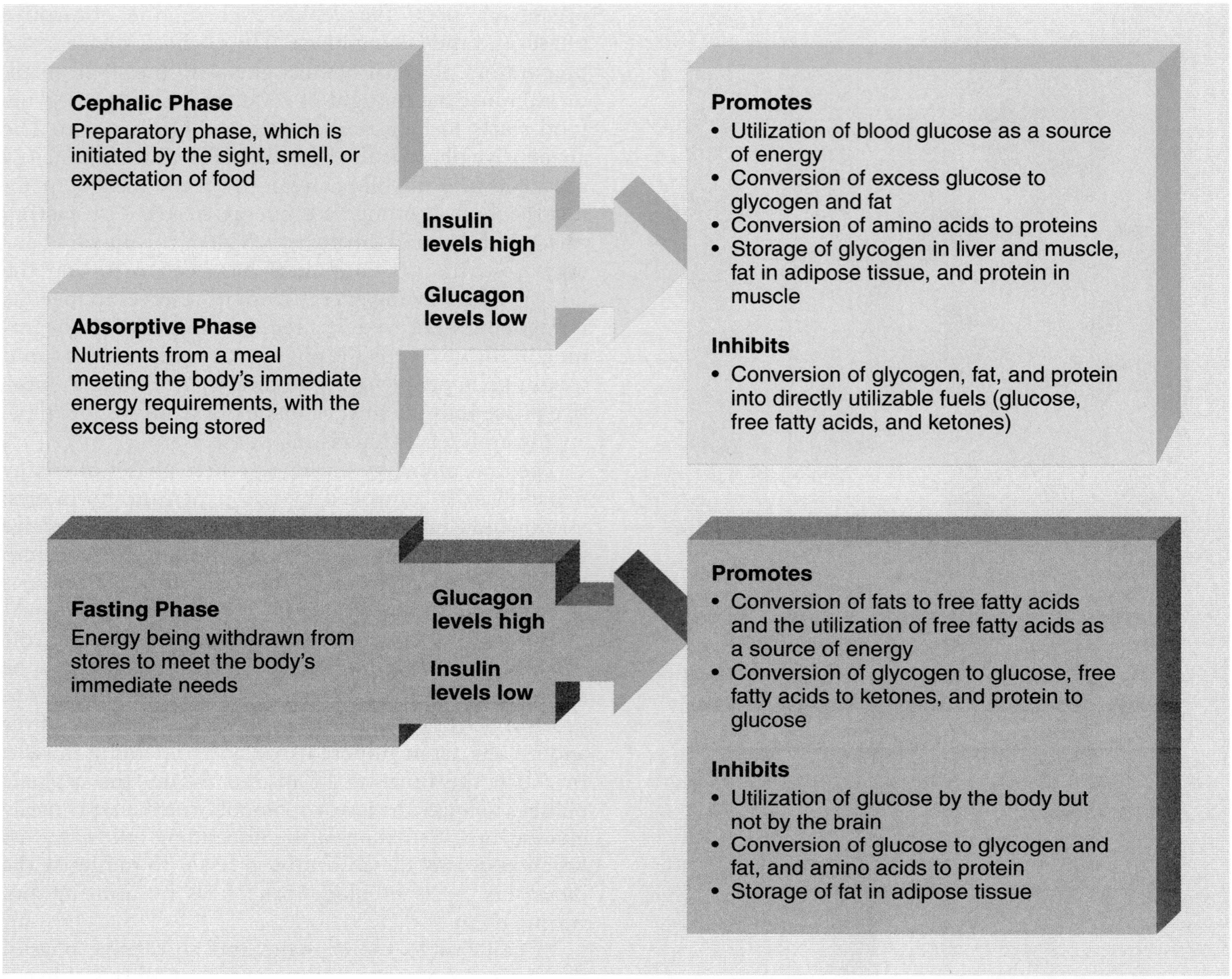

Cephalic Phase
Preparatory phase, which is initiated by the sight, smell, or expectation of food

Absorptive Phase
Nutrients from a meal meeting the body's immediate energy requirements, with the excess being stored

Insulin levels high

Glucagon levels low

Promotes
- Utilization of blood glucose as a source of energy
- Conversion of excess glucose to glycogen and fat
- Conversion of amino acids to proteins
- Storage of glycogen in liver and muscle, fat in adipose tissue, and protein in muscle

Inhibits
- Conversion of glycogen, fat, and protein into directly utilizable fuels (glucose, free fatty acids, and ketones)

Fasting Phase
Energy being withdrawn from stores to meet the body's immediate needs

Glucagon levels high

Insulin levels low

Promotes
- Conversion of fats to free fatty acids and the utilization of free fatty acids as a source of energy
- Conversion of glycogen to glucose, free fatty acids to ketones, and protein to glucose

Inhibits
- Utilization of glucose by the body but not by the brain
- Conversion of glucose to glycogen and fat, and amino acids to protein
- Storage of fat in adipose tissue

FIGURE 12.3 The major events associated with the three phases of energy metabolism: the cephalic, absorptive, and fasting phases.

12.2

Theories of Hunger and Eating: Set Points versus Positive Incentives

One of the main difficulties I have in teaching the fundamentals of hunger, eating, and body weight regulation is the **set-point assumption**. Although it dominates most people's thinking about hunger and eating (Assanand, Pinel, & Lehman, 1998a, 1998b), whether they realize it or not, it is inconsistent with the bulk of the evidence. What exactly is the set-point assumption?

Set-Point Assumption

Most people attribute *hunger* (the motivation to eat) to the presence of an energy deficit, and they view eating as the means by which the energy resources of the body are returned to their optimal level—that is, to the *energy set point*. Figure 12.4 summarizes this set-point assumption. After a *meal* (a bout of eating), a person's energy resources are assumed to be near their set point and to decline thereafter as the body uses energy to fuel its physiological processes. When the level of the body's energy resources falls far enough below the set point, a person becomes motivated by hunger to initiate another meal. The meal continues, according to the set-point assumption, until the energy level

FIGURE 12.4 The energy set-point view that is the basis of many people's thinking about hunger and eating.

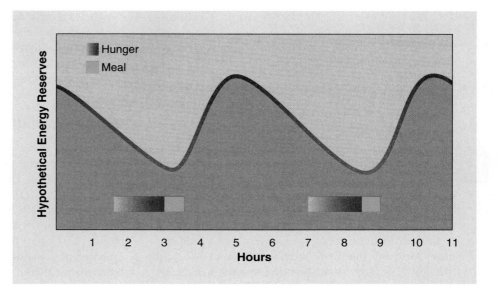

returns to its set point and the person feels *satiated* (no longer hungry).

Set-point models assume that hunger and eating work in much the same way as a thermostat-regulated heating system in a cool climate. The heater increases the house temperature until it reaches its set point (the thermostat setting). The heater then shuts off, and the temperature of the house gradually declines until it becomes low enough to turn the heater back on. All set-point systems have three components: a set-point mechanism, a detector mechanism, and an effector mechanism. The *set-point mechanism* defines the set point, the *detector mechanism* detects deviations from the set point, and the *effector mechanism* acts to eliminate the deviations. For example, the set-point, detector, and effector mechanisms of a heating system are the thermostat, the thermometer, and the heater, respectively.

All set-point systems are **negative feedback systems**—systems in which feedback from changes in one direction elicit compensatory effects in the opposite direction. Negative feedback systems are common in mammals because they act to maintain **homeostasis**—a stable internal environment—which is critical for mammals' survival (see Wenning, 1999). Set-point systems combine negative feedback with a set point to keep an internal environment fixed at the prescribed point. Set-point systems seemed necessary when the adult human brain was assumed to be immutable: Because the brain couldn't change, energy resources had to be highly regulated. However, we now know that the adult human brain is plastic and capable of considerable adaptation. Thus, there is no longer a logical imperative for the set-point regulation of eating. Throughout this chapter, you will need to put aside your preconceptions and base your thinking about hunger and eating entirely on the empirical evidence.

Glucostatic and Lipostatic Set–Point Theories of Hunger and Eating

In the 1940s and 1950s, researchers working under the assumption that eating is regulated by some type of set-point system speculated about the nature of the regulation. Several researchers suggested that eating is regulated by a system that is designed to maintain a blood glucose set point—the idea being that we become hungry when our blood glucose levels drop significantly below their set point and that we become satiated when eating returns our blood glucose levels to their set point. The various versions of this theory are collectively referred to as the **glucostatic theory**. It seemed to make good sense that the main purpose of eating is to defend a blood glucose set point, because glucose is the brain's primary fuel.

The **lipostatic theory** is another set-point theory that was proposed in various forms in the 1940s and 1950s. According to this theory, every person has a set point for body fat, and deviations from this set point produce compensatory adjustments in the level of eating that return levels of body fat to their set point. The most frequently cited support for the theory is the fact that the body weights of adults stay relatively constant.

The glucostatic and lipostatic theories were viewed as complementary, not mutually exclusive. The glucostatic theory was thought to account for meal initiation and termination, whereas the lipostatic theory was thought to account for long-term regulation. Thus, the dominant view in the 1950s was that eating is regulated by the interaction between two set-point systems: a short-term glucostatic system and a long-term lipostatic system. The simplicity of these 1950s theories is appealing. Remarkably, they are still being presented as the latest word in some textbooks; perhaps you have encountered them.

Problems with Set–Point Theories of Hunger and Eating

Set-point theories of hunger and eating have several serious weaknesses (see de Castro & Plunkett, 2002). You have already learned one fact that undermines these theories: There is an epidemic of obesity and overweight,

Thinking Creatively which should not occur if eating is regulated by a set point. Let's look at three more major weaknesses of set-point theories of hunger and eating.

- First, set-point theories of hunger and eating are inconsistent with basic eating-related evolutionary pressures as we understand them. The major eating-related problem faced by our ancestors was the incon-*Evolutionary Perspective* sistency and unpredictability of the food supply. Thus, in order to survive, it was important for them to eat large quantities of good food when it was available so that calories could be banked in the form of body fat. Any ancestor—human or otherwise—that stopped feeling hungry as soon as immediate energy needs were met would not have survived the first hard winter or prolonged drought. For any warm-blooded species to survive under natural conditions, it needs a hunger and eating system that prevents energy deficits, rather than one that merely responds to them once they have developed. From this perspective, it is difficult to imagine how a set-point hunger and feeding system could have evolved in mammals (see Pinel, Assanand, & Lehman, 2000).

- Second, major predictions of the set-point theories of hunger and eating have not been confirmed. Early studies seemed to support the set-point theories by showing that large reductions in body fat, produced by starvation, or large reductions in blood glucose, produced by insulin injections, induce increases in eating in laboratory animals. The problem is that reductions in blood glucose of the magnitude needed to reliably induce eating rarely occur naturally. Indeed, as you have already learned in this chapter, about 65% of U.S. adults have a significant excess of fat deposits when they begin a meal. Conversely, efforts to reduce meal size by having subjects consume a high-calorie drink before eating have been largely unsuccessful; indeed, beliefs about the caloric content of a premeal drink often influence the size of a subsequent meal more than does its actual caloric content (see Lowe, 1993).

- Third, set-point theories of hunger and eating are deficient because they fail to recognize the major influences on hunger and eating of such important factors as taste, learning, and social influences. To convince yourself of the importance of these factors, pause for a minute and imagine the sight, smell, and taste of your favorite food. Perhaps it is a succulent morsel of lobster meat covered with melted garlic butter, a piece of chocolate cheesecake, or a plate of sizzling homemade french fries. Are you starting to feel a bit hungry? If the homemade french fries—my personal weakness—were sitting in front of you right now, wouldn't you reach out and have one, or maybe the whole plateful? Have you not on occasion felt discomfort after a large main course, only to polish off a substantial dessert? The usual positive answers to these questions lead unavoidably to the conclusion that hunger and eating are not rigidly controlled by deviations from energy set points.

Positive–Incentive Perspective

The inability of set-point theories to account for the basic phenomena of eating and hunger led to the development of an alternative theoretical perspective (see Berridge, 2004). The central assertion of this perspective, commonly referred to as **positive-incentive theory**, is that humans and other animals are not normally driven to eat by internal energy deficits but are drawn to eat by the anticipated pleasure of eating—the anticipated pleasure of a behavior is called its **positive-incentive value** (see Bolles, 1980; Booth, 1981; Collier, 1980; Rolls, 1981; Toates, 1981). There are several different positive-incentive theories, and I refer generally to all of them as the *positive-incentive perspective*.

The major tenet of the positive-incentive perspective on eating is that eating is controlled in much the same way as sexual behavior: We engage in sexual *Evolutionary Perspective* behavior not because we have an internal deficit, but because we have evolved to crave it. The evolutionary pressures of unexpected food shortages have shaped us and all other warm-blooded animals, who need a continuous supply of energy to maintain their body temperatures, to take advantage of good food when it is present and eat it. According to the positive-incentive perspective, it is the presence of good food, or the anticipation of it, that normally makes us hungry, not an energy deficit.

According to the positive-incentive perspective, the degree of hunger you feel at any particular time depends on the interaction of all the factors that influence the positive-incentive value of eating (see Palmiter, 2007). These include the following: the flavor of the food you are likely to consume, what you have learned about the effects of this food either from eating it previously or from other people, the amount of time since you last ate, the type and quantity of food in your gut, whether or not other people are present and eating, whether or not your blood glucose levels are within the normal range. This partial list illustrates one strength of the positive-incentive perspective. Unlike set-point theories, positive-incentive theories do not single out one factor as the major determinant of hunger and ignore the others. Instead, they acknowledge that many factors interact to determine a person's hunger at any time, and they suggest that this interaction occurs through the influence of these various factors on the positive-incentive value of eating (see Cabanac, 1971).

In this section, you learned that most people think about hunger and eating in terms of energy set points and were introduced to an alternative way of thinking—the positive-incentive perspective. Which way is correct? If you are like most people, you have an attachment to familiar ways of thinking and a resistance to new ones. Try to put this tendency aside and base your views about this important issue entirely on the evidence.

You have already learned about some of the major weaknesses of strict set-point theories of hunger and eating. The next section describes some of the things that biopsychological research has taught us about hunger and eating. As you progress through the section, notice the superiority of the positive-incentive theories over set-point theories in accounting for the basic facts.

12.3
Factors That Determine What, When, and How Much We Eat

This section describes major factors that commonly determine what we eat, when we eat, and how much we eat. Notice that energy deficits are not included among these factors. Although major energy deficits clearly increase hunger and eating, they are not a common factor in the eating behavior of people like us, who live in food-replete societies. Although you may believe that your body is short of energy just before a meal, it is not. This misconception is one that is addressed in this section. Also, notice how research on nonhumans has played an important role in furthering understanding of human eating.

Factors That Determine What We Eat

Certain tastes have a high positive-incentive value for virtually all members of a species. For example, most humans have a special fondness for sweet, fatty, and salty tastes. This species-typical pattern of human taste preferences is adaptive because in nature sweet and fatty tastes are typically characteristic of high-energy foods that are rich in vitamins and minerals, and salty tastes are characteristic of sodium-rich foods. In contrast, bitter tastes, for which most humans have an aversion, are often associated with toxins. Superimposed on our species-typical taste preferences and aversions, each of us has the ability to learn specific taste preferences and aversions (see Rozin & Shulkin, 1990).

Learned Taste Preferences and Aversions   Animals learn to prefer tastes that are followed by an infusion of calories, and they learn to avoid tastes that are followed by illness (e.g., Baker & Booth, 1989; Lucas & Sclafani, 1989; Sclafani, 1990). In addition, humans and other animals learn what to eat from their conspecifics. For example,

rats learn to prefer flavors that they experience in mother's milk and those that they smell on the breath of other rats (see Galef, 1995, 1996; Galef, Whishkin, & Bielavska, 1997). Similarly, in humans, many food preferences are culturally specific—for example, in some cultures, various nontoxic insects are considered to be a delicacy. Galef and Wright (1995) have shown that rats reared in groups, rather than in isolation, are more likely to learn to eat a healthy diet.

Learning to Eat Vitamins and Minerals   How do animals select a diet that provides all of the vitamins and minerals they need? To answer this question, researchers have studied how dietary deficiencies influence diet selection. Two patterns of results have emerged: one for sodium and one for the other essential vitamins and minerals. When an animal is deficient in sodium, it develops an immediate and compelling preference for the taste of sodium salt (see Rowland, 1990). In contrast, an animal that is deficient in some vitamin or mineral other than sodium must learn to consume foods that are rich in the missing nutrient by experiencing their positive effects; this is because vitamins and minerals other than sodium normally have no detectable taste in food. For example, rats maintained on a diet deficient in *thiamine* (vitamin B$_1$) develop an aversion to the taste of that diet; and if they are offered two new diets, one deficient in thiamine and one rich in thiamine, they often develop a preference for the taste of the thiamine-rich diet over the ensuing days, as it becomes associated with improved health.

If we, like rats, are capable of learning to select diets that are rich in the vitamins and minerals we need, why are dietary deficiencies so prevalent in our society? One reason is that, in order to maximize profits, manufacturers produce foods that have the tastes we prefer but lack many of the nutrients we need to maintain our health. (Even rats prefer chocolate chip cookies to nutritionally complete rat chow.) The second reason is illustrated by the classic study of Harris and associates (1933). When thiamine-deficient rats were offered two new diets, one with thiamine and one without, almost all of them learned to eat the complete diet and avoid the deficient one. However, when they were offered ten new diets, only one of which contained the badly needed thiamine, few developed a preference for the complete diet. The number of different substances, both nutritious and not, consumed each day by most people in industrialized societies is immense, and this makes it difficult, if not impossible, for their bodies to learn which foods are beneficial and which are not.

There is not much about nutrition in this chapter: Although it is critically important to eat a nutritious diet, nutrition seems to have little direct effect on our feelings of hunger. However, while I am on the topic, I would like to direct you to a good source of information about nutrition that could have a positive effect on your health: Some popular books on nutrition are dangerous, and even governments, inordinately influenced by economic considerations and special-interest groups, often do not provide the best nutritional advice (see Nestle, 2003). For sound research-based advice on nutrition, check out an article by Willett and Stampfer (2003) and the book on which it is based, *Eat, Drink, and Be Healthy* by Willett, Skerrett, and Giovannucci (2001).

Factors That Influence When We Eat

Collier and his colleagues (see Collier, 1986) found that most mammals choose to eat many small meals (snacks) each day if they have ready access to a continuous supply of food. Only when there are physical costs involved in initiating meals—for example, having to travel a considerable distance—does an animal opt for a few large meals.

The number of times humans eat each day is influenced by cultural norms, work schedules, family routines, personal preferences, wealth, and a variety of other factors. However, in contrast to the usual mammalian preference, most people, particularly those living in family groups, tend to eat a few large meals each day at regular times. Interestingly, each person's regular mealtimes are the very same times at which that person is likely to feel most hungry; in fact, many people experience attacks of malaise (headache, nausea, and an inability to concentrate) when they miss a regularly scheduled meal.

Premeal Hunger   I am sure that you have experienced attacks of premeal hunger. Subjectively, they seem to provide compelling support for set-point theories. Your body seems to be crying out: "I need more energy. I cannot function without it. Please feed me." But things are not always the way they seem. Woods has straightened out the confusion (see Woods, 1991; Woods & Ramsay, 2000; Woods & Strubbe, 1994).

According to Woods, the key to understanding hunger is to appreciate that eating meals stresses the body. Before a meal, the body's energy reserves are in reasonable homeostatic balance; then, as a meal is consumed, there is a homeostasis-disturbing influx of fuels into the bloodstream. The body does what it can to defend its homeostasis. At the first indication that a person will soon be eating—for example, when the usual mealtime approaches—the body enters the cephalic phase and takes steps to soften the impact of the impending homeostasis-disturbing influx by releasing insulin into the blood and thus reducing blood glucose. Woods's message is that the strong, unpleasant feelings of hunger that you may experience at mealtimes are not cries from your body for food; they are the sensations of your body's preparations for the expected homeostasis-disturbing meal. Mealtime

hunger is caused by the expectation of food, not by an energy deficit.

As a high school student, I ate lunch at exactly 12:05 every day and was overwhelmed by hunger as the time approaches. Now, my eating schedule is different, and I never experience noontime hunger pangs; I now get hungry just before the time at which I usually eat. Have you had a similar experience?

Pavlovian Conditioning of Hunger   In a classic series of Pavlovian conditioning experiments on laboratory rats, Weingarten (1983, 1984, 1985) provided strong support for the view that hunger is often caused by the expectation of food, not by an energy deficit. During the conditioning phase of one of his experiments, Weingarten presented rats with six meals per day at irregular intervals, and he signaled the impending delivery of each meal with a buzzer-and-light conditional stimulus. This conditioning procedure was continued for 11 days. Throughout the ensuing test phase of the experiment, the food was continuously available. Despite the fact that the subjects were never deprived during the test phase, the rats started to eat each time the buzzer and light were presented—even if they had recently completed a meal.

Factors That Influence How Much We Eat

The motivational state that causes us to stop eating a meal when there is food remaining is **satiety**. Satiety mechanisms play a major role in determining how much we eat.

Satiety Signals   As you will learn in the next section of the chapter, food in the gut and glucose entering the blood can induce satiety signals, which inhibit subsequent consumption. These signals depend on both the volume and the **nutritive density** (calories per unit volume) of the food.

The effects of nutritive density have been demonstrated in studies in which laboratory rats have been maintained on a single diet. Once a stable baseline of consumption has been established, the nutritive density of the diet is changed. Some rats learn to adjust the volume of food they consume to keep their caloric intake and body weights relatively stable. However, there are major limits to this adjustment: Rats rarely increase their intake sufficiently to maintain their body weights if the nutritive density of their conventional laboratory feed is reduced by more than 50% or if there are major changes in the diet's palatability.

Sham Eating   The study of **sham eating** indicates that satiety signals from the gut or blood are not necessary to terminate a meal. In sham-eating experiments, food is chewed and swallowed by the subject; but rather than

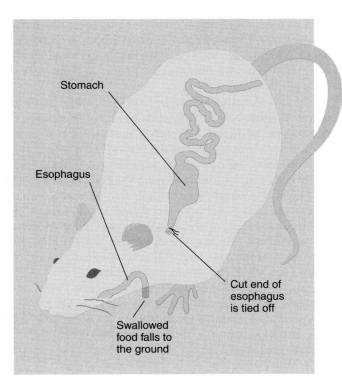

FIGURE 12.5   The sham-eating preparation.

passing down the subject's esophagus into the stomach, it passes out of the body through an implanted tube (see Figure 12.5).

Because sham eating adds no energy to the body, set-point theories predict that all sham-eaten meals should be huge. But this is not the case. Weingarten and Kulikovsky (1989) sham fed rats one of two differently flavored diets: one that the rats had naturally eaten many times before and one that they had never eaten before. The first sham meal of the rats that had previously eaten the diet was the same size as the previously eaten meals of that diet; then, on ensuing days they began to sham eat more and more (see Figure 12.6). In contrast, the rats that were presented with the unfamiliar diet

FIGURE 12.6   Change in the magnitude of sham eating over repeated sham-eating trials. The rats in one group sham ate the same diet they had eaten before the sham-eating phase; the rats in another group sham ate a diet different from the one they had previously eaten. (Based on Weingarten, 1990.)

sham ate large quantities right from the start. Weingarten and Kulikovsky concluded that the amount we eat is influenced largely by our previous experience with the particular food's physiological effects, not by the immediate effect of the food on the body.

Appetizer Effect and Satiety   The next time you attend a dinner party, you may experience a major weakness of the set-point theory of satiety. If appetizers are served, you will notice that small amounts of food  consumed before a meal actually increase hunger rather than reducing it. This is the **appetizer effect**. Presumably, it occurs because the consumption of a small amount of food is particularly effective in eliciting cephalic-phase responses.

Serving Size and Satiety   Many experiments have shown that the amount of consumption is influenced by serving size (Geier, Rozin, & Doros, 2006). The larger the servings, the more we tend to eat. There is even evidence that we tend to eat more when we eat with larger spoons.

Social Influences and Satiety   Feelings of satiety may also depend on whether we are eating alone or with others. Redd and de Castro (1992) found that their subjects consumed 60% more when eating with others. Laboratory rats also eat substantially more when fed in groups.

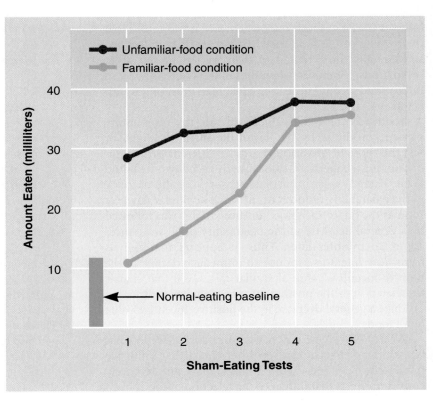

In humans, social factors have also been shown to reduce consumption. Many people eat less than they would like in order to achieve their society's ideal of slenderness, and others refrain from eating large amounts in front of others so as not to appear gluttonous. Unfortunately, in our culture, females are influenced by such pressures more than males are, and, as you will learn later in the chapter, some develop serious eating disorders as a result.

Sensory-Specific Satiety The number of different tastes available at each meal has a major effect on meal size. For example, the effect of offering a laboratory rat a varied diet of highly palatable foods—a **cafeteria diet**—is dramatic. Adults rats that were offered bread and chocolate in addition to their usual laboratory diet increased their average intake of calories by 84%, and after 120 days they had increased their average body weights by 49% (Rogers & Blundell, 1980). The spectacular effects of cafeteria diets on consumption and body weight clearly run counter to the idea that satiety is rigidly controlled by internal energy set points.

The effect on meal size of cafeteria diets results from the fact that satiety is to a large degree sensory-specific. As you eat one food, the positive-incentive value of all foods declines slightly, but the positive-incentive value of that particular food plummets. As a result, you soon become satiated on that food and stop eating it. However, if another food is offered to you, you will often begin eating again.

In one study of **sensory-specific satiety** (Rolls et al., 1981), human subjects were asked to rate the palatability of eight different foods, and then they ate a meal of one of them. After the meal, they were asked to rate the palatability of the eight foods once again, and it was found that their rating of the food they had just eaten had declined substantially more than had their ratings of the other seven foods. Moreover, when the subjects were offered an unexpected second meal, they consumed most of it unless it was the same as the first.

Booth (1981) asked subjects to rate the momentary pleasure produced by the flavor, the smell, the sight, or just the thought of various foods at different times after consuming a large, high-calorie, high-carbohydrate liquid meal. There was an immediate sensory-specific decrease in the palatability of foods of the same or similar flavor as soon as the liquid meal was consumed. This was followed by a general decrease in the palatability of all substances about 30 minutes later. Thus, it appears that signals from taste receptors produce an immediate decline in the positive-incentive value of similar tastes and that signals associated with the postingestive consequences of eating produce a general decrease in the positive-incentive value of all foods.

Rolls (1990) suggested that sensory-specific satiety has two kinds of effects: relatively brief effects that influence the selection of foods within a single meal, and relatively enduring effects that influence the selection of foods from meal to meal. Some foods seem to be relatively immune to long-lasting sensory-specific satiety; foods such as rice, bread, potatoes, sweets, and green salads can be eaten almost every day with only a slight decline in their palatability (Rolls, 1986).

The phenomenon of sensory-specific satiety has two adaptive consequences. First, it encourages the consumption of a varied diet. If there were no sensory-specific satiety, a person would tend to eat her or his preferred food and nothing else, and the result would be malnutrition. Second, sensory-specific satiety encourages animals that have access to a variety of foods to eat a lot; an animal that has eaten its fill of one food will often begin eating again if it encounters a different one (Raynor & Epstein, 2001). This encourages animals to take full advantage of times of abundance, which are all too rare in nature.

Evolutionary Perspective

This section has introduced you to several important properties of hunger and eating. How many support the set-point assumption, and how many are inconsistent with it?

Thinking Creatively

Scan Your Brain

Are you ready to move on to the discussion of the physiology of hunger and satiety in the following section? Find out by completing the following sentences with the most appropriate terms. The correct answers are provided at the end of the exercise. Before proceeding, review material related to your incorrect answers and omissions.

1. The primary function of the _____ is to serve as a storage reservoir for undigested food.
2. Most of the absorption of nutrients into the body takes place through the wall of the _____, or upper intestine.
3. The phase of energy metabolism that is triggered by the expectation of food is the _____ phase.
4. During the absorptive phase, the pancreas releases a great deal of _____ into the bloodstream.
5. During the fasting phase, the primary fuels of the body are _____.
6. During the fasting phase, the primary fuel of the brain is _____.
7. The three components of a set-point system are a set-point mechanism, a detector, and an _____.
8. The theory that hunger and satiety are regulated by a blood glucose set point is the _____ theory.
9. Evidence suggests that hunger is greatly influenced by the current _____ value of food.

10. Most humans have a preference for sweet, fatty, and _____ tastes.

11. There are two mechanisms by which we learn to eat diets containing essential vitamins and minerals: one mechanism for _____ and another mechanism for the rest.

12. Satiety that is specific to the particular foods that produce it is called _____ satiety.

Scan Your Brain answers: (1) stomach, (2) duodenum, (3) cephalic, (4) insulin, (5) free fatty acids, (6) glucose, (7) effector, (8) glucostatic, (9) positive-incentive, (10) salty, (11) sodium, (12) sensory-specific.

12.4
Physiological Research on Hunger and Satiety

Now that you have been introduced to set-point theories, the positive-incentive perspective, and some basic factors that affect why, when, and how much we eat, this section introduces you to five prominent lines of research on the physiology of hunger and satiety.

Role of Blood Glucose Levels in Hunger and Satiety

As I have already explained, efforts to link blood glucose levels to eating have been largely unsuccessful. However, there was a renewed interest in the role of glucose in the regulation of eating in the 1990s, following the development of methods of continually monitoring blood glucose levels. In the classic experiment of Campfield and Smith (1990), rats were housed individually, with free access to a mixed diet and water, and their blood glucose levels were continually monitored via a chronic intravenous catheter (i.e., a hypodermic needle located in a vein). In this situation, baseline blood glucose levels rarely fluctuated more than 2%. However, about 10 minutes before a meal was initiated, the levels suddenly dropped about 8% (see Figure 12.7).

Do the observed reductions in blood glucose before a meal lend support to the glucostatic theory of hunger? I think not, for five reasons:

● It is a simple matter to construct a situation in which drops in blood glucose levels do not precede eating (e.g., Strubbe & Steffens, 1977)—for example, by unexpectedly serving a food with a high positive-incentive value.

● The usual premeal decreases in blood glucose seem to be a response to the intention to start eating, not the other way round. The premeal decreases in blood glucose are typically preceded by increases in blood insulin levels, which indicates that the decreases do not reflect gradually declining energy reserves but are actively produced by an increase in blood levels of insulin (see Figure 12.7).

● If an expected meal is not served, blood glucose levels soon return to their previous homeostatic level.

● The glucose levels in the extracellular fluids that surround CNS neurons stay relatively constant, even when blood glucose levels drop (see Seeley & Woods, 2003).

● Injections of insulin do not reliably induce eating unless the injections are sufficiently great to reduce blood glucose levels by 50% (see Rowland, 1981), and large premeal infusions of glucose do not suppress eating (see Geiselman, 1987).

Myth of Hypothalamic Hunger and Satiety Centers

In the 1950s, experiments on rats seemed to suggest that eating behavior is controlled by two different regions of the hypothalamus: satiety by the **ventromedial**

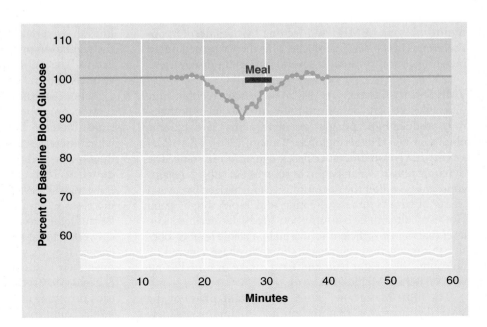

FIGURE 12.7 The meal-related changes in blood glucose levels observed by Campfield and Smith (1990).

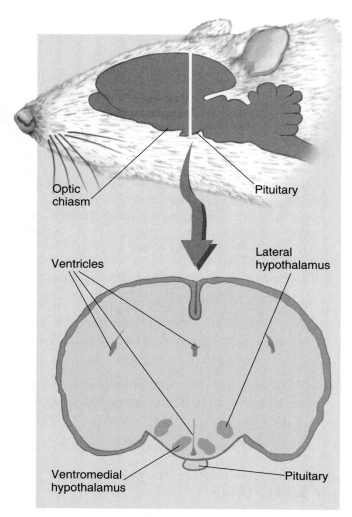

FIGURE 12.8 The locations in the rat brain of the ventromedial hypothalamus and the lateral hypothalamus.

hypothalamus (VMH) and feeding by the **lateral hypothalamus (LH)**—see Figure 12.8. This theory turned out to be wrong, but it stimulated several important discoveries.

VMH Satiety Center   In 1940, it was discovered that large bilateral electrolytic lesions to the ventromedial hypothalamus produce **hyperphagia** (excessive eating) and extreme obesity in rats (Hetherington & Ranson, 1940). This *VMH syndrome* has two different phases: dynamic and static. The **dynamic phase**, which begins as soon as the subject regains consciousness after the operation, is characterized by several weeks of grossly excessive eating and rapid weight gain. However, after that, consumption gradually declines to a level that is just sufficient to maintain a stable level of obesity; this marks the beginning of the **static phase**. Figure 12.9 illustrates the weight gain and food intake of an adult rat with bilateral VMH lesions.

The most important feature of the static phase of the VMH syndrome is that the animal maintains its new

body weight. If a rat in the static phase is deprived of food until it has lost a substantial amount of weight, it will regain the lost weight once the deprivation ends; conversely, if it is made to gain weight by forced feeding, it will lose the excess weight once the forced feeding is curtailed.

Paradoxically, despite their prodigious levels of consumption, VMH-lesioned rats in some ways seem less hungry than unlesioned controls. Although VMH-lesioned rats eat much more than normal rats when palatable food is readily available, they are less willing to work for it (Teitelbaum, 1957) or to consume it if it is slightly unpalatable (Miller, Bailey, & Stevenson, 1950). Weingarten, Chang, and Jarvie (1983) showed that the finicky eating of VMH-lesioned rats is a consequence of their obesity, not a primary effect of their lesion; they are no less likely to consume unpalatable food than are unlesioned rats of equal obesity.

LH Feeding Center   In 1951, Anand and Brobeck reported that bilateral electrolytic lesions to the *lateral hypothalamus* produce **aphagia**—a complete cessation of eating. Even rats that were first made hyperphagic by VMH lesions were rendered aphagic by the addition of LH lesions. Anand and Brobeck concluded that the lateral region of the hypothalamus is a feeding center. Teitelbaum and Epstein (1962) subsequently discovered two important features of the *LH syndrome*. First, they found that the aphagia was accompanied by **adipsia**—a complete cessation of drinking. Second, they found that LH-lesioned rats partially recover if they are kept alive by tube feeding. First, they begin to eat wet, palatable foods, such as chocolate chip cookies soaked in milk, and eventually they will eat dry food pellets if water is concurrently available.

Reinterpretation of the Effects of VMH and LH Lesions   The theory that the VMH is a satiety center crumbled in the face of two lines of evidence. One of these lines showed that the primary role of the hypothalamus is the regulation of energy metabolism, not the regulation of eating. The initial interpretation was that VMH-lesioned animals become obese because they overeat; however, the evidence suggests the converse—that they overeat because they become obese. Bilateral VMH lesions increase blood insulin levels, which increases **lipogenesis** (the production of body fat) and decreases **lipolysis** (the breakdown of body fat to utilizable forms of energy)—see Powley et al. (1980). Both are likely to be the result of the increases in insulin levels that occur following the lesion. Because the calories ingested by VMH-lesioned rats are converted to fat at a high rate, the rats must keep eating to ensure that they have enough calories in their blood to meet their immediate energy requirements (e.g., Hustvedt & Løvø, 1972); they are like misers who run to the bank each time they make a bit of money and deposit it in a savings account from which withdrawals cannot be made.

Thinking Creatively

FIGURE 12.9 Postoperative hyper-phagia and obesity in a rat with bilateral VMH lesions. (Based on Teitelbaum, 1961.)

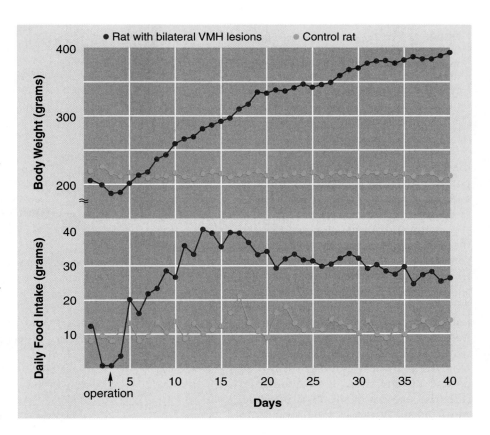

The second line of evidence that undermined the theory of a VMH satiety center has shown that many of the effects of VMH lesions are not attributable to VMH damage. A large fiber bundle, the *ventral noradrenergic bundle*, courses past the VMH and is thus inevitably damaged by large electrolytic VMH lesions; in particular, fibers that project from the nearby **paraventricular nuclei** of the hypothalamus are damaged (see Figure 12.10). Bilateral lesions of the noradrenergic bundle (e.g., Gold et al., 1977) or the paraventricular nuclei (Leibowitz, Hammer, & Chang, 1981) produce hyperphagia and obesity, just as VMH lesions do.

Most of the evidence against the notion that the LH is a feeding cen-

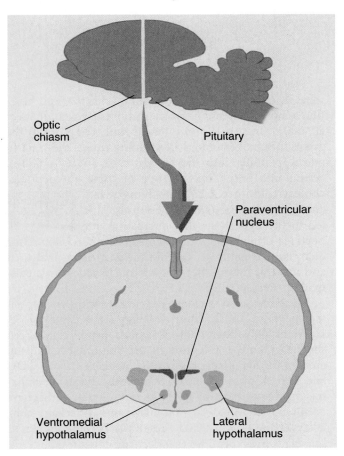

FIGURE 12.10 Location of the paraventricular nucleus in the rat hypothalamus. Note that the section through the hypothalamus is slightly different than the one in Figure 12.8.

ter has come from a thorough analysis of the effects of bilateral LH lesions. Early research focused exclusively on the aphagia and adipsia that are produced by LH lesions, but subsequent research has shown that LH lesions produce a wide range of severe motor disturbances and a general lack of responsiveness to sensory input (of which food and drink are but two examples). Consequently, the idea that the LH is a center specifically dedicated to feeding no longer warrants serious consideration.

Role of the Gastrointestinal Tract in Satiety

One of the most influential early studies of hunger was published by Cannon and Washburn in 1912. It was a perfect collaboration: Cannon had the ideas, and Washburn had the ability to swallow a balloon. First, Washburn swallowed an empty balloon tied to the end of a thin tube. Then, Cannon pumped some air into the balloon and connected the end of the tube to a water-filled glass U-tube so that Washburn's stomach contractions produced a momentary increase in the level of the water at

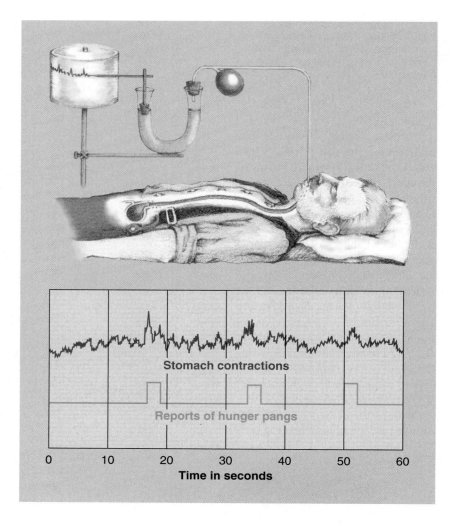

FIGURE 12.11 The system developed by Cannon and Washburn in 1912 for measuring stomach contractions. They found that large stomach contractions were related to pangs of hunger.

Stomach contractions

Reports of hunger pangs

0   10   20   30   40   50   60
Time in seconds

the other end of the U-tube. Washburn reported a "pang" of hunger each time that a large stomach contraction was recorded (see Figure 12.11).

Cannon and Washburn's finding led to the theory that hunger is the feeling of contractions caused by an empty stomach, whereas satiety is the feeling of stomach distention. However, support for this theory and interest in the role of the gastrointestinal tract in hunger and satiety quickly waned with the discovery that human patients whose stomach had been surgically removed and whose esophagus had been hooked up directly to their **duodenum** (the first segment of the small intestine, which normally carries food away from the stomach) continued to report feelings of hunger and satiety and continued to maintain their normal body weight by eating more meals of smaller size.

In the 1980s, there was a resurgence of interest in the role of the gastrointestinal tract in eating. It was stimulated by a series of experiments that indicated that the gastrointestinal tract is the source of satiety signals. For example, Koopmans (1981) transplanted an extra stomach and length of intestine into rats and then joined the major arteries and veins of the implants to the recipients' circulatory systems (see Figure 12.12). Koopmans found that food injected into the transplanted stomach and kept there by a noose around the *pyloric sphincter* decreased eating in proportion to both its caloric content and volume. Because the transplanted stomach had no functional nerves, the gastrointestinal satiety signal had to be reaching the brain through the blood. And because nutrients are not absorbed from the stomach, the bloodborne satiety signal could not have been a nutrient. It had to be some chemical or chemicals that were released from the stomach in response to the caloric value and volume of the food—which leads us nicely into the next subsection.

Hunger and Satiety Peptides

Soon after the discovery that the stomach and other parts of the gastrointestinal tract release chemical signals to the brain, evidence began to accumulate that these chemicals

were *peptides*, short chains of amino acids that can function as hormones and neurotransmitters (see Fukuhara et al., 2005). Ingested food interacts with receptors in the gastrointestinal tract and in so doing causes the tract to release peptides into the bloodstream. In 1973, Gibbs, Young, and Smith injected one of these gut peptides, **cholecystokinin (CCK)**, into hungry rats and found that they ate smaller meals. This led to the hypothesis that circulating gut peptides provide the brain with information about the quantity and nature of food in the gastrointestinal tract and that this information plays a role in satiety (see Badman & Flier, 2005; Flier, 2006).

There has been considerable support for the hypothesis that peptides can function as satiety signals (see Gao & Horvath, 2007; Ritter, 2004). Several gut peptides have been shown to bind to receptors in the brain, particularly in areas of the hypothalamus involved in energy metabolism, and a dozen or so (e.g., CCK, bombesin, glucagon, alpha-melanocyte-stimulating hormone, and somatostatin) have been reported to reduce food intake (see Batterham et al., 2006; Zhang et al., 2005). These have become known as *satiety peptides* (peptides that decrease appetite).

Evolutionary Perspective

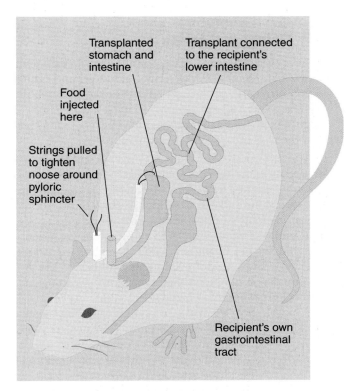

FIGURE 12.12 Transplantation of an extra stomach and length of intestine in a rat. Koopmans (1981) implanted an extra stomach and length of intestine in each of his experimental subjects. He then connected the major blood vessels of the implanted stomachs to the circulatory systems of the recipients. Food injected into the extra stomach and kept there by a noose around the pyloric sphincter decreased eating in proportion to its volume and caloric value.

In studying the appetite-reducing effects of peptides, researchers had to rule out the possibility that these effects are not merely the consequence of illness (see Moran, 2004). Indeed, there is evidence that one peptide in particular, CCK, induces illness: CCK administered to rats after they have eaten an unfamiliar substance induces a *conditioned taste aversion* for that substance, and CCK induces nausea in human subjects. However, CCK reduces appetite and eating at doses substantially below those that are required to induce taste aversion in rats, and thus it qualifies as a legitimate satiety peptide.

Several *hunger peptides* (peptides that increase appetite) have also been discovered. These peptides tend to be synthesized in the brain, particularly in the hypothalamus. The most widely studied of these are neuropeptide Y, galanin, orexin-A, and ghrelin (e.g., Baird, Gray, & Fischer, 2006; Olszewski, Schiöth & Levine, 2008; Williams et al., 2004).

The discovery of the hunger and satiety peptides has had two major effects on the search for the neural mechanisms of hunger and satiety. First, the sheer number of these hunger and satiety peptides indicates that the neural system that controls eating likely reacts to many different signals (Nogueiras & Tschöp, 2005; Schwartz & Azzara, 2004), not just to one or two (e.g., not just to glucose and fat). Second, the discovery that many of the hunger and satiety peptides have receptors in the hypothalamus has renewed interest in the role of the hypothalamus in hunger and eating (Gao & Horvath, 2007; Lam, Schwartz, & Rossetti, 2006; Luquet et al., 2005). This interest was further stimulated by the discovery that microinjection of gut peptides into some sites in the hypothalamus can have major effects on eating. Still, there is a general acceptance that hypothalamic circuits are only one part of a much larger system (see Berthoud & Morrison, 2008; Cone, 2005).

Serotonin and Satiety

The monoaminergic neurotransmitter serotonin is another chemical that plays a role in satiety. The initial evidence for this role came from a line of research in rats. In these studies, serotonin-produced satiety was found to have three major properties (see Blundell & Halford, 1998):

Evolutionary Perspective

- It caused the rats to resist the powerful attraction of highly palatable cafeteria diets.
- It reduced the amount of food that was consumed during each meal rather than reducing the number of meals (see Clifton, 2000).
- It was associated with a shift in food preferences away from fatty foods.

This profile of effects suggested that serotonin might be useful in combating obesity in humans. Indeed, serotonin agonists (e.g., fenfluramine, dexfenfluramine, fluoxetine) have been shown to reduce hunger, eating, and body weight under some conditions (see Blundell & Halford, 1998). Later in this chapter, you will learn about the use of serotonin to treat human obesity (see De Vry & Schreiber, 2000).

Prader-Willi Syndrome: Patients with Insatiable Hunger

Prader-Willi syndrome could prove critical in the discovery of the neural mechanisms of hunger and satiety (Goldstone, 2004). Individuals with **Prader-Willi syndrome,** which results from an accident of chromosomal replication, experience insatiable hunger, little or no satiety, and an exceptionally slow metabolism. In short, the Prader-Willi patient acts as though he or she is starving. Other common physical and neurological symptoms include weak muscles, small hands and feet, feeding difficulties in infancy, tantrums, compulsivity, and skin picking. If untreated, most patients become extremely obese, and they often die in early adulthood from diabetes, heart disease, or other obesity-related disorders. Some have even died from gorging until their stomachs split open. Fortunately, Miss A. was diagnosed in infancy and received excellent care, which kept her from becoming obese (Martin et al., 1998).

Prader-Willi Syndrome: The Case of Miss A.

Miss A. was born with little muscle tone. Because her sucking reflex was so weak, she was tube fed. By the time she was 2 years old, her *hypotonia* (below-normal muscle tone) had resolved itself, but a number of characteristic deformities and developmental delays began to appear.

At $3^{1}/_{2}$ years of age, Miss A. suddenly began to display a voracious appetite and quickly gained weight. Fortunately, her family maintained her on a low-calorie diet and kept all food locked away.

Miss A. is moderately retarded, and she suffers from psychiatric problems. Her major problem is her tendency to have tantrums any time anything changes in her environment (e.g., a substitute teacher at school). Thanks largely to her family and pediatrician, she has received excellent care, which has minimized the complications that arise with Prader-Willi syndrome—most notably those related to obesity and its pathological effects.

Although the study of Prader-Willi syndrome has yet to provide any direct evidence about the neural mechanisms of hunger and eating, there has been a marked surge in its investigation. This increase has been stimulated by the recent identification of the genetic cause of the condition: an accident of reproduction that deletes or disrupts a section of chromosome 15 coming from the father. This information has provided clues about genetic factors in appetite.

12.5
Body Weight Regulation: Set Points versus Settling Points

One strength of set-point theories of eating is that they explain body weight regulation. You have already learned that set-point theories are largely inconsistent with the facts of eating, but how well do they account for the regulation of body weight? Certainly, many people in our culture believe that body weight is regulated by a body-fat set point (Assanand, Pinel, & Lehman, 1998a, 1998b). They believe that when fat deposits are below a person's set point, a person becomes hungrier and eats more, which results in a return of body-fat levels to that person's set point; and, conversely, they believe that when fat deposits are above a person's set point, a person becomes less hungry and eats less, which results in a return of body-fat levels to their set point.

Set-Point Assumptions about Body Weight and Eating

You have already learned that set-point theories do a poor job of explaining the characteristics of hunger and eating. Do they do a better job of accounting for the facts of body weight regulation? Let's begin by looking at three lines of evidence that challenge fundamental aspects of many set-point theories of body weight regulation.

Variability of Body Weight   The set-point model was expressly designed to explain why adult body weights remain constant. Indeed, a set-point mechanism should make it virtually impossible for an adult to gain or lose large amounts of weight. Yet, many adults experience large and lasting changes in body weight (see Booth, 2004). Moreover, set-point thinking crumbles in the face of the epidemic of obesity that is currently sweeping fast-food societies (Rosenheck, 2008).

Set-point theories of body weight regulation suggest that the best method of maintaining a constant body weight is to eat each time there is a motivation to eat, because, according to the theory, the main function of hunger is to defend the set point. However, many people avoid obesity only by resisting their urges to eat.

Set Points and Health   One implication of set-point theories of body weight regulation is that each person's set point is optimal for that person's health—or at least not incompatible with good health. This is why popular psychologists commonly advise people to "listen to the wisdom of their bodies" and eat as much as they need to satisfy their hunger. Experimental results indicate that this common prescription for good health could not be further from the truth.

Two kinds of evidence suggest that typical *ad libitum* (free-feeding) levels of consumption are unhealthy (see Brownell & Rodin, 1994). First are the results of studies of humans who consume fewer calories than others. For example, people living on the Japanese island of Okinawa seemed to eat so few calories that their eating habits became a concern of health officials. When the health officials took a closer look, here is what they found (see Kagawa, 1978). Adult Okinawans were found to consume, on average, 20% fewer calories than other adult Japanese, and Okinawan school children were found to consume 38% fewer calories than recommended by public health officials. It was somewhat surprising then that rates of morbidity and mortality and of all aging-related diseases were found to be substantially lower in Okinawa than in other parts of Japan, a country in which overall levels of caloric intake and obesity are far below Western norms. For example, the death rates from stroke, cancer, and heart disease in Okinawa were only 59%, 69%, and 59%, respectively, of those in the rest of Japan. Indeed, the proportion of Okinawans living to be over 100 years of age was up to 40 times greater than that of inhabitants of various other regions of Japan.

The Okinawan study and the other studies that have reported major health benefits in humans who eat less (e.g., Manson et al., 1995; Meyer et al., 2006; Walford & Walford, 1994) are not controlled experiments; therefore, they must be interpreted with caution. For example, perhaps it is not simply the consumption of fewer calories that leads to health and longevity; perhaps in some cultures people who eat less tend to eat healthier diets.

Thinking Creatively

Controlled experimental demonstrations in over a dozen different mammalian species, including monkeys (see Coleman et al., 2009), of the beneficial effects of calorie restriction constitute the second kind of evidence that *ad libitum* levels of consumption are unhealthy. Fortunately, the results of such controlled experiments do not present the same problems of interpretation as do the findings of the Okinawa study and other similar correlational studies in humans. In typical *calorie-restriction experiments*, one group of subjects is allowed to eat as much as they choose, while other groups of subjects have their caloric intake of the same diets substantially reduced (by between 25% and 65% in various studies). Results of such experiments have been remarkably consistent (see Bucci, 1992; Masoro, 1988; Weindruch, 1996; Weindruch & Walford, 1988): In experiment after experiment, substantial reductions in the caloric intake of balanced diets have improved numerous indices of health and increased longevity. For example, in one experiment (Weindruch et al., 1986), groups of mice had their caloric intake of a well-balanced commercial diet reduced by either 25%, 55%, or 65% after weaning. All levels of dietary restriction substantially improved health and increased longevity, but the benefits were greatest in the mice whose intake was reduced the most. Those mice that consumed the least had the lowest incidence of cancer, the best immune responses, and the greatest maximum life span—they lived 67% longer than mice that ate as much as they liked. Evidence suggests that dietary restriction can have beneficial effects even if it is not initiated until later in life (Mair et al., 2003; Vaupel, Carey, & Christensen, 2003).

Evolutionary Perspective

Evolutionary Perspective

One important point about the results of the calorie-restriction experiments is that the health benefits of the restricted diets may not be entirely attributable to loss of body fat (see Weindruch, 1996). In some dietary restriction studies, the health of subjects has improved even if they did not reduce their body fat, and there are often no significant correlations between amount of weight loss and improvements in health. This suggests excessive energy consumption, independent of fat accumulation, may accelerate aging with all its attendant health problems (Lane, Ingram, & Roth, 2002; Prolla & Mattson, 2001).

Remarkably, there is evidence that dietary restriction can be used to treat some neurological conditions. Caloric restriction has been shown to reduce seizure susceptibility in human epileptics (see Maalouf, Rho, & Mattson, 2008) and to improve memory in the elderly (Witte et al., 2009). Please stop and think about the implications of all these findings about calorie restriction. How much do you eat?

Thinking Creatively

Regulation of Body Weight by Changes in the Efficiency of Energy Utilization   Implicit in many set-point theories is the premise that body weight is largely a function of how much a person eats. Of course, how much someone eats plays a role in his or her body weight, but it is now clear that the body controls its fat levels, to a large degree, by changing the efficiency with which it uses energy. As a person's level of body fat declines, that person starts to use energy resources more efficiently, which limits further weight loss (see Martin, White, & Hulsey, 1991); conversely, weight gain is limited by a progressive decrease in the efficiency of energy utilization. Rothwell and Stock (1982) created a group of obese rats by maintaining them on a cafeteria diet, and they found that the resting level of energy expenditure in these obese rats was 45% greater than in control rats.

This point is illustrated by the progressively declining effectiveness of weight-loss programs. Initially, low-calorie diets produce substantial weight loss. But the rate of weight loss diminishes with each successive week on the diet, until an equilibrium is achieved and little or no further weight loss occurs. Most dieters are familiar with this disappointing trend. A similar effect occurs with weight-gain programs (see Figure 12.13 on page 316).

The mechanism by which the body adjusts the efficiency of its energy utilization in response to its levels of body fat has been termed **diet-induced thermogenesis**. Increases in the levels of body fat produce increases in body temperature, which require additional energy to maintain them—and decreases in the level of body fat have the opposite effects (see Lazar, 2008).

There are major differences among humans both in **basal metabolic rate** (the rate at which energy is utilized to maintain bodily processes when resting) and in the ability to adjust the metabolic rate in response to changes in the levels of body fat. We all know people who remain slim even though they eat gluttonously. However, the research on calorie-restricted diets suggests that these people may not eat with impunity: There may be a health cost to pay for overeating even in the absence of obesity.

Set Points and Settling Points in Weight Control

The theory that eating is part of a system designed to defend a body-fat set point has long had its critics (see

FIGURE 12.13 The diminishing effects on body weight of a low-calorie diet and a high-calorie diet.

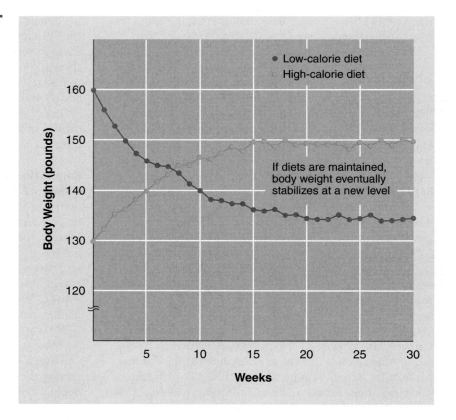

Booth, Fuller, & Lewis, 1981; Wirtshafter & Davis, 1977), but for many years their arguments were largely ignored and the set-point assumption ruled. This situation has been changing: Several prominent reviews of research on hunger and weight regulation generally acknowledge that a strict set-point model cannot account for the facts of weight regulation, and they argue for a more flexible model (see Berthoud, 2002; Mercer & Speakman, 2001; Woods et al., 2000). Because the body-fat set-point model still dominates the thinking of many people, I want to review the main advantages of an alternative and more flexible regulatory model: the settling-point model. Can you change your thinking?

Thinking Creatively

According to the settling-point model, body weight tends to drift around a natural **settling point**—the level at which the various factors that influence body weight achieve an equilibrium. The idea is that as body-fat levels increase, changes occur that tend to limit further increases until a balance is achieved between all factors that encourage weight gain and all those that discourage it.

The settling-point model provides a loose kind of homeostatic regulation, without a set-point mechanism or mechanisms to return body weight to a set point. According to the settling-point model, body weight remains stable as long as there are no long-term changes in the factors that influence it; and if there are such changes, their impact is limited by negative feedback. In the settling-point model, the negative feedback merely limits further changes in the same direction, whereas in the set-point model, negative feedback triggers a return to the set point. A neuron's resting potential is a well-known biological settling point—see Chapter 4.

The seductiveness of the set-point mechanism is attributable in no small part to the existence of the thermostat model, which provides a vivid means of thinking about it. Figure 12.14 presents an analogy I like to use to think about the settling-point mechanism. I call it the **leaky-barrel model**: (1) The amount of water entering the hose is analogous to the amount of food available to the subject; (2) the water pressure at the nozzle is analogous to the positive-incentive value of the available food; (3) the amount of water entering the barrel is analogous to the amount of

energy consumed; (4) the water level in the barrel is analogous to the level of body fat; (5) the amount of water leaking from the barrel is analogous to the amount of energy being expended; and (6) the weight of the barrel on the hose is analogous to the strength of the satiety signal.

The main advantage of the settling-point model of body weight regulation over the body-fat set-point model is that it is more consistent with the data. Another advantage is that in those cases in which both models make the same prediction, the settling-point model does so more parsimoniously—that is, with a simpler mechanism that requires fewer assumptions. Let's use the leaky-barrel analogy to see how the two models account for four key facts of weight regulation.

● Body weight remains relatively constant in many adult animals. On the basis of this fact, it has been argued that body fat must be regulated around a set point. However, constant body weight does not require, or even imply, a set point. Consider the leaky-barrel model. As water from the tap begins to fill the barrel, the weight of the water in the barrel increases. This increases the amount of water leaking out of the barrel and decreases the amount of water entering the barrel by increasing the pressure of the barrel on the hose. Eventually, this system settles into an equilibrium where the water level stays constant; but because this level is neither predetermined nor actively defended, it is a settling point, not a set point.

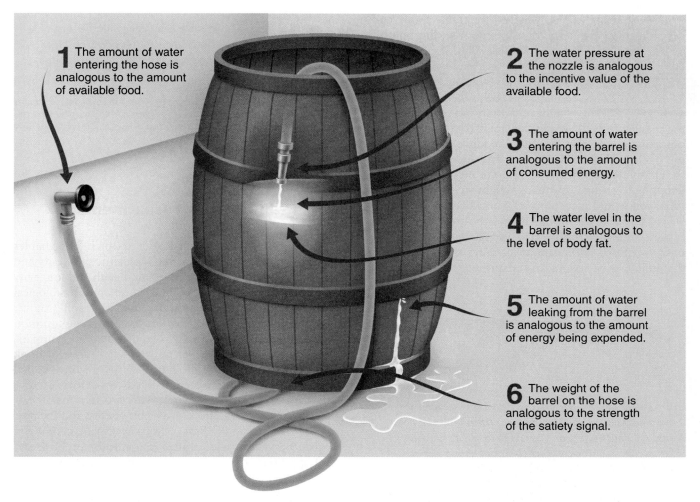

1 The amount of water entering the hose is analogous to the amount of available food.

2 The water pressure at the nozzle is analogous to the incentive value of the available food.

3 The amount of water entering the barrel is analogous to the amount of consumed energy.

4 The water level in the barrel is analogous to the level of body fat.

5 The amount of water leaking from the barrel is analogous to the amount of energy being expended.

6 The weight of the barrel on the hose is analogous to the strength of the satiety signal.

FIGURE 12.14   The leaky-barrel model: a settling-point model of eating and body weight homeostasis.

- Many adult animals experience enduring changes in body weight. Set-point systems are designed to maintain internal constancy in the face of fluctuations of the external environment. Thus, the fact that many adult animals experience long-term changes in body weight is a strong argument against the set-point model. In contrast, the settling-point model predicts that when there is an enduring change in one of the parameters that affect body weight—for example, a major increase in the positive-incentive value of available food—body weight will drift to a new settling point.

- If a subject's intake of food is reduced, metabolic changes that limit the loss of weight occur; the opposite happens when the subject overeats. This fact is often cited as evidence for set-point regulation of body weight; however, because the metabolic changes merely limit further weight changes rather than eliminating those that have occurred, they are more consistent with a settling-point model. For example, when water intake in the leaky-barrel model is reduced, the water

level in the barrel begins to drop; but the drop is limited by a decrease in leakage and an increase in inflow attributable to the falling water pressure in the barrel. Eventually, a new settling point is achieved, but the reduction in water level is not as great as one might expect because of the loss-limiting changes.

- After an individual has lost a substantial amount of weight (by dieting, exercise, or the surgical removal of fat), there is a tendency for the original weight to be regained once the subject returns to the previous eating- and energy-related lifestyle. Although this finding is often offered as irrefutable evidence of a body-weight set point, the settling-point model readily accounts for it. When the water level in the leaky-barrel model is reduced—by temporarily decreasing input (dieting), by temporarily increasing output (exercising), or by scooping out some of the water (surgical removal of fat)— only a temporary drop in the settling point is produced. When the original conditions are reinstated, the water level inexorably drifts back to the original settling point.

Does it really matter whether we think about body weight regulation in terms of set points or settling points—or is making such a distinction just splitting hairs? It certainly matters to biopsychologists: Understanding that

Thinking Creatively

body weight is regulated by a settling-point system helps them better understand, and more accurately predict, the changes in body weight that are likely to occur in various situations; it also indicates the kinds of physiological mechanisms that are likely to mediate these changes. And it should matter to you. If the set-point model is correct, attempting to change your body weight would be a waste of time; you would inevitably be drawn back to your body-weight set point. On the other hand, the leaky-barrel model suggests that it is possible to permanently change your body weight by permanently changing any of the factors that influence energy intake and output.

Scan Your Brain

Are you ready to move on to the final two sections of the chapter, which deal with eating disorders? This is a good place to pause and scan your brain to see if you understand the physiological mechanisms of eating and weight regulation. Complete the following sentences by filling in the blanks. The correct answers are provided at the end of the exercise. Before proceeding, review material related to your incorrect answers and omissions.

1. The expectation of a meal normally stimulates the release of _____ into the blood, which reduces blood glucose.
2. In the 1950s, the _____ hypothalamus was thought to be a satiety center.
3. A complete cessation of eating is called _____.
4. _____ is the breakdown of body fat to create usable forms of energy.
5. The classic study of Washburn and Cannon was the perfect collaboration: Cannon had the ideas, and Washburn could swallow a _____.
6. CCK is a gut peptide that is thought to be a _____ peptide.
7. _____ is the monoaminergic neurotransmitter that seems to play a role in satiety.
8. Okinawans eat less and live _____.
9. Experimental studies of _____ have shown that typical *ad libitum* (free-feeding) levels of consumption are unhealthy in many mammalian species.
10. As an individual grows fatter, further weight gain is minimized by diet-induced _____.

11. _____ models are more consistent with the facts of body-weight regulation than are set-point models.
12. _____ are to set points as leaky barrels are to settling points.

Scan Your Brain answers: (1) insulin, (2) ventromedial, (3) aphagia, (4) Lipolysis, (5) balloon, (6) satiety, (7) Serotonin, (8) longer, (9) calorie restriction, (10) thermogenesis, (11) Settling-point, (12) Thermostats.

12.6

Human Obesity: Causes, Mechanisms, and Treatments

This is an important point in this chapter. The chapter opened by describing the current epidemic of obesity and overweight and its adverse effects on health and longevity and then went on to discuss behavioral and physiological factors that influence eating and weight. Most importantly, as the chapter progressed, you learned that some common beliefs about eating and weight regulation are incompatible with the evidence, and you were challenged to think about eating and weight regulation in unconventional ways that are more consistent with current evidence. Now, the chapter completes the circle with two sections on eating disorders: This section focuses on obesity, and the next covers anorexia and bulimia. I hope that by this point you realize that obesity is currently a major health problem and will appreciate the relevance of what you are learning to your personal life and the lives of your loved ones.

Who Needs to Be Concerned about Obesity?

Almost everyone needs to be concerned about the problem of obesity. If you are currently overweight, the reason for concern is obvious: The relation between obesity and poor health has been repeatedly documented (see Eilat-Adar, Eldar, & Goldbourt, 2005; Ferrucci & Alley, 2007; Flegal et al., 2007; Hjartåker et al., 2005; Stevens, McClain, & Truesdale, 2006). Moreover, some studies have shown that even individuals who are only a bit overweight run a greater risk of developing health problems (Adams et al., 2006; Byers, 2006; Jee et al., 2006), as do obese individuals who manage to keep their blood pressure and blood cholesterol at normal levels (Yan et al., 2006). And the risk is not only to one's own health: Obese women are at increased risk of having infants with health problems (Nohr et al., 2007).

Even if you are currently slim, there is cause for concern about the problem of obesity. The incidence of obesity is so high that it is almost certain to be a problem for somebody you care about. Furthermore, because weight tends to increase substantially with age, many people who are slim as youths develop serious weight problems as they age.

There is cause for special concern for the next generation. Because rates of obesity are increasing in most parts of the world (Rosenheck, 2008; Sofsian, 2007), public health officials are concerned about how they are going to handle the growing problem. For example, it has been estimated that over one-third of the children born in the United States in 2000 will eventually develop diabetes, and 10% of these will develop related life-threatening conditions (see Haslam, Sattar, & Lean, 2006; Olshansky et al., 2005).

Why Is There an Epidemic of Obesity?

Let's begin our analysis of obesity by considering the pressures that are likely to have led to the evolution of our eating and weight-regulation systems (see Flier & Maratos-Flier, 2007; Lazar, 2005; Pinel et al., 2000). During the course of evolution, inconsistent food supplies were one of the main threats to survival. As a result, the fittest individuals were those who preferred high-calorie foods, ate to capacity when food was available, stored as many excess calories as possible in the form of body fat, and used their stores of calories as efficiently as possible. Individuals who did not have these characteristics were unlikely to survive a food shortage, and so these characteristics were passed on to future generations.

Evolutionary Perspective

The development of numerous cultural practices and beliefs that promote consumption has augmented the effects of evolution. For example, in my culture, it is commonly believed that one should eat three meals per day at regular times, whether one is hungry or not; that food should be the focus of most social gatherings; that meals should be served in courses of progressively increasing palatability; and that salt, sweets (e.g., sugar), and fats (e.g., butter or cream) should be added to foods to improve their flavor and thus increase their consumption.

Each of us possesses an eating and weight-regulation system that evolved to deal effectively with periodic food shortages, and many of us live in cultures whose eating-related practices evolved for the same purpose. However, our current environment differs from our "natural" environment in critical food-related ways. We live in an environment in which an endless variety of foods of the highest positive-incentive and caloric value are readily and continuously available. The consequence is an appallingly high level of consumption.

Why Do Some People Become Obese While Others Do Not?

Why do some people become obese while others living under the same obesity-promoting conditions do not? At a superficial level, the answer is obvious: Those who are obese are those whose energy intake has exceeded their energy output; those who are slim are those whose energy intake has not exceeded their energy output (see Nestle, 2007). Although this answer provides little insight, it does serve to emphasize that two kinds of individual differences play a role in obesity: those that lead to differences in energy input and those that lead to differences in energy output.

Differences in Consumption   There are many factors that lead some people to eat more than others who have comparable access to food. For example, some people consume more energy because they have strong preferences for the taste of high-calorie foods (see Blundell & Finlayson, 2004; Epstein et al., 2007); some consume more because they were raised in families and/or cultures that promote excessive eating; and some consume more because they have particularly large cephalic-phase responses to the sight or smell of food (Rodin, 1985).

👁 **Watch**
Eating and the Brain
www.mypsychlab.com

Differences in Energy Expenditure   With respect to energy output, people differ markedly from one another in the degree to which they can dissipate excess consumed energy. The most obvious difference is that people differ substantially in the amount of exercise they get; however, there are others. You have already learned about two of them: differences in *basal metabolic rate* and in the ability to react to fat increases by *diet-induced thermogenesis*. The third factor is called **NEAT**, or *nonexercise activity thermogenesis*, which is generated by activities such as fidgeting and the maintenance of posture and muscle tone (Ravussin & Danforth, 1999) and can play a small role in dissipating excess energy (Levine, Eberhardt, & Jensen, 1999; Ravussin, 2005).

Genetic Differences   Given the number of factors that can influence food consumption and energy metabolism, it is not surprising that many genes can influence body weight. Indeed, over 100 human chromosome loci (regions) have already been linked to obesity (see Fischer et al., 2009; Rankinen et al., 2006). However, because body weight is influenced by so many genes, it is proving difficult to understand how their interactions with one another and with experience contribute to obesity in healthy people. Although it is proving difficult to unravel the various genetic factors that influence variations in body weight among the healthy, single gene mutations have been linked to pathological conditions that involve obesity. You will encounter an example of such a condition later in this section.

Why Are Weight-Loss Programs Typically Ineffective?

Figure 12.15 describes the course of the typical weight-loss program. Most weight-loss programs are unsuccessful in the sense that, as predicted by the settling-point model, most of the lost weight is regained once the dieter

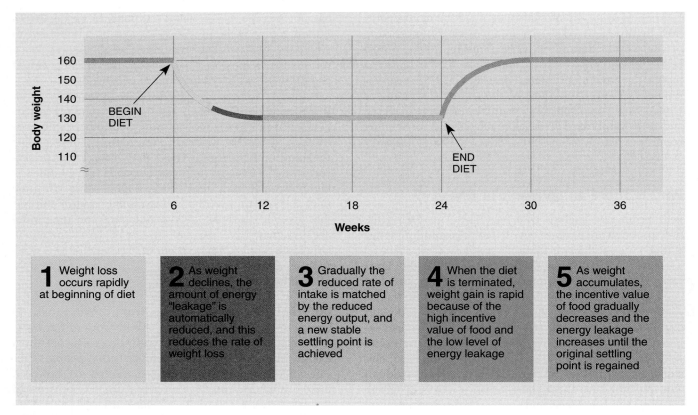

1 Weight loss occurs rapidly at beginning of diet

2 As weight declines, the amount of energy "leakage" is automatically reduced, and this reduces the rate of weight loss

3 Gradually the reduced rate of intake is matched by the reduced energy output, and a new stable settling point is achieved

4 When the diet is terminated, weight gain is rapid because of the high incentive value of food and the low level of energy leakage

5 As weight accumulates, the incentive value of food gradually decreases and the energy leakage increases until the original settling point is regained

FIGURE 12.15   The five stages of a typical weight-loss program.

stops following the program and the original conditions are reestablished. The key to permanent weight loss is a permanent lifestyle change.

Exercise has many health-promoting effects; however, despite the general belief that exercise is the most effective method of losing weight, several studies have shown that it often contributes little to weight loss (e.g., Sweeney et al., 1993). One reason is that physical exercise normally accounts for only a small proportion of total energy expenditure: About 80% of the energy you expend is used to maintain the resting physiological processes of your body and to digest your food (Calles-Escandon & Horton, 1992). Another reason is that our bodies are efficient machines, burning only a small number of calories during a typical workout. Moreover, after exercise, many people feel free to consume extra drinks and foods that contain more calories than the relatively small number that were expended during the exercise.

Leptin and the Regulation of Body Fat

Fat is more than a passive storehouse of energy; it actively releases a peptide hormone called **leptin**. The discovery of leptin has been extremely influential (see Elmquist & Flier, 2004). The following three subsections describe (1) the discovery of leptin, (2) how its discovery has fueled the development of a new approach to the treatment

of human obesity, and (3) how the understanding that leptin (and insulin) are feedback signals led to the discovery of a hypothalamic nucleus that plays an important role in the regulation of body fat.

Obese Mice and the Discovery of Leptin   In 1950, a spontaneous genetic mutation occurred in the mouse colony being maintained in the Jackson Laboratory at Bar Harbor, Maine. The mutant mice were *homozygous* for the gene (ob), and they were grossly obese, weighing up to three times as much as typical mice. These mutant mice are commonly referred to as **ob/ob mice**. See Figure 12.16.

Ob/ob mice eat more than control mice; they convert calories to fat more efficiently; and they use their calories more efficiently. Coleman (1979) hypothesized that ob/ob mice lack a critical hormone that normally inhibits fat production and maintenance.

Evolutionary Perspective

In 1994, Friedman and his colleagues characterized and cloned the gene that is mutated in ob/ob mice (Zhang et al., 1994). They found that this gene is *expressed* only in fat cells, and they characterized the protein that it normally encodes, a peptide hormone that they named leptin. Because of their mutation, ob/ob mice lack leptin. This finding led to an exciting hypothesis: Perhaps leptin is a negative feedback signal that is normally released from fat

FIGURE 12.16 An ob/ob mouse and a control mouse.

stores to decrease appetite and increase fat metabolism. Could leptin be administered to obese humans to reverse the current epidemic of obesity?

Leptin, Insulin, and the Arcuate Melanocortin System There was great fanfare when leptin was discovered. However, it was not the first peptide hormone to be discovered that seems to function as a negative feedback signal in the regulation of body fat (see Schwartz, 2000; Woods, 2004). More than 25 years ago, Woods and colleagues (1979) suggested that the pancreatic peptide hormone insulin serves such a function.

At first, the suggestion that insulin serves as a negative feedback signal for body fat regulation was viewed with skepticism. After all, how could the level of insulin in the body, which goes up and then comes back down to normal following each meal, provide the brain with information about gradually changing levels of body fat? It turns out that insulin does not readily penetrate the blood–brain barrier, and its levels in the brain were found to stay relatively stable—indeed, high levels of glucose are toxic to neurons (Tomlinson & Gardiner, 2008). The following findings supported the hypothesis that insulin serves as a negative feedback signal in the regulation of body fat:

- Brain levels of insulin were found to be positively correlated with levels of body fat (Seeley et al., 1996).
- Receptors for insulin were found in the brain (Baura et al., 1993).
- Infusions of insulin into the brains of laboratory animals were found to reduce eating and body weight (Campfield et al., 1995; Chavez, Seeley, & Woods, 1995).

Why are there two fat feedback signals? One reason may be that leptin levels are more closely correlated with **subcutaneous fat** (fat stored under the skin), whereas insulin levels are more closely correlated with **visceral fat** (fat stored around the internal organs of the body cavity)—see Hug & Lodish (2005). Thus, each fat signal provides different information. Visceral fat is more common in males than females and poses the greater threat to health (Wajchenberg, 2000). Insulin, but not leptin, is also involved in glucose regulation (see Schwartz & Porte, 2005).

The discovery that leptin and insulin are signals that provide information to the brain about fat levels in the body provided a means for discovering the neural circuits that participate in fat regulation. Receptors for both peptide hormones are located in many parts of the nervous system, but most are in the hypothalamus, particularly in one area of the hypothalamus: the **arcuate nucleus**.

A closer look at the distribution of leptin and insulin receptors in the arcuate nucleus indicated that these receptors are not randomly distributed throughout the nucleus. They are located in two classes of neurons: neurons that release **neuropeptide Y** (the gut hunger peptide that you read about earlier in the chapter), and neurons that release **melanocortins**, a class of peptides that includes the gut satiety peptide *α-melanocyte-stimulating hormone* (alpha-melanocyte-stimulating hormone). Attention has been mostly focused on the melanocortin-releasing neurons in the arcuate nucleus (often referred to as the **melanocortin system**) because injections of α-melanocyte-stimulating hormone have been shown to suppress eating and promote weight loss (see Horvath, 2005; Seeley & Woods, 2003). It seems, however, that the melanocortin system is only a minor component of a much larger system: Elimination of leptin receptors in the melanocortin system produces only a slight weight gain (see Münzberg & Myers, 2005).

Leptin as a Treatment for Human Obesity The early studies of leptin seemed to confirm the hypothesis that it could function as an effective treatment for obesity. Receptors for leptin were found in the brain, and injecting it into ob/ob mice reduced both their eating and their body fat (see Seeley & Woods, 2003). All that remained was to prove leptin's effectiveness in human patients.

However, when research on leptin turned from ob/ob mice to obese humans, the program ran into two major snags. First, obese humans—unlike ob/ob mice—were found to have high, rather than low, levels of leptin (see Münzberg & Myers, 2005). Second, injections of leptin did not reduce either the eating or the body fat of obese humans (see Heymsfield et al., 1999).

Why the actions of leptin are different in humans and ob/ob mice has yet to be explained. Nevertheless, efforts to use leptin in the treatment of human obesity have not been a total failure. Although few obese humans have a genetic mutation to the ob gene, leptin is a panacea for those few who do. Consider the following case.

Clinical Implications

The Case of the Child with No Leptin

The patient was of normal weight at birth, but her weight soon began to increase at an excessive rate. She demanded food continually and was disruptive when denied food. As a result of her extreme obesity, deformities of her legs developed, and surgery was required.

She was 9 when she was referred for treatment. At this point, she weighed 94.4 kilograms (about 210 pounds), and her weight was still increasing at an alarming rate. She was found to be homozygous for the ob gene and had no detectable leptin. Thus, leptin therapy was commenced.

The leptin therapy immediately curtailed the weight gain. She began to eat less, and she lost weight steadily over the 12-month period of the study, a total of 16.5 kilograms (about 36 pounds), almost all in the form of fat. There were no obvious side effects (Farooqi et al., 1999).

Treatment of Obesity

Because obesity is such a severe health problem, there have been many efforts to develop an effective treatment. Some of these—such as the leptin treatment you just read about—have worked for a few, but the problem of obesity continues to grow. The following two subsections discuss two treatments that are at different stages of development: serotonergic agonists and gastric surgery.

Serotonergic Agonists   Because—as you have already learned—serotonin agonists have been shown to reduce food consumption in both human and nonhuman subjects, they have considerable potential in the treatment of obesity (Halford & Blundell, 2000a). Serotonin agonists seem to act by a mechanism different from that for leptin and insulin, which produce long-term satiety signals based on fat stores. Serotonin agonists seem to increase short-term satiety signals associated with the consumption of a meal (Halford & Blundell, 2000b).

Serotonin agonists have been found in various studies of obese patients to reduce the following: the urge to eat high-calorie foods, the consumption of fat, the subjective inten- | *Clinical Implications* | sity of hunger, the size of meals, the number of between-meal snacks, and bingeing. Because of this extremely positive profile of effects and the severity of the obesity problem, serotonin agonists (fenfluramine and dexfenfluramine) were rushed into clinical use. However, they were subsequently withdrawn from the market because chronic use was found to be associated with heart disease in a small, but significant, number of users. Currently, the search is on for serotonergic weight-loss medications that do not have dangerous side effects.

Gastric Surgery   Cases of extreme obesity sometimes warrant extreme treatment. **Gastric bypass** is a surgical treatment for extreme obesity that involves short-circuiting the normal path of food through the digestive tract so that its absorption is reduced. The first gastric bypass was done in 1967, and it is currently the most commonly prescribed surgical treatment for extreme obesity. An alternative is the **adjustable gastric band procedure,** which involves surgically positioning a hollow silicone band around the stomach to reduce the flow of food through it; the circumference of the band can be adjusted by injecting saline into the band through a port that is implanted in the skin. One advantage of the gastric band over the gastric bypass is that the band can readily be removed.

The gastric bypass and adjustable gastric band are illustrated in Figure 12.17. A meta-analysis of studies comparing the two procedures found both to be highly effective (Maggard et al., 2005). However, neither procedure is effective unless patients change their eating habits.

12.7
Anorexia and Bulimia Nervosa

In contrast to obesity, **anorexia nervosa** is a disorder of underconsumption (see Södersten, Bergh, & Zandian, 2006). Anorexics eat so little that they experience health-threatening weight loss; and despite their emaciated appearance, they often perceive themselves as fat (see Benning- | *Clinical Implications* | hoven et al., 2006). Anorexia nervosa is a serious condition; In approximately 10% of diagnosed cases, complications from starvation result in death (Birmingham et al., 2005), and there is a high rate of suicide among anorexics (Pompili et al., 2004).

Anorexia nervosa is related to bulimia nervosa. **Bulimia nervosa** is a disorder characterized by periods of not eating interrupted by *bingeing* (eating huge amounts of food in short periods of time) followed by efforts to immediately eliminate the consumed calories from the body by voluntary *purging* (vomiting); by excessive use of laxatives, enemas, or diuretics; or by extreme exercise. Bulimics may be obese or of normal weight. If they are underweight, they are diagnosed as *bingeing anorexics.*

Relation between Anorexia and Bulimia

Are anorexia nervosa and bulimia nervosa really different disorders, as current convention dictates? The answer to this question depends on one's perspective. From the perspective of a physician, it is important to distinguish between these disorders because starvation pro- | *Thinking Creatively* | duces different health problems than does repeated bingeing and purging.

Gastric Bypass

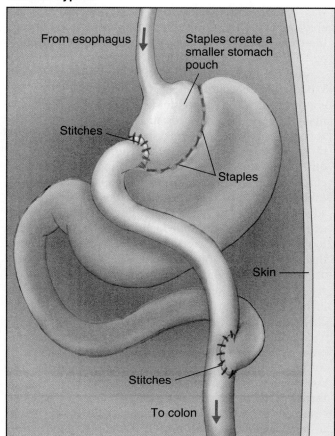

Adjustable Gastric Band

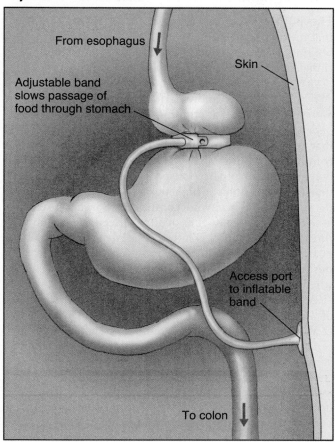

FIGURE 12.17 Two surgical methods for treating extreme obesity: gastric bypass and adjustable gastric band. The gastric band can be tightened by injecting saline into the access port implanted just beneath the skin.

For example, anorexics often require treatment for reduced metabolism, *bradycardia* (slow heart rate), *hypotension* (low blood pressure), *hypothermia* (low body temperature), and *anemia* (deficiency of red blood cells) (Miller et al., 2005). In contrast, bulimics often require treatment for irritation and inflammation of the esophagus, vitamin and mineral deficiencies, electrolyte imbalance, dehydration, and acid reflux.

Although anorexia and bulimia nervosa may seem like very different disorders from a physician's perspective, scientists often find it more appropriate to view them as variations of the same disorder. According to this view, both anorexia and bulimia begin with an obsession about body image and slimness and extreme efforts to lose weight. Both anorexics and bulimics attempt to lose weight by strict dieting, but bulimics are less capable of controlling their appetites and thus enter into a cycle of starvation, bingeing, and purging (see Russell, 1979). The following are other similarities that support the view that anorexia and bulimia are variants of the same disorder (see Kaye et al., 2005):

- Both anorexics and bulimics tend to have distorted body images, seeing themselves as much fatter and

less attractive than they are in reality (see Grant et al., 2002).

- In practice, many patients seem to straddle the two diagnoses and cannot readily be assigned to one or the other categories and many patients flip-flop between the two diagnoses as their circumstances change (Lask & Bryant-Waugh, 2000; Santonastaso et al., 2006; Tenconi et al., 2006).

- Anorexia and bulimia show the same pattern of distribution in the population. Although their overall incidence in the population is low (lifetime incidence estimates for American adults are 0.6% and 1.0% for anorexia and bulimia, respectively; Hudson et al., 2007), both conditions occur more commonly among educated females in affluent cultural groups (Lindberg & Hjern, 2003).

- Both anorexia and bulimia are highly correlated with obsessive-compulsive disorder and depression (Kaye et al., 2004; O'Brien & Vincent, 2003).

- Neither disorder responds well to existing therapies. Short-term improvements are common, but relapse is usual (see Södersten et al., 2006).

Anorexia and Positive Incentives

The positive-incentive perspective on eating suggests that the decline in eating that defines both anorexia (and bulimia) is likely a consequence of a corresponding decline in the positive-incentive value of food. However, the

Thinking Creatively

positive-incentive value of food for anorexia patients has received little attention—in part, because anorexic patients often display substantial interest in food. The fact that many anorexic patients are obsessed with food—continually talking about it, thinking about it, and preparing it for others (Crisp, 1983)—seems to suggest that food still holds a high positive-incentive value for them. However, to avoid confusion, it is necessary to keep in mind that the positive-incentive value of *interacting* with food is not necessarily the same as the positive-incentive value of *eating* food—and it is the positive-incentive value of eating food that is critical when considering anorexia nervosa.

A few studies have examined the positive-incentive value of various tastes in anorexic patients (see, e.g., Drewnowski et al., 1987; Roefs et al., 2006; Sunday & Halmi, 1990). In general, these studies have found that the positive-incentive value of various tastes is lower in anorexic patients than in control participants. However, these studies grossly underestimate the importance of reductions in the positive-incentive value of food in the etiology of anorexia nervosa, because the anorexic participants and the normal-weight control participants were not matched for weight—such matching is not practical.

We can get some insight into the effects of starvation on the positive-incentive value of food by studying starvation. That starvation normally triggers a radical increase in the positive-incentive value of food has been best documented by the descriptions and behavior of participants voluntarily undergoing experimental semistarvation. When asked how it felt to starve, one participant replied:

> I wait for mealtime. When it comes I eat slowly and make the food last as long as possible. The menu never gets monotonous even if it is the same each day or is of poor quality. It is food and all food tastes good. Even dirty crusts of bread in the street look appetizing. (Keys et al., 1950, p. 852)

Anorexia Nervosa: A Hypothesis

The dominance of set-point theories in research into the regulation of hunger and eating has resulted in widespread inattention to one of the major puzzles of anorexia: Why does the adaptive massive increase in the positive-incentive value of eating that occurs in victims of starvation not occur in starving anorexics? Under conditions of starvation, the positive-incentive value of eating normally increases to such high levels that it is difficult to imagine how anybody who was starving—no matter how

controlled, rigid, obsessive, and motivated that person was—could refrain from eating in the presence of palatable food. Why this protective mechanism is not activated in severe anorexics is a pressing question about the etiology of anorexia nervosa.

I believe that part of the answer lies in the research of Woods and his colleagues on the aversive physiological effects of meals. At the beginning of meals, people are normally in reasonably homeostatic balance, and this homeostasis is disrupted by the sudden infusion of calories. The

Thinking Creatively

other part of the answer lies in the finding that the aversive effects of meals are much greater in people who have been eating little (Brooks & Melnik, 1995). Meals, which produce adverse, but tolerable, effects in healthy individuals, may be extremely aversive for individuals who have undergone food deprivation. Evidence for the extremely noxious effects that eating meals has on starving humans is found in the reactions of World War II concentration camp victims to refeeding—many were rendered ill and some were even killed by the food given to them by their liberators (Keys et al., 1950; see also Soloman & Kirby, 1990).

So why do severe anorexics not experience a massive increase in the positive-incentive value of eating, similar to the increase experienced by other starving individuals? The answer may be *meals*—meals forced on these patients as a result of the misconception of our society that meals are

👁 **Watch**
Anorexia
www.mypsychlab.com

the healthy way to eat. Each meal consumed by an anorexic may produce a variety of conditioned taste aversions that reduce the motivation to eat. This hypothesis needs to be addressed because of its implication for treatment: Anorexic patients—or anybody else who is severely undernourished—should not be encouraged, or even permitted, to eat meals. They should be fed—or infused with—small amounts of food intermittently throughout the day.

I have described the preceding hypothesis to show you the value of the new ideas that you have encountered in this chapter: The major test of a new theory is whether it leads

Thinking Creatively

to innovative hypotheses. A while ago, as I was perusing an article on global famine and malnutrition, I noticed an intriguing comment: One of the clinical complications that results from feeding meals to famine victims is anorexia (Blackburn, 2001). What do you make of this?

The Case of the Anorexic Student

In a society in which obesity is the main disorder of consumption, anorexics are out of step. People who are struggling to eat less have difficulty understanding those who have to struggle to eat. Still, when you stare anorexia in the face, it is difficult not to be touched by it.

Clinical Implications

She began by telling me how much she had been enjoying the course and how sorry she was to be dropping out of the university. She was articulate and personable, and her grades were high—very high. Her problem was anorexia; she weighed only 82 pounds, and she was about to be hospitalized.

"But don't you want to eat?" I asked naively. "Don't you see that your plan to go to medical school will go up in smoke if you don't eat?"

"Of course I want to eat. I know I am terribly thin— my friends tell me I am. Believe me, I know this is wrecking my life. I try to eat, but I just can't force myself. In a strange way, I am pleased with my thinness."

She was upset, and I was embarrassed by my insensitivity. "It's too bad you're dropping out of the course before we cover the chapter on eating," I said, groping for safer ground.

"Oh, I've read it already," she responded. "It's the first chapter I looked at. It had quite an effect on me; a lot of things started to make more sense. The bit about positive incentives and learning was really good. I think my problem began when eating started to lose its positive-incentive value for me—in my mind, I kind of associated eating with being fat and all the boyfriend problems I was having. This made it easy to diet, but every once in a while I would get hungry and binge, or my parents would force me to eat a big meal. I would eat so much that I would feel ill. So I would put my finger down my throat and make myself throw up. This kept me from gaining weight, but I think it also taught my body to associate my favorite foods with illness— kind of a conditioned taste aversion. What do you think of my theory?"

Her insightfulness impressed me; it made me feel all the more sorry that she was going to discontinue her studies. After a lengthy chat, she got up to leave, and I walked her to the door of my office. I wished her luck and made her promise to come back for a visit. I never saw her again, but the image of her emaciated body walking down the hallway from my office has stayed with me.

Themes Revisited

Three of the book's four themes played prominent roles in this chapter. The thinking creatively theme was prevalent as you were challenged to critically evaluate your own beliefs and ambiguous research findings, to consider the scientific implications of your own experiences, and to think in new ways about phenomena with major personal and clinical implications. The chapter ended by using these new ideas to develop a potentially important hypothesis about the etiology of anorexia nervosa. Because of its emphasis on thinking, this chapter is my personal favorite.

Thinking Creatively

Both aspects of the evolutionary perspective theme were emphasized repeatedly. First, you saw how thinking about hunger and eating from an evolutionary perspective leads to important insights. Second, you saw how controlled research on nonhuman species has contributed to our current understanding of human hunger and eating.

Evolutionary Perspective

Finally, the clinical implications theme pervaded the chapter, but it was featured in the cases of the man who forgot not to eat, the child with Prader-Willi syndrome, the child with no leptin, and the anorexic student.

Clinical Implications

Think about It

1. Set-point theories suggest that attempts at permanent weight loss are a waste of time. On the basis of what you have learned in this chapter, design an effective and permanent weight-loss program.
2. Most of the eating-related health problems of people in our society occur because the conditions in which we live are different from those in which our species evolved. Discuss.
3. On the basis of what you have learned in this chapter, develop a feeding program for laboratory rats that would lead to obesity. Compare this program with the eating habits prevalent in your culture.
4. What causes anorexia nervosa? Summarize the evidence that supports your view.
5. Given the weight of evidence, why is the set-point theory of hunger and eating so prevalent?

Key Terms

Set point (p. 299)

12.1 Digestion, Energy Storage, and Energy Utilization

Digestion (p. 299)
Lipids (p. 300)
Amino acids (p. 300)
Glucose (p. 300)
Cephalic phase (p. 301)
Absorptive phase (p. 301)
Fasting phase (p. 301)
Insulin (p. 301)
Glucagon (p. 301)
Gluconeogenesis (p. 301)
Free fatty acids (p. 301)
Ketones (p. 301)

12.2 Theories of Hunger and Eating: Set Points versus Positive Incentives

Set-point assumption (p. 302)
Negative feedback systems (p. 303)

Homeostasis (p. 303)
Glucostatic theory (p. 303)
Lipostatic theory (p. 303)
Positive-incentive theory (p. 304)
Positive-incentive value (p. 304)

12.3 Factors That Determine What, When, and How Much We Eat

Satiety (p. 306)
Nutritive density (p. 306)
Sham eating (p. 306)
Appetizer effect (p. 307)
Cafeteria diet (p. 308)
Sensory-specific satiety (p. 308)

12.4 Physiological Research on Hunger and Satiety

Ventromedial hypothalamus (VMH) (p. 309)
Lateral hypothalamus (LH) (p. 310)
Hyperphagia (p. 310)

Dynamic phase (p. 310)
Static phase (p. 310)
Aphagia (p. 310)
Adipsia (p. 310)
Lipogenesis (p. 310)
Lipolysis (p. 310)
Paraventricular nuclei (p. 311)
Duodenum (p. 312)
Cholecystokinin (CCK) (p. 312)
Prader-Willi syndrome (p. 313)

12.5 Body Weight Regulation: Set Points versus Settling Points

Diet-induced thermogenesis (p. 315)
Basal metabolic rate (p. 315)
Settling point (p. 316)
Leaky-barrel model (p. 316)

12.6 Human Obesity: Causes, Mechanisms, and Treatments

NEAT (p. 319)
Leptin (p. 320)

Ob/ob mice (p. 320)
Subcutaneous fat (p. 321)
Visceral fat (p. 321)
Arcuate nucleus (p. 321)
Neuropeptide Y (p. 321)
Melanocortins (p. 321)
Melanocortin system (p. 321)
Gastric bypass (p, 322)
Adjustable gastric band procedure (p. 322)

12.7 Anorexia and Bulimia Nervosa

Anorexia nervosa (p. 322)
Bulimia nervosa (p. 322)

✔•–Quick Review   Test your comprehension of the chapter with this brief practice test. You can find the answers to these questions as well as more practice tests, activities, and other study resources at www.mypsychlab.com.

1. The phase of energy metabolism that often begins with the sight, the smell, or even the thought of food is the
   a. luteal phase.
   b. absorptive phase.
   c. cephalic phase.
   d. fasting phase.
   e. none of the above

2. The ventromedial hypothalamus (VH) was once believed to be
   a. part of the hippocampus.
   b. a satiety center.
   c. a hunger center.
   d. static.
   e. dynamic.

3. Patients with Prader-Willi syndrome suffer from
   a. anorexia nervosa.
   b. bulimia.
   c. an inability to digest fats.
   d. insatiable hunger.
   e. lack of memory for eating.

4. In comparison to obese people, slim people tend to
   a. have longer life expectancies.
   b. be healthier.
   c. be less efficient in their use of body energy.
   d. all of the above
   e. both a and b

5. Body fat releases a hormone called
   a. leptin.
   b. glucagon.
   c. insulin.
   d. glycogen.
   e. serotonin.

13 Hormones and Sex
What's Wrong with the Mamawawa?

This chapter is about hormones and sex, a topic that some regard as unfit for conversation but that fascinates many others. Perhaps the topic of hormones and sex is so fascinating because we are intrigued by the fact that our sex is so greatly influenced by the secretions of a small pair of glands. Because we each think of our gender as fundamental and immutable, it is a bit disturbing to think that it could be altered with a few surgical snips and some hormone injections. And there is something intriguing about the idea that our sex lives might be enhanced by the application of a few hormones. For whatever reason, the topic of hormones and sex is always a hit with my students. Some remarkable things await you in this chapter; let's go directly to them.

Men-Are-Men-and-Women-Are-Women Assumption

Many students bring a piece of excess baggage to the topic of hormones and sex: the men-are-men-and-women-are-women assumption—or "mamawawa." This assumption is seductive; it seems so right that we are continually drawn to it without considering alternative views. Unfortunately, it is fundamentally flawed.

The men-are-men-and-women-are-women assumption is the tendency to think about femaleness and maleness as discrete, mutually exclusive, opposite categories. In thinking about hormones and sex, this general attitude leads one to assume that females have female sex hormones that give them female bodies and make them do "female" things, and that males have male sex hormones that give them male bodies and make them do opposite "male" things. Despite

the fact that this approach to hormones and sex is inconsistent with the evidence, its simplicity, symmetry, and comfortable social implications draw us to it. That's why this chapter grapples with it throughout. In so doing, this chapter encourages you to think about hormones and sex in 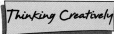 new ways that are more consistent with the evidence.

Developmental and Activational Effects of Sex Hormones

Before we begin discussing hormones and sex, you need to know that hormones influence sex in two fundamentally different ways (see Phoenix, 2008): (1) by influencing the development from conception to sexual maturity of the anatomical, physiological, and behavioral characteristics that distinguish one as female or male; and (2) by activating the reproduction-related behavior of sexually mature adults. Both the *developmental* (also called *organizational*) and *activational* effects of sex hormones are discussed in different sections of this chapter. Although the distinction between the developmental and activational effects of sex hormones is not always as clear as it was once assumed to be—for example, because the brain continues to develop into the late teens, adolescent hormone surges can have both effects—the distinction is still useful (Cohen-Bendahan, van de Beek, & Berenbaum, 2005).

13.1
Neuroendocrine System

This section introduces the general principles of neuroendocrine function. It introduces these principles by focusing on the glands and hormones that are directly involved in sexual development and behavior.

The endocrine glands are illustrated in Figure 13.1. By convention, only the organs whose primary function appears to be the release of hormones are referred to as *endocrine glands*. However, other organs (e.g., the

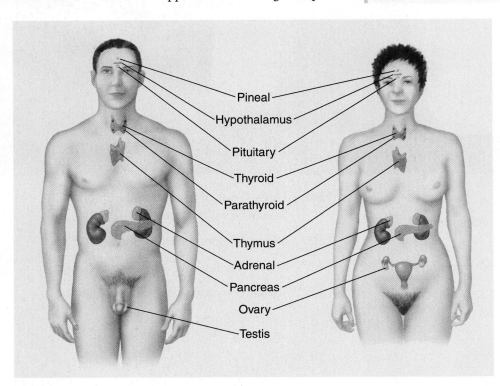

Pineal
Hypothalamus
Pituitary
Thyroid
Parathyroid
Thymus
Adrenal
Pancreas
Ovary
Testis

FIGURE 13.1 The endocrine glands.

stomach, liver, and intestine) and body fat also release hormones into general circulation (see Chapter 12), and they are thus, strictly speaking, also part of the endocrine system.

Glands

There are two types of glands: exocrine glands and endocrine glands. **Exocrine glands** (e.g., sweat glands) release their chemicals into ducts, which carry them to their targets, mostly on the surface of the body. **Endocrine glands** (ductless glands) release their chemicals, which are called **hormones**, directly into the circulatory system. Once released by an endocrine gland, a hormone travels via the circulatory system until it reaches the targets on which it normally exerts its effect (e.g., other endocrine glands or sites in the nervous system).

Gonads

Central to any discussion of hormones and sex are the **gonads**—the male **testes** (pronounced TEST-eez) and the female **ovaries** (see Figure 13.1). As you learned in Chapter 2, the primary function of the testes and ovaries is the production of *sperm cells* and *ova*, respectively. After **copulation** (sexual intercourse), a single sperm cell may *fertilize* an *ovum* to form one cell called a **zygote**, which contains all of the information necessary for the normal growth of a complete adult organism in its natural environment (see Primakoff & Myles, 2002). With the exception of ova and sperm cells, each cell of the human body has 23 pairs of chromosomes. In contrast, the ova and sperm cells contain only half that number, one member of each of the 23 pairs. Thus, when a sperm cell fertilizes an ovum, the resulting zygote ends up with the full complement of 23 pairs of chromosomes, one of each pair from the father and one of each pair from the mother.

Of particular interest in the context of this chapter is the pair of chromosomes called the **sex chromosomes**, so named because they contain the genetic programs that direct sexual development. The cells of females have two large sex chromosomes, called *X chromosomes*. In males, one sex chromosome is an X chromosome, and the other is called a *Y chromosome*. Consequently, the sex chromosome of every ovum is an X chromosome, whereas half the sperm cells have X chromosomes and half have Y chromosomes. Your sex with all its social, economic, and personal ramifications was determined by which of your father's sperm cells won the dash to your mother's ovum. If a sperm cell with an X sex chromosome won, you are a female; if one with a Y sex chromosome won, you are a male.

You might reasonably assume that X chromosomes are X-shaped and Y chromosomes are Y-shaped, but this is incorrect. Once a chromosome has duplicated, the two products remain joined at one point, producing an X shape. This is true of all chromosomes, including Y chromosomes. Because the Y chromosome is much smaller than

the X chromosome, early investigators failed to discern one small arm and thus saw a Y. In humans, Y-chromosome genes encode only 27 proteins; in comparison, about 1,500 proteins are encoded by X-chromosome genes (see Arnold, 2004).

Writing this section reminded me of my seventh-grade basketball team, the "Nads." The name puzzled our teacher because it was not at all like the names usually favored by pubescent boys—names such as the "Avengers," the "Marauders," and the "Vikings." Her puzzlement ended abruptly at our first game as our fans began to chant their support. You guessed it: "Go Nads, Go! Go Nads, Go!" My 14-year-old spotted-faced teammates and I considered this to be humor of the most mature and sophisticated sort. The teacher didn't.

Classes of Hormones

Vertebrate hormones fall into one of three classes: (1) amino acid derivatives, (2) peptides and proteins, and (3) steroids. **Amino acid derivative hormones** are hormones that are synthesized in a few simple steps from an amino acid molecule; an example is *epinephrine*, which is released from the *adrenal medulla* and synthesized from *tyrosine*. **Peptide hormones** and **protein hormones** are chains of amino acids—peptide hormones are short chains, and protein hormones are long chains. **Steroid hormones** are hormones that are synthesized from *cholesterol*, a type of fat molecule.

The hormones that influence sexual development and the activation of adult sexual behavior (i.e., the sex hormones) are all steroid hormones. Most other hormones produce their effects by binding to receptors in cell membranes. Steroid hormones can influence cells in this fashion; however, because they are small and fat-soluble, they can readily penetrate cell membranes and often affect cells in a second way. Once inside a cell, the steroid molecules can bind to receptors in the cytoplasm or nucleus and, by so doing, directly influence gene expression (amino acid derivative hormones and peptide hormones affect gene expression less commonly and by less direct mechanisms). Consequently, of all the hormones, steroid hormones tend to have the most diverse and long-lasting effects on cellular function (Brown, 1994).

Sex Steroids

The gonads do more than create sperm and egg cells; they also produce and release steroid hormones. Most people are surprised to learn that the testes and ovaries release the very same hormones. The two main classes of gonadal hormones are **androgens** and **estrogens; testosterone** is the most common androgen, and **estradiol** is the most common estrogen. The fact that adult ovaries tend to release more estrogens than they do androgens and that adult testes release more androgens than they do estrogens has led to the common, but misleading, practice of referring to androgens as "the *male* sex hormones" and to estrogens as "the *female* sex hormones." This practice should be avoided because of its men-are-men-and-women-are-women implication that androgens produce maleness and estrogens produce femaleness. They don't.

The ovaries and testes also release a third class of steroid hormones called **progestins**. The most common progestin is **progesterone**, which in women prepares the uterus and the breasts for pregnancy. Its function in men is unclear.

Because the primary function of the **adrenal cortex**—the outer layer of the *adrenal glands* (see Figure 13.1)—is the regulation of glucose and salt levels in the blood, it is not generally thought of as a sex gland. However, in addition to its principal steroid hormones, it does release small amounts of all of the sex steroids that are released by the gonads.

Hormones of the Pituitary

The pituitary gland is frequently referred to as the *master gland* because most of its hormones are tropic hormones. *Tropic hormones* are hormones whose primary function is to influence the release of hormones from other glands (*tropic* means "able to stimulate or change something"). For example, **gonadotropin** is a pituitary tropic hormone that travels through the circulatory system to the gonads, where it stimulates the release of gonadal hormones.

The pituitary gland is really two glands, the posterior pituitary and the anterior pituitary, which fuse during the

course of embryological development. The **posterior pituitary** develops from a small outgrowth of hypothalamic tissue that eventually comes to dangle from the *hypothalamus* on the end of the **pituitary stalk** (see Figure 13.2). In contrast, the **anterior pituitary** begins as part of the same embryonic tissue that eventually develops into the roof of the mouth; during the course of development, it pinches off and migrates upward to assume its position next to the posterior pituitary. It is the anterior pituitary that releases tropic hormones; thus, it is the anterior pituitary in particular, rather than the pituitary in general, that qualifies as the master gland.

Female Gonadal Hormone Levels Are Cyclic; Male Gonadal Hormone Levels Are Steady

Although men and women possess the same hormones, these hormones are not present at the same levels, and they do not necessarily perform the same functions. The major difference between the endocrine function of women and men is that in women the levels of gonadal and gonadotropic hormones go through a cycle that repeats itself every 28 days or so. It is these more-or-less regular hormone fluctuations that control the female **menstrual cycle**. In contrast, human males are, from a neuroendocrine perspective, rather dull creatures; males' levels of gonadal and gonadotropic hormones change little from day to day.

Because the anterior pituitary is the master gland, many early scientists assumed that an inherent difference between the male and female anterior pituitary was the basis for the difference in male and female patterns of gonadotropic and gonadal hormone release. However, this hypothesis was discounted by a series of clever transplant studies conducted by Geoffrey Harris in the 1950s (see Raisman, 1997). In these studies, a cycling pituitary removed from a mature female rat became a steady-state pituitary when transplanted at the appropriate site in a male, and a steady-state pituitary removed from a mature male rat began to cycle once transplanted into a female. What these studies established was that anterior pituitaries are not inherently female (cyclical) or male (steady-state); their patterns

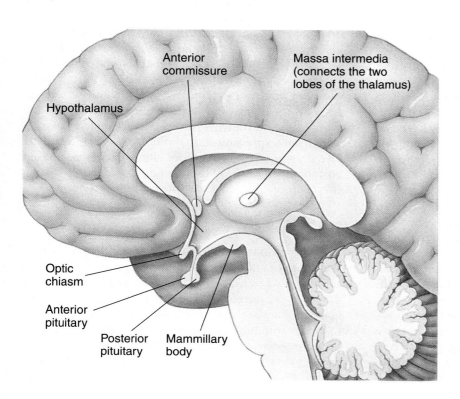

FIGURE 13.2 A midline view of the posterior and anterior pituitary and surrounding structures.

of hormone release are controlled by some other part of the body. The master gland seemed to have its own master. Where was it?

Neural Control of the Pituitary

The nervous system was implicated in the control of the anterior pituitary by behavioral research on birds and other animals that breed only during a specific time of the year. It was found that the seasonal variations in the light–dark cycle triggered many of the breeding-related changes in hormone release. If the lighting conditions under which the animals lived were reversed, for example, by having the animals transported across the equator, the breeding seasons were also reversed. Somehow, visual input to the nervous system was controlling the release of tropic hormones from the anterior pituitary.

The search for the particular neural structure that controlled the anterior pituitary turned, naturally enough, to the hypothalamus, the structure from which the pituitary is suspended. Hypothalamic stimulation and lesion experiments quickly established that the hypothalamus is the regulator of the anterior pituitary, but how the hypothalamus carries out this role remained a mystery. You see, the

anterior pituitary, unlike the posterior pituitary, receives no neural input whatsoever from the hypothalamus, or from any other neural structure (see Figure 13.3).

Control of the Anterior and Posterior Pituitary by the Hypothalamus

There are two different mechanisms by which the hypothalamus controls the pituitary: one for the posterior pituitary and one for the anterior pituitary. The two major hormones of the posterior pituitary, **vasopressin** and **oxytocin**, are peptide hormones that are synthesized in the cell bodies of neurons in the **paraventricular nuclei** and **supraoptic nuclei** on each side of the hypothalamus (see Figure 13.3 and Appendix VI). They are then transported along the axons of these neurons to their terminals in the posterior pituitary and are stored there until the arrival of action potentials causes them to be released into the bloodstream. (Neurons that release hormones into general circulation are called *neurosecretory cells*.) Oxytocin stimulates contractions of the uterus during labor and the ejection of milk during suckling. Vasopressin (also called *antidiuretic hormone*) facilitates the reabsorption of water by the kidneys.

The means by which the hypothalamus controls the release of hormones from the neuron-free anterior pituitary was more difficult to explain. Harris (1955) suggested that the release of hormones from the anterior pituitary was itself regulated by hormones released from the hypothalamus. Two findings provided early support for this hypothesis. The first was the discovery of a vascular network, the **hypothalamopituitary portal system**, that seemed well suited to the task of carrying hormones from the hypothalamus to the anterior pituitary. As Figure 13.4 on page 332 illustrates, a network of hypothalamic capillaries feeds a bundle of portal veins that carries blood down the pituitary stalk into another network of capillaries in the anterior pituitary. (A *portal vein* is a vein that connects one capillary network with another.) The second finding was the discovery that cutting the portal veins of the pituitary stalk disrupts the release of anterior pituitary hormones until the damaged veins regenerate (Harris, 1955).

Discovery of Hypothalamic Releasing Hormones

It was hypothesized that the release of each anterior pituitary hormone is controlled by a different hypothalamic hormone. The hypothalamic hormones that were thought to stimulate the release of an anterior pituitary hormone were referred to as **releasing hormones**; those thought to inhibit the release of an anterior pituitary hormone were referred to as **release-inhibiting factors**.

Efforts to isolate the putative (hypothesized) hypothalamic releasing and inhibitory factors led to a major breakthrough in the late 1960s. Guillemin and his colleagues isolated **thyrotropin-releasing hormone** from

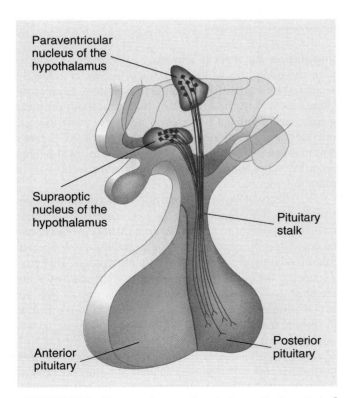

FIGURE 13.3 The neural connections between the hypothalamus and the pituitary. All neural input to the pituitary goes to the posterior pituitary; the anterior pituitary has no neural connections.

Paraventricular nucleus of the hypothalamus

Supraoptic nucleus of the hypothalamus

Anterior pituitary

Pituitary stalk

Posterior pituitary

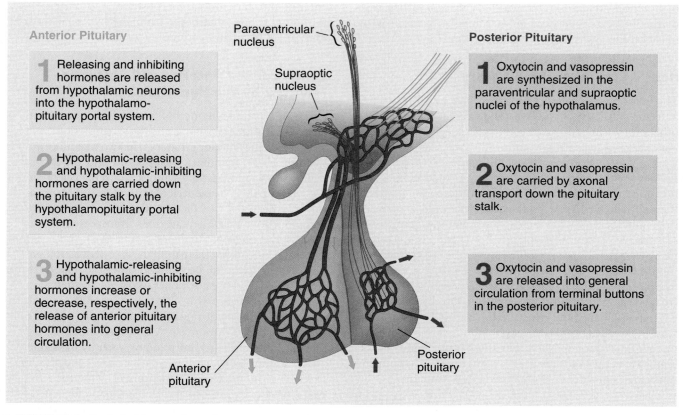

FIGURE 13.4 Control of the anterior and posterior pituitary by the hypothalamus.

the hypothalamus of sheep, and Schally and his colleagues isolated the same hormone from the hypothalamus of pigs.

 Thyrotropin-releasing hormone triggers the release of **thyrotropin** from the anterior pituitary, which in turn stimulates the release of hormones from the *thyroid gland*. For their efforts, Guillemin and Schally were awarded Nobel Prizes in 1977.

Schally's and Guillemin's isolation of thyrotropin-releasing hormone confirmed that hypothalamic releasing hormones control the release of hormones from the anterior pituitary and thus provided the major impetus for the isolation and synthesis of several other releasing hormones. Of direct relevance to the study of sex hormones was the subsequent isolation of **gonadotropin-releasing hormone** by Schally and his group (Schally, Kastin, & Arimura, 1971). This releasing hormone stimulates the release of both of the anterior pituitary's gonadotropins: **follicle-stimulating hormone (FSH)** and **luteinizing hormone (LH)**. All hypothalamic releasing hormones, like all tropic hormones, have proven to be peptides.

Regulation of Hormone Levels

Hormone release is regulated by three different kinds of signals: signals from the nervous system, signals from hormones, and signals from nonhormonal chemicals in the blood.

Regulation by Neural Signals All endocrine glands, with the exception of the anterior pituitary, are directly regulated by signals from the nervous system. Endocrine glands located in the brain (i.e., the pituitary and pineal glands) are regulated by cerebral neurons; those located outside the CNS are innervated by the *autonomic nervous system*—usually by both the *sympathetic* and *parasympathetic* branches, which often have opposite effects on hormone release.

The effects of experience on hormone release are usually mediated by signals from the nervous system. It is extremely important to remember that hormone release can be regulated by experience—for example, many species that breed only in the spring are often prepared for reproduction by the release of sex hormones triggered by the increasing daily duration of daylight. This means that an explanation of any behavioral phenomenon in terms of a hormonal mechanism does not necessarily rule out an explanation in terms 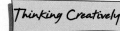 of an experiential mechanism. Indeed, hormonal and experiential explanations may merely be different aspects of the same hypothetical mechanism.

Regulation by Hormonal Signals The hormones themselves also influence hormone release. You have already learned, for example, that the tropic hormones of the anterior pituitary influence the release of hormones from their respective target glands. However, the regulation of endocrine function by the anterior pituitary is not a one-way street. Circulating hormones often provide feedback to the very structures that influence their release: the pituitary gland, the hypothalamus, and other sites in the brain. The function of most hormonal feedback is the maintenance of stable blood levels of the hormones. Thus, high gonadal hormone levels usually have effects on the hypothalamus and pituitary that decrease subsequent gonadal hormone release, and low levels usually have effects that increase hormone release.

Regulation by Nonhormonal Chemicals   Circulating chemicals other than hormones can play a role in regulating hormone levels. Glucose, calcium, and sodium levels in the blood all influence the release of particular hormones. For example, you learned in Chapter 12 that increases in blood glucose increase the release of *insulin* from the *pancreas*, and insulin, in turn, reduces blood glucose levels.

Pulsatile Hormone Release

Hormones tend to be released in pulses (see Armstrong et al., 2009; Khadra & Li, 2006); they are discharged several times per day in large surges, which typically last no more than a few minutes. Hormone levels in the blood are regulated by changes in the frequency and duration of the hormone pulses. One consequence of **pulsatile hormone release** is that there are often large minute-to-minute fluctuations in the levels of circulating hormones (e.g., Koolhaas, Schuurman, & Wierpkema, 1980). Accordingly, when the pattern of human male gonadal hormone release is referred to as "steady," it means that there are no major systematic changes in circulating gonadal hormone levels from day to day, not that the levels never vary.

Summary Model of Gonadal Endocrine Regulation

Figure 13.5 is a summary model of the regulation of gonadal hormones. According to this model, the brain controls the release of gonadotropin-releasing hormone from the hypothalamus into the hypothalamo-pituitary portal system, which carries it to the anterior pituitary. In the anterior pituitary, the gonadotropin-releasing hormone stimulates the release of gonadotropin, which is carried by the circulatory system to the gonads. In response to the gonadotropin, the

gonads release androgens, estrogens, and progestins, which feed back into the pituitary and hypothalamus to regulate subsequent gonadal hormone release.

Armed with this general perspective of neuroendocrine function, you are ready to consider how gonadal hormones direct sexual development and activate adult sexual behavior.

13.2
Hormones and Sexual Development of the Body

You have undoubtedly noticed that humans are *dimorphic*—that is, they come in two standard models: female and male. This section describes how the development of female and male bodily characteristics is directed by hormones.

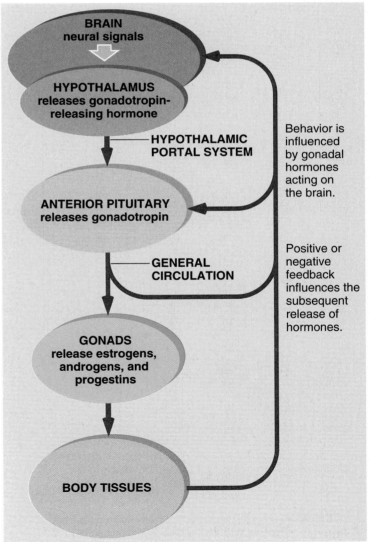

FIGURE 13.5   A summary model of the regulation of gonadal hormones.

Sexual differentiation in mammals begins at fertilization with the production of one of two different kinds of zygotes: either one with an XX (female) pair of sex chromosomes or one with an XY (male) pair. It is the genetic information on the sex chromosomes that normally determines whether development will occur along female or male lines. But be cautious here: Do not fall into the seductive embrace of the men-are-men-and-women-are-women assumption. Do not begin by assuming that there are two parallel but opposite genetic programs for sexual development, one for female development and one for male development. As you are about to learn, sexual development unfolds according to an entirely different principle, one that males who still stubbornly cling to notions of male preeminence find unsettling. This principle is that we are all genetically programmed to develop female bodies; genetic males develop male bodies only because their fundamentally female program of development is overruled.

Thinking Creatively

Fetal Hormones and Development of Reproductive Organs

Gonads   Figure 13.6 illustrates the structure of the gonads as they appear 6 weeks after fertilization. Notice

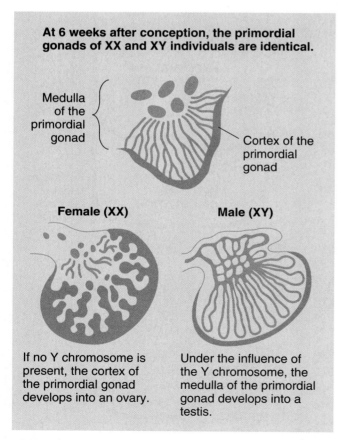

At 6 weeks after conception, the primordial gonads of XX and XY individuals are identical.

Medulla of the primordial gonad

Cortex of the primordial gonad

Female (XX)

Male (XY)

If no Y chromosome is present, the cortex of the primordial gonad develops into an ovary.

Under the influence of the Y chromosome, the medulla of the primordial gonad develops into a testis.

FIGURE 13.6   The development of an ovary and a testis from the cortex and the medulla, respectively, of the primordial gonadal structure that is present 6 weeks after conception.

that at this stage of development, each fetus, regardless of its genetic sex, has the same pair of gonadal structures, called *primordial gonads* (*primordial* means "existing at the beginning"). Each primordial gonad has an outer covering, or *cortex*, which has the potential to develop into an ovary; and each has an internal core, or *medulla*, which has the potential to develop into a testis.

Six weeks after conception, the **Sry gene** on the Y chromosome of the male triggers the synthesis of **Sry protein** (see Arnold, 2004; Wu et al., 2009), and this protein causes the medulla of each primordial gonad to grow and to develop into a testis. There is no female counterpart of Sry protein; in the absence of Sry protein, the cortical cells of the primordial gonads automatically develop into ovaries. Accordingly, if Sry protein is injected into a genetic female fetus 6 weeks after conception, the result is a genetic female with testes; or if drugs that block the effects of Sry protein are injected into a male fetus, the result is a genetic male with ovaries. Such "mixed-sex" individuals expose in a dramatic fashion the weakness of mamawawa thinking (thinking of "male" and "female" as mutually exclusive, opposite categories).

Internal Reproductive Ducts   Six weeks after fertilization, both males and females have two complete sets of reproductive ducts. They have a male **Wolffian system**, which has the capacity to develop into the male reproductive ducts (e.g., the *seminal vesicles*, which hold the fluid in which sperm cells are ejaculated; and the *vas deferens*, through which the sperm cells travel to the seminal vesicles). And they have a female **Müllerian system**, which has the capacity to develop into the female ducts (e.g., the *uterus;* the upper part of the *vagina;* and the *fallopian tubes*, through which ova travel from the ovaries to the uterus, where they can be fertilized).

In the third month of male fetal development, the testes secrete testosterone and **Müllerian-inhibiting substance.** As Figure 13.7 illustrates, the testosterone stimulates the development of the Wolffian system, and the Müllerian-inhibiting substance causes the Müllerian system to degenerate and the testes to descend into the **scrotum**—the sac that holds the testes outside the body cavity. Because it is testosterone—not the sex chromosomes—that triggers Wolffian development, genetic females who are injected with testosterone during the appropriate fetal period develop male reproductive ducts along with their female ones.

The differentiation of the internal ducts of the female reproductive system (see Figure 13.7) is not under the control of ovarian hormones; the ovaries are almost completely inactive during fetal development. The development of the Müllerian system occurs in any fetus that is not exposed to testicular hormones during the critical fetal period. Accordingly, normal female fetuses, ovariectomized female fetuses, and orchidectomized male fetuses all develop female reproductive ducts (Jost, 1972). **Ovariectomy** is the removal of the ovaries, and **orchidectomy** is the removal of the testes

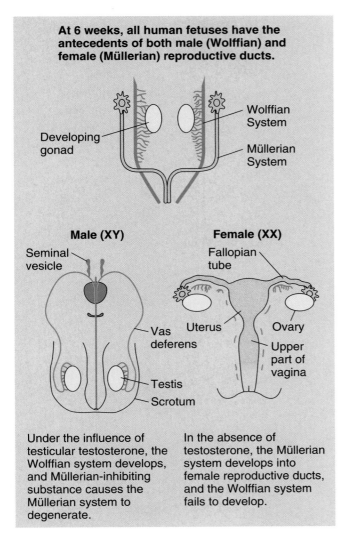

At 6 weeks, all human fetuses have the antecedents of both male (Wolffian) and female (Müllerian) reproductive ducts.

Developing gonad

Wolffian System

Müllerian System

Male (XY)

Seminal vesicle

Vas deferens

Testis

Scrotum

Female (XX)

Fallopian tube

Uterus

Ovary

Upper part of vagina

Under the influence of testicular testosterone, the Wolffian system develops, and Müllerian-inhibiting substance causes the Müllerian system to degenerate.

In the absence of testosterone, the Müllerian system develops into female reproductive ducts, and the Wolffian system fails to develop.

FIGURE 13.7 The development of the internal ducts of the male and female reproductive systems from the Wolffian and Müllerian systems, respectively.

(the Greek word *orchis* means "testicle"). **Gonadectomy**, or *castration*, is the surgical removal of gonads—either ovaries or testes.

External Reproductive Organs There is a basic difference between the differentiation of the external reproductive organs and the differentiation of the internal reproductive organs (i.e., the gonads and reproductive ducts). As you have just read, every normal fetus develops separate precursors for the male (medulla) and female (cortex) gonads and for the male (Wolffian system) and female (Müllerian system) reproductive ducts; then, only one set, male or female, develops. In contrast, both male and female **genitals**—external reproductive organs—develop from the same precursor. This

⊙→ Simulate Differentiating the External Genitals: The Penis and the Vagina
www.mypsychlab.com

bipotential precursor and its subsequent differentiation are illustrated in Figure 13.8 on page 336.

In the second month of pregnancy, the bipotential precursor of the external reproductive organs consists of four parts: the glans, the urethral folds, the lateral bodies, and the labioscrotal swellings. Then it begins to differentiate. The *glans* grows into the head of the *penis* in the male or the *clitoris* in the female; the *urethral folds* fuse in the male or enlarge to become the *labia minora* in the female; the *lateral bodies* form the shaft of the penis in the male or the hood of the clitoris in the female; and the *labioscrotal swellings* form the *scrotum* in the male or the *labia majora* in the female.

Like the development of the internal reproductive ducts, the development of the external genitals is controlled by the presence or absence of testosterone. If testosterone is present at the appropriate stage of fetal development, male external genitals develop from the bipotential precursor; if testosterone is not present, development of the external genitals proceeds along female lines.

Puberty: Hormones and Development of Secondary Sex Characteristics

During childhood, levels of circulating gonadal hormones are low, reproductive organs are immature, and males and females differ little in general appearance. This period of developmental quiescence ends abruptly with the onset of *puberty*—the transitional period between childhood and adulthood during which fertility is achieved, the adolescent growth spurt occurs, and the secondary sex characteristics develop. **Secondary sex characteristics** are those features other than the reproductive organs that distinguish sexually mature men and women. The body changes that occur during puberty are illustrated in Figure 13.9 on page 337.

Puberty is associated with an increase in the release of hormones by the anterior pituitary (see Grumbach, 2002). The increase in the release of **growth hormone**—the only anterior pituitary hormone that does not have a gland as its primary target—acts directly on bone and muscle tissue to produce the pubertal growth spurt. Increases in the release of gonadotropic hormone and **adrenocorticotropic hormone** cause the gonads and adrenal cortex to increase their release of gonadal and adrenal hormones, which in turn initiate the maturation of the genitals and the development of secondary sex characteristics.

The general principle guiding normal pubertal sexual maturation is a simple one: In pubertal males, androgen levels are higher than estrogen levels, and masculinization is the result; in pubertal females, the estrogens predominate, and the result is feminization. Individuals castrated prior to puberty do not become sexually mature unless they receive replacement injections of androgens or estrogens.

But even during puberty, its only period of relevance, the men-are-men-and-women-are-women assumption stumbles badly. You see, **androstenedione**, an androgen

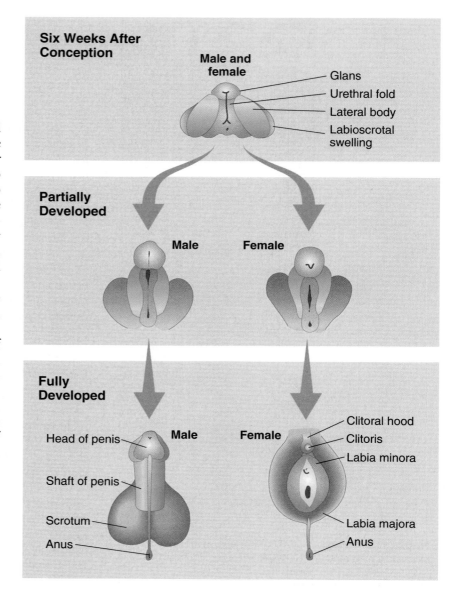

that is released primarily by the adrenal cortex, is normally responsible for the growth of pubic hair and *axillary hair* (underarm hair) in females. It is hard to take seriously the practice of referring to androgens as "male hormones" when one of them is responsible for the development of the female pattern of pubic hair growth. The male pattern is a pyramid, and the female pattern is an inverted pyramid (see Figure 13.9).

Do you remember how old you were when you started to go through puberty? In most North American and European countries, puberty begins at about 10.5 years of age for girls and 11.5 years for boys. I am sure you would have been unhappy if you had not started puberty until you were 15 or 16, but this was the norm in North America and Europe just a century and a half ago. Presumably, this acceleration of puberty has resulted from improvements in dietary, medical, and socioeconomic conditions.

13.3
Hormones and Sexual Development of Brain and Behavior

Biopsychologists have been particularly interested in the effects of hormones on the sexual differentiation of the brain and the effects of brain differences on behavior. This section reveals how seminal studies conducted in the 1930s generated theories that have gradually morphed, under the influence of subsequent research, into our current views. But first, let's take a quick look at the differences between male and female brains.

Sex Differences in the Brain

The brains of men and women may look the same on casual inspection, and it may be politically correct to believe that they are—but they are not. The brains of men tend to be about 15% larger than those of women, and many other anatomical differences between average male and female brains have been documented. There are statistically

significant sex differences in the volumes of various nuclei and fiber tracts, in the numbers and types of neural and glial cells that compose various structures, and in the numbers and types of synapses that connect the cells in various structures. *Sexual dimorphisms* (male–female structural differences) of the brain are typically studied in nonhuman mammals, but many have also been documented in humans (see Arnold, 2003; Cahill, 2005, 2006; de Vries & Södersten, 2009).

Let's begin with the first functional sex difference to be identified in mammalian brains. It set the stage for everything that followed.

First Discovery of a Sex Difference in Mammalian Brain Function The first attempts to discover sex differences in the mammalian brain focused on the factors that control the development of the steady and cyclic patterns

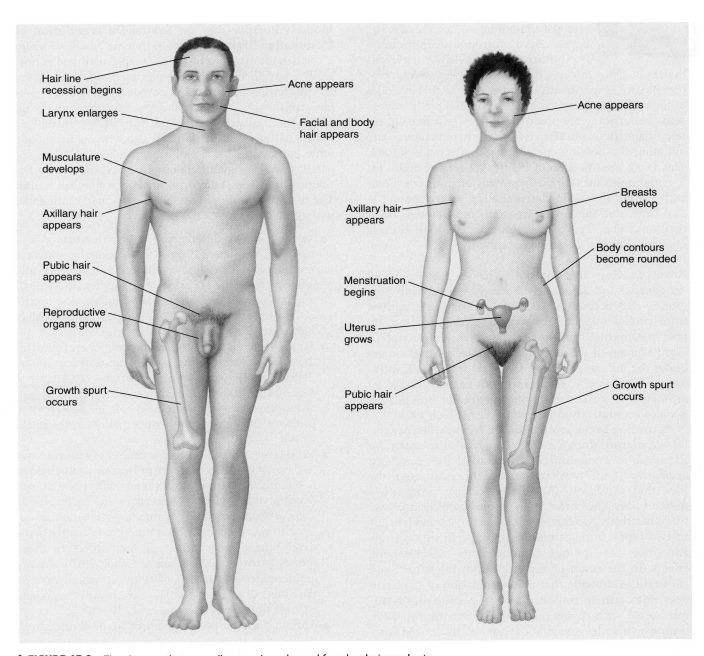

Hair line recession begins

Acne appears

Larynx enlarges

Facial and body hair appears

Musculature develops

Axillary hair appears

Pubic hair appears

Reproductive organs grow

Growth spurt occurs

Acne appears

Axillary hair appears

Breasts develop

Body contours become rounded

Menstruation begins

Uterus grows

Pubic hair appears

Growth spurt occurs

FIGURE 13.9 The changes that normally occur in males and females during puberty.

Evolutionary Perspective of gonadotropin release in males and females, respectively. The seminal experiments were conducted by Pfeiffer in 1936. In his experiments, some neonatal rats (males and females) were gonadectomized and some were not, and some received gonad transplants (ovaries or testes) and some did not.

Remarkably, Pfeiffer found that gonadectomizing neonatal rats of either genetic sex caused them to develop into adults with the female cyclic pattern of gonadotropin release. In contrast, transplantation of testes into gonad-

ectomized or intact female neonatal rats caused them to develop into adults with the steady male pattern of gonadotropin release. Transplantation of ovaries had no effect on the pattern of hormone release. Pfeiffer concluded that the female cyclic pattern of gonadotropin release develops unless the preprogrammed female cyclicity is overridden by testosterone during perinatal development (see Harris & Levine, 1965).

Pfeiffer incorrectly concluded that the presence or absence of testicular hormones in neonatal rats influenced the development of the pituitary because he was not

Neuroplasticity aware of something we know today: The release of gonadotropins from the anterior pituitary is controlled by the hypothalamus. Once this was discovered, it became apparent that Pfeiffer's experiments had provided the first evidence of the role of *perinatal* (around the time of birth) androgens in overriding the preprogrammed cyclic female pattern of gonadotropin release from the hypothalamus and initiating the development of the steady male pattern. This 1960s modification of Pfeiffer's theory of brain differentiation to include the hypothalamus was consistent with the facts of brain differentiation as understood at that time, but subsequent research necessitated major revisions. The first of these major revisions became known as the aromatization hypothesis.

Aromatization Hypothesis

What is aromatization? All gonadal and adrenal sex hormones are steroid hormones, and because all steroid hormones are derived from cholesterol, they have similar structures and are readily converted from one to the other. For example, a slight change to the testosterone molecule that occurs under the influence of the *enzyme* (a protein that influences a biochemical reaction without participating in it) **aromatase** converts testosterone to estradiol. This process is called **aromatization** (see Balthazart & Ball, 1998).

According to the **aromatization hypothesis**, perinatal testosterone does not directly masculinize the brain; the brain is masculinized by estradiol that has been aromatized from perinatal testosterone. Although the idea that estradiol—the alleged female hormone—masculinizes the brain may seem counterintuitive, there is strong evidence for it. Most of the evidence is of two types, both coming from experiments on rats and mice: (1) findings demonstrating masculinizing effects on the brain of early estradiol injections, and (2) findings showing that masculinization of the brain does not occur in response to testosterone that is administered with agents that block aromatization or in response to androgens that cannot be aromatized (e.g., dihydrotestosterone).

How do genetic females of species whose brains are masculinized by estradiol keep from being masculinized by their mothers' estradiol, which circulates through the fetal blood supply? Alpha fetoprotein is the answer. **Alpha fetoprotein** is present in the blood of rats during the perinatal period, and it deactivates circulating estradiol by binding to it (Bakker et al., 2006; Bakker & Baum, 2007; De Mees et al., 2006). How, then, does estradiol masculinize the brain of the male fetus in the presence of the deactivating effects of alpha fetoprotein? Because testosterone is immune to alpha fetoprotein, it can travel unaffected from the testes to the brain cells where it is converted to estradiol. Estradiol is not broken down in the brain because alpha fetoprotein does not readily penetrate the blood–brain barrier.

Modern Perspectives on Sexual Differentiation of Mammalian Brains

The view that the female program is the default program of brain development and is normally overridden in genetic males by perinatal exposure to testosterone aromatized to estradiol remained the preeminent theory of the sexual differentiation of the brain as long as research focused on the rat hypothalamus. Once studies of brain differentiation began to include other parts of the brain and other species, it became apparent that no single mechanism can account for the development of sexual dimorphisms of mammalian brains. The following findings have been particularly influential in shaping current views:

- Various sexual differences in brain structure and function have been found to develop by different mechanisms; for example, aromatase is found in only a few areas of the rat brain (e.g., the hypothalamus), and it is only in these areas that aromatization is critical for testosterone's masculinizing effects (see Ball & Balthazart, 2006; Balthazart & Ball, 2006).
- Sexual differences in the brain have been found to develop by different mechanisms in different mammalian species (see McCarthy, Wright, & Schwartz, 2009); for example, aromatization plays a less prominent role in primates than in rats and mice (see Zuloaga et al., 2008). *Evolutionary Perspective*
- Various sex differences in the brain have been found to develop at different stages of development (Bakker & Baum, 2007); for example, many differences do not develop until puberty (Ahmed et al., 2008; Sisk & Zehr, 2005), a possibility ignored by early theories.
- Sex chromosomes have been found to influence brain development independent of their effect on hormones (Arnold, 2009; Jazon & Cahill, 2010); for example, different patterns of gene expression exist in the brains of male and female mice before the gonads become functional (Dewing et al., 2003).
- Although the female program of brain development had been thought to proceed normally in the absence of gonadal steroids, recent evidence suggests that estradiol plays an active role; knockout mice without the gene that forms estradiol receptors do not display a normal female pattern of brain development (see Bakker & Baum, 2007).

In short, there is overwhelming evidence that various sexual differences in mammalian brains emerge at different stages of development under different genetic and hormonal influences (see Wagner, 2006). Although the conventional view that a female program of development is the default does an excellent job of explaining differentiation of the reproductive organs, it falters badly when it comes to differentiation of the brain.

In studying the many sexual differences of mammalian brains, it is easy to lose sight of the main point: We still do not understand how any of the anatomical differences that have been identified influence behavior.

Perinatal Hormones and Behavioral Development

In view of the fact that perinatal hormones influence the development of the brain, it should come as no surprise that they also influence the development of behavior. Much of the research on perinatal hormones and behavioral development was conducted before the discoveries about brain development that we have just considered. Consequently, most of the studies have been based on the idea of a female default program that can be overridden by testosterone and have assessed the effects of perinatal testosterone exposure on reproductive behaviors in laboratory animals.

Phoenix and colleagues (1959) were among the first to demonstrate that the perinatal injection of testosterone **masculinizes** and **defeminizes** a genetic female's adult copulatory behavior. First, they injected pregnant guinea pigs with testosterone. Then, when the litters were born, the researchers ovariectomized the female offspring. Finally, when these ovariectomized female guinea pigs reached maturity, the researchers injected them with testosterone and assessed their copulatory behavior. Phoenix and his colleagues found that the females that had been exposed to perinatal testosterone displayed more male-like mounting behavior in response to testosterone injections in adulthood than did adult females that had not been exposed to perinatal testosterone. And when, as adults, the female guinea pigs were injected with progesterone and estradiol and mounted by males, they displayed less **lordosis**—the intromission-facilitating arched-back posture that signals female rodent receptivity.

In a study complementary to that of Phoenix and colleagues, Grady, Phoenix, and Young (1965) found that the lack of early exposure of male rats to testosterone both **feminizes** and **demasculinizes** their copulatory behavior as adults. Male rats castrated shortly after birth failed to display the normal male copulatory pattern of mounting, **intromission** (penis insertion), and **ejaculation** (ejection of sperm) when they were treated with testosterone and given access to a sexually receptive female; and when they were injected with estrogen and progesterone as adults, they exhibited more lordosis than did uncastrated controls.

The aromatization of perinatal testosterone to estradiol seems to be important for both the defeminization and the masculinization of rodent copulatory behavior (Goy & McEwen, 1980; Shapiro, Levine, & Adler, 1980). In contrast, that aromatization does not seem to be critical for these effects in monkeys (Wallen, 2005).

When it comes to the effects of perinatal testosterone on behavioral development, timing is critical. The ability of single injections of testosterone to masculinize and

defeminize the rat brain seems to be restricted to the first 11 days after birth.

Because much of the research on hormones and behavioral development has focused on the copulatory act, we know less about the role of hormones in the development of **proceptive behaviors** (solicitation behaviors) and in the development of gender-related behaviors that are not directly related to reproduction. However, perinatal testosterone has been reported to disrupt the proceptive hopping, darting, and ear wiggling of receptive female rats; to increase the aggressiveness of female mice; to disrupt the maternal behavior of female rats; and to increase rough social play in female monkeys and rats.

Ethical considerations prohibit experimental studies of the developmental effects of hormones on human development. However, there have been many correlational studies of clinical cases and of ostensibly healthy individuals who received abnormal prenatal exposure to androgens (due to their own pathology or to drugs taken by their mothers). The results have been far from impressive. Cohen-Bendahan, van de Beek, and Berenbaum (2005) reviewed the extensive research literature and concluded that, despite many inconsistencies, the weight of evidence indicated that prenatal androgen exposure contributes to the differences in interests, spatial ability, and aggressiveness typically observed between men and women. However, there was no convincing evidence that differences in prenatal androgen exposure contribute to behavioral differences observed *among* women or *among* men.

Before you finish this subsection, I want to clarify an important point. If you are like many of my students, you may be wondering why biopsychologists who study the development of male–female behavioral differences always measure *masculinization* separately from *defeminization* and *feminization* separately from *demasculinization*. If you think that masculinization and defeminization are the same thing and that feminization and demasculinization are the same thing, you have likely fallen into the trap of the men-are-men-and-women-are-women assumption—that is, into the trap of thinking of maleness and femaleness as discrete, mutually exclusive, opposite categories. In fact, male behaviors and female behaviors can co-exist in the same individual, and they do not necessarily change in opposite directions if the individual receives physiological treatment such as hormones or brain lesions. For example, "male" behaviors (e.g., mounting receptive females) have been observed in the females of many different mammalian species, and "female" behaviors (e.g., lordosis) have been observed in males (see Dulac & Kimchi, 2007). And, lesions in medial preoptic areas have been shown to abolish male reproductive behaviors in both male and female rats, without affecting female behaviors (Singer, 1968). Think about this idea carefully, it plays an important role in later sections of the chapter.

Scan Your Brain

Before you proceed to a consideration of three cases of exceptional human sexual development, scan your brain to see whether you understand the basics of normal sexual development. Fill in the blanks in the following sentences. The correct answers are provided at the end of the exercise. Review material related to your errors and omissions before proceeding.

1. Six weeks after conception, the Sry gene on the Y chromosome of the human male triggers the production of _____.
2. In the absence of the Sry protein, the cortical cells of the primordial gonads develop into _____.
3. In the third month of male fetal development, the testes secrete testosterone and _____ substance.
4. The hormonal factor that triggers the development of the human Müllerian system is the lack of _____ around the third month of fetal development.
5. The scrotum and the _____ develop from the same bipotential precursor.
6. The female pattern of cyclic _____ release from the anterior pituitary develops in adulthood unless androgens are present in the body during the perinatal period.
7. It has been hypothesized that perinatal testosterone must first be changed to estradiol before it can masculinize the male rat brain. This is called the _____ hypothesis.
8. _____ is normally responsible for pubic and axillary hair growth in human females during puberty.
9. Girls usually begin puberty _____ boys do.
10. The simplistic, seductive, but incorrect assumption that sexual differentiation occurs because male and female sex hormones trigger programs of development that are parallel but opposite to one another has been termed the _____.

Scan Your Brain answers: (1) Sry protein, (2) ovaries, (3) Müllerian-inhibiting, (4) androgens (or testosterone), (5) labia majora, (6) gonadotropin, (7) aromatization, (8) Androstenedione, (9) before, (10) mamawawa.

13.4

Three Cases of Exceptional Human Sexual Development

This section discusses three cases of abnormal sexual development. I am sure you will be intrigued by these three cases, but that is not the only reason I have chosen to present them. My main reason is expressed by a proverb: The exception proves the rule. Most people think this proverb means that the exception "proves" the rule in the sense that it establishes its truth, but this is clearly wrong: The truth of a rule is challenged by, not confirmed by, exceptions to it. The word *proof* comes from the Latin *probare*, which means "to test"—as in *proving ground* or printer's *proof*—and this is the sense in which it is used in the proverb. Hence, the proverb means that the explanation of exceptional cases is a major challenge for any theory.

So far in this chapter, you have learned the "rules" according to which hormones seem to influence normal sexual development. Now, three exceptional cases are offered to prove (to test) these rules.

The Case of Anne S., the Woman Who Wasn't

Anne S., an attractive 26-year-old female, sought treatment for two sex-related disorders: lack of menstruation and pain during sexual intercourse (Jones & Park, 1971). She sought help because she and her husband of 4 years had been trying without success to have children, and she correctly surmised that her lack of a menstrual cycle was part of the problem. A physical examination revealed that Anne was a healthy young woman. Her only readily apparent peculiarity was the sparseness and fineness of her pubic and axillary hair. Examination of her external genitals revealed no abnormalities; however, there were some problems with her internal genitals. Her vagina was only 4 centimeters long, and her uterus was underdeveloped.

Clinical Implications

At the start of this chapter, I said that you would encounter some remarkable things, and the diagnosis of Anne's case certainly qualifies as one of them. Anne's doctors concluded that her sex chromosomes were those of a man. No, this is not a misprint; they concluded that Anne, the attractive young housewife, had the genes of a genetic male. Three lines of evidence supported their diagnosis. First, analysis of cells scraped from the inside of Anne's mouth revealed that they were of the male XY type. Second, a tiny incision in Anne's abdomen, which enabled Anne's physicians to look inside, revealed a pair of internalized testes but no ovaries. Finally, hormone tests revealed that Anne's hormone levels were those of a male.

Anne suffers from complete **androgenic insensitivity syndrome**; all her symptoms stem from a mutation to the androgen receptor gene that rendered her androgen receptors totally unresponsive (see Fink et al., 1999; Goldstein, 2000). Complete androgen insensitivity is rare, occurring in about 5 of 100,000 male births.

During development, Anne's testes released normal amounts of androgens for a male, but her body could not respond to them because of the mutation to her androgen receptor gene; and thus, her development proceeded as if no androgens had been released. Her external genitals, her brain, and her behavior developed along female lines, without the effects of androgens to override the female program, and her testes could not descend from her body cavity with no scrotum for them to descend into. Furthermore, Anne did not develop normal internal female reproductive ducts because, like other genetic males, her testes released Müllerian-inhibiting substance; that is why her vagina was short and her uterus undeveloped. At puberty, Anne's testes released enough estrogens to feminize her body in the absence of the counteracting effects of androgens; however, adrenal androstenedione was not able to stimulate the growth of pubic and axillary hair.

Although the samples are small, patients with complete androgen insensitivity have been found to be comparable to genetic females. All aspects of their behavior that have been studied—including gender identity, sexual orientation, interests, and cognitive abilities—have been found to be typically female (see Cohen-Bendahan, van de Beek, & Berenbaum, 2005).

An interesting issue of medical ethics is raised by the androgenic insensitivity syndrome. Many people believe that physicians should always disclose all relevant findings to their patients. If you were Anne's physician, would you tell her that she is a genetic male? Would you tell her husband? Her doctor did not. Anne's vagina was surgically enlarged, she was counseled to consider adoption, and, as far as I know, she is still happily married and unaware of her genetic sex. On the other hand, I have heard from several women who suffer from partial androgenic insensitivity, and they recommended full disclosure. They had faced a variety of sexual ambiguities throughout their lives, and learning the cause helped them.

The Case of the Little Girl Who Grew into a Boy

The patient—let's call her Elaine—sought treatment in 1972. Elaine was born with somewhat ambiguous external genitals, but she was raised by her parents as a girl without incident, until the onset of puberty, when she suddenly began to develop male secondary sex characteristics. This was extremely distressing. Her treatment had two aspects: surgical and hormonal. Surgical treatment was used to increase the size of her vagina and decrease the size of her clitoris; hormonal treatment was used to suppress androgen release so that her own estrogen could feminize her body. Following treatment, Elaine developed into an attractive young woman—narrow hips and a husky voice being the only signs of her brush with masculinity. Fifteen years later, she was married and enjoying a normal sex life (Money & Ehrhardt, 1972).

Clinical Implications

Elaine suffered from adrenogenital syndrome, which is the most common disorder of sexual development, affecting about 1 in 10,000. **Adrenogenital syndrome** is caused by **congenital adrenal hyperplasia**—a congenital deficiency in the release of the hormone *cortisol* from the adrenal cortex, which results in compensatory adrenal hyperactivity and the excessive release of adrenal androgens. This has little effect on the development of males, other than accelerating the onset of puberty, but it has major effects on the development of genetic females. Females who suffer from the adrenogenital syndrome are usually born with an enlarged clitoris and partially fused labia. Their gonads and internal ducts are usually normal because the adrenal androgens are released too late to stimulate the development of the Wolffian system.

👁 **Watch** Intersexuals
www.mypsychlab.com

Most female cases of adrenogenital syndrome are diagnosed at birth. In such cases, the abnormalities of the external genitals are immediately corrected, and cortisol is administered to reduce the levels of circulating adrenal androgens. Following early treatment, adrenogenital females grow up to be physically normal except that the onset of menstruation is likely to be later than normal. This makes them good subjects for studies of the effects of fetal androgen exposure on psychosexual development.

Adrenogenital teenage girls who have received early treatment tend to display more tomboyishness, greater strength, and more aggression than most teenage girls, and they tend to prefer boys' clothes and toys, play mainly with boys, and daydream about future careers rather than motherhood (e.g., Collaer et al., 2008; Hines, 2003; Matthews et al., 2009). However, it is important not to lose sight of the fact that many teenage girls display similar characteristics—and why not? Accordingly, the behavior of treated adrenogenital females, although tending toward the masculine, is usually within the range considered to be normal female behavior by the current standards of our culture.

The most interesting questions about the development of females with adrenogenital syndrome concern their romantic and sexual preferences as adults. They seem to lag behind normal females in dating and marriage—perhaps because of the delayed onset of their menstrual cycle. Most are heterosexual, although a few studies have found an increased tendency for these women to express interest in bisexuality or homosexuality and a tendency to be less involved in heterosexual relationships (see Gooren, 2006). Complicating the situation further is the fact that these slight differences may not be direct consequences of early androgen exposure but arise from the fact that some adrenogenital girls have ambiguous genitalia and other male characteristics (e.g., body hair), which may result in different experiential influences.

Thinking Creatively

Prior to the development of cortisol therapy in 1950, genetic females with adrenogenital syndrome were left untreated. Some were raised as boys and some as girls, but the direction of their pubertal development was unpredictable. In some cases, adrenal androgens predominated and masculinized their bodies; in others, ovarian estrogens predominated and feminized their bodies. Thus, some who were raised as boys were transformed at puberty into women

((• **Listen** Psychology in the News: Sexuality and Gender **www.mypsychlab.com**

and some who were raised as girls were transformed into men, with devastating emotional consequences.

The Case of the Twin Who Lost His Penis

One of the most famous cases in the literature on sexual development is that of a male identical twin whose penis was accidentally destroyed during circumcision at the age of 7 months. Because there was no satisfactory way of surgically replacing the lost penis, a respected expert in such matters, John Money, recommended that the boy be castrated, that an artificial vagina be created, that the boy be raised as a girl, and that estrogen be administered at puberty to feminize the body. After a great deal of consideration and anguish, the parents followed Money's advice.

Money's (1975) report of this case of **ablatio penis** has been influential. It has been seen by some as the ultimate test of the *nature–nurture controversy* (see Chapter 2)

Clinical Implications

with respect to the development of sexual identity and behavior. It seemed to pit the masculinizing effects of male genes and male prenatal hormones against the effects of being reared as a female. And the availability of a genetically identical control subject, the twin brother, made the case all the more interesting. According to Money, the outcome of this case strongly supported the *social-learning theory* of sexual identity. Money reported in 1975, when the patient was 12, that "she" had developed as a normal female, thus confirming his prediction that being gonadectomized, having the genitals surgically altered, and being raised as a girl would override the masculinizing effects of male genes and early androgens.

A long-term follow-up study published by impartial experts tells an entirely different story (Diamond & Sigmundson, 1997). Despite having female genitalia and being treated as a female, John/Joan developed along male lines. Apparently, the organ that determines the course of psychosocial development is the brain, not the genitals (Reiner, 1997). The following paraphrases from Diamond and Sigmundson's report give you a glimpse of John/Joan's life:

> From a very early age, Joan tended to act in a masculine way. She preferred boys' activities and games and displayed little interest in dolls, sewing, or other conventional female activities. When she was four, she was watching her father shave and her mother put on lipstick,

and she began to put shaving cream on her face. When she was told to put makeup on like her mother, she said, "No, I don't want no makeup, I want to shave."

> "Things happened very early. As a child, I began to see that I felt different about a lot of things than I was supposed to. I suspected I was a boy from the second grade on."

Despite the absence of a penis, Joan often tried to urinate while standing, and she would sometimes go to the boys' lavatory.

Joan was attractive as a girl, but as soon as she moved or talked her masculinity became apparent. She was teased incessantly by the other girls, and she often retaliated violently, which resulted in her expulsion from school.

Joan was put on an estrogen regimen at the age of 12 but rebelled against it. She did not want to feminize; she hated her developing breasts and refused to wear a bra.

At 14, Joan decided to live as a male and switched to John. At that time, John's father tearfully revealed John's entire early history to him. "All of a sudden everything clicked. For the first time I understood who and what I was."

John requested androgen treatment, a *mastectomy* (surgical removal of breasts), and *phaloplasty* (surgical creation of a penis). He became a handsome and popular young man. He married at the age of 25 and adopted his wife's children. He is strictly heterosexual.

John's ability to ejaculate and experience orgasm returned following his androgen treatments. However, his early castration permanently eliminated his reproductive capacity.

"John" remained bitter about his early treatment and his inability to produce offspring. To save others from his experience, he cooperated in writing his biography, *As Nature Made Him* (Colapinto, 2000). His real name was David Reimer (see Figure 13.10). David never recovered from his emotional scars. On May 4, 2004, he committed suicide.

David Reimer's case suggests that the clinical practice of surgically modifying a person's sex at birth should be curtailed. Any such irrevocable treatments should await early puberty and the emergence of the patient's sexual identity and sexual attraction. At that stage, a compatible course of treatment can be selected.

Do the Exceptional Cases Prove the Rule?

Do current theories of hormones and sexual development pass the test of the three preceding cases of exceptional sexual development? In my view, the answer is "yes." Although *Thinking Creatively* current theories do not supply all of the answers, especially when it comes to brain dimorphisms and behavior, they have contributed greatly to the understanding of exceptional patterns of sexual differentiation of the body.

FIGURE 13.10 David Reimer, the twin whose penis was accidentally destroyed.

For centuries, cases of abnormal sexual development have befuddled scholars, but now, armed with a basic understanding of the role of hormones in sexual development, they have been able to make sense of some of the most puzzling of such cases. Moreover, the study of sexual development has pointed the way to effective treatments. Judge these contributions for yourself by comparing your current understanding of these three cases with the understanding that you would have had if you had encountered them before beginning this chapter.

Notice one more thing about the three cases: Each of the three subjects was male in some respects and female in others. Accordingly, each case is a serious challenge to the men-are-men-and-women-are-women assumption: Male and female are not opposite, mutually exclusive categories.

13.5

Effects of Gonadal Hormones on Adults

Once an individual reaches sexual maturity, gonadal hormones begin to play a role in activating reproductive behavior. These activational effects are the focus of the first two parts of this section. They deal with the role of hormones in activating the reproduction-related behavior of men and women, respectively. The third part of this section deals with anabolic steroids, and the fourth describes the neuroprotective effects of estradiol.

Male Reproduction–Related Behavior and Testosterone

The important role played by gonadal hormones in the activation of male sexual behavior is clearly demonstrated by the asexualizing effects of orchidectomy. Bremer (1959) reviewed the cases of 157 orchidectomized Norwegians. Many had committed sex-related offenses and had agreed to castration to reduce the length of their prison terms.

Two important generalizations can be drawn from Bremer's study. The first is that orchidectomy leads to a reduction in sexual interest and behavior; the second is that the rate and the degree of the loss are variable. About half the men became completely asexual within a few weeks of the operation; others quickly lost their ability to

achieve an erection but continued to experience some sexual interest and pleasure; and a few continued to copulate successfully, although somewhat less enthusiastically, for the duration of the study. There were also body changes: a reduction of hair on the trunk, extremities, and face; the deposition of fat on the hips and chest; a softening of the skin; and a reduction in strength.

Of the 102 sex offenders in Bremer's study, only 3 were reconvicted of sex offenses. Accordingly, he recommended castration as an effective treatment of last resort for male sex offenders.

Why do some men remain sexually active for months after orchidectomy, despite the fact that testicular hormones are cleared from their bodies within days? It has been suggested that adrenal androgens may play some role in the maintenance of sexual activity in some castrated men, but there is no direct evidence for this hypothesis.

Orchidectomy removes, in one fell swoop—or, to put it more precisely, in two fell swoops—a pair of glands that release many hormones. Because testosterone is the major testicular hormone, the major symptoms of orchidectomy have been generally attributed to the loss of testosterone, rather than to the loss of some other testicular hormone or to some nonhormonal consequence of the surgery. The therapeutic effects of **replacement injections** of testosterone have confirmed this assumption.

The Case of the Man Who Lost and Regained His Manhood

The very first case report of the effects of testosterone replacement therapy concerned an unfortunate 38-year-old World War I veteran, who was castrated in 1918 at the age of 19 by a shell fragment that removed his testes but left his penis undamaged.

Clinical Implications

His body was soft; it was as if he had almost no muscles at all; his hips had grown wider and his shoulders seemed narrower than when he was a soldier. He had very little drive. . . .

Just the same this veteran had married, in 1924, and you'd wonder why, because the doctors had told him he would surely be **impotent** [unable to achieve an erection]. . . . he made some attempts at sexual intercourse "for his wife's satisfaction" but he confessed that he had been unable to satisfy her at all. . . .

Dr. Foss began injecting it [testosterone] into the feeble muscles of the castrated man. . . .

After the fifth injection, erections were rapid and prolonged. . . . But that wasn't all. During twelve weeks of treatment he had gained eighteen pounds, and all his clothes had become too small. . . . testosterone had resurrected a broken man to a manhood he had lost forever. (de Kruif, 1945, pp. 97–100)

Since this first clinical trial, testosterone has breathed sexuality into the lives of many men. Testosterone does not, however, eliminate the *sterility* (inability to reproduce) of males who lack functional testes.

The fact that testosterone is necessary for male sexual behavior has led to two widespread assumptions: (1) that the level of a man's sexuality is a function of the amount of testosterone he has in his blood, and (2) that a man's sex drive can be increased by increasing his testosterone levels. Both assumptions are incorrect. Sex drive and testosterone levels are uncorrelated in healthy men, and testosterone injections do not increase their sex drive.

Evolutionary Perspective

It seems that each healthy male has far more testosterone than is required to activate the neural circuits that produce his sexual behavior and that having more than the minimum is of no advantage in this respect (Sherwin, 1988). A classic experiment by Grunt and Young (1952) clearly illustrates this point.

First, Grunt and Young rated the sexual behavior of each of the male guinea pigs in their experiment. Then,

on the basis of the ratings, the researchers divided the male guinea pigs into three experimental groups: low, medium, and high sex drive. Following castration, the sexual behavior of all of the guinea pigs fell to negligible levels within a few weeks (see Figure 13.11), but it recovered after the initiation of a series of testosterone replacement injections. The important point is that although each subject received the same, very large replacement injections of testosterone, the injections simply returned each to its previous level of copulatory activity. The conclusion is clear: With respect to the effects of testosterone on sexual behavior, more is not necessarily better.

Dihydrotestosterone, a nonaromatizable androgen, restores the copulatory behavior of castrated male primates (e.g., Davidson, Kwan, & Greenleaf, 1982); however, it fails to restore the copulatory behavior of castrated male rodents (see MacLusky & Naftolin, 1981). These findings indicate that the restoration of copulatory behavior by testosterone occurs by different mechanisms in rodents and primates: It appears to be a direct effect of testosterone in primates, but appears to be produced by estradiol aromatized from testosterone in rodents (see Ball & Balthazart, 2006).

Female Reproduction-Related Behavior and Gonadal Hormones

Sexually mature female rats and guinea pigs display 4-day cycles of gonadal hormone release. There is a gradual increase in the secretion of estrogens by the developing *follicle* (ovarian structure in which eggs mature) in the

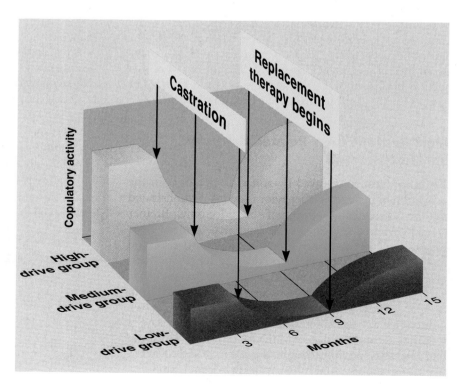

FIGURE 13.11 The sexual behavior of male guinea pigs with low, medium, and high sex drive. Sexual behavior was disrupted by castration and returned to its original level by very large replacement injections of testosterone. (Based on Grunt & Young, 1952.)

Evolutionary
Perspective

2 days prior to ovulation, followed by a sudden surge in progesterone as the egg is released. These surges of estrogens and progesterone initiate **estrus**—a period of 12 to 18 hours during which the female is *fertile, receptive* (likely to assume the lordosis posture when mounted), *proceptive* (likely to engage in behaviors that serve to attract the male), and *sexually attractive* (smelling of chemicals that attract males).

The close relation between the cycle of hormone release and the **estrous cycle**—the cycle of sexual receptivity—in female rats and guinea pigs and in many other mammalian species suggests that female sexual behavior in these species is under hormonal control. The effects of ovariectomy confirm this conclusion; ovariectomy of female rats and guinea pigs produces a rapid decline of both proceptive and receptive behaviors. Furthermore, estrus can be induced in ovariectomized rats and guinea pigs by an injection of estradiol followed about a day and a half later by an injection of progesterone.

Women are different from female rats, guinea pigs, and other mammals when it comes to the hormonal control of their sexual behavior: Female primates are the only female mammals that are motivated to copulate during periods of nonfertility (Ziegler, 2007). Moreover, ovariectomy has surprisingly little direct effect on either their sexual motivation or their sexual behavior (e.g., Martin, Roberts, & Clayton, 1980). Other than sterility, the major consequence of ovariectomy in women is a decrease in vaginal lubrication.

Numerous studies have investigated the role of estradiol in the sex drive of women by relating various measures of their sexual interest and activity to phases of their menstrual cycles. The results of this research are difficult to interpret. Some women do report that their sex drive is related to their menstrual cycles, and many studies have reported statistically significant correlations. The confusion arises because many studies have found no significant correlations, and because many different patterns of correction have been reported (see Regan, 1996; Sanders & Bancroft, 1982). No single pattern has emerged that characterizes fluctuations in human female sexual motivation. Paradoxically, there is evidence that the sex drive of women is under the control of androgens (the so-called male sex hormones), not estrogens (see Davis & Tran, 2001; Sherwin, 1988). Apparently, enough androgens are released from the human adrenal glands to maintain the sexual motivation of women even after their ovaries have been removed. Support for the theory that androgens control human female sexuality has come from three sources:

- In experiments with nonhuman female primates, replacement injections of testosterone, but not estradiol, increased the proceptivity of ovariectomized and adrenalectomized rhesus monkeys (see Everitt & Herbert, 1972; Everitt, Herbert, & Hamer, 1971).

- In correlational studies of healthy women, various measures of sexual motivation have been shown to correlate with testosterone levels but not with estradiol levels (see Bancroft et al., 1983; Morris et al., 1987).

- In clinical studies of women following ovariectomy and adrenalectomy or menopause, replacement injections of testosterone, but not of estradiol, rekindled the patients' sexual motivation (see de Paula et al., 2007; Sherwin, Gelfand, & Brender, 1985).

This research has led to the development of a testosterone skin patch for the treatment of low sex drive in women. The patch has been shown to be effective for women who have lost their sex drive following radical hysterectomy (Buster et al., 2005), Although a few studies have reported positive correlations between blood testosterone levels and the strength of sex drive in women (e.g., Turna et al., 2004), most women with low sex drive do not have low blood levels of testosterone (Davis et al., 2005; Gerber et al., 2005). Thus, the testosterone skin patch is unlikely to help most women with libido problems.

Clinical Implications

Thinking Creatively

Although neither the sexual motivation nor the sexual activity of women has found to be linked to their menstrual cycles, the type of men they prefer may be. Several studies have shown that women prefer masculine faces more on their fertile days than on their nonfertile days (e.g., Gangestad, Thornhill, & Garver-Apgar, 2005; Penton-Voak & Perrett, 2000).

Anabolic Steroid Abuse

Anabolic steroids are steroids, such as testosterone, that have *anabolic* (growth-promoting) effects. Testosterone itself is not very useful as an anabolic drug because it is broken down soon after injection and because it has undesirable side effects. Chemists have managed to synthesize a number of potent anabolic steroids that are long-acting, but they have not managed to synthesize one that does not have side effects.

According some experts, we are currently in the midst of an epidemic of anabolic steroid abuse. Many competitive athletes and bodybuilders are self-administering appallingly large doses, and many others use them for cosmetic purposes. Because steroids are illegal, estimates of the numbers who use them are likely underestimates. Still, the results of some surveys have been disturbing: For example, a survey by the U.S. Centers for Disease Control and Prevention (Eaton et al., 2005) found that almost 5% of high school students had been steroid users.

Clinical Implications

Effects of Anabolic Steroids on Athletic Performance

Do anabolic steroids really increase the muscularity and strength of the athletes who use them? Surprisingly, the

FIGURE 13.12 An athlete who used anabolic steroids to augment his training program.

early scientific evidence was inconsistent (see Yesalis & Bahrke, 1995), even though many athletes and coaches believe that it is impossible to compete successfully at the highest levels of their sports without an anabolic steroid boost. The failure of the early experiments to confirm the benefits that had been experienced by many athletes likely results from two shortcomings of the research. First, the early experimental studies tended to use doses of steroids smaller than those used by athletes and for shorter periods of time. Second, the early studies were often conducted on volunteers who were not involved in intense training. However, despite the inconsistent experimental evidence, the results achieved by numerous individual steroid users, such as the man pictured in Figure 13.12, are convincing.

Physiological Effects of Anabolic Steroids There is general agreement (see Maravelias et al., 2005; Yesalis & Bahrke, 1995) that people who take high doses of anabolic steroids risk side effects. In men, the negative feedback from high levels of anabolic steroids reduces gonadotropin release; this leads to a reduction in testicular activity, which can result in *testicular atrophy* (wasting away of the testes) and sterility. *Gynecomastia* (breast growth in men) can also occur, presumably as the result of the aromatization of anabolic steroids to estrogens.

In women, anabolic steroids can produce *amenorrhea* (cessation of menstruation), sterility, *hirsutism* (excessive growth of body hair), growth of the clitoris, development of a masculine body shape, baldness, shrinking of the breasts, and deepening and coarsening of the voice. Unfortunately, some of the masculinizing effects of anabolic steroids on women appear to be irreversible.

Both men and women who use anabolic steroids can suffer muscle spasms, muscle pains, blood in the urine, acne, general swelling from the retention of water, bleeding of the tongue, nausea, vomiting, and a variety of psychotic behaviors, including fits of depression and anger (Maravelias et al., 2005). Oral anabolic steroids produce cancerous liver tumors.

One controlled evaluation of the effects of exposure to anabolic steroids was conducted in adult male mice. Adult male mice were exposed for 6 months to a cocktail of four anabolic steroids at relative levels comparable to those used by human athletes (Bronson & Matherne, 1997). *Evolutionary Perspective* None of the mice died during the period of steroid exposure; however, by 20 months of age (6 months after termination of the steroid exposure), 52% of the steroid-exposed mice had died, whereas only 12% of the controls had died.

There are two general points of concern about the adverse health consequences of anabolic steroids: First, the use of anabolic steroids in puberty, before developmental programs of sexual differentiation are complete, is particularly risky (see Farrell & McGinnis, 2003). Second, many of the adverse effects of anabolic steroids may take years to be manifested—steroid users who experience few immediate adverse effects may pay the price later.

Behavioral Effects of Anabolic Steroids Other than those focusing on athletic performance, which you have just read about, few studies have systematically investigated the effects of anabolic steroids on behavior. However, *Thinking Creatively* because of the similarity between anabolic steroids and testosterone, there has been some suggestion that anaboic steroid use might increase aggressive and sexual behaviors. Let's take a brief look at the evidence.

Evidence that anabolic steroid use increases aggression comes almost entirely from the claims of steroid users. These anecdotal reports are unconvincing for three reasons:

- Because of the general belief that testosterone causes aggression, reports of aggressive behavior in male steroid users might be a consequence of expectation.
- Many individuals who use steroids (e.g., professional fighters or football players) are likely to have been aggressive before they started treatment.
- Aggressive behavior might be an indirect consequence of increased size and muscularity.

In one experimental assessment of the effects of anabolic steroids on aggression, Pope, Kouri, and Hudson (2000) administered either testosterone or placebo injections in a double-blind study of 53 men. Each volunteer completed tests of aggression and kept a daily aggression-related diary. Pope and colleagues found increases in aggression in only a few of the volunteers.

Although their similarity to testosterone suggests that steroids might increase sexual motivation, there is no evidence of such an effect. On the contrary, there are several anecdotal reports of the disruptive effects of anabolic steroids on human male copulatory behavior, and controlled experiments have shown that anabolic steroids disrupt the copulatory behavior of both male and female rodents (see Clark & Henderson, 2003).

Neuroprotective Effects of Estradiol

Although estradiol is best known for its sex-related organizational and activational effects, this hormone also can reduce the brain damage associated with stroke and various neurodegenerative disorders (see De Butte-Smith et al., 2009). For example, Yang and colleagues (2003) showed that estradiol administered to rats just before, during, or just after the induction of *cerebral hypoxia* (reduction of oxygen to the brain) reduced subsequent brain damage (see Chapter 10).

Estradiol has been shown to have several neurotrophic effects that might account for its neuroprotective properties (see Chapter 10). It has been shown to reduce inflammation, encourage axonal regeneration, promote synaptogenesis (see Stein & Hoffman, 2003; Zhang et al., 2004), and increase adult neurogenesis (see Chapter 10). Injection of estradiol initially increases the number of new neurons created in the dentate gyri of the hippocampuses of adult female rats and then, about 48 hours later, there is a period of reduced neurogenesis (see Galea et al., 2006; Ormerod, Falconer, & Galea, 2003). As well as increasing adult neurogenesis, estradiol increases the survival rate of the new neurons (see Galea et al., 2006; Ormerod & Galea, 2001b).

The discovery of estradiol's neuroprotective properties has created a lot of excitement among neuroscientists. These properties may account for women's greater longevity and their lower incidence of several common neuropsychological disorders, such as Parkinson's disease. They may also explain the decline in memory and some other cognitive deficits experienced by postmenopausal women (see Bisagno, Bowman, & Luine, 2003; Gandy, 2003).

Several studies have assessed the ability of estrogen treatments to reduce the cognitive deficits experienced by postmenopausal women. The results of some studies have been encouraging, but others have observed either no benefit or an increase in cognitive deficits (see Blaustein, 2008; Frick, 2009). Two suggestions have been made for improving the effectiveness of estradiol therapy: First, Sherwin (2007) pointed out that such therapy appears to be effective in both humans and non-humans only if the estradiol treatment is commenced at menopause or shortly thereafter. Second, Marriott and Wenk (2004) argued that the chronically high doses that have been administered to postmenopausal women are unnatural and potentially toxic; they recommend instead that estradiol therapy should mimic the natural cycle of estradiol levels in premenopausal women.

Thinking Creatively

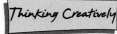

You encountered many clinical problems in the preceding two sections of the chapter. Do you remember them? Write the name of the appropriate condition or syndrome in each blank, based on the clues provided. The answers appear at the end of the exercise. Before proceeding, review material related to your errors and omissions.

Name of condition or syndrome	Clues
1. _____	Genetic male, sparse pubic hair, short vagina
2. _____	Congenital adrenal hyperplasia, elevated androgen levels
3. _____	David Reimer, destruction of penis
4. _____	Castrated males, gonadectomized males
5. _____	Castrated females, gonadectomized females
6. _____	Unable to achieve erection
7. _____	Anabolic steroids, breasts on men
8. _____	Anabolic steroids, cessation of menstruation
9. _____	Anabolic steroids, excessive body hair
10. _____	Reduction of oxygen to brain, effects can be reduced by estradiol

Scan Your Brain answers: (1) androgenic insensitivity syndrome, (2) adrenogenital syndrome, (3) ablatio penis, (4) orchidectomized, (5) ovariectomized, (6) impotent, (7) gynecomastia, (8) amenorrhea, (9) hirsutism, (10) cerebral hypoxia.

13.6

Neural Mechanisms of Sexual Behavior

Major differences among cultures in sexual practices and preferences indicate that the control of human sexual behavior involves the highest levels of the nervous system (e.g., association cortex), and this point is reinforced by controlled demonstrations of the major role played by experience in the sexual behaviors of nonhuman animals (see Woodson, 2002; Woodson & Balleine, 2002; Woodson, Balleine, & Gorski, 2002). Nevertheless, research on the neural mechanisms of sexual behavior has focused almost exclusively on hypothalamic circuits. Consequently, I am forced to do the same here: When it comes to the study of the neural regulation of sexual behavior, the hypothalamus is virtually the only game in town.

Why has research on the neural mechanisms of sexual behavior focused almost exclusively on hypothalamic circuits? There are three obvious reasons: First, because of the difficulty of studying the neural mechanisms of complex human sexual behaviors, researchers have focused on the relatively simple, controllable copulatory behaviors (e.g., ejaculation, mounting, and lordosis) of laboratory animals (see Agmo & Ellingsen, 2003), which tend to be controlled by the hypothalamus. Second, because the hypothalamus controls gonadotropin release, it was the obvious place to look for sexually dimorphic structures and circuits that might control copulation. And third, early studies confirmed that the hypothalamus does play a major role in sexual behavior, and this finding led subsequent neuroscientific research on sexual behavior to focus on that brain structure.

Structural Differences between the Male and Female Hypothalamus

You have already learned that the male hypothalamus and the female hypothalamus are functionally different in their control of anterior pituitary hormones (steady versus cyclic release, respectively). In the 1970s, structural differences between the male and female hypothalamus were discovered in rats (Raisman & Field, 1971). Most notably, Gorski and his colleagues (1978) discovered a nucleus in the **medial preoptic area** of the rat hypothalamus that was several times larger in males (see Figure 13.13). They called this nucleus the **sexually dimorphic nucleus**.

Evolutionary Perspective

At birth, the sexually dimorphic nuclei of male and female rats are the same size. In the first few days after birth, the male sexually dimorphic nuclei grow at a high rate and the female sexually dimorphic nuclei do not. The growth of the male sexually dimorphic nuclei is normally triggered by estradiol, which has been aromatized from testosterone (see McEwen, 1987). Accordingly, castrating day-old

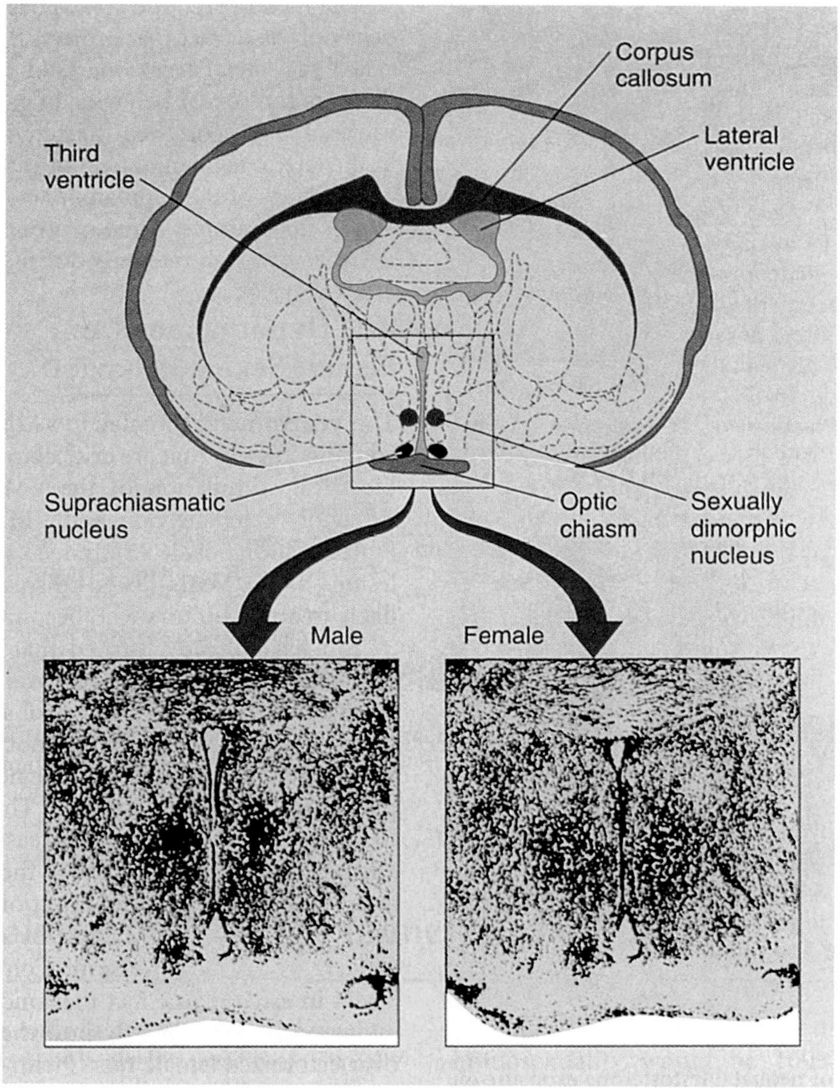

FIGURE 13.13 Nissl-stained coronal sections through the preoptic area of male and female rats. The sexually dimorphic nuclei are larger in male rats than in female rats. (Based on Gorski et al., 1978.)

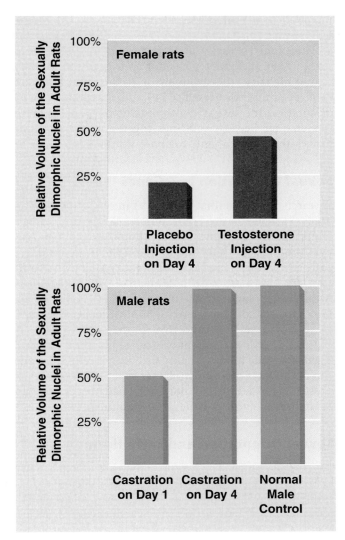

FIGURE 13.14 The effects of neonatal testosterone exposure on the size of the sexually dimorphic nuclei in male and female adult rats. (Based on Gorski, 1980.)

Since the discovery of the sexually dimorphic nuclei in rats, other sex differences in hypothalamic anatomy have been identified in rats and in other species (see Swaab & Hofman, 1995; Witelson, 1991). In humans, for example, there are nuclei in the preoptic (Swaab & Fliers, 1985), suprachiasmatic (Swaab et al., 1994), and anterior (Allen et al., 1989) regions of the hypothalamus that differ in men and women.

Hypothalamus and Male Sexual Behavior

The medial preoptic area (which includes the sexually dimorphic nucleus) is one area of the hypothalamus that plays a key role in male sexual behavior (see Dominguez & Hull, 2005). Destruction of the entire area abolishes sexual behavior in the males of all mammalian species that have been studied (see Hull et al., 1999). In contrast, medial preoptic area lesions do not eliminate the female sexual behaviors of females, but they do eliminate the male sexual behaviors (e.g., mounting) that are often observed in females (Singer, 1968). Thus, bilateral medial preoptic lesions appear to abolish male copulatory behavior in both sexes. Conversely, electrical stimulation of the medial preoptic area elicits copulatory behavior in male rats (Malsbury, 1971; Rodriguez-Manzo et al., 2000), and copulatory behavior can be reinstated in castrated male rats by medial preoptic implants of testosterone (Davidson, 1980).

The medial preoptic circuits that control male sexual behavior appear to be dopaminergic (see Dominguez & Hull, 2005; Lagoda et al., 2004). Dopamine agonists microinjected into the medial preoptic area facilitate male sexual behavior, whereas dopamine agonists block it.

It is not clear why males with medial preoptic lesions stop copulating. One possibility is that the lesions disrupt the ability of males to copulate; another is that the lesions reduce the motivation of the males to engage in sexual behavior. The evidence is mixed, but it favors the hypothesis that the medial preoptic area is involved in the motivational aspects of male sexual behavior (Paredes, 2003).

The medial preoptic area appears to control male sexual behavior via a tract that projects to an area of the midbrain called the *lateral tegmental field*. Destruction of this tract disrupts the sexual behavior of male rats (Brackett & Edwards, 1984). Moreover, the activity of individual neurons in the lateral tegmental field of male rats is often correlated with aspects of the copulatory act (Shimura & Shimokochi, 1990); for example, some neurons in the lateral tegmental field fire at a high rate only during intromission.

Hypothalamus and Female Sexual Behavior

The **ventromedial nucleus (VMN)** of the rat hypothalamus contains circuits that appear to be critical for female sexual behavior. Female rats with bilateral lesions of the

(but not 4-day-old) male rats significantly reduces the size of their sexually dimorphic nuclei as adults, whereas injecting neonatal (newborn) female rats with testosterone significantly increases the size of theirs (Gorski, 1980)—see Figure 13.14. Although the overall size of the sexually dimorphic nucleus diminishes only slightly in male rats that are castrated in adulthood, specific areas of the nucleus do display significant degeneration (Bloch & Gorski, 1988).

The size of a male rat's sexually dimorphic nucleus is correlated with the rat's testosterone levels and aspects of its sexual activity (Anderson et al., 1986). However, bilateral lesions of the sexually dimorphic nucleus have only slight disruptive effects on male rat sexual behavior (e.g., De Jonge et al., 1989; Turkenburg et al., 1988), and the specific function of this nucleus is unclear.

VMN do not display lordosis, and they are likely to attack suitors who become too ardent.

You have already learned that an injection of progesterone brings into estrus an ovariectomized female rat that received an injection of estradiol about 36 hours before. Because the progesterone by itself does not induce estrus, the estradiol must in some way prime the nervous system so that the progesterone can exert its effect. This priming effect appears to be mediated by the large increase in the number of *progesterone receptors* that occurs in the VMN and surrounding area following an estradiol injection (Blaustein et al., 1988); the estradiol exerts this effect by entering VMN cells and influencing gene expression. Confirming the role of the VMN in estrus is the fact that microinjections of estradiol and progesterone directly into the VMN induce estrus in ovariectomized female rats (Pleim & Barfield, 1988).

The influence of the VMN on the sexual behavior of female rats appears to be mediated by a tract that descends to the *periaqueductal gray (PAG)* of the tegmentum. Destruction of this tract eliminates female sexual behavior (Hennessey et al., 1990), as do lesions of the PAG itself (Sakuma & Pfaff, 1979).

In conclusion, although many parts of the brain play a role in sexual behavior, much of the research has focused on the role of the hypothalamus in the copulatory behavior of rats. Several areas of the hypothalamus influence this copulatory behavior, and several hypothalamic nuclei are sexually dimorphic in rats, but the medial preoptic area and the ventromedial nucleus are two of the most widely studied. Male rat sexual behavior is influenced by a tract that runs from the medial preoptic area to the lateral tegmental field, and female rat sexual behavior is influenced by a tract that runs from the ventromedial nucleus to the periaqueductal gray (see Figure 13.15).

members of the other sex), some are **homosexual** (sexually attracted to members of the same sex), and some are **bisexual** (sexually attracted to members of both sexes). Also, the chapter has not addressed the topic of **sexual identity** (the sex, male or female, that a person believes himself or herself to be). A discussion of sexual orientation and sexual identity is a fitting conclusion to this chapter because it brings together the exception-proves-the-rule and anti-mamawawa themes.

Sexual Orientation and Genes

Research has shown that differences in sexual orientation have a genetic basis. For example, Bailey and Pillard (1991) studied a group of male homosexuals who had twin brothers, and they found that 52% of the monozygotic twin brothers and 22% of the dizygotic twin brothers were homosexual. In a comparable study of female twins by the same group of researchers (Bailey et al., 1993), the concordance rates for homosexuality were 48% for monozygotic twins and 16% for dizygotic twins.

Considerable excitement was created by the claim that a gene for male homosexuality had been localized on one end of the X chromosome (Hamer et al., 1993). However, subsequent research has not confirmed this claim (see Mustanski, Chivers, & Bailey, 2002; Rahman, 2005).

Sexual Orientation and Early Hormones

Many people mistakenly assume that homosexuals have lower levels of sex hormones. They don't: Heterosexuals and homosexuals do not differ in their levels of circulating hormones. Moreover, orchidectomy reduces the sexual behavior of both heterosexual and homosexual males, but it

13.7
Sexual Orientation and Sexual Identity

So far, this chapter has not addressed the topic of sexual orientation. As you know, some people are **heterosexual** (sexually attracted to

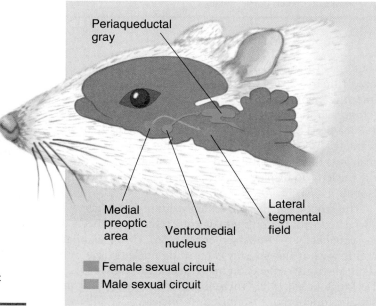

FIGURE 13.15 The hypothalamus-tegmentum circuits that play a role in female and male sexual behavior in rats.

does not redirect it; and replacement injections simply re-activate the preferences that existed prior to surgery.

Many people also assume that sexual preference is a matter of choice. It isn't: People discover their sexual preferences; they don't choose them. Sexual preferences seem to develop very early, and a child's first indication of the direction of sexual attraction usually does not change as he or she matures. Could perinatal hormone exposure be the early event that shapes sexual orientation?

Because experiments involving levels of perinatal hormone exposure are not feasible with humans, efforts to determine whether perinatal hormone levels influence the development of sexual orientation have focused on nonhuman species. A consistent pattern of findings has emerged. In those species that have been studied (e.g., rats, hamsters, ferrets, pigs, zebra finches, and dogs), it has not been uncommon to see males engaging in female sexual behavior, being mounted by other males; nor has it been uncommon to see females engaging in male sexual behavior, mounting other females. However, because the defining feature of sexual orientation is sexual preference, the key studies have examined the effect of early hormone exposure on the sex *Evolutionary Perspective* of preferred sexual partners. In general, the perinatal castration of males has increased their preference as adults for male sex partners; similarly, prenatal testosterone exposure in females has increased their preference as adults for female sex partners (see Baum et al., 1990; Henley, Nunez, & Clemens, 2009; Hrabovszky & Hutson, 2002).

On the one hand, we need to exercise prudence in drawing conclusions about the development of sexual preferences in humans based on the results of experiments on laboratory species; it would be a mistake to ignore the profound cognitive and emotional components of human sexuality, which have no counterpart in laboratory animals. On the other hand, it would also be a mistake to think that a pattern of results that runs so consistently through so many mammalian species has no relevance to humans (Swaab, 2004).

In addition, there are some indications that perinatal hormones do influence sexual orientation in humans—although the evidence is sparse (see Diamond, 2009). Support comes from the quasiexperimental study of Ehrhardt and her colleagues (1985). They interviewed adult women whose mothers had been exposed to *diethylstilbestrol* (a synthetic estrogen) during pregnancy. The women's responses indicated that they were significantly more sexually attracted to women than was a group of matched controls. Ehrhardt and her colleagues concluded that perinatal estrogen exposure does encourage homosexuality and bisexuality in women but that its effect is relatively weak: The sexual behavior of all but 1 of the 30 participants was primarily heterosexual.

One promising line of research on sexual orientation focuses on the **fraternal birth order effect**, the finding that the probability of a man's being homosexual increases as a function of the number of older brothers he has

(Blanchard, 2004; Blanchard & Lippa, 2007). A recent study of blended families (families in which biologically related siblings were raised with adopted siblings or step-siblings) found that the effect is related to the number of boys previously born to the mother, not the number of boys one is reared with (Bogaert, 2007). The effect is quite large: The probability of a male's being homosexual increases by 33.3% for every older brother he has (see Puts, Jordan, & Breedlove, 2006), and an estimated 15% of gay men can attribute their homosexuality to the fraternal birth order effect (Cantor et al., 2002). The **maternal immune hypothesis** has been proposed to explain the fraternal birth order effect; this hypothesis is that some mothers become progressively more immune to masculinizing hormones in male fetuses (see Blanchard, 2004), and a mother's immune system might deactivate masculinizing hormones in her younger sons.

What Triggers the Development of Sexual Attraction?

The evidence indicates that most girls and boys living in Western countries experience their first feelings of sexual attraction at about 10 years of age, whether they are heterosexual or homosexual (see Quinsey, 2003). This finding is at odds with the usual assumption that sexual interest is triggered by puberty, which, as you have learned, currently tends to occur at 10.5 years of age in girls and at 11.5 years in boys.

McClintock and Herdt (1996) have suggested that the emergence of sexual attraction may be stimulated by adrenal cortex steroids. Unlike gonadal maturation, adrenal maturation occurs at about the age of 10.

👁 **Watch** Adolescence: Identity and Role Development and Sexual Orientation **www.mypsychlab.com**

Is There a Difference in the Brains of Homosexuals and Heterosexuals?

The brains of homosexuals and heterosexuals must differ in some way, but how? Many studies have attempted to identify neuroanatomical, neuropsychological, neurophysiological, and hormonal response differences between homosexuals and heterosexuals.

In a highly publicized study, LeVay (1991) found that the structure of one hypothalamic nucleus in male homosexuals was intermediate between that in female heterosexuals and that in male heterosexuals. This study has not been consistently replicated, however. Indeed, no reliable difference between the brains of heterosexuals and homosexuals has yet been discovered (see Rahman, 2005).

Sexual Identity

Sexual identity is the sex, male or female, that a person believes himself or herself to be. Usually, sexual identity coincides with a person's anatomical sex, but not always.

Transsexualism is a condition of sexual identity in which an individual believes that he or she is trapped in a body of the other sex. To put it mildly, the transsexual faces a bizarre conflict: "I am a woman (or man) trapped in the body of a man (or woman). Help!" It is important to appreciate the desperation of these individuals; they do not merely think that life might be better if their gender were different. Although many transsexuals do seek *surgical sexual reassignment* (surgery to change their sex), the desperation is better indicated by how some of them dealt with their problem before surgical sexual reassignment was an option: Some biological males (psychological females) attempted self-castration, and others consumed copious quantities of estrogen-containing face creams in order to feminize their bodies.

● Watch Transsexuality
www.mypsychlab.com

How does surgical sexual reassignment work? I will describe the male-to-female procedure. The female-to-male procedure is much more complex (because a penis must be created) and far less satisfactory (for example, because a surgically created penis has no erectile potential), and male-to-female sexual reassignment is three times more prevalent.

Clinical Implications

The first step in male-to-female reassignment is psychiatric assessment to establish that the candidate for surgery is a true transsexual. Once accepted for surgical re-assignment, each transsexual receives in-depth counseling to prepare for the difficulties that will ensue. If the candidate is still interested in re-assignment after counseling, estrogen administration is initiated to feminize the body; the hormone regimen continues for life to maintain the changes. Then, comes the surgery. The penis and testes are surgically removed, and female external genitalia and a vagina are constructed—the vagina is lined with skin and nerves from the former penis so that it will have sensory nerve endings that will respond to sexual stimulation. Finally, some patients have cosmetic surgery to feminize the face (e.g., to reduce the size of the Adam's apple). Generally, the adjustment of transsexuals after surgical sexual reassignment is good.

The causes of transsexualism are unknown. Transsexualism was once thought to be a product of social learning, that is, of inappropriate child-rearing practices (e.g., mothers dressing their little boys in dresses). The occasional case that is consistent with this view can be found, but in most cases, there is no obvious cause (see Diamond, 2009; Swaab, 2004). One of the major difficulties in identifying the causes and mechanisms of transsexualism is that there is no comparable syndrome in nonhumans (Baum, 2006).

Evolutionary Perspective

Independence of Sexual Orientation and Sexual Identity

To complete this chapter, I would like to remind you of two of its main themes and show you how useful they are in thinking about one of the puzzles of human sexuality. One of the two themes is that the exception proves the rule: that a powerful test of any theory is its ability to explain exceptional cases. The second is that the mamawawa is seriously flawed: We have seen that men and women are similar in some ways (Hyde, 2005) and different in others (Cahill, 2006), but they are certainly not opposites, and their programs of development are neither parallel nor opposite.

Thinking Creatively

Here, I want to focus on the puzzling fact that sexual attraction, sexual identity, and body type are sometimes unrelated. For example, consider transsexuals: They, by definition, have the body type of one sex and the sexual identity of the other sex, but the orientation of their sexual attraction is an independent matter. Some transsexuals with a male body type are sexually attracted to females, others are sexually attracted to males, and others are sexually attracted to neither—and this is not changed by sexual reassignment (see Van Goozen et al., 2002). Also, it is important to realize that a particular sex-related trait in an individual can lie at midpoint between the female and male norms.

Obviously, the mere existence of homosexuality and transsexualism is a challenge to the mamawawa, the assumption that males and females belong to distinct and opposite categories. Many people tend to think of "femaleness" and "maleness" as being at opposite ends of a continuum, with a few abnormal cases somewhere between the two. Perhaps this is how you tend to think. However, the fact that body type, sexual orientation, and sexual identity are often independent constitutes a serious attack on any assumption that femaleness and maleness lie at opposite ends of a single scale. Clearly, femaleness or maleness is a combination of many different attributes (e.g., body type, sexual orientation, and sexual identity), each of which can develop quite independently. This is a real puzzle for many people, including scientists, but what you have already learned in this chapter suggests a solution.

Thinking Creatively

Think back to the section on brain differentiation. Until recently, it was assumed that the differentiation of the human brain into its usual female and male forms occurred through a single testosterone-based mechanism. However, a different notion has developed from recent evidence. Now, it is clear that male and female brains differ in many ways and that the differences develop at different times and by different mechanisms. If you keep this developmental principle in mind, you will have no difficulty understanding how it is possible for some individuals to be female in some ways and male in others, and to lie between the two norms in still others.

This analysis exemplifies a point I make many times in this book. The study of biopsychology often has important personal and social implications: The search for the neural basis of a behavior frequently provides us with a greater understanding of that behavior. I hope that you now have a greater understanding of, and acceptance of, differences in human sexuality.

Themes Revisited

Three of the book's four major themes were repeatedly emphasized in this chapter: the evolutionary perspective, clinical implications, and thinking creatively themes.

The evolutionary perspective theme was pervasive. It received frequent attention because most experimental studies of hormones and sex have been conducted in nonhuman species. The other major source of information about hormones and sex has been the study of human clinical cases, which is why the clinical implications theme was prominent in the cases of the woman who wasn't, the little girl who grew into a boy, the twin who lost his penis, and the man who lost and regained his manhood.

The thinking creatively theme was emphasized throughout the chapter because conventional ways of thinking about hormones and sex have often been at odds with the results of biopsychological research. If you are now better able to resist the seductive appeal of the men-are-men-and-women-are-women assumption, you are a more broadminded and understanding  person than when you began this chapter. I hope you have gained an abiding appreciation of the fact that maleness and femaleness are multidimensional and, at times, ambiguous variations of each other.

The fourth major theme of the book, neuroplasticity, arose during the discussions of the effects of hormones on the development of sex differences in the brain and of the neurotrophic effects of estradiol.

Think about It

1. Over the last century and a half, the onset of puberty has changed from age 15 or 16 to age 10 or 11, but there has been no corresponding acceleration in psychological and intellectual development. Precocious puberty is like a loaded gun in the hand of a child. Discuss.

2. Do you think adult sex-change operations should be permitted? Should they be permitted in preadolescents? Explain and supply evidence.

3. What should be done about the current epidemic of anabolic steroid abuse? Would you make the same recommendation if a safe anabolic steroid were developed? If a safe drug that would dramatically improve your memory were developed, would you take it?

4. Heterosexuality cannot be understood without studying homosexuality. Discuss.

5. What treatment should be given to infants born with ambiguous external genitals? Why?

6. Sexual orientation, sexual identity, and body type are not always related. Discuss.

Key Terms

13.1 Neuroendocrine System

Exocrine glands (p. 329)
Endocrine glands (p. 329)
Hormones (p. 329)
Gonads (p. 329)
Testes (p. 329)
Ovaries (p. 329)
Copulation (p. 329)
Zygote (p. 329)
Sex chromosomes (p. 329)
Amino acid derivative
    hormones (p. 329)
Peptide hormones (p. 329)
Protein hormones (p. 329)
Steroid hormones (p. 329)
Androgens (p. 329)

Estrogens (p. 329)
Testosterone (p. 329)
Estradiol (p. 329)
Progestins (p. 330)
Progesterone (p. 330)
Adrenal cortex (p. 330)
Gonadotropin (p. 330)
Posterior pituitary (p. 330)
Pituitary stalk (p. 330)
Anterior pituitary
    (p. 330)
Menstrual cycle (p. 330)
Vasopressin (p. 331)
Oxytocin (p. 331)
Paraventricular nuclei
    (p. 331)
Supraoptic nuclei (p. 331)

Hypothalamopituitary portal
    system (p. 331)
Releasing hormones
    (p. 331)
Release-inhibiting factors
    (p. 331)
Thyrotropin-releasing hormone
    (p. 331)
Thyrotropin (p. 332)
Gonadotropin-releasing
    hormone (p. 332)
Follicle-stimulating hormone
    (FSH) (p. 332)
Luteinizing hormone (LH)
    (p. 332)
Pulsatile hormone release
    (p. 333)

13.2 Hormones and Sexual Development of the Body

Sry gene (p. 334)
Sry protein (p. 334)
Wolffian system (p. 334)
Müllerian system (p. 334)
Müllerian-inhibiting substance
    (p. 334)
Scrotum (p. 334)
Ovariectomy (p. 334)
Orchidectomy (p. 334)
Gonadectomy (p. 335)
Genitals (p. 335)
Secondary sex characteristics
    (p. 335)
Growth hormone (p. 335)

Adrenocorticotropic hormone
(p. 335)
Androstenedione
(p. 335)

13.3 Hormones and Sexual Development of Brain and Behavior

Aromatase (p. 338)
Aromatization (p. 338)
Aromatization hypothesis
(p. 338)
Alpha fetoprotein (p. 338)
Masculinizes (p. 339)
Defeminizes (p. 339)
Lordosis (p. 339)

Feminizes (p. 339)
Demasculinizes (p. 339)
Intromission (p. 339)
Ejaculation (p. 339)
Proceptive behaviors (p. 339)

13.4 Three Cases of Exceptional Human Sexual Development

Androgenic insensitivity
syndrome (p. 340)
Adrenogenital syndrome (p. 341)
Congenital adrenal hyperplasia
(p. 341)
Ablatio penis (p. 342)

13.5 Effects of Gonadal Hormones on Adults

Replacement injections (p. 343)
Impotent (p. 344)
Estrus (p. 345)
Estrous cycle (p. 345)
Anabolic steroids (p. 345)

13.6 Neural Mechanisms of Sexual Behavior

Medial preoptic area (p. 348)
Sexually dimorphic nucleus
(p. 348)

Ventromedial nucleus (VMN)
(p. 349)

13.7 Sexual Orientation and Sexual Identity

Heterosexual (p. 350)
Homosexual (p. 350)
Bisexual (p. 350)
Sexual identity (p. 350)
Fraternal birth order effect
(p. 351)
Maternal immune hypothesis
(p. 351)
Transsexualism (p. 352)

✓●⌐**Quick Review** Test your comprehension of the chapter with this brief practice test. You can find the answers to these questions as well as more practice tests, activities, and other study resources at www.mypsychlab.com.

1. The ovaries and testes are
   a. zygotes.
   b. exocrine glands.
   c. gonads.
   d. both a and c
   e. both b and c

2. Gonadotropin is released by the
   a. anterior pituitary.
   b. posterior pituitary.
   c. hypothalamus.
   d. gonads.
   e. adrenal cortex.

3. Releasing hormones are released by the
   a. anterior pituitary.
   b. posterior pituitary.
   c. hypothalamus.
   d. gonads.
   e. adrenal cortex.

4. Which term refers specifically to the surgical removal of the testes?
   a. orchidectomy
   b. castration
   c. gonadectomy
   d. ovariectomy
   e. both b and c

5. Adrenogenital syndrome typically has severe consequences for
   a. rodents but not primates.
   b. Caucasians but not other ethnic groups.
   c. girls but not boys.
   d. boys but not girls.
   e. men but not women.

14

Sleep, Dreaming, and Circadian Rhythms

How Much Do You Need to Sleep?

Most of us have a fondness for eating and sex—the two highly esteemed motivated behaviors discussed in Chapter 12 and 13. But the amount of time devoted to these behaviors by even the most amorous gourmands pales in comparison to the amount of time spent sleeping: Most of us will sleep for well over 175,000 hours in our lifetimes. This extraordinary commitment of time implies that sleep fulfills a critical biological function. But what is it? And what about dreaming: Why do we spend so much time dreaming? And why do we tend to get sleepy at about the same time every day? Answers to these questions await you in this chapter.

Almost every time I lecture about sleep, somebody asks "How much sleep do we need?" Each time, I provide the same unsatisfying answer: I explain that there are two fundamentally different answers to this question, but neither has emerged a clear winner.

Watch Sleep: How Much?
www.mypsychlab.com

One answer stresses the presumed health-promoting and recuperative powers of sleep and suggests that people need as much sleep as they can comfortably get—the usual prescription being at least 8 hours per night. The other answer is that many of us sleep more than we need to and are consequently sleeping part of our life away. Just think how your life could change if you slept 5 hours per night instead of 8. You would have an extra 21 waking hours each week, a mind-boggling 10,952 hours each decade.

As I prepared to write this chapter, I began to think of the personal implications of the idea that we get more sleep than we need. That is when I decided to do something a bit unconventional. I am going to participate in a sleep-reduction experiment—by trying to get no more than 5 hours of sleep per night—11:00 P.M. to 4:00 A.M.—until this chapter is written. As I begin, I am excited by the prospect of having more time to write, but a little worried that this extra time might cost me too dearly.

Thinking Creatively

It is now the next day—4:50 Saturday morning to be exact—and I am just sitting down to write. There was a party last night, and I didn't make it to bed by 11:00; but considering that I slept for only 3 hours and 35 minutes, I feel quite good. I wonder what I will feel like later in the day. In any case, I will report my experiences to you at the end of the chapter.

The following case study challenges several common beliefs about sleep. Ponder its implications before proceeding to the body of the chapter.

The Case of the Woman Who Wouldn't Sleep

Miss M . . . is a busy lady who finds her ration of twenty-three hours of wakefulness still insufficient for her needs.

Even though she is now retired she is still busy in the community, helping sick friends whenever requested. She is an active painter and . . . writer. Although she becomes tired physically, when she needs to sit down to rest her legs, she does not ever report feeling sleepy. During the night she sits on her bed . . . reading, writing, crocheting or painting. At about 2:00 A.M. she falls asleep without any preceding drowsiness often while still holding a book in her hands. When she wakes about an hour later, she feels as wide awake as ever. . . .

We invited her along to the laboratory. She came willingly but on the first evening we hit our first snag. She announced that she did not sleep at all if she had interesting things to do, and by her reckoning a visit to a university sleep laboratory counted as very interesting. Moreover, for the first time in years, she had someone to talk to for the whole of the night. So we talked.

In the morning we broke into shifts so that some could sleep while at least one person stayed with her and entertained her during the next day. The second night was a repeat performance of the first night. . . .

In the end we prevailed upon her to allow us to apply EEG electrodes and to leave her sitting comfortably on the bed in the bedroom. She had promised that she would co-operate by not resisting sleep although she claimed not to be especially tired. . . . At approximately 1:30 A.M., the EEG record showed the first signs of sleep even though . . . she was still sitting with the book in her hands. . . .

The only substantial difference between her sleep and what we might have expected. . . was that it was of short duration. . . . [After 99 minutes], she had no further interest in sleep and asked to . . . join our company again.

("The Case of the Woman Who Wouldn't Sleep," from *The Sleep Instinct* by R. Meddis. Copyright © 1977, Routledge & Kegan Paul, London, pp. 42–44. Reprinted by permission of the Taylor & Francis Group.)

14.1
Stages of Sleep

Many changes occur in the body during sleep. This section introduces you to the major ones.

Three Standard Psychophysiological Measures of Sleep

There are major changes in the human EEG during the course of a night's sleep. Although the EEG waves that accompany sleep are generally high-voltage and slow, there are periods throughout the night that are dominated by low-voltage, fast waves similar to those in nonsleeping individuals. In the 1950s, it was discovered that *rapid eye movements (REMs)* occur under the closed eyelids of sleepers during these periods of low-voltage, fast EEG activity. And in 1962,

Berger and Oswald discovered that there is also a loss of electromyographic activity in the neck muscles during these same sleep periods. Subsequently, the **electroencephalogram (EEG)**, the **electrooculogram (EOG)**, and the neck **electromyogram (EMG)** became the three standard psychophysiological bases for defining stages of sleep.

Figure 14.1 depicts a volunteer participating in a sleep experiment. A participant's first night of sleep in a laboratory is often fitful. That's why the usual practice is to have each participant sleep several nights in the laboratory before commencing a sleep study. The disturbance of sleep observed during the first night in a sleep laboratory is called the *first-night phenomenon*. It is well known to graders of introductory psychology examinations because of the creative definitions of it that are offered by students who forget that it is a sleep-related, rather than a sex-related, phenomenon.

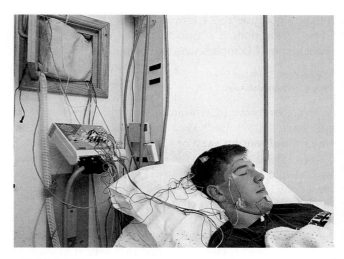

FIGURE 14.1   A participant in a sleep experiment.

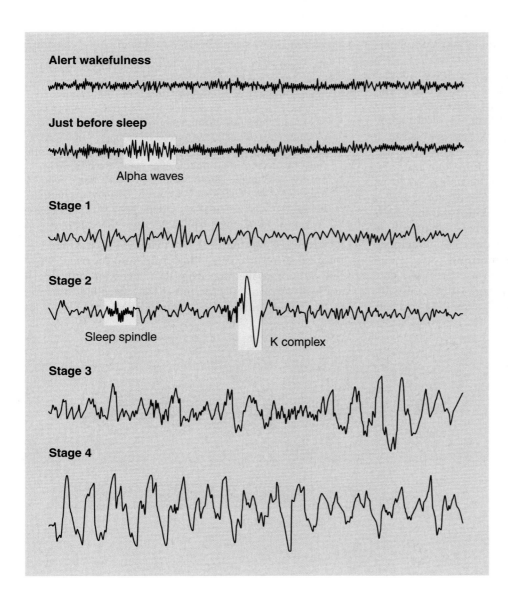

FIGURE 14.2 The EEG of alert wakefulness, the EEG that precedes sleep onset, and the four stages of sleep EEG. Each trace is about 10 seconds long.

Four Stages of Sleep EEG

There are four stages of sleep EEG: stage 1, stage 2, stage 3, and stage 4. Examples of these are presented in Figure 14.2.

After the eyes are shut and a person prepares to go to sleep, **alpha waves**—waxing and waning bursts of 8- to 12-Hz EEG waves—begin to punctuate the low-voltage, high-frequency waves of alert wakefulness. Then, as the person falls asleep, there is a sudden transition to a period of stage 1 sleep EEG. The stage 1 sleep EEG is a low-voltage, high-frequency signal that is similar to, but slower than, that of alert wakefulness.

There is a gradual increase in EEG voltage and a decrease in EEG frequency as the person progresses from stage 1 sleep through stages 2, 3, and 4. Accordingly, the stage 2 sleep EEG has a slightly higher amplitude and a lower frequency than the stage 1 EEG; in addition, it is punctuated by two characteristic wave forms: K complexes and sleep spindles. Each *K complex* is a single large negative wave (upward deflection) followed immediately by a single large positive wave (downward deflection)—see Cash and colleagues (2009). Each *sleep spindle* is a 1- to 2-second waxing and waning burst of 12- to 14-Hz waves. The stage 3 sleep EEG is defined by the occasional presence of **delta waves**—the largest and slowest EEG waves, with a frequency of 1 to 2 Hz—whereas the stage 4 sleep EEG is defined by a predominance of delta waves.

Watch Stages of Sleep
www.mypsychlab.com

Once sleepers reach stage 4 EEG sleep, they stay there for a time, and then they retreat back through the stages of sleep to stage 1. However, when they return to stage 1, things are not at all the same as they were the first time through. The first period of stage 1 EEG during a night's sleep (**initial stage 1 EEG**) is not marked by any striking electromyographic or electrooculographic changes, whereas subsequent periods of stage 1 sleep EEG (**emergent stage 1 EEG**) are accompanied by REMs and by a loss of tone in the muscles of the body core.

After the first cycle of sleep EEG—from initial stage 1 to stage 4 and back to emergent stage 1—the rest of the night is spent going back and forth through the stages. Figure 14.3 illustrates the EEG cycles of a typical night's sleep and the close relation between emergent stage 1 sleep, REMs, and the loss of tone in core muscles. Notice that each cycle tends to be about 90 minutes long and that, as the night progresses, more and more time is spent in emergent stage 1 sleep, and less and less time is spent in the other stages, particularly stage 4. Notice also that there are brief periods during the night when the person is awake, although he or she usually does not remember these periods of wakefulness in the morning.

Let's pause here to get some sleep-stage terms straight. The sleep associated with emergent stage 1 EEG is usually called **REM sleep** (pronounced "rehm"), after the associated rapid eye movements; whereas all other stages of sleep together are called *NREM sleep* (non-REM sleep). Stages 3 and 4 together are referred to as **slow-wave sleep (SWS)**, after the delta waves that characterize them.

REMs, loss of core-muscle tone, and a low-amplitude, high-frequency EEG are not the only physiological correlates of REM sleep. Cerebral activity (e.g., oxygen consumption, blood flow, and neural firing) increases to waking levels in many brain structures, and there is a general increase in the variability of autonomic nervous system activity (e.g., in blood pressure, pulse, and respiration). Also, the muscles of the extremities occasionally twitch, and there is often some degree of penile erection in males.

REM Sleep and Dreaming

Nathaniel Kleitman's laboratory was an exciting place in 1953. REM sleep had just been discovered, and Kleitman and his students were driven by the fascinating implication of the discovery. With the exception of the loss of tone in the core muscles, all of the other measures suggested that REM sleep episodes were emotion-charged. Could REM sleep be the physiological correlate of dreaming? Could it provide researchers with a window into the subjective inner world of dreams? The researchers began by waking a few sleepers in the middle of REM episodes:

> The vivid recall that could be elicited in the middle of the night when a subject was awakened while his eyes were moving rapidly was nothing short of miraculous. It [seemed to open] . . . an exciting new world to the subjects whose only previous dream memories had been the vague morning-after recall. Now, instead of perhaps some fleeting glimpse into the dream world each night, the subjects could be tuned into the middle of as many as ten or twelve dreams every night. (From *Some Must Watch While Some Must Sleep* by William C. Dement, Portable Stanford Books, Stanford Alumni Association, Stanford University, 1978, p. 37. Used by permission of William C. Dement.)

Strong support for the theory that REM sleep is the physiological correlate of dreaming came from the observation that 80% of awakenings from REM sleep but only 7% of awakenings from NREM (non-REM) sleep led to dream recall. The dreams recalled from NREM sleep tended to

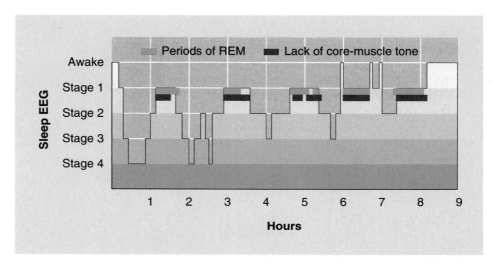

FIGURE 14.3 The course of EEG stages during a typical night's sleep and the relation of emergent stage 1 EEG to REMs and lack of tone in core muscles.

be isolated experiences (e.g., "I was falling"), while those associated with REM sleep tended to take the form of stories, or narratives. The phenomenon of dreaming, which for centuries had been the subject of wild speculation, was finally rendered accessible to scientific investigation.

((•● **Listen** Dreaming
www.mypsychlab.com

Testing Common Beliefs about Dreaming

The high correlation between REM sleep and dream recall provided an opportunity to test some common beliefs about dreaming. The following five beliefs were among the first to be addressed:

- Many people believe that external stimuli can become incorporated into their dreams. Dement and Wolpert (1958) sprayed water on sleeping volunteeers after they had been in REM sleep for a few minutes, and then awakened them a few seconds later. In 14 of 33 cases, the water was incorporated into the dream report. The following narrative was reported by one participant who had been dreaming that he was acting in a play:

  ● **Watch** Lucid Dreaming
  www.mypsychlab.com

  > I was walking behind the leading lady when she suddenly collapsed and water was dripping on her. I ran over to her and water was dripping on my back and head. The roof was leaking. . . . I looked up and there was a hole in the roof. I dragged her over to the side of the stage and began pulling the curtains. Then I woke up. (p. 550)

- Some people believe that dreams last only an instant, but research suggests that dreams run on "real time." In one study (Dement & Kleitman, 1957), volunteers were awakened 5 or 15 minutes after the beginning of a REM episode and asked to decide on the basis of the duration of the events in their dreams whether they had been dreaming for 5 or 15 minutes. They were correct in 92 of 111 cases.
- Some people claim that they do not dream. However, these people have just as much REM sleep as normal dreamers. Moreover, they report dreams if they are awakened during REM episodes (Goodenough et al., 1959), although they do so less frequently than do normal dreamers.
- Penile erections are commonly assumed to be indicative of dreams with sexual content. However, erections are no more complete during dreams with frank sexual content than during those without it (Karacan et al., 1966). Even babies have REM-related penile erections.
- Many people believe that sleeptalking (*somniloquy*) and sleepwalking (*somnambulism*) occur only during dreaming. This is not so (see Dyken, Yamada, & Lin-Dyken, 2001). Sleeptalking has no special association with REM sleep—it can occur during any stage but often occurs

during a transition to wakefulness. Sleepwalking usually occurs during stage 3 or 4 sleep, and it never occurs during dreaming, when core muscles tend to be totally relaxed (Usui et al., 2007). There is no proven treatment for sleepwalking (Harris & Grunstein, 2008).

Interpretation of Dreams

Sigmund Freud believed that dreams are triggered by unacceptable repressed wishes, often of a sexual nature. He argued that because dreams represent unacceptable wishes, the dreams we experience (our *manifest dreams*) are merely disguised versions of our real dreams (our *latent dreams*): He hypothesized an unconscious censor that disguises and subtracts information from our real dreams so that we can endure them. Freud thus concluded that one of the keys to understanding people and dealing with their psychological problems is to expose the meaning of their latent dreams through the interpretation of their manifest dreams.

There is no convincing evidence for the Freudian theory of dreams; indeed, the brain science of the 1890s, which served as its foundation, is now obsolete. Yet many people accept the notion that dreams bubble up from a troubled subconscious and that they represent repressed thoughts and wishes.

The modern alternative to the Freudian theory of dreams is Hobson's (1989) activation-synthesis theory (see Eiser, 2005). It is based on the observation that, during REM sleep, many brain-stem circuits become active and bombard the cerebral cortex with neural signals. The essence of the **activation-synthesis theory** is that the information supplied to the cortex during REM sleep is largely random and that the resulting dream is the cortex's effort to make sense of these random signals.

14.2
Why Do We Sleep, and Why Do We Sleep When We Do?

Now that you have been introduced to the properties of sleep and its various stages, the focus of this chapter shifts to a consideration of two fundamental questions about sleep: Why do we sleep? And why do we sleep when we do?

Two kinds of theories for sleep have been proposed: *recuperation theories* and *adaptation theories*. The differences between these two theoretical approaches are revealed by the answers they offer to the two fundamental questions about sleep.

The essence of **recuperation theories of sleep** is that being awake disrupts the *homeostasis* (internal physiological stability) of the body in some way and sleep is required to restore it. Various recuperation theories differ in terms of the particular physiological disruption they propose as the trigger for sleep—for example, it is commonly believed that the function of sleep is to restore energy levels. However, regardless of the particular function postulated by restoration theories of sleep, they all imply that sleepiness is triggered by a deviation from homeostasis caused by wakefulness and that sleep is terminated by a return to homeostasis.

The essence of **adaptation theories of sleep** is that sleep is not a reaction to the disruptive effects of being awake but the result of an internal 24-hour timing mechanism—that is, we humans are programmed to sleep at night regardless of what happens to us during the day. According to these theories, we have evolved to sleep at night because sleep protects us from accident and predation during the night. (Remember that humans evolved long before the advent of artificial lighting.)

Adaptation theories of sleep focus more on when we sleep than on the function of sleep. Some of these theories even propose that sleep plays no role in the efficient physiological functioning of the body. According to these theories, early humans had enough time to get their eating, drinking, and reproducing out of the way during the daytime, and their strong motivation to sleep at night evolved to conserve their energy resources and to make them less susceptible to mishap (e.g., predation) in the dark (Rattenborg, Martinez-Gonzales, & Lesku, 2009; Siegel, 2009). Adaptation theories suggest that sleep is like reproductive behavior in the sense that we are highly motivated to engage in it, but we don't need it to stay healthy.

Comparative Analysis of Sleep

Sleep has been studied in only a small number of species, but the evidence so far suggests that most mammals and birds sleep. Furthermore, the sleep of mammals and birds, like ours, is characterized by high-amplitude, low-frequency EEG waves punctuated by periods of low-amplitude, high-frequency waves (see Siegel, 2008). The evidence for sleep in amphibians, reptiles, fish, and insects is less clear: Some display periods of inactivity and unresponsiveness, but the relation of these periods to mammalian sleep has not been established (see Siegel, 2008; Zimmerman et al., 2008). Table 14.1 gives the average number of hours per day that various mammalian species spend sleeping.

The comparative investigation of sleep has led to several important conclusions. Let's consider four of these.

First, the fact that most mammals and birds sleep suggests that sleep serves some important physiological function, rather than merely protecting animals from mishap and conserving energy. The evidence is strongest in species that are at increased risk of predation when they sleep (e.g., antelopes) and in species that have evolved complex mechanisms that enable them to sleep.

For example, some marine mammals, such as dolphins, sleep with only half of their brain at a time so that the other half can control resurfacing for air (see Rattenborg, Amlaner, & Lima, 2000). It is against the logic of natural selection for some animals to risk predation while sleeping and for others to have evolved complex mechanisms to permit them to sleep, unless sleep itself serves some critical function.

Second, the fact that most mammals and birds sleep suggests that the primary function of sleep is not some special, higher-order human function. For example, suggestions that sleep helps humans reprogram our complex brains or that it permits some kind of emotional release to maintain our mental health are improbable in view of the comparative evidence.

Third, the large between-species differences in sleep time suggest that although sleep may be essential for survival, it is not necessarily needed in large quantities (refer to Table 14.1). Horses and many other animals get by quite nicely on 2 or 3 hours of sleep per day. Moreover, it is important to realize that the sleep patterns of mammals and birds in their natural environments can vary substantially from their patterns in captivity, which is where they are typically studied (see Horne, 2009). For example, some animals that sleep a great deal in captivity sleep little in the wild when food is in short supply or during periods of migration (Siegel, 2008).

TABLE 14.1   Average Number of Hours Slept per Day by Various Mammalian Species

Mammalian Species	Hours of Sleep per Day
Giant sloth	20
Opossum, brown bat	19
Giant armadillo	18
Owl monkey, nine-banded armadillo	17
Arctic ground squirrel	16
Tree shrew	15
Cat, golden hamster	14
Mouse, rat, gray wolf, ground squirrel	13
Arctic fox, chinchilla, gorilla, raccoon	12
Mountain beaver	11
Jaguar, vervet monkey, hedgehog	10
Rhesus monkey, chimpanzee, baboon, red fox	9
Human, rabbit, guinea pig, pig	8
Gray seal, gray hyrax, Brazilian tapir	6
Tree hyrax, rock hyrax	5
Cow, goat, elephant, donkey, sheep	3
Roe deer, horse	2

FIGURE 14.4 After gorging themselves on a kill, African lions often sleep almost continuously for 2 or 3 days. And where do they sleep? Anywhere they want!

Fourth, many studies have tried to identify some characteristic that identifies various species as long sleepers or short sleepers. Why do cats tend to sleep about 14 hours a day and horses only about 2? Under the influence of recuperation theories, researchers have focused on energy-related factors in their efforts. However, there is no strong relationship between a species' sleep time and its level of activity, its body size, or its body temperature (see Siegel, 2005). The fact that giant sloths sleep 20 hours per day is a strong argument against the theory that sleep is a compensatory reaction to energy expenditure—similarly, energy expenditure has been shown to have little effect on subsequent sleep in humans (Driver & Taylor, 2000; Youngstedt & Kline, 2006). In contrast, adaptation theories correctly predict that the daily sleep time of each species is related to how vulnerable it is while it is asleep and how much time it must spend each day to feed itself and to take care of its other survival requirements. For example, zebras must graze almost continuously to get enough to eat and are extremely vulnerable to predatory attack when they are asleep—and they sleep only about 2 hours per day. In contrast, African lions often sleep more or less continuously for 2 or 3 days after they have gorged themselves on a kill. Figure 14.4 says it all.

14.3
Effects of Sleep Deprivation

One way to identify the functions of sleep is to determine what happens when a person is deprived of sleep. This section begins with a cautionary note about the interpretation of the effects of sleep deprivation, a description of

the predictions that recuperation theories make about sleep deprivation, and two classic case studies of sleep deprivation. Then, it summarizes the results of sleep-deprivation research.

Interpretation of the Effects of Sleep Deprivation: The Stress Problem

I am sure that you have experienced the negative effects of sleep loss. When you sleep substantially less than you are used to, the next day you feel out of sorts and unable to function as well as you usually do. Although such experiences of sleep deprivation are compelling, you need to be cautious in interpreting them. In Western cultures, most people who sleep little or irregularly do so because they are under extreme stress (e.g., from illness, excessive work, shift work, drugs, or examinations), which could have adverse effects independent of any sleep loss. Even when sleep deprivation studies are conducted on healthy volunteers in controlled laboratory environments, stress can be a contributing factor because many of the volunteers will find the sleep-deprivation procedure itself stressful. Because it is difficult to separate the effects of sleep loss from the effects of stressful conditions that may have induced the loss, results of sleep-deprivation studies must be interpreted with particular caution.

Unfortunately, many studies of sleep deprivation, particularly those that are discussed in the popular media, do not control for stress. For example, almost weekly I read an article in my local newspaper decrying the effects of sleep loss in the general population. It will point out that many people who are pressured by the demands of their work schedule sleep little and experience a variety of health and accident problems. There is a place for this kind of research because it identifies a problem that requires public attention; however, because the low levels of sleep are hopelessly confounded with high levels of stress, many sleep-deprivation studies tell us little about the functions of sleep and how much we need.

Predictions of Recuperation Theories about Sleep Deprivation

Because recuperation theories of sleep are based on the premise that sleep is a response to the accumulation of some debilitating effect of wakefulness, they make the following three predictions about sleep deprivation:

- Long periods of wakefulness will produce physiological and behavioral disturbances.
- These disturbances will grow steadily worse as the sleep deprivation continues.
- After a period of deprivation has ended, much of the missed sleep will be regained.

Have these predictions been confirmed?

Two Classic Sleep–Deprivation Case Studies

Let's look at two widely cited sleep-deprivation case studies. First is the study of a group of sleep-deprived students, described by Kleitman (1963); second is the case of Randy Gardner, described by Dement (1978).

The Case of the Sleep–Deprived Students

While there were differences in the many subjective experiences of the sleep-evading persons, there were several features common to most.... [D]uring the first night the subject did not feel very tired or sleepy. He could read or study or do laboratory work, without much attention from the watcher, but usually felt an attack of drowsiness between 3 A.M. and 6 A.M. ... Next morning the subject felt well, except for a slight malaise which always appeared on sitting down and resting for any length of time. However, if he occupied himself with his ordinary daily tasks, he was likely to forget having spent a sleepless night. During the second night . . . reading or study was next to impossible because sitting quietly was conducive to even greater sleepiness. As during the first night, there came a 2–3 hour period in the early hours of the morning when the desire for sleep was almost overpowering. . . . Later in the morning the sleepiness diminished once more, and the subject could perform routine laboratory work, as usual. It was not safe for him to sit down, however, without danger of falling asleep, particularly if he attended lectures. . . .

  The third night resembled the second, and the fourth day was like the third. . . . At the end of that time the individual was as sleepy as he was likely to be. Those who continued to stay awake experienced the wavelike increase and decrease in sleepiness with the greatest drowsiness at about the same time every night. (Kleitman, 1963, pp. 220–221)

The Case of Randy Gardner

As part of a 1965 science fair project, Randy Gardner and two classmates, who were entrusted with keeping him awake, planned to break the then world record of 260 hours of consecutive wakefulness. Dement read about the project in the newspaper and, seeing an opportunity to collect some important data, joined the team, much to the comfort of Randy's worried parents. Randy proved to be a friendly and cooperative subject, although he did complain vigorously when his team would not permit him to close his eyes for more than a few seconds at a time. However, in no sense could Randy's behavior be considered abnormal or disturbed. Near the end of his vigil, Randy held a press conference attended by reporters and television crews from all over the United States, and he conducted

himself impeccably. When asked how he had managed to stay awake for 11 days, he replied politely, "It's just mind over matter." Randy went to sleep exactly 264 hours and 12 minutes after his alarm clock had awakened him 11 days before. And how long did he sleep? Only 14 hours the first night, and thereafter he returned to his usual 8-hour schedule. Although it may seem amazing that Randy did not have to sleep longer to "catch up" on his lost sleep, the lack of substantial recovery sleep is typical of such cases.

(From *Some Must Watch While Some Must Sleep* by William C. Dement, Portable Stanford Books, Stanford Alumni Association, Stanford University, 1978, pp. 38–39. Used by permission of William C. Dement.)

Experimental Studies of Sleep Deprivation in Humans

Since the first studies of sleep deprivation by Dement and Kleitman in the mid-20th century, there have been hundreds of studies assessing the effects on humans of sleep-deprivation schedules ranging from a slightly reduced amount of sleep during one night to total sleep deprivation for several nights (see Durmer & Dinges, 2005). The studies have assessed the effects of these schedules on many different measures of sleepiness, mood, cognition, motor performance, physiological function, and even molecular function (see Cirelli, 2006).

  Even moderate amounts of sleep deprivation—for example, sleeping 3 or 4 hours less than normal for one night—have been found to have three consistent effects. First, sleep-deprived individuals display an increase in sleepiness: They report being more sleepy, and they fall asleep more quickly if given the opportunity. Second, sleep-deprived individuals display negative affect on various written tests of mood. And third, they perform poorly on tests of vigilance, such as watching a computer screen and responding when a moving light flickers.

  The effects of sleep deprivation on complex cognitive functions have been less consistent (see Drummond et al., 2004). Consequently, researchers have preferred to assess performance on the simple, dull, monotonous tasks most sensitive to the effects of sleep deprivation (see Harrison & Horne, 2000). Nevertheless, a growing number of studies have been able to demonstrate disruption of the performance of complex cognitive tasks by sleep deprivation (Blagrove, Alexander, & Horne, 2006; Durmer & Dinges, 2005; Killgore, Balkin, & Wesensten, 2006; Nilsson et al., 2005) although a substantial amount of sleep deprivation (e.g., 24 hours) has often been required to produce consistent disruption (e.g., Killgore, Balkin, & Wesensten, 2006; Strangman et al., 2005).

  The disruptive impact of sleep deprivation on cognitive function has been clarified by the discovery that only some cognitive functions are susceptible. Many early studies of the effect of sleep deprivation on cognitive function used tests of logical deduction or critical thinking, and performance on these has proved to be largely immune to the disruptive effects of sleep loss. In contrast, performance on tests of **executive function** (cognitive abilities that appear to depend on the *prefrontal cortex*) has proven much more susceptible (see Nilsson et al., 2005). Executive function includes innovative thinking, lateral thinking, insightful thinking, and assimilating new information to update plans and strategies.

The adverse effects of sleep deprivation on physical performance have been surprisingly inconsistent considering the general belief that a good night's sleep is essential for optimal motor performance. Only a few measures tend to be affected, even after lengthy periods of deprivation (see Van Helder & Radomski, 1989).

Sleep deprivation has been found to have a variety of physiological consequences such as reduced body temperature, increases in blood pressure, decreases in some aspects of immune function, hormonal changes, and metabolic changes (e.g., Dinges et al., 1994; Kato et al., 2000; Knutson et al., 2007; Ogawa et al., 2003). The problem is that there is little evidence that these changes have any consequences for health or performance. For example, the fact that a decline in immune function was discovered in sleep-deprived volunteers does not necessarily mean that they would be more susceptible to infection—the immune system is extremely complicated and a decline in one aspect can be compensated for by other changes. This is why I want to single out a study by Cohen and colleagues (2009) for commendation: Rather than studying immune function, these researchers focused directly on susceptibility to infection and illness. They exposed 153 healthy volunteers to a cold virus. Those who reported sleeping less than 8 hours a night were not less likely to become infected, but they were more likely to develop cold symptoms. Although this is only a correlational study (see Chapter 1) and thus cannot directly implicate sleep duration as the causal factor, experimental studies of sleep and infectious disease need to follow this example and directly measure susceptibility to infection and illness.

After 2 or 3 days of continuous sleep deprivation, most study participants experience microsleeps, unless they are in a laboratory environment where the microsleeps can be interrupted as soon as they begin. **Microsleeps** are brief periods of sleep, typically about 2 or 3 seconds long, during which the eyelids droop and the subjects become less responsive to external stimuli, even though they remain sitting or standing. Microsleeps disrupt performance on tests of vigilance, but such performance deficits also occur in sleep-deprived individuals who are not experiencing microsleeps (Ferrara, De Gennaro, & Bertini, 1999).

It is useful to compare the effects of sleep deprivation with those of deprivation of the motivated behaviors

discussed in Chapters 12 and 13. If people were deprived of the opportunity to eat or engage in sexual activity, 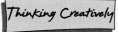 the effects would be severe and unavoidable: In the first case, starvation and death would ensue; in the second, there would be a total loss of reproductive capacity. Despite our powerful drive to sleep, the effects of sleep deprivation tend to be subtle, selective, and variable. This is puzzling. Another puzzling thing is that performance deficits observed after extended periods of sleep deprivation disappear so readily—for example, in one study, 4 hours of sleep eliminated the performance deficits produced by 64 hours of sleep deprivation (Rosa, Bonnett, & Warm, 2007).

Sleep-Deprivation Studies with Laboratory Animals

The **carousel apparatus** (see Figure 14.5) has been used to deprive rats of sleep. Two rats, an experimental rat and its *yoked control*, are placed in separate chambers of the apparatus. Each time the EEG activity of the experimental rat indicates that it is sleeping, the disk, which serves as the floor of half of both chambers, starts to slowly rotate. As a result, if the sleeping experimental rat does not awaken immediately, it gets shoved off the disk into a shallow pool of water. The yoked control is exposed to exactly the same pattern of disk rotations; but if it is not sleeping, it can easily avoid getting dunked by walking in the direction opposite to the direction of disk rotation. The experimental rats typically died after about 12 days,

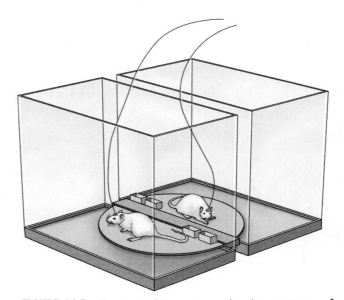

FIGURE 14.5   The carousel apparatus used to deprive an experimental rat of sleep while a yoked control rat is exposed to the same number and pattern of disk rotations. The disk on which both rats rest rotates every time the experimental rat has a sleep EEG. If the sleeping rat does not awaken immediately, it is deposited in the water. (Based on Rechtschaffen et al., 1983.)

while the yoked controls stayed reasonably healthy (see Rechtschaffen & Bergmann, 1995).

The fact that humans and rats have been sleep-deprived by other means for similar periods of time without dire consequences argues for caution in interpreting the results of the carousel sleep-deprivation experiments (see Rial et al., 2007; Siegel, 2009). It may be that repeatedly being awakened by this apparatus kills the experimental rats not because it keeps them from sleeping but because it is stressful. This interpretation is consistent with the pathological problems in the experimental rats that were revealed by postmortem examination: swollen adrenal glands, gastric ulcers, and internal bleeding.

You have already encountered many examples in this book of the value of the comparative approach. However, sleep deprivation may be one phenomenon that cannot be productively studied in nonhumans because of the unavoidable confounding effects of extreme stress (see Benington & Heller, 1999; D'Almeida et al., 1997; Horne, 2000).

Thinking Creatively

REM-Sleep Deprivation

Because of its association with dreaming, REM sleep has been the subject of intensive investigation. In an effort to reveal the particular functions of REM sleep, sleep researchers have specifically deprived sleeping volunteers of REM sleep by waking them up each time a bout of REM sleep begins.

REM-sleep deprivation has been shown to have two consistent effects (see Figure 14.6). First, following REM-sleep deprivation, participants display a *REM rebound;* that is, they have more than their usual amount of REM sleep for the first two or three nights (Brunner et al., 1990). Second, with each successive night of deprivation, there is a greater tendency for participants to initiate REM sequences. Thus, as REM-sleep deprivation proceeds, participants have to be awakened more and more frequently to keep them from accumulating significant amounts of REM sleep. For example, during the first night of REM-sleep deprivation in one experiment (Webb & Agnew, 1967), the participants had to be awakened 17 times to keep them from having extended periods of REM sleep; but during the seventh night of deprivation, they had to be awakened 67 times.

The compensatory increase in REM sleep following a period of REM-sleep deprivation suggests that the amount of REM sleep is regulated separately from the amount of slow-wave sleep and that REM sleep serves a special function. This finding, coupled with the array of interesting physiological and psychological events that define REM sleep, has led to much speculation about its function.

Considerable attention has focused on the potential role of REM sleep in strengthening explicit memory (see Chapter 11). Many reviewers of the literature on this topic have treated the positive effect of REM sleep on the storage of existing memories as well established, and researchers have moved on to study the memory-promoting effects of other stages of sleep (e.g. Deak & Stickgold, 2010; Rasch & Born, 2008; Stickgold & Walker, 2007) and the physiological mechanisms of these memory-promoting effects (e.g., Rasch et al., 2007). However, two eminent sleep researchers,

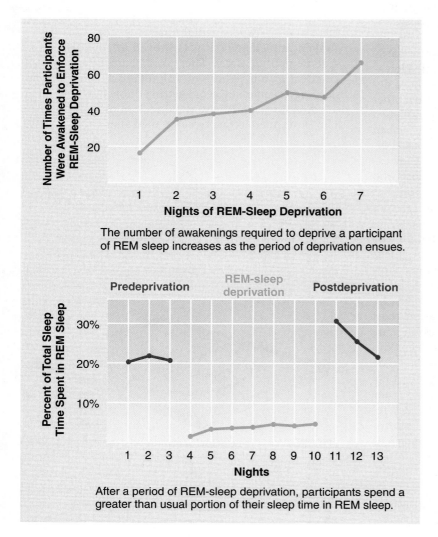

The number of awakenings required to deprive a participant of REM sleep increases as the period of deprivation ensues.

After a period of REM-sleep deprivation, participants spend a greater than usual portion of their sleep time in REM sleep.

FIGURE 14.6 The two effects of REM-sleep deprivation.

Robert Vertes and Jerome Seigel (2005), have argued that the evidence that REM sleep strengthens memory is unconvincing (see Vertes, 2004). They point out, for example, that numerous studies failing to support a *mnemonic* (pertaining to memory) function of REM sleep have been ignored. They also question why the many patients who have taken antidepressant drugs that block REM sleep experience no obvious memory problems, even if they have taken the drugs for months or even years. In one study, pharmacologically blocking REM sleep in human volunteers did not disrupt consolidation of verbal memories, and it actually improved the consolidation of the memory of motor tasks (Rasch et al., 2008)

The *default theory* of REM sleep is a different approach (Horne, 2000). According to this theory, it is difficult to stay continuously in NREM sleep, so the brain periodically switches to one of two other states. If there is any immediate bodily need to take care of (e.g., eating or drinking), the brain switches to wakefulness; if there are no immediate needs, it switches to the default state—REM sleep. According to the default theory, REM sleep and wakefulness are similar states, but REM sleep is more adaptive when there are no immediate bodily needs. Indirect support for this theory comes from the many similarities between REM sleep and wakefulness.

A study by Nykamp and colleagues (1998) supported the default theory of REM sleep. The researchers awakened young adults every time they entered REM sleep, but instead of letting them go back to sleep immediately, substituted a 15-minute period of wakefulness for each lost REM period. Under these conditions, the participants, unlike the controls, were not tired the next day, despite getting only 5 hours of sleep, and they displayed no REM rebound. In other words, there seemed to be no need for REM sleep if periods of wakefulness were substituted for it. This finding has been replicated in rats (Oniani & Lortkipanidze, 2003), and it is consistent with the finding that as antidepressants reduce REM sleep, the number of nighttime awakenings increases (see Horne, 2000).

Sleep Deprivation Increases the Efficiency of Sleep

One of the most important findings of human sleep-deprivation research is that individuals who are deprived of sleep become more efficient sleepers (see Elmenhorst et al., 2008). In particular, their sleep has a higher proportion of slow-wave sleep (stages 3 and 4), which seems to serve the main restorative function. Because this is such an important finding, let's look at six major pieces of evidence that support it:

- Although people regain only a small proportion of their total lost sleep after a period of sleep deprivation, they regain most of their lost stage 4 sleep (e.g.,

Borbély et al., 1981; De Gennaro, Ferrara, & Bertini, 2000; Lucidi et al., 1997).
- After sleep deprivation, the slow-wave sleep EEG of humans is characterized by an even higher proportion than usual of slow waves (Aeschbach et al., 1996; Borbély, 1981; Borbély et al., 1981).
- People who sleep 6 hours or less per night normally get as much slow-wave sleep as people who sleep 8 hours or more (e.g., Jones & Oswald, 1966; Webb & Agnew, 1970).
- If individuals take a nap in the morning after a full night's sleep, their naptime EEG shows few slow waves, and the nap does not reduce the duration of the following night's sleep (e.g., Åkerstedt & Gillberg, 1981; Hume & Mills, 1977; Karacan et al., 1970).
- People who gradually reduce their usual sleep time get less stage 1 and stage 2 sleep, but the duration of their slow-wave sleep remains about the same as before (Mullaney et al., 1977; Webb & Agnew, 1975).
- Repeatedly waking individuals during REM sleep produces little increase in the sleepiness they experience the next day, whereas repeatedly waking individuals during slow-wave sleep has major effects (Nykamp et al., 1998).

The fact that sleep becomes more efficient in people who sleep less means that conventional sleep-deprivation studies are virtually useless for discovering how much sleep people need. Certainly, our bodies respond negatively when we get less sleep than we are used to getting. However, the negative consequences of sleep loss in inefficient sleepers does not indicate whether the lost sleep was really needed. The true need for sleep can be assessed only by experiments in which sleep is regularly reduced for many weeks, to give the participants the opportunity to adapt to getting less sleep by maximizing their sleep efficiency. Only when people are sleeping at their maximum efficiency is it possible to determine how much sleep they really need. Such sleep-reduction studies are discussed later in the chapter, but please pause here to think about this point—it is extremely important, and it is totally consistent with the growing appreciation of the plasticity and adaptiveness of the adult mammalian brain.

Thinking Creatively

Neuroplasticity

This is an appropriate time, here at the end of the section on sleep deprivation, for me to file a brief progress report. It has now been 2 weeks since I began my 5-hours-per-night sleep schedule. Generally, things are going well. My progress on this chapter has been faster than usual. I am not having any difficulty getting up on time or getting my work done, but I am finding that it takes a major effort to stay awake in the evening. If I try to read or watch a bit of television after 10:30, I experience microsleeps. My so-called friends delight in making sure that my transgressions are quickly interrupted.

Scan Your Brain

Before continuing with this chapter, scan your brain by completing the following exercise to make sure you understand the fundamentals of sleep. The correct answers appear at the end of the exercise. Before proceeding, review material related to your errors and omissions.

1. The three standard psychophysiological measures of sleep are the EEG, the EMG, and the _____.
2. Stage 4 sleep EEG is characterized by a predominance of _____ waves.
3. _____ stage 1 EEG is accompanied by neither REM nor loss of core-muscle tone.
4. Dreaming occurs predominantly during _____ sleep.
5. The modern alternative to Freud's theory of dreaming is Hobson's _____ theory.
6. There are two fundamentally different kinds of theories of sleep: recuperation theories and _____ theories.
7. The effects of sleep deprivation are often difficult to study because they are often confounded by _____.
8. Convincing evidence that REM-sleep deprivation does not produce severe memory problems comes from the study of patients taking certain _____ drugs.
9. After a lengthy period of sleep deprivation (e.g., several days), a person's first night of sleep is only slightly longer than usual, but it contains a much higher proportion of _____ waves.
10. _____ sleep in particular, rather than sleep in general, appears to play the major restorative role.

Scan Your Brain answers: (1) EOG, (2) delta, (3) Initial, (4) REM, (5) activation-synthesis, (6) adaptation, (7) stress, (8) antidepressant, (9) slow (or delta), (10) Slow-wave (or stage 3 and 4).

14.4

Circadian Sleep Cycles

The world in which we live cycles from light to dark and back again once every 24 hours, and most surface-dwelling species have adapted to this regular change in their environment with a variety of **circadian rhythms** (see Foster & Kreitzman, 2004; *circadian* means "lasting about a day"). For example, most species display a regular circadian sleep–wake cycle. Humans take advantage of the light of day to take care of their biological needs, and then they sleep for much of the night; in contrast, *nocturnal animals*, such as rats, sleep for much of the day and stay awake at night.

Although the sleep–wake cycle is the most obvious circadian rhythm, it is difficult to find a physiological, biochemical, or behavioral process in animals that does not display some measure of circadian rhythmicity (Gillette & Sejnowski, 2005). Each day, our bodies adjust themselves in a variety of ways to meet the demands of the two environments in which we live: light and dark.

Our circadian cycles are kept on their once-every-24-hours schedule by temporal cues in the environment. The most important of these cues for the regulation of mammalian circadian rhythms is the daily cycle of light and dark. Environmental cues, such as the light–dark cycle, that can *entrain* (control the timing of) circadian rhythms are called **zeitgebers** (pronounced "ZITE-gay-bers"), a German word that means "time givers." In controlled laboratory environments, it is possible to lengthen or shorten circadian cycles somewhat by adjusting the duration of the light–dark cycle; for example, when exposed to alternating 11.5-hour periods of light and 11.5-hour periods of dark, subjects' circadian cycles begin to conform to a 23-hour day. In a world without 24-hour cycles of light and dark, other *zeitgebers* can entrain circadian cycles. For example, the circadian sleep–wake cycles of hamsters living in continuous darkness or in continuous light can be entrained by regular daily bouts of social interaction, hoarding, eating, or exercise (see Mistlberger et al., 1996; Sinclair & Mistlberger, 1997). Hamsters display particularly clear circadian cycles and thus are frequent subjects of research on circadian rhythms.

Free-Running Circadian Sleep–Wake Cycles

What happens to sleep–wake cycles and other circadian rhythms in an environment that is devoid of *zeitgebers*? Remarkably, under conditions in which there are absolutely no temporal cues, humans and other animals maintain all of their circadian rhythms. Circadian rhythms in constant environments are said to be **free-running rhythms**, and their duration is called the **free-running period**. Free-running periods vary in length from subject to subject, are of relatively constant duration within a given subject, and are usually longer than 24 hours—about 24.2 hours is typical in humans living under constant moderate illumination (see Czeizler et al., 1999). It seems that we all have an internal *biological clock* that habitually runs a little slow unless it is entrained by time-related cues in the environment.

A typical free-running circadian sleep–wake cycle is illustrated in Figure 14.7. Notice its regularity. Without any external cues, this man fell asleep at intervals of approximately 25.3 hours for an entire month. The regularity of free-running sleep–wake cycles despite variations in physical and mental activity provides support for the dominance of circadian factors over recuperative factors in the regulation of sleep.

Free-running circadian cycles do not have to be learned. Even rats that are born and raised in an unchanging

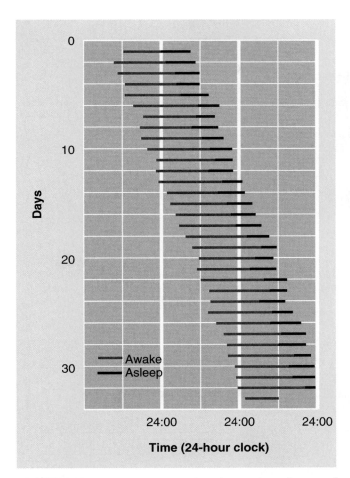

FIGURE 14.7 A free-running circadian sleep–wake cycle 25.3 hours in duration. Despite living in an unchanging environment with no time cues, the man went to sleep each day approximately 1.3 hours later than he had the day before. (Based on Wever, 1979, p. 30.)

laboratory environment (in continuous light or in continuous darkness) display regular free-running sleep–wake cycles that are slightly longer than 24 hours (Richter, 1971).

Many animals display a circadian cycle of body temperature that is related to their circadian sleep–wake cycle: They tend to sleep during the falling phase of their circadian body temperature cycle and awaken during its rising phase. However, when subjects are housed in constant laboratory environments, their sleep–wake and body temperature cycles sometimes break away from one another. This phenomenon is called **internal desynchronization** (see De La Iglesia, Cambras, & Díez-Noguera, 2008). For example, in one human volunteer, the free-running periods of *both* the sleep–wake and body temperature cycles were initially 25.7 hours; then, for some unknown reason, there was an increase in the free-running period of the sleep–wake cycle to 33.4 hours

and a decrease in the free-running period of the body temperature cycle to 25.1 hours. The potential for the simultaneous existence of two different free-running periods suggests that there is more than one circadian timing mechanism, and that sleep is not causally related to the decreases in body temperature that are normally associated with it.

There is another point about free-running circadian sleep–wake cycles that is incompatible with recuperation theories of sleep. On occasions when subjects stay awake longer than usual, the following sleep time is shorter rather than longer (Wever, 1979). Humans and other animals are programmed to have sleep–wake cycles of approximately 24 hours; hence, the more wakefulness there is during a cycle, the less time there is for sleep.

Jet Lag and Shift Work

People in modern industrialized societies are faced with two different disruptions of circadian rhythmicity: jet lag and shift work. **Jet lag** occurs when the *zeitgebers* that control the phases of various circadian rhythms are accelerated during east-bound flights (*phase advances*) or decelerated during west-bound flights (*phase delays*). In *shift work*, the *zeitgebers* stay the same, but workers are forced to adjust their natural sleep–wake cycles in order to meet the demands of changing work schedules. Both of these disruptions produce sleep disturbances, fatigue, general malaise, and deficits on tests of physical and cognitive function. The disturbances can last for many days; for example, it typically takes about 10 days to completely adjust to the phase advance of 10.5 hours that one experiences on a Tokyo-to-Boston flight.

What can be done to reduce the disruptive effects of jet lag and shift work? Two behavioral approaches have been proposed for the reduction of jet lag. One is gradually shifting one's sleep–wake cycle in the days prior to the flight. The other is administering treatments after the flight that promote the required shift in the circadian rhythm. For example, exposure to intense light early in the morning following an east-bound flight accelerates adaptation to the phase advance. Similarly, the results of a study of hamsters (Mrosovsky & Salmon, 1987) suggest that a good workout early in the morning of the first day after an east-bound flight might accelerate adaptation to the phase advance; hamsters that engaged in one 3-hour bout of wheel running 7 hours before their usual period of activity adapted quickly to an 8-hour advance in their light–dark cycle (see Figure 14.8 on page 368).

Companies that employ shift workers have had success in improving the productivity and job satisfaction of those workers by scheduling phase delays rather than phase advances; whenever possible, shift workers are transferred from their current schedule to one that begins later in the day (see Driscoll, Grunstein, & Rogers, 2007). It is much more difficult to go to sleep 4 hours earlier and

FIGURE 14.8   A period of forced exercise accelerates adaptation to an 8-hour phase advance in the circadian light–dark cycle. Daily activity is shown in red; periods of darkness are shown in black; and the period of forced exercise is shown in green. (Based on Mrosovsky & Salmon, 1987.)

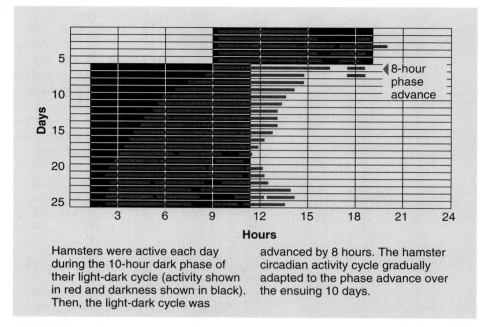

Hamsters were active each day during the 10-hour dark phase of their light-dark cycle (activity shown in red and darkness shown in black). Then, the light-dark cycle was advanced by 8 hours. The hamster circadian activity cycle gradually adapted to the phase advance over the ensuing 10 days.

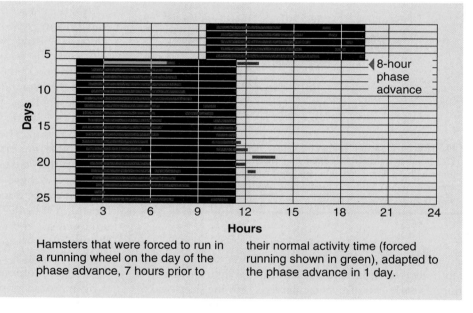

Hamsters that were forced to run in a running wheel on the day of the phase advance, 7 hours prior to their normal activity time (forced running shown in green), adapted to the phase advance in 1 day.

get up 4 hours earlier (a phase advance) than it is to go to sleep 4 hours later and get up 4 hours later (a phase delay). This is also why east-bound flights tend to be more problematic for travelers than west-bound flights.

A Circadian Clock in the Suprachiasmatic Nuclei

The fact that circadian sleep–wake cycles persist in the absence of temporal cues from the environment indicates that the physiological systems that regulate sleep are controlled by an internal timing mechanism—the **circadian clock**.

The first breakthrough in the search for the circadian clock was Richter's 1967 discovery that large medial hypothalamic lesions disrupt circadian cycles of eating, drinking, and activity in rats. Next, specific lesions of the **suprachiasmatic nuclei (SCN)** of the medial hypothalamus were shown to disrupt various circadian cycles, including sleep–wake cycles. Although SCN lesions do not greatly affect the amount of time mammals spend sleeping, they do abolish the circadian periodicity of sleep cycles. Further support for the conclusion that the suprachiasmatic nuclei contain a circadian timing mechanism comes from the observation that the nuclei display circadian cycles of electrical, metabolic, and biochemical activity that can be entrained by the light–dark cycle (see Mistlberger, 2005; Saper et al., 2005).

If there was any lingering doubt about the location of the circadian clock, it was eliminated by the brilliant experiment of Ralph and his colleagues (1990). They removed the SCN from the fetuses of a strain of mutant hamsters that had an abnormally short (20-hour) free-running sleep–wake cycle. Then, they transplanted the SCN into normal adult hamsters whose free-running sleep–wake cycles of 25 hours had been abolished by SCN lesions. These transplants restored free-running sleep–wake cycles in the recipients; but, remarkably, the cycles were about 20 hours long rather than the original 25 hours. Transplants in the other direction—that is, from normal hamster fetuses to SCN-lesioned adult mutants—had the complementary effect: They restored free-running sleep–wake cycles that were about 25 hours long rather than the original 20 hours.

Although the suprachiasmatic nuclei are unquestionably the major circadian clocks in mammals, they are not the only ones (e.g., Tosini et al., 2008). Three lines of experiments, largely conducted in the 1980s and 1990s, pointed to the existence of other circadian timing mechanisms:

- Under certain conditions, bilateral SCN lesions have been shown to leave some circadian rhythms unaffected while abolishing others.
- Bilateral SCN lesions do not eliminate the ability of all environmental stimuli to entrain circadian rhythms; for example, SCN lesions can block entrainment by light but not by food or water availability.
- Just like suprachiasmatic neurons, cells from other parts of the body display free-running circadian cycles of activity when maintained in tissue culture.

Neural Mechanisms of Entrainment

How does the 24-hour light–dark cycle entrain the sleep–wake cycle and other circadian rhythms? To answer this question, researchers began at the obvious starting point: the eyes (see Morin & Allen, 2006). They tried to identify and track the specific neurons that left the eyes and carried the information about light and dark that entrained the biological clock. Cutting the *optic nerves* before they reached the *optic chiasm* eliminated the ability of the light–dark cycle to entrain circadian rhythms; however, when the *optic tracts* were cut at the point where they left the optic chiasm, the ability of the light–dark cycle to entrain circadian rhythms was unaffected. As Figure 14.9 illustrates, these two findings indicated that visual axons critical for the entrainment of circadian rhythms branch off from the optic nerve in the vicinity of the optic chiasm. This finding led to the discovery of the *retinohypothalamic tracts*, which leave the optic chiasm and project to the adjacent suprachiasmatic nuclei.

Surprisingly, although the retinohypothalamic tracts mediate the ability of light to entrain photoreceptors, neither rods nor cones are necessary for the entrainment. The mystery photoreceptors have

proven to be neurons, a rare type of *retinal ganglion cells* with distinctive functional properties (see Berson, 2003; Hattar et al., 2002). During the course of evolution, these photoreceptors have sacrificed the ability to respond quickly and briefly to rapid changes of light in favor of the ability to respond consistently to slowly changing levels of background illumination. Their photopigment is **melanopsin** (Hankins, Peirson, & Foster, 2007; Panda et al., 2005).

Genetics of Circadian Rhythms

An important breakthrough in the study of circadian rhythms came in 1988 when routine screening of a shipment of hamsters revealed that some of them had abnormally short 20-hour free-running circadian rhythms. Subsequent breeding experiments showed that the abnormality was the result of a genetic mutation, and the gene that was mutated was named **tau** (Ralph & Menaker, 1988).

Although tau was the first mammalian circadian gene to be identified, it was not the first to have its molecular structure characterized. This honor went to *clock*, a mammalian circadian gene discovered in mice. The structure

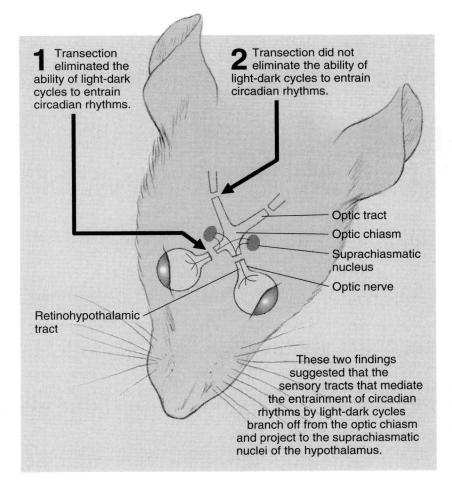

1 Transection eliminated the ability of light-dark cycles to entrain circadian rhythms.

2 Transection did not eliminate the ability of light-dark cycles to entrain circadian rhythms.

Optic tract
Optic chiasm
Suprachiasmatic nucleus
Optic nerve

Retinohypothalamic tract

These two findings suggested that the sensory tracts that mediate the entrainment of circadian rhythms by light-dark cycles branch off from the optic chiasm and project to the suprachiasmatic nuclei of the hypothalamus.

FIGURE 14.9 The discovery of the retinohypothalamic tracts. Neurons from each retina project to both suprachiasmatic nuclei.

of the clock gene was characterized in 1997, and that of the tau gene was characterized in 2000 (Lowrey et al., 2000). The molecular structures of several other mammalian circadian genes have now been specified (see Morse & Sassone-Corsi, 2002).

The identification of circadian genes has led to three important discoveries:

● The same or similar circadian genes have been found in many species of different evolutionary ages (e.g., bacteria, flies, fish, frogs, mice, and humans), indicating that circadian genes evolved early in evolutionary history and have been conserved in various descendant species (see Cirelli, 2009).

● Once the circadian genes were discovered, the fundamental molecular mechanism of circadian rhythms was quickly clarified. The key mechanism seems to be gene expression, that is, the transcription of proteins by the circadian genes displays a circadian cycle (see Dunlap, 2006; Hardin, 2006; Meyer, Saez, & Young, 2006).

● The identification of circadian genes provided a more direct method of exploring the circadian timing capacities of parts of the body other than the SCN. Molecular circadian timing mechanisms similar to those in the SCN exist in most cells of the body (see Green & Menaker, 2003; Hastings, Reddy, & Maywood, 2003; Yamaguchi et al., 2003). Although most cells contain circadian timing mechanisms, these cellular clocks are normally entrained by neural and hormonal signals from the SCN.

14.5
Four Areas of the Brain Involved in Sleep

You have just learned about the neural structures involved in controlling the circadian timing of sleep. This section describes four areas of the brain that are directly involved in producing or reducing sleep. You will learn more about their effects in the later section on sleep disorders.

Two Areas of the Hypothalamus Involved in Sleep

It is remarkable that two areas of the brain that are involved in the regulation of sleep were discovered early in the 20th century, long before the advent of modern behavioral neuroscience. The discovery was made by Baron Constantin von Economo, a Viennese neurologist (see Saper, Scammell, & Lu, 2005).

The Case of Constantin von Economo, the Insightful Neurologist

During World War I, the world was swept by a serious viral infection of the brain: *encephalitis lethargica*. Many of its victims slept almost continuously. Baron Constantin von Economo discovered that the brains of deceased victims who had problems with excessive sleep all had damage in the *posterior hypothalamus* and adjacent parts of the midbrain. He then turned his attention to the brains of a small group of victims of encephalitis lethargica who had had the opposite sleep-related problem: In contrast to most victims, they had difficulty sleeping. He found that the brains of the deceased victims in this minority always had damage in the *anterior hypothalamus* and adjacent parts of the basal forebrain. On the basis of these clinical observations, von Economo concluded that the posterior hypothalamus promotes wakefulness, whereas the anterior hypothalamus promotes sleep.

Since von Economo's discovery of the involvement of the posterior hypothalamus and the anterior hypothalamus in human wakefulness and sleep, respectively, that involvement has been confirmed by lesion and recording studies in experimental animals (see Szymusiak, Gvilia, & McGinty, 2007; Szymusiak & McGinty, 2008). The locations of the posterior and anterior hypothalamus are shown in Figure 14.10.

Reticular Formation and Sleep

Another area involved in sleep was discovered through the comparison of the effects of two different brain-stem transections in cats. First, in 1936, Bremer severed the brain stems of cats between their *inferior colliculi* and *superior colliculi* in order to disconnect their forebrains from ascending sensory input (see Figure 14.11). This surgical preparation is called a **cerveau isolé preparation** (pronounced "ser-VOE ees-o-LAY"—literally, "isolated forebrain").

Bremer found that the cortical EEG of the isolated cat forebrains was indicative of almost continuous slow-wave sleep. Only when strong visual or olfactory stimuli were presented (the cerveau isolé has intact visual and olfactory input) could the continuous high-amplitude, slow-wave activity be changed to a **desynchronized EEG**—a low-amplitude, high-frequency EEG. However, this arousing effect barely outlasted the stimuli.

Next, for comparison purposes, Bremer (1937) *transected* (cut through) the brain stems of a different group of cats. These transections were located in the caudal brain stem, and thus, they disconnected the brain from the rest of the nervous system (see Figure 14.11). This experimental

FIGURE 14.10 Two regions of the brain involved in sleep. The anterior hypothalamus and adjacent basal forebrain are thought to promote sleep; the posterior hypothalamus and adjacent midbrain are thought to promote wakefulness.

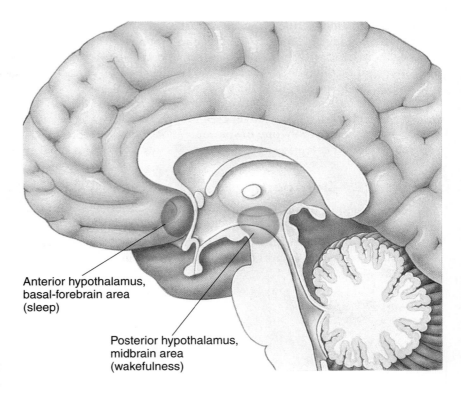

Anterior hypothalamus, basal-forebrain area (sleep)

Posterior hypothalamus, midbrain area (wakefulness)

preparation is called the **encéphale isolé preparation** (pronounced "on-say-FELL ees-o-LAY").

Although it cut most of the same sensory fibers as the cerveau isolé transection, the encéphale isolé transection did not disrupt the normal cycle of sleep EEG and wakefulness EEG. This suggested that a structure for maintaining wakefulness was located somewhere in the brain stem between the two transections.

Later, two important findings suggested that this wakefulness structure in the brain stem was the *reticular formation*. First, it was shown that partial transections at the cerveau isolé level

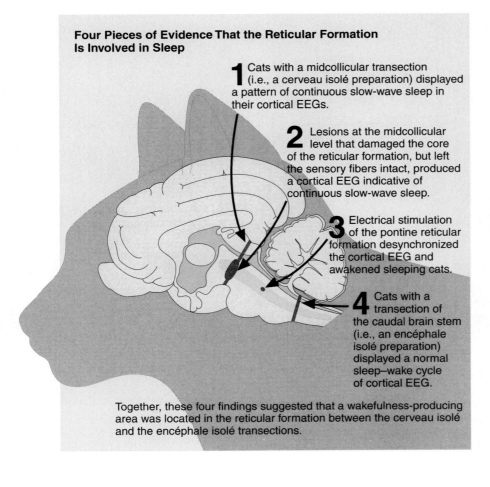

Four Pieces of Evidence That the Reticular Formation Is Involved in Sleep

1 Cats with a midcollicular transection (i.e., a cerveau isolé preparation) displayed a pattern of continuous slow-wave sleep in their cortical EEGs.

2 Lesions at the midcollicular level that damaged the core of the reticular formation, but left the sensory fibers intact, produced a cortical EEG indicative of continuous slow-wave sleep.

3 Electrical stimulation of the pontine reticular formation desynchronized the cortical EEG and awakened sleeping cats.

4 Cats with a transection of the caudal brain stem (i.e., an encéphale isolé preparation) displayed a normal sleep–wake cycle of cortical EEG.

Together, these four findings suggested that a wakefulness-producing area was located in the reticular formation between the cerveau isolé and the encéphale isolé transections.

disrupted normal sleep–wake cycles of cortical EEG only when they severed the reticular formation core of the brain stem; when the partial transections were restricted to more lateral areas, which contain the ascending sensory tracts, they had little effect on the cortical EEG (Lindsey, Bowden, & Magoun, 1949). Second, it was shown that electrical stimulation of the reticular formation of sleeping cats awakened them and produced a lengthy period of EEG desynchronization (Moruzzi & Magoun, 1949).

In 1949, Moruzzi and Magoun considered these four findings together: (1) the effects on cortical EEG of the cerveau isolé preparation, (2) the effects on cortical EEG of the encéphale isolé preparation, (3) the effects of reticular formation lesions, and (4) the effects on

FIGURE 14.11 Four pieces of evidence that the reticular formation is involved in sleep.

sleep of stimulation of the reticular formation. From these four key findings, Moruzzi and Magoun proposed that low levels of activity in the reticular formation produce sleep and that high levels produce wakefulness (see McCarley, 2007). Indeed, this theory is so widely accepted that the reticular formation is commonly referred to as the **reticular activating system,** even though maintaining wakefulness is only one of the functions of the many nuclei that it comprises.

Reticular REM-Sleep Nuclei

The fourth area of the brain that is involved in sleep controls REM sleep and is included in the brain area I have just described—it is part of the caudal reticular formation. It makes sense that an area of the brain involved in

maintaining wakefulness would also be involved in the production of REM sleep because of the similarities between the two states. Indeed, REM sleep is controlled by a variety of nuclei scattered throughout the caudal reticular formation. Each site is responsible for controlling one of the major indices of REM sleep (Datta & MacLean, 2007; Siegel, 1983; Vertes, 1983)—a site for the reduction of core-muscle tone, a site for EEG desynchronization, a site for rapid eye movements, and so on. The approximate location in the caudal brain stem of each of these REM-sleep nuclei is illustrated in Figure 14.12.

Please think for a moment about the broad implications of these various REM-sleep nuclei. In thinking about the brain mechanisms of behavior, many people assume that if there is one name for a behavior, there must be a single structure for it in the brain: In other words, they

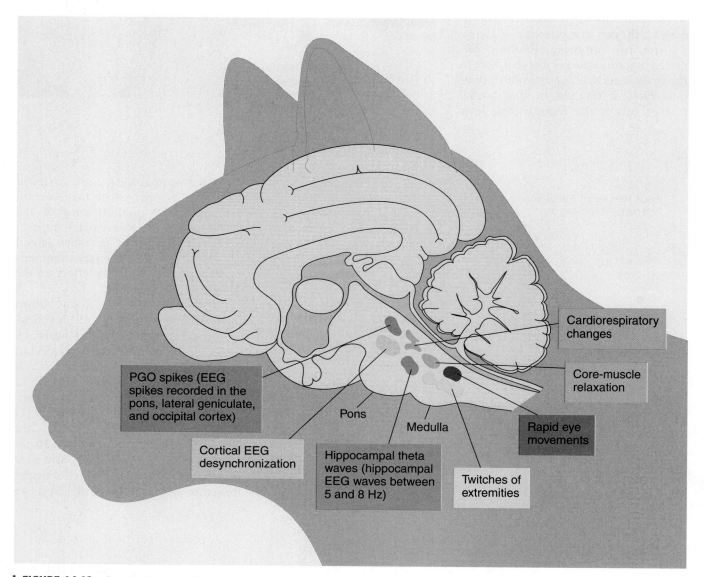

FIGURE 14.12 A sagittal section of the brain stem of the cat illustrating the areas that control the various physiological indices of REM sleep. (Based on Vertes, 1983.)

Thinking Creatively assume that evolutionary pressures have acted to shape the human brain according to our current language and theories. Here we see the weakness of this assumption: The brain is organized along different principles, and REM sleep occurs only when a network of independent structures becomes active together. Relevant to this is the fact that the physiological changes that go together to define REM sleep sometimes break apart and go their separate ways—and the same is true of the changes that define slow-wave sleep. For example, during REM-sleep deprivation, penile erections, which normally occur during REM sleep, begin to occur during slow-wave sleep. And during total sleep deprivation, slow waves, which normally occur only during slow-wave sleep, begin to occur during wakefulness. This suggests that REM sleep, slow-wave sleep, and wakefulness are not each controlled by a single mechanism. Each state seems to result from the interaction of several mechanisms that are capable under certain conditions of operating independently of one another.

Scan Your Brain

Before continuing with this chapter, scan your brain by completing the following exercise to make sure you understand the fundamentals of sleep. The correct answers appear at the end of the exercise. Before proceeding, review material related to your errors and omissions.

1. _____ means lasting about one day.
2. Free-running rhythms are those that occur in environments devoid of _____.
3. The major circadian clock seems to be located in the _____ nuclei of the hypothalamus.
4. The _____ tracts conduct information about light–dark cycles to the circadian clock in the SCN.
5. The first mammalian circadian gene to have its structure characterized was _____.
6. Patients with damage to the _____ hypothalamus and adjacent basal ganglia often have difficulty sleeping.
7. Damage to the _____ hypothalamus and adjacent areas of the midbrain often cause excessive sleepiness.
8. The low-amplitude high-frequency EEG of wakefulness is said to be _____.
9. In Bremer's classic study, the _____ preparation displayed an EEG characteristic of continuous sleep.
10. The indices of REM sleep are controlled by a variety of nuclei located in the caudal _____.

Scan Your Brain answers: (1) Circadian, (2) zeitgebers, (3) suprachiasmatic, (4) retinohypothalamic, (5) clock, (6) anterior, (7) posterior, (8) desynchronized, (9) encéphale isolé, (10) reticular formation.

14.6
Drugs That Affect Sleep

Most drugs that influence sleep fall into two different classes: hypnotic and antihypnotic. **Hypnotic drugs** are drugs that increase sleep; **antihypnotic drugs** are drugs that reduce sleep. A third class of sleep-influencing drugs comprises those that influence its circadian rhythmicity; the main drug of this class is **melatonin**.

⦿ Watch Nightsleep
www.mypsychlab.com

Hypnotic Drugs

The **benzodiazepines** (e.g., Valium and Librium) were developed and tested for the treatment of anxiety, yet they are the most commonly prescribed hypnotic medications. In the short term, they increase drowsiness, decrease the time it takes to fall asleep, reduce the number of awakenings during a night's sleep, and increase total sleep time (Krystal, 2008). Thus, they can be effective in the treatment of occasional difficulties in sleeping.

Clinical Implications

Although benzodiazepines can be effective therapeutic hypnotic agents in the short term, their prescription for the treatment of chronic sleep difficulties, though common, is ill-advised (Riemann & Perlis, 2008). Five complications are associated with the chronic use of benzodiazepines as hypnotic agents:

- Tolerance develops to the hypnotic effects of benzodiazepines; thus, patients must take larger and larger doses to maintain the drugs' efficacy and often become addicted.
- Cessation of benzodiazepine therapy after chronic use causes *insomnia* (sleeplessness), which can exacerbate the very problem that the benzodiazepines were intended to correct.
- Benzodiazepines distort the normal pattern of sleep; they increase the duration of stage 2 sleep, while actually decreasing the duration of stage 4 and of REM sleep.
- Benzodiazepines lead to next-day drowsiness (Ware, 2008) and increase the incidence of traffic accidents (Gustavsen et al., 2008).
- Most troubling is that chronic use of benzodiazepines has been shown to substantially reduce life expectancy (see Siegel, 2010).

Evidence that the raphé nuclei, which are serotonergic, play a role in sleep suggested that serotonergic drugs might be effective hypnotics. Efforts to demonstrate the hypnotic effects of such drugs have focused on **5-hydroxytryptophan (5-HTP)**—the precursor of serotonin—because 5-HTP, but not serotonin, readily passes through the blood–brain barrier. Injections of 5-HTP do reverse the insomnia produced in both cats and rats by the serotonin antagonist PCPA; however, they appear to be of no therapeutic benefit in the treatment of human insomnia (see Borbély, 1983).

Antihypnotic Drugs

The mechanisms of the following three classes of anti-hypnotic drugs are well understood: *cocaine-derived stimulants, amphetamine-derived stimulants,* and *tricyclic antidepressants.* The drugs in these three classes seem to promote wakefulness by boosting the activity of cate-cholamines (norepinephrine, epinephrine, and dopamine)— by increasing their release into synapses, by blocking their  reuptake from synapses, or both. The antihypnotic mechanisms of two other stimulant drugs, codeine and modafinil, are less well understood.

The regular use of antihypnotic drugs is risky. Antihyp-notics tend to produce a variety of adverse side effects, such as loss of appetite, anxiety, tremor, addiction, and distur-bance of normal sleep patterns. Moreover, they may mask the pathology that is causing the excessive sleepiness.

Melatonin

Melatonin is a hormone that is synthesized from the neu-rotransmitter serotonin in the **pineal gland** (see Moore, 1996). The pineal gland is an inconspicuous gland that

René Descartes, whose dualistic philosophy was discussed in Chapter 2, once believed to be the seat of the soul. The pineal gland is located on the midline of the brain just ventral to the rear portion of the corpus callosum (see Figure 14.13).

The pineal gland has important functions in birds, reptiles, amphibians, and fish (see Cassone, 1990). The pineal gland of these species has inherent timing properties and regulates circadian rhythms and seasonal changes in reproduc-tive behavior through its release of melatonin. In humans and other mammals, however, the functions of the pineal gland and melatonin are not as apparent.

In humans and other mammals, circulating levels of melatonin display circadian rhythms under control of the suprachiasmatic nuclei (see Gillette & McArthur, 1996), with the highest levels being associated with darkness and sleep (see Foulkes et al., 1997). On the basis of this correla-tion, it has long been assumed that melatonin plays a role in promoting sleep or in regulating its timing in mammals.

In order to put the facts about melatonin in perspec-tive, it is important to keep one significant point firmly in mind. In adult mammals, pinealectomy and the conse-quent elimination of melatonin appear to have little ef-fect. The pineal gland plays a role in the development of mammalian sexual maturity, but its functions after pu-berty are not at all obvious.

Does *exogenous* (externally produced) melatonin im-prove sleep, as widely believed? The evidence is mixed

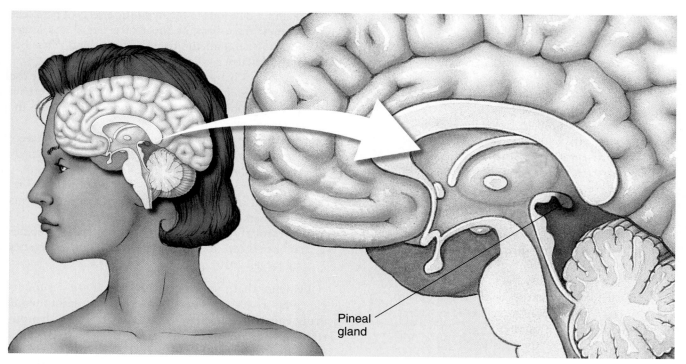

FIGURE 14.13 The location of the pineal gland, the source of melatonin.

(see van den Heuvel et al., 2005). However, a *meta-analysis* (a combined analysis of results of more than one study) of 17 studies indicated that exogenous melatonin has a slight, but statistically significant, *soporific* (sleep-promoting) effect (Brzezinski et al., 2005).

In contrast to the controversy over the soporific effects of exogenous melatonin in mammals, there is good evidence that it can shift the timing of mammalian circadian cycles. Indeed, several researchers have argued that melatonin is better classified as a **chronobiotic** (a substance that adjusts the timing of internal biological rhythms) than as a soporific (see Scheer & Czeisler, 2005). Arendt and Skene (2005) have argued that administration of melatonin in the evening increases sleep by accelerating the start of the nocturnal phase of the circadian rhythm and that administration at dawn increases sleep by delaying the end of the nocturnal phase.

Exogenous melatonin has been shown to have a therapeutic potential in the treatment of two types of sleep problems (see Arendt & Skene, 2005). Melatonin before bedtime has been shown to improve the sleep of those insomniacs who are melatonin-deficient and of blind people who have sleep problems attributable to the lack of the synchronizing effects of the light–dark cycle. Melatonin's effectiveness in the treatment of other sleep disorders remains controversial.

14.7

Sleep Disorders

Many sleep disorders fall into one of two complementary categories: insomnia and hypersomnia. **Insomnia** includes all disorders of initiating and maintaining sleep, whereas **hypersomnia** includes disorders of excessive sleep or sleepiness. A third major class of sleep disorders includes all those disorders that are specifically related to REM-sleep dysfunction. Ironically, both insomnia and hypersomnia are common symptoms of depression and other mood disorders (Kaplan & Harvey, 2009).

In various surveys, approximately 30% of respondents report significant sleep-related problems. However, it is important to recognize that complaints of sleep problems often come from people whose sleep appears normal in laboratory sleep tests. For example, many people normally sleep 6 hours or less a night and seem to do well sleeping that amount, but they are pressured by their doctors, their friends, and their own expectations to sleep more (e.g., at least 8 hours). As a result, they spend more time in bed than they should and have difficulty getting to sleep. Often, the anxiety associated with their inability to sleep more makes it even more difficult for them to sleep (see Espie, 2002). Such patients can often be helped by counseling that convinces them to go to bed only when they are very sleepy (see Anch et al., 1988). Others with disturbed sleep have more serious problems (see Mahowald & Schenck, 2005).

Insomnia

Many cases of insomnia are **iatrogenic** (physician-created)—in large part because sleeping pills (i.e., benzodiazepines), which are usually prescribed by physicians, are a major cause of insomnia. At first, hypnotic drugs may be effective in increasing sleep, but soon the patient may become trapped in a rising spiral of drug use, as *tolerance* to the drug develops and progressively more of it is required to produce its original hypnotic effect. Soon, the patient cannot stop taking the drug without running the risk of experiencing *withdrawal symptoms*, which include insomnia. The case of Mr. B. illustrates this problem.

Mr. B., the Case of Iatrogenic Insomnia

Mr. B. was studying for a civil service exam, the outcome of which would affect his entire future. He was terribly worried about the test and found it difficult to get to sleep at night. Feeling that the sleep loss was affecting his ability to study, he consulted his physician. . . . His doctor prescribed a moderate dose of barbiturate at bedtime, and Mr. B. found that this medication was very effective . . . for the first several nights. After about a week, he began having trouble sleeping again and decided to take two sleeping pills each night. Twice more the cycle was repeated, until on the night before the exam he was taking four times as many pills as his doctor had prescribed. The next night, with the pressure off, Mr. B. took no medication. He had tremendous difficulty falling asleep, and when he did, his sleep was terribly disrupted. . . . Mr. B. now decided that he had a serious case of insomnia, and returned to his sleeping pill habit. By the time he consulted our clinic several years later, he was taking approximately 1,000 mg sodium amytal every night, and his sleep was more disturbed than ever. . . . Patients may go on for years and years—from one sleeping pill to another—never realizing that their troubles are caused by the pills.

("Mr. B., the Case of Iatrogenic Insomnia," from *Some Must Watch While Some Must Sleep* by William C. Dement, Portable Stanford Books, Stanford Alumni Association, Stanford University, 1978, p. 80. Used by permission of William C. Dement.)

In one study, insomniacs claimed to take an average of 1 hour to fall asleep and to sleep an average of only 4.5 hours per night; but when they were tested in a sleep laboratory, they were found to have an average *sleep latency* (time to fall asleep) of only 15 minutes and an average nightly sleep duration of 6.5 hours. It used to be common medical practice to assume that people who claimed to suffer from insomnia but slept more than 6.5 hours per night were neurotic. However, this practice stopped when some of those diagnosed as *neurotic pseudoinsomniacs* were subsequently found to be suffering from sleep apnea, nocturnal myoclonus, or other sleep-disturbing problems. Insomnia is not necessarily a problem of too little sleep; it is often a problem of too little undisturbed sleep (Bonnet & Arand, 2002; Stepanski et al., 1987).

One of the most effective treatments for insomnia is *sleep restriction therapy* (Morin, Kowatch, & O'Shanick, 1990): First, the amount of time that an insomniac is allowed to spend in bed is substantially reduced. Then, after a period of sleep restriction, the amount of time spent in bed is gradually increased in small increments, as long as sleep latency remains in the normal range. Even severe insomniacs can benefit from this treatment.

Some cases of insomnia have specific medical causes; **sleep apnea** is one such cause. The patient with sleep apnea stops breathing many times each night. Each time, the patient awakens, begins to breathe again, and drifts back to sleep. Sleep apnea usually leads to a sense of having slept poorly and is thus often diagnosed as insomnia. However, some patients are totally unaware of their multiple awakenings and instead complain of excessive sleepiness during the day, which can lead to a diagnosis of *hypersomnia* (Stepanski et al., 1984).

Sleep apnea disorders are of two types: (1) *obstructive sleep apnea* results from obstruction of the respiratory passages by muscle spasms or *atonia* (lack of muscle tone) and often occurs in individuals who are vigorous snorers; (2) *central sleep apnea* results from the failure of the central nervous system to stimulate respiration (Banno & Kryger, 2007). Sleep apnea is more common in males, in the overweight, and in the elderly (Villaneuva et al., 2005).

Two other specific causes of insomnia are related to the legs: periodic limb movement disorder and restless legs syndrome. **Periodic limb movement disorder** is disorder characterized by periodic, involuntary movements of the limbs, often involving twitches of the legs during sleep. Most patients suffering from this disorder complain of poor sleep and daytime sleepiness but are unaware of the nature of their problem. In contrast, people with **restless legs syndrome** are all too aware of their problem. They complain of a hard-to-describe tension or uneasiness in their legs that keeps them from falling asleep. Once established, both of these disorders are chronic (see Garcia-Borreguero et al., 2006). Much more research into their treatment is needed, although some dopamine agonists can be effective (Ferini-Strambi et al., 2008; Hornyak et al., 2006).

Hypersomnia

Narcolepsy is the most widely studied disorder of hypersomnia. It occurs in about 1 out of 2,000 individuals (Ohayon, 2008) and has two prominent symptoms (see

Nishino, 2007). First, narcoleptics experience severe daytime sleepiness and repeated, brief (10- to 15-minute) daytime sleep episodes. Narcoleptics typically sleep only about an hour per day more than average; it is the inappropriateness of their sleep episodes that most clearly defines their condition. Most of us occasionally fall asleep on the beach, in front of the television, or in that most soporific of all daytime sites—the large, stuffy, dimly lit lecture hall. But narcoleptics fall asleep in the middle of a conversation, while eating, while scuba diving, or even while making love.

The second prominent symptom of narcolepsy is cataplexy (Houghton, Scammell, & Thorpy, 2004). **Cataplexy** is characterized by recurring losses of muscle tone during wakefulness, often triggered by an emotional experience. In its mild form, it may simply force the patient to sit down for a few seconds until it passes. In its extreme form, the patient drops to the ground as if shot and remains there for a minute or two, fully conscious.

In addition to the two prominent symptoms of narcolepsy (daytime sleep attacks and cataplexy), narcoleptics often experience two other symptoms: sleep paralysis and hypnagogic hallucinations. **Sleep paralysis** is the inability to move (paralysis) just as one is falling asleep or waking up. **Hypnagogic hallucinations** are dreamlike experiences during wakefulness. Many healthy people occasionally experience sleep paralysis and hypnagogic hallucinations. Have you experienced them?

Three lines of evidence suggested to early researchers that narcolepsy results from an abnormality in the mechanisms that trigger REM sleep. First, unlike normal people, narcoleptics often go directly into REM sleep when they fall asleep. Second and third, as you have already learned, narcoleptics often experience dreamlike states and loss of muscle tone during wakefulness.

Some of the most exciting current research on the neural mechanisms of sleep in general and narcolepsy in particular began with the study of a strain of narcoleptic dogs. After 10 years of studying the genetics of these narcoleptic dogs, Lin and colleagues (1999) finally isolated the gene that causes the disorder. The gene encodes a receptor protein that binds to a neuropeptide called **orexin** (sometimes called *hypocretin*), which exists in two forms: orexin-A and orexin-B (see Sakurai, 2005). Although discovery of the orexin gene has drawn attention to genetic factors in narcolepsy, the concordance rate between identical twins is only about 25% (Raizen, Mason, & Pack, 2006).

Several studies have documented reduced levels of orexin in the cerebrospinal fluid of living narcoleptics and in the brains of deceased narcoleptics (see Nishino & Kanbayashi, 2005). Also, the number of orexin-releasing neurons has been found to be reduced in the brains of narcoleptics (e.g., Peyron et al., 2000; Thannickal et al., 2000).

Where is orexin synthesized in the brain? Orexin is synthesized by neurons in the region of the hypothalamus that has been linked to the promotion of wakefulness: the posterior hypothalamus (mainly its lateral regions). The orexin-producing neurons project diffusely throughout the brain, but they show many connections with neurons of the other wakefulness-promoting area of the brain: the reticular formation. Currently, there is considerable interest in understanding the role of the orexin circuits in normal sleep–wake cycles (see Sakurai, 2007; Siegel, 2004).

Narcolepsy has traditionally been treated with stimulants (e.g., amphetamine, methylphenidate), but these have substantial addiction potential and produce many undesirable side effects. The antihypnotic stimulant modafinil has been shown to be effective in the treatment of narcolepsy, and antidepressants can be effective against cataplexy (Thorpy, 2007).

REM-Sleep–Related Disorders

Several sleep disorders are specific to REM sleep; these are classified as *REM-sleep–related disorders*. Even narcolepsy, which is usually classified as a hypersomnic disorder, can reasonably be considered to be a REM-sleep–related disorder—for reasons you have just encountered.

Occasionally, patients who have little or no REM sleep are discovered. Although this disorder is rare, it is important because of its theoretical implications. Lavie and others (1984) described a patient who had suffered a brain injury that presumably involved damage to the REM-sleep controllers in the caudal reticular formation. The most important finding of this case study was that the patient did not appear to be adversely affected by his lack of REM sleep. After receiving his injury, he completed high school, college, and law school and established a thriving law practice.

Some patients experience REM sleep without core-muscle atonia. It has been suggested that the function of REM-sleep atonia is to prevent the acting out of dreams. This theory receives support from case studies of people who suffer from this disorder—case studies such as the following one.

The Case of the Sleeper Who Ran Over Tackle

I was a halfback playing football, and after the quarterback received the ball from the center he lateraled it sideways to me and I'm supposed to go around end and cut back over tackle and—this is very vivid—as I cut back over tackle there is this big 280-pound tackle waiting, so I, according to football rules, was to give him my shoulder and bounce him out of the way. . . . [W]hen I came to I was standing in front of our dresser and I had [gotten up out of bed and run and] knocked lamps, mirrors and everything off the dresser, hit my head against the wall and my knee against the dresser. (Schenck et al., 1986, p. 294)

Presumably, REM sleep without atonia is caused by damage to the nucleus magnocellularis or to an interruption

of its output. The **nucleus magnocellularis** is a structure of the caudal reticular formation that controls muscle relaxation during REM sleep. In normal dogs, it is active only during REM sleep; in narcoleptic dogs, it is also active during their catalectic attacks.

Evolutionary Perspective

14.8
Effects of Long-Term Sleep Reduction

When people sleep less than they are used to sleeping, they do not feel or function well. I am sure that you have experienced these effects. But what do they mean? Most people—nonexperts and experts alike—believe that the adverse effects of sleep loss indicate that we need the sleep we typically get. However, there is an alternative interpretation, one that is consistent with the now acknowledged plasticity of the adult human brain. Perhaps the brain needs a small amount of sleep each day but will sleep much more under ideal conditions because of sleep's high positive incentive value. The brain then slowly adapts to the amount of sleep it is getting—even though this amount may be far more than it needs—and is disturbed when there is a sudden reduction.

Neuroplasticity

Fortunately, there are ways to determine which of these two interpretations of the effects of sleep loss is correct. The key is to study individuals who sleep little, either because they have always done so or because they have purposefully reduced their sleep times. If people need at least 8 hours of sleep each night, short sleepers should be suffering from a variety of health and performance problems. Before I summarize the results of this key research, I want to emphasize one point: Because they are so time-consuming, few studies of long-term sleep patterns have been conducted, and some of those that have been conducted are not sufficiently thorough. Nevertheless, there have been enough of them for a clear pattern of results to have emerged. I think they will surprise you.

This final section begins with a comparison of short and long sleepers. Then, it discusses two kinds of long-term sleep-reduction studies: studies in which volunteers reduced the amount they slept each night and studies in which volunteers reduced their sleep by restricting it to naps. Next comes a discussion of studies that have examined the relation between sleep duration and health. Finally, I relate my own experience of long-term sleep reduction

Differences between Short and Long Sleepers

Numerous studies have compared short sleepers (those who sleep 6 hours or less per night) and long sleepers (those who sleep 8 hours or more per night). I focus here on the 2004 study of Fichten and colleagues because it is the most thorough. The study had three strong features:

- It included a large sample (239) of adult short sleepers and long sleepers.
- It compared short and long sleepers in terms of 48 different measures, including daytime sleepiness, daytime naps, regularity of sleep times, busyness, regularity of meal times, stress, anxiety, depression, life satisfaction, and worrying.
- Before the study began, the researchers carefully screened out volunteers who were ill or under various kinds of stress or pressure; thus, the study was conducted with a group of healthy volunteers who slept the amount that they felt was right for them.

The findings of Fichten and colleagues are nicely captured by the title of their paper, "Long sleepers sleep more and short sleepers sleep less." In other words, other than the differences in sleep time, there were no differences between the two groups on any of the other measures—no indication that the short sleepers were suffering in any way from their shorter sleep time. Fichten and colleagues report that these results are consistent with most previous comparisons of short and long sleepers (e.g., Monk et al., 2001), except for a few studies that did not screen out subjects who slept little because they were under pressure (e.g., from worry, illness, or a demanding work schedule). Those studies did report some negative characteristics in the short-sleep group, which likely reflected the stress experienced by some in that group.

Long-Term Reduction of Nightly Sleep

Are short sleepers able to live happy productive lives because they are genetically predisposed to be short sleepers, or is it possible for average people to adapt to a short sleep schedule? There have been only two published studies in which healthy volunteers have reduced their nightly sleep for several weeks or longer. In one (Webb & Agnew, 1974), a group of 16 volunteers slept for only 5.5 hours per night for 60 days, with only one detectable deficit on an extensive battery of mood, medical, and performance tests: a slight deficit on a test of auditory vigilance.

In the other systematic study of long-term nightly sleep reduction (Friedman et al., 1977; Mullaney et al., 1977), 8 volunteeers reduced their nightly sleep by 30 minutes every 2 weeks until they reached 6.5 hours per night, then by 30 minutes every 3 weeks until they reached 5 hours, and then by 30 minutes every 4 weeks thereafter. After a participant indicated a lack of desire to reduce sleep further, the person spent 1 month sleeping the shortest duration of nightly sleep that had been achieved, then 2 months sleeping the shortest duration plus 30 minutes. Finally, each participant slept however long was preferred each night for 1 year. The minimum duration of nightly sleep achieved during this experiment was 5.5 hours for 2 participants, 5.0 hours for 4 participants, and an impressive 4.5 hours for 2 participants. In each participant, a reduction in sleep time was associated with an increase in sleep efficiency: a decrease in the amount of time it took to fall asleep after going to bed, a decrease in the number of nighttime awakenings, and an increase in the proportion of stage 4 sleep. After the participants had reduced their sleep to 6 hours per night, they began to experience daytime sleepiness, and this became a problem as sleep time was further reduced. Nevertheless, there were no deficits on any of the mood, medical, or performance tests administered throughout the experiment. The most encouraging result was that an unexpected follow-up 1 year later found that all participants were sleeping less than they had previously—between 7 and 18 hours less each week—with no excessive sleepiness.

Long-Term Sleep Reduction by Napping

Most mammals and human infants display **polyphasic sleep cycles**; that is, they regularly sleep more than once per day. In contrast, most adult humans display **monophasic sleep cycles**; that is, they sleep once per day. Nevertheless, most adult humans do display polyphasic cycles of sleepiness, with periods of sleepiness occurring in late afternoon and late morning (Stampi, 1992a). Have you ever experienced them?

Do adult humans need to sleep in one continuous period per day, or can they sleep effectively in several naps as human infants and other mammals do? Which of the two sleep patterns is more efficient? Research has shown that naps have recuperative powers out of proportion with their brevity (e.g., Milner & Cote, 2008; Smith et al., 2007), suggesting that polyphasic sleep might be particularly efficient.

Interest in the value of polyphasic sleep was stimulated by the legend that Leonardo da Vinci managed to generate a steady stream of artistic and engineering accomplishments during his life by napping for 15 minutes every 4 hours, thereby limiting his sleep to 1.5 hours per day. As unbelievable as this sleep schedule may seem, it has been replicated in several experiments (see Stampi, 1992b). Here are the main findings of these truly mind-boggling experiments: First, participants required a long time, several weeks, to adapt to a polyphasic sleep schedule. Second, once adapted to polyphasic sleep, participants were content and displayed no deficits on the performance tests they were given. Third, Leonardo's 4-hour schedule works quite well, but in unstructured working situations (e.g., around-the-world solo sailboat races), individuals often vary the duration of the cycle without feeling negative consequences. Fourth, most people display a strong preference for particular sleep durations (e.g., 25 minutes) and refrain from sleeping too little, which leaves them unrefreshed, or too much, which leaves them groggy for several minutes when they awake—an effect called **sleep inertia** (e.g., Fushimi & Hayashi, 2008; Ikeda & Hayashi, 2008; Wertz et al., 2006). Fifth, when individuals first adopt a polyphasic sleep cycle, most of their sleep is slow-wave sleep, but eventually they return to a mix of REM and slow-wave sleep.

The following are the words of artist Giancarlo Sbragia, who adopted Leonardo's purported sleep schedule:

> This schedule was difficult to follow at the beginning. . . . It took about 3 wk to get used to it. But I soon reached a point at which I felt a natural propensity for sleeping at this rate, and it turned out to be a thrilling and exciting experience.
> . . . How beautiful my life became: I discovered dawns, I discovered silence, and concentration. I had more time for studying and reading—far more than I did before. I had more time for myself, for painting, and for developing my career. (Sbragia, 1992, p. 181)

Effects of Shorter Sleep Times on Health

For decades, it was believed that sleeping 8 hours or more per night is ideal for promoting optimal health and longevity. Then, a series of large-scale epidemiological studies conducted in both the United States and Japan challenged this belief (e.g., Ayas et al., 2003; Kripke et al., 2002; Patel et al., 2003; Tamakoshi & Ohno, 2004). These studies did not include participants who were a potential source of bias, for example, people who slept little *Thinking Creatively*

because they were ill, depressed, or under stress. The studies started with a sample of healthy volunteers and followed their health for several years.

The results of these studies are remarkably uniform (Kripke, 2004). Figure 14.14 presents data from Tamakoshi and Ohno (2004), who followed 104,010 volunteers for 10 years. You will immediately see that sleeping 8 hours per night is not the healthy ideal that it has been assumed to be: The fewest deaths occurred among people sleeping between 5 and 7 hours per night, far fewer than among those who slept 8 hours. You should be aware that other studies that are not as careful in excluding volunteers who sleep little because of stress or ill health do find more problems associated with short sleep (see Cappuccio et al., 2008), but any such finding is likely an artifact of pre-existing ill health or stress, which is more prevalent among short sleepers. *Clinical Implications*

Because these epidemiological data are correlational, it is important not to interpret them causally (see Grandner & Drummond, 2007; Stamatakis & Punjabi, 2007; Youngstedt & Kripke, 2004). They do not prove that sleeping 8 or more hours a night causes health problems: Perhaps there is something about people who sleep 8 hours or more per night that leads them to die sooner than people who sleep less. Thus, these studies do not prove that reducing your sleep will cause you to live longer—although some experts are advocating sleep reduction as a means of improving health (e.g., Youngstedt & Kripke, 2004). These studies do, however, provide strong evidence that sleeping less than 8 hours is not the risk to life and health that it is often made out to be. *Thinking Creatively*

Long-Term Sleep Reduction: A Personal Case Study

I began this chapter 4 weeks ago with both zeal and trepidation. I was fascinated by the idea that I could wring 2 or 3 extra hours of living out of each day by sleeping less, and I hoped that adhering to a sleep-reduction program while writing about

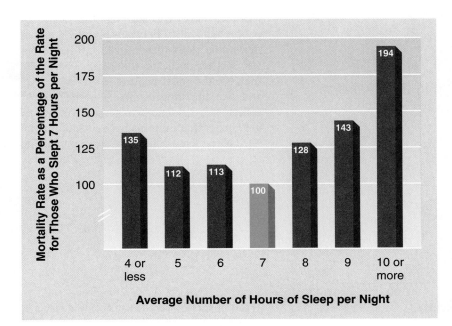

FIGURE 14.14 The mortality rates associated with different amounts of sleep, based on 104,010 volunteers followed over 10 years. The mortality rate at 7 hours of sleep per night has been arbitrarily set at 100%, and the other mortality rates are presented in relation to it. (Based on Tamakoshi & Ohno, *Sleep* 2004, 27(1): 51–4.)

sleep would create an enthusiasm for the subject that would color my writing and be passed on to you. I began with a positive attitude because I was aware of the relevant evidence; still I was more than a little concerned about the negative effect that reducing my sleep by 3 hours per night might have on me and my writing.

The Case of the Author Who Reduced His Sleep

Rather than using the gradual stepwise reduction method of Friedman and his colleagues, I jumped directly into my 5-hours-per-night sleep schedule. This proved to be less difficult than you might think. I took advantage of a trip to the East Coast from my home on the West Coast to reset my circadian clock. While I was in the East, I got up at 7:00 A.M., which is 4:00 A.M. on the West Coast, and I just kept on the same schedule when I got home. I decided to add my extra waking hours to the beginning of my day rather than to the end so there would be no temptation for me to waste them—there are not too many distractions around this university at 5:00 A.M.

Figure 14.15 is a record of my sleep times for the 4-week period that it took me to write a first draft of this chapter. I didn't quite meet my goal of sleeping less than 5 hours every night, but I didn't miss by much: My overall mean was 5.05 hours per night. Notice that in the last week, there was a tendency for my circadian clock to run a bit slow; I began sleeping in until 4:30 A.M. and staying up until 11:30 P.M.

What were the positives and negatives of my experience? The main positive was the added time to do things: Having an extra 21 hours per week was wonderful. Furthermore, because my daily routine was out of synchrony with everybody else's, I spent little time sitting in traffic. The only negative of the experience was sleepiness. It was no problem during the day, when I was active. However, staying awake during the last hour before I went to bed—an hour during which I usually engaged in sedentary activities, such as reading—was at times a problem. This is when I became personally familiar with the phenomenon of microsleeps, and it was then that I required some assistance in order to stay awake. Each night of sleep became a highly satisfying but all too brief experience.

I began this chapter with this question: How much sleep do we need? Then, I gave you my best professorial it-could-be-this, it-could-be-that answer. However, that was a month ago. Now, after experiencing sleep reduction firsthand and reviewing the evidence yet again, I am less inclined toward wishy-washiness on the topic of sleep. The fact that most committed subjects who are active during the day can reduce their sleep to about 5.5 hours per night without great difficulty or major adverse consequences suggested to me that the answer is 5.5 hours of sleep. But that was before I learned about polyphasic sleep schedules. Now, I must revise my estimate downward.

Conclusion

In this section, you have learned that many people sleep little with no apparent ill effects and that people who are average sleepers can reduce their sleep time substantially, again with no apparent ill effects. You also learned that the health of people who sleep between 5 and 7 hours a night does not suffer; indeed, epidemiological studies indicate that they are the most healthy *Thinking Creatively* and live the longest. Together, this evidence challenges the widely held belief that humans have a fundamental need for at least 8 hours of sleep per night.

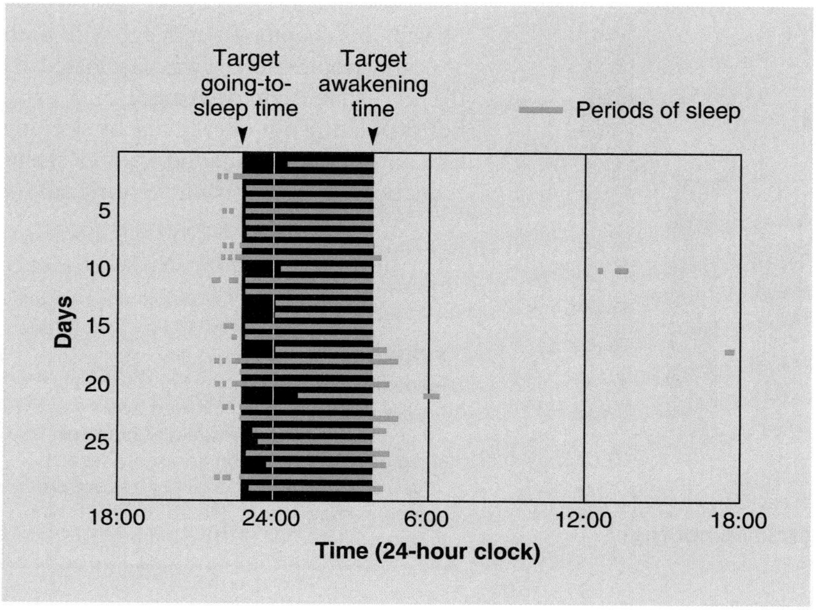

FIGURE 14.15 Sleep record of Pinel during a 4-week sleep-reduction program.

Themes Revisited

The thinking creatively theme pervaded this chapter. The major purpose of the chapter was to encourage you to re-evaluate conventional ideas about  sleep in the light of relevant evidence. Has this chapter changed your thinking about sleep? Writing it changed mine.

The evolutionary perspective theme also played a prominent role in this chapter. You learned how thinking about the adaptive function of sleep and comparing sleep in different species have led to interesting insights. Also, you saw how research into the physiology and genetics of sleep has been conducted on nonhuman species.

The clinical implications theme received emphasis in the section on sleep disorders. Perhaps most exciting and interesting were the recent breakthroughs in the understanding of the genetics and physiology of narcolepsy.

Finally, the neuroplasticity theme arose in a fundamental way. The fact that the adult human brain has the capacity to change and adapt raises the possibility that it might successfully adapt to a consistent long-term schedule of sleep that is of shorter duration than most people currently choose.

Think about It

1. Do you think your life could be improved by changing when or how long you sleep each day? In what ways? What negative effects do you think such changes might have on you?
2. Some people like to stay up late, some people like to get up early, others like to do both, and still others like to do neither. Design a sleep-reduction program that is tailored to your own preferences and lifestyle and that is consistent with the research literature on circadian cycles and sleep deprivation. The program should produce the greatest benefits for you with the least discomfort.
3. How has reading about sleep research changed your views about sleep? Give three specific examples.
4. Given the evidence that the long-term use of benzodiazepines actually contributes to the problems of insomnia, why are they so commonly prescribed for its treatment?
5. Your friend tells you that everybody needs 8 hours of sleep per night; she points out that every time she stays up late to study, she feels lousy the next day. What evidence would you provide to convince her that she does not need 8 hours of sleep per night?

Key Terms

14.1 Stages of Sleep

Electroencephalogram (EEG) (p. 357)
Electrooculogram (EOG) (p. 357)
Electromyogram (EMG) (p. 357)
Alpha waves (p. 357)
Delta waves (p. 358)
Initial stage 1 EEG (p. 358)
Emergent stage 1 EEG (p. 358)
REM sleep (p. 358)
Slow-wave sleep (SWS) (p. 358)
Activation-synthesis theory (p. 359)

14.2 Why Do We Sleep, and Why Do We Sleep When We Do?

Recuperation theories of sleep (p. 359)
Adaptation theories of sleep (p. 360)

14.3 Effects of Sleep Deprivation

Executive function (p. 363)
Microsleeps (p. 363)
Carousel apparatus (p. 363)

14.4 Circadian Sleep Cycles

Circadian rhythms (p. 366)
Zeitgebers (p. 366)
Free-running rhythms (p. 366)
Free-running period (p. 366)
Internal desynchronization (p. 367)
Jet lag (p. 367)
Circadian clock (p. 368)
Suprachiasmatic nuclei (SCN) (p. 368)
Melanopsin (p. 369)
Tau (p. 369)

14.5 Four Areas of the Brain Involved in Sleep

Cerveau isolé preparation (p. 370)
Desynchronized EEG (p. 370)
Encéphale isolé preparation (p. 371)
Reticular activating system (p. 372)

14.6 Drugs That Affect Sleep

Hypnotic drugs (p. 373)
Antihypnotic drugs (p. 373)
Melatonin (p. 373)
Benzodiazepines (p. 373)
5-Hydroxytryptophan (5-HTP) (p. 373)
Pineal gland (p. 374)
Chronobiotic (p. 375)

14.7 Sleep Disorders

Insomnia (p. 375)

Hypersomnia (p. 375)
Iatrogenic (p. 375)
Sleep apnea (p. 376)
Periodic limb movement disorder (p. 376)
Restless legs syndrome (p. 376)
Narcolepsy (p. 376)
Cataplexy (p. 376)
Sleep paralysis (p. 376)
Hypnagogic hallucinations (p. 376)
Orexin (p. 376)
Nucleus magnocellularis (p. 377)

14.8 Effects of Long-Term Sleep Reduction

Polyphasic sleep cycles (p. 378)
Monophasic sleep cycles (p. 378)
Sleep inertia (p. 379)

✔•⎯Quick Review   Test your comprehension of the chapter with this brief practice test. You can find the answers to these questions as well as more practice tests, activities, and other study resources at www.mypsychlab.com.

1. In which stage of sleep do delta waves predominate?

   a. initial stage 1
   b. emergent stage 1
   c. stage 2
   d. stage 3
   e. stage 4

2. The results of many sleep-deprivation studies are difficult to interpret because of the confounding effects of

   a. sex.
   b. dreaming.
   c. shift work.
   d. memory loss.
   e. stress.

3. The carousel apparatus has been used to

   a. entertain sleep-deprived volunteers.
   b. synchronize zeitgebers.
   c. synchronize circadian rhythms.
   d. deprive rodents of sleep.
   e. block microsleeps in sleep-deprived humans.

4. Dreaming occurs during

   a. initial stage 1 sleep.
   b. stage 2 sleep.
   c. stage 3 sleep.
   d. stage 4 sleep.
   e. none of the above

5. Circadian rhythms without zeitgebers are said to be

   a. entrained.
   b. free-running.
   c. desynchronized.
   d. internal.
   e. pathological.

15 Drug Addiction and the Brain's Reward Circuits

Chemicals That Harm with Pleasure

Drug addiction is a serious problem in most parts of the world. For example, in the United States alone, over 60 million people are addicted to nicotine, alcohol, or both; 5.5 million are addicted to illegal drugs; and many millions more are addicted to prescription drugs. Pause for a moment and think about the sheer magnitude of the problem represented by such figures—hundreds of millions of addicted people worldwide. The incidence of drug addiction is so high that it is almost certain that you, or somebody dear to you, will be adversely affected by drugs.

This chapter introduces you to some basic **pharmacological** (pertaining to the scientific study of drugs) principles and concepts, compares the effects of five common addictive drugs, and reviews the research on the neural mechanisms of addiction. You likely already have strong views about drug addiction; thus, as you progress through this chapter, it is particularly important that you do not let your thinking be clouded by preconceptions. In particular, it is important that you do not fall into the trap of assuming that a drug's legal status has much to say about its safety. You will be less likely to assume that legal drugs are safe and illegal drugs are dangerous if you remember that most laws governing drug abuse in various parts of the world were enacted in the early part of the 20th century, long before there was any scientific research on the topic.

Thinking Creatively

The Case of the Drugged High School Teachers

People's tendency to equate drug legality with drug safety was recently conveyed to me in a particularly ironic fashion: I was invited to address a convention of high school teachers on the topic of drug abuse. When I arrived at the convention center to give my talk, I was escorted to a special suite, where I was encouraged to join the executive committee in a round of drug taking—the drug being a special high-proof single-malt whiskey. Later, the irony of the situation had its full impact. As I stepped to the podium under the influence of a psychoactive drug (the whiskey), I looked out through the haze of cigarette smoke at an audience of educators who had invited me to speak to them because they were concerned about the unhealthy impact of drugs on their students. The welcoming applause gradually gave way to the melodic tinkling of ice cubes in liquor glasses, and I began. They did not like what I had to say.

15.1
Basic Principles of Drug Action

This section focuses on the basic principles of drug action, with an emphasis on **psychoactive drugs**—drugs that influence subjective experience and behavior by acting on the nervous system.

Drug Administration and Absorption

Drugs are usually administered in one of four ways: by oral ingestion, by injection, by inhalation, or by absorption through the mucous membranes of the nose, mouth, or rectum. The route of administration influences the rate at which and the degree to which the drug reaches its sites of action in the body.

Oral Ingestion   The oral route is the preferred route of administration for many drugs. Once they are swallowed, drugs dissolve in the fluids of the stomach and are carried to the intestine, where they are absorbed into the bloodstream. However, some drugs readily pass through the stomach wall (e.g., alcohol), and these take effect sooner because they do not have to reach the intestine to be absorbed. Drugs that are not readily absorbed from the digestive tract or that are broken down into inactive *metabolites* (breakdown products of the body's chemical reactions) before they can be absorbed must be taken by some other route.

The two main advantages of the oral route of administration over other routes are its ease and relative safety. Its main disadvantage is its unpredictability: Absorption from the digestive tract into the bloodstream can be greatly influenced by such difficult-to-gauge factors as the amount and type of food in the stomach.

Injection   Drug injection is common in medical practice because the effects of injected drugs are strong, fast, and predictable. Drug injections are typically made *subcutaneously (SC)*, into the fatty tissue just beneath the skin; *intramuscularly (IM)*, into the large muscles; or *intravenously (IV)*, directly into veins at points where they run just beneath the skin. Many addicts prefer the intravenous route because the bloodstream delivers the drug directly to the brain. However, the speed and directness of the intravenous route are mixed blessings; after an intravenous injection, there is little or no opportunity to counteract the effects of an overdose, an impurity, or an allergic reaction. Furthermore, many addicts develop scar tissue, infections, and collapsed veins at the few sites on their bodies where there are large accessible veins.

Inhalation   Some drugs can be absorbed into the bloodstream through the rich network of capillaries in the lungs. Many anesthetics are typically administered by *inhalation*, as are tobacco and marijuana. The two main shortcomings of this route are that it is difficult to precisely regulate the dose of inhaled drugs, and many substances damage the lungs if they are inhaled chronically.

Absorption through Mucous Membranes   Some drugs can be administered through the mucous membranes of the nose, mouth, and rectum. Cocaine, for example, is commonly self-administered through the nasal membranes (snorted)—but not without damaging them.

Drug Penetration of the Central Nervous System

Once a drug enters the bloodstream, it is carried in the blood to the blood vessels of the central nervous system. Fortunately, a protective filter, the *blood–brain barrier*, makes it difficult for many potentially dangerous blood-borne chemicals to pass from the blood vessels of the CNS into its neurons.

Mechanisms of Drug Action

Psychoactive drugs influence the nervous system in many ways (see Koob & Bloom, 1988). Some drugs (e.g., alcohol and many of the general anesthetics) act diffusely on neural membranes throughout the CNS. Others act in a more specific way: by binding to particular synaptic receptors; by influencing the synthesis, transport, release, or deactivation of particular neurotransmitters; or by influencing the chain of chemical reactions elicited in postsynaptic neurons by the activation of their receptors (see Chapter 4).

Drug Metabolism and Elimination

The actions of most drugs are terminated by enzymes synthesized by the *liver*. These liver enzymes stimulate the conversion of active drugs to nonactive forms—a process referred to as **drug metabolism**. In many cases, drug metabolism eliminates a drug's ability to pass through lipid membranes of cells so that it can no longer penetrate the blood–brain barrier. In addition, small amounts of some psychoactive drugs are passed from the body in urine, sweat, feces, breath, and mother's milk.

Drug Tolerance

Drug tolerance is a state of decreased sensitivity to a drug that develops as a result of exposure to it. Drug tolerance can be demonstrated in two ways: by showing that a given dose of the drug has less effect than it had before drug exposure or by showing that it takes more of the drug to produce the same effect. In essence, what this means is that drug tolerance is a shift in the *dose-response curve* (a graph of the magnitude of the effect of different doses of the drug) to the right (see Figure 15.1).

There are three important points to remember about the specificity of drug tolerance.

- One drug can produce tolerance to other drugs that act by the same mechanism; this is known as **cross tolerance**.
- Drug tolerance often develops to some effects of a drug but not to others. Failure to understand this second point can have tragic conse- 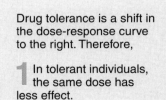 quences for people who think that because they have become tolerant to some effects of a drug (e.g., to the nauseating effects of alcohol or tobacco), they are tolerant to all of them. In fact, tolerance may develop to some effects of a drug while sensitivity to other effects of the same drug increases. Increasing sensitivity to a drug is called **drug sensitization** (Robinson, 1991).
- Drug tolerance is not a unitary phenomenon; that is, there is no single mechanism that underlies all examples of it (Littleton, 2001). When a drug is administered at doses that affect nervous system function, many kinds of adaptive changes can occur to reduce its effects.

Two categories of changes underlie drug tolerance: metabolic and functional. Drug tolerance that results from changes that reduce the amount of the drug getting to its sites of action is called **metabolic tolerance**. Drug tolerance that results from changes that reduce the reactivity of the sites of action to the drug is called **functional tolerance**.

Tolerance to psychoactive drugs is largely functional. Functional tolerance to psychoactive drugs can result from several different types of adaptive neural changes (see Treistman & Martin, 2009). For example, exposure to a psychoactive drug can reduce the number of receptors for it, decrease the efficiency with which it binds to existing receptors, or diminish the impact of receptor binding on the activity

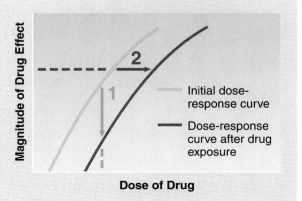

Drug tolerance is a shift in the dose-response curve to the right. Therefore,

1 In tolerant individuals, the same dose has less effect.

2 In tolerant individuals, a greater dose is required to produce the same effect.

Magnitude of Drug Effect

Dose of Drug

Initial dose-response curve

Dose-response curve after drug exposure

FIGURE 15.1 Drug tolerance: A shift in the dose-response curve to the right as a result of exposure to the drug.

of the cell. At least some of these adaptive neural changes are caused by epigenetic mechanisms that affect gene expression (Wang et al., 2007).

Drug Withdrawal Effects and Physical Dependence

After significant amounts of a drug have been in the body for a period of time (e.g., several days), its sudden elimination can trigger an adverse physiological reaction called a **withdrawal syndrome**. The effects of drug withdrawal are virtually always opposite to the initial effects of the drug. For example, the withdrawal of anticonvulsant drugs often triggers convulsions, and the withdrawal of sleeping pills often produces insomnia. Individuals who suffer withdrawal reactions when they stop taking a drug are said to be **physically dependent** on that drug.

Clinical Implications

The fact that withdrawal effects are frequently opposite to the initial effects of the drug suggests that withdrawal effects may be produced by the same neural changes that produce drug tolerance (see Figure 15.2). According to this theory, exposure to a drug produces compensatory changes in the nervous system that offset the drug's effects and produce tolerance. Then, when the drug is eliminated from the body, these compensatory neural changes, without the drug to offset them, manifest themselves as withdrawal symptoms opposite to the initial effects of the drug.

The severity of withdrawal symptoms depends on the particular drug in question, on the duration and degree of the preceding drug exposure, and on the speed with which the drug is eliminated from the body. In general, longer exposure to greater doses followed by more rapid elimination produces greater withdrawal effects.

Addiction: What Is It?

Addicts are habitual drug users, but not all habitual drug users are addicts. **Addicts** are those habitual drug users who continue to use a drug despite its adverse effects on their health and social life, and despite their repeated efforts to stop using it (see Volkow & Li, 2004).

The greatest confusion about the nature of drug addiction concerns its relation to physical dependence. Many people equate the two: They see drug addicts as people who are trapped on a merry-go-round of drug taking, withdrawal symptoms, and further drug taking to combat the withdrawal symptoms. Although appealing in its simplicity, this conception of drug addiction is inconsistent with the evidence. Addicts sometimes take drugs to prevent or alleviate their withdrawal symptoms (Baker et al., 2006), but this is not the major motivating factor in their addiction. If it were, drug addicts could be easily cured by hospitalizing them for a few days, until their withdrawal symptoms subsided. However, most addicts renew their drug taking even after months of enforced abstinence. This is an important issue, and it will be revisited later in this chapter.

Thinking Creatively

It may have occurred to you, given the foregoing definition of addiction, that drugs are not the only substances to which humans are commonly addicted.

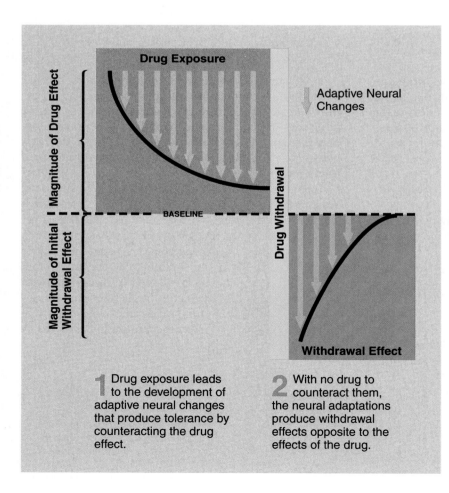

1 Drug exposure leads to the development of adaptive neural changes that produce tolerance by counteracting the drug effect.

2 With no drug to counteract them, the neural adaptations produce withdrawal effects opposite to the effects of the drug.

FIGURE 15.2 The relation between drug tolerance and withdrawal effects. The same adaptive neurophysiological changes that develop in response to drug exposure and produce drug tolerance manifest themselves as withdrawal effects once the drug is removed. As the neurophysiological changes develop, tolerance increases; as they subside, the severity of the withdrawal effects decreases.

Indeed, people who risk their health by continually bingeing on high-calorie foods, risk their family life by repeated illicit sex, or risk their economic stability through compulsive gambling clearly satisfy the definition of an addict (Johnson & Kenny, 2010; Pelchat, 2009; Volkow & Wise, 2005). Although this chapter focuses on drug addiction, food, sex, and gambling addictions may be based on the same neural mechanisms.

15.2

Role of Learning in Drug Tolerance

An important line of psychopharmacologic research has shown that learning plays a major role in drug tolerance. In addition to contributing to our understanding of drug tolerance, this research has established that efforts to understand the effects of psychoactive drugs without considering the experience and behavior of the subjects can provide only partial answers.

Research on the role of learning in drug tolerance has focused on two phenomena: contingent drug tolerance and conditioned drug tolerance. These two phenomena are discussed in the following subsections.

Contingent Drug Tolerance

Contingent drug tolerance refers to demonstrations that tolerance develops only to drug effects that are actually experienced. Most studies of contingent drug tolerance employ the **before-and-after design**. In before-and-after experiments, two groups of subjects receive the same series of drug injections and the same series of repeated tests, but the subjects in one group receive the drug before each test of the series and those in the other group receive the drug after each test. At the end of the experiment, all subjects receive the same dose of the drug followed by the test so that the degree to which the drug disrupts test performance in the two groups can be compared.

My colleagues and I (Pinel, Mana, & Kim, 1989) used the before-and-after design to study contingent tolerance to the anticonvulsant effect of alcohol. In one study, two groups of rats received exactly the same regimen of alcohol injections: one injection every 2 days for the duration of the experiment. During the tolerance development phase, the rats in one group received each alcohol injection 1 hour before a mild convulsive amygdala stimulation so that the anticonvulsant effect of the alcohol could be experienced on each trial. The rats in the other group received their injections 1 hour after each convulsive stimulation so that the anticonvulsant effect could not be experienced. At the end of the experiment, all of the subjects received a test injection of alcohol, followed 1 hour later by a convulsive stimulation so that the amount of tolerance to the anticonvulsant effect of alcohol could be compared in the two groups. As Figure 15.3 illustrates, the rats that received alcohol on each trial before a convulsive stimulation became almost totally tolerant to alcohol's anticonvulsant effect, whereas those that received the same injections and stimulations in the reverse order developed no tolerance whatsoever to alcohol's anticonvulsant effect. Contingent drug tolerance has been demonstrated for many other drug effects in many species, including humans (see Poulos & Cappell, 1991; Wolgin & Jakubow, 2003).

Evolutionary Perspective

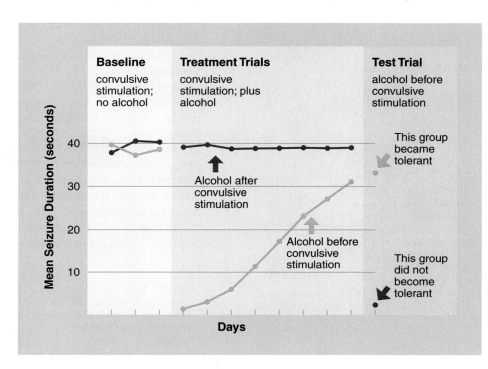

Baseline
convulsive stimulation; no alcohol

Treatment Trials
convulsive stimulation; plus alcohol

Test Trial
alcohol before convulsive stimulation

This group became tolerant

Alcohol after convulsive stimulation

Alcohol before convulsive stimulation

This group did not become tolerant

Mean Seizure Duration (seconds)

Days

FIGURE 15.3 Contingent tolerance to the anticonvulsant effect of alcohol. The rats that received alcohol on each trial *before* a convulsive stimulation became tolerant to its anticonvulsant effect; those that received the same injections *after* a convulsive stimulation on each trial did not become tolerant. (Based on Pinel et al., 1989.)

Conditioned Drug Tolerance

Whereas studies of contingent drug tolerance focus on what subjects do while they are under the influence of drugs, studies of conditioned drug tolerance focus on the situations in which drugs are taken. **Conditioned drug tolerance** refers to demonstrations that tolerance effects are maximally expressed only when a drug is administered in the same situation in which it has previously been administered (see McDonald & Siegel, 2004; Mitchell, Basbaum, & Fields, 2000; Weise-Kelley & Siegel, 2001).

((•• **Listen** The Effects of Drugs and Alcohol www.mypsychlab.com

In one demonstration of conditioned drug tolerance (Crowell, Hinson, & Siegel, 1981), two groups of rats received 20 alcohol and 20 saline injections in an alternating sequence, 1 injection every other day. The only difference between the two groups was that the rats in one group received all 20 alcohol injections in a distinctive test room and the 20 saline injections in their colony room, while the rats in the other group received the alcohol in the colony room and the saline in the distinctive test room. At the end of the injection period, the tolerance of all rats to the *hypothermic* (temperature-reducing) effects of alcohol was assessed in both environments. As Figure 15.4 illustrates, tolerance was observed only when the rats were injected in the environment that had previously been paired with alcohol administration. There have been dozens of other demonstrations of the *situational speci-*

Evolutionary Perspective

ficity of drug tolerance: The effect is large, reliable, and general.

The situational specificity of drug tolerance led Siegel and his colleagues to propose that addicts may be particularly susceptible to the lethal effects of a drug *overdose* when the drug is administered in a new context. Their hypothesis is that addicts become tolerant when they repeatedly self-administer their drug in the same environment and, as a result, begin taking larger and larger doses to counteract the diminution of drug effects. Then, if the addict administers the usual massive dose in an unusual situation, tolerance effects are not present to counteract the effects of the drug, and there is a greater risk of death from overdose. In support of this hypothesis, Siegel and colleagues (1982) found that many more heroin-tolerant rats died following a high dose of heroin administered in a novel environment than died in the usual injection environment. (Heroin, as you will learn later in the chapter, kills by suppressing respiration.)

Siegel views each incidence of drug administration as a Pavlovian conditioning trial in which various environmental stimuli (e.g., bars, washrooms, needles, or other addicts) that regularly predict the administration of the drug are conditional stimuli and the drug effects are unconditional stimuli. The central assumption of the theory is that conditional stimuli that predict drug administration come to elicit conditional responses opposite to the unconditional effects of the drug. Siegel has termed these hypothetical opposing conditional responses **conditioned compensatory responses**. The theory is that as the stimuli that repeatedly predict the effects of a drug come to elicit greater and greater conditioned compensatory responses, they increasingly counteract the unconditional effects of the drug and produce situationally specific tolerance.

Alert readers will have recognized the relation between Siegel's theory of drug tolerance and Woods's theory of meal-time hunger, *Thinking Creatively* which you learned about in Chapter 12. Stimuli that predict the homeostasis-disrupting effects of meals trigger conditioned compensatory responses to minimize

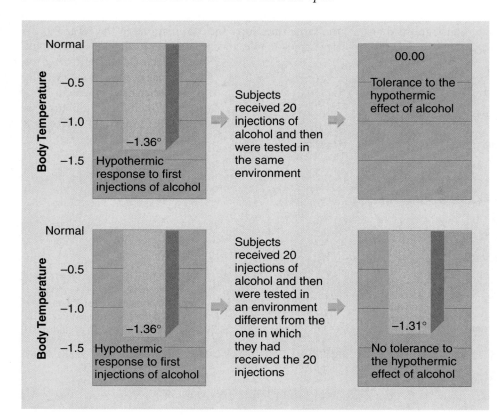

FIGURE 15.4 The situational specificity of tolerance to the hypothermic effects of alcohol in rats. (Based on Crowell et al., 1981.)

a meal's disruptive effects in the same way that stimuli that predict the homeostasis-disturbing effects of a drug trigger conditioned compensatory responses to minimize the drug's disruptive effects.

Most demonstrations of conditioned drug tolerance have employed **exteroceptive stimuli** (external, public stimuli, such as the drug-administration environment) as the conditional stimuli. However, **interoceptive stimuli** (internal, private stimuli) are just as effective in this role. For example, both the feelings produced by the drug-taking ritual and the first mild effects of the drug experienced soon after administration can, through conditioning, come to reduce the full impact of a drug (Siegel, 2005). This point about interoceptive stimuli is important because it indicates that just thinking about a drug can evoke conditioned compensatory responses.

Although tolerance develops to many drug effects, sometimes the opposite occurs, that is, drug sensitization. *Drug sensitization*, like drug tolerance, can be situationally specific (see Arvanitogiannis, Sullivan, & Amir, 2000). For example, Anagnostaras and Robinson (1996) demonstrated the situational specificity of sensitization to the motor stimulant effects of amphetamine. They found that 10 amphetamine injections, 1 every 3 or 4 days, greatly increased the ability of amphetamine to activate the motor activity of rats—but only when the rats were injected and tested in the same environment in which they had experienced the previous amphetamine injections.

Drug withdrawal effects and conditioned compensatory responses are similar: They are both responses that are opposite to the unconditioned effect of the drug. The difference is that drug withdrawal effects are produced by elimination of the drug from the body, whereas conditioned compensatory responses are elicited by drug-predictive cues in the absence of the drug. In complex, real-life situations, it is often difficult to tell them apart.

Thinking about Drug Conditioning

In any situation in which drugs are repeatedly administered, conditioned effects are inevitable. That is why it is particularly important to understand them. However, most theories of drug conditioning have a serious problem: They have difficulty predicting the direction of the conditioned effects. For example, Siegel's conditioned compensatory response theory predicts that conditioned drug effects will always be opposite to the unconditioned effects of the drug, but there are many documented instances in which conditional stimuli elicit responses similar to those of the drug.

Ramsay and Woods (1997) contend that much of the confusion about conditioned drug effects stems from a misunderstanding of Pavlovian conditioning. In particular,

Thinking Creatively they criticize the common assumption that the unconditional stimulus in a drug-tolerance experiment is the

drug and that the unconditional response is whatever change in physiology or behavior the experimenter happens to be recording. They argue instead that the unconditional stimulus (i.e., the stimulus to which the subject reflexively reacts) is the disruption of neural functioning that has been directly produced by the drug, and that the unconditional responses are the various neurally mediated compensatory reactions to the unconditional stimulus.

This change in perspective makes a big difference. For example, in the previously described alcohol tolerance experiment by Crowell and colleagues (1981), alcohol was designated as the unconditional stimulus and the resulting hypothermia as the unconditional response. Instead, Ramsay and Woods would argue that the unconditional stimulus was the hypothermia directly produced by the exposure to alcohol, whereas the compensatory changes that tended to counteract the reductions in body temperature were the unconditional responses. The important point about all of this is that once one determines the unconditional stimulus and unconditional response, it is easy to predict the direction of the conditional response in any drug-conditioning experiment: The conditional response is always similar to the unconditional response.

15.3

Five Commonly Abused Drugs

This section focuses on the hazards of chronic use of five commonly abused drugs: tobacco, alcohol, marijuana, cocaine, and the opiates.

Tobacco

When a cigarette is smoked, **nicotine**—the major psychoactive ingredient of tobacco—and some 4,000 other chemicals, collectively referred to as *tar*, are absorbed through the lungs. Nicotine acts on nicotinic cholinergic receptors in the brain (see Benowitz, 2008). Tobacco is the

leading preventable cause of death in Western countries. In the United States, it contributes to 400,000 premature deaths a year—about 1 in every 5 deaths (U.S. Centers for Disease Control and Prevention, 2008b).

Because considerable tolerance develops to some of the immediate effects of tobacco, the effects of smoking a cigarette on nonsmokers and smokers can be quite different. Nonsmokers often respond to a few puffs of a cigarette with various combinations of nausea, vomiting, coughing, sweating, abdominal cramps, dizziness, flushing, and diarrhea. In contrast, smokers report that they are more relaxed, more alert, and less hungry after a cigarette.

There is no question that heavy smokers are drug addicts in every sense of the word (see Hogg & Bertrand, 2004). Can you think of any other psychoactive drug that is self-administered almost continually—even while the addict is walking along the street? The compulsive drug craving, which is the major defining feature of addiction, is readily apparent in any habitual smoker who has run out of cigarettes or who is forced by circumstance to refrain from smoking for several hours. Furthermore, habitual smokers who stop smoking experience a variety of withdrawal effects, such as depression, anxiety, restlessness, irritability, constipation, and difficulties in sleeping and concentrating.

About 70% of all people who experiment with smoking become addicted—this figure compares unfavorably with 10% for alcohol and 30% for heroin. Moreover, nicotine addiction typically develops quickly, within a few weeks (Di Franza, 2008), and only about 20% of all attempts to stop smoking are successful for 2 years or more. Twin studies (Lerman et al., 1999; True et al., 1999) confirm that nicotine addiction, like other addictions, has a major genetic component. The heritability estimate is about 65%.

The consequences of long-term tobacco use are alarming. **Smoker's syndrome** is characterized by chest pain, labored breathing, wheezing, coughing, and a heightened susceptibility to infections of the respiratory tract.

Clinical Implications

Chronic smokers are highly susceptible to a variety of potentially lethal lung disorders, including pneumonia, *bronchitis* (chronic inflammation of the bronchioles of the lungs), *emphysema* (loss of elasticity of the lung from chronic irritation), and lung cancer. Although the increased risk of lung cancer receives the greatest publicity, smoking also increases the risk of cancer of the larynx (voice box), mouth, esophagus, kidneys, pancreas, bladder, and stomach. Smokers also run a

👁 **Watch** Smoking Damage
www.mypsychlab.com

greater risk of developing a variety of cardiovascular diseases, which may culminate in heart attack or stroke.

Many smokers claim that they smoke despite the adverse effects because smoking reduces tension. However, smokers are actually more tense than nonsmokers: Their levels of tension are reasonably normal while they are smoking, but they increase markedly between cigarettes.

Thus, the apparent relaxant effect of smoking merely reflects the temporary reversal of the stress caused by the smoker's addiction (see Parrott, 1999). Consistent with this finding is the fact that smokers are more prone than nonsmokers to experience panic attacks (Zvolensky & Bernstein, 2005).

Sufferers from Buerger's disease provide a shocking illustration of the addictive power of nicotine. In **Buerger's disease**—which occurs in about 15 of 100,000 individuals, mostly in male smokers— the blood vessels, especially those supplying the legs, become constricted.

> If a patient with this condition continues to smoke, gangrene may eventually set in. First a few toes may have to be amputated, then the foot at the ankle, then the leg at the knee, and ultimately at the hip. Somewhere along this gruesome progression gangrene may also attack the other leg. Patients are strongly advised that if they will only stop smoking, it is virtually certain that the otherwise inexorable march of gangrene up the legs will be curbed. Yet surgeons report that it is not at all uncommon to find a patient with Buerger's disease vigorously puffing away in his hospital bed following a second or third amputation operation. (Brecher, 1972, pp. 215–216)

The adverse effects of tobacco smoke are unfortunately not restricted to those who smoke. Individuals who live or work with smokers are more likely to develop heart disease and cancer than those who don't. Even the unborn are vulnerable (see Huizink & Mulder, 2006; Thompson, Levitt, & Stanwood, 2009). Nicotine is a **teratogen** (an agent that can disturb the normal development of the fetus): Smoking during pregnancy increases the likelihood of miscarriage, stillbirth, and

👁 **Watch** Prenatal Smoking
www.mypsychlab.com

early death of the child. And the levels of nicotine in the blood of breastfed infants are often as great as those in the blood of their smoking mothers.

If you or a loved one is a cigarette smoker, some recent findings provide both good news and bad news (see West, 2007). First the bad news: Treatments for nicotine addiction are only marginally effective—nicotine patches have been shown to help some in the short term (Shiffman & Ferguson, 2008). The good news: Many people do stop smoking, and they experience major health benefits. For example, smokers who manage to stop smoking before the age of 30 live almost as long as people who have never smoked (Doll et al., 2004).

Alcohol

Alcohol is involved in over 3% of all deaths in the United States, including deaths from birth defects, ill health, accidents, and violence (see Mokdad et al., 2004). Approximately 13 million Americans are heavy users, and about 80,000 die each year from alcohol-

👁 **Watch** Alcoholism
www.mypsychlab.com

related diseases and accidents (U.S. Centers for Disease Control and Prevention, 2008a).

Because alcohol molecules are small and soluble in both fat and water, they invade all parts of the body. Alcohol is classified as a **depressant** because at moderate-to-high doses it depresses neural firing; however, at low doses it can stimulate neural firing and facilitate social interaction. Alcohol addiction has a major genetic component (McGue, 1999): Heritability estimates are about 55%, and several genes associated with alcoholism have been identified (Nurnberger & Bierut, 2007).

 Watch Genetic Predisposition to Alcoholism
www.mypsychlab.com

With moderate doses, the alcohol drinker experiences various degrees of cognitive, perceptual, verbal, and motor impairment, as well as a loss of control that can lead to a variety of socially unacceptable actions. High doses result in unconsciousness; and if blood levels reach 0.5%, there is a risk of death from respiratory depression. The telltale red facial flush of alcohol intoxication is produced by the dilation of blood vessels in the skin; this dilation increases the amount of heat that is lost from the blood to the air and leads to a decrease in body temperature (*hypothermia*). Alcohol is also a *diuretic*; that is, it increases the production of urine by the kidneys.

Alcohol, like many addictive drugs, produces both tolerance and physical dependence. The livers of heavy drinkers metabolize alcohol more quickly than do the livers of nondrinkers, but this increase in metabolic efficiency contributes only slightly to overall alcohol tolerance; most alcohol tolerance is functional. Alcohol withdrawal often produces a mild syndrome of headache, nausea, vomiting, and tremulousness, which is euphemistically referred to as a *hangover*.

 Watch Alcohol Withdrawal
www.mypsychlab.com

A full-blown alcohol withdrawal syndrome comprises three phases (see De Witte et al., 2003). The first phase begins about 5 or 6 hours after the cessation of a long bout of heavy drinking and is characterized by severe tremors, agitation, headache, nausea, vomiting, abdominal cramps, profuse sweating, and sometimes hallucinations. The defining feature of the second phase, which typically occurs between 15 and 30 hours after cessation of drinking,

Clinical Implications

is convulsive activity. The third phase, which usually begins a day or two after the cessation of drinking and lasts for 3 or 4 days, is called **delirium tremens (DTs)**. The DTs are characterized by disturbing hallucinations, bizarre delusions, agitation, confusion, *hyperthermia* (high body temperature), and *tachycardia* (rapid heartbeat). The convulsions and the DTs produced by alcohol withdrawal can be lethal.

Alcohol attacks almost every tissue in the body (see Anderson et al., 1993). Chronic alcohol consumption produces extensive brain damage. This damage is produced both directly (see Mechtcheriakov et al., 2007) and indirectly. For example, you learned in Chapter 1 that alcohol indirectly causes **Korsakoff's syndrome** (a neuropsychological disorder characterized by memory loss, sensory and motor dysfunction, and, in its advanced stages, severe dementia) by inducing thiamine deficiency, and it also indirectly causes brain damage by increasing susceptibility to stroke (Rehm, 2006). Alcohol affects the brain function of drinkers in other ways, as well. For example, it reduces the flow of calcium ions into neurons by acting on ion channels; it interferes with the function of second messengers inside neurons; it disrupts GABAergic and glutaminergic transmission; and it triggers apoptosis (see Farber & Olney, 2003; Ikonomidou et al., 2000).

Chronic alcohol consumption also causes extensive scarring, or **cirrhosis**, of the liver, which is the major cause of death among heavy alcohol users. Alcohol erodes the muscles of the heart and thus increases the risk of heart attack. It irritates the lining of the digestive tract and, in so doing, increases the risk of oral and liver cancer, stomach ulcers, *pancreatitis* (inflammation of the pancreas), and *gastritis* (inflammation of the stomach). And not to be forgotten is the carnage that alcohol produces from accidents on our roads, in our homes, in our workplaces, and at recreational sites—in the United States, over 20,000 people die each year in alcohol-related traffic accidents alone.

Many people assume that the adverse effects of alcohol occur only in people who drink a lot—they tend to define "a lot" as "much more than they themselves consume." But they are wrong. Several large-scale studies have shown that even low-to-moderate regular drinking (a drink or two per day) is associated with elevated levels of most cancers, including breast, prostate, ovary, and skin cancer (Allen et al., 2009; Bagnardi et al., 2001; Benedetti, Parent, & Siemiatycki, 2009).

Like nicotine, alcohol readily penetrates the placental membrane and acts as a teratogen. The result is that the offspring of mothers who consume substantial quantities of alcohol during pregnancy can develop **fetal alcohol syndrome (FAS)**—see Calhoun and Warren (2007). The FAS child suffers from some or all of the following symptoms: brain damage, mental retardation, poor coordination, poor muscle tone, low birth weight, retarded growth, and/or physical deformity. Because alcohol can disrupt brain development in so many ways (e.g., by disrupting neurotrophic support, by disrupting the production of cell-adhesion molecules, or by disrupting normal patterns of apoptosis), there is no time during pregnancy when alcohol consumption is safe (see Farber & Olney, 2003; Guerri, 2002). Moreover, there seems to be no safe amount. Although full-blown FAS is rarely seen in the babies of mothers who never had more than one drink a day during pregnancy, children of mothers who drank only moderately while pregnant are sometimes found to have a variety of cognitive problems, even though they are not diagnosed with FAS (see Korkman, Kettunen, & Autti-Ramo, 2003).

There is no cure for alcoholism; however, disulfiram (Antabuse) can help reduce alcohol consumption under certain conditions. **Disulfiram** is a drug that interferes with the metabolism of alcohol and produces an accumulation in the bloodstream of *acetaldehyde* (one of alcohol's breakdown products). High levels of acetaldehyde produce flushing, dizziness, headache, vomiting, and difficulty breathing; thus, a person who is medicated with disulfiram cannot drink much alcohol without feeling ill. Unfortunately, disulfiram is not a cure for alcoholism because alcoholics simply stop taking it when they return to drinking alcohol. However, treatment with disulfiram can be useful in curtailing alcohol consumption in hospital or outpatient environments, where patients take the medication each day under supervision (Brewer, 2007).

One of the most widely publicized findings about alcohol is that moderate drinking reduces the risk of coronary heart disease. This conclusion is based on the finding that the incidence of coronary heart disease is less among moderate drinkers than among abstainers. You learned in Chapter 1 about the difficulty in basing causal interpretations on correlational data, and researchers worked diligently to identify and rule out factors other than the alcohol that might protect moderate drinkers from coronary heart disease. They seemed to rule out every other possibility. However, a thoughtful new analysis has led to a different conclusion. Let me explain. In a culture in which alcohol

Thinking Creatively

consumption is the norm, any large group of abstainers will always include some people who have stopped drinking because they are ill—perhaps this is why abstainers have more heart attacks than moderate drinkers. This hypothesis was tested by including in a meta-analysis only those studies that used an abstainers control group consisting of individuals who had never consumed alcohol. This meta-analysis indicated that alcohol in moderate amounts does not prevent coronary heart disease; that is, moderate drinkers did not suffer less coronary heart disease than lifelong abstainers (Fillmore et al., 2006; Stockwell et al., 2007).

Marijuana

Marijuana is the name commonly given to the dried leaves and flowers of **Cannabis sativa**—the common hemp plant. Approximately 2 million Americans have used marijuana in the last month. The usual mode of

consumption is to smoke these leaves in a *joint* (a cigarette of marijuana) or a pipe; but marijuana is also effective when ingested orally, if first baked into an oil-rich substrate, such as a chocolate brownie, to promote absorption from the gastrointestinal tract.

The psychoactive effects of marijuana are largely attributable to a constituent called **THC** (delta-9-tetrahydrocannabinol). However, marijuana contains over 80 *cannabinoids* (chemicals of the same chemical class as THC), which may also be psychoactive. Most of the cannabinoids are found in a sticky resin covering the leaves and flowers of the plant; this resin can be extracted and dried to form a dark corklike material called **hashish**. Hashish can be further processed into an extremely potent product called *hash oil*.

Written records of marijuana use go back 6,000 years in China, where its stems were used to make rope, its seeds were used as a grain, and its leaves and flowers were used for their psychoactive and medicinal effects. In the Middle Ages, cannabis cultivation spread into Europe, where it was grown primarily for the manufacture of rope. During the period of European imperialism, rope was in high demand for sailing vessels, and the American colonies responded to this demand by growing cannabis as a cash crop. George Washington was one of the more notable cannabis growers.

The practice of smoking the leaves of *Cannabis sativa* and the word *marijuana* itself seem to have been introduced to the southern United States in the early part of the 20th century. In 1926, an article appeared in a New Orleans newspaper exposing the "menace of marijuana," and soon similar stories were appearing in newspapers all over the United States claiming that marijuana turns people into violent, drug-crazed criminals. The misrepresentation of the effects of marijuana by the news media led to the rapid enactment of laws against the drug. In many states, marijuana was legally classified a **narcotic** (a legal term generally used to refer to opiates), and punishment for its use was dealt out accordingly. Marijuana bears no resemblance to opiate narcotics.

Popularization of marijuana smoking among the middle and upper classes in the 1960s stimulated a massive program of research. One of the difficulties in studying the effects of marijuana is that they are subtle, difficult to measure, and greatly influenced by the social situation:

At low, usual "social" doses, the intoxicated individual may experience an increased sense of well-being: initial restlessness and hilarity followed by a dreamy, carefree state of relaxation; alteration of sensory perceptions including expansion of space and time; and a more vivid sense of touch, sight, smell, taste, and sound; a feeling of hunger, especially a craving for sweets; and subtle changes in thought formation and expression. To an unknowing observer, an individual in this state of consciousness would not appear noticeably different. (National Commission on Marijuana and Drug Abuse, 1972, p. 68)

Although the effects of typical social doses of marijuana are subtle, high doses do impair psychological functioning. At high doses, short-term memory is impaired, and the ability to carry out tasks involving multiple steps to reach a specific goal declines. Speech becomes slurred, and meaningful conversation becomes difficult. A sense of unreality, emotional intensification, sensory distortion, feelings of paranoia, and motor impairment are also common.

● Watch Marijuana and Performance
www.mypsychlab.com

Some people do become addicted to marijuana, but its addiction potential is low. Most people who use marijuana do so only occasionally, with only about 10% of them using daily; moreover, most people who try marijuana do so in their teens and curtail their use by their 30s or 40s (see Room et al., 2010). Tolerance to marijuana develops during periods of sustained use; however, obvious withdrawal symptoms (e.g., nausea, diarrhea, sweating, chills, tremor, sleep disturbance) are rare, except in contrived laboratory situations in which massive oral doses are administered.

What are the health hazards of marijuana use? Two have been documented. First, the few marijuana smokers who do smoke it regularly for long periods (estimated to be about 10%) tend to develop respiratory problems (see Aldington et al., 2008; Brambilla & Colonna, 2008; Tetrault et al., 2007): cough, bronchitis, and asthma. Second, because marijuana produces *tachycardia* (elevated heart rate), single large doses can trigger heart attacks in susceptible individuals who have previously suffered a heart attack.

Clinical Implications

Although many people believe that marijuana causes brain damage, almost all efforts to document brain damage in marijuana users have proven negative. The one exception is an MRI study by Yücel and colleagues (2008), who studied the brains of 15 men who had an extremely high level of marijuana exposure—at least 5 joints per day for almost 20 years. These men had hippocampuses and amygdalae with reduced volumes. Because this finding is the single positive report in a sea of negative findings, it

needs to be replicated. Furthermore, because the finding is correlational, it cannot prove that extremely high doses of marijuana can cause brain damage—one can just as easily conclude that brain damage predisposes individuals to pathological patterns of marijuana use.

Because it has been difficult to directly document brain damage in marijuana users, many studies have taken an indirect approach: They have attempted to document permanent memory loss in marijuana users—the assumption being that such loss would be indicative of brain damage. Many studies have documented memory deficits in marijuana users, but these deficits tend to be acute effects associated with marijuana intoxication that disappear after a few weeks of abstinence. Indeed, there seems to be a general consensus that marijuana use is not associated with substantial permanent memory problems (see Grant et al., 2003; Jager et al., 2006; Iversen, 2005). There have been reports (see Medina et al., 2007) that people who become heavy marijuana users in adolescence display memory and other cognitive deficits; however, Pope and colleagues (2003) found that adolescents with lower verbal intelligence scores are more likely to become heavy marijuana users, which likely accounts for the poorer cognitive performance.

Several correlational studies have found that heavy marijuana users are more likely to be be diagnosed with schizophrenia (see Arseneault et al., 2004). The best of these studies followed a group of Swedish males for 25 years (Zammit et al., 2002); after some of the obvious confounds had been controlled, there was a higher incidence of schizophrenia among heavy marijuana users. This correlation has led some to conclude that marijuana causes schizophrenia, but, as you know, correlational evidence cannot prove causation. In this case, it is also possible that youths in the early developmental stages of schizophrenia have a particular attraction and/or susceptibility to marijuana; however, more research is required to understand the causal factors involved in this correlation (see Pollack & Reurer, 2007). In the meantime, individuals with a history of schizophrenia in their families should avoid marijuana.

Thinking Creatively

THC has been shown to have several therapeutic effects (see Karanian & Bahr, 2006). Since the early 1990s, it has been widely used to suppress nausea and vomiting in cancer patients and to stimulate the appetite of AIDS patients (see DiMarzo & Matias, 2005). THC has also been shown to block seizures; to dilate the bronchioles of asthmatics; to decrease the severity of *glaucoma* (a disorder characterized by an increase in the pressure of the fluid inside the eye); and to reduce anxiety, some kinds of pain, and the symptoms of multiple sclerosis (Agarwal et al., 2007; Nicoll & Alger, 2004; Page et al., 2003). Medical use of THC does not appear to be associated with adverse side effects (Degenhardt & Hall, 2008; Wang et al., 2008).

Research on THC changed irrevocably in the early 1990s with the discovery of two receptors for it in the brain: CB_1 and CB_2. CB_1 turned out to be the most prevalent

G-protein–linked receptor in the brain (see Chapter 4); CB_2 is found in the brain stem and in the cells of the immune system (see Van Sickle et al., 2005). But why are there THC receptors in the brain? They could hardly have evolved to mediate the effects of marijuana smoking. This puzzle was quickly solved with the discovery of a class of endogenous cannabinoid neurotransmitters: the endocannabinoids (see Harkany, Mackie, & Doherty, 2008). The first endocannabinoid neurotransmitter to be isolated and characterized was named **anandamide**, from a word that means "internal bliss" (see Nicoll & Alger, 2004).

I cannot end this discussion of marijuana (*Cannabis sativa*) without telling you the following story:

> You can imagine how surprised I was when my colleague went to his back door, opened it, and yelled, "Sativa, here Sativa, dinner time."
>
> "What was that you called your dog?" I asked as he returned to his beer.
>
> "Sativa," he said. "The kids picked the name. I think they learned about it at school; a Greek goddess or something. Pretty, isn't it? And catchy too: Every kid on the street seems to remember her name."
>
> "Yes," I said. "Very pretty."

Cocaine and Other Stimulants

Stimulants are drugs whose primary effect is to produce general increases in neural and behavioral activity. Although stimulants all have a similar profile of effects, they differ greatly in their potency. Coca-Cola is a mild commercial stimulant preparation consumed by many people around the world. Today, its stimulant action is attributable to *caffeine*, but when it was first introduced, "the pause that refreshes" packed a real wallop in the form of small amounts of cocaine. **Cocaine** and its derivatives are the most commonly abused stimulants, and thus they are the focus of this discussion.

Cocaine is prepared from the leaves of the coca bush, which grows primarily in Peru and Bolivia. For centuries, a crude extract called *coca paste* has been made directly from the leaves and eaten. Today, it is more common to treat the coca paste and extract *cocaine hydrochloride*, the nefarious white powder that is referred to simply as *cocaine* and typically consumed by snorting or by injection. Cocaine hydrochloride may be converted to its base form by boiling it in a solution of baking soda until the water has evaporated. The impure residue of this process is **crack**, which is a potent, cheap, smokable form of cocaine. However, because crack is impure, variable, and consumed by smoking, it is difficult to study, and most research on cocaine derivatives has thus focused on pure cocaine hydrochloride. Approximately 36 million Americans have used cocaine or crack (Substance Abuse and Mental Health Services Administration [SAMSHA], 2009).

👁 **Watch** Cocaine
www.mypsychlab.com

Cocaine hydrochloride is an effective local anesthetic and was once widely prescribed as such until it was supplanted by synthetic analogues such as *procaine* and *lidocaine*. It is not, however, cocaine's anesthetic actions that are of interest to users. People eat, smoke, snort, or inject cocaine or its derivatives in order to experience its psychological effects. Users report being swept by a wave of well-being; they feel self-confident, alert, energetic, friendly, outgoing, fidgety, and talkative; and they have less than their usual desire for food and sleep.

Cocaine addicts tend to go on so-called **cocaine sprees**, binges in which extremely high levels of intake are maintained for periods of a day or two. During a cocaine spree, users become increasingly tolerant to the euphoria-producing effects of cocaine. Accordingly, larger and larger doses are often administered. The spree usually ends when the cocaine is gone or when it begins to have serious toxic effects. The effects of cocaine sprees include sleeplessness, tremors, nausea, hyperthermia, and psychotic behavior, which is called **cocaine psychosis** and has often been mistakenly diagnosed as *paranoid schizophrenia*. During cocaine sprees, there is a risk of loss of consciousness, seizures, respiratory arrest,

Clinical Implications

heart attack, or stroke (Kokkinos & Levine, 1993). Although tolerance develops to most effects of cocaine (e.g., to the euphoria), repeated cocaine exposure sensitizes subjects (i.e., makes them even more responsive) to its motor and convulsive effects (see Robinson & Berridge, 1993). The withdrawal effects triggered by abrupt termination of a cocaine spree are relatively mild. Common cocaine withdrawal symptoms include a negative mood swing and insomnia.

Cocaine and its various derivatives are not the only commonly abused stimulants. **Amphetamine** (speed) and its relatives also present major health problems. Amphetamine has been in wide illicit use since the 1960s. It is usually consumed orally in the potent form called

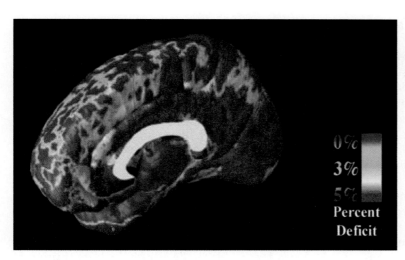

FIGURE 15.5 Structural MRIs have revealed widespread loss of cortical volume in methamphetamine users. Red indicates the areas of greatest loss. (From Thompson et al., 2004.)

d-amphetamine (dextroamphetamine). The effects of *d*-amphetamine are comparable to those of cocaine; for example, it produces a syndrome of psychosis called *amphetamine psychosis*.

In the 1990s, *d*-amphetamine was supplanted as the favored amphetamine-like drug by several more potent relatives. One is *methamphetamine*, or "meth" (see Cho, 1990), which is commonly used in its even more potent, smokable, crystalline form (ice or crystal). Another potent relative of amphetamine is *3,4-methylenedioxy-methamphetamine* (MDMA, or ecstasy), which is taken orally (see Baylen & Rosenberg, 2006).

The primary mechanism by which cocaine and its derivatives exert their effects is the blockade of **dopamine transporters**, molecules in the presynaptic membrane that normally remove dopamine from synapses and transfer it back into presynaptic neurons. Other stimulants increase the release of monoamines into synapses (Sulzer et al., 2005).

Do stimulants have long-term adverse effects on the health of habitual users? There is mounting evidence that they do. Users of MDMA have deficits in the performance of various neuropsychological tests; they have deficiencies in various measures of dopaminergic and serotonergic

Clinical Implications

function; and functional brain imaging during tests of executive functioning, inhibitory control, and decision making often reveals abnormalities in many areas of the cortex and limbic system (see Aron & Paulus, 2007; Baicy & London, 2007; Chang et al., 2007; Volz, Fleckenstein, & Hanson, 2007). The strongest evidence that methamphetamine damages the brain comes from a structural MRI study that found decreases in volume of various parts of the brains of persons who had used methamphetamine for an average of 10 years (Thompson et al.,

2004)—the reductions in cortical volume are illustrated in Figure 15.5. Controlled experiments on nonhumans have confirmed the adverse effects of stimulants on brain function (see McCann & Ricaurte, 2004).

Although research on the health hazards of stimulants has focused on brain pathology, there is also evidence of heart pathology—many methamphetamine-dependent patients have been found to have electrocardiographic abnormalities (Haning & Goebert, 2007). Also, many behavioral, neurological, and cardiovascular problems have been observed in infants born to mothers who have used stimulants while pregnant (see Harvey, 2004).

The Opiates: Heroin and Morphine

Opium—the dried form of sap exuded by the seed pods of the opium poppy—has several psychoactive ingredients. Most notable are **morphine** and **codeine**, its weaker relative. Morphine, codeine, and other drugs that have similar structures or effects are commonly referred to as **opiates**. The opiates exert their effects by binding to receptors whose normal function is to bind to endogenous opiates. The endogenous opiate neurotransmitters that bind to such receptors are of two classes: *endorphins* and *enkephalins* (see Chapter 4).

The opiates have a Jekyll-and-Hyde character. On their Dr. Jekyll side, the opiates are effective as **analgesics** (painkillers; see Watkins et al., 2005); they are also extremely effective in the treatment of cough and diarrhea. But, unfortunately, the kindly Dr. Jekyll brings with him the evil Mr. Hyde—the risk of addiction.

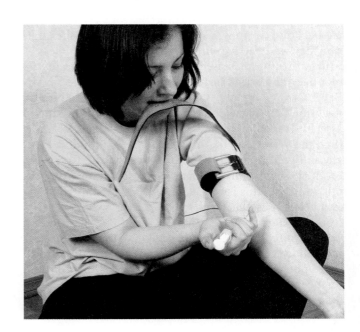

The practice of eating opium spread from the Middle East sometime before 4000 B.C. Three historic events fanned the flame of opiate addiction. First, in 1644, the Emperor of China banned tobacco smoking, and this contributed to a gradual increase in opium smoking in China, spurred on by the smuggling of opium into China by the British East India Company. Because smoking opium has a greater effect on the brain than does eating it, many more people became addicted. Second, morphine, the most potent constituent of opium, was isolated in 1803, and it became available commercially in the 1830s. Third, the hypodermic needle was invented in 1856, and soon the injured were introduced to morphine through a needle.

Until the early part of the 20th century, opium was available legally in many parts of the world, including Europe and North America. Indeed, opium was an ingredient in cakes, candies, and wines, as well as in a variety of over-the-counter medicinal offerings. Opium potions such as *laudanum* (a very popular mixture of opium and alcohol), *Godfrey's Cordial*, and *Dalby's Carminative* were very popular. (The word *carminative* should win first prize for making a sow's ear at least sound like a silk purse: A carminative is a drug that expels gas from the digestive tract, thereby reducing stomach cramps and flatulence. *Flatulence* is the obvious pick for second prize.) There were even over-the-counter opium potions just for baby—such as *Mrs. Winslow's Soothing Syrup* and the aptly labeled *Street's Infant Quietness*. Although pure morphine required a prescription at the time, physicians prescribed it for so many different maladies that morphine addiction was common among those who could afford a doctor.

The **Harrison Narcotics Act**, passed in 1914, made it illegal to sell or use opium, morphine, or cocaine in the United States—although morphine and its analogues are still legally prescribed for their medicinal properties. However, the act did not include the semisynthetic opiate **heroin**. Heroin was synthesized in 1870 by the addition of two acetyl groups to the morphine molecule, which greatly increased its ability to penetrate the blood–brain barrier. In 1898, heroin was marketed by the Bayer Drug Company; it was freely available without prescription and was widely advertised as a superior kind of aspirin. Tests showed that it was a more potent analgesic than morphine and that it was less likely to induce nausea and vomiting. Moreover, the Bayer Drug Company, on the basis of flimsy evidence, claimed that heroin was not addictive; this is why it was not covered by the Harrison Narcotics Act. The consequence of omitting heroin from the Harrison Narcotics Act was that opiate addicts in the United States, forbidden by law to use opium or morphine, turned to the readily available and much more potent heroin—and the flames of addiction were further fanned. In 1924, the U.S. Congress made it illegal for anybody to possess, sell, or use heroin. Unfortunately, the laws enacted to stamp out opiate addiction in the United States have been far from successful: An esti-mated 136,000 Americans currently use heroin (National Survey on Drug Use and Health, 2005), and organized crime flourishes on the proceeds.

The effect of opiates most valued by addicts is the *rush* that follows intravenous injection. The *heroin rush* is a wave of intense abdominal, orgasmic pleasure that evolves into a state of serene, drowsy euphoria. Many opiate users, drawn by these pleasurable effects, begin to use the drug more and more frequently. Then, once they reach a point where they keep themselves drugged much of the time, tolerance and physical dependence develop and contribute to the problem. Opiate tolerance encourages addicts to progress to higher doses, to more potent drugs (e.g., heroin), and to more direct routes of administration (e.g., IV injection); and physical dependence adds to the already high motivation to take the drug.

The classic opiate withdrawal syndrome usually begins 6 to 12 hours after the last dose. The first withdrawal sign is typically an increase in restlessness; the addict begins to pace and fidget. Watering eyes, running nose, yawning, and sweating are also common during the early stages of opiate withdrawal. Then, the addict often falls into a fitful sleep, which typically lasts for several hours. Once the person wakes up, the original symptoms may be joined in extreme cases by chills, shivering, profuse sweating, gooseflesh, nausea, vomiting, diarrhea, cramps, dilated pupils, tremor, and muscle pains and spasms. The gooseflesh skin and leg spasms of the opiate withdrawal syndrome are the basis for the expressions "going cold turkey" and "kicking the habit." The symptoms of opiate withdrawal are typically most severe in the second or third day after the last injection, and by the seventh day they have all but disappeared. In short, opiate withdrawal is about as serious as a bad case of the flu:

> Opiate withdrawal is probably one of the most misunderstood aspects of drug use. This is largely because of the image of withdrawal that has been portrayed in the movies and popular literature for many years.... Few addicts . . . take enough drug to cause the . . . severe withdrawal symptoms that are shown in the movies. Even in its most severe form, however, opiate withdrawal is not as dangerous or terrifying as withdrawal from barbiturates or alcohol. (McKim, 1986, p. 199)

Although opiates are highly addictive, the direct health hazards of chronic exposure are surprisingly minor. The main direct risks are constipation, pupil constriction, menstrual irregularity, and reduced libido (sex drive). Many opiate addicts have taken pure heroin or morphine for years with no serious ill effects. In fact, opiate addiction is more prevalent among doctors, nurses, and dentists than among other professionals (e.g., Brewster, 1986):

Clinical Implications

> An individual tolerant to and dependent upon an opiate who is socially or financially capable of obtaining an adequate supply of good quality drug, sterile syringes and needles, and other paraphernalia may maintain his or her

proper social and occupational functions, remain in fairly good health, and suffer little serious incapacitation as a result of the dependence. (Julien, 1981, p. 117)

One such individual was Dr. William Steward Halsted, one of the founders of Johns Hopkins Medical School and one of the most brilliant surgeons of his day . . . known as "the father of modern surgery." And yet, during his career he was addicted to morphine, a fact that he was able to keep secret from all but his closest friends. In fact, the only time his habit caused him any trouble was when he was attempting to reduce his dosage. (McKim, 1986, p. 197)

Most medical risks of opiate addiction are indirect—that is, not entirely attributable to the drug itself. Many of the medical risks arise out of the battle between the relentless addictive power of opiates and the attempts of governments to eradicate addiction by making opiates *Thinking Creatively* illegal. The opiate addicts who cannot give up their habit—treatment programs report success rates of only about 10%—are caught in the middle. Because most opiate addicts must purchase their morphine and heroin from illicit dealers at greatly inflated prices, those who are not wealthy become trapped in a life of poverty and petty crime. They are poor, they are undernourished, they receive poor medical care, they are often driven to prostitution, and they run great risk of contracting AIDS and other infections (e.g., hepatitis, syphilis, and gonorrhea) from unsafe sex and unsterile needles. Moreover, they never know for sure what they are injecting: Some street drugs are poorly processed, and virtually all have been *cut* (stretched by the addition of some similar-appearing substance) to some unknown degree.

Death from heroin overdose is a serious problem—high doses of heroin kill by suppressing breathing (Megarbane et al., 2005). However, death from heroin overdose is not well understood. The following are three points of confusion:

- Medical examiners often attribute death to heroin overdose without assessing blood levels of heroin. Careful toxicological analysis at autopsy often reveals that this diagnosis is questionable (Poulin, Stein, & Butt, 2000). In many cases, the deceased have low levels of heroin in the blood and high levels of other CNS depressants such as alcohol and benzodiazepines. In short, many so-called heroin overdose deaths appear to be a product of drug interaction (Darke et al., 2000; Darke & Zador, 1996; Mirakbari, 2004).

- Some deaths from heroin overdose are a consequence of its legal status. Because addicts are forced to buy their drugs from criminals, they never know for sure what they are buying. Reports of death from heroin overdose occur when a shipment of heroin hits the street that has been cut by a toxic substance or when the heroin is more pure than normal (Darke et al.,1999; Mcgregor et al.,1998).

- In the United States, deaths from opiate overdose have increased precipitously in the last few years, and many people attribute this increase to heroin.

However, the sharp increase is almost entirely due to legal synthetic opioid analgesics such as Oxycontin and Lorcet (Manchikanti, 2007; Paulozzi, Budnitz, & Xi, 2006).

The primary treatment for heroin addiction in most countries is methadone. Ironically, methadone is itself an opiate with many of the same adverse effects as heroin. However, because methadone produces less pleasure than heroin, the strategy has been to block heroin withdrawal effects with methadone and then maintain addicts on methadone until they can be weaned from it. Methadone replacement has been shown to improve the success rate of some treatment programs, but its adverse effects and the high drop-out rates from such programs are problematic (see Zador, 2007). *Buprenorphine* is an alternative treatment for heroin addiction. Buprenorphine has a high and long-lasting affinity for opiate receptors and thus blocks the effects on the brain of other opiates, without producing powerful euphoria. Studies suggest that it is as effective as methadone (see Davids & Gaspar, 2004; Gerra et al., 2004).

In 1994, the Swiss government took an alternative approach to the problem of heroin addiction—despite substantial opposition from the Swiss public. It established a series of clinics in which, as part of a total treatment package, Swiss heroin addicts could receive heroin injections from a physician for a small fee. The Swiss government wisely funded a major research program to evaluate the clinics (see Gschwend et al., 2002). The results have been uniformly positive. Once they had a reliable source of heroin, most addicts gave up their crim- *Clinical Implications* inal lifestyles, and their health improved once they were exposed to the specialized medical and counseling staff at the clinics. Many addicts returned to their family and jobs, and many opted to reduce or curtail their heroin use. As a result, addicts are no longer a presence in Swiss streets and parks; drug-related crime has substantially declined,and the physical and social well-being of the addicts has greatly improved. Furthermore, the number of new heroin addicts has declined, apparently because once addiction becomes treated as an illness, it becomes less cool (see Brehmer & Iten, 2001; De Preux, Dubois-Arber, & Zobel, 2004; Gschwend et al., 2003; Nordt & Stohler, 2006; Rehm et al., 2001).

These positive results have led to the establishment of similar experimental programs in other countries (e.g., Canada, Norway, Netherlands, and Germany) with similar success (see Skeie et al., 2008; Yan, 2009). Furthermore, safe injection facilities have managed to reduce the spread of infection and death from heroin overdose in many cities (e.g., Milloy et al., 2008). Given the unqualified success of such pro- *Thinking Creatively* grams in dealing with the drug problem, it is interesting to consider why some governments have not adopted them (see Fischer et al., 2007). What do you think?

Comparison of the Hazards of Tobacco, Alcohol, Marijuana, Cocaine, and Heroin

One way of comparing the adverse effects of tobacco, alcohol, marijuana, cocaine, and heroin is to compare the

Thinking Creatively

prevalence of their use in society as a whole. In terms of this criterion, it is clear that tobacco and alcohol have a greater negative impact than do marijuana, cocaine, and heroin (see Figure 15.6). Another method of comparison is one based on death rates: Tobacco has been implicated in the deaths of approximately 400,000 Americans per year; alcohol, in approximately 80,000 per year; and all other drugs combined, in about 25,000 per year.

But what about the individual drug user? Who is taking greater health risks: the cigarette smoker, the alcohol drinker, the marijuana smoker, the cocaine user, or the heroin user? You now have the information to answer this question. Complete the Scan Your Brain, which will help you appreciate the positive impact that studying 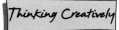 *Thinking Creatively* biopsychology is having on your understanding of important issues. Would you have ranked the health risks of these drugs in the same way before you began this chapter? How have the laws, or lack thereof, influenced the hazards associated with the five drugs?

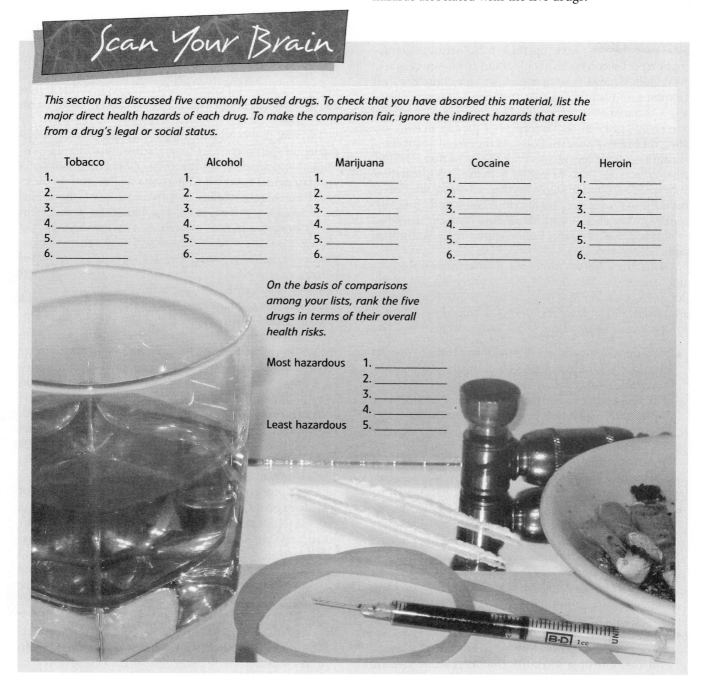

Scan Your Brain

This section has discussed five commonly abused drugs. To check that you have absorbed this material, list the major direct health hazards of each drug. To make the comparison fair, ignore the indirect hazards that result from a drug's legal or social status.

Tobacco	Alcohol	Marijuana	Cocaine	Heroin
1. _____	1. _____	1. _____	1. _____	1. _____
2. _____	2. _____	2. _____	2. _____	2. _____
3. _____	3. _____	3. _____	3. _____	3. _____
4. _____	4. _____	4. _____	4. _____	4. _____
5. _____	5. _____	5. _____	5. _____	5. _____
6. _____	6. _____	6. _____	6. _____	6. _____

On the basis of comparisons among your lists, rank the five drugs in terms of their overall health risks.

Most hazardous   1. _____
                 2. _____
                 3. _____
                 4. _____
Least hazardous 5. _____

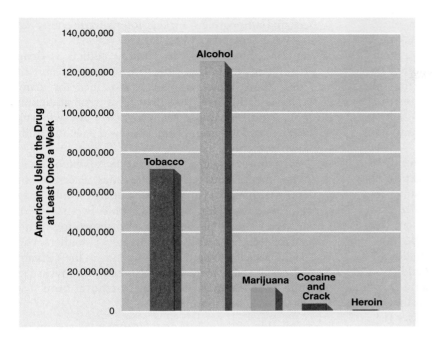

who are no longer experiencing withdrawal symptoms.

The failure of detoxification as a treatment for addiction is not surprising, for two reasons. First, some highly addictive drugs, such as cocaine and amphetamines, do not produce severe withdrawal distress (see Gawin, 1991). Second, the pattern of drug taking routinely displayed by many addicts involves an alternating cycle of binges and detoxification (Mello & Mendelson, 1972). There are a variety of reasons for this pattern of drug use. For example, some addicts adopt it because weekend binges are compatible with their work schedules, others adopt it because they do not have enough money to use drugs continuously, others have it forced on them because their binges often land them in jail, and others have it forced on them by their repeated unsuccessful efforts to shake their habit. However, whether detoxification is by choice or necessity, it does not stop addicts from renewing their drug-taking habits (see Leshner, 1997).

As a result of these problems with physical-dependence theories of addiction, a different approach began to predominate in the 1970s and 1980s (see Higgins, Heil, & Lussier, 2004). This approach was based on the assumption that most addicts take drugs not to escape or to avoid the unpleasant consequences of withdrawal, but rather to obtain the drugs' positive effects. Theories of addiction based on this premise are called **positive-incentive theories of addiction**. They hold that the primary factor in most cases of addiction is the craving for the positive-incentive (expected pleasure-producing) properties of the drug.

There is no question that physical dependence does play a role in addiction: Addicts do sometimes consume the drug to alleviate their withdrawal symptoms. However, most researchers now assume that the primary factor in addiction is the drugs' *hedonic* (pleasurable) effects (see Cardinal & Everitt, 2004; Everitt, Dickinson, & Robbins, 2001). All drugs with addiction potential have some pleasurable effects for users.

From Pleasure to Compulsion: Incentive-Sensitization Theory

To be useful, positive-incentive theories of drug addiction need to offer explanations for two puzzling aspects of drug addiction. First, they must explain why there is often such a big difference between the hedonic value of drug taking and the positive-incentive value of drug taking.

Biopsychological Approaches to Theories of Addiction

This section of the chapter introduces two diametrically different ways of thinking about an addiction: Are addicts driven to take drugs by an internal need, or are they drawn to take drugs by the anticipated positive effects? I am sure you will recognize, after having read the preceding chapters, that this is the same fundamental question that has been the focus of biopsychological research on the motivation to eat and sleep.

Physical-Dependence and Positive-Incentive Perspectives of Addiction

Early attempts to explain the phenomenon of drug addiction attributed it to physical dependence. According to various **physical-dependence theories of addiction**, physical dependence traps addicts in a vicious circle of drug taking and withdrawal symptoms. The idea was that drug users whose intake has reached a level sufficient to induce physical dependence are driven by their withdrawal symptoms to self-administer the drug each time they attempt to curtail their intake.

Early drug addiction treatment programs were based on the physical-dependence perspective. They attempted to break the vicious circle of drug taking by gradually withdrawing drugs from addicts in a hospital environment. Unfortunately, once discharged, almost all detoxified addicts return to their former drug-taking habits—**detoxified addicts** are addicts who have no drugs in their bodies and

Positive-incentive value refers specifically to the anticipated pleasure associated with an action (e.g., taking a drug), whereas **hedonic value** refers to the amount of pleasure that is actually experienced. Addicts often report a huge discrepancy between them: Although they are compulsively driven to take their drug by its positive-incentive value (i.e., by the anticipated pleasure), taking the drug is often not as pleasurable as it once was (see Ahmed, 2004; Redish, 2004).

The second challenge faced by positive-incentive theories of drug addiction is that they must explain the process that transforms a drug user into a drug addict. Many people periodically use addictive drugs and experience their hedonic effects without becoming addicted to them (see Everitt & Robbins, 2005; Kreek et al., 2005). What transforms some drug users into compulsive users, or addicts?

The **incentive-sensitization theory** of drug addiction meets these two challenges (see Berridge, Robinson, & Aldridge, 2009). The central tenet of this theory is that the positive-incentive value of addictive drugs increases (i.e., is sensitized) with drug use (see Miles et al., 2004). Robinson and Berridge (2003) have suggested that in addiction-prone individuals, the use of a drug sensitizes the drug's positive-incentive value, thus rendering such individuals highly motivated to seek and consume the drug. A key point of Robinson and Berridge's incentive-sensitization theory is that it isn't the pleasure (liking) of taking the drug that is the basis of addiction; it is the anticipated pleasure (wanting) of drug taking (i.e., the drug's positive-incentive value). Initially, a drug's positive-incentive value is closely tied to its pleasurable effects; but tolerance often develops to the pleasurable effects, whereas the addict's wanting for the drug is sensitized. Thus, in chronic addicts, the positive-incentive value of the drug is often out of proportion with the pleasure actually derived from it: Many addicts are miserable, their lives are in ruins, and the drug effects are not that great anymore; but they crave the drug more than ever.

Relapse and Its Causes

The most difficult problem in treating drug addicts is not getting them to stop using their drug. The main problem is preventing those who stop from relapsing. The propensity to **relapse** (to return to one's drug taking habit after a period of voluntary abstinence), even after a long period of voluntary abstinence, is a hallmark of addiction. Thus, understanding the causes of relapse is one key to understanding addiction and its treatment.

Three fundamentally different causes of relapse in drug addicts have been identified (see Shaham & Hope, 2005):

- Many therapists and patients point to stress as a major factor in relapse. The impact of stress on drug taking was illustrated in a dramatic fashion by the marked increases in cigarette and alcohol consumption that occurred among New Yorkers following the terrorist attacks of September 11, 2001.
- Another cause of relapse in drug addicts is **drug priming** (a single exposure to the formerly abused drug). Many addicts who have abstained for many weeks, and thus feel that they have their addiction under control, sample their formerly abused drug just once and are immediately plunged back into full-blown addiction.
- A third cause of relapse in drug addicts is exposure to environmental cues (e.g., people, times, places, or objects) that have previously been associated with drug taking (see Concklin, 2006; Di Ciano & Everitt, 2003). Such environmental cues have been shown to precipitate relapse. The fact that the many U.S. soldiers who became addicted to heroin while fighting in the Vietnam War easily shed their addiction when they returned home has been attributed to their removal from that drug-associated environment.

Explanation of the effects of environmental cues on relapse is related to our discussion of conditioned drug tolerance earlier in the chapter (see Kauer & Malenka, 2007). You may recall that cues that predict drug exposure come to elicit conditioned compensatory responses through a Pavlovian conditioning mechanism, and because conditioned compensatory responses are usually opposite to the original drug effects, they produce tolerance. The point here is that these same conditioned compensatory responses seem to increase craving in abstinent drug addicts and, in so doing, trigger relapse. Moreover, because interoceptive cues have been shown to function as conditional stimuli in conditioned tolerance experiments, they can also induce craving—that is why just thinking about drugs is enough to induce craving and relapse. Because susceptibility to relapse is a defining feature of drug addicts, conditioned drug responses play a major role in most modern theories of drug addiction (see Day & Carelli, 2007; Hellemans, Dickinson, & Everitt, 2006; Hyman, Malenka, & Nestler, 2006).

So far in this chapter, you have been introduced to the principles of drug action, the role of learning in drug tolerance, five common addictive drugs, and theories of drug addiction. This is a good place to pause and reinforce what you have learned. In each blank, write the appropriate term. The correct answers are provided at the end of the exercise. Review material related to your errors and omissions before proceeding.

1. Drugs that affect the nervous system and behavior are called _____ drugs.
2. The most dangerous route of drug administration is _____ injection.
3. Drug tolerance is of two different types: metabolic and _____.
4. An individual who displays a withdrawal syndrome when intake of a drug is curtailed is said to be _____ on that drug.
5. The before-and-after design is used to study _____ drug tolerance.
6. The fact that drug tolerance is often _____ suggests that Pavlovian conditioning plays a major role in addiction.
7. _____ disease provides a compelling illustration of nicotine's addictive power.
8. Convulsions and hyperthermia are symptoms of withdrawal from _____.
9. Anandamide was the first endogenous _____ to be identified.
10. Cocaine sprees can produce cocaine psychosis, a syndrome that is similar to paranoid _____.
11. Morphine and codeine are constituents of _____.
12. _____ is a semisynthetic opiate that penetrates the blood–brain barrier more effectively than morphine.
13. _____ heroin addicts were among the first to legally receive heroin injections from a physician for a small fee.
14. Many current theories of addiction focus on the _____ of addictive drugs.

Scan Your Brain answers: (1) psychoactive, (2) intravenous (or IV), (3) functional, (4) physically dependent, (5) contingent, (6) situationally specific, (7) Buerger's, (8) alcohol, (9) cannabinoid, (10) schizophrenia, (11) opium, (12) Heroin, (13) Swiss, (14) positive-incentive value.

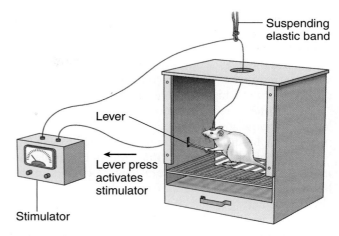

FIGURE 15.7 A rat pressing a lever to obtain rewarding brain stimulation.

Olds and Milner (1954), the discoverers of intracranial self-stimulation, argued that the specific brain sites that mediate self-stimulation are those that normally mediate the pleasurable effects of natural rewards (i.e., food, water, and sex). Accordingly, researchers studied the self-stimulation of various brain sites in order to map the neural circuits that mediate the experience of pleasure.

Fundamental Characteristics of Intracranial Self-Stimulation

It was initially assumed that intracranial self-stimulation was a unitary phenomenon—that is, that its fundamental properties were the same regardless of the site of stimulation. Most early studies of intracranial self-stimulation involved septal or lateral hypothalamic stimulation because the rates of self-stimulation from these sites are spectacularly high: Rats typically press a lever thousands of times per hour for stimulation of these sites, stopping only when they become exhausted. However, self-stimulation of many other brain structures has been documented.

Early studies of intracranial self-stimulation suggested that lever pressing for brain stimulation was fundamentally different from lever pressing for natural reinforcers such as food or water. Two puzzling observations contributed to this view. First, despite their extremely high response rates, many rats stopped pressing the self-stimulation lever *Evolutionary Perspective* almost immediately when the current delivery mechanism was shut off. This finding was puzzling because high rates of operant responding are generally assumed to indicate that the reinforcer is particularly pleasurable, whereas rapid rates of extinction are usually assumed to indicate that it is not. Would you stop pressing a lever that had been delivering $100 bills the first few times that a press did not produce one? Second, experienced self-stimulators often did not recommence lever pressing

15.5
Intracranial Self-Stimulation and the Pleasure Centers of the Brain

Rats, humans, and many other species will administer brief bursts of weak electrical stimulation to specific sites in their own brains (see Figure 15.7). This phenomenon is known as **intracranial self-stimulation (ICSS)**, and the brain sites capable of mediating the phenomenon are often called *pleasure centers*. When research on addiction turned to positive incentives in the 1970s and 1980s, what had been learned about the neural mechanisms of pleasure from studying intracranial self-stimulation served as a starting point for the study of the neural mechanisms of addiction.

when they were returned to the apparatus after being briefly removed from it. In such cases, the rats had to be **primed** to get them going again: The experimenter simply pressed the lever a couple of times, to deliver a few free stimulations, and the hesitant rat immediately began to self-stimulate at a high rate once again.

These differences between lever pressing for rewarding lateral hypothalamic or septal stimulation and lever pressing for food or water seemed to discredit Olds and Milner's original theory that intracranial self-stimulation involves the activation of natural reward circuits in the brain. However, several lines of research indicate that the circuits mediating intracranial self-stimulation are natural reward circuits. Let's consider three of these.

First, brain stimulation through electrodes that mediate self-stimulation often elicits a natural motivated behavior such as eating, drinking, or copulation in the presence of the appropriate goal object. Second, producing increases in natural motivation (for example, by food or water deprivation, by hormone injections, or by the presence of prey objects) often increases self-stimulation rates.

The third point is a bit more complex: It became clear that differences between the situations in which the rewarding effects of brain stimulation and those of natural rewards were usually studied contribute to the impression that these effects are qualitatively different. For example, comparisons between lever pressing for food and lever pressing for brain stimulation are usually confounded by the fact that subjects pressing for brain stimulation are nondeprived and by the fact that the lever press delivers the reward directly and immediately. In contrast, in studies of lever pressing for natural rewards, subjects are often

Thinking Creatively

deprived, and they press a lever for a food pellet or a drop of water, which they must then approach and consume to experience the rewarding effects. This point was illustrated by a clever experiment (Panksepp & Trowill, 1967) that compared lever pressing for brain stimulation and lever pressing for a natural reinforcer in a situation in which the usual confounds were absent. In the absence of the confounds, some of the major differences between lever pressing for food and lever pressing for brain stimulation disappeared. When nondeprived rats pressed a lever to inject a small quantity of chocolate milk directly into their mouths through an intraoral tube, they behaved remarkably like self-stimulating rats: They quickly learned to press the lever, they pressed at high rates, they extinguished quickly, and some even had to be primed.

Mesotelencephalic Dopamine System and Intracranial Self-Stimulation

The mesotelencephalic dopamine system plays an important role in intracranial self-stimulation. The **mesotelencephalic dopamine system** is a system of dopaminergic neurons that projects from the mesencephalon (the midbrain) into various regions of the telencephalon. As Figure 15.8 indicates, the neurons that compose the mesotelencephalic dopamine system have their cell bodies in two midbrain nuclei—the **substantia nigra** and the **ventral tegmental area**. Their axons project to a variety of telencephalic sites, including specific regions of the prefrontal neocortex, the limbic cortex, the olfactory tubercle, the amygdala, the septum, the dorsal striatum, and, in particular, the **nucleus accumbens** (nucleus of the ventral striatum)—see Zahm, 2000.

Most of the axons of dopaminergic neurons that have their cell bodies in the substantia nigra project to the dorsal striatum; this component of the mesotelencephalic dopamine system is called the *nigrostriatal pathway*. It is degeneration in this pathway that is associated with Parkinson's disease.

Most of the axons of dopaminergic neurons that have their cell bodies in the ventral tegmental area project to various cortical and limbic sites. This component of the mesotelencephalic

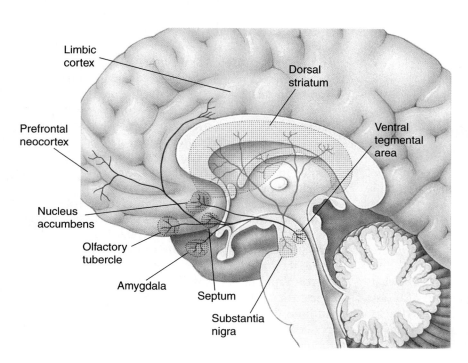

FIGURE 15.8   The mesotelencephalic dopamine system in the human brain, consisting of the nigrostriatal pathway (green) and the mesocorticolimbic pathway (red). (Based on Klivington, 1992.)

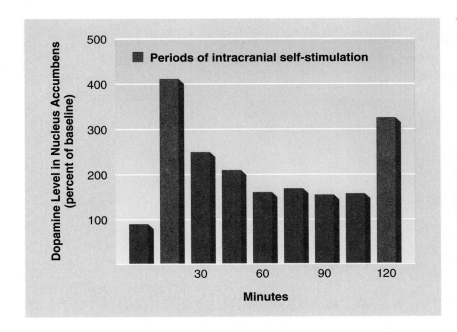

dopamine system is called the *mesocorticolimbic pathway*. Although there is some intermingling of the neurons between these two dopaminergic pathways, it is the particular neurons that project from the ventral tegmental area to the nucleus accumbens that have been most frequently implicated in the rewarding effects of brain stimulation, natural rewards, and addictive drugs.

Several pieces of evidence have supported the view that the mesocorticolimbic pathway of the mesotelencephalic dopamine system plays an important role in mediating intracranial self-stimulation. The following are four of them:

- Many of the brain sites at which self-stimulation occurs are part of the mesotelencephalic dopamine system.
- Intracranial self-stimulation is often associated with an increase in dopamine release in the mesocorticolimbic pathway (Hernandez et al., 2006). See Figure 15.9.
- Dopamine agonists tend to increase intracranial self-stimulation, and dopamine antagonists tend to decrease it.
- Lesions of the mesocorticolimbic pathway tend to disrupt intracranial self-stimulation.

15.6

Early Studies of Brain Mechanisms of Addiction: Dopamine

The positive-incentive value of drug taking had been implicated in addiction, and the experience of pleasure had been linked to the mesocorticolimbic pathway. It was natural, therefore, that the first sustained efforts to discover the neural

mechanisms of drug addiction should focus on the mesocorticolimbic pathway.

In considering the neural mechanisms of drug addiction, it is important to appreciate that specific brain mechanisms could not *Thinking Creatively* possibly have evolved for the purpose of mediating addiction—drug addiction is not adaptive. Thus, the key to understanding the neural *Evolutionary Perspective* mechanisms of addiction lies in understanding natural motivational mechanisms and how they are co-opted and warped by addictive drugs (Nesse & Berridge, 1997).

Two Key Methods for Measuring Drug–Produced Reinforcement in Laboratory Animals

Most of the research on the neural mechanisms of addiction has been conducted in nonhumans. Because of the presumed role of the positive-incentive value of drugs in addiction, methods used to measure the rewarding effects of drugs in the nonhuman subjects have played a key role in this research. Two such methods have played particularly important roles: the drug self-administration paradigm and the conditioned place-preference paradigm (see Aguilar, Rodríguez-Arias, & Miñarro, 2008; Sanchis-Segura & Spanagel, 2006). They are illustrated in Figure 15.10.

In the **drug self-administration paradigm**, laboratory rats or primates press a lever to inject drugs into themselves through implanted *cannulas* (thin tubes). They readily learn to self-administer intravenous injections of drugs to which humans become addicted. Furthermore, once they have learned to self-administer an addictive drug, their drug taking often mimics in major respects the drug taking of human addicts (Deroche-Gamonet, Belin, & Piazza, 2004; Louk Vanderschuren & Everitt, 2004; Robinson, 2004). Studies in which microinjections have been self-administered directly into particular brain structures have proved particularly enlightening.

In the **conditioned place-preference paradigm**, rats repeatedly receive a drug in one compartment (the *drug compartment*) of a two-compartment box. Then, during the test phase, the drug-free rat is placed in the box, and the

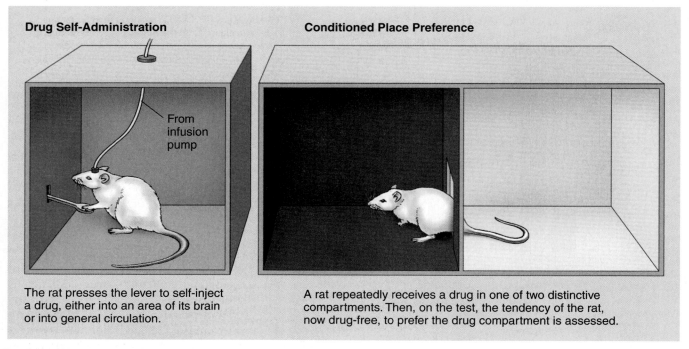

Drug Self-Administration

From infusion pump

The rat presses the lever to self-inject a drug, either into an area of its brain or into general circulation.

Conditioned Place Preference

A rat repeatedly receives a drug in one of two distinctive compartments. Then, on the test, the tendency of the rat, now drug-free, to prefer the drug compartment is assessed.

FIGURE 15.10 Two behavioral paradigms that are used extensively in the study of the neural mechanisms of addiction: the drug self-administration paradigm and the conditioned place-preference paradigm.

proportion of time it spends in the drug compartment, as opposed to the equal-sized but distinctive *control compartment*, is measured. Rats usually prefer the drug compartment over the control compartment when the drug compartment has been associated with the effects of drugs to which humans become addicted. The main advantage of the conditioned place-preference paradigm is that the subjects are tested while they are drug-free, which means that the measure of the incentive value of a drug is not confounded by other effects the drug might have on behavior.

Early Evidence of the Involvement of Dopamine in Drug Addiction

In the 1970s, following much research on the role of dopamine in intracranial self-stimulation, experiments began to implicate dopamine in the rewarding effects of natural reinforcers and addictive drugs. For example, in rats, dopamine antagonists blocked the self-administration of, or the conditioned preference for, several different addictive drugs; and they reduced the reinforcing effects of food. These findings suggested that dopamine signaled something akin to reward value or pleasure.

Evolutionary Perspective

The Nucleus Accumbens and Drug Addiction

Once evidence had accumulated linking dopamine to natural reinforcers and drug-induced reward, investigators began to explore particular sites in the mesocorticolimbic dopamine pathway by conducting experiments on laboratory animals. Their findings soon focused attention on the nucleus accumbens. Events occurring in the nucleus accumbens and dopaminergic input to it from the ventral tegmental area appeared to be most clearly related to the experience of reward and pleasure.

The following are four kinds of findings from research on laboratory animals that focused attention on the nucleus accumbens (see Deadwyler et al., 2004; Nestler, 2005; Pierce & Kumaresan, 2006):

Evolutionary Perspective

- Laboratory animals self-administered microinjections of addictive drugs (e.g., cocaine, amphetamine, and morphine) directly into the nucleus accumbens.
- Microinjections of addictive drugs into the nucleus accumbens produced a conditioned place preference for the compartment in which they were administered.
- Lesions to either the nucleus accumbens or the ventral tegmental area blocked the self-administration of drugs into general circulation or the development of drug-associated conditioned place preferences.
- Both the self-administration of addictive drugs and the experience of natural reinforcers were found to be associated with elevated levels of extracellular dopamine in the nucleus accumbens.

Support for the Involvement of Dopamine in Addiction: Evidence from Imaging Human Brains

With the development of brain-imaging techniques for measuring dopamine in human brains, considerable evidence began to emerge that dopamine is involved in human reward in general and human addiction in particular (see O'Doherty, 2004; Volkow et al., 2004). One of the strongest of the early brain-imaging studies linking dopamine to addiction was published by Volkow and colleagues (1997). They administered various doses of radioactively labeled cocaine to addicts and asked the addicts to rate the resulting "high." They also used positron emission tomography (PET) to measure the degree to which the labeled cocaine bound to dopamine transporters. As you learned earlier in this chapter, cocaine has its agonistic effects on dopamine by binding to these transporters, blocking reuptake, and thus increasing extracellular dopamine levels. The intensity of the "highs" experienced by the addicts was correlated with the degree to which cocaine bound to the dopamine transporters—no high at all was experienced unless the drug bound to 50% of the dopamine transporters.

Brain-imaging studies have also indicated that the nucleus accumbens plays an important role in mediating the rewarding effects of addictive behavior. For example, in one study, healthy (i.e., nonaddicted) human subjects were given an IV injection of amphetamine (Drevets et al., 2001). As dopamine levels in the nucleus accumbens increased in response to the amphetamine injection, the subjects reported a parallel increase in their experience of euphoria.

In general, brain-imaging studies have shown that dopamine function is markedly diminished in human addicts—see Figure 15.11 (Volkow et al., 2009). However, when addicts are exposed to their drug or to stimuli associated with their drug, the nucleus accumbens and some of the other parts of the mesocorticolimbic dopamine pathway tend to become hyperactive.

Dopamine Release in the Nucleus Accumbens: What Is Its Function?

As you have just learned, substantial evidence links dopamine release, particularly in the nucleus accumbens, to the rewarding effects of addictive drugs and other reinforcers (see Kelley, 2004; Nestler & Malenka, 2004). But, reward is a complex process, with many different psychological components (see Berridge & Robinson, 2003): What exactly is the role in reward of dopamine release in the nucleus accumbens?

Several studies have found increases in extracellular dopamine levels in the nucleus accumbens following the presentation of a natural reward (e.g., food), rewarding brain stimulation (Hernandez et al., 2007), or an addictive

Cocaine

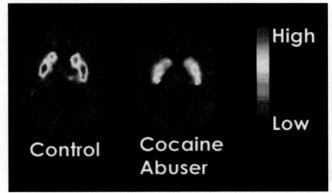

Methamphetamine

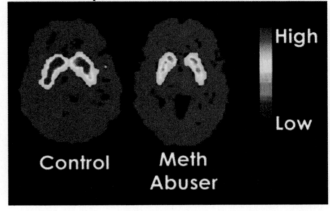

FIGURE 15.11 Chronic use of cocaine and methamphetamine reduces binding of radioactive tracers to D_2 receptors in the striatum. (From Volkow et al., 2009.)

drug (see Joseph, Datla, & Young, 2003; Ungless, 2004). Even stronger evidence for the idea that increased dopamine levels in the nucleus accumbens are related to the *experience* of reward came from the finding that ventral tegmental neurons, which release their dopamine into the nucleus accumbens, fire in response to a stimulus at a rate proportional to its reward value. Other studies have suggested that dopamine released in the nucleus accumbens is related to the *expectation* of reward, rather than to its experience. For example, some studies have shown that neutral stimuli that signal the impending delivery of a reward (e.g., food or an addictive drug) can trigger dopamine release in the nucleus accumbens (e.g., Fiorino, Coury, & Phillips, 1997; Weiss et al., 2000).

A third theory about dopamine release in the nucleus accumbens encompasses and extends the experience-of-reward and expectation-of-reward theories (Caplin & Dean, 2008). The theory was proposed by Tobler, Forillo, and Schultz (2005). They found that dopaminergic neurons with their cell bodies in the ventral tegmental area fire at a rate related to the value of the reward. When the expected reward was delivered, there was no change in firing rate; when a greater than expected reward was delivered, firing increased; and when a less than expected reward was delivered, firing decreased. Thus, dopamine release in the nucleus accumbens reflected both the experience and expectation of reward, but not in a straightforward fashion: It seemed to reflect discrepancies between expected and actual rewards (see Fiorillo, Newsome, & Schultz, 2008; Schultz, 2007).

15.7
Current Approaches to Brain Mechanisms of Addiction

The previous three sections of the chapter have brought us from the beginnings of research on the brain mechanisms of addiction to current research, which will be discussed in this section. Figure 15.12 summarizes the major shifts in thinking about the brain mechanisms of addiction that have occurred over time.

Figure 15.12 shows that two lines of thinking about the brain mechanisms of addiction both had their origins in classic research on drug tolerance and physical dependence. One line developed into physical-dependence theories of addiction, which though appealing in their simplicity, proved to be inconsistent with the evidence, and these inconsistencies led to the emergence of positive-incentive theories. In turn, the positive-incentive approach to addiction, in combination with research on dopamine and pleasure centers in the brain, led to a focus on the mesocorticolimbic pathway and the mechanisms of reward. The second line of thinking about the brain mechanisms of addiction also began with early research on drug tolerance and physical dependence. This line moved ahead with the discovery that drug-associated cues come to elicit conditioned compensatory responses through a Pavlovian conditioning mechanism and that these conditioned responses are largely responsible for drug tolerance. This finding gained further prominence when researchers discovered that conditioned responses elicited by drug-associated cues were major factors in drug craving and relapse.

These two lines of research together have shaped modern thinking about the brain mechanisms of addiction, but this is not the end of the story (see Koob, 2006). In this, the final section of the chapter, you will learn about issues that are the focus of current research on the brain mechanisms of addiction and about new areas of the brain that have been linked to drug addiction.

Current Issues in Addiction Research

The early decades of research on addiction and its neural mechanisms clarified a number of issues about addiction, but it raised others that are the focus of current research. The following are four of these issues:

Addiction Is Psychologically Complex Studies of addicted patients have found that drug addicts differ psychologically from healthy controls in a variety of ways. Drug addicts have been found to make poor decisions, to engage in excessive risk taking, and to have deficits in self control (see Baler & Volkow,

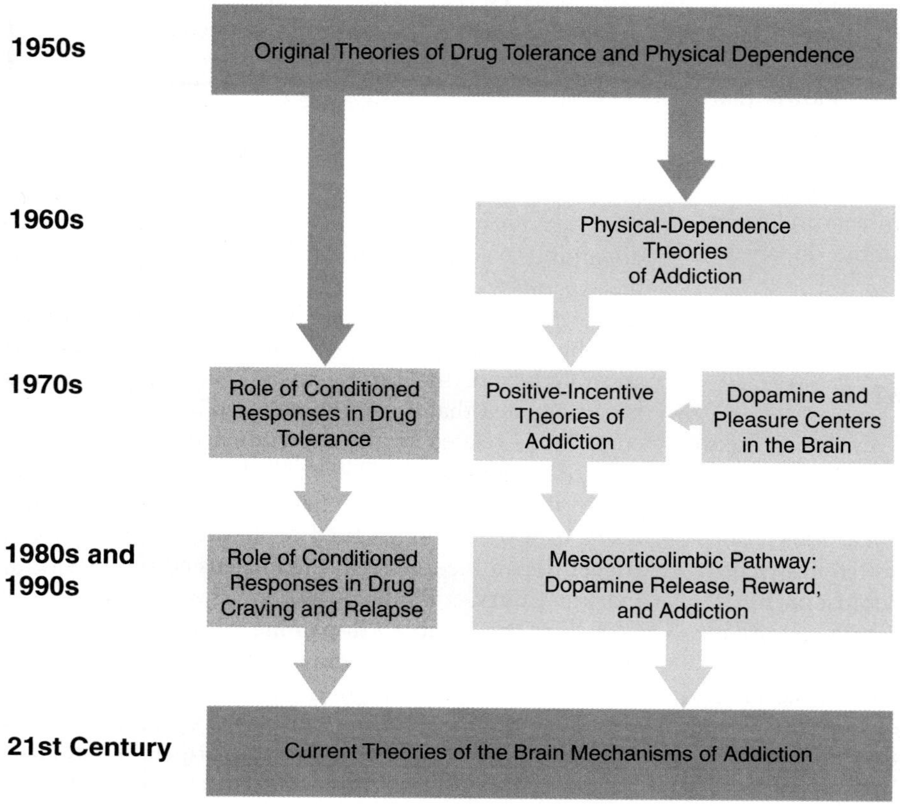

FIGURE 15.12 Historic influences that shaped current thinking about the brain mechanisms of addiction.

2006; Diekof, Falkai, & Gruber, 2008; Yucel & Lubman, 2007). There is a growing appreciation for the fact that efforts to develop theories of drug addiction and effective treatments for it must take these psychological differences into account.

Addiction Is a Disturbance of Decision Making   There has been an increasing appreciation that the primary symptom of addiction is a disturbance of decision making: Why do addicts decide to engage in harmful behaviors? This has had two beneficial effects: It has led investigators who study drug addiction to consider research on decision making from other fields (e.g., economics and social psychology), and it has led investigators in these other fields to consider research on drug addiction (e.g., Cardinal & Everitt, 2004; Lieberman & Eisenberger, 2009; Sanfey, 2007).

Addiction Is Not Limited to Drugs   There has been a growing consensus that drug addiction is a specific expression of a more general problem and that other behaviors exhibit the defining feature of drug addiction: the inability to refrain from a behavior despite its adverse effects. A lot of attention has recently been paid to overeating as an addiction because of its major adverse health consequences (e.g., Di Chiara & Bassareo, 2007; Trinko et al., 2007), but compulsive gambling, compulsive sexual behavior, kleptomania (compulsive shoplifting), and compulsive shopping also seem to share some brain mechanisms with drug addiction (see Grant, Brewer, & Potenza, 2006; Tanabe et al., 2007).

Addiction Involves Many Neurotransmitters   The evidence implicating dopamine in addiction is diverse and substantial, but it is not possible for any complex behavior to be the product of a single neurotransmitter. Some evidence has pointed to a role for glutamate in addiction (Kalivas, 2004)—of particular interest are prefrontal glutaminergic neurons that project into the nucleus accumbens. Also of interest to researchers are endogenous opioids, norepinephrine, GABA, and endocannabinoids (see Koob, 2006; Weinshenker & Schroeder, 2007).

Brain Structures That Mediate Addiction: The Current View

Although researchers do not totally agree about the neural mechanisms of drug addiction, there seems to be an emerging consensus that areas of the brain other than the nucleus accumbens are involved in its three stages (see Everitt & Robbins, 2005; Koob, 2006): (1) initial drug taking, (2) the change to craving and compulsive drug taking, and (3) relapse.

Initial Drug Taking   Initial taking of potentially addictive drugs is thought to be mediated in much the same way

as any pleasurable activity, with the mesocorticolimbic pathway—in particular, the nucleus accumbens—playing a key role. But the nucleus accumbens does not act alone; its interactions with three other areas of the brain are thought to be important. The prefrontal lobes are thought to play a major role in the decision to take a drug (Grace et al., 2007); the hippocampus and related structures are assumed to provide information about previous relevant experiences; and the amygdala is thought to coordinate the positive or negative emotional reactions to the drug taking.

Change to Craving and Compulsive Drug Taking   The repeated consumption of an addictive drug brings major changes in the motivation of the developing addict. Drug taking develops into a habit and then to a compulsion; that is, despite its numerous adverse effects, drug taking starts to dominate the addict's life. Earlier in this chapter, you learned that this change has been described as an increase in the positive-incentive value of taking the drug that occurs in the absence of any increase in its hedonic effects. It is not yet known why this change occurs and why it occurs in some drug takers but not in others. The change may be a direct neural response to the repeated experience of drug-induced pleasure; it could be a product of the myriad conditioned responses to drug-associated cues (Cardinal & Everitt, 2004; Kenny, 2007); or, more likely, it could be a product of both of these influences.

Several changes in the brain's responses appear to contribute to the development of addiction. First, there is a change in how the striatum reacts to drugs and drug-associated cues. As addiction develops, striatal control of addiction spreads from the nucleus accumbens (i.e., the ventral striatum) to the dorsal striatum, an area that is known to play a role in habit formation and retention (see Chapter 11). Also, at the same time, the role of the prefrontal cortex in controlling drug-related behaviors apparently declines, and stress *Neuroplasticity* circuits in the hypothalamus (see Chapter 17) begin to interact with the dorsal striatum. In essence, the development of addiction is a pathological neuroplastic response that some people show with repeated drug taking (see Kalivas, 2005; Koob & Le Moal, 2005).

Relapse   As you learned in an earlier section of this chapter, three factors are known to trigger relapse in abstinent addicts: priming doses of the drug, drug associated cues, and stress. Each cause of relapse appears to be mediated by interaction of a different brain structure with the striatum. Evidence from research on drug self-administration in laboratory animals suggests that the prefrontal cortex mediates priming-induced relapse, the amygdala mediates conditional cue-induced relapse, and the hypothalamus mediates stress-induced relapse.

15.8

A Noteworthy Case of Addiction

To illustrate in a more personal way some of the things you have learned about addiction, this chapter concludes with a case study of one addict. The addict was Sigmund Freud, a man of great significance to psychology.

Freud's case is particularly important for two reasons. First, it shows that nobody, no matter how powerful their intellect, is immune to the addictive effects of drugs. Second, it allows comparisons between the two drugs of addiction with which Freud had problems.

● Watch Dr. Freud and the Self-Administration Paradigm **www.mypsychlab.com**

The Case of Sigmund Freud

In 1883, a German army physician prescribed cocaine, which had recently been isolated, to Bavarian soldiers to help them deal with the demands of military maneuvers. When Freud read about this, he decided to procure some of the drug.

In addition to taking cocaine himself, Freud pressed it on his friends and associates, both for themselves and for their patients. He even sent some to his fiancée. In short, by today's standards, Freud was a public menace.

Freud's famous essay "Song of Praise" was about cocaine and was published in July 1884. Freud wrote in such glowing terms about his own personal experiences with cocaine that he created a wave of interest in the drug. But within a year, there was a critical reaction to Freud's premature advocacy of the drug. As evidence accumulated that cocaine was highly addictive and produced a psychosis-like state at high doses, so too did published criticisms of Freud.

Freud continued to praise cocaine until the summer of 1887, but soon thereafter he suddenly stopped all use of cocaine—both personally and professionally. Despite the fact that he had used cocaine for 3 years, he seems to have had no difficulty stopping.

Some 7 years later, in 1894, when Freud was 38, his physician and close friend ordered him to stop smoking because it was causing a heart arrhythmia. Freud was a heavy smoker; he smoked approximately 20 cigars per day.

Freud did stop smoking, but 7 weeks later he started again. On another occasion, Freud stopped for 14 months, but at the age of 58, he was still smoking 20 cigars a day—and still struggling against his addiction. He wrote to friends that smoking was adversely affecting his heart and making it difficult for him to work . . . yet he kept smoking.

In 1923, at the age of 67, Freud developed sores in his mouth. They were cancerous. When he was recovering from oral surgery, he wrote to a friend that smoking was the cause of his cancer . . . yet he kept smoking.

In addition to the cancer, Freud began to experience severe heart pains (tobacco angina) whenever he smoked . . . still he kept smoking.

At 73, Freud was hospitalized for his heart condition and stopped smoking. He made an immediate recovery. But 23 days later, he started to smoke again.

In 1936, at the age of 79, Freud was experiencing more heart trouble, and he had had 33 operations to deal with his recurring oral cancer. His jaw had been entirely removed and replaced by an artificial one. He was in constant pain, and he could swallow, chew, and talk only with difficulty . . . yet he kept smoking.

Freud died of cancer in 1939 (see Sheth, Bhagwate, & Sharma, 2005).

Themes Revisited

Two of this book's themes—thinking creatively and clinical implications—received strong emphasis in this chapter because they are integral to its major objective: to sharpen your thinking about the effects of addiction on people's health. You were repeatedly challenged to think about drug addiction in ways that may have been new to you but are more consistent with the evidence.

The evolutionary perspective theme was also highlighted frequently in this chapter, largely because of the nature of biopsychological research into drug addiction. Because of the risks associated with the administration of addictive drugs and the direct manipulation of brain structures, the majority of biopsychological studies of drug addiction involve nonhumans—mostly rats and monkeys. Also, in studying the neural mechanisms of addiction, there is a need to maintain an evolutionary perspective. It is important not to lose sight of the fact that brain mechanisms did not evolve to support addiction; they evolved to serve natural adaptive functions and have somehow been co-opted by addictive drugs.

Although the neuroplasticity theme pervades this chapter, the neuroplasticity tag appeared only once—where the text explains that the development of addiction is a pathological neuroplastic response. The main puzzle in research on addiction is how the brain of an occasional drug user is transformed into the brain of an addict.

Neuroplasticity

Think about It

1. There are many misconceptions about drug addiction. Describe three. What factors contribute to these misconceptions? In what ways is the evidence about drug addiction often misrepresented?

2. A doctor who had been a morphine user for many years was found dead of an overdose at a holiday resort. She appeared to have been in good health, and no foul play was suspected. Explain how conditioned tolerance may have contributed to her death.

3. If you had an opportunity to redraft current laws related to drug use in light of what you have learned in this chapter, what changes would you make? Do you think that all drugs, including nicotine and alcohol, should be illegal? Explain.

4. Speculate: How might recent advances in the study of the mesotelencephalic dopamine system eventually lead to effective treatments?

5. Does somebody you love use a hard drug such as nicotine or alcohol? What should you do?

6. One of my purposes in writing this chapter was to provide you with an alternative way of thinking about drug addiction, one that might benefit you. Imagine my dismay when I received an e-mail message suggesting that this chapter was making things worse for addicts. According to this message, discussion of addiction induces craving in addicts who have stopped taking drugs, thus encouraging them to recommence their drug taking. Discuss this point, and consider its implications for the design of antidrug campaigns.

Key Terms

Pharmacological (p. 384)

15.1 Basic Principles of Drug Action

Psychoactive drugs (p. 384)
Drug metabolism (p. 385)
Drug tolerance (p. 385)
Cross tolerance (p. 385)
Drug sensitization (p. 385)
Metabolic tolerance (p. 385)
Functional tolerance (p. 385)
Withdrawal syndrome (p. 386)
Physically dependent (p. 386)
Addicts (p. 386)

15.2 Role of Learning in Drug Tolerance

Contingent drug tolerance (p. 387)
Before-and-after design (p. 387)
Conditioned drug tolerance (p. 388)
Conditioned compensatory responses (p. 388)
Exteroceptive stimuli (p. 389)

Interoceptive stimuli (p. 389)

15.3 Five Commonly Abused Drugs

Nicotine (p. 389)
Smoker's syndrome (p. 390)
Buerger's disease (p. 390)
Teratogen (p. 390)
Depressant (p. 391)
Delirium tremens (DTs) (p. 391)
Korsakoff's syndrome (p. 391)
Cirrhosis (p. 391)
Fetal alcohol syndrome (FAS) (p. 391)
Disulfiram (p. 392)
Cannabis sativa (p. 392)
THC (p. 392)
Hashish (p. 392)
Narcotic (p. 392)
Anandamide (p. 394)
Stimulants (p. 394)
Cocaine (p. 394)

Crack (p. 394)
Cocaine sprees (p. 394)
Cocaine psychosis (p. 394)
Amphetamine (p. 394)
Dopamine transporters (p. 395)
Opium (p. 395)
Morphine (p. 395)
Codeine (p. 395)
Opiates (p. 395)
Analgesics (p. 395)
Harrison Narcotics Act (p. 396)
Heroin (p. 396)

15.4 Biopsychological Approaches to Theories of Addiction

Physical-dependence theories of addiction (p. 399)
Detoxified addicts (p. 399)
Positive-incentive theories of addiction (p. 399)
Positive-incentive value (p. 400)
Hedonic value (p. 400)
Incentive-sensitization theory (p. 400)

Relapse (p. 400)
Drug priming (p. 400)

15.5 Intracranial Self-Stimulation and the Pleasure Centers of the Brain

Intracranial self-stimulation (ICSS) (p. 401)
Primed (p. 402)
Mesotelencephalic dopamine system (p. 402)
Substantia nigra (p. 402)
Ventral tegmental area (p. 402)
Nucleus accumbens (p. 402)

15.6 Early Studies of Brain Mechanisms of Addiction: Dopamine

Drug self-administration paradigm (p. 403)
Conditioned place-preference paradigm (p. 403)

✓● **Quick Review**   Test your comprehension of the chapter with this brief practice test. You can find the answers to these questions as well as more practice tests, activities, and other study resources at www.mypsychlab.com.

1. Tolerance to psychoactive drugs is largely
   a. nonexistent.
   b. metabolic.
   c. functional.
   d. sensitization.
   e. cross tolerance.

2. Which drug is thought to lead to about 400,000 deaths each year in the United States alone?
   a. heroin
   b. cocaine
   c. alcohol
   d. nicotine
   e. marijuana

3. Delirium tremens can be produced by withdrawal from
   a. heroin.
   b. morphine.
   c. alcohol.
   d. amphetamines.
   e. both a and b

4. Animals that have been previously trained to press a lever to deliver rewarding electrical stimulation to their own brains will often not begin pressing unless they have been
   a. primed.
   b. extinguished.
   c. fed.
   d. frightened.
   e. punished.

5. A method of measuring drug-produced reinforcement or pleasure in laboratory animals is the
   a. drug self-administration paradigm.
   b. conditioned place-preference paradigm.
   c. conditioned tolerance paradigm.
   d. all of the above
   e. both a and b

16 Lateralization, Language, and the Split Brain

The Left Brain and the Right Brain of Language

With the exception of a few midline orifices, we humans have two of almost everything—one on the left and one on the right. Even the brain, which most people view as the unitary and indivisible basis of self, reflects this general principle of bilateral duplication. In its upper reaches, the brain comprises two structures, the left and right cerebral hemispheres, which are entirely separate except for the **cerebral commissures** connecting them. The fundamental duality of the human forebrain and the locations of the cerebral commissures are illustrated in Figure 16.1.

👁 Watch The Forebrain; The Midbrain; The Hindbrain www.mypsychlab.com

Although the left and right hemispheres are similar in appearance, there are major differences between them in function. This chapter is about these differences, a topic commonly referred to as **lateralization of function**. The study of **split-brain patients**—patients whose left and right hemispheres have been separated by **commissurotomy**—is a major focus of discussion. Another focus is the cortical localization of language abilities in the left hemisphere; language abilities are the most highly lateralized of all cognitive abilities.

You will learn in this chapter that your left and right hemispheres have different abilities and that they have the capacity to function independently—to have different thoughts, memories, and emotions. Thus, this chapter will challenge the concept you have of yourself as a unitary being. I hope you both enjoy it.

16.1
Cerebral Lateralization of Function: Introduction

In 1836, Marc Dax, an unknown country doctor, presented a short report at a medical society meeting in France. It was his first and only scientific presentation. Dax was struck by the fact that of the 40 or so brain-damaged patients with speech problems whom he had seen during his career, not a single one had damage restricted to the right hemisphere. His report aroused little interest, and Dax died the following year unaware that he had anticipated one of the most important areas of modern neuropsychological research.

Discovery of the Specific Contributions of Left–Hemisphere Damage to Aphasia and Apraxia

One reason Dax's important paper had so little impact was that most of his contemporaries believed that the brain acted as a whole and that specific functions could

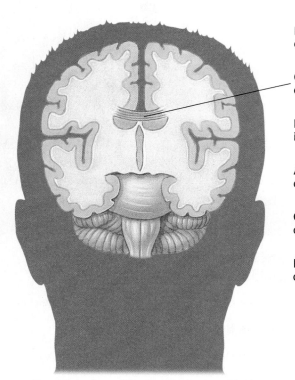

Frontal section of the human brain, which illustrates the fundamental duality of the human forebrain.

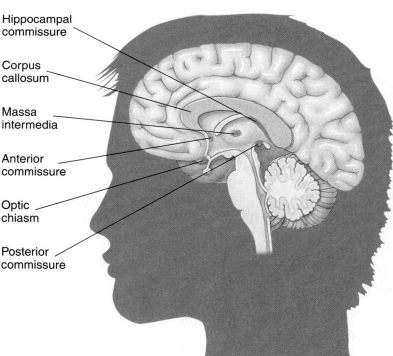

Hippocampal commissure
Corpus callosum
Massa intermedia
Anterior commissure
Optic chiasm
Posterior commissure

Midsagittal section of the human brain, which illustrates the corpus callosum and other commissures.

FIGURE 16.1 The cerebral hemispheres and cerebral commissures.

not be attributed to particular parts of it. This view began to change 25 years later, when Paul Broca reported his postmortem examination of two aphasic patients. **Aphasia** is a brain-damage–produced deficit in the ability to produce or comprehend language.

Both of Broca's patients had a left-hemisphere lesion that involved an area in the frontal cortex just in front of the face area of the primary motor cortex. Broca at first did not realize that there was a relation between aphasia and the side of the brain damage; he had not heard of Dax's report. However, by 1864, Broca had performed postmortem examinations on seven more aphasic patients, and he was struck by the fact that, like the first two, they all had damage to the *inferior prefrontal cortex* of the left hemisphere—which by then had become known as **Broca's area** (see Figure 16.2).

Clinical Implications

In the early 1900s, another example of *cerebral lateralization of function* was discovered. Hugo-Karl Liepmann found that **apraxia**, like aphasia, is almost always associated with left-hemisphere damage, despite the fact that its symptoms are *bilateral* (involving both sides of the body). Apraxic patients have difficulty performing movements when asked to perform them out of context, even though they often have no difficulty performing the same movements when they are not thinking about doing so.

The combined impact of the evidence that the left hemisphere plays a special role in both language and voluntary movement led to the concept of *cerebral dominance*.

According to this concept, one hemisphere—usually the left—assumes the dominant role in the control of all complex behavioral and cognitive processes, and the other plays only a minor role. This concept led to the practice of referring to the left hemisphere as the **dominant hemisphere** and the right hemisphere as the **minor hemisphere**.

Clinical Implications

Tests of Cerebral Lateralization

Early research on the cerebral lateralization of function compared the effects of left-hemisphere and right-hemisphere lesions. Now, however, other techniques are also used for this purpose. The sodium amytal test, the dichotic listening test, and functional brain imaging are three of them.

Sodium Amytal Test   The **sodium amytal test** of language lateralization (Wada, 1949) is often given to patients prior to neurosurgery. The neurosurgeon uses the results of the test to plan the surgery; every effort is made to avoid damaging areas of the cortex that are likely to be involved in language. The sodium amytal test involves the injection of a small amount of sodium amytal into the carotid artery on one side of the neck. The injection anesthetizes the hemisphere on that side for a few minutes, thus allowing the capacities of the other hemisphere to be assessed. During the test, the patient is asked to recite well-known series (e.g., letters of the alphabet, days of the week, months of the year) and to name pictures of common objects. Then, an injection is administered to the other side, and the test is repeated. When the hemisphere that is specialized for speech, usually the left hemisphere, is anesthetized, the patient is rendered completely mute for a minute or two; and once the ability to talk returns, there are errors of serial order and naming. In contrast, when the minor speech hemisphere, usually the right, is anesthetized, mutism often does not occur at all, and errors are few.

Dichotic Listening Test   Unlike the sodium amytal test, the **dichotic listening test** is noninvasive; thus, it can be administered to healthy individuals. In the standard dichotic listening test (Kimura, 1961), three pairs of spoken digits are presented through earphones; the digits of each pair are presented simultaneously, one to each ear. For example, a person might hear the sequence 3, 9, 2 through one ear and at the same time 1, 6, 4 through the other. The person is then asked to report all of the digits. Kimura found that most people report slightly more of the digits presented to the right ear than the left, which is indicative of left-hemisphere specialization for language. In contrast, Kimura found that all the patients who had been identified by the sodium amytal test as having right-hemisphere specialization for language performed better with the left ear than the right.

Why does the superior ear on the dichotic listening test indicate the language specialization of the contralateral hemisphere? Kimura argued that although the sounds from

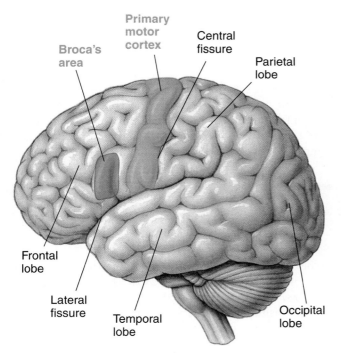

FIGURE 16.2   The location of Broca's area: in the inferior left prefrontal cortex, just anterior to the face area of the left primary motor cortex.

each ear are projected to both hemispheres, the contralateral connections are stronger and take precedence when two different sounds are simultaneously competing for access to the same cortical auditory centers.

Functional Brain Imaging Lateralization of function has also been studied using functional brain-imaging techniques. While a volunteer engages in some activity, such as reading, the activity of the brain is monitored by positron emission tomography (PET) or functional magnetic resonance imaging (fMRI). On language tests, functional brain-imaging techniques typically reveal far greater activity in the left hemisphere than in the right hemisphere (see Martin, 2003).

Discovery of the Relation between Speech Laterality and Handedness

Two early large-scale lesion studies clarified the relation between the cerebral lateralization of speech and handedness. One study involved military personnel who suffered brain damage in World War II (Russell & Espir, 1961), and the other focused on neurological patients who underwent unilateral excisions for the treatment of neurological disorders (Penfield & Roberts, 1959). In both studies, approximately 60% of **dextrals** (right-handers) with left-hemisphere lesions and 2% of those with right-hemisphere lesions were diagnosed as aphasic; the comparable figures for **sinestrals** (left-handers) were about 30% and 24%, respectively. These results indicate that the left hemisphere is dominant for language-related abilities in almost all dextrals and in the majority of sinestrals. Consequently, sinestrals are more variable (less predictable) than dextrals with respect to their hemisphere of language lateralization.

Results of the sodium amytal test have confirmed the relation between handedness and language lateralization that was first observed in early lesion studies. For example, Milner (1974) found that almost all right-handed patients without early left-hemisphere damage had left-hemisphere specialization for speech (92%), that most left-handed and ambidextrous patients without early left-hemisphere damage had left-hemisphere specialization for speech (69%), and that early left-hemisphere damage decreased left-hemisphere specialization for speech in left-handed and ambidextrous patients (30%).

In interpreting Milner's figures, it is important to remember that sodium amytal tests are administered only to people who are experiencing brain dysfunction, that early brain damage can cause the lateralization of speech to shift to the other hemisphere (see Maratsos & Matheny, 1994; Stiles, 1998), and that many more people have left-hemisphere specialization for speech to start with. Considered together, these points suggest that Milner's findings likely underestimate the proportion of people with left-

hemisphere specialization for speech among healthy members of the general population.

Sex Differences in Brain Lateralization

Interest in the possibility that the brains of females and males differ in their degree of lateralization was stimulated by McGlone's (1977, 1980) studies of unilateral stroke victims. McGlone found that male victims of unilateral strokes were three times more likely to suffer from aphasia than female victims. She found that male victims of left-hemisphere strokes had deficits on the Wechsler Adult Intelligence Scale (WAIS) verbal subtests, whereas male victims of right-hemisphere strokes had deficits on the WAIS performance subtests. In contrast, in female victims, there were no significant differences between the disruptive effects of left- and right-hemisphere unilateral strokes on performance on the WAIS. On the basis of these three findings, McGlone concluded that the brains of males are more lateralized than the brains of females.

McGlone's hypothesis of a sex difference in brain lateralization has been widely embraced, and it has been used to explain almost every imaginable behavioral difference between the sexes. But support for McGlone's hypothesis has been mixed. Some researchers have failed to confirm her report of a sex difference in the effects of unilateral brain lesions (see Inglis & Lawson, 1982). In addition, although a few functional brain-imaging studies have suggested that females, more than males, use both hemispheres in the performance of language-related tasks (e.g., Jaeger et al., 1998; Kansaku, Yamaura, & Kitazawa, 2000), a meta-analysis of 14 functional brain-imaging studies did not find a significant effect of sex on language lateralization (Sommer et al., 2004).

In this introductory section, you have been introduced to early research on the lateralization of function, and you learned about four methods of studying cerebral lateralization of function: comparing the effects of unilateral left- and right-hemisphere brain lesions, the sodium amytal test, the dichotic listening test, and functional brain imaging. The next section focuses on a fifth method.

16.2
The Split Brain

In the early 1950s, the **corpus callosum**—the largest cerebral commissure—constituted a paradox of major proportions. Its size, an estimated 200 million axons, and its central position, right between the two cerebral hemispheres, implied that it performed an extremely important function; yet research in the 1930s and 1940s seemed to suggest that it did nothing at all. The corpus callosum had been cut in monkeys and in several other laboratory species, but the animals seemed no different after the surgery than they had been before. Similarly, human patients who were born without a corpus callosum or had it damaged seemed quite normal (see Paul et al., 2007). In the early 1950s, Roger Sperry, whom you may remember for the eye-rotation experiments described in Chapter 9, and his colleagues were intrigued by this paradox.

Groundbreaking Experiment of Myers and Sperry

The solution to the puzzle of the corpus callosum was provided in 1953 by an experiment on cats by Myers and Sperry. The experiment made two astounding theoretical points. First, it showed that one function of the corpus callosum is to transfer learned information from one hemisphere to the other. Second, it showed that when the corpus callosum is cut, each hemisphere can function independently; each split-brain cat appeared to have two brains. If you find the thought of a cat with two brains provocative, you will almost certainly be bowled over by similar observations about split-brain humans. But I am getting ahead of myself. Let's first consider the research on cats.

Evolutionary Perspective

⊙ **Watch** The Visual Cortex
www.mypsychlab.com

In their experiment, Myers and Sperry trained cats to perform a simple visual discrimination. On each trial, each cat was confronted by two panels, one with a circle on it and one with a square on it. The relative positions of the circle and square (right or left) were varied randomly from trial to trial, and the cats had to learn which symbol to press in order to get a food reward. Myers and Sperry correctly surmised that the key to split-brain research was to develop procedures for teaching and testing one hemisphere at a time. Figure 16.3 illustrates the method they used to isolate visual-discrimination learning in one hemisphere of the cats. There are two routes by which visual information can cross from one eye to the contralateral hemisphere: via the corpus callosum or via the optic chiasm. Accordingly, in their

key experimental group, Myers and Sperry *transected* (cut completely through) both the optic chiasm and the corpus callosum of each cat and put a patch on one eye. This restricted all incoming visual information to the hemisphere ipsilateral to the uncovered eye.

The results of Myers and Sperry's experiment are illustrated in Figure 16.4 on page 416. In the first phase of the study, all cats learned the task with a patch on one eye. The cats in the key experimental group (those with both the optic chiasm and the corpus callosum transected) learned the simple discrimination as rapidly as did unlesioned control cats or control cats with either the corpus callosum or the optic chiasm transected, despite the fact that cutting the optic chiasm produced a **scotoma**—an area of blindness—involving the entire medial half of each retina. This result suggested that one hemisphere working alone can learn simple tasks as rapidly as two hemispheres working together.

More surprising were the results of the second phase of Myers and Sperry's experiment, during which the patch was transferred to each cat's other eye. The transfer of the patch had no effect on the performance of the intact control cats or of the control cats with either the optic chiasm or the corpus callosum transected; these subjects continued to perform the task with close to 100% accuracy. In contrast, transferring the eye patch had a devastating effect on the performance of the experimental cats. In effect, it blindfolded the hemisphere that had originally learned the task and tested the knowledge of the other hemisphere, which had been blindfolded during initial training. When the patch was transferred, the performance of the experimental cats dropped immediately to baseline (i.e., to 50% correct); and then the cats relearned

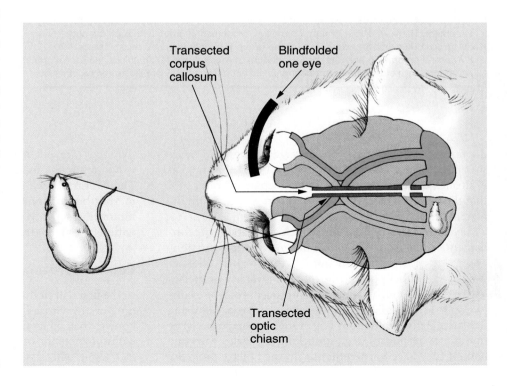

FIGURE 16.3 Restricting visual information to one hemisphere in cats. To restrict visual information to one hemisphere, Myers and Sperry (1) cut the corpus callosum, (2) cut the optic chiasm, and (3) blindfolded one eye. This restricted the visual information to the hemisphere ipsilateral to the uncovered eye.

Transected corpus callosum

Blindfolded one eye

Transected optic chiasm

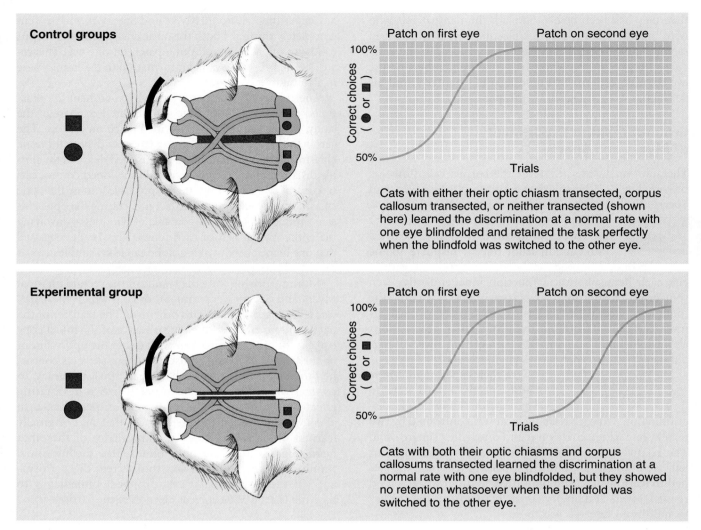

FIGURE 16.4 Schematic illustration of Myers and Sperry's (1953) groundbreaking split-brain experiment. There were four groups: (1) the key experimental group with both the optic chiasm and corpus callosum transected, (2) a control group with only the optic chiasm transected, (3) a control group with only the corpus callosum transected, and (4) an unlesioned control group. The performance of the three control groups did not differ, so they are illustrated together here.

the task with no savings whatsoever, as if they had never seen it before. Myers and Sperry concluded that the cat brain has the capacity to act as two separate brains and that the function of the corpus callosum is to transmit information between them.

Myers and Sperry's startling conclusions about the fundamental duality of the cat brain and the information-transfer function of the corpus callosum have been confirmed in a variety of species with a variety of test procedures. For example, split-brain monkeys cannot perform tasks requiring fine tactual discriminations (e.g., rough versus smooth) or fine motor responses (e.g., unlocking a puzzle) with one hand if they have learned them with the other—provided

Evolutionary Perspective

that they are not allowed to watch their hands, which would allow the information to enter both hemispheres. There is no transfer of fine tactual and motor information in split-brain monkeys because the somatosensory and motor fibers involved in fine sensory and motor discriminations are all contralateral.

Commissurotomy in Human Epileptics

In the first half of the 20th century, when the normal function of the corpus callosum was still a mystery, it was known that epileptic discharges often spread from one hemisphere to the other through the corpus callosum. This fact, along with the fact that cutting the corpus callosum had proven in numerous studies to have no obvious effect on performance outside the contrived conditions of Sperry's laboratory, led two neurosurgeons, Vogel and Bogen, to initiate a program of commissurotomy for the treatment of severe intractable cases of epilepsy—despite the fact that a previous similar attempt had failed, presumably because of incomplete transections (Van Wagenen & Herren, 1940). The rationale underlying therapeutic commissurotomy—which typically involves transecting the corpus callosum and leaving the smaller commissures intact—was that the severity of the patient's convulsions might be reduced if the discharges could be limited to the hemisphere of their origin. The therapeutic benefits of commissurotomy turned out to be even greater than anticipated: Despite the fact that commissurotomy is performed in only the most severe cases, many commissurotomized patients do not experience another major convulsion.

Clinical Implications

Evaluation of the neuropsychological status of Vogel and Bogen's split-brain patients was conducted by Sperry and his associate Gazzaniga, and this work was a major factor in Sperry's being awarded a Nobel Prize in 1981 (see Table 1.1). Sperry and Gazzaniga began by developing a battery of tests based on the same methodological strategy that had proved so informative in their studies of laboratory animals: delivering information to one hemisphere while keeping it out of the other (see Gazzaniga, 2005).

They could not use the same visual-discrimination procedure that had been used in studies of split-brain laboratory animals (i.e., cutting the optic chiasm and blindfolding one eye) because cutting the optic chiasm produces a scotoma. Instead, they employed the procedure illustrated in Figure 16.5. Each split-brain patient was asked to fixate on the center of a display screen; then, visual stimuli were flashed onto the left or right side of the screen for 0.1 second. The 0.1-second exposure time was long enough for the subjects to perceive the stimuli but short enough to preclude the confounding effects of eye movement. All stimuli thus presented in the left visual field were transmitted to the right visual cortex, and all stimuli thus presented in the right visual field were transmitted to the left visual cortex.

Fine tactual and motor tasks were performed by each hand under a ledge. This procedure was used so that the nonperforming hemisphere, that is, the ipsilateral hemisphere, could not monitor the performance via the visual system.

The results of the tests on split-brain patients have confirmed the findings in split-brain laboratory animals in one major respect, but not in another. Like split-brain laboratory

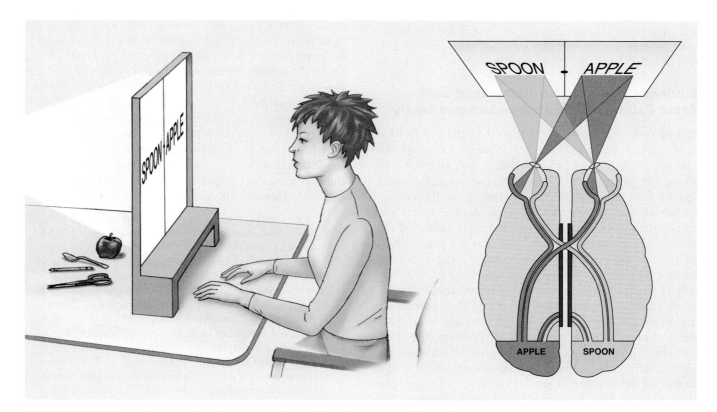

FIGURE 16.5 The testing procedure that was used to evaluate the neuropsychological status of split-brain patients. Visual input goes from each visual field to the contralateral hemisphere; fine tactile input goes from each hand to the contralateral hemisphere; and each hemisphere controls the fine motor movements of the contralateral hand.

animals, human split-brain patients seem to have in some respects two independent brains, each with its own stream of consciousness, abilities, memories, and emotions (e.g., Gazzaniga, 1967; Gazzaniga & Sperry, 1967; Sperry, 1964). But unlike the hemispheres of split-brain laboratory animals, the hemispheres of split-brain patients are far from equal in their ability to perform certain tasks. Most notably, the left hemisphere of most split-brain patients is capable of speech, whereas the right hemisphere is not.

Before I recount some of the key results of the tests on split-brain humans, let me give you some advice. Some students become confused by the results of these tests because their tendency to think of the human brain as a single unitary organ is deeply engrained. If you become confused, think of each split-brain patient as two separate individuals: Ms. or Mr. Right Hemisphere, who understands a few simple instructions but cannot speak, who receives sensory information from the left visual field and left hand, and who controls the fine motor responses of the left hand; and Ms. or Mr. Left Hemisphere, who is verbally adept, who receives sensory information from the right visual field and right hand, and who controls the fine motor responses of the right hand. In everyday life, the behavior of split-brain patients is reasonably normal because their two brains go through life together and acquire much of the same information; however, in the neuropsychological laboratory, major discrepancies in what the two hemispheres learn can be created. As you are about to find out, this situation has some interesting consequences.

Thinking Creatively

Evidence That the Hemispheres of Split-Brain Patients Can Function Independently

If a picture of an apple were flashed in the right visual field of a split-brain patient, the left hemisphere could do one of two things to indicate that it had received and stored the information. Because it is the hemisphere that speaks, the left hemisphere could simply tell the experimenter that it saw a picture of an apple. Or the patient could reach under the ledge with the right hand, feel the test objects that are there, and pick out the apple. Similarly, if the apple were presented to the left hemisphere by being placed in the patient's right hand, the left hemisphere could indicate to the experimenter that it was an apple either by saying so or by putting the apple down and picking out another apple with the right hand from the test objects under the ledge. If, however, the non-speaking right hemisphere were asked to indicate the identity of an object that had previously been presented to the left hemisphere, it could not do so. Although objects that have been presented to the left hemisphere can be accurately identified with the right hand, performance is no better than chance with the left hand.

When test objects are presented to the right hemisphere either visually (in the left visual field) or tactu-

ally (in the left hand), the pattern of responses is entirely different. A split-brain patient asked to name an object flashed in the left visual field is likely to claim that nothing appeared on the screen. (Remember that it is the left hemisphere who is talking and the right hemisphere who has seen the stimulus.) A patient asked to name an object placed in the left hand is usually aware that something is there, presumably because of the crude tactual information carried by ipsilateral somatosensory fibers, but is unable to say what it is (see Fabri et al., 2001). Amazingly, all the while the patient is claiming (i.e., all the while the left hemisphere is claiming) the inability to identify a test object presented in the left visual field or left hand, the left hand (i.e., the right hemisphere) can identify the correct object. Imagine how confused the patient must become when, in trial after trial, the left hand can feel an object and then fetch another just like it from a collection of test items under the ledge, while the left hemisphere is vehemently claiming that it does not know the identity of the test object.

Cross-Cuing

Although the two hemispheres of a split-brain patient have no means of direct neural communication, they can communicate neurally via indirect pathways through the brain stem. They can also communicate with each other by an external route, by a process called **cross-cuing**. An example of cross-cuing occurred during a series of tests designed to determine whether the left hemisphere could respond to colors presented in the left visual field. To test this possibility, a red or a green stimulus was presented in the left visual field, and the split-brain patient was asked to verbally report the color: red or green. At first, the patient performed at a chance level on this task (50% correct); but after a time, performance improved appreciably, thus suggesting that the color information was somehow being transferred over neural pathways from the right hemisphere to the left. However, this proved not to be the case:

> We soon caught on to the strategy the patient used. If a red light was flashed and the patient by chance guessed red, he would stick with that answer. If the flashed light was red, and the patient by chance guessed green, he would frown, shake his head and then say, "Oh no, I meant red." What was happening was that the right hemisphere saw the red light and heard the left hemisphere make the guess "green." Knowing that the answer was wrong, the right hemisphere precipitated a frown and a shake of the head, which in turn cued in the left hemisphere to the fact that the answer was wrong and that it had better correct itself! . . . The realization that the neurological patient has various strategies at his command emphasizes how difficult it is to obtain a clear neurological description of a human being with brain damage. (Reprinted with permission from "The Split Brain in Man" from *Scientific American*, Aug. 1967, 24–29. Copyright © 1967 Scientific American a division of Nature America, Inc. All rights reserved.)

Doing Two Things at Once

In many of the classes I teach, there is a student who fits the following stereotype: He sits—or rather sprawls—near the back of the class; and despite good grades, he tries to create the impression that he is above it all by making sarcastic comments. I am sure you recognize him. Such a student inadvertently triggered an interesting discussion in one of my classes. His comment went something like this: "If getting my brain cut in two could create two separate brains, perhaps I should get it done so that I could study for two different exams at the same time."

The question raised by this comment is a good one. If the two hemispheres of a split-brain patient are capable of independent functioning, then they should be able to do two different things at the same time—in this case, learn two different things at the same time. Can they? Indeed they can. For example, in one test, two different visual stimuli appeared simultaneously on the test screen—let's say a pencil in the left visual field and an orange in the right visual field. The split-brain patient was asked to simultaneously reach into two bags—one with each hand—and grasp in each hand the object that was on the screen. After grasping the objects, but before withdrawing them, the patient was asked to tell the experimenter what was in the two hands; the patient (i.e., the left hemisphere) replied, "Two oranges." Much to the bewilderment of the verbal left hemisphere, when the hands were withdrawn, there was an orange in the right hand and a pencil in the left. The two hemispheres of the split-brain patient had learned two different things at exactly the same time.

In another test in which two visual stimuli were presented simultaneously—again, let's say a pencil to the left visual field and an orange to the right—the split-brain patient was asked to pick up the presented object from an assortment of objects on a table, this time in full view. As the right hand reached out to pick up the orange under the direction of the left hemisphere, the right hemisphere saw what was happening and thought an error was being made (remember that the right hemisphere saw a pencil). On some trials, the right hemisphere dealt with this problem in the only way that it could: The left hand shot out, grabbed the right hand away from the orange, and redirected it to the pencil. This response is called the **helping-hand phenomenon**.

The special ability of split brains to do two things at once has also been demonstrated on tests of attention. Each hemisphere of split-brain patients appears to be

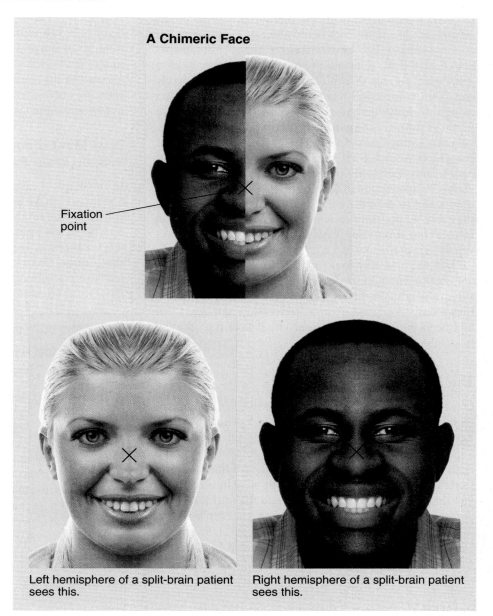

A Chimeric Face

Fixation point

Left hemisphere of a split-brain patient sees this.

Right hemisphere of a split-brain patient sees this.

FIGURE 16.6   The chimeric figures test. The left hemisphere of a split-brain patient sees a single normal face that is a completed version of the half face on the right. At the same time, the right hemisphere sees a single normal face that is a completed version of the half face on the left.

able to maintain an independent focus of attention (see Gazzaniga, 2005). This leads to an ironic pattern of results: Split-brain patients can search for, and identify, a visual target item in an array of similar items more quickly than healthy controls can (Luck et al., 1989)—presumably because the two split hemispheres are conducting two independent searches.

Yet another example of the split brain's special ability to do two things at once involves the phenomenon of **visual completion**. As you may recall from Chapter 6, individuals with scotomas are often unaware of them because their brains have the capacity to fill them in (to complete them) by using information from the surrounding areas of the visual field. In a sense, each hemisphere of a split-brain patient is a subject with a scotoma covering the entire ipsilateral visual field. The ability of the hemispheres of a split-brain patient to simultaneously and independently engage in completion has been demonstrated in studies using the **chimeric figures test**—named after *Chimera*, a mythical monster composed of parts of different animals. Levy, Trevarthen, and Sperry (1972) flashed photographs composed of fused-together half-faces of two different people onto the center of a screen in front of split-brain patients—see Figure 16.6 on page 419. The patients were then asked to describe what they saw or to indicate what they saw by pointing to it in a series of photographs of intact faces. Amazingly, each patient (i.e., each left hemisphere) reported seeing a complete, bilaterally symmetrical face, even when asked such leading questions as "Did you notice anything peculiar about what you just saw?" When the patients were asked to describe what they saw, they usually described a completed version of the half that had been presented to the right visual field (i.e., the left hemisphere).

The Z Lens

Once it was firmly established that the two hemispheres of each split-brain patient can function independently, it became clear that the study of split-brain patients provided a unique opportunity to compare the abilities of left and right hemispheres. However, early studies of the lateralization of function in split-brain patients were limited by the fact that visual stimuli requiring more than 0.1 second to perceive could not be studied using the conventional method for restricting visual input to one hemisphere. This methodological barrier was eliminated by Zaidel in 1975. Zaidel developed a lens, called the Z lens, that limits visual input to one hemisphere of split-brain patients while they scan complex visual material such as the pages of a book. As Figure 16.7 on page 421 illustrates, the Z lens is a contact lens that is opaque on one side (left or right). Because it moves with the eye, it permits visual input to enter only one hemisphere, irrespective of eye movement. Zaidel used the Z lens to compare the ability of the left and

right hemispheres of split-brain patients to perform on various tests.

The usefulness of the Z lens is not restricted to purely visual tests. For example, it has been used to compare the ability of the left and right hemispheres to comprehend speech. Because each ear projects to both hemispheres, it is not possible to present spoken words to only one hemisphere. Thus, to assess the ability of a hemisphere to comprehend spoken words or sentences, Zaidel presented them to both ears, and then he asked the split-brain patients to pick the correct answer or to perform the correct response under the direction of visual input to only that hemisphere. For example, to test the ability of the right hemisphere to understand oral commands, the patients were given an oral instruction (such as "Put the green square under the red circle"), and then the right hemisphere's ability to comprehend the direction was tested by allowing only the right hemisphere to observe the colored tokens while the task was being completed.

Dual Mental Functioning and Conflict in Split–Brain Patients

In most split-brain patients, the right hemisphere does not seem to have a strong will of its own; the left hemisphere seems to control most everyday activities. However, in a few split-brain patients, the right hemisphere takes a more active role in controlling behavior, and in these cases, there can be serious conflicts between the left and right hemispheres. One patient (let's call him Peter) was such a case.

The Case of Peter, the Split–Brain Patient Tormented by Conflict

At the age of 8, Peter began to suffer from complex partial seizures. Antiepileptic medication was ineffective, and at 20, he received a commissurotomy, which greatly improved his condition but did not completely block his seizures. A *Clinical Implications* sodium amytal test administered prior to surgery showed that he was left-hemisphere dominant for language.

Following surgery, Peter, unlike most other split-brain patients, was not able to respond with the left side of his body to verbal input. When asked to make whole-body movements (e.g., "Stand like a boxer") or movements of the left side of his body (e.g., "Touch your left ear with your left hand"), he could not respond correctly. Apparently, his left hemisphere could not, or would not, control the left side of his body via ipsilateral fibers. During such tests, Peter—or, more specifically, Peter's left hemisphere—often remarked that he hated the left side of his body.

The independent, obstinate, and sometimes mischievous behavior of Peter's right hemisphere often caused him (his left hemisphere) considerable frustration. He (his left hemisphere) complained that his left hand would turn off television shows that he was enjoying, that his left leg would not always walk in the intended direction, and that his left arm would sometimes perform embarrassing, socially unacceptable acts (e.g., striking a relative).

In the laboratory, he (his left hemisphere) sometimes became angry with his left hand, swearing at it, striking it, and trying to force it with his right hand to do what he (his left hemisphere) wanted. In these cases, his left hand usually resisted his right hand and kept performing as directed by his right hemisphere. In these instances, it was always clear that the right hemisphere was behaving with intent and understanding and that the left hemisphere had no clue why the despised left hand was doing what it was doing (Joseph, 1988).

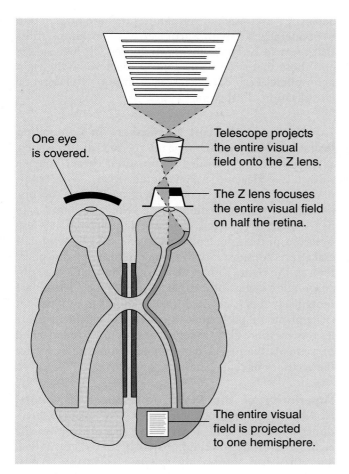

One eye is covered.

Telescope projects the entire visual field onto the Z lens.

The Z lens focuses the entire visual field on half the retina.

The entire visual field is projected to one hemisphere.

FIGURE 16.7 The Z lens, which was developed by Zaidel to study functional asymmetry in split-brain patients. It is a contact lens that is opaque on one side (left or right), so that visual input reaches only one hemisphere.

Independence of Split Hemispheres: Current Perspective

Discussions of split-brain patients tend to focus on the many examples of complete functional independence, as I have done here. These examples are not only intriguing but important, because they demonstrate major differences from normal integrated brain function. However, it is important not to lose sight of the fact that surgically separated hemispheres retain the ability to interact via the brain stem and, in some cases, do function together. The hemispheres of split-brain patients are more likely to perform independently on some types of tests than on others, but sometimes two patients performing on the same test may differ in their hemispheric independence (see Wolford, Miller, & Gazzaniga, 2004).

The classic study of Sperry, Zaidel, and Zaidel (1979) provided early evidence that some types of information are more likely to be shared between split hemispheres. These researchers used the Z lens to assess the behavioral reactions of the right hemispheres of split-brain patients to various emotion-charged images: photographs of relatives; of pets; of themselves; and of political, historical, and religious figures and emblems. The patients' behavioral reactions were emotionally appropriate, thus indicating that right hemispheres are capable of emotional expression. In addition, there was an unexpected finding: The emotional content of images presented to the right hemisphere was reflected in the patients' speech as well as in their nonverbal behavior. This suggested that emotional information was somehow being passed from the right to the verbal left hemisphere of the split-brain patients. The ability of emotional reactions, but not visual information, to be readily passed from the right hemisphere to the left hemisphere created a bizarre situation. A patient's left hemisphere often reacted with the appropriate emotional verbal response to an image that had been presented to the right hemisphere, even though it did not know what the image was.

Consider the following remarkable exchange (paraphrased from Sperry, Zaidel, & Zaidel, 1979, pp. 161–162). The patient's right hemisphere was presented with an array of photos, and the patient was asked if one was familiar. He pointed to the photo of his aunt.

> *Experimenter:* "Is this a neutral, a thumbs-up, or a thumbs-down person?"
> *Patient:* With a smile, he made a thumbs-up sign and said, "This is a happy person."
> *Experimenter:* "Do you know him personally?"
> *Patient:* "Oh, it's not a him, it's a her."
> *Experimenter:* "Is she an entertainment personality or an historical figure?"
> *Patient:* "No, just . . ."
> *Experimenter:* "Someone you know personally?"

PATIENT: He traced something with his left index finger on the back of his right hand, and then he exclaimed, "My aunt, my Aunt Edie."

EXPERIMENTER: "How do you know?"

PATIENT: "By the E on the back of my hand."

Another factor that has been shown to contribute substantially to the hemispheric independence of split-brain patients is task difficulty (Weissman & Banich, 2000). As tasks become more difficult, they are more likely to involve both hemispheres of split-brain patients. It appears that simple tasks are best processed in one hemisphere, the hemisphere specialized for the specific activity, but complex tasks require the cognitive power of both hemispheres. This is an important finding for two reasons. First, it complicates the interpretation of functional brain-imaging studies of lateralization of function: If tasks are difficult, both hemispheres may show substantial activity, even though one hemisphere is specialized for the performance of the task. Second, it explains why the elderly often display less lateralization of function: As neural resources decline, it may become necessary to involve both hemispheres in most tasks.

16.3
Differences between Left and Right Hemispheres

So far in this chapter, you have learned about five methods of studying cerebral lateralization of function: unilateral lesions, the sodium amytal test, the dichotic listening test, functional brain imaging, and studies of split-brain patients. This section takes a look at some of the major functional differences between the left and right cerebral hemispheres that have been discovered using these methods. Because the verbal and motor abilities of the left hemisphere are readily apparent (see Beeman & Chiarello, 1998; Reuter-Lorenz & Miller, 1998), most research on the lateralization of function has focused on uncovering the special abilities of the right hemisphere.

Before I introduce you to some of the differences between the left and right hemispheres, I need to clear up a common misconception: For many *Thinking Creatively* functions, there are no substantial differences between the hemispheres; and when functional differences do exist, these tend to be slight biases in favor of one hemisphere or the other—not absolute differences (see Brown & Kosslyn, 1993). Disregarding these facts, the popular media inevitably portray left–right cerebral differences as absolute. As a result, it is widely believed that various abilities reside exclusively in one hemisphere or the other. For example, it is widely believed that the left hemisphere has exclusive control over language and the right hemisphere has exclusive control over emotion and creativity.

Language-related abilities provide a particularly good illustration of the fact that lateralization of function is statistical rather than absolute. Language is the most lateralized of all cognitive abilities. Yet, even in this most extreme case, lateralization is far from total; there is substantial language-related activity in the right hemisphere. For example, on the dichotic listening test, people who are left-hemisphere dominant for language tend to identify more digits with the right ear than the left ear, but this right-ear advantage is only 55% to 45%. Furthermore, the right hemispheres of most left-hemisphere dominant split-brain patients can understand many spoken or written words and simple sentences (see Baynes & Gazzaniga, 1997; Zaidel, 1987)—the language abilities of the right hemispheres tend to be comparable to those of a preschool child (Gazzaniga, 1998).

Examples of Cerebral Lateralization of Function

Table 16.1 lists some of the abilities that are often found to be lateralized. They are arranged in two columns: those that seem to be controlled more by the left hemisphere and those that seem to be controlled more by the right hemisphere. Let's consider several examples of cerebral lateralization of function.

Superiority of the Left Hemisphere in Controlling Ipsilateral Movement   One unexpected left-hemisphere specialization was revealed by functional brain-imaging studies (see Haaland & Harrington, 1996). When complex, cognitively driven movements are made by one hand, most of the activation is observed in the *contralateral* hemisphere, as expected. However, some activation is also observed in the *ipsilateral* hemisphere, and these ipsilateral effects are substantially greater in the left hemisphere than in the right (Kim et al., 1993). Consistent with this observation is the finding that left-hemisphere lesions are more likely than right-hemisphere lesions to produce ipsilateral motor problems—for example, left-hemisphere lesions are more likely to reduce the accuracy of left-hand movements than right-hemisphere lesions are to reduce the accuracy of right-hand movements.

Superiority of the Right Hemisphere in Spatial Ability   In a classic early study, Levy (1969) placed a three-dimensional block of a particular shape in either the right hand or the left hand of split-brain patients. Then, after they had thoroughly *palpated* (tactually investigated) it, she asked them to point to the two-dimensional test stimulus that best represented what the three-dimensional block would look like if it were made of cardboard and unfolded. She found a right-hemisphere

TABLE 16.1 Abilities That Display Cerebral Lateralization of Function

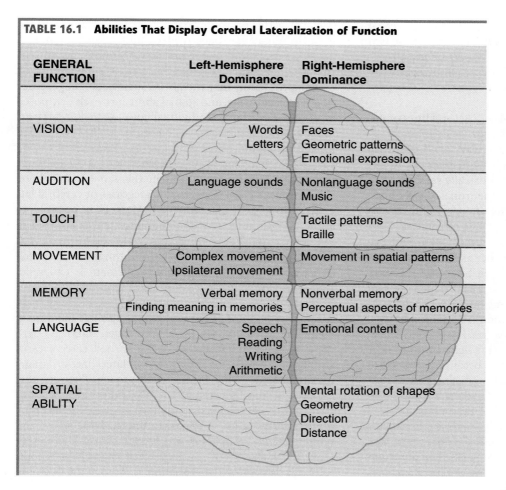

GENERAL FUNCTION	Left-Hemisphere Dominance	Right-Hemisphere Dominance
VISION	Words Letters	Faces Geometric patterns Emotional expression
AUDITION	Language sounds	Nonlanguage sounds Music
TOUCH		Tactile patterns Braille
MOVEMENT	Complex movement Ipsilateral movement	Movement in spatial patterns
MEMORY	Verbal memory Finding meaning in memories	Nonverbal memory Perceptual aspects of memories
LANGUAGE	Speech Reading Writing Arithmetic	Emotional content
SPATIAL ABILITY		Mental rotation of shapes Geometry Direction Distance

Superior Musical Ability of the Right Hemisphere Kimura (1964) compared the performance of 20 right-handers on the standard digit version of the dichotic listening test with their performance on a version of the test involving the dichotic presentation of melodies. In the melody version of the test, Kimura simultaneously played two different melodies—one to each ear—and then asked the participants to identify the two they had just heard from four that were subsequently played to them through both ears. The right ear (i.e., the left hemisphere) was superior in the perception of digits, whereas the left ear (i.e., the right hemisphere) was superior in the perception of melodies. This is consistent with the observation that right temporal lobe lesions are more likely to disrupt music discriminations than are left temporal lobe lesions.

Hemispheric Differences in Memory Early studies of the lateralization of cognitive function were premised on the assumption that particular cognitive abilities reside in one or the other of the two hemispheres. However, the results of research have led to an alternative way of thinking: The two hemispheres have similar abilities that tend to be expressed in different ways. The study of the lateralization of memory was one of the first areas of research on cerebral lateralization to lead to this modification in thinking. You see, both the left and right hemispheres have the ability to perform on tests of memory, but the left hemisphere is better on some tests, whereas the right hemisphere is better on others.

There are two approaches to studying the cerebral lateralization of memory. One approach is to try to link particular memory processes with particular hemispheres—for example, it has been argued that the left hemisphere is specialized for encoding episodic memory (see Chapter 11). The other approach (e.g., Wolford, Miller, & Gazzaniga, 2004) is to link the memory processes of each hemisphere to specific materials rather than to specific processes. In general, the left hemisphere has been found to play the greater role in memory for verbal material, whereas the right hemisphere has been

superiority on this task, and she found that the two hemispheres seemed to go about the task in different ways. The performance of the left hand and right hemisphere was rapid and silent, whereas the performance of the right hand and left hemisphere was hesitant and often accompanied by a running verbal commentary that was difficult for the patients to inhibit. Levy concluded that the right hemisphere is superior to the left at spatial tasks. This conclusion has been frequently confirmed (e.g., Funnell, Corballis, & Gazzaniga, 1999; Kaiser et al., 2000), and it is consistent with the finding that disorders of spatial perception (e.g., contralateral neglect—see Chapters 7 and 8) tend to be associated with right-hemisphere damage.

Specialization of the Right Hemisphere for Emotion

According to the old concept of left-hemisphere dominance, the right hemisphere is uninvolved in emotion. This presumption has been proven false. Indeed, analysis of the effects of unilateral brain lesions indicates that the right hemisphere is superior to the left at performing on some tests of emotion—for example, in accurately identifying facial expressions of emotion (Bowers et al., 1985).

found to play the greater role in memory for nonverbal material (e.g., Kelley et al., 2002). Whichever of these two approaches proves more fruitful, they represent an advance over the tendency to think that memory is totally lateralized to one hemisphere.

The Left–Hemisphere Interpreter   Several lines of evidence suggest that the left and right hemispheres approach cognitive tasks in different ways. The cognitive approach that is typical of the left hemisphere is attributed to a mechanism that is metaphorically referred to as the **interpreter**—a hypothetical neuronal mechanism that continuously assesses patterns of events and tries to make sense of them.

The following experiment illustrates the kind of evidence that supports the existence of a left-hemisphere interpreter. The left and right hemispheres of split-brain patients were tested separately. The task was to guess which of two lights—top or bottom—would come on next. The top light came on 80% of the time in a random sequence, but the subjects were not given this information. Intact control participants quickly discovered that the top light came on more often than the bottom one; however, because they tried to figure out the nonexistent rule that predicted the exact sequence, they were correct only 68% of the time—even though they could have scored 80% if they always selected the top light. The left hemispheres of the split-brain patients performed on this test like intact controls: They attempted to find deeper meaning and as a result performed poorly. In contrast, the right hemispheres, like intact rats or pigeons, did not try to interpret the events and readily learned to maximize their correct responses by always selecting the top light (see Metcalfe, Funnell, & Gazzaniga, 1995; Roser & Gazzaniga, 2004).

What Is Lateralized—Broad Clusters of Abilities or Individual Cognitive Processes?

You have undoubtedly encountered information about cerebral lateralization of function before. You have most likely heard or read that the left hemisphere is the logical language hemisphere and the right hemisphere is the emotional spatial hemisphere. Some of you may even believe that we can all be classified as left-hemisphere or right-hemisphere people. Information like that in Table 16.1 summarizes the results of many studies, and thus it serves a useful function if not taken too literally. The problem is that such information is almost always taken too literally by those unfamiliar with the complexities of the relevant research literature. Let me explain.

Early theories of cerebral laterality tended to ascribe complex clusters of mental abilities to one hemisphere or the other. The left hemisphere tended to perform better on language tests, so it was presumed to be dominant for language-related abilities; the right hemisphere tended to

perform better on some spatial tests, so it was presumed to be dominant for space-related abilities; and so on. Perhaps this was a reasonable first step, but now the consensus among researchers is that this conclusion is simplistic.

The problem is that categories such as language, emotion, musical ability, and spatial ability are each composed of dozens of different individual cognitive activities, and there is no reason to assume that all those activities associated with a general English label (e.g., spatial ability) will necessarily be lateralized in the same hemisphere. The inappropriateness of broad categories of cerebral lateralization has been confirmed. How is it possible to argue that all language-related abilities are lateralized in the left hemisphere, when the right hemisphere has proved superior in perceiving the intonation of speech and the identity of the speaker (Beeman & Chiarello, 1998)? Indeed, notable exceptions to all broad categories of cerebral lateralization have emerged (see Baas, Aleman, & Kahn, 2004; Josse & Tzourio-Mazoyer, 2004; Tervaniemi & Hugdahl, 2003; Vogel, Bowers, & Vogel, 2003).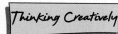

Thinking Creatively

As a result of the evidence that broad categories of abilities are not the units of cerebral lateralization, many researchers are taking a different approach. They are basing their studies of cerebral lateralization on the work of cognitive psychologists, who have broken down complex cognitive tasks—such as reading, judging space, and remembering—into their *constituent cognitive processes.* Once the laterality of the individual cognitive elements has been determined, it is possible to predict the laterality of cognitive tasks based on the specific cognitive elements that compose them.

Anatomical Asymmetries of the Brain

Many anatomical differences between the hemispheres have been documented. These differences appear to result from interhemispheric differences in gene expression (Sun et al., 2005).

Most of the research on anatomical asymmetries has focused on areas of cortex that are important for language. Three of these areas are the frontal operculum, the planum temporale, and Heschl's gyrus. The **frontal operculum** is the area of frontal lobe cortex that lies just in front of the face area of the primary motor cortex; in the left hemisphere, it is the location of Broca's area. The planum temporale and Heschl's gyrus are areas of temporal lobe cortex (see Figure 16.8). The **planum temporale** lies in the posterior region of the lateral fissure; it is thought to play a role in the comprehension of language and is often referred to as *Wernicke's area.* **Heschl's gyrus** is located in the lateral fissure just anterior to the planum temporale in the temporal lobe; it is the location of primary auditory cortex.

Documenting anatomical asymmetries in the frontal operculum, planum temporale, and Heschl's gyrus has

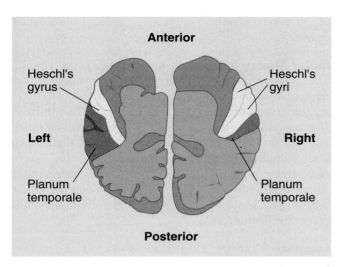

FIGURE 16.8 Two language areas of the cerebral cortex that display neuroanatomical asymmetry: the planum temporale (Wernicke's area) and Heschl's gyrus (primary auditory cortex).

proven to be far from simple. The main difficulties are that there are no clear boundaries around these cortical areas and no consensus on how best to define them (see Davis et al., 2008; Friederici, 2009; Keller et al., 2009). In general, the planum temporale does seem to be larger in the left hemisphere, but in only about 65% of human brains (Geschwind & Levitsky, 1968). In contrast, the cortex of Heschl's gyrus seems to be larger in the right hemisphere, primarily because there are often two Heschl's gyri in the right hemisphere and only one in the left. The laterality of the frontal operculum is even less clear. The area of the frontal operculum that is visible on the surface of the brain tends to be larger in the right hemisphere; but when the cortex buried within sulci of the frontal operculum is considered, there tends to be a greater volume of frontal

operculum cortex in the left hemisphere (Falzi, Perrone, & Vignolo, 1982).

Thus, of these three language areas, only the planum temporale displays the expected left-hemisphere size advantage. Complicating things further is that there is no evidence that the tendency for the planum temporale to be larger in the left hemisphere is related to the tendency for the left hemisphere to be dominant for language. People with well-developed planum temporale asymmetries do not tend to have more lateralized language functions (see Dos Santos Sequeira et al., 2006; Eckert et al., 2006; Jäncke & Steinmetz, 2003). In addition, similar asymmetries of the planum temporale have been reported in nonhumans (see Dehaene-Lambertz, Hertz-Pannier, & Dubois, 2006).

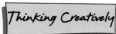

One particularly interesting study of planum temporale asymmetry was conducted by Schlaug and colleagues (1995). They used structural magnetic resonance imaging (MRI) to measure the asymmetry of the planum temporale and relate it to the presence of *perfect pitch* (the ability to identify the pitch of individual musical notes). The planum temporale tended to be larger in the left hemisphere in musicians with perfect pitch than in nonmusicians or in musicians without perfect pitch (see Figure 16.9).

Most studies of anatomical asymmetries of the brain have compared the sizes of particular gross structures in the left and right hemispheres. However, some anatomists have studied differences in cellular-level structure between corresponding areas of the two hemispheres that have been found to differ in function (see Gazzaniga, 2000; Hutsler & Galuske, 2003). One such study was conducted by Galuske and colleagues (2000), who compared organization of the neurons in

FIGURE 16.9 The anatomical asymmetry detected in the planum temporale of musicians by magnetic resonance imaging. In most people, the planum temporale is larger in the left hemisphere than in the right; this difference was found to be greater in musicians with perfect pitch than in either musicians without perfect pitch or controls. (Based on Schlaug et al., 1995.)

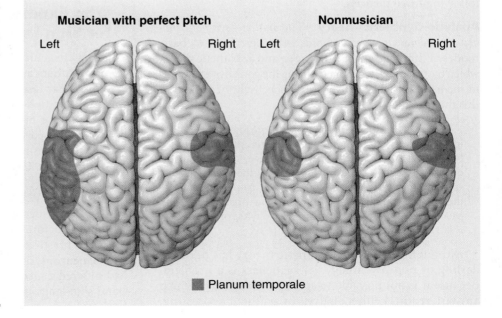

part of Wernicke's area with organization of the neurons in the same part of the right hemisphere. They found that the areas in both hemispheres are organized into regularly spaced columns of interconnected neurons and that the columns are interconnected by medium-range axons. The columns are the same diameter in both hemispheres, but they are about 20% farther apart in the left hemisphere and are interconnected by longer axons. The particular way in which the columns of neurons are organized in Wernicke's area may be an adaptation for the processing of language signals.

16.4
Evolutionary Perspective of Cerebral Lateralization and Language

You have already seen in this chapter how the discussion of cerebral lateralization inevitably leads to a discussion of language: Language is the most lateralized cognitive function. This section considers the evolution of cerebral lateralization and then the evolution of language.

Theories of the Evolution of Cerebral Lateralization

Many theories have been proposed to explain why cerebral lateralization of function evolved. Most are based on the same general premise: that it is advantageous for areas of the brain that perform similar functions to be located in the same hemisphere. However, each theory of cerebral asymmetry postulates a different fundamental distinction between left and right hemisphere function. Consider the following three theories.

Analytic–Synthetic Theory   The *analytic–synthetic theory of cerebral asymmetry* holds that there are two basic modes of thinking, an analytic mode and a synthetic mode, which have become segregated during the course of evolution in the left and right hemispheres, respectively. According to this theory,

> . . . the left hemisphere operates in a more logical, analytical, computerlike fashion, analyzing stimulus information input sequentially and abstracting the relevant details, to which it attaches verbal labels; the right hemisphere is primarily a synthesizer, more concerned with the overall stimulus configuration, and organizes and processes information in terms of gestalts, or wholes. (Harris, 1978, p. 463)

Although the analytic–synthetic theory has been the darling of pop psychology, its vagueness is a problem. Because it is not possible to specify the degree to which any task requires either analytic or synthetic processing,

it has been difficult to subject the theory to empirical tests.

Motor Theory   According to a second theory of cerebral asymmetry (see Kimura, 1979), the *motor theory of cerebral asymmetry*, the left hemisphere is specialized not for the control of speech specifically but for the control of fine movements, of which speech is only one category. Support for this theory comes from reports that lesions that produce aphasia often produce other motor deficits (see Serrien, Ivry, & Swinnen, 2006). One shortcoming of the motor theory of cerebral asymmetry is that it does not suggest why motor function became lateralized in the first place (see Beaton, 2003).

Linguistic Theory   A third theory of cerebral asymmetry, the *linguistic theory of cerebral asymmetry*, posits that the primary role of the left hemisphere is language; this is in contrast to the analytic–synthetic and motor theories, which view language as a secondary specialization residing in the left hemisphere because of that hemisphere's primary specialization for analytic thought and skilled motor activity, respectively.

The linguistic theory of cerebral asymmetry is based to a large degree on the study of deaf people who use *American Sign Language* (a sign language with a structure similar that of to spoken language) and who suffer unilateral brain damage (see Hickok, Bellugi, & Klima, 2001). The fact that left-hemisphere damage can disrupt the use of sign language but not pantomime gestures, as occurred in the case of W.L., suggests that the fundamental specialization of the left hemisphere may be language.

The Case of W.L., the Man Who Experienced Aphasia for Sign Language

W.L. is a congenitally deaf, right-handed male who grew up using American Sign Language. W.L. had a history of cardiovascular disease; and 7 months prior to testing, he was admitted to hospital complaining of right-side weakness and motor problems. A CT scan revealed a large left frontotemporoparietal stroke. At that time, W.L.'s wife noticed that he was making many uncharacteristic errors in signing and was having difficulty understanding the signs of others.

Fortunately, W.L.'s neuropsychologists managed to obtain a 2-hour videotape of an interview with him recorded 10 months before his stroke, which served as a valuable source of prestroke performance measures. Formal poststroke neuropsychological testing confirmed that W.L. had suffered a specific loss in his ability to use and understand sign language. The fact that he could produce and understand complex pantomime gestures suggested that his sign-language aphasia was not the result of motor or sensory deficits, and the results of cognitive tests suggested that it was not the result of general cognitive deficits (Corina et al., 1992).

When Did Cerebral Lateralization Evolve?

Until recently, cerebral lateralization had been assumed to be an exclusive feature of the hominin brain. For example, one version of the motor theory of cerebral asymmetry is that left-hemisphere specialization for motor control evolved in early hominins in response to their use of tools, and then the capacity for vocal language subsequently evolved in the left hemisphere because of its greater motor dexterity. However, there is evidence of lateralization of function in many vertebrates that evolved long before we humans did (see Hopkins & Cantalupo, 2008; Hopkins, Russell, & Cantalupo, 2007). Indeed, it has been suggested that lateralization of function may have been present in its basic form when vertebrates emerged about 500 million years ago (MacNeilage, Rogers, & Vallortigara, 2009).

Evolutionary Perspective

Right-handedness may have evolved from a preference for the use of the right side of the body for feeding—such a right-sided preference has been demonstrated in species of all five classes of vertebrates (fishes, reptiles, birds, amphibians, and mammals). Then, once hands evolved, those species with hands (i.e., species of monkeys and apes) displayed a right-hand preference for feeding and other complex responses (see MacNeilage et al., 2009).

A left-hemisphere specialization for communication is also present in species that existed prior to human evolution. For example, you learned in Chapter 2 that the left hemisphere plays the dominant role in birdsong, and the left hemispheres of dogs and monkeys have been found to be dominant in the perception of conspecific calls.

What Are the Survival Advantages of Cerebral Lateralization?

The discovery of examples of cerebral lateralization in species from all five vertebrate classes indicates that cerebral lateralization must have survival advantages: But what are they? There seem to be two fundamental advantages. First, in some cases, it may be more efficient for the neurons performing a particular function to be concentrated in one hemisphere. For example, in most cases, it is advantageous to have one highly skilled hand rather than having two moderately skilled hands. Second, in some cases, two different kinds of cognitive processes may be more readily performed simultaneously if they are lateralized to different hemispheres (see MacNeilage et al., 2009).

Once the control of some abilities becomes lateralized, this may make the lateralization of more abilities advantageous. There may be situations in which there is an advantage to having the control of one ability lateralized in the hemisphere of another ability. For example, the motor theory of lateralization suggests that language became lateralized in the left hemisphere because fine motor control was already lateralized there.

Evolution of Human Language

Human communication is different from the communication of other species. Human language is a system allowing a virtually limitless number of ideas to be expressed by combining a finite set of elements (Hauser, Chomsky, & Fitch, 2005; Wargo, 2008). Other species do have language of sorts, but it can't compare with human language. For example, monkeys have distinct warning calls for different threats, but they do not combine the calls to express new ideas. Also, birds and whales sing complex songs, but creative recombination of the sounds to express new ideas does not occur in these animals either.

Language has been called a human instinct because it is so readily and universally learned by infants. At 10 months of age, infants say little, but 30-month-old infants speak in complete sentences and use over 500 words (Golinkoff & Hirsh-Pasek, 2006). Moreover, over this same 20-month period, the plastic infant brain reorganizes itself to learn its parents' languages. At 10 months, human infants can distinguish the sounds of all human languages, but by 30 months, they can readily discriminate only those sounds that compose the languages to which they have been exposed (Kraus & Banai, 2007). Once the ability to discriminate particular speech sounds is lost, it is difficult to regain, which is one reason why adults usually have difficulty learning to speak new languages without an accent.

Neuroplasticity

Words do not leave fossils, and thus, insights into the evolution of human language can be obtained only through the comparative study of existing species. Naturally enough, researchers interested in the evolution of human language turned first to the vocal communications of our primate relatives.

Vocal Communication in Nonhuman Primates As you have just learned, no other species has a language that can compare with human language. However, each nonhuman primate species has a variety of calls, each with a specific meaning that is understood by conspecifics. Moreover, the calls are not simply reflexive reactions to particular situations: They are dependent on the social context. For example, vervet monkeys do not make alarm calls unless other vervet monkeys are nearby, and the calls are most likely to be made if the nearby vervets are relatives (Cheney & Seyfarth, 2005). And chimpanzees vary the screams they produce during aggressive encounters depending on the severity of the encounter, their role in it, and which other chimpanzees can hear them (Slocombe & Zuberbühler, 2007; Zuberbühler, 2005).

A consistent pattern has emerged from studies of nonhuman vocal language: There is typically a substantial difference between vocal production and auditory comprehension. Even the most vocal nonhumans can produce a relatively few calls, yet they are capable of interpreting a wide range of other sounds in their environments. This suggests that the ability of nonhumans to produce vocal language may be limited, not by their inability to interpret sounds, but by their inability to exert fine motor control over their voices—only humans have this ability. It also suggests that human language may have evolved from a competence in comprehension already existing in our primate ancestors.

Motor Theory of Speech Perception   The **motor theory of speech perception** proposes that the perception of speech depends on the words activating the same neural circuits in the motor system that would be activated if the listener said the words (see Scott, McGettigan, & Eisner, 2009). General support for this theory has come from the discovery that just thinking about performing a particular action often activates the same areas of the brain as performing the action and from the discovery of *mirror neurons* (see Chapter 8), motor cortex neurons that fire when particular responses are either performed or observed (see Fogassi & Ferrari, 2007).

It was reasonable for Broca to believe that Broca's area played a specific role in language expression (speech): After all, Broca's area is part of the left premotor cortex. However, the main thesis of the motor theory of speech perception is that motor cortex plays a role in language comprehension (see Andres, Olivier, & Badets, 2008; Hagoort & Levelt, 2009; Sahin et al., 2009). Indeed, many functional brain-imaging studies have revealed activity in primary or secondary motor cortex during language tests that do not involve language expression (i.e., speaking or writing). Scott, McGettigan, and Eisner (2009) compiled and evaluated the results of studies that recorded activity in motor cortex during speech perception and concluded that the motor cortex is particularly active during conversational exchanges.

Gestural Language   Because only humans are capable of a high degree of motor control over their vocal apparatus, language in nonhuman primates might be mainly gestural, rather than vocal. To test this hypothesis, Pollick and de Waal (2007) compared the gestures and the vocalizations of chimpanzees. They found a highly nuanced vocabulary of hand gestures that were used in many situations and in various combinations. In short, the chimpanzees' gestures were much more like human language than were their vocalizations. Could primate gestures

have been a critical stage in the evolution of human language (Corballis, 2003)?

Scan Your Brain

The chapter now switches its focus to the cerebral mechanisms of language and language disorders. This is a good point for you to review what you have learned about cerebral lateralization by filling in the blanks in the following sentences. The correct answers are provided at the end of the exercise. Be sure to review material related to your errors and omissions before proceeding.

1. The cerebral _____ connect the two hemispheres.
2. Left-hemisphere damage plays a special role in both aphasia and _____.
3. Cortex of the left inferior prefrontal lobe became known as _____.
4. One common test of language lateralization is invasive; it involves injecting _____ into the carotid artery.
5. Some evidence suggests that the brains of males are _____ lateralized than the brains of females.
6. The _____ is the largest cerebral commissure.
7. _____ received a Nobel Prize for his research on split-brain patients.
8. Commissurotomy is an effective treatment for severe cases of _____.
9. The two hemispheres of a split-brain patient can communicate via an external route; such external communication has been termed _____.
10. Damage to the _____ hemisphere is more likely to produce ipsilateral motor problems.
11. Traditionally, musical ability, spatial ability, and _____ have been viewed as right-hemisphere specializations.
12. A neural mechanism metaphorically referred to as the *interpreter* is assumed to reside in the _____ hemisphere.
13. Because broad categories of abilities do not appear to be the units of cerebral lateralization, researchers have turned to studying the laterality of _____ cognitive processes.
14. Three common theories of cerebral asymmetry are the analytic–synthetic theory, the motor theory, and the _____ theory.
15. Broca believed that Broca's area is an area of speech production, but there is now strong evidence that Broca's area and other areas of motor cortex also play a role in language _____.

Scan Your Brain answers: (1) commissures, (2) apraxia, (3) Broca's area, (4) sodium amytal, (5) more, (6) corpus callosum, (7) Sperry, (8) epilepsy, (9) cross-cuing, (10) left, (11) emotions, (12) left, (13) constituent, (14) linguistic, (15) comprehension.

Cortical Localization of Language: Wernicke–Geschwind Model

This section focuses on the cerebral localization of language. In contrast to language lateralization, which refers to the relative control of language-related functions by the left and right hemispheres, *language localization* refers to the location within the hemispheres of the circuits that participate in language-related activities.

Like most introductions to language localization, the following discussion begins with the *Wernicke-Geschwind model*, the predominant theory of language localization. Because most of the research on the localization of language has been conducted and interpreted within the context of this model, reading about the localization of language without a basic understanding of the Wernicke-Geschwind model would be like watching a game of chess without knowing the rules.

Historical Antecedents of the Wernicke–Geschwind Model

The history of the localization of language and the history of the lateralization of function began at the same point, with Broca's assertion that a small area (Broca's area) in the inferior portion of the left prefrontal cortex is the center for speech production. Broca hypothesized that programs of articulation are stored within this area and that speech is produced when these programs activate the adjacent area of the precentral gyrus, which controls the muscles of the face and oral cavity. According to Broca, damage restricted to Broca's area should disrupt speech production without producing deficits in language comprehension.

The next major event in the study of the cerebral localization of language occurred in 1874, when Carl Wernicke

Clinical Implications

(pronounced "VER-ni-key") concluded on the basis of 10 clinical cases that there is a language area in the left temporal lobe just posterior to the primary auditory cortex (i.e., in the left planum temporale). This second language area, which Wernicke argued was the cortical area of language comprehension, subsequently became known as **Wernicke's area**.

Wernicke suggested that selective lesions of Broca's area produce a syndrome of aphasia whose symptoms are primarily **expressive**—characterized by normal comprehension of both written and spoken language and by speech that retains its meaningfulness despite being slow, labored, disjointed, and poorly articulated. This hypothetical form of aphasia became known as **Broca's aphasia**. In contrast, Wernicke suggested that selective lesions of Wernicke's area produce a syndrome of aphasia whose

deficits are primarily **receptive**—characterized by poor comprehension of both written and spoken language and speech that is meaningless but still retains the superficial structure, rhythm, and intonation of normal speech. This hypothetical form of aphasia became known as **Wernicke's aphasia**, and the normal-sounding but nonsensical speech of Wernicke's aphasia became known as *word salad*.

The following are examples of the kinds of speech that are presumed to be associated with selective damage to Broca's and Wernicke's areas (Geschwind, 1979, p. 183):

> **Broca's aphasia**: A patient who was asked about a dental appointment replied haltingly and indistinctly: "Yes . . . Monday . . . Dad and Dick . . . Wednesday nine o'clock . . . 10 o'clock . . . doctors . . . and . . . teeth."
>
> **Wernicke's aphasia**: A patient who was asked to describe a picture that showed two boys stealing cookies reported smoothly: "Mother is away here working her work to get her better, but when she's looking the two boys looking in the other part. She's working another time."

Wernicke reasoned that damage to the pathway connecting Broca's and Wernicke's areas—the **arcuate fasciculus**—would produce a third type of aphasia, one he called **conduction aphasia**. He contended that comprehension and spontaneous speech would be largely intact in patients with damage to the arcuate fasciculus but that they would have difficulty repeating words they had just heard.

The left **angular gyrus**—the area of left temporal and parietal cortex just posterior to Wernicke's area—is another cortical area that has been implicated in language. Its role in language was recognized in 1892 by neurologist Joseph Jules Dejerine on the basis of the postmortem examination of one special patient. The patient suffered from **alexia** (the inability to read) and **agraphia** (the inability to write). What made this case special was that the alexia and agraphia were exceptionally pure: Although the patient could not read or write, he had no difficulty speaking or understanding speech. Dejerine's postmortem examination revealed damage in the pathways connecting the visual cortex with the left angular gyrus. He concluded that the left angular gyrus is responsible for comprehending language-related visual input, which is received directly from the adjacent left visual cortex and indirectly from the right visual cortex via the corpus callosum.

During the era of Broca, Wernicke, and Dejerine, many influential scholars (e.g., Freud, Head, and Marie) opposed their attempts to localize various language-related abilities to specific neocortical areas. In fact, advocates of the holistic approach to brain function gradually gained the upper hand, and interest in the cerebral localization of language waned. However, in the mid-1960s, Norman Geschwind (1970) revived the old localizationist ideas of

FIGURE 16.10 The seven components of the Wernicke-Geschwind model.

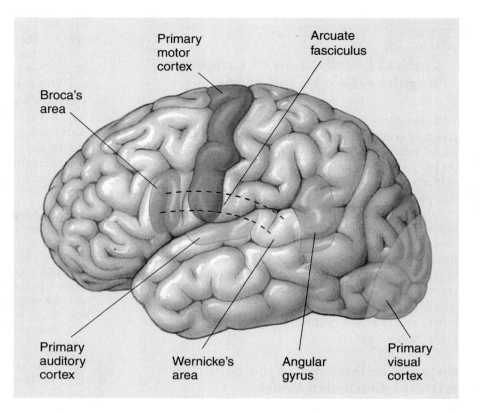

Broca, Wernicke, and Dejerine, added some new data and insightful interpretation, and melded the mix into a powerful theory: the Wernicke-Geschwind model.

The Wernicke–Geschwind Model

The following are the seven components of the **Wernicke-Geschwind model**:

○▶ **Simulate** The Wernicke-Geschwind Model of Language **www.mypsychlab.com**

primary visual cortex, angular gyrus, primary auditory cortex, Wernicke's area, arcuate fasciculus, Broca's area, and primary motor cortex—all of which are in the left hemisphere. They are shown in Figure 16.10.

The following two examples illustrate how the Wernicke-Geschwind model is presumed to work (see Figure 16.11). First, when you are having a conversation, the auditory signals triggered by the speech of the other person are received by your primary auditory cortex and conducted to Wernicke's area, where they are comprehended. If a response is in order, Wernicke's area generates the neural representation of the thought underlying the reply, and it is transmitted to Broca's area via the left arcuate fasciculus. In Broca's area, this signal activates the appropriate program of articulation that drives the appropriate neurons of your primary motor cortex and ultimately your muscles of articulation. Second, when you are reading aloud, the signal received by your primary visual cortex is transmitted to your left angular gyrus, which translates the visual form of the word into its auditory code and transmits it to Wernicke's area for comprehension. Wernicke's area then triggers the appropriate responses in your arcuate fasciculus, Broca's area, and motor cortex, respectively, to elicit the appropriate speech sounds.

are provided at the end of the exercise. Review material related to your errors and omissions before proceeding.

According to the Wernicke-Geschwind model, the following seven areas of the left cerebral cortex play a role in language-related activities:

1. The _____ gyrus translates the visual form of a read word into an auditory code.
2. The _____ cortex controls the muscles of articulation.
3. The _____ cortex perceives the written word.
4. _____ area is the center for language comprehension.
5. The _____ cortex perceives the spoken word.
6. _____ area contains the programs of articulation.
7. The left _____ carries signals from Wernicke's area to Broca's area.

Scan Your Brain answers: (1) angular, (2) primary motor, (3) primary visual, (4) Wernicke's, (5) primary auditory, (6) Broca's, (7) arcuate fasciculus.

Scan Your Brain

Before proceeding to the following evaluation of the Wernicke-Geschwind model, scan your brain to confirm that you understand its fundamentals. The correct answers

16.6

Wernicke–Geschwind Model: The Evidence

Unless you are reading this text from back to front, you should have read the preceding description of the Wernicke-Geschwind model with some degree of skepticism. By this

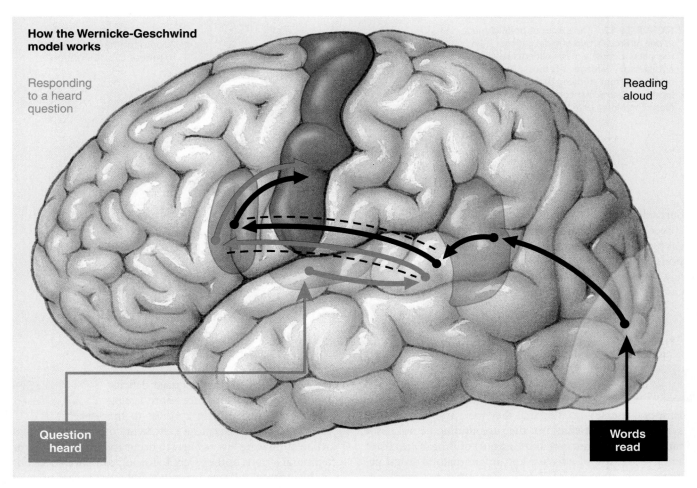

How the Wernicke-Geschwind model works

Responding to a heard question

Reading aloud

Question heard

Words read

FIGURE 16.11   How the Wernicke-Geschwind model works in a person who is responding to a heard question and reading aloud. The hypothetical circuit that allows the person to respond to heard questions is in green; the hypothetical circuit that allows the person to read aloud is in black.

point in the text, you will almost certainly recognize that any model of a complex cognitive process that involves a few localized neocortical centers joined in a serial fashion by a few arrows is sure to have major shortcomings, and you will appreciate that the neocortex is not divided into neat compartments whose cognitive functions conform to vague concepts such as language comprehension, speech motor programs, and conversion of written language to auditory language (see Thompson-Schill, Bedny, & Goldberg, 2005). Initial skepticism aside, the ultimate test of a theory's validity is the degree to which its predictions are consistent with the empirical evidence.

Thinking Creatively

Before we examine this evidence, I want to emphasize one point. The Wernicke-Geschwind model was initially based on case studies of aphasic patients with strokes, tumors, and penetrating brain injuries. Damage in such cases is often diffuse, and it inevitably encroaches on subcortical nerve fibers that connect the lesion site to other areas of the brain (see Bogen & Bogen, 1976). For example, Figure 16.12 on page 432 shows the extent of the cortical damage in one of Broca's two original cases (see Mohr, 1976)—the damage is so diffuse that the case provides little evidence that Broca's area plays a role in speech.

Effects of Cortical Damage on Language Abilities

In view of the fact that the Wernicke-Geschwind model grew out of the study of patients with cortical damage, it is appropriate to begin evaluating it by assessing its ability to predict the language-related deficits produced by damage to various parts of the cortex.

The study of patients in whom discrete areas of cortex have been surgically removed has been particularly informative about the cortical localization of language, because the location and extent of these patients' lesions can be

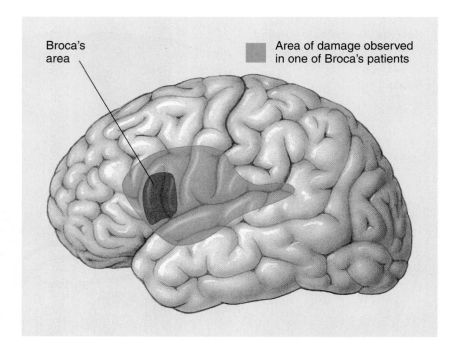

FIGURE 16.12 The extent of brain damage in one of Broca's two original patients. Like this patient, most aphasic patients have diffuse brain damage. It is thus difficult to determine from studying them the precise location of particular cortical language areas. (Based on Mohr, 1976.)

Broca's area

Area of damage observed in one of Broca's patients

derived with reasonable accuracy from the surgeons' reports. The study of neurosurgical patients has not confirmed the predictions of the Wernicke-Geschwind model by any stretch of the imagination. See the six cases summarized in Figure 16.13.

Surgery that destroys all of Broca's area but little surrounding tissue typically has no lasting effects on speech (Penfield & Roberts, 1959; Rasmussen & Milner, 1975; Zangwill, 1975). *Clinical Implications* Some speech problems were observed after the removal of Broca's area, but their temporal course suggested that they were products of postsurgical *edema* (swelling) in the surrounding neural tissue rather than of the *excision* (cutting out) of Broca's area itself. Prior to the use of effective anti-inflammatory drugs, patients with excisions of Broca's area often regained consciousness with their language abilities fully intact only to have serious language-related problems develop over the next few hours and then subside in the following weeks. Similarly, permanent speech difficulties were not produced by discrete surgical lesions to the arcuate fasciculus, and permanent alexia and agraphia were not produced by surgical lesions restricted to the cortex of the angular gyrus (Rasmussen & Milner, 1975).

The consequences of surgical removal of Wernicke's area are less well documented; surgeons have been hesitant to remove it in light of Wernicke's dire predictions. Nevertheless, in some cases, a good portion of Wernicke's area has been removed without lasting language-related deficits (e.g., Ojemann, 1979; Penfield & Roberts, 1959).

Hécaen and Angelergues (1964) published the first large-scale study of accident or disease-related brain damage and aphasia. They rated the articulation, fluency, comprehension, naming ability, ability to repeat spoken sentences, reading, and writing of 214 right-handed patients with left-hemisphere damage. The extent and location of the damage in each case were estimated by either postmortem examination or visual inspection during subsequent surgery.

Hécaen and Angelergues found that small lesions to Broca's area seldom produced lasting language deficits and that small lesions restricted to Wernicke's area did not always produce lasting language deficits. *Clinical Implications* Larger lesions did produce more lasting language deficits; but in contrast to the predictions of the Wernicke-Geschwind model, problems of articulation were just as likely to occur following parietal or temporal lesions as they were following comparable lesions in the vicinity of Broca's area. It is noteworthy that none of the 214 patients studied by Hécaen & Angelergues displayed syndromes of aphasia that were either totally expressive (Broca's aphasia) or totally receptive (Wernicke's aphasia).

Since their development in the 1970s, CT and MRI techniques have been used extensively to analyze the brain damage associated with aphasia. Several large studies have assessed the CT and structural MRI scans of aphasic patients with accidental or disease-related brain damage *Clinical Implications* (e.g., Alexander, 1989; Damasio, 1989; Mazzocchi & Vignolo, 1979; Naeser et al., 1981). In confirming and extending the results of earlier studies, they have not been kind to the Wernicke-Geschwind model. The following have been the major findings:

- No aphasic patients have damage restricted to Broca's area or Wernicke's area.
- Aphasic patients almost always have significant damage to subcortical white matter.
- Large anterior lesions are more likely to produce expressive symptoms, whereas large posterior lesions are more likely to produce receptive symptoms.
- **Global aphasia** (a severe disruption of all language-related abilities) is usually related to massive lesions of anterior cortex, posterior cortex, and underlying white matter.

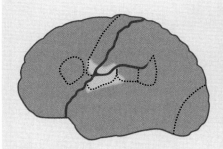

Case J.M. No speech difficulties for 2 days after his surgery, but by Day 3 he was almost totally aphasic; 18 days after his operation he had no difficulty in spontaneous speech, naming, or reading, but his spelling and writing were poor.

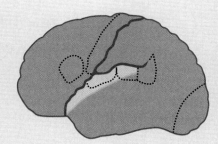

Case H.N. After his operation, he had a slight difficulty in spontaneous speech, but 4 days later he was unable to speak; 23 days after surgery, there were minor deficits in spontaneous speech, naming, and reading aloud, and a marked difficulty in oral calculation.

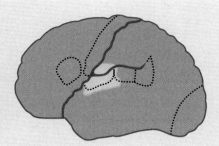

Case J.C. There were no immediate speech problems; 18 hours after his operation he became completely aphasic, but 21 days after surgery, only mild aphasia remained.

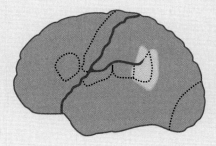

Case P.R. He had no immediate speech difficulties; 2 days after his operation, he had some language-related problems, but they cleared up.

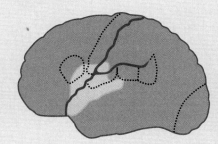

Case D.H. This operation was done in two stages; following completion of the second stage, no speech-related problems were reported.

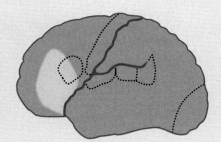

Case A.D. He had no language-related problems after his operation, except for a slight deficit in silent reading and writing.

FIGURE 16.13   The lack of permanent disruption of language-related abilities after surgical excision of the classic Wernicke-Geschwind language areas. (Based on Penfield & Roberts, 1959.)

● Aphasic patients sometimes have brain damage that does not encroach on the Wernicke-Geschwind areas; aphasic patients with damage only to the medial frontal lobe, subcortical white matter, basal ganglia, or thalamus have been reported.

In summary, large-scale, objective studies of the relationship between language deficits and brain damage—whether utilizing autopsy, direct observation during surgery, or brain scan—have not confirmed the major predictions of the Wernicke-Geschwind model. Has the model been treated more favorably by studies of electrical brain stimulation?

Effects of Electrical Stimulation to the Cortex on Language Abilities

The first large-scale electrical brain-stimulation studies of humans were conducted by Wilder Penfield and his colleagues in the 1940s at the Montreal Neurological Institute (see Feindel, 1986). One purpose of the studies was to map the language areas of each patient's brain so that tissue involved in language could be avoided during the surgery. The mapping was done by assessing the responses of conscious patients, who were under local anesthetic, to stimulation applied to various points on the cortical surface. The description of the effects of each stimulation was dictated to a stenographer—this was before the days of tape recorders—and then a tiny numbered card was dropped on the stimulation site for subsequent photography.

Figure 16.14 on page 434 illustrates the responses to stimulation of a 37-year-old right-handed epileptic patient. He had started to have seizures about 3 months after receiving a blow to the head; at the time of his operation, in 1948, he had been suffering from seizures for 6 years, despite

Clinical Implications

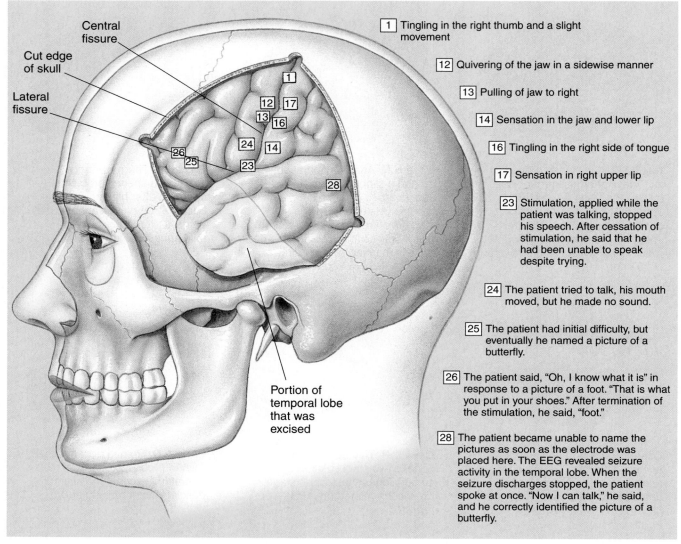

Central fissure

Cut edge of skull

Lateral fissure

Portion of temporal lobe that was excised

1 Tingling in the right thumb and a slight movement

12 Quivering of the jaw in a sidewise manner

13 Pulling of jaw to right

14 Sensation in the jaw and lower lip

16 Tingling in the right side of tongue

17 Sensation in right upper lip

23 Stimulation, applied while the patient was talking, stopped his speech. After cessation of stimulation, he said that he had been unable to speak despite trying.

24 The patient tried to talk, his mouth moved, but he made no sound.

25 The patient had initial difficulty, but eventually he named a picture of a butterfly.

26 The patient said, "Oh, I know what it is" in response to a picture of a foot. "That is what you put in your shoes." After termination of the stimulation, he said, "foot."

28 The patient became unable to name the pictures as soon as the electrode was placed here. The EEG revealed seizure activity in the temporal lobe. When the seizure discharges stopped, the patient spoke at once. "Now I can talk," he said, and he correctly identified the picture of a butterfly.

FIGURE 16.14 The responses of the left hemisphere of a 37-year-old epileptic to electrical stimulation. Numbered cards were placed on the brain during surgery to mark the sites where brain stimulation had been applied. (Based on Penfield & Roberts, 1959.)

efforts to control them with medication. In considering his responses, remember that the cortex just posterior to the central fissure is primary somatosensory cortex and that the cortex just anterior to the central fissure is primary motor cortex.

Because electrical stimulation of the cortex is much more localized than a brain lesion, it has been a useful method for testing predictions of the Wernicke-Geschwind model. Penfield and Roberts (1959) published the first large-scale study of the effects of cortical stimulation on speech. They found that sites at which stimulation blocked or disrupted speech in conscious neurosurgical patients were scattered throughout a large expanse of frontal, temporal, and parietal cortex, rather than being restricted to the Wernicke-Geschwind language areas (see Figure 16.15).

They also found no tendency for particular kinds of speech disturbances to be elicited from particular areas of the cortex: Sites at which stimulation produced disturbances of pronunciation, confusion of counting, inability to name objects, or misnaming of objects were pretty much intermingled. Right-hemisphere stimulation almost never disrupted speech.

Ojemann and his colleagues (see Ojemann, 1983) assessed naming, reading of simple sentences, short-term verbal memory, ability to mimic movements of the face and mouth, and ability to recognize phonemes during cortical stimulation. A **phoneme** is the smallest unit of sound that distinguishes various words in a language; the pronunciation of each phoneme varies slightly, depending on the sounds next to it. The following are the

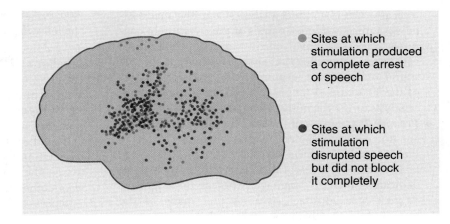

FIGURE 16.15 The wide distribution of left hemisphere sites where cortical stimulation either blocked speech or disrupted it. (Based on Penfield & Roberts, 1959.)

● Sites at which stimulation produced a complete arrest of speech

● Sites at which stimulation disrupted speech but did not block it completely

findings of Ojemann and colleagues related to the Wernicke-Geschwind model:

- Areas of cortex at which stimulation could disrupt language extended far beyond the boundaries of the Wernicke-Geschwind language areas.
- Each of the language tests was disrupted by stimulation at widely scattered sites.
- There were major differences among the subjects in the organization of language abilities (see McDermott, Watson, & Ojemann, 2005).

Because the disruptive effects of stimulation at a particular site were frequently quite specific (i.e., disrupting only a single test), Ojemann suggested that the language cortex is organized like a *mosaic*, with the discrete columns of tissue that perform a particular function widely distributed throughout the language areas of cortex.

Current Status of the Wernicke–Geschwind Model

Evidence from studies of brain damage and from observations of electrical stimulation of the brain has supported the Wernicke-Geschwind model in two general respects. First, the evidence has confirmed that Broca's and Wernicke's areas play important roles in language; many aphasics have diffuse cortical damage that involves one or both of these areas. Second, there is a tendency for aphasias associated with anterior damage to involve deficits that are more expressive and those associated with posterior damage to involve deficits that are more receptive. However, other observations have not confirmed predictions of the Wernicke-Geschwind model:

- Damage restricted to the boundaries of the Wernicke-Geschwind cortical areas often has little lasting effect on the use of language—aphasia is typically associated with widespread damage.
- Brain damage that does not include any of the Wernicke-Geschwind areas can produce aphasia.
- Broca's and Wernicke's aphasias rarely exist in the pure forms implied by the Wernicke-Geschwind model; aphasia virtually always involves both expressive and receptive symptoms (see Benson, 1985).
- There are major differences in the locations of cortical language areas in different individuals (e.g., Casey, 2002; Schlaggar et al., 2002).

Despite these problems, the Wernicke-Geschwind model has been an extremely important theory. It guided the study and clinical diagnosis of aphasia for more than four decades. Indeed, clinical neuropsychologists still use *Broca's aphasia* and *Wernicke's aphasia* as diagnostic categories, but with an understanding that the syndromes are much less selective and the precipitating damage much more diffuse and variable than implied by the model (Alexander, 1997; Hickok & Poeppel, 2007; Poeppel & Monahan, 2008). Because of the lack of empirical support for its major predictions, the Wernicke-Geschwind model has been largely abandoned by researchers, but it is still prominent in the classroom and clinic.

16.7
Cognitive Neuroscience of Language

The *cognitive neuroscience approach*, which currently dominates research on language, is the focus of the final two sections of this chapter. Three premises define the cognitive neuroscience approach to language and differentiate it from the Wenicke-Geschwind model.

- *Premise 1:* The use of language is mediated by activity in all the areas of the brain that participate in the cognitive processes involved in the particular language-related behavior. The Wernicke-Geschwind model theorized that particular areas of the brain involved in language were each dedicated to a complex process such as speech, comprehension, or reading. But cognitive neuroscience research is based on the premise that each of these processes is the combination of several *constituent cognitive processes*, which may be organized separately in different parts of the brain (Neville & Bavelier, 1998). Accordingly, the specific constituent cognitive processes, not the general Wernicke-Geschwind processes, appear to be the appropriate level at which to conduct analysis. Cognitive neuroscientists typically divide analysis of the constituent cognitive processes involved in language into three categories: **phonological analysis** (analysis of the sound of language), **grammatical analysis** (analysis of the structure of language), and **semantic analysis** (analysis of the meaning of language).

- *Premise 2:* The areas of the brain involved in language are not dedicated solely to that purpose (Nobre & Plunkett, 1997). In the Wernicke-Geschwind model, large areas of left cerebral cortex were thought to be dedicated solely to language, whereas the cognitive neuroscience approach assumes that many of the constituent cognitive processes involved in language also play roles in other behaviors (see Bischoff-Grethe et al., 2000). For example, some of the areas of the brain that participate in short-term memory and visual pattern recognition are clearly involved in reading.
- *Premise 3:* Because many of the areas of the brain that perform specific language functions are also parts of other functional systems, these areas are likely to be small, widely distributed, and specialized (Neville & Bavelier, 1998). In contrast, the language areas of the Wernicke-Geschwind model are assumed to be large, circumscribed, and homogeneous.

In addition to these three premises, the methodology of the cognitive neuroscience approach to language distinguishes it from previous approaches. The Wernicke-Geschwind model rested heavily on the analysis of brain-damaged patients, whereas researchers using the cognitive neuroscience approach also employ an array of other techniques—most notably, functional brain imaging—in studying the localization of language in healthy volunteers.

It is important to remember that functional brain-imaging studies cannot prove causation. There is a tendency to assume that brain activity recorded during a particular cognitive process plays a causal role in that process, but, as you learned in Chapter 1, correlation cannot prove causation. For example, substantial right-hemisphere activity is virtually always recorded during various language-related cognitive tasks, and it is tempting to assume that this activity is thus crucial to language-related cognitions. However, lesions of the right hemisphere are only rarely associated with lasting language-related deficits (see Hickok et al., 2008).

Functional Brain Imaging and the Localization of Language

Functional brain-imaging techniques have revolutionized the study of the localization of language. In the last decade, there have been numerous PET and fMRI studies of volunteers engaging in various language-related tasks (see Martin,

2005; Nakamura et al., 2005; Schlaggar & Church, 2009). The following two have been particularly influential.

Bavelier's fMRI Study of Reading   Bavelier and colleagues (1997) used fMRI to measure the brain activity of healthy volunteers while they read silently. The methodology of these researchers was noteworthy in two respects. First, they used a particularly sensitive fMRI machine that allowed them to identify areas of activity with more accuracy than in most previous studies and without having to average the scores of several participants (see Chapter 5). Second, they recorded brain activity during the reading of sentences—rather than during simpler, controllable, unnatural tasks (e.g., listening to individual words) as in most functional brain-imaging studies of language.

The volunteers in Bavelier and colleagues' study viewed sentences displayed on a screen. Interposed between periods of silent reading were control periods, during which the participants were presented with strings of consonants. The differences in cortical activity during the reading and control periods served as the basis for determining those areas of cortical activity associated with reading. Because of the computing power required for the detailed analyses, only the lateral cortical surfaces were monitored.

Let's begin by considering the findings obtained for individual participants on individual trials, before any averaging took place. Three important points emerged from this analysis:

- The areas of activity were patchy; that is, they were tiny areas of activity separated by areas of inactivity.
- The patches of activity were variable; that is, the areas of activity differed from participant to participant and even from trial to trial in the same participant.
- Although some activity was observed in the classic Wernicke-Geschwind areas, it was widespread over the lateral surfaces of the brain.

Even though Bavelier and colleagues' fMRI machine was sensitive enough to render averaging unnecessary, they did average their data in the usual way to illustrate its misleading effects. Figure 16.16 illustrates the reading-related increases of activity averaged over all the trials and participants in the study by Bavelier and colleagues—as they are typically reported. The averaging creates the false impression that large, homogeneous expanses of tissue were active during reading, whereas the patches of activity induced on any given trial comprised only between 5% and 10% of the illustrated areas. Still, two points are clear from the averaged data: First, although there was significant activity in the right hemisphere, there was far more activity in the left hemisphere; second, the activity extended far beyond those areas predicted by the Wernicke-Geschwind model to be involved in silent reading (e.g., activity in Broca's area and the primary motor cortex would not have been predicted).

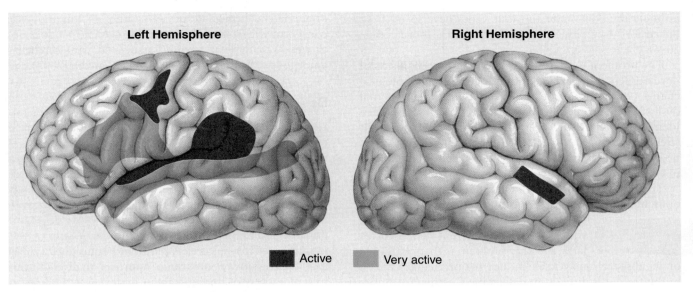

Left Hemisphere   **Right Hemisphere**

■ Active   ■ Very active

FIGURE 16.16 The areas in which reading-associated increases in activity were observed in the fMRI study of Bavelier and colleagues (1997). These maps were derived by averaging the scores of all participants, each of whom displayed patchy increases of activity in 5–10% of the indicated areas on any particular trial.

Damasio's PET Study of Naming The objective of the study of Damasio and colleagues (1996) was to look selectively at the temporal-lobe activity involved in naming objects within particular categories. PET activity was recorded from the left temporal lobes of healthy volunteers while they named images presented on a screen. The images were of three different types: famous faces, animals, and tools. To get a specific measure of the temporal-lobe activity involved in naming, the researchers subtracted from the activity recorded during this task the activity recorded while the volunteers judged the orientation of the images. The researchers limited the PET images to the left temporal lobe to permit a more fine-grained PET analysis.

Naming objects activated areas of the left temporal lobe outside the classic Wernicke's language area. Remarkably, the precise area that was activated by the naming depended on the category: Famous faces, animals, and tools each activated a slightly different area. In general, the areas for naming famous faces, animals, and tools are arrayed from anterior to posterior along the middle portions of the left temporal lobe.

Other functional brain-imaging studies have confirmed category-specific encoding of words in the left temporal lobe (see Binder et al., 2005; Brambati et al., 2006). Moreover, aphasic patients who have naming difficulties that are specific to famous faces, animals, or tools have been shown to have damage in one of the three areas of the left temporal lobe identified by Damasio and colleagues.

16.8
Cognitive Neuroscience of Dyslexia

This final section of the chapter looks at how the cognitive neuroscience approach to language views dyslexia, one of the major subjects of cognitive neuroscience research.

Dyslexia is a pathological difficulty in reading, one that does not result from general visual, motor, or intellectual deficits. There are two fundamentally different types of dyslexias: **developmental dyslexias**, which become apparent when a child is learning to read; and **acquired dyslexias**, which are caused by brain damage in individuals who were already capable of reading. Developmental dyslexia is a widespread problem. Estimates of the overall incidence of developmental dyslexia among English-speaking children range from 5.3% to 11.8%, depending on the criteria that are employed to define dyslexia, and the incidence is two to three times higher among boys than among girls (Katusic et al., 2001). In contrast, acquired dyslexias are relatively rare.

Developmental Dyslexia: Causes and Neural Mechanisms

Because developmental dyslexia is far more common and its causes are less obvious, most research on dyslexia has focused on this form. There is a major genetic component to developmental dyslexia. The disorder has a heritability estimate of about 50%, and four genes have so far been linked to it (see Fisher & Francks, 2006; Galaburda et al., 2006).

The problem in identifying the neural mechanisms of developmental dyslexia is not in discovering pathological changes in the brains of individuals with developmental dyslexia. The problem is that so many changes have been identified that it has been difficult to sort them out (Eckert & Leonard, 2003; Roach & Hogben, 2004). As yet, no single kind of brain pathology has been found to occur in all cases of developmental dyslexia.

The task of identifying the neural correlates of developmental dyslexia is further complicated by the fact that the disorder occurs in various forms, *Neuroplasticity* which likely have different neural correlates. Another problem is that reading—or its absence—may itself induce major changes in the brain; thus, it is difficult to determine whether an abnormality in the brain of a person with developmental dyslexia is likely to be the cause or the result of the disorder (see Price & Mechelli, 2005).

Many researchers who study the neural mechanisms of dyslexia have studied one kind of brain pathology and have tried to attribute developmental dyslexia to it. For example, developmental dyslexia has been attributed to attentional and other sensorimotor deficits caused by damage to neural circuits linked to the *magnocellular layers* (see Chapter 6) of the *lateral geniculate nuclei* (e.g., Stevens & Neville, 2006). Although many patients with dyslexia do experience a variety of subtle visual, auditory, and motor deficits (Wilmer et al., 2004), many do not (see Roach & Hogben, 2004). Moreover, even when visual, auditory, or motor deficits are present in dyslexic patients, they do not account for all aspects of the disorder. As a result, there is now widespread agreement that dyslexia results from a disturbance of *phonological processing* (the representation and comprehension of speech sounds)—not a disturbance of sensorimotor functioning (see Ahissar, 2007; Shaywitz, Mody, & Shaywitz, 2006; Tallal & Gaab, 2006).

Ramus (2004) has proposed a theory that is compatible with much of what is known about the developmental dyslexias and about associated errors in neurodevelopment. Ramus agrees with the view that *Neuroplasticity* deficits in phonological processing are the defining features of acquired dyslexia. The strength of his theory is that it explains why phonological deficits often occur in conjunction with a variety of subtle sensorimotor deficits that can exacerbate the condition. Ramus argues that the first stage in the development of dyslexia is the occurrence of developmental errors in auditory areas around the lateral fissure—these errors in neural development have been well documented, and one of the genes associated with dyslexia controls neuronal migration (see Galaburda et al., 2006). In some individuals, the cortical abnormalities extend back to and affect areas in the thalamus (e.g., the magnocellular layers

of the lateral geniculate nuclei). Although this model is consistent with most of the available evidence, it does not explain why dyslexia is often associated with cerebellar damage (e.g., Justus, 2004; Nicolson & Fawcett, 2007).

Developmental Dyslexia and Culture

Although it is established that developmental dyslexia is associated with abnormalities of brain function, it was once considered to be a psychological rather than a neural disorder. Why? Because for many years, those whose thinking was affected by the physiology-or-psychology dichotomy (see Chapter 2) assumed that developmental dyslexia could not 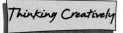 possibly be a brain disorder because it is influenced by culture. Paulesu and colleagues (2001) used the cognitive neuroscience approach to discredit this misguided way of thinking about dyslexia.

Paulesu and colleagues were intrigued by the finding that about twice as many English speakers as Italian speakers are diagnosed as dyslexic. They reasoned that this difference resulted from differences in the complexity of the two languages. English consists of 40 phonemes, which can be spelled, by one count, in 1120 different ways. In contrast, Italian is composed of 25 phonemes, which can be spelled in 33 different ways.

Paulesu and colleagues (2000) began by comparing PET activity in the brains of normal English-speaking and Italian-speaking adults as they read. These researchers hypothesized that since the cognitive demands of reading aloud are different for Italian and English speakers, their volunteers might use different parts of their brains while reading. That is exactly what the researchers found. Although the same general areas were active during reading in both groups, Italian readers displayed more activity in the left superior temporal lobe, whereas English readers displayed more activity in the left inferior temporal and frontal lobes.

Next, Paulesu and colleagues (2001) turned their attention to Italian and English readers with developmental dyslexia. Despite the fact that the Italian dyslexics had less severe reading problems, both groups of dyslexics displayed the same pattern of abnormal PET activity when reading: less than normal reading-related activity in the posterior region of the temporal lobe. Thus, although dyslexia can manifest itself differently in people who speak different languages, the underlying neural pathology appears to be the same. Clearly, the fact that developmental dyslexia is influenced by cultural factors does not preclude the involvement of neural mechanisms.

Cognitive Neuroscience of Deep and Surface Dyslexia

Cognitive psychologists have long recognized that reading aloud can be accomplished in two entirely different ways.

One is by a **lexical procedure**, which is based on specific stored information that has been acquired about written words: The reader simply looks at the word, recognizes it, and says it. The other way reading can be accomplished is by a **phonetic procedure**: The reader looks at the word, recognizes the letters, sounds them out, and says the word. The lexical procedure dominates in the reading of familiar words; the phonetic procedure dominates in the reading of unfamiliar words.

This simple cognitive analysis of reading aloud has proven useful in understanding the symptoms of two kinds of dyslexia resulting from brain damage (see Crisp & Ralph, 2006): *surface dyslexia* and *deep dyslexia*. (Two similar types of developmental dyslexia are also observed, but they tend to be less severe.)

In cases of **surface dyslexia**, patients have lost their ability to pronounce words based on their specific memories of the words (i.e., they have lost the *lexical procedure*), but they can still apply rules of pronunciation in their reading (i.e., they can still use the *phonetic procedure*). Accordingly, they retain their ability to pronounce words whose pronunciation is consistent with common rules (e.g., *fish, river,* and *glass*) and their ability to pronounce nonwords according to common rules of pronunciation (e.g., *spleemer* and *twipple*); but they have great difficulty pronouncing words that do not follow common rules of pronunciation (e.g., *have, lose,* and *steak*). The errors they make often involve the misapplication of common rules of pronunciation; for example, *have, lose,* and *steak* are typically pronounced as if they rhymed with *cave, hose,* and *beak*.

In cases of **deep dyslexia** (also called *phonological dyslexia*), patients have lost their ability to apply rules of pronunciation in their reading (i.e., they have lost the *phonetic procedure*), but they can still pronounce familiar words based on their specific memories of them (i.e., they can still use the *lexical procedure*). Accordingly, they are completely incapable of pronouncing nonwords and have difficulty pronouncing uncommon words and words whose meaning is abstract. In attempting to pronounce words, patients with deep dyslexia try to react to them by using various lexical strategies, such as responding to the overall look of the word, the meaning of the word, or the derivation of the word. This leads to a characteristic pattern of errors. A patient with deep dyslexia might say "quill" for *quail* (responding to the overall look of the word), "hen" for *chicken* (responding to the meaning of the word), or "wise" for *wisdom* (responding to the derivation of the word).

I used to have difficulty keeping these two types of acquired dyslexia straight. Now I remember which is which by reminding myself that surface dyslexics have difficulty reacting to the overall shape of the word, which is metaphorically more superficial (less deep), rather than a problem in applying rules of pronunciation, which is experienced by deep dyslexics.

Clinical Implications

Where are the lexical and phonetic procedures performed in the brain? Much of the research attempting to answer this question has focused on the study of deep dyslexia. Deep dyslexics most often have extensive damage to the left-hemisphere language areas, suggesting that the disrupted phonetic procedure is widely distributed in the frontal and temporal areas of the left hemisphere. But which part of the brain maintains the lexical procedure in deep dyslexics? There have been two theories, both of which have received some support. One theory is that the surviving lexical abilities of deep dyslexics are mediated by activity in surviving parts of the left-hemisphere language areas. Evidence for this theory comes from the observation of neural activity in the surviving regions during reading (Laine et al., 2000; Price et al., 1998). The other theory is that the surviving lexical abilities of deep dyslexics are mediated by activity in the right hemisphere. The following remarkable case study provides support for this theory.

The Case of N.I., the Woman Who Read with Her Right Hemisphere

Prior to the onset of her illness, N.I. was a healthy girl. At the age of 13, she began to experience periods of aphasia, and several weeks later, she suffered a generalized convulsion. She subsequently had many convulsions, and her speech and motor abilities deteriorated badly. CT scans indicated ischemic brain damage to the left hemisphere.

Two years after the onset of her disorder, N.I. was experiencing continual seizures and blindness in her right visual field, and there was no meaningful movement or perception in her right limbs. In an attempt to relieve these symptoms, a total left **hemispherectomy** was performed; that is, her left hemisphere was totally removed. Her seizures were totally arrested by this surgery.

The reading performance of N.I. is poor, but she displays a pattern of retained abilities strikingly similar to those displayed by deep dyslexics or split-brain patients reading with their right hemispheres. For example, she recognizes letters but is totally incapable of translating them into sounds; she can read concrete familiar words; she cannot pronounce even simple nonsense words (e.g., *neg*); and her reading errors indicate that she is reading on the basis of the meaning and appearance of words rather than by translating letters into sounds (e.g., when presented with the word *fruit*, she responded, "Juice . . . it's apples and pears and . . . fruit"). In other words, she suffers from a severe case of deep dyslexia (Patterson, Vargha-Khadem, & Polkey, 1989).

The case of N.I. completes the circle: The chapter began with a discussion of lateralization of function, and the case of N.I. concludes it on the same note.

Themes Revisited

In positioning the themes tabs throughout this chapter, I learned something. I learned why this is one of my favorite chapters: This chapter contributes most to developing the themes of the book. Indeed, several passages in this chapter are directly relevant to more than one of the themes, which made placing the tabs difficult for me.

The clinical implications theme is the most prevalent because much of what we know about the lateralization of *Clinical Implications* function and the localization of language in the brain comes from the study of neuropsychological patients.

Because lateralization of function and language localization are often covered by the popular media, they have become integrated into pop culture, and many widely *Thinking Creatively* held ideas about these subjects are overly simplistic. In this chapter, thinking creatively tabs mark aspects

of laterality and language that require unconventional ways of thinking.

Evolutionary analysis has not played a major role in the study of the localization of language, largely *Evolutionary Perspective* because humans are the only species with well-developed language. However, it has played a key role in trying to understand why cerebral lateralization of function evolved in the first place, and the major breakthrough in understanding the split-brain phenomenon came from comparative research.

The neuroplasticity theme arose during the discussion of the effect of early brain damage on speech lateralization and during the discussion of *Neuroplasticity* the neurodevelopmental bases of developmental dyslexia. Damage to the brain always triggers a series of neuroplastic changes, which can complicate the study of the behavioral effects of the damage.

Think about It

1. Design an experiment to show that it is possible for a split-brain student to study for an English exam and a geometry exam at the same time by using a Z lens.

2. The decision to perform commissurotomies on epileptic patients turned out to be a good one; the decision to perform prefrontal lobotomies on mental patients (see Chapter 1) turned out to be a bad one. Was this just the luck of the draw? Discuss.

3. Design an fMRI study to identify the areas of the brain involved in comprehending speech.

4. Why do you think cerebral lateralization of function evolved?

5. Do chimpanzees have language?

Key Terms

Cerebral commissures (p. 412)
Lateralization of function (p. 412)
Split-brain patients (p. 412)
Commissurotomy (p. 412)

16.1 Cerebral Lateralization of Function: Introduction

Aphasia (p. 413)
Broca's area (p. 413)
Apraxia (p. 413)
Dominant hemisphere (p. 413)
Minor hemisphere (p. 413)
Sodium amytal test (p. 413)
Dichotic listening test (p. 413)
Dextrals (p. 414)
Sinestrals (p. 414)

16.2 The Split Brain

Corpus callosum (p. 414)
Scotoma (p. 415)

Cross-cuing (p. 418)
Helping-hand phenomenon (p. 419)
Visual completion (p. 420)
Chimeric figures test (p. 420)
Z lens (p. 420)

16.3 Differences between Left and Right Hemispheres

Interpreter (p. 424)
Frontal operculum (p. 424)
Planum temporale (p. 424)
Heschl's gyrus (p. 424)

16.4 Evolutionary Perspective of Cerebral Lateralization and Language

Motor theory of speech perception (p. 428)

16.5 Cortical Localization of Language: Wernicke-Geschwind Model

Wernicke's area (p. 429)
Expressive (p. 429)
Broca's aphasia (p. 429)
Receptive (p. 429)
Wernicke's aphasia (p. 429)
Arcuate fasciculus (p. 429)
Conduction aphasia (p. 429)
Angular gyrus (p. 429)
Alexia (p. 429)
Agraphia (p. 429)
Wernicke-Geschwind model (p. 430)

16.6 Wernicke-Geschwind Model: The Evidence

Global aphasia (p. 432)
Phoneme (p. 434)

16.7 Cognitive Neuroscience of Language

Phonological analysis (p. 436)
Grammatical analysis (p. 436)
Semantic analysis (p. 436)

16.8 Cognitive Neuroscience of Dyslexia

Dyslexia (p. 437)
Developmental dyslexias (p. 437)
Acquired dyslexias (p. 437)
Lexical procedure (p. 439)
Phonetic procedure (p. 439)
Surface dyslexia (p. 439)
Deep dyslexia (p. 439)
Hemispherectomy (p. 439)

1. Which of the following is associated with left-hemisphere damage?
    a. apraxia
    b. aphasia
    c. contralateral neglect
    d. all of the above
    e. both a and b

2. The largest cerebral commissure in humans is the
    a. optic chiasm.
    b. anterior commissure.
    c. scotoma.
    d. commissura grandus.
    e. none of the above

3. Who won a Nobel Prize for his research on split-brain patients?
    a. Moniz
    b. Sperry
    c. Wada
    d. Myers
    e. Penfield

4. Broca's area is located in the cortex of the left hemisphere in an area known as
    a. the planum temporal.
    b. Heschl's gyrus.
    c. the frontal operculum.
    d. both a and b
    e. none of the above

5. According to the Wernicke-Geschwind model, which structure transfers information from the area of comprehension to the area of speech production?
    a. arcuate fasciculus
    b. angular gyrus
    c. Wernicke's area
    d. Broca's area
    e. both c and d

17 Biopsychology of Emotion, Stress, and Health
Fear, the Dark Side of Emotion

This chapter about the biopsychology of emotion, stress, and health begins with a historical introduction to the biopsychology of emotion and then focuses in the next two sections on the dark end of the emotional spectrum: fear. Biopsychological research on emotions has concentrated on fear not because biopsychologists are a scary bunch, but because fear has three important qualities: It is the easiest emotion to infer from behavior in various species; it plays an important adaptive function in motivating the avoidance of threatening situations; and chronic fear induces stress. In the final two sections of the chapter, you will learn how stress increases susceptibility to illness and how some brain structures have been implicated in human emotion.

17.1

Biopsychology of Emotion: Introduction

To introduce the biopsychology of emotion, this section reviews several classic early discoveries and then discusses the role of the autonomic nervous system in emotional experience and the facial expression of emotion.

Early Landmarks in the Biopsychological Investigation of Emotion

This subsection describes, in chronological sequence, six early landmarks in the biopsychological investigation of emotion. It begins with the 1848 case of Phineas Gage.

The Mind–Blowing Case of Phineas Gage

In 1848, Phineas Gage, a 25-year-old construction foreman for the Rutland and Burlington Railroad, was the victim of a tragic accident. In order to lay new tracks, the terrain had to be leveled, and Gage was in charge of the blasting. His task involved drilling holes in the rock, pouring some gun powder into each hole, covering it with sand, and tamping the material down with a large tamping iron before detonating it with a fuse. On the fateful day, the gunpowder exploded while Gage was tamping it, launching the 3-cm-thick, 90-cm-long tamping iron through his face, skull, and brain and out the other side.

Amazingly, Gage survived his accident, but he survived it a changed man. Before the accident, Gage had been a responsible, intelligent, socially well-adapted person, who was well liked by his friends and fellow workers. Once recovered, he appeared to be as able-bodied and intellectually capable as before, but his personality and emotional life had totally changed. Formerly a religious, respectful, reliable man,

Gage became irreverent and impulsive. In particular, his abundant profanity offended many. He became so unreliable and undependable that he soon lost his job, and was never again able to hold a responsible position.

Gage became itinerant, roaming the country for a dozen years until his death in San Francisco. His bizarre accident and apparently successful recovery made headlines around the world, but his death went largely unnoticed and unacknowledged.

Gage was buried next to the offending tamping iron. Five years later, neurologist John Harlow was granted permission from Gage's family to exhume the body and tamping iron to study them. Since then, Gage's skull and the tamping iron have been on display in the Warren Anatomical Medical Museum at Harvard University.

In 1994, Damasio and her colleagues brought the power of computerized reconstruction to bear on Gage's classic case. They began by taking an X-ray of the skull and measuring it precisely, paying particular attention to the position of the entry and exit holes. From these measurements, they reconstructed the accident and determined the likely region of Gage's brain damage (see Figure 17.1). It was apparent that the damage to Gage's

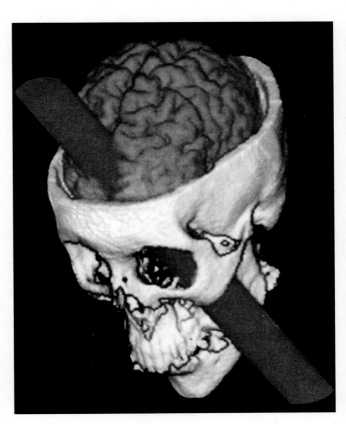

FIGURE 17.1 A reconstruction of the brain injury of Phineas Gage. The damage focused on the medial prefrontal lobes. (Based on Damasio et al., 1994.)

brain affected both medial prefrontal lobes, which we now know are involved in planning and emotion (see Machado & Bachevalier, 2006; Vogt, 2005).

Darwin's Theory of the Evolution of Emotion   The first major event in the study of the biopsychology of emotion was the publication in 1872 of Darwin's book *The Expression of Emotions in Man and Animals*. In it, Darwin argued, largely on the basis of anecdotal evidence, that particular emotional responses, such as human facial expressions, tend to accompany the same emotional states in all members of a species.

Evolutionary Perspective

Darwin believed that expressions of emotion, like other behaviors, are products of evolution; he therefore tried to understand them by comparing them in different species. From such interspecies comparisons, Darwin developed a theory of the evolution of emotional expression that was composed of three main ideas:

- Expressions of emotion evolve from behaviors that indicate what an animal is likely to do next.
- If the signals provided by such behaviors benefit the animal that displays them, they will evolve in ways that enhance their communicative function, and their original function may be lost.
- Opposite messages are often signaled by opposite movements and postures, an idea called the *principle of antithesis*.

Consider how Darwin's theory accounts for the evolution of *threat displays*. Originally, facing one's enemies, rising up, and exposing one's weapons were the components of the early stages of combat. But once enemies began to recognize these behaviors as signals of impending aggression, a survival advantage accrued to attackers that could communicate their aggression most effectively and intimidate their victims without actually fighting. As a result, elaborate threat displays evolved, and actual combat declined.

To be most effective, signals of aggression and submission must be clearly distinguishable; thus, they tended to evolve in opposite directions. For example, gulls signal aggression by pointing their beaks at one another and submission by pointing their beaks away from one another; primates signal aggression by staring and submission by averting their gaze. Figure 17.2 reproduces the woodcuts Darwin used in his 1872 book to illustrate this principle of antithesis in dogs.

James-Lange and Cannon-Bard Theories   The first physiological theory of emotion was proposed independently by James and Lange in 1884. According to the **James-Lange theory**, emotion-inducing sensory stimuli are received and interpreted by the cortex, which triggers changes in the visceral organs via the autonomic nervous system and in the skeletal muscles via the somatic nervous system. Then, the autonomic and somatic responses trigger the experience of emotion in the brain. In effect,

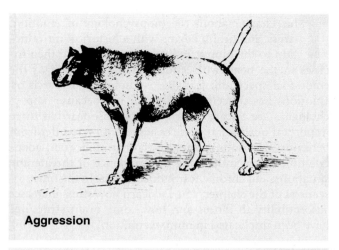

Aggression

Submission

FIGURE 17.2   Two woodcuts from Darwin's 1872 book, *The Expression of Emotions in Man and Animals*, that he used to illustrate the principle of antithesis. The aggressive posture of dogs features ears forward, back up, hair up, and tail up; the submissive posture features ears back, back down, hair down, and tail down.

what the James-Lange theory did was to reverse the usual common-sense way of thinking about the causal relation between the experience of emotion and its expression. James and Lange argued that the autonomic activity and behavior that are triggered by the emotional event (e.g., rapid heartbeat and running away) produce the feeling of emotion, not vice versa.

Around 1915, Cannon proposed an alternative to the James-Lange theory of emotion, and it was subsequently extended and promoted by Bard. According to the **Cannon-Bard theory**, emotional stimuli have two independent excitatory effects: They excite both the feeling of emotion in the brain and the expression of emotion in the autonomic and somatic nervous systems. That is, the Cannon-Bard theory, in contrast to the James-Lange theory, views emotional experience and emotional expression as parallel processes that have no direct causal relation.

The James-Lange and Cannon-Bard theories make different predictions about the role of feedback from autonomic and somatic nervous system activity in emotional experience. According to the James-Lange theory, emotional experience depends entirely on feedback from autonomic and somatic nervous system activity; according to the Cannon-Bard theory, emotional experience is totally independent of such feedback. Both extreme positions have proved to be incorrect. On the one hand, it seems that the autonomic and somatic feedback is not necessary for the experience of emotion: Human patients whose autonomic and somatic feedback has been largely eliminated by a broken neck are capable of a full range of emotional experiences (e.g., Lowe & Carroll, 1985). On the other hand, there have been numerous reports—some of which you will soon encounter—that autonomic and somatic responses to emotional stimuli can influence emotional experience.

Failure to find unqualified support for either the James-Lange or the Cannon-Bard theory led to the modern biopsychological view. According to this view, each of the three principal factors in an emotional response—the perception of the emotion-inducing stimulus, the autonomic and somatic responses to the stimulus, and the experience of the emotion—can influence the other two (see Figure 17.3).

Sham Rage In the late 1920s, Bard (1929) discovered that **decorticate** cats—cats whose cortex has been removed—respond aggressively to the slightest provocation: After a light touch, they arch their backs, erect their hair, growl, hiss, and expose their teeth.

The aggressive responses of decorticate animals are abnormal in two respects: They are inappropriately severe, and they are not directed at particular targets. Bard referred to the exaggerated, poorly directed aggressive responses of decorticate animals as **sham rage**.

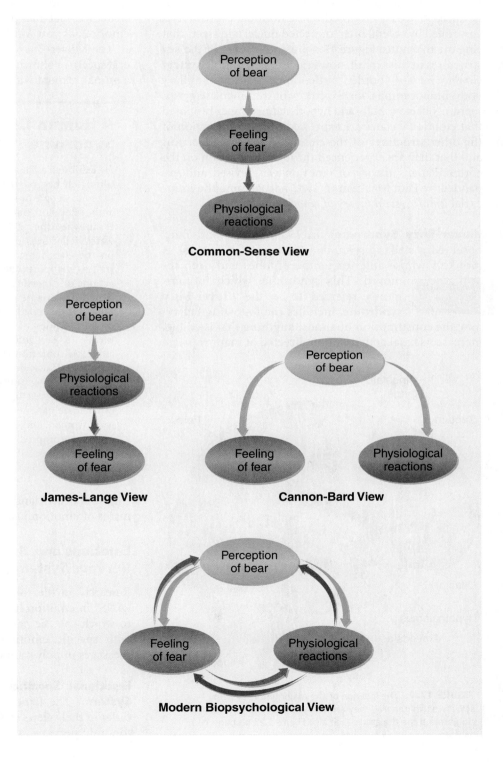

FIGURE 17.3 Four ways of thinking about the relations among the perception of emotion-inducing stimuli, the autonomic and somatic responses to the stimuli, and the emotional experience.

Sham rage can be elicited in cats whose cerebral hemispheres have been removed down to, but not including, the hypothalamus; but it cannot be elicited if the hypothalamus is also removed. On the basis of this observation, Bard concluded that the hypothalamus is critical for the expression of aggressive responses and that the function of the cortex is to inhibit and direct these responses.

Limbic System and Emotion In 1937, Papez (pronounced "Payps") proposed that emotional expression is controlled by several interconnected nuclei and tracts that ring the thalamus. Figure 17.4 illustrates some of the key structures in this circuit, now known as the **limbic system** (*limbic* means "border"): the amygdala, mammillary body, hippocampus, fornix, cortex of the cingulate gyrus, septum, olfactory bulb, and hypothalamus. Papez proposed that emotional states are expressed through the action of the other structures of the circuit on the hypothalamus and that they are experienced through their action on the cortex. Papez's theory of emotion was revised and expanded by Paul MacLean in 1952 and became the influential *limbic system theory of emotion.*

Kluver-Bucy Syndrome In 1939, Kluver and Bucy observed a striking *syndrome* (pattern of behavior) in monkeys whose anterior temporal lobes had been removed. This syndrome, which is commonly referred to as the **Kluver-Bucy syndrome**, includes the following behaviors: the consumption of almost anything that is edible, increased sexual activity often directed at inappropriate

Evolutionary Perspective

objects, a tendency to repeatedly investigate familiar objects, a tendency to investigate objects with the mouth, and a lack of fear. Monkeys that could not be handled before surgery were transformed by bilateral anterior temporal lobectomy into tame subjects that showed no fear whatsoever—even in response to snakes, which terrify normal monkeys. In primates, most of the symptoms of the Kluver-Bucy syndrome appear to result from damage to the **amygdala** (see Phelps, 2006), a structure that has played a major role in research on emotion, as you will learn later in this chapter.

The Kluver-Bucy syndrome has been observed in several species. Following is a description of the syndrome in a human patient with a brain infection.

A Human Case of Kluver-Bucy Syndrome

He exhibited a flat affect, and although originally restless, ultimately became remarkably placid. He appeared indifferent to people or situations. He spent much time gazing at the television, but never learned to turn it on; when the set was off, he tended to watch reflections of others in the room on the glass screen. On occasion he became facetious, smiling inappropriately and mimicking the gestures and actions of others. Once initiating an imitative series, he would perseverate copying all movements made by another for extended periods of time. . . . He engaged in oral exploration of all objects within his grasp, appearing unable to gain information via tactile or visual means alone. All objects that he could lift were placed in his mouth and sucked or chewed. . . .

Clinical Implications

Although vigorously heterosexual prior to his illness, he was observed in hospital to make advances toward other male patients. . . . [H]e never made advances toward women, and, in fact, his apparent reversal of sexual polarity prompted his fiancée to sever their relationship. (Marlowe, Mancall, & Thomas, 1985, pp. 55–56)

The six early landmarks in the study of brain mechanisms of emotion just reviewed are listed in Table 17.1.

Emotions and the Autonomic Nervous System

Research on the role of the autonomic nervous system (ANS) in emotion has focused on two issues: the degree to which specific patterns of ANS activity are associated with specific emotions and the effectiveness of ANS measures in polygraphy (lie detection).

Emotional Specificity of the Autonomic Nervous System The James-Lange and Cannon-Bard theories differ in their views of the emotional specificity of the autonomic nervous system. The James-Lange theory says

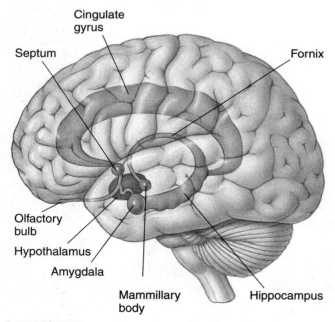

Cingulate gyrus

Septum

Fornix

Olfactory bulb

Hypothalamus

Amygdala

Mammillary body

Hippocampus

FIGURE 17.4 The location of the major limbic system structures. In general, they are arrayed near the midline in a ring around the thalamus. (See also Figure 3.28 on page 70.)

TABLE 17.1	Biopsychological Investigation of Emotion: Six Early Landmarks
Event	*Date*
Case of Phineas Gage	1848
Darwin's theory of the evolution of emotion	1872
James-Lange and Cannon-Bard theories	about 1900
Discovery of sham rage	1929
Discovery of Kluver-Bucy syndrome	1939
Limbic system theory of emotion	1952

that different emotional stimuli induce different patterns of ANS activity and that these different patterns produce different emotional experiences. In contrast, the Cannon-Bard theory claims that all emotional stimuli produce the same general pattern of sympathetic activation, which prepares the organism for action (i.e., increased heart rate, increased blood pressure, pupil dilation, increased flow of blood to the muscles, increased respiration, and increased release of epinephrine and norepinephrine from the adrenal medulla).

The experimental evidence suggests that the specificity of ANS reactions lies somewhere between the extremes of total specificity and total generality (Levenson, 1994). There is ample evidence that not all emotions are associated with the same pattern of ANS activity (see Ax, 1955); however, there is insufficient evidence to make a strong case for the view that each emotion is characterized by a different pattern of ANS activity.

Polygraphy **Polygraphy** is a method of interrogation that employs autonomic nervous system indexes of emotion to infer the truthfulness of the subject's responses. Polygraph tests administered by skilled examiners can be useful additions to normal interrogation procedures, but they are far from infallible.

The main problem in evaluating the effectiveness of polygraphy is that it is rarely possible in real-life situations to know for certain whether a suspect is guilty or innocent. Consequently, many studies of polygraphy have employed the *mock-crime procedure*: Volunteer subjects participate in a mock crime and are then subjected to a polygraph test by an examiner who is unaware of their "guilt" or "innocence." The usual interrogation method is the **control-question technique**, in which the physiological response to the target question (e.g., "Did you steal that purse?") is compared with the physiological responses to control questions whose answers are known (e.g., "Have you ever been in jail before?"). The assumption is that lying will be associated with greater sympathetic activation. The average success rate in various mock-crime studies using the control-question technique is about 80%.

Despite being commonly referred to as *lie detection*, polygraphy detects emotions, not lies. Consequently, it is less likely to successfully identify lies in real life than in experiments. In real-life situations, questions such as "Did you steal that purse?" are likely to elicit a reaction from all suspects, regardless of their guilt or innocence, making it difficult to detect deception. The **guilty-knowledge technique** circumvents this problem. In order to use this technique, the polygrapher must have a piece of information concerning the crime that would be known only to the guilty person. Rather than attempting to catch the suspect in a lie, the polygrapher simply assesses the suspect's reaction to a list of actual and contrived details of the crime. Innocent suspects, because they have no knowledge of the crime, react to all such details in the same way; the guilty react differentially.

In the classic study of the guilty-knowledge technique (Lykken, 1959), subjects waited until the occupant of an office went to the washroom. Then, they entered her office, stole her purse from her desk, removed the money, and left the purse in a locker. The critical part of the interrogation went something like this: "Where do you think we found the purse? In the washroom? . . . In a locker? . . . Hanging on a coat rack? . . ." Even though electrodermal activity was the only measure of ANS activity used in this study, 88% of the mock criminals were correctly identified; more importantly, none of the innocent subjects was judged guilty—see MacLaren (2001) for a review.

Emotions and Facial Expression

Ekman and his colleagues have been preeminent in the study of facial expression (see Ekman, 2003). They began in the 1960s by analyzing hundreds of films and photographs of people experiencing various real emotions. From these, they compiled an atlas of the facial expressions that are normally associated with different emotions (Ekman & Friesen, 1975). For example, to produce the facial expression for surprise, models were instructed to pull their brows upward so as to wrinkle their forehead, to open their eyes wide so as to reveal white above the iris, to slacken the muscles around their mouth, and to drop their jaw. Try it.

Universality of Facial Expression Several studies have found that people of different cultures make similar facial expressions in similar situations and that they can correctly identify the emotional significance of facial expressions displayed by people from cultures other than their own. The most convincing of these studies was a study of the members of an isolated New Guinea tribe who had had little or no contact with the outside world (Ekman & Friesen, 1971). However, some studies have identified some subtle cultural differences in facial expressions (see Russell, Bachorowski, & Fernandez-Dols, 2003). Remarkably, human facial expressions are similar in many respects to those of our primate relatives (see Parr, Waller, & Fugate, 2005; Parr, Waller, & Vick, 2007).

Primary Facial Expressions Ekman and Friesen concluded that the facial expressions of the following six emotions are primary: surprise, anger, sadness, disgust, fear, and happiness (however, see Tracy & Robins, 2004). They further concluded that all other facial expressions of genuine emotion are composed of predictable mixtures of these six primaries. Figure 17.5 illustrates these six primary facial expressions and the combination of two of them to form a nonprimary expression.

Facial Feedback Hypothesis Is there any truth to the old idea that putting on a happy face can make you feel better? Research suggests that there is (see Adelmann & Zajonc, 1989). The hypothesis that our facial expressions influence our emotional experience is called the **facial feedback hypothesis**. In a test of the facial feedback hypothesis, Rutledge and Hupka (1985) instructed subjects to assume one of two patterns of facial contractions while they viewed a series of slides; the patterns corresponded to happy or angry faces, although the subjects were unaware of that. The subjects reported that the slides made them feel more happy and less angry

when they were making happy faces, and less happy and more angry when they were making angry faces (see Figure 17.6).

Voluntary Control of Facial Expression Because we can exert voluntary control over our facial muscles, it is possible to inhibit true facial expressions and to substitute false ones. There are many reasons for choosing to put on a false facial expression. Some of them are positive (e.g., putting on a false smile to reassure a worried friend), and some are negative (e.g., putting on a false smile to disguise a lie). In either case, it is difficult to fool an expert.

There are two ways of distinguishing true expressions from false ones (Ekman, 1985). First, *microexpressions* (brief facial expressions) of the real emotion often break through the false one (Porter & ten Brinke, 2008). Such microexpressions last only about 0.05 second, but with practice they can be detected without the aid of slow-motion photography. Second, there are often subtle differences between genuine facial expressions and false ones that can be detected by skilled observers.

The most widely studied difference between a genuine and a false facial expression was first described by the French anatomist Duchenne in 1862. Duchenne said that the smile of enjoyment could be distinguished from

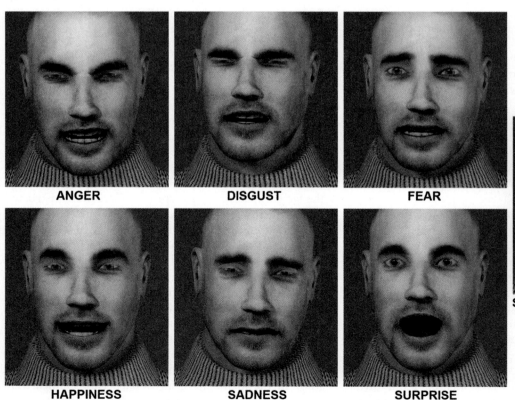

ANGER   DISGUST   FEAR

HAPPINESS   SADNESS   SURPRISE

SAD EYES / HAPPY MOUTH

FIGURE 17.5 Ekman's six primary facial expressions of emotion, and one combination facial expression. (Generously supplied by Kyung Jae Lee and Stephen DiPaola of the iVizLab, Simon Fraser University. The expressions were created in video game character style using FaceFx 3D software, which allows DiPaola and Lee to create and control facial expressions of emotion in stills and animated sequences; see ivizlab.sfu.ca).

FIGURE 17.6 The effects of facial expression on the experience of emotion. Participants reported feeling more happy and less angry when they viewed slides while making a happy face, and less happy and more angry when they viewed slides while making an angry face. (Based on Rutledge & Hupka, 1985.)

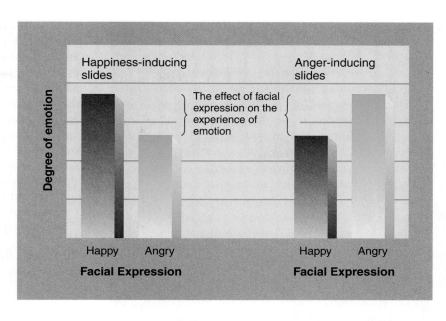

deliberately produced smiles by consideration of the two facial muscles that are contracted during genuine smiles: *orbicularis oculi*, which encircles the eye and pulls the skin from the cheeks and forehead toward the eyeball, and *zygomaticus major*, which pulls the lip corners up (see Figure 17.7). According to Duchenne, the zygomaticus major can be controlled voluntarily, whereas the orbicularis oculi is normally contracted only by genuine pleasure. Thus, inertia of the orbicularis oculi in smiling

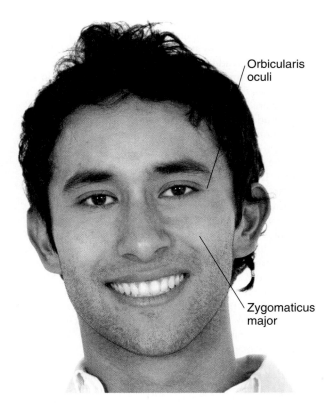

Orbicularis oculi

Zygomaticus major

FIGURE 17.7 A fake smile. The orbicularis oculi and the zygomaticus major are two muscles that contract during genuine (Duchenne) smiles. Because the lateral portion of the orbicularis oculi is difficult for most people to contract voluntarily, fake smiles usually lack this component. This young man is faking a smile for the camera. Look at his eyes.

unmasks a false friend—a fact you would do well to remember. Ekman named the genuine smile the **Duchenne smile** (see Ekman & Davidson, 1993).

Facial Expressions: Current Perspectives Ekman's work on facial expression began before video recording became commonplace. Now, video recordings provide almost unlimited access to natural facial expressions made in response to real-life situations. As a result, it is now clear that Ekman's six primary facial expressions of emotion rarely occur in pure form—they are ideals with many subtle variations. Also, the existence of other primary emotions has been recognized. For example, Ekman (1992) agrees that there is evidence for adding contempt and embarrassment to his original six.

Have you noticed that only one of the eight primary emotions, happiness, has a positive emotional valence? (*Emotional valence* refers to the general positive or negative character of an emotion.) This imbalance has led some to hypothesize that all positive emotions may share the same facial expression. The research on pride by Tracy

Check It Out

Experiencing Facial Feedback

Why don't you try the facial feedback hypothesis? Pull your eyebrows down and together; raise your upper eyelids and tighten your lower eyelids, and narrow your lips and press them together. Now, hold this expression for a few seconds. If it makes you feel slightly angry and uncomfortable, you have just experienced the effect of facial feedback.

FIGURE 17.8 An expression of pride. (Reproduced with permission of Jessica Tracy, Department of Psychology, University of British Columbia.)

and Robins (2004, 2007a, 2007b) argues against this view. The expression of pride is readily identified by individuals of various cultures, cannot be created from a mixture of other primary expressions, and involves postural as well as facial components (Tracy & Robins, 2007a). Pride is expressed through a small smile, with the head tilted back slightly and the hands on the hips, raised above the head, or clenched in fists with the arms crossed on the chest—see Figure 17.8.

17.2

Fear, Defense, and Aggression

Most biopsychological research on emotion has focused on fear and defensive behaviors. **Fear** is the emotional reaction to threat; it is the motivating force for defensive behaviors. **Defensive behaviors** are behaviors whose primary function is to protect the organism from threat or harm. In contrast, **aggressive behaviors** are behaviors whose primary function is to threaten or harm.

⊙➤ **Simulate** Coping Strategies and Their Effect; Defense Mechanisms
www.mypsychlab.com

Although one purpose of this section is to discuss fear, defense, and aggression, it has another important purpose: to explain a common problem faced by biopsychologists and the way in which those who conduct research in this particular area have managed to circumvent it. Barrett (2006) pointed out that progress in the study of the neu-

ral basis of emotion has been limited because neuroscientists have often been guided by unsubstantiated cultural assumptions about emotion: Because we have words such as *fear*, *happiness*, and *anger* in our language, scientists have often assumed that these emotions exist as entities in the brain, and they have searched for them—often with little success. The following lines of research on fear, defense, and aggression illustrate how biopsychologists can overcome the problem of vague, subjective, everyday concepts by basing their search for neural mechanisms on the thorough descriptions of relevant behaviors and the environments in which they occur, and on the putative adaptive functions of such behaviors (see Barrett & Wager, 2006; Panksepp, 2007).

Types of Aggressive and Defensive Behaviors

Considerable progress in the understanding of aggressive and defensive behaviors has come from the research of Blanchard and Blanchard (see 1989, 1990) on the *colony-intruder model of aggression and defense* in rats. Blanchard and Blanchard *Evolutionary Perspective* have derived rich descriptions of rat intraspecific aggressive and defensive behaviors by studying the interactions between the **alpha male**—the dominant male—of an established mixed-sex colony and a small male intruder:

> The alpha male usually bites the intruder [on the back], and the intruder runs away. The alpha chases after it, and after one or two additional [back] bites, the intruder stops running and turns to face its attacker. It rears up on its hind legs, using its forelimbs to push off the alpha. . . . However, rather than standing nose to nose with the "boxing" intruder, the attacking rat abruptly moves to a lateral orientation, with the long axis of its body perpendicular to the front of the defending rat. . . . It moves sideways toward the intruder, crowding and sometimes pushing it off balance. If the defending rat stands solid against this "lateral attack" movement, the alpha may make a quick lunge forward and around the defender's body to bite at its back. In response to such a lunge, the defender usually pivots on its hind feet, in the same direction as the attacker is moving, continuing its frontal orientation to the attacker. If the defending rat moves quickly enough, no bite will be made. (From "Affect and Aggression: An Animal Model Applied to Human Behavior," by D. C. Blanchard and R. J. Blanchard, in *Advances in the Study of Aggression*, Vol. 1, 1984, edited by D. C. Blanchard and R. J. Blanchard. San Diego: Academic Press. Copyright © 1984 by Academic Press. Reprinted by permission of Elsevier.)

Another excellent illustration of how careful observation of behavior has led to improved understanding of aggressive and defensive behaviors is provided by the study of Pellis and colleagues (1988) of cats. They began by videotaping interactions between cats and mice. They found that different cats reacted to mice in different ways: Some were efficient mouse killers, some reacted defensively, and some seemed to play with the mice. Careful analysis of the "play" sequences led to two important conclusions. The first conclusion was that, in contrast to the common belief, cats do not play with their prey; the cats that appeared to be playing with the mice were simply vacillating between attack and defense. The second conclusion was that one can best understand each cat's interactions with mice by locating the interactions on a linear scale, with total aggressiveness at one end, total defensiveness at the other, and various proportions of the two in between.

Pellis and colleagues tested their conclusions by reducing the defensiveness of the cats with an antianxiety drug. As predicted, the drug moved each cat along the scale toward more efficient killing. Cats that avoided mice before the injection "played with" them after the injection, those that "played with" them before the injection killed them after the injection, and those that killed them before the injection killed them more quickly after the injection.

Based on the numerous detailed descriptions of rat aggressive and defensive behaviors provided by the Blanchards and other biopsychologists who have followed their example, most researchers now distinguish among different categories of such behaviors. These categories of rat aggressive and defensive behaviors are based on three criteria: (1) their *topography* (form), (2) the situations that elicit them, and (3) their apparent function. Several of these categories are described in Table 17.2 (see also Blanchard et al., 2001; Dielenberg & McGregor, 2001; Kavaliers & Choleris, 2001).

The analysis of aggressive and defensive behaviors has led to the development of the **target-site concept**—the idea that the aggressive and defensive behaviors of an animal are often designed to attack specific sites on the body of another animal while protecting specific sites on its own. For example, the behavior of a socially aggressive rat (e.g., lateral attack) appears to be designed to deliver bites to the defending rat's back and to protect its own face, the likely target of a defensive attack. Conversely, most of the maneuvers of the defending rat (e.g., boxing and pivoting) appear to be designed to protect the target site on its back.

TABLE 17.2	Categories of Aggressive and Defensive Behaviors in Rats	
Aggressive Behaviors	*Predatory Aggression*	The stalking and killing of members of other species for the purpose of eating them. Rats kill prey, such as mice and frogs, by delivering bites to the back of the neck.
	Social Aggression	Unprovoked aggressive behavior that is directed at a *conspecific* (member of the same species) for the purpose of establishing, altering, or maintaining a social hierarchy. In mammals, social aggression occurs primarily among males. In rats, it is characterized by piloerection, lateral attack, and bites directed at the defender's back.
Defensive Behaviors	*Intraspecific Defense*	Defense against social aggression. In rats, it is characterized by freezing and flight and by various behaviors, such as boxing, that are specifically designed to protect the back from bites.
	Defensive Attacks	Attacks that are launched by animals when they are cornered by threatening members of their own or other species. In rats, they include lunging, shrieking, and biting attacks that are usually directed at the face of the attacker.
	Freezing and Flight	Responses that many animals use to avoid attack. For example, if a human approaches a wild rat, it will often freeze until the human penetrates its safety zone, whereupon it will explode into flight.
	Maternal Defensive Behaviors	The behaviors by which mothers protect their young. Despite their defensive function, they are similar to male social aggression in appearance.
	Risk Assessment	Behaviors that are performed by animals in order to obtain specific information that helps them defend themselves more effectively. For example, rats that have been chased by a cat into their burrow do not emerge until they have spent considerable time at the entrance scanning the surrounding environment.
	Defensive Burying	Rats and other rodents spray sand and dirt ahead with their forepaws to bury dangerous objects in their environment, to drive off predators, and to construct barriers in burrows.

The discovery that aggressive and defensive behaviors occur in a variety of stereotypical species-common forms was the necessary first step in the identification of their neural bases. Because the different categories of aggressive and defensive behaviors are mediated by different neural circuits, little progress was made in identifying these circuits before the categories were delineated. For example, the lateral septum was once believed to inhibit all aggression, because lateral septal lesions rendered laboratory rats notoriously difficult to handle—the behavior of the lesioned rats was commonly referred to as *septal aggression* or *septal rage*. However, we now know that lateral septal lesions do not increase aggression: Rats with lateral septal lesions do not initiate attacks at an experimenter unless they are threatened.

Aggression and Testosterone

The fact that social aggression in many species occurs more commonly among males than among females is usually explained with reference to the organizational and activational effects of testosterone. The brief period of testosterone release that occurs around birth in genetic males is thought to organize their nervous systems along masculine lines and hence to create the potential for male patterns of social aggression to be activated by the high testosterone levels that are present after puberty. These organizational and activational effects have been demonstrated in some mammalian species. For example, neonatal castration of male mice eliminates the ability of testosterone injections to induce social aggression in adulthood, and adult castration eliminates social aggression in males that do not receive testosterone replacement injections. Unfortunately, research on testosterone and aggression in other species has not been so straightforward (see Wingfield, 2005).

 Soma and his colleagues have reviewed the extensive comparative research literature on testosterone and aggression (Demas et al., 2005; Soma, 2006). Here are their major conclusions:

- Testosterone increases social aggression in the males of many species; aggression is largely abolished by castration in these same species.
- In some species, castration has no effect on social aggression; in still others, castration reduces aggression during the breeding season but not at other times.
- The relation between aggression and testosterone levels is difficult to interpret because engaging in aggressive activity can itself increase testosterone levels—for example, just playing with a gun increased the testosterone levels of male college students (Klinesmith, Kasser, & McAndrew, 2006).
- The blood level of testosterone, which is the only measure used in many studies, is not the best measure.

What matters more are the testosterone levels in the relevant areas of the brain. Although studies focusing on brain levels of testosterone are rare, it has been shown that testosterone can be synthesized in particular brain sites and not in others.

It is unlikely that humans are an exception to the usual involvement of testosterone in mammalian social aggression. However, the evidence is far from clear. In human males, aggressive behavior does not increase at puberty as testosterone levels in the blood increase; aggressive behavior is not eliminated by castration; and it is not increased by testosterone injections that elevate blood levels of testosterone. A few studies have found that violent male criminals and aggressive male athletes tend to have higher testosterone levels than normal (see Bernhardt, 1997); however, this correlation may indicate that aggressive encounters increase testosterone, rather than vice versa.

The lack of strong evidence of the involvement of testosterone in human aggression could mean that hormonal and neural regulation of aggression in humans differs from that 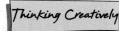 in many other mammalian species. Or, it could mean that the research on human aggression and testosterone is flawed. For example, human studies are typically based on blood levels of testosterone, often inferred from levels in saliva, rather than on brain levels. Also, the researchers who study human aggression have often failed to appreciate the difference between social aggression, which is related to testosterone in many species, and defensive attack, which is not. Most aggressive outbursts in humans are overreactions to real or perceived threat, and thus they are more appropriately viewed as defensive attack, not social aggression.

17.3
Neural Mechanisms of Fear Conditioning

Much of what we know about the neural mechanisms of fear has come from the study of fear conditioning (Olsson & Phelps, 2007). **Fear conditioning** is the establishment of fear in response to a previously neutral stimulus (the *conditional stimulus*) by presenting it, usually several times, before the delivery of an aversive stimulus (the *unconditional stimulus*).

In the usual fear-conditioning experiment, the subject, often a rat, hears a tone (conditional stimulus) and then receives a mild electric shock to its feet (unconditional stimulus). After several pairings of the tone and the shock, the rat responds to the tone with a variety of defensive behaviors (e.g., freezing and increased susceptibility to startle) and sympathetic nervous system responses (e.g., increased heart rate and blood pressure). LeDoux and his colleagues have mapped the neural mechanism that mediates this form of auditory fear conditioning (see Schafe & LeDoux, 2004).

Amygdala and Fear Conditioning

LeDoux and his colleagues began their search for the neural mechanisms of auditory fear conditioning by making lesions in the auditory pathways of rats. They found that bilateral lesions to the *medial geniculate nucleus* (the auditory relay nucleus of the thalamus) blocked fear conditioning to a tone, but bilateral lesions to the auditory cortex did not. This indicated that for auditory fear conditioning to occur, it is necessary for signals elicited by the tone to reach the medial geniculate nucleus but not the auditory cortex. It also indicated that a pathway from the medial geniculate nucleus to a structure other than the auditory cortex plays a key role in fear conditioning. This pathway proved to be the pathway from the medial geniculate nucleus to the amygdala. Lesions of the amygdala, like lesions of the medial geniculate nucleus, blocked fear conditioning. The amygdala receives input from all sensory systems, and it is believed to be the structure in which the emotional significance of sensory signals is learned and retained.

Several pathways (see Balleine & Killcross, 2006; LaBar, 2007) carry signals from the amygdala to brain-stem structures that control the various emotional responses. For example, a pathway to the periaqueductal gray of the midbrain elicits appropriate defensive responses (see Bandler & Shipley, 1994), whereas another pathway to the lateral hypothalamus elicits appropriate sympathetic responses.

The fact that auditory cortex lesions do not disrupt fear conditioning to simple tones does not mean that the auditory cortex is not involved in auditory fear conditioning. There are two pathways from the medial geniculate nucleus to the amygdala: the direct one, which you have already learned about, and an indirect one that projects via the auditory cortex (Romanski & LeDoux, 1992). Both routes are capable of mediating fear conditioning to simple sounds; if only one is destroyed, conditioning progresses normally. However, only the cortical route is capable of mediating fear conditioning to complex sounds (Jarrell et al., 1987).

Figure 17.9 illustrates the circuit of the brain that is thought to mediate fear conditioning to auditory conditional stimuli (see LeDoux, 1994). Sound signals from the medial geniculate nucleus of the thalamus reach the amygdala either directly or via the auditory cortex. The amygdala assesses the emotional significance of the sound on the basis of previous encounters with it, and then the amygdala activates the appropriate response circuits—for example, behavioral circuits in the periaqueductal gray and sympathetic circuits in the hypothalamus.

Contextual Fear Conditioning and the Hippocampus

Environments, or *contexts*, in which fear-inducing stimuli are encountered can themselves come to elicit fear. For example, if you repeatedly encountered a bear on a particular trail in the forest, the trail itself would elicit fear in you. The process by which benign contexts come to elicit fear through their association with fear-inducing stimuli is called **contextual fear conditioning**.

Contextual fear conditioning has been produced in the laboratory in two ways. First, it has been produced by the conventional fear-conditioning procedure, which we just discussed. For example, if a rat repeatedly receives an electric shock following a conditional stimulus, such as a tone, the rat will become fearful of the conditional context (the test chamber) as well as the tone. Second, contextual fear conditioning has been produced by delivering aversive stimuli in a

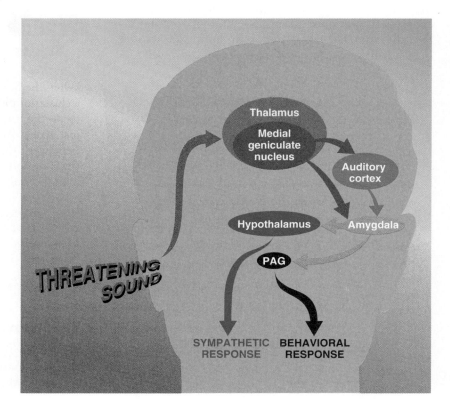

FIGURE 17.9 The structures that are thought to mediate the sympathetic and behavioral responses conditioned to an auditory conditional stimulus.

particular context in the absence of any other conditional stimulus. For example, if a rat receives shocks in a distinctive test chamber, the rat will become fearful of that chamber.

In view of the fact that the **hippocampus** plays a key role in memory for spatial location, it is reasonable to expect that it would be involved in contextual fear conditioning. This seems to be the case (see Antoniadis & McDonald, 2000). Bilateral hippocampal lesions block the subsequent development of a fear response to the context without blocking the development of a fear response to the explicit conditional stimulus (e.g., a tone).

Amygdala Complex and Fear Conditioning

The preceding discussion has probably left you with the impression that the amygdala is a single brain structure; it isn't. It is actually a cluster of many nuclei, often referred to as the *amygdala complex*. The amygdala is composed of a dozen or so major nuclei, which are themselves divided into subnuclei. Each of these subnuclei is structurally distinct, has different connections (see LeDoux, 2000b), and is thus likely to have different functions.

The study of fear conditioning provides a compelling demonstration of the inadvisability of assuming that the amygdala is a single structure. Evidence has been accumulating that it is actually the **lateral nucleus of the amygdala**—not the entire amygdala—that is critically involved in the acquisition, storage, and expression of conditioned fear (see Kim & Jung, 2006; Maren & Quirk, 2004). Both the prefrontal cortex and the hippocampus project to the lateral nucleus of the amygdala: The **prefrontal cortex** is thought to act on the lateral nucleus of the amygdala to suppress conditioned fear, and the hippocampus is thought to interact with that part of the amygdala to mediate learning about the context of fear-related events.

Scan Your Brain

This chapter is about to change direction: The remaining two sections focus on the effects of stress on health and on the neural mechanisms of human emotion. This is a good point for you to scan your brain to see whether it has retained the introductory material on emotion and fear. Fill in each of the following blanks with the most appropriate term. The correct answers are provided at the end of the exercise. Before continuing, review material related to your errors and omissions.

1. The theory that the subjective experience of emotion is triggered by ANS responses is called the _____ theory.

2. The pattern of aggressive responses observed in decorticate animals is called _____.
3. Between the amygdala and the fornix in the limbic system is the _____.
4. A Duchenne smile, but not a false smile, involves contraction of the _____.
5. Aggression directed by the alpha male of a colony at a male intruder is called _____ aggression.
6. The usual target site of rat defensive attacks is the _____ of the attacking rat.
7. Testosterone increases _____ aggression in rats.
8. In humans, most violent outbursts that are labeled as aggression are more appropriately viewed as _____ attacks.
9. The establishing of a fear response to a previously neutral stimulus, such as a tone, is accomplished by fear _____.
10. In the typical auditory fear-conditioning experiment, the _____ is a tone.
11. Auditory fear conditioning to simple tones depends on a pathway from the _____ to the amygdala.
12. Unlike auditory fear conditioning to simple tones, fear conditioning to complex sounds involves the _____.
13. The prefrontal cortex is thought to act on the _____ of the amygdala to inhibit conditioned fear.

Scan Your Brain answers: (1) James-Lange, (2) sham rage, (3) hippocampus, (4) orbicularis oculi, (5) social, (6) face, (7) social, (8) defensive, (9) conditioning, (10) conditional stimulus, (11) medial geniculate nucleus, (12) auditory cortex, (13) lateral nucleus.

17.4
Stress and Health

When the body is exposed to harm or threat, the result is a cluster of physiological changes that is generally referred to as the *stress response*—or just **stress**. All **stressors** (experiences that induce the stress response) produce the same core pattern of physiological changes, whether psychological (e.g., dismay at the loss of one's job) or physical (e.g., long-term exposure to cold). However, it is *chronic psychological stress* that has been most frequently implicated in ill health (see Kiecolt-Glaser et al., 2002; Natelson, 2004), which is the focus of this section.

((•● **Listen** Stress and Suicide **www.mypsychlab.com**

◉ **Watch** Gender Differences in Stress Vulnerability **www.mypsychlab.com**

The Stress Response

Hans Selye (pronounced "SELL-yay") first described the stress response in the 1950s, and he emphasized its dual nature. In the short term, it produces adaptive changes that help the animal respond to the stressor (e.g., mobilization of energy resources); in the long term, however, it produces changes that are maladaptive (e.g., enlarged adrenal glands)—see de Kloet, Joëls, and Holsboer (2005).

Selye attributed the stress response to the activation of the *anterior-pituitary adrenal-cortex system*. He concluded that stressors acting on neural circuits stimulate the release of **adrenocorticotropic hormone (ACTH)** from the anterior pituitary, that the ACTH in turn triggers the release of **glucocorticoids** from the **adrenal cortex**, and that the glucocorticoids produce many of the components of the stress response (see Erickson, Drevets, & Schulkin, 2003; Schulkin, Morgan, & Rosen, 2005). The level of circulating glucocorticoids is the most commonly employed physiological measure of stress.

Selye largely ignored the contributions of the sympathetic nervous system to the stress response. However, stressors activate the sympathetic nervous system, thereby increasing the amounts of epinephrine and norepinephrine released from the **adrenal medulla**. Most modern theories of stress acknowledge the roles of both the anterior-pituitary adrenal-cortex system and the sympathetic-nervous-system adrenal-medulla system (see Gunnar & Quevedo, 2007; Ulrich-Lai & Herman, 2009). Figure 17.10 illustrates the two-system view.

The major feature of Selye's landmark theory is its assertion that both physical and psychological stressors induce the same general stress response. This assertion has proven to be partly correct. There is good

Clinical Implications

evidence that all kinds of common psychological stressors—such as losing a job, taking a final exam, or ending a relationship—act like physical stressors. However, Selye's contention that there is only one stress response has proven to be a gross simplification. Stress responses are complex and varied, with the exact response depending on the stressor, its timing, the nature of the stressed person, and how the stressed person reacts to the stressor (e.g., Joëls & Baram, 2009; Miller, Chen, & Zhou, 2007; Smith, 2006). For example, in a study of women awaiting surgery for possible breast cancer, the levels of stress were lower in those who had convinced themselves that they could not possibly have cancer, that their prayers were certain to be answered, or that it was counterproductive to worry (Katz et al., 1970).

In the 1990s, there was an important advance in the understanding of the stress response (see Fleshner & Laudenslager, 2004). It was discovered that brief stressors produce physiological reactions that participate in the body's inflammatory responses. Most notably, it was found that brief stressors produced an increase in blood levels of **cytokines**, a group of peptide hormones that are released by many cells and participate in a variety of physiological and immunological responses, causing inflammation and fever. The cytokines are now classified with the adrenal hormones as major stress hormones.

Animal Models of Stress

Most of the early research on stress was conducted with nonhumans, and even today most lines of stress research begin with controlled experiments involving nonhumans before moving to correlational studies of humans. Early stress research on nonhumans tended to involve extreme forms of stress such as repeated exposure to electric shock or long periods of physical restraint. There are two problems with this kind of research. First is the problem of ethics. Any research that involves creating stressful situations is going to be controversial, but many of the early stress studies were "over the top" and would not be permitted today in many countries. The second problem is that studies that use extreme, unnatural forms of stress are often of questionable scientific value. Responses to extreme stress tend to mask normal variations in the stress response, and it is difficult to relate the results of such studies to common human stressors (see Fleshner & Laudenslager, 2004; Koolhass, de Boer, & Buwalda, 2006).

Evolutionary Perspective

◉ **Watch** Can Yoga or Meditation Help You Relax? **www.mypsychlab.com**

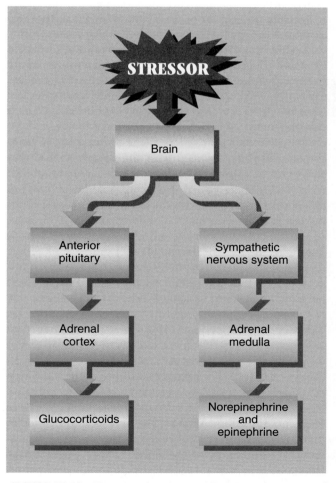

FIGURE 17.10 The two-system view of the stress response.

Some animal models of stress involve the study of threat from *conspecifics* (members of the same species). Virtually all mammals—particularly males—experience threats from conspecifics at certain points in their lives. When conspecific threat becomes an enduring feature of daily life, the result is **subordination stress** (see Berton et al., 2006; Sapolsky, 2005). Subordination stress is most readily studied in social species that form stable *dominance hierarchies* (pecking orders; see Chapter 2). What do you think happens to subordinate male rodents who are continually attacked by more dominant males? They are more likely to attack juveniles, and they have small testes, shorter life spans, lower blood levels of testosterone, and higher blood levels of glucocorticoids.

Psychosomatic Disorders: The Case of Gastric Ulcers

◉ Watch Fighting Stress
www.mypsychlab.com

Interest in pathological effects of stress has increased as researchers have identified more and more **psychosomatic disorders** (medical disorders that have psychological causes). So many adverse effects of stress on health (e.g., in heart disease, asthma, and skin disorders) have been documented that it is now more reasonable to think of most, if not all, medical disorders as psychosomatic (Miller & Blackwell, 2006; Wargo, 2007).

Gastric ulcers were one of the first medical disorders to be classified as psychosomatic. **Gastric ulcers** are painful lesions to the lining of the stomach and duodenum, which in extreme cases can be life-threatening. About 500,000 new cases are reported each year in the United States.

Clinical Implications

The view of gastric ulcers as the prototypical psychosomatic disorder changed with the discovery that they seemed to be caused by bacteria. It was claimed that bacteria (*Helicobacter pylori*) are responsible for all cases of gastric ulcers except those caused by nonsteroidal anti-inflammatory agents such as aspirin (Blaser, 1996). This seemed to rule out stress as a causal factor, but a consideration of the evidence suggests otherwise.

There is no denying that *H. pylori* damage the stomach wall or that antibiotic treatment of gastric ulcers helps many sufferers. The facts do, however, suggest that *H. pylori* infection alone is insufficient to produce the disorder in most people. Although it is true that most patients with gastric ulcers display signs of *H. pylori* infection, so too do 75% of healthy individuals. Also, although it is true that antibiotics improve the condition of many patients with gastric ulcers, so do psychological treatments—and they do it without reducing signs of *H. pylori* infection. Apparently, there is another factor that increases the susceptibility of the stomach wall to damage from *H. pylori*, and this factor appears to be stress. Gastric ulcers occur more commonly in people living in stressful situations, and stressors produce gastric ulcers in laboratory animals.

Psychoneuroimmunology: Stress, the Immune System, and the Brain

A major change in the study of psychosomatic disorders came in the 1970s with the discovery that stress can increase susceptibility to infectious diseases. Up to that point, infectious diseases had been regarded as "strictly physical": The discovery that stress can increase susceptibility to infection led in the early 1980s to the emergence of a new field of research: **psychoneuroimmunology**—the study of interactions among psychological factors, the nervous system, and the immune system (see Fleshner & Laudenslager, 2004). Psychoneuroimmunological research is the focus of this subsection. Let's begin with an introduction to the immune system.

Clinical Implications

Immune System Microorganisms of every description revel in the warm, damp, nutritive climate of your body. Your **immune system** keeps your body from being overwhelmed by these invaders, but, before it can take any action against an invading microorganism, the immune system must have some way of distinguishing foreign cells from body cells. That is the function of **antigens**—protein molecules on the surface of a cell that identify it as native or foreign.

There are two divisions of the mammalian immune system: the innate immune system and the adaptive immune system (see Iwasaki & Medzhitov, 2010; O'Neill, 2005). The **innate immune system** is the first line of defense. It acts near entry points to the body and attacks general classes of molecules produced by a variety of **pathogens** (disease-causing agents). If the general innate immune system fails to destroy a pathogen, it is dealt with by the specific adaptive immune system. The **adaptive immune system** mounts a targeted attack by binding to the antigens on foreign cells and destroying them or marking them for destruction by other cells. An important feature of the adaptive immune system is that it has a memory; once particular pathogens have been recognized and destroyed, they are promptly eliminated if they invade again (see Littman & Singh, 2007; Reiner, Sallusto, & Lanzavecchia, 2007). The memory of the adaptive immune system is the mechanism that gives vaccinations their *prophylactic* (preventive) effect—**vaccination** involves administering a weakened form of a virus so that if the virus later invades, the adaptive immune system is prepared to act against it. For example, smallpox has been largely eradicated by programs of vaccination with the weaker virus of its largely benign relative, cowpox. The process of creating immunity through vaccination is termed **immunization**.

Until recently, most immunological research has focused on the adaptive immune system; however, the discovery of the role of cytokines in the innate immune system stimulated interest in that system (O'Neill, 2005).

The innate immune system is activated by **toll-like receptors** (receptors that are similar in structure to a receptor called *toll*, which had previously been discovered in fruit flies). Various kinds of **phagocytes** (cells, such as macrophages and microglia, that destroy and ingest pathogens) have toll-like receptors in their membranes, and upon binding to pathogens, the toll-like receptors trigger two responses (see Kettenmann, 2006; Pocock & Kettenmann, 2007). First, the phagocytes destroy and consume the pathogens (see Deretic & Klionsky, 2008)—in Figure 17.11, you see a **macrophage** about to engage in **phagocytosis** (the destruction and ingestion of foreign matter) of a bacterium. Then, the phagocytes release cytokines, which trigger an inflammatory response that results in swelling and redness at sites of local infection and produces the fever, body aches, and other flulike symptoms that often accompany infections. Cytokines also attract more phagocytes from the blood into the infected area.

Another important effect of the cytokines is that they activate lymphocytes. **Lymphocytes** are cells of the adaptive immune system, specialized white blood cells that are produced in *bone marrow* and the *thymus gland* and are stored in the lymphatic system until they have been activated (see Terszowski et al., 2006; von Boehmer, 2006).

There are many kinds of lymphocytes, but they are considered to be of two general types: T lymphocytes and B lymphocytes. Each is involved in a different adaptive immune reaction. **Cell-mediated immunity** is directed by **T cells** (T lymphocytes); **antibody-mediated immunity** is directed by **B cells** (B lymphocytes)—see Figure 17.12 on page 458.

The cell-mediated immune reaction begins when a macrophage ingests a foreign microorganism. The macrophage then displays the microorganism's antigens on the surface of its cell membrane, and this display attracts T cells. Each T cell has two kinds of receptors on its surface, one for molecules that are normally found on the surface of macrophages and other body cells, and one for a specific foreign antigen. There are millions of different receptors for foreign antigens on T cells, but there is only one kind on each T cell, and there are only a few T cells with each kind of receptor. Once a T cell with a receptor for the foreign antigen binds to the surface of an infected macrophage, a series of reactions is initiated (Grakoui et al., 1999; Malissen, 1999). Among these reactions is the multiplication of the bound T cell, creating more T cells with the specific receptor necessary to destroy all invaders that contain the target antigens and all body cells that have been infected by the invaders.

The antibody-mediated immune reaction begins when a B cell binds to a foreign antigen for which it contains an appropriate receptor. This causes the B cell to multiply and to synthesize a lethal form of its receptor molecules. These lethal receptor molecules, called **antibodies**, are released into the intracellular fluid, where they bind to the foreign antigens and destroy or deactivate the microorganisms that possess them. Memory B cells for the specific antigen are also produced during the process; these cells have a long life and accelerate antibody-mediated immunity if there is a subsequent infection by the same microorganism (see Ahmed & Gray, 1996).

The recent discovery of **T-reg cells** (regulatory T cells) is potentially of major clinical significance (Fehervari & Sakaguchi, 2006). Sometimes T cells start to attack the body's own tissue, mistaking it for a pathogen, thus producing **autoimmune diseases**—for example, T cells sometimes attack the body's own myelin, causing *multiple sclerosis* (see Chapter 10). T-reg cells combat autoimmune diseases by identifying and destroying T cells that engage in such attacks.

Researchers had been puzzled by the complex and highly coordinated nature of immune reactions. Then, high-resolution microscopic images of immune cells revealed connections between them that looked like synapses (see Biber et al., 2007; Davis, 2006).

What Effect Does Stress Have on Immune Function: Disruptive or Beneficial?

It is widely believed that the main effect of stress on immune function is disruptive. I am sure that you have heard this from family members, friends, and even physicians. But is this true?

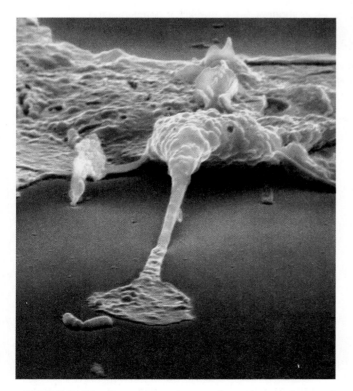

FIGURE 17.11 Phagocytosis: A macrophage about to destroy and ingest a bacterium.

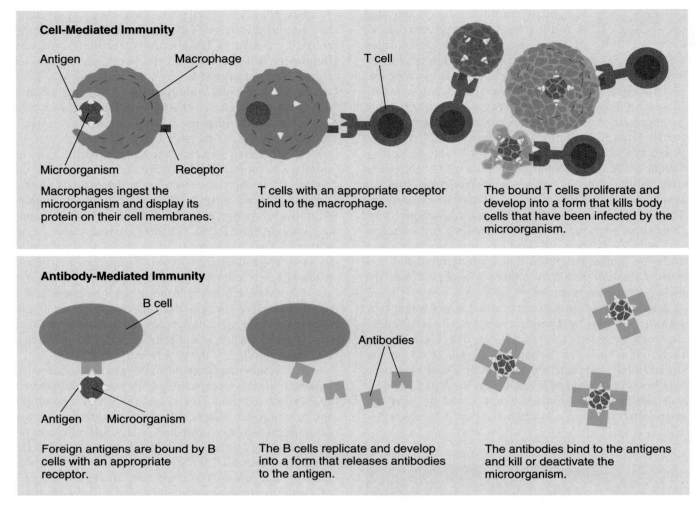

FIGURE 17.12 Two adaptive barriers to infection: Cell-mediated immunity and antibody-mediated immunity. In cell-mediated immunity, microorganisms or body cells that they have invaded are killed by T cells; in antibody-mediated immunity, microorganisms are killed by antibodies produced by B cells.

One of the logical problems with the view that stress always disrupts immune function is that it is inconsistent with the principles of evolution (see Segerstrom, 2007). Virtually every individual organism encounters many stressors during the course of its life, and it is difficult to see how a maladaptive response to stress, such as a disruption of immune function, could have evolved—or could have survived if it had been created by a genetic accident or as a *spandrel* (a nonadaptive by-product of an adaptive evolutionary change; see Chapter 2).

Two events have helped clarify the relation between stress and immune function. The first was the *meta-analysis* of Segerstrom and Miller (2004), which reviewed about 300 previous studies of stress and immune function. Segerstrom and Miller found that the effects of stress on immune function depended on the kind of stress. They found that acute (brief) stressors (i.e., those lasting

less than 100 minutes, such as public speaking, an athletic competition, or a musical performance) actually led to improvements in immune function. Not surprisingly, the improvements in immune function following acute stress occurred mainly in the innate immune system, whose components can be marshaled quickly. In contrast, chronic (long-lasting) stressors, such as caring for an ill relative or experiencing a period of unemployment, adversely affected the adaptive immune system. Stress that disrupts health or other aspects of functioning is called *distress*, and stress that improves health or other aspects of functioning is called *eustress*.

The second event that has helped clarify the relation between stress and immune function was the discovery of the bidirectional role played by the cytokines in the innate immune system. Short-term cytokine-induced inflammatory responses help the body combat infection, whereas long-term cytokine release is associated with a variety of adverse health consequences (Robles, Glaser, & Kiecolt-Glaser, 2005). This finding provided an explanation of the pattern of results discovered by Segerstrom and Miller's meta-analysis.

How Does Stress Influence Immune Function?   The mechanisms by which stress influences immune function have been difficult to specify because there are so many possibilities (see Dustin & Colman, 2002). Stress produces widespread changes in the body through its effects on the anterior-pituitary adrenal-cortex system and the sympathetic-nervous-system adrenal-medulla system, and there are innumerable mechanisms by which these systems can influence immune function. For example, both T cells and B cells have receptors for glucocorticoids; and lymphocytes have receptors for epinephrine, norepinephrine, and glucocorticoids. In addition, many of the neuropeptides that are released by neurons are also released by cells of the immune system. Conversely, cytokines, originally thought to be produced only by cells of the immune system, have been found to be produced in the nervous system (Salzet, Vieau, & Day, 2000). In short, the physiological mechanisms by which the nervous system and the immune system can interact are innumerable.

It is important to appreciate that there are behavioral routes by which stress can affect immune function. For example, people under severe stress often change their diet, exercise, sleep, and drug use, any of which could influence immune function. Also, the behavior of a stressed or ill person can produce stress and illness in others. For example, Wolf and colleagues (2007) found that stress in mothers aggravates asthmatic symptoms in their children; conversely, asthma in the children increases measures of stress in their mothers.

Does Stress Affect Susceptibility to Infectious Disease?   You have just learned that stress influences immune function. Most people assume that this means that  stress increases susceptibility to infectious diseases. But it doesn't, and it is important that you understand why. There are three reasons why decreases in immune function may not be reflected in an increased incidence of infectious disease:

- The immune system seems to have many redundant components; thus, disruption of one of them may have little or no effect on vulnerability to infection.
- Stress-produced changes in immune function may be too short-lived to have substantial effects on the probability of infection.
- Declines in some aspects of immune function may induce compensatory increases in others.

It has proven difficult to show unequivocally that stress causes increases in susceptibility to infectious diseases in humans. One reason for this difficulty is that only correlational

studies are possible. Numerous studies have reported *positive* correlations between stress and ill health in human subjects; for example, students in one study reported more respiratory infections during final exams (Glaser et al., 1987). However, interpretation of such correlations is never straightforward: Subjects may report more illness during times of stress because they expect to be more ill, because their experience of illness during times of stress is more unpleasant, or because the stress changed their behavior in ways that increased their susceptibility to infection.

Despite the difficulties of proving a causal link between stress and susceptibility to infectious disease in humans, the evidence for such a link is strong. Three basic types of evidence, when considered together, are almost persuasive:

- Correlational studies in humans—as you have just learned—have found correlations between stress levels and numerous measures of health.
- Controlled experiments conducted with laboratory animals show that stress can increase susceptibility to infectious disease in these species.
- Partially controlled studies conducted with humans, which are rare for ethical reasons, have added greatly to the weight of evidence.

One of the first partially controlled studies demonstrating stress-induced increases in the susceptibility of humans to infectious disease was conducted by Cohen and colleagues (1991). Using questionnaires, they assessed psychological stress levels in 394 healthy participants. Then, each participant randomly received saline nasal drops that contained a respiratory virus or only the saline. Then, all of the participants were quarantined until the end of the study. A higher proportion of those participants who scored highly on the stress scales developed colds.

Early Experience of Stress

Early exposure to severe stress can have a variety of adverse effects on subsequent development. Children subjected to maltreatment or other forms of severe stress display a variety of brain and endocrine system abnormalities (Evans & Kim, 2007; Teicher et al., 2003). As you will learn in Chapter 18, some psychiatric disorders are thought to result from an interaction between an inherited susceptibility to a disorder and early exposure to severe stress. Also, early exposure to stress often increases the intensity of subsequent stress responses (e.g., increases the subsequent release of glucocorticoids in response to stressors).

It is important to understand that the developmental period during which early stress can adversely affect neural and endocrine development begins before birth. Many experiments have demonstrated the adverse effects of prenatal stress in laboratory animals; pregnant females have been exposed to stressors, and the adverse effects of that exposure on their offspring have subsequently been documented (Maccari et al., 2003; Weinstock, 2008).

One particularly interesting line of research on the role of early experience in the development of the stress response began with the observation that handling of rat pups by researchers for a few minutes per day during the first few weeks of the rats' lives has a variety of salutary (health-promoting) effects (see Sapolsky, 1997). The majority of these effects seemed to result from a decrease in the magnitude of the handled pups' responses to stressful events. As adults, rats that were handled as pups displayed smaller increases in circulating glucocorticoids in response to stressors (see Francis & Meaney, 1999). It seemed remarkable that a few hours of handling early in life could have such a significant and lasting effect. In fact, evidence supports an alternative interpretation.

Liu and colleagues (1997) found that handled rat pups are groomed (licked) more by their mothers, and they hypothesized that the salutary effects of the early handling resulted from the extra grooming, rather than from the handling itself. They confirmed this hypothesis by showing that unhandled rat pups that received a lot of grooming from their mothers developed the same profile of increased glucocorticoid release that was observed in handled pups.

The research on the effects of early grooming on the subsequent neural and behavioral development of rat pups is important because it is a particularly well-documented case of epigenetic transmission of a trait (see Parent et al., 2005). I hope you remember from earlier chapters that traits can be passed from parents to offspring by epigenetic mechanisms—literally, **epigenetic** means "not of the genes." In this case, the female rat pups of fearful, poor grooming mothers or foster mothers are themselves fearful, poor grooming mothers as adults.

Thinking Creatively

In contrast, early separation of rat pups from their mothers seems to have effects opposite to those of high levels of early grooming (see Cirulli, Berry, & Alleva, 2003; Pryce & Feldon, 2003; Rhees, Lephart, & Eliason, 2001). As adults, rats that are separated from their mothers in infancy display elevated behavioral and hormonal responses to stress.

Stress and the Hippocampus

Exposure to stress affects the structure and function of the brain in a variety of ways (see Rodrigues, LeDoux, & Sapolsky, 2009). However, the hippocampus is particularly susceptible to stress-induced effects. The reason for this susceptibility may be the particularly dense population of glucocorticoid receptors in the hippocampus (see McEwen, 2004).

Neuroplasticity

Stress has been shown to reduce dendritic branching in the hippocampus, to reduce adult neurogenesis in the

hippocampus, to modify the structure of some hippocampal synapses, and to disrupt the performance of hippocampus-dependent tasks (see Sandi, 2004). These effects of stress on the hippocampus appear to be mediated by elevated glucocorticoid levels: They can be induced by **corticosterone** (a major glucocorticoid) and can be blocked by **adrenalectomy** (surgical removal of the adrenal glands)—see Brummelte, Pawluski, and Galea (2006), Gould (2004), and McEwen (2004).

Scan Your Brain

Pause here to review the preceding section on stress and health. Fill in each of the following blanks. The correct answers are provided at the end of the exercise. Review material related to your errors or omissions before proceeding.

1. Glucocorticoids are released from the _____ as part of the stress response.
2. Stressors increase the release of epinephrine and norepinephrine from the _____.
3. Brief stressors trigger the release of _____, which participate in the body's inflammatory responses.
4. When threats from conspecifics become an enduring feature of daily life, the result is _____.
5. Gastric ulcers have been associated with *H. pylori* infection, but it seems likely that _____ is another causal factor in their development.
6. The study of the interactions among psychological factors, the nervous system, and the immune system is called _____.
7. There are two immune systems: the _____ immune system and the adaptive immune system.
8. Disease-causing agents are known as _____.
9. Lymphocytes participate in two immune reactions: _____ and antibody-mediated.
10. T cells and B cells are involved in cell-mediated and _____ immune reactions, respectively.
11. Rat pups groomed intensely by their mothers display decreased _____ release from the adrenal cortex in response to stressors in adulthood.
12. Corticosterone is a _____.
13. In laboratory animals, stress has been shown to reduce adult neurogenesis in the _____.

Scan Your Brain answers: (1) adrenal cortex, (2) adrenal medulla, (3) cytokines, (4) subordination stress, (5) stress, (6) psychoneuroimmunology, (7) innate, (8) pathogens, (9) cell-mediated, (10) antibody-mediated, (11) glucocorticoid, (12) glucocorticoid, (13) hippocampus.

Brain Mechanisms of Human Emotion

This final section of the chapter deals with the brain mechanisms of human emotion. We still do not know how the brain controls the experience or expression of emotion, or how the brain interprets emotion in others, but progress has been made. Each of the following subsections illustrates an area of progress.

Cognitive Neuroscience of Emotion

Cognitive neuroscience is currently the dominant approach being used to study the brain mechanisms of human emotion. There have been many functional brain-imaging studies of people experiencing or imagining emotions or watching others experiencing them. These studies have established three points that have advanced our understanding of the brain mechanisms of emotion in fundamental ways (see Bastiaansen, Thioux, & Keysers, 2009; Niedenthal, 2007):

- Brain activity associated with each human emotion is diffuse—there is not a center for each emotion. Think "mosaic," not "center," for locations of brain mechanisms of emotion.
- There is virtually always activity in motor and sensory cortices when a person experiences an emotion or empathizes with a person experiencing an emotion (see Figure 17.13).
- Very similar patterns of brain activity tend to be recorded when a person experiences an emotion, imagines that emotion, or sees somebody else experience that emotion.

These three fundamental findings are influencing how researchers are thinking about the neural mechanisms of emotion. For example, the activity observed in sensory and motor cortex during the experience of human emotions is now believed to be an important part of the mechanism by which the emotions are experienced. The re-experiencing of related patterns of motor, autonomic, and sensory neural activity during emotional experiences is generally referred to as the *embodiment of emotions* (see Niedenthal, 2007).

These three fundamental findings may also help explain the remarkable ability of humans to grasp the emotional states of others (see Bastiaansen et al., 2009). You may recall that *mirror neurons*, which have been identified in nonhuman primates, are neurons that fire when a specific response is performed by a subject or the subject watches the response being performed. The discovery that certain patterns of brain activity are observed on fMRI scans when individuals experience an emotion or watch somebody else experience the same emotion suggests that a **mirror-like system** might be the basis for human empathy (Fabbri-Destro & Rizzolatti, 2008; Iacoboni, 2009; Keysers & Gazzola, 2009; Oberman & Ramachandran, 2007).

Amygdala and Human Emotion

You have already learned that the amygdala plays an important role in fear conditioning in rats. Numerous functional brain-imaging studies have found the amygdala to be involved in human emotions—particularly in fear and other negative emotions (see Adolphs, 2008; Cheng et al., 2006; Sergerie, Lepage, & Armony, 2006). Furthermore, the amygdala appears to be involved in only some aspects of human fear. It seems to be more involved in the perception of fear in others

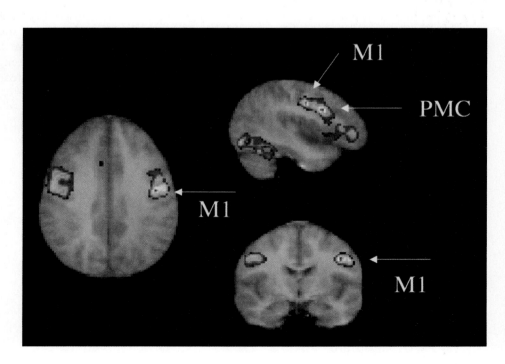

FIGURE 17.13 Horizontal, sagittal, and coronal functional MRIs show areas of increased activity in the primary motor cortex (M1) and the premotor cortex (PMC) when volunteers watched facial expressions of emotion. The same areas were active when the volunteers made the expressions themselves. (From Carr et al., 2003.)

than in its experience. The following case illustrates these points.

The Case of S.P., the Woman Who Couldn't Perceive Fear

Clinical Implications

At the age of 48, S.P. had her right amygdala and adjacent tissues removed for the treatment of epilepsy. Because her left amygdala had been damaged, she in effect had a bilateral amygdalar lesion.

Following her surgery, S.P. had an above average I.Q., and her perceptual abilities were generally normal. Of particular relevance was the fact that she had no difficulty in identifying faces or extracting information from them (e.g., information about age or gender). However, S.P. did have a severe postsurgical deficit in recognizing facial expressions of fear and less striking deficits in recognizing facial expressions of disgust, sadness, and happiness.

In contrast, S.P. had no difficulty specifying which emotion would go with particular sentences. Also, she had no difficulty using facial expressions upon request to express various emotions (Anderson & Phelps, 2000). This case is consistent with previous reports that the human amygdala is specifically involved in perceiving facial expressions of emotion, particularly of fear (e.g., Broks et al., 1998; Calder et al., 1996).

The case of S.P. is similar to reported cases of Urbach-Wiethe disease (see Aggleton & Young, 2000). **Urbach-Wiethe disease** is a genetic disorder that often results in *calcification* (hardening by conversion to calcium carbonate, the main component of bone) of the amygdala and surrounding anterior medial temporal-lobe structures in both hemispheres (see Figure 17.14). One Urbach-Wiethe patient with bilateral amygdalar damage was found to have lost the ability to recognize facial expressions

of fear (Adolphs, 2006). Indeed, she could not describe fear-inducing situations or produce fearful expressions, although she had no difficulty on tests involving other emotions. Although recent research has focused on the role of the amygdala in the recognition of negative facial expressions, patients with Urbach-Wiethe disease sometimes have difficulty recognizing other complex visual stimuli (Adolphs & Tranel, 1999).

Medial Prefrontal Lobes and Human Emotion

Emotion and cognition are often studied independently, but it is now believed that they are better studied as components of the same system (Phelps, 2004, 2006; Beer, Knight, & Esposito, 2006). The medial portions of the prefrontal lobes (including the medial portions of the orbitofrontal cortex and cingulate cortex) are the sites of emotion-cognition interaction that have received the most attention. Functional brain-imaging studies have found evidence of activity in the medial prefrontal lobes when emotional reactions are being cognitively suppressed or re-evaluated (see Quirk & Beer, 2006). Most of the studies of medial prefrontal lobe activity employ suppression paradigms or reappraisal paradigms. In studies that use **suppression paradigms**, participants are directed to inhibit their emotional reactions to unpleasant films or pictures; in studies that use **reappraisal paradigms**, participants are instructed to reinterpret a picture to change their emotional reaction to it. The medial prefrontal lobes are active when both of these paradigms are used, and they seem to exert their cognitive control of emotion by interacting with the amygdala (see Holland & Gallagher, 2004; Quirk & Beer, 2006).

Many theories of the specific functions of the medial prefrontal lobes have been proposed. The medial prefrontal lobes have been hypothesized to monitor the difference between outcome and expectancy (Potts et al., 2006), to respond to personal choices that result in losses (Gehring & Willoughby, 2002), to predict the likelihood of error (Brown & Braver, 2005), to guide behavior based on previous actions and outcomes (Kennerly et al.,

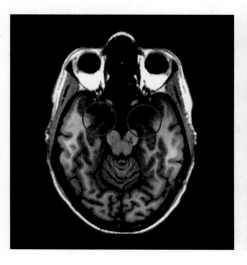
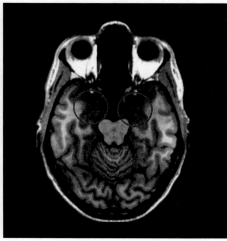

FIGURE 17.14 Bilateral calcification of the amygdalae in twins with Urbach-Wiethe disease. The red circles indicate the areas of calcification. (From Hurlemann et al., 2007.)

2006), and to respond to social rejection (Somerville, Heatherton, & Kelley, 2006). Which hypothesis is correct? Perhaps all are; the medial prefrontal lobes are large and complex, and they likely perform many functions. This point was made by Kawasaki and colleagues (2005).

Kawasaki and colleagues used microelectrodes to record from 267 neurons in the anterior cingulate cortices of four patients prior to surgery. They assessed the activity of the neurons when the patients viewed photographs with emotional content. Of these 267 neurons, 56 responded most strongly and consistently to negative emotional content. This confirms previous research linking the medial prefrontal lobes with negative emotional reactions, but it also shows that not all neurons in the area perform the same function—neurons directly involved in emotional processing appear to be sparse and widely distributed in the human brain.

Clinical Implications

Lateralization of Emotion

There is considerable evidence that emotional functions are lateralized, that is, that the left and right cerebral hemispheres are specialized to perform different emotional functions (e.g., Kim et al., 2004; Shaw et al., 2005)—as you learned in Chapter 16. This evidence has led to several theories of the cerebral lateralization of emotion (see Demaree et al., 2005); the following are the two most prominent:

- The *right-hemisphere model* of the cerebral lateralization of emotion holds that the right hemisphere is specialized for all aspects of emotional processing: perception, expression, and experience of emotion.
- The *valence model* proposes that the right hemisphere is specialized for processing negative emotion and the left hemisphere is specialized for processing positive emotion.

Most studies of the cerebral lateralization of emotion have employed functional brain-imaging methods, and the results have been complex and variable. Wager and colleagues (2003) performed a meta-analysis of the data from 65 such studies. Which of the theories does the analysis support?

The main conclusion of Wager and colleagues was that the current theories of lateralization of emotion are too general from a neuroanatomical perspective. Overall comparisons between left and right hemispheres revealed no interhemispheric differences in either the amount of emotional processing or the valence of the emotions being processed. However, when the comparisons were conducted on a structure-by-structure basis, they revealed substantial evidence of lateralization of emotional processing. Some kinds of emotional processing were lateralized to the left hemisphere in certain structures and to the right in others. Functional brain-imaging studies of emotion have commonly observed lateralization in the amygdalae—more activity is often observed in the left amygdala (Baas, Aleman, & Kahn, 2004). Clearly, neither the right-hemisphere model nor the valence model of the lateralization of emotion is supported by the evidence. The models are too general.

Another approach to studying the lateralization of emotions is based on observing the asymmetry of facial expressions. In most people, each facial expression begins on the left side of the face and, when fully expressed, is more pronounced there—which implies right-hemisphere dominance for facial expressions (see Figure 17.15). Remarkably, the same asymmetry of facial expressions has been documented in monkeys (Hauser, 1993). A right-hemisphere dominance for the recognition of facial expressions has also been demonstrated—people base their judgments of facial expression more on the right side of an observed face (Coolican et al., 2007).

Individual Differences in the Neural Mechanisms of Emotion

In general, more complex brain functions tend to show more individual differences in cerebral localization. For example, in Chapter 16, you learned that the cortical localization of language

FIGURE 17.15 The asymmetry of facial expressions. Notice that the expressions are more obvious on the left side of two well-known faces: those of Mona Lisa and Albert Einstein. The Einstein face is actually that of a robot that has been programmed to make natural facial expressions.

(Right-hand image from Tingfan Wu, Nicholas J. Butko, Paul Ruvulo, Marian S. Bartlett, Javier R. Movellan, "Learning to Make Facial Expressions," *devlrn*, pp.1-6, 2009 IEEE 8th International Conference on Development and Learning, 2009. © 2009 IEEE.)

processes varies substantially from person to person. Nevertheless, few studies of the neural mechanisms of emotion have focused on individual differences (see Leppänen & Nelson, 2009; Samanez-Larkin et al., 2008). Let's consider two studies and one notorious case that have a bearing on this issue.

First, Adolphs and colleagues (1999) tested the ability of nine neuropsychological patients with bilateral amygdalar damage to correctly identify facial expressions of emotion. As others had reported, these researchers found that the group of patients as a whole had difficulty identifying facial expressions of fear. However, there were substantial differences among the patients: Some also had difficulty identifying other negative emotions, and two had no deficits whatsoever in identifying facial expressions of emotions. Remarkably, structural MRIs revealed that both of these latter two patients had total bilateral amygdalar lesions.

Second, Canli and colleagues (2002) used functional MRIs to compare the reactions of healthy participants who scored high on *extraversion* with those of healthy participants who scored high on *neuroticism*. These personality dimensions were selected because of their relation to emotion—people high on the extraversion scale have a tendency toward positive emotional reaction; people high on the neuroticism scale have a tendency toward negative emotional reaction. Although all the participants displayed increased activity in the amygdala when viewing fearful faces, only the extraverts displayed increased amygdalar activity when viewing happy faces.

The following case study ends the chapter by emphasizing the point that the brain mechanisms of emotion differ from person to person. Fortunately, the reactions of Charles Whitman to brain damage are atypical.

The Case of Charles Whitman, the Texas Tower Sniper

After having lunch with his wife and his mother, Charles Whitman went home and typed a letter of farewell—perhaps as an explanation for what would soon happen.

He stated in his letter that he was having many compelling and bizarre ideas. Psychiatric care had been no help. He asked that his brain be autopsied after he was through; he was sure that they would find the problem.

By all reports, Whitman had been a good person. An Eagle Scout at 12 and a high school graduate at 17, he then enlisted in the Marine Corps, where he established himself as an expert marksman. After his discharge, he entered the University of Texas to study architectural engineering.

Nevertheless, in the evening of August 1, 1966, Whitman killed his wife and mother. He professed love for both of them, but he did not want them to face the aftermath of what was to follow.

The next morning, at about 11:30, Whitman went to the Tower of the University of Texas, carrying six guns, ammunition, several knives, food, and water. He clubbed the receptionist to death and shot four more people on his way to the observation deck. Once on the deck, he opened fire on people crossing the campus and on nearby streets. He was deadly, killing people as far as 300 meters away—people who assumed they were out of range.

At 1:24 that afternoon, the police fought their way to the platform and shot Whitman to death. All told, 17 people, including Whitman, had been killed, and another 31 had been wounded (Helmer, 1986).

An autopsy was conducted. Whitman was correct: They found a walnut-sized tumor in his right amygdala.

Themes Revisited

All four of the book's themes were prevalent in this chapter. The clinical implications theme appeared frequently, both because brain-damaged patients have taught us much about the neural mechanisms of emotion and because emotions have a major impact on health. The evolutionary perspective theme also occurred frequently because comparative research and the consideration of evolutionary pressures have also had a major impact on current thinking about the biopsychology of emotion.

The thinking creatively theme appeared where the text encouraged you to think in unconventional ways about the relation between testosterone and human aggression, the interpretation of reports of correlations between stress and ill health, and the possibility of susceptibility to stress being passed from generation to generation by maternal care.

Neuroplasticity was the major theme of the discussion of the effects of stress on the hippocampus.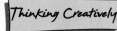

Think about It

1. With practice, you could become an expert in the production and recognition of facial expressions. How could you earn a living with these skills?
2. Does the target-site concept have any relevance to human aggression, defense, and play fighting?
3. Genes are not the only means by which behavioral tendencies can be passed from generation to generation. Discuss, with reference to maternal care and susceptibility to stress.
4. It is misleading to think of the amygdala as a single structure. Discuss.
5. Evidence suggests that emotion is a right-hemisphere phenomenon. Discuss.
6. Research on emotion has focused on fear. Why?

Key Terms

17.1 Biopsychology of Emotion: Introduction

James-Lange theory (p. 444)
Cannon-Bard theory (p. 444)
Decorticate (p. 445)
Sham rage (p. 445)
Limbic system (p. 446)
Kluver-Bucy syndrome (p. 446)
Amygdala (p. 446)
Polygraphy (p. 447)
Control-question technique (p. 447)
Guilty-knowledge technique (p. 447)
Facial feedback hypothesis (p. 448)
Duchenne smile (p. 449)

17.2 Fear, Defense, and Aggression

Fear (p. 450)
Defensive behaviors (p. 450)

Aggressive behaviors (p. 450)
Alpha male (p. 450)
Target-site concept (p. 451)

17.3 Neural Mechanisms of Fear Conditioning

Fear conditioning (p. 452)
Contextual fear conditioning (p. 453)
Hippocampus (p. 454)
Lateral nucleus of the amygdala (p. 454)
Prefrontal cortex (p. 454)

17.4 Stress and Health

Stress (p. 454)
Stressors (p. 454)
Adrenocorticotropic hormone (ACTH) (p. 455)
Glucocorticoids (p. 455)

Adrenal cortex (p. 455)
Adrenal medulla (p. 455)
Cytokines (p. 455)
Subordination stress (p. 456)
Psychosomatic disorder (p. 456)
Gastric ulcers (p. 456)
Psychoneuroimmunology (p. 456)
Immune system (p. 456)
Antigens (p. 456)
Innate immune system (p. 456)
Pathogens (p. 456)
Adaptive immune system (p. 456)
Vaccination (p. 456)
Immunization (p. 456)
Toll-like receptors (p. 457)
Phagocytes (p. 457)
Macrophage (p. 457)
Phagocytosis (p. 457)
Lymphocytes (p. 457)

Cell-mediated immunity (p. 457)
T cells (p. 457)
Antibody-mediated immunity (p. 457)
B cells (p. 457)
Antibodies (p. 457)
T-reg cells (p. 457)
Autoimmune diseases (p. 457)
Epigenetic (p. 460)
Corticosterone (p. 460)
Adrenalectomy (p. 460)

17.5 Brain Mechanisms of Human Emotion

Mirror-like system (p. 461)
Urbach-Wiethe disease (p. 462)
Suppression paradigm (p. 462)
Reappraisal paradigm (p. 462)

✓ **Quick Review** Test your comprehension of the chapter with this brief practice test. You can find the answers to these questions as well as more practice tests, activities, and other study resources at www.mypsychlab.com.

1. Sham rage was first observed in
   a. Papez's circuit.
   b. wild rats.
   c. decorticate cats.
   d. monkeys with no limbic system.
   e. patients with Kluver-Bucy syndrome.

2. When an alpha male rat attacks a submissive male intruder, he
   a. directs his attack at the intruder's face.
   b. directs his attack at the intruder's back.
   c. moves toward the "boxing" intruder with a lateral (sideways) attack.
   d. both a and c
   e. both b and c

3. A genuine smile
   a. involves the orbicularis oculi.

   b. involves the zygomaticus major.
   c. is called a *Duchenne smile*.
   d. all of the above
   e. both a and c

4. The most commonly used measure of stress is
   a. the level of glucocorticoids circulating in the blood.
   b. blood pressure.
   c. heart rate.
   d. the release of glucocorticoids from the pituitary.
   e. both b and c

5. T cells and B cells are
   a. phagocytes.
   b. lymphocytes.
   c. antibodies.
   d. antigens.
   e. macrophages.

18 Biopsychology of Psychiatric Disorders
The Brain Unhinged

This chapter is about the biopsychology of **psychiatric disorders** (disorders of psychological function sufficiently severe to require treatment). One of the main difficulties in studying or treating psychiatric disorders is that they are difficult to diagnose. The psychiatrist or clinical psychologist must first decide whether a patient's psychological function is pathological or merely an extreme of normal human variation: For example, does a patient with a poor memory suffer from a pathological condition or is he merely a healthy person with a poor memory? If a patient is judged to be suffering from a psychiatric disorder, then the particular disorder must be diagnosed. Because we cannot identify the specific brain pathology associated with various disorders, their diagnosis usually rests entirely on the patient's symptoms. The diagnosis is guided by the **DSM-IV-TR** (the current edition of the *Diagnostic and Statistical Manual* of the American Psychiatric Association). There are two main difficulties in diagnosing particular psychiatric disorders: (1) patients suffering the same disorder often display different symptoms, and (2) patients suffering from different disorders often display many of the same symptoms. Consequently, experts often disagree on the diagnosis of particular cases, and the guidelines provided by the DSM change with each new edition.

⊙➤ **Simulate** The Axes of the DSM
www.mypsychlab.com

This chapter begins with discussions of four psychiatric disorders: schizophrenia, affective (emotional) disorders, anxiety disorders, and Tourette syndrome. It ends with a description of how new *psychotherapeutic* drugs are developed and tested.

18.1

Schizophrenia

Schizophrenia means "the splitting of psychic functions." The term was coined in the early years of the 20th century to describe what was assumed at that time to be the primary symptom of the disorder: the breakdown of integration among emotion, thought, and action.

Schizophrenia is the disease that is most commonly associated with the concept of madness. It attacks about 1% of individuals of all races and cultural groups, typically beginning in adolescence or early adulthood. Schizophrenia occurs in many forms, but the case of Lena introduces you to some of its common features.

The Case of Lena, the Catatonic Schizophrenic

Lena's mother was hospitalized with schizophrenia when Lena was 2. She died in the hospital under peculiar circumstances, a suspected suicide. As a child, Lena displayed periods of hyperactivity; as an adolescent, she was viewed by others as odd. Although she enjoyed her classes and got good grades, she seldom established relationships with her fellow students. Lena rarely dated. However, she married her husband only a few months after meeting him. He was a quiet man who tried to avoid fuss or stress at all costs and who was attracted to Lena because she was quiet and withdrawn.

Clinical Implications

Shortly after their marriage, Lena's husband noticed that Lena was becoming even more withdrawn. She would sit for hours barely moving a muscle. He also found her having lengthy discussions with nonexistent persons.

About 2 years after he first noticed her odd behavior, Lena's husband found her sitting on the floor in an odd posture staring into space. She was totally unresponsive. When he tried to move her, Lena displayed *waxy flexibility*—that is, she reacted like a mannequin, not resisting movement but holding her new position until she was moved again. At that point, he took her to the hospital, where her disorder was immediately diagnosed as *stuporous catatonic schizophrenia* (schizophrenia characterized by long periods of immobility and waxy flexibility).

In the hospital, Lena displayed a speech pattern that is exhibited by many schizophrenics: echolalia (vocalized repetition of some or all of what has just been heard).

> DOCTOR: How are you feeling today?
> LENA: I am feeling today, feeling the feelings today.
> DOCTOR: Are you still hearing the voices?
> LENA: Am I still hearing the voices, voices?
> (Meyer & Salmon, 1988)

What Is Schizophrenia?

The major difficulty in studying and treating schizophrenia is accurately defining it (Heinrichs, 2005; Krueger & Markon, 2006). Its symptoms are complex and diverse; they overlap greatly with those of other psychiatric disorders and frequently change during the progression of the disorder. Also, various neurological disorders (e.g., complex partial epilepsy; see Chapter 10) have symptoms that might suggest a diagnosis of schizophrenia. In recognition of the fact that the current definition of schizophrenia likely includes several different brain diseases, some experts prefer to use the plural form to refer to this disorder: the *schizophrenias* (Wong & Van Tol, 2003).

Clinical Implications

The following are some symptoms of schizophrenia, although none of them appears in all cases. In an effort to categorize cases of schizophrenia so that they can be studied and treated more effectively, it is common practice to consider **positive symptoms** (symptoms that seem to represent an excess or distortion of normal function) separately from **negative symptoms** (symptoms that seem to represent a reduction or loss of normal function).

Positive Symptoms

- *Delusions.* Delusions of being controlled (e.g., "Martians are making me steal"), delusions of persecution (e.g., "My mother is poisoning me"), or delusions of grandeur (e.g., "Tiger Woods admires my backswing").
- *Hallucinations.* Imaginary voices making critical comments or telling patients what to do.
- *Inappropriate affect.* Failure to react with the appropriate emotion to positive or negative events.
- *Incoherent speech or thought.* Illogical thinking, echolalia, peculiar associations among ideas, belief in supernatural forces.
- *Odd behavior.* Difficulty performing everyday tasks, lack of personal hygiene, talking in rhymes, *catatonia* (remaining motionless, often in awkward positions for long periods).

Negative Symptoms

- *Affective flattening.* Reduction or absence of emotional expression.
- *Alogia.* Reduction or absence of speech.
- *Avolition.* Reduction or absence of motivation.
- *Anhedonia.* Inability to experience pleasure.

The recurrence of any two of these symptoms for 1 month is sufficient for the diagnosis of schizophrenia (Tamminga & Holcomb, 2005; Walker et al., 2004). Only one symptom is necessary if the symptom is a delusion that is particularly bizarre or an hallucination that includes voices.

Causal Factors in Schizophrenia

In the first half of the 20th century, the cloak of mystery began to be removed from mental illness by a series of studies that established schizophrenia's genetic basis (see Walker & Tessner, 2008). First, it was discovered that although only 1% of the population develops schizophrenia, the probability of schizophrenia's occurring in a close biological relative (i.e., a parent, child, or sibling) of a schizophrenic patient is about 10%, even if the relative was adopted shortly after birth by a healthy family (e.g., Kendler & Gruenberg, 1984; Rosenthal et al., 1980). Then, it was discovered that the concordance rates for schizophrenia are higher in identical twins (45%) than in fraternal twins (10%)—see Holzman and Matthyse (1990) and Kallman (1946). Finally, adoption studies have shown that the risk of schizophrenia is increased by the presence of the disorder in biological parents but not by its presence in adoptive parents (Gottesman & Shields, 1982).

Clinical Implications

((•● **Listen** Heritability and Onset of Schizophrenia Disorder **www.mypsychlab.com**

The fact that the concordance rate for schizophrenia in identical twins is substantially less than 100% suggests that differences in experience have a significant effect on the development of schizophrenia. The current view is that some people inherit a potential for schizophrenia, which may or may not be activated by experience. Supporting this view is a recent comparison of the offspring of a large sample of identical twins who were themselves discordant for schizophrenia (i.e., one had the disorder and one did not); the incidence of schizophrenia was as great in the offspring of the twin without schizophrenia as in the offspring of the twin with schizophrenia (Gottesman & Bertelsen, 1989).

It is clear that schizophrenia has multiple causes. Several different genes have been linked to the disorder (see Hall et al., 2009; O'Tuathaigh et al., 2007; Walsh et al., 2008). However, the mechanisms by which these genes contribute to schizophrenia have yet to be determined. Also, a variety of early experiential factors have been implicated in the development of schizophrenia—for example, birth complications, early infections, autoimmune reactions, toxins, traumatic injury, and stress. These early experiences are thought to alter the normal course of neurodevelopment, leading to schizophrenia in individuals who have a genetic susceptibility (see Jarskog, Miyamoto, & Lieberman, 2007;

Lenzenweger, 2006). Supporting this neurodevelopmental theory of schizophrenia is the study of two 20th-century famines: the Nazi-induced Dutch famine of 1944–1945 and the Chinese famine of 1959–1961. Fetuses whose pregnant mothers suffered in those famines were more likely to develop schizophrenia as adults (Kyle & Pichard, 2006; McClellan, Susser, & King, 2006; St. Clair et al., 2005).

Discovery of the First Antischizophrenic Drugs

The first major breakthrough in the study of the biochemistry of schizophrenia was the accidental discovery in the early 1950s of the first antischizophrenic drug, **chlorpromazine**. Chlorpromazine was developed by a French drug company as an antihistamine. Then, in 1950, a French surgeon noticed that chlorpromazine given prior to surgery to counteract swelling had a calming effect on some of his patients, and he suggested that it might have a calming effect on difficult-to-handle psychotic patients. His suggestion proved to be incorrect, but the research it triggered led to the discovery that chlorpromazine alleviates schizophrenic symptoms: Agitated patients with schizophrenia were calmed by chlorpromazine, and emotionally blunted patients with schizophrenia were activated by it. Don't get the idea that chlorpromazine cures schizophrenia. It doesn't. But it often reduces the severity of schizophrenic symptoms enough to allow institutionalized patients to be discharged.

Shortly after the antischizophrenic action of chlorpromazine was first documented, an American psychiatrist became interested in reports that the snakeroot plant had long been used in India for the treatment of mental illness. He gave **reserpine**—the active ingredient of the snakeroot plant—to his patients with schizophrenia and confirmed its antischizophrenic action. Reserpine is no longer used in the treatment of schizophrenia because it produces a dangerous decline in blood pressure at the doses needed for the treatment.

Although the chemical structures of chlorpromazine and reserpine are dissimilar, their antischizophrenic effects are similar in two major respects. First, the antischizophrenic effect of both drugs is manifested only after a patient has been medicated for 2 or 3 weeks. Second, the onset of this antischizophrenic effect is usually associated with motor effects similar to the symptoms of Parkinson's disease: tremors at rest, muscular rigidity, and a general decrease in voluntary movement. These similarities suggested to researchers that chlorpromazine and reserpine were acting through the same mechanism, one that was related to Parkinson's disease.

Dopamine Theory of Schizophrenia

Paradoxically, the next major breakthrough in the study of schizophrenia came from research on Parkinson's disease. In 1960, it was reported that the *striatums* (caudates plus putamens) of persons dying of Parkinson's disease had been depleted of dopamine (Ehringer & Hornykiewicz, 1960). This finding suggested that a disruption of dopaminergic transmission might produce Parkinson's disease, and, because of the relation between symptoms of Parkinson's disease and the antischizophrenic effects of chlorpromazine and reserpine, that antischizophrenic drug effects might be produced in the same way. Thus was born the *dopamine theory of schizophrenia*—the theory that schizophrenia is caused by too much dopamine and, conversely, that antischizophrenic drugs exert their effects by decreasing dopamine levels.

Lending instant support to the dopamine theory of schizophrenia were two already well-established facts. First, the antischizophrenic drug reserpine was known to deplete the brain of dopamine and other monoamines by breaking down the synaptic vesicles in which these neurotransmitters are stored. Second, drugs such as amphetamine and cocaine, which can trigger schizophrenic episodes in healthy subjects, were known to increase the extracellular levels of dopamine and other monoamines in the brain.

An important step in the evolution of the dopamine theory of schizophrenia came in 1963, when Carlsson and Lindqvist assessed the effects of chlorpromazine on extracellular levels of dopamine and its *metabolites* (substances that are created by the breakdown of another substance in cells). Although they expected to find that chlorpromazine, like reserpine, depletes the brain of dopamine, they didn't. The extracellular levels of dopamine were unchanged by chlorpromazine, and the extracellular levels of its metabolites were increased. The researchers concluded that both chlorpromazine and reserpine antagonize transmission at dopamine synapses but that they do it in different ways: reserpine by depleting the brain of dopamine and chlorpromazine by binding to dopamine receptors. Carlsson and Lindqvist argued that chlorpromazine is a *receptor blocker* at dopamine synapses—that is, that it binds to dopamine receptors without activating them and, in so doing, keeps dopamine from activating them (see Figure 18.1 on page 470). We now know that many psychoactive drugs are receptor blockers, but chlorpromazine was the first to be identified as such.

Carlsson and Lindqvist further postulated that the lack of activity at postsynaptic dopamine receptors sent a feedback signal to the presynaptic cells that increased their release of dopamine, which was broken down in the synapses. This explained why dopaminergic activity was reduced while extracellular levels of dopamine stayed about the same and extracellular levels of its metabolites were increased. Carlsson and Lindqvist's findings led to an important revision of the dopamine theory of schizophrenia:

Clinical Implications

Clinical Implications

◉➔ Simulate The Dopamine Theory of Schizophrenia **www.mypsychlab.com**

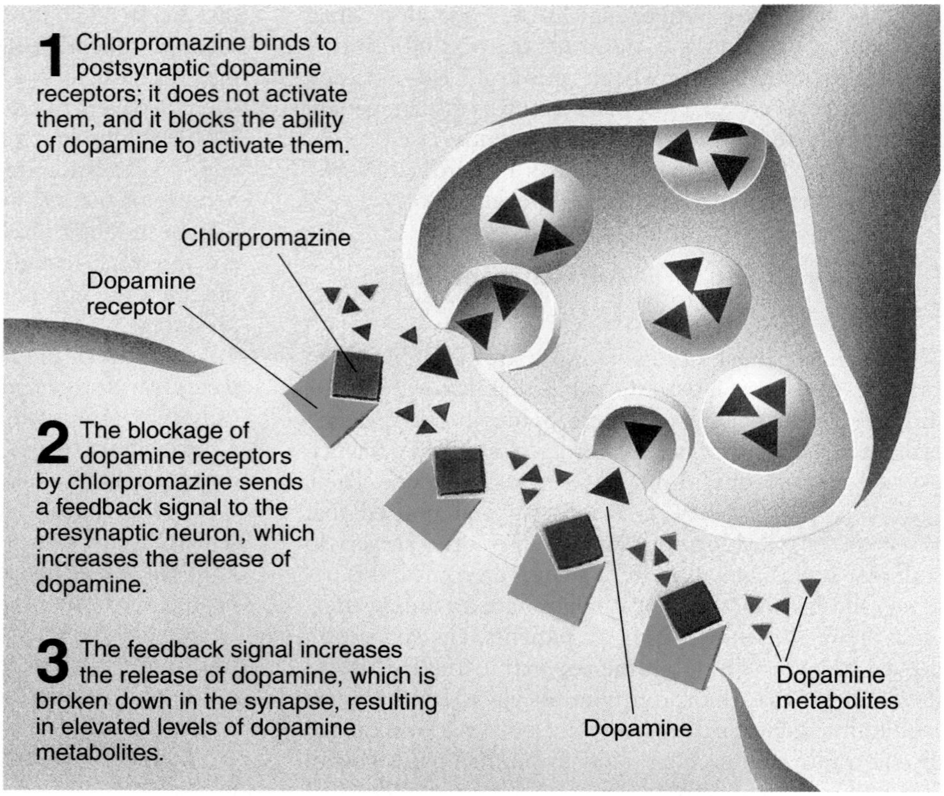

FIGURE 18.1 Chlorpromazine is a receptor blocker at dopamine synapses. Chlorpromazine was the first receptor blocker to be identified, and its discovery changed psychopharmacology.

1 Chlorpromazine binds to postsynaptic dopamine receptors; it does not activate them, and it blocks the ability of dopamine to activate them.

Chlorpromazine

Dopamine receptor

2 The blockage of dopamine receptors by chlorpromazine sends a feedback signal to the presynaptic neuron, which increases the release of dopamine.

3 The feedback signal increases the release of dopamine, which is broken down in the synapse, resulting in elevated levels of dopamine metabolites.

Dopamine

Dopamine metabolites

Rather than high dopamine levels, the main factor in schizophrenia was presumed to be high levels of activity at dopamine receptors.

In the mid-1970s, Snyder and his colleagues (Creese, Burt, & Snyder, 1976) assessed the degree to which the various antischizophrenic drugs that had been developed by that time bind to dopamine receptors. First, they added radioactively labeled dopamine to samples of dopamine-receptor–rich neural membrane obtained from calf striatums. Then, they rinsed away the unbound dopamine molecules from the samples and measured the amount of radioactivity left in them to obtain a measure of the number of dopamine receptors. Next, in other samples, they measured each drug's ability to block the binding of radioactive dopamine to the sample, the assumption being that the drugs with a high affinity for dopamine receptors would leave fewer sites available for the dopamine. In general, they found that chlorpromazine and the other effective antischizophrenic drugs had a high affinity for dopamine receptors, whereas ineffective antischizophrenic drugs had a low affinity. There were, however, several major exceptions, one of them being haloperidol. Although **haloperidol** was one of the most potent antischizophrenic drugs of its day, it had a relatively low affinity for dopamine receptors.

A solution to the haloperidol puzzle came with the discovery that dopamine binds to more than one receptor subtype—five have been identified (Hartmann & Civelli, 1997). It turns out that chlorpromazine and other antischizophrenic drugs in the same chemical class (the **phenothiazines**) all bind effectively to both D_1 and D_2 receptors, whereas haloperidol and the other antischizophrenic drugs in its chemical class (the **butyrophenones**) all bind effectively to D_2 receptors but not to D_1 receptors.

This discovery of the selective binding of butyrophenones to D_2 receptors led to an important revision in the dopamine theory of schizophrenia. It suggested that schizophrenia is caused by hyperactivity specifically at D_2 receptors, rather than at dopamine receptors in general. Snyder and his colleagues (see Snyder, 1978) subsequently

confirmed that the degree to which **neuroleptics**—antischizophrenic drugs—bind to D_2 receptors is highly correlated with their effectiveness in suppressing schizophrenic symptoms (see Figure 18.2). For example, the butyrophenone *spiroperidol* had the greatest affinity for

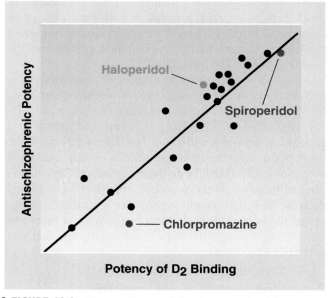

FIGURE 18.2 The positive correlation between the ability of various neuroleptics to bind to D_2 receptors and their clinical potency. (Based on Snyder, 1978.)

TABLE 18.1   The Key Events That Led to the Development and Refinement of the Dopamine Theory of Schizophrenia

Early 1950s	The antischizophrenic effects of both chlorpromazine and reserpine were documented and related to their parkinsonian side effects.
Late 1950s	The brains of recently deceased Parkinson's patients were found to be depleted of dopamine.
Early 1960s	It was hypothesized that schizophrenia was associated with excessive activity at dopaminergic synapses.
1960s and early 1970s	Chlorpromazine and other clinically effective neuroleptics were found to act as receptor blockers at dopamine synapses.
Mid-1970s	The affinity of neuroleptics for dopamine receptors was found to be only roughly correlated with their antischizophrenic potency.
Late 1970s	The binding of existing antischizophrenic drugs to D_2 receptors was found to be highly correlated with their antischizophrenic potency.
1980s and 1990s	It became clear that the D_2 version of the dopamine theory of schizophrenia cannot account for all of the research findings.

D_2 receptors and the most potent antischizophrenic effect.

Although the evidence implicating D_2 receptors in schizophrenia is strong, it has become apparent that the D_2 version of the dopamine theory of schizophrenia cannot explain several key findings. Appreciation of these limitations has led to the current version of the theory. This version holds that excessive activity at D_2 receptors is a factor in the disorder but that there are other, as yet unidentified, causal factors. The major events in the development of the dopamine theory are summarized in Table 18.1.

Neural Basis of Schizophrenia: Limitations of the Dopamine Theory

The following are four key discoveries about the neural bases of schizophrenia that cannot be explained by the D_2 version of the dopamine theory. These discoveries suggest that although overactivity at D_2 receptors plays a role in schizophrenia, other important factors are yet to be identified.

Receptors Other Than D_2 Receptors Are Involved in Schizophrenia   Neurotransmitters other than dopamine have been linked to schizophrenia, including glutamate (Javitt & Coyle, 2004; Tuominen, Tiihonen, & Wahlbeck, 2005), GABA (Lewis, Hashimoto, & Volk, 2005), and serotonin (Sawa & Snyder, 2002). Much of the evidence implicating serotonin and glutamate comes from the study of *hallucinogenic drugs* such as *lysergic acid diethylamide* (LSD) and *phencyclidine* (PCP). Both LSD and PCP produce psychological symptoms similar to those of schizophrenia by acting on serotonergic and glutaminergic transmission (see González-Maeso & Sealfon, 2009).

Compelling evidence that D_2 receptors are not the sole mechanism underlying schizophrenia came from

the development of *atypical neuroleptics* (antischizophrenic drugs that are not primarily D_2 receptor blockers). For example, **clozapine**, the first atypical neuroleptic for the treatment of schizophrenia, has an affinity for D_1 receptors, D_4 receptors, and several serotonin receptors, but only a slight affinity for D_2 receptors.

Clozapine has some promising therapeutic properties. It is often effective in treating patients with schizophrenia who have not responded to *typical neuroleptics*, and it does not produce parkinsonian side effects. Unfortunately, the therapeutic utility of clozapine is limited because it produces a severe blood disorder in some patients who use it (see Wong & Van Tol, 2003); however, a number of other atypical neuroleptics are in wide use.

It Takes Weeks of Neuroleptic Therapy to Alleviate Schizophrenic Symptoms   As you have already learned, it takes several weeks of neuroleptic therapy to alleviate schizophrenic symptoms. However, neuroleptics effectively block activity at D_2 receptors within hours. This time difference indicates that the blockage of D_2 receptors is not the *Neuroplasticity* specific mechanism by which the neuroleptics have their therapeutic effect. It appears that blocking D_2 receptors triggers some slow-developing compensatory change in the brain that is the key factor in the therapeutic effect.

Schizophrenia Is Associated with Widespread Brain Damage   With the development of neuroimaging techniques in the 1960s, reports of brain pathology in patients with schizophrenia accumulated rapidly. The first generation of studies reported enlarged ventricles and fissures (see Figure 18.3 on page 472), which indicated reduced brain size. Subsequent studies focused on specific cortical areas and subcortical structures. A recent meta-analysis reported reduced volume in 50 different brain regions (see

FIGURE 18.3 Brain scans of a schizophrenic patient and his nonschizophrenic identical (monozygotic) twin. Notice the enlarged ventricles (i.e., reduced brain volume) in the schizophrenic brain.

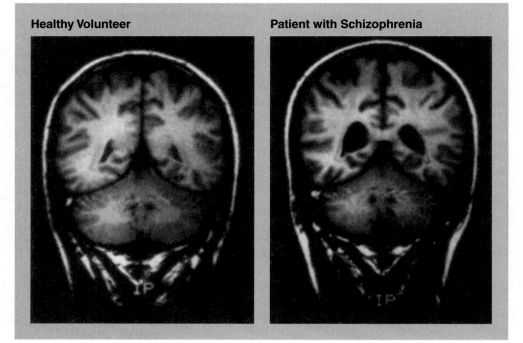

Healthy Volunteer   Patient with Schizophrenia

Kubicki et al., 2007). Schizophrenia-related brain damage occurred in both gray and white matter, and was most consistently observed in the temporal lobes (see Honea et al., 2005). Figure 18.4 illustrates the amount of gray matter loss documented by structural MRIs in various cortical areas of a group of teenagers with schizophrenia (Thompson et al., 2001). Similarly, postmortem studies of schizophrenic brains have found widespread neuron loss and abnormalities of neuron structure and circuitry in many parts of the brain (see Walker et al., 2004).

Two things about the pattern of brain damage observed in many patients with schizophrenia are problematic for the dopamine theory: One is that there is little evidence of specific structural damage to dopaminergic circuits (see Egan & Weinberger, 1997; Nopoulis et al., 2001); the other is that the dopamine theory provides no rationale for the diffuse pattern of brain damage that is typically observed.

Because schizophrenia is believed to be a neurodevelopmental disorder, many studies have assessed the development of brain damage in patients with schizophrenia. Three important findings have emerged—see the meta-analyses of Steen and colleagues (2006) and Vita and colleagues (2006):

- Extensive brain damage exists when patients first seek medical treatment and have their first brain scan.
- Subsequent brain scans reveal that the brain damage has continued to develop.
- Damage to different areas of the brain develops at different rates (Vidal et al., 2006).

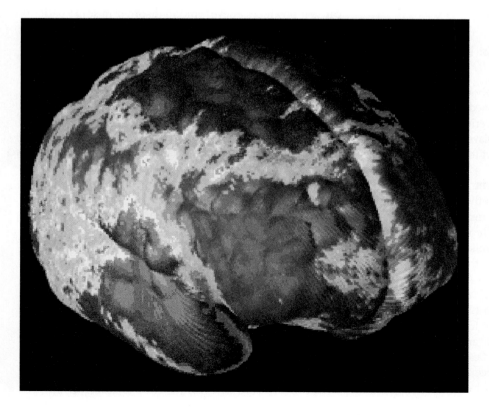

FIGURE 18.4 Structural MRIs reveal cortical loss in adolescent patients with schizophrenia. Here dark blue indicates normal areas, and red indicates areas of greatest tissue loss. (From Thompson et al., 2001.)

Neuroleptics Are Only Marginally Effective If the D_2 version of the dopamine theory of schizophrenia were correct, then typical neuroleptics should be effective treatments for all cases of schizophrenia. They aren't. It is important to remember that just because a drug has been proven to have statistically significant beneficial effects, that doesn't mean that these effects are large.

To put this point in perspective, consider the results of a meta-analysis that re-evaluated all of the studies that had compared the effectiveness of the neuroleptic chlorpromazine to that of a placebo (Adams et al., 2005). The analysis revealed that there were clear beneficial effects only when chlorpromazine was administered for more than 6 months and that, even then, only about 1 in 7 patients was helped substantially.

Neuroleptics are also marginally effective in the sense that even when they do have beneficial effects, they tend to act on only some symptoms of schizophrenia—contradicting the dopamine theory, which predicts that neuroleptics should improve the entire condition. In general, neuroleptics are much more effective in treating positive symptoms of schizophrenia than they are in treating negative symptoms (Murphy et al., 2006).

Patients often stop taking typical neuroleptics because of the motor side effects. There was great optimism when the atypical neuroleptics were introduced because they did not produce the movement disturbances associated with typical neuroleptics. However, it has become apparent that the atypical neuroleptics have their own array of side effects, including diabetes, weight gain, and problems with fat regulation (e.g., Melkersson & Dahl, 2004). As a result, according to another meta-analysis, patients are just as likely to refuse to keep taking atypical neuroleptics as typical neuroleptics (Leucht et al., 2004). Indeed, in one 18-month comparison of patients who had been prescribed atypical or typical neuroleptics, the overall dropout rate was 74% (Lieberman et al., 2005).

18.2
Affective Disorders: Depression and Mania

All of us have experienced depression. Depression is a normal reaction to grievous loss such as the loss of a loved one, the loss of self-esteem, or the loss of health. However, there are people whose tendency toward depression is out of proportion. These people repeatedly fall into the depths of despair and experience **anhedonia** (loss of the capacity to experience pleasure), often for no apparent reason. Their depression can be so extreme that it is almost impossible for them to meet the essential requirements of their daily lives—to keep a job, to maintain social contacts, to eat, or even to maintain an acceptable level of personal hygiene.

Sleep disturbances and thoughts of suicide are common. When this condition lasts for more than 2 weeks, these people are said to be suffering from **clinical depression**, or **major depressive disorder**. The case of P.S. introduces you to some of the main features of clinical depression.

◉ Watch Depression
www.mypsychlab.com

The Case of P.S., the Weeping Widow

Clinical Implications

P.S. was a 57-year-old widow and mother of four. She was generally cheerful and friendly and known for her meticulous care of her home and children. She took great pride in having reared her children by herself following the death of her husband 14 years earlier.

For no apparent reason, her life began to change. She suddenly appeared more fatigued, less cheerful, and more lackadaisical about her housework. Over the ensuing weeks, she stopped going to church and cancelled all of her regular social engagements, including the weekly family dinner, which she routinely hosted. She started to spend all her time sleeping or rocking back and forth and sobbing in her favorite chair. She wasn't eating, bathing, or changing her clothes. And her house was rapidly becoming a garbage dump. She woke up every morning at about 3:00 A.M. and was unable to get back to sleep.

Things got so bad that her two children who were still living with her called her oldest son for advice. He drove from the nearby town where he lived. What he found reminded him of an episode about 10 years earlier, when his mother had attempted suicide by slitting her wrists.

At the hospital, P.S. answered few questions. She cried throughout the admission interview and sat rocking in her chair wringing her hands and rolling her head up towards the ceiling. When asked to explain what was bothering her, she just shook her head no.

She was placed on a regimen of antidepressant medication. Several weeks later, she was discharged, much improved (Spitzer et al., 1983).

Major Categories of Affective Disorders

Depression is an **affective disorder** (any psychiatric disorder characterized by disturbances of mood or emotion). **Mania**, another affective disorder, is in some respects the opposite of depression; it is characterized by overconfidence, impulsivity, distractibility, and high energy. Affective disorders are also commonly known as **mood disorders**.

Clinical Implications

⊙→ **Simulate** Recognizing Mood Disorders **www.mypsychlab.com**

During periods of mild mania, people are talkative, energetic, impulsive, positive, and very confident. In this state, they can be very effective at certain jobs and can be great fun to be with. But when mania becomes extreme, it is a serious clinical problem. When mania is full-blown, the person often awakens in a state of unbridled enthusiasm, with an outflow of incessant chatter that careens nonstop from topic to topic. No task is too difficult. No goal is unattainable. This confidence and grandiosity, coupled with high energy, distractibility, and a leap-before-you-look impulsiveness, result in a continual series of disasters. Mania often leaves behind it a trail of unfinished projects, unpaid bills, and broken relationships.

Many depressive patients experience periods of mania. Those who do are said to suffer from **bipolar affective disorder**. Those depressive patients who do not experience periods of mania are said to suffer from **unipolar affective disorder**. Depression is often further divided into two categories. Depression triggered by a negative experience (e.g., the death of a friend, the loss of a job) is called **reactive depression**; depression with no apparent cause is called **endogenous depression**.

⊙ **Watch** Bipolar Disorder **www.mypsychlab.com**

⊙→ **Simulate** Bipolar Quiz **www.mypsychlab.com**

((⊙• **Listen** Famous People with Bipolar Disorder **www.mypsychlab.com**

In most countries, the probability of suffering from clinical depression during one's lifetime is about 10%. Women tend to be diagnosed with unipolar affective disorder about twice as frequently as men (see Altemus, 2006; Bale, 2006; Hyde, Mezulis, & Abramson, 2008)— but the reason for this is unclear. There is no sex difference in the incidence of bipolar affective disorder. A high rate of suicide among the clinically depressed is well documented, and the lifetime rate that is often reported is 15%, which is based on a 1970 meta-analysis (Guze & Robins, 1970). Because this high figure seemed to be inconsistent with their experiences at the Mayo Clinic, Bostwick and Pankratz (2000) re-examined the 1970 meta-analysis and spotted a major methodological flaw. They then conducted a more accurate meta-analysis and found that the lifetime risk of suicide in an individual diagnosed with clinical depression is about 5%.

Affective disorders attack children, adolescents, and adults. Indeed, it has been suggested that bipolar affective disorders that first appear in childhood, adolescence, and adulthood may constitute three different subgroups with different symptoms and causes (Carlson & Meyer, 2006). In adults, affective disorders are associated with heart disease (Frasure-Smith & Lespérance, 2005; Miller & Blackwell, 2006); in adult women, these disorders are associated with bone loss (Eskandari et al., 2007).

Causal Factors in Affective Disorders

Genetic factors contribute to differences among people in the development of affective disorders. Twin studies of affective disorders suggest a concordance rate of about 60% for identical twins and 15% for fraternal twins, whether they are reared together or apart. Although there are many exceptions, there is a tendency for affected twins to suffer from the same type of disorder, unipolar or bipolar; and the concordance rates for bipolar disorders tend to be higher than those for unipolar disorders. No particular gene has been linked to affective disorders (see Berton & Nestler, 2006).

Clinical Implications

Most of the research on the causal role of experience in affective disorders has focused on the role of stress in the etiology of depression. Indeed, depression is often described as a stress-related disorder (Bale, 2005). However, good evidence linking stress to affective disorders is sparse. Several studies have shown that stressful experiences can trigger attacks in people already suffering from depression (e.g., Brown, 1993), but there is little evidence showing that stress can increase the susceptibility to affective disorders in the healthy—even childhood sexual abuse appears to have only a slight effect on the development of depression in adulthood (see Nelson et al., 2002). Rather than depression, extreme stress tends to produce *posttraumatic stress disorder*, which you will learn more about later in this chapter.

There are two affective disorders whose cause is more apparent because of the timing of the attacks. One is **seasonal affective disorder (SAD)**, in which attacks of depression and lethargy typically recur every winter (Lam et al., 2006). Two lines of evidence suggest that the attacks are triggered by the reduction in sunlight. One is that the incidence of the disorder is higher in the northern United States (9%) than in Florida (1.5%), where the winter days are longer and brighter (Modell et al., 2005). The other is that *light therapy* is often effective in reducing the symptoms (Lam et al., 2006). The second affective disorder with an obvious cause is **postpartum depression**, the intense, sustained depression experienced by some women after they give birth. The diagnosis of postpartum depression requires that the depression last for at least 1 month. It normally lasts no longer than 3 months. Although estimates vary, the disorder seems to develop following about 10% of deliveries.

Discovery of Antidepressant Drugs

Four major classes of drugs have been used in the treatment of affective disorders (see Berton & Nestler, 2006): monoamine oxidase inhibitors, tricyclic antidepressants, selective monoamine-reuptake inhibitors, and mood stabilizers.

Monoamine Oxidase Inhibitors   **Iproniazid**, the first antidepressant drug, was originally developed for the treatment of tuberculosis, for which it proved to be a dismal flop. However, interest in the antidepressant potential of the drug was kindled by the observation that it left patients with tuberculosis less concerned about their disorder. As a result, iproniazid was tested on a mixed group of psychiatric patients and seemed to act against clinical depression. It was first marketed as an antidepressant drug in 1957.

Iproniazid is a monoamine agonist; it increases the levels of monoamines (e.g., norepinephrine and serotonin) by inhibiting the activity of *monoamine oxidase (MAO)*, the enzyme that breaks down monoamine neurotransmit-

ters in the *cytoplasm* (cellular fluid) of the neuron. **MAO inhibitors** have several side effects; the most dangerous is known as the **cheese effect** (see Youdim, Edmonson, & Tipton, 2006). Foods such as cheese, wine, and pickles contain an amine called *tyramine*, which is a potent elevator of blood pressure. Normally, these foods have little effect on blood pressure, because tyramine is rapidly metabolized in the liver by MAO. However, people who take MAO inhibitors and consume tyramine-rich foods run the risk of strokes caused by surges in blood pressure.

Tricyclic Antidepressants   The **tricyclic antidepressants** are so named because of their antidepressant action and because their chemical structures include three rings of atoms. **Imipramine**, the first tricyclic antidepressant, was initially thought to be an antischizophrenic drug. However, when its effects on a mixed sample of psychiatric patients were assessed, it had no effect against schizophrenia but seemed to help some depressed patients. Tricyclic antidepressants block the reuptake of both serotonin and norepinephrine, thus increasing their levels in the brain. They are a safer alternative to MAO inhibitors.

Selective Monoamine–Reuptake Inhibitors   In the late 1980s, a new class of drugs—the selective serotonin-reuptake inhibitors—was introduced for treating clinical depression. **Selective serotonin-reuptake inhibitors (SSRIs)** are serotonin agonists that exert agonistic effects by blocking the reuptake of serotonin from synapses—see Figure 18.5.

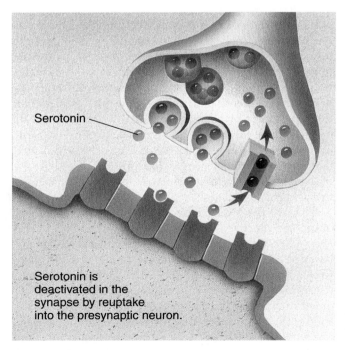

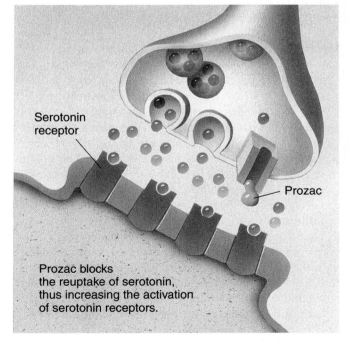

FIGURE 18.5   Blocking of serotonin reuptake by fluoxetine (Prozac).

Fluoxetine, which is marketed as **Prozac**, was the first SSRI to be developed. Now there are many more (e.g., Paxil, Zoloft, Luvox, Remeron). Prozac's structure is a slight variation of that of imipramine and other tricyclic antidepressants; in fact, Prozac is no more effective than imipramine in treating depression. Nevertheless, it was immediately embraced by the psychiatric community and has been prescribed in many millions of cases. The remarkable popularity of Prozac and other SSRIs is attributable to two things: First, they have few side effects; second, it is claimed that they act against a wide range of psychological disorders in addition to depression. Because SSRIs are so effective against disorders that were once considered to be the exclusive province of psychotherapy (e.g., lack of self-esteem, fear of failure, excessive sensitivity to criticism, and inability to experience pleasure), they have had a major impact on psychiatry and clinical psychology.

In 2003, there was an indication that SSRIs might increase suicide rates, and U.S. and European public health agencies issued warnings. However, increased suicide rates in SSRI users were not observed in subsequent studies (e.g., Bridge et al., 2007; Simon et al., 2006; Tiihonen, 2006). As a result of the premature warning from governmental agencies, SSRI prescriptions decreased, and suicide rates increased markedly between 2003 and 2005 (Gibbons et al., 2007).

The success of the SSRIs spawned the introduction of a similar class of drugs, the *selective norepinephrine-reuptake inhibitors (SNRIs)*. These (e.g., Reboxetine) have proven to be as effective as the SSRIs in the treatment of depression. Also used against depression are drugs (e.g., Wellbutrin, Effexor) that block the reuptake of more than one monoamine neurotransmitter.

Mood Stabilizers   Antidepressant drugs have a major drawback: They often act against depression in bipolar patients by triggering bouts of mania. This led to the development of a new class of drugs for the treatment of bipolar affective disorders—the mood stabilizers. **Mood stabilizers** are drugs that act against depression without increasing mania or, conversely, act against mania without increasing depression (Bourin & Prica, 2007). The mechanism by which mood stabilizers work is unknown (see Gould & Einat, 2007), but for some reason these drugs are also effective in the treatment of epilepsy.

Lithium, a simple metallic ion, was the first drug found to act as a mood stabilizer. The discovery of lithium's anti-mania action is yet another important pharmacological breakthrough that occurred by accident. John Cade, an Australian psychiatrist, mixed the urine of manic patients with lithium to form a soluble salt; then he injected the salt into a group of guinea pigs to see if it would induce mania. As a control, he injected lithium into another group. Instead of inducing mania, the urine solution seemed to calm the guinea pigs; and because the lithium control injections

had the same effect, Cade concluded that lithium, not uric acid, was the calming agent. In retrospect, Cade's conclusion was incredibly foolish. We now know that at the doses he used, lithium salts produce extreme nausea. To Cade's untrained eye, his subjects' inactivity may have looked like calmness. But the subjects weren't calm; they were sick. In any case, flushed with what he thought was the success of his guinea pig experiments, in 1954 Cade tried lithium on a group of 10 manic patients, and it seemed to have a therapeutic effect. There was little immediate reaction to Cade's report—few scientists were impressed by his scientific credentials, and few drug companies were interested in spending millions of dollars to evaluate the therapeutic potential of a metallic ion that could not be protected by a patent. It was not until the late 1960s that lithium was found to treat mania without increasing depression.

Effectiveness of Drugs in the Treatment of Affective Disorders   About 2 billion dollars is spent in the United States each year by the Medicaid program on antidepressants (Chen et al., 2008). But how effective are antidepressants? Numerous studies have evaluated the effectiveness of antidepressant drugs against unipolar affective disorder. Hollon, Thase, and Markowitz (2002) compared the efficacy of the various pharmacological treatments for depression. The results were about the same for MAO inhibitors, tricyclic antidepressants, and selective monoamine-reuptake inhibitors: About 50% of clinically depressed patients improved. This rate seems quite good; however, the control group showed a 25% rate of improvement, so only 25% of the depressed group were actually helped by the antidepressants.

In 2008, the news about the benefits of antidepressants got a lot worse. Kirsch and colleagues (2008) conducted a meta-analysis of all the evaluation studies of the four most commonly prescribed antidepressants that were on record with the U.S. Food and Drug Administration. They found that the antidepressants did not produce improvements that were significantly greater than those produced by placebos in patients who were mildly or moderately depressed. In the severely depressed, the improvements were slight. The meta-analysis indicated that antidepressants work, but mostly through the placebo effect—overall, the placebos were 82% as effective as the actual drugs.

The effectiveness of mood stabilizers in the treatment of bipolar affective disorder was reviewed by Bourin and Prica (2007), who concluded that the ideal mood stabilizer does not exist—an ideal mood stabilizer would effectively stop attacks of depression or mania without triggering the alternative condition and then keep subsequent attacks from recurring. Although the evidence is sparse, they concluded that lithium and *carbamazepine* (an anti-epileptic drug) are best for treating mania, and *lamotrigine* (another anti-epileptic drug) is best for treating depression.

Brain Pathology and Affective Disorders

Numerous MRI studies of the brains of bipolar patients have been published. Reductions in overall brain size (e.g., Frazier et al., 2005) and in the size of many different brain structures (e.g., amygdala, cingulate cortex, or prefrontal cortex) have been reported (e.g., Savitz & Drevets, 2009; Delbello et al., 2004). However, the pattern of results is inconsistent: A particular brain structure found to be reduced in size in some studies is often found to be of normal size in others (McDonald et al., 2004). This suggests that not all patients diagnosed as having bipolar affective disorder by current criteria suffer from the same disorder (see Fountoulakis et al., 2008).

However, there are two structures that are found to be abnormal in many structural and functional brain-imaging studies of affective disorders: the amygdala (e.g., Gotlib & Hamilton, 2008; Savitz & Drevets, 2009) and the anterior cingulate cortex (Fountoulakis et al., 2008; Greicius et al., 2007). Even the connections between these two structures appear to be disturbed in patients with affective disorders (e.g., Matthews et al., 2009; Wang et al., 2009). Figure 18.6 illustrates the loss of tissue in the anterior cingulate cortex and the amygdala in a group of healthy volunteers who are genetically disposed to developing depression.

Theories of Depression

The search for the neural mechanisms of affective disorders has focused on clinical depression. However, the fact that depression and mania often occur in the same patients—that is, in those with bipolar affective disorder—suggests that the mechanisms of the two are closely related. None of the prominent theories of depression deals adequately with its relation to mania.

Monoamine Theory of Depression One prominent theory of clinical depression is the *monoamine theory*. The monoamine theory of depression holds that depression is associated with underactivity at serotonergic and noradrenergic synapses. The theory is largely based on the fact that monoamine oxidase inhibitors, tricyclic antidepressants, selective serotonin-reuptake inhibitors, and selective norepinephrine-reuptake inhibitors are all agonists of serotonin, norepinephrine, or both.

Other support for the monoamine theory of depression has been provided by autopsy studies (see Nemeroff, 1998). Norepinephrine and serotonin receptors have been found to be more numerous in the brains of deceased clinically depressed individuals who had not received pharmacological treatment. This implicates a deficit in monoamine release: When an insufficient amount of a neurotransmitter is released at a synapse, there are usually compensatory increases in the number of receptors for that neurotransmitter—a process called **up-regulation**.

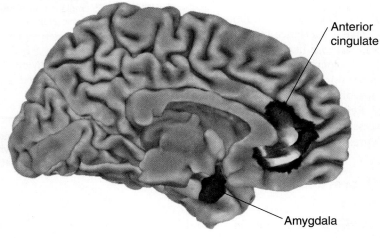

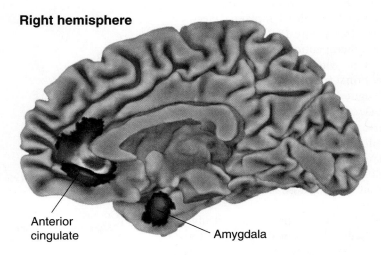

FIGURE 18.6 Structural MRIs of healthy volunteers with a genetic predisposition to developing depression reveal cell loss in the anterior cingulate and the amygdala.

Overall, support for the monoamine theory of depression is weak. Although the theory is based on the fact that monoamine agonists are used to treat depressed patients, few depressed patients benefit substantially from them.

Diathesis–Stress Model of Depression A second theory of depression is the *diathesis–stress model*. According to this theory (see Gillespie & Nemeroff, 2007; Pariante & Lightman, 2008), some people inherit a **diathesis** (a genetic susceptibility), which is incapable of initiating the disorder by itself. The central idea of the diathesis–stress model is that if susceptible individuals are exposed to stress early in life, their systems become permanently sensitized, and they overreact to mild stressors for the rest of their lives.

Support for the diathesis–stress model of depression is largely indirect: It is based on the finding that depressed people tend to release more stress hormones (see Gillespie & Nemeroff, 2007). For example, depressed individuals synthesize more hypothalamic *corticotropin-releasing hormone* and release more *adrenocorticotropic hormone* from the anterior pituitary and more *glucocorticoids* from the adrenal cortex.

Although there has been some evidence that individuals who have experienced early stress (e.g., sexual abuse) are more likely to suffer from depression later (see Carpenter et al., 2004; Nelson et al., 2002), this evidence is not convincing because it is based on the recollections of patients, which cannot be confirmed (Monroe & Reid, 2008). Moreover, there is no evidence of excessive early stress in the lives of the majority of depressed patients.

Treatment of Depression with Brain Stimulation

Perhaps the most exciting recent advance in the study of affective disorders was the demonstration that chronic brain stimulation through an implanted electrode (see Figure 18.7) has a significant therapeutic effect in depressed patients who had repeatedly failed to respond to conventional treatments.

Clinical Implications

Lozano and colleagues (2008) implanted the tip of a stimulation electrode into an area of the white matter of the anterior cingulate gyrus just ventral to the anterior end of the corpus callosum (see Figure 18.8). The stimulator, which was implanted under the skin, delivered continual pulses of electrical stimulation that could not be detected by the patients.

Considering that the 20 patients in this study had been selected because they repeatedly failed to

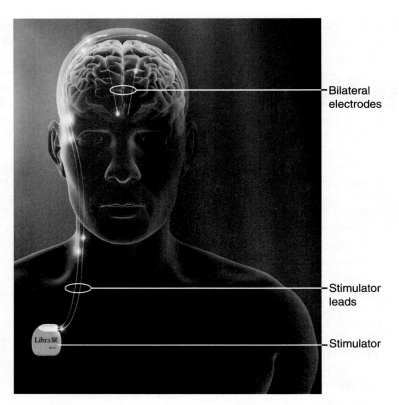

FIGURE 18.7 Implantation of bilateral anterior cingulate electrodes and stimulator for chronic deep brain stimulation for the treatment of depression.

respond to conventional treatments, the results were striking: 60% showed substantial improvements, 35% were

FIGURE 18.8 The site in the anterior cingulate gyrus at which chronic brain stimulation to subcortical white matter alleviated symptoms in treatment-resistant depressed patients.

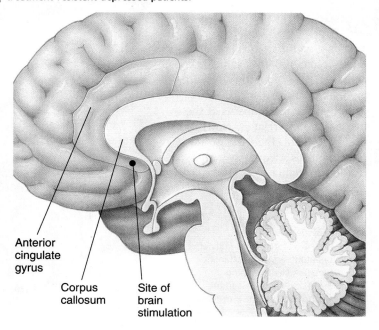

largely symptom free, and the improvements were maintained for at least 1 year (the duration of the study). However, a careful double-blind evaluation of the effectiveness of the procedure must be performed before it can be recommended for wider use.

18.3
Anxiety Disorders

Anxiety—chronic fear that persists in the absence of any direct threat—is a common psychological correlate of stress. Anxiety is adaptive if it motivates effective coping behaviors; however, when it becomes so severe that it disrupts normal functioning, it is referred to as an **anxiety disorder**. All anxiety disorders are associated with feelings of anxiety (e.g., fear, worry, despondency) and with a variety of physiological stress reactions—for example, *tachycardia* (rapid heartbeat), *hypertension* (high blood pressure), nausea, breathing difficulty, sleep disturbances, and high glucocorticoid levels.

⊙ **Watch** Clinical Anxiety
www.mypsychlab.com

Anxiety disorders are the most prevalent of all psychiatric disorders. A meta-analysis of 46 studies found that about 17% of people suffer from an anxiety disorder at some point in their lives and that the incidence seems to be about twice as great in females as in males (Somers et al., 2006). M.R., a woman who was afraid to leave her home, suffered from one type of anxiety disorder.

The Case of M.R., the Woman Who Was Afraid to Go Out

M.R. was a 35-year-old woman who developed a pathological fear of leaving home. The onset of her problem was sudden. Following an argument with her husband, she went out to mail a letter and cool off, but before she could accomplish her task, she was overwhelmed by dizziness and fear. She immediately struggled back to her house and rarely left it again, for about 2 years. Then, she gradually started to improve.

Clinical Implications

Her recovery was abruptly curtailed, however, by the death of her sister and another argument with her husband. Following the argument, she tried to go shopping, panicked, and had to be escorted home by a stranger. Following that episode, she was not able to leave her house by herself without experiencing an anxiety attack. Shortly after leaving home by herself, she would feel dizzy and sweaty, and her heart would start to pound; at that point, she would flee home to avoid a full-blown panic attack.

Although M.R. could manage to go out if she was escorted by her husband or one of her children, she felt anxious the entire time. Even with an escort, she was terrified of crowds—crowded stores, restaurants, or movie theaters were out of the question.

Five Classes of Anxiety Disorders

There are five major classes of anxiety disorders:

Clinical Implications

- **Generalized anxiety disorders** are characterized by stress responses and extreme feelings of anxiety that occur in the absence of any obvious precipitating stimulus.
- **Phobic anxiety disorders** are similar to generalized anxiety disorders except that they are triggered by exposure to particular objects (e.g., birds, spiders) or situations (e.g., crowds, darkness). M.R., the woman who was afraid to go out, suffered from a common phobic anxiety disorder: agoraphobia. **Agoraphobia** is the pathological fear of public places and open spaces.
- **Panic disorders** are characterized by rapid-onset attacks of extreme fear and severe symptoms of stress (e.g., choking, heart palpitations, shortness of breath); they are often components of generalized anxiety and phobic disorders, but they also occur as separate disorders.
- **Obsessive-compulsive disorders** are characterized by frequently recurring, uncontrollable, anxiety-producing thoughts (obsessions) and impulses (compulsions). Responding to them— for example, by repeated compulsive hand washing—is a means of dissipating the anxiety associated with them.

⊙ **Watch** Obsessive-Compulsive Disorder
www.mypsychlab.com

- **Posttraumatic stress disorder** is a persistent pattern of psychological distress following exposure to extreme stress, such as war or being the victim of sexual assault (McNally, 2003; McNally, Bryant, & Ehlers, 2003; Newport & Nemeroff, 2000).

Etiology of Anxiety Disorders

Because anxiety disorders are often triggered by identifiable stressful events and because the anxiety is often focused on particular objects or situations, the role of experience in shaping the disorder is often apparent (see Anagnostaras, Craske, & Fanselow, 1999). For example, in addition to having agoraphobia, M.R. was obsessed by her health—particularly by high blood pressure, although hers was in the normal range. The fact that both her grandfather and her father suffered from high blood pressure and died of heart attacks clearly shaped this component of her disorder.

Clinical Implications

Like other psychiatric disorders, anxiety disorders have a significant genetic component—heritability estimates range from 30% to 40% in various studies (Leonardo & Hen, 2006). The concordance rates for various anxiety disorders are substantially higher for identical twins than for fraternal twins. However, the timing and focus of anxiety disorders often reflect the particular experiences of the patient (see Gross & Hen, 2004). No specific genes have yet been linked to anxiety disorders (Gordon & Hen, 2004).

Pharmacological Treatment of Anxiety Disorders

Three categories of drugs are effective against anxiety disorders: benzodiazepines, serotonin agonists, and antidepressants.

Clinical Implications

Benzodiazepines   Benzodiazepines such as *chlordiazepoxide* (Librium) and *diazepam* (Valium) are widely prescribed for the treatment of anxiety disorders. They are also prescribed as *hypnotics* (sleep-inducing drugs), anticonvulsants, and muscle relaxants. Indeed, benzodiazepines are the most widely prescribed psychoactive drugs; approximately 10% of adult North Americans are currently taking them. The benzodiazepines have several adverse side effects: sedation, *ataxia* (disruption of motor activity), tremor, nausea, and a withdrawal reaction that includes rebound anxiety. Another serious problem with benzodiazepines is that they are highly addictive. Consequently, they should be prescribed for only short-term use (see Gray & McNaughton, 2000). The behavioral effects of benzodiazepines are thought to be mediated by their agonistic action on $GABA_A$ receptors.

Serotonin Agonists   The serotonin agonist *buspirone* is widely used in the treatment of anxiety disorders. Buspirone appears to have selective agonist effects at one subtype of serotonin receptor, the 5-HT_{1A} receptor. Its mechanism of action is not totally understood, but it does not function as an SSRI. The main advantage of buspirone over the benzodiazepines is its specificity: It produces *anxiolytic* (anti-anxiety) effects without producing ataxia, muscle relaxation, and sedation, the common side effects of the benzodiazepines. Buspirone does, however, have other side effects (e.g., dizziness, nausea, headache, and insomnia).

Antidepressant Drugs   One of the complications in studying both anxiety disorders and depression is their **comorbidity** (their tendency to occur together in the same individual)—in one study of patients with unipolar or bipolar affective disorder, over half had also been previously diagnosed with an anxiety disorder (Simon et al., 2004). The comorbidity is thought to exist because both disorders involve a heightened emotional response to stress (Morilak & Frazer, 2004). Consistent with the comorbid relationship are the observations that antidepressants, such as the SSRIs, are often effective against anxiety disorders, and **anxiolytic drugs** (anti-anxiety drugs) are often effective against depression.

Animal Models of Anxiety

Animal models have played an important role in the study of anxiety and in the assessment of the anxiolytic potential of new drugs (see Gray & McNaughton, 2000; Green, 1991; Treit, 1985). A weakness of these models is that they typically involve animal defensive behaviors, the implicit assumption being that defensive behaviors are motivated by fear and that fear and anxiety are similar states (see McNaughton & Zangrossi, 2008). Three animal behaviors that model anxiety are elevated-plus-maze performance, defensive burying, and risk assessment.

In the **elevated-plus-maze test**, rats are placed on a four-armed plus-sign-shaped maze that is 50 centimeters above the floor. Two arms have sides and two arms have no sides, and the measure of anxiety is the proportion of time the rats spend in the enclosed arms, rather than venturing onto the exposed arms (see Pellow et al., 1985).

In the **defensive-burying test** (see Figure 5.27), rats are shocked by a wire-wrapped wooden dowel mounted on the wall of a familiar test chamber. The measure of anxiety is the amount of time the rats spend spraying bedding material from the floor of the chamber at the source of the shock with forward thrusting movements of their head and forepaws (see Treit et al., 1993).

In the **risk-assessment test**, after a single brief exposure to a cat on the surface of a laboratory burrow system, rats flee to their burrows and freeze. Then, they engage in a variety of risk-assessment behaviors (e.g., scanning the surface from the mouth of the burrow or exploring the surface in a cautious stretched posture) before their behavior returns to normal (see Blanchard, Blanchard, & Rodgers, 1991; Blanchard et al., 1990). The measures of anxiety are the amounts of time that the rats spend in freezing and in risk assessment.

The elevated-plus-maze, defensive-burying, and risk-

Thinking Creatively

assessment tests of anxiety have all been validated by demonstrations that benzodiazepines reduce the various indices of anxiety used in the tests, whereas nonanxiolytic drugs usually do not. However, a potential problem with this line of evidence stems from the fact that many cases of anxiety do not respond well to benzodiazepine therapy. Therefore, existing animal models of anxiety may be models of benzodiazepine-sensitive anxiety rather than of anxiety in general, and thus the models may not be sensitive to anxiolytic drugs that act by a different (i.e., a non-GABAergic) mechanism. For example, the serotonin agonist buspirone does not have a reliable anxiolytic effect on subjects performing on the elevated-plus-maze test.

Neural Bases of Anxiety Disorders

Like current theories of the neural bases of schizophrenia and depression, current theories of the neural bases of anxiety disorders rest heavily on the analysis of therapeutic drug effects. The fact that many anxiolytic drugs are agonists at either $GABA_A$ receptors (e.g., the benzodiazepines) or serotonin receptors (e.g., buspirone, Prozac, and Paxil) has focused attention on the possible role in anxiety disorders of deficits in both GABAergic and serotonergic transmission.

There is substantial overlap between the brain structures involved in affective and anxiety disorders. Indeed, the amygdala and the anterior cingulate cortex, which you have just learned have been implicated in affective disorders, have also been implicated in anxiety disorders. This is hardly surprising given the comorbidity of affective and anxiety disorders, the effectiveness of some drugs (e.g., SSRIs) against both, and the fact that both are primarily disturbances of emotion.

Although the amygdala and the anterior cingulate cortex have been implicated in both affective and anxiety disorders, the patterns of evidence differ. With affective disorders, you have already seen that there seems to be shrinkage to these structures; however, with anxiety disorders, there appears to be no gross damage. Most of the evidence linking the amygdala and the anterior cingulate cortex to anxiety disorders has come from functional brain-imaging studies in which abnormal activity in these areas has been recorded during the performance of various emotional tasks (see Bishop, 2007; Kim & Whalen, 2009; Melcher, Falkai, & Gruber, 2008; Nitschke et al., 2009). For example, Figure 18.9 illustrates increased functional MRI activity in the amygdalae of patients with spider phobias when they viewed photographs of spiders.

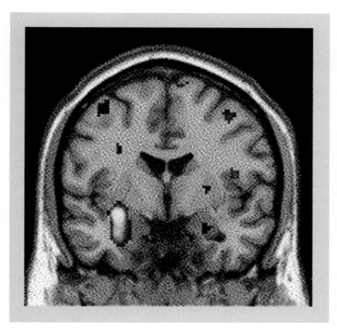

FIGURE 18.9 Functional brain image taken at the moment when patients with spider phobia were presented with spiders. Increased activity was recorded in the amygdala. (From Schienle et al., 2005.)

18.4
Tourette Syndrome

Tourette syndrome is the last of the four psychiatric disorders discussed in this chapter. It differs from the first three (schizophrenia, affective disorders, and anxiety) in the specificity of its effects. And they are as interesting as they are specific. The case of R.G. introduces you to Tourette syndrome.

The Case of R.G.—Barking Mad

When R.G. was 15, he developed *tics* (involuntary, repetitive, stereotyped movements or vocalizations). For the first week, his tics took the form of involuntary blinking, but after that they started to involve other parts of the body, particularly his arms and legs.

Clinical Implications

R.G. and his family were religious, so it was particularly distressing when his tics became verbal. He began to curse repeatedly and involuntarily. Involuntary cursing is a common symptom of Tourette syndrome and of several other psychiatric and neurological disorders (Van Lancker & Cummings, 1999). R.G. also started to bark like a dog. Finally, he developed echolalia: When his mother said, "Dinner is ready," he responded, "Is ready, is ready."

Prior to the onset of R.G.'s symptoms, he was an A student, apparently happy and with an outgoing, engaging personality. Once his symptoms developed, he was jeered at, imitated, and ridiculed by his schoolmates. He responded by becoming anxious, depressed, and withdrawn. His grades plummeted.

Once R.G. was taken to a psychiatrist by his parents, his condition was readily diagnosed—the symptoms of Tourette syndrome are unmistakable. Medication eliminated 99% of his symptoms, and once his disorder was explained to him and he realized he was not mad, he resumed his former outgoing manner (Spitzer et al., 1983).

Many people with Tourette syndrome experience no symptoms other than tics. Thus, if their friends, family members, and colleagues are understanding and supportive, these people can live happy, productive lives—for example, Tim Howard (shown in the photo) is goalkeeper for both the American national team and Everton Football Club, a team in the English Premier Division.

What Is Tourette Syndrome?

Tourette syndrome is a disorder of **tics** (involuntary, repetitive, stereotyped movements or vocalizations) (Gerard & Peterson, 2003). It typically begins early in life—usually in childhood or early adolescence—with simple motor tics, such as eye blinking or head movements, but the symptoms tend to become more complex and severe as the patient grows older. Common complex motor tics include hitting, touching objects, squatting, hopping, twirling, and sometimes even making lewd gestures. Common verbal tics include inarticulate sounds (e.g., barking, coughing, grunting), *coprolalia* (uttering obscenities), *echolalia* (repetition of another's words), and *palilalia* (repetition of one's own words). The symptoms usually reach a peak after a few years, and they often gradually subside as the patient matures.

Tourette syndrome develops in approximately 0.7% of the population and is three times more frequent in males than in females. There is a major genetic component: Concordance rates are 55% for identical twins and 8% for fraternal twins (see Pauls, 2001).

Some patients with Tourette syndrome also display signs of *attention-deficit/hyperactivity disorder*, obsessive-compulsive disorder, or both (Sheppard et al., 1999). For example, R.G. was obsessed by odd numbers and refused to sit in even-numbered seats.

Although the tics of Tourette syndrome are involuntary, they can be temporarily suppressed with concentration and effort by the patient. The effect of suppression has been widely misunderstood. A study published in 2004 by Marcks and colleagues found that 77% of medical professionals believed that tic suppression is inevitably followed by a *rebound* (that the tics become even worse following a period of suppression). However, this has proven not to be the case (Himle & Woods, 2005; Meidinger et al., 2005)—see Figure 18.10.

Imagine how difficult it would be to get on with your life if you suffered from an extreme form of Tourette syndrome—for example if you frequently made obscene gestures and barked like a dog. No matter how polite, intelligent, and kind you were inside, not many people would be willing to socialize with, or employ, you (see Kushner, 1999).

Neuropathology of Tourette Syndrome

Because Tourette syndrome is a well-defined disorder with clearly observable symptoms, its neuropathology is more amenable to study than that of the other disorders that you have encountered in this chapter. Still, there are impediments: for example, the lack of an animal model and the lack of a strong link to any particular gene. The greatest difficulties in studying Tourette syndrome are attributable to the fact that symptoms often subside as people age. Because Tourette patients are rarely under care for the syndrome when they die, few postmortem studies of

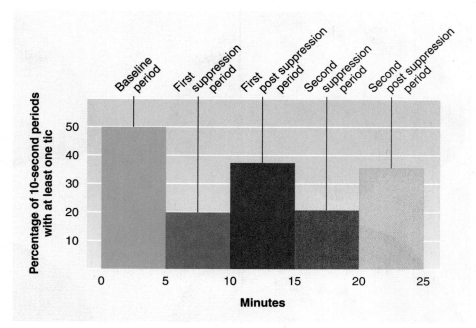

FIGURE 18.10 No rebound effect above baseline was observed following periods of tic suppression by children with Tourette syndrome. (Based on Himle & Woods, 2005.)

Tourette syndrome have been conducted. Consequently, the study of the disorder's neuropathology is based almost exclusively on brain-imaging studies, which are difficult to conduct because of the requirement that the patients remain motionless. As a result, many brain-imaging studies of Tourette syndrome have focused on adult patients in whom the tics have largely subsided. This makes it difficult for researchers to identify the neural mechanisms of tic production (see Gerard & Peterson, 2003).

Most research on the cerebral pathology associated with Tourette syndrome has focused on the caudate. Patients with this disorder tend to have smaller caudate nuclei, and when they suppress their tics, fMRI activity is recorded in both prefrontal cortex and caudate nuclei (see Albin & Mink, 2006; Gerard & Peterson, 2003). Presumably, the decision to suppress the tics comes from the prefrontal cortex, which initiates the suppression by acting on the caudate nuclei. Although most studies of the neuropathology associated with Tourette syndrome have focused on the caudate nuclei, the brain damage appears to be more widespread. A recent MRI study of children with Tourette syndrome (Sowell et al., 2008) documented thinning in sensorimotor cortex that was particularly prominent in the areas that controlled the face, mouth, and *larynx* (voice box)—see Figure 18.11 on page 484.

Treatment of Tourette Syndrome

Although tics are the defining feature of Tourette syndrome, treatment typically begins by focusing on other aspects of the disorder. First, the patient, family members, friends, and 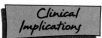 teachers are educated about the nature of the syndrome. Second, the treatment focuses on the ancillary emotional problems (e.g., anxiety and depression). Once these first two steps have been taken, attention turns to treating the symptoms.

The tics of Tourette syndrome are usually treated with *neuroleptics* (the D_2 receptor blockers that are used in the treatment of schizophrenia). Neuroleptics can reduce tics by about 70%, but in practice their benefits are only modest because patients often refuse, or are not allowed by their parents, to take them because of the adverse side effects (e.g., weight gain, fatigue, and dry mouth).

The success of D_2 receptor blockers in blocking Tourette tics is consistent with the current hypothesis that the disorder is related to an abnormality of the basal ganglia–thalamus–cortex feedback circuit. The efficacy of these drugs especially implicates the *striatum* (caudate plus putamen), which is the target of many of the dopaminergic projections into the basal ganglia. The binding of extracellular D_2 receptors was found to be reduced in the brains of a group of adult Tourette patients who had never taken D_2-receptor-blocking medication (Gilbert et al., 2006).

P.H. is a scientist who counsels Tourette patients and their families. He also has Tourette syndrome, which provides him with a useful perspective.

The Case of P.H., the Neuroscientist with Tourette Syndrome

Tourette syndrome has been P.H.'s problem for more than three decades (Hollenbeck, 2001). Taking advantage of his position as a medical school faculty member, he regularly offers a series of lectures on the topic. Along with students, many other Tourette patients and their families are attracted to his lectures.

Encounters with Tourette patients of his own generation taught P.H. a real lesson. He was astounded to learn that most of them did not have his thick skin. About half of them were still receiving treatment for psychological wounds inflicted during childhood.

For the most part, these patients' deep-rooted pain and anxiety did not result from the tics themselves. They derived

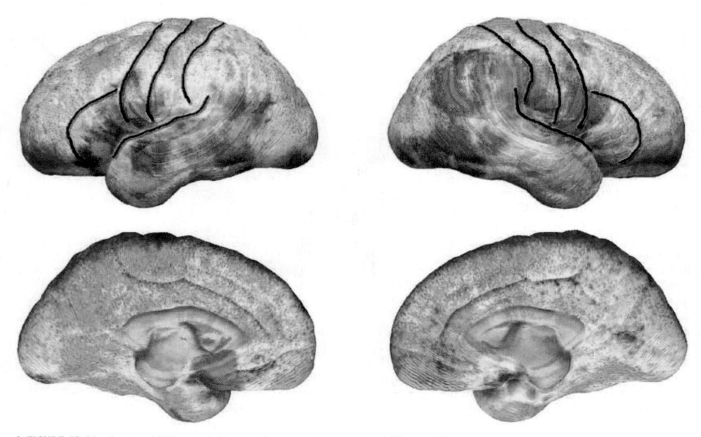

FIGURE 18.11 Structural MRIs reveal thinning of sensorimotor cortex in children with Tourette syndrome. Red and yellow indicate large areas of cortical loss. (From Sowell et al., 2008.)

from being ridiculed and tormented by others and from the self-righteous advice repeatedly offered by well-meaning "clods." The malfunction may be in a patient's striatum, but in reality this is more a disorder of the on-looker than of the patient.

Last year, I received an e-mail from a professor of biological sciences at Purdue University. He came across this text because it was being used in his department's behavioral neurobiology course. He thanked me for my coverage of Tourette syndrome but said that he found the case study "a bit eerie." The message began with "From one case study to another," and it ended "All the best, P.H."

18.5
Clinical Trials: Development of New Psychotherapeutic Drugs

Almost daily, there are news reports of exciting discoveries that appear to be pointing to effective new therapeutic drugs or treatments for psychiatric disorders. But most

often, the promise does not materialize. For example, almost 50 years after the revolution in molecular biology began, not a single form of gene therapy is yet in widespread use. The reason is that the journey of a drug or other medical treatment from promising basic research to useful reality is excruciatingly complex, time-consuming, and expensive. Research designed to translate basic scientific discoveries into effective clinical treatments is called **translational research**.

So far, the chapter has focused on early drug discoveries and their role in the development of theories of psychiatric dysfunction. In the early years, the development of psychotherapeutic drugs was largely a hit-or-miss process. New drugs were tested on patient populations with little justification and then quickly marketed to an unsuspecting public, often before it was discovered that they were ineffective for their original purpose.

Things have changed. The testing of experimental drugs on human subjects and their subsequent release for sale are now strictly regulated by government agencies.

The process of gaining permission from the government to market a new psychotherapeutic drug begins with the synthesis of the drug, the development of procedures for synthesizing it economically, and the collection

TABLE 18.2 Phases of Drug Development

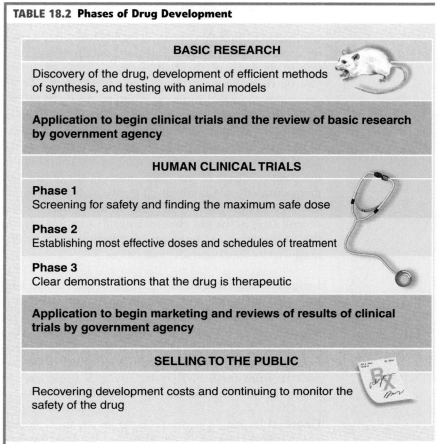

BASIC RESEARCH

Discovery of the drug, development of efficient methods of synthesis, and testing with animal models

Application to begin clinical trials and the review of basic research by government agency

HUMAN CLINICAL TRIALS

Phase 1
Screening for safety and finding the maximum safe dose

Phase 2
Establishing most effective doses and schedules of treatment

Phase 3
Clear demonstrations that the drug is therapeutic

Application to begin marketing and reviews of results of clinical trials by government agency

SELLING TO THE PUBLIC

Recovering development costs and continuing to monitor the safety of the drug

Source: Adapted from Zivin (2000).

(2) establishing the testing protocol, and (3) final testing (see Zivin, 2000). The average duration and cost of each phase, based on estimates by Adams and Brantner (2010), are summarized in Figure 18.12 on page 486.

Screening for Safety The purpose of the first phase of a clinical trial is to determine whether the drug is safe for human use and, if it is, to determine how much of the drug can be tolerated. Administering the drug to humans for the first time is always a risky process because there is no way of knowing for certain how they will respond. The subjects in phase 1 are typically healthy paid volunteers. Phase 1 clinical trials always begin with tiny injections, which are gradually increased as the tests proceed. The reactions of the subjects are meticulously monitored, and if strong adverse reactions are observed, phase 1 is curtailed.

Establishing the Testing Protocol
The purpose of the second phase of a clinical trial is to establish the *protocol* (the conditions) under which the final tests are likely to provide a clear result. For example, in phase 2, researchers hope to discover which doses are likely to be therapeutically effective, how frequently they should be administered, how long they need to be administered to have a therapeutic effect, what benefits are likely to occur, and which patients are likely to be helped. Phase 2 tests are conducted on patients suffering from the target disorder; the tests usually include *placebo-control groups* (groups of patients who receive a control substance rather than the drug), and their designs are *double-blind*—that is, the tests are conducted so that neither the patients nor the physicians interacting with them know which treatment (drug or placebo) each patient has received.

of evidence from nonhuman subjects showing that the drug is likely safe for human consumption and has potential therapeutic benefits. These initial steps usually take a long time—at least 5 years—and only if the evidence is sufficiently promising is permission granted to proceed to clinical trials. **Clinical trials** are studies conducted on human volunteers to assess the therapeutic efficacy of an untested drug or other treatment. This process is summarized in Table 18.2.

This final section of the chapter focuses on the process of conducting clinical trials. But before we begin, I want to emphasize the critical role played by research on nonhuman subjects in the development of effective therapeutic drugs for human patients (see Chapter 1). Without a solid foundation of comparative translational research, it is extremely difficult to gain government permission to begin clinical trials.

Evolutionary Perspective

Clinical Trials: The Three Phases

Once approval has been obtained from the appropriate government agencies, clinical trials of a new drug with therapeutic potential can commence. Clinical trials are conducted in three separate phases: (1) screening for safety,

Clinical Implications

Final Testing Phase 3 of a clinical trial is typically a double-blind, placebo-control study on large numbers—often, many thousands—of patients suffering from the target disorder. The design of the phase 3 tests is based on the results of phase 2 so that the final tests are likely to demonstrate positive therapeutic effects, if these exist. The first test of the final phase is often not conclusive, but if it is promising, a second test based on a redesigned protocol may be conducted. In most cases, two independent successful tests are required to convince government regulatory agencies. A successful test is one in which the beneficial effects are substantially greater than the adverse side effects.

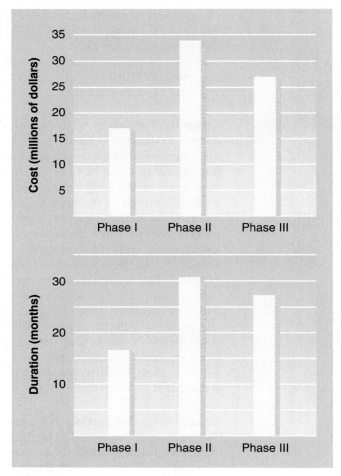

FIGURE 18.12 The cost and duration of the three phases of drug testing on human volunteers.

Controversial Aspects of Clinical Trials

The clinical trial process is not without controversy, as is clear from the following major focuses of criticism and debate (Zivin, 2000).

Requirement for Double-Blind Design and Placebo Controls In most clinical trials, patients are assigned to drug or placebo groups randomly and do not know for sure which treatment they are receiving (see Woods et al., 2001). Thus, some patients whose only hope for recovery may be the latest experimental treatment will, without knowing it, receive the placebo. Drug companies and government agencies concede that this is true, but they argue that there can be no convincing evidence that the experimental treatment is effective until a double-blind, placebo-control trial is complete. Because psychiatric disorders often improve after a placebo, a double-blind, placebo-control procedure is essential in the evaluation of any psychotherapeutic drug.

The Need for Active Placebos Conventional wisdom has been that the double-blind, placebo-control procedure is the perfect control procedure to establish the effectiveness of new drugs, but it isn't (see Salamone, 2000). Here is a new way to think about the double-blind placebo-control procedure. At therapeutic doses, many drugs have side effects that are obvious to people taking them, and thus the participants in double-blind, placebo-control studies who receive the drug can be certain that they are not in the placebo group. This knowledge may greatly contribute to the positive effects of the drug, independent of any real therapeutic effect. Accordingly, it is now widely recognized that an active placebo is better than an inert placebo as the control drug. **Active placebos** are control drugs that have no therapeutic effect but produce side effects similar to those produced by the drug under evaluation.

Length of Time Required Patients desperately seeking new treatments are frustrated by the amount of time needed for clinical trials (see Figure 18.12). Therefore, researchers, drug companies, and government agencies are striving to speed up the evaluation process, without sacrificing the quality of the procedures designed to protect patients from ineffective treatments. It is imperative to strike the right compromise (see Benderly, 2007).

Financial Issues The drug companies pay the scientists, physicians, technicians, assistants, and patients involved in drug trials. Considering the millions these companies spend and the fact that only about 22% of the candidate drugs entering phase 1 testing ever gain final approval (see Figure 18.13), it should come as no surprise that the companies are anxious to recoup their costs. In view of this pressure, many have questioned the impartiality of those conducting and reporting the trials (see Fisher, 2006). The scientists themselves have often complained that the sponsoring drug company makes them sign an agreement that prohibits them from publishing or discussing negative findings without the company's consent.

Another financial issue is profitability—drug companies seldom develop drugs to treat rare disorders because such treatments will not be profitable. Drugs for which the market is too small for them to be profitable are called **orphan drugs**. Governments in Europe and North America have passed laws intended to promote the development of orphan drugs (see Maeder, 2003). Also, the massive costs of clinical trials have contributed to a **translational bottleneck** (Hyman & Fenton, 2003)—only a small proportion of potentially valuable ideas or treatments receive funding for translational research.

Targets of Psychopharmacology Hyman and Fenton (2003) have argued that a major impediment to the development of effective psychotherapeutic drugs is that the effort is often aimed at curing disease entities as currently

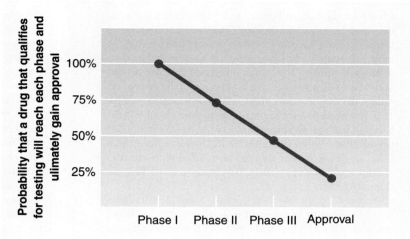

FIGURE 18.13 The probabilities that a drug that qualifies for testing in humans will reach each phase of testing and ultimately gain approval. Only 22% of the drugs that initially qualify for testing eventually gain approval.

conceived—for example, as defined by DSM-IV-TR. The current characterizations of various psychiatric disorders are the best they can be, based on existing evidence; however, it is clear that most psychiatric disorders, as currently conceived, are likely clusters of disorders, each with a different pattern of brain pathology. Thus, effective new drugs are likely to benefit only a proportion of those patients who have been given a particular diagnosis, and thus their effectiveness might go unrecognized. Hyman and Fenton recommend that drug development should focus on treating specific, readily monitored symptoms, rather than on treating general psychiatric disorders as currently conceived.

Thinking Creatively

Effectiveness of Clinical Trials

Despite the controversy that surrounds the clinical trial process, there is no question that it works.

> A long, dismal history tells of charlatans who make unfounded promises and take advantage of people at the time when they are least able to care for themselves. The clinical trial process is the most objective method ever devised to assess the efficacy of a treatment. It is expensive and slow, and in need of constant refinements, and oversight, but the process is trustworthy. (Zivin, 2000, p. 75)

Certainly, the clinical trial process is far from perfect. For example, concerns about the ethics of randomized double-blind, placebo-control studies are often warranted. Still, the vast majority of those in the medical and research professions accept that these studies are the essential critical test of any new therapy. This is particularly true of psychotherapeutic drugs because psychiatric disorders often respond to placebo treatments and because assessment of their severity is subjective and can be greatly influenced by the expectations of the therapist.

Everybody agrees that clinical trials are too expensive and take too long. But one expert responds to this concern in the following way: Clinical trials can be trustworthy, fast, or cheap; but in any one trial, only two of the three are possible (Zivin, 2000). Think about it.

It is important to realize that every clinical trial is carefully monitored as it is being conducted. Any time the results warrant it, changes to the research protocol are made to reduce costs and the time required to deliver an effective treatment to patients in need—particularly to the patients in the placebo-control group.

Conclusion

The chapter, and indeed the book, ends with the case of S.B., who suffers from bipolar affective disorder. S.B.'s case is appropriate here because S.B. benefited greatly from the clinical trial process and because S.B.'s case demonstrates the value of a biopsychological education that stresses independent thinking and the importance of taking responsibility for one's own health. You see, S.B. took a course similar to the one that you are currently taking, and the things that he learned in the course enabled him to steer his own treatment to a positive outcome.

The Case of S.B., the Biopsychology Student Who Took Control

I met S.B. when he was a third-year undergraduate. S.B. is a quiet, pleasant, shy person; he has an unassuming manner, but he is kind, knowledgeable, and intelligent, with broad interests. For example, I was surprised to learn that he was a skilled artist, interested in medical illustration, so we chatted at length about the illustrations in this book.

I was delighted to discover that S.B.'s grades confirmed my positive impression of him. In addition, it soon became apparent that he had a real "touch" for research, so I invited him to become my graduate student. He accepted and was truly exceptional. As you can tell, I am very proud of him.

S.B. is now going to describe his case to you in his own words. I wanted to tell you a bit about him myself so that you would have a clear picture of his situation. As you are about to discover, S.B.'s view of himself, obscured by a black cloud of depression, bears little relation to reality.

> As an undergraduate student, I suffered from depression. Although my medication improved things somewhat, I still

felt stupid, disliked, and persecuted. There were some positives in my undergraduate years. Dr. Pinel was very good to me, and I liked his course. He always emphasized that the most important part of his course was learning to be an independent thinker, and I was impressed by how he had been able to diagnose his own brain tumor. I did not appreciate at the time just how important these lessons would be.

A few months after beginning graduate school, my depression became so severe that I could not function. My psychiatrist advised me to take a leave of absence, which I did. I returned a few months later, filled with antipsychotics and antidepressants, barely capable of keeping things together.

You can appreciate how pleased I was 2 years later, when I started to snap out of it. It occurred to me that I was feeling better than I had ever felt. My productivity and creativity increased. I read, wrote, and drew, and new ideas for experiments flooded into my head. Things were going so well that I found that I was sleeping only 2 or 3 hours a night, and my brain was so energized that my friends sometimes begged me to talk more slowly so that they could follow what I was saying.

But my euphoria soon came to an abrupt end. I was still energetic and creative, but the content of my ideas changed. My consciousness was again dominated by feelings of inferiority, stupidity, and persecution. Thoughts of suicide were a constant companion. As a last resort, I called my psychiatrist, and when she saw me, she immediately had me committed. My diagnosis was bipolar affective disorder with a mixed episode. As she explained, *mixed episodes* are transition states between mania and depression and are associated with particularly high suicide rates.

Heavily sedated, I slept for much of the first week. When I came out of my stupor, two resident psychiatrists informed me that I would be placed on a mood stabilizer and would likely have to take it for the rest of my life. Two things made me feel uncomfortable about this. First, many patients in the ward were taking this drug, and they seemed like bloated zombies; second, my physicians seemed to know less about this drug and its mechanisms than I did. So I requested that they give me access to the hospital library so I could learn about my disorder and drug.

I was amazed by what I found. The drug favored by the residents had been shown several months before to be no more effective in the long-term treatment of bipolar affective disorder 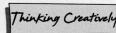 than an active placebo. Moreover, a new drug that had recently cleared clinical trials was proving to be effective with fewer side effects.

When I confronted my psychiatrists with this evidence, they were surprised and agreed to prescribe the new drug. Today, I am feeling well and am finishing graduate school. I still find it difficult to believe that I had enough nerve to question my physicians and prescribe for myself. I never imagined that the lessons learned from this book would have such a positive impact on my life. I am glad that I could tell you my story.

(Used by permission of Steven Barnes.)

Themes Revisited

This entire chapter focused on psychiatric disorders, so it should come as no surprise that the clinical implications theme was predominant. Nevertheless, the other three major themes of this book also received coverage.

The thinking creatively theme arose during the discussions of the following ideas: the variability of the brain pathology associated with bipolar affective disorder suggests that it may comprise several different disorders; animal models of anxiety may be models of benzodiazepine effects; active placebos are needed to establish the clinical efficacy of psychotherapeutic drugs; and scientists should try to focus on treatments for specific measurable symptoms rather than general diseases as currently conceived.

The evolutionary perspective theme came up twice: in the discussions of animal models of anxiety and of the important role played by research on nonhuman subjects in gaining official clearance to commence human clinical trials.

The neuroplasticity theme was discussed explicitly only once, when it was pointed out that the therapeutic effect of neuroleptic drugs seems to result from a neuroplastic response to D_2 receptor blockage.

Think about It

1. Blunders often play an important role in scientific progress. Discuss this with respect to the development of drugs for the treatment of psychiatric disorders.
2. The mechanism by which a disorder is alleviated is not necessarily opposite to the mechanism by which it was caused. Discuss this with respect to the evidence supporting current theories of schizophrenia, depression, and anxiety.
3. Discuss the diathesis–stress model of depression. Design an experiment to test the model.

4. Judge people by what they do, not by what they say. Discuss this recommendation with respect to Tourette syndrome.
5. Tourette syndrome is a disorder of onlookers. Explain and discuss.

6. Clinical trials are no more than excessive government bureaucracy. The prescription of drugs should be left entirely to the discretion of physicians. Discuss.

Key Terms

Psychiatric disorders (p. 467)
DSM-IV-TR (p. 467)

18.1 Schizophrenia

Positive symptoms (p. 468)
Negative symptoms (p. 468)
Chlorpromazine (p. 469)
Reserpine (p. 469)
Haloperidol (p. 470)
Phenothiazines (p. 470)
Butyrophenones (p. 470)
Neuroleptics (p. 470)
Clozapine (p. 471)

18.2 Affective Disorders: Depression and Mania

Clinical depression (major depressive disorder) (p. 473)

Affective disorders (p. 474)
Mania (p. 474)
Mood disorder (p. 474)
Bipolar affective disorder (p. 474)
Unipolar affective disorder (p. 474)
Reactive depression (p. 474)
Endogenous depression (p. 474)
Seasonal affective disorder (SAD) (p. 474)
Postpartum depression (p. 475)
Iproniazid (p. 475)
MAO inhibitors (p. 475)
Cheese effect (p. 475)
Tricyclic antidepressants (p. 475)
Imipramine (p. 475)

Selective serotonin-reuptake inhibitors (SSRIs) (p. 475)
Prozac (p. 476)
Mood stabilizers (p. 476)
Lithium (p. 476)
Up-regulation (p. 477)
Diathesis (p. 478)

18.3 Anxiety Disorders

Anxiety (p. 479)
Anxiety disorder (p. 479)
Generalized anxiety disorders (p. 479)
Phobic anxiety disorders (p. 479)
Agoraphobia (p. 479)
Panic disorders (p. 479)
Obsessive-compulsive disorders (p. 479)
Posttraumatic stress disorder (p. 480)

Benzodiazepines (p. 480)
Comorbidity (p. 480)
Anxiolytic drugs (p. 480)
Elevated-plus-maze test (p. 480)
Defensive-burying test (p. 480)
Risk-assessment test (p. 480)

18.4 Tourette Syndrome

Tics (p. 482)

18.5 Clinical Trials: Development of New Psychotherapeutic Drugs

Translational research (p. 484)
Clinical trials (p. 485)
Active placebos (p. 486)
Orphan drugs (p. 486)
Translational bottleneck (p. 486)

✔•Quick Review Test your comprehension of the chapter with this brief practice test. You can find the answers to these questions as well as more practice tests, activities, and other study resources at www.mypsychlab.com.

1. The first antischizophrenic drug was
   a. clozapine.
   b. iproniazid.
   c. chlorpromazine.
   d. imipramine.
   e. haloperidol.

2. When clinical depression alternates with periods of mania, the disorder is termed
   a. major depressive disorder.
   b. bipolar affective disorder.
   c. endogenous depression.
   d. reactive depression.
   e. postpartum depression.

3. Lithium is classified as a
   a. mood stabilizer.
   b. monoamine oxidase inhibitor.
   c. tricyclic antidepressant.
   d. SSRI.
   e. both a and c

4. Because the tics of Tourette syndrome seem to be associated with activity in the caudate nuclei, they have been treated with
   a. D_2 receptor blockers.
   b. tricyclic antidepressants.
   c. SSRIs.
   d. benzodiazepines.
   e. phenobarbital.

5. Research designed to develop effective clinical treatments from basic scientific discoveries is termed
   a. phase 1 research.
   b. phase 2 research.
   c. phase 3 research.
   d. translational research.
   e. orphan research.

Epilogue

I feel relieved to be finishing this edition of *Biopsychology*, which I began almost 2 years ago, and I am excited by the prospect of being able to speak to so many students like you through this edition and MyPsychLab. You must also feel relieved to be finishing this book; still, I hope that you feel a tiny bit of regret that our time together is over.

Like good friends, we have shared good times and bad. We have shared the fun and wonder of Rhonda, the dexterous cashier; the Nads basketball team; people who rarely sleep; the "mamawawa"; split brains; and brain transplantation. But we have also been touched by many personal tragedies: for example, the victims of Alzheimer's disease and MPTP poisoning; Jimmie G.; H.M.; the man who mistook his wife for a hat; Professor P., the biopsychologist who experienced brain surgery from the other side of the knife; and his student, S.B., who guided the treatment of his own disease. Thank you for allowing me to share *Biopsychology* with you. I hope you have found it to be an enriching experience.

Right now, I am sitting at my desk looking out over my garden and the Pacific Ocean. There has just been a summer shower, and the air smells and feels particularly fresh as the setting sun sends rays of light through the trees. It is the evening of Tuesday August 10, 2010. Where and when are you?

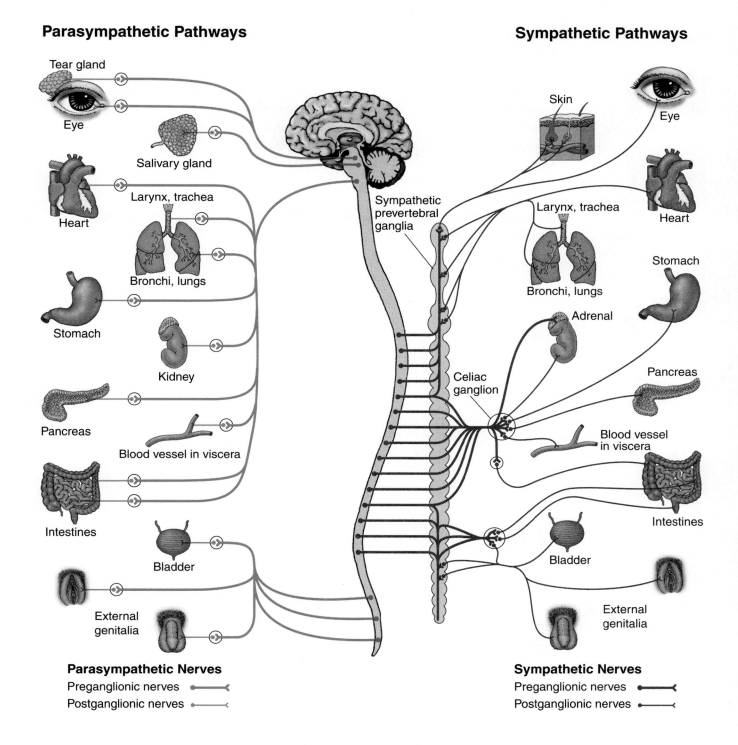

Parasympathetic Pathways

Tear gland

Eye

Salivary gland

Heart

Larynx, trachea

Bronchi, lungs

Stomach

Kidney

Pancreas

Blood vessel in viscera

Intestines

Bladder

External genitalia

Sympathetic Pathways

Skin

Eye

Heart

Sympathetic prevertebral ganglia

Larynx, trachea

Bronchi, lungs

Adrenal

Stomach

Pancreas

Celiac ganglion

Blood vessel in viscera

Intestines

Bladder

External genitalia

Parasympathetic Nerves

Preganglionic nerves

Postganglionic nerves

Sympathetic Nerves

Preganglionic nerves

Postganglionic nerves

Appendix II
Some Functions of Sympathetic and Parasympathetic Neurons

Organ	Sympathetic Effect	Parasympathetic Effect
Salivary gland	Decreases secretion	Increases secretion
Heart	Increases heart rate	Decreases heart rate
Blood vessels	Constricts blood vessels in most organs	Dilates blood vessels in a few organs
Penis	Ejaculation	Erection
Iris radial muscles	Dilates pupils	No effect
Iris sphincter muscles	No effect	Constricts pupils
Tear gland	No effect	Stimulates secretion
Sweat gland	Stimulates secretion	No effect
Stomach and intestine	No effect	Stimulates secretion
Lungs	Dilates bronchioles; inhibits mucous secretion	Constricts bronchioles; stimulates mucous secretion
Arrector pili muscles	Erects hair and creates gooseflesh	No effect

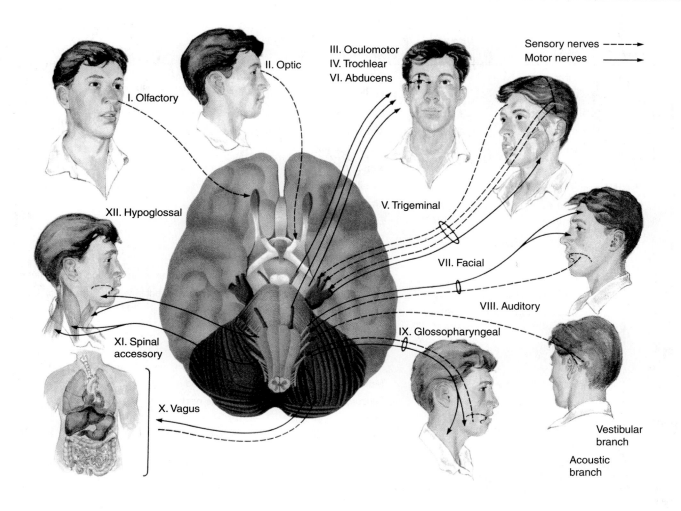

I. Olfactory

II. Optic

III. Oculomotor
IV. Trochlear
VI. Abducens

V. Trigeminal

Sensory nerves ⤍
Motor nerves →

XII. Hypoglossal

VII. Facial

VIII. Auditory

IX. Glossopharyngeal

XI. Spinal accessory

X. Vagus

Vestibular branch

Acoustic branch

Appendix IV
Functions of the Cranial Nerves

Number	Name	General Function	Specific Functions
I	Olfactory	Sensory	Smell
II	Optic	Sensory	Vision
III	Oculomotor	Motor	Eye movement and pupillary constriction
		Sensory	Sensory signals from certain eye muscles
IV	Trochlear	Motor	Eye movement
		Sensory	Sensory signals from certain eye muscles
V	Trigeminal	Sensory	Facial sensations
		Motor	Chewing
VI	Abducens	Motor	Eye movement
		Sensory	Sensory signals from certain eye muscles
VII	Facial	Sensory	Taste from anterior two-thirds of tongue
		Motor	Facial expression, secretion of tears, salivation, cranial blood vessel dilation
VIII	Auditory-Vestibular	Sensory	Audition; sensory signals from the organs of balance in the inner ear
IX	Glossopharyngeal	Sensory	Taste from posterior third of tongue
		Motor	Salivation, swallowing
X	Vagus	Sensory	Sensations from abdominal and thoracic organs
		Motor	Control over abdominal and thoracic organs and muscles of the throat
XI	Spinal Accessory	Motor	Movement of neck, shoulders, and head
		Sensory	Sensory signals from muscles of the neck
XII	Hypoglossal	Motor	Tongue movements
		Sensory	Sensory signals from tongue muscles

Note: Some authors describe cranial nerves III, IV, VI, XI, and XII as purely motor. However, each of these cranial nerves contains a small proportion of sensory fibers that conduct information from receptors to the brain. This sensory information is necessary for directing the respective cranial nerve's motor responses. See the discussion of sensory feedback in Chapter 8.

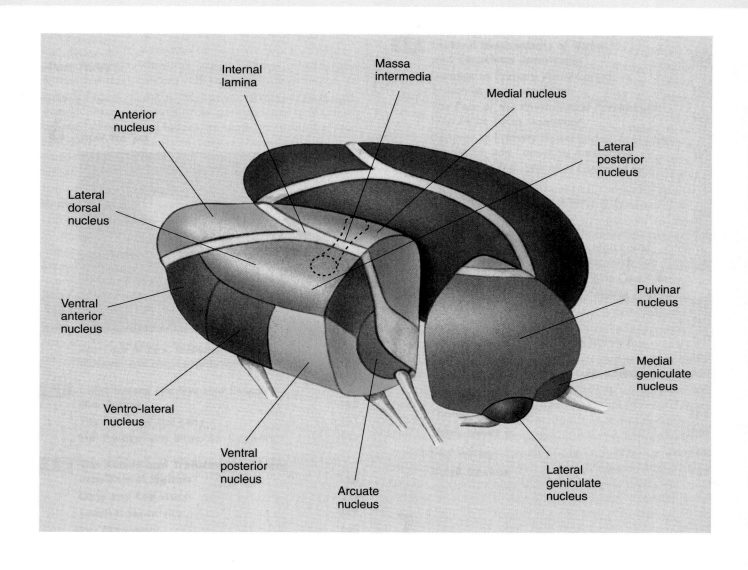

Appendix VI

Nuclei of the Hypothalamus

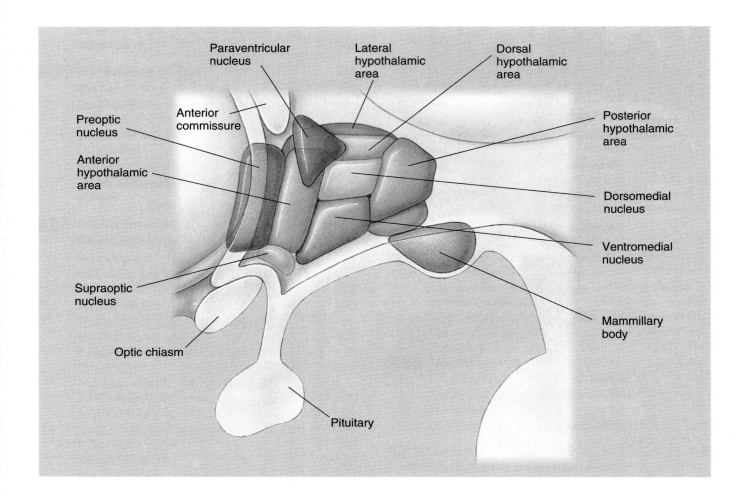

Ablatio penis. Accidental destruction of the penis.

Absolute refractory period. A brief period (typically 1 to 2 milliseconds) after the initiation of an action potential during which it is impossible to elicit another action potential in the same neuron.

Absorption spectrum. A graph of the ability of a substance to absorb light of different wavelengths.

Absorptive phase. The metabolic phase during which the body is operating on the energy from a recently consumed meal and is storing the excess as body fat, glycogen, and proteins.

Accommodation. The process of adjusting the configuration of the lenses to bring images into focus on the retina.

Acetylcholine. A neurotransmitter that is created by the addition of an acetyl group to a choline molecule.

Acetylcholinesterase. The enzyme that breaks down the neurotransmitter acetylcholine.

Acquired dyslexias. Dyslexias caused by brain damage in people already capable of reading.

Action potential (AP). A massive momentary reversal of a neuron's membrane potential from about −70 mV to about +50 mV.

Activation-synthesis theory. The theory that dream content reflects the cerebral cortex's inherent tendency to make sense of, and give form to, the random signals it receives from the brain stem during REM sleep.

Active placebos. Control drugs that have no therapeutic effect but produce side effects similar to those produced by the drug under evaluation in a clinical trial.

Acuity. The ability to see the details of objects.

Adaptation theories of sleep. Theories of sleep based on the premise that sleep evolved to protect organisms from predation and accidents and to conserve their energy, rather than to fulfill some particular physiological need.

Adaptive immune system. The division of the immune system that mounts targeted attacks on foreign pathogens by binding to antigens in their cell membranes.

Addicts. Habitual drug users who continue to use a drug despite its adverse effects on their health and social life and despite their repeated efforts to stop using it.

Adipsia. Complete cessation of drinking.

Adjustable gastric band procedure. A surgical procedure for treating extreme obesity in which an adjustable band is implanted around the stomach to reduce the flow of food.

Adrenal cortex. The outer layer of the adrenal glands, which releases glucocorticoids in response to stressors, as well as steroid hormones in small amounts.

Adrenal medulla. The core of each adrenal gland, which releases epinephrine and norepinephrine in response to stressors.

Adrenalectomy. Surgical removal of the adrenal glands.

Adrenocorticotropic hormone (ACTH). The anterior pituitary hormone that triggers the release of gonadal and adrenal hormones from the adrenal cortices.

Adrenogenital syndrome. A sexual developmental disorder in which high levels of adrenal androgens, resulting from congenital adrenal hyperplasia, masculinize the bodies of genetic females.

Affective disorders. Depression and mania.

Afferent nerves. Nerves that carry sensory signals to the central nervous system; sensory nerves.

Ageusia. The inability to taste.

Aggregation. The alignment of cells within different areas of the embryo during development to form various structures.

Aggressive behaviors. Behaviors whose primary function is to threaten or harm other organisms.

Agnosia. An inability to consciously recognize sensory stimuli of a particular class that is not attributable to a sensory deficit or to verbal or intellectual impairment.

Agonists. Drugs that facilitate the effects of a particular neurotransmitter.

Agoraphobia. Pathological fear of public places and open spaces.

Agraphia. A specific inability to write; one that does not result from general visual, motor, or intellectual deficits.

Akinetopsia. A deficiency in the ability to perceive motion, which often results from damage to the dorsal visual pathway.

Alexia. A specific inability to read; one that does not result from general visual, motor, or intellectual deficits.

Alleles. The two genes that control the same trait.

All-or-none responses. Responses that are not graded, that either occur to their full extent or not at all.

Alpha fetoprotein. A protein that is present in the blood of many mammals during the perinatal period and that deactivates circulating estradiol by binding to it.

Alpha male. The dominant male of a colony.

Alpha waves. Regular, 8- to 12-per-second, high-amplitude EEG waves that typically occur during relaxed wakefulness and just before falling asleep.

Alternative splicing. A mechanism by which the actions of individual genes can be controlled or "edited" so that one gene can produce two or more proteins.

Alzheimer's disease. The major cause of dementia in old age, characterized by neurofibrillary tangles, amyloid plaques, and neuron loss.

Amacrine cells. A type of retinal neurons whose specialized function is lateral communication.

Amino acid derivative hormones. Hormones that are synthesized in a few simple steps from amino acids.

Amino acid neurotransmitters. A class of small-molecule neurotransmitters, which includes the amino acids glutamate and GABA.

Amino acids. The building blocks and breakdown products of proteins.

Amnesia. Any pathological loss of memory.

Amphetamine. A stimulant drug whose effects are similar to those of cocaine.

Amphibians. Species that spend their larval phase in water and their adult phase on land.

Amygdala. A structure in the anterior temporal lobe, just anterior to the hippocampus; plays a role in emotion.

Amyloid. A protein that is normally present in small amounts in the human brain but is a major constituent of the numerous plaques in the brains of Alzheimer's patients.

Anabolic steroids. Steroid drugs that are similar to testosterone and have powerful anabolic (growth-promoting) effects.

Analgesics. Drugs that reduce pain.

Analogous. Having a similar structure because of convergent evolution (e.g., a bird's wing and a bee's wing are analogous).

Anandamide. The first endogenous endocannabinoid to be discovered and characterized.

Androgenic insensitivity syndrome. The developmental disorder of genetic males in which a mutation to the androgen receptor gene renders the androgen receptors defective and causes the development of a female body.

Androgens. The class of steroid hormones that includes testosterone.

Androstenedione. The adrenal androgen that triggers the growth of pubic and axillary hair in human females.

Aneurysm. A pathological balloonlike dilation that forms in the wall of a blood vessel at a point where the elasticity of the vessel wall is defective.

Angular gyrus. The gyrus of the posterior cortex at the boundary between the temporal and parietal lobes, which in the left hemisphere is thought to play a role in reading.

Anorexia nervosa. An eating disorder that is characterized by a pathological fear of obesity and that results in health-threatening weight loss.

Anosmia. The inability to smell.

Anosognosia. The common failure of neuropsychological patients to recognize their own symptoms.

Antagonistic muscles. Pairs of muscles that act in opposition.

Antagonists. Drugs that inhibit the effects of a particular neurotransmitter.

Anterior. Toward the nose end of a vertebrate.

Anterior cingulate cortex. The cortex of the anterior cingulate gyrus, which is involved in the emotional reaction to painful stimulation.

Anterior pituitary. The part of the pituitary gland that releases tropic hormones.

Anterograde amnesia. Loss of memory for events occurring after the amnesia-inducing brain injury.

Anterograde degeneration. The degeneration of the distal segment of a cut axon.

Anterolateral system. The division of the somatosensory system that ascends in the anterolateral portion of spinal white matter and carries signals related to pain and temperature.

Antibodies. Proteins that bind specifically to antigens on the surface of invading micro-organisms and in so doing promote the destruction of the micro-organisms.

Antibody-mediated immunity. The immune reaction by which B cells destroy invading micro-organisms.

Antidromic conduction. Axonal conduction opposite to the normal direction; conduction from axon terminals back toward the cell body.

Antigens. Proteins on the surface of cells that identify them as native or foreign.

Antihypnotic drugs. Sleep-reducing drugs.

Anxiety. Chronic fear that persists in the absence of any direct threat.

Anxiety disorder. Anxiety that is so extreme and so pervasive that it disrupts normal functioning.

Anxiolytics. Anti-anxiety drugs.

Aphagia. Complete cessation of eating.

Aphasia. A brain-damage–produced deficit in the ability to use or comprehend language.

Apoptosis. Cell death that is actively induced by genetic programs; programmed cell death.

Appetizer effect. The increase in hunger that is produced by the consumption of small amounts of palatable food.

Applied research. Research that is intended to bring about some direct benefit to humankind.

Apraxia. A disorder in which patients have great difficulty performing movements when asked to do so out of context but can readily perform them spontaneously in natural situations.

Arachnoid membrane. The meninx that is located between the dura mater and the pia mater and has the appearance of a gauzelike spiderweb.

Arcuate fasciculus. The major neural pathway between Broca's area and Wernicke's area.

Arcuate nucleus. A nucleus of the hypothalamus that contains high concentrations of both leptin receptors and insulin receptors.

Aromatase. An enzyme that promotes the conversion of testosterone to estradiol.

Aromatization. The chemical process by which testosterone is converted to estradiol.

Aromatization hypothesis. The hypothesis that the brain is masculinized by estradiol that is produced from perinatal testosterone through a process called *aromatization*.

Arteriosclerosis. A condition in which blood vessels are blocked by the accumulation of fat deposits on their walls.

Asomatognosia. A deficiency in the awareness of parts of one's own body that is typically produced by damage to the parietal lobe.

Aspartate. An amino acid neurotransmitter that is a constituent of many of the proteins that we eat.

Asperger's syndrome. A mild form of autism in which cognitive and linguistic functions are well preserved.

Aspiration. A lesion technique in which tissue is drawn off by suction through the fine tip of a glass pipette.

Association cortex. Any area of the cortex that receives input from more than one sensory system.

Astereognosia. An inability to recognize objects by touch that is not attributable to a simple sensory deficit or to general intellectual impairment.

Astrocytes. Large, star-shaped glial cells that play a role in the passage of chemicals from the blood into CNS neurons and perform several other important functions that are not yet well understood.

Ataxia. Loss of motor coordination.

Atropine. A receptor blocker that exerts an antagonistic effect at muscarinic receptors.

Auditory nerve. The branch of cranial nerve VIII that carries auditory signals from the hair cells in the basilar membrane.

Autism. A neurodevelopmental disorder characterized by (1) a reduced ability to interpret the emotions and intentions of others, (2) a reduced capacity for social interaction and communication, and (3) a preoccupation with a single subject or activity.

Autism spectrum disorders. A group of related neurodevelopmental disorders typically characterized by preoccupation with a particular object or activity and deficits in communication and understanding the feelings of others.

Autoimmune diseases. Diseases (e.g., multiple sclerosis) that arise when the immune system begins to attack healthy body cells as if they were foreign pathogens.

Autonomic nervous system (ANS). The part of the peripheral nervous system that participates in the regulation of the body's internal environment.

Autoradiography. The technique of photographically developing brain slices that have been exposed to a radioactively labeled substance such as 2-DG so that regions of high uptake are visible.

Autoreceptors. A type of metabotropic receptor located on the presynaptic membrane and sensitive to a neuron's own neurotransmitter.

Autosomal chromosomes. Chromosomes that come in matched pairs; in mammals, all of the chromosomes except the sex chromosomes are autosomal.

Axon hillock. The conical structure at the junction between the axon and cell body.

B cells. B lymphocytes; lymphocytes that manufacture antibodies against antigens they encounter.

Basal forebrain. A midline area of the forebrain, which is located just in front of and above the hypothalamus and is the brain's main source of acetylcholine.

Basal ganglia. A collection of subcortical nuclei (e.g., striatum and globus pallidus) that have important motor functions.

Basal metabolic rate. The rate at which an individual utilizes energy to maintain bodily processes.

Basilar membrane. The membrane of the organ of Corti in which the hair cell receptors are embedded.

Before-and-after design. The experimental design used to demonstrate contingent drug tolerance; the experimental group receives the drug before each of a series of behavioral tests and the control group receives the drug after each test.

Behavioral paradigm. A single set of procedures developed for the investigation of a particular behavioral phenomenon.

Benign tumors. Tumors that are surgically removable with little risk of further growth in the body.

Benzodiazepines. A class of GABA agonists with anxiolytic, sedative, and anticonvulsant properties; drugs such as chlordiazepoxide (Librium) and diazepam (Valium).

Between-subjects design. An experimental design in which a different group of subjects is tested under each condition.

Betz cells. Large pyramidal neurons of the primary motor cortex that synapse directly on motor neurons in the lower regions of the spinal cord.

Bilateral medial temporal lobectomy. The removal of the medial portions of both temporal lobes, including the hippocampus, the amygdala, and the adjacent cortex.

Binocular. Involving both eyes.

Binocular disparity. The difference in the position of the retinal image of the same object on the two retinas.

Biopsychology. The scientific study of the biology of behavior.

Bipolar affective disorder. A disorder of emotion in which the patient experiences periods of mania interspersed with periods of depression.

Bipolar cells. Bipolar neurons that form the middle layer of the retina.

Bipolar neuron. A neuron with two processes extending from its cell body.

Bisexual. Sexually attracted to members of both sexes.

Blind spot. The area on the retina where the bundle of axons of the retinal ganglion cells penetrate the receptor layer and leave the eye as the optic nerve.

Blindsight. The ability of some patients who are blind as a consequence of cortical damage to unconsciously see some aspects of their visual environments.

Blobs. Peglike, cytochrome oxidase–rich, dual-opponent color columns.

Blood–brain barrier. The mechanism that keeps certain toxic substances in the blood from passing into brain tissue.

BOLD signal. A blood-oxygen-level-dependent signal, which is recorded by fMRI and is related to the level of neural firing.

Botox. *Botulinium toxin*, which blocks release of acetylcholine at neuromuscular junctions and is used cosmetically to treat wrinkles.

Brainbow. A neuroanatomical technique that involves inserting various mutations of the green fluorescent protein gene into neural tissue so that different neurons fluoresce in different colors.

Brain–gut peptides. One of the five classes of neuropeptide transmitters; consists of those first discovered in the gut.

Brain stem. The part of the brain on which the cerebral hemispheres rest; in general, it regulates reflex activities that are critical for survival (e.g., heart rate and respiration).

Bregma. The point on the surface of the skull where two of the major sutures intersect, commonly used as a reference point in stereotaxic surgery on rodents.

Broca's aphasia. A hypothetical disorder of speech production with no associated deficits in language comprehension.

Broca's area. The area of the inferior prefrontal cortex of the left hemisphere hypothesized by Broca to be the center of speech production.

Buerger's disease. A condition in which the blood vessels, especially those supplying the legs, are constricted whenever nicotine enters the bloodstream, the ultimate result being gangrene and amputation.

Bulimia nervosa. An eating disorder that is characterized by recurring cycles of fasting, bingeing, and purging without dangerous weight loss.

Butyrophenones. A class of antischizophrenic drugs that bind primarily to D_2 receptors.

CA1 subfield. The region of the hippocampus that is commonly damaged by cerebral ischemia.

Cafeteria diet. A diet offered to experimental animals that is composed of a wide variety of palatable foods.

Cannabis sativa. The common hemp plant, which is the source of marijuana.

Cannon-Bard theory. The theory that emotional experience and emotional expression are parallel processes that have no direct causal relation.

Cannula. A fine, hollow tube that is implanted in the body for the purpose of introducing or extracting substances.

Carbon monoxide. A soluble-gas neurotransmitter.

Carousel apparatus. An apparatus used to study the effects of sleep deprivation in laboratory rats.

Cartesian dualism. The philosophical position of René Descartes, who argued that the universe is composed of two elements: physical matter and the human mind.

Case studies. Studies that focus on a single case, or subject.

Cataplexy. A disorder that is characterized by recurring losses of muscle tone during wakefulness and is often seen in cases of narcolepsy.

Catecholamines. The three monoamine neurotransmitters that are synthesized from the amino acid tyrosine: dopamine, epinephrine, and norepinephrine.

Caudate. The tail-like structure that is part of the striatum.

Cell-adhesion molecules (CAMs). Molecules on the surface of cells that have the ability to recognize specific molecules on the surface of other cells and bind to them.

Cell-mediated immunity. The immune reaction by which T cells destroy invading micro-organisms.

Central canal. The small CSF-filled channel that runs the length of the spinal cord.

Central fissure. The large fissure that separates the frontal lobe from the parietal lobe.

Central nervous system (CNS). The portion of the nervous system within the skull and spine.

Central sensorimotor programs. Patterns of activity that are programmed into the sensorimotor system.

Cephalic phase. The metabolic phase during which the body prepares for food that is about to be absorbed.

Cerebellum. The metencephalic structure that has been shown to mediate the retention of Pavlovian eyeblink conditioning.

Cerebral angiography. A contrast X-ray technique for visualizing the cerebral circulatory system by infusing a radio-translucent dye into a cerebral artery.

Cerebral commissures. Tracts that connect the left and right cerebral hemispheres.

Cerebral cortex. The layer of neural tissue covering the cerebral hemispheres of humans and other mammals.

Cerebral dialysis. A method for recording changes in brain chemistry in behaving animals in which a fine tube with a short semipermeable section is implanted in the brain, and extracellular neurochemicals are continuously drawn off for analysis.

Cerebral hemorrhage. Bleeding in the brain.

Cerebral ischemia. An interruption of the blood supply to an area of the brain; a common cause of medial temporal lobe amnesia.

Cerebral ventricles. The four CSF-filled internal chambers of the brain: the two lateral ventricles, the third ventricle, and the fourth ventricle.

Cerebrospinal fluid (CSF). The colorless fluid that fills the subarachnoid space, the central canal, and the cerebral ventricles.

Cerebrum. The portion of the brain that sits on the brain stem; in general, it plays a role in complex adaptive processes (e.g., learning, perception, and motivation).

Cerveau isolé preparation. An experimental preparation in which the forebrain is disconnected from the rest of the brain by a midcollicular transection.

Change blindness. The difficulty perceiving major changes to unattended-to parts of a visual image when the changes are introduced during brief interruptions in the presentation of the image.

Cheese effect. The surges in blood pressure that occur when individuals taking MAO inhibitors consume tyramine-rich foods, such as cheese.

Chemoaffinity hypothesis. The hypothesis that growing axons are attracted to the correct targets by different chemicals released by the target sites.

Chimeric figures test. A test of visual completion in split-brain subjects that uses pictures composed of the left and right halves of two different faces.

Chlorpromazine. The first antischizophrenic drug.

Cholecystokinin (CCK). A peptide that is released by the gastrointestinal tract and is thought to function as a satiety signal.

Chordates. Animals with dorsal nerve cords.

Choroid plexuses. The networks of capillaries that protrude into the ventricles from the pia mater and continuously produce cerebrospinal fluid.

Chromosomes. Threadlike structures in the cell nucleus that contain the genes; each chromosome is a DNA molecule.

Chronobiotic. A substance that influences the timing of internal biological rhythms.

Ciliary muscles. The eye muscles that control the shape of the lenses.

Cingulate cortex. The cortex of the cingulate gyri, which are located on the medial surfaces of the frontal lobes.

Cingulate gyri. Large gyri located on the medial surfaces of the frontal lobes, just superior to the corpus callosum.

Cingulate motor areas. Two small areas of secondary motor cortex located in the cortex of the cingulate gyrus of each hemisphere.

Circadian clock. An internal timing mechanism that is capable of maintaining daily cycles of physiological functions, even when there are no temporal cues from the environment.

Circadian rhythms. Diurnal (daily) cycles of body functions.

Cirrhosis. Scarring, typically of the liver.

Clinical. Pertaining to illness or treatment.

Clinical depression (major depressive disorder). Depression that is so severe that it is difficult for the patient to meet the essential requirements of daily life.

Clinical trials. Studies conducted on human subjects to assess the therapeutic efficacy of an untested drug or other treatment.

Clozapine. An atypical neuroleptic that is used to treat schizophrenia, does not produce Parkinsonian side effects, and does not have a high affinity for D_2 receptors.

Cocaine. A potent catecholamine agonist and stimulant that is highly addictive.

Cocaine psychosis. Psychotic behavior observed during a cocaine spree, similar in many respects to paranoid schizophrenia.

Cocaine sprees. Binges of cocaine use.

Cochlea. The long, coiled tube in the inner ear that is filled with fluid and contains the organ of Corti and its auditory receptors.

Cocktail-party phenomenon. The ability to unconsciously monitor the contents of one conversation while consciously focusing on another.

Cocontraction. The simultaneous contraction of antagonistic muscles.

Codeine. A relatively weak psychoactive ingredient of opium.

Codon. A group of three consecutive nucleotide bases on a DNA or messenger RNA strand; each codon specifies the particular amino acid that is to be added to an amino acid chain during protein synthesis.

Coexistence. The presence of more than one neurotransmitter in the same neuron.

Cognition. Higher intellectual processes such as thought, memory, attention, and complex perceptual processes.

Cognitive map theory. The theory that the main function of the hippocampus is to store memories of spatial location.

Cognitive neuroscience. A division of biopsychology that focuses on the use of functional brain imaging to study the neural bases of human cognition.

Collateral sprouting. The growth of axon branches from mature neurons, usually to postsynaptic sites abandoned by adjacent axons that have degenerated.

Colony–intruder paradigm. A paradigm for the study of aggressive and defensive behaviors in male rats; a small male intruder rat is placed in an established colony in order to study the aggressive responses of the colony's alpha male and the defensive responses of the intruder.

Color constancy. The tendency of an object to appear the same color even when the wavelengths of light that it reflects change.

Columnar organization. The functional organization of the neocortex in vertical columns; the cells in each column form a mini-circuit that performs a single function.

Commissurotomy. Surgical severing of the cerebral commissures.

Comorbidity. The tendency for two or more diseases to occur together in the same individual.

Comparative approach. The study of biological processes by comparing different species—usually from the evolutionary perspective.

Comparative psychology. The division of biopsychology that studies the evolution, genetics, and adaptiveness of behavior, often by using the comparative approach.

Complementary colors. Pairs of colors that produce white or gray when combined in equal measure; every color has a complementary color.

Completion. The visual system's automatic use of information obtained from receptors around the blind spot, or scotoma, to create a perception of the missing portion of the retinal image.

Complex cells. Neurons in the visual cortex that respond optimally to straight-edge stimuli in a certain orientation in any part of their receptive field.

Complex partial seizures. Seizures that are characterized by various complex psychological phenomena and are thought to result from temporal lobe discharges.

Component theory. The theory that the relative amount of activity produced in three different classes of cones by light determines its perceived color (also called *trichromatic theory*).

Computed tomography (CT). A computer-assisted X-ray procedure that can be used to visualize the brain and other internal structures of the living body.

Concussion. Disturbance of consciousness following a blow to the head with no cerebral bleeding or obvious structural damage.

Conditioned compensatory responses. Physiological responses opposite to the effects of a drug that are thought to be elicited by stimuli that are regularly associated with experiencing the drug effects.

Conditioned defensive burying. The burial of a source of aversive stimulation by rodents.

Conditioned drug tolerance. Tolerance effects that are maximally expressed only when a drug is administered in the situation in which it has previously been administered.

Conditioned place-preference paradigm. A test that assesses a laboratory animal's preference for environments in which it has previously experienced drug effects.

Conditioned taste aversion. An avoidance response developed by animals to the taste of food whose consumption has been followed by illness.

Conduction aphasia. Aphasia that is thought to result from damage to the neural pathway between Broca's area and Wernicke's area.

Cones. The visual receptors in the retina that mediate high acuity color vision in good lighting.

Confounded variable. An unintended difference between the conditions of an experiment that could have affected the dependent variable.

Congenital. Present at birth.

Congenital adrenal hyperplasia. A congenital deficiency in the release of cortisol from the adrenal cortex, which leads to the excessive release of adrenal androgens.

Conscious awareness. The ability to perceive one's experiences; typically inferred from the ability to verbally describe them.

Conspecifics. Members of the same species.

Constituent cognitive processes. Simple cognitive processes that combine to produce complex cognitive processes and that are assumed to be mediated by neural activity in particular parts of the brain.

Contextual fear conditioning. The process by which benign contexts (situations) come to elicit fear through their association with fear-inducing stimuli.

Contingent drug tolerance. Drug tolerance that develops as a reaction to the experience of the effects of drugs rather than to drug exposure alone.

Contralateral. Projecting from one side of the body to the other.

Contralateral neglect. A disturbance of the patient's ability to respond to visual, auditory, and somatosensory stimuli on the side of the body opposite to a site of brain damage, usually the left side of the body following damage to the right parietal lobe.

Contrast enhancement. The intensification of the perception of edges.

Contrast X-ray techniques. X-ray techniques that involve the injection into one compartment of the body a substance that absorbs X-rays either less than or more than the surrounding tissue.

Contrecoup injuries. Contusions that occur on the side of the brain opposite to the side of a blow.

"Control of behavior" versus "conscious perception" theory. The theory that the dorsal stream mediates behavioral interactions with objects and the ventral stream mediates conscious perception of objects.

Control-question technique. A lie-detection interrogation method in which the polygrapher compares the physiological responses to target questions with the responses to control questions.

Contusions. Closed-head injuries that involve damage to the cerebral circulatory system, which produces internal hemorrhaging.

Convergent evolution. The evolution in unrelated species of similar solutions to the same environmental demands.

Converging operations. The use of several research approaches to solve a single problem.

Convolutions. Folds on the surface of the cerebral hemispheres.

Convulsions. Motor seizures.

Coolidge effect. The fact that a copulating male who becomes incapable of continuing to copulate with one sex partner can often recommence copulating with a new sex partner.

Copulation. Sexual intercourse.

Corpus callosum. The largest cerebral commissure.

Corticosterone. The predominant glucocorticoid in humans.

Crack. A potent, cheap, smokable form of cocaine.

Cranial nerves. The 12 pairs of nerves extending from the brain (e.g., the optic nerves, the olfactory nerves, and the vagus nerves).

Critical period. A period during development in which a particular experience must occur for it to influence the course of subsequent development.

Critical thinking. Carefully assessing the strength of the evidence presented to support an idea.

Cross-cuing. Nonneural communication between hemispheres that have been separated by commissurotomy.

Cross section. Section cut at a right angle to any long, narrow structure of the CNS.

Cross tolerance. Tolerance to the effects of one drug that develops as the result of exposure to another drug that acts by the same mechanism.

Cryogenic blockade. The temporary elimination of neural activity in an area of the brain by cooling the area with a cryoprobe.

Cytochrome oxidase. An enzyme present in particularly high concentrations in the mitochondria of dual-opponent color cells of the visual cortex.

Cytokines. A group of peptide hormones that are released by many cells and participate in a variety of physiological and immunological responses, causing inflammation and fever.

Decorticate. Lacking a cortex.

Decussate. To cross over to the other side of the brain.

Deep brain stimulation. A treatment for neurological disorders (such as Parkinson's disease) that involves the delivery of continuous, low-level electrical stimulation to particular brain structures through implanted electrodes.

Deep dyslexia. A reading disorder in which the phonetic procedure is disrupted while the lexical procedure is not.

Default mode. The pattern of brain activity that is associated with relaxed wakefulness, when an individual is not focused on the external world.

Default mode network. The network of brain structures that tends to be active when the brain is in default mode.

Defeminizes. Suppresses or disrupts female characteristics.

Defensive behaviors. Behaviors whose primary function is protection from threat or harm.

Defensive burying test. An animal model of anxiety; anxious rats bury objects that generate anxiety.

Delayed nonmatching-to-sample test. A test in which the subject is presented with an unfamiliar sample object and then, after a delay, is presented with a choice between the sample object and an unfamiliar object, where the correct choice is the unfamiliar object.

Delirium tremens (DTs). The phase of alcohol withdrawal syndrome characterized by hallucinations, delusions, agitation, confusion, hyperthermia, and tachycardia.

Delta waves. The largest and slowest EEG waves.

Demasculinizes. Suppresses or disrupts male characteristics.

Dementia. General intellectual deterioration.

Dendritic spines. Tiny nodules of various shapes that are located on the surfaces of many dendrites and are the sites of most excitatory synapses in the mature mammalian brain.

2-Deoxyglucose (2-DG). A substance similar to glucose that is taken up by active neurons in the brain and accumulates in them because, unlike glucose, it cannot be metabolized.

Deoxyribonucleic acid (DNA). The double-stranded, coiled molecule of genetic material; a chromosome.

Dependent variable. The variable measured by the experimenter to assess the effect of the independent variable.

Depolarize. To decrease the resting membrane potential.

Deprenyl. A monoamine agonist that has been shown to retard the development of Parkinson's disease.

Depressant. A drug that depresses neural activity.

Dermatome. An area of the body that is innervated by the left and right dorsal roots of one segment of the spinal cord.

Desynchronized EEG. Low-amplitude, high-frequency EEG.

Detoxified addicts. Addicts who have none of the drug to which they are addicted in their body and who are no longer experiencing withdrawal symptoms.

Developmental dyslexias. Dyslexias that become apparent when a child tries to learn to read but has little success.

Dextrals. Right-handers.

Diathesis. A genetic susceptibility to a disorder, as in the diathesis–stress model of depression.

Dichotic listening test. A test of language lateralization in which two different sequences of three spoken digits are presented simultaneously, one to each ear, and the subject is asked to report all of the digits heard.

Dichotomous traits. Traits that occur in one form or the other, never in combination.

Diencephalon. One of the five major divisions of the brain; it is composed of the thalamus and hypothalamus.

Diet-induced thermogenesis. The homeostasis-defending increases in body temperature that are associated with increases in body fat.

Digestion. The process by which food is broken down and absorbed through the lining of the gastrointestinal tract.

Digit span. The longest sequence of random digits that can be repeated correctly 50% of the time—most people have a digit span of 7.

Directed synapses. Synapses at which the site of neurotransmitter release and the site of neurotransmitter reception are in close proximity.

Distal. Farther from the central core of the body (e.g., the wrists are distal to the elbows).

Distal segment. The segment of a cut axon between the cut and the axon terminals.

Disulfiram. A drug that blocks the normal metabolism of alcohol and has been used in the treatment of alcoholism.

Dizygotic twins. Fraternal twins; twins that develop from two zygotes and thus tend to be as genetically similar as any pair of siblings.

Dominant hemisphere. A term used in the past to refer to the left hemisphere, based on the incorrect assumption that the left hemisphere is dominant in all complex behavioral and cognitive activities.

Dominant trait. The trait of a dichotomous pair that is expressed in the phenotypes of heterozygous individuals.

L-Dopa. The chemical precursor of dopamine, which is used in the treatment of Parkinson's disease.

Dopamine. One of the three catecholamine neurotransmitters; dopaminergic neurons are damaged in Parkinson's disease.

Dopamine transporters. Molecules in the presynaptic membrane of dopaminergic neurons that attract dopamine molecules in the synaptic cleft and deposit them back inside the neuron.

Dorsal. Toward the surface of the back of a vertebrate or toward the top of the head.

Dorsal-column medial-lemniscus system. The division of the somatosensory system that ascends in the dorsal portion of the spinal white matter and carries signals related to touch and proprioception.

Dorsal columns. The somatosensory tracts that ascend in the dorsal portion of the spinal cord white matter.

Dorsal horns. The two dorsal arms of the spinal gray matter.

Dorsal root ganglia. Structures just outside the spinal cord that are composed of the cell bodies of dorsal root axons.

Dorsal stream. The group of visual pathways that flows from the primary visual cortex to the dorsal prestriate cortex to the posterior parietal cortex; according to one theory, its function is the control of visually guided behavior.

Dorsolateral corticorubrospinal tract. The descending motor tract that synapses in the red nucleus of the midbrain, decussates, and descends in the dorsolateral spinal white matter.

Dorsolateral corticospinal tract. The motor tract that leaves the primary motor cortex, descends to the medullary pyramids, decussates, and then descends in the contralateral dorsolateral spinal white matter.

Dorsolateral prefrontal association cortex. The area of the prefrontal association cortex that plays a role in the evaluation of external stimuli and the initiation of complex voluntary motor responses.

Down syndrome. A disorder associated with the presence of an extra chromosome 21, resulting in disfigurement and mental retardation.

Drug metabolism. The conversion of a drug from its active form to a nonactive form.

Drug priming. A single exposure to a formerly abused drug.

Drug self-administration paradigm. A test of the addictive potential of drugs in which laboratory animals can inject drugs into themselves by pressing a lever.

Drug sensitization. An increase in the sensitivity to a drug effect that develops as the result of exposure to the drug.

Drug tolerance. A state of decreased sensitivity to a drug that develops as a result of exposure to the drug.

DSM-IV-TR. The current edition of the *Diagnostic and Statistical Manual of Mental Disorders,* produced by the American Psychiatric Association.

Dual-opponent color cells. Neurons that respond to the differences in the wavelengths of light stimulating adjacent areas of their receptive field.

Duchenne smile. A genuine smile, one that includes contraction of the facial muscles called the *orbicularis oculi.*

Duodenum. The upper portion of the intestine through which most of the glucose and amino acids are absorbed into the bloodstream.

Duplexity theory. The theory that cones and rods mediate photopic and scotopic vision, respectively.

Dura mater. The tough outer meninx.

Dynamic contraction. Contraction of a muscle that causes the muscle to shorten.

Dynamic phase. The first phase of the VMH syndrome, characterized by grossly excessive eating and rapid weight gain.

Dyslexia. A pathological difficulty in reading, one that does not result from general visual, motor, or intellectual deficits.

Efferent nerves. Nerves that carry motor signals from the central nervous system to the skeletal muscles or internal organs.

Ejaculate. To eject sperm from the penis.

Ejaculation. Ejection of sperm.

Electrocardiogram (ECG or EKG). A recording of the electrical signals associated with heartbeats.

Electroconvulsive shock (ECS). An intense, brief, diffuse, seizure-inducing current administered to the brain via large electrodes attached to the scalp.

Electroencephalogram (EEG). A measure of the gross electrical activity of the brain, commonly recorded through scalp electrodes.

Electroencephalography. A technique for recording the gross electrical activity of the brain through disk-shaped electrodes, which in humans are usually taped to the surface of the scalp.

Electromyogram (EMG). A measure of the electrical activity of muscles.

Electromyography. A procedure for measuring muscle tension by recording the gross electrical discharges of muscles.

Electron microscopy. A neuroanatomical technique used to study the fine details of cellular structure.

Electrooculogram (EOG). A measure of eye movement.

Electrooculography. A technique for recording eye movements through electrodes placed around the eye.

Elevated plus maze. An apparatus for recording defensiveness or anxiety in rats by assessing their tendency to avoid the two open arms of a plus-sign–shaped maze mounted some distance above the floor of a lab.

Elevated-plus-maze test. An animal model of anxiety; anxious rats tend to stay in the enclosed arms of the maze rather than venturing onto the open arms.

Embolism. The blockage of blood flow in a smaller blood vessel by a plug that was formed in a larger blood vessel and carried by the bloodstream to the smaller one.

Emergent stage 1 EEG. All periods of stage 1 sleep EEG except initial stage 1; each is associated with REMs.

Encapsulated tumors. Tumors that grow within their own membrane.

Encéphale isolé preparation. An experimental preparation in which the brain is separated from the rest of the nervous system by a transection of the caudal brain stem.

Encephalitis. The inflammation associated with brain infection.

Endocannabinoids. A class of unconventional neurotransmitters that are chemically similar to the active components of marijuana.

Endocrine glands. Ductless glands that release chemicals called hormones directly into the circulatory system.

Endogenous. Naturally occurring in the body (e.g., endogenous opioids).

Endogenous depression. Depression that occurs with no apparent cause.

Endorphins. A class of endogenous opioids.

Engram. A change in the brain that stores a memory.

Enhancers. Stretches of DNA that control the rate of expression of target genes.

Enkephalins. The first class of endogenous opioids to be discovered.

Enriched environments. Laboratory environments designed to promote cognitive and physical activity by providing opportunities for a greater variety of sensory and motor experiences than available in conventional laboratory environments; commonly used to study the effects of experience on development in rats and mice.

Entorhinal cortex. The portion of the rhinal cortex within the rhinal fissure.

Enzymatic degradation. The breakdown of chemicals by enzymes—one of the two mechanisms for deactivating released neurotransmitters.

Enzymes. Proteins that stimulate or inhibit biochemical reactions without being affected by them.

Epidemiology. The study of the factors that influence the distribution of a disease in the general population.

Epigenetics. A field of research that focuses on factors that influence the expression of genes.

Epigenetic. Not of the genes; refers to nongenetic means by which traits are passed from parents to offspring.

Epilepsy. A neurological disorder characterized by spontaneously recurring seizures.

Epileptic auras. Psychological symptoms that precede the onset of a convulsion.

Epileptogenesis. Development of epilepsy.

Epinephrine. One of the three catecholamine neurotransmitters.

Episodic memories. Explicit memories for the particular events and experiences of one's life.

Estradiol. The most common estrogen.

Estrogens. The class of steroid hormones that are released in large amounts by the ovaries; an example is estradiol.

Estrous cycle. The cycle of sexual receptivity displayed by many female mammals.

Estrus. The portion of the estrous cycle characterized by proceptivity, sexual receptivity, and fertility (*estrus* is a noun and *estrous* an adjective).

Ethological research. The study of animal behavior in its natural environment.

Ethology. The study of the behavior of animals in their natural environments.

Event-related potentials (ERPs). The EEG waves that regularly accompany certain psychological events.

Evolutionary perspective. The approach that focuses on the environmental pressures that likely led to the evolution of the characteristics (e.g., of brain and behavior) of current species.

Evolve. To undergo gradual orderly change.

Exaptation. A characteristic that evolved because it performed one function but was later co-opted to perform another.

Excitatory postsynaptic potentials (EPSPs). Graded post-synaptic depolarizations, which increase the likelihood that an action potential will be generated.

Executive function. A collection of cognitive abilities (e.g., planning, insightful thinking, and reference memory) that appear to depend on the prefrontal cortex.

Exocrine glands. Glands that release chemicals into ducts that carry them to targets, mostly on the surface of the body.

Exocytosis. The process of releasing a neurotransmitter.

Explicit memories. Conscious memories.

Expressive. Pertaining to the generation of language; that is, pertaining to writing or talking.

Extensors. Muscles that act to straighten or extend a joint.

Exteroceptive sensory systems. The five sensory systems that interpret stimuli from outside the body: vision, hearing, touch, smell, and taste.

Exteroceptive stimuli. Stimuli that arise from outside the body.

Facial feedback hypothesis. The hypothesis that our facial expressions can influence how we feel.

Far-field potentials. EEG signals recorded in attenuated form at the scalp because they originate far away—for example, in the brain stem.

Fasciculation. The tendency of developing axons to grow along the paths established by preceding axons.

Fasting phase. The metabolic phase that begins when energy from the preceding meal is no longer sufficient to meet the immediate needs of the body and during which energy is extracted from fat and glycogen stores.

Fear. The emotional reaction that is normally elicited by the presence or expectation of threatening stimuli.

Fear conditioning. Establishing fear of a previously neutral conditional stimulus by pairing it with an aversive unconditional stimulus.

Feminizes. Enhances or produces female characteristics.

Fetal alcohol syndrome (FAS). A syndrome produced by prenatal exposure to alcohol and characterized by brain damage, mental retardation, poor coordination, poor muscle tone, low birth weight, retarded growth, and/or physical deformity.

Fissures. The large furrows in a convoluted cortex.

Fitness. According to Darwin, the ability of an organism to survive and contribute its genes to the next generation.

Fixational eye movements. Involuntary movements of the eyes (tremor, drifts, and saccades) that occur when a person tries to fix his or her gaze on (i.e., stare at) a point.

Flavor. The combined impression of taste and smell.

Flexors. Muscles that act to bend or flex a joint.

Follicle-stimulating hormone (FSH). The gonadotropic hormone that stimulates development of ovarian follicles.

Fornix. The major tract of the limbic system; it connects the hippocampus with the septum and mammillary bodies.

Fourier analysis. A mathematical procedure for breaking down a complex wave form (e.g., an EEG signal) into component sine waves of varying frequency.

Fovea. The central indentation of the retina, which is specialized for high-acuity vision.

Fraternal birth order effect. The finding that the probability of a man's being homosexual increases as a function of the number of older brothers he has.

Free fatty acids. The main source of the body's energy during the fasting phase; released from adipose tissue in response to high levels of glucagon.

Free nerve endings. Neuron endings that lack specialized structures on them and that detect cutaneous pain and changes in temperature.

Free-running period. The duration of one cycle of a free-running rhythm.

Free-running rhythms. Circadian rhythms that do not depend on environmental cues to keep them on a regular schedule.

Frontal eye field. A small area of prefrontal cortex that controls eye movements.

Frontal lobe. The most anterior of the four cerebral lobes.

Frontal operculum. The area of prefrontal cortex that in the left hemisphere is the location of Broca's area.

Frontal sections. Any slices of brain tissue cut in a plane that is parallel to the face; also termed *coronal sections*.

Functional MRI (fMRI). A magnetic resonance imaging technique for inferring brain activity by measuring increased oxygen flow into particular areas.

Functional segregation. Organization into different areas, each of which performs a different function; for example, in sensory systems, different areas of secondary and association cortex analyze different aspects of the same sensory stimulus.

Functional tolerance. Tolerance resulting from a reduction in the reactivity of the nervous system (or other sites of action) to a drug.

G proteins. Proteins that are located inside neurons (and some other cells) and are attached to metabotropic receptors in the cell membrane.

GABA (gamma-aminobutyric acid). The amino acid neurotransmitter that is synthesized from glutamate; the most prevalent inhibitory neurotransmitter in the mammalian nervous system.

Gametes. Egg cells and sperm cells.

Ganglia. Clusters of neuronal cell bodies in the peripheral nervous system (singular *ganglion*).

Gap junctions. Narrow spaces between adjacent neurons that are bridged by fine tubular channels containing cytoplasm, through which electrical signals and small molecules can pass readily.

Gastric bypass. A surgical procedure for treating extreme obesity in which the intestine is cut and connected to the upper portion of the stomach, which is isolated from the rest of the stomach by a row of staples.

Gastric ulcers. Painful lesions to the lining of the stomach or duodenum.

Gate-control theory. The theory that signals descending from the brain can activate neural gating circuits in the spinal cord to block incoming pain signals.

Gene. A unit of inheritance; for example, the section of a chromosome that controls the synthesis of one protein.

Gene expression. The production of the protein specified by a particular gene.

Gene knockout techniques. Procedures for creating organisms that lack a particular gene.

Gene replacement techniques. Procedures for creating organisms in which a particular gene has been replaced with another.

General paresis. The insanity and intellectual deterioration resulting from syphilitic infection.

Generalizability. The degree to which the results of a study can be applied to other individuals or situations.

Generalized anxiety disorders. Anxiety disorders that are not precipitated by any obvious event.

Generalized seizures. Seizures that involve the entire brain.

Genetic recombination. The meiotic process by which pairs of chromosomes cross over one another at random points, break apart, and exchange genes.

Genitals. The external reproductive organs.

Genotype. The traits that an organism can pass on to its offspring through its genetic material.

Glial cells. Several classes of nonneural cells of the nervous system, whose important contributions to nervous system function are just starting to be understood.

Glia-mediated migration. One of two major modes of neural migration during development, by which immature neurons move out from the central canal along radial glial cells.

Global amnesia. Amnesia for information presented in all sensory modalities.

Global aphasia. Severe disruption of all language-related abilities.

Globus pallidus. One of the basal ganglia; it is located between the putamen and thalamus.

Glucagon. A pancreatic hormone that promotes the release of free fatty acids from adipose tissue, their conversion to ketones, and the use of both as sources of energy.

Glucocorticoids. Steroid hormones that are released from the adrenal cortex in response to stressors.

Gluconeogenesis. The process by which protein is converted to glucose.

Glucose. A simple sugar that is the breakdown product of complex carbohydrates; it is the body's primary, directly utilizable source of energy.

Glucostatic theory. The theory that eating is controlled by deviations from a hypothetical blood glucose set point.

Glutamate. The brain's most prevalent excitatory neurotransmitter, whose excessive release causes much of the brain damage resulting from cerebral ischemia.

Glycine. An amino acid neurotransmitter that is a constituent of many of the proteins that we eat.

Golgi complex. Structures in the cell bodies and terminal buttons of neurons that package neurotransmitters and other molecules in vesicles.

Golgi stain. A neural stain that completely darkens a few of the neurons in each slice of tissue, thereby revealing their silhouettes.

Golgi tendon organs. Receptors that are embedded in tendons and are sensitive to the amount of tension in the skeletal muscles to which their tendons are attached.

Gonadectomy. The surgical removal of the gonads (testes or ovaries); castration.

Gonadotropin. The pituitary tropic hormone that stimulates the release of hormones from the gonads.

Gonadotropin-releasing hormone. The hypothalamic releasing hormone that controls the release of the two gonadotropic hormones from the anterior pituitary.

Gonads. The testes and the ovaries.

Graded responses. Responses whose magnitude is indicative of the magnitude of the stimuli that induce them.

Grammatical analysis. Analysis of the structure of language.

Grand mal seizure. A seizure whose symptoms are loss of consciousness, loss of equilibrium, and a violent tonic-clonic convulsion.

Gray matter. Portions of the nervous system that are gray because they are composed largely of neural cell bodies and unmyelinated interneurons.

Green fluorescent protein (GFP). A protein that is found in some species of jellyfish and that fluoresces when exposed to blue light; thus, inserting GFP genes into neurons allows researchers to visualize the neurons.

Grid cells. Entorhinal neurons that have multiple, evenly spaced place fields.

Growth cone. Amoebalike structure at the tip of each growing axon or dendrite that guides growth to the appropriate target.

Growth hormone. The anterior pituitary hormone that acts directly on bone and muscle tissue to produce the pubertal growth spurt.

Guilty-knowledge technique. A lie-detection method in which the polygrapher records autonomic nervous system responses to a list of control and crime-related information known only to the guilty person and the examiner.

Gyri. The cortical ridges that are located between fissures or sulci.

Hair cells. The receptors of the auditory system.

Haloperidol. A butyrophenone that was used as an antischizophrenic drug.

Harrison Narcotics Act. The act, passed in 1914, that made it illegal to sell or use opium, morphine, or cocaine in the United States.

Hashish. Dark corklike material extracted from the resin on the leaves and flowers of *Cannabis sativa*.

Hedonic value. The amount of pleasure produced by an action.

Helping-hand phenomenon. The redirection of one hand of a split-brain patient by the other hand.

Hematoma. A bruise.

Hemianopsic. Having a scotoma that covers half of the visual field.

Hemispherectomy. The removal of one cerebral hemisphere.

Heritability estimate. A numerical estimate of the proportion of variability that occurred in a particular trait in a particular study and that resulted from the genetic variation among the subjects in that study.

Heroin. A powerful semisynthetic opiate.

Heschl's gyrus. The temporal lobe gyrus that is the location of primary auditory cortex.

Heterosexual. Sexually attracted to members of the other sex.

Heterozygous. Possessing two different genes for a particular trait.

Hierarchical organization. Organization into a series of levels that can be ranked with respect to one another; for example, primary cortex, secondary cortex, and association cortex perform progressively more detailed analyses.

Hippocampus. A structure of the medial temporal lobes that plays a role in memory for spatial location.

Homeostasis. The stability of an organism's constant internal environment.

Hominins. The family of primates that includes *Homo sapiens* (humans), *Homo erectus*, and *Australopithecus*.

Homologous. Having a similar structure because of a common evolutionary origin (e.g., a human's arm and a bird's wing are homologous).

Homosexual. Sexually attracted to members of the same sex.

Homozygous. Possessing two identical genes for a particular trait.

Horizontal cells. Type of retinal neurons whose specialized function is lateral communication.

Horizontal sections. Any slices of brain tissue cut in a plane that is parallel to the top of the brain.

Hormones. Chemicals released by the endocrine system directly into the circulatory system.

Human genome project. The international research effort to construct a detailed map of the human chromosomes.

Huntingtin. Dominant gene that is mutated in cases of Huntington's disease.

Huntingtin protein. Protein whose synthesis is controlled by the huntingtin gene and is thus abnormal in individuals with Huntington's disease.

Huntington's disease. A progressive terminal disorder of motor and intellectual function that is produced in adulthood by a dominant gene.

5-Hydroxytryptophan (5-HTP). The precursor of serotonin.

Hyperphagia. Excessive eating.

Hyperpolarize. To increase the resting membrane potential.

Hypersomnia. Disorders characterized by excessive sleep or sleepiness.

Hypertension. Chronically high blood pressure.

Hypnagogic hallucinations. Dreamlike experiences that occur during wakefulness.

Hypnotic drugs. Sleep-promoting drugs.

Hypothalamic peptides. One of the five classes of neuropeptide transmitters; it consists of those first identified as hormones released by the hypothalamus.

Hypothalamopituitary portal system. The vascular network that carries hormones from the hypothalamus to the anterior pituitary.

Hypothalamus. The diencephalic structure that sits just below the anterior portion of the thalamus; it plays a role in motivated behaviors, in part by controlling the pituitary gland.

Hypoxia. Shortage of oxygen supply to tissue—for example, to the brain.

Iatrogenic. Physician-created.

Imipramine. The first tricyclic antidepressant drug.

Immune system. The system that protects the body against infectious micro-organisms.

Immunization. The process of creating immunity through vaccination.

Immunocytochemistry. A procedure for locating particular proteins in the brain by labeling their antibodies with a dye or radioactive element and then exposing slices of brain tissue to the labeled antibodies.

Implicit memories. Memories that are expressed by improved performance without conscious recall or recognition.

Impotent. Unable to achieve a penile erection.

In situ hybridization. A technique for locating particular proteins in the brain; molecules that bind to the mRNA that directs the synthesis of the target protein are synthesized and labeled, and brain slices are exposed to them.

Incentive-sensitization theory. Theory that addictions develop when drug use sensitizes the neural circuits mediating wanting of the drug—not necessarily liking for the drug.

Incomplete-pictures test. A test of memory measuring the improved ability to identify fragmented figures that have been previously observed.

Independent variable. The difference between experimental conditions that is arranged by the experimenter.

Infantile amnesia. The normal inability to recall events from early childhood.

Indolamines. The class of monoamine neurotransmitters that are synthesized from tryptophan; serotonin is the only member of this class found in the mammalian nervous system.

Inferior. Toward the bottom of the primate head or brain.

Inferior colliculi. The structures of the tectum that receive auditory input from the superior olives.

Inferotemporal cortex. The cortex of the inferior temporal lobe, in which is located an area of secondary visual cortex that is involved in object recognition.

Infiltrating tumors. Tumors that grow diffusely through surrounding tissue.

Inhibitory postsynaptic potentials (IPSPs). Graded post-synaptic hyperpolarizations, which decrease the likelihood that an action potential will be generated.

Initial stage 1 EEG. The period of the stage 1 EEG that occurs at the onset of sleep; it is not associated with REM.

Innate immune system. The immune system's first line of defense; it acts near entry points to the body and attacks generic classes of molecules produced by a variety of pathogens.

Inside-out pattern. The pattern of cortical development in which orderly waves of tangential migrations progress systematically from deeper to more superficial layers.

Insomnia. Disorders of initiating and maintaining sleep.

Instinctive behaviors. Behaviors that occur in all like members of a species, even when there seems to have been no opportunity for them to have been learned.

Instructive experiences. Particular experiences that influence the direction of a genetic program of development.

Insulin. A pancreatic hormone that facilitates the entry of glucose into cells and the conversion of bloodborne fuels to forms that can be stored.

Integration. Adding or combining a number of individual signals into one overall signal.

Internal desynchronization. The cycling on different schedules of the free-running circadian rhythms of two different processes.

Interneurons. Neurons with short axons or no axons at all, whose function is to integrate neural activity within a single brain structure.

Interoceptive stimuli. Stimuli that arise from inside the body.

Interpreter. A hypothetical mechanism that is assumed to reside in the left hemisphere and that continuously assesses patterns of events and tries to make sense of them.

Intracranial self-stimulation (ICSS). The repeated performance of a response that delivers electrical stimulation to certain sites in the animal's brain.

Intrafusal motor neuron. A motor neuron that innervates an intrafusal muscle.

Intrafusal muscle. A threadlike muscle that adjusts the tension on a muscle spindle.

Intromission. Insertion of the penis into the vagina.

Ion channels. Pores in neural membranes through which specific ions pass.

Ionotropic receptors. Receptors that are associated with ligand-activated ion channels.

Ions. Positively or negatively charged particles.

Iproniazid. The first antidepressant drug; a monoamine oxidase inhibitor.

Ipsilateral. On the same side of the body.

Isometric contraction. Contraction of a muscle that increases the force of its pull but does not shorten the muscle.

James-Lange theory. The theory that emotional experience results from the brain's perception of the pattern of autonomic and somatic nervous system responses elicited by emotion-inducing sensory stimuli.

Jet lag. The adverse effects on body function of the acceleration of zeitgebers during east-bound flights or their deceleration during west-bound flights.

Ketones. Breakdown products of free fatty acids that are used by muscles as a source of energy during the fasting phase.

Kindling phenomenon. The progressive development and intensification of convulsions elicited by a series of periodic low-intensity brain stimulations—most commonly by daily electrical stimulations to the amygdala.

Kluver-Bucy syndrome. The syndrome of behavioral changes (e.g., lack of fear and hypersexuality) that is induced in primates by bilateral damage to the anterior temporal lobes.

Korsakoff's syndrome. A neuropsychological disorder that is common in alcoholics and whose primary symptom is severe memory loss.

Lateral. Away from the midline of the body of a vertebrate, toward the body's lateral surfaces.

Lateral fissure. The large fissure that separates the temporal lobe from the frontal lobe.

Lateral geniculate nuclei. The six-layered thalamic structures that receive input from the retinas and transmit their output to the primary visual cortex.

Lateral hypothalamus (LH). The area of the hypothalamus once thought to be the feeding center.

Lateral inhibition. Inhibition of adjacent neurons or receptors in a topographic array.

Lateral nucleus of the amygdala. The nucleus of the amygdala that plays the major role in the acquisition, storage, and expression of conditioned fear.

Lateralization of function. The unequal representation of various psychological functions in the two hemispheres of the brain.

Leaky-barrel model. A settling-point model of body-fat regulation.

Learning. The brain's ability to change in response to experience.

Leptin. A protein normally synthesized in fat cells; it is thought to act as a negative feedback fat signal, reducing consumption.

Leucotome. Any of the various surgical devices used for performing lobotomies—*leucotomy* is another word for lobotomy.

Lewy bodies. Clumps of proteins observed in the surviving dopaminergic neurons of Parkinson's patients.

Lexical procedure. A procedure for reading aloud that is based on specific stored information acquired about written words.

Ligand. A molecule that binds to another molecule; neurotransmitters are ligands of their receptors.

Limbic system. A collection of interconnected nuclei and tracts that borders the thalamus and is widely assumed to play a role in emotion.

Lipids. Fats.

Lipogenesis. The production of body fat.

Lipolysis. The breakdown of body fat.

Lipostatic theory. The theory that eating is controlled by deviations from a hypothetical body-fat set point.

Lithium. A metallic ion that is used in the treatment of bipolar affective disorder.

Lobectomy. An operation in which a lobe, or a major part of one, is removed from the brain.

Lobotomy. An operation in which a lobe, or a major part of one, is separated from the rest of the brain by a large cut but is not removed.

Longitudinal fissure. The large fissure that separates the two cerebral hemispheres.

Long-term memory. Memory for experiences that endures after the experiences are no longer the focus of attention.

Long-term potentiation (LTP). The enduring facilitation of synaptic transmission that occurs following activation of synapses by high-intensity, high-frequency stimulation of the presynaptic neurons.

Lordosis. The arched-back, rump-up, tail-to-the-side posture of female rodent sexual receptivity, which serves to facilitate intromission.

Lordosis quotient. The proportion of mounts that elicit lordosis.

Luteinizing hormone (LH). The gonadotropic hormone that causes the developing ovum to be released from its follicle.

Lymphocytes. Specialized white blood cells that are produced in bone marrow and play important roles in the body's immune reactions.

Macrophage. A large phagocyte that plays a role in cell-mediated immunity.

Magnetic resonance imaging (MRI). A procedure in which high-resolution images of the structures of the living brain are constructed from the measurement of waves that hydrogen atoms emit when they are activated by radio-frequency waves in a magnetic field.

Magnetoencephalography (MEG). A technique for recording changes produced in magnetic fields on the surface of the scalp by changes in underlying patterns of neural activity.

Magnocellular layers. The layers of the lateral geniculate nuclei that are composed of neurons with large cell bodies; the bottom two layers (also called *M layers*).

Malignant tumors. Tumors that may continue to grow in the body even after attempted surgical removal.

Mammals. Species whose young are fed from mammary glands.

Mammillary bodies. The pair of spherical nuclei that are located on the inferior surface of the posterior hypothalamus.

Mania. An affective disorder in which the patient is overconfident, impulsive, distractible, and highly energetic.

MAO inhibitors. Antidepressant drugs that increase the level of monoamine neurotransmitters by inhibiting the action of monoamine oxidase.

Masculinizes. Enhances or produces male characteristics.

Massa intermedia. The neural structure that is located in the third ventricle and connects the two lobes of the thalamus.

Maternal immune hypothesis. The hypothesis that mothers become progressively more immune to some masculinizing hormone in their male fetuses; proposed to explain the fraternal birth order effect.

Medial. Toward the midline of the body of a vertebrate.

Medial diencephalic amnesia. Amnesia that is associated with damage to the medial diencephalon (e.g., Korsakoff's amnesia).

Medial dorsal nuclei. The thalamic relay nuclei of the olfactory system.

Medial geniculate nuclei. The auditory thalamic nuclei that receive input from the inferior colliculi and project to primary auditory cortex.

Medial lemniscus. The somatosensory pathway between the dorsal column nuclei and the ventral posterior nucleus of the thalamus.

Medial preoptic area. The area of the hypothalamus that includes the sexually dimorphic nuclei and that plays a key role in the control of male sexual behavior.

Medial temporal lobe amnesia. Amnesia associated with bilateral damage to the medial temporal lobes; its major feature is anterograde amnesia for explicit memories in combination with preserved intellectual functioning.

Mediodorsal nuclei. A pair of medial diencephalic nuclei in the thalamus, damage to which is thought to be responsible for many of the memory deficits associated with Korsakoff's syndrome.

Medulla. The myelencephalon.

Meiosis. The process of cell division that produces cells (e.g., egg cells and sperm cells) with half the chromosomes of the parent cell.

Melanocortin system. Neurons in the arcuate nucleus that regulate α-melanocyte-stimulating hormone and in so doing play a role in satiety.

Melanocortins. A class of neuropeptides that includes α-melanocyte-stimulating hormone, which suppresses eating.

Melanopsin. Photopigment found in retinal cells that respond to changes in background illumination and play a role in synchronizing circadian rhythms.

Melatonin. A hormone that is synthesized from serotonin in the pineal gland and influences the circadian rhythm of sleep.

Membrane potential. The difference in electrical charge between the inside and the outside of a cell.

Memory. The brain's ability to store and access the learned effects of experiences.

Memory consolidation. The transfer of short-term memories to long-term storage.

Meninges. The three protective membranes that cover the brain and spinal cord (singular *meninx*).

Meningiomas. Tumors that grow between the meninges.

Meningitis. Inflammation of the meninges, usually caused by bacterial infection.

Menstrual cycle. The hormone-regulated cycle in women of follicle growth, egg release, buildup of the uterus lining, and menstruation.

Mesencephalon. One of the five major divisions of the brain; it is composed of the tectum and tegmentum.

Mesoderm layer. The middle of the three cell layers in the developing embryo.

Mesotelencephalic dopamine system. The ascending projections of dopamine-releasing neurons from the substantia nigra and ventral tegmental area of the mesencephalon (midbrain) into various regions of the telencephalon.

Messenger RNA. A strand of RNA that is transcribed from DNA and carries the genetic code out of the cell nucleus to direct the synthesis of a protein.

Metabolic tolerance. Tolerance that results from a reduction in the amount of a drug getting to its sites of action.

Metabotropic receptors. Receptors that are associated with signal proteins and G proteins.

Metastatic tumors. Tumors that originate in one organ and spread to another.

Metencephalon. One of the five major divisions of the brain; it includes the pons and cerebellum.

Microelectrodes. Extremely fine recording electrodes, which are used for intracellular recording.

Microglia. Glial cells that respond to injury or disease by engulfing cellular debris and triggering inflammatory responses.

MicroRNAs. Short strands of RNA that have been found to have major effects on gene expression.

Microsleeps. Brief periods of sleep that occur in sleep-deprived subjects while they remain sitting or standing.

Migration. The movement of cells from their site of creation in the ventricular zone of the neural tube to their ultimate location in the mature nervous system.

Minor hemisphere. A term used in the past to refer to the right hemisphere, based on the incorrect assumption that the left hemisphere is dominant.

Mirror neurons. Neurons that fire both when a person makes a particular movement and when the person observes somebody else making the same movement.

Mirror-like system. Areas of the cortex that are active both when a person performs a particular response and when the person perceives somebody else performing the same response.

Miscellaneous peptides. One of the five classes of neuropeptide transmitters; it consists of those that don't fit into the other four classes.

Mitochondria. The energy-generating, DNA-containing structures in each cell's cytoplasm.

Mitosis. The process of cell division that produces cells with the same number of chromosomes as the parent cell.

Monoallelic expression. A mechanism of gene expression that inactivates one gene of a pair of alleles and allows the other gene of the pair to be expressed.

Monoamine neurotransmitters. Small-molecule neurotransmitters that are synthesized from monoamines and comprise two classes: catecholamines and indolamines.

Monocular. Involving only one eye.

Monogamy. A pattern of mate bonding in which one male and one female form an enduring bond.

Monophasic sleep cycles. Sleep cycles that regularly involve only one period of sleep per day, typically at night.

Monozygotic twins. Identical twins; twins that develop from the same zygote and are thus genetically identical.

Mood disorders. Affective disorders.

Mood stabilizer. A drug that blocks the rapid transition between depression and mania.

Morgan's Canon. The rule that the simplest possible interpretation for a behavioral observation should be given precedence.

Morphine. The major psychoactive ingredient in opium.

Morris water maze. A pool of milky water that has a goal platform invisible just beneath its surface and is used to study the ability of rats to learn spatial locations.

Morris water maze test. A widely used test of spatial memory in which rats must learn to swim directly to a platform hidden just beneath the surface of a circular pool of murky water.

Motor end-plate. The receptive area on a muscle fiber at a neuromuscular junction.

Motor equivalence. The ability of the sensorimotor system to carry out the same basic movement in different ways that involve different muscles.

Motor homunculus. The somatotopic map of the human primary motor cortex.

Motor pool. All of the motor neurons that innervate the fibers of a given muscle.

Motor theory of speech perception. The theory that the perception of speech involves activation of the same areas of the brain that are involved in the production of speech.

Motor units. A single motor neuron and all of the skeletal muscle fibers that are innervated by it.

MPTP. A neurotoxin that produces a disorder in primates that is similar to Parkinson's disease.

Müllerian-inhibiting substance. The testicular hormone that causes the precursor of the female reproductive ducts (the Müllerian system) to degenerate and the testes to descend.

Müllerian system. The embryonic precursor of the female reproductive ducts.

Multiple sclerosis (MS). A progressive disease that attacks the myelin of axons in the CNS.

Multiple-trace theory. Theory that memories are encoded in a distributed fashion throughout the hippocampus and other brain structures for as long as the memories exist.

Multiplier effect. A mechanism by which the behavioral effects of a gene are increased because the gene promotes the choice of experiences that have the same behavioral effects.

Multipolar neuron. A neuron with more than two processes extending from its cell body.

Multipotent. Capable of developing into a limited number of types of mature body cell.

Mumby box. An apparatus that is used in a rat version of the delayed nonmatching-to-sample test.

Muscle spindles. Receptors that are embedded in skeletal muscle tissue and are sensitive to changes in muscle length.

Mutations. Accidental alterations in individual genes that arise during chromosome duplication.

Myelencephalon. The most posterior of the five major divisions of the brain; also called the medulla.

Myelin. A fatty insulating substance found in the extensions of glial cells.

Myelin sheaths. Coverings on the axons of some CNS neurons that are rich in myelin and increase the speed and efficiency of axonal conduction.

Narcolepsy. A disorder in the hypersomnia category that is characterized by repeated, brief daytime sleep attacks and cataplexy.

Narcotic. A legal category of drugs, mostly opiates.

Nasal hemiretina. The half of each retina next to the nose.

Natural selection. The idea that heritable traits that are associated with high rates of survival and reproduction are preferentially passed on to future generations.

Nature–nurture issue. The debate about the relative contributions of nature (genes) and nurture (experience) to the behavioral capacities of individuals.

NEAT. Nonexercise activity thermogenesis, which is generated by activities such as fidgeting and the maintenance of posture and muscle tone.

Necrosis. Passive cell death, which is characterized by inflammation.

Negative feedback systems. Systems in which feedback from changes in one direction elicit compensatory effects in the opposite direction.

Negative symptoms. Symptoms of schizophrenia that involve the loss of a normal behavior or ability (e.g., anhedonia).

Neocortex. Six-layered cerebral cortex of relatively recent evolution; it constitutes 90% of human cerebral cortex.

Neoplasm. Tumor; literally, "new growth."

Nerve growth factor (NGF). A neurotrophin that attracts the growing axons of the sympathetic nervous system and promotes their survival.

Nerves. Bundles of axons in the peripheral nervous system.

Neural crest. The structure that is formed by cells breaking off from the neural groove during the formation of the neural tube and that develops into the peripheral nervous system.

Neural plate. A small patch of ectodermal tissue on the dorsal surface of the vertebrate embryo, from which the neural groove, the neural tube, and, ultimately, the mature nervous system develop.

Neural proliferation. The rapid increase in the number of neurons that follows the formation of the neural tube.

Neural regeneration. The regrowth of damaged neurons.

Neural tube. The tube that is formed in the vertebrate embryo when the edges of the neural groove fuse and that develops into the central nervous system.

Neuroanatomy. The study of the structure of the nervous system.

Neurochemistry. The study of the chemical bases of neural activity.

Neuroendocrinology. The study of the interactions between the nervous system and the endocrine system.

Neurogenesis. The growth of new neurons.

Neuroleptics. Drugs that alleviate schizophrenic symptoms.

Neuromuscular junctions. The synapses of a motor neuron on a muscle.

Neurons. Cells of the nervous system that are specialized for the reception, conduction, and transmission of electrochemical signals.

Neuropathic pain. Severe chronic pain in the absence of a recognizable painful stimulus.

Neuropathology. The study of nervous system disorders.

Neuropeptide transmitters. Peptides that function as neurotransmitters, of which about 100 have been identified; also called *neuropeptides*.

Neuropeptide Y. A neuropeptide that is released both in the gut and by neurons, particularly those in the arcuate nucleus; its release is associated with hunger.

Neuropharmacology. The study of the effects of drugs on neural activity.

Neurophysiology. The study of the functions and activities of the nervous system.

Neuropsychology. The division of biopsychology that studies the psychological effects of brain damage in human patients.

Neuroscience. The scientific study of the nervous system.

Neurotoxins. Neural poisons.

Neurotrophins. Chemicals that are supplied to developing neurons by their targets and that promote their survival.

Nicotine. The major psychoactive ingredient of tobacco.

Nigrostriatal pathway. The pathway along which axons from the substantia nigra project to the striatum.

Nissl stain. A neural stain that has an affinity for structures in neuron cell bodies.

Nitric oxide. A soluble-gas neurotransmitter.

NMDA (N-methyl-D-aspartate) receptors. Glutamate receptors that play key roles in the development of stroke-induced brain damage and long-term potentiation at glutaminergic synapses.

Nodes of Ranvier. The gaps between adjacent myelin segments on an axon.

Nondirected synapses. Synapses at which the site of neurotransmitter release and the site of neurotransmitter reception are not close together.

Nootropics (smart drugs). Drugs that purportedly improve memory.

Norepinephrine. One of the three catecholamine neurotransmitters.

Nuclei. The DNA-containing structures of cells; also, clusters of neuronal cell bodies in the central nervous system (singular *nucleus*).

Nucleotide bases. A class of chemical substances that includes adenine, thymine, guanine, and cytosine—the constituents of the genetic code.

Nucleus accumbens. Nucleus of the ventral striatum and a major terminal of the mesocorticolimbic dopamine pathway.

Nucleus magnocellularis. The nucleus of the caudal reticular formation that promotes relaxation of the core muscles during REM sleep and during cataplectic attacks.

Nutritive density. Calories per unit volume of a food.

Ob/ob mice. Mice that are homozygous for the mutant ob gene; their body fat produces no leptin, and they become very obese.

Obsessive-compulsive disorders. Anxiety disorders characterized by recurring uncontrollable, anxiety-producing thoughts and impulses.

Occipital lobe. The most posterior of the four cerebral lobes; its function is primarily visual.

Off-center cells. Visual neurons that respond to lights shone in the center of their receptive fields with "off" firing and to lights shone in the periphery of their fields with "on" firing.

Olfactory bulbs. The first cranial nerves, whose output goes primarily to the amygdala and piriform cortex.

Olfactory glomeruli. Discrete clusters of neurons on the surface of the olfactory bulbs; each neuron in a particular cluster contains the same type of receptor protein.

Olfactory mucosa. The mucous membrane that lines the upper nasal passages and contains the olfactory receptor cells.

Oligodendrocytes. Glial cells that myelinate axons of the central nervous system; also known as *oligodendroglia*.

On-center cells. Visual neurons that respond to lights shone in the center of their receptive fields with "on" firing and to lights shone in the periphery of their fields with "off" firing.

Ontogeny. The development of individuals over their life span.

Open-field test. A method for recording and scoring the general activity of an animal in a large, barren chamber.

Operant conditioning paradigm. A paradigm in which the rate of a particular voluntary response is increased by reinforcement or decreased by punishment.

Opiates. Morphine, codeine, heroin, and other chemicals with similar structures or effects.

Opioid peptides. One of the five classes of neuropeptide transmitters; it consists of those with a structure similar to the active ingredients of opium.

Opium. The sap that exudes from the seed pods of the opium poppy.

Opponent-process theory. The theory that a visual receptor or a neuron signals one color when it responds in one way (e.g., by increasing its firing rate) and signals the complementary color when it responds in the opposite way (e.g., by decreasing its firing rate).

Optic chiasm. The X-shaped structure on the inferior surface of the diencephalon; the point where the optic nerves decussate.

Optic tectum. The main destination of retinal ganglion cells in lower vertebrates.

Orbitofrontal cortex. The cortex of the inferior frontal lobes, adjacent to the orbits, which receives olfactory input from the thalamus.

Orchidectomy. The removal of the testes.

Orexin. A neuropeptide that has been implicated in narcolepsy in dogs and in knockout mice.

Organ of Corti. The auditory receptor organ, comprising the basilar membrane, the hair cells, and the tectorial membrane.

Orphan drugs. Drugs for which the market is too small for the necessary developmental research to be profitable.

Orthodromic conduction. Axonal conduction in the normal direction—from the cell body toward the terminal buttons.

Ossicles. The three small bones of the middle ear: the malleus, the incus, and the stapes.

Oval window. The membrane that transfers vibrations from the ossicles to the fluid of the cochlea.

Ovariectomy. The removal of the ovaries.

Ovaries. The female gonads.

Oxytocin. One of the two major peptide hormones of the posterior pituitary, which in females stimulates contractions of the uterus during labor and the ejection of milk during suckling.

Pacinian corpuscles. The largest and most deeply positioned cutaneous receptors, which are sensitive to sudden displacements of the skin.

Paired-image subtraction technique. The use of PET or fMRI to locate constituent cognitive processes in the brain by producing an image of the difference in brain activity associated with two cognitive tasks that differ in terms of a single constituent cognitive process.

Panic disorders. Anxiety disorders characterized by recurring rapid-onset attacks of extreme fear and severe symptoms of stress (choking, heart palpitations, and shortness of breath).

Parallel processing. The simultaneous analysis of a signal in different ways by the multiple parallel pathways of a neural network.

Parasympathetic nerves. Those motor nerves of the autonomic nervous system that project from the brain (as components of cranial nerves) or from the sacral region of the spinal cord.

Paraventricular nuclei. Hypothalamic nuclei that play a role in eating and synthesize hormones released by the posterior pituitary.

Parietal lobe. One of the four cerebral lobes; it is located just posterior to the central fissure.

Parkinson's disease. A movement disorder that is associated with degeneration of dopaminergic neurons in the nigrostriatal pathway.

Partial seizures. Seizures that do not involve the entire brain.

Parvocellular layers. The layers of the lateral geniculate nuclei that are composed of neurons with small cell bodies; the top four layers (also called *P layers*).

Patellar tendon reflex. The stretch reflex that is elicited when the patellar tendon is struck.

Pathogens. Disease-causing agents.

Pavlovian conditioning paradigm. A paradigm in which the experimenter pairs an initially neutral stimulus (conditional stimulus) with a stimulus (unconditional stimulus) that elicits a reflexive response (unconditional response); after several pairings, the neutral stimulus elicits a response (conditional response).

Penumbra. The area of brain tissue around an infarct, in which the degree of damage can vary.

Peptide hormones. Hormones that are short chains of amino acids.

Perception. The higher-order process of integrating, recognizing, and interpreting complex patterns of sensations.

Periaqueductal gray (PAG). The gray matter around the cerebral aqueduct, which contains opiate receptors and activates a descending analgesia circuit.

Perimetry test. The procedure used to map scotomas.

Periodic limb movement disorder. Recurrent involuntary movements of the limbs during sleep; a major cause of insomnia.

Peripheral nervous system (PNS). The portion of the nervous system outside the skull and spine.

Perirhinal cortex. The portion of the rhinal cortex around the rhinal fissure.

Permissive experiences. Particular experiences that are necessary for a particular genetic program to be manifested.

Perseveration. The tendency to continue making a formerly correct response that is currently incorrect.

Petit mal seizure. A generalized seizure that is characterized by a disruption of consciousness and a 3-per-second spike-and-wave EEG discharge.

Phagocytes. Cells, such as macrophages and microglia, that destroy and ingest pathogens.

Phagocytosis. The destruction and ingestion of foreign matter by cells of the immune system.

Phantom limb. The vivid perception that an amputated limb still exists.

Pharmacological. Pertaining to the scientific study of drugs.

Phenothiazines. A class of antischizophrenic drugs that bind effectively to both D_1 and D_2 receptors.

Phenotype. An organism's observable traits.

Phenylketonuria (PKU). A neurological disorder whose symptoms are vomiting, seizures, hyperactivity, hyperirritability, mental retardation, brain damage, and high levels of phenylpyruvic acid in the urine.

Phenylpyruvic acid. A substance that is found in abnormally high concentrations in the urine of those suffering from phenylketonuria.

Pheromones. Chemicals that are released by an animal and elicit through their odor specific patterns of behavior in its conspecifics.

Phobic anxiety disorders. Anxiety disorders characterized by extreme, largely irrational fears of specific objects or situations.

Phoneme. The smallest unit of sound that distinguishes among various words in a language.

Phonetic procedure. A procedure for reading aloud that involves the recognition of letters and the application of a language's rules of pronunciation.

Phonological analysis. Analysis of the sound of language.

Photopic spectral sensitivity curve. The graph of the sensitivity of cone-mediated vision to different wavelengths of light.

Photopic vision. Cone-mediated vision, which predominates when lighting is good.

Phylogeny. The evolutionary development of species.

Physical-dependence theories of addiction. Theories holding that the main factor that motivates drug addicts to keep taking drugs is the prevention or termination of withdrawal symptoms.

Physically dependent. Being in a state in which the discontinuation of drug taking will induce withdrawal reactions.

Physiological psychology. The division of biopsychology that studies the neural mechanisms of behavior through direct manipulation of the brains of nonhuman animal subjects in controlled experiments.

Pia mater. The delicate, innermost meninx.

Pineal gland. The endocrine gland that is the human body's sole source of melatonin.

Pioneer growth cones. The first growth cones to travel along a particular route in the developing nervous system.

Piriform cortex. An area of medial temporal cortex that is adjacent to the amygdala and that receives direct olfactory input.

Pituitary gland. The gland that dangles from, and is controlled by, the hypothalamus.

Pituitary peptides. One of the five classes of neuropeptide transmitters; it consists of those first identified as hormones released by the pituitary.

Pituitary stalk. The structure connecting the hypothalamus and the pituitary gland.

Place cells. Neurons that develop place fields—that is, that respond only when the subject is in a particular place in a familiar test environment.

Planum temporale. An area of temporal lobe cortex that lies in the posterior region of the lateral fissure and, in the left hemisphere, roughly corresponds to Wernicke's area.

Plethysmography. Any technique for measuring changes in the volume of blood in a part of the body.

Polyandry. A pattern of mate bonding in which one female bonds with more than one male.

Polygraphy. A method of interrogation in which autonomic nervous system indexes of emotion are used to infer the truthfulness of the responses.

Polygyny. A pattern of mate bonding in which one male bonds with more than one female; the most prevalent pattern of mate bonding in mammals.

Polyphasic sleep cycles. Sleep cycles that regularly involve more than one period of sleep per day.

Pons. The metencephalic structure that creates a bulge on the ventral surface of the brain stem.

Positive symptoms. Symptoms of schizophrenia that involve the production of an abnormal behavior or trait (e.g., hallucinations).

Positive-incentive theories of addiction. Theories holding that the primary factor in most cases of addiction is a craving for the pleasure-producing properties of drugs.

Positive-incentive theory. The idea that behaviors (e.g., eating and drinking) are motivated by their anticipated pleasurable effects.

Positive-incentive value. The anticipated pleasure associated with a particular action, such as taking a drug.

Positron emission tomography (PET). A technique for visualizing brain activity, usually by measuring the accumulation of radioactive 2-deoxyglucose (2-DG) or radioactive water in the various areas of the brain.

Postcentral gyrus. The gyrus located just posterior to the central fissure; its function is primarily somatosensory.

Posterior. Toward the tail end of a vertebrate or toward the back of the head.

Posterior parietal (association) cortex. An area of association cortex that receives input from the visual, auditory, and somatosensory systems and is involved in the perception of spatial location and guidance of voluntary behavior.

Posterior pituitary. The part of the pituitary gland that contains the terminals of hypothalamic neurons.

Postpartum depression. Clinical depression that some women suffer after they have given birth.

Posttraumatic amnesia (PTA). Amnesia produced by a nonpenetrating head injury (a blow to the head that does not penetrate the skull).

Posttraumatic stress disorder. An anxiety disorder exhibited as a persistent pattern of psychological distress that follows a period of exposure to extreme stress.

Prader-Willi syndrome. A neurodevelopmental disorder that is characterized by insatiable appetite and exceptionally slow metabolism.

Precentral gyrus. The gyrus located just anterior to the central fissure; its function is primarily motor.

Prefrontal cortex. The areas of frontal cortex that are anterior to the frontal motor areas.

Prefrontal lobotomy. A surgical procedure in which the connections between the prefrontal lobes and the rest of the brain are cut, as a treatment for mental illness.

Premotor cortex. The area of secondary motor cortex that lies between the supplementary motor area and the lateral fissure.

Prestriate cortex. The band of tissue in the occipital lobe that surrounds the primary visual cortex and contains areas of secondary visual cortex.

Primary motor cortex. The cortex of the precentral gyrus, which is the major point of departure for motor signals descending from the cerebral cortex into lower levels of the sensorimotor system.

Primary sensory cortex. An area of sensory cortex that receives most of its input directly from the thalamic relay nuclei of one sensory system.

Primary visual cortex. The area of the cortex that receives direct input from the lateral geniculate nuclei (also called *striate cortex*).

Primates. One of 14 different orders of mammals; there are five families of primates: prosimians, New-World monkeys, Old-World monkeys, apes, and hominids.

Primed. Induced to resume self-stimulation by the delivery of a few free stimulations.

Proceptive behaviors. Behaviors that solicit the sexual advances of members of the other sex.

Progesterone. A progestin that prepares the uterus and breasts for pregnancy.

Progestins. The class of steroid hormones that includes progesterone.

Prosopagnosia. Visual agnosia for faces.

Protein hormones. Hormones that are long chains of amino acids.

Proteins. Long chains of amino acids.

Proximal. Nearer the central core of the body (e.g., the elbows are proximal to the wrists).

Proximal segment. The segment of a cut axon between the cut and the cell body.

Prozac. The trade name of fluoxetine, the first selective serotonin-reuptake inhibitor developed for treating depression.

Psychiatric disorder. A disorder of psychological function sufficiently severe to require treatment by a psychiatrist or clinical psychologist.

Psychoactive drugs. Drugs that influence subjective experience and behavior by acting on the nervous system.

Psychoneuroimmunology. The study of interactions among psychological factors, the nervous system, and the immune system.

Psychopharmacology. The division of biopsychology that studies the effects of drugs on the brain and behavior.

Psychophysiology. The division of biopsychology that studies the relation between physiological activity and psychological processes in human subjects by noninvasive methods.

Psychosomatic disorder. Any physical disorder that can be caused or exacerbated by stress.

Psychosurgery. Any brain surgery performed for the treatment of a psychological problem (e.g., prefrontal lobotomy).

P300 wave. The positive EEG wave that usually occurs about 300 milliseconds after a momentary stimulus that has meaning for the subject.

Pulsatile hormone release. The typical pattern of hormone release, which occurs in large surges several times a day.

Punch-drunk syndrome. The dementia and cerebral scarring that result from repeated concussions.

Pure research. Research motivated primarily by the curiosity of the researcher and done solely for the purpose of acquiring knowledge.

Purkinje effect. In intense light, red and yellow wavelengths look brighter than blue or green wavelengths of equal intensity; in dim light, blue and green wavelengths look brighter than red and yellow wavelengths of equal intensity.

Putamen. A structure that is joined to the caudate by a series of fiber bridges; together the putamen and caudate compose the striatum.

Pyramidal cell layer. The major layer of cell bodies in the hippocampus.

Pyramidal cells. Large multipolar cortical neurons with a pyramid-shaped cell body, an apical dendrite, and a very long axon.

Quasiexperimental studies. Studies of groups of subjects who have been exposed to the conditions of interest in the real world; such studies have the appearance of experiments but are not true experiments because potential confounded variables have not been controlled.

Radial arm maze. A maze in which several arms radiate out from a central starting chamber, commonly used to study spatial learning in rats.

Radial arm maze test. A widely used test of rats' spatial ability in which the same arms are baited on each trial, and the rats must learn to visit only the baited arms only one time on each trial.

Radial glial cells. Glial cells that exist in the neural tube only during the period of neural migration and that form a network along which radial migration occurs.

Radial migration. Movement of cells in the developing neural tube from the ventricular zone in a straight line outward toward the tube's outer wall.

Reactive depression. Depression that is triggered by a negative experience.

Reappraisal paradigm. An experimental method for studying emotion; subjects are asked to reinterpret a film or photo to change their emotional reaction to it while their brain activity is recorded.

Receptive. Pertaining to the comprehension of language and speech.

Receptive field. The area of the visual field within which it is possible for the appropriate stimulus to influence the firing of a visual neuron.

Receptor blockers. Antagonistic drugs that bind to postsynaptic receptors without activating them and block the access of the usual neurotransmitter.

Receptor subtypes. The different types of receptors to which a particular neurotransmitter can bind.

Receptors. Cells that are specialized to receive chemical, mechanical, or radiant signals from the environment; also proteins that contain binding sites for particular neurotransmitters.

Recessive trait. The trait of a dichotomous pair that is not expressed in the phenotype of heterozygous individuals.

Reciprocal innervation. The principle of spinal cord circuitry that causes a muscle to automatically relax when a muscle that is antagonistic to it contracts.

Recuperation theories of sleep. Theories based on the premise that being awake disturbs the body's homeostasis and the function of sleep is to restore it.

Recurrent collateral inhibition. The inhibition of a neuron that is produced by its own activity via a collateral branch of its axon and an inhibitory interneuron.

Red nucleus. A motor structure that is located in the tectum of the mesencephalon.

Reference memory. Memory for the general principles and skills that are required to perform a task.

Relapse. To return to a diseased state after a period of improvement (e.g., to return to addictive drug taking after a period of voluntary abstinence).

Relative refractory period. A period after the absolute refractory period during which a higher-than-normal amount of stimulation is necessary to make a neuron fire.

Release-inhibiting factors. Hypothalamic hormones that inhibit the release of hormones from the anterior pituitary.

Releasing hormones. Hypothalamic hormones that stimulate the release of hormones from the anterior pituitary.

REM sleep. The stage of sleep characterized by rapid eye movements, loss of core muscle tone, and emergent stage 1 EEG.

Remote memory. Memory for events of the distant past.

Repetition priming tests. Tests of implicit memory; in one example, a list of words is presented, then fragments of the original words are presented and the subject is asked to complete them.

Replacement injections. Injections of a hormone whose natural release has been curtailed by the removal of the gland that normally releases it.

Replication. The process by which the DNA molecule duplicates itself.

Reserpine. The first monoamine antagonist to be used in the treatment of schizophrenia; the active ingredient of the snakeroot plant.

Response-chunking hypothesis. The idea that practice combines the central sensorimotor programs that control individual responses into programs that control sequences of responses (chunks of behavior).

Resting potential. The steady membrane potential of a neuron at rest, usually about −70 mV.

Restless legs syndrome. Tension or uneasiness in the legs that is particularly prevalent at bedtime and is a major cause of insomnia.

Reticular activating system. The hypothetical arousal system in the reticular formation.

Reticular formation. A complex network of nuclei in the core of the brain stem that contains, among other things, motor programs that regulate complex species-common movements such as walking and swimming.

Retina-geniculate-striate pathway. The major visual pathway from each retina to the striate cortex (primary visual cortex) via the lateral geniculate nuclei of the thalamus.

Retinal ganglion cells. Retinal neurons whose axons leave the eyeball and form the optic nerve.

Retinex theory. Land's theory that the color of an object is determined by its reflectance, which the visual system calculates by comparing the ability of adjacent surfaces to reflect short, medium, and long wavelengths.

Retinotopic. Organized, like the primary visual cortex, according to a map of the retina.

Retrograde amnesia. Loss of memory for events or information learned before the amnesia-inducing brain injury.

Retrograde degeneration. Degeneration of the proximal segment of a cut axon.

Reuptake. The drawing back into the terminal button of neurotransmitter molecules after their release into the synapse; the more common of the two mechanisms for deactivating a released neurotransmitter.

Rhinal cortex. An area of medial temporal cortex adjacent to the amygdala and hippocampus.

Rhodopsin. The photopigment of rods.

Ribonucleic acid (RNA). A molecule that is similar to DNA except that it has the nucleotide base uracil and a phosphate and ribose backbone.

Ribosome. A structure in the cell's cytoplasm that translates the genetic code from strands of messenger RNA.

Risk-assessment test. An animal model of anxiety; anxious rats carefully observe an open environment before venturing into it.

Rods. The visual receptors in the retina that mediate achromatic, low-acuity vision under dim light.

Saccades. The rapid movements of the eyes between fixations.

Sagittal sections. Any slices of brain tissue cut in a plane that is parallel to the side of the brain.

Saltatory conduction. Conduction of an action potential from one node of Ranvier to the next along a myelinated axon.

Satiety. The motivational state that terminates a meal when there is food remaining.

Savants. Intellectually handicapped individuals who nevertheless display amazing and specific cognitive or artistic abilities; savant abilities are sometimes associated with autism.

Schwann cells. The glial cells that compose the myelin sheaths of PNS axons and promote their regeneration.

Scientific inference. The logical process by which observable events are used to infer the properties of unobservable events.

Scotoma. An area of blindness produced by damage to, or disruption of, an area of the visual system.

Scotopic spectral sensitivity curve. The graph of the sensitivity of rod-mediated vision to different wavelengths of light.

Scotopic vision. Rod-mediated vision, which predominates in dim light.

Scrotum. The sac that holds the male testes outside the body cavity.

Seasonal affective disorder (SAD). Type of affective disorder in which attacks of depression and lethargy typically occur every winter, presumably triggered by the reduction in sunlight.

Second messenger. A chemical synthesized in a neuron in response to the binding of a neurotransmitter to a metabotropic receptor in its cell membrane.

Secondary motor cortex. Areas of the cerebral cortex that receive much of their input from association cortex and send much of their output to primary motor cortex.

Secondary sensory cortex. Areas of sensory cortex that receive most of their input from the primary sensory cortex of one sensory system or from other areas of secondary cortex of the same system.

Secondary sex characteristics. Body features, other than the reproductive organs, that distinguish men from women.

Secondary visual cortex. Areas of cerebral cortex that receive most of their input from primary visual cortex.

Selective attention. The ability to focus on a small subset of the multitude of stimuli that are being received at any one time.

Selective serotonin-reuptake inhibitors (SSRIs). Class of drugs that exert agonistic effects by blocking the reuptake of serotonin from synapses; used to treat depression.

Self-stimulation paradigm. A paradigm in which animals press a lever to administer reinforcing electrical stimulation to their own brains.

Semantic analysis. Analysis of the meaning of language.

Semantic memories. Explicit memories for general facts and knowledge.

Semicircular canals. The receptive organs of the vestibular system.

Sensation. The process of detecting the presence of stimuli.

Sensitive period. The period during the development of a particular trait, usually early in life, when a particular experience is likely to change the course of that development.

Sensitivity. In vision, the ability to detect the presence of dimly lit objects.

Sensorimotor phase. The second of the two phases of bird-song development, during which juvenile birds progress from subsongs to adult songs.

Sensory evoked potential. A change in the electrical activity of the brain (e.g., in the cortical EEG) that is elicited by the momentary presentation of a sensory stimulus.

Sensory feedback. Sensory signals that are produced by a response and are often used to guide the continuation of the response.

Sensory phase. The first of the two phases of birdsong development, during which young birds do not sing but form memories of the adult songs they hear.

Sensory relay nuclei. Those nuclei of the thalamus whose main function is to relay sensory signals to the appropriate areas of cortex.

Sensory-specific satiety. The fact that the consumption of a particular food produces increased satiety for foods of the same taste than for other foods.

Septum. A midline nucleus of the limbic system, located near the anterior tip of the cingulate cortex.

Serotonin. An indolamine neurotransmitter; the only member of this class of monoamine neurotransmitters found in the mammalian nervous system.

Set point. The value of a physiological parameter that is maintained constantly by physiological or behavioral mechanisms; for example, the body's energy resources are often assumed to be maintained at a constant optimal level by compensatory changes in hunger.

Set-point assumption. The assumption that hunger is typically triggered by the decline of the body's energy reserves below their set point.

Settling point. The point at which various factors that influence the level of some regulated function (such as body weight) achieve an equilibrium.

Sex chromosomes. The pair of chromosomes that determine an individual's sex: XX for a female and XY for a male.

Sex-linked traits. Traits that are influenced by genes on the sex chromosomes.

Sexual identity. The sex (male or female) that a person feels himself or herself to be.

Sexually dimorphic nucleus. The nucleus in the medial preoptic area of rats that is larger in males than in females.

Sham eating. The experimental protocol in which an animal chews and swallows food, which immediately exits its body through a tube implanted in its esophagus.

Sham rage. The exaggerated, poorly directed aggressive responses of decorticate animals.

Short-term memory. Memories (e.g., recall of a phone number) that are stored only until a person stops focusing on them—typically assessed with the digit-span test.

Signal averaging. A method of increasing the signal-to-noise ratio by reducing background noise.

Simple cells. Neurons in the visual cortex that respond maximally to straight-edge stimuli in a certain position and orientation.

Simple partial seizures. Partial seizures in which the symptoms are primarily sensory or motor or both.

Simultanagnosia. A disorder characterized by the inability to attend to more than one thing at a time.

Sinestrals. Left-handers.

Skeletal muscle (extrafusal muscle). Striated muscle that is attached to the skeleton and is usually under voluntary control.

Skin conductance level (SCL). The steady level of skin conductance associated with a particular situation.

Skin conductance response (SCR). The transient change in skin conductance associated with a brief experience.

Sleep apnea. A condition in which sleep is repeatedly disturbed by momentary interruptions in breathing.

Sleep inertia. The unpleasant feeling of grogginess that is sometimes experienced for a few minutes after awakening.

Sleep paralysis. A sleep disorder characterized by the inability to move (paralysis) just as a person is falling asleep or waking up.

Slow-wave sleep (SWS). Stages 3 and 4 of sleep, which are characterized by the largest and slowest EEG waves.

Smoker's syndrome. The chest pain, labored breathing, wheezing, coughing, and heightened susceptibility to infections of the respiratory tract commonly observed in tobacco smokers.

Sodium amytal test. A test involving the anesthetization of first one cerebral hemisphere and then the other to determine which hemisphere plays the dominant role in language.

Sodium–potassium pumps. Active transport mechanisms that pump Na^+ ions out of neurons and K^+ ions in.

Solitary nucleus. The medullary relay nucleus of the gustatory system.

Soluble-gas neurotransmitters. A class of unconventional neurotransmitters that includes nitric oxide and carbon monoxide.

Somal translocation. One of two major modes of neural migration, in which an extension grows out from the undeveloped neuron and draws the cell body up into it.

Somatic nervous system (SNS). The part of the peripheral nervous system that interacts with the external environment.

Somatosensory homunculus. The somatotopic map that corresponds to the primary somatosensory cortex.

Somatotopic. Organized, like the primary somatosensory cortex, according to a map of the surface of the body.

Spandrels. Nonadaptive characteristics that evolve because they are related to evolutionary changes that are adaptive.

Spatial resolution. Ability of a recording technique to detect differences in spatial location (e.g., to pinpoint a location in the brain).

Spatial summation. The integration of signals that occur at different sites on the neuron's membrane.

Species. A group of organisms that is reproductively isolated from other organisms; the members of one species cannot produce fertile offspring by mating with members of other species.

Species-common behaviors. Behaviors that are displayed in the same manner by virtually all like members of a species.

Spindle afferent neurons. Neurons that carry signals from muscle spindles into the spinal cord via the dorsal root.

Split-brain patients. Commissurotomized patients.

Sry gene. A gene on the Y chromosome that triggers the release of Sry protein, which in turn stimulates the development of testes.

Sry protein. A protein that causes the medulla of each primordial gonad to develop into a testis.

Standard consolidation theory. Theory that memories are temporarily stored in the hippocampus until they can be transferred to a more stable cortical storage system.

Static phase. The second phase of the VMH syndrome, during which the grossly obese animal maintains a stable level of obesity.

Stellate cells. Small star-shaped cortical interneurons.

Stem cells. Developing cells that have the capacity for self-renewal and the potential to develop into various types of mature cells.

Stereognosis. The process of identifying objects by touch.

Stereotaxic atlas. A series of maps representing the three-dimensional structure of the brain that is used to determine coordinates for stereotaxic surgery.

Stereotaxic instrument. A device for performing stereotaxic surgery, composed of two parts: a head holder and an electrode holder.

Steroid hormones. Hormones that are synthesized from cholesterol.

Stimulants. Drugs that produce general increases in neural and behavioral activity.

Stress. The physiological response to physical or psychological threat.

Stressors. Experiences that induce the stress response.

Stretch reflex. A reflexive counteracting reaction to an unanticipated external stretching force on a muscle.

Striatum. A structure of the basal ganglia that is the terminal of the dopaminergic nigrostriatal pathway and is damaged in Parkinson's patients; it seems to play a role in memory for consistent relationships between stimuli and responses in multiple-trial tasks.

Strokes. Sudden-onset cerebrovascular disorders that cause brain damage.

Subarachnoid space. The space beneath the arachnoid membrane, which contains many large blood vessels and cerebrospinal fluid.

Subcutaneous fat. Body fat that is stored under the skin and is positively correlated with leptin levels.

Subordination stress. Stress experienced by animals, typically males, that are continually attacked by higher-ranking conspecifics.

Substantia nigra. The midbrain nucleus whose neurons project via the nigrostriatal pathway to the striatum of the basal ganglia; it is part of the mesotelencephalic dopamine system and degenerates in cases of Parkinson's disease.

Subthalamic nucleus. A nucleus that lies just below the thalamus and is connected to the basal ganglia; deep brain stimulation applied to this site has been used to treat Parkinson's disease.

Sulci. Small furrows in a convoluted cortex.

Superior. Toward the top of the primate head.

Superior colliculi. Two of the four nuclei that compose the tectum; they receive major visual input.

Superior olives. Medullary nuclei that play a role in sound localization.

Superior temporal gyrus. The large gyrus of the temporal lobe adjacent to the lateral fissure; the location of auditory cortex.

Supplementary motor area. The area of secondary motor cortex that is within and adjacent to the longitudinal fissure.

Suppression paradigm. An experimental method for studying emotion; subjects are asked to inhibit their emotional reactions to unpleasant films or photos while their brain activity is recorded.

Suprachiasmatic nuclei (SCN). Nuclei of the medial hypothalamus that control the circadian cycles of various body functions.

Supraoptic nuclei. Hypothalamic nuclei in which the hormones of the posterior pituitary are synthesized.

Surface dyslexia. A reading disorder in which the lexical procedure is disrupted while the phonetic procedure is not.

Surface interpolation. The process by which the visual system perceives large surfaces, by extracting information about edges and from it, inferring the appearance of adjacent surfaces.

Sympathetic nerves. Those motor nerves of the autonomic nervous system that project from the CNS in the lumbar and thoracic areas of the spinal cord.

Synaptic vesicles. Small spherical membranes that store neurotransmitter molecules and release them into the synaptic cleft.

Synaptogenesis. The formation of new synapses.

Synergistic muscles. Pairs of muscles whose contraction produces a movement in the same direction.

T cells. T lymphocytes; lymphocytes that bind to foreign micro-organisms and cells that contain them and, in so doing, destroy them.

Tangential migration. Movement of cells in the developing neural tube in a direction parallel to the tube's walls.

Tardive dyskinesia (TD). A motor disorder that results from chronic use of certain antipsychotic drugs.

Target-site concept. The idea that aggressive and defensive behaviors of an animal are often designed to attack specific sites on the body of another animal while protecting specific sites on its own.

Taste buds. Clusters of taste receptors found on the tongue and in parts of the oral cavity.

Tau. The first circadian gene to be identified in mammals.

Tectorial membrane. The cochlear membrane that rests on the hair cells.

Tectum. The "roof," or dorsal surface, of the mesencephalon; it includes the superior and inferior colliculi.

Tegmentum. The ventral division of the mesencephalon; it includes part of the reticular system, substantia nigra, and red nucleus.

Telencephalon. The most superior of the brain's five major divisions; also called the *cerebral hemispheres.*

Temporal hemiretina. The half of each retina next to the temple.

Temporal lobe. One of the four major cerebral lobes; it lies adjacent to the temples and contains the hippocampus and amygdala.

Temporal resolution. Ability of a recording technique to detect differences in time (i.e., to pinpoint when an event occurred).

Temporal summation. The integration of neural signals that occur at different times at the same synapse.

Teratogen. A drug or other chemical that causes birth defects.

Testes. The male gonads.

Testosterone. The most common androgen.

Thalamus. The large two-lobed diencephalic structure that constitutes the anterior end of the brain stem; many of its nuclei are sensory relay nuclei that project to the cortex.

THC. Delta-9-tetrahydrocannabinol, the main psychoactive constituent of marijuana.

Thigmotaxic. Tending to stay near the walls of an open space such as a test chamber.

Thinking creatively. Thinking in productive, unconventional ways that are consistent with the evidence.

3-per-second spike-and-wave discharge. The characteristic EEG pattern of the petit mal seizure.

Threshold of excitation. The level of depolarization necessary to generate an action potential, usually about −65 mV.

Thrombosis. The blockage of blood flow by a plug (a thrombus) at the site of its formation.

Thyrotropin. The anterior pituitary hormone that stimulates the release of hormones from the thyroid gland.

Thyrotropin-releasing hormone. The hypothalamic hormone that stimulates the release of thyrotropin from the anterior pituitary.

Tics. Involuntary, repetitive, stereotyped movements or vocalizations; the defining feature of Tourette syndrome.

Tinnitus. Ringing in the ears.

Token test. A preliminary test for language-related deficits that involves following verbal instructions to touch or move tokens of different shapes, sizes, and colors.

Toll-like receptors. Receptors found in the cell membranes of many cells of the innate immune system; they trigger phagocytosis and inflammatory responses.

Tonotopic. Organized, like the primary auditory cortex, according to the frequency of sound.

Topographic gradient hypothesis. The hypothesis that axonal growth is guided by the relative position of the cell bodies on intersecting gradients, rather than by point-to-point coding of neural connections.

Totipotent. Capable of developing into any type of mature body cell.

Toxic psychosis. A chronic psychiatric disorder produced by exposure to a neurotoxin.

Tracts. Bundles of axons in the central nervous system.

Transcranial magnetic stimulation (TMS). A technique for disrupting the activity in an area of the cortex by creating a magnetic field under a coil postitioned next to the skull; the effect of the disruption on cognition is assessed to clarify the function of the affected area of cortex.

Transcription factors. Proteins that bind to DNA and influence the expression of particular genes.

Transduction. The conversion of one form of energy to another.

Transfer RNA. Molecules of RNA that carry amino acids to ribosomes during protein synthesis; each kind of amino acid is carried by a different kind of transfer RNA molecule.

Transgenic. Containing the genes of another species, which have been implanted there for research purposes.

Transgenic mice. Mice into which the genetic material of another species has been introduced.

Translational bottleneck. A barrier keeping promising ideas and treatments from becoming the focus of translational research; largely created by the massive cost of such research.

Translational research. Research designed to translate basic scientific discoveries into effective applications (e.g., into clinical treatments).

Transneuronal degeneration. Degeneration of a neuron caused by damage to another neuron to which it is linked by a synapse.

Transorbital lobotomy. A prefrontal lobotomy performed with a cutting instrument inserted through the eye socket.

Transporters. Mechanisms in the membrane of a cell that actively transport ions or molecules across the membrane.

Transsexualism. A disorder of sexual identity in which the individual believes that he or she is trapped in a body of the other sex.

T-reg cells. Regulatory T cells; they protect the body from autoimmune disease by identifying and destroying T cells that engage in autoimmune activity.

Tricyclic antidepressant drugs, or tricyclic antidepressants. Drugs with an antidepressant action and a three-ring molecular structure; they selectively suppress REM sleep.

True-breeding lines. Breeding lines in which interbred members always produce offspring with the same trait, generation after generation.

Tumor (neoplasm). A mass of cells that grows independently of the rest of the body.

Tympanic membrane. The eardrum.

Unipolar affective disorder. A disorder of emotion in which a patient experiences depression but no periods of mania.

Unipolar neuron. A neuron with one process extending from its cell body.

Up-regulation. An increase in the number of receptors for a neurotransmitter in response to decreased release of that neurotransmitter.

Urbach-Wiethe disease. A genetic disorder that often results in the calcification of the amygdala and surrounding brain structures.

Vaccination. Administering a weakened form of a virus so that if the virus later invades, the adaptive immune system is prepared to deal with it.

Vasopressin. One of the two major peptide hormones of the posterior pituitary; it facilitates reabsorption of water by kidneys and is thus also called *antidiuretic hormone*.

Ventral. Toward the chest surface of a vertebrate or toward the bottom of the head.

Ventral horns. The two ventral arms of the spinal gray matter.

Ventral posterior nucleus. A thalamic relay nucleus in both the somatosensory and gustatory systems.

Ventral stream. The group of visual pathways that flows from the primary visual cortex to the ventral prestriate cortex to the inferotemporal cortex; according to one theory, its function is conscious visual perception.

Ventral tegmental area. The midbrain nucleus of the mesotelencephalic dopamine system that is a major source of the mesoscorticolimbic pathway.

Ventricular zone. The region adjacent to the ventricle in the developing neural tube; the zone where neural proliferation occurs.

Ventromedial cortico-brainstem-spinal tract. The indirect ventromedial motor pathway, which descends bilaterally from the primary motor cortex to several interconnected brain stem motor structures and then descends in the ventromedial portions of the spinal cord.

Ventromedial corticospinal tract. The direct ventromedial motor pathway, which descends ipsilaterally from the primary motor cortex directly into the ventromedial areas of the spinal white matter.

Ventromedial hypothalamus (VMH). The area of the hypothalamus that was once thought to contain the satiety center.

Ventromedial nucleus (VMN). A hypothalamic nucleus that is thought to be involved in female sexual behavior.

Vertebrates. Chordates that possess spinal bones.

Vestibular nucleus. The brain stem nucleus that receives information about balance from receptors in the semicircular canals.

Vestibular system. The sensory system that detects changes in the direction and intensity of head movements and that contributes to the maintenance of balance through its output to the motor system.

Visceral fat. Body fat that is stored in the organs of the body and is positively correlated with insulin levels.

Visual agnosia. A failure to recognize visual stimuli that is not attributable to sensory, verbal, or intellectual impairment.

Visual association cortex. Areas of cerebral cortex that receive input from areas of secondary visual cortex as well as from secondary cortex of other sensory systems.

Visual completion. The completion or filling in of a scotoma by the brain.

Voltage-activated ion channels. Ion channels that open and close in response to changes in the level of the membrane potential.

Wechsler Adult Intelligence Scale (WAIS). A widely used test of general intelligence that includes 11 subtests.

Wernicke-Geschwind model. An influential model of cortical language localization in the left hemisphere.

Wernicke's aphasia. A hypothetical disorder of language comprehension with no associated deficits in speech production.

Wernicke's area. The area of the left temporal cortex hypothesized by Wernicke to be the center of language comprehension.

"Where" versus "what" theory. The theory that the dorsal stream mediates the perception of where things are and the ventral stream mediates the perception of what things are.

White matter. Portions of the nervous system that are white because they are composed largely of myelinated axons.

Williams syndrome. A neurodevelopmental disorder characterized by severe mental retardation, accompanied by preserved language and social skills.

Wisconsin Card Sorting Test. A neuropsychological test that evaluates a patient's ability to remember that previously learned rules of behavior are no longer effective and to learn to respond to new rules.

Withdrawal reflex. The reflexive withdrawal of a limb when it comes in contact with a painful stimulus.

Withdrawal syndrome. The illness brought on by the elimination from the body of a drug on which the person is physically dependent.

Within-subjects design. An experimental design in which the same subjects are tested under each condition.

Wolffian system. The embryonic precursor of the male reproductive ducts.

Working memory. Temporary memory necessary for the successful performance of a task on which one is currently working.

Z lens. A contact lens that is opaque on one side (left or right) and thus allows visual input to enter only one hemisphere of a split-brain subject, irrespective of eye movement.

Zeitgebers. Environmental cues, such as the light–dark cycle, that entrain circadian rhythms.

Zeitgeist. The general intellectual climate of a culture.

Zygote. The cell formed from the amalgamation of a sperm cell and an ovum.

References

Abbott, N. J., Rönnbäck, L., & Hannson, E. (2006). Astrocyte–endothelial interactions at the blood-brain barrier. *Nature Reviews Neuroscience, 7,* 41–53.

Abelson, P., & Kennedy, D. (2004). The obesity epidemic. *Science, 304,* 1413.

Abou-Sleiman, P. M., Muqit, M. M. K., & Wood, N. W. (2006). Expanding insights of mitochondrial dysfunction in Parkinson's disease. *Nature Reviews Neuroscience, 7,* 207–219.

Abraham, W. C. (2006). Memory maintenance: The changing nature of neural mechanisms. *Current Directions in Psychological Science, 15,* 5–8.

Abraham, W. C. (2008). Metaplasticity: Tuning synapses and networks for plasticity. *Nature Reviews Neuroscience, 9,* 387–399.

Acker, W., Ron, M. A., Lishman, W. A., & Shaw, G. K. (1984). A multivariate analysis of psychological, clinical and CT scanning measures in detoxified chronic alcoholics. *British Journal of Addiction, 79,* 293–301.

Adams, C. E., Rathbone, J., Thornley, B., Clarke, M., Borrill, J., Wahlbeck, K., & Awad, A. G. (2005). Chlorpromazine for schizophrenia: A Cochrane systematic review of 50 years of randomized controlled trials. *BMC Medicine, 3,* 15–25.

Adams, C. P., and Brantner, V. V. (2010). Spending on new drug development. *Health Economics, 19,* 130–141.

Adams, K. F., Schatzkin, A., Harris, T. B., Kipnis, V., Mouw, T., Ballard-Barbash, R., . . . Leitzmann, M. F. (2006). Overweight, obesity, and mortality in a large prospective cohort of persons 50 to 71 years old. *New England Journal of Medicine, 355,* 763–778.

Adelmann, P. K., & Zajonc, R. B. (1989). Facial efference and the experience of emotion. *Annual Review of Psychology, 40,* 249–280.

Adlard, P. A., Perreau, V. M., Pop, V., & Cotman, C. W. (2005). Voluntary exercise decreases amyloid load in a transgenic model of Alzheimer's diseases. *Journal of Neuroscience, 25,* 4217–4221.

Adolphs, R. (2006). Perception and emotion: How we recognize facial expressions. *Current Directions in Psychological Science, 15,* 222–226.

Adolphs, R. (2008). Fear, faces, and the human amygdala. *Current Opinion in Neurobiology, 18,* 166–172.

Adolphs, R., & Tranel, D. (1999). Preferences for visual stimuli following amygdala damage. *Journal of Cognitive Neuroscience, 11,* 610–616.

Adolphs, R., Tranel, D., Hamann, S., Young, A. W., Calder, A. J., Phelps, E. A., . . . Damasio, A.R. (1999). Recognition of facial emotion in nine individuals with bilateral amygdala damage. *Neuropsychologia, 37,* 1111–1117.

Aeschbach, D., Cajochen, C., Landolt, H., & Borbely, A. A. (1996). Homeostatic sleep regulation in habitual short sleepers and long sleepers. *American Journal of Regulatory, Integrative and Comparative Physiology, 270,* R41–R53.

Agarwal, N., Pacher, P., Tegeder, I., Amaya, F., Constantin, C. E., Brenner, G. J., . . . Kuner, R. (2007). Cannabinoids mediate analgesia largely via peripheral type 1 cannabinoid receptors in nociceptors. *Nature Neuroscience, 10,* 870–879.

Aggleton, J. P., & Young, A. W. (2000). The enigma of the amygdala: On its contribution to human emotion. In R. D. Lane & L. Nadel (Eds.), *Cognitive neuroscience of emotion* (pp. 106–128). New York, NY: Oxford University Press.

Aglioti, S., Smania, N., & Peru, A. (1999). Frames of reference for mapping tactile stimuli in brain-damaged patients. *Journal of Cognitive Neuroscience, 11,* 67–79.

Agmo, A., & Ellingsen, E. (2003). Relevance of non-human animal studies to the understanding of human sexuality. *Scandinavian Journal of Psychology, 44,* 293–301.

Agnew, N., & Demas, M. (1998, September). Preserving the Laetoli footprints. *Scientific American, 279,* 46–55.

Aguilar, M. A., Rodríguez-Arias, M., & Miñarro, J. (2008). Neurobiological mechanisms of the reinstatement of drug-conditioned place preference. *Brain Research Reviews, 59,* 253–277.

Ahissar, M. (2007). Dyslexia and the anchoring-deficit hypothesis. *Trends in Cognitive Sciences, 11,* 458–465.

Ahmed, E. I., Zehr, J. L., Schulz, K. M., Lorenz, B. H., DonCarlos, L. L., & Sisk, C. L. (2008). Pubertal hormones modulate the addition of new cells to sexually dimorphic brain regions. *Nature Neuroscience, 11,* 995–997.

Ahmed, O. J., & Mehta, M. M. (2009). The hippocampal rate code: Anatomy, physiology and theory. *Trends in Neurosciences, 32,* 329–338.

Ahmed, R., & Gray, D. (1996). Immunological memory and protective immunity: Understanding their relation. *Science, 272,* 54–60.

Ahmed, S. H. (2004). Addiction as compulsive reward prediction. *Science, 306,* 1901–1902.

Åkerstedt, R., & Gillberg, M. (1981). The circadian variation of experimentally displaced sleep. *Sleep, 4,* 159–169.

Akins, M. R., & Biederer, T. (2006). Cell-cell interactions in synaptogenesis. *Current Opinion in Neurobiology, 16,* 83–89.

Albasser, M. M., Davies, M., Futter, J. E., & Aggleton, J. P. (2009). Magnitude of the object recognition deficit associated with perirhinal cortex damage in rats: Effects of varying the lesion extent and the duration of the sample period. *Behavioral Neuroscience, 123,* 115–124.

Albin, R. L., & Mink, J. W. (2006). Recent advances in Tourette syndrome research. *Trends in Neurosciences, 29,* 175–182.

Albright, T. D., Kandel, E. R., & Posner, M. I. (2000). Cognitive neuroscience. *Current Opinion in Neurobiology, 10,* 612–624.

Aldington, S., Harwood, M., Cox, B., Weatherall, M., Beckert, L., Hansell, A., . . . Beasley, R. (2008). Cannabis use and risk of lung cancer: A case-control study. *European Respiratory Journal, 31,* 280–286.

Alexander, M. P. (1989). Clinical-anatomical correlations of aphasia following predominantly subcortical lesions. In H. Goodglass (Ed.), *Handbook of neuropsychology* (Vol. II, Pt. 2, pp. 47–66). New York, NY: Elsevier.

Alexander, M. P. (1997). Aphasia: Clinical and anatomic aspects. In T. E. Feinberg & M. J. Farah, (Eds.), *Behavioral neurology and neuropsychology* (pp. 133–149). New York, NY: McGraw-Hill.

Alle, H., & Geiger, J. R. P. (2006). Combined analog and action potential coding in hippocampal mossy fibers. *Science, 311,* 1290–1293.

Alle, H., & Geiger, J. R. P. (2008). Analog signalling in mammalian cortical axons. *Current Opinion in Neurobiology, 18,* 314–320.

Allen, E. A., Pasley, B. N., Duong, T., & Freeman, R. D. (2007). Transcranial magnetic stimulation elicits coupled neural and hemodynamic consequences. *Science, 317,* 1918–1921.

Allen, J. S., Tranel, D., Bruss, J., & Damasio, H. (2006). Correlations between regional brain volumes and memory in anoxia. *Journal of Clinical and Experimental Neuropsychology, 28,* 457–476.

Allen, L. S., Hines, M., Shryne, J. E., & Gorski, R. A. (1989). Two sexually dimorphic cell groups in the human brain. *Journal of Neuroscience, 9,* 497–506.

Allen, N. E., Beral, V., Casabonne, D., Kan, S. W., Reeves, G. K., Brown, A., & Green, J. (2009). Moderate alcohol intake and cancer incidence in women. *Journal of the National Cancer Institute, 101,* 296–305.

Allen, N. J., & Barres, B. A. (2005). Signaling between glia and neurons: Focus on synaptic plasticity. *Current Opinion in Neurobiology, 15,* 542–548.

Allsop, T. E., & Fazakerley, J. K. (2000). Altruistic cell suicide and the specialized case of the virus-infected nervous system. *Trends in Neurosciences, 23,* 284–290.

Altemus, M. (2006). Sex differences in depression and anxiety disorders: Potential biological determinants. *Hormones and Behavior, 50,* 534–538.

Alvarez, V. A., & Sabatini, B. L. (2007). Anatomical and physiological plasticity of dendritic spines. *Annual Review of Neuroscience, 30*, 79–97.

Amaral, D. G., Schumann, C. M., & Nordahl, C. W. (2008). Neuroanatomy of autism. *Trends in Neurosciences, 31*, 137–145.

Amariglio, N., Hirschberg, A., Scheithauer, B. W., Cohen, Y., Lowenthal, R., Trakhtenbrot, L., . . . Rechavi, G. (2009). Donor-derived brain tumor following neural stem cell transplantation in an ataxia telangiectasia patient. *PLoS Medicine, 6*: e1000029. doi:10.1371/journal.pmed.1000029

Ambrose, S. H. (2001). Paleolithic technology and human evolution. *Science, 291*, 1748–1753.

Amedi, A., Merabet, L. B., Bermpohl, F., & Pascual-Leone, A. (2005). The occipital cortex in the blind. *Current Directions in Psychological Science, 14*, 306–311.

Amso, D., & Casey, B. J. (2006). Beyond what happens when. *Current Directions in Psychological Science, 15*, 24–29.

Anagnostaras, S. G., Craske, M. G., & Fanselow, M. S. (1999). Anxiety: At the intersection of genes and experience. *Nature Neuroscience, 2*, 780–782.

Anagnostaras, S. G., & Robinson, T. E. (1996). Sensitization to the psychomotor stimulant effects of amphetamine: Modulation by associative learning. *Behavioral Neuroscience, 110*(6), 1397–1414.

Anand, B. K., & Brobeck, J. R. (1951). Localization of a "feeding center" in the hypothalamus of the rat. *Proceedings of the Society for Experimental Biology and Medicine, 77*, 323–324.

Anch, A. M., Browman, C. P., Mitler, M. M., & Walsh, J. K. (1988). *Sleep: A scientific perspective.* Englewood Cliffs, NJ: Prentice Hall.

Anderson, A. K., & Phelps, E. A. (2000). Expression without recognition: Contributions of the human amygdala to emotional communication. *Psychological Science, 11*, 106–111.

Anderson, P., Cremona, A., Paton, A., Turner, C., & Wallace, P. (1993). The risk of alcohol. *Addiction, 88*, 1493–1508.

Anderson, R. H., Fleming, D. E., Rhees, R. W., & Kinghorn, E. (1986). Relationships between sexual activity, plasma testosterone, and the volume of the sexually dimorphic nucleus of the preoptic area in prenatally stressed and non-stressed rats. *Brain Research, 370*, 1–10.

Andres, M., Olivier, E., & Badets, A. (2008). Actions, words, and numbers: A motor contribution to semantic processing? *Current Directions in Psychological Science, 17*, 313–317.

Anikovich, M. V., Sinitsyn, A. A., Hoffecker, J. F., Holliday, V. T., Popov, V. V., Lisitsyn, S. N., . . . Praslov, N. D. (2007). Early upper Paleolithic in eastern Europe and implications for the dispersal of modern humans. *Science, 315*, 223–226.

Antle, M. C., & Silver, R. (2005). Orchestrating time: Arrangements of the brain circadian clock. *Trends in Neurosciences, 28*, 145–151.

Antoniadis, E. A., & McDonald, R. J. (2000). Amygdala, hippocampus, and discriminative fear conditioning to context. *Behavioural Brain Research, 108*, 1–19.

Antonini, A., & Stryker, M. P. (1993). Rapid remodeling of axonal arbors in the visual cortex. *Science, 260*, 1819–1821.

Antshel, K. M., & Waisbren, S. E. (2003). Timing is everything: Executive functions in children exposed to elevated levels of phenylalanine. *Neuropsychology, 17*, 458–468.

Apkarian, A. V. (2008). Pain perception in relation to emotional learning. *Current Opinion in Neurobiology, 18*, 464–468.

Apps, R., & Garwicz, M. (2005). Anatomical and physiological foundations of cerebellar information processing. *Nature Reviews Neuroscience, 6*, 297–311.

Apps, R., & Hawkes, R. (2009). Cerebellar cortical organization: A one-map hypothesis. *Nature Reviews Neuroscience, 10*, 670–681.

Araújo, S. J., & Tear, G. (2003). Axon guidance mechanisms and molecules: Lessons from invertebrates. *Nature Reviews Neuroscience, 4*, 910–922.

Arendt, J., & Skene, D. J. (2005). Melatonin as a chronobiotic. *Sleep Medicine Reviews, 9*, 25–39.

Arieli, A., Sterkin, A., Grinvald, A., & Aertsen, A. (1996). Dynamics of ongoing activity: Explanation of the large variability in evoked cortical responses. *Science, 273*, 1868–1870.

Arikkath, J., & Reichardt, L. F. (2008). Cadherins and catenins at synapses: Roles in synaptogenesis and synaptic plasticity. *Trends in Neurosciences, 31*, 487–494.

Armstrong, C. M. (2007). Life among the axons. *Annual Review of Physiology, 69*, 1–18.

Armstrong, S. P., Caunt, C. J., Fowkes, R. C., Tsaneva-Atanasova, K., & McArdle, C. A. (2009). Pulsatile and sustained gonadotropin-releasing hormone (GnRH) receptor signaling: Does the Ca^{2+}/NFAT signaling pathway decode GnRH pulse frequency? *Journal of Biological Chemistry, 284*, 35746–35757.

Arnold, A. P. (2003). The gender of the voice within: The neural origin of sex differences in the brain. *Current Opinion in Neurobiology, 13*, 759–764.

Arnold, A. P. (2004). Sex chromosomes and brain gender. *Nature Reviews Neuroscience, 5*, 701–708.

Arnold, A. P. (2009). The organizational-activational hypothesis as the foundation for a unified theory of sexual differentiation of all mammalian tissues. *Hormones and Behavior, 55*, 570–578.

Arnold, A. P., Xu, J., Grisham, W., Chen, X., Kim, Y. H., & Itoh, Y. (2004). Sex chromosomes and brain sexual differentiation [Review]. *Endocrinology, 145*, 1057–1062.

Arnold, S. R. (2009, January 25). Obesity statistics—the facts presented by modern research on obesity. *EzineArticles.com.* Retrieved from http://ezinearticles.com/?Obesity-Statistics—The-Facts-Presented-by-Modern-Research-on-Obesity&id=191960

Aron, J. L., & Paulus, M. P. (2007). Location, location: Using functional magnetic resonance imaging to pinpoint brain differences relevant to stimulant use. *Addiction, 102*, 33–43.

Aronov, D., Andalman, A. S., & Fee, M. S. (2008). A specialized forebrain circuit for vocal babbling in the juvenile songbird. *Science, 320*, 630–634.

Arseneault, L., Cannon, M., Witton, J., & Murray, R. M. (2004). Causal association between cannabis and psychosis: Examination of the evidence. *British Journal of Psychiatry, 184*, 110–117.

Arvanitogiannis, A., Sullivan, J., & Amir, S. (2000). Time acts as a conditioned stimulus to control behavioral sensitization to amphetamine in rats. *Neuroscience, 101*, 1–3.

Ascherio, A., & Munger, K. (2008). Epidemiology of multiple sclerosis: From risk factors to prevention. *Seminars in Neurology, 28*, 17–28.

Ashe, J., Lungu, O. V., Basford, A. T., & Lu, X. (2006). Cortical control of motor sequences. *Current Opinion in Neurobiology, 16*, 213–221.

Assanand, S., Pinel, J. P. J., & Lehman, D. R. (1998a). Personal theories of hunger and eating. *Journal of Applied Social Psychology, 28*, 998–1015.

Assanand, S., Pinel, J. P. J., & Lehman, D. R. (1998b). Teaching theories of hunger and eating: Overcoming students' misconceptions. *Teaching Psychology, 25*, 44–46.

Ast, G. (April, 2005). The alternative genome. *Scientific American, 292*, 58–65.

Au, E., & Fishell, G. (2006). Adult cortical neurogenesis: Nuanced, negligible or nonexistent? *Nature Neuroscience, 9*, 1086–1088.

Auld, V. J. (2001). Why didn't the glia cross the road? *Trends in Neurosciences, 24*, 309–311.

Autism Genome Project. (2007). Mapping autism risk loci using genetic linkage and chromosomal rearrangements. *Nature Genetics, 39*, 319–328.

Ax, A. F. (1955). The physiological differentiation between fear and anger in humans. *Psychosomatic Medicine, 15*, 433–442.

Ayas, N. T., White, D. P., Manson, J. E., Stampfer, M. J., Speizer, F. E., Malhotra, A., & Hu, F. B. (2003). A prospective study of sleep duration and coronary heart disease in women. *Archives of Internal Medicine, 163*, 205–209.

Azevedo, F. A., Carvalho, L. R., Grinberg, L. T., Farfel, J. M., Ferretti, R. E., Leite, R. E., . . . Herculano-Houzel, S. (2009). Equal numbers of neuronal and nonneuronal cells make the human brain an isometrically scaled-up primate brain. *Journal of Comparative Neurology, 513*, 532–541.

Baas, D., Aleman, A., & Kahn, R. S. (2004). Lateralization of amygdala activation: A systematic review of functional neuroimaging studies. *Brain Research Reviews, 45*, 96–103.

Bachevalier, J., & Loveland, K. A. (2006). The orbitofrontal-amygdala circuit and self-regulation of social-emotional behavior in autism. *Neuroscience and Biobehavioural Reviews, 30*, 97–117.

Badman, M. K., & Flier, J. S. (2005). The gut and energy balance: Visceral allies in the obesity wars. *Science, 307*, 1909–1914.

Bagnardi, V., Blangiardo, M., La Vecchia, C., & Corrao, G. (2001). Alcohol consumption and the risk of cancer: A meta-analysis. *Alcohol Research & Health, 25*, 263–270.

Baicy, K., & London, E. D. (2007). Corticolimbic dysregulation and chronic methamphetamine abuse. *Addiction, 102*, 5–15.

Bailey, J. M., Pillard, R. C., Neale, M. C., & Agyei, Y. (1993). Heritable factors influence sexual orientation in women. *Archives of General Psychiatry, 50*, 217–223.

Bailey, M. J., & Pillard, R. C. (1991). A genetic study of male sexual orientation. *Archives of General Psychiatry, 48*, 1089–1096.

Bains, J. S., & Oliet, S. H. R. (2007). Glia: They make your memories stick! *Trends in Neurosciences, 30*, 418–424.

Bair, W. (2005). Visual receptive field organization. *Current Opinion in Neurobiology, 15*, 459–464.

Baird, J-P., Gray, N. E., & Fischer, S. G. (2006). Effects of neuropeptide Y on feeding microstructure: Dissociation of appetitive and consummatory actions. *Behavioral Neuroscience, 120*, 937–951.

Baker, B. J., & Booth, D. A. (1989). Preference conditioning by concurrent diets with delayed proportional reinforcement. *Physiology & Behavior, 46*, 585–590.

Baker, T. B., Japuntich, S. J., Hogle, J. M., McCarthy, D. E., & Curtin, J. J. (2006). Pharmacologic and behavioral withdrawal from addictive drugs. *Current Directions in Psychological Science, 15*, 232–236.

Bakker, J., & Baum, M. J. (2007). Role for estradiol in female-typical brain and behavioral sexual differentiation. *Frontiers in Neuroendocrinology, 29*, 1–16.

Bakker J., De Mees, C., Douhard, Q., Balthazart, J., Gabant, P., Szpirer, J., & Szpirer, C. (2006). Alpha-fetoprotein protects the developing female mouse brain from masculinization and defeminization by estrogens. *Nature Neuroscience, 9*, 220–226.

Bale, T. L. (2005). Sensitivity to stress: Dysregulation of CRF pathways and disease development. *Hormones and Behavior, 48*, 1–10.

Bale, T. L. (2006). Stress sensitivity and the development of affective disorders. *Hormones and Behavior, 50*, 529–533.

Baler, R. D., & Volkow, N. D. (2006). Drug addiction: The neurobiology of disrupted self-control. *Trends in Molecular Medicine, 12*, 559–566.

Ball, G. F., & Balthazart, J. (2006). Androgen metabolism and the activation of male sexual behavior: It's more complicated than you think! *Hormones and Behavior, 49*, 1–3.

Ballatore, C., Lee, V. M.-Y., & Trojanowski, J. Q. (2007). Tau-mediated neurodegeneration in Alzheimer's disease and related disorders. *Nature Reviews Neuroscience, 8*, 663–672.

Balleine, B. W., & Killcross, S. (2006). Parallel incentive processing: An integrated view of amygdala function. *Trends in Neurosciences, 29*, 272–279.

Balthazart, J., & Ball, G. F. (1998). New insights into the regulation and function of brain estrogen synthase (aromatase). *Trends in Neurosciences, 21*, 243–249.

Balthazart, J., & Ball, G. F. (2006). Is brain estradiol a hormone or a neurotransmitter? *Trends in Neurosciences, 29*, 241–249.

Balu, D. T., & Lucki, I. (2008). Adult hippocampal neurogenesis: Regulation, functional implications, and contribution to disease pathology. *Neuroscience and Biobehavioral Reviews, 33*, 232–252.

Bancroft, J., Sanders, D., Davidson, D., & Warner, P. (1983). Mood, sexuality, hormones and the menstrual cycle: III. Sexuality and the role of androgens. *Psychosomatic Medicine, 45*, 509–516.

Bandler, R., & Shipley, M. T. (1994). Columnar organization in the midbrain periaqueductal gray: Modules for emotional expression? *Trends in Neurosciences, 17*, 379–389.

Banerjee, S., & Bhat, M. A. (2007). Neuron-glial interactions in blood-brain barrier formation. *Annual Review of Neuroscience, 30*, 235–258.

Bankiewicz, K. S., Plunkett, R. J., Jaconowitz, D. M., Porrino, L., di Porzio, U., London, W. T., . . . Oldfield, E. H. (1990). The effect of fetal mesencephalon implants on primate MPTP-induced Parkinsonism: Histochemical and behavioral studies. *Journal of Neuroscience, 72*, 231–244.

Bannerman, D. M., Rawlins, J. N. P., McHugh, S. B., Deacon, R. M. J., Yee, B. K., Bast, T., . . . Feldon, J. (2004). Regional dissociations within the hippocampus—memory and anxiety. *Neuroscience and Biobehavioural Reviews, 28*, 273–283.

Banno, K., & Kryger, M. H. (2007). Sleep apnea: Clinical investigations in humans. *Sleep Medicine, 8*, 400–426.

Bard P. (1929). The central representation of the sympathetic system. *Archives of Neurology and Psychiatry, 22*, 230–246.

Baron-Cohen, S., & Belmonte, M. K. (2005). Autism: A window onto the development of the social and the analytic brain. *Annual Review of Psychology, 28*, 109–126.

Barrett, L. F. (2006). Are emotions natural kinds? *Perspectives on Psychological Science, 1*, 28–58.

Barrett, L. F., & Wager, R. D. (2006). The structure of emotion: Evidence from neuroimaging studies. *Current Directions in Psychological Science, 15*, 79–83.

Barros, L. F., Porras, O. H., & Bittner, C. X. (2005). Why glucose transport in the brain matters for PET. *Trends in Neurosciences, 3*, 117–119.

Barry, C., Hayman, R., Burgess, N., & Jeffery, K. J. (2007). Experience-dependent rescaling of entorhinal grids. *Nature Neuroscience, 10*, 682–684.

Bartels, A., Logothetis, N K., & Moutoussis, K. (2008). fMRI and its interpretation: An illustration on directional selectivity in area V5/MT. *Trends in Neurosciences, 31*, 444–453.

Basbaum, A. I., & Fields, H. L. (1978). Endogenous pain control mechanisms: Review and hypothesis. *Annals of Neurology, 4*, 451–462.

Basbaum, A. I., & Julius, D. (June, 2006). Control. *Scientific American, 294*, 61–67.

Bashir, Z. I., & Collingridge, G. L. (1992). Synaptic plasticity: Longterm potentiation in the hippocampus. *Current Opinion in Neurobiology, 2*, 328–335.

Bastiaansen, J. A. C. J., Thioux, M., & Keysers, C. (2009). Evidence for mirror systems in emotions. *Philosophical Transactions of the Royal Society B, 364*, 2391–2404.

Bastian, A. J. (2006). Learning to predict the future: The cerebellum adapts feedforward movement control. *Current Opinion in Neurobiology, 16*, 645–649.

Batterham, R. L., Heffron, H., Kapoor, S., Chivers, J. E., Chandarana, K., Herzog, H., . . . Withers, D. J. (2006). Critical role for peptide YY in protein-mediated satiation and body-weight regulation. *Cell Metabolism, 4*, 223–233.

Baum, M. J. (2006). Mammalian animal models of psychosexual differentiation: When is "translation" to the human situation possible? *Hormones and Behavior, 50*, 579–588.

Baum, M. J., Erskine, M. S., Kornberg, E., & Weaver, C. E. (1990). Prenatal and neonatal testosterone exposure interact to affect differentiation of sexual behavior and partner preference in female ferrets. *Behavioral Neuroscience, 104*, 183–198.

Baura, G., Foster, D., Porte, D., Kahn, S. E., Bergman, R. N., Cobelli, C., & Schwartz, M. W. (1993). Saturable transport of insulin from plasma into the central nervous system of dogs in vivo: A mechanism for regulated insulin delivery to the brain. *Journal of Clinical Investigations, 92*, 1824–1830.

Bavelier, D., Corina, D., Jessard, P., Padmanabhan, S., Clark, V. P., Karni, A., et al. (1997). Sentence reading: A functional MRI study at 4 tesla. *Journal of Cognitive Neuroscience, 9*, 664–686.

Baxter, M. G., & Chiba, A. A. (1999). Cognitive functions of the basal forebrain. *Current Opinion in Neurobiology, 9*, 178–183.

Baylen, C. A., & Rosenberg, H. (2006). A review of the acute subjective effects of MDMA/ecstasy. *Addiction, 101*, 933–947.

Baynes, K., & Gazzaniga, M. S. (1997). Callosal disconnection. In T. E. Feinberg & M. J. Farah (Eds.), *Behavioral neurology and neuropsychology* (pp. 419–425). New York, NY: McGraw-Hill.

Bays, P. M., & Husain, M. (2008). Dynamic shifts of limited working memory resources in human vision. *Science, 321*, 851–854.

Beal, M. F., & Ferrante, R. J. (2004). Experimental therapeutics in transgenic mouse models of Huntington's disease. *Nature Reviews Neuroscience, 5*, 373–384.

Bean, B. P. (2007). The action potential in mammalian central neurons. *Nature Reviews Neuroscience, 8,* 451–465.

Beaton, A. A. (2003). The nature and determinants of handedness. In K. Hugdahl and R. J. Davidson (Eds.), *The asymmetrical brain* (pp. 105–158). Cambridge, MA: Bradford Books/MIT Press.

Bechara, A., Tranel, D., Damasio, H., Adolphs, R., Rockland, C., & Damasio, A. R. (1995). Double dissociation of conditioning and declarative knowledge relative to the amygdala and hippocampus in humans. *Science, 269,* 1115–1118.

Beckers, G., & Hömberg, V. (1992). Cerebral visual motion blindness: Transitory akinetopsia induced by transcranial magnetic stimulation of human area V5. *Proceedings. Biological Sciences/The Royal Society, 249,* 173–178.

Beckers, G., & Zeki, S. (1995). The consequences of inactivating areas V1 and V5 on visual motion perception. *Brain, 118,* 49–60.

Beeman, M. J., & Chiarello, C. (1998). Complementary right- and left-hemisphere language comprehension. *Current Directions in Psychological Science, 7,* 2–8.

Beer, J. S., Knight, R. T., & Esposito, M. (2006). Controlling the integration of emotion and cognition: The role of frontal cortex in distinguishing helpful from hurtful emotional information. *Psychological Science, 17,* 448–453.

Begun, D. R. (2004). The earliest hominins—is less more? *Science, 303,* 1478–1480.

Behl, C. (2002). Oestrogen as a neuroprotective hormone. *Nature Reviews Neuroscience, 3,* 433–442.

Behrensmeyer, A. K. (2006). Climate change and human evolution. *Science, 311,* 476–478.

Behrmann, M., Avidan, G., Marotta, J. J., & Kimchi, R. (2005). Detailed exploration of face-related processing in congenital prosopagnosia: 1. Behavioral findings. *Journal of Cognitive Neuroscience, 17,* 1130–1149.

Bell, C. C., Han, V., & Sawtell, N. B. (2008). Cerebellum-like structures and their implications for cerebellar function. *Annual Review of Neuroscience, 31,* 1–24.

Ben-Ari, Y. (2008). Neuro-archeology: Pre-symptomatic architecture and signature of neurological disorders. *Trends in Neurosciences, 31,* 626–636.

Benderly, B. L. (2007, October). Experimental drugs on trial. *Scientific American, 297,* 93–99.

Bendor, D., & Wang, S. (2005). The neuronal representation of pitch in primate auditory cortex. *Nature, 436,* 1161–1165.

Bendor, D., & Wang, S. (2006). Cortical representations of pitch in monkeys and humans. *Current Opinion in Neurobiology, 16,* 391–399.

Benedetti, A., Parent, M., & Siemiatycki, J. (2009). Lifetime consumption of alcoholic beverages and risk of 13 types of cancer in men: Results from a case-control study in Montreal. *Cancer Detection and Prevention, 32,* 352–362.

Benedict, R. H. B., & Bobholz, J. H. (2007). Multiple sclerosis. *Seminars in Neurology, 27,* 78–85.

Benington, J. H., & Heller, H. C. (1999). Implications of sleep deprivation experiments for our understanding of sleep homeostasis. *Sleep, 22,* 1033–1043.

Benjamin, L. T., Jr. (2003). Behavioral science and Nobel Prize: A history. *American Journal of Psychology, 58,* 731–741.

Bennett, M. V. L. (2000). Electrical synapses, a personal perspective (or history). *Brain Research Reviews, 32,* 16–28.

Bennett, M. V. L., Contreras, J. E., Bukauskas, F. F., & Saez, J. C. (2003). New roles for astrocytes: Gap junction hemichannels have something to communicate. *Trends in Neurosciences, 26,* 610–617.

Benninghoven, D., Jürgens, E., Mohr, A., Heberlein, I., Kunzendorf, S., & Jantschek, G. (2006). Different changes of body-images in patients with anorexia or bulimia nervosa during inpatient psychosomatic treatment. *European Eating Disorders Review, 14,* 88–96.

Benowitz, N. L. (2008). Clinical pharmacology of nicotine: Implications for understanding, preventing, and treating tobacco addiction. *Clinical Pharmacology and Therapeutics, 83,* 531–541.

Bensafi, M., Zelano, C., Johnson, B., Mainland, J., Khan, R., & Sobel, N. (2004). Olfaction: From sniff to percept. In M. S. Gazzaniga (Ed.), *The cognitive neurosciences* (pp. 259–280). Cambridge, MA: MIT Press.

Benson, D. F. (1985). Aphasia. In K. M. Heilman & E. Valenstein (Eds.), *Clinical neuropsychology* (pp. 17–47). New York, NY: Oxford University Press.

Benton, A. L. (1994). Neuropsychological assessment. *Annual Review of Psychology, 45,* 1–23.

Berger, R. J., & Oswald, I. (1962). Effects of sleep deprivation on behaviour, subsequent sleep, and dreaming. *Journal of Mental Science, 106,* 457–465.

Bergman, T. J., Beehner, J. C., Cheney, D. L., & Seyfarth, R. M. (2003). Hierarchical classification by rank and kinship in baboons. *Science, 302,* 1234–1236.

Berkovic, S. F., Mulley, J. C., Scheffer, I. E., & Petrou, S. (2006). Human epilepsies: Interaction of genetic and acquired factors. *Trends in Neurosciences, 29,* 391–397.

Bernard, C., Anderson, A., Becker, A., Poolos, N. P., Beck, H., & Johnston, D. (2004). Acquired dendritic channelopathy in temporal lobe epilepsy. *Science, 305,* 532–535.

Bernhardt, P. C. (1997). Influences of serotonin and testosterone in aggression and dominance: Convergence with social psychology. *Current Directions in Psychological Science, 6,* 44–48.

Bernstein, I. L., & Webster, M. M. (1980). Learned taste aversion in humans. *Physiology & Behavior, 25,* 363–366.

Berridge, K. C. (2004). Motivation concepts in behavioral neuroscience. *Physiology & Behavior, 81,* 179–209.

Berridge, K. C., & Robinson, T. E. (2003). Parsing reward. *Trends in Neurosciences, 26,* 507–513.

Berridge, K. C., Robinson, T. E., & Aldridge, J. W. (2009). Dissecting components of reward: "Liking," "wanting," and learning. *Current Opinion in Pharmacology, 9,* 65–73.

Berson, D. M. (2003). Strange vision: Ganglion cells as circadian photoreceptors. *Trends in Neurosciences, 26,* 314–320.

Berthoud, H.-R. (2002). Multiple neural systems controlling food intake and body weight. *Neuroscience and Biobehavioural Reviews, 26,* 393–428.

Berthoud, H-R., & Morrison, C. (2008). The brain, appetite, and obesity. *Annual Review of Psychology, 59,* 55–92.

Berti, A., Bottini, G., Gandola, M., Pia, L., Smania, N., Stracciari, A., . . . Paulesu, E. (2005). Shared cortical anatomy for motor awareness and motor control. *Science, 309,* 488–491.

Berton, O., & Nestler, E. J. (2006). New approaches to antidepressant drug discovery: Beyond monoamines. *Nature Reviews Neuroscience, 7,* 137–151.

Berton, O., McClund, C. A., DiLeone, R. J., Krishnan, V., Renthal, W., Russo, S. J., . . . Nestler, E. J. (2006). Essential role of BDNF in the mesolimbic dopamine pathway in social defeat stress. *Science, 311,* 864–868.

Bertram, L., & Tanzi, R. E. (2008). Thirty years of Alzheimer's disease genetics: The implications of systematic meta-analyses. *Nature Reviews Neuroscience, 9,* 768–778.

Bestmann, S. (2007). The physiological basis of transcranial magnetic stimulation. *Trends in Cognitive Sciences, 12,* 81–83.

Bhardwaj, R. D., Curtis, M. A., Spalding, K. L., Buchholz, B. A., Fink, D., Björk-Eriksson, T., . . . Frisén, J. (2006). Neocortical neurogenesis in humans is restricted to development. *Proceedings of the National Academy of Sciences, USA,* 12564–12568.

Bi, G.-Q., & Poo, M.-M. (2001). Synaptic modification by correlated activity: Hebb's postulate revisited. *Annual Review of Neuroscience, 24,* 139–166.

Biber, K., Neumann, H., Inoue, K., & Boddeke, H. W. G. M. (2007). Neuronal "on" and "off" signals control microglia. *Trends in Neurosciences, 30,* 596–602.

Biedenkapp, J. C., & Rudy, J. W. (2004). Context memories and reactivation: Constraints on the reconsolidation hypothesis. *Behavioral Neuroscience, 118,* 956–964.

Biglan, K. M., & Ravina, B. (2007). Neuroprotection in Parkinson's disease: An elusive goal. *Seminars in Neurology, 27,* 106–112.

Billock, V. A., & Tsou, B. H. (2004). What do catastrophic visual binding failures look like? *Trends in Neurosciences, 27,* 84–89.

Billock, V. A., & Tsou, B. H. (2010, February). Seeing forbidden colors. *Scientific American, 302,* 72–77.

Binder, J. R., Westbury, C. F., McKiernan, K. A., Possing, E. T., & Medler, D. A. (2005). Distinct brain systems for processing concrete and abstract concepts. *Journal of Cognitive Neuroscience, 17,* 905–917.

Bird, A. (2007). Perceptions of epigenetics. *Nature, 447,* 396–398.

Bird, C. M., & Burgess, N. (2008). The hippocampus and memory: Insights from spatial processing. *Nature Reviews Neuroscience, 9,* 182–194.

Birmingham, C. L., Su, J., Hlynsky, J. A., Goldner, E. M., & Gao, M. (2005). The mortality rate from anorexia nervosa. *International Journal of Eating Disorders, 38,* 143–146.

Bisagno, V., Bowman, R., & Luine, V. (2003). Functional aspects of estrogen neuroprotection. *Endocrine, 21,* 33–41.

Bischoff-Grethe, A., Proper, S. M., Mao, H., Daniels, K. A., & Berns, G. S. (2000). Conscious and unconscious processing of nonverbal predictability in Wernicke's area. *Journal of Neuroscience, 20,* 1975–1981.

Bishop, D. V. M. (1999). An innate basis for language? *Science, 286,* 2283–2355.

Bishop, S. J. (2007). Neurocognitive mechanisms of anxiety: An integrative account. *Trends in Cognitive Sciences, 11,* 307–316.

Björklund, A., & Lindvall, O. (2000). Cell replacement therapies for central nervous system disorders. *Nature, 3,* 537–544.

Black, J. E., Isaacs, K. R., Anderson, B. J., Alcantara, A. A., & Greenhough, W. T. (1990). Learning causes synaptogenesis, whereas motor activity causes angiogenesis, in cerebellar cortex of adult rats. *Proceedings of the National Academy of Science, USA, 87,* 5568–5572.

Blackburn, G. L. (2001). Pasteur's quadrant and malnutrition. *Nature, 409,* 397–401.

Blagrove, M., Alexander, C., & Horne, J. A. (2006). The effects of chronic sleep reduction on the performance of cognitive tasks sensitive to sleep deprivation. *Applied Cognitive Psychology, 9,* 21–40.

Blakemore, S-J. (2008). The social brain in adolescence. *Nature Reviews Neuroscience, 9,* 267–277.

Blanchard, D. C., & Blanchard, R. J. (1984). Affect and aggression: An animal model applied to human behavior. In D. C. Blanchard & R. J. Blanchard (Eds.), *Advances in the study of aggression* (pp. 1–62). Orlando, FL: Academic Press.

Blanchard, D. C., & Blanchard, R. J. (1988). Ethoexperimental approaches to the biology of emotion. *Annual Review of Psychology, 39,* 43–68.

Blanchard, D. C., & Blanchard, R. J. (1990). Behavioral correlates of chronic dominance-subordination relationships of male rats in a seminatural situation. *Neuroscience and Biobehavioural Reviews, 14,* 455–462.

Blanchard, D. C., Blanchard, R. J., & Rodgers, R. J. (1991). Risk assessment and animal models of anxiety. In J. Olivier, J. Mos, & J. L. Slangen (Eds.), *Animal models in psycho-pharmacology* (pp. 117–134). Basel, Switzerland: Birkhauser Verlag.

Blanchard, D. C., Blanchard, R. J., Tom, P., & Rodgers, R. J. (1990). Diazepam changes risk assessment in an anxiety/defense test battery. *Psychopharmacology, 101,* 511–518.

Blanchard, D. C., Hynd, A. L., Minke, K. A., Minemoto, T., & Blanchard, R. J. (2001). Human defensive behaviors to threat scenarios show parallels to fear and anxiety-related defense patterns of non-human mammals. *Neuroscience and Biobehavioural Reviews, 25,* 761–770.

Blanchard, R. (2004). Quantitative and theoretical analyses of the relation between older brothers and homosexuality in men. *Journal of Theoretical Biology, 230,* 173–187.

Blanchard, R. J., & Blanchard, D. C. (1989). Anti-predator defensive behaviors in a visible burrow system. *Journal of Comparative Psychology, 103,* 70–82.

Blanchard, R., & Lippa, R. A. (2007). Birth order, sibling sex ratio, handedness, and sexual orientation of male and female participants in a BBC internet research project. *Archives of Sexual Behavior, 36,* 163–176.

Blanke, O., Landis, T., Safran, A. B., & Seeck, M. (2002). Direction-specific motion blindness induced by focal stimulation of human extrastriate cortex. *European Journal of Neuroscience, 15,* 2043–2048.

Blankenship, A. G., & Feller, M. B. (2010). Mechanisms underlying spontaneous patterned activity in developing neural circuits. *Nature Reviews Neuroscience, 11,* 18–29.

Blaser, M. J. (1996, February). The bacteria behind ulcers. *Scientific American, 275,* 104–107.

Blaustein, J. D. (2008). Neuroendocrine regulation of feminine sexual behavior: Lessons from rodent models and thoughts about humans. *Annual Review of Psychology, 59,* 93–118.

Blaustein, J. D., King, J. C., Toft, D. O., & Turcotte, J. (1988). Immunocytochemical localization of estrogen-induced progestin receptors in guinea pig brain. *Brain Research, 474,* 1–15.

Bliss, T. V. P., & Lømø, T. (1973). Long-lasting potentiation of synaptic transmission in the dentate area of the anaesthetized rabbit following stimulation of the perforant path. *Journal of Physiology, 232,* 331–356.

Bliss, T., & Schoepfer, R. (2004). Controlling the ups and downs of synaptic strength. *Science, 304,* 973–974.

Bloch, G. J., & Gorski, R. A. (1988). Cytoarchitectonic analysis of the SDN-POA of the intact and gonadectomized rat. *Journal of Comparative Neurology, 275,* 604–612.

Bloch, G. J., & Mills, R. (1995). Prepubertal testosterone treatment of neonatally gonadectomized male rats: Defeminization and masculinization of behavioral and endocrine function in adulthood. *Neuroscience and Behavioural Reviews, 19,* 187–200.

Bloch, G. J., Mills, R., & Gale, S. (1995). Prepubertal testosterone treatment of female rats: Defeminization of behavioral and endocrine function in adulthood. *Neuroscience and Biobehavioural Reviews, 19,* 177–186.

Bloom, P., & Weisberg, D. S. (2007). Childhood origins of adult resistance to science. *Science, 316,* 996–997.

Bloomfield, S. A., & Völgyi, B. (2009). The diverse functional roles and regulation of neuronal gap junctions in the retina. *Nature Reviews Neuroscience, 10,* 495–506.

Blundell, J. E., & Finlayson, G. (2004). Is susceptibility to weight gain characterized by homeostatic or hedonic risk factors for overconsumption? *Physiology & Behavior, 82,* 21–25.

Blundell, J. E., & Halford, J. C. G. (1998). Serotonin and appetite regulation. *CNS Drugs, 9,* 473–495.

Boehning, D., & Snyder, S. H. (2003). Novel neural modulators. *Annual Review of Neuroscience, 26,* 105–131.

Bogaert, A. F. (2006). Biological versus nonbiological older brothers and men's sexual orientation. *Proceedings of the National Academy of Sciences, USA, 103,* 10771–10774.

Bogen, J. G., & Bogen, G. M. (1976). Wernicke's region—where is it? *Annals of the New York Academy of Science, 280,* 834–843.

Bolles, R. C. (1980). Some functionalistic thought about regulation. In F. M. Toates & T. R. Halliday (Eds.), *Analysis of motivational processes* (pp. 63–75). London, England: Academic Press.

Bölte, S., Hubl, D., Feineis-Matthews, S., Prvulovic, D., Dierks, T., & Poustka, F. (2006). Facial affect recognition training in autism: Can we animate the fusiform gyrus? *Behavioral Neuroscience, 120,* 211–216.

Bolton, M. M., & Eroglu, C. (2009). Look who is weaving the neural web: Glial control of synapse formation. *Current Opinion in Neurobiology, 19,* 491–497.

Bonnel, A., Mottron, L., Peretz, I., Trudel, M., Gallun, E., & Bonnel, A. M. (2003). Enhanced pitch sensitivity in individuals with autism: A signal detection analysis. *Journal of Cognitive Neuroscience, 15,* 226–235.

Bonnet, M. H., & Arand, D. L. (2003). Clinical effects of sleep fragmentation versus sleep deprivation. *Sleep Medicine Reviews, 7,* 297–310.

Booth, D. A. (1981). The physiology of appetite. *British Medical Bulletin, 37,* 135–140.

Booth, D. A. (2004). How observations on oneself can be scientific. *Behavioral and Brain Sciences, 27,* 262–263.

Booth, D. A., Fuller, J., & Lewis, V. (1981). Human control of body weight: Cognitive or physiological? Some energy-related perceptions and misperceptions. In L. A. Cioffi (Ed.), *The body weight regulatory system: Normal and disturbed systems* (pp. 305–314). New York, NY: Raven Press.

Booth, M. (2004). *Cannabis: A history.* London, England: Bantam Books.

Borbély, A. A. (1981). The sleep process: Circadian and homeostatic aspects. *Advances in Physiological Sciences, 18,* 85–91.

Borbély, A. A. (1983). Pharmacological approaches to sleep regulation. In A. R. Mayes (Ed.), *Sleep mechanisms and functions in humans and animals* (pp. 232–261). Wokingham, England: Van Nostrand Reinhold.

Borbély, A. A., Baumann, F., Brandeis, D., Strauch, I., & Lehmann, D. (1981). Sleep deprivation: Effect on sleep stages and EEG power density in man. *Electroencephalography and Clinical Neurophysiology, 51*, 483–493.

Bossy-Wetzel, E., Petrilli, A., & Knott, A. B. (2008). Mutant huntingtin and mitochondrial dysfunction. *Trends in Neurosciences, 31*, 609–616.

Bostwick, J. M., & Pankratz, V. S. (2000). Affective disorders and suicide risk: A reexamination. *American Journal of Psychiatry, 157*, 1925–1932.

Botly, L. C. P., & De Rosa, E. (2008). A cross-species investigation of acetylcholine, attention, and feature binding. *Psychological Science, 19*, 1185–1193.

Bottjer, D. J. (2005, August). The early evolution of animals. *Scientific American, 293*, 43–47.

Bouchard, T. J., Jr. (1998). Genetic and environmental influences on adult intelligence and special mental abilities. *Human Biology, 70*, 257–279.

Bouchard, T. J., Jr., & Pedersen, N. (1998). Twins reared apart: Nature's double experiment. In E. L. Grigorenko & S. Scarr (Eds.), *On the way to individuality: Current methodological issues in behavioral genetics.* Commack, NY: Nova Science.

Boulenguez, P., & Vinay, L. (2009). Strategies to restore motor functions after spinal cord injury. *Current Opinion in Neurobiology, 19*, 587–600.

Bourgeron, T. (2009). A synaptic trek to autism. *Current Opinion in Neurobiology, 19*, 231–234.

Bourin, M., & Prica, C. (2007). The role of mood stabilisers in the treatment of the depressive facet of bipolar disorder. *Neuroscience and Biobehavioural Reviews, 31*, 963–975.

Bowers, D., Bauer, R. M., Coslett, H. B., & Heilman, K. M. (1985). Processing of face by patients with unilateral hemisphere lesions. I. Dissociations between judgements of facial affect and facial identity. *Brain and Cognition, 4*, 258–272.

Bowler, P. J. (2009). Darwin's originality. *Science, 323*, 223–226.

Boyden, E. S., Katoh, A., & Raymond, J. L. (2004). Cerebellum-dependent learning: The role of multiple plasticity mechanisms. *Annual Review of Neuroscience, 27*, 581–609.

Brackett, N. L., & Edwards, D. A. (1984). Medial preoptic connections with the midbrain tegmentum are essential for male sexual behavior. *Physiology & Behavior, 32*, 79–84.

Brambati, S. M., Myers, D., Wilson, A., Rankin, K. P., Allison, S. C., Rosen, H. J., . . . Gorno-Tempini, M. L. (2006). The anatomy of category-specific object naming in neurodegenerative diseases. *Journal of Cognitive Neuroscience, 18*, 1644–1653.

Brambilla, C., & Colonna, M. (2008). Cannabis: The next villain on the lung cancer battlefield? *European Respiratory Journal, 31*, 227–228.

Branch, G., & Scott, E. C. (2009, January). The latest face of creationism. *Scientific American, 300*, 92–99.

Brass, M., Ullsperger, M., Knoesche, T. R., von Cramon, D. Y., & Phillips, N. A. (2005). Who comes first? The role of the prefrontal and parietal cortex in cognitive control. *Journal of Cognitive Neuroscience, 17*, 1367–1375.

Brecher, E. M. (1972). *Licit and illicit drugs.* Boston, MA: Little, Brown.

Brehmer, C., & Iten, P. X. (2001). Medical prescription of heroin to chronic heroin addicts in Switzerland—a review. *Forensic Science International, 121*, 23–26.

Breitner, J. C. S. (1990). Life table methods and assessment of familial risk in Alzheimer's disease. *Archives of General Psychiatry, 47*, 395–396.

Bremer, F. (1936). Nouvelles recherches sur le mécanisme du sommeil. *Comptes rendus de la Société de Biologie, 22*, 460–464.

Bremer, F. L. (1937). L'activité cérébrale au cours du sommeil et de la narcose. Contribution à l'étude du mécanisme du sommeil. *Bulletin de l'Académie Royale de Belgique, 4*, 68–86.

Bremer, J. (1959). *Asexualization.* New York, NY: Macmillan.

Brenowitz, E. A. (2004). Plasticity of the adult avian song control system. *Annals of the New York Academy of Sciences, 1016*, 749–777.

Brewer, C. (2007). Controlled trials of antabuse in alcoholism: The importance of supervision and adequate dosage. *Acta Psychiatrica Scandinavica, 86*, 51–58.

Brewster, J. M. (1986). Prevalence of alcohol and other drug problems among physicians. *Journal of the American Medical Association, 255*, 1913–1920.

Bridge, J. A., Iyengar, S., Salary, C. B., Barbe, R. P., Birmaher, B., Pincus, H. A., . . . Brent, D. A. (2007). Clinical response and risk for reported suicidal ideation and suicide attempts in pediatric antidepressant treatment. *Journal of the American Medical Association, 297*, 1683–1696.

Britten, K. H. (2008). Mechanisms of self-motion perception. *Annual Review of Neuroscience, 31*, 389–410.

Brittle, E. R., & Waters, M. G. (2000). ER-to-Golgi traffic—this bud's for you. *Science, 289*, 403–448.

Brivanlou, A. H., Gage, F. H., Jaenisch, R., Jessell, T., Melton, D., & Rossant, J. (2003). Setting standards for human embryonic stem cells. *Science, 300*, 913–916.

Brochier, T., & Umiltà, M. A. (2007). Cortical control of grasp in non-human primates. *Current Opinion in Neurobiology, 17*, 637–643.

Brody, D. L., & Holtzman, D. M. (2008). Active and passive immunotherapy for neurodegenerative disorders. *Annual Review of Neuroscience, 31*, 175–193.

Broks, P., Young, A. W., Maratos, E. J., Coffey, P. J., Calder, A. J., Isaac, C. L., . . . Hadley, D. (1998). Face processing impairments after encephalitis: Amygdala damage and recognition of fear. *Neuropsychologia, 36*, 59–70.

Bronson, F. H., & Matherne, C. M. (1997). Exposure to anabolic-androgenic steroids shortens life span of male mice. *Medicine and Science in Sports and Exercise, 29*, 615–619.

Brooks, M. J., & Melnik, G. (1995). The refeeding syndrome: An approach to understanding its complications and preventing its occurrence. *Pharmacotherapy, 15*, 713–726.

Brooks, S. P., & Dunnett, S. B. (2009). Tests to assess motor phenotype in mice: A user's guide. *Nature Reviews Neuroscience, 10*, 519–528.

Brooks, V. B. (1986). *The neural basis of motor control.* New York, NY: Oxford University Press.

Brown, G. W. (1993). The role of life events in the aetiology of depressive and anxiety disorders. In S. C. Stanford & S. Salmon (Eds.) *Stress: From synapse to syndrome* (pp. 23–50). San Diego, CA: Academic Press.

Brown, H. D., & Kosslyn, S. M. (1993). Cerebral lateralization. *Current Opinion in Neurobiology, 3*, 183–186.

Brown, J. W., & Braver, T. S. (2005). Learned predictions of error likelihood in the anterior cingulate cortex. *Science, 307*, 1118–1121.

Brown, M. W., & Aggleton, J. P. (2001). Recognition memory: What are the roles of the perirhinal cortex and hippocampus? *Nature Reviews Neuroscience, 2*, 51–61.

Brown, R. E. (1994). *An introduction to neuroendocrinology.* Cambridge, England: Cambridge University Press.

Brown, R. E., & Milner, P. M. (2003). The legacy of Donald Hebb: More than the Hebb synapse. *Nature Reviews Neuroscience, 4*, 1013–1019.

Brownell, K. D., & Rodin, J. (1994). The dieting maelstrom: Is it possible and advisable to lose weight? *American Psychologist, 49*, 781–791.

Brummelte, S., Pawluski, J. L., & Galea, L. A. M. (2006). High post-partum levels of corticosterone given to dams influence postnatal hippocampal cell proliferation and behavior of offspring: A model of post-partum stress and possible depression. *Hormones and Behavior, 50*, 370–382.

Brundin, P., Li, J-Y., Holton, J. L., Lindvall, O., & Revesz, T. (2008). Research in motion: The enigma of Parkinson's disease pathology spread. *Nature Reviews Neuroscience, 9*, 741–745.

Brunner, D. P., Dijk, D-J., Tobler, I., & Borbély, A. A. (1990). Effect of partial sleep deprivation on sleep stages and EEG power spectra: Evidence for non-REM and REM sleep homeostaşis. *Electroencephalography and Clinical Neurophysiology, 75*, 492–499.

Bruno, R. M., & Sakmann, B. (2006). Cortex is driven by weak but synchronously active thalamocortical synapses. *Science, 312*, 1622–1627.

Brzezinski, A., Vangel, M. G., Wurtman, R. J., Norrie, G., Zhdanova, I., Ben-Shushan, A., & Ford, I. (2005). Effects of exogenous melatonin on sleep: A meta-analysis. *Sleep Medicine Reviews, 9*, 41–50.

Bucci, T. J. (1992). Dietary restriction: Why all the interest? An overview. *Laboratory Animal, 21*, 29–34.

Buckley, M. J., & Gaffan, D. (1998). Perirhinal cortex ablation impairs visual object identification. *Journal of Neuroscience, 18*(6), 2268–2275.

Bulkin, D. A., & Groh, J. M. (2006). Seeing sounds: Visual and auditory interactions in the brain. *Current Opinion in Neurobiology, 16*, 415–419.

Bunge, S. A., & Zelazo, P. D. (2006). A brain-based account of the development of rule use in childhood. *Current Directions in Psychological Science, 15,* 118–121.

Bunin, M. A., & Wightman, R. M. (1999). Paracrine neurotransmission in the CNS: Involvement of 5-HT. *Trends in Neurosciences, 22,* 377–382.

Buonomano, D. V., & Merzenich, M. M. (1998). Cortical plasticity: From synapses to maps. *Annual Reviews of Neuroscience, 21,* 149–186.

Burbach, J. P. H., & van der Zwaag, B. (2008). Contact in the genetics of autism and schizophrenia. *Trends in Neurosciences, 32,* 69–72.

Burkhardt, R. W., Jr. (2005). *Konrad Lorenz, Niko Tinbergen, and the founding of ethology.* Chicago, IL: University of Chicago Press.

Buss, D. M. (1992). Mate preference mechanisms: Consequences for partner choice and intrasexual competition. In J. M. Barkow, L. Cosmides, & J. Tooby (Eds.), *The adapted mind* (pp. 249–265). New York, NY: Oxford University Press.

Bussey, T. J., & Saksida, L. M. (2005). Object memory and perception in the medial temporal lobe: An alternative approach. *Current Opinion in Neurobiology, 15,* 730–737.

Bussey, T. J., Warburton, E. C., Aggleton, J. P., & Muir, J. L. (1998). Fornix lesions can facilitate acquisition of the transverse patterning task: A challenge for "configural" theories of hippocampal function. *Journal of Neuroscience, 18*(4), 1622–1631.

Buster, J. E., Kingsberg, S. A., Aguirre, O., Brown, C., Breaux, J. G., Buch, A., . . . Casson, P. (2005). Testosterone patch for low sexual desire in surgically menopausal women: A randomized trial. *Obstetrics & Gynecology, 105,* 944–952.

Butters, N., & Delis, D. C. (1995). Clinical assessment of memory disorders in amnesia and dementia. *Annual Review of Psychology, 46,* 493–523.

Buzsáki, G. (2005). Similar is different in hippocampal networks. *Science, 309,* 568–569.

Byers, T. (2006). Overweight and mortality among baby boomers—now we're getting personal. *New England Journal of Medicine, 355,* 758–760.

Bystron, I., Blakemore, C., & Rakic, P. (2008). Development of the human cerebral cortex: Boulder committee revisited. *Nature Reviews Neuroscience, 9,* 110–122.

Cabanac, M. (1971). Physiological role of pleasure. *Science, 173,* 1103–1107.

Cabeza, R., & Kingston, A. (2002). Cognitive neuroimaging for all! *Trends in Neurosciences, 25*(5), 275.

Cacioppo, J. T., & Decety, J. (2009). What are the brain mechanisms on which psychological processes are based? *Perspectives on Psychological Science, 4,* 10–18.

Cafferty, W. B. J., McGee, A. W., & Strittmatter, S. M. (2008). Axonal growth therapeutics: Regeneration or sprouting or plasticity? *Trends in Neurosciences, 31,* 215–220.

Caggiula, A. R. (1970). Analysis of the copulation-reward properties of posterior hypothalamic stimulation in male rats. *Journal of Comparative and Physiological Psychology, 70,* 399–412.

Cahill, L. (2006). Why sex matters for neuroscience. *Nature Reviews Neuroscience, 7,* 477–484.

Cahill, L. (May, 2005). His brain, her brain. *Scientific American, 292,* 40–47.

Cain, D. P. (1997). LTP, NMDA, genes and learning. *Current Opinion in Neurobiology, 7,* 235–242.

Calder, A. J., Young, A. W., Rowland, D., Perrett, D. I., Hodges, J. R., & Etcoff, N. L. (1996). Facial emotion recognition after bilateral amygdala damage: Differentially severe impairment of fear. *Cognitive Neuropsychology, 13,* 699–745.

Calhoun, F., & Warren, K. (2007). Fetal alcohol syndrome: Historical perspectives. *Neuroscience and Biobehavioral Reviews, 31,* 168–171.

Calles-Escandon, J., & Horton, E. S. (1992). The thermogenic role of exercise in the treatment of morbid obesity: A critical evaluation. *American Journal of Clinical Nutrition, 55,* 533S–537S.

Calmon, G., Roberts, N., Eldridge, P., & Thirion, J-P. (1998). Automatic quantification of changes in the volume of brain structures. *Lecture Notes in Computer Science, 1496,* 761–769.

Calne, S., Schoenberg, B., Martin, W., Uitti, J., Spencer, P., & Calne, D. B. (1987). Familial Parkinson's disease: Possible role of environmental factors. *Canadian Journal of Neurological Sciences, 14,* 303–305.

Calton, J. L., & Taube, J. S. (2009). Where am I and how will I get there from here? A role for the posterior parietal cortex in the integration of spatial information and route planning. *Neurobiology of Learning and Memory, 91,* 186–196.

Cameron, H. A., Woolley, C. S., McEwen, B. S., & Gould, E. (1993). Differentiation of newly born neurons and glia in the dentate gyrus of the adult rat. *Neuroscience, 56,* 337–344.

Campbell, K., & Gotz, M. (2002). Radial glia: Multi-purpose cells for vertebrate brain development. *Trends in Neurosciences, 25,* 235–238.

Campfield, L. A., & Smith, F. J. (1990). Transient declines in blood glucose signal meal initiation. *International Journal of Obesity, 14* (Suppl. 3), 15–33.

Campfield, L. A., Smith, F. J., Gulsez, Y., Devos, R., & Burn, P. (1995). Mouse Ob protein: Evidence for a peripheral signal linking adiposity and central neural networks. *Science, 269,* 546–550.

Canli, T., Sivers, H., Whitfield, S. L., Gotlib, I. H., & Gabrieli, J. D. E. (2002). Amygdala response to happy faces as a function of extraversion. *Science, 296,* 2191.

Cannon, W. B., & Washburn, A. L. (1912). An explanation of hunger. *American Journal of Physiology, 29,* 441–454.

Cantor, J. M., Blanchard, R., Paterson, A. D., & Bogaert, A. F. (2004). How many gay men owe their sexual orientation to fraternal birth order? *Archives of Sexual Behavior, 31,* 63–71.

Capaday, C. (2002). The special nature of human walking and its neural control. *Trends in Neurosciences, 25,* 370–376.

Caplin, A., & Dean, M. (2008). Axiomatic methods, dopamine and reward prediction error. *Current Opinion in Neurobiology, 18,* 197–202.

Caporael, L. R. (2001). Evolutionary psychology: Toward a unifying theory and a hybrid science. *Annual Review of Psychology, 52,* 607–628.

Cappuccio, F. P., D'Elia, L., Strazzullo, P., & Miller, M. A. (2008). Sleep duration and all-cause mortality: A systematic review and meta-analysis of prospective studies. *Sleep, 33,* 585–592.

Cardinal, R. N., & Everitt, B. J. (2004). Neural and psychological mechanisms underlying appetitive learning: Links to drug addiction. *Current Opinion in Neurobiology, 14,* 156–162.

Carlson, G. A., & Meyer, S. E. (2006). Phenomenology and diagnosis of bipolar disorder in children, adolescents, and adults: Complexities and developmental issues. *Development and Psychopathology, 18,* 939.

Carlsson, A., & Lindqvist, M. (1963). Effect of chlorpromazine or haloperidol on formation of 3-methoxytyramine and normetanephrine in mouse brains. *Acta Pharmacologica et Toxicologica, 20,* 140–144.

Carlsson, K., Petrovic, P., Skare, S., Petersson, K. M., & Ingvar, M. (2000). Tickling expectations: Neural processing in anticipation of a sensory stimulus. *Journal of Cognitive Neuroscience, 12,* 691–703.

Carpenter, L. L., Tyrka, A. R., McDougle, C. J., Malison, R. T., Owens, M. J., Nemeroff, C. B., & Price, L. H. (2004). Cerebrospinal fluid corticotropin-releasing factor and perceived early-life stress in depressed patients and healthy control subjects. *Neuropsychopharmacology, 29,* 777–784.

Carr, L., Iacoboni, M., Dubeau, M-C., Mazziotta, J. C., & Lenzi, G. L. (2003). Neural mechanisms of empathy in humans: A relay from neural systems for imitation to limbic areas. *Proceedings of the National Academy of Sciences, USA, 100,* 5497–5502.

Carroll, S. B., Prud'homme, B., & Gompel, N. (2008, May). Regulating evolution. *Scientific American, 298,* 61–67.

Carson, R. A., & Rothstein, M. A. (1999). *Behavioral genetics: The clash of culture and biology.* New York, NY: Worth.

Caselli, R. J. (1997). Tactile agnosia and disorders of tactile perception. In T. E. Feinberg & M. J. Farah (Eds.), *Behavioral neurology and neuropsychology* (pp. 277–288). New York, NY: McGraw-Hill.

Casey, B. J. (2002). Windows into the human brain. *Science, 296,* 1408–1409.

Casey, B. J., Giedd, J. N., & Thomas, K. M. (2000). Structural and functional brain development and its relation to cognitive development. *Biological Psychology, 54,* 241–257.

Casey, B. J., Tottenham, N., Liston, C., & Durston, S. (2005). Imaging the developing brain: What have we learned about cognitive development? *Trends in Cognitive Sciences, 9,* 104–110.

Cash, S. S., Halgren, E., Dehghani, N., Rossetti, A. O., Thesen, T., Wang, C., . . . Ulbert, I. (2009). The human K-complex represents an isolated cortical down-state. *Science, 324,* 1084–1087.

Cassone, V. M. (1990). Effects of melatonin on vertebrate circadian systems. *Trends in Neurosciences, 13*(11), 457–467.

Cermakian, N., & Sassone-Corsi, P. (2002). Environmental stimulus perception and control of circadian clocks. *Current Opinion in Neurobiology, 12*, 359–365.

Chalfie, M., Tu, Y., Euskirchen, G., Ward, W., & Prasher, D. (1994). Green fluorescent protein as a marker for gene expression. *Science, 263*, 802–805.

Chalupa, L. M., & Huberman, A. D. (2004). A new perspective on the role of activity in the development of eye-specific retinogeniculate projections. In M. S. Gazzaniga (Ed.), *The cognitive neurosciences* (pp. 85–92). Cambridge, MA: MIT Press.

Chan, C. S., Gertler, T. S., & Surmeier, D. J. (2009). Calcium homeostasis, selective vulnerability and Parkinson's disease. *Trends in Neurosciences, 32*, 249–256.

Chang, L., Alicata, D., Ernst, T., & Volkow, N. (2007). Structural and metabolic brain changes in the striatum associated with methamphetamine abuse. *Addiction, 102*, 16–32.

Changeux, J. P., & Edelstein, S. J. (2005). Allosteric mechanisms of signal transduction. *Science, 308*, 1424–1428.

Chao, D. L., Ma, L., & Shen, K. (2009). Transient cell-cell interactions in neural circuit formation. *Nature Reviews Neuroscience, 10*, 262–271.

Chatterjee, S., & Callaway, E. M. (2003). Parallel colour-opponent pathways to primary visual cortex. *Nature, 426*, 668–671.

Chavez, M., Seeley, R. J., & Woods, S. C. (1995). A comparison between the effects of intraventricular insulin and intraperitoneal LiCl on three measures sensitive to emetic agents. *Behavioral Neuroscience, 109*, 547–550.

Chen, E., Hanson, M. D., Paterson, L. Q., Griffin, M. J., Walker, H. A., & Miller, G. E. (2006). Socioeconomic status and inflammatory processes in childhood asthma: The role of psychological stress. *Journal of Allergy and Clinical Immunology, 117*, 1014–1020.

Chen, S-Y., & Cheng, H-J. (2009). Functions of axon guidance molecules in synapse formation. *Current Opinion in Neurobiology, 19*, 471–478.

Chen, W. R., Midtgaard, J., & Shepherd, G. M. (1997). Forward and backward propagation of dendritic impulses and their synaptic control in mitral cells. *Science, 278*, 463–466.

Chen, Y., Grossman, E. D., Bidwell, L. C., Yurgelun-Todd, D., Gruber, S. A., Levy, D. L., . . . Holzman, P. S. (2008). Differential activation patterns of occipital and prefrontal cortices during motion processing: Evidence from normal and schizophrenic brains. *Cognitive, Affective, & Biobehavioral Neuroscience, 8*, 293–303.

Chen, Y., Kelton, C. M., Jing, Y., Guo, J. J., Li, X., & Patel, N. C. (2008). Utilization, price, and spending trends for antidepressants in the US medicaid program. *Research in Social and Administrative Pharmacy, 4*, 244–257.

Chen, Z-L., Yu, W-M., & Strickland, S. (2007). Peripheral regeneration. *Annual Review of Neuroscience, 30*, 209–233.

Cheney, D. L., & Seyfarth, R. M. (2005). Constraints and preadaptations in the earliest stages of language evolution. *Linguistic Review, 22*, 135–159.

Cheng, D. T., Knight, D. C., Sjith, C. N., & Helmstetter, F. J. (2006). Human amygdala activity during the expression of fear responses. *Behavioral Neuroscience, 120*, 1187–1195.

Cheng, J., Cao, Y., & Olson, L. (1996). Spinal cord repair in adult paraplegic rats: Partial restoration of hind limb function. *Science, 273*, 510–513.

Chih, B., Engelman, H., & Scheiffele, P. (2005). Control of excitatory and inhibitory synapse formation by neuroligins. *Science, 307*, 1324–1328.

Chimpanzee Sequencing and Analysis Consortium. (2005). Initial sequence of the chimpanzee genome and comparison with the human genome. *Nature, 437*, 69–87.

Chizhikou, V. V., & Millen, K. J. (2004). Mechanisms of roof plate formation in the vertebrate CNS. *Nature Reviews Neuroscience, 5*, 808–812.

Chklovskii, D. B., & Koulakov, A. A. (2004). Maps in the brain: What can we learn from them? *Annual Review of Neuroscience, 27*, 369–392.

Cho, A. K. (1990). Ice: A new dosage form of an old drug. *Science, 249*, 631–634.

Chorover, S. L., & Schiller, P. H. (1965). Short-term retrograde amnesia in rats. *Journal of Comparative and Physiological Psychology, 59*, 73–78.

Chotard, C., & Salecker, I. (2004). Neurons and glia: Team players in axon guidance. *Trends in Neurosciences, 27*, 655–661.

Churchland, P. C. (2002). Self representation in nervous systems. *Science, 296*, 308–310.

Cirelli, C. (2006). Cellular consequences of sleep deprivation in the brain. *Sleep Medicine Reviews, 10*, 307–321.

Cirelli, C. (2009). The genetic and molecular regulation of sleep: From fruit flies to humans. *Nature Reviews Neuroscience, 10*, 549–560.

Cirulli, F., Berry, A., & Alleva, E. (2003). Early disruption of the mother–infant relationship: Effects on brain plasticity and implications for psychopathology. *Neuroscience and Biobehavioural Reviews, 27*, 73–82.

Citron, M. (2004). Strategies for disease modification in Alzheimer's disease. *Nature Reviews Neuroscience, 5*, 677–685.

Clandinin, T. R., & Feldheim, D. A. (2009). Making a visual map: Mechanisms and molecules. *Current Opinion in Neurobiology, 19*, 174–180.

Clark, A. S., & Henderson, L. P. (2003). Behavioral and physiological responses to anabolic-androgenic steroids. *Neuroscience and Biobehavioural Reviews, 27*, 413–436.

Clayton, N. S. (2001). Hippocampal growth and maintenance depend on food-caching experience in juvenile mountain chickadees (Poecile gambeli). *Behavioral Neuroscience, 115*, 614–625.

Clifton, P. G. (2000). Meal patterning in rodents: Psychopharmacological and neuroanatomical studies. *Neuroscience and Biobehavioural Reviews, 24*, 213–222.

Cohen, I., Navarro, V., Clemenceau, S., Baulac, M., & Miles, R. (2002). On the origin of interictal activity in human temporal lobe epilepsy in vitro. *Science, 298*, 1418–1421.

Cohen, J. (2007). Relative differences: The myth of 1%. *Science, 316*, 1836.

Cohen, S., Doyle, W. J., Alper, C. M., Janicki-Deverts, D., & Turner, R. B. (2009). Sleep habits and susceptibility to the common cold. *Archives of Internal Medicine, 169*, 62–67.

Cohen, S., Tyrrell, D. A. J., & Smith, A. P. (1991). Psychological stress and susceptibility to the common cold. *New England Journal of Medicine, 325*, 606–612.

Cohen, Y. E., & Knudsen, E. I. (1999). Maps versus clusters: Different representations of auditory space in the midbrain and forebrain. *Trends in Neuroscience, 22*, 128–135.

Cohen, Y. E., Russ, B. E., & Gifford, G. W. (2005). Auditory processing in the posterior parietal cortex. *Behavioral and Cognitive Neuroscience Reviews, 4*, 218–231.

Cohen-Bendahan, C. C. C., van de Beek, C., & Berenbaum, S. A. (2005). Prenatal sex hormone effects on child and adult sex-typed behavior: Methods and findings. *Neuroscience and Biobehavioural Reviews, 29*, 353–384.

Colapinto, J. (2000). *As nature made him: The boy who was raised as a girl*. New York, NY: HarperCollins.

Cole, M. W., Yeung, N., Freiwald, W. A., & Botvinick, M. (2009). Cingulate cortex: Diverging data from humans and monkeys. *Trends in Neurosciences, 32*, 566–574.

Coleman, D. L. (1979). Obesity genes: Beneficial effects in heterozygous mice. *Science, 203*, 663–665.

Coleman, M. (2005). Axon degeneration mechanisms: Commonality amid diversity. *Nature Reviews Neuroscience, 6*, 889–898.

Coleman, R. J., Anderson, R. M., Johnson, S. C., Kastman, E. K., Kosmatka, K. J., Beasley, T. M., . . . Weindruch, R. (2009). Caloric restriction delays disease onset and mortality in rhesus monkeys. *Science, 325*, 201–204.

Collaer, M. L., Brook, C. G. D., Conway, G. S., Hindmarsh, P. C., & Hines, M. (2008). Motor development in individuals with congenital adrenal hyperplasia: Strength, targeting, and fine motor skill. *Psychoneuroendocrinology, 34*, 249–258.

Collie, A., & Maruff, P. (2000). The neuropsychology of preclinical Alzheimer's disease and mild cognitive impairment. *Neuroscience and Biobehavioural Reviews, 24*, 365–374.

Collier, G. (1986). The dialogue between the house economist and the resident physiologist. *Nutrition and Behavior, 3*, 9–26.

Collier, G. H. (1980). An ecological analysis of motivation. In F. M. Toates & T. R. Halliday (Eds.), *Analysis of motivational processes* (pp. 125–151). London, England: Academic Press.

Collin, T., Marty, A., & Llano, I. (2005). Presynaptic calcium stores and synaptic transmission. *Current Opinion in Neurobiology, 15,* 275–281.

Collingridge, G. L., Isaac, J. T. R., & Wang, Y. T. (2004). Receptor trafficking and synaptic plasticity. *Nature Reviews Neuroscience, 5,* 952–962.

Colombo, M., & Broadbent, N. (2000). Is the avian hippocampus a functional homologue of the mammalian hippocampus? *Neuroscience and Biobehavioural Reviews, 24,* 465–484.

Colvin, M. K., Dunbar, K., & Grafman, J. (2001). The effects of frontal lobe lesions on goal achievement in the water jug task. *Journal of Cognitive Neuroscience, 13,* 1129–1147.

Concklin, C. A. (2006). Environments as cues to smoke: Implications for human extinction-based research and treatment. *Experimental and Clinical Psychopharmacology, 14,* 12–19.

Cone, R. D. (2005). Anatomy and regulation of the central melanocortin system. *Nature Neuroscience, 8,* 571–578.

Conners, B. W., & Long, M. A. (2004). Electrical synapses in the mammalian brain. *Annual Review of Neuroscience, 27,* 393–418.

Connolly, J. D., Andersen, R. A., & Goodale, M. A. (2003). fMRI evidence for a "parietal reach region" in the human brain. *Experimental Brain Research, 153,* 140–145.

Connor, C. E. (2006). Attention: Beyond neural response increases. *Nature Neuroscience, 9,* 1083–1084.

Coolen, M., & Bally-Cuif, L. (2009). MicroRNAs in brain development and physiology. *Current Opinion in Neurobiology, 19,* 461–470.

Coolican, J., Eskes, G. A., McMullen, P. A., & Lecky, E. (2007). Perceptual biases in processing facial identity and emotion. *Brain and Cognition, 66,* 176–187.

Cooper, J. A. (2008). A mechanism for inside-out lamination in the neocortex. *Trends in Neurosciences, 31,* 113–119.

Cooper, R. M., & Zubek, J. P. (1958). Effects of enriched and restricted early environments on the learning ability of bright and dull rats. *Canadian Journal of Psychology, 12,* 159–164.

Cooper, S. J. (2005). Donald O. Hebb's synapse and learning rule: A history and commentary. *Neuroscience and Biobehavioural Reviews, 28,* 851–874.

Corballis, M. C. (2003). From mouth to hand: Gesture, speech, and the evolution of right-handedness. *Behavioral and Brain Sciences, 26,* 199–208.

Corbetta, M., Miezin, F. M., Dobmeyer, S., Shulman, G. L., & Petersen, S. E. (1990). Attentional modulation of neural processing of shape, color, and velocity in humans. *Science, 248,* 1556–1559.

Corina, D. P., Poizner, H., Bellugi, U., Feinberg, T., Dowd, D., & O'Grady-Batch, L. (1992). Dissociation between linguistic and nonlinguistic gestural systems: A case for compositionality. *Brain and Language, 43,* 414–447.

Corkin, S. (1968). Acquisition of motor skill after bilateral medial temporal-lobe excision. *Neuropsychologia, 6,* 255–265.

Corkin, S. (2002). What's new with the amnesic patient H.M.? *Nature Reviews Neuroscience, 3,* 153–160.

Corkin, S., Milner, B., & Rasmussen, R. (1970). Somatosensory thresholds. *Archives of Neurology, 23,* 41–59.

Corsi, P. I. (1991). *The enchanted loom: Chapters in the history of neuroscience.* New York, NY: Oxford University Press.

Coslett, H. B. (1997). Acquired dyslexia. In T. E. Feinberg & M. J. Farah (Eds.), *Behavioral neurology and neuropsychology* (pp. 209–218). New York, NY: McGraw-Hill.

Cossart, R., Bernard, C., & Ben-Ari, Y. (2005). Multiple facets of GABAergic neurons and synapses: Multiple fates of GABA signaling in epilepsies. *Trends in Neurosciences, 28,* 108–114.

Costafreda, S. G., Brammer, M. J., David, A. S., & Fu, C. H. Y. (2007). Predictors of amygdala activation during the processing of emotional stimuli: A meta-analysis of 385 PET and fMRI studies. *Brain Research Reviews, 58,* 57–70.

Cotman, C. W., Berchtold, N. C., & Christie, L.-A. (2007). Exercise builds brain health: Key roles of growth factor cascades and inflammation. *Trends in Neurosciences, 30,* 464–472.

Courchesne, E., & Pierce, K. (2005). Why the frontal cortex in autism might be talking only to itself: Local over-connectivity but long-distance disconnection. *Current Opinion in Neurobiology, 15,* 225–230.

Courtine, G., Song, B., Roy, R. R., Zhong, H., Herrmann, J. E., Ao, Y., . . . Sofroniew, M. V. (2008). Recovery of supraspinal control of stepping via indirect propriospinal relay connections after spinal cord injury. *Nature Medicine, 14,* 69–74.

Cowan, W. M. (1979, September). The development of the brain. *Scientific American, 241,* 113–133.

Cox, J. J., Reimann, F., Nicholas, A. K., Thornton, G., Roberts, E., Springell, K., . . . Woods, C. G. (2006). An SCN9A channelopathy causes congenital inability to experience pain. *Nature, 444,* 894–898.

Craelius, W. (2002). The bionic man: Restoring mobility. *Science, 295,* 1018–1019.

Craig, A. D. (2003). Pain mechanisms: Labeled lines versus convergence in central processing. *Annual Review of Neuroscience, 26,* 1–30.

Creese, I., Burt, D. R., & Snyder, S. H. (1976). Dopamine receptor binding predicts clinical and pharmacological potencies of anti-schizophrenic drugs. *Science, 192,* 481–483.

Crisp, A. H. (1983). Some aspects of the psychopathology of anorexia nervosa. In P. L. Darby, P. E. Garfinkel, D. M. Garner, & D. V. Coscina (Eds.), *Anorexia nervosa: Recent developments in research* (pp. 15–28). New York, NY: Alan R. Liss.

Crisp, J., & Ralph, M. A. L. (2006). Unlocking the nature of the phonological-deep dyslexia continuum: The keys to reading aloud are in phonology and semantics. *Journal of Cognitive Neuroscience, 18,* 348–362.

Crowell, C. R., Hinson, R. E., & Siegel, S. (1981). The role of conditional drug responses in tolerance to the hypothermic effects of ethanol. *Psychopharmacology, 73,* 51–54.

Crunelli, V., & Leresche, N. (2002). Childhood absence epilepsy: Genes, channels, neurons and networks. *Nature Reviews Neuroscience, 3,* 371–382.

Cudeiro, J., & Sillito, A. M. (2006). Looking back: Corticothalamic feedback and early visual processing. *Trends in Neurosciences, 29,* 298–306.

Culbertson, F. M. (1997). Depression and gender. *American Psychologist, 52,* 25–31.

Culham, J. C., & Kanwisher, N. G. (2001). Neuroimaging of cognitive functions in human parietal cortex. *Current Opinion in Neurobiology, 11,* 157–163.

Culham, J. C., & Valyear, K. F. (2006). Human parietal cortex in action. *Current Opinion in Neurobiology, 16,* 205–212.

Curcio, G., Ferrara, M., & De Gennaro, L. (2006). Sleep loss, learning capacity and academic performance. *Sleep Medicine Reviews, 10,* 323–337.

Curran, T., & Schacter, D. L. (1997). Amnesia: Cognitive neuropsychological aspects. In T. E. Feinberg & M. J. Farah (Eds.), *Behavioral neurology and neuropsychology* (pp. 463–472) New York, NY: McGraw-Hill.

Curtis, M. A., Kam, M., Nannmark, U., Anderson, M. F., Axel, M. Z., Wikkelso, C., . . . Eriksson, P. S. (2007). Human neuroblasts migrate to the olfactory bulb via a lateral ventricular extension. *Science, 315,* 1243–1249.

Curtiss, S. (1977). *Genie: A psycholinguistic study of a modern-day "wild child."* New York, NY: Academic Press

Czeisler, C. A., Duffy, J. F., Shanahan, T. L., Brown, E. N., Mitchell, J. F., Rimmer, D. W., . . . Kronauer, R. E. (1999). Stability, precision, and near-24-hour period of the human circadian pacemaker. *Science, 284,* 2177–2181.

Dacey, D. (2004). Origins of perception: Retinal ganglion cell diversity and the creation of parallel visual pathways. In M. S. Gazzaniga (Ed.), *The cognitive neurosciences* (pp. 281–302). Cambridge, MA: MIT Press.

Daeschler, E. B., Shubin, N. H., & Jenkins, F. A., Jr. (2006). A Devonian tetrapod-like fish and the evolution of the tetrapod body plan. *Nature, 440,* 757–763.

Dallos, P. (2008). Cochlear amplification, outer hair cells and prestin. *Current Opinion in Neurobiology, 18,* 370–376.

D'Almeida, V., Hipólide, D. C., Azzalis, L. A., Lobo, L. L., Junqueira, V. B. C., & Tufik, S. (1997). Absence of oxidative stress following paradoxical sleep deprivation in rats. *Neuroscience Letters, 235,* 25–28.

Dalton, K. M., Nacewicz, B. M., Johnstone, T., Schaefer, H. S., Gernsbacher, M. A., Goldsmith, H. H., . . . Davidson, R. J. (2005). Gaze fixation and the neural circuitry of face processing in autism. *Nature Neuroscience, 8,* 519–526.

Dalva, M. B., McClelland, A. C., & Kayser, M. S. (2007). Cell adhesion molecules: Signalling functions at the synapse. *Nature Reviews Neuroscience, 8*, 206–220.

Daly, M., & Wilson, M. (1983). *Sex, evolution, and behavior.* Boston, MA: Allard Grant Press.

Damasio, A. R. (1999, December). How the brain creates the mind. *Scientific American, 281*, 112–117.

Damasio, H. (1989). Neuroimaging contributions to the understanding of aphasia. In F. Boller & J. Grafman (Eds.), *Handbook of neuropsychology* (Vol. 2, pp. 3–46). New York, NY: Elsevier.

Damasio, H., Grabowski, T., Frank, R., Galaburda, A. M., & Damasio, A. R. (1994). The return of Phineas Gage: Clues about the brain from the skull of a famous patient. *Science, 264*, 1102–1105.

Damasio, H., Grabowski, T. J., Tranel, D., Hichwa, R. D., & Damasio, A. R. (1996). A neural basis for lexical retrieval. *Nature, 380*, 499–505.

Danckert, J., & Rossetti, Y. (2005). Blindsight in action: What can the different sub-types of blindsight tell us about the control of visually guided actions? *Neuroscience and Biobehavioural Reviews, 29*, 1035–1046.

D'Angelo, E., & De Zeeuw, C. I. (2008). Timing and plasticity in the cerebellum: Focus on the granular layer. *Trends in Neurosciences, 32*, 30–40.

Dapretto, M., Davies, M. S., Pfeifer, J. H., Scott, A. A., Sigman, M., Bookheimer, S. Y., & Iacoboni, M. (2006). Understanding emotions in others: Mirror neuron dysfunction in children with autism spectrum disorders. *Nature Neuroscience, 9*, 28–30.

Darian-Smith, I., Burman, K., & Darian-Smith, C. (1999). Parallel pathways mediating manual dexterity in the macaque. *Experimental Brain Research, 128*, 101–108.

Darke, S., Hall, W., Weatherburn, D., & Lind, B. (1998). Fluctuations in heroin purity and the incidence of fatal heroin overdose. *Drug and Alcohol Dependence, 54*, 155–161.

Darke, S., Ross, J., Zador, D., & Sunjic, S. (2000). Heroin-related deaths in New South Wales, Australia, 1992–1996. *Drug and Alcohol Dependence, 60*, 141–150.

Darke, S., & Zador, D. (1996). Fatal heroin "overdose": A review. *Addiction, 91*, 1765–1772.

Darlison, M. G., & Richter, D. (1999). Multiple genes for neuropeptides and their receptors: Co-evolution and physiology. *Trends in Neurosciences, 22*, 81–88.

Datta, S., & MacLean, R. R. (2007). Neurobiological mechanisms for the regulation of mammalian sleep-wake behavior: Reinterpretation of historical evidence and inclusion of contemporary cellular and molecular evidence. *Neuroscience and Biobehavioral Reviews, 31*, 775–824.

Davids, E., & Gaspar, M. (2004). Buprenorphine in the treatment of opioid dependence. *European Neuropsychopharmacology, 14*, 209–216.

Davidson, A. G., Chan, V., O'Dell, R., & Schieber, M. H. (2007). Rapid changes in throughput from single motor cortex neurons to muscle activity. *Science, 318*, 1934–1937.

Davidson, J. M. (1980). Hormones and sexual behavior in the male. In D. T. Krieger & J. C. Hughes (Eds.), *Neuroendocrinology* (pp. 232–238). Sunderland, MA: Sinauer Associates.

Davidson, J. M., Kwan, M., & Greenleaf, W. J. (1982). Hormonal replacement and sexuality in men. *Clinics in Endocrinology and Metabolism, 11*, 599–623.

Davis, C., Kleinman, J. T., Newhart, M., Gingis, L., Pawlak, M., & Hillis, A. E. (2008). Speech and language functions that require a functioning Broca's area. *Brain and Language, 105*, 50–58.

Davis, D. M. (2006, February). Intrigue at the immune synapse. *Scientific American, 294*, 50–55.

Davis, S. R., Davison, S. L., Donath, S., & Bell, R. J. (2005). Circulating androgen levels and self-reported sexual function in women. *Journal of the American Medical Association, 294*, 91–96.

Davis, S. R., & Tran, J. (2001). Testosterone influences libido and well being in women. *Trends in Endocrinology & Metabolism, 12*, 33–37.

Dawson, M., Soulières, I., Gernsbacher, M. A., & Mottron, L. (2007). The level and nature of autistic intelligence. *Psychological Science, 18*, 657–662.

Day, J. J., & Carelli, R. M. (2007). The nucleus accumbens and Pavlovian reward learning. *Neuroscientist, 13*, 148–159.

Deadwyler, S. A., Hayashizaki, S., Cheer, J., & Hampson, R. E. (2004). Reward, memory and substance abuse: Functional neuronal circuits in the nucleus accumbens. *Neuroscience and Biobehavioural Reviews, 27*, 703–711.

Deak, M. C., & Stickgold, R. (2010). Sleep and cognition. *Wiley Interdisciplinary Review: Cognitive Science.* doi:10.1002/wcs.52

Dean, C., & Dresbach, T. (2006). Neuroligins and neurexins: Linking cell adhesion, synapse formation and cognitive function. *Trends in Neurosciences, 29*, 21–29.

Debanne, D. (2004). Information processing in the axon. *Nature Reviews Neuroscience, 5*, 304–316.

De Beaumont, L., Théoret, H., Mongeon, D., Messier, J., Leclerc, S., Tremblay, S., . . . Lassonde, M. (2009). Brain function decline in healthy retired athletes who sustained their last sports concussion in early adulthood. *Brain, 132*, 695–708.

De Butte-Smith, M., Gulinello, M., Zukin, R. S., & Etgen, A. M. (2009). Chronic estradiol treatment increases CA1 cell survival but does not improve visual or spatial recognition memory after global ischemia in middle-aged female rats. *Hormones and Behavior, 55*, 442–453.

de Castro, J. M., & Plunkett, S. (2002). A general model of intake regulation. *Neuroscience and Biobehavioural Reviews, 26*, 581–595.

Degenhardt, L., & Hall, W. D. (2008). The adverse effects of cannabinoids: Implications for use of medical marijuana. *Canadian Medical Association Journal, 178*, 1685–1686.

De Gennaro, L., Ferrara, M., & Bertini, M. (2000). Muscle twitch activity during REM sleep: Effect of sleep deprivation and relation with rapid eye movement activity. *Psychobiology, 28*, 432–436.

de Haan, M., Mishkin, M., Baldeweg, T., & Vargha-Khadem, F. (2006). Human memory development and its dysfunction after early hippocampal injury. *Trends in Neurosciences, 29*, 374–380.

Dehaene-Lambertz, G., Hertz-Pannier, L., & Dubois, J. (2006). Nature and nurture in language acquisition: Anatomical and functional brain-imaging studies in infants. *Trends in Neurosciences, 29*, 367–373.

De Jager, P. L., & Hafler, D. A. (2007). New therapeutic approaches for multiple sclerosis. *Annual Review of Medicine, 58*, 417–432.

De Jonge, F. H., Louwerse, A. L., Ooms, M. P., Evers, P., Endert, E., & van de Poll, N. E. (1989). Lesions of the SDN-POA inhibit sexual behavior of male Wistar rats. *Brain Research Bulletin, 23*, 483–492.

de Kloet, E. R., Joëls, M., & Holsboer, F. (2005). Stress and the brain: From adaptation to disease. *Nature Reviews Neuroscience, 6*, 463–475.

de Kruif, P. (1945). *The male hormone.* New York, NY: Harcourt, Brace.

De La Iglesia, H. O., Cambras, T., & Díez-Noguera, A. (2008). Circadian internal desynchronization: Lessons from a rat. *Sleep and Biological Rhythms, 6*, 76–83.

de Lange, F. P., Hagoort, P., & Toni, I. (2005). Neural topography and content of movement representations. *Journal of Cognitive Neuroscience, 17*, 97–112.

Delbello, M. P., Zimmerman, M. E., Mills, N. P., Getz, G. E., & Strakowski, S. M. (2004). Magnetic resonance imaging analysis of amygdala and other subcortical brain regions in adolescents with bipolar disorder. *Bipolar Disorders, 6*, 43–52.

Delgado-García, J. M., & Gruart, A. (2006). Building new motor responses: Eyelid conditioning revisited. *Trends in Neurosciences, 29*, 330–338.

Demakis, G. J. (2003). A meta-analytic review of the sensitivity of the Wisconsin card sorting test to frontal and lateralized frontal brain damage. *Neuropsychology, 17*, 255–264.

Demaree, H. A., Everhart, E., Youngstrom, E. A., & Harrison, D. W. (2005). Brain lateralization of emotional processing: Historical roots and a future incorporating "dominance." *Behavioral and Cognitive Neuroscience Reviews, 4*, 2–20.

Demas, G. E., Cooper, M. A., Albers, H. E., & Soma, K. K. (2005). Novel mechanisms underlying neuroendocrine regulation of aggression: A synthesis of rodent, avian, and primate studies. In J. D. Blaustein (Ed.), *Behavioral neurochemistry and neuroendocrinology* (pp. 2–25). New York, NY: Kluwer Press.

De Mees, C., Bakker, J., Szpirer, J., & Szpirer, C. (2006). Alpha-fetoprotein: From a diagnostic biomarker to a key role in female fertility. *Biomarker Insights, 1*, 82–85.

Dement, W. C. (1978). *Some must watch while some must sleep.* New York, NY: Norton.

Dement, W. C., & Kleitman, N. (1957). The relation of eye movement during sleep to dream activity: An objective method for the study of dreaming. *Journal of Experimental Psychology, 53,* 339–346.

Dement, W. C., & Wolpert, E. A. (1958). The relation of eye movements, body motility and external stimuli to dream content. *Journal of Experimental Psychology, 55,* 543–553.

Demers, G., Griffin, G., De Vroey, G., Haywood, J. R., Zurlo, J., & Bédard, M. (2006). Harmonization of animal care and use guidance. *Science, 312,* 700–701.

de Paula, F. J., Soares, J. M., Jr., Haidar, M. A., de Lima, G. R., & Baracat, E. C. (2007). The benefits of androgens combined with hormone replacement therapy regarding to patients with postmenopausal sexual symptoms. *Maturitas, 56,* 69–77.

Deppmann, C. D., Mihalas, S., Shama, N., Lonze, B. E., Niebur, E., & Ginty, D. D. (2008). A model for neuronal competition during development. *Science, 320,* 369–373.

De Preux, E., Dubois-Arber, F., & Zobel, F. (2004). Current trends in illegal drug use and drug related health problems in Switzerland. *Swiss Medical Weekly, 134,* 313–321.

De Renzi, E. (1997). Visuospatial and constructional disorders. In T. E. Feinberg & M. J. Farah (Eds.), *Behavioral neurology and neuropsychology* (pp. 297–308). New York, NY: McGraw-Hill.

Deretic, V., & Klionsky, D. J. (2008, May). How cells clean house. *Scientific American,* 74–81.

Deroche-Gamonet, V., Belin, D., & Piazza, P. V. (2004). Evidence for addiction-like behavior in the rat. *Science, 305,* 1014–1017.

Desmurget, M., Reilly, K. T., Richard, N., Szathmari, A., Mottolese, C., & Sirigu, A. (2009). Movement intention after parietal cortex stimulation in humans. *Science, 324,* 811–813.

Dess, N. K., & Chapman, C. D. (1998). "Humans and animals"? On saying what we mean. *Psychological Science, 9*(2), 156–157.

De Valois, R. L., Cottaris, N. P., Elfar, S. D., Mahon, L. E., & Wilson, J. A. (2000). Some transformations of color information from lateral geniculate nucleus to striate cortex. *Proceedings of the National Academy of Science, USA, 97,* 4997–5002.

de Vries, G. J., & Södersten, P. (2009). Sex differences in the brain: The relation between structure and function. *Hormones and Behavior, 55,* 589–596.

De Vry, J., & Schreiber, R. (2000). Effects of selected serotonin 5-HT1 and 5-HT2 receptor agonists on feeding behavior: Possible mechanisms of action. *Neuroscience and Biobehavioural Reviews, 24,* 341–353.

de Waal, F. B. M. (1999, December). The end of nature versus nurture. *Scientific American, 281,* 94–99.

Dewing, P., Shi, T., Horvath, S., & Vilain, E. (2003). Sexually dimorphic gene expression in mouse brain precedes gonadal differentiation. *Molecular Brain Research, 118,* 82–90.

De Witte, P., Pinto, E., Ansseau, M., & Verbanck, P. (2003). Alcohol and withdrawal: From animal research to clinical issues. *Neuroscience and Biobehavioural Reviews, 27,* 189–197.

Dewsbury, D. A. (1988). The comparative psychology of monogamy. In D. W. Leger (Ed.), *Comparative perspectives in modern psychology: Nebraska Symposium on Motivation* (Vol. 35, pp. 1–50). Lincoln: University of Nebraska Press.

Dewsbury, D. A. (1991). Psychobiology. *American Psychologist, 46,* 198–205.

De Zeeuw, C., & Yeo, C. H. (2005). Time and tide in cerebellar memory formation. *Current Opinion in Neurobiology, 15,* 667–674.

Diamond, A. (1985). Development of the ability to use recall to guide action, as indicated by infants' performance on AB. *Child Development, 56,* 868–883.

Diamond, A. (1991). Neuropsychological insights into the meaning of object concept development. In S. Carey & R. Gelman (Eds.), *The epigenesis of mind: Essays on biology and cognition* (pp. 67–110). Hillsdale, NJ: Lawrence Erlbaum.

Diamond, J. (2004). The astonishing micropygmies. *Science, 306,* 2047–2048.

Diamond, M. (2009). Clinical implications of the organizational and activational effects of hormones. *Hormones and Behavior, 55,* 621–632.

Diamond, M., & Sigmundson, H. K. (1997). Sex reassignment at birth: Long-term review and clinical implications. *Archives of Pediatric and Adolescent Medicine, 151,* 298–304.

DiCarlo, J. J., & Johnson, K. O. (2000). Spatial and temporal structure of receptive fields in primate somatosensory area 3b: Effects of stimulus scanning direction and orientation. *Journal of Neuroscience, 20,* 495–510.

Dicarlo, J. J., Johnson, K. O., & Hsaio, S. S. (1998). Structure of receptive fields in area 3b of primary somatosensory cortex in the alert monkey. *Journal of Neuroscience, 18,* 2626–2645.

Di Chiara, G., & Bassareo, V. (2007). Reward system and addiction: What dopamine does and doesn't do. *Current Opinion in Pharmacology, 7,* 69–76.

Di Ciano, P., & Everitt, B. J. (2003). Differential control over drug-seeking behavior by drug-associated conditioned reinforcers and discriminative stimuli predictive of drug availability. *Behavioral Neuroscience, 117,* 952–960.

DiCicco-Bloom, E. (2006). Neuron, know thy neighbor. *Science, 311,* 1560–1562.

Diekhof, E. K., Falkai, P., & Gruber, O. (2008). Functional neuroimaging of reward processing and decision making: A review of aberrant motivational and affective processing in addiction and mood disorders. *Brain Research Reviews, 59,* 164–184.

Dielenberg, R. A., & McGregor, I. S. (2001). Defensive behavior in rats towards predatory odors: A review. *Neuroscience and Biobehavioural Reviews, 25,* 597–609.

Dietz, V. (2002). Do human bipeds use quadrupedal coordination? *Trends in Neurosciences, 25,* 462–467.

Di Franza, J. R. (2008, May). Hooked from the first cigarette. *Scientific American, 298,* 82–87.

Dillehay, T. D., Rossen, J., Andres, T. C., & Williams, D. E. (2007). Preceramic adoption of peanut, squash, and cotton in northern Peru. *Science, 316,* 1890–1894.

DiMarzo, V., & Matias, I. (2005). Endocannabinoid control of food intake and energy balance. *Nature Neuroscience, 8,* 585–589.

Dinges, D. F., Douglas, S. D., Zaugg, L., Campbell, D. E., McMann, J. M., Whitehouse, W. G., . . . Orne, M. T. (1994). Leukocytosis and natural killer cell function parallel neurobehavioral fatigue induced by 64 hours of sleep deprivation. *Journal of Clinical Investigation, 93,* 1930–1939.

Dinges, D. F., Pack, F., Williams, K., Gillen, K. A., Powell, J. W., Ott, G. E., et al. (1997). Cumulative sleepiness, mood disturbance, and psychomotor vigilance performance decrements during a week of sleep restricted to 4–5 hours per night. *Sleep, 20,* 267–277.

Dirnagl, U., Simon, R. P., & Hallenbeck, J. M. (2003). Ischemic tolerance and endogenous neuroprotection. *Trends in Neurosciences, 24,* 248–254.

Dobbs, D. (2005, April). Fact or phrenology? *Scientific American Mind, 16,* 25–31.

Dobelle, W. H., Mladejovsky, M. G., & Girvin, J. P. (1974). Artificial vision for the blind: Electrical stimulation of visual cortex offers hope for a functional prosthesis. *Science, 183,* 440–444.

Döbrössy, M., & Dunnett, S. B. (2001). The influence of environment and experience on neural grafts. *Nature Reviews Neuroscience, 2,* 871–909.

Dodick, D. W., & Gargus, J. J. (2008, August). Why migraines strike. *Scientific American, 299,* 56–63.

Doetsch, F., & Hen, R. (2005). Young and excitable: The function of new neurons in the adult mammalian brain. *Current Opinion in Neurobiology, 15,* 121–128.

Dominguez, J. M., & Hull, E. M. (2005). Dopamine, the medial preoptic area, and male sexual behavior. *Physiology and Behavior, 86,* 356–368.

Dos Santos Sequeira, S., Woemer, W., Walter, C., Kreuder, F., Lueken, U., Westerhausen, R., . . . Wittling, W. (2006). Handedness, dichotic-listening ear advantage, and gender effects on planum temporale asymmetry: A volumetric investigation using structural magnetic resonance imaging. *Neuropsychologia, 44,* 622–636.

Doty, R. L. (2001). Olfaction. *Annual Review of Psychology, 52,* 423–452.

Douglas, R. J., & Martin, K. A. C. (2004). Neuronal circuits of the neocortex. *Annual Review of Neuroscience, 27,* 419–451.

Doupe, A. J. (1993). A neural circuit specialized for vocal learning. *Current Opinion in Neurobiology, 3*, 104–110.

Doupe, A. J., Solis, M. M., Boettiger, C. A., & Hessler, N. A. (2004). Birdsong: Hearing in the service of vocal learning. In M. S. Gazzaniga (Ed.), *The cognitive neurosciences* (pp. 245–258). Cambridge, MA: MIT Press.

Doya, K. (2000). Complementary roles of basal ganglia and cerebellum in learning and motor control. *Current Opinion in Neurobiology, 10*, 732–739.

Drachman, D. A., & Arbit, J. (1966). Memory and the hippocampal complex. *Archives of Neurology, 15*, 52–61.

Drevets, W. C., Gautier, C., Price, J. C., Kupfer, D. J., Kinahan, P. E., Grace, A. A., . . . Mathis, C. A. (2001). Amphetamine-induced dopamine release in human ventral striatum correlates with euphoria. *Biological Psychiatry, 49*, 81–96.

Drew, T., Jiang, W., & Widajewicz, W. (2002). Contributions of the motor cortex to the control of the hindlimbs during locomotion in the cat. *Brain Research Reviews, 40*, 178–191.

Drewnowski, A., Halmi, K. A., Pierce, B., Gibbs, J., & Smith, G. P. (1987). Taste and eating disorders. *American Journal of Clinical Nutrition, 46*, 442–450.

Driscoll, T. R., Grunstein, R. R., & Rogers, N. L. (2007). A systematic review of the neurobehavioural and physiological effects of shiftwork systems. *Sleep Medicine Reviews, 11*, 179–194.

Driver, J., Vuilleumier, P., & Husain, M. (2004). Spatial neglect and extinction. In M. S. Gazzaniga (Ed.), *The cognitive neurosciences* (pp. 589–606). Cambridge, MA: MIT Press.

Drummey, A. B., & Newcombe, N. (1995). Remembering versus knowing the past: Children's explicit and implicit memories for pictures. *Journal of Experimental Child Psychology, 59*, 540–565.

Drummond, S. P. A., Brown, G. G., Salamat, J. S., & Gillin, J. C. (2004). Increasing task difficulty facilitates the cerebral compensatory response to total sleep deprivation. *Sleep, 27*, 445–451.

Duan, X., Kang, E., Liu, C. Y., Ming, G-L., & Song, H. (2008). Development of neural stem cell in the adult brain. *Current Opinion in Neurobiology, 18*, 108–115.

Duchaine, B., Cosmides, L., & Tooby, J. (2001). Evolutionary psychology and the brain. *Current Opinion in Neurobiology, 11*, 225–230.

Duchaine, B., & Nakayama, K. (2005). Dissociations of face and object recognition in developmental prosopagnosia. *Journal of Cognitive Neuroscience, 17*, 249–261.

Duchaine, B. C., & Nakayama, K. (2006). Developmental prosopagnosia: A window to content-specific face processing. *Current Opinion in Neurobiology, 16*, 166–173.

Dulac, C., & Kimchi, T. (2007). Neural mechanisms underlying sex-specific behaviors in vertebrates. *Current Opinion in Neurobiology, 17*, 675–683.

Dully, H., & Fleming, C. (2007). *My lobotomy: A memoir.* New York, NY: Crown.

Dunbar, R. (2003). Evolution of the social brain. *Science, 302*, 1160–1161.

Dunlap, J. C. (2006). Running a clock requires quality time together. *Science, 311*, 184–186.

Dunnett, S. B., Björklund, A., & Lindvall, O. (2001). Cell therapy in Parkinson's disease—stop or go? *Nature Reviews Neuroscience, 2*, 365–368.

Durmer, J. S., & Dinges, D. F. (2005). Neurocognitive consequences of sleep deprivation. *Seminars in Neurology, 25*, 117–129.

Dusart, I., Ghoumari, A., Wehrle, R., Morel, M. P., Bouslama-Oueghlani, L., Camand, E., & Sotelo, C. (2005). Cell death and axon regeneration of Purkinje cells after axotomy: Challenges of classical hypotheses of axon regeneration. *Brain Research Reviews, 49*, 300–316.

Dustin, M. L., & Colman, D. R. (2002). Neural and immunological synaptic relations. *Science, 298*, 785–789.

Duva, C. A., Kornecook, T. J., & Pinel, J. P. J. (2000). Animal models of medial temporal lobe amnesia: The myth of the hippocampus. In M. Haug & R. E. Whalen (Eds.), *Animal models of human emotion and cognition* (pp. 197–214). Washington, DC: American Psychological Association.

Dyken, M. E., Yamada, T., & Lin-Dyken, D. C. (2001). Abnormal behaviors in sleep. *Acta Neurologica Taiwanica, 10*, 241–247.

Eacker, S. M., Dawson, T. M., & Dawson, V. L. (2009). Understanding microRNAs in neurodegeneration. *Nature Reviews Neuroscience, 10*, 837–841.

Eacott, M. J., Machin, P. E., & Gaffan, E. A. (2001). Elemental and configural visual discrimination learning following lesions to perirhinal cortex in the rat. *Behavioural Brain Research, 124*, 55–70.

Eagleman, D. M. (2001). Visual illusions and neurobiology. *Nature Reviews Neuroscience, 2*, 920–926.

Ecker, A. S., Berens, P., Keliris, G. A., Bethge, M., Logothetis, N. K., & Tolias, A. S. (2010). Decorrelated neuronal firing in cortical microcircuits. *Science, 327*, 584–586.

Eckert, M. A., & Leonard, C. M. (2003). Developmental disorders: Dyslexia. In K. Hugdahl and R. J. Davidson (Eds.), *The asymmetrical brain* (pp. 651–679). Cambridge, MA: Bradford/MIT Press.

Eckert, M. A., Leonard, C. M., Possing, E. T., & Binder, J. R. (2006). Uncoupled leftward asymmetries for planum morphology and functional language processing. *Brain and Language, 98*, 102–111.

Edgar, J. M., & Nave, K-A. (2009). The role of CNS glia in preserving axon function. *Current Opinion in Neurobiology, 19*, 498–504.

Egan, J. F. (2004). Down syndrome births in the United States from 1989 to 2001. *American Journal of Obstetrics and Gynecology, 191*, 1044–1048.

Egan, M. F., & Weinberger, D. R. (1997). Neurobiology of schizophrenia. *Current Opinion in Neurobiology, 7*, 701–707.

Eggermont, J. J., & Roberts, L. E. (2004). The neuroscience of tinnitus. *Trends in Neurosciences, 27*, 676–682.

Ehrhardt, A. A., Meyer-Bahlburg, H. F. L., Rosen, L. R., Feldman, J. F., Veridiano, N. P., Zimmerman, I., & McEwen, B. S. (1985). Sexual orientation after prenatal exposure to exogenous estrogen. *Archives of Sexual Behavior, 14*, 57–77.

Ehringer, H., & Hornykiewicz, O. (1960). Verteilung von Noradrenalin und Dopamin (3-Hydroxytyramin) im gehirn des Menschen und ihr Verhalten bei Erkrankungen des Extrapyramidalen Systems. *Klinische Wochenschrift, 38*, 1236–1239.

Ehrlich, I., Humeau, Y., Grenier, F., Ciocchi, S., Herry, C., & Lüthi, A. (2009). Amygdala inhibitory circuits and the control of fear memory. *Neuron, 62*, 757–771.

Eichenbaum, H. (1996). Learning from LTP: A comment on recent attempts to identify cellular and molecular mechanisms of memory. *Learning & Memory, 3*, 61–73.

Eichenbaum, H. (1999). Conscious awareness, memory and the hippocampus. *Nature, 2*, 775–776.

Eichenbaum, H., Yonelinas, A. P., & Ranganath, C. (2007). The medial temporal lobe and object recognition memory. *Annual Review of Neuroscience, 30*, 123–152.

Eigsti, I. M., & Shapiro, T. (2004). A systems neuroscience approach to autism: Biological, cognitive, and clinical perspectives. *Mental Retardation and Developmental Disabilities Research Reviews, 9*, 205–215.

Eilat-Adar, S., Eldar, M., & Goldbourt, U. (2005). Association of intentional changes in body weight with coronary heart disease event rates in overweight subjects who have an additional coronary risk factor. *American Journal of Epidemiology, 161*, 352–358.

Eisener-Dorman, A. F., Lawrence, D. A., & Bolivar, V. J. (2008). Cautionary insights on knockout mouse studies: The gene or not the gene? *Brain, Behavior, and Immunity, 23*, 318–324.

Eiser, A. S. (2005). Physiology and psychology of dreams. *Seminars in Neurology, 25*, 97–105.

Ekman, P. (1985). *Telling lies.* New York, NY: Norton.

Ekman, P. (1992). An argument for basic emotions. *Cognition and Emotion, 6*, 169–200.

Ekman, P. (2003). *Emotions revealed: Recognizing faces and feelings to improve communication and emotional life.* New York, NY: Times/Holt.

Ekman, P., & Davidson, R. J. (1993). Voluntary smiling changes regional brain activity. *Psychological Science, 4*, 342–345.

Ekman, P., & Friesen, W. V. (1971). Constants across cultures in the face and emotion. *Journal of Personality and Social Psychology, 17*, 124–129.

Ekman, P., & Friesen, W. V. (1975). *Unmasking the face: A guide to recognizing emotions from facial clues.* Englewood Cliffs, NJ: Prentice-Hall.

Ekstrom, A., Suthana, N., Millet, D., Fried, I., & Bookheimer, S. (2009). Correlation between BOLD fMRI and theta-band local field potentials in the human hippocampal area. *Journal of Neurophysiology, 10*, 2668–2678.

Elbert, T., Pantev, C., Wienbruch, C., Rockstroh, B., & Taub, E. (1995). Increased cortical representation of the fingers of the left hand in string players. *Science, 270,* 305–307.

Elbert, T., & Rockstroh, B. (2004). Reorganization of human cerebral cortex: The range of changes following use and injury. *Neuroscientist, 10,* 129–141.

Elbert, T., Sterr, A., Rockstroh, B., Pantev, C., Müller, M. M., & Taub, E. (2002). Expansion of the tonotopic area in the auditory cortex of the blind. *Journal of Neuroscience, 22,* 9941–9944.

Elefteriades, J. A. (2005, August). Beating a sudden killer. *Scientific American, 293,* 64–71.

Elias, L. A. B., & Kriegstein, A. R. (2008). Gap junctions: Multifaceted regulators of embryonic cortical development. *Trends in Neurosciences, 31,* 243–250.

Elibol, B., Söylemezoglu, F., Ünal, I., Fujii, M., Hirt, L., Huang, P. L., . . . Dalkara, T. (2001). Nitric oxide is involved in ischemia-induced apoptosis in brain: A study in neuronal nitric oxide synthase null mice. *Neuroscience, 105,* 79–86.

Eling, P., Derckx, K., & Maes, R. (2008). On the historical and conceptual background of the Wisconsin Card Sorting Test. *Brain and Cognition, 67,* 247–253.

Ellenbogen, J. M., Payne, J. D., & Stickgold, R. (2006). The role of sleep in declarative memory consolidation: Passive, permissive, active or none? *Current Opinion in Neurobiology, 16,* 716–722.

Ellis, L., & Ames, M. A. (1987). Neurohormonal functioning and sexual orientation: A theory of homosexuality-heterosexuality. *Psychological Bulletin, 101,* 233–258.

Elmenhorst, E-M., Elmenhorst, D., Luks, N., Maass, H., Vejvoda, M., & Samel, A. (2008). Partial sleep deprivation: Impact on the architecture and quality of sleep. *Sleep Medicine, 9,* 840–850.

Elmquist, J. K., & Flier, J. S. (2004). The fat-brain axis enters a new dimension. *Science, 304,* 63–64.

Engel, S. A. (1999). Using neuroimaging to measure mental representations: Finding color-opponent neurons in visual cortex. *Current Directions in Psychological Science, 8,* 23–27.

Epstein, L. H., Temple, J. L., Neaderhiser, B. J., Salis, R. J., Erbe, R. W., & Leddy, J. J. (2007). Food reinforcement, the dopamine D_2 receptor genotype, and energy intake in obese and nonobese humans. *Behavioral Neuroscience, 121,* 877–886.

Erickson, K., Drevets, W., & Schulkin, J. (2003). Glucocorticoid regulation of diverse cognitive functions in normal and pathological emotional states. *Neuroscience and Biobehavioural Reviews, 27,* 233–246.

Erikkson, P. S., Perfilieva, E., Björk-Eriksson, T., Alborn, A., Nordberg, C., Peterson, D. A., & Gage, F. H. (1998). Neurogenesis in the adult human hippocampus. *Nature Medicine, 4,* 1313–1317.

Eskandari, F., Martinez, P. E., Torvik, S., Phillips, T. M., Sternberg, E. M., Mistry, S., . . . Cizza, G. (2007). Low bone mass in premenopausal women with depression. *Archives of Internal Medicine, 167,* 2329–2336.

Espie, C. A. (2002). Insomnia: Conceptual issues in the development, persistence, and treatment of sleep disorder in adults. *Annual Review of Neuroscience, 53,* 215–243.

Evans, G. W., & Kim, P. (2007). Childhood poverty and health: Cumulative risk exposure and stress dysregulation. *Psychological Science, 18,* 953–957.

Everitt, B. J., Dickinson, A., & Robbins, T. W. (2001). The neuropsychological basis of addictive behaviour. *Brain Research Reviews, 36,* 129–138.

Everitt, B. J., & Herbert, J. (1972). Hormonal correlates of sexual behavior in sub-human primates. *Danish Medical Bulletin, 19,* 246–258.

Everitt, B. J., Herbert, J., & Hamer, J. D. (1971). Sexual receptivity of bilaterally adrenalectomized female rhesus monkeys. *Physiology & Behavior, 8,* 409–415.

Everitt, B. J., & Robbins, T. W. (1997). Central cholinergic systems and cognition. *Annual Review of Psychology, 48,* 649–684.

Everitt, B. J., & Robbins, T. W. (2005). Neural systems of reinforcement for drug addiction: From actions to habits to compulsion. *Nature Neuroscience, 8,* 1481–1489.

Fabbri-Destro, M., & Rizzolatti, G. (2008). Mirror neurons and mirror systems in monkeys and humans. *Physiology, 23,* 171–179.

Fabbro, F., Tavano, A., Corti, S., Bresolin, N., De Fabritiis, P., & Borgatti, R. (2004). Long-term neuropsychological deficits after cerebellar infarctions in two young adult twins. *Neuropsychologia, 42,* 536–545.

Fabri, M., Polonara, G., Del Pesce, M., Quattrini, A., Salvolini, U., & Manzoni, T. (2001). Posterior corpus callosum and interhemispheric transfer of somatosensory information: An fMRI and neuropsychological study of a partially callosotomized patient. *Journal of Cognitive Neuroscience, 13,* 1071–1079.

Fadiga, L., Craighero, L., & Olivier, E. (2005). Human motor cortex excitability during the perception of others' action. *Current Opinion in Neurobiology, 15,* 213–218.

Fagiolini, M., Jensen, C. L., & Champagne, F. A. (2009). Epigenetic influences on brain development and plasticity. *Current Opinion in Neurobiology, 19,* 207–212.

Falzi, G., Perrone, P., & Vignolo, L. A. (1982). Right-left asymmetry in anterior speech region. *Archives of Neurology, 39,* 239–240.

Farah, M. J. (1990). *Visual agnosia: Disorders of object recognition and what they tell us about normal vision.* Cambridge, MA: MIT Press.

Farah, M. J., & Murphy, N. (2009). Neuroscience and the soul [Letter to editor]. *Science, 323,* 1168.

Farber, N. B., & Olney, J. W. (2003). Drugs of abuse that cause developing neurons to commit suicide. *Developmental Brain Research, 147,* 37–45.

Farmer, J., Zhao, X., Van Praag, H., Wodke, K., Gage, F. H., & Christie, B. R. (2004). Effects of voluntary exercise on synaptic plasticity and gene expression in the dentate gyrus of adult male Sprague-Dawley rats in vivo. *Neuroscience, 124,* 71–79.

Farooqi, I. S., Jebb, S. A., Langmack, G., Lawrence, E., Cheetham, C. H., Prentice, A. M., . . . O'Rahilly, S. (1999). Effects of recombinant leptin therapy in a child with congenital leptin deficiency. *New England Journal of Medicine, 341,* 879–884.

Farrell, S. F., & McGinnis, M. Y. (2003). Effects of pubertal anabolic-androgenic steroid (AAS) administration on reproductive and aggressive behaviors in male rats. *Behavioral Neuroscience, 117,* 904–911.

Fava, G. A., Ruini, C., Rafanelli, C., Finos, L., Conti, S., & Grandi, S. (2004). Six-year outcome of cognitive behavior therapy for prevention of recurrent depression. *American Journal of Psychiatry, 161,* 1872–1876.

Fehervari, Z., & Sakaguchi, S. (2006, October). Peacekeepers of the immune system. *Scientific American, 295,* 57–63.

Feindel, W. (1986). Electrical stimulation of the brain during surgery for epilepsy—historical highlights. In G. P. Varkey (Ed.), *Anesthetic considerations for craniotomy in awake patients* (pp. 75–87). Boston, MA: Little, Brown.

Feldman, D. E., & Brecht, M. (2005). Map plasticity in somatosensory cortex. *Science, 310,* 810–815.

Feller, M. B., & Scanziani, M. (2005). A precritical period for plasticity in visual cortex. *Current Opinion in Neurobiology, 15,* 94–100.

Felsen, G., & Dan, Y. (2006). A natural approach to studying vision. *Nature Neuroscience, 8,* 1643–1646.

Feng, A. S., & Ratnam, R. (2000). Neural basis of hearing in real-world situations. *Annual Review of Psychology, 51,* 699–725.

Fenton, A. A. (2007). Where am I? *Science, 315,* 947–949.

Fentress, J. C. (1973). Development of grooming in mice with amputated forelimbs. *Science, 179,* 704–705.

Ferini-Strambi, L., Aarskog, D., Partinen, M., Chaudhuri, K. R., Sohr, M., Verri, D., & Albrecht, S. (2008). Effect of pramipexole on RLS symptoms and sleep: A randomized, double-blind, placebo-controlled trial. *Sleep Medicine, 9,* 874–881.

Fernald, R. D. (2000). Evolution of eyes. *Current Opinion in Neurobiology, 10,* 444–450.

Ferrara, M., De Gennaro, L., & Bertini, M. (1999). The effects of slow-wave sleep (SWS) deprivation and time of night on behavioral performance upon awakening. *Physiology & Behavior, 68,* 55–61.

Ferrucci, L., & Alley, D. (2007). Obesity, disability, and mortality. *Archives of Internal Medicine, 167,* 750–751.

ffrench-Constant, C., Colognato, H., & Franklin, R. J. M. (2004). The mysteries of myelin unwrapped. *Science, 304,* 688–703.

Fichten, C. S., Libman, E., Creti, L., Bailes, S., & Sabourin, S. (2004). Long sleepers sleep more and short sleepers sleep less: A comparison of older adults who sleep well. *Behavioral Sleep Medicine, 2*, 2–23.

Field, G. D., & Chichilnisky, E. J. (2007). Information processing in the primate retina: Circuitry and coding. *Annual Review of Neuroscience, 30*, 1–30.

Fields, H. (2004). State-dependent opioid control of pain. *Nature Reviews Neuroscience, 5*, 565–575.

Fields, R. D. (2004, April). The other half of the brain. *Scientific American, 290*, 55–61.

Fields, R. D. (2005, February). Making memories stick. *Scientific American, 292*, 75–80.

Fields, R. D. (2008a, March). White matter. *Scientific American, 298*, 42–49.

Fields, R. D. (2008b). White matter in learning, cognition and psychiatric disorders. *Trends in Neurosciences, 31*, 361–370.

Fields, R. D. (2008c, March). White matter matters. *Scientific American, 298*, 54–61.

Fields, R. D. (2009, November). New culprits in chronic pain. *Scientific American, 301*, 50–57.

Fields, R. D., & Burnstock, G. (2006). Purinergic signalling in neuron-glia interactions. *Nature Reviews Neuroscience, 7*, 423–435.

Fields, R. D., & Stevens-Graham, B. (2002). New insights into neuronglia communication. *Science, 298*, 556–562.

Fillion, T. J., & Blass, E. M. (1986). Infantile experience with suckling odors determines adult sexual behavior in male rats. *Science, 231*, 729–731.

Fillmore, K. M., Kerr, W. C., Stockwell, T., Chikritzhs, T., & Bostrom, A. (2006). Moderate alcohol use and reduced mortality risk: Systematic error in prospective studies. *Addiction Research and Therapy, 14*, 101–132.

Fink, G., Sumner, B., Rosie, R., Wilson, H., & McQueen, J. (1999). Androgen actions on central serotonin neurotransmission: Relevance for mood, mental state and memory. *Behavioural Brain Research, 105*, 53–68.

Finlay, B. L., & Darlington, R. B. (1995). Linked regularities in the development and evolution of mammalian brains. *Science, 268*, 1578–1584.

Finn, R. (1991, June). Different minds. *Discover*, 55–58.

Fiorillo, C. D., Newsome, W. T., & Schultz, W. (2008). The temporal precision of reward prediction in dopamine neurons. *Nature Neuroscience, 11*, 966–973.

Fiorino, D. F., Coury, A., & Phillips, A. G. (1997). Dynamic changes in nucleus accumbens dopamine efflux during the Coolidge effect in male rats. *Journal of Neuroscience, 17*, 4849–4855.

Fischer, J., Koch, L., Emmerling, C., Vierkotten, J., Peters, T., Brüning, J. C., & Rüther, U. (2009). Inactivation of the Fto gene protects from obesity. *Nature, 458*, 894–898.

Fisher, C. B. (2006). Clinical trials results databases: Unanswered questions. *Science, 311*, 180–181.

Fisher, S. E., & Francks, C. (2006). Genes, cognition and dyslexia: Learning to read the genome. *Trends in Cognitive Sciences, 10*, 249–257.

Fitzpatrick, S. M., & Rothman, D. L. (2000). Meeting report: Transcranial magnetic stimulation and studies of human cognition. *Journal of Cognitive Neuroscience, 12*, 704–709.

Flanagan, J. G. (2006). Neural map specification by gradients. *Current Opinion in Neurobiology, 16*, 59–66.

Flavell, S. W., & Greenberg, M. E. (2008). Signaling mechanisms linking neuronal activity to gene expression and plasticity of the nervous system. *Annual Review of Neuroscience, 31*, 563–590.

Flegal, K. M., Graubard, B. I., Williamson, D. F., & Gail, M. H. (2007). Cause-specific excess deaths associated with underweight, overweight, and obesity. *Journal of the American Medical Association, 298*, 2028–2037.

Fleshner, M., & Laudenslager, M. L. (2004). Psychoneuroimmunology: Then and now. *Behavioral and Cognitive Neuroscience Reviews, 3*, 114–130.

Flier, J. S. (2006). Regulating energy balance: The substrate strikes back. *Science, 312*, 861–864.

Flier, J. S., & Maratos-Flier, E. (2007, September). What fuels fat. *Scientific American, 297*, 72–81.

Flor, H., Nikolajsen, L., & Jensen, T. S. (2006). Phantom limb pain: A case of maladaptive CNS plasticity? *Nature Reviews Neuroscience, 7*, 873–881.

Floresco, S. B., Seamans, J. K., & Phillips, A. G. (1997). Selective roles for hippocampal, prefrontal cortical, and ventral striatal circuits in radial-arm maze tasks with or without a delay. *Journal of Neuroscience, 17*(5), 1880–1890.

Fogassi, L., & Ferrari, P. F. (2007). Mirror neurons and the evolution of embodied language. *Current Directions in Psychological Science, 16*, 136–141.

Fogassi, L., & Luppino, G. (2005). Motor functions of the parietal lobe. *Current Opinion in Neurobiology, 15*, 626–631.

Förster, E., Ahao, S., & Frotscher, M. (2006). Laminating the hippocampus. *Nature Reviews Neuroscience, 7*, 259–267.

Foster, R. G., & Kreitzman, L. (2004). *Rhythms of life*. London: Profiles.

Foulkes, N. S., Borjigin, J., Snyder, S. H., & Sassone-Corsi, P. (1997). Rhythmic transcription: The molecular basis of circadian melatonin synthesis. *Trends in Neurosciences, 20*, 487–492.

Fountoulakis, K. N., Giannakopoulos, P., Kövari, E., & Bouras, C. (2008). Assessing the role of cingulate cortex in bipolar disorder: Neuropathological, structural and functional imaging data. *Brain Research Reviews, 59*, 9–21.

Fowler, C. D., Liu, Y., & Wang, Z. (2007). Estrogen and adult neurogenesis in the amygdala and hypothalamus. *Brain Research Reviews, 57*, 342–351.

Fox, K., Glazewski, S., & Schulze, S. (2000). Plasticity and stability of somatosensory maps in thalamus and cortex. *Current Opinion in Neurobiology, 10*, 494–497.

Francis, D. D., & Meaney, M. J. (1999). Maternal care and the development of stress responses. *Current Opinion in Neurobiology, 9*, 128–134.

Frank, M. J., Samanta, J., Moustafa, A. A., & Sherman, S. J. (2007). Hold your horses: Impulsivity, deep brain stimulation, and medication in Parkinsonism. *Science, 318*, 1309–1312.

Frankland, P. W., & Miller, F. D. (2008). Regenerating your senses: Multiple roles for neurogenesis in the adult brain. *Nature Neuroscience, 11*, 1124–1126.

Franklin, R. J. M., & ffrench-Constant, C. (2008). Remyelination in the CNS: From biology to therapy. *Nature Reviews Neuroscience, 9*, 839–855.

Frasure-Smith, N., & Lespérance, F. (2005). Depression and coronary heart disease: Complex synergism of mind, body, and environment. *Current Directions in Psychological Science, 14*, 39–43.

Frazier, J. A., Chiu, S., Breeze, J. L., Makris, N., Lange, N., Kennedy, D. N., . . . Biederman, J. (2005). Structural brain magnetic resonance imaging of limbic and thalamic volumes in pediatric bipolar disorder. *American Journal of Psychiatry, 162*, 1256–1265.

Freeman, M. R. (2006). Sculpting the nervous system: Glial control of neuronal development. *Current Opinion in Neurobiology, 16*, 119–125.

Freund, H. J. (2003). Somatosensory and motor disturbances in patients with parietal lobe lesions. *Advances in Neurology, 93*, 179–193.

Frick, K. M. (2009). Estrogens and age-related memory decline in rodents: What have we learned and where do we go from here? *Hormones and Behavior, 55*, 2–23.

Friederici, A. D. (2009). Pathways to language: Fiber tracts in the human brain. *Trends in Cognitive Sciences, 13*, 175–181.

Friedman, D., Cycowicz, Y. M., & Gaeta, H. (2001). The novelty P3: An event-related brain potential (ERP) sign of the brain's evaluation of novelty. *Neuroscience and Biobehavioural Reviews, 25*, 355–373.

Friedman, J., Globus, G., Huntley, A., Mullaney, D., Naitoh, P., & Johnson, L. (1977). Performance and mood during and after gradual sleep reduction. *Psychophysiology, 14*, 245–250.

Frigon, A., & Rossignol, S. (2008). Adaptive changes of the locomotor pattern and cutaneous reflexes during locomotion studied in the same cats before and after spinalization. *Journal of Physiology, 12*, 2927–2945.

Fritz, J. B., Elhilali, M., David, S. V., & Shamma, S. A. (2007). Auditory attention—focusing the searchlight on sound. *Current Opinion in Neurobiology, 17*, 437–455.

Fuerst, P. G., & Burgess, R. W. (2009). Adhesion molecules in establishing retinal circuitry. *Current Opinion in Neurobiology, 19*, 389–394.

Fujioka, M., Okuchi, K., Hiramatsu, K-I., Sakaki, T., Sakaguchi, S., & Ishii, Y. (1997). Specific changes in human brain after hypoglycemic injury. *Stroke, 28*, 584–587.

Fukuhara, A., Matsuda, M., Nishizawa, M., Segawa, K., Tanaka, M., Kishi-moto, K., . . . Shimomura, I. (2005). Visfatin: A protein secreted by visceral fat that mimics the effects of insulin. *Science, 307*, 426–430.

Funnell, M. G., Corballis, P. M., & Gazzaniga, M. S. (1999). A deficit in perceptual matching in the left hemisphere of a callostomy patient. *Neuropsychologia, 37*, 1143–1154.

Fushimi, A., & Hayashi, M. (2008). Pattern of slow-wave sleep in afternoon naps. *Sleep and Biological Rhythms, 6*, 187–189.

Fuster, J. M. (2000). The prefrontal cortex of the primate: A synopsis. *Psychobiology, 28*, 125–131.

Gabrieli, J. D. E., Corkin, S., Mickel, S. F., & Growdon, J. H. (1993). Intact acquisition and long-term retention of mirror-tracing skill in Alzheimer's disease and in global amnesia. *Behavioral Neuroscience, 107*, 899–910.

Gaffan, D. (1974). Recognition impaired and association intact in the memory of monkeys after transection of the fornix. *Journal of Comparative and Physiological Psychology, 86*, 1100–1109.

Galaburda, A. M., LoTurco, J., Ramus, F., Fitch, R. H., & Rosen, G. D. (2006). From genes to behavior in developmental dyslexia. *Nature Neuroscience, 9*, 1213–1217.

Galea, L. A. M., Spritzer, M. D., Barker, J. M., & Pawluski, J. L. (2006). Gonadal hormone modulation of hippocampal neurogenesis in the adult. *Hippocampus, 16*, 225–232.

Galef, B. G. (1989). Laboratory studies of naturally-occurring feeding behaviors: Pitfalls, progress and problems in ethoexperimental analysis. In R. J. Blanchard, P. F. Brain, D. C. Blanchard, & S. Parmigiani (Eds.), *Ethoexperimental approaches to the study of behavior* (pp. 51–77). Dordrecht, Netherlands: Kluwer Academic.

Galef, B. G. (1995). Food selection: Problems in understanding how we choose foods to eat. *Neuroscience and Biobehavioural Reviews, 20*, 67–73.

Galef, B. G. (1996). Social enhancement of food preferences in Norway rats: A brief review. In C. M. Heyes & B. G. Galef, Jr. (Eds.), *Social learning in animals: The roots of culture* (pp. 49–64). New York, NY: Academic Press.

Galef, B. G., Whishkin, E. E., & Bielavska, E. (1997). Interaction with demonstrator rats changes observer rats' affective responses to flavors. *Journal of Comparative Psychology, 111*, 393–398.

Galef, B. G., & Wright, T. J. (1995). Groups of naive rats learn to select nutritionally adequate foods faster than do isolated naive rats. *Animal Behavior, 49*, 403–409.

Gallup, G. G., Jr. (1983). Toward a comparative psychology of mind. In R. L. Mellgren (Ed.), *Animal cognition and behavior* (pp. 473–505). New York, NY: North-Holland.

Galluzzi, L., Blomgren, K., & Kroemer, G. (2009). Mitochondrial membrane permeabilization in neuronal injury. *Nature Reviews Neuroscience, 10*, 481–494.

Galuske, R. A. W., Schlote, W., Bratzke, H., & Singer, W. (2000). Interhemispheric asymmetries of the modular structure in human temporal cortex. *Science, 289*, 1946–1949.

Gandy, S. (2003). Estrogen and neurodegeneration. *Neurochemical Research, 28*, 1003–1008.

Ganel, T., Tanzer, M., & Goodale, M. A. (2008). A double dissociation between action and perception in the context of visual illusions. *Psychological Science, 19*, 221–225.

Gangestad, S. W., Thornhill, R., & Garver-Apgar, C. E. (2005). Adaptations to ovulation: Implications for sexual and social behavior. *Current Directions in Psychological Science, 14*, 312–316.

Gao, Q., & Horvath, T. L. (2007). Neurobiology of feeding and energy expenditure. *Annual Review of Neuroscience, 30*, 367–398.

Garcia, J., & Koelling, R. A. (1966). Relation of cue to consequence in avoidance learning. *Psychonomic Science, 4*, 123–124.

Garcia-Borreguero, D., Egatz, R., Winkelmann, J., & Berger, K. (2006). Epidemiology of restless legs syndrome: The current status. *Sleep Medicine Reviews, 10*, 153–167.

Gatz, M. (2007). Genetics, dementia, and the elderly. *Current Directions in Psychological Science, 16*, 123–127.

Gawin, F. H. (1991). Cocaine addiction: Psychology and neurophysiology. *Science, 251*, 1580–1586.

Gazzaniga, M. S. (1967, August). The split brain in man. *Scientific American, 217*, 24–29.

Gazzaniga, M. S. (1998, July). The split brain revisited. *Scientific American, 278*, 51–55.

Gazzaniga, M. S. (2000). Regional differences in cortical organization. *Science, 289*, 1887–1888.

Gazzaniga, M. S. (2005). Forty-five years of split-brain research and still going strong. *Nature Reviews Neuroscience, 6*, 653–659.

Gazzaniga, M. S., & Sperry, R. W. (1967). Language after section of the cerebral commissure. *Brain, 90*, 131–148.

Ge, W-P., Yang, X-J., Zhang, Z., Wang, H-K., Shen, W., Deng, Q-D., & Duan, S. (2006). Long-term potentiation of neuron-glia synapses mediated by Ca^{2+}-permeable AMPA receptors. *Science, 312*, 1533–1537.

Gebhart, G. F. (2004). Descending modulation of pain. *Neuroscience and Biobehavioural Reviews, 27*, 729–737.

Geddes, J. R., Carney, S. M., Davies, C., Furukawa, T. A., Kupfer, D. J., Frank, E., & Goodwin, G. M. (2003). Relapse prevention with antidepressant drug treatment in depressive disorders: A systematic review. *The Lancet, 361*, 653–661.

Gegenfurtner, K. R., & Kiper, D. C. (2003). Color vision. *Annual Review of Neuroscience, 26*, 181–206.

Gehring, W. J., & Willoughby, A. R. (2002). The medial frontal cortex and the rapid processing of monetary gains and losses. *Science, 295*, 2279–2282.

Geier, A. B., Rozin, P., & Doros, G. (2006). A new heuristic that helps explain the effect of portion size on food intake. *Psychological Science, 17*, 521–525.

Geiselman, P. J. (1987). Carbohydrates do not always produce satiety: An explanation of the appetite- and hunger-stimulating effects of hexoses. *Progress in Psychobiology and Physiological Psychology, 12*, 1–46.

Geng, J. J., & Behrmann, M. (2002). Probability cuing of target location facilitates visual search implicitly in normal participants and patients with hemispatial neglect. *Psychological Science, 13*, 520–525.

Georgopoulos, A. P. (1991). Higher order motor control. *Annual Review of Neuroscience, 14*, 361–377.

Gerard, E., & Peterson, B. S. (2003). Developmental processes and brain imaging studies in Tourette syndrome. *Journal of Psychosomatic Research, 55*, 13–22.

Gerber, J. R., Johnson, J. V., Bunn, J. Y., & O'Brien, S. L. (2005). A longitudinal study of the effects of free testosterone and other psychosocial variables on sexual function during the natural traverse of menopause. *Fertility and Sterility, 83*, 643–648.

Gerlai, R., & Clayton, N. S. (1999). Analyzing hippocampal function in transgenic mice: An ethological perspective. *Trends in Neurosciences, 22*, 47–51.

Gerra, G., Borella, F., Zaimovic, A., Moi, G., Bussandri, M., Bubici, C., & Bertacca, S. (2004). Buprenorphine versus methadone for opioid dependence: Predictor variables for treatment outcome. *Drug and Alcohol Dependence, 75*, 37–45.

Gerstein, M., & Zheng, D. (2006, August). The real life of pseudogenes. *Scientific American, 295*, 49–55.

Geschwind, N. (1970). The organization of language and the brain. *Science, 170*, 940–944.

Geschwind, N. (1979, September). Specializations of the human brain. *Scientific American, 241*, 180–199.

Geschwind, N., & Levitsky, W. (1968). Human brain: Left-right asymmetries in temporal speech region. *Science, 161*, 186–187.

Gevins, A., Leong, H., Smith, M. E., Le, J., & Du, R. (1995). Mapping cognitive brain function with modern high-resolution electroencephalography. *Trends in Neurosciences, 18*, 429–436.

Ghashghaei, H. T., Lai, C., & Anton, E. S. (2007). Neuronal migration in the adult brain: Are we there yet? *Nature Reviews Neuroscience, 8*, 141–151.

Giardino, N. D., Friedman, S. D., & Dager, S. R. (2007). Anxiety, respiration, and cerebral blood flow: Implications for functional brain imaging. *Comprehensive Psychiatry, 48*, 103–112.

Giaume, C., Koulakoff, A., Roux, L., Holcman, D., & Rouach, N. (2010). Astroglia networks: A step further in neuroglial and gliovascular interactions. *Nature Reviews Neuroscience, 11*, 87–99.

Gibbons, R. D., Brown, C. H., Hur, K., Marcus, S. M., Bhaumik, D. K., Erkens, J. A., . . . Mann, J. J. (2007). Early evidence on the effects of regulators' suicidality warnings on SSRI prescriptions and suicide in children and adolescents. *American Journal of Psychiatry, 164,* 1356–1363.

Gibbs, J., Young, R. C., & Smith, G. P. (1973). Cholecystokinin decreases food intake in rats. *Journal of Comparative and Physiological Psychology, 84,* 488–495.

Gilbert, C. D., & Wiesel, T. N. (1992). Receptive field dynamics in adult primary visual cortex. *Nature, 356,* 150–152.

Gilbert, D. L., Christian, B. T., Gelfand, M. J., Shi, B., Mantil, J., & Sallee, F. R. (2006). Altered mesolimbocortical and thalamic dopamine in Tourette syndrome. *Neurology, 67,* 1695–1697.

Gilbertson, T., Damak, S., & Margolskee, R. F. (2000). The molecular physiology of taste transduction. *Current Opinion in Neurobiology, 10,* 519–527.

Gillespie, C. F., & Nemeroff, C. B. (2007). Corticotropin-releasing factor and the psychobiology of early-life stress. *Current Directions in Psychological Science, 16,* 87–89.

Gillette, M. U., & McArthur, A. J. (1996). Circadian actions of melatonin at the suprachiasmatic nucleus. *Behavioural Brain Research, 73,* 135–139.

Gillette, M. U., & Sejnowski, T. J. (2005). Biological clocks coordinately keep life on time. *Science, 309,* 1196–1198.

Gilliam, T. C., Gusella, J. F., & Lehrach, H. (1987). Molecular genetic strategies to investigate Huntington's disease. *Advances in Neurology, 48,* 17–29.

Gimelbrant, A., Hutchinson, J. N., Thompson, B. R., & Chess, A. (2007). Widespread monoallelic expression of human autosomes. *Science, 318,* 1136–1140.

Glaser, R., Rice, J., Sheridan, J., Fertel, R., Stout, J., Speicher, C., . . . Kiecolt-Glaser, J. (1987). Stress-related immune suppression: Health implications. *Brain, Behavior, and Immunity, 1,* 7–20.

Glickfeld, L. L., & Scanziani, M. (2005). Self-administering cannabinoids. *Trends in Neurosciences, 28,* 341–343.

Glickstein, M. (2000). How are visual areas of the brain connected to motor areas for the sensory guidance of movement? *Trends in Neurosciences, 23,* 613–617.

Goddard, G. V., McIntyre, D. C., & Leech, C. K. (1969). A permanent change in brain function resulting from daily electrical stimulation. *Experimental Neurology, 25,* 295–330.

Goebel, T. (2007). The missing years for modern humans. *Science, 315,* 194–196.

Goebel, T., Waters, M. R., & O'Rourke, D. H. (2008). The late Pleistocene dispersal of modern humans in the Americas. *Science, 319,* 1497–1502.

Goedert, M. (1993). Tau protein and the neurofibrillary pathology of Alzheimer's disease. *Trends in Neurosciences, 16,* 460–465.

Goense, J. B. M., & Logothetis, N. K. (2008). Neurophysiology of the BOLD fMRI signal in awake monkeys. *Current Biology, 18,* 631–640.

Gogolla, N., Galimberti, I., & Caroni, P. (2007). Structural plasticity of axon terminals in the adult. *Current Opinion in Neurobiology, 17,* 516–524.

Gold, P. E., Cahill, L., & Wenk, G. L. (2002). Ginkgo biloba: A cognitive enhancer? *Psychological Science in the Public Interest, 3,* 2–11.

Gold, R. M., Jones, A. P., Sawchenko, P. E., & Kapatos, G. (1977). Paraventricular area: Critical focus of a longitudinal neurocircuitry mediating food intake. *Physiology & Behavior, 18,* 1111–1119.

Goldberg, A. M., & Hartung, T. (2006, January). Protecting more than animals. *Scientific American, 294,* 84–91.

Goldberg, J. L., & Barres, B. A. (2000). The relationship between neuronal survival and regeneration. *Annual Review of Neuroscience, 23,* 576–612.

Goldman, S. A., & Nottebohm, F. (1983). Neuronal production, migration, and differentiation in a vocal control nucleus of the adult female canary brain. *Proceedings of the National Academy of Sciences, USA, 80,* 2390–2394.

Goldstein, I. (2000, August). Male sexual circuitry. *Scientific American, 283,* 70–75.

Goldstone, A. P. (2004). Prader-Willi syndrome: Advances in genetics, pathophysiology and treatment. *Trends in Endocrinology and Metabolism, 15,* 12–20.

Golinkoff, R. M., & Hirsh-Pasek, K. (2006). Baby wordsmith: From associationist to social sophisticate. *Current Directions in Psychological Science, 15,* 30–33.

Gollin, E. S. (1960). Developmental studies of visual recognition of incomplete objects. *Perceptual Motor Skills, 11,* 289–298.

Gollnick, P. D., & Hodgson, D. R. (1986). The identification of fiber types in skeletal muscle: A continual dilemma. *Exercise and Sport Sciences Reviews, 14,* 81–104.

Gomi, H. (2008). Implicit online corrections of reaching movements. *Current Opinion in Neurobiology, 18,* 558–564.

González-Maeso, J., & Sealfon, S. C. (2009). Psychedelics and schizophrenia. *Trends in Neurosciences, 32,* 225–232.

Good, M. A., Barnes, P., Staal, V., McGregor, A., & Honey, R. C. (2007). Context- but not familiarity-dependent forms of object recognition are impaired following excitotoxic hippocampal lesions in rats. *Behavioral Neuroscience, 121,* 218–223.

Goodale, M. A. (2004). Perceiving the world and grasping it: Dissociations between conscious and unconscious visual processing. In M. Gazzaniga (Ed.), *The cognitive neurosciences* (pp. 1159–1172). Cambridge, MA: MIT Press.

Goodale, M. A., & Milner, A. D. (1992). Separate visual pathways for perception and action. *Trends in Neurosciences, 15,* 20–25.

Goodale, M. A., & Milner, A. D. (2004). *Sight unseen: An exploration of conscious and unconscious perception.* Oxford, England: Oxford University Press.

Goodale, M. A., & Westwood, D. A. (2004). An evolving view of duplex vision: Separate but interacting cortical pathways for perception and action. *Current Opinion in Neurobiology, 14,* 203–211.

Goodenough, D. R., Shapiro, A., Holden, M., & Steinschriber, L. (1959). A comparison of "dreamers" and "nondreamers": Eye movements, electroencephalograms, and the recall of dreams. *Journal of Abnormal and Social Psychology, 59,* 295–303.

Gooren, L. (2006). The biology of human psychosexual differentiation. *Hormones and Behavior, 50,* 589–601.

Gordon, J. A., & Hen, R. (2004). Genetic approaches to the study of anxiety. *Annual Review of Neuroscience, 27,* 193–222.

Gorski, R. A. (1980). Sexual differentiation in the brain. In D. T. Krieger & J. C. Hughes (Eds.), *Neuroendocrinology* (pp. 215–222). Sunderland, MA: Sinauer.

Gorski, R. A., Gordon, J. H., Shryne, J. E., & Southam, A. M. (1978). Evidence for a morphological sex difference within the medial preoptic area of the rat brain. *Brain Research, 148,* 333–346.

Gotlib, I. H., & Hamilton, J. P. (2008). Neuroimaging and depression: Current status and unresolved issues. *Current Directions in Psychological Science, 17,* 159–163.

Gottesman, I. I., & Bertelsen, A. (1989). Confirming unexpressed genotypes for schizophrenia. *Archives of General Psychiatry, 46,* 867–872.

Gottesman, I. I., & Hanson, D. R. (2005). Human development: Biological and genetic processes. *Annual Review of Psychology, 56,* 263–286.

Gottesman, I. I., & Shields, J. (1982). *Schizophrenia: The epigenetic puzzle.* Cambridge, England: Cambridge University Press.

Gottfried, J. A. (2009). Function follows form: Ecological constraints on odor codes and olfactory percepts. *Current Opinion in Neurobiology, 19,* 422–429.

Gottfried, J. A., & Zald, D. H. (2005). On the scent of human olfactory orbitofrontal cortex: Meta-analysis and comparison to non-human primates. *Brain Research Reviews, 50,* 287–304.

Gottlieb, R. A. (2000). Role of mitochondria in apoptosis. *Critical Reviews in Eukaryotic Gene Expression, 10,* 231–239.

Götz, J., & Ittner, L. M. (2008). Animal models of Alzheimer's disease and frontotemporal dementia. *Nature Reviews Neuroscience, 9,* 532–544.

Gougoux, F., Zatorre, R. J., Lassonde, M., Voss, P., & Lepore, F. (2005). A functional neuroimaging study of sound localization: Visual cortex activity predicts performance in early-blind individuals. *PloS Biology, 3,* 324–333.

Gould, E. (2004). Stress, deprivation, and adult neurogenesis. In M. S. Gazzaniga (Ed.), *The cognitive neurosciences* (pp. 139–148). Cambridge, MA: MIT Press.

Gould, E., Reeves, A. J., Graziano, M. S. A., & Gross, C. G. (1999). Neurogenesis in the neocortex of adult primates. *Science, 286*, 548–552.

Gould, T. D., & Einat, H. (2007). Animal models of bipolar disorder and mood stabilizer efficacy: A critical need for improvement. *Neuroscience and Biobehavioral Reviews, 31*, 825–831.

Goy, R. W., & McEwen, B. S. (1980). *Sexual differentiation of the brain.* Cambridge, MA: MIT Press.

Grace, A. A., Floresco, S. B., Goto, Y., & Lodge, D. J. (2007). Regulation of firing of dopaminergic neurons and control of goal-directed behaviors. *Trends in Neurosciences, 30*, 220–227.

Grady, K. L., Phoenix, C. H., & Young, W. C. (1965). Role of the developing rat testis in differentiation of the neural tissues mediating mating behavior. *Journal of Comparative and Physiological Psychology, 59*, 176–182.

Grakoui, A., Bromley, S. K., Sumen, C., Davis, M. M., Shaw, A. S., Allen, P. M., & Dustin, M. L. (1999). The immunological synapse: A molecular machine controlling T cell activation. *Science, 285*, 221–227.

Grandner, M. A., & Drummond, S. P. A. (2007). Who are the long sleepers? Towards an understanding of the mortality relationship. *Sleep Medicine Reviews, 11*, 341–360.

Grant, I., Gonzalez, R., Carey, C. L., Natarajan, L., & Wolfson, T. (2003). Nonacute (residual) neurocognitive effects of cannabis use: A meta-analytic study. *Journal of the International Neuropsychological Society, 9*, 679–689.

Grant, J. E., Brewer, J. A., & Potenza, M. N. (2006). The neurobiology of substance and behavioral addictions. *CNS Spectrums, 11*, 924–930.

Grant, J. E., Kim, S. W., & Eckert, E. D. (2002). Body dysmorphic disorder in patients with anorexia nervosa: Prevalence, clinical features and delusionality of body image. *International Journal of Eating Disorders, 32*, 291–300.

Grant, P. R. (1991, October). Natural selection and Darwin's finches. *Scientific American, 265*, 82–87.

Gray, J. A., & McNaughton, N. (2000). *The neuropsychology of anxiety.* London: Oxford University Press.

Graybiel, A. M. (2005). The basal ganglia: Learning new tricks and loving it. *Current Opinion in Neurobiology, 15*, 638–644.

Graybiel, A. M. (2008). Habits, rituals, and the evaluative brain. *Nature Reviews Neuroscience, 31*, 359–387.

Graybiel, A. M., & Saka, E. (2004). The basal ganglia and the control of action. In M. S. Gazzaniga (Ed.), *The cognitive neurosciences* (pp. 495–510). Cambridge, MA: MIT Press.

Graziano, M. S. A. (2009). *The intelligent movement machine: An ethological perspective on the primate motor system.* New York, NY: Oxford University Press.

Green, C. B., & Menaker, M. (2003). Clock on the brain. *Science, 301*, 319–320.

Green, S. (1991). Benzodiazepines, putative anxiolytics and animal models of anxiety. *Trends in Neurosciences, 14*, 101–103.

Greene, P. E., Fahn, S., Tsai, W. Y., Winfield, H., Dillon, S., Kao, R., et al. (1999). Double-blind controlled trial of human embryonic dopaminergic tissue transplants in advanced Parkinson's disease: Long-term unblinded follow-up phase. *Neurology, 52*(Suppl. 2).

Greenspan, R. J. (2004). E pluribus unum, ex uno plura: Quantitative and single-gene perspectives on the study of behavior. *Annual Review of Neuroscience, 27*, 79–105.

Greer, D. M. (2006). Mechanisms of injury in hypoxic-ischemic encephalopathy: Implications to therapy. *Seminars in Neurology, 26*, 373–379.

Greicius, M. D., Flores, B. H., Menon, V., Glover, G. H., Solvason, H. B., Kenna, H., . . . Schatzberg, A. F. (2007). Resting-state functional connectivity in major depression: Abnormally increased contributions from subgenual cingulate cortex and thalamus. *Biological Psychiatry, 62*, 429–437.

Griffiths, T. D., Warren, J. D., Scott, S. K., Nelken, I., & King, A. J. (2004). Cortical processing of complex sound: A way forward? *Trends in Neurosciences, 27*, 181–185.

Grillner, S. (1985). Neurobiological bases of rhythmic motor acts in vertebrates. *Science, 228*, 143–149.

Grillner, S., & Dickinson, M. (2002). Motor systems. *Current Opinion in Neurobiology, 12*, 629–632.

Grillner, S., & Jessell, T. M. (2009). Measured motion: Searching for simplicity in spinal locomotor networks. *Current Opinion in Neurobiology, 19*, 572–586.

Grill-Spector, K., & Mallach, R. (2004). The human visual cortex. *Annual Review of Neuroscience, 27*, 649–677.

Grill-Spector, K., Sayres, R., & Ress, D. (2006). High-resolution imaging reveals highly selective nonface clusters in the fusiform face area. *Nature Neuroscience, 9*, 1177–1185.

Grimson, W. E. L., Kikinis, R., Jolesz, F. A., & Black, P. M. (1999, June). Virtual-reality technology. *Scientific American, 280*, 63–69.

Grine, F. E., Bailey, R. M., Harvati, K., Nathan, R. P., Morris, A. G., Henderson, G. M., . . . Pike, A. W. G. (2007). Late Pleistocene human skull from Hofmeyr, South Africa, and modern human origins. *Science, 315*, 226–229.

Gross, C., & Hen, R. (2004). The developmental origins of anxiety. *Nature Reviews Neuroscience, 5*, 545–552.

Gross, C. G., Moore, T., & Rodman, H. R. (2004). Visually guided behavior after V1 lesions in young and adult monkeys and its relation to blindsight in humans. *Progress in Brain Research, 144*, 279–294.

Grossman, E., Donnelly, M., Price, R., Pickens, D., Morgan, V., Neighbor, G., & Blake, R. (2000). Brain areas involved in perception of biological motion. *Journal of Cognitive Neuroscience, 12*, 711–720.

Grubbe, M. S., & Thompson, I. D. (2004). The influence of early experience on the development of sensory systems. *Current Opinion in Neurobiology, 14*, 503–512.

Grumbach, M. M. (2002). The neuroendocrinology of human puberty revisited. *Hormone Research, 57*, S2–S14.

Grunt, J. A., & Young, W. C. (1952). Differential reactivity of individuals and the response of the male guinea pig to testosterone propionate. *Endocrinology, 51*, 237–248.

Gschwend, P., Rehm, J., Lezzi, S., Blattler, R., Steffen, T., Gutzwiller, F., & Uchtenhagen, A. (2002). Development of a monitoring system for heroin-assisted substitution treatment in Switzerland. *Sozial und Praventivmedizin, 47*, 33–38.

Guerri, C. (2002). Mechanisms involved in central nervous system dysfunctions induced by prenatal ethanol exposure. *Neurotoxicity Research, 4*, 327–335.

Guerrini, R., Dobyns, W. B., & Barkovich, A. J. (2008). Abnormal development of the human cerebral cortex: Genetics, functional consequences and treatment options. *Trends in Neurosciences, 31*, 154–162.

Guillamón, A., & Segovia, S. (1996). Sexual dimorphism in the CNS and the role of steroids. In T. W. Stone (Ed.), *CNS neurotransmitters and neuromodulators. Neuroactive steroids* (pp. 127–152). Boca Raton: CRC Press.

Gunnar, M., & Quevedo, K. (2007). The neurobiology of stress and development. *Annual Review of Psychology, 58*, 145–173.

Guo, Y., & Udin, S. B. (2000). The development of abnormal axon trajectories after rotation of one eye in *Xenopus. Journal of Neuroscience, 20*, 4189–4197.

Gustavsen, I., Bramness, J. G., Skurtveit, S., Engeland, A., Neutel, I., & Mørland, J. (2008). Road traffic accident risk related to prescriptions of the hypnotics zopiclone, zolpidem, flunitrazepam and nitrazepam. *Sleep Medicine, 9*, 818–822.

Gutfreund, Y., Zheng, W., & Knudsen, E. I. (2002). Gated visual input to the central auditory system. *Science, 297*, 1556–1559.

Guthrie, S. (2007). Patterning and axon guidance of cranial motor neurons. *Nature Reviews Neuroscience, 8*, 859–871.

Guttinger, F., Gschwend, P., Schulte, B., Rehm, J., & Uchtenhagen, A. (2003). Evaluating long-term effects of heroin-assisted treatment: The results of a 6-year follow-up. *European Addiction Research, 9*, 73–79.

Guyenet, P. G. (2006). The sympathetic control of blood pressure. *Nature Reviews Neuroscience, 7*, 335–346.

Guze, S. B., & Robins, E. (1970) Suicide and primary affective disorders. *British Journal of Psychiatry, 117*, 437–438.

Haaland, K. Y., & Harrington, D. L. (1996). Hemispheric asymmetry of movement. *Current Opinion in Neurobiology, 6*, 796–800.

Haase, G., Pettmann, B., Raoul, C., & Henderson, C. E. (2008). Signaling by death receptors in the nervous system. *Current Opinion in Neurobiology, 18,* 284–291.

Hackett, T. A., & Kaas, J. H. (2004). Auditory cortex in primates: Functional subdivisions and processing streams. In M. S. Gazzaniga (Ed.), *The cognitive neurosciences* (pp. 215–232). Cambridge, MA: MIT Press.

Haffenden, A. M., & Goodale, M. A. (1998). The effect of pictorial illusion on prehension and perception. *Journal of Cognitive Neuroscience, 10,* 122–136.

Hagg, T. (2006). Molecular regulation of adult CNS neurogenesis: An integrated view. *Trends in Neurosciences, 28,* 589–595.

Haggard, P. (2008). Human volition: Towards a neuroscience of will. *Nature Reviews Neuroscience, 9,* 934–946.

Hagoort, P., & Levelt, W. J. M. (2009). The speaking brain. *Science, 326,* 372–374.

Haist, F., Bowden, G. J., & Mao, H. (2001). Consolidation of human memory over decades revealed by functional magnetic resonance imaging. *Nature Neuroscience, 4,* 1057–1058.

Halford, J. C. G., & Blundell, J. E. (2000a). Pharmacology of appetite suppression. In E. Jucker (Ed.), *Progress in drug research* (Vol. 54, pp. 25–58). Basel, Switzerland: Birkhäuser, Verlag.

Halford, J. C. G., & Blundell, J. E. (2000b). Separate systems for serotonin and leptin in appetite control. *Annals of Medicine, 32,* 222–232.

Hall, J., Romaniuk, L., McIntosh, A. M., Steele, J. D., Johnstone, E. C., & Lawrie, S. M. (2009). Associative learning and the genetics of schizophrenia. *Trends in Neurosciences, 32,* 359–365.

Hamer, D. H., Hu, S., Magnuson, V. L., Hu, N., & Pattatucci, A. M. L. (1993). A linkage between DNA markers on the chromosome and male sexual orientation. *Science, 261,* 321–327.

Hammond, P. H., Merton, P. A., & Sutton, G. G. (1956). Nervous gradation of muscular contraction. *British Medical Bulletin, 12,* 214–218.

Hampton, R. R., & Murray, E. A. (2002). Learning of discriminations is impaired, but generalization to altered views is intact, in monkeys (*Macaca mulatta*) with perirhinal cortex removal. *Behavioral Neuroscience, 116,* 363–377.

Hampton, R. R., & Schwartz, B. L. (2004). Episodic memory in nonhumans: What, and where, is when? *Current Opinion in Neurobiology, 14,* 192–197.

Haning, W., & Goebert, D. (2007). Electrocardiographic abnormalities in methamphetamine abusers. *Addiction, 102*(Suppl. 1), 70–75.

Hankins, M. W., Peirson, S. N., & Foster, R. G. (2007). Melanopsin: An exciting photopigment. *Trends in Neurosciences, 31,* 27–36.

Hanson, G. R., Bunsey, M. D., & Riccio, D. C. (2002). The effects of pretraining and reminder treatments on retrograde amnesia in rats: Comparison of lesions to the fornix or perirhinal and entorhinal cortices. *Neurobiology of Learning & Memory, 78,* 365–378.

Happé, F., Ronald, A., & Plomin, R. (2006). Time to give up on a single explanation for autism. *Nature Neuroscience, 9,* 1218–1219.

Hardin, P. E. (2006). Essential and expendable features of the circadian timekeeping mechanism. *Current Opinion in Neurobiology, 16,* 686–692.

Hardy, J., & Selkoe, D. J. (2002). The amyloid hypothesis of Alzheimer's disease: Progress and problems on the road to therapeutics. *Science, 297,* 353–356.

Harkany, T., Mackie, K., & Doherty, P. (2008). Wiring and firing neuronal networks: Endocannabinoids take center stage. *Current Opinion in Neurobiology, 18,* 338–345.

Harris, G. W. (1955). *Neural control of the pituitary gland.* London, England: Edward Arnold.

Harris, G. W., & Levine, S. (1965). Sexual differentiation of the brain and its experimental control. *Journal of Physiology, 181,* 379–400.

Harris, L. J. (1978). Sex differences in spatial ability: Possible environmental, genetic, and neurological factors. In M. Kinsbourne (Ed.), *Asymmetrical function of the brain* (p. 463). Cambridge, England: Cambridge University Press.

Harris, L. J., Clay, J., Hargreaves, F. J., & Ward, A. (1933). Appetite and choice of diet: The ability of the vitamin B deficient rat to discriminate between diets containing and lacking the vitamin. *Proceedings of the Royal Society of London (B), 113,* 161–190.

Harris, M., & Grunstein, R. R. (2008). Treatments for somnambulism in adults: Assessing the evidence. *Sleep Medicine Reviews, 13,* 295–297.

Harrison, Y., & Horne, J. A. (2000). The impact of sleep deprivation on decision making—a review. *Journal of Experimental Psychology—Applied, 6,* 236–249.

Hartmann, D. S., & Civelli, O. (1997). Dopamine receptor diversity: Molecular and pharmacological perspectives. *Progress in Drug Research, 48,* 173–194.

Harvey, J. A. (2004). Cocaine effects on the developing brain: Current status. *Neuroscience and Biobehavioral Reviews, 27,* 751–764.

Harvey, P. H., & Krebs, J. R. (1990). Comparing brains. *Science, 249,* 140–145.

Haslam, D., Sattar, N., & Lean, M. (2006). Obesity—time to wake up. *British Medical Journal, 333,* 640–642.

Hastings, M. H., Reddy, A. B., & Maywood, E. S. (2003). A clockwork web: Circadian timing in brain and periphery, in health and disease. *Nature Reviews Neuroscience, 4,* 649–661.

Hata, Y., & Stryker, M. P. (1994). Control of thalamocortical afferent rearrangement by postsynaptic activity in developing visual cortex. *Science, 265,* 1732–1735.

Hattar, S., Liao, H.-W., Takao, M., Berson, D. M., & Yau, K.-W. (2002). Melanopsin-containing retinal ganglion cells: Architecture, projections, and intrinsic photosensitivity. *Science, 295,* 1065–1070.

Hattar, S., Lucas, R. J., Mrosovsky, N., Thompson, S., Douglas, R. H., Hankins, M. W., . . . Yau, K.-W. (2003). Melanopsin and rod-cone photoreceptive systems account for all major accessory visual functions in mice. *Nature, 424,* 75–81.

Hatten, M. E. (2002). New directions in neural migration. *Science, 297,* 1660–1665.

Hauser, M. D. (1993). Right hemisphere dominance for the production of facial expression in monkeys. *Science, 261,* 475–478.

Hauser, M. D., Chomsky, N., & Fitch, W. T. (2005). The faculty of language: What is it, who has it, and how did it evolve? *Science, 298,* 1569–1579.

Haut, S. R., & Shinnar, S. (2008). Considerations in the treatment of a first unprovoked seizure. *Seminars in Neurology, 28,* 289–296.

Haxby, J. V. (2006). Fine structure in representations of faces and objects. *Nature Neuroscience, 9,* 1084–1085.

Hayne, H. (2003). Infant memory development: Implications for childhood amnesia. *Developmental Review, 24,* 33–73.

Haynes, J-D., & Rees, G. (2006). Decoding mental states from brain activity in humans. *Nature Reviews Neuroscience, 7,* 523–534.

Hebb, D. O. (1949). *The organization of behavior.* New York, NY: Wiley.

Hébert, S. S., & De Strooper, B. (2009). Alterations of the microRNA network cause neurodegenerative disease. *Trends in Neurosciences, 32,* 199–206.

Hécaen, H., & Angelergues, R. (1964). Localization of symptoms in aphasia. In A. V. S. de Reuck & M. O'Connor (Eds.), *CIBA foundation symposium on the disorders of language* (pp. 222–256). London: Churchill Press.

Heffner, H. E., & Heffner, R. S. (2003). Audition. In S. Davis (Ed.), *Handbook of research methods in experimental psychology* (pp. 413–440). Oxford, England: Blackwell.

Heffner, H. E., & Masterton, R. B. (1990). Sound localization in mammals: Brainstem mechanisms. In M. Berkley & W. Stebbins (Eds.), *Comparative perception, Vol. I: Discrimination.* New York, NY: Wiley.

Heilman, K. M., Watson, R. T., & Rothi, L. J. G. (1997). Disorders of skilled movements: Limb apraxia. In T. E. Feinberg & M. J. Farah (Eds.), *Behavioral neurology and neuropsychology* (pp. 227–236). New York, NY: McGraw-Hill.

Heinrichs, R. W. (2005). The primacy of cognition in schizophrenia. *American Psychologist, 60,* 229–242.

Hellemans, K. G. C., Dickinson, A., & Everitt, B. J. (2006). Motivational control of heroin seeking by conditioned stimuli associated with withdrawal and heroin taking by rats. *Behavioral Neuroscience, 120,* 103–114.

Helmer, W. J. (1986, February). The madman in the tower. *Texas Monthly.*

Henke, K., Kroll, N. E. A., Behniea, H., Amaral, D. G., Miller, M. B., Rafal, R., & Gazzaniga, M. S. (1999). Memory lost and regained following bilateral hippocampal damage. *Journal of Cognitive Neuroscience, 11,* 682–697.

Henley, C. L., Nunez, A. A., & Clemens, L. G. (2009). Estrogen treatment during development alters adult partner preference and reproductive behavior in female laboratory rats. *Hormones and Behavior, 55,* 68–75.

Hennessey, A. C., Camak, L., Gordon, F., & Edwards, D. A. (1990). Connections between the pontine central gray and the ventromedial hypothalamus are essential for lordosis in female rats. *Behavioral Neuroscience, 104,* 477–488.

Hensch, T. K. (2004). Critical period regulation. *Annual Review of Neuroscience, 27,* 549–579.

Herman, R., He, J., D'Luzansky, S., Willis, W., & Dilli, S. (2002). Spinal cord stimulation facilitates functional walking in a chronic, incomplete spinal cord injured. *Spinal Cord, 2,* 65–68.

Hermanowicz, N. (2007). Drug therapy for Parkinson's disease. *Seminars in Neurology, 27,* 97–105.

Hernandez, G., Haines, E., Rajabi, H., Stewart, J., Arvanitogiannis, A., & Shizgal, P. (2007). Predictable and unpredictable rewards produce similar changes in dopamine tone. *Behavioral Neuroscience, 121,* 887–895.

Hernandez, G., Hamdani, S., Rajabi, H., Conover, K., Stewart, J., Arvanitogiannis, A., & Shizgal, P. (2006). Prolonged rewarding stimulation of the rat medial forebrain bundle: Neurochemical and behavioral consequences. *Behavioral Neuroscience, 120,* 888–904.

Herrmann, E., Call, J., Hernández-Lloreda, M. V., Hare, B., & Tomasello, M. (2007). Humans have evolved specialized skills of social cognition: The cultural intelligence hypothesis. *Science, 317,* 1360–1366.

Hertzog, C., Kramer, A. F., Wilson, R. S., & Lindenberger, U. (2008). Enrichment effects on adult cognitive development: Can the functional capacity of older adults be preserved or enhanced? *Psychological Science in the Public Interest, 9,* 1–65.

Hertz-Picciotto, I., & Delwiche, L. (2009). The rise in autism and the role of age at diagnosis. *Epidemiology, 20,* 84–90.

Hestrin, S., & Galarreta, M. (2005). Electrical synapses define networks of neocortical GABAergic neurons. *Trends in Neurosciences, 28,* 304–309.

Hetherington, A. W., & Ranson, S. W. (1940). Hypothalamic lesions and adiposity in the rat. *Anatomical Record, 78,* 149–172.

Heuser, J. E., Reese, T. S., Dennis, M. J., Jan, Y., Jan, L., & Evans, L. (1979). Synaptic vesicle exocytosis captured by quick freezing and correlated with quantal transmitter release. *Journal of Cell Biology, 81,* 275–300.

Heuss, C., & Gerber, U. (2000). G-protein–independent signaling by G-protein–coupled receptors. *Trends in Neurosciences, 23,* 469–474.

Heymsfield, S. B., Greenberg, A. S., Fujioka, K., Dixon, R. M., Kushner, R., Hunt, T., . . . McCamish, M. (1999). Recombinant leptin for weight loss in obese and lean adults. *Journal of the American Medical Association, 282,* 1568–1575.

Hickok, G., Bellugi, U., & Klima, E. S. (2001, June). Sign language in the brain: How does the human brain process language? New studies of deaf signers hint at an answer. *Scientific American, 284,* 58–65.

Hickok, G., Okada, K., Barr, W., Pa, J., Rogalsky, C., Donnelly, K., . . . Grant, A. (2008). Bilateral capacity for speech sound processing in auditory comprehension: Evidence from Wada procedures. *Brain and Language, 107,* 179–184.

Hickok, G., & Poeppel, D. (2007). The cortical organization of speech processing. *Nature Reviews Neuroscience, 8,* 393–402.

Higgins, S. T., Heil, S. H., & Lussier, J. P. (2004). Clinical implications of reinforcement as a determinant of substance use disorders. *Annual Review of Psychology, 55,* 431–461.

Higgs, S., Williamson, A. C., Rotshtein, P., & Humphreys, G. W. (2008). Sensory-specific satiety is intact in amnesics who eat multiple meals. *Psychological Science, 19,* 623–628.

Hilgetag, C. C., & Barbas, H. (2009, February). Sculpting the brain. *Scientific American, 300,* 66–71.

Himle, M. B., & Woods, D. W. (2005). An experimental evaluation of tic suppression and the tic rebound effect. *Behaviour Research and Therapy, 43,* 1443–1451.

Himmelbach, M., & Karnath, H-O. (2005). Dorsal and ventral stream interaction: Contributions from optic ataxia. *Journal of Cognitive Neuroscience, 17,* 632–640.

Hines, M. (2003). Sex steroids and human behavior: Prenatal androgen exposure and sex-typical play behavior in children. *Annals of the New York Academy of Sciences, 1007,* 272–282.

Hirsch, J. A., & Martinez, L. M. (2006). Circuits that build visual cortical receptive fields. *Trends in Neurosciences, 29,* 30–39.

Hjartåker, A., Adami, H-O., Lund, E., & Weiderpass, E. (2005). Body mass index and mortality in a prospectively studied cohort of Scandinavian women: The Women's Lifestyle and Health Cohort Study. *European Journal of Epidemiology, 20,* 747–754.

Hobson, J. A. (1989). *Sleep.* New York, NY: Scientific American Library.

Hobson, J. A., & Pace-Schott, E. F. (2002). The cognitive neuroscience of sleep: Neuronal systems, consciousness and learning. *Nature Reviews Neuroscience, 3,* 679–693.

Hockfield, S., & Kalb, R. G. (1993). Activity-dependent structural changes during neuronal development. *Current Opinion in Neurobiology, 3,* 87–92.

Hofer, S. B., Mrsic-Flogel, T. D., Bonhoeffer, T., & Hübener, M. (2005). Prior experience enhances plasticity in adult visual cortex. *Nature Neuroscience, 9,* 127–132.

Hofer, S. B., Mrsic-Flogel, T. D., Bonhoeffer, T., & Hübener, M. (2006). Lifelong learning: Ocular dominance plasticity in mouse visual cortex. *Current Opinion in Neurobiology, 16,* 451–459.

Hogg, R. C., & Bertrand, D. (2004). What genes tell us about nicotine addiction. *Science, 306,* 983–986.

Holdsworth, S. J., & Bammer, R. (2008). Magnetic resonance imaging techniques: fMRI, DWI, and PWI. *Seminars in Neurology, 28,* 395–406.

Holland, L. Z. (2009). Chordate roots of the vertebrate nervous system: Expanding the molecular toolkit. *Nature Reviews Neuroscience, 10,* 736–746.

Holland, P. C., & Gallagher, M. (2004). Amygdala-frontal interactions and reward expectancy. *Current Opinion in Neurobiology, 14,* 148–155.

Hollenbeck, P. J. (2001). Insight and hindsight into Tourette syndrome. In D. J. Cohen, C. G. Goetz, & J. Jankovic (Eds.), *Tourette syndrome* (pp. 363–367). Philadelphia, PA: Lippincott Williams & Wilkins.

Hollon, S. D., Thase, M. E., & Markowitz, J. C. (2002). Treatment and prevention of depression. *Psychological Science in the Public Interest, 3,* 39–77.

Holtmaat, A., & Svoboda, K. (2009). Experience-dependent structural synaptic plasticity in the mammalian brain. *Nature Reviews Neuroscience, 10,* 647–658.

Holzman, P. S., & Matthyse, S. (1990). The genetics of schizophrenia: A review. *Current Directions in Psychological Science, 1,* 279–286.

Honea, R., Crow, T. J., Passingham, D., & Mackay, C. E. (2005). Regional deficits in brain volume in schizophrenia: A meta-analysis of voxel-based morphometry studies. *American Journal of Psychiatry, 162,* 2233–2245.

Hopkins, W. D., & Cantalupo, C. (2008). Theoretical speculations on the evolutionary origins of hemispheric specialization. *Current Directions in Psychological Science, 17,* 233–237.

Hopkins, W. D., Russell, J. L., & Cantalupo, C. (2007). Neuroanatomical correlates of handedness for tool use in chimpanzees (*Pan troglodytes*). *Psychological Science, 18,* 971–977.

Hoppenbrouwers, S. S., Schutter, D. J. L. G., Fitzgerald, P. B., Chen, R., & Daskalakis, Z. J. (2008). The role of the cerebellum in the pathophysiology and treatment of neuropsychiatric disorders: A review. *Brain Research Reviews, 59,* 185–200.

Horgan, J. (2005, October). The forgotten era of brain chips. *Scientific American, 293,* 166–173.

Horne, J. (2009). REM sleep, energy balance and "optimal foraging." *Neuroscience and Biobehavioral Reviews, 33,* 466–474.

Hornyak, M., Feige, B., Riemann, D., & Voderholzer, U. (2006). Periodic leg movements in sleep and periodic limb movement disorder: Prevalence, clinical significance and treatment. *Sleep Medicine Reviews, 10,* 169–177.

Horton, J. (2009). Akinetopsia from nefazodone toxicity. *American Journal of Ophthalmology, 128,* 530–531.

Horton, J. C., & Sinich, L. C. (2004). A new foundation for the visual cortical hierarchy. In M. Gazzaniga (Ed.), *The cognitive neurosciences* (pp. 233–244). Cambridge, MA: MIT Press.

Horvath, T. L. (2005). The hardship of obesity: A soft-wired hypothalamus. *Nature Neuroscience, 8*, 561–565.

Hoshi, E., & Tanji, J. (2007). Distinctions between dorsal and ventral premotor areas: Anatomical connectivity and functional properties. *Current Opinion in Neurobiology, 17*, 234–242.

Houghton, W. C., Scammell, T. E., & Thorpy, M. (2004). Pharmacotherapy for cataplexy. *Sleep Medicine Reviews, 8*, 355–366.

Howald, C., Merla, G., Digilio, M. C., Amenta, S., Lyle, R., Deutsch, S., . . . Reymond, A. (2006). Two high throughput technologies to detect segmental aneuploidies identify new Williams-Beuren syndrome patients with atypical deletions. *Journal of Medical Genetics, 43*, 266–273.

Hrabovszky, Z., & Hutson, J. M. (2002). Androgen imprinting of the brain in animal models and humans with intersex disorders: Review and recommendations. *The Journal of Urology, 168*, 2142–2148.

Hromádka, T., & Zador, A. M. (2009). Representations in auditory cortex. *Current Opinion in Neurobiology, 19*, 430–433.

Hsiao, K., Chapman, P., Nilsen, S., Eckman, C., Harigaya, Y., Younkin, S., . . . Cole, G. (1996). Correlative memory deficits, Aβ elevation, and amyloid plaques in transgenic mice. *Science, 274*, 99–102.

Hsiao, S. (2008). Central mechanisms of tactile shape perception. *Current Opinion in Neurobiology, 18*, 418–424.

Huang, L., Treisman, A., & Pashler, H. (2007). Characterizing the limits of human visual awareness. *Science, 317*, 823–825.

Hubel, D. H., & Wiesel, T. N. (1979, September). Brain mechanisms of vision. *Scientific American, 241*, 150–162.

Hubel, D. H., & Wiesel, T. N. (2004). *Brain and visual perception: The story of a 25-year collaboration.* New York: Oxford University Press.

Hubel, D. H., Wiesel, T. N., & LeVay, S. (1977). Plasticity of ocular dominance columns in the monkey striate cortex. *Philosophical Transactions of the Royal Society of London, 278*, 377–409.

Huberman, A. D., Feller, M. B., & Chapman, B. (2008). Mechanisms underlying development of visual maps and receptive fields. *Annual Review of Neuroscience, 31*, 479–509.

Hudson, J. I., Hiripi, E., Pope, H. G., & Kessler, R. C. (2007). The prevalence and correlates of eating disorders in the National Comorbidity Survey Replication. *Biological Psychiatry, 61*, 348–358.

Huey, E. D., Krueger, F., & Grafman, J. (2006). Representations in the human prefrontal cortex. *Current Directions in Psychological Science, 15*, 167–171.

Huffman, M. A., Nahallage, C. A. D., & Leca, J-B. (2008). Cultured monkeys. *Current Directions in Psychological Science, 17*, 410–414.

Hug, C., & Lodish, H. F. (2005). Visfatin: A new adipokine. *Science, 307*, 366–367.

Huguenard, J. R. (2000). Reliability of axonal propagation: The spike doesn't stop here. *Proceedings of the National Academy of Science, U.S.A., 97*, 9349–9350.

Huijbregts, S. C. J., de Sonneville, L. M. J., van Spronsen, F. J., Licht, R., & Sergeant, J. A. (2002). The neuropsychological profile of early and continuously treated phenylketonuria: Orienting, vigilance, and maintenance versus manipulation-functions of working memory. *Neuroscience and Biobehavioural Reviews, 26*, 697–712.

Huizink, A. C., & Mulder, E. J. H. (2006). Maternal smoking, drinking or cannabis use during pregnancy and neurobehavioral and cognitive functioning in human offspring. *Neuroscience and Biobehavioural Reviews, 30*, 24–41.

Hull, E. M., Lorrain, D. S., Du, J., Matuszewich, L., Lumley, L. A., Putnam, S. K., & Moses, J. (1999). Hormone-neurotransmitter interaction in the control of sexual behavior. *Behavioural Brain Research, 105*, 105–116.

Hume, K. I., & Mills, J. N. (1977). Rhythms of REM and slow wave sleep in subjects living on abnormal time schedules. *Waking and Sleeping, 1*, 291–296.

Hurlbert, A. (2003). Colour vision: Primary visual cortex shows its influence. *Current Biology, 13*, 270–272.

Hurlbert, A., & Wolf, K. (2004). Color contrast: A contributory mechanism to color constancy. *Progress in Brain Research, 144*, 147–160.

Hurlemann, R., Wagner, M., Hawellek, B., Reich, H., Pieperhoff, P., Amunts, K., . . . Dolan, R. J. (2007). Amygdala control of emotion-induced forgetting and remembering: Evidence from Urbach-Wiethe disease. *Neuropsychologia, 45*, 877–884.

Husain, M., & Nachev, P. (2006). Space and the parietal cortex. *Trends in Cognitive Science, 11*, 30–36.

Hussain, N. K., & Sheng, M. (2005). Making synapses: A balancing act. *Science, 307*, 1207–1208.

Hustvedt, B. E., & Løvø, A. (1972). Correlation between hyperinsulinemia and hyperphagia in rats with ventromedial hypothalamic lesions. *Acta Physiologica Scandinavica, 84*, 29–33.

Hutsler, J., & Galuske, R. A. W. (2003). Hemispheric asymmetries in cerebral cortical networks. *Trends in Neurosciences, 26*, 429–435.

Huxley, A. (2002). From overshoot to voltage clamp. *Trends in Neurosciences, 25*(11), 553–558.

Hyde, J. S. (2005). The gender similarities hypothesis. *American Psychologist, 60*, 581–592.

Hyde, J. S., Mezulis, A. H., & Abramson, L. Y. (2008). The ABCs of depression: Integrating affective, biological, and cognitive models to explain the emergence of the gender difference in depression. *Psychological Review, 115*, 291–313.

Hyman, S. E., & Fenton, W. S. (2003). What are the right targets for psychopharmacology? *Science, 299*, 350–351.

Hyman, S. E., Malenka, R. C., & Nestler, E. J. (2006). Neural mechanisms of addiction: The role of reward-related learning. *Annual Review of Neuroscience, 29*, 565–598.

Iacoboni, M. (2005). Neural mechanisms of imitation. *Current Opinion in Neurobiology, 15*, 632–637.

Iacoboni, M. (2009). Imitation, empathy, and mirror neurons. *Annual Review of Psychology, 60*, 653–670.

Iacono, W. G., & Koenig, W. G. R. (1983). Features that distinguish the smooth-pursuit eye-tracking performance of schizophrenic, affective-disorder, and normal individuals. *Journal of Abnormal Psychology, 92*, 29–41.

Iino, M., Goto, K., Kakegawa, W., Okado, H., Sudo, M., Ishiuchi, S., . . . Ozawa, S. (2001). Glia-synapse interaction through calcium-permeable AMPA receptors in Bergmann glia. *Science, 292*, 926–929.

Ikeda, H., & Hayashi, M. (2008). Electroencephalogram activity before self-awakening. *Sleep and Biological Rhythms, 6*, 256–259.

Ikeda, H., Heinke, B., Ruscheweyh, R., & Sandkuhler, J. (2003). Synaptic plasticity in spinal lamina I projection neurons that mediate hyperalgesia. *Science, 299*, 1237–1240.

Ikonomidou, C., Bittigau, P., Ishimaru, M. J., Wozniak, D. F., Koch, C., Geenz, K., et al. (2000). Ethanol-induced apoptotic neurodegeneration and fetal alcohol syndrome. *Science, 287*, 1056–1060.

Illert, M., & Kümmel, H. (1999). Reflex pathways from large muscle spindle afferents and recurrent axon collaterals to motoneurones of wrist and digit muscles: A comparison in cats, monkeys and humans. *Experimental Brain Research, 128*, 13–19.

Imai, T., & Sakano, H. (2007). Roles of odorant receptors in projecting axons in the mouse olfactory system. *Current Opinion in Neurobiology, 17*, 507–515.

Imai, T., Yamazaki, T., Kobayakawa, R., Kobayakawa, K., Abe, T., Suzuki, M., & Sakano, H. (2009). Pre-target axon sorting establishes the neural map topography. *Science, 325*, 585–590.

Inestrosa, N. C., & Arenas, E. (2010). Emerging roles of Wnts in the adult nervous system. *Nature Reviews Neuroscience, 11*, 77–86.

Inglis, J., & Lawson, J. S. (1982). A meta-analysis of sex differences in the effects of unilateral brain damage on intelligence test results. *Canadian Journal of Psychology, 36*, 670–683.

Innocenti, G. M., & Price, D. J. (2005). Exuberance in the development of cortical networks. *Nature Reviews Neuroscience, 6*, 955–965.

Intriligator, J. M., Xie, R., & Barton, J. J. S. (2002). Blindsight modulation of motion perception. *Journal of Cognitive Neuroscience, 14*, 1174–1183.

Iversen, L. (2005). Long-term effects of exposure to cannabis. *Current Opinion in Pharmacology, 5*, 69–72.

Iwamura, Y. (1998). Hierarchical somatosensory processing. *Current Opinion in Neurobiology, 8*, 522–528.

Iwaniuk, A. N., & Whishaw, I. Q. (2000). On the origin of skilled forelimb movements. *Trends in Neurosciences, 23*, 372–376.

Iwasaki, A., & Medzhitov, R. (2010). Regulation of adaptive immunity by the innate immune system. *Science, 327*, 291–295.

Jääskeläinen, I. P., Ahveninen, J., Belliveau, J. W., Raij, T., & Sams, M. (2007). Short-term plasticity in auditory cognition. *Trends in Neurosciences, 30*, 653–661.

Jackson, P. L., & Decety, J. (2004). Motor cognition: A new paradigm to study self-other interactions. *Current Opinion in Neurobiology, 14*, 259–263.

Jacob, T. C., Moss, S. J., & Jurd, R. (2008). GABA_A receptor trafficking and its role in the dynamic modulation of neuronal inhibition. *Nature Reviews Neuroscience, 9*, 331–343.

Jacobs, G. H., & Nathans, J. (2009, April). The evolution of primate color vision. *Scientific American, 300*, 56–63.

Jacobs, G. H., Williams, G. A., Cahill, H., & Nathans, J. (2007). Emergence of novel color vision in mice engineered to express a human cone photopigment. *Science, 315*, 1723–1725.

Jacobson, G. A., Rokni, D., & Yarom, Y. (2008). A model of the olivo-cerebellar system as a temporal pattern generator. *Trends in Neurosciences, 31*, 617–625.

Jaeger, J. J., Lockwood, A. H., Van Valin, R. D., Kemmerer, D. L., Murphy, B. W., & Wack, D. S. (1998). Sex differences in brain regions activated by grammatical and reading tasks. *NeuroReport, 9*, 2803–2807.

Jaeger, J-J., & Marivaux, L. (2005). Shaking the earliest branches of anthropoid primate evolution. *Science, 310*, 244–245.

Jager, G., Kahn, R. S., Van Den Brink, W., Van Ree, J. M., & Ramsey, N. F. (2006). Long-term effects of frequent cannabis use on working memory and attention: An fMRI study. *Psychopharmacology, 185*, 358–368.

James, L. E., & MacKay, D. G. (2001). H. M., word knowledge, and aging: Support for a new theory of long-term retrograde amnesia. *Current Directions in Psychological Science, 12*, 485–492.

Jameson, K. A., Highnote, S. M., & Wasserman, L. M. (2001). Richer color experience in observers with multiple photopigment opsin genes. *Psychonomic Bulletin & Review, 8*(2), 244–261.

Janardhan, V., & Qureshi, A. I. (2004). Mechanisms of ischemic brain injury. *Current Cardiology Reports, 6*, 117–123.

Jäncke, L., & Steinmetz, H. (2003). Anatomical brain asymmetries and their relevance for functional asymmetries. In K. Hugdahl & R. J. Davidson (Eds.), *The asymmetrical brain* (pp. 187–229). Cambridge, MA: Bradford Books/MIT Press.

Jarrell, T. W., Gentile, C. G., Romanski, L. M., McCabe, P. M., & Schneiderman, N. (1987). Involvement of cortical and thalamic auditory regions in retention of differential bradycardia conditioning to acoustic conditioned stimuli in rabbits. *Brain Research, 412*, 285–294.

Jarskog, L. F., Miyamoto, S., & Lieberman, J. A. (2007). Schizophrenia: New pathological insights and therapies. *Annual Review of Medicine, 58*, 49–61.

Jasny, B. R., Kelner, K. L., & Pennisi, P. (2008). From genes to social behavior. *Science, 322*, 891.

Javitt, D. C., & Coyle, J. T. (2004, January). Decoding schizophrenia. A fuller understanding of signaling in the brain of people with this disorder offers new hope for improved therapy. *Scientific American, 290*, 48–55.

Jax, S. A., Buxbaum, L. J., & Moll, A. D. (2006). Deficits in movement planning and intrinsic coordinate control in ideomotor apraxia. *Journal of Cognitive Neuroscience, 18*, 2063–2076.

Jazin, E., & Cahill, L. (2010). Sex differences in molecular neuroscience: From fruit flies to humans. *Nature Reviews Neuroscience, 11*, 9–16.

Jeannerod, M., Arbib, M. A., Rizzolatti, G., & Sakarta, H. (1995). Grasping objects: The cortical mechanisms of visuomotor transformation. *Trends in Neurosciences, 18*(7), 314–327.

Jee, S. H., Sull, J. W., Park, J., Lee, S-Y., Ohrr, H., Guallar, E., & Samet, J. M. (2006). Body-mass index and mortality in Korean men and women. *New England Journal of Medicine, 355*, 779–787.

Jeffery, K. J. (2007). Self-localization and the entorhinal-hippocampal system. *Current Opinion in Neurobiology, 17*, 684–691.

Jegalian, K., & Lahn, B. T. (2001, February). Why the Y is so weird. *Scientific American, 284*, 56–61.

Jenkins, I. H., Brooks, D. J., Bixon, P. D., Frackowiak, R. S. J., & Passingham, R. E. (1994). Motor sequence learning: A study with positron emission tomography. *Journal of Neuroscience, 14*(6), 3775–3790.

Jenner, P. (2008). Molecular mechanisms of L-DOPA-induced dyskinesia. *Nature Reviews Neuroscience, 9*, 665–677.

Jessen, K. R., & Mirsky, R. (2005). The origin and development of glial cells in peripheral nerves. *Nature Reviews Neuroscience, 6*, 671–682.

Joëls, M., & Baram, T. Z. (2009). The neuro-symphony of stress. *Nature Reviews Neuroscience, 10*, 459–466.

Johansson, R. S., & Flanagan, J. R. (2009). Coding and use of tactile signals from the fingertips in object manipulation tasks. *Nature Reviews Neuroscience, 10*, 345–359.

Johnson, K. O. (2001). The roles and functions of cutaneous mechanoreceptors. *Current Opinion in Neurobiology, 11*, 455–461.

Johnson, M. H. (2001). Functional brain development in humans. *Nature Reviews Neuroscience, 2*, 475–483.

Johnson, P. M., & Kenny, P. J. (2010). Dopamine D2 receptors in addiction-like reward dysfunction and compulsive eating in obese rats. *Nature Neuroscience, 13*, 635–641.

Johnston, T. D. (1987). The persistence of dichotomies in the study of behavioral development. *Developmental Review, 7*, 149–182.

Jones, H. S., & Oswald, I. (1966). Two cases of healthy insomnia. *Electroencephalography and Clinical Neurophysiology, 24*, 378–380.

Jones, H. W., & Park, I. J. (1971). A classification of special problems in sex differentiation. In D. Bergsma (Ed.), *The clinical delineation of birth defects. Part X: The endocrine system* (pp. 113–121). Baltimore, MD: Williams and Wilkins.

Jordan, H., Reis, J. E., Hoffman, J. E., & Landau, B. (2002). Intact perception of biological motion in the face of profound spatial deficits: Williams syndrome. *Psychological Science, 13*, 162–167.

Joseph, M. H., Datla, K., & Young, A. M. J. (2003). The interpretation of the measurement of nucleus accumbens dopamine by in vivo dialysis: The kick, the craving or the cognition? *Neuroscience and Biobehavioural Reviews, 27*, 527–541.

Joseph, R. (1988). Dual mental functioning in a split-brain patient. *Journal of Clinical Psychology, 44*, 771–779.

Joshua, M., Adler, A., & Bergman, H. (2009). The dynamics of dopamine in control of motor behavior. *Current Opinion in Neurobiology, 19*, 615–620.

Josse, G., & Tzourio-Mazoyer, N. (2004). Hemispheric specialization for language. *Brain Research Reviews, 44*, 1–12.

Jost, A. (1972). A new look at the mechanisms controlling sex differentiation in mammals. *Johns Hopkins Medical Journal, 130*, 38–53.

Jourdain, P., Bergersen, L. H., Bhaukaurally, K., Bezzi, P., Santello, M., Domercq, M., . . . Volterra, A. (2007). Glutamate exocytosis from astrocytes controls synaptic strength. *Nature Neuroscience, 10*, 331–339.

Julien, R. M. (1981). *A primer of drug action*. San Francisco, CA: W. H. Freeman.

Justus, T. (2004). The cerebellum and English grammatical morphology: Evidence from production, comprehension, and grammaticality judgments. *Journal of Cognitive Neuroscience, 16*, 1115–1130.

Juusola, M., French, A. S., Uusitalo, R. O., & Weckström, M. (1996). Information processing by graded-potential transmission through tonically active synapses. *Trends in Neurosciences, 19*, 292–297.

Kaas, J. H., & Collins, C. E. (2001). The organization of sensory cortex. *Current Opinion in Neurobiology, 11*, 498–504.

Kaas, J. H., Krubtzer, L. A., Chino, Y. M., Langston, A. L., Polley, E. H., & Blair, N. (1990). Reorganization of retinotopic cortical maps in adult mammals after lesions of the retina. *Science, 248*, 229–231.

Kaas, J. H., Nelson, R. J., Sur, M., & Merzenich, M. M. (1981). Organization of somatosensory cortex in primates. In F. O. Schmitt, F. G. Worden, G. Adelman, & S. G. Dennis (Eds.), *The organization of the cerebral cortex* (pp. 237–261). Cambridge, MA: MIT Press.

Kaessmann, H., & Pääbo, S. (2002). The genetical history of humans and the great apes. *Journal of Internal Medicine, 251*, 1–18.

Kagan, J., & Baird, A. (2004). Brain and behavioral development during childhood. In M. S. Gazzaniga (Ed.), *The cognitive neurosciences* (pp. 93–103). Cambridge, MA: MIT Press.

Kagawa, Y. (1978). Impact of Westernization on the nutrition of Japanese: Changes in physique, cancer, longevity, and centenarians. *Preventive Medicine, 7*, 205–217.

Kaiser, J., Lutzenberger, W., Preissl, H., Ackermann, H., & Birbaumer, N. (2000). Right-hemisphere dominance for the processing of sound-source lateralization. *Journal of Neuroscience, 20*, 6631–6639.

Kalaria, R. N. (2001). Advances in molecular genetics and pathology of cerebrovascular disorders. *Trends in Neurosciences, 24*, 392–400.

Kalil, K., & Dent, E. W. (2005). Touch and go: Guidance cues signal to the growth cone cytoskeleton. *Current Opinion in Neurobiology, 15*, 521–526.

Kalil, R. E. (1989, December). Synapse formation in the developing brain. *Scientific American, 261*, 76–85.

Kalivas, P. W. (2004). Glutamate systems in cocaine addiction. *Current Opinion in Pharmacology, 4*, 23–29.

Kalivas, P. W. (2005). How do we determine which drug-induced neuroplastic changes are important? *Nature Neuroscience, 8*, 1440–1441.

Kallman, F. J. (1946). The genetic theory of schizophrenia: An analysis of 691 schizophrenic twin index families. *American Journal of Psychiatry, 103*, 309–322.

Kalynchuk, L. E. (2000). Long-term amygdala kindling in rats as a model for the study of interictal emotionality in temporal lobe epilepsy. *Neuroscience and Biobehavioural Reviews, 24*, 691–704.

Kalynchuk, L. E., Pinel, J. P. J., Treit, D., & Kippin, T. E. (1997). Changes in emotional behavior produced by long-term amygdala kindling in rats. *Biological Psychiatry, 41*, 438–451.

Kandel, E. R. (2001). The molecular biology of memory storage: A dialogue between genes and synapses. *Science, 294*, 1030–1038.

Kandel, E. R., & Squire, L. R. (2000). Neuroscience: Breaking down barriers to the study of brain and mind. *Science, 290*, 1113–1120.

Kansaku, K., Yamaura, A., & Kitazawa, S. (2000). Sex differences in lateralization revealed in the posterior language areas. *Cerebral Cortex, 10*, 866–872.

Kantarci, O. H. (2008). Genetics and natural history of multiple sclerosis. *Seminars in Neurology, 28*, 7–16.

Kanwisher, N. (2006). What's in a face? *Science, 311*, 617–618.

Kaplan, K. A., & Harvey, A. G. (2009). Hypersomnia across mood disorders: A review and synthesis. *Sleep Medicine Reviews, 13*, 275–285.

Kapur, N. (1997). *Injured brains of medical minds.* Oxford, England: Oxford University Press.

Karacan, I., Goodenough, D. R., Shapiro, A., & Starker, S. (1966). Erection cycle during sleep in relation to dream anxiety. *Archives of General Psychiatry, 15*, 183–189.

Karacan, I., Williams, R. L., Finley, W. W., & Hursch, C. J. (1970). The effects of naps on nocturnal sleep: Influence on the need for stage-1 REM and stage-4 sleep. *Biological Psychiatry, 2*, 391–399.

Karanian, D. A., & Bahr, B. A. (2006). Cannabinoid drugs and enhancement of endocannabinoid responses: Strategies for a wide array of disease states. *Current Molecular Medicine, 6*, 677–684.

Kato, M., Phillips, B. G., Sigurdsson, G., Narkiewicz, K., Pesek, C. A., & Somers, V. K. (2000). Effects of sleep deprivation on neural circulatory control. *Hypertension, 35*, 1173–1175.

Katsumata, N., Kuroiwa, T., Ishibashi, S., Li, S., Endo, S., & Ohno, K. (2006). Heterogeneous hyperactivity and distribution of ischemic lesions after focal cerebral ischemia in Mongolian gerbils. *Neuropathology, 26*, 283–292.

Katusic, S. K., Colligan, R. C., Barbaresi, W. J., Schaid, D. J., & Jacobsen, S. J. (2001). Incidence of reading disability in a population-based birth cohort, 1976–1982, Rochester, Minn. *Mayo Clinic Proceedings, 76*, 1081–1092.

Katz, J. L., Ackman, P., Rothwax, Y., Sachar, E. J., Weiner, H., Hellman, L., & Gallagher, T. F. (1970). Psychoendocrine aspects of cancer of the breast. *Psychosomatic Medicine, 32*, 1–18.

Kauer, J. A., & Malenka, R. C. (2007). Synaptic plasticity and addiction. *Nature Neuroscience Reviews, 8*, 844–858.

Kaut, K. P., & Bunsey, M. D. (2001). The effects of lesions to the rat hippocampus or rhinal cortex on olfactory and spatial memory: Retrograde and anterograde findings. *Cognitive, Affective, & Behavioral Neuroscience, 1*, 270–286.

Kavaliers, M., & Choleris, E. (2001). Antipredator responses and defensive behavior: Ecological and ethological approaches for the neurosciences. *Neuroscience and Biobehavioural Reviews, 25*, 577–586.

Kawasaki, H., Adolphs, R., Oya, H., Kovach, C., Damasio, H., Kaufman, O., & Howard, M. (2005). Analysis of single-unit responses to emotional scenes in human ventromedial prefrontal cortex. *Journal of Cognitive Neuroscience, 17*, 1509–1518.

Kaye, W. H., Bailer, U. F., Frank, G. K., Wagner, A., & Henry, S. E. (2005). Brain imaging of serotonin after recovery from anorexia and bulimia nervosa. *Physiology & Behavior, 86*, 15–17.

Kaye, W. H., Bulik, C. M., Thornton, L., Barbarich, N., Masters, K., & the Price Foundation Collaborative Group. (2004). Comorbidity of anxiety disorders with anorexia and bulimia nervosa. *American Journal of Psychiatry, 161*, 2215–2221.

Kayser, C., Körding, K. P., & König, P. (2004). Processing of complex stimuli and natural scenes in the visual cortex. *Current Opinion in Neurobiology, 14*, 468–473.

Keirstead, H. S. (2005). Stem cells for the treatment of myelin loss. *Trends in Neurosciences, 28*, 677–683.

Keller, A., & Vosshall, L. B. (2008). Better smelling through genetics: Mammalian odor perception. *Current Opinion in Neurobiology, 18*, 364–369.

Keller, S. S., Crow, T., Foundas, A., Amunts, K., & Roberts, N. (2009). Broca's area: Nomenclature, anatomy, typology and asymmetry. *Brain and Language, 109*, 29–48.

Kelley, A. E. (2004). Ventral striatal control of appetitive motivation: Role in ingestive behavior and reward-related learning. *Neuroscience and Biobehavioural Reviews, 27*, 765–776.

Kelley, M. W. (2006). Regulation of cell fate in the sensory epithelia of the inner ear. *Nature Reviews Neuroscience, 7*, 837–849.

Kelley, W. M., Ojemann, J. G., Wetzel, R. D., Derdeyn, C. P., Moran, C. J., Cross, D. T., . . . Petersen, S. E. (2002). Wada testing reveals frontal lateralization for the memorization of words and faces. *Journal of Cognitive Neuroscience, 14*, 116–125.

Kelly, M. C., & Chen, P. (2009). Development of form and function in the mammalian cochlea. *Current Opinion in Neurobiology, 19*, 395–401.

Keltner, D., Kring, A. M., & Bonanno, G. A. (1999). Fleeting signs of the course of life: Facial expression and personal adjustment. *Current Directions in Psychological Science, 8*, 18–22.

Kemp, A., & Manahan-Vaughan, D. (2006). Hippocampal long-term depression: Master or minion in declarative memory processes? *Trends in Neurosciences, 30*, 111–118.

Kempermann, G., & Gage, F. H. (1999, May). New nerve cells for the adult brain. *Scientific American, 282*, 48–53.

Kempermann, G., Jessberger, S., Steiner, B., & Kronenberg, G. (2004). Milestones of neuronal development in the adult hippocampus. *Trends in Neurosciences, 27*, 447–452.

Kempermann, G., Wiskott, L., & Gage, F. H. (2004). Functional significance of adult neurogenesis. *Current Opinion in Neurobiology, 14*, 186–191.

Kenakin, T. (2005, October). New bull's-eyes for drugs. *Scientific American, 293*, 51–57.

Kendler, K. S., & Gruenberg, A. M. (1984). An independent analysis of the Danish adoption study of schizophrenia: VI. The relationship between psychiatric disorders as defined by DSM-III in the relatives and adoptees. *Archives of General Psychiatry, 41*, 555–564.

Kennerly, S. W., Walton, M. E., Behrens, T. E. J., Buckley, M. J., & Rushworth, M. F. S. (2006). Optimal decision making and the anterior cingulate cortex. *Nature Neuroscience, 9*, 940–947.

Kenny, P. J. (2007). Brain reward systems and compulsive drug use. *Trends in Pharmacological Science, 28*, 135–141.

Kenrick, D. T. (2001). Evolutionary psychology, cognitive science, and dynamical systems: Building an integrative paradigm. *Current Directions in Psychological Science, 10*(1), 13–17.

Kentridge, R. W., Heywood, C. A., & Weiskrantz, L. (1997). Residual vision in multiple retinal locations within a scotoma: Implications for blindsight. *Journal of Cognitive Neuroscience, 9,* 191–202.

Kessels, R. P. C., De Haan, E. H. F., Kappelle, L. J., & Postma, A. (2001). Varieties of human spatial memory: A meta-analysis on the effects of hippocampal lesions. *Brain Research Reviews, 35,* 295–303.

Kettenmann, H. (2006). Triggering the brain's pathology sensor. *Nature Neuroscience, 9,* 1463–1464.

Kettenmann, H., & Verkharatsky, A. (2008). Neuroglia: The 150 years after. *Trends in Neurosciences, 31,* 653–659.

Keys, A., Broz, J., Henschel, A., Mickelsen, O., & Taylor H. L. (1950). *The biology of human starvation.* Minneapolis: University of Minnesota Press.

Keysers, C., & Gazzola, V. (2009). Expanding the mirror: Vicarious activity for actions, emotions, and sensations. *Current Opinion in Neurobiology, 19,* 666–671.

Khadra, A., & Li, Y-X. (2006). A model for the pulsatile secretion of gonadotropin-releasing hormone from synchronized hypothalamic neurons. *Biophysical Journal, 91,* 74–83.

Kiecolt-Glaser, J. K., McGuire, L., Robles, T. F., & Glaser, R. (2002). Emotions, morbidity, and mortality: New perspectives from psychoneuroimmunology. *Annual Review of Psychology, 53,* 83–107.

Kiehl, K. A., Liddle, P. F., Smith, A. M., Mendrek, A., Forster, B. B., & Hare, R. D. (1999). Neural pathways involved in the processing of concrete and abstract words. *Human Brain Mapping, 7,* 225–233.

Kiehn, O. (2006). Locomotor circuits in the mammalian spinal cord. *Annual Review of Neuroscience, 29,* 279–306.

Killacky, H. P. (1995). Evolution of the human brain: A neuroanatomical perspective. In M. S. Gazzaniga (Ed.), *The cognitive neurosciences* (pp. 1243–1253). Cambridge, MA: MIT Press.

Killgore, W. D. S., Balkin, T. J., & Wesensten, N. J. (2006). Impaired decision making following 49 h of sleep deprivation. *Journal of Sleep Research, 15,* 7–13.

Kim, H., Somerville, L. H., Johnstone, T., Polis, S., Alexander, A. L., Shin, L. M., & Whalen, P. J. (2004). Contextual modulation of amygdala responsivity to surprised faces. *Journal of Cognitive Neuroscience, 16,* 1730–1745.

Kim, J. J., & Jung, M. W. (2006). Neural circuits and mechanisms involved in Pavlovian fear conditioning: A critical review. *Neuroscience and Biobehavioural Reviews, 30,* 188–202.

Kim, M. J., & Whalen, P.J. (2009). The structural integrity of an amygdala-prefrontal pathway predicts trait anxiety. *Journal of Neuroscience, 29,* 11614–11618.

Kim, S-G., Ashe, J., Hendrich, K., Ellermann, J. M., Merkle, H., Ugurbil, K., & Georgopoulos, A. P. (1993). Functional magnetic resonance imaging of motor cortex: Hemispheric asymmetry and handedness. *Science, 261,* 615–617.

Kimberg, D. Y., D'Esposito, M., & Farah, M. J. (1998). Cognitive functions in the prefrontal cortex—working memory and executive control. *Current Directions in Psychological Science, 6,* 185–192.

Kimble, G. A. (1989). Psychology from the standpoint of a generalist. *American Psychologist, 44,* 491–499.

Kimura, D. (1961). Some effects of temporal-lobe damage on auditory perception. *Canadian Journal of Psychology, 15,* 156–165.

Kimura, D. (1964). Left-right differences in the perception of melodies. *Quarterly Journal of Experimental Psychology, 16,* 355–358.

Kimura, D. (1973, March). The asymmetry of the human brain. *Scientific American, 228,* 70–78.

Kimura, D. (1979). Neuromotor mechanisms in the evolution of human communication. In H. E. Steklis & M. J. Raleigh (Eds.), *Neurobiology of social communication in primates* (pp. 197–219). New York, NY: Academic Press.

Kind, P. C., & Neumann, P. E. (2001). Plasticity: Downstream of glumate. *Trends in Neurosciences, 24,* 553–555.

King, A. J., Schnupp, J. W. H., & Thompson, I. D. (1998). Signals from the superficial layers of the superior colliculus enable the development of the auditory space map in the deeper layers. *Journal of Neuroscience, 18,* 9394–9408.

Kingsley, D. M. (2009, January). From atoms to traits. *Scientific American, 300,* 52–59.

Kirsch, I., Deacon, B. J., Huedo-Medina, T. B., Scoboria, A., Moore, T. J., & Johnson, B. T. (2008). Initial severity and antidepressant benefits: A meta-analysis of data submitted to the Food and Drug Administration. *PloS Medicine, 5,* e45.

Klawans, H. L. (1990). *Newton's madness: Further tales of clinical neurology.* New York: Harper & Row.

Klein, C., & Schlossmacher, M. G. (2006). The genetics of Parkinson disease: Implications for neurological care. *Nature Clinical Practice Neurology, 2,* 136–146.

Klein, C. A. (2008). The metastasis cascade. *Science, 321,* 1785–1787.

Klein, D. A., & Walsh, T. B. (2004). Eating disorders: Clinical features and pathophysiology. *Physiology & Behavior, 81,* 359–374.

Kleiner-Fisman, G., Fisman, D. N., Sime, E., Saint-Cyr, J. A., Lozano, A. M., & Lang, A. E. (2003). Long-term follow up of bilateral deep brain stimulation of the subthalamic nucleus in patients with advanced Parkinson disease. *Journal of Neuroscience, 99,* 489–495.

Kleinman, J. T., Newhart, M., Davis, C., Heidler-Gary, J., Gottesman, R. F., & Hillis, A. E. (2007). Right hemispatial neglect: Frequency and characterization following acute left hemisphere stroke. *Brain and Cognition, 64,* 50–59.

Kleitman, N. (1963). *Sleep and wakefulness.* Chicago, IL: University of Chicago Press.

Klinesmith, J., Kasser, T., & McAndrew, F. T. (2006). Guns, testosterone, and aggression: An experimental test of a mediational hypothesis. *Psychological Science, 17,* 568–571.

Klivington, K. A. (Ed.). (1992). *Gehirn und Geist.* Heidelberg, Germany: Spektrum Akademischer Verlag.

Kluver, H., & Bucy, P. C. (1939). Preliminary analysis of the temporal lobes in monkeys. *Archives of Neurology and Psychiatry, 42,* 979–1000.

Knoblick, G., & Sebanz, N. (2006). The social nature of perception and action. *Current Directions in Psychological Science, 15,* 99–104.

Knowlton, B. J., Mangels, J. A., & Squire, L. R. (1996). A neostriatal habit learning system in humans. *Science, 273,* 1399–1402.

Knudsen, E. I. (2007). Fundamental components of attention. *Annual Review of Neuroscience, 30,* 57–78.

Knudsen, E. I., & Brainard, M. S. (1991). Visual instruction of the neural map of auditory space in the developing optic tectum. *Science, 253,* 85–87.

Knutson, K. L., Spiegel, K., Penev, P., & Cauter, E. V. (2007). The metabolic consequences of sleep deprivation. *Sleep Medicine Research, 11,* 163–178.

Koekkoek, S. K. E., Hulscher, H. C., Dortland, B. R., Hensbroek, R. A., Elgersma, Y., Ruigrok, T. J. H., & De Zeeuw, C. I. (2003). Cerebellar LTD and learning-dependent timing of conditioned eyelid responses. *Science, 301,* 1736–1739.

Kokkinos, J., & Levine, S. R. (1993). Stroke. *Neurologic Complications of Drug and Alcohol Abuse, 11,* 577–590.

Kolb, B., Morshead, C., Gonzalez, C., Kim, M., Gregg, C., Shingo, T., & Weiss, S. (2006). Growth factor-stimulated generation of new cortical tissue and functional recovery after stroke damage to the motor cortex of rats. *Journal of Cerebral Blood Flow and Metabolism, 27,* 983–997.

Kolb, B., & Whishaw, I. Q. (1990). *Fundamentals of human neuropsychology* (3rd ed.). New York, NY: Freeman.

Komatsu, H. (2006). The neural mechanisms of perceptual filling-in. *Nature, 7,* 220–231.

Konen, C. S., & Kastner, S. (2008). Two hierarchically organized neural systems for object information in human visual cortex. *Nature Neuroscience, 11,* 224–231.

König, P., & Verschure, P. F. M. J. (2002). Neurons in action. *Science, 296,* 1817–1818.

Konishi, M. (2003). Coding of auditory space. *Annual Review of Neuroscience, 26,* 31–35.

Koob, G. F. (2006). The neurobiology of addiction: A neuro-adaptational view relevant for diagnosis. *Addiction, 101,* 23–30.

Koob, G. F., & Bloom, F. E. (1988). Cellular and molecular mechanisms of drug dependence. *Science, 242,* 715–723.

Koob, G. F., & Le Moal, M. (2005). Plasticity of reward neurocircuitry and the "dark side" of drug addiction. *Nature Neuroscience, 8,* 1442–1444.

Koolhass, J. M., de Boer, S. F., & Buwalda, B. (2006). Stress and adaptation: Toward ecologically relevant animal models. *Current Directions in Psychological Science, 15*, 109–112.

Koolhaas, J. M., Schuurman, T., & Wierpkema, P. R. (1980). The organization of intraspecific agonistic behaviour in the rat. *Progress in Neurobiology, 15*, 247–268.

Koopmans, H. S. (1981). The role of the gastrointestinal tract in the satiation of hunger. In L. A. Cioffi, W. B. T. James, & T. B. Van Italie (Eds.), *The body weight regulatory system: Normal and disturbed mechanisms* (pp. 45–55). New York, NY: Raven Press.

Kopec, C., & Malinow, R. (2006). Matters of size. *Science, 314*, 1554–1555.

Kopelman, P. G. (2000). Obesity as a medical problem. *Nature, 404*, 635–648.

Korkman, M., Kettunen, S., & Autti-Ramo, I. (2003). Neurocognitive impairment in early adolescence following prenatal alcohol exposure of varying duration. *Neuropsychology, Development, and Cognition: Section C, Child Neuropsychology, 9*, 117–128.

Kornack, D. R., & Rakic, P. (1999). Continuation of neurogenesis in the hippocampus of the adult macaque monkey. *Proceedings of the National Academy of Sciences, USA, 96*, 5768–5773.

Kornell, N. (2009). Metacognition in humans and animals. *Current Directions in Psychological Science, 18*, 11–15.

Kosik, K. S. (2009). MicroRNAs tell an evo-devo story. *Nature Reviews Neuroscience, 10*, 754–759.

Kosslyn, S. M., & Andersen, R. A. (1992). *Frontiers in cognitive neuroscience.* Cambridge, MA: MIT Press.

Kosslyn, S. M., Ganis, G., & Thompson, W. L. (2001). Neural foundations of imagery. *Nature Reviews Neuroscience, 2*, 635–642.

Koutalos, Y., & Yau, K-W. (1993). A rich complexity emerges in phototransduction. *Current Opinion in Neurobiology, 3*, 513–519.

Kraus, N., & Banai, K. (2007) Auditory-processing malleability: Focus on language and music. *Current Directions on Psychological Science, 16*, 105–110.

Kreek, M. J., Nielsen, D. A., Butelman, E. R., & LaForge, K. S. (2005). Genetic influences on impulsivity, risk taking, stress responsivity and vulnerability to drug abuse and addiction. *Nature Neuroscience, 8*, 1450–1457.

Kreitzer, A. C. (2009). Physiology and pharmacology of striatal neurons. *Annual Review of Neuroscience, 32*, 33–55.

Krieglstein, J. (1997). Mechanisms of neuroprotective drug actions. *Clinical Neuroscience, 4*, 184–193.

Kriegstein, A., & Alvarez-Buylla, A. (2009). The glial nature of embryonic and adult neural stem cells. *Annual Review of Neuroscience, 32*, 149–184.

Kriegstein, A. R., & Noctor, S. C. (2004). Patterns of neuronal migration in the embryonic cortex. *Trends in Neurosciences, 27*, 392–399.

Kring, A. M. (1999). Emotion in schizophrenia: Old mystery, new understanding. *Current Directions in Psychological Science, 8*, 160–163.

Kripke, D. F. (2004). Do we sleep too much? *Sleep, 27*, 13–14.

Kripke, D. F., Garfinkel, L., Wingard, D. L., Klauber, M. R., & Marler, M. R. (2002). Mortality associated with sleep duration and insomnia. *Archives of General Psychiatry, 59*, 131–136.

Krueger, R. F., & Markon, K. E. (2006). Understanding psychopathology: Melding behavior genetics, personality, and quantitative psychology to develop an empirically based model. *Current Directions in Psychological Science, 15*, 113–117.

Krystal, A. D. (2008). A compendium of placebo-controlled trials of the risks/benefits of pharmacological treatments for insomnia: The empirical basis for U.S. clinical practice. *Sleep Medicine Reviews, 13*, 265–274.

Kubicki, M., McCarley, R., Westin, C. F., Park, H. J., Maier, S., Kikinis, R., Jolesz, . . . Shenton, M. E. (2007). A review of diffusion tensor imaging studies in schizophrenia. *Journal of Psychiatric Research, 41*, 15–30.

Kubie, J. L., Fenton, A., Novikov, N., Touretzky, D., & Muller, R. U. (2007). Changes in goal selection induced by cue conflicts are in register with predictions from changes in place cell field locations. *Behavioral Neuroscience, 121*, 751–763.

Kübler, A., Dixon, V., & Garavan, H. (2006). Automaticity and reestablishment of executive control—an fMRI study. *Journal of Cognitive Neuroscience, 18*, 1331–1342.

Kullman, D. M., & Lamsa, K. P. (2007). Long-term synaptic plasticity in hippocampal interneurons. *Nature Reviews Neuroscience, 8*, 687–699.

Kumar, A., Godwin, J. W., Gates, P. B., Garza-Garcia, A. A., & Brockes, J. P. (2007). Molecular basis for the nerve dependence of limb regeneration in an adult vertebrate. *Science, 318*, 772–777.

Kurbat, M. A., & Farah, M. J. (1998). Is the category-specific deficit for living things spurious? *Journal of Cognitive Neuroscience, 10*, 355–361.

Kushner, H. I. (1999). *A cursing brain? The histories of Tourette syndrome.* Cambridge, MA: Harvard University Press.

Kyle, U. G., & Pichard, C. (2006). The Dutch famine of 1944–1945: A pathophysiological model of long-term consequences of wasting disease. *Current Opinion in Clinical Nutrition and Metabolic Care, 9*, 388–394.

LaBar, K. S. (2007). Beyond fear: Emotional memory mechanisms in the human brain. *Current Directions in Psychological Science, 16*, 173–177.

LaBar, K. S., & Cabeza, R. (2006). Cognitive neuroscience of emotional memory. *Nature Reviews Neuroscience, 7*, 54–64.

Laeng, B., & Caviness, V. S. (2001). Prosopagnosia as a deficit in encoding curved surface. *Journal of Cognitive Neuroscience, 13*, 556–576.

Lagoda, G., Muschamp, J. W., Vigdorchik, A., & Hull, E. M. (2004). A nitric oxide synthesis inhibitor in the medial preoptic area inhibits copulation and stimulus sensitization in male rats. *Behavioral Neuroscience, 118*, 1317–1323.

Lai, K-O., & Ip, N. Y. (2009). Synapse development and plasticity: Roles of ephrin/Eph receptor signaling. *Current Opinion in Neurobiology, 19*, 275–283.

Laine, M., Salmelin, R., Helenius, P., & Marttila, R. (2000). Brain activation during reading in deep dyslexia: An MEG study. *Journal of Cognitive Neuroscience, 12*, 622–634.

Lam, R. W., Levitt, A. J., Levitan, R. D., Enns, M. W., Morehouse, R., Michalak, E. E., & Tam, E. M. (2006). The Can-SAD study: A randomized controlled trial of the effectiveness of light therapy and fluoxetine in patients with winter seasonal affective disorder. *American Journal of Psychiatry, 163*, 805–812.

Lam, T. K. T., Schwartz, G. J., & Rossetti, L. (2006). Hypothalamic sensing of fatty acids. *Nature Neuroscience, 8*, 579–584.

Lamb, T. D., Collin, S. P., & Pugh, E. N. (2007). Evolution of the vertebrate eye: Opsins, photoreceptors, retina and eye cup. *Nature Reviews Neuroscience, 8*, 960–975.

Lamsa, K. P., Heeroma, J. H., Somogyi, P., Rusakov, D. A., & Kullman, D. M. (2007). Anti-Hebbian long-term potentiation in the hippocampal feedback inhibitory circuit. *Science, 315*, 1262–1266.

Land, E. H. (1977, April). The retinex theory of color vision. *Scientific American, 237*, 108–128.

Landau, B., & Lakusta, L. (2009). Spatial representation across species: Geometry, language, and maps. *Current Opinion in Neurobiology, 19*, 1–8.

Lane, M. A., Ingram, D. K., & Roth, G. S. (2002, August). The serious search for an anti-aging pill. *Scientific American, 287*, 36–41.

Langston, J. W. (1985). MPTP and Parkinson's disease. *Trends in Neurosciences, 8*, 79–83.

Langston, J. W. (1986). MPTP-induced Parkinsonism: How good a model is it? In S. Fahn, C. P. Marsden, P. Jenner, & P. Teychenne (Eds.), *Recent developments in Parkinson's disease* (pp. 119–126). New York, NY: Raven Press.

Lansbury, P. T., Jr. (2006). Improving synaptic function in a mouse model of AD. *Cell, 126*, 655–657.

Lanza, R., & Rosenthal, N. (2004, June). The stem cell challenge. *Scientific American, 290*, 93–99.

Lashley, K. S. (1941). Patterns of cerebral integration indicated by the scotomas of migraine. *Archives of Neurology and Psychiatry, 46*, 331–339.

Lask, B., & Bryant-Waugh, R. (Eds.) (2000). *Anorexia nervosa and related eating disorders in childhood and adolescence.* Hove, England: Psychology Press.

Laughlin, S. B., & Sejnowski, T. J. (2003). Communication in some neuronal networks. *Science, 301*, 1870–1874.

Lavie, P., Pratt, H., Scharf, B., Peled, R., & Brown, J. (1984). Localized pontine lesion: Nearly total absence of REM sleep. *Neurology, 34*, 1118–1120.

Lawrence, D. G., & Kuypers, H. G. J. M. (1968a). The functional organization of the motor system in the monkey: I. The effects of bilateral pyramidal lesions. *Brain, 91*, 1–14.

Lawrence, D. G., & Kuypers, H. G. J. M. (1968b). The functional organization of the motor system in the monkey: II. The effects of lesions of the descending brain-stem pathways. *Brain, 91*, 15–36.

Lazar, M. A. (2005). How obesity causes diabetes: Not a tall tale. *Science, 307*, 373–375.

Lazar, M. A. (2008). How now, brown fat? *Science, 321*, 1048–1049.

Lazarov, O., Robinson, J., Tang, Y-P., Hairston, I. S., Korade-Mirnics, Z., Lee, V. M-Y., . . . Sisodia, S. S. (2005). Environmental enrichment reduces Aβ levels and amyloid deposition in transgenic mice. *Cell, 120*, 701–713.

Learney, K., Van Wart, A., & Sur, M. (2009). Intrinsic patterning and experience-dependent mechanisms that generate eye-specific projections and binocular circuits in the visual pathway. *Current Opinion in Neurobiology, 19*, 181–187.

Lebedev, M. A., & Nicolelis, M. A. L. (2006). Brain-machine interfaces: Past, present and future. *Trends in Neurosciences, 29*, 536–546.

LeDoux, J. E. (1994, June). Emotion, memory and the brain. *Scientific American, 270*, 50–57.

LeDoux, J. E. (2000). Emotion circuits in the brain. *Annual Review of Neuroscience, 23*, 155–184.

Lee, H. W., Hong, S. B., Seo, D. W., Tae, W. S., & Hong, S. C. (2000). Mapping of functional organization in human visual cortex. *Neurology, 54*, 849–854.

Lee, J. C. (2009). Reconsolidation: Maintaining memory relevance. *Trends in Neurosciences, 32*, 413–420.

Lee, L. J., Hughes, T. R., & Frey, B. J. (2006). How many new genes are there? *Science, 3*, 1709.

Lee, V. M. (2001). Tauists and βaptists united—well, almost! *Science, 293*, 1446–1495.

Leibowitz, S. F., Hammer, N. J., & Chang, K. (1981). Hypothalamic paraventricular nucleus lesions produce overeating and obesity in the rat. *Physiology & Behavior, 27*, 1031–1040.

Lennoff, H. M., Wang, P. P., Greenberg, F., & Bellugi, U. (1997, December). Williams syndrome and the brain. *Scientific American, 279*, 68–73.

Lennox, W. G. (1960). *Epilepsy and related disorders.* Boston, MA: Little, Brown.

Lenroot, R. K., & Giedd, J. N. (2006). Brain development in children and adolescents: Insights from anatomical magnetic resonance imaging. *Neuroscience and Biobehavioural Reviews, 30*, 718–729.

Lenzenweger, M. F. (2006). Schizotypy: An organizing framework for schizophrenia research. *Current Directions in Psychological Science, 15*, 162–166.

Leonardo, E. D., & Hen, R. (2006). Genetics of affective and anxiety disorders. *Annual Review of Psychology, 57*, 117–137.

Leppänen, J. M., & Nelson, C. A. (2009). Tuning the developing brain to social signals of emotions. *Nature Reviews Neuroscience, 10*, 37–47.

Lerman, C., Caporaso, N. E., Audrain, J., Main, D., Bowman, E. D., Lockshin, B., . . . Shields, P. G. (1999). Evidence suggesting the role of specific genetic factors in cigarette smoking. *Health Psychology, 18*, 14–20.

Leshner, A. I. (1997). Addiction is a brain disease, and it matters. *Science, 278*, 45–46.

Lester, G. L. L., & Gorzalka, B. B. (1988). Effect of novel and familiar mating partners on the duration of sexual receptivity in the female hamster. *Behavioral and Neural Biology, 49*, 398–405.

Leucht, S., Barnes, T. R., Kissling, W., Engel, R. R., Correll, C., & Kane, J. M. (2004). Relapse prevention in schizophrenia with new-generation antipsychotics: A systematic review and exploratory meta-analysis of randomized, controlled trials. *American Journal of Psychiatry, 160*, 1209–1222.

Leung, L. S., Ma, J., & McLachlan, R. S. (2000). Behaviors induced or disrupted by complex partial seizures. *Neuroscience and Biobehavioural Reviews, 24*, 763–775.

Leutgeb, J. K., Leutgeb, S., Moser, M-B., & Moser, E. I. (2007). Pattern separation in the dentate gyrus and CA3 of the hippocampus. *Science, 315*, 961–966.

Leutgeb, S., Leutgeb, J. K., Treves, A., Moser, M-B., & Moser, E. I. (2004). Distinct ensemble codes in hippocampal areas CA3 and CA1. *Science, 305*, 1295–1298.

LeVay, S. (1991). A difference in hypothalamic structure between heterosexual and homosexual men. *Science, 253*, 1034–1037.

Levenson, R. W. (1994). The search for autonomic specificity. In P. Ekman & R. J. Davidson (Eds.), *The nature of emotion: Fundamental questions* (pp. 252–257). New York, NY: Oxford University Press.

Levi-Montalcini, R. (1952). Effects of mouse motor transplantation on the nervous system. *Annals of the New York Academy of Sciences, 55*, 330–344.

Levi-Montalcini, R. (1975). NGF: An uncharted route. In F. G. Worden, J. P. Swazey, & G. Adelman (Eds.), *The neurosciences: Paths of discovery* (pp. 245–265). Cambridge, MA: MIT Press.

Levin, H. S. (1989). Memory deficit after closed-head injury. *Journal of Clinical and Experimental Neuropsychology, 12*, 129–153.

Levin, H. S., Papanicolaou, A., & Eisenberg, H. M. (1984). Observations on amnesia after non-missile head injury. In L. R. Squire & N. Butters (Eds.), *Neuropsychology of memory* (pp. 247–257). New York: Guilford Press.

Levine, J. A., Eberhardt, N. L., & Jensen, M. D. (1999). Role of nonexercise activity thermogenesis in resistance to fat gain in humans. *Science, 283*, 212–214.

Levine, M. S., Cepeda, C., Hickey, M. A., Fleming, S. M., & Chesselet, M-F. (2004). Genetic mouse models of Huntington's and Parkinson's diseases: Illuminating but imperfect. *Trends in Neurosciences, 27*, 691–697.

Levitt, J. B. (2001). Function following form. *Science, 292*, 232–234.

Levy, J. (1969). Possible basis for the evolution of lateral specialization of the human brain. *Nature, 224*, 614–615.

Levy, J., Trevarthen, C., & Sperry, R. W. (1972). Perception of bilateral chimeric figures following hemispheric deconnection. *Brain, 95*, 61–78.

Lewis, D. A., Hashimoto, T., & Volk, D. W. (2005). Cortical inhibitory neurons and schizophrenia. *Nature Reviews Neuroscience, 6*, 312–324.

Lezak, M. D. (1997). Principles of neuropsychological assessment. In T. E. Feinberg & M. J. Farah (Eds.), *Behavioral neurology and neuropsychology* (pp. 43–54). New York, NY: McGraw-Hill.

Li, J-Y., Christophersen, N. S., Hall, V., Soulet, D., & Brundin, P. (2008). Critical issues of clinical human embryonic stem cell therapy for brain repair. *Trends in Neurosciences, 31*, 146–153.

Li, Q., Lee, J-A., & Black, D. L. (2007). Neuronal regulation of alternative pre-mRNA splicing. *Nature Reviews Neuroscience, 8*, 819–831.

Libby, P. (2002, May). Atherosclerosis: The new view. *Scientific American, 286*(5), 47–55.

Lieberman, J. A., Stroup, T. S., McEvoy, J. P., Swartz, M. S., Rosenheck, R. A., Perkins, D. O., . . . Hsiao, J. K. (2005). Effectiveness of antipsychotic drugs in patients with chronic schizophrenia. *New England Journal of Medicine, 353*, 1209–1223.

Lieberman, M. D., & Eisenberger, N. I. (2009). Pains and pleasures of social life. *Science, 323*, 890–891.

Lien, W-H., Klezovitch, O., Fernandez, T. E., Delrow, J., & Vasioukhin, V. (2006). AlphaE-catenin controls cerebral cortical size by regulating the hedgehog signaling pathway. *Science, 311*, 1609–1612.

Lin, L., Faraco, J., Li, R., Kadotani, H., Rogers, W., Lin, X., . . . Mignot, E. (1999). The sleep disorder canine narcolepsy is caused by a mutation in the hypocretin (orexin) receptor 2 gene. *Cell, 98*, 365–376.

Lindberg, L., & Hjern, A. (2003). Risk factors for anorexia nervosa: A national cohort study. *International Journal of Eating Disorders, 34*, 397–408.

Lindner, A., Haarmeier, T., Erb, M., Grodd, W., & Thier, P. (2006). Cerebrocerebellar circuits for the perceptual cancellation of eye-movement–induced retinal image motion. *Journal of Cognitive Neuroscience, 18*, 1899–1912.

Lindsay, P. H., & Norman, D. A. (1977). *Human information processing* (2nd ed.). New York, NY: Academic Press.

Lindsey, D. B., Bowden, J., & Magoun, H. W. (1949). Effect upon the EEG of acute injury to the brain stem activating system. *Electroencephalography and Clinical Neurophysiology, 1*, 475–486.

Lindwall, C., Fothergill, T., & Richards, L. J. (2007). Commissure formation in the mammalian forebrain. *Current Opinion in Neurobiology, 17,* 3–14.

Lisman, J. (2003). Long-term potentiation: Outstanding questions and attempted synthesis. *Philosophical Transactions of the Royal Society of London, 358,* 829–842.

Lisman, J., Lichtman, J. W., & Sanes, J. R. (2003). LTP: Perils and progress. *Nature Reviews Neuroscience, 4,* 926–929.

Lisman, J. E., Raghavachari, S., & Tsien, R W. (2007). The sequence of events that underlie quantal transmission at central glutamatergic synapses. *Nature Reviews Neuroscience, 8,* 597–609.

Littleton, J. (2001). Receptor regulation as a unitary mechanism for drug tolerance and physical dependence—not quite as simple as it seemed! *Addiction, 96,* 87–101.

Littman, D. R., & Singh, H. (2007). Asymmetry and immune memory. *Science, 315,* 1673–1674.

Liu, D., Dorio, J., Tannenbaum, B., Caldji, C., Francis, D., Freedman, A., . . . Meaney, M. J. (1997). Maternal care, hippocampal glucocorticoid receptors, and hypothalamic-pituitary-adrenal responses to stress. *Science, 277,* 1659–1662.

Liu, G., & Roo, Y. (2004). Neuronal migration in the brain. In M. S. Gazzaniga (Ed.), *The cognitive neurosciences* (pp. 51–68). Cambridge, MA: MIT Press.

Liu, L., Wong, T. P., Pozza, M. F., Lingenhoehl, K., Wang, Y., Sheng, M., . . . Wang, Y. T. (2004). Role of NMDA receptor subtypes in governing the direction of hippocampal synaptic plasticity. *Science, 304,* 1021–1024.

Livet, J., Weissman, T. A., Kang, H., Draft, R. W., Lu, J., Bennis, R. A., . . . Lichtman, J. W. (2007). Transgenic strategies for combinatorial expression of fluorescent proteins in the nervous system. *Nature, 450,* 56–62.

Livingstone, M. S., & Hubel, D. H. (1984). Anatomy and physiology of a color system in the primate visual cortex. *Journal of Neuroscience, 4,* 309–356.

Livingstone, M. S., & Hubel, D. S. (1988). Segregation of form, color, movement, and depth: Anatomy, physiology, and perception. *Science, 240,* 740–749.

Livingstone, M. S., & Tsao, D. Y. (1999). Receptive fields of disparity-selective neurons in macaque striate cortex. *Nature Reviews Neuroscience, 2,* 825–832.

Lledo, P-M., Alonso, M., & Grubb, M. S. (2006). Adult neurogenesis and functional plasticity in neuronal circuits. *Nature Reviews Neuroscience, 7,* 179–193.

Lledo, P-M., & Saghatelyan, A. (2006). Integrating new neurons into the adult olfactory bulb: Joining the network, life-death decisions, and the effects of sensory experience. *Trends in Neurosciences, 28,* 248–254.

Lleras, A., & Moore, C. M. (2006). What you see is what you get. *Psychological Science, 17,* 876–881.

Lo, E. H., Dalkara, T., & Moskowitz, M. A. (2003). Mechanisms, challenges and opportunities in stroke. *Nature Reviews Neuroscience, 4,* 399–415.

Logothetis, N. K., & Sheinberg, D. L. (1996). Visual object recognition. *Annual Review of Neuroscience, 19,* 577–621.

Lombardino, A. J., & Nottebohm, F. (2000). Age at deafening affects the stability of learned song in adult male zebra finches. *Journal of Neuroscience, 20,* 5054–5064.

Lomber, S. G., & Malhotra, S. (2008). Double dissociation of "what" and "where" processing in auditory cortex. *Nature Neuroscience, 11,* 609–616.

London, S. E., & Clayton, D. F. (2008). Functional identification of sensory mechanisms required for developmental song learning. *Nature Neuroscience, 11,* 579–586.

Lorincz, M. T. (2006). Clinical implications of Parkinson's disease genetics. *Seminars in Neurology, 26,* 492–498.

Löscher, W., & Potschka, H. (2005). Drug resistance in brain diseases and the role of drug efflux transporters. *Nature Reviews Neuroscience, 6,* 591–602.

Louk Vanderschuren, J. M. J., & Everitt, B. J. (2004). Drug seeking becomes compulsive after prolonged cocaine self-administration. *Science, 305,* 1017–1019.

Lovejoy, C. O., Suwa, G., Simpson, S. W., Matternes, J. H., & White, T. D. (2009). The great divides: *Ardipithecus ramidus* reveals the postcrania of our last common ancestors with African apes. *Science, 326,* 100–106.

Low, L. K., & Cheng, H-J. (2005). A little nip and tuck: Axon refinement during development and axonal injury. *Current Opinion in Neurobiology, 15,* 549–556.

Lowe, J., & Carroll, D. (1985). The effects of spinal injury on the intensity of emotional experience. *The British Journal of Clinical Psychology, 24,* 135–136.

Lowe, M. R. (1993). The effects of dieting on eating behavior: A three-factor model. *Psychological Bulletin, 114,* 100–121.

Lowrey, P. L., Shimomura, K., Antoch, M. P., Yamazaki, S., Zemenides, P. D., Ralph, M. R., . . . Takahashi, J. S. (2000). Positional syntenic cloning and functional characterization of the mammalian circadian mutation tau. *Science, 288,* 483–491.

Lozano, A. M., & Kalia, S. K. (2005, July). New movement in Parkinson's. *Scientific American, 293,* 68–75.

Lozano, A. M., Mayberg, H. S., Giacobbe, P., Hamani, C., Craddock, R. C., & Kennedy, S. H. (2008). Subcallosal cingulate gyrus deep brain stimulation for treatment-resistant depression. *Biological Psychiatry, 64,* 461–467.

Luauté, J., Halligan, P., Rode, G., Rossetti, Y., & Boisson D. (2006). Visuospatial neglect: A systematic review of current interventions and their effectiveness. *Neuroscience and Biobehavioural Reviews, 30,* 961–982.

Lucas F., & Sclafani, A. (1989). Flavor preferences conditioned by intragastric fat infusions in rats. *Physiology & Behavior, 46,* 403–412.

Lucidi, F., Devoto, A., Violani, C., Mastracci, P., & Bertini, M. (1997). Effects of different sleep duration on delta sleep in recovery nights. *Psychophysiology, 34,* 227–233.

Luck, S. J., Hillyard, S. A., Mangoun, G. R., & Gazzaniga, M. S. (1989). Independent hemispheric attentional systems mediate visual search in split-brain patients. *Nature, 343,* 543–545.

Ludwig, M., & Leng, G. (2006). Dendritic peptide release and peptide-dependent behaviours. *Nature Reviews Neuroscience, 7,* 126–136.

Ludwig, M., & Pittman, Q. J. (2003). Talking back: Dendritic neurotransmitter release. *Trends in Neurosciences, 26,* 255–261.

Lumpkin, E. A., & Bautista, D. M. (2005). Feeling the pressure in mammalian somatosensation. *Current Opinion in Neurobiology, 15,* 382–388.

Luo, L., & O'Leary, D. D. M. (2005). Axon retraction and degeneration in development and disease. *Annual Review of Neuroscience, 28,* 127–156.

Luquet, S., Perez, F. A., Hnasko, T. S., & Palmiter, R. D. (2005). NPy/AgRP neurons are essential for feeding in adult mice but can be ablated in neonates. *Science, 310,* 683–685.

Lykken, D. T. (1959). The GSR in the detection of guilt. *Journal of Applied Psychology, 43,* 385–388.

Lynch, M. A. (2004). Long-term potentiation and memory. *Physiological Review, 84,* 87–136.

Lyons, D. E., Santos, L. R., & Keil, F. C. (2006). Reflections of other minds: How primate social cognition can inform the function of mirror neurons. *Current Opinion in Neurobiology, 16,* 230–234.

Maalouf, M., Rho, J. M., & Mattson, M. P. (2008). The neuroprotective properties of caloric restriction, the ketogenic diet, and ketone bodies. *Brain Research Reviews, 59,* 293–315.

Macaluso, E., & Driver, J. (2005). Multisensory spatial interactions: A window onto functional integration in the human brain. *Trends in Neurosciences, 28,* 259–266.

Macaluso, E., Driver, J., & Frith, C. D. (2003). Multimodal spatial representations engaged in human parietal cortex during both saccadic and manual spatial orienting. *Current Biology, 13,* 990–999.

Maccari, S., Darnaudery, M., Morley-Fletcher, S., Zuena, A. R., Cinque, C., & Van Reeth, O. (2003). Prenatal stress and long-term consequences: Implications of glucocorticoid hormones. *Neuroscience and Biobehavioural Reviews, 27,* 119–127.

Macchi, G. (1989). Anatomical substrate of emotional reactions. In F. Boller & J. Grafman (Eds.), *Handbook of neuropsychology* (Vol. 3, pp. 283–304). New York, NY: Elsevier.

MacDougall-Shackleton, S. A., & Ball, G. F. (1999). Comparative studies of sex differences in the song-control system of songbirds. *Trends in Neurosciences, 22*(10), 432–436.

Machado, C. J., & Bachevalier, J. (2006). The impact of selective amygdala, orbital frontal cortex, or hippocampal formation lesions on established social relationships in rhesus monkeys (*Macaca mulatta*). *Behavioral Neuroscience, 120,* 761–786.

MacLaren, V. V. (2001). A qualitative review of the guilty knowledge test. *Journal of Applied Psychology, 86,* 674–683.

MacLean, P. D. (1952). Some psychiatric implications of physiological studies on frontotemporal portion of limbic system (visceral brain). *Electroencephalography and Clinical Neurophysiology, 4,* 407–418.

MacLusky, N. J., & Naftolin, F. (1981). Sexual differentiation of the central nervous system. *Science, 211,* 1294–1302.

MacNeilage, P. F., Rogers, L. J., & Vallortigara, G. (2009, July). Origins of the left and right brain. *Scientific American, 301,* 60–67.

Maddox, W. T., Aparicio, P., Marchant, N. L., & Ivry, R. B. (2005). Rule-based category learning is impaired in patients with Parkinson's disease but not in patients with cerebellar disorders. *Journal of Cognitive Neuroscience, 17,* 707–723.

Maeder, T. (2003, May). The orphan drug backlash. *Scientific American, 288,* 81–87.

Maggard, M. A., Shugarman, L. R., Suttorp, M., Maglione, M., Sugerman, H. J., Livingston, E. H., . . . Shekelle, P. G. (2005). Meta-analysis: Surgical treatment of obesity. *Annals of Internal Medicine, 142,* 547–559.

Maguire, E. A., Frith, C. D., Burgess, N., Donnett, J. G., & O'Keefe, J. (1998). Knowing where things are: Parahippocampal involvement in encoding object locations in virtual large-scale space. *Journal of Cognitive Neuroscience, 19,* 61–76.

Mahowald, M. W., & Schenck, C. H. (2005). Insights from studying human sleep disorders. *Nature, 437,* 1279–1285.

Mair, W., Goymer, P., Pletcher, S. D., & Partridge, L. (2003). Demography of dietary restriction and death in *Drosophila. Science, 301,* 1731–1733.

Majdan, M., & Shatz, C. J. (2006). Effects of visual experience on activity-dependent gene regulation in cortex. *Nature Neuroscience, 9,* 650–659.

Majewska, A. K., & Sur, M. (2006). Plasticity and specificity of cortical processing networks. *Trends in Neurosciences, 29,* 323–329.

Malenka, R. C. (2003). The long-term potential of LTP. *Nature Reviews Neuroscience, 4,* 923–926.

Malissen, B. (1999). Dancing the immunological two-step. *Science, 285,* 207–208.

Malsbury, C. W. (1971). Facilitation of male rat copulatory behavior by electrical stimulation of the medial preoptic area. *Physiology & Behavior, 7,* 797–805.

Manchikanti, L. (2007). National drug control policy and prescription drug abuse: Facts and fallacies. *Pain Physician, 10,* 399–424.

Manson, J. E., Willett, W. C., Stampfer, M. J., Colditz, G. A., Hunter, D. J., Hankinson, S. E., . . . Speizer, F. E. (1995). Body weight and mortality among women. *New England Journal of Medicine, 333,* 677–685.

Maratsos, M., & Matheny, L. (1994). Language specificity and elasticity: Brain and clinical syndrome studies. *Annual Review of Psychology, 45,* 487–516.

Maravelias, C., Dona, A., Stefanidou, M., & Spiliopoulou, C. (2005). Adverse effects of anabolic steroids in athletes: A constant threat. *Toxicology Letters, 158,* 167–175.

Marcks, B. A., Woods, D. A., Teng, E. J., & Twohig, M. P. (2004). What do those who know, know? Investigating providers' knowledge about Tourette's syndrome and its treatment. *Cognitive and Behavioral Practice, 11,* 298–305.

Maren, S., & Quirk, G. J. (2004). Neuronal signaling of fear memory. *Nature Reviews Neuroscience, 5,* 844–852.

Maries, E., Dass, B., Collier, T. J., Kordower, J. H., & Steece-Collier, K. (2003). The role of alpha-synuclein in Parkinson's disease: Insights from animal models. *Nature Reviews Neuroscience, 4,* 727–738.

Marin, O., & Rubenstein, J. L. R. (2003). Cell migration in the forebrain. *Annual Review of Neuroscience, 26,* 441–483.

Marin, O., Valdeolmillos, M., & Moya, F. (2006). Neurons in motion: Same principles for different shapes? *Trends in Neurosciences, 29,* 656–661.

Marin, O., Yaron, A., Bagri, A., Tessier-Lavigne, M., & Rubenstein, J. L. R. (2001). Sorting of striatal and cortical interneurons regulated by semaphoring-neuropilin interactions. *Science, 293,* 872–875.

Mark, V. H., Ervin, F. R., & Yakolev, P. I. (1962). The treatment of pain by stereotaxic methods. First International Symposium on Stereoencephalotomy, Philadelphia, 1961. *Confina Neurologica, 22,* 238–245.

Markram, H., Toledo-Rodriguez, M., Wang, Y., Gupta, A., Silberberg, G., & Wu, C. (2004). Interneurons of the neocortical inhibitory system. *Nature Reviews Neuroscience, 5,* 793–807.

Marlowe, W. B., Mancall, E. L., & Thomas, J. J. (1985). Complete Kluver-Bucy syndrome in man. *Cortex, 11,* 53–59.

Marriott, L. K., & Wenk, G. L. (2004). Neurobiological consequences of long-term estrogen therapy. *Current Directions in Psychological Science, 13,* 173–176.

Martin, A. (2006). The representation of object concepts in the brain. *Annual Review of Psychology, 58,* 25–45.

Martin, A., State, M., Koenig, K., Schultz, R., Dykens, E. M., Cassidy, S. B., & Leckman, J. F. (1998). Prader-Willi syndrome. *American Journal of Psychiatry, 155,* 1265–1272.

Martin, B. J. (1981). Effect of sleep deprivation on tolerance of prolonged exercise. *European Journal of Applied Physiology, 47,* 345–354.

Martin, J. B. (1987). Molecular genetics: Applications to the clinical neurosciences. *Science, 238,* 765–772.

Martin, R. C. (2003). Language processing: Functional organization and neuroanatomical basis. *Annual Review of Psychology, 54,* 55–89.

Martin, R. C. (2005). Components of short-term memory and their relation to language processing. *Current Directions in Psychological Science, 14,* 204–208.

Martin, R. J., White, B. D., & Hulsey, M. G. (1991). The regulation of body weight. *American Scientist, 79,* 528–541.

Martin, R. L., Roberts, W. V., & Clayton, P. J. (1980). Psychiatric status after a one-year prospective follow-up. *Journal of the American Medical Association, 244,* 350–353.

Martinez, L. M., Wang, Q., Reid, R. C., Pillai, C., Alonso, J.-M., Sommer, F. T., & Hirsch, J. A. (2005). Receptive field structure varies with layer in the primary visual cortex. *Nature Neuroscience, 8,* 372–379.

Martinez-Conde, S., Macknik, S. L., & Hubel, D. H. (2004). The role of fixational eye movements in visual perception. *Nature Reviews Neuroscience, 5,* 229–240.

Mashour, G. A., Walker, E. E., & Martuza, R. L. (2005). Psychosurgery: Past, present, and future. *Brain Research Reviews, 48,* 409–419.

Masland, R. H. (2001). Neuronal diversity in the retina. *Current Opinion in Neurobiology, 11,* 431–436.

Mason, M. F., Norton, M. I., Van Horn, J. D., Wegner, D. M., Grafton, S. T., & Macrae, C. N. (2007) Wandering minds: The default network and stimulus-independent thought. *Science, 315,* 393–395.

Masoro, E. J. (1988). Food restriction in rodents: An evaluation of its role in the study of aging [Minireview]. *Journal of Gerontology, 43,* 59–64.

Massey, P. V., & Bashir, Z. I. (2007). Long-term depression: Multiple forms and implications for brain function. *Trends in Neurosciences, 30,* 176–184.

Matsui, K., & Jahr, C. E. (2006). Exocytosis unbound. *Current Opinion in Neurobiology, 16,* 305–311.

Matsumoto, K., & Tanaka, K. (2004). Conflict and cognitive control. *Science, 303,* 969–970.

Matthews, G. A., Fane, B. A., Conway, G. S., Brook, C. G. D., & Hines, M. (2009). Personality and congenital adrenal hyperplasia: Possible effects of prenatal androgen exposure. *Hormones and Behavior, 55,* 285–291.

Matthews, P. M., Honey, G. D., & Bullmore, E. T. (2006). Applications of fMRI in translational medicine and clinical practice. *Nature, 7,* 732–744.

Matthews, S., Strigo, I., Simmons, A., Yang, T., & Paulus, M. (2009). Decreased functional coupling of the amygdala and supragenual cingulate is related to increased depression in unmedicated individuals with current major depressive disorder. *Journal of Affective Disorders, 111,* 13–20.

Matzinger, P. (2002). The danger model: A renewed sense of self. *Science, 296,* 301–305.

Mauch, D. H., Nagler, K., Schumacher, S., Goritz, C., Muller, E.-C., Otto, A., & Pfrieger, F. W. (2001). CNS synaptogenesis promoted by glia-derived cholesterol. *Science, 294,* 1354–1357.

Mayr, E. (2000, July). Darwin's influence on modern thought. *Scientific American, 283,* 70–83.

Mazzocchi, F., & Vignolo, L. A. (1979). Localisation of lesions in aphasia: Clinical–CT scan correlations in stroke patients. *Cortex, 15*, 627–654.

McCann, T. S. (1981). Aggression and sexual activity of male southern elephant seals, *Mirounga leonina. Journal of Zoology, 195*, 295–310.

McCann, U. D., & Ricaurte, G. A. (2004). Amphetamine neurotoxicity: Accomplishments and remaining challenges. *Neuroscience and Behavioural Reviews, 27*, 821–826.

McCarley, R. W. (2007). Neurobiology of REM and NREM sleep. *Sleep Medicine, 8*, 302–330.

McCarthy, M. M., Auger, A. P., & Perrot-Sinal, T. S. (2002). Getting excited about GABA and sex differences in the brain. *Trends in Neurosciences, 25*, 307–313.

McCarthy, M. M., Wright, C. L., & Schwartz, J. M. (2009). New tricks by an old dogma: Mechanisms of the organizational/activational hypothesis of steroid-mediated sexual differentiation of brain and behavior. *Hormones and Behavior, 55*, 655–665.

McClellan, J. M., Susser, E., & King, M-C. (2006). Maternal famine, de novo mutations, and schizophrenia. *Journal of the American Medical Association, 296*, 582–594.

McClintock, M. K., & Herdt, G. (1996). Rethinking puberty: The development of sexual attraction. *Current Directions in Psychological Sciences, 5*, 178–183.

McConkey, E. H., & Varki, A. (2005). Thoughts on the future of great ape research. *Science, 309*, 1499–1501.

McCoy, P. A., Huang, H-S., & Philpot, B. D. (2009). Advances in understanding visual cortex plasticity. *Current Opinion in Neurobiology, 19*, 298–304.

McDaniel, M. A., Maier, S. F., & Einstein, G. O. (2002). "Brain-specific" nutrients: A memory cure? *Psychological Science in the Public Interest, 3*, 12–38.

McDermott, K. B., Watson, J. M., & Ojemann, J. G. (2005). Presurgical language mapping. *Current Directions in Psychological Science, 14*, 291–295.

McDonald, C., Zanelli, J., Rabe-Hesketh, S., Ellison-Wright, I., Sham, P., Kalidindi, S., . . . Kennedy, N. (2004). Meta-analysis of magnetic resonance imaging brain morphometry studies in bipolar disorder. *Biological Psychiatry, 56*, 411–417.

McDonald, J. W., Liu, X-Z., Qu, Y., Liu, S., Mickey, S. K., Turetsky, D., . . . Choi, D. W. (1999). Transplanted embryonic stem cells survive, differentiate and promote recovery in injured rat spinal cord. *Nature Medicine, 5*, 1410–1412.

McDonald, R. J., & White, N. M. (1993). Triple dissociation of memory systems: Hippocampus, amygdala, and dorsal striatum. *Behavioral and Neural Biology, 59*, 107–119.

McDonald, R. V., & Siegel, S. (2004). Intra-administration associations and withdrawal symptoms: Morphine-elicited morphine withdrawal. *Experimental and Clinical Psychopharmacology, 12*, 3–11.

McEwen, B. S. (1983). Gonadal steroid influences on brain development and sexual differentiation. In R. O. Greep (Ed.), *Reproductive physiology IV*. Baltimore: University Park Press.

McEwen, B. S. (1987). Sexual differentiation. In G. Adelman (Ed.), *Encyclopedia of neuroscience* (Vol. II, pp. 1086–1088). Boston, MA: Birkhäuser.

McEwen, B. S. (2004). How sex and stress hormones regulate the structural and functional plasticity of the hippocampus. In M. S. Gazzaniga (Ed.), *The cognitive neurosciences* (pp. 171–182). Cambridge, MA: MIT Press.

McGaugh, J. L. (2002). Memory consolidation and the amygdala: A systems perspective. *Trends in Neurosciences, 25*, 456–461.

McGlone, J. (1977). Sex differences in the cerebral organization of verbal functions in patients with unilateral brain lesions. *Brain, 100*, 775–793.

McGlone, J. (1980). Sex differences in human brain asymmetry: A critical survey. *Behavioral and Brain Sciences, 3*, 215–263.

McGregor, C., Darke, S., Ali, R., & Christie, P. (1998). Experience of non-fatal overdose among heroin users in Adelaide, Australia: Circumstances and risk perceptions. *Addiction, 93*, 701–711.

McGue, M. (1999). The behavioral genetics of alcoholism. *Current Directions in Psychological Science, 8*, 109–115.

McGuffin, P., Riley, B., & Plomin, R. (2001). Toward behavioral genomics. *Science, 291*, 1232–1249.

McHaffie, J. G., Stanford, T. R., Stein, B. E., Coizet, V., & Redgrave, P. (2005). Subcortical loops through the basal ganglia. *Trends in Neurosciences, 28*, 401–407.

McKim, W. A. (1986). *Drugs and behavior: An introduction to behavioral pharmacology*. Englewood Cliffs, NJ: Prentice-Hall.

McLaughlin, T., Hindges, R., & O'Leary, D. D. M. (2003). Regulation of axial patterning of the retina and its topographic mapping in the brain. *Current Opinion in Neurobiology, 13*, 57–69.

McLaughlin, T., & O'Leary, D. D. M. (2005). Molecular gradients and development of retinotopic maps. *Annual Review of Neuroscience, 28*, 327–355.

McNally, R. J. (2003). Progress and controversy in the study of posttraumatic stress disorder. *Annual Review of Psychology, 54*, 229–252.

McNally, R. J., Bryant, R. A., & Ehlers, A. (2003). Does early psychological intervention promote recovery from posttraumatic stress? *Psychological Science in the Public Interest, 4*, 45–79.

McNaughton, N., & Zangrossi, H. H., Jr. (2008). Theoretical approaches to the modeling of anxiety in animals. In R. J. Blanchard, D. C. Blanchard, G. Griebel, & D. J. Nutt (Eds.), *Handbook of anxiety and fear*, Vol. 17 (pp. 11–27). Oxford, England: Elsevier.

Meaney, M. J., & Szyf, M. (2005). Maternal care as a model for experience-dependent chromatin plasticity? *Trends in Neurosciences, 28*, 456–463.

Mechtcheriakov, S., Brenneis, C., Koppelstaetter, F., Schocke, M., & Marksteiner, J. (2007). A widespread distinct pattern of cerebral atrophy in patients with alcohol addiction revealed by voxel-based morphometry. *Journal of Neurology, Neurosurgery, and Psychiatry, 8*, 610–614.

Meddis, R. (1977). *The sleep instinct*. London, England: Routledge & Kegan Paul.

Medina, J. F., Repa, J. C., Mauk, M. D., & LeDoux, J. E. (2002). Parallels between cerebellum- and amygdala-dependent conditioning. *Nature Reviews Neuroscience, 3*, 122–131.

Medina, K. L., Hanson, K. L., Schweinsburg, A. D., Cohen-Zion, M., Nagel, B. J., & Tapert, S. F. (2007). Neuropsychological functioning in adolescent marijuana users: Subtle deficits detectable after a month of abstinence. *Journal of the International Neuropsychological Society, 13*, 807–820.

Medzhitov, R., & Janeway, C. A., Jr. (2002). Decoding the patterns of self and nonself by the innate immune system. *Science, 296*, 298–301.

Megarbane, B., Buisine, A., Jacobs, F., Resiere, D., Guerrier, G., Delerme, S., . . . Baud, F. J. (2005). Heroin, methadone, and buprenorphine overdoses: Prospective comparative assessment of conditions, severity, and treatment. *Clinical Toxicology, 43*, 431–432.

Meidinger, A. L., Miltenberger, R. G., Himle, M., Omvig, M., Trainor, C., & Crosby, R. (2005). An investigation of tic suppression and the rebound effect in Tourette's disorder. *Behavior Modification, 29*, 716–745.

Mel, B. W. (2002). What the synapse tells the neuron. *Science, 295*, 1845–1846.

Melcher, T., Falkai, P., & Gruber, O. (2008). Functional abnormalities in psychiatric disorders: Neural mechanisms to detect and resolve cognitive conflict and interference. *Brain Research Reviews, 59*, 96–124.

Melkersson, K., & Dahl, M. L. (2004). Adverse metabolic effects associated with atypical antipsychotics: Literature review and clinical implications. *Drugs, 64*, 701–723.

Mello, N. K., & Mendelson, J. H. (1972). Drinking patterns during work-contingent and noncontingent alcohol acquisition. *Psychosomatic Medicine, 34*, 139–165.

Melzack, R. (1992, April). Phantom limbs. *Scientific American, 266*, 120–126.

Melzack, R., & Wall, P. D. (1982). *The challenge of pain*. London, England: Penguin.

Meney, I., Waterhouse, J., Atkinson, G., Reilly, T., & Davenne, D. (1988). The effect of one night's sleep deprivation on temperature, mood, and physical performance in subjects with different amounts of habitual physical activity. *Chronobiology International, 15*, 349–363.

Merabet, L. B., Rizzo, J. F., Amedi, A., Somers, D. C., & Pascual-Leone, A. (2005). What blindness can tell us about seeing again: Merging neuroplasticity and neuroprostheses. *Nature Reviews Neuroscience, 6*, 71–77.

Mercer, J. G., & Speakman, J. R. (2001). Hypothalamic neuropeptide mechanisms for regulating energy balance: From rodent models to human obesity. *Neuroscience and Biobehavioural Reviews, 25*, 101–116.

Metcalfe, J., Funnell, M., & Gazzaniga, M. S. (1995). Right hemisphere memory superiority: Studies of a split-brain patient. *Psychological Science, 6*, 157–165.

Meurnier, M., Murray, E. A., Bachevalier, J., & Mishkin, M. (1990). Effects of perirhinal cortical lesions on visual recognition memory in rhesus monkeys. *Society for Neuroscience Abstracts, 17*, 337.

Meyer, P., Saez, L., & Young, M. W. (2006). PER-TIM interactions in living Drosophila cells: An interval timer for the circadian clock. *Science, 311*, 226–229.

Meyer, R. G., & Salmon, P. (1988). *Abnormal psychology* (2nd ed.). Boston, MA: Allyn & Bacon.

Meyer, T. E., Kovács, S. J., Ehsani, A. A., Klein, S., Holloszy, J. O., & Fontana, L. (2006). Long-term caloric restriction ameliorates the decline in diastolic function in humans. *Journal of the American College of Cardiology, 47*, 398–402.

Meyer-Lindenberg, A., Mervis, C. B., & Berman, K. F. (2006). Neural mechanisms in Williams syndrome: A unique window to genetic influences on cognition and behaviour. *Nature Reviews Neuroscience, 7*, 380–393.

Miguel-Aliaga, I., & Thor, S. (2009). Programmed cell death in the nervous system—a programmed cell fate? *Current Opinion in Neurobiology, 19*, 127–133.

Miles, F. J., Everitt, B. J., Dalley, J. W., & Dickinson, A. (2004). Conditioned activity and instrumental reinforcement following long-term oral consumption of cocaine by rats. *Behavioral Neuroscience, 118*, 1331–1339.

Miller, G. E., & Blackwell, E. (2006). Turning up the heat: Inflammation as a mechanism linking chronic stress, depression and heart disease. *Current Directions in Psychological Science, 15*, 269–272.

Miller, G. E., & Chen, E. (2006). Life stress and diminished expression of genes encoding glucocorticoid receptor and β_2-adrenergic receptor in children with asthma. *Proceedings of the National Academy of Sciences, USA, 103*, 5496–5501.

Miller, G. E., Chen, E., & Zhou, E. S. (2007). If it goes up, must it come down: Chronic stress and the hypothalamic pituitary-adrenocortical axis in humans. *Psychological Bulletin, 133*, 25–45.

Miller, G. L., & Knudsen, E. I. (1999). Early visual experience shapes the representation of auditory space in the forebrain gaze fields of the barn owl. *Journal of Neuroscience, 19*, 2326–2336.

Miller, K. K., Grinspoon, S. K., Ciampa, J., Hier, J., Herzog, D., & Klibanski, A. (2005). Medical findings in outpatients with anorexia nervosa. *Archives of Internal Medicine, 165*, 561–566.

Miller, N. E., Bailey, C. J., & Stevenson, J. A. F. (1950). Decreased "hunger" but increased food intake resulting from hypothalamic lesions. *Science, 112*, 256–259.

Miller, T. M., & Cleveland, D. W. (2005). Treating neurodegenerative diseases with antibiotics. *Science, 307*, 361–362.

Milloy, M-J. S., Kerr, T., Tyndall, M., Montaner, J., & Wood, E. (2008). Estimated drug overdose deaths averted by North America's first medically supervised safer injection facility. *PLoS ONE, 3*, e3351. doi:10.1371/journal.pone.0003351

Milner, A. D., & Goodale, M. A. (1993). Visual pathways to perception and action. *Progress in Brain Research, 95*, 317–337.

Milner, B. (1965). Memory disturbances after bilateral hippocampal lesions. In P. Milner & S. Glickman (Eds.), *Cognitive processes and the brain* (pp. 104–105). Princeton, NJ: D. Van Nostrand.

Milner, B. (1971). Interhemispheric differences in the localization of psychological processes in man. *British Medical Bulletin, 27*, 272–277.

Milner, B. (1974). Hemispheric specialization: Scope and limits. In F. O. Schmitt & F. G. Worden (Eds.), *The neurosciences: Third study program* (pp. 75–89). Cambridge, MA: MIT Press.

Milner, B., Corkin, S., & Teuber, H. L. (1968). Further analysis of the hippocampal amnesic syndrome: 14-year follow-up study of H.M. *Neuropsychologia, 6*, 317–338.

Milner, C. E., & Cote, K. A. (2008). A dose-response investigation of the benefits of napping in healthy young, middle-aged, and older adults. *Sleep and Biological Rhythms, 6*, 2–15.

Milner, P. M. (1993, January). The mind and Donald O. Hebb. *Scientific American, 268*, 124–129.

Mindell, D. P. (2009, January). Evolution in the everyday world. *Scientific American, 300*, 82–89.

Ming, G-l., & Song, H. (2005). Adult neurogenesis in the mammalian central nervous system. *Annual Review of Neuroscience, 28*, 223–250.

Mirakbari, S. M. (2004). Heroin overdose as cause of death: Truth or myth. *Australian Journal of Forensic Sciences, 36*, 73–78.

Mishkin, M., & Appenzeller, T. (1987, June). The anatomy of memory. *Scientific American, 256*, 80–89.

Mishkin, M., & Delacour, J. (1975). An analysis of short-term visual memory in the monkey. *Journal of Experimental Psychology: Animal Behavior Processes, 1*, 326–334.

Mistlberger, R. E. (2005). Circadian regulation of sleep in mammals: Role of the suprachiasmatic nucleus. *Brain Research Reviews, 49*, 429–454.

Mistlberger, R. E., de Groot, M. H. M., Bossert, J. M., & Marchant, E. G. (1996). Discrimination of circadian phase in intact and suprachiasmatic nuclei-ablated rats. *Brain Research, 739*, 12–18.

Mitchell, J. M., Basbaum, A. I., & Fields, H. L. (2000). A locus and mechanism of action for associative morphine tolerance. *Nature Neuroscience, 3*, 47–53.

Mitchell, K. J., Johnson, M. K., Raye, C. L., & Greene, E. J. (2004). Prefrontal cortex activity associated with source monitoring in a working memory task. *Journal of Cognitive Neuroscience, 16*, 921–934.

Miyamichi, K., & Luo, L. (2009). Brain wiring by presorting axons. *Science, 325*, 544–545.

Miyashita, Y. (2004). Cognitive memory: Cellular and network machineries and their top-down control. *Science, 306*, 435–440.

Modell, J., Rosenthal, N. E., Harriett, A. E., Krishen, A., Asgharian, A., Foster, V. J., . . . Wightman, D. S. (2005). Seasonal affective disorder and its prevention by anticipatory treatment with bupropion SL. *Biological Psychiatry, 58*, 658–667.

Moeller, S., Freiwald, W. A., & Tsao, D. Y. (2008). Patches with links: A united system for processing faces in the macaque temporal lobe. *Science, 320*, 1355–1359.

Moffitt, T. E., Caspi, A., & Rutter, M. (2006). Measured gene-environment interactions in psychology. *Perspectives on Psychological Science, 1*, 5–27.

Mohr, J. P. (1976). Broca's area and Broca's aphasia. In H. Whitaker & H. A. Whitaker (Eds.), *Studies in neurolinguistics* (Vol. 1, pp. 201–235). New York, NY: Academic Press.

Mokdad, A. H., Marks, J. S., Stroup, D. F., & Gerberding, J. L. (2004). Actual causes of death in the United States, 2000. *The Journal of the American Medical Association, 291*, 1238–1245.

Molday, R. S., & Hsu, Y-T. (1995). The cGMP-gated channel of photoreceptor cells: Its structural properties and role in phototransduction. *Behavioral and Brain Sciences, 18*, 441–451.

Molyneaux, B. J., Arlotta, P., Menezes, J. R. L., & Macklis, J. D. (2007). Neuronal subtype specification in the cerebral cortex. *Nature Reviews Neuroscience, 8*, 427–437.

Money, J. (1975). Ablatio penis: Normal male infant sex-reassigned as a girl. *Archives of Sexual Behavior, 4*(1), 65–71.

Money, J., & Ehrhardt, A. A. (1972). *Man & woman, boy & girl.* Baltimore, MD: Johns Hopkins University Press.

Monk, T. H., Buysse, D. J., Welsh, D. K., Kennedy, K. S., & Rose, L. R. (2001). A sleep diary and questionnaire study of naturally short sleepers. *Journal of Sleep Research, 10*, 173–179.

Monroe, S. M., & Reid, M. W. (2008). Gene-environment interactions in depression research. *Psychological Science, 19*, 947–956.

Montgomery, E. B., & Gale, J. T. (2008). Mechanisms of action of deep brain stimulation (DBS). *Neuroscience and Biobehavioral Reviews, 32*, 388–407.

Montoro, R. J., & Yuste, R. (2004). Gap junctions in developing neocortex: A review. *Brain Research Reviews, 47*, 216–226.

Mooney, R. (2009). Neurobiology of song learning. *Current Opinion in Neurobiology, 19*, 654–660.

Moore, D. J., West, A. B., Dawson, V. L., & Dawson, T. M. (2005). Molecular pathophysiology of Parkinson's disease. *Annual Review of Neuroscience, 28*, 57–87.

Moore, R. Y. (1996). Neural control of the pineal gland. *Behavioural Brain Research, 73*, 125–130.

Moore, T. (2006). The neurobiology of visual attention: Finding sources. *Current Opinion in Neurobiology, 16*, 159–165.

Moran, J., & Desimone, R. (1985). Selective attention gates visual processing in the extrastriate cortex. *Science, 229*, 782–784.

Moran, M. H. (2004). Gut peptides in the control of food intake: 30 years of ideas. *Physiology & Behavior, 82*, 175–180.

Morgan, T. H., Sturtevant, A. H., Muller, H. J., & Bridges, C. B. (1915). *The mechanism of Mendelian heredity.* New York, NY: Holt.

Morilak, D. A., & Frazer, A. (2004). Antidepressants and brain monoaminergic system: A dimensional approach to understanding their behavioural effects in depression and anxiety disorders. *International Journal of Neuropsychopharmacology, 7*, 193–218.

Morimoto, K., Fahnestock, M., & Racine, R. J. (2004). Kindling and status epilepticus models of epilepsy: Rewiring the brain. *Progress in Neurobiology, 73*, 1–60.

Morin, C. M., Kowatch, R. A., & O'Shanick, G. (1990). Sleep restriction for the inpatient treatment of insomnia. *Sleep, 13*, 183–186.

Morin, L. P., & Allen, C. N. (2006). The circadian visual system. *Brain Research Reviews, 51*, 1–61.

Morishima, Y., Akaishi, R., Yamada, Y., Okuda, J., Toma, K., & Sakai, K. (2009). Task-selective signal transmission from prefrontal cortex in visual selective attention. *Nature Neuroscience, 12*, 85–91.

Morishita, H., & Hensch, T. K. (2008). Critical period revisited: Impact on vision. *Current Opinion in Neurobiology, 18*, 101–107.

Morris, M. K., Bowers, D., Chatterjee, A., & Heilman, K. M. (1992). Amnesia following a discrete basal forebrain lesion. *Brain, 115*, 1827–1847.

Morris, N. M., Udry, J. R., Khan-Dawood, F., & Dawood, M. Y. (1987). Marital sex frequency and midcycle female testosterone. *Archives of Sexual Behavior, 16*, 27–37.

Morris, R. G. M. (1981). Spatial localization does not require the presence of local cues. *Learning and Motivation, 12*, 239–260.

Morris, R. G. M., Moser, E. I., Riedel, G., Martin, S. J., Sandin, J., Day, M., & O'Carroll, C. O. (2003). Elements of a neurobiological theory of the hippocampus: The role of activity-dependent synaptic plasticity in memory. *Philosophical Transactions of the Royal Society of London, 358*, 773–786.

Morse, D., & Sassone-Corsi, P. (2002). Time after time: Inputs to and outputs from the mammalian circadian oscillators. *Trends in Neurosciences, 25*, 632–637.

Mort, D. J., Malhotra, P., Mannan, S. K., Rorden, C., Pambakian, A., Kennard, C., & Husain, M. (2003). The anatomy of visual neglect. *Brain, 126*, 1986–1997.

Moruzzi, G., & Magoun, H. W. (1949). Brain stem reticular formation and activation of the EEG. *Electroencephalography and Clinical Neurophysiology, 1*, 455–473.

Moscovitch, M., Nadel, L., Winocur, G., Gilboa, A., & Rosenbaum, R. S. (2006). The cognitive neuroscience of remote episodic, semantic and spatial memory. *Current Opinion in Neurobiology, 16*, 179–190.

Moser, E. I., Kropff, E., & Moser, M-B. (2008). Place cells, grid cells, and the brain's spatial representation system. *Annual Review of Neuroscience, 31*, 69–89.

Mountcastle, V. B., & Powell, T. P. S. (1959). Neural mechanisms subserving cutaneous sensibility with special references to the role of afferent inhibition in sensory perception and discrimination. *Bulletin of Johns Hopkins Hospital, 105*, 201–232.

Mrosovsky, N., & Salmon, P. A. (1987). A behavioral method for accelerating re-entrainment of rhythms to new light-dark cycles. *Nature, 330*, 372–373.

Mudher, A., & Lovestone, S. (2002). Alzheimer's disease—do tauists and baptists finally shake hands? *Trends in Neurosciences, 25*, 22–26.

Mühlnickel, W., Elbert, T., Taub, E., & Flor, H. (1998). Reorganization of auditory cortex in tinnitus. *Proceedings of the National Academy of Sciences, USA, 95*, 10340–10343.

Mulette-Gillman, O. A., Cohen, Y. E., & Groh, J. M. (2005). Eye-centered, head-centered, and complex coding of visual and auditory targets in the intraparietal sulcus. *Journal of Neurophysiology, 94*, 2331–2352.

Mullaney, D. J., Johnson, L. C., Naitoh, P., Friedman, J. K., & Globus, G. G. (1977). Sleep during and after gradual sleep reduction. *Psychophysiology, 14*, 237–244.

Mumby, D. G. (2001). Perspectives on object-recognition memory following hippocampal damage: Lessons from studies in rats. *Behavioural Brain Research, 14*, 159–181.

Mumby, D. G., Cameli, L., & Glenn, M. J. (1999). Impaired allocentric spatial working memory and intact retrograde memory after thalamic damage caused by thiamine deficiency in rats. *Behavioral Neuroscience, 113*, 42–50.

Mumby, D. G., & Pinel, J. P. J. (1994). Rhinal cortex lesions impair object recognition in rats. *Behavioral Neuroscience, 108*, 11–18.

Mumby, D. G., Pinel, J. P. J., & Wood, E. R. (1989). Nonrecurring items delayed nonmatching-to-sample in rats: A new paradigm for testing nonspatial working memory. *Psychobiology, 18*, 321–326.

Mumby, D. G., Wood, E. R., Duva, C. A., Kornecook, T. J., Pinel, J. P. J., & Phillips, A. G. (1996). Ischemia-induced object-recognition deficits in rats are attenuated by hippocampal ablation before or soon after ischemia. *Behavioral Neuroscience, 110*(2), 266–281.

Mumby, D. G., Wood, E. R., & Pinel, J. P. J. (1992). Object-recognition memory is only mildly impaired in rats with lesions of the hippocampus and amygdala. *Psychobiology, 20*, 18–27.

Muneoka, K., Han, M., & Gardiner, D. M. (2008, April). Regrowing human limbs. *Scientific American*, 56–63.

Munoz, D. P. (2006). Stabilizing the visual world. *Nature Neuroscience, 9*, 1467–1468.

Muñoz-Sanjuán, I., & Brivanlou, A. H. (2002). Neural induction, the default model and embryonic stem cells. *Nature Reviews Neuroscience, 3*, 271–280.

Münte, T. F., Altenmüller, E., & Jänke, L. (2002). The musician's brain as a model of neuroplasticity. *Nature Reviews Neuroscience, 3*, 473–478.

Münzberg, H., & Myers, M. G. (2005). Molecular and anatomical determinants of central leptin resistance. *Nature Neuroscience, 8*, 566–570.

Murai, K. K., & Pasquale, E. B. (2008). Axons seek neighborly advice. *Science, 320*, 185–186.

Murphy, B. P., Chung, Y. C., Park, T. W., & McGorry, P. D. (2006). Pharmacological treatment of primary negative symptoms in schizophrenia: A systematic review. *Schizophrenia Research, 88*, 5–25.

Murphy, M. R., & Schneider, G. E. (1970). Olfactory bulb removal eliminates mating behavior in the male golden hamster. *Science, 157*, 302–304.

Murphy, T. H., & Corbett, D. (2009). Plasticity during stroke recovery: From synapse to behaviour. *Nature Reviews Neuroscience, 10*, 861–872.

Murray, E. A. (1996). What have ablation studies told us about the neural substrates of stimulus memory? *Seminars in the Neurosciences, 8*, 13–22.

Murray, E. A., Bussey, T. J., and Saksida, L. M. (2007). Visual perception and memory: A new view of medial temporal lobe function in primates and rodents. *Annual Review of Neuroscience, 30*, 99–122.

Murray, E. A., & Richmond, B. J. (2001). Role of perirhinal cortex in object perception, memory, and associations. *Current Opinion in Neurobiology, 11*, 188–193.

Mustanski, B. S., Chivers, M. L., & Bailey, J. M. (2002). A critical review of recent biological research on human sexual orientation. *Annual Review of Sex Research, 13*, 89–140.

Myers, R. E., & Sperry, R. W. (1953). Interocular transfer of a visual form discrimination habit in cats after section of the optic chiasma and corpus callosum. *Anatomical Record, 115*, 351–352.

Nachev, P., Kennard, C., & Husain, M. (2008). Functional role of the supplementary and pre-supplementary motor areas. *Nature Reviews Neuroscience, 9*, 856–869.

Nadarajah, B., & Parnavelas, J. G. (2002). Modes of neuronal migration in the developing cerebral cortex. *Nature Reviews Neuroscience, 3*, 423–432.

Nadel, L., & Moscovitch, M. (1997). Memory consolidation, retrograde amnesia and the hippocampal complex. *Current Opinion in Neurobiology, 7*, 217–227.

Nader, K., & Hardt, O. (2009). A single standard for memory: The case for reconsolidation. *Nature Reviews Neuroscience, 10*, 224–234.

Nader, K., Schafe, G. E., & LeDoux, J. E. (2000). Fear memories require protein synthesis in the amygdala for reconsolidation after retrieval. *Nature, 406*, 722–726.

Naeser, M. A., Hayward, R. W., Laughlin, S. A., & Zatz, L. M. (1981). Quantitative CT scan studies in aphasia. *Brain and Language, 12,* 140–164.

Nagy, J. I., Dudek, F. E., & Rash, J. E. (2004). Update on connexins and gap junctions in neurons and glia in the mammalian nervous system. *Brain Research Reviews, 47,* 191–215.

Nagy, Z., Westerberg, H., & Klinsberg, T. (2004). Maturation of white matter is associated with the development of cognitive functions during childhood. *Journal of Cognitive Neuroscience, 16,* 1227–1233.

Nakahara, K., Hayashi, T., Konishi, S., & Miyashita, Y. (2002). Functional MRI of macaque monkeys performing a cognitive set-shifting task. *Science, 295,* 1532–1536.

Nakamura, K., Dehaene, S., Jobert, A., Le Bihan, D., & Kouider, S. (2005). Subliminal convergence of *kanji* and *kana* words: Further evidence for functional parcellation of the posterior temporal cortex in visual word perception. *Journal of Cognitive Neuroscience, 17,* 954–968.

Nambu, A. (2008). Seven problems on the basal ganglia. *Current Opinion in Neurobiology, 18,* 595–604.

Nassi, J. J., & Calloway, E. M. (2009). Parallel processing strategies of the primate visual system. *Nature Reviews Neuroscience, 10,* 360–372.

Nassi, J. J., Lyon, D. C., & Callaway, E. M. (2006). The parvocellular LGN provides a robust disynaptic input to the visual motion area MT. *Neuron, 50,* 319–327.

Natelson, B. H. (2004). Stress, hormones and disease. *Physiology & Behavior, 82,* 139–143.

National Commission on Marijuana and Drug Abuse. (1972). *Marijuana: A signal of misunderstanding.* New York, NY: New American Library.

Nau, R., & Brück, W. (2002). Neuronal injury in bacterial meningitis: Mechanisms and implications for therapy. *Trends in Neurosciences, 25,* 38–45.

Naya, Y., Yoshida, M., & Miyashita, Y. (2001). Backward spreading of memory-retrieval signal in the primate temporal cortex. *Science, 291,* 661–664.

Neary, J. T., & Zimmerman, H. (2009). Trophic functions of nucleotides in the central nervous system. *Trends in Neurosciences, 32,* 189–198.

Nee, D. E., Berman, M. G., Moore, K. S., & Jonides, J. (2008). Neuroscientific evidence about the distinction between short- and long-term memory. *Current Directions in Psychological Science, 17,* 102–106.

Nelken, I. (2008). Processing of complex sounds in the auditory system. *Current Opinion in Neurobiology, 18,* 413–417.

Nelson, E. C., Heath, A. C., Madden, P. A. F., Cooper, M. L., Dinwiddie, S. H., Bucholz, . . . Martin, N. G. (2002). Association between self-reported childhood sexual abuse and adverse psychosocial outcomes. *Archives of General Psychiatry, 59,* 139–145.

Nelson, S. B., Sugino, K., & Hempel, C. M. (2006). The problem of neuronal cell types: A physiological genomics approach. *Trends in Neurosciences, 29,* 339–345.

Nemeroff, C. B. (1998, June). The neurobiology of depression. *Scientific American, 278,* 42–49.

Nesse, R. M., & Berridge, K. C. (1997). Psychoactive drug use in evolutionary perspective. *Science, 278,* 63–66.

Nestle, M. (2003). The ironic politics of obesity. *Science, 299,* 781.

Nestle, M. (2007, September). Eating made simple. *Scientific American, 297,* 60–69.

Nestler, E. J. (2005). Is there a common molecular pathway for addiction? *Nature Reviews Neuroscience, 8,* 1445–1449.

Nestler, E. J., Barrot, M., DiLeone, R. J., Eisch, A. J., Gold, S. J., & Monteggia, L. M. (2002). Neurobiology of depression. *Neuron, 34,* 13–25.

Nestler, E. J., & Malenka, R. C. (2004, March). The addicted brain. *Scientific American, 290,* 78–85.

Netter, F. H. (1962). *The CIBA collection of medical illustrations. Vol. 1, The nervous system.* New York, NY: CIBA.

Neves, S. R., Ram, P. T., & Iyengar, R. (2002). G protein pathways. *Science, 296,* 1636–1639.

Neville, H., & Bavelier, D. (1998). Neural organization and plasticity of language. *Current Opinion in Neurobiology, 8,* 254–258.

Newcombe, N., & Fox, N. (1994). Infantile amnesia: Through a glass darkly. *Child Development, 65,* 31–40.

Newcombe, N. S., Drummey, A. B., Fox, N. A., Lie, E., & Ottinger Alberts, W. (2000). Remembering early childhood: How much, how, and why (or why not). *Current Directions in Psychological Science, 9,* 55–58.

Newman, E. A. (2003). New roles for astrocytes: Regulation of synaptic transmission. *Trends in Neurosciences, 26,* 536–542.

Newport, D. J., & Nemeroff, C. B. (2000). Neurobiology of posttraumatic stress disorder. *Current Opinion in Neurobiology, 10,* 211–218.

Newsom-Davis, J., & Vincent, A. (1991). Antibody-mediated neurological disease. *Current Opinion in Neurobiology, 1,* 430–435.

Nicolesis, M. A. L., & Chapin, J. K. (October, 2002). People with nerve or limb injuries may one day be able to command wheelchairs, prosthetics and even paralyzed arms and legs by "thinking them through" the motions. *Scientific American, 287,* 47–53.

Nicolelis, M. A. L., & Ribeiro, S. (2006, December). Seeking the neural code. *Scientific American, 295,* 71–77.

Nicoll, R. A., & Alger, B. E. (December, 2004). The brain's own marijuana. *Scientific American, 291,* 68–75.

Nicolson, R. I., & Fawcett, A. J. (2007). Procedural learning difficulties: Reuniting the developmental disorders? *Trends in Neurosciences, 30,* 135–141.

Niedenthal, P. M. (2007). Embodying emotion. *Science, 316,* 1002–1005.

Nielsen, J. B. (2002). Motoneuronal drive during human walking. *Brain Research Reviews, 40,* 192–201.

Nijhawan, D., Honarpour, N., & Wang, X. (2000). Apoptosis in neural development and disease. *Annual Review of Neuroscience, 23,* 73–89.

Nilsson, J. P., Söderström, M., Karlsson, A. U., Lekander, M., Åkerstedt, T., Lindroth, N. E., & Axelsson, J. (2005). Less effective executive functioning after one night's sleep deprivation. *Journal of Sleep Research, 14,* 1–6.

Nimmerjahn, A., Kirchhoff, F., & Helmchen, F. (2005). Resting microglial cells are highly dynamic surveillants of brain parenchyma in vivo. *Science, 308,* 1314–1318.

Nishimura, Y., Onoe, H., Morichika, Y., Perfiliev, S., Tsukada, H., & Isa, T. (2007). Time-dependent central compensatory mechanisms of finger dexterity after spinal cord injury. *Science, 318,* 1150–1156.

Nishino, S. (2007). Clinical and neurobiological aspects of narcolepsy. *Sleep Medicine, 8,* 373–399.

Nishino, S., & Kanbayashi, T. (2005). Symptomatic narcolepsy, cataplexy and hypersomnia, and their implications in the hypothalamic hypocretin/orexin system. *Sleep Medicine Reviews, 9,* 269–310.

Nithianantharajah, J., & Hannan, A. J. (2006). Enriched environments, experience-dependent plasticity and disorders of the nervous system. *Nature Reviews Neuroscience, 7,* 697–709.

Nitschke, J. B., Sarinopoulos, I., Oathes, D. J., Johnstone, T., Whalen, P. J., Davidson, R. J., & Kalin, N. H. (2009). Anticipatory activation in the amygdala and anterior cingulate in generalized anxiety disorder and prediction of treatment response. *American Journal of Psychiatry, 166,* 302–310.

Nobre, A. C., & Plunkett, K. (1997). The neural system of language: Structure and development. *Current Opinion in Neurobiology, 7,* 262–268.

Nogueiras, R., & Tschöp, M. (2005). Separation of conjoined hormones yields appetite rivals. *Science, 310,* 985–987.

Nohr, E. A., Vaeth, M., Bech, B. H., Henriksen, T. B., Cnattingius, S., & Olsen, J. (2007). Maternal obesity and neonatal mortality according to subtypes of preterm birth. *Obstetrics and Gynecology, 110,* 1082–1090.

Nopoulis, P. C., Ceilley, J. W., Gailis, E. A., & Andreasen, N. C. (2001). An MRI study of midbrain morphology in patients with schizophrenia: Relationship to psychosis, neuroleptics, and cerebellar neural circuitry. *Biological Psychiatry, 49,* 13–19.

Nordt, C., & Stohler, R. (2006). Incidence of heroin use in Zurich, Switzerland: A treatment case register analysis. *Lancet, 367,* 1830–1834.

Norman, G., & Eacott, M. J. (2005). Dissociable effects of lesions to the perirhinal cortex and the postrhinal cortex on memory for context and objects in rats. *Behavioral Neuroscience, 119,* 557–566.

Nudo, R. J. (2006). Mechanisms for recovery of motor function following cortical damage. *Current Opinion in Neurobiology, 16,* 638–644.

Nudo, R. J., Jenkins, W. M., & Merzenich, M. M. (1996). Repetitive microstimulation alters the cortical representation of movements in adult rats. *Somatosensory Motor Research, 7,* 463–483.

Nurnberger, J. I., & Bierut, L. J. (2007, April). Seeking the connections: Alcoholism and our genes. *Scientific American, 296,* 46–53.

Nusser, Z. (2009). Variability in the subcellular distribution of ion channels increases neuronal diversity. *Trends in Neurosciences, 32,* 267–274.

Nykamp, K., Rosenthal, L., Folkerts, M., Roehrs, T., Guido, P., & Roth, T. (1998). The effects of REM sleep deprivation on the level of sleepiness/alertness. *Sleep, 21,* 609–614.

O'Brien, K. M., & Vincent, N. K. (2003). Psychiatric co-morbidity in anorexia and bulimia nervosa: Nature, prevalence, and causal relationships. *Clinical Psychology Review, 23,* 57–74.

Oberman, L. M., & Ramachandran, V. S. (2007). The simulating social mind: The role of the mirror neuron system and simulation in the social and communicative deficits of autism spectrum disorders. *Psychological Bulletin, 133,* 310–327.

O'Doherty, J. P. (2004). Reward representations and reward-related learning in the human brain: Insights from neuroimaging. *Current Opinion in Neurobiology, 14,* 769–776.

O'Donnell, M., Chance, R. K., & Bashaw, G. J. (2009). Axon growth and guidance: Receptor regulation and signal transduction. *Annual Review of Neuroscience, 32,* 383–412.

Ogawa, Y., Kanbayashi, T., Saito, Y., Takahashi, Y., Kitajima, T., Takahashi, K., . . . Shimizu, T. (2003). Total sleep deprivation elevates blood pressure through arterial baroreflex resetting: A study with microneurographic technique. *Sleep, 26,* 986–989.

Ohayon, M. M. (2008). From wakefulness to excessive sleepiness: What we know and still need to know. *Sleep Medicine Reviews, 12,* 129–141.

Ohbayashi, M., Ohki, K., & Miyashita, Y. (2003). Conversion of working memory to motor sequence in the monkey premotor cortex. *Science, 301,* 233–236.

Ohlsson, R. (2007). Widespread monoallelic expression. *Science, 318,* 1077–1078.

Ohtaki, H., Mori, S., Nakamachi, T., Dohi, K., Yin, L., Endo, S., . . . Shioda, S. (2003). Evaluation of neuronal cell death after a new global ischemia model in infant mice. *Acta Neurochirurgica Suppl., 86,* 97–100.

Ohzawa, I. (1998). Mechanisms of stereoscopic vision: The disparity energy model. *Current Opinion in Neurobiology, 8,* 509–515.

Ojemann, G. A. (1979). Individual variability in cortical localization of language. *Journal of Neurosurgery, 50,* 164–169.

Ojemann, G. A. (1983). Brain organization for language from the perspective of electrical stimulation mapping. *Behavioral and Brain Sciences, 2,* 189–230.

Okamoto-Barth, S., Call, J., & Tomasello, M. (2007). Great apes' understanding of other individuals' line of sight. *Psychological Science, 18,* 462–468.

Okano, H., & Temple, S. (2009). Cell types to order: Temporal specification of CNS stem cells. *Current Opinion in Neurobiology, 19,* 112–119.

O'Keefe, J. (1993). Hippocampus, theta, and spatial memory. *Current Opinion in Neurobiology, 3,* 917–924.

O'Keefe, J., & Nadel, L. (1978). *The hippocampus as a cognitive map.* Oxford, England: Clarendon Press.

O'Keefe, J., & Speakman, A. (1987). Single unit activity in the rat hippocampus during a spatial memory task. *Experimental Brain Research, 68,* 1–27.

Olds, J., & Milner, P. (1954). Positive reinforcement produced by electrical stimulation of septal area and other regions of rat brain. *Journal of Comparative and Physiological Psychology, 47,* 419–427.

O'Leary, C. M. (2004). Fetal alcohol syndrome: Diagnosis, epidemiology, and developmental outcomes. *Journal of Paediatrics and Child Health, 40,* 2–7.

Olshansky, S. J., Passaro, D. J., Hershow, R. C., Layden, J., Carnes, B. A., Brody, J., . . . Ludwig, D. S. (2005). A potential decline in life expectancy in the United States in the 21st century. *New England Journal of Medicine, 352,* 1138–1145.

Olson, C. R. (2003). Brain representations of object-centered space in monkeys and humans. *Annual Review of Neuroscience, 26,* 331–354.

Olsson, A., & Phelps, E. A. (2007). Social learning of fear. *Nature Neuroscience, 10,* 1095–1102.

Olszewski, P. K., Schiöth, H. B., & Levine, A. S. (2008). Ghrelin in the CNS: From hunger to a rewarding and memorable meal? *Brain Research Reviews, 58,* 160–170.

Olton, D. S., & Samuelson, R. J. (1976). Remembrance of places: Spatial memory in rats. *Journal of Experimental Psychology: Animal Behavior Processes, 2,* 97–116.

O'Neill, L. A. J. (2005, January). Immunity's early-warning system. *Scientific American, 292,* 38–45.

Oniani, T., & Lortkipanidze, N. (2003). Effect of paradoxical sleep deprivation on the learning and memory. *Neurobiology of Sleep-Wakefulness Cycle, 3,* 9–43.

Ormerod, B. K., & Galea, L. A. M. (2001). Mechanism and function of adult neurogenesis. In C. A. Shaw & J. C. McEachern (Eds.), *Towards a theory of neuroplasticity* (pp. 85–100). Philadelphia, PA: Taylor & Francis.

Ormerod, B. K., Falconer, E. M., & Galea, L. A. M. (2003). N-methyl-D-aspartate receptor activity and estradiol: Separate regulation of cell proliferation in the dentate gyrus of adult female meadow vole. *Journal of Endocrinology, 179,* 155–163.

Orr, H. A. (2009, January). Testing natural selection. *Scientific American, 300,* 44–50.

Orr, H. T., & Zoghbi, H. Y. (2007). Trinucleotide repeat disorders. *Annual Review of Neuroscience, 30,* 575–621.

Orser, B. A. (2007, June). Lifting the fog around anesthesia. *Scientific American, 296,* 54–61.

O'Tuathaigh, C. M. P., Babovic, D., O'Meara, G., Clifford, J. J., Croke, D. T., & Waddington, J. L. (2007). Susceptibility genes for schizophrenia: Characterisation of mutant mouse models at the level of phenotypic behaviour. *Neuroscience and Biobehavioral Reviews, 31,* 60–78.

Overmier, J. B., & Murison, R. (1997). Animal models reveal the "psych" in the psychosomatics of peptic ulcers. *Current Directions in Psychological Science, 6,* 180–184.

Pääbo, S. (1995). The Y chromosome and the origin of all of us (men). *Science, 268,* 1141–1142.

Page, S. A., Verhoef, M. J., Stebbins, R. A., Metz, L. M., & Levy, J. C. (2003). Cannabis use as described by people with multiple sclerosis. *Canadian Journal of Neurological Sciences, 30,* 201–205.

Palmer, L. M., & Stuart, G. J. (2006). Site of action potential initiation in layer 5 pyramidal neurons. *Journal of Neuroscience, 26,* 1854–1863.

Palmiter, R. D. (2007). Is dopamine a physiologically relevant mediator of feeding behavior? *Trends in Neurosciences, 30,* 375–381.

Pancrazio, J. J., & Peckham, P. H. (2009). Neuroprosthetic devices: How far are we from recovering movements in paralyzed patients? *Expert Review of Neurotherapeutics, 9,* 427–430.

Panda, S., Nayak, S. K., Campo, B., Walker, J. R., Hogenesch, J. B., & Jegla, T. (2005). Illumination of the melanopsin signaling pathway. *Science, 307,* 600–604.

Panksepp, J. (2003). Feeling the pain of social loss. *Science, 302,* 237–239.

Panksepp, J. (2007). Neurologizing the psychology of affect: How appraisal-based constructivism and basic emotion theory can coexist. *Perspectives on Psychological Science, 2,* 281–296.

Panksepp, J., & Trowill, J. A. (1967). Intraoral self-injection: II. The simulation of self-stimulation phenomena with a conventional reward. *Psychonomic Science, 9,* 407–408.

Paredes, R. G. (2003). Medial preoptic area/anterior hypothalamus and sexual motivation. *Scandinavian Journal of Psychology, 44,* 203–212.

Parent, C., Zhang, T-Y., Caldji, C., Bagot, R., Champagne, F. A., Pruessner, J., & Meaney, M. J. (2005). Maternal care and individual differences in defensive responses. *Current Directions in Psychological Science, 14,* 229–233.

Pariante, C. M., & Lightman, S. L. (2008). The HPA axis in major depression: Classical theories and new developments. *Trends in Neurosciences, 31,* 464–468.

Parker, A. J. (2007). Binocular depth perception and the cerebral cortex. *Nature Reviews Neuroscience, 8,* 379–391.

Parker, S. T., Mitchell, R. W., & Boccia, M. L. (1994). *Self-awareness in animals and humans: Developmental perspectives.* New York, NY: Cambridge University Press.

Parr, L. A., Waller, B. M., & Fugate, J. (2005). Emotional communication in primates: Implications for neurobiology. *Current Opinion in Neurobiology, 15,* 716–720.

Parr, L. A., Waller, B. M., & Vick, S. J. (2007). New developments in understanding emotional facial signals in chimpanzees. *Current Directions in Psychological Science, 16,* 117–122.

Parrott, A. C. (1999). Does cigarette smoking cause stress? *American Psychologist, 54,* 817–820.

Pascalis, O., & Kelly, D. J. (2009). The origins of face processing in humans. *Perspectives on Psychological Science, 4,* 200–209.

Pascual, O., Casper, K. B., Kubera, C., Zhang, J., Revilla-Sanchez, R., Sul, J-Y., . . . Haydon, P. G. (2005). Astrocytic purinergic signaling coordinates synaptic networks. *Science, 310,* 113–116.

Pascual-Leone, A., Amedi, A., Fregni, F., & Merabet, L. B. (2005). The plastic human brain cortex. *Annual Review of Neuroscience, 28,* 377–401.

Pascual-Leone, A., Walsh, V., & Rothwell, J. (2000). Transcranial magnetic stimulation in cognitive neuroscience—virtual lesion, chronometry, and functional connectivity. *Current Opinion in Neurobiology, 10,* 232–237.

Passingham, R. (2009). How good is the macaque monkey model of the human brain? *Current Opinion in Neurobiology, 19,* 6–11.

Pasternak, T., & Greenlee, M. W. (2005). Working memory in primate sensory systems. *Nature Reviews Neuroscience, 6,* 97–107.

Patel, S. R., Ayas, N. T., White, D. P., Speizer, F. E., Stampfer, M. J., & Hu, F. B. (2003). A prospective study of sleep duration and mortality risk in women. *Sleep, 26,* A184.

Paterson, S. J., Brown, J. H., Gsödl, M. K., Johnson, M. H., & Karmiloff-Smith, A. (1999). Cognitive modularity and genetic disorders. *Science, 286,* 2355–2358.

Patil, P. G. (2009). Introduction: Advances in brain-machine interfaces. *Neurosurgical Focus, 27,* E1.

Patterson, K., Nestor, P. J., & Rogers, T. T. (2007). Where do you know what you know? The representation of semantic knowledge in the human brain. *Nature Reviews Neuroscience, 8,* 976–987.

Patterson, K., & Ralph, M. A. L. (1999). Selective disorders of reading? *Current Opinion in Neurobiology, 9,* 235–239.

Patterson, K., Vargha-Khadem, F., & Polkey, C. E. (1989). Reading with one hemisphere. *Brain, 112,* 39–63.

Paul, L. K., Brown, W. S., Adolphs, R., Tyszka, J. M., Richards, L. J., Mukherjee, P., & Sherr, E. H. (2007). Agenesis of the corpus callosum: Genetic, developmental and functional aspects of connectivity. *Nature Reviews Neuroscience, 8,* 287–299.

Paulesu, E., Démonet, J.-F., Fazio, F., McCrory, E., Chanoine, V., Brunswick, N., . . . Frith, U. (2001). Dyslexia: Cultural diversity and biological unity. *Science, 291,* 2165–2167.

Paulesu, E., McCrory, E., Fazio, F., Menoncello, L., Brunswick, N., Cappa, S. F., . . . Frith, U. (2000). A cultural effect on brain function. *Nature Neuroscience, 3,* 91–96.

Paulozzi, L. J., Budnitz, D. S., & Xi, Y. (2006). Increasing deaths from opioid analgesics in the United States. *Pharmacoepidemiology and Drug Safety, 15,* 618–627.

Pauls, D. L. (2001). Update on the genetics of Tourette syndrome. In D. J. Cohen, C. G. Goetz, & J. Jankovic (Eds.), *Tourette syndrome* (pp. 281–293). Philadelphia, PA: Lippincott Williams & Wilkins.

Payne, B. R., & Lomber, S. G. (2001). Reconstructing functional systems after lesions of cerebral cortex. *Nature Reviews Neuroscience, 2,* 911–919.

Paz, R., Pelletier, J. G., Bauer, E. P., & Paré, D. (2006). Emotional enhancement of memory via amygdala-driven facilitation of rhinal interactions. *Nature Neuroscience, 9,* 1321–1329.

Peissig, J. J., & Tarr, M. J. (2007). Visual object recognition: Do we know more now than we did 20 years ago? *Annual Review of Psychology, 56,* 75–96.

Pelchat, M. L. (2009). Food addiction in humans. *Journal of Nutrition, 139,* 620–622.

Pellis, S. M., O'Brien, D. P., Pellis, V. C., Teitelbaum, P., Wolgin, D. L., & Kennedy, S. (1988). Escalation of feline predation along a gradient from avoidance through "play" to killing. *Behavioral Neuroscience, 102,* 760–777.

Pellow, S., Chopin, P., File, S. E., & Briley, M. (1985). Validation of open:closed arm entries in an elevated plus-maze as a measure of anxiety in the rat. *Journal of Neuroscience Methods, 14,* 149–167.

Penfield, W., & Boldrey, E. (1937). Somatic motor and sensory representations in cerebral cortex of man as studied by electrical stimulation. *Brain, 60,* 389–443.

Penfield, W., & Evans, J. (1935). The frontal lobe in man: A clinical study of maximum removals. *Brain, 58,* 115–133.

Penfield, W., & Rasmussen, T. (1950). *The cerebral cortex of man: A clinical study of the localization of function.* New York, NY: Macmillan.

Penfield, W., & Roberts, L. (1959). *Speech and brain mechanisms.* Princeton, NJ: Princeton University Press.

Penton-Voak, I. S., & Perrett, D. I. (2000). Female preferences for male faces changes cyclically—further evidence. *Evolution and Human Behavior, 21,* 39–48.

Perea, G., Navarrete, M., & Araque, A. (2009). Tripartite synapses: Astrocytes process and control synaptic information. *Trends in Neurosciences, 32,* 421–431.

Persico, A. M., & Bourgeron, T. (2006). Searching for ways out of the autism maze: Genetic, epigenetic and environmental clues. *Trends in Neurosciences, 29,* 349–358.

Peters, M., & Brooke, J. (1998). Conduction velocity in muscle and cutaneous afferents in humans. *Journal of Motor Behavior, 30,* 285–287.

Petersen, S. E., Fox, P. T., Posner, M. I., Mintun, M., & Raichle, M. E. (1988). Positron emission tomographic studies of the cortical anatomy of single-word processing. *Nature, 331,* 585–589.

Peyron, C., Faraco, J., Rogers, W., Ripley, B., Overeem, S., Charnay, Y., . . . Mignot, E. (2000). A mutation in a case of early onset narcolepsy and a generalized absence of hypocretin peptides in human narcoleptic brains. *Nature Medicine, 6,* 991–997.

Pezawas, L., Meyer-Lindenberg, A., Drabant, E. M., Verchinski, B. A., Munoz, K. E., Kolachana, B. S., . . . Weinberger, D. R. (2005). 5-HTTLPR polymorphism impacts human cingulate-amygdala interactions: A genetic susceptibility mechanism for depression. *Nature Neuroscience, 8,* 828–834.

Pfeiffer, C. A. (1936). Sexual differences of the hypophyses and their determination by the gonads. *American Journal of Anatomy, 58,* 195–225.

Pfreiger, F. W. (2002). Role of glia in synapse development. *Current Opinion in Neurobiology, 12,* 486–490.

Phelps, E. A. (2004). The human amygdala and awareness: Interactions between emotion and cognition. In M. S. Gazzaniga (Ed.), *The cognitive neurosciences* (pp. 1005–1016). Cambridge, MA: MIT Press.

Phelps, E. A. (2006). Emotion and cognition: Insights from studies of the human amygdala. *Annual Review of Psychology, 57,* 27–53.

Phelps, M. E., & Mazziotta, J. (1985). Positron tomography: Human brain function and biochemistry. *Science, 228,* 804.

Phillips, A. G., Coury, A., Fiorino, D., LePiane, F. G., Brown, E., & Fibiger, H. C. (1992). Self-stimulation of the ventral tegmental area enhances dopamine release in the nucleus accumbens: A microdialysis study. *Annals of the New York Academy of Sciences, 654,* 199–206.

Phoenix, C. H., Goy, R. W., Gerall, A. A., & Young, W. C. (1959). Organizing action of prenatally administered testosterone proprionate on the tissues mediating mating behavior in the female guinea pig. *Endocrinology, 65,* 369–382.

Pierce, R. C., & Kumaresan, V. (2006). The mesolimbic dopamine system: The final common pathway for the reinforcing effect of drugs of abuse? *Neuroscience and Biobehavioural Reviews, 30,* 215–238.

Pietrobon, D., & Striessnig, J. (2003). Neurobiology of migraine. *Nature Reviews Neuroscience, 4,* 386–398.

Pietropaolo, S., Feldon, J., & Yee, B. K. (2008). Age-dependent phenotypic characteristics of a triple transgenic mouse model of Alzheimer disease. *Behavioral Neuroscience, 122,* 733–747.

Pinel, J. P. J. (1969). A short gradient of ECS-produced amnesia in a one-trial appetitive learning situation. *Journal of Comparative and Physiological Psychology, 68,* 650–655.

Pinel, J. P. J., Assanand, S., & Lehman, D. R. (2000). Hunger, eating, and ill health. *American Psychologist, 55,* 1105–1116.

Pinel, J. P. J., & Mana, M. J. (1989). Adaptive interactions of rats with dangerous inanimate objects: Support for a cognitive theory of defensive behavior. In R. J. Blanchard, P. F. Brain, D. C. Blanchard, & S. Parmigiani

(Eds.), *Ethoexperimental approaches to the study of behavior* (pp. 137–150). Dordrecht, The Netherlands: Kluwer Academic Publishers.

Pinel, J. P. J., Mana, M. J., & Kim, C. K. (1989). Effect-dependent tolerance to ethanol's anticonvulsant effect on kindled seizures. In R. J. Porter, R. H. Mattson, J. A. Cramer, & I. Diamond (Eds.), *Alcohol and seizures: Basic mechanisms and clinical implications* (pp. 115–125). Philadelphia, PA: F. A. Davis.

Pinel, J. P. J., & Treit, D. (1978). Burying as a defensive response in rats. *Journal of Comparative and Physiological Psychology, 92,* 708–712.

Piomelli, D., Astarita, G., & Rapaka, R. (2007). A neuroscientist's guide to lipidomics. *Nature Reviews Neuroscience, 8,* 743–754.

Pisella, L., & Mattingley, J. B. (2004). The contribution of spatial remapping impairments to unilateral visual neglect. *Neuroscience and Biobehavioural Reviews, 28,* 181–200.

Placzek, M., & Briscoe, J. (2005). The floor plate: Multiple cells, multiple signals. *Nature Reviews Neuroscience, 6,* 230–240.

Platt, M. L., & Spelke, E. S. (2009). What can developmental and comparative cognitive neuroscience tell us about the adult human brain? *Current Opinion in Neurobiology, 19,* 1–5.

Pleim, E. T., & Barfield, R. J. (1988). Progesterone versus estrogen facilitation of female sexual behavior by intracranial administration to female rats. *Hormones and Behavior, 22,* 150–159.

Ploegh, H. L. (1998). Viral strategies of immune evasion. *Science, 280,* 248–252.

Plomin, R., & DeFries, J. C. (1998, May). The genetics of cognitive abilities and disabilities. *Scientific American, 278,* 62–69.

Plomin, R., DeFries, J. C., McGuffin, P., & Craig, I. W. (2002). *Behavioral genetics in the postgenomic era.* Washington, DC: American Psychological Association.

Plomin, R., & McGuffin, P. (2005). Psychopathology in the postgenomic era. *Annual Review of Psychology, 54,* 205–228.

Plomin, R., & Neiderhiser, J. M. (1992). Genetics and experience. *Current Directions in Psychological Science, 1,* 160–163.

Pluchino, S., Zanotti, L., Deleidi, M., & Marino, G. (2005). Neural stem cells and their use as therapeutic tools in neurological disorders. *Brain Research Reviews, 48,* 211–219.

Pocock, J. M., & Kettenmann, H. (2007). Neurotransmitter receptors on microglia. *Trends in Neurosciences, 30,* 527–535.

Poeppel, D., & Monahan, P. J. (2008). Speech perception: Cognitive foundations and cortical implementation. *Current Directions in Psychological Science, 17,* 80–85.

Poldrack, R. A. (2008). The role of fMRI in cognitive neuroscience: Where do we stand? *Current Opinion in Neurobiology, 18,* 223–227.

Poliak, S., & Peles, E. (2003). The local differentiation of myelinated axons at nodes of Ranvier. *Nature, 4,* 968–980.

Poliakoff, E., & Smith-Spark, J. H. (2008). Everyday cognitive failures and memory problems in Parkinson's patients without dementia. *Brain and Cognition, 67,* 340–350.

Polivy, J., & Herman, P. C. (2002). Causes of eating disorders. *Annual Review of Neuroscience, 53,* 187–213.

Pollack, H. A., & Reuter, P. (2007). The implications of recent findings on the link between cannabis and psychosis. *Addiction, 102,* 173–176.

Pollard, K. S. (2009, May). What makes us human? *Scientific American, 300,* 44–49.

Polleux, F., Ince-Dunn, G., & Gosh, A. (2007). Transcriptional regulation of vertebrate axon guidance and synapse formation. *Nature Reviews Neuroscience, 8,* 331–340.

Pompili, M., Mancinelli, I., Girardi, P., Ruberto, A., & Tatarelli, R. (2004). Suicide in anorexia nervosa: A meta-analysis. *International Journal of Eating Disorders, 36,* 99–103.

Pons, T. P., Garraghty, P. E., Ommaya, A. K., Kaas, J. H., Taub, E., & Mishkin, M. (1991). Massive cortical reorganization after sensory deafferentation in adult macaques. *Science, 252,* 1857–1860.

Pope, H. G., Gruber, A. J., Hudson, J. I., Cohane, G., Huestis, M. A., & Yurgelun-Todd, D. (2003). Early-onset cannabis use and cognitive deficits: What is the nature of the association? *Drug and Alcohol Dependence, 69,* 303–310.

Pope, H. G., Jr., Kouri, E. M., & Hudson, J. I. (2000). Effects of supraphysiologic doses of testosterone on mood and aggression in normal men. *Archives of General Psychiatry, 57,* 133–140.

Poppele, R., & Bosco, G. (2003). Sophisticated spinal contributions to motor control. *Trends in Neurosciences, 26,* 269–276.

Porter, R., & Lemon, R. N. (1993). Corticospinal function and voluntary movement. *Monographs of the Physiological Society, No. 45.* Oxford, England: Oxford University Press.

Porter, S., & ten Brinke, L. (2008). Reading between the lies: Identifying concealed and falsified emotions in universal facial expressions. *Psychological Science, 19,* 508–514.

Posner, M. I., & Raichle, M. E. (1994). *Images of the mind.* New York, NY: Scientific American Library.

Postle, B. R., Corkin, S., & Growdon, J. H. (1996). Intact implicit memory for novel patterns in Alzheimer's disease. *Learning & Memory, 3,* 305–312.

Potts, G. F., Martin, L. E., Burton, P., & Montague, P. R. (2006). When things are better or worse than expected: The medial frontal cortex and allocation of processing resources. *Journal of Cognitive Neuroscience, 18,* 1112–1119.

Poucet, B., & Save, E. (2005). Attractors in memory. *Science, 308,* 799–800.

Pouget, A., & Driver, J. (2000). Relating unilateral neglect to the neural coding of space. *Current Opinion in Neurobiology, 10,* 242–249.

Poulin, C., Stein, J., & Butt, J. (2000). Surveillance of drug overdose deaths using medical examiner data. *Chronic Diseases in Canada, 19.* Retrieved from http://www.phas-aspc.gc.ca/publicat/cdic-mcc/19-4/f_e.html

Poulos, C. X., & Cappell, H. (1991). Homeostatic theory of drug tolerance: A general model of physiological adaptation. *Psychological Review, 98,* 390–408.

Pourtois, G., Schwartz, S., Seghier, M. L., Lazeyras, F., & Vuilleumier, P. (2005). Portraits or people? Distinct representation of face identity in the human visual cortex. *Journal of Cognitive Neuroscience, 17,* 1043–1057.

Powley, T. L., Opsahl, C. A., Cox, J. E., & Weingarten, H. P. (1980). The role of the hypothalamus in energy homeostasis. In P. J. Morgane & J. Panksepp (Eds.), *Handbook of the hypothalamus, 3A: Behavioral studies of the hypothalamus* (pp. 211–298). New York, NY: Marcel Dekker.

Preti, A. (2007). New developments in the pharmacotherapy of cocaine abuse. *Addiction Biology, 12,* 133–151.

Price, C. J., Howard, D., Patterson, K., Warburton, E. A., Friston, K. J., & Frackowiak, R. S. J. (1998). A functional neuroimaging description of two deep dyslexic patients. *Journal of Cognitive Neuroscience, 10,* 303–315.

Price, C. J., & Mechelli, A. (2005). Reading and reading disturbance. *Current Opinion in Neurobiology, 15,* 231–238.

Price, D. D. (2000). Psychological and neural mechanisms of the affective dimension of pain. *Science, 288,* 1769–1772.

Primakoff, P., & Myles, D. G. (2002). Penetration, adhesion, and fusion in mammalian sperm-egg interaction. *Science, 296,* 2183–2185.

Prolla, T. A., & Mattson, M. P. (2001). Molecular mechanisms of brain aging and neurodegenerative disorders: Lessons from dietary restriction. *Trends in Neurosciences, 24,* S21–S31.

Provine, R. R. (2005). Yawning. *American Scientist, 93,* 532–539.

Pryce, C. R., & Feldon, J. (2003). Long-term neurobehavioural impact of the postnatal environment in rats: Manipulations, effects, and mediating mechanisms. *Neuroscience and Biobehavioural Reviews, 27,* 57–71.

Purves, D., Augustine, G. J., Fitzpatrick, D., Hall, W. C., LaMantia, A-S., McNamara, J. O., & Williams, S. M. (2004). *Neuroscience* (3rd ed.) Sunderland, MA: Sinauer Associates.

Pusey, A., Williams, J., & Goodall, J. (1997). The influence of dominance rank on the reproductive success of female chimpanzees. *Science, 277,* 828–830.

Puts, D. A., Jordan, C. L., & Breedlove, S. M. (2006). O brother, where art thou? The fraternal birth-order effect on male sexual orientation. *Proceedings of the National Academy of Sciences, USA, 103,* 10531–10532.

Quammen, D. (2004, May). Was Darwin wrong? No. The evidence for evolution is overwhelming. *National Geographic, 206,* 2–35.

Quinsey, V. L. (2003). The etiology of anomalous sexual preferences in men. *Annals of the New York Academy of Sciences, 989,* 105–117.

Quirk, G. J., & Beer, J. S. (2006). Prefrontal involvement in the regulation of emotion: Convergence of rat and human studies. *Current Opinion in Neurobiology, 16,* 723–727.

Rademakers, R., & Rovelet-Lecrux, A. (2009). Recent insights into the molecular genetics of dementia. *Trends in Neurosciences, 32,* 451–461.

Rahman, Q. (2005). The neurodevelopment of human sexual orientation. *Neuroscience and Biobehavioural Reviews, 29,* 1057–1066.

Raichle, M. E. (2006). The brain's dark energy. *Science, 314,* 1249–1250.

Raichle, M. E. (2008). A brief history of human brain mapping. *Trends in Neurosciences, 32,* 118–126.

Raichle, M. E. (2010, March). The brain's dark energy. *Scientific American, 302,* 44–49.

Raichle, M. E., & Mintun, M. A. (2006). Brain work and brain imaging. *Annual Review of Neuroscience, 29,* 449–476.

Rainville, P. (2002). Brain mechanisms of pain affect and pain modulation. *Current Opinion in Neurobiology, 12,* 195–204.

Raisman, G. (1997). An urge to explain the incomprehensible: Geoffrey Harris and the discovery of the neural control of the pituitary gland. *Annual Review of Neuroscience, 20,* 533–566.

Raisman, G., & Field, P. M. (1971). Sexual dimorphism in the neuropil of the preoptic area of the rat and its dependence on neonatal androgens. *Brain Research, 54,* 1–29.

Raisman, G., & Li, Y. (2007). Repair of neural pathways by olfactory ensheathing cells. *Nature Reviews Neuroscience, 8,* 312–319.

Raizen, D. M., Mason, T. B. A., & Pack, A. I. (2006). Genetic basis for sleep regulation and sleep disorders. *Seminars in Neurology, 5,* 467–483.

Rakic, P. (1979). Genetic and epigenetic determinants of local neuronal circuits in the mammalian central nervous system. In F. O. Schmitt & F. G. Worden (Eds.), *The neurosciences: Fourth study program.* Cambridge, MA: MIT Press.

Rakic, P. (2006). No more cortical neurons for you. *Science, 313,* 928–929.

Rakic, P. (2009). Evolution of the neocortex: a perspective from developmental biology. *Nature Reviews Neuroscience, 10,* 724–735.

Rakic, P., Ayoub, A. E., Breunig, J. J., & Dominguez, M. H. (2009). Decision by division: Making cortical maps. *Trends in Neurosciences, 32,* 291–301.

Ralph, M. R., Foster, T. G., Davis, F. C., & Menaker, M. (1990). Transplanted suprachiasmatic nucleus determines circadian period. *Science, 247,* 975–978.

Ralph, M. R., & Menaker, M. (1988). A mutation of the circadian system in golden hamsters. *Science, 241,* 1225–1227.

Ramachandran, V. S., & Blakeslee, S. (1998). *Phantoms in the brain.* New York, NY: Morrow.

Ramachandran, V. S., & Oberman, L. M. (November, 2006). Broken mirrors, a theory of autism. *Scientific American, 295,* 63–69.

Ramachandran, V. S., & Rogers-Ramachandran, D. (2000). Phantom limbs and neuronal plasticity. *Archives of Neurology, 57,* 317–320.

Ramagopalan, S. V., Dyment, D. A., & Ebers, G. C. (2008). Genetic epidemiology: The use of old and new tools for multiple sclerosis. *Trends in Neurosciences, 31,* 645–652.

Ramagopalan, S. V., Maugeri, N. J., Handunnetthi, L., Lincoln, M. R., Orton, S-M., Dyment, D. A., . . . Knight, J. C. (2009). Expression of the multiple sclerosis-associated MHC class II allele HLA-DRB$_{1*1501}$ is regulated by vitamin D. *PLoS Genetics, 5:* e1000369. doi:10.1371/journal.pgen .1000369

Ramnani, N. (2006). The primate cortico-cerebellar system: Anatomy and function. *Nature Reviews Neuroscience, 7,* 511–521.

Ramsay, D. S., & Woods, S. C. (1997). Biological consequences of drug administration: Implications for acute and chronic tolerance. *Psychological Review, 104*(1), 170–193.

Ramus, F. (2004). Neurobiology of dyslexia: A reinterpretation of the data. *Trends in Neurosciences, 27,* 720–726.

Rankinen, T., Zuberi, A., Chagnon, Y. C., Weisnagel, S. J., Argyropoulos, G., Walts, B., . . . Bouchard, C. (2006). The human obesity gene map: The 2005 update. *Obesity, 14,* 529–644.

Rao, S. C., Rainer, G., & Miller, E. K. (1997). Integration of what and where in the primate prefrontal cortex. *Science, 276,* 821–824.

Rapport, R. (2005). *Nerve endings: The discovery of the synapse.* New York, NY: Norton.

Rasch, B., & Born, J. (2008). Reactivation and consolidation of memory during sleep. *Current Directions in Psychological Science, 17,* 188–192.

Rasch, B., Büchel, C., Gais, S., & Born, J. (2007). Odor cues during slow-wave sleep prompt declarative memory consolidation. *Science, 315,* 1426–1429.

Rasch, B., Pommer, J., Diekelmann, S., & Born, J. (2008). Pharmacological REM sleep suppression paradoxically improves rather than impairs skill memory. *Nature Neuroscience, 12,* 396–397.

Rash, B. G., & Groves, E. A. (2006). Area and layer patterning in the developing cerebral cortex. *Current Opinion in Neurobiology, 16,* 25–34.

Rasmussen, T., & Milner, B. (1975). Clinical and surgical studies of the cerebral speech areas in man. In K. J. Zulch, O. Creutzfeldt, & G. C. Galbraith (Eds.), *Cerebral localization* (pp. 238–257). New York, NY: Springer-Verlag.

Ratliff, F. (1972, June). Contour and contrast. *Scientific American, 226,* 90–101.

Rattenborg, N. C., Amlaner, C. J., & Lima, S. L. (2000). Behavioral, neurophysiological and evolutionary perspectives on unihemispheric sleep. *Neuroscience and Biobehavioural Reviews, 24,* 817–842.

Rattenborg, N. C., Martinez-Gonzales, D., & Lesku, J. A. (2009). Avian sleep homeostasis: Convergent evolution of complex brains, cognition and sleep functions in mammals and birds. *Neuroscience and Biobehavioral Reviews, 33,* 253–270.

Ravizza, S. M., & Ivry, R. B. (2001). Comparison of the basal ganglia and cerebellum in shifting attention. *Journal of Cognitive Neuroscience, 13,* 285–297.

Ravussin, E. (2005). A NEAT way to control weight? *Science, 307,* 530–531.

Ravussin, E., & Danforth, E., Jr. (1999). Beyond sloth—physical activity and weight gain. *Science, 283,* 184–185.

Raymond, C. R. (2007). LTP forms 1, 2 and 3: Different mechanisms for the "long" in long-term potentiation. *Trends in Neurosciences, 30,* 168–175.

Raynor, H. A., & Epstein, L. H. (2001). Dietary variety, energy regulation, and obesity. *Psychological Bulletin, 127,* 325–341.

Reber, P. J., Knowlton, B. J., & Squire, L. R. (1996). Dissociable properties of memory system: Differences in the flexibility of declarative and nondeclarative knowledge. *Behavioral Neuroscience, 110*(5), 861–871.

Recanzone, G. H. (2004). Acoustic stimulus processing and multimodal interactions in primates. In M. S. Gazzaniga (Ed.), *The cognitive neurosciences* (pp. 359–368). Cambridge, MA: MIT Press.

Recanzone, G. H., & Sutter, M. L. (2008). The biological basis of audition. *Annual Review of Psychology, 59,* 119–142.

Rechtschaffen, A., & Bergmann, B. M. (1995). Sleep deprivation in the rat by the disk-over-water method. *Behavioural Brain Research, 69,* 55–63.

Rechtschaffen, A., Gilliland, M. A., Bergmann, B. M., & Winter, J. B. (1983). Physiological correlates of prolonged sleep deprivation in rats. *Science, 221,* 182–184.

Redd, M., & de Castro, J. M. (1992). Social facilitation of eating: Effects of social instruction on food intake. *Physiology & Behavior, 52,* 749–754.

Reddy, L., & Kanwisher, N. (2006). Coding of visual objects in the ventral stream. *Current Opinion in Neurobiology, 16,* 408–414.

Redish, A. D. (2004). Addiction as a computational process gone awry. *Science, 306,* 1944–1948.

Rees, G., Kreiman, G., & Koch, C. (2002). Neural correlates of consciousness in humans. *Nature Reviews Neuroscience, 3,* 261–270.

Rees, G., Russell, C., Frith, C. D., & Driver, J. (1999). Inattentional blindness versus inattentional amnesia for fixated but ignored words. *Science, 286,* 2504–2507.

Regan, P. C. (1996). Rhythms of desire: The association between menstrual cycle phases and female sexual desire. *Canadian Journal of Human Sexuality, 5*(3), 145–156.

Rehm, J. (2006). Alcohol consumption, stroke and public health—an overlooked relation? *Addiction, 101,* 1679–1681.

Rehm, J., Gschwend, P., Steffen, T., Gutzwiller, F., Dobler-Mikola, A., & Uchtenhagen, A. (2001). Feasibility, safety, and efficacy of injectable heroin prescription for refractory opioid addicts: A follow-up study. *Lancet, 358,* 1417–1423.

Reichling, D. B., & Levine, J. D. (2009). Critical role of nociceptor plasticity in chronic pain. *Trends in Neurosciences, 32,* 611–618.

Reid, R. C., & Alonso, J-M. (1996). The processing and encoding of information in the visual cortex. *Current Opinion in Neurobiology, 6,* 475–480.

Reijmers, L. G., Perkins, B. L., Matsuo, N., & Mayford, M. (2007). Localization of a stable neural correlate of associative memory. *Science, 317,* 1230–1233.

Reinarman, C., Cohen, P. D., & Kaal, H. L. (2004). The limited relevance of drug policy: Cannabis in Amsterdam and in San Francisco. *American Journal of Public Health, 94,* 836–842.

Reiner, S. L., Sallusto, F., & Lanzavecchia, A. (2007). Division of labor with a workforce of one: Challenges in specifying effector and memory T cell fate. *Science, 317,* 622–625.

Reiner, W. (1997). To be male or female—that is the question. *Archives of Pediatrics and Adolescent Medicine, 151,* 224–225.

Ren, T., & Gillespie, P. G. (2007). A mechanism for active hearing. *Current Opinion in Neurobiology, 17,* 498–503.

Rensberger, B. (2000). The nature of evidence. *Science, 289,* 61.

Represa, A., & Ben-Ari, Y. (2006). Trophic actions of GABA on neuronal development. *Trends in Neurosciences, 28,* 279–283.

Reuter-Lorenz, P. A., & Cappell, K. A. (2008). Neurocognitive aging and the compensation hypothesis. *Current Directions in Psychological Science, 17,* 177–182.

Reuter-Lorenz, P. A., & Miller, A. C. (1998). The cognitive neuroscience of human laterality: Lessons from the bisected brain. *Current Directions in Psychological Science, 7,* 15–20.

Revusky, S. H., & Garcia, J. (1970). Learned associations over long delays. In G. H. Bower & J. T. Spence (Eds.), *The psychology of learning and motivation* (Vol. 4, pp. 1–85). New York, NY: Academic Press.

Reynolds, D. V. (1969). Surgery in the rat during electrical analgesia induced by focal brain stimulation. *Science, 164,* 444–445.

Rhees, R. W., Lephart, E. D., & Eliason, D. (2001). Effects of maternal separation during early postnatal development on male sexual behavior and female reproductive function. *Behavioural Brain Research, 123,* 1–10.

Rhinn, M., Picker, A., & Brand, M. (2006). Global and local mechanisms of forebrain and midbrain patterning. *Current Opinion in Neurobiology, 16,* 5–12.

Rhodes, G., Byatt, G., Michie, P. T., & Puce, A. (2004). Is the fusiform face area specialized for faces, individuation, or expert individuation? *Journal of Cognitive Neuroscience, 16,* 189–203.

Rial, R. V., Nicolau, M. C., Gamundi, A., Akaârir, M., Aparicio, S., Garau, C., . . . Estaban, S. (2007). The trivial function of sleep. *Sleep Medicine Reviews, 11,* 311–325.

Riccio, D. C., Millin, P. M., & Gisquet-Verrier, P. (2003). Retrograde amnesia: Forgetting back. *Current Directions in Psychological Science, 12,* 41–44.

Richards, W. (1971, May). The fortification illusions of migraines. *Scientific American, 224,* 89–97.

Richer, F., Martinez, M., Cohen, H., & Saint-Hilaire, J.-M. (1991). Visual motion perception from stimulation of the human medial parieto-occipital cortex. *Experimental Brain Research, 87,* 649–652.

Richter, C. P. (1967). Sleep and activity: Their relation to the 24-hour clock. *Proceedings of the Association for Research on Nervous and Mental Disorders, 45,* 8–27.

Richter, C. P. (1971). Inborn nature of the rat's 24-hour clock. *Journal of Comparative and Physiological Psychology, 75,* 1–14.

Riemann, D., & Perlis, M. L. (2008). The treatments of chronic insomnia: A review of benzodiazepine receptor agonists and psychological and behavioral therapies. *Sleep Medicine Reviews, 13,* 205–214.

Rijntjes, M., Dettmers, C., Büchel, C., Kiebel, S., Frackowiak, R. S. J., & Weiller, C. (1999). A blueprint for movement: Functional and anatomical representations in the human motor system. *Journal of Neuroscience, 19,* 8043–8048.

Ritter, R. C. (2004). Gastrointestinal mechanisms of satiation for food. *Physiology & Behavior, 81,* 249–273.

Rizzolatti, G., & Fabbri-Destro, M. (2008). The mirror system and its role in social cognition. *Current Opinion in Neurobiology, 18,* 179–184.

Rizzolatti, G., Fogassi, L., & Gallese, V. (2002). Motor and cognitive functions of the ventral premotor cortex. *Current Opinion in Neurobiology, 12,* 149–154.

Rizzolatti, G., Fogassi, L., & Gallese, V. (2006, November). Mirrors in the mind. *Scientific American, 295,* 54–61.

Rizzoli, S. O., & Betz, W. (2004). The structural organization of the readily releasable pool of synaptic vesicles. *Science, 303,* 2037–2039.

Rizzoli, S. O., & Betz, W. (2005). Synaptic vesicle pools. *Nature Reviews Neuroscience, 6,* 57–69.

Roach, N. W., & Hogben, J. H. (2004). Attentional modulation of visual processing in adult dyslexia: A spatial-cuing deficit. *Psychological Science, 15,* 650–654.

Robinson, G. E., Fernald, R. D., & Clayton, D. F. (2008). Genes and social behavior. *Science, 322,* 896–900.

Robinson, T. E. (1991). Persistent sensitizing effects of drugs on brain dopamine systems and behavior: Implications for addiction and relapse. In J. Barchas & S. Korenman (Eds.), *The biological basis of substance abuse and its therapy.* New York, NY: Oxford University Press.

Robinson, T. E. (2004). Addicted rats. *Science, 305,* 951–953.

Robinson, T. E., & Berridge, K. C. (1993). The neural basis of drug craving: An incentive-sensitization theory of addiction. *Brain Research Reviews, 18,* 247–291.

Robinson, T. E., & Justice, J. B. (Eds.). (1991). *Microdialysis in the neurosciences. Vol. 7, Techniques in the neural and behavioral sciences.* Amsterdam, Netherlands: Elsevier.

Robles, T. F., Glaser, R., & Kiecolt-Glaser, J. K. (2005). Out of balance: A new look at chronic stress, depression, and immunity. *Current Directions in Psychological Science, 14,* 111–115.

Rodier, P. M. (2000, February). The early origins of autism. *Scientific American, 284,* 56–63.

Rodin, J. (1985). Insulin levels, hunger, and food intake: An example of feedback loops in body weight regulation. *Health Psychology, 4,* 1–24.

Rodrigues, S. M., LeDoux, J. E., & Sapolsky, R. M. (2009). The influence of stress hormones on fear circuitry. *Annual Review of Neuroscience, 32,* 289–313.

Rodriguez, M., Llanos, C., Gonzalez, S., & Sabate, M. (2008). How similar are motor imagery and movement? *Behavioral Neuroscience, 122,* 910–916.

Rodríguez-Manzo, G., Pellicer, F., Larsson, K., & Fernández-Guasti, A. (2000). Stimulation of the medial preoptic area facilitates sexual behavior but does not reverse sexual satiation. *Behavioral Neuroscience, 114,* 553–560.

Roe, A. W., Pallas, S. L., Hahm, J-O., & Sur, M. (1990). A map of visual space induced in primary auditory cortex. *Science, 250,* 818–820.

Roefs, A., Werrij, M. Q., Smulders, F. T. Y., & Jansen, A. (2006). The value of indirect measures for assessing food preferences in abnormal eating. *Journal für Verbraucherschutz und Lebensmittelsicherheit, 1,* 180–186.

Rogawski, M. A., & Löscher, W. (2004). The neurobiology of antiepileptic drugs. *Nature Reviews Neuroscience, 5,* 553–564.

Rogers, P. J., & Blundell, J. E. (1980). Investigation of food selection and meal parameters during the development of dietary induced obesity. *Appetite, 1,* 85–88.

Rolls, B. J. (1986). Sensory-specific satiety. *Nutrition Reviews, 44,* 93–101.

Rolls, B. J. (1990). The role of sensory-specific satiety in food intake and food selection. In E. D. Capaldi & T. L. Powley (Eds.), *Taste, experience, and feeding* (pp. 28–42). Washington, DC: American Psychological Association.

Rolls, B. J., Rolls, E. T., Rowe, E. A., & Sweeney, K. (1981). Sensory specific satiety in man. *Physiology & Behavior, 27,* 137–142.

Rolls, E. T. (1981). Central nervous mechanisms related to feeding and appetite. *British Medical Bulletin, 37,* 131–134.

Rolls, E. T., Robertson, R. G., & Georges-François, P. (1995). The representation of space in the primate hippocampus. *Society for Neuroscience Abstracts, 21,* 1492.

Romanski, L. M., & Averbeck, B. B. (2009). The primate cortical auditory system and neural representation of conspecific vocalizations. *Annual Review of Neuroscience, 32,* 315–346.

Romanski, L. M., & LeDoux, J. E. (1992). Equipotentiality of thalamo-cortico-amygdala projections as auditory conditioned stimulus pathways. *Journal of Neuroscience, 12,* 4501–4509.

Roozendaal, B., McEwen, B. S., & Chattarji, S. (2009). Stress, memory, and the amygdala. *Nature Reviews Neuroscience, 10,* 423–433.

Rorden, C., & Karnath, H-O. (2004). Using human brain lesions to infer function: A relic from a past era in the fMRI age? *Nature Reviews Neuroscience, 5,* 813–819.

Rosa, M. G. P., Tweedale, R., & Elston, G. N. (2000). Visual responses of neurons in the middle temporal area of New World monkeys after lesions of striate cortex. *Journal of Neuroscience, 20,* 5552–5563.

Rosa, R. R., Bonnet, M. H., & Warm, J. S. (2007). Recovery of performance during sleep following sleep deprivation. *Psychophysiology, 20,* 152–159.

Rose, S. P. R. (2002). "Smart drugs": Do they work? Are they ethical? Will they be legal? *Nature Reviews Neuroscience, 3,* 975–979.

Rosenheck, R. (2008). Fast food consumption and increased caloric intake: A systematic review of a trajectory towards weight gain and obesity risk. *Obesity Reviews, 9,* 535–547.

Rosenthal, D., Wender, P. H., Kety, S. S., Welner, J., & Schulsinger, F. (1980). The adopted-away offspring of schizophrenics. *American Journal of Psychiatry, 128,* 87–91.

Roser, M., & Gazzaniga, M. S. (2004). Automatic brains—interpretive minds. *Current Directions in Psychological Science, 13,* 56–59.

Rossini, P. M., Martino, G., Narici, L., Pasquarelli, A., Peresson, M., Pizzella, V., . . . Romani, G. L. (1994). Short-term brain "plasticity" in humans: Transient finger representation changes in sensory cortex somatotopy following ischemic anesthesia. *Brain Research, 642,* 169–177.

Rothwell, J. C., Traub, M. M., Day, B. L., Obeso, J. A., Thomas, P. K., & Marsden, C. D. (1982). Manual motor performance in a deafferented man. *Brain, 105,* 515–542.

Rothwell, N. J., & Stock, M. J. (1982). Energy expenditure derived from measurements of oxygen consumption and energy balance in hyperphagic, "cafeteria"-fed rats. *Journal of Physiology, 324,* 59–60.

Rouach, N., Koulakoff, A., Abudara, V., Willecke, K., & Giaume, C. (2008). Astroglial metabolic networks sustain hippocampal synaptic transmission. *Science, 322,* 1551–1555.

Round, J., & Stein, E. (2007). Netrin signaling leading to directed growth cone steering. *Current Opinion in Neurobiology, 17,* 15–21.

Rowe, J. B., Toni, I., Josephs, O., Frackowiak, R. S., & Pasingham, R. E. (2000). The prefrontal cortex: Response selection or maintenance within working memory? *Science, 288,* 1656–1660.

Rowitch, D. H. (2004). Glial specification in the vertebrate neural tube. *Annual Review of Neuroscience, 28,* 251–274.

Rowland, N. (1981). Glucoregulatory feeding in cats. *Physiology & Behavior, 26,* 901–903.

Rowland, N. E. (1990). Sodium appetite. In E. D. Capaldi & T. L. Powley (Eds.), *Taste, experience, and feeding* (pp. 94–104). Washington, DC: American Psychological Association.

Rozin, P., Dow, S., Moscovitch, M., & Rajaram, S. (1998). What causes humans to begin and end a meal? A role for memory for what has been eaten, as evidenced by a study of multiple meal eating in amnesic patients. *Psychological Science, 9,* 392–396.

Rozin, P. N., & Schulkin, J. (1990). Food selection. In E. M. Stricker (Ed.), *Handbook of behavioral neurobiology* (pp. 297–328). New York, NY: Plenum Press.

Rubin, D. C. (2006). The basic-systems model of episodic memory. *Perspectives on Psychological Science, 1,* 277–311.

Ruby, N. F., Brennan, T. J., Xie, X., Cao, V., Franken, P., Heller, H. C., & O'Hara, B. F. (2002). Role of melanopsin in circadian responses to light. *Science, 298,* 2211–2213.

Rudy, J. W., & Sutherland, R. J. (1992). Configural and elemental associations and the memory coherence problem. *Journal of Cognitive Neuroscience, 4,* 208–216.

Russell, G. (1979). Bulimia nervosa: An ominous variant of anorexia nervosa. *Psychological Medicine, 9,* 429–448.

Russell, J. A., Bachorowski, J.-A., & Fernandez-Dols, J.-M. (2003). Facial and vocal expressions of emotion. *Annual Review of Psychology, 54,* 329–349.

Russell, W. R., & Espir, M. I. E. (1961). *Traumatic aphasia—a study of aphasia in war wounds of the brain.* London, England: Oxford University Press.

Rutledge, L. L., & Hupka, R. B. (1985). The facial feedback hypothesis: Methodological concerns and new supporting evidence. *Motivation and Emotion, 9,* 219–240.

Rutter, M. L. (1997). Nature-nurture integration: The example of antisocial behavior. *American Psychologist, 52,* 390–398.

Rymer, R. (1993). *Genie.* New York, NY: HarperCollins

Ryugo, D. K., Kretzmer, E. A., & Niparko, J. K. (2005). Restoration of auditory nerve synapses in cats by cochlear implants. *Science, 310,* 1490–1492.

Saalmann, Y. B., Pigarev, I. N., & Vidyasagar, T. R. (2007). Neural mechanisms of visual attention: How top-down feedback highlights relevant locations. *Science, 316,* 1612–1615.

Sack, A. T. (2006). Transcranial magnetic stimulation, causal structure–function mapping and networks of functional relevance. *Current Opinion in Neurobiology, 16,* 593–599.

Sacks, O. (1985). *The man who mistook his wife for a hat and other clinical tales.* New York, NY: Summit Books.

Saffran, E. M. (1997). Aphasia: Cognitive neuropsychological aspects. In T. E. Feinberg & M. J. Farah (Eds.), *Behavioral neurology and neuropsychology* (pp. 151–166). New York, NY: McGraw-Hill.

Sahin, N. T., Pinker, S., Cash, S. S., Schomer, D., & Halgren, E. (2009). Sequential processing of lexical, grammatical, and phonological information within Broca's area. *Science, 326,* 445–450.

Sakuma, Y., & Pfaff, D. W. (1979). Mesencephalic mechanisms for the integration of female reproductive behavior in the rat. *American Journal of Physiology, 237,* 285–290.

Sakurai, T. (2005). Roles of orexin/hypocretin in regulation of sleep/wakefulness and energy homeostasis. *Sleep Medicine Reviews, 9,* 231–241.

Sakurai, T. (2007). The neural circuit of orexin (hypocretin): Maintaining sleep and wakefulness. *Nature Reviews Neuroscience, 8,* 171–181.

Salamone, J. D. (2000). A critique of recent studies on placebo effects of antidepressants: Importance of research on active placebos. *Psychopharmacology, 152,* 1–6.

Sale, A., Berardi, N., & Maffei, L. (2008). Enrich the environment to empower the brain. *Trends in Neurosciences, 32,* 233–239.

Salmon, D. P., & Bondi, M. W. (2009). Neuropsychological assessment of dementia. *Annual Review of Psychology, 60,* 257–282.

Salzer, J. L. (2002). Nodes of Ranvier come of age. *Trends in Neurosciences, 25*(1), 2–5.

Salzet, M., Vieau, D., & Day, R. (2000). Crosstalk between nervous and immune systems through the animal kingdom: Focus on opioids. *Trends in Neurosciences, 23,* 550–555.

Samanez-Larkin, G. R., Hollon, N. G., Carstensen, L. L., & Knutson, B. (2008). Individual differences in insular sensitivity during loss anticipation predict avoidance learning. *Psychological Science, 19,* 320–323.

Sanchis-Segura, C., & Spanagel, R. (2006). Behavioural assessment of drug reinforcement and addictive features in rodents: An overview. *Addiction Biology, 11,* 2–38.

Sanders, D., & Bancroft, J. (1982). Hormones and the sexuality of women—the menstrual cycle. *Clinics in Endocrinology and Metabolism, 11,* 639–659.

Sandi, C. (2004). Stress, cognitive impairment and cell adhesion molecules. *Nature Reviews Neuroscience, 5,* 917–930.

Sanes, D. H., & Bao, S. (2009). Tuning up the developing auditory CNS. *Current Opinion in Neurobiology, 19,* 188–199.

Sanes, J. N. (2003). Neocortical mechanisms in motor learning. *Current Opinion in Neurobiology, 13,* 225–231.

Sanes, J. N., Donoghue, J. P., Thangaraj, V., Edelman, R. R., & Warach, S. (1995). Shared neural substrates controlling hand movements in human motor cortex. *Science, 268,* 1775–1777.

Sanes, J. N., Suner, S., & Donoghue, J. P. (1990). Dynamic organization of primary motor cortex output to target muscles in adult rats. I. Long-term patterns of reorganization following motor or mixed peripheral nerve lesions. *Experimental Brain Research, 79,* 479–491.

Sanfey, A. G. (2007). Decision neuroscience: New directions in studies of judgement and decision making. *Current Directions in Psychological Science, 16*, 151–155.

Santonastaso, P., Zanetti, T., De Antoni, C., Tenconi, E., & Favaro, A. (2006). Anorexia nervosa patients with a prior history of bulimia nervosa. *Comprehensive Psychiatry, 47*, 519–522.

Saper, C. B., Lu, J., Chou, T. C., & Gooley, J. (2005). The hypothalamic integrator for circadian rhythms. *Trends in Neurosciences, 28*, 152–157.

Saper, C. B., Scammell, T. E., & Lu, J. (2005). Hypothalamic regulation of sleep and circadian rhythms. *Nature, 437*, 1257–1263.

Sapolsky, R. M. (1997). The importance of a well-groomed child. *Science, 277*, 1620–1622.

Sapolsky, R. M. (2005). The influence of social hierarchy on primate health. *Science, 308*, 648–652.

Särkämö, T., Tervaniemi, M., Laitinen, S., Forsblom, A., Soinila, S., Mikkonen, M., . . . Hietanen, M. (2008). Music listening enhances cognitive recovery and mood after middle cerebral artery stroke. *Brain, 131*, 866–876.

Savic, I. (2002). Imaging of brain activation by odorants in humans. *Current Opinion in Neurobiology, 12*, 455–461.

Savill, J., Gregory, C., & Haslett, C. (2003). Eat me or die. *Science, 302*, 1516–1517.

Savitz, J., & Drevets, W. C. (2009). Bipolar and major depressive disorder: Neuroimaging the developmental-degenerative divide. *Neuroscience and Biobehavioral Reviews, 33*, 699–771.

Sawa, A., & Snyder, S. H. (2002). Schizophrenia: Diverse approaches to a complex disease. *Science, 296*, 692–695.

Sawada, H., & Shimohama, S. (2000). Neuroprotective effects of estradiol in mesencephalic dopaminergic neurons. *Neuroscience and Biobehavioural Reviews, 24*, 143–147.

Sawamoto, K., Wichterle, H., Gonzalez-Perez, O., Cholfin, J. A., Yamada, M., Spassky, N., . . . Alvarez-Buylla, A. (2006). New neurons follow the flow of cerebrospinal fluid in the adult brain. *Science, 311*, 629–632.

Sawle, G. V., & Myers, R. (1993) The role of positron emission tomography in the assessment of human neurotransplantation. *Trends in Neurosciences, 16*, 172–176.

Saxena, A. (2003). Issues in newborn screening. *Genetic Testing, 7*, 131–134.

Sbragia, G. (1992). Leonardo da Vinci and ultrashort sleep: Personal experience of an eclectic artist. In C. Stampi (Ed.), *Why we nap: Evolution, chronobiology, and functions of polyphasic and ultra-short sleep*. Boston, MA: Birkhäuser.

Schacter, D. L., Dobbins, I. G., & Schnyer, D. M. (2004). Specificity of priming: A cognitive neuroscience perspective. *Nature Reviews Neuroscience, 5*, 853–861.

Schafe, G. E., & Le Doux, J. E. (2004). The neural basis of fear. In M. S. Gazzaniga (Ed.), *The cognitive neurosciences* (pp. 987–1003). Cambridge, MA: MIT Press.

Schally, A. V., Kastin, A. J., & Arimura, A. (1971). Hypothalamic folliclestimulating hormone (FSH) and luteinizing hormone (LH)–regulating hormone: Structure, physiology, and clinical studies. *Fertility and Sterility, 22*, 703–721.

Schärli, H., Harman, A. M., & Hogben, J. H. (1999a). Blindsight in subjects with homonymous visual field defects. *Journal of Cognitive Neuroscience, 11*, 52–66.

Schärli, H., Harman, A. M., & Hogben, J. H. (1999b). Residual vision in a subject with damaged visual cortex. *Journal of Cognitive Neuroscience, 11*, 502–510.

Scheer, F. A. J. L., & Czeisler, C. A. (2005). Melatonin, sleep, and circadian rhythms. *Sleep Medicine Reviews, 9*, 5–9.

Scheiber, M. H. (1999). Somatotopic gradients in the distributed organization of the human primary motor cortex hand area: Evidence from small infarcts. *Experimental Brain Research, 128*, 139–148.

Scheiber, M. H., & Poliakov, A. V. (1998). Partial inactivation of the primary motor cortex hand area: Effects on individuated finger movements. *Journal of Neuroscience, 18*, 9038–9054.

Schenck, C. H., Bundlie, S. R., Ettinger, M. G., & Mahowald, M. W. (1986). Chronic behavioral disorders of human REM sleep: A new category of parasomnia. *Sleep, 9*, 293–308.

Schendel, K., & Robertson, L. C. (2004). Reaching out to see: Arm position can attenuate human visual loss. *Journal of Cognitive Neuroscience, 16*, 935–943.

Schenk, T., & Zihl, J. (1997). Visual motion perception after brain damage: I. Deficits in global motion perception. *Neuropsychologia, 35*, 1289–1297.

Scherberger, H. (2009). Neural control of motor prostheses. *Current Opinion in Neurobiology, 19*, 629–633.

Schienle, A., Schäfer, A., Walter, B., Stark, R., & Vaitl, D. (2005). Brain activation of spider phobics towards disorder-relevant, generally disgust- and fear-inducing pictures. *Neuroscience Letters, 388*, 1–6.

Schiffman, S. S., & Erickson, R. P. (1980). The issue of primary tastes versus a taste continuum. *Neuroscience and Biobehavioural Reviews, 4*, 109–117.

Schlaggar, B. L., Brown, T. T., Lugar, H. M., Visscher, K. M., Miezin, F. M., & Petersen, S. E. (2002). Functional neuroanatomical differences between adults and school-age children in the processing of single words. *Science, 296*, 1476–1479.

Schlaggar, B. L., & Church, J. A. (2009). Functional neuroimaging insights into the development of skilled reading. *Current Directions in Psychological Science, 18*, 21–26.

Schlaug, G., Jäncke, L., Huang, Y., & Steinmetz, H. (1995). In vivo evidence of structural brain asymmetry in musicians. *Science, 267*, 699–701.

Schmitt, D. P., & Pilcher, J. J. (2004). Evaluating evidence of psychological adaptation. *Psychological Science, 15*, 643–649.

Schneggenburger, R., & Neher, E. (2005). Presynaptic calcium and control of vesicle fusion. *Current Opinion in Neurobiology, 15*, 266–274.

Schoppa, N. E. (2009). Making scents out of how olfactory neurons are ordered in space. *Nature Neuroscience, 12*, 103–104.

Schratt, G. (2009). MicroRNAs at the synapse. *Nature Reviews Neuroscience, 10*, 842–849.

Schreiner, C. E. (1992). Functional organization of the auditory cortex: Maps and mechanisms. *Current Opinion in Neurobiology, 2*, 516–521.

Schreiner, C. E., Read, H. L., & Sutter, M. L. (2000). Modular organization of frequency integration in primary auditory cortex. *Annual Review of Neuroscience, 23*, 501–529.

Schroeder, C. E., & Foxe, J. (2005). Multisensory contributions to low-level, "unisensory" processing. *Current Opinion in Neurobiology, 15*, 454–458.

Schulkin, J. (2007). Autism and the amygdala: An endocrine hypothesis. *Brain and Cognition, 65*, 87–99.

Schulkin, J., Morgan, M. A., & Rosen, J. B. (2005). A neuroendocrine mechanism for sustaining fear. *Trends in Neurosciences, 28*, 629–635.

Schultz, W. (2002). Getting formal with dopamine and reward. *Neuron, 36*, 241–263.

Schultz, W. (2007a). Behavioral dopamine signals. *Trends in Neurosciences, 30*, 203–210.

Schultz, W. (2007b). Multiple dopamine functions at different time courses. *Annual Review of Neuroscience, 30*, 259–288.

Schultz, W., Tremblay, L., & Hollerman, J. R. (2003). Changes in behavior-related neuronal activity in the striatum during learning. *Trends in Neurosciences, 26*, 321–328.

Schwartz, A. B. (2004). Cortical neural prosthetics. *Annual Review of Neuroscience, 27*, 487–507.

Schwartz, G. J., & Azzara, A. V. (2004). Sensory neurobiological analysis of neuropeptide modulation of meal size. *Physiology & Behavior, 82*, 81–87.

Schwartz, M. W. (2000). Staying slim with insulin in mind. *Science, 289*, 2066–2067.

Schwartz, M. W., & Porte, D. (2005). Diabetes, obesity, and the brain. *Science, 307*, 375–379.

Schweinsburg, A. D., Brown, S. A., & Tapert, S. F. (2008). The influence of marijuana use on neurocognitive functioning in adolescents. *Current Drug Abuse Reviews, 1*, 99–111.

Schweizer, F. E., & Ryan, T. A. (2006). The synaptic vesicle: Cycle of exocytosis and endocytosis. *Current Opinion in Neurobiology, 16*, 298–304.

Sclafani, A. (1990). Nutritionally based learned flavor preferences in rats. In E. D. Capaldi & T. L. Powley (Eds.), *Taste, experience, and feeding* (pp. 139–156). Washington, DC: American Psychological Association.

Scott, K. (2004). The sweet and the bitter of mammalian taste. *Current Opinion in Neurobiology, 14,* 423–427.

Scott, S. K. (2005). Auditory processing—speech, space and auditory objects. *Current Opinion in Neurobiology, 15,* 197–201.

Scott, S. K., McGettigan, C., & Eisner, F. (2009). A little more conversation, a little less action—candidate roles for the motor cortex in speech production. *Nature Reviews Neuroscience, 10,* 295–302.

Scoville, W. B., & Milner, B. (1957). Loss of recent memory after bilateral hippocampal lesions. *Journal of Neurology, Neurosurgery and Psychiatry, 20,* 11–21.

Seaberg, R. M., & van der Kooy, D. (2003). Stem and progenitor cells: The premature desertion of rigorous definitions. *Trends in Neurosciences, 26,* 125–131.

Searle, L. V. (1949). The organization of hereditary maze-brightness and maze-dullness. *Genetic Psychology Monographs, 39,* 279–325.

Seeley, R. J., van Dijk, G., Campfield, L. A., Smith, F. J., Nelligan, J. A., Bell, S. M., et al. (1996). The effect of intraventricular administration of leptin (Ob protein) on food intake and body weight in the rat. *Hormone and Metabolic Research, 28,* 664–668.

Seeley, R. J., & Woods, S. C. (2003). Monitoring of stored and available fuel by the CNS: Implication for obesity. *Nature Reviews Neuroscience, 4,* 901–909.

Segal, M. (2005). Dendritic spines and long-term plasticity. *Nature Reviews Neuroscience, 6,* 277–284.

Segerstrom, S. C. (2007). Stress, energy, and immunity: An ecological view. *Current Directions in Psychological Science, 16,* 326–330.

Segerstrom, S. C., & Miller, G. E. (2004). Psychological stress and the human immune system: A meta-analytic study of 30 years of inquiry. *Psychological Bulletin, 130,* 601–630.

Segovia, S., Guillamón, A., del Cerro, M. C. R., Ortega, E., Pérez-Laso, C., Rodriguez-Zafra, M., & Beyer, C. (1999). The development of brain sex differences: A multisignaling process. *Behavioural Brain Research, 105,* 69–80.

Seifert, G., Schilling, K., & Steinhäuser, C. (2006). Astrocyte dysfunction in neurological disorders: A molecular perspective. *Nature Reviews Neuroscience, 7,* 194–206.

Seitz, R. J., Roland, P. E., Bohm, C., Greitz, T., & Stone-Elanders, S. (1990). Motor learning in man: A positron emission tomographic study. *NeuroReport, 1,* 17–20.

Selkoe, D. J. (1991, November). Amyloid protein and Alzheimer's. *Scientific American, 265,* 68–78.

Selkoe, D. J. (2002). Alzheimer's disease is a synaptic failure. *Science, 298,* 789–791.

Sener, R. N. (2003). Phenylkenonuria: Diffusion magnetic resonance imaging and proton magnetic resonance spectroscopy. *Computer Assisted Tomography, 27,* 541–543.

Sergerie, K., Lepage, M., & Armony, J. L. (2006). A process-specific functional dissociation of the amygdala in emotional memory. *Journal of Cognitive Neuroscience, 18,* 1359–1367.

Serrien, D. J., Ivry, R. B., & Swinnen, S. P. (2006). Dynamics of hemispheric specialization and integration in the context of motor control. *Nature Reviews Neuroscience, 7,* 160–166.

Servos, P., Engel, S. A., Gati, J., & Menon, R. (1999). fMRI evidence for an inverted face representation in human somatosensory cortex. *NeuroReport, 10*(7), 1393–1395.

Sewards, T. V., & Sewards, M. A. (2001). Cortical association areas in the gustatory system. *Neuroscience and Biobehavioural Reviews, 25,* 395–407.

Shaham, Y., & Hope, B. T. (2005). The role of neuroadaptations in relapse to drug seeking. *Nature Neuroscience, 8,* 1437–1439.

Shapely, R., & Hawken, M. (2002). Neural mechanisms for color perception in the primary visual cortex. *Current Opinion in Neurobiology, 12,* 426–432.

Shapiro, B. H., Levine, D. C., & Adler, N. T. (1980). The testicular feminized rat: A naturally occurring model of androgen independent brain masculinization. *Science, 209,* 418–420.

Shapiro, L., Love, J., & Colman, D. R. (2007). Adhesion molecules in the nervous system: Structural insights into function and diversity. *Annual Review of Neuroscience, 30,* 451–474.

Shapiro, M. L., Kennedy, P. J., & Ferbinteanu, J. (2006). Representing episodes in the mammalian brain. *Current Opinion in Neurobiology, 16,* 701–709.

Shaw, P., Bramham, J., Lawrence, E. J., Morris, R., Baron-Cohen, S., & David, A. S. (2005). Differential effects of lesions of the amygdala and prefrontal cortex on recognizing facial expressions of complex emotions. *Journal of Cognitive Neuroscience, 17,* 1410–1419.

Shaywitz, S. E., Mody, M., & Shaywitz, B. A. (2006). Neural mechanisms in dyslexia. *Current Directions in Psychological Science, 15,* 278–281.

Shelton, J. F., Tancredi, D. J., & Hertz-Picciotto, I. (2010). Independent and dependent contributions of advanced maternal and paternal ages to autism risk. *Autism Research, 3,* 30–39.

Shen, S., Lang, B., Nakamoto, C., Zhang, F., Pu, J., Kuan, S-L., . . . St. Clair, D. (2008). Schizophrenia-related neural and behavioral phenotypes in transgenic mice expressing truncated *disc1. Journal of Neuroscience, 28,* 10893–10904.

Sheng, M., & Kim, M. J. (2002). Postsynaptic signaling and plasticity mechanisms. *Science, 298,* 776–780.

Shepherd, G. M., & Erulkar, S. D. (1997). Centenary of the synapse: From Sherrington to the molecular biology of the synapse and beyond. *Trends in Neurosciences, 20,* 385–392.

Sheppard, D. M., Bradshaw, J. L., Purcell, R., & Pantelis, C. (1999). Tourette's and comorbid syndromes: Obsessive compulsive and attention deficit hyperactivity disorder. A common etiology? *Clinical Psychology Review, 19,* 531–552.

Sherman, D. L., & Brophy, P. J. (2005). Mechanisms of axon ensheathment and myelin growth. *Nature Reviews Neuroscience, 6,* 683–690.

Sherman, M. (2007). The thalamus is more than just a relay. *Current Opinion in Neurobiology, 17,* 417–422.

Sherry, D. F., & Vaccarino, A. L. (1989). Hippocampus and memory for food caches in black-capped chickadees. *Behavioral Neuroscience, 103,* 308–318.

Sherwin, B. B. (1985). Changes in sexual behavior as a function of plasma sex steroid levels in post-menopausal women. *Maturitas, 7,* 225–233.

Sherwin, B. B. (1988). A comparative analysis of the role of androgen in human male and female sexual behavior: Behavioral specificity, critical thresholds, and sensitivity. *Psychobiology, 16,* 416–425.

Sherwin, B. B., Gelfand, M. M., & Brender, W. (1985). Androgen enhances sexual motivation in females: A prospective cross-over study of sex steroid administration in the surgical menopause. *Psychosomatic Medicine, 47,* 339–351.

Sheth, D. N., Bhagwate, M. R., & Sharma, N. (2005). Curious clicks—Sigmund Freud. *Journal of Postgraduate Medicine, 51,* 240–241.

Shiffman, S., & Ferguson, S. G. (2008). Nicotine patch therapy prior to quitting smoking: A meta-analysis. *Addiction, 103,* 557–563.

Shimomura, O., Johnson, F., & Saiga, Y. (1962). Extraction, purification and properties of aequorin, a bioluminescent protein from the luminous hydromedusan, *Aequorea. Journal of Cellular and Comparative Physiology, 59,* 223–239.

Shimura, T., & Shimokochi, M. (1990). Involvement of the lateral mesencephalic tegmentum in copulatory behavior of male rats: Neuron activity in freely moving animals. *Neuroscience Research, 9,* 173–183.

Shin, J. C., & Ivry, R. B. (2003). Spatial and temporal sequence learning in patients with Parkinson's disease or cerebellar lesions. *Journal of Cognitive Neuroscience, 15,* 1232–1243.

Shipp, S., de Jong, B. M., Zihl, J., Frackowiak, R. S., & Zeki, S. (1994). The brain activity related to residual motion vision in a patient with bilateral lesions of V5. *Brain, 117,* 1023–1038.

Shmuel, A., & Leopold, D. A. (2009). Neuronal correlates of spontaneous fluctuations in fMRI signal in monkey visual cortex: Implications for functional connectivity at rest. *Human Brain Mapping, 29,* 751–761.

Shors, T. J. (2009, March). Saving new brain cells. *Scientific American, 300,* 47–52.

Shors, T. J., & Matzel, L. D. (1997). Long-term potentiation: What's learning got to do with it? *Behavioral and Brain Sciences, 20,* 597–655.

Shubin, N. H. (2009, January). This old body. *Scientific American, 300,* 64–67.

Shubin, N. H., Daeschler, E. B., & Jenkins, F. A., Jr. (2006). The pectoral fin of *Tiktaalik roseae* and the origin of the tetrapod limb. *Nature, 440,* 764–771.

Siegel, J. M. (1983). A behavioral approach to the analysis of reticular formation unit activity. In T. E. Robinson (Ed.), *Behavioral approaches to brain research* (pp. 94–116). New York, NY: Oxford University Press.

Siegel, J. M. (2004). Hypocretin (orexin): Role in normal behavior and neuropathology. *Annual Review of Psychology, 55,* 125–148.

Siegel, J. M. (2005). Clues to the functions of mammalian sleep. *Nature 437,* 1264–1271.

Siegel, J. M. (2008). Do all animals sleep? *Trends in Neurosciences, 31,* 208–213.

Siegel, J. M. (2009). Sleep viewed as a state of adaptive inactivity. *Nature Reviews Neuroscience, 10,* 747–752.

Siegel, S. (2005). Drug tolerance, drug addiction, and drug anticipation. *Current Directions in Psychological Science, 14,* 296–300.

Siegel, S., Hinson, R. E., Krank, M. D., & McCully, J. (1982). Heroin "overdose" death: Contribution of drug-associated environmental cues. *Science, 216,* 436–437.

Silk, J. B., Alberts, S. C., & Altmann, J. (2003). Social bonds of female baboons enhance infant survival. *Science, 302,* 1231–1233.

Sillito, A. M., Cudeiro, J., & Jones, H. E. (2006). Always returning: Feedback and sensory processing in visual cortex and thalamus. *Trends in Neurosciences, 29,* 307–316.

Simon, G. E., Savarino, J., Operskalski, B., & Wang, P. S. (2006). Suicide risk during antidepressant treatment. *American Journal of Psychiatry, 163,* 41–47.

Simon, N. M., Otto, M. W., Wisniewski, S. R., Fossey, M., Sagduyu, K., Frank, E., . . . Pollack, M. H. (2004). Anxiety disorder comorbidity in bipolar disorder patients: Data from the first 500 participants in the Systematic Treatment Enhancement Program for Bipolar Disorder (STEP-BD). *American Journal of Psychiatry, 161,* 2222–2229.

Simon, S. A., de Araujo, I. E., Gutierrez, R., & Nicolelis, M. A. L. (2006). The neural mechanisms of gustation: A distributed processing code. *Nature Reviews Neuroscience, 7,* 890–901.

Simons, D. J., & Ambinder, M. (2005). Change blindness: Theory and consequences. *Current Directions in Psychological Science, 14,* 44–48.

Simons, D. J., & Rensink, R. A. (2005). Change blindness: Past, present, and future. *Trends in Cognitive Science, 9,* 16–20.

Sinclair, S. V., & Mistlberger, R. E. (1997). Scheduled activity reorganizes circadian phase of Syrian hamsters under full and skeleton photoperiods. *Behavioural Brain Research, 87,* 127–137.

Singer, J. J. (1968). Hypothalamic control of male and female sexual behavior in female rats. *Journal of Comparative and Physiological Psychology, 66,* 738–742.

Singh, J., Hallmayer, J., & Illes, J. (2007). Interacting and paradoxical forces in neuroscience and society. *Nature Reviews Neuroscience, 8,* 153–160.

Sirigu, A., & Duhamel, J. R. (2001). Motor and visual imagery as two complementary but neurally dissociable mental processes. *Journal of Cognitive Neuroscience, 13,* 910–919.

Sisk, C. L., & Zehr, J. L. (2005). Pubertal hormones organize the adolescent brain and behavior. *Frontiers in Neuroendocrinology, 26,* 163–174.

Skeie, I., Brekke, M., Lindbæk, M., & Waal, H. (2008). Somatic health among heroin addicts before and during opioid maintenance treatment: A retrospective cohort study. *BMC Public Health, 8.* doi:10.1186/1471-2458-8-43

Sladek, J. R., Jr., Redmond, D. E., Jr., Collier, T. J., Haber, S. N., Elsworth, J. D., Deutch, A. Y., & Roth, R. H. (1987). Transplantation of fetal dopamine neurons in primate brain reverses MPTP induced Parkinsonism. In F. J. Seil, E. Herbert, & B. M. Carlson (Eds.), *Progress in brain research* (Vol. 71, pp. 309–323). New York, NY: Elsevier.

Slocombe, K. E., & Zuberbühler, K. (2007). Chimpanzees modify recruitment screams as a function of audience composition. *Proceedings of the National Academy of Sciences, USA, 104,* 17228–17233.

Sluder, G., & McCollum, D. (2000). The mad ways of meiosis. *Science, 289,* 254–255.

Smith, E. E. (2000). Neural bases of human working memory. *Current Directions in Psychological Science, 9,* 45–49.

Smith, S. S., Kilby, S., Jorgensen, G., & Douglas, J. A. (2007). Napping and night-shift work: Effects of a short nap on psychomotor vigilance and subjective sleepiness in health workers. *Sleep and Biological Rhythms, 5,* 117–125.

Smith, T. W. (2006). Personality as risk and resilience in physical health. *Current Directions in Psychological Science, 15,* 227–231.

Smith, Y., Raju, D. V., Pare, J-F., & Sidibe, M. (2004). The thalamostriatal system: A highly specific network of the basal ganglia circuitry. *Trends in Neurosciences, 27,* 520–527.

Snyder, S. H. (1978). Neuroleptic drugs and neurotransmitter receptors. *Journal of Clinical and Experimental Psychiatry, 133,* 21–31.

Södersten, P., Bergh, C., & Zandian, M. (2006). Understanding eating disorders. *Hormones and Behavior, 50,* 572–578.

Sofroniew, M. V., Howe, C. L., & Mobley, W. C. (2001). Nerve growth factor signaling, neuroprotection, and neural repair. *Annual Review of Neuroscience, 24,* 1217–1281.

Sofsian, D. (2007, January 5). Obesity statistics. *EzineArticles.com.* Retrieved from http://ezinearticles. com/?Obesity-Statistics&id=405478

Soloman, S. M., & Kirby, D. F. (1990). The refeeding syndrome: A review. *Journal of Parenteral and Enteral Nutrition, 14,* 90–97.

Solstad, T., Boccara, C. N., Kropff, E., Moser, M-B., & Moser, E. I. (2008). Representation of geometric borders in the entorhinal cortex. *Science, 322,* 1865–1868.

Soma, K. K. (2006). Testosterone and aggression: Berthold, birds and beyond. *Journal of Neuroendocrinology, 18,* 543–551.

Somers, J. M., Goldner, E. M., Waraich, P., & Hsu, L. (2006). Prevalence and incidence studies of anxiety disorders: A systematic review of the literature. *Canadian Journal of Psychiatry, 51,* 100–112.

Somerville, L. H., Heatherton, T. F., & Kelley, W. M. (2006). Anterior cingulate cortex responds differentially to expectancy violation and social rejection. *Nature Neuroscience, 9,* 1007–1008.

Sommer, E. C., Aleman, A., Bouma, A., & Kahn, R. S. (2004). Do women really have more bilateral language representation than men? A meta-analysis of functional imaging studies. *Brain, 127,* 1845–1852.

Sommer, M. A., & Wurtz, R. H. (2008). Brain circuits for the internal monitoring of movements. *Annual Review of Neuroscience, 31,* 317–338.

Soucy, E. R., Albeanu, D. F., Fantana, A. L., Murthy, V. N., & Meister, M. (2009). Precision and diversity in an odor map on the olfactory bulb. *Nature Neuroscience, 12,* 210–220.

Sowell, E. R., Kan, E., Yoshii, J., Thompson, P. M., Bansal, R., Xu, D., . . . Peterson, B. S. (2008). Thinning of sensorimotor cortices in children with Tourette syndrome. *Nature Neuroscience, 11,* 637–639.

Spadoni, A. D., McGee, C. L., Fryer, S. L., & Riley, E. P. (2007). Neuroimaging and fetal alcohol spectrum disorders. *Neuroscience and Biobehavioral Reviews, 31,* 239–245.

Spector, A. C., & Glendinning, J. I. (2009). Linking peripheral taste processes to behavior. *Current Opinion in Neurobiology, 19,* 370–377.

Spector, A. C., & Travers, S. P. (2005). The representation of taste quality in the mammalian nervous system. *Behavioral and Cognitive Neuroscience Reviews, 4,* 143–191.

Spence, I., Wong, P., Rusau, M., & Rastegar, N. (2006). How color enhances visual memory for natural scenes. *Psychological Science, 17,* 1–6.

Sperry, R. W. (1963). Chemoaffinity in the orderly growth of nerve fiber patterns and connections. *Proceedings of the National Academy of Sciences, USA, 50,* 703–710.

Sperry, R. W. (1964, January). The great cerebral commissure. *Scientific American, 210,* 42–52.

Sperry, R. W., Zaidel, E., & Zaidel, D. (1979). Self recognition and social awareness in the deconnected minor hemisphere. *Neuropsychologia, 17,* 153–166.

Spires-Jones, T. L., Stoothoff, W. H., de Calignon, A., Jones, P. B., & Hyman, B. T. (2009). Tau pathophysiology in neurodegeneration: A tangled issue. *Trends in Neurosciences, 32,* 150–159.

Spitzer, R. L., Skodol, A. E., Gibbon, M., & Williams, J. B. W. (1983). *Psychopathology: A case book.* New York, NY: McGraw-Hill.

Spoor, F., Leakey, M. G., Gathogo, P. N., Brown, F. H., Antón, S. C., McDougall, I., . . . Leakey, L. N. (2007). Implications of new early *Homo* fossils from Ileret, east of Lake Turkana, Kenya. *Nature, 448,* 688–691.

Spradling, A. C., & Zheng, Y. (2007). The mother of all stem cells? *Science, 315,* 469–470.

Spruston, N. (2008). Pyramidal neurons: Dendritic structure and synaptic integration. *Nature Reviews Neuroscience, 9,* 206–221.

Squire, L. R. (1987). *Memory and brain.* New York, NY: Oxford University Press.

Squire, L. R., Amaral, D. G., Zola-Morgan, S., Kritchevsky, M., & Press, G. (1989). Description of brain injury in the amnesic patient N.A. based on magnetic resonance imaging. *Experimental Neurology, 105,* 23–35.

Squire, L. R., & Bayley, P. J. (2007). The neuroscience of remote memory. *Current Opinion in Neurobiology, 17,* 185–196.

Squire, L. R., Clark, R. E., & Knowlton, B. J. (2001). Retrograde amnesia. *Hippocampus, 11,* 50–55.

Squire, L. R., Slater, P. C., & Chace, P. M. (1975). Retrograde amnesia: Temporal gradient in very long term memory following electroconvulsive therapy. *Science, 187,* 77–79.

Squire, L. R., & Spanis, C. W. (1984). Long gradient of retrograde amnesia in mice: Continuity with the findings in humans. *Behavioral Neuroscience, 98,* 345–348.

Squire, L. R., & Zola-Morgan, S. (1985). The neuropsychology of memory: New links between humans and experimental animals. *Annals of the New York Academy of Sciences, 444,* 137–149.

Squire, L. R. & Zola-Morgan, S. (1991). The medial temporal lobe memory system. *Science, 253,* 1380–1386.

St. Clair, D., Xu, M., Wang, P., Yu, Y., Fang, Y., Zhang, X., . . . He, L. (2005). Rates of adult schizophrenia following prenatal exposure to the Chinese famine of of 1959–1961. *Journal of the American Medical Association, 294,* 557–562.

St. George-Hyslop, P. H. (2000, December). Piecing together Alzheimer's. *Scientific American, 283,* 76–83.

Stamatakis, K. A., & Punjabi, N. M. (2007). Long sleep duration: A risk to health or a marker of risk? *Sleep Medicine Reviews, 11,* 337–339.

Stampi, C. (1992a). Evolution, chronobiology, and functions of polyphasic and ultrashort sleep: Main issues. In C. Stampi (Ed.), *Why we nap: Evolution, chronobiology, and functions of polyphasic and ultrashort sleep.* Boston, MA: Birkhaüser.

Stampi, C. (Ed.). (1992b). *Why we nap: Evolution, chronobiology, and functions of polyphasic and ultrashort sleep.* Boston, MA: Birkhaüser.

Steen, R. G., Mull, C., McClure, R., Hamer, R. M., & Lieberman, J. A. (2006). Brain volume in first-episode schizophrenia: Systematic review and meta-analysis of magnetic resonance imaging studies. *British Journal of Psychiatry, 188,* 510–518.

Stein, D. G. (2001). Brain damage, sex hormones and recovery: A new role for progesterone and estrogen? *Trends in Neurosciences, 24,* 386–391.

Stein, D. G., & Hoffman, S. W. (2003). Estrogen and progesterone as neuroprotective agents in the treatments of acute brain injuries. *Pediatric Rehabilitation, 6,* 13–22.

Stepanski, E., Lamphere, J., Badia, P., Zorick, F., & Roth, T. (1984). Sleep fragmentation and daytime sleepiness. *Sleep, 7,* 18–26.

Stepanski, E., Lamphere, J., Roehrs, T., Zorick, F., & Roth, T. (1987). Experimental sleep fragmentation in normal subjects. *International Journal of Neuroscience, 33,* 207–214.

Stephan, B. C. M., & Caine, D. (2009). Aberrant pattern of scanning in prosopagnosia reflects impaired face processing. *Brain and Cognition, 69,* 262–268.

Steven, M. S., & Blakemore, C. (2004). Cortical plasticity in the adult human brain. In M. S. Gazzaniga (Ed.), *The cognitive neurosciences* (pp. 1243–1254). Cambridge, MA: MIT Press.

Stevens, C., & Neville, H. (2006). Neuroplasticity as a double-edged sword: Deaf enhancements and dyslexic deficits in motion processing. *Journal of Cognitive Neuroscience, 18,* 701–714.

Stevens, J., McClain, J. E., & Truesdale, K. P. (2006). Commentary: Obesity claims and controversies. *International Journal of Epidemiology, 35,* 77–78.

Stickgold, R. (2005). Sleep-dependent memory consolidation. *Nature, 437,* 1272–1278.

Stickgold, R., & Walker, M. P. (2005). Memory consolidation and reconsolidation: What is the role of sleep? *Trends in Neurosciences, 28,* 408–415.

Stickgold, R., & Walker, M. P. (2007). Sleep-dependent memory consolidation and reconsolidation. *Sleep Medicine, 8,* 331–343.

Stiles, J. (1998). The effects of early focal brain injury on lateralization of cognitive function. *Current Directions in Psychological Science, 7,* 21–26.

Stix, G. (2008, July). Traces of a distant past. *Scientific American, 299,* 56–63.

Stockwell, T., Chikritzhs, T., Bostrom, A., Fillmore, K., Kerr, W., Rehm, J., & Taylor, B. (2007). Alcohol-caused mortality in Australia and Canada: Scenario analyses using different assumptions about cardiac benefit. *Journal of Studies on Alcohol and Drugs, 14,* 101–132.

Stocum, D. L. (2007). Acceptable nAGging. *Science, 318,* 755–756.

Stoeckli, E. T. (2006). Longitudinal axon guidance. *Current Opinion in Neurobiology, 16,* 35–39.

Storandt, M. (2008). Cognitive deficits in the early stages of Alzheimer's disease. *Current Directions in Psychological Science, 17,* 198–202.

Strangman, G., Thompson, J. H., Strauss, M. M., Marshburn, T. H., & Sutton, J. P. (2005). Functional brain imaging of a complex navigation task following one night of total sleep deprivation: A preliminary study. *Journal of Sleep Research, 14,* 369–375.

Strathdee, S. A., & Pollini, R. A. (2007). A 21st-century Lazarus: The role of safer injection sites in harm reduction and recovery. *Addiction, 102,* 848–849.

Strick, P. L. (2004). Basal ganglia and cerebellar circuits with the cerebral cortex. In M. S. Gazzaniga (Ed.), *The cognitive neurosciences* (pp. 453–462). Cambridge, MA: MIT Press.

Strick, P. L., Dum, R. P., & Fiez, J. A. (2009). Cerebellum and nonmotor function. *Annual Review of Neuroscience, 32,* 413–434.

Strickland, D., & Bertoni, J. M. (2004). Parkinson's prevalence estimated by a state registry. *Movement Disorders, 19,* 318–323.

Strømme, P., Bjornstad, P. G., & Ramstad, K. (2002). Prevalence estimation of Williams syndrome. *Journal of Child Neurology, 17*(4), 269–271.

Strub, R. L., & Black, F. W. (1997). The mental status exam. In T. E. Feinberg & M. J. Farah (Eds.), *Behavioral neurology and neuropsychology* (pp. 25–42). New York, NY: McGraw-Hill.

Strubbe, J. H., & Steffens, A. B. (1997). Blood glucose levels in portal and peripheral circulation and their relation to food intake in the rat. *Physiology & Behavior, 19,* 303–307.

Stuss, D. T., & Alexander, M. P. (2005). Does damage to the frontal lobes produce impairment in memory? *Current Directions in Psychological Science, 14,* 84–88.

Stuss, D. T., & Levine, B. (2002). Adult clinical neuropsychology: Lessons from studies of the frontal lobes. *Annual Review of Psychology, 53,* 401–433.

Südhof, T. C. (2004). The synaptic vesicle cycle. *Annual Review of Neuroscience, 27,* 509–547.

Sugita, M., & Shipa, Y. (2005). Genetic tracing shows segregation of taste neuronal circuitries for bitter and sweet. *Science, 309,* 781–785.

Sullivan, E. V., & Marsh, L. (2003). Hippocampal volume deficits in alcoholic Korsakoff's syndrome. *Neurology, 61,* 1716–1719.

Sulzer, D. (2007). Multiple hit hypotheses for dopamine neuron loss in Parkinson's disease. *Trends in Neurosciences, 30,* 244–250.

Sulzer, D., Sonders, M. S., Poulsen, N. W., & Galli, A. (2005). Mechanisms of neurotransmitter release by amphetamines: A review. *Progress in Neurobiology, 75,* 406–433.

Sun, T., Patoine, C., Abu-Khalil, A., Visvader, J., Sum, E., Cherry, T. J., . . . Walsh C. A. (2005). Early asymmetry of gene transcription in embryonic human left and right cerebral cortex. *Science, 308,* 1794–1798.

Sunday, S. R., & Halmi, K. A. (1990). Taste perceptions and hedonicas in eating disorders. *Physiology & Behavior, 48,* 587–594.

Sur, M., & Rubenstein, J. L. R. (2005). Patterning and plasticity of the cerebral cortex. *Science, 310,* 805–810.

Surmeier, D. J., Mercer, J. N., & Chan, C. S. (2005). Autonomous pacemakers in the basal ganglia: Who needs excitatory synapses anyway? *Current Opinion in Neurobiology, 15,* 312–218.

Surmeier, D. J., Plotkin, J., & Shen, W. (2009). Dopamine and synaptic plasticity in dorsal striatal circuits controlling action selection. *Current Opinion in Neurobiology, 19,* 621–628.

Suwanwela, N., & Koroshetz, W. J. (2007). Acute ischemic stroke: Overview of recent therapeutic developments. *Annual Review of Medicine, 58,* 89–106.

Suzuki, W. A., & Clayton, N. S. (2000). The hippocampus and memory: A comparative and ethological perspective. *Current Opinion in Neurobiology, 10,* 768–773.

Swaab, D. F. (2004). Sexual differentiation of the human brain: Relevance for gender identity, transsexualism and sexual orientation. *Gynecological Endocrinology, 19,* 301–312.

Swaab, D. F., & Fliers, E. (1985). A sexually dimorphic nucleus in the human brain. *Science, 188,* 1112–1115.

Swaab, D. F., & Hofman, M. A. (1995). Sexual differentiation of the human hypothalamus in relation to gender and sexual orientation. *Trends in Neurosciences, 18,* 264–270.

Swaab, D. F., Zhou, J. N., Ehlhart, T., & Hofman, M. A. (1994). Development of vasoactive intestinal polypeptide neurons in the human suprachiasmatic nucleus in relation to birth and sex. *Developmental Brain Research, 79,* 249–259.

Swanson, L. W. (2000). What is the brain? *Trends in Neurosciences, 23,* 519–527.

Swanson, L. W., & Petrovich, G. D. (1998). What is the amygdala? *Trends in Neurosciences, 21,* 323–331.

Sweeney, M. E., Hill, P. A., Baney, R., & DiGirolamo, M. (1993). Severe vs. moderate energy restriction with and without exercise in the treatment of obesity: Efficiency of weight loss. *American Journal of Clinical Nutrition, 57,* 127–134.

Swinnen, S. P. (2002). Intermanual coordination: From behavioral principles to neural-network interactions. *Nature Reviews Neuroscience, 3,* 350–361.

Syntichaki, P., & Tavernarakis, N. (2003). The biochemistry of neuronal necrosis: Rogue biology? *Nature Reviews Neuroscience, 4,* 672–684.

Szabadics, J., Varga, C., Molnár, G., Oláh, S., Barzó, P., & Tamás, G. (2006). Excitatory effect of GABAergic axo-axonic cells in cortical microcircuits. *Science, 311,* 233–235.

Szymusiak, R., Gvilia, I., & McGinty, D. (2007). Hypothalamic control of sleep. *Sleep Medicine, 8,* 291–301.

Szymusiak, R., & McGinty, D. (2008). Hypothalamic regulation of sleep and arousal. *Annals of the New York Academy of Sciences, 1129,* 275–286.

Takeichi, M. (2007). The cadherin superfamily in neuronal connections and interactions. *Nature Reviews Neuroscience, 8,* 11–20.

Tallal, P., & Gaab, N. (2006). Dynamic auditory processing, musical experience and language development. *Trends in Neurosciences, 29,* 382–390.

Tamakoshi, A., & Ohno, Y. (2004). Self-reported sleep duration as a predictor of all-cause mortality: Results from the JACC study, Japan. *Sleep, 27,* 51–54.

Tamminga, C. A., & Holcomb, H. H. (2005). Phenotype of schizophrenia: A review and formulation. *Molecular Psychiatry, 10,* 27–39.

Tanabe, J., Thompson, L., Claus, E., Dalwani, M., Hutchison, K., & Banich, M. T. (2007). Prefrontal cortex activity is reduced in gambling and nongambling substance users during decision-making. *Human Brain Mapping, 28*(12), 1276–1286.

Tanaka, E. M., & Ferretti, P. (2009). Considering the evolution of regeneration in the central nervous system. *Nature Reviews Neuroscience, 10,* 713–723.

Tanji, J., & Hoshi, E. (2001). Behavioral planning in the prefrontal cortex. *Current Opinion in Neurobiology, 11,* 164–170.

Tank, D. W., Sugimori, M., Connoer, J. A., & Llinás, R. R. (1998). Spatially resolved calcium dynamics of mammalian Purkinje cells in cerebellar slice. *Science, 242,* 7733–7777.

Tattersall, I., & Matternes, J. H. (2000, January). Once we were not alone. *Scientific American, 282,* 56–62.

Taub, E., Uswatte, G., & Elbert, T. (2002). New treatments in neurorehabilitation founded on basic research. *Nature Reviews Neuroscience, 3,* 228–236.

Taylor, D. M., Tillery, S. I. H., & Schwartz, A. B. (2002). Direct cortical control of 3D neuroprosthetic devices. *Science, 296,* 1829–1832.

Taylor, J. R., Elsworth, J. D., Roth, J. R., Sladek, J. R., Jr., & Redmond, D. E., Jr. (1990). Cognitive and motor deficits in the acquisition of an object retrieval/detour task in MPTP-treated monkeys. *Brain, 113,* 617–637.

Taylor, R., & Forge, A. (2005). Life after deaf for hair cells? *Science, 307,* 1056–1058.

Teicher, M. H., Andersen, S. L., Polcari, A., Anderson, C. M., Navalta, C. P., & Kim, D. M. (2003). The neurobiological consequences of early stress and childhood maltreatment. *Neuroscience and Biobehavioural Reviews, 27,* 33–44.

Teitelbaum, P. (1957). Random and food-directed activity in hyperphagic and normal rats. *Journal of Comparative and Physiological Psychology, 50,* 486–490.

Teitelbaum, P. (1961). Disturbances in feeding and drinking behavior after hypothalamic lesions. In M. R. Jones (Ed.), *Nebraska symposium on motivation* (pp. 39–69). Lincoln: University of Nebraska Press.

Teitelbaum, P., & Epstein, A. N. (1962). The lateral hypothalamic syndrome: Recovery of feeding and drinking after lateral hypothalamic lesions. *Psychological Review, 69,* 74–90.

Tenconi, E., Lunardi, N., Zanetti, T., Santonastaso, P., & Favaro, A. (2006). Predictors of binge eating in restrictive anorexia nervosa patients in Italy. *Journal of Nervous & Mental Disease, 194,* 712–715.

Terszowski, G., Müller, S. M., Bleul, C. C., Blum, C., Schirmbeck, R., Reimann, J., Du Pasquier, L., Amagai, T., Boehm, T., & Rodewald, H-R. (2006). Evidence for a functional second thymus in mice. *Science, 312,* 284–287.

Tervaniemi, M., & Hugdahl, K. (2003). Lateralization of auditory-cortex functions. *Brain Research Reviews, 43,* 231–246.

Tetrud, J. W., & Langston, J. W. (1989). The effect of deprenyl (Selegiline) on the natural history of Parkinson's disease. *Science, 245,* 519–522.

Teuber, H.-L. (1975). Recovery of function after brain injury in man. In *Outcomes of severe damage to the nervous system.* CIBA Foundation Symposium 34. Amsterdam, Netherlands: Elsevier North-Holland.

Teuber, H.-L., Battersby, W. S., & Bender, M. B. (1960). Recovery of function after brain injury in man. In *Outcomes of severe damage to the nervous system.* CIBA Foundation Symposium 34. Amsterdam, Netherlands: Elsevier North-Holland.

Teuber, H.-L., Milner, B., & Vaughan, H. G., Jr. (1968). Persistent anterograde amnesia after stab wound of the basal brain. *Neuropsychologia, 6,* 267–282.

Thach, W. T., & Bastian, A. J. (2004). Role of the cerebellum in the control and adaptation of gait in health and disease. *Progress in Brain Research, 143,* 353–366.

Thannickal, T. C., Moore, R. Y., Nienhaus, R., Ramanathan, L., Gulyani, S., Aldrich, M., . . . Siegel, J. M. (2000). Reduced number of hypocretin neurons in human narcolepsy. *Neuron, 27,* 469–474.

Thomas, M. S. C., & Johnson, M. H. (2008). New advances in understanding sensitive periods in brain development. *Current Directions in Psychological Science, 17,* 1–5.

Thompson, B. L., Levitt, P., & Stanwood, G. D. (2009). Prenatal exposure to drugs: Effects on brain development and implications for policy and education. *Nature Reviews Neuroscience, 10,* 303–312.

Thompson, P. M., Vidal, C., Giedd, J. N., Gochman, P., Blumenthal, J., Nicolson, R., . . . Rapoport, J. L. (2001). Mapping adolescent brain change reveals dynamic wave of accelerated gray matter loss in very early-onset schizophrenia. *Proceedings of the National Academy of Sciences, USA, 98,* 11650–11655.

Thompson, R. F. (2005). In search of memory traces. *Annual Review of Psychology, 56,* 1–23.

Thompson-Schill, S. L., Bedny, M., & Goldberg, R. F. (2005). The frontal lobes and the regulation of mental activity. *Current Opinion in Neurobiology, 15,* 219–224.

Thorpy, M. (2007). Therapeutic advances in narcolepsy. *Sleep Medicine, 8,* 427–440.

Tiihonen, J., Lönnqvist, J., Wahlbeck, K., Klaukka, T., Tanskanen, A., & Haukka, J. (2006). Antidepressants and the risk of suicide, attempted suicide, and overall mortality in a nationwide cohort. *Archives of General Psychiatry, 63,* 1358–1367.

Toates, F. M. (1981). The control of ingestive behaviour by internal and external stimuli—a theoretical review. *Appetite, 2,* 35–50.

Tobler, P. N., Forillo, C. D., & Schultz, W. (2005). Adaptive coding of reward value by dopamine neurons. *Science, 307,* 1642–1645.

Toga, A. W., & Thompson, P. M. (2005). Genetics of brain structure and intelligence. *Annual Review of Neuroscience, 28,* 1–23.

Toga, A. W., Thompson, P. M., & Sowell, E. R. (2006). Mapping brain maturation. *Trends in Neurosciences, 29,* 148–159.

Toledano, R., & Gil-Nagel, A. (2008). Adverse effects of antiepileptic drugs. *Seminars in Neurology, 28,* 317–327.

Tomlinson, D. R., & Gardiner, N. J. (2008). Glucose neurotoxicity. *Nature Reviews Neuroscience, 9,* 36–45.

Tong, F. (2003). Primary visual cortex and visual awareness. *Nature Reviews Neuroscience, 4,* 219–229.

Toni, N., Laplagne, D. A., Zhao, C., Lombardi, G., Ribak, C. E., Gage, F. H., & Schinder, A. F. (2008). Neurons born in the adult dentate gyrus form functional synapses with target cells. *Nature Neuroscience, 11,* 901–907.

Tootell, R. B. H., Dale, A. M., Sereno, M. I., & Malach, R. (1996). New images from human visual cortex. *Trends in Neurosciences, 19,* 481–489.

Tosini, G., Pozdeyev, N., Sakamoto, K., & Iuvone, P. M. (2008). The circadian clock system in the mammalian retina. *Bioessays, 30,* 624–633.

Tracey, I. (2005). Nociceptive processing in the human brain. *Current Opinion in Neurobiology, 15,* 478–487.

Tracy, J. L., & Robins, R. W. (2004). Show your pride: Evidence for a discrete emotion expression. *Psychological Science, 15,* 194–197.

Tracy, J. L., & Robins, R. W. (2007a). Emerging insights into the nature and function of pride. *Current Directions in Psychological Science, 16,* 147–150.

Tracy, J. L., & Robins, R. W. (2007b). The psychological structure of pride: A tale of two facets. *Journal of Personality and Social Psychology, 92,* 506–525.

Tramontin, A. D., & Brenowitz, E. A. (2000). Seasonal plasticity in the adult brain. *Trends in Neurosciences, 23*(6), 251–258.

Tramontin, A. D., Hartman, V. N., & Brenowitz, E. A. (2000). Breeding conditions induce rapid and sequential growth in adult avian song control circuits: A model of seasonal plasticity in the brain. *Journal of Neuroscience, 20*(2), 854–861.

Tranel, D., & Damasio, A. R. (1985). Knowledge without awareness: An autonomic index of facial recognition by prosopagnosics. *Science, 228,* 1453–1454.

Trapp, B. D., & Nave, K-A. (2008). Multiple sclerosis: An immune or neurodegenerative disease? *Annual Review of Neurosciences, 31,* 247–269.

Treffert, D. A., & Christensen, D. D. (December, 2005). Inside the mind of a savant. *Scientific American, 293,* 109–113.

Treffert, D. A., & Wallace, G. L. (2002, June). Islands of genius. *Scientific American, 286,* 76–85.

Treistman, S. N., & Martin, G. E. (2009). BK channels: Mediators and models for alcohol tolerance. *Trends in Neurosciences, 32,* 629–636.

Treit, D. (1985). Animal models for the study of anti-anxiety agents: A review. *Neuroscience and Biobehavioural Reviews, 9,* 203–222.

Treit, D. (1987). RO 15-1788, CGS 8216, picrotoxin, pentylenetetrazol: Do they antagonize anxiolytic drug effects through an anxiogenic action? *Brain Research Bulletin, 19,* 401–405.

Treit, D., Robinson, A., Rotzinger, S., & Pesold, C. (1993). Anxiolytic effects of serotonergic interventions in the shock-probe burying test and the elevated plus-maze test. *Behavioural Brain Research, 54,* 23–34.

Tresch, M. C., Saltiel, P., d'Avella, A., & Bizzi, E. (2002). Coordination and localization in spinal motor systems. *Brain Research Reviews, 40,* 66–79.

Trinko, R., Sears, R. M., Guarnieri, D. J., & Dileone, R. J. (2007). Neural mechanisms underlying obesity and drug addiction. *Physiology & Behavior, 91*(5), 499–505.

Trivers, R. L. (1972). Parental investment and sexual selection. In B. Campbell (Ed.), *Sexual selection and the descent of man* (pp. 136–179). Chicago, IL: Aldine.

Trommershäuser, J., Glimcher, P. W., & Gegenfurtner, K. R. (2009). Visual processing, learning and feedback in the primate eye movement system. *Trends in Neurosciences, 32,* 583–590.

Tronson, N. C., & Taylor, J. R. (2007). Molecular mechanisms of memory reconsolidation. *Nature Reviews Neuroscience, 8,* 262–275.

Trottier, G., Srivastava, L., & Walker, C. D. (1999). Etiology of infantile autism: A review of recent advances in genetic and neurobiological research. *Journal of Psychiatry and Neuroscience, 24,* 103–115.

True, W. R., Xian, H., Scherrer, J. F., Madden, P. A. F., Bucholz, K. K., Heath, A. C., . . . Tsuang, M. (1999). Common genetic vulnerability for nicotine and alcohol dependence in men. *Archives of General Psychiatry, 56,* 655–661.

Tryon, R. C. (1934). Individual differences. In F. A. Moss (Ed.), *Comparative psychology* (pp. 409–448). New York, NY: Prentice-Hall.

Tsao, D. Y., Freiwald, W. A., Tootell, R. B. H., & Livingstone, M. S. (2006). A cortical region consisting entirely of face-selective cells. *Science, 311,* 670–674.

Tsao, D. Y., & Livingstone, M. S. (2008). Mechanisms of face perception. *Annual Review of Neuroscience, 31,* 411–437.

Tsien, R. (1998). The green fluorescent protein. *Annual Review of Biochemistry, 67,* 509–544.

Tsivilis, D., Vann, S. D., Denby, C., Roberts, N., Mayes, A. R., Montaldi, D., & Aggleton, J. P. (2008). A disproportionate role for the fornix and mammillary bodies in recall versus recognition memory. *Nature Neuroscience, 11,* 834–842.

Tsodyks, M., Kenet, T., Grinvald, A., & Arieli, A. (1999). Linking spontaneous activity of single cortical neurons and the underlying functional architecture. *Science, 286,* 1943–1946.

Tsunozaki, M., & Bautista, D. M. (2009). Mammalian somatosensory mechanotransduction. *Current Opinion in Neurobiology, 19,* 362–369.

Tulving, E. (2002). Episodic memory: From mind to brain. *Annual Review of Neuroscience, 53,* 1–25.

Tuominen, H. J., Tiihonen, J., & Wahlbeck, K. (2005). Glutaminergic drugs for schizophrenia: A systematic review and meta-analysis. *Schizophrenia Research, 72,* 225–234.

Turella, L., Pierno, A. C., Tubaldi, F., & Catsiello, U. (2009). Mirror neurons in humans: Consisting or confounding evidence? *Brain & Language, 108,* 10–21.

Turkenburg, J. L., Swaab, D. F., Endert, E., Louwerse, A. L., & van de Poll, N. E. (1988). Effects of lesions of the sexually dimorphic nucleus on sexual behavior of testosterone-treated female Wistar rats. *Brain Research Bulletin, 329,* 195–203.

Turkheimer, E. (2000). Three laws of behavior genetics and what they mean. *Current Directions in Psychological Science, 9*(5), 160–164.

Turkheimer, E., Haley, A., Waldron, M., D'Onofrio, B., & Gottesman, I. (2003). Socioeconomic status modifies heritability of IQ in young children. *Psychological Science, 14*(6), 623–628.

Turna, B., Apaydin, E., Semerci, B., Altay, B., Cikili, N., & Nazli, O. (2004). Women with low libido: Correlation of decreased androgen levels with female sexual function index. *International Journal of Impotence Research, 17,* 148–153.

Turner, R. S. (2006). Alzheimer's disease. *Seminars in Neurology, 26,* 499–506.

Tuxhorn, I., & Kotagal, P. (2008). Classification. *Seminars in Neurology, 28,* 277–288.

Tzingounis, A. V., & Wadiche, J. I. (2007). Glutamate transporters: Confining runaway excitation by shaping synaptic transmission. *Nature Reviews Neuroscience, 8,* 935-947.

Uc, E. Y., & Follett, K. A. (2007). Deep brain stimulation in movement disorders. *Seminars in Neurology, 27,* 170–182.

Ulrich, R. E. (1991). Commentary: Animal rights, animal wrongs and the question of balance. *Psychological Science, 2,* 197–201.

Ulrich-Lai, Y. M., & Herman, J. P. (2009). Neural regulation of endocrine and autonomic stress responses. *Nature Reviews Neuroscience, 10,* 397–408.

Ungerleider, L. G., & Haxby, J. V. (1994). "What" and "where" in the human brain. *Current Opinion in Neurobiology, 4,* 157–165.

Ungerleider, L. G., & Mishkin, M. (1982). Two cortical visual systems. In D. J. Ingle, M. A. Goodale, & R. J. W. Mansfield (Eds.), *Analysis of visual behavior* (pp. 549–586). Cambridge, MA: MIT Press.

Ungless, M. A. (2004). Dopamine: The salient issue. *Trends in Neurosciences, 27,* 702–706.

U.S. Centers for Disease Control and Prevention (CDC). (2008a). *Alcohol-related disease impact (ARDI)*. Atlanta, GA: Author.

U.S. Centers for Disease Control and Prevention (CDC). (2008b). Smoking and tobacco use—fact sheet: Health effects of cigarette smoking. Atlanta, GA: Author.

U.S. Department of Health and Human Services, Substance Abuse and Mental Health Services Administration. (2006). *2006 National Survey on Drug Use and Health*. Retrieved from http://www.oas.samhsa.gov/nsduhLatest.htm

Usui, A., Matsushita, Y., Kitahara, Y., Sakamoto, R., Watanabe, T., & Motohashi, N. (2007). Two cases of young adults sleepwalking. *Sleep and Biological Rhythms, 5*, 291–293.

Vaina, L. M., Cowey, A., Eskew, R. T., Jr., LeMay, M., & Kemper, T. (2001). Regional cerebral correlates of global motion perception: Evidence from unilateral cerebral brain damage. *Brain, 124*, 310–321.

Valenstein, E. S. (1973). *Brain control*. New York, NY: John Wiley & Sons.

Valenstein, E. S. (1980). *The psychosurgery debate: Scientific, legal, and ethical perspectives*. San Francisco, CA: W. H. Freeman.

Valenstein, E. S. (1986). *Great and desperate cures: The rise and decline of psychosurgery and other radical treatments for mental illness*. New York, NY: Basic Books.

van den Heuvel, C. J., Ferguson, S. A., Macchi, M. M., & Dawson, D. (2005). Melatonin as a hypnotic: Con. *Sleep Medicine Reviews, 7*, 71–80.

van den Pol, A. N. (2006). Viral infections in the developing and mature brain. *Trends in Neurosciences, 29*, 398–406.

Vanderhaegen, P., & Polleux, F. (2004). Developmental mechanisms patterning thalamocortical projections: Intrinsic, extrinsic and in between. *Trends in Neurosciences, 27*, 385–391.

Van der Linden, A., Van Meir, V., Boumans, T., Poirier, C., & Balthazart, J. (2009). MRI in small brains displaying extensive plasticity. *Trends in Neurosciences, 32*, 257–266.

Vanegas, H., & Schaible, H-G. (2004). Descending control of persistent pain: Inhibitory or facilitatory? *Brain Research Reviews, 46*, 295–309.

Van Essen, D. C., Anderson, C. H., & Felleman, D. J. (1992). Information processing in the primate visual system: An integrated systems perspective. *Science, 255*, 419–423.

Van Goozen, S. H. M., Slabbekoorn, D., Gooren, L. J. G., Sanders, G., & Cohen-Kettenis, P. T. (2002). Organizing and activating effects of sex hormones in homosexual transsexuals. *Behavioral Neuroscience, 116*, 982–988.

van Groen, T., Puurunen, K., Mäki, H-M., Sivenius, J., & Jolkkonen, J. (2005). Transformation of diffuse β-amyloid precursor protein and β-amyloid deposits to plaques in the thalamus after transient occlusion of the middle cerebral artery in rats. *Stroke, 36*, 1551.

VanHelder, T., & Radomski, M. W. (1989). Sleep deprivation and the effect on exercise performance. *Sports Medicine, 7*, 235–247.

Van Lancker, D., & Cummings, J. L. (1999). Expletives: Neurolinguistic and neurobehavioral perspectives on swearing. *Brain Research Reviews, 31*, 83–104.

Vann, S. D., & Aggleton, J. P. (2004). The mammillary bodies: Two memory systems in one? *Nature Reviews Neuroscience, 5*, 35–43.

Van Praag, H., Christie, B. R., Sejnowski, T. J., & Gage, F. H. (1999). Running enhances neurogenesis, learning, and long-term potentiation in mice. *Proceedings of the National Academy of Sciences, USA, 19*, 13427–13431.

Van Praag, H., Schinder, A. F., Christie, B. R., Toni, N., Palmer, T. D., & Gage, F. H. (2002). Functional neurogenesis in the adult hippocampus. *Nature, 415*, 1030–1034.

Van Praag, H., Shubert, T., Zhao, C., & Gage, F. H. (2005). Exercise enhances learning and hippocampal neurogenesis in aged mice. *Journal of Neuroscience, 25*, 8680–8685.

Van Praag, H., Zhao, X., & Gage, F. H. (2004). Neurogenesis in the adult mammalian brain. In M. S. Gazzaniga (Ed.), *The cognitive neurosciences* (pp. 127–159). Cambridge, MA: MIT Press.

Van Sickle, M. D., Duncan, M., Kingsley, P. J., Mouhihate, A., Urbani, P., Mackie, K., . . . Sharkey, K. A. (2005). Identification and functional characterization of brainstem cannabinoid CB2 receptors. *Science, 310*, 329–332.

Van Vleet, T. M., & Robertson, L. C. (2006). Cross-modal interactions in time and space: Auditory influence on visual attention in hemispatial neglect. *Journal of Cognitive Neuroscience, 18*, 1368–1379.

Van Wagenen, W. P., & Herren, R. Y. (1940). Surgical division of commissural pathways in the corpus callosum: Relation to spread of an epileptic attack. *Archives of Neurology and Psychiatry, 44*, 740–759.

Vargha-Khadem, F., Gadian, D. G., Watkins, K. E., Connelly, A., van Paesschen, W., & Mishkin, M. (1997). Differential effects of early hippocampal pathology on episodic and semantic memory. *Science, 277*, 376–380.

Vargas, M. E., & Barres, B. A. (2007). Why is Wallerian degeneration in the CNS so slow? *Annual Review of Neuroscience, 30*, 153–179.

Vaupel, J. W., Carey, J. R., & Christensen, K. (2003). It's never too late. *Science, 301*, 1679–1681.

Vertes, R. P. (1983). Brainstem control of the events of REM sleep. *Progress in Neurobiology, 22*, 241–288.

Vertes, R. P. (2004). Memory consolidation in sleep: Dream or reality? *Neuron, 44*, 135–148.

Vertes, R. P., & Eastman, K. E. (2000). The case against memory consolidation in REM sleep. *Behavioral and Brain Sciences, 23*, 867–876.

Vertes, R. P., & Seigel, J. M. (2005). Time for the sleep community to take a critical look at the purported role of sleep in memory processing. *Sleep, 28*, 1228–1231.

Vicario, D. S. (1991). Neural mechanisms of vocal production in songbirds. *Current Opinion in Neurobiology, 1*, 595–600.

Vidal, C. N., Rapoport, J. L., Hayashi, K. M., Geaga, J. A., McLemore, L. E., Alaghband Y., . . . Thompson, P. M. (2006). Dynamically spreading frontal and cingulate deficits mapped in adolescents with schizophrenia. *Archives of General Psychiatry, 63*, 25–34.

Villanueva, A. T. C., Buchanan, P. R., Yee, B. J., & Grunstein, R. R. (2005). Ethnicity and obstructive sleep apnoea. *Sleep Medicine Reviews, 9*, 419–436.

Vita, A., De Peri, L., Silenzi, C., & Dieci, M. (2006). Brain morphology in first-episode schizophrenia: A meta-analysis of quantitative magnetic resonance imaging studies. *Schizophrenia Research, 82*, 75–88.

Vogel, J. J., Bowers, C. A., & Vogel, D. S. (2003). Cerebral lateralization of spatial abilities: A meta-analysis. *Brain and Cognition, 52*, 197–204.

Vogt, B. A. (2005). Pain and emotion interactions in subregions of the cingulate gyrus. *Nature Reviews Neuroscience, 6*, 533–544.

Volkow, N. D., Fowler, J. S., Wang, G. J., & Swanson, J. M. (2004). Dopamine in drug abuse and addiction: Results from imaging studies and treatment implications. *Molecular Psychiatry, 9*, 557–569.

Volkow, N. D., Fowler, J. S., Wang, G. J., Baler, R., & Telang, F. (2009). Imaging dopamine's role in drug abuse and addiction. *Neuropharmacology, 56*, 3–8.

Volkow, N. D., & Li, T-K. (2004). Drug addiction: The neurobiology of behaviour gone awry. *Nature Reviews Neuroscience, 5*, 963–970.

Volkow, N. D., Wang, G. J., Fischman, M. W., Foltin, R. W., Fowler, J. S., Abumrad, N. N., . . . Shea, C. E. (1997). Relationship between subjective effects of cocaine and dopamine transporter occupancy. *Nature, 386*, 827–830.

Volkow, N. D., & Wise, R. A. (2005). How can drug addiction help us understand obesity? *Nature Neuroscience, 8*, 555–560.

Volz, R. J., Fleckenstein, A. E., & Hanson, G. R. (2007). Methamphetamine-induced alterations in monoamine transport: Implications for neurotoxicity, neuroprotection and treatment. *Addiction, 102*, 44–48.

von Boehmer, H. (2006). Thoracic thymus, exclusive no longer. *Science, 312*, 206–207.

Vuilleumier, P., Schwartz, S., Clarke, K., Husain, M., & Driver, J. (2002). Testing memory for unseen visual stimuli in patients with extinction and spatial neglect. *Journal of Cognitive Neuroscience, 14*, 875–886.

Wada, J. A. (1949). A new method for the determination of the side of cerebral speech dominance. *Igaku to Seibutsugaku, 14*, 221–222.

Wager, T. D. (2005). The neural bases of placebo effects in pain. *Current Directions in Psychological Science, 14*, 175–179.

Wager, T. D., Phan, K. L., Liberzon, I., & Taylor, S. F. (2003). Valence, gender, and lateralization of functional brain anatomy in emotion: A meta-analysis of findings from neuroimaging. *Neuroimage, 19*, 513–531.

Wagner, C. K. (2006). The many faces of progesterone: A role in adult and developing male brain. *Frontiers in Neuroendocrinology, 27,* 340–359.

Wagner, T., Valero-Cabre, A., & Pascual-Leone, A. (2007). Noninvasive human brain stimulation. *Annual Review of Biomedical Engineering, 9,* 527–565.

Wahlgren, N. G., & Ahmed, N. (2004). Neuroprotection in cerebral ischaemia: Facts and fancies—the need for new approaches. *Cerebrovascular Diseases, 17,* 153–166.

Waites, C. L., Craig, A. M., & Garner, C. C. (2005). Mechanisms of vertebrate synaptogenesis. *Annual Review of Neuroscience, 28,* 251–274.

Wajchenberg, B. L. (2000). Subcutaneous and visceral adipose tissue: Their relation to the metabolic syndrome. *Endocrinology Review, 21,* 697–738.

Wald, G. (1964). The receptors of human color vision. *Science, 145,* 1007–1016.

Walford, R. L., & Walford, L. (1994) The anti-aging plan. New York, NY: Four Walls Eight Windows.

Walker, E., Kestler, L., Bollini, A., & Hochman, K. M. (2004). Schizophrenia: Etiology and course. *Annual Review of Psychology, 55,* 401–430.

Walker, E., & Tessner, K. (2008). Schizophrenia. *Perspectives on Psychological Science, 3,* 30–37.

Wall, T. C., Brumfield, C. G., Cliver, S. P., Hou, J., Ashworth, C. S., & Norris, M. J. (2003). Does early discharge with nurse home visits affect adequacy of newborn metabolic screening? *The Journal of Pediatrics, 143,* 213–218.

Wallace, D. C. (1997, August). Mitochondrial DNA in aging and disease. *Scientific American, 277,* 40–47.

Wallen, K. (2005). Hormonal influences on sexually differentiated behavior in nonhuman primates. *Frontiers in Neuroendocrinology, 26,* 7–26.

Wallentin, M. (2009). Putative sex differences in verbal abilities and language cortex: A critical review. *Brain and Language, 108,* 175–183.

Walsh, T., McClellan, J. M., McCarthy, S. E., Addington, A. M., Pierce, S. B., Cooper, G. M., . . . Sebat, J. (2008). Rare structural variants disrupt multiple genes in neurodevelopmental pathways in schizophrenia. *Science, 320,* 539–543.

Wan, H., Aggleton, J. P., & Brown, M. W. (1999). Different contributions of the hippocampus and perirhinal cortex to recognition memory. *Journal of Neuroscience, 19,* 1142–1148.

Wang, F., Kalmar, J., He, Y., Jackowski, M., Chepenik, L., Edmiston, E., . . . Jones, M. (2009). Functional and structural connectivity between the perigenual anterior cingulate and amygdala in bipolar disorder. *Biological Psychiatry, 66,* 516–521.

Wang, M. M. (2006). Genetics of ischemic stroke: Future clinical applications. *Seminars in Neurology, 26,* 523–530.

Wang, T., Collet, J-P., Shapiro, S., & Ware, M. A. (2008). Adverse effects of medical cannabinoids: A systematic review. *Canadian Medical Association Journal, 178,* 1669–1678.

Wang, Y., Krishnan, H. R., Ghezzi, A., Yin, J. C. P., & Atkinson, N. S. (2007). Drug-induced epigenetic changes produce drug tolerance. *PLoS Biology, 5,* e265. doi:10.1371/journal.pbio.0050265

Ware, J. C. (2008). Will sleeping pills ever wake us up? *Sleep Medicine, 9,* 811–812.

Wargo, E. (2007). Understanding the have-knots: The role of stress in just about everything. *Association for Psychological Science, 20,* 18–23.

Wargo, E. (2008). Talk to the hand: New insights into the evolution of language and gesture. *Association for Psychological Science, 21,* 16–22.

Warneken, F., Hare, B., Melis, A. P., Hanus, D., & Tomasello, M. (2007). Spontaneous altruism by chimpanzees and young children. *Public Library of Science: Biology, 5,* 1414–1420.

Wässle, H. (2004). Parallel processing in the mammalian retina. *Nature Reviews Neuroscience, 5,* 747–767.

Watkins, L. R., Hutchinson, M. R., Johnston, I. N., & Maier, S. F. (2005). Glia: Novel counter-regulators of opioid analgesia. *Trends in Neurosciences, 28,* 661–669.

Watson, J. B. (1930). *Behaviorism.* New York, NY: Norton.

Waxham, M. N. (1999). Neurotransmitter receptors. In M. J. Zigmond, F. E. Bloom, S. C. Landis, J. L. Roberts, & L. R. Squire (Eds.), *Fundamental neuroscience* (pp. 235–268). New York, NY: Academic Press.

Webb, W. B., & Agnew, H. W. (1967). Sleep cycling within the twenty-four hour period. *Journal of Experimental Psychology, 74,* 167–169.

Webb, W. B., & Agnew, H. W. (1970). Sleep stage characteristics of long and short sleepers. *Science, 163,* 146–147.

Webb, W. B., & Agnew, H. W. (1974). The effects of a chronic limitation of sleep length. *Psychophysiology, 11,* 265–274.

Webb, W. B., & Agnew, H. W. (1975). The effects on subsequent sleep of an acute restriction of sleep length. *Psychophysiology, 12,* 367–370.

Wechsler-Reya, R., & Scott, M. P. (2001). The developmental biology of brain tumors. *Annual Review of Neuroscience, 24,* 385–428.

Weiller, C., & Rijntjes, M. (1999). Learning, plasticity, and recovery in the central nervous system. *Experimental Brain Research, 128,* 134–138.

Weinberger, N. M. (2004). Specific long-term memory traces in primary auditory cortex. *Nature Reviews Neuroscience, 5,* 279–290.

Weindruch, R. (1996, January). Caloric restriction and aging. *Scientific American, 274,* 46–52.

Weindruch, R., & Walford, R. L. (1988). *The retardation of aging and disease by dietary restriction.* Springfield, IL: Charles C. Thomas.

Weindruch, R., Walford, R. L., Fligiel, S., & Guthrie, D. (1986). The retardation of aging in mice by dietary restriction: Longevity, cancer, immunity, and lifetime energy intake. *Journal of Nutrition, 116,* 641–654.

Weingarten, H. P. (1983). Conditioned cues elicit feeding in sated rats: A role for learning in meal initiation. *Science, 220,* 431–433.

Weingarten, H. P. (1984). Meal initiation controlled by learned cues: Basic behavioral properties. *Appetite, 5,* 147–158.

Weingarten, H. P. (1985). Stimulus control of eating: Implications for a two-factor theory of hunger. *Appetite, 6,* 387–401.

Weingarten, H. P. (1990). Learning, homeostasis, and the control of feeding behavior. In E. D. Capaldi & T. L. Powley (Eds.), *Taste, experience, and feeding* (pp. 14–27). Washington, DC: American Psychological Association.

Weingarten, H. P., Chang, P. K., & Jarvie, K. R. (1983). Reactivity of normal and VMH-lesion rats to quinine-adulterated foods: Negative evidence for negative finickiness. *Behavioral Neuroscience, 97,* 221–233.

Weingarten H. P., & Kulikovsky, O. T. (1989). Taste-to-postingestive consequence conditioning: Is the rise in sham feeding with repeated experience a learning phenomenon? *Physiology & Behavior, 45,* 471–476.

Weinshenker, D., & Schroeder, J. P. (2007). There and back again: A tale of norepinephrine and drug addiction. *Neuropsychopharmacology, 32,* 1433–1451.

Weinstock, M. (2008). The long-term behavioural consequences of prenatal stress. *Neuroscience and Biobehavioural Reviews, 32,* 1073–1086.

Weise-Kelley, L., & Siegel, S. (2001). Self-administration cues as signals: Drug self-administration and tolerance. *Journal of Experimental Psychology, 27,* 125–136.

Weiskrantz, L. (2004). Roots of blindsight. *Progress in Brain Research, 144,* 229–241.

Weiskrantz, L., Warrington, E. K., Sanders, M. D., & Marshall, J. (1974). Visual capacity in the hemianopic field following a restricted occipital ablation. *Brain, 97,* 709–728.

Weiss, F., Maldonado-Vlaar, C. S., Parsons, L. H., Kerr, T. M., Smith, D. L., & Ben-Shahar, O. (2000). Control of cocaine-seeking behavior by drug-associated stimuli in rats: Effects on recovery of extinguished operant-responding and extracellular dopamine levels in amygdala and nucleus accumbens. *Proceedings of the National Academy of Sciences, USA, 97,* 4321–4326.

Weissman, D. H., & Banich, M. T. (2000). The cerebral hemispheres cooperate to perform complex but not simple tasks. *Neuropsychology, 14,* 41–59.

Weissman, M. M., & Olfson, M. (1995). Depression in women: Implications for health care research. *Science, 269,* 799–801.

Wen, Z., & Zheng, J. O. (2006). Directional guidance of nerve growth cones. *Current Opinion in Neurobiology, 16,* 52–58.

Wenning, A. (1999). Sensing effectors make sense. *Trends in Neurosciences, 22,* 550–555.

Werblin, F., & Roska, B. (2007, April). The movies in our eyes. *Scientific American, 296,* 73–79.

Wertz, A. T., Wright, K. P., Jr., Ronda, J. M., & Czeisler, C. A. (2006). Effects of sleep inertia on cognition. *Journal of the American Medical Association, 295*, 163.

West, A. E., Griffith, E. C., & Greenberg, M. E. (2002). Regulation of transcription factors by neuronal activity. *Nature Reviews Neuroscience, 3*, 921–931.

West, M. J., Slomianka, L., & Gunderson, H. J. G. (1991). Unbiased stereological estimation of the total number of neurons in the subdivisions of the rat hippocampus using the optical fractionator. *Anatomical Record, 231*, 482–497.

West, R. (2007). The clinical significance of "small" effects of smoking cessation treatments. *Addiction, 102*, 506–509.

Wever, R. A. (1979). *The circadian system of man.* Seewiesen-Andechs, Germany: Max-Planck-Institut für Verhaltensphysiologie.

White, N. M. (1997). Mnemonic functions of the basal ganglia. *Current Opinion in Neurobiology, 7*, 164–169.

White, T. D., Asfaw, B., Beyene, Y., Haile-Selassie, Y., Lovejoy, C. O., Suwa, G., & WoldeGabriel, G. (2009). *Ardipithecus ramidus* and the paleobiology of early hominids. *Science, 326*, 75–86.

Whiten, A., & Boesch, C. (2001, January). The cultures of chimpanzees. *Scientific American, 284*, 61–67.

Wickelgren, W. A. (1968). Sparing of short-term memory in an amnesic patient: Implications for strength theory of memory. *Neuropsychologia, 6*, 31–45.

Widaman, K. F. (2009). Phenylketonuria in children and mothers. *Current Directions in Psychological Science, 18*, 48–52.

Wiech, K., Ploner, M., & Tracey, I. (2008). Neurocognitive aspects of pain perception. *Trends in Cognitive Science, 12*, 306–313.

Wig, G. S., Grafton, S. T., Demos, K. E., & Kelley, W. M. (2005). Reductions in neural activity underlie behavioral components of repetition priming. *Nature Neuroscience, 8*, 1228–1233.

Wilkinson, L. S., Davies, W., & Isles, A. R. (2007). Genomic imprinting effects on brain development and function. *Nature, 8*, 832–843.

Willet, W. C., Skerrett, P. J., & Giovannucci, E. L. (2001). *Eat, drink, and be healthy: The Harvard Medical School guide to healthy eating.* New York, NY: Simon & Schuster.

Willett, W. C., & Stampfer, M. J. (2003, January). Rebuild the food pyramid. *Scientific American, 288*, 65–71.

Williams, G., Cai, X. J., Elliot, J. C., & Harrold, J. A. (2004). Anabolic neuropeptides. *Physiology & Behavior, 81*, 211–222.

Williams, M. (1970). *Brain damage and the mind.* Baltimore, MD: Penguin Books.

Williams, S. R., & Stuart, G. J. (2002). Dependence of EPSP efficacy on synapse location in neocortical pyramidal neurons. *Science, 295*, 1907–1910.

Williams, S. R., & Stuart, G. J. (2003). Role of dendritic synapse location in the control of action potential output. *Trends in Neurosciences, 26*(3), 147–154.

Willis, J. H. (2009). Origin of species in overdrive. *Science, 323*, 350–351.

Willoughby, R. E. (2007, April). A cure for rabies? *Scientific American*, 89–95.

Wills, T. J., Lever, C., Cacucci, F., Burgess, N., & O'Keefe, J. (2005). Attractor dynamics in the hippocampal representation of the local environment. *Science, 308*, 873–876.

Wilmer, J. B., Richardson, A. J., Chen, Y., & Stein, J. F. (2004). Two visual motion processing deficits in developmental dyslexia associated with different reading skills deficits. *Journal of Cognitive Neuroscience, 16*, 528–540.

Wilson, B. A. (1998). Recovery of cognitive functions following nonprogressive brain injury. *Current Opinion in Neurobiology, 8*, 281–287.

Wilson, M. A., & McNaughton, B. L. (1993). Dynamics of the hippocampal ensemble code for space. *Science, 261*, 1055–1058.

Wilson, R. I. (2008). Neural and behavioral mechanisms of olfactory perception. *Current Opinion in Neurobiology, 18*, 408–412.

Wilson, R. I., & Mainen, Z. F. (2006). Early events in olfactory processing. *Annual Review of Neuroscience, 29*, 163–201.

Wiltstein, S. (1995, October 26). Quarry KO'd by dementia. *Vancouver Sun,* p. 135.

Wingerchuk, D. M. (2008). Current evidence and therapeutic strategies for multiple sclerosis. *Seminars in Neurology, 28*, 56–68.

Wingfield, A., Tun, P. A., & McCoy, S. L. (2005). Hearing loss in older adulthood: What it is and how it interacts with cognitive performance. *American Psychological Society, 14*, 144–148.

Wingfield, J. C. (2005). A continuing saga: The role of testosterone in aggression. *Hormones and Behavior, 48*, 253–255.

Winkler, C., Kirik, D., & Björklund, A. (2004). Cell transplantation in Parkinson's disease: How can we make it work? *Trends in Neurosciences, 28*, 87–92.

Wintink, A. J., Young, N. A., Davis, A. C., Gregus, A., & Kalynchuck, L. E. (2003). Kindling-induced emotional behavior in male and female rats. *Behavioral Neuroscience, 117*, 632–640.

Wirtshafter, D., & Davis, J. D. (1977). Set points, settling points, and the control of body weight. *Physiology & Behavior, 19*, 75–78.

Wise, P. M., Dubal, D. B., Wilson, M. E., Shane, W. R., Böttner, M., & Rosewell, K. L. (2001). Estradiol is a protective factor in the adult and aging brain: Understanding of mechanisms derived from in vivo and in vitro studies. *Brain Research Review, 37*, 313–319.

Wise, S. P. (2008). Forward frontal fields: Phylogeny and fundamental function. *Trends in Neurosciences, 31*, 599–608.

Witelson, S. F. (1991). Neural sexual mosaicism: Sexual differentiation of the human temporo-parietal region for functional asymmetry. *Psychoneuroendocrinology, 16*, 133–153.

Witte, A. V., Fobker, M., Gellner, R., Knecht, S., & Flöel, A. (2009). Caloric restriction improves memory in elderly humans. *Proceedings of the National Academy of Sciences, USA, 106*, 1255–1260.

Wolf, J. M., Walker, D. A., Cochrane, K. L., & Chen, E. (2007, May). *Parental perceived stress influences inflammatory markers in children with asthma.* Paper delivered at the Western Psychological Association Convention, Vancouver, BC.

Wolfe, M. S. (2006, May). Shutting down Alzheimer's. *Scientific American, 294*, 73–79.

Wolford, G., Miller, M. B., & Gazzaniga, M. S. (2004). Split decisions. In M. S. Gazzaniga (Ed.), *The cognitive neurosciences* (pp. 1189–1199). Cambridge, MA: MIT Press.

Wolgin, D. L., & Jakubow, J. J. (2003). Tolerance to amphetamine hypophagia: A microstructural analysis of licking behavior in the rat. *Behavioral Neuroscience, 117*, 95–104.

Wolpaw, J. R., & Tennissen, A. (2001). Activity-dependent spinal cord plasticity in health and disease. *Annual Review of Neuroscience, 24*, 807–843.

Wommelsdorf, T., Anton-Erxleben, K., Pieper, F., & Treue, S. (2006). Dynamic shifts of visual receptive fields in cortical area MT by spatial attention. *Nature Neuroscience, 9*, 1156–1160.

Wonders, C. P., & Anderson, S. A. (2006). The origin and specification of cortical interneurons. *Nature Reviews Neuroscience, 7*, 687–696.

Wong, A. H. C., & Van Tol, H. M. (2003). Schizophrenia: From phenomenology to neurobiology. *Neuroscience and Biobehavioural Reviews, 27*, 269–306.

Wong, K. (2005a, February). The littlest human. *Scientific American, 292*, 56–65.

Wong, K. (2005b, June). The morning of the modern mind. *Scientific American, 292*, 86–95.

Wong, W. T., & Wong, R. O. L. (2000). Rapid dendritic movements during synapse formation and rearrangement. *Current Opinion in Neurobiology, 10*, 118–124.

Wood, E. R., Mumby, D. G., Pinel, J. P. J., & Phillips, A. G. (1993). Impaired object recognition memory in rats following ischemia-induced damage to the hippocampus. *Behavioral Neuroscience, 107*, 51–62.

Wood, J. N., Glynn, D. D., Phillips, B. C., & Hauser, M. D. (2007). The perception of rational, goal-directed action in nonhuman primates. *Science, 317*, 1402–1405.

Woodlee, M. T., & Schallert, T. (2006). The impact of motor activity and inactivity on the brain. *Current Directions in Psychological Science, 15*, 203–206.

Woodruff-Pak, D. S. (1993). Eyeblink classical conditioning in H.M.: Delay and trace paradigms. *Behavioral Neuroscience, 107*, 911–925.

Woods, S. C. (1991). The eating paradox: How we tolerate food. *Psychological Review, 98,* 488–505.

Woods, S. C. (2004). Lessons in the interactions of hormones and ingestive behavior. *Physiology & Behavior, 82*(1), 187–190.

Woods, S. C., Lotter, E. C., McKay, L. D., & Porte, D., Jr. (1979). Chronic intracerebroventricular infusion of insulin reduces food intake and body weight of baboons. *Nature, 282,* 503–505.

Woods, S. C., & Ramsay, D. S. (2000). Pavlovian influences over food and drug intake. *Behavioural Brain Research, 110,* 175–182.

Woods, S. C., Schwartz, M. W., Baskin, D. G., & Seeley, R. J. (2000). Food intake and the regulation of body weight. *Annual Review of Neuroscience, 51,* 255–277.

Woods, S. C., & Strubbe, J. H. (1994). The psychology of meals. *Psychonomic Bulletin & Review, 1,* 141–155.

Woods, S. W., Stolar, M., Sernyak, M. J., & Charney, D. S. (2001). Consistency of atypical antipsychotic superiority to placebo in recent clinical trials. *Biological Psychiatry, 49,* 64–70.

Woodson, J. C. (2002). Including "learned sexuality" in the organization of sexual behavior. *Neuroscience and Biobehavioural Reviews, 26,* 69–80.

Woodson, J. C., & Balleine, B. W. (2002). An assessment of factors contributing to instrumental performance for sexual reward in the rat. *Quarterly Journal of Experimental Psychology, 55*(B), 75–88.

Woodson, J. C., Balleine, B. W., & Gorski, R. A. (2002). Sexual experience interacts with steroid exposure to shape the partner preferences of rats. *Hormones and Behavior, 42,* 148–157.

Woodson, J. C., & Gorski, R. A. (2000). Structural sex differences in the mammalian brain: Reconsidering the male/female dichotomy. In A. Matsumoto (Ed.), *Sexual differentiation of the brain* (pp. 229–255). Boca Raton, FL: CRC Press.

Wray, G. A., & Babbitt, C. C. (2008). Enhancing gene regulation. *Science, 321,* 1300–1302.

Wu, J. B., Chen, K., Li, Y., Lau, Y-F. C., & Shih, J. C. (2009). Regulation of monoamine oxidase A by the SRY gene on the Y chromosome. *FASEB Journal, 23,* 4029–4038.

Wüst, S., Kasten, E., & Sabel, A. (2002). Blindsight after optic nerve injury indicates functionality of spared fibers. *Journal of Cognitive Neuroscience, 14,* 243–253.

Xu, D., Bureau, Y., McIntyre, D. C., Nicholson, D. W., Liston, P., Zhu, Y., . . . Robertson, G. S. (1999). Attenuation of ischemia-induced cellular and behavioral deficits by X chromosome–linked inhibitor of apoptosis protein overexpression in the rat hippocampus. *Journal of Neuroscience, 19,* 5026–5033.

Yaari, R., & Corey-Bloom, J. (2007). Alzheimer's disease. *Seminars in Neurology, 27,* 32–41.

Yabuta, N. H., Sawatari, A., & Callaway, E. M. (2001). Two functional channels from primary visual cortex to dorsal visual cortical areas. *Science, 292,* 297–301.

Yamaguchi, S., Isejima, H., Matsuo, T., Okura, R., Yagita, K., Kobayashi, M., & Okamura, H. (2003). Synchronization of cellular clocks in the suprachiasmatic nucleus. *Science, 302,* 1408–1412.

Yan, J. (2009). Heroin-assisted treatment helps some patients. *Psychiatric News, 44,* 16.

Yan, L. L., Davighas, M. L., Liu, K., Stamler, J., Wang, R., Pirzada, A., . . . Greenland, P. (2006). Midlife body mass index and hospitalization and mortality in older age. *Journal of the American Medical Association, 295,* 190–198.

Yang, S. H., Liu, R., Wu, S. S., & Simpkins, J. W. (2003). The use of estrogens and related compounds in the treatment of damage from cerebral ischemia. *Annals of the New York Academy of Sciences, 1007,* 101–107.

Yantis, S. (2008). The neural basis of selective attention: Cortical sources and targets of attentional modulation. *Current Directions in Psychological Science, 17,* 86–90.

Yesalis, C. E., & Bahrke, M. S. (1995). Anabolic-androgenic steroids: Current issues. *Sports Medicine, 19,* 326–340.

Yiu, G., & He, Z. (2006). Glial inhibition of CNS axon regeneration. *Nature Reviews Neuroscience, 7,* 617–627.

Youdim, M. B. H., Edmonson, D., & Tipton, K. F. (2006). The therapeutic potential of monoamine oxidase inhibitors. *Nature Reviews Neuroscience, 7,* 295–309.

Youngstedt, S. D., & Kline, C. E. (2006). Epidemiology of exercise and sleep. *Sleep and Biological Rhythms, 4,* 215–221.

Youngstedt, S. D., & Kripke, D. F. (2004). Long sleep and mortality: Rationale for sleep restriction. *Sleep Medicine Reviews, 8,* 159–174.

Yücel, M., & Lubman, D. I. (2007). Neurocognitive and neuroimaging evidence of behavioural dysregulation in human drug addiction: Implications for diagnosis, treatment and prevention. *Drug and Alcohol Review, 26,* 33–39.

Yücel, M., Solowij, N., Respondek, C., Whittle, S., Fornito, A., Pantelis, C., & Lubman, D. I. (2008). Regional brain abnormalities associated with long-term heavy cannabis use. *Archives of General Psychiatry, 65,* 694–701.

Yurgelun-Todd, D. (2007). Emotional and cognitive changes during adolescence. *Current Opinion in Neurobiology, 17,* 251–257.

Zador, D. (2007). Methadone maintenance: Making it better. *Addiction, 102,* 350–351.

Zahm, D. S. (2000). An integrative neuroanatomical perspective on some subcortical substrates of adaptive responding with emphasis on the nucleus accumbens. *Neuroscience and Biobehavioural Reviews, 24,* 85–105.

Zaidel, E. (1975). A technique for presenting lateralized visual input with prolonged exposure. *Vision Research, 15,* 283–289.

Zaidel, E. (1987). Language in the disconnected right hemisphere. In G. Adelman (Ed.), *Encyclopedia of neuroscience* (pp. 563–564). Cambridge, MA: Birkhäuser.

Zammit, S., Allebeck, P., Andreasson, S., Lundberg, I., & Lewis, G. (2002). Self reported cannabis use as a risk factor for schizophrenia in Swedish conscripts of 1969: Historical cohort study. *British Medical Journal, 325,* 1199.

Zangwill, O. L. (1975). Excision of Broca's area without persistent aphasia. In K. J. Zulch, O. Creutzfeldt, & G. C. Galbraith (Eds.), *Cerebral localization* (pp. 258–263). New York, NY: Springer-Verlag.

Zeigler, T. E. (2007). Female sexual motivation during non-fertile periods: A primate phenomenon. *Hormones and Behavior, 51,* 1–2.

Zeki, S. M. (1993a). *A vision of the brain.* Oxford, England: Blackwell Scientific.

Zeki, S. M. (1993b). The visual association cortex. *Current Opinion in Neurobiology, 3,* 155–159.

Zhang, J. V., Ren, P-G., Avsian-Kretchmer, O., Luo, C-W., Rauch, R., Klein, C., & Hsueh, A. J. W. (2005). Obestatin, a peptide encoded by the ghrelin gene, opposes ghrelin's effects on food intake. *Science, 310,* 996–999.

Zhang, L., Nair, A., Krady, K., Corpe, C., Bonneau, R. H., Simpson, I. A., & Vannucci, S. J. (2004). Estrogen stimulates microglia and brain recovery from hypoxia-ischemia in normoglycemic but not diabetic female mice. *Journal of Clinical Investigations, 113,* 85–95.

Zhang, N., Yacoub, E., Zhu, X-H., Ugurbil, K., & Chen, W. (2009). Linearity of blood-oxygenation-level dependent signal at microvasculature. *NeuroImage, 48,* 313–318.

Zhang, Y., Proenca, R., Maffie, M., Barone, M., Leopold, L., & Friedman, J. M. (1994). Positional cloning of the mouse obese gene and its human homologue. *Nature, 372,* 425–432.

Zhou, C., Wen, Z-X., Wang, Z-P., Guo, X., Shi, D-M., Zuo, H-C., & Xie, Z-P. (2003). Green fluorescent protein labeled mapping of neural stem cells migrating towards damaged areas in the adult central nervous system. *Cell Biology International, 27,* 943–945.

Zhuo, M. (2008). Cortical excitation and chronic pain. *Trends in Neurosciences, 31,* 199–207.

Zilles, K., & Amunts, K. (2010). Centenary of Brodmann's map—conception and fate. *Nature Reviews Neuroscience, 11,* 139–145.

Zimitat, C., Kril, J., Harper, C. G., & Nixon, P. F. (1990). Progression of neurological disease in thiamin-deficient rats is enhanced by ethanol. *Alcohol, 7,* 493–501.

Zimmer, C. (2008). Isolated tribe gives clues to the origins of syphilis. *Science, 319,* 272.

Zimmer, C. (2008, June). What is a species? *Scientific American, 298*, 72–88.

Zimmerman, J. E., Naidoo, N., Raizen, D. M., & Pack, A. I. (2008). Conservation of sleep: Insights from non-mammalian model systems. *Trends in Neurosciences, 31*, 371–376.

Zivin, J. A. (2000, April). Understanding clinical trials. *Scientific American, 282*, 69–75.

Zoghbi, H. Y. (2003). Postnatal neurodevelopmental disorders: Meeting at the synapse? *Science, 302*, 826–830.

Zola-Morgan, S., Squire, L. R., & Amaral, D. G. (1986). Human amnesia and the medial temporal region: Enduring memory impairment following a bilateral lesion limited to field CA1 of the hippocampus. *Journal of Neuroscience, 6*, 2950–2967.

Zola-Morgan, S. M., Squire, L. R., Amaral, D. G., & Suzuki, W. A. (1989). Lesions of perirhinal and parahippocampal cortex that spare the amygdala and hippocampal formation produce severe memory impairment. *Journal of Neuroscience, 9*, 4355–4370.

Zola-Morgan, S., Squire, L. R., & Mishkin, M. (1982). The neuroanatomy of amnesia: Amygdala-hippocampus versus temporal stem. *Science, 218*, 1337–1339.

Zola-Morgan, S., Squire, L. R., Rempel, N. L., Clower, R. P., & Amaral, D. G. (1992). Enduring memory impairment in monkeys after ischemic damage to the hippocampus. *Journal of Neuroscience, 12*, 2582–2596.

Zou, D-J., Chesler, A., & Firestein, S. (2009). How the olfactory bulb got its glomeruli: A just so story? *Nature Reviews Neuroscience, 10*, 611–618.

Zou, Y. (2004). Wnt signaling in axon guidance. *Trends in Neurosciences, 27*, 529–532.

Zou, Z., & Buck, L. B. (2006). Combinatorial effects of odorant mixes in olfactory cortex. *Science, 311*, 1477–1481.

Zuberbühler, K. (2005). The phylogenetic roots of language: Evidence from primate communication and cognition. *Current Directions in Psychological Science, 14*, 126–130.

Zufall, F., & Leinders-Zufall, T. (2007). Mammalian pheromone sensing. *Current Opinion in Neurobiology, 17*, 483–489.

Zuloaga, D. G., Puts, D. A., Jordan, C. L., & Breedlove, S. M. (2008). The role of androgen receptors in the masculinization of brain and behavior: What we've learned from the testicular feminization mutation. *Hormones and Behavior, 53*, 613–626.

Zur, D., & Ullman, S. (2003). Filling-in of retinal scotomas. *Vision Research, 43*, 971–982.

Zvolensky, M. J., & Bernstein, A. (2005). Cigarette smoking and panic psychopathology. *Current Directions in Psychological Science, 14*, 301–305.

Credits

Photo Credits

p. iv left, © Lee Cohen/CORBIS; **p. iv right,** Corbis/Super Stock Royalty Free; **p. v left,** © Wang Yuguo/Xinhua Press/Corbis; **p. v right,** © Dennis Hallinan/Alamy; **p. vi,** © The Photolibrary Wales/Alamy; **p. vii left,** Massimo Pizzotti/Photographer's Choice/Getty Images; **p. vii right,** Corbis Royalty Free; **p. viii,** © Sean Sprague/The Image Works; **p. ix left,** Exactostock/SuperStock Royalty Free; **p. ix right,** PHANIE/Photo Researchers, Inc.; **p. x,** Marc Grimberg/The Image Bank/Getty Images; **p. xi,** © Cultura/Alamy; **p. xii left,** © Radius/SuperStock Royalty Free; **p. xii right,** Justin Guariglia/National Geographic/Getty Images; **p. xiii,** © Exactostock/SuperStock Royalty Free; **p. xiv,** P Deliss/Godong/Corbis; **p. xv,** Mike Powell, Getty Images; **p. xvi,** © Matthieu Spohn/ès Photography/Corbis; **p. 1,** © Lee Cohen/CORBIS; **Fig. 1.1,** UHB Trust/Stone/Getty Images; **Fig. 1.2,** Bettmann/Corbis; **Fig. 1.5,** Courtesy of Todd Handy, Department of Psychology, University of British Columbia; **p. 20,** Corbis/Super Stock Royalty Free; **Fig. 2.2,** Donna Bierschwale, Courtesy of the New Iberia Research Center; **Fig. 2.5,** Dale and Marion Zimmerman/Animals Animals; **Fig. 2.6,** University of Chicago; **Fig. 2.7,** ape: Kevin Schafer/Peter Arnold; man: Suza Scalora/Photodisc/Getty Images Royalty Free; prosimian: A. Comoost/Peter Arnold; NW monkey: Erwin and Peggy Bauer/Bruce Coleman; OW monkey: Erwin and Peggy Bauer/Bruce Coleman; **Fig. 2.10,** LEALISA WESTERHOFF/AFP/Getty Images; **Fig. 2.11,** John Reader/Photo Researchers, Inc.; **Fig. 2.14,** Eastcott/Momatiuk/Stone/Getty Images; **Fig. 2.16,** David M. Phillips/Photo Researchers, Inc.; **Fig. 2.23,** Original by Arturo Alverez-Buylla, Illustration kindly provided by *Trends in Neuroscience*; **Fig. 2.25,** Thomas Wanstall/The Image Works; **p. 50,** © Wang Yuguo/Xinhua Press/Corbis; **Fig. 3.10,** Courtesy of T. Chan-Ling; **Fig. 3.11,** Ed Reschke/Peter Arnold; **Fig. 3.12,** Courtesy of Carl Ernst and Brian Christie, Department of Psychology, University of British Columbia; **Fig. 3.13,** Courtesy of Jerold J. M. Chun, M.D., Ph.D.; **Fig. 3.31,** Courtesy of Miles Herkenham, Unit of Functional Neuroanatomy, National Institute of Mental Health, Bethesda, MD; **p. 75,** © Dennis Hallinan/Alamy; **Fig. 4.11,** J.E. Heuser et al./The Rockefeller University Press; **Fig. 4.15,** Floyd E. Bloom, M.D./The Scripps Research Institute; **Fig. 4.20,** Image Source/Corbis Royalty Free; **p. 101,** © The Photolibrary Wales/Alamy; **Fig. 5.1,** Science Photo Library/Photo Researchers, Inc.; **Fig. 5.3,** Scott Camazine/Photo Researchers, Inc.; **Fig. 5.4,** Bruce Foster and Robert Hare/The University of British Columbia; **Fig. 5.5,** Courtesy of Neil Roberts, University of Liverpool; **Fig. 5.6,** Courtesy of Drs. Michael E. Phelps and John Mazziotta, UCLA School of Medicine; **Fig. 5.7,** Courtesy of Kent Kiehl and Peter Liddle, Department of Psychiatry, University of British Columbia; **Fig. 5.18,** Courtesy of Rod Cooper, Department of Psychology, University of Calgary; **Fig. 5.19,** Courtesy of Mark Klitenick and Chris Fibiger, Department of Psychiatry, University of British Columbia; **Fig. 5.20,** Courtesy of Ningning Guo and Chris Fibiger, Department of Psychiatry, University of British Columbia; **Fig. 5.21,** Nematode Caenorhabditis elegans from the cover of *Science* Vol. 263, no. 5148, 11 February 1994. Photo: Martin Chalfie. Reprinted with permission from AAAS. Photo provided by Martin Chalfie, Columbia University; **Fig. 5.22,** Image provided by Weissman, Livet, Sanes, Lichtman, Harvard University. Livet et al., 2007 *Nature* 450, 56–62; **Fig. 5.24,** PET scans courtesy of Marcus Raichle, Mallinckrodt Institute of Radiology, Washington University Medical Center; **Fig. 5.27,** Courtesy of Jack Wong; **p. 131,** Massimo Pizzotti/Photographer's Choice/Getty Images; **Fig. 6.3,** Alamy Images Royalty Free; **p. 135,** woman: Jorgen Schytte/Peter Arnold; owl: Michael Fairchild/Peter Arnold; lion: John Cancalosi/Peter Arnold;

impala: Gerhard Jaegle/Das Photoarchiv/Peter Arnold; bird: Cyril Laubscher/Dorling Kindersley Media Library; squirrel: Colin Varndell/Nature Picture Library; **Fig. 6.6,** Science Photo Library/Photo Researchers, Inc.; **Fig. 6.7,** Ralph C. Eagle/Photo Researchers, Inc.; **p. 164,** Corbis Royalty Free; **Fig. 7.9, left,** AJ Photo/Photo Researchers, Inc.; **Fig. 7.9, right,** AP Images; **p. 179,** Kateland Photo; **Fig. 7.17,** Roman Soumar/Corbis; **Fig. 7.20,** Omicron/Photo Researchers, Inc.; **Fig. 7.22,** Courtesy of James Enns, Department of Psychology; **p. 191,** © Sean Sprague/The Image Works; **p. 207,** Science Photo Library/Photo Researchers, Inc.; **p. 219,** Exactostock/SuperStock Royalty Free; **Fig. 9.4,** Courtesy Naweed I. Syed, Ph.D., Departments of Anatomy and Medical Physiology, University of Calgary; **Fig. 9.9,** Doug Goodman/Photo Researchers; **Fig. 9.11,** Courtesy of Carl Ernst and Brian Christie, Department of Psychology, University of British Columbia; **Fig. 9.12,** AP Images; **Fig. 9.14,** Courtesy of Williams Syndrome Foundation, Inc.; **p. 240,** PHANIE/Photo Researchers, Inc.; **Fig. 10.1,** Courtesy of Kenneth Berry, Head of Neuropathology, Vancouver General Hospital; **Fig. 10.2,** Custom Medical Stock Photography; **Fig. 10.3,** Courtesy of Dr. John P. J. Pinel; **Fig. 10.4,** Volker Steger/Peter Arnold; **Fig. 10.6,** © 1996 Scott Camazine; **Fig. 10.7,** Courtesy of Kenneth E. Salyer, Director, International Cranial Institute; **Fig. 10.8,** New York Public Library/Photo Researchers, Inc.; **Fig. 10.11,** James Stevens/Photo Researchers, Inc.; **Fig. 10.12,** Cecil Fox/Photo Researchers, Inc.; **Fig. 10.21,** Courtesy of Carl Ernst and Brian Christie, Department of Psychology, University of British Columbia; **Fig. 10.22,** Carolyn A. McKeone/Photo Researchers, Inc.; **p. 268,** Marc Grimberg/The Image Bank/Getty Images; **Fig. 11.6,** Kevin R. Morris/Corbis; **Fig. 11.16,** Richard R. Hansen/Photo Researchers, Inc.; **Fig. 11.20,** Both images courtesy of Tank et al., 1988; **p. 298,** © Cultura/Alamy; **Fig. 12.2,** David © Richard Reinauer/Color-Pic; **Fig. 12.17,** Jackson Laboratory; **p. 327,** © Radius/SuperStock Royalty Free; **Fig. 13.10,** CP/Winnipeg Free Press; **Fig. 13.12,** Woodfin Camp & Associates; **p. 355,** Justin Guariglia/National Geographic/Getty Images; **Fig. 14.1,** Hank Morgan/Photo Researchers, Inc.; **Fig. 14.4,** Animals Animals/Earth Scenes; **p. 383,** © Exactostock/SuperStock Royalty Free; **p. 389,** Digital Vision/Getty Images Royalty Free; **p. 391,** Michael Blann/Lifesize/Getty Images Royalty Free; **p. 392,** Creasource/Corbis; **p. 394,** © new photo service/Alamy Royalty Free; **p. 395,** siamionau pavel/Shutterstock; **Fig. 15.5,** Thompson et al., 2004, *Journal of Neuroscience,* 24, pgs. 6028–6036; **p. 398,** Courtesy of Maggie Edwards; **Fig. 15.11,** Reprinted from *Neuropharmacology,* Vol. 56 / N.D. Volkow, J.S. Fowler, G.J. Wang, R. Baler, F. Telang, Imaging dopamine's role in drug abuse and addiction, p. 12, © 2009, with permission from Elsevier; **p. 411,** © P Deliss/Godong/Corbis; **Fig. 16.6,** woman: EDHAR/Shutterstock; man: Jason Stitt/Shutterstock; **p. 442,** Mike Powell/Getty Images; **Fig. 17.1,** From: Damasia H., Grabowski T., Frank R., Galaburda, A.M., Damasio A.R.: The Return of Phineas Gage: Clues about the brain from a famous patient. *Science,* 264: 1102–1105, 1994. Department of Neurology and Image Analysis Facility, University of Iowa; **Fig. 17.5,** Kyung and DiPaola of the iViz Lab, Simon Fraser University, ivizlab.sfu.ca; **Fig. 17.7,** PhotosToGo; **Fig. 17.8,** Courtesy of Jessica Tracy, Department of Psychology, University of British Columbia; **Fig. 17.11,** Lennert Nillson/Albert Bonniers Forlag AB; **Fig. 17.13,** Carr et al., 2003, *Neural Mechanisms of Empathy,* 100, pgs. 5497–5502. Copyright 2003, National Academy of Sciences, U.S.A.; **Fig. 17.14,** Reprinted from *Neuropsychologia,* Vol. 45, René Hurlemann, Michael Wagner, Barbara Hawellek, Harald Reich, Peter Pieperhoff, Katrin Amunts, Ana-Maria Oros-Peusquens, Nadim J. Shah, Wolfgang Maier, Raymond J. Dolan, Amygdala control of emotion-induced forgetting and remembering:

Evidence from Urbach-Wiethe disease, p. 879, Copyright 2007, with permission from Elsevier; **Fig. 17.15, left,** © Dennis Hallinan/Alamy; **Fig. 17.15, right,** Tingfan Wu, Nicholas J. Butko, Paul Ruvulo, Marian S. Bartlett, Javier R. Movellan, "Learning to Make Facial Expressions," devlrn, pp. 1–6, 2009 IEEE 8th International Conference on Development and Learning, 2009. © 2009 IEEE; **p. 466,** © Matthieu Spohn/ès Photography/Corbis; **p. 467,** © Enigma/Alamy; **Fig. 18.3,** Dr. E Fuller Torrey and Dr. Weinberger/National Institute of Medical Health; **Fig. 18.4,** Paul Thompson/UCLA School of Medicine; **p. 473,** Konstantin Sutyagin/Shutterstock; **Fig. 18.6,** Reprinted by permission from Macmillan Publishers Ltd: *Nature Neuroscience* Vol. 8, Lukas Pezawas, Andreas Meyer-Lindenberg, Emily M Drabant, Beth A Verchinski, Karen E Munoz et al., 5-HTTLPR polymorphism impacts human cingulate-amygdala interactions: a genetic susceptibility mechanism for depression, pp. 828–834, copyright 2005; **Fig. 18.7,** Illustration provided by St. Jude Medical; **p. 479,** Andrew Lever/Shutterstock; **Fig. 18.9,** Reprinted from *Neuroscience Letters*, Vol. 388, Anne Schienle, Axel Schäfer, Bertram Walter, Rudolf Stark, Dieter Vaitl, Brain activation of spider phobics towards disorder-relevant, generally disgust- and fear-inducing pictures, pp. 1–6, Copyright 2005, with permission from Elsevier; **p. 482,** Laurence Griffiths/Getty Images; **Fig. 18.11,** Reprinted by permission from Macmillan Publishers Ltd: *Nature Neuroscience* Vol. 11, Elizabeth R Sowell, Eric Kan, June Yoshii, Paul M Thompson, Ravi Bansal et al., Thinning of sensorimotor cortices in children with Tourette syndrome, pp. 637–639, copyright 2008.

Illustration Credits

Figs. 1.4, 1.6, 1.7, 1.8, 1.9, 2.1, 2.17, 2.18, 2.19, 3.1, 3.3, 3.14, 3.23, 3.28, 3.29, 4.9, 4.10, 4.12, 4.13, 4.14, 4.18, 4.19, 5.2, 5.13, 5.14, 5.16, 5.23, 5.26, 6.28, 6.29, 7.7, 7.8, 7.15, 7.19, 8.2, 8.3, 8.4, 8.5, 8.11, 8.12, 8.13, 8.14, 8.15, 8.16, 8.17, 8.18, 9.2, 9.3, 9.13, 10.23, 11.2, 11.3, 11.9, 11.12, 11.13, 12.5, 12.12, 12.14, 13.8, 14.5, 14.9, 14.10, 14.13, 15.7, 15.8, 15.10, 16.1, 16.3, 16.4, 16.5, 16.7, 16.11, 16.16, Appendix I, unnumbered figure on p. 135, Frank Forney. **Other illustrations,** Modern Graphics, Inc., Omegatype Typography, Inc.; Schneck-DePippo Graphics; Williams C. Ober and Claire W. Garrison; Illustrious Interactive; Mark Leftowitz; Celadon Digital Studios; Academy Artworks; Leo Harrington; Adrienne Lehmann; Gale Mueller.

Name Index

Subject Index